AA002374

2013 IEEE 5th International Nanoelectronics Conference

(INEC 2013)

Singapore
2 – 4 January 2013

IEEE Catalog Number: CFP13625-PRT
ISBN: 978-1-4673-4840-9

Copyright © 2013 by the Institute of Electrical and Electronic Engineers, Inc
All Rights Reserved

Copyright and Reprint Permissions: Abstracting is permitted with credit to the source. Libraries are permitted to photocopy beyond the limit of U.S. copyright law for private use of patrons those articles in this volume that carry a code at the bottom of the first page, provided the per-copy fee indicated in the code is paid through Copyright Clearance Center, 222 Rosewood Drive, Danvers, MA 01923.

For other copying, reprint or republication permission, write to IEEE Copyrights Manager, IEEE Service Center, 445 Hoes Lane, Piscataway, NJ 08854. All rights reserved.

***This publication is a representation of what appears in the IEEE Digital Libraries. Some format issues inherent in the e-media version may also appear in this print version.**

IEEE Catalog Number: CFP13625-PRT
ISBN 13:   978-1-4673-4840-9
ISSN: 2159-3523

Additional Copies of This Publication Are Available From:

Curran Associates, Inc
57 Morehouse Lane
Red Hook, NY 12571 USA
Phone: (845) 758-0400
Fax: (845) 758-2633
E-mail: curran@proceedings.com
Web: www.proceedings.com

2013 IEEE 5th International Nanoelectronics Conference

(INEC 2013)

Singapore
2-4 January 2013

IEEE Catalog Number: CFP13625-POD
ISBN: 978-1-46734-840-9

Proceedings of the

2013 IEEE 5th International Nanoelectronics Conference (INEC)

2 – 4 January 2013
Resorts World Sentosa, Singapore

Technical support & inquiries
Research Publishing Services
t:+65-6492 1137; f:+65-6747 4355
e:enquiries@rpsonline.com.sg

Conference Sponsors

**IEEE
Nanotech Chapter**

Technical Sponsors

Platinum Sponsors

General Chair's Message

On behalf of the conference executive committee, I would like to welcome you to the 5th IEEE International Nanoelectronics Conference. It is my pleasure to have this conference returning to Singapore, the birth place of this conference. IEEE INEC was first held in Singapore in 2006, followed by Shanghai 2008, Hong Kong 2010 and Taiwan 2011. Since its inception, it has continued to become an important conference on nanoelectronics linking academics and engineers in industry.

The theme of this conference is **SUSTAINABLE NANOELECTRONICS**, aiming in nanoelectronics for the future. This conference also aims to identify the paths between fundamental research and potential electronics, photonics and nano-science applications. To meet these aims, the Technical Committee has put together an exciting and comprehensive programme covering the four areas in: (1) Nano-Fabrication, (2) Nano-Electronics, (3) Nano-Photonics and (4) Nano-Sciences. We are also honored to have four prominent keynote speakers: Prof. Kang L. Wang (University of California, CA, USA), Prof. Yoong Ahm Kim (Shinshu University, Nagano, Japan), Prof. Hongqi Xu (Lund University, Sweden) and Dr. Dennis L. Polla (IARPA, USA) to share with us their latest works. I would like to thank the delegates who have chosen this conference for their paper submissions. After the review process, a total of 230 papers have been selected to be presented at the conference covering the four parallel tracks. This includes 58 invited papers, 79 regular papers and 93 poster papers. I am also pleased to announce that for the first time in this conference, we will give out the Best Oral Presentation and the Best Poster award. We have also introduced an Invited talks by Young Scientist program to encourage young professionals to participate and contribute to this conference.

IEEE INEC 2013 is held at the beautiful and exciting Resorts World Sentosa. This is one of the latest additions to the Singapore's beautiful sceneries where it boasts of having the unique Maritime Experiential Museum & Aquarium and Universal Studios Singapore equipped with the latest Battlestar Galactica® and Transformers 3D® Rides. I hope that you will enjoy your stay and have a good time exploring the many fun and exciting places in this wonderful resort.

Finally, I would like to take this opportunity to thank the conference sponsors IEEE Singapore Section and IEEE Nanotech Chapter, and the technical sponsors Nanyang Technological University and IEEE Electron Device Society for their support in this conference. I would also like to thank our financial sponsors and exhibitors for their financial support in this conference. Last but not least, my sincere thanks to my colleagues in the Organizing Committee and the International Advisory Committee who have contributed their time and effort to this conference.

Wish you all have an enjoyable and fruitful time at this conference and a Happy 2013!

NG Geok Ing
General Chair
Nanyang Technological University

Committees

General Chair	**Ng Geok Ing (EEE)** *Nanyang Technological University*
Founding Chair	**Tan Cher Ming (EEE)** *Nanyang Technological University*
Vice General Chair	**Wang Hong (EEE)** *Nanyang Technological University*
Finance Chair	**Radhakrishnan K (EEE)** *Nanyang Technological University*
Local Arrangement Chair	**Terrence Wong (EEE)** *Nanyang Technological University*
Technical Programme Chair	**Zhang Dao Hua (EEE)** *Nanyang Technological University*
Technical Programme Co-Chair	**Wang Qijie** *Nanyang Technological University*
	Sivashankar Krishnamoorthy *Institute of Material Research Engineering*
	Sun Changqing *Nanyang Technological University*
	Su Haibin (MSE) *Nanyang Technological University*
Poster Session Chair	**Zheng Lianxi (MAE)** *Nanyang Technological University*
Poster Session Co-Chair	**Xiong Qihua (EEE/SMPS)** *Nanyang Technological University*
Publications Chair	**Miao Jianmin (MAE)** *Nanyang Technological University*
Publicity Chair	**Daniel Chua (NUS)** *National University of Singapore*
Publicity Co-Chair	**Wei Jun (SIMTECH)** *Singapore Institute of manufacturing Technology*
Sponsorship & Exhibition Chair	**Sun Xiaowei (EEE)** *Nanyang Technological University*
Sponsorship & Exhibition Co-Chair	**Goutam Kumar DALAPATI (IMRE)** *Institute of Material Research Engineering*

International Advisory Committee

AMARATUNGA Gehan
University of Cambridge, UK

Andrew T. S. WEE
National University of Singapore

BOEY Freddy
Nanyang Technological University, Singapore

CARTY Arthur
Waterloo Institute of Nanotechnology, Canada

CHU Paul
Hong Kong City University, Hong Kong

CHUA Soo Jin
National University of Singapore

HOR Tzi Sum Andy
Institute of Materials Research and Engineering (IMRE), Singpapore

HUANG Wei
Nanjing Univ of Post and Telecommunications, China

JAGADISH C.
Australian National University, Australia

KAM Chan Hin
Nanyang Technological University, Singapore

LAI Chao-Sung
Chang Gung University, Taiwan

LI Shushen
Institute of Semiconductors, Chinese Academy of Sciences, China

LIOU J. J.
University of Central Florida, USA

MASAHARA Meishoku
National Institute of Advanced Industrial Science and Technology, Japan

MORINOBU Endo
Shinshu University, Japan

RAO Ramgopal
IIT Bombay, India

ROSSEL Christophe
IBM Research GmbH, Germnay

SHUR Michael
Rensselaer Polytechnic Institute, USA

TAN Cher Ming (Founding Chair)
Nanyang Technological University, Singapore

TAY Beng Kang
Nanyang Technological University, Singapore

TEO Kie Leong
National University of Singapore

Keynote Speakers

Speaker **Kang L. Wang**, *University of California, USA*

 **Kang L. Wang** received his BS degree from National Cheng Kung University, Taiwan (1964) and Ph.D. (1970) degrees from the Massachusetts Institute of Technology. In 1970 to 1972 he was the Assistant Professor at MIT. From 1972 to 1979, he worked at th e General Electric Corporate Research a nd Development Center as a physicist/engineer. In 1979 he joined the Electrical Engineering Department of t he University of California, Los Angeles (UCLA), where he is the endowed Raytheon Professor of Physical Science and Electronics. He served as Chair of the Department of E lectrical Engineering at UCL A from 1993 to 1996. His research activities include semiconductor nano devices, spintronics/ferromagnetic materials and devices, nano science and technology; Molecular Beam Epitaxy and nano-epitaxy of hetero-structures. He has held more than 20 patents and publishe d over 400 pa pers. He received many awards, including SIA Unive rsity Research Award (2009); IBM Faculty Award; Guggenheim Fellow; IEEE Fellow; TSMC Honor Lectureship Award; Honoris Causa at Politechnico University, Torino, Italy; Semiconductor Research Corporation Inventor Awards; European Material Research Society Meeting Best pape r award; the Sem iconductor Research Corporation Technical Excellence Achievement Award.

He serves on the editorial board of the Encyclopedia of Nanoscience and Nanotechnology TM (American Scientific publishers). He currently also se rves as t he Director of Marco Focus Center on Functional Engineered Nano Architectonics – FEN A, an interdisciplinary Research Center, funded by Semiconductor Industry Association and Department of Defense to address the challenge of low-dissipation information processing technology beyond scaled CMOS. The Center involves 14 universities across the nation with 43 participating faculty members. He also leads Western Institute of Nan oelectronics (WIN) – a coord inated multi-project spintronics Research Institute. WIN is funded by NRI, Intel and the State of California and is a single largest university spintronics research institute in US. He was also the Dean of Engineering from 2000 to 2002 at the Hong Kong University of Science and Technology.

Title	Recent Studies on Carbon Nanotubes and Graphenes in Shinshu University
Speaker	Yoong Ahm Kim, *Shinshu University, Nagano, Japan*

Abstract

Nanocarbons including graphene and carbon nanotubes have attracted lots of at tention from various fields of scientists because they exhi bited extraordinary physical and c hemical properties due to their intrinsi c nano-sized feature. It should be noted that lots of nanocarbons -derived products are already in use and their viability strongly depend on their commercialization. However, at present, the biggest hurdles in nanocarbon business are considered to be their limited end uses. Thus, it is now urgently needed to speed up their current applications as well as find out their new end uses.

In this talk, I, first, describe the industrial usages of carbon nanotubes as functional filler lithium ion secondary batteries. Then, I focus on a super rubber sealant that is able to withstand temperatures up to 260oC and pressures as high as 239 MPa b y incorporating surface-modified nanotubes into fluorine rubber. Thus our rubber sealant will contribute to a revolutionary enhancement in the oil recovery efficiency from the current 35% to more than 70% by excavating previously inaccessible depo sits. Then, I will descri be the phys ics of double walled carbon nanotubes that have been less explored in nanotube. Finally, I'd like to tell you our recent results on boron doped gra phenes and carbon nanotubes.

Biography

Yoong Ahm Kim is Associate Professor at Shinshu University (Nagano, Japan), where he started as Assistant Professor in 2002. His research interests include: (a) Synthesis, ultra-high purification and dispersion of CNTs, (b) CNTs-based transparent conductive film, (c) Study of the optical properties of DWCNTs, (d) Fab rication of high-quality graphene and (e) Applications of nanocarbons in nanocomposites, nanofibers and energy storage systems. Prof Kim is a h ighly active and prolific researcher, as he has already published 170 papers and 12 book chapters. Many of his publications have been made with top-researchers in the field (e.g., Profs. Morinobu Endo, Mildred Dresselhaus, Riichiro Saito, and Mauricio Terrones). His contributions were highly cited (citation = 5,335) with H-index of 38, and have been presented in more than 30 international conferences with many invited talks. The train ing component of Prof. Kim is also impressing as he has supervised so far 44 students at various levels (21 BS, 16 MS and 6 PhD). He has chaired sessions in many nanocarbon related conferences and has served on the technical program of t he prestigious International Carbon Conference in 2008. He has been "Carbon Top Reviewer" in 2008 and 2011.

Title	Building Future Electronics and Optoelectronics with Semiconductor Nanostructures
Speaker	Hongqi Xu, *Lund University, Sweden*

Abstract

Epitaxially grown semiconductor nanostructures are emerging materials systems with potential applications in the future nanoelectronics and optoelectronics. Semiconductor nanostructures, such as nanowires and nanowire heterostructures, possess variety of novel structural and physics properties which offer great opportunities for the development of the next generation information, light emitting and photovoltaic devices. However, growth of microelectronic compatible, high crystalline quality semiconductor nanostructures required for this development, is still facing great scientific and technical challenges. In this talk, the state of the art research and technology development in epitaxially grown semiconductor nanowires and nanowire heterostructures, as well as novel devices made from these semiconductor nanostructures, will be presented. After the current fabrication technology of semiconductor nanowires and nanowire heterostructures is briefly reviewed, novel structural and physics properties of nanowires and nanowire heterostructures will be presented and discussed. The talk will then focus on the presence of a few examples of nanoelectronic and optoelectronic devices made based on these novel properties of nanowires and nanowire heterostructures. Especially, a wrap-gate technology for high speed nanowire transistors and a p-n junction free technology for photovoltaic and light emitting devices will be reported. The talk will be concluded by a perspective view of future nanowire-based information, light emitting and photovoltaic devices.

Biography

Hongqi Xu is currently a Professor at Lund University and a Professor at Peking University. He received the B.S. degree in applied physics from the Dalian University of Technology, China, in 1982 and the Ph.D. degree in theoretical physics from Lund University, Lund, Sweden, in 1991. From 1991 to 1993, he was a Postdoctoral Fellow at Linköping University. From 1993 to 1995, he was a Research Associate at Lund University, where he became an Assistant Professor in 1995, a Docent during 1995, an Associate Professor in 2001, and a Full Professor in 2003. He was also appointed Chair Professor at Peking University in 2010.

His recent research interests include electronic structures and photonic properties of semiconductors nanostructures, electron transport in low-dimensional systems, mesoscopic physics and devices, nanoelectronics, optoelectronic and solar-cell devices, quantum transport phenomena in nanostructures. He has authored or coauthored more than 150 refereed papers in scientific journals and has made about 300 contributions in scientific conferences and workshops.

Title	Intelligent Nanosystems
Speaker	Dennis L. Polla, *Intelligence Advanced Research Projects Agency, USA* Michael B. Wolfson, *System Planning Corporation, USA*

Abstract

The integrated circuit industry has been steadily investing in technologies to fabricate at ever-smaller nodes, and is predicted to achieve sub-10 nm feature sizes within a decade. This inexorable push by a US$300B industry creates opportunities in other semiconductor device disciplines to realizenew, powerful,self-contained, ultra-miniature systems capable of autonomously sensing, transducing, processing, and communicating information. We refer to this new opportunity as enabling "Intelligent Nanosystems."

Realization of such nanosystems will enable unique capabilities and benefits, but require overcoming significant challenges. This talk focuses on technical challenges of the four key nano-basedsub-systems components that comprise a fully-integrated, self-containedintelligent nanosystem-on-a chip: 1) N/MEMSsensors, 2) mixed signal circuits, 3) communication linkages, and 4) on-chip power.Many classical sensing methods benefit greatly from scalingsize to nanometer dimensions, and new modalities become available at quantum-scale dimensions. Most properties of transistors benefit from smaller node sizes, but analog performance often suffers. Nanometer-scale features put a lower limit on the wavelength of communication methods, requiring the use of THz and optical transduction. Power is a serious concern, as it scales to smaller system volumes. New approaches to minimizing standby power dissipation will be needed.

The practical integration of these four essential nano-scaled components into a compact chip stack represents a significant technology integration challenge.The Intelligence Research Projects Activity (IARPA) has begun an investment in this future electronics direction including forming strategic worldwide partnerships among industry, government, and academia. Two case examples of aggressive dimensional scaling in both physical and biochemical sensing will be presented as a new pathway forward in enabling future intelligent nanosystems.

Biography

Dr. Dennis L. Polla currently serves as the Deputy Director Safe and Secure Operations for the Intelligence Advanced Research Projects Agency (IARPA). He previously has served as a Program Manager at the Defense Advanced Research Projects Agency (DARPA) from 2004-2011 and has previously held faculty positions in multiple departments at the University of California Berkeley, Yale University , and the University Minnesota where he currently the W.R. Sweatt Chair in the Management of Technology.

He received B.S. degrees in electrical engineering and physics, the M.S. degree in electrical engineering, and the electrical engineering E.E. degree from the Massachusetts Institute of Technology, Cambridge. He received the Ph.D. degree in electrical engineering and the MBA degree from the University of California, Berkeley.

At IARPA Dr. Polla manages a portfolio of cybersecurity, microelectronics, and quantum computing programs for the Office of the Director of National Intelligence.

Invited Speakers

ALEXANDER B. Smetana
Nanoink Inc., USA

Abdel-Aziz El Mel
Université de Mons, Belgium

ARBIOL Jordi
ICREA at Institut de Ciència de Materials de Barcelona, ICMAB-CSIC

CHAN Che-Ting
The Hong Kong University of Science and Technology, China

CHELLAPPAN Vijila
Institute of Materials Research and Engineering, Singapore

CHE-SHENG Chung
National Taiwan University of Science and Technology, Taiwan

CZYSZANOWSKI Tomasz
Lodz University of Technology, Poland

DEGIRON Aloyse
CNRS/ Université Paris-Sud 11, France

FENG Yuanping
National University of Singapore, Singapore

FU Yunyi
Beijing University

Golap Kalita
Nagoya Institute of Technology, Japan

GUISBIERS Gregory
University of Mons, Belgium

Hara Kenji
Hokkaido University, Japan

HE Jr-Hau
National Taiwan University

HONG Minghui
National University of Singapore, Singapore

HORNG Ray-Hua
National Chung Hsing University, Taiwan

Hu Junhui
Nanjing Univ. of Aeronaut. & Astronaut, China

KIM Beomjoon
Institute of Industrial Science, Japan

KULKARNI G. U.
Jawaharlal Nehru Center for Advanced Scientific Research, India

LAI Chao-Sung
Chang Gung University, Taiwan

LEE Ming-Kwei
Chung Yuan Christian University, Taiwan

LI Chaoyang
Kochi University of Technology, Japan

LI Erping
IHPC, Singapore

LI Sean
University New South Wales, Australia

LI Tao
Nanjing University, China

LIU Huiyun
University College London, UK

MA Jianguo
Tianjin University, China

MAI Liqiang
Wuhan University of Technology, China

MAIKAP S.
Chang Gung University, Taiwan

MALLICK Govind
Army Research Laboratory, USA

MEHTA B. R.
IIT Delhi, India

MILNE William I.
Cambridge University, UK

MOHAN H.K.S.V
Nanyang Technological University, Singapore

NAZIR Kherani
University of Toronto, Canada

NOJEH Alireza
The University of British Columbia, Canada

OKANO Ken
International Christian University, Japan

OSTRIKOV Kostya
CSIRO, Australia

PENG Huisheng
Fudan University, China

Philippe Coquet
Institut d'électronique de Microélectronique et de Nanotechnologie (IEMN), France

QIHUA Xiong
Nayang Technological University, Singapore

RAO Ram Gopal
IIT Bombay, India

RAY Samit K.
IIT Kharagpur, India

SAMUKAWA Seiji
Tohoku University, Japan

Satheesh Krishnamurthy
The Open University, UK

SUN Handong
Nanyang Technological University, Singapore

TANEMURA Masaki
Nagoya Institute of Technology, Japan

TEO Hang Tong Edwin
Nanyang Technological University, Singapore

TEO Kie Leong
National University of Singapore, Singapore

TSENG Tseung-Yuen
National Chiao Tung University, Taiwan

VETRONE Fiorenzo
Institut National de la Recherche Scientifique, Canada

WEIMIN Chen
Linköping University, Sweden

WOLF Heiko
IBM Research Laboratory, Switzerland

Wu Yongling Linda
SIMTech, Singapore

YE Peide (Peter)
Purdue University, USA

YEO Yee Chia
National University of Singapore, Singapore

YU Siu Fung
City University of Hong Kong, China

ZHANG Ke-Qin
Suchow University, China

ZHANG Qinyuan
South China University, China

ZHAO Jianhua
Institute of Semiconductors Chinese Academy of Sciences, China

Technical Program

Session 1A	Nanofabrication for Electronic Devices I
Chairs	Sivashankar Krishnamoorthy and Goutam Dalapati

ZnO Nanowires Lift-Off From Silicon Substrate Embedded in Flexible Films 1
Ray-Hua Horng, Hung-I Lin and Dong-Sing Wuu

The Influence of Titanium Nitride Barrier Layer on the Properties of CNT Bundles 4
Chin Chong Yap, Dunlin Tan, Christophe Brun, Hong Li, Edwin Hang Tong Teo, Baillargeat Dominique and Beng Kang Tay

Low Temperature ISSG Oxidation and Its Application in SSRW for 20nm and Below Semiconductor Devices 7
Weihua Tong, Really Kim, Zhao Lun, Henry Lim and Sung Kim

Nanoscale Tri Gate MOSFET for Ultra Low Power Applications Using High-k Dielectrics 12
D. Nirmal, P. Vijay Kumar, Doreen Joy, Binola K Jebalin and N. Mohan Kumar

Session 1B	Advanced CMOS Devices I
Chairs	Che-Sheng Chung and Ye Peide

Near Flatband Mode Currents Evaluated by One-Particle Self-Consistent Calculations 20
Che-Sheng Chung and Sheng-Lyang Jang

Optimizing Nanoscale MOSFET Architecture for Low Power Analog/RF Applications 22
Dipankar Ghosh, Mukta Singh Parihar and Abhinav Kranti

High Performance Gadolinium Oxide Nanocrystal Memory with Optimized Charge Storage and Blocking Dielectric Thickness 24
Chih-Ting Lin, Chi-Feng Chang, Yu-Ren Yen, Chin-Hsiang Liao, Po-Wei Huang and Jer-Chyi Wang

Threshold Voltage of Nanoscale Si Gate-All-Around MOSFET: Short-Channel, Quantum, and Volume Effects 27
Min-Chul Sun, Hyun Woo Kim, Sang Wan Kim, Jung Han Lee, Hyungjin Kim and Byung-Gook Park

Session 1C	Nanophotonics I
Chairs	Degiron Aloyse and Chan Cheting

Enhancement of Angular CCT by Hybrid Phosphor Structure in White Light-Emitting Diodes 30
Kuo-Ju Chen, Hsin-Chu Chen, Min-Hsiung Shih, Chao Hsun Wang, Shih-Hsuan Chien, Hsin-Han Tsai, Chien-Chung Lin and Hao-Chung Kuo

Cu-Insulator-Si Hybrid Plasmonic Waveguide Based CMOS-Compatible Nanophotonic Devices 33
Shiyang Zhu, G. Q. Lo and D. L. Kwong

Session 1D	Nano Physics and Chemistry
Chairs	Su Haibin and Fan Hongjin

Electrical Probing of Multi-Ions Solution by Using Graphene-Based Sensor 37
Kuan-I Ho, Jia-Hong Liao, Chi-Hsien Huang, Ching-Yuan Su and Chao-Sung Lai

ZnO Nanorods Based Ultra Sensitive and Selective Explosive Sensor 40
Rashi Nathawat, M. Patel, P. Ray, N. A. Gilda, M. S. Vinchurkar and V. Ramgopal Rao

Patterning Metallic Films to Enhance Plasmonic Modes for mRNA Detection 43
Ping Bai, Lin Wu and Er Ping Li

xii

Session 2A	Investigating Properties at the Nanoscale
Chairs	Chao-Sung Lai and Gregory Guisbiers

Fabrication of Highly Ordered Hollow Oxide Nanostructures Based on Nanoscale Kirkendall Effect and Ostwald Ripening 46
Abdel-Aziz El Mel, Marie Buffière, Pierre-Yves Tessier, Wei Xu, Ke Du, Chang-Hwan Choi, Stephanos Konstantinidis, Carla Bittencourt and Rony Snyders

Microstructure and Piezoelectric Properties of c-BN Nano-Films Deposited on Si by RF Sputtering for Piezoelectric Devices 49
Wang Fang, Yang Baohe, Wei Jun and Zhang Kailiang

Investigation of the Dynamics of Carbon Nanotube Deposition in Dielectrophoresis 52
Ali Kashefian Naieni and Alireza Nojeh

Investigation of Rock-Salt CrTe Thin Film Grown by Molecular Beam Epitaxy Toward Half-Metal 56
Lu Hui and Teo Kie Leong

Session 2B	Advanced CMOS Devices II
Chairs	Ma JG and Abhinav Kranti

Status Overview: Fabrication, Characterization and Modeling of Flexible RF/Microwave Nanoelectronics 59
Guoxuan Qin, Tianhao Cai, Zhenqiang Ma and Jianguo Ma

Characterization of MIM Diodes Based on Nb/ Nb_2O_5 61
Islam E. Hashem, Nadia H. Rafat and Ezzeldin A. Soliman

UTBB with Ground-Plane Dopant-Segregated Schottky Barrier SOI MOSFET for Thermally Efficient Low- Variability Nanoscale CMOS Circuits 65
S. Qureshi and Ganesh C. Patil

A Study on Gate-All-Around (GAA) Polycrystalline Silicon Channel SONOS Flash Memory 69
Joo Yun Seo, Sang-Ho Lee, Yoon Kim, Se Hwan Park, Wandong Kim, Do-Bin Kim and Byung-Gook Park

Single Transistor Latch Phenomena in Junctionless Nanotransistors 72
Mukta Singh Parihar, Dipankar Ghosh, G. Alastair Armstrong and Abhinav Kranti

Session 2C	Nanophotonics II
Chairs	Czyszanowski Tomasz and Chen Rui

Modeling and Analysis of 1.3 μm InAs/GaAs Self-Assembled Quantum Dot Lasers with Rate Equation 74
C. Y. Liu, H. Wang and Q. Q. Meng

Session 2D	Nano Fabrication Synthesis
Chairs	Teo Hang Tong Edwin and Mallick Govind

Formation of Thick Textured Carbon Film Using Filtered Cathodic Vacuum Arc Technique 78
Naiyun Xu, Siu Hon Tsang, Edwin Hang Tong Teo and Beng Kang Tay

Effects of Carbon Loading on the Performance of Functionalized Carbon Nanotube Polymer Heat Sink for Ultra High Power Light-Emitting Diode 81
S. H. Chen, C. M. Tan, E. Tan and J. Kong

Session 3A	Novel Nanofabrication Strategies I
Chairs	Beomjoon Kim and Alireza Nojeh

Session 3B	Advanced CMOS Devices III
Chairs	Y. Y. Chia and Sunghun Jung

A Neural Network Approach to Classify Inversion Regions of High Mobility Ultralong Channel Single Walled Carbon Nanotube Field-Effect Transistors for Sensing Applications 85
S. V. Hari Krishna, Jianing An and Lianxi Zheng

Impact of Stress Induced by Stressors on Hot Carrier Reliability of Strained nMOSFETs 89
H. W. Hsu, H. S. Huang, S. Y. Chen, M. C. Wang, K. C. Li, K. C. Lin and C. H. Liu

Capture Cross Section Analysis of Four-level Random-Telegraph-Noise in Gate-Induced Drain Leakage Current 91
Seulki Park, Sungwon Yoo and Hyungcheol Shin

Self-Consistent Quasi Static CV Characterization of In_xGa_{1-x} Sb Buried Channel n-MOSFET 94
Muhammad Shaffatul Islam, Md. Nur Kutubul Alam and Md. Rafiqul Islam

Investigation of Conduction Mechanism in $Ti/Si_3N_4/p$-Si Stacked RRAM 97
Sunghun Jung, Sungjun Kim, Jeong-Hoon Oh, Kyung-Chang Ryoo, Jong-Ho Lee, Hyungcheol Shin and Byung-Gook Park

Session 3C	Nanophotonics Materials and Structures I
Chair	Xiong Qihua

Experimental Demonstration of Fano Resonance in Microfabricated Phononic Crystal Resonators Based on Two–Dimensional Silicon Slab 100
Nan Wang, Fu-Li Hsiao, J. M. Tsai, Moorthi Palaniapan, Dim-Lee Kwong and Chengkuo Lee

ZnS/ZnO/ZnS Core/Shell/Shell Hollow Spheres and its Optical Properties 103
Yi-Ting Hung and Shih-Shou Lo

Session 3D	Nano Technology Characterization and Application
Chairs	Peng Huisheng and Ke-Qin Zhang

Selenium Surface Energy Determination From Size-Dependent Considerations 105
G. Guisbiers, S. Arscott, M. Gaudet, A. Belfiore and R. Snyders

Facile Nanometer Thick Native Oxide Based Passivation of Silicon for High Efficiency Photovoltaics 110
Nazir P. Kherani and Zahidur R. Chowdhury

Various Size Images Mapping Technique to Analyze Trap-Assisted Non-Radiative Recombination Mechanism Using Cathodo- and Electro- Luminescences Measurement in GaN-Based LEDs 112
Euyhwan Park, Garam Kim, Janghyun Kim, Donghoon Kang, Joong-Kon Son and Byung-Gook Park

The Temperature Dependent TCAD and SPICE Modeling of AlGaN/GaN HEMTs 115
Li Yuan, Weizhu Wang, Kean Boon Lee, Haifeng Sun, Susai Lawrence Selvaraj, Xing Zhou and Guo-Qiang Lo

Session 4A	Tools for Fabrication, Measurement and Quality Control
Chairs	Vijila Chellappan and Jordi Arbiol

SOI CMOS Integrated Zinc Oxide Nanowire for Toluene Detection 119
S. Santra, P. K. Guha, S. K. Ray, F. Udrea and J. W. Gardner

Critical Thickness of Diamond-Like Carbon Study Using X-Ray Photoelectron Spectroscopy Depth Profiling 122
Nattaporn Khamnaulthong, Krisda Siangchaew and Pichet Limsuwan

Impact of Source Pupil Shapes on Process Windows in EUV Lithography 124
Hung-Fei Kuo

Session 4B	Advanced CMOS Devices IV
Chairs	H. K. S. V. Mohan and Jiandong Ye

Technology Options for Reducing Contact Resistances in Nanoscale Metal-Oxide-Semiconductor Field-Effect Transistors 128
Yee-Chia Yeo

Impact of Atomic-Scale Structural Design on Ultra-Short Channel (3 nm) MOSFETs 132
Shinji Migita, Yukinori Morita, Meishoku Masahara and Hiroyuki Ota

Improved Resistance Memory Characteristics and Switching Mechanism Using TiN Electrode on TaO_x/W Structure 136
A. Prakash and S. Maikap

Session 4C	Nanophotonics III
Chairs	Mohsen Rahmani and Zhang Qinyuan

A Single-Walled Carbon Nanotube Wall Paper as an Absorber for Simultaneously Achieving Passively Mode-Locked and Q-Switched Yb-Doped Fiber Lasers 139
Xiaohui Li, Yonggang Wang, Yishan Wang, Qijie Wang, Wei Zhao, Yongzhe Zhang, Xia Ya and Ying Zhang

Simulation of Grading Double Hetero-Junction Non-Polar InGaN Solar Cell 143
Hsun-Wen Wang, Pei-Chen Yu, Hau-Vei Han, Chien-Chung Lin, Hao-Chung Kuo and Shiuan-Huei Lin

Optical Properties of Si/SiO2 and GaAs/AlOx Sub-Wavelength HCG Mirrors on GaAs Substrate and an Impact of Structural Imperfections on Their Performance 146
Marcin Gebski, Maciej Dems, Jian Chen, Wang Qijie, Zhang Dao Hua and Tomasz Czyszanowski

Enhance Current Density and Light Trapping Effect in A-Si Thin Film Solar Cells by Flexible Textured PDMS Film 150
H. V. Han, H. C. Chen, Y. L. Tsai, C. C. Lin, H. C. Kuo and P. Yu

Session 4D	Nano Technology Procession
Chairs	Gregory Guisbiers and Philippe Coquet

A Sustainable Approach to Individualized Disease Treatment: The Engineering of a Multiple Use MEMS Drug Delivery Device 153
Danny Jian Hang Tng, Peiyi Song, Rui Hu, Guimiao Lin and Ken-Tye Yong

Microfluidic Device Optimization for Cell Growth 157
Vibha Jayaraj, Pramod. P. Wangikar and Sameer Jadhav

Poster 1	Nanofabrication (NF)

Stoichiometric Amorphous Hydrogenated Silicon Carbide Thin Film Synthesis Using DC-Saddle Plasma Enhanced Chemical Vapour Deposition 160
Behzad Jazizadeh Karimi, Ali B. Alamin Dow and Nazir P. Kherani

Realization and Application of Nanometer E-Beam Lithography System 164
Wei Shuhua, Dai Lan and Zhang Jing

Oxide Thin Film Transistors with Ink-Jet Printed In-Ga-Zn Oxide Channel Layer and ITO/IZO Source/Drain Contacts 168
Y. Wang, X. W. Sun, S. W. Liu, A. K. K. Kyaw and J. L. Zhao

Enhancement of Nanoelectronic Sensor Performance with Microfluidic Device 172
Kyungsup Han, Yong-Jin Yoon, Jack Sheng Kee and Mi Kyoung Park

A Novel Density Control of Carbon Nanotubes by Partial Oxidation of Catalyst Metal and its Field Emission Enhancement 175
Chuan-Ping Juan, Chia-Tsung Chang, Jyh-Liang Wang and Chuan-Chou Hwang

Magnetic and Leakage Current Properties of $Bi_{1-x} Gd_x FeO_3$ Thin Films 178
Ming-Cheng Kao, Hone-Zern Chen and San-Lin Young

A Novel Synthesis Approach of Gold Nanoparticles by Amino Acids 181
Cheng-Sheng Wu, Hong-Huei Huang, Fu-Ken Liu and Ching-Chich Leu

The Studies of Poly (Amino Acid) Assisted Synthesis of Gold Nanoparticles 183
Ying-Hui Hsu, Jen-Hau Yeh, Jeng-Shiung Jan and Ching-Chich Leu

Memory property of APTMS-Mediated Au-SiO2 Core-Shell Nanocrystal Memory 185
Sheng-Fu Huang, An-Ching Hsiao, Fu-Ken Liu and Ching-Chich Leu

Optimization of Al-doped Zinc Oxide Grown on Sapphire Using Dual-Plasma-Enhanced Metal Organic Chemical Vapor Deposition for InGaN/GaN Light-emitting Diodes 187
Po-Hsun Lei, Chia-Ming Hsu, Chia-Te Lin, Yu-Siang Fan and Sheng-Jhan Ye

Characterization and Photocurrent Study of Al, Mg-Doped ZnO Transparent Thin Films Prepared by Sol-Gel Route 189
W. Techitdheera, K. Chongsri and W. Pecharapa

Titanium Dioxide/Vanadium Oxide Nanocomposites Synthesized Via Sonochemical and Hydrothermal Process for Energy Storage Application 193
C. Kahattha, W. Techitdheera, N. Vittayakorn and W. Pecharapa

Poster 2	Nanoelectronics (NE)

In$_x$Ga$_{1-x}$Sb n-Channel MOSFET: Effect of Interface States on C-V Characteristics 197
Muhammad Shaffatul Islam, Md. Nur Kutubul Alam and Md. Rafiqul Islam

Growth Control of CuO Nanowires on Copper Thin Films: Toward the Development of pn Nanojunction Arrays 201
A. A. El Mel, M. Buffière, N. Bouts, E. Gautron, C. Bittencourt, P. Guttmann, P. Y. Tessier, S. Konstantinidis and R. Snyders

Growth, Structure and Optical Properties of GaSb Quantum Dot by LPE Technique 203
F. Qiu, Y. Zhang, Y.F. Lv, J. H. Guo, G. J. Hu, Sun, H. Y. Deng, S. H. Hu, N. Dai, Q. D. Zhuang, M. Yin, A. Krier and Z. Zhao

Design and Growth Optimization by Dual Ion Beam Sputtering of ZnO-Based High-Efficiency Multiple Quantum Well Green Light Emitting Diode 205
Sushil Kumar Pandey, Saurabh Kumar Pandey and Shaibal Mukherjee

Accurate Numerical Model for Surface Scattering, Grain Boundary Scattering and Anomalous Skin Effect of Copper Wires 209
Elhameh Abbaspour, Reza Sarvari, Alireza Akbarzadeh and Melika Rostami

Conduction Mechanism of Single-Electron Transistors Fabricated by Field-Emission-Induced Electromigration 211
Shunsuke Akimoto, Ryutaro Suda, Mitsuki Ito, Masazumi Ando and Jun-ichi Shirakashi

Structural and Magnetic Properties of Fe$_2$CrSi Heusler Alloy and Tunneling MagnetoResistance of Its Magnetic Tunneling Junctions 215
Yu-Pu Wang, Jin-Jun Qiu, Hui Lu, Qi-Jia Yap, Wen-Hong Wang, Gu-Chang Han, Duc-The Ngo and Kie-Leong Teo

The Effect of Intrinsic Defects on Resistive Switching Based on p-n Heterojunction 219
K. Zheng, X. W. Sun and K. L. Teo

Ink-Jet Printed In-Ga-Zn Oxide Nonvolatile TFT Memory Utilizing Silicon Nanocrystals Embedded in SiO$_2$ Gate Dielectric 222
Y. Wang, T. P. Chen, X. W. Sun, J. I. Wong, H. Y. Yang and J. L. Zhao

Ultimate Performance Projection of Ballistic III-V Ultra-Thin-Body MOSFET 226
Yan Guo, Kai-Tak Lam, Yee-Chia Yeo and Gengchiau Liang

Fully CMOS Compatible 1T1R Integration of Vertical Nanopillar GAA Transistor and Oxide Based RRAM Cell for High Density Nonvolatile Memory Application 228
Z. Fang, X. P. Wang, B. B. Weng, Z. X. Chen, A. Kamath, G. Q. Lo and D. L. Kwong

Ultra-low Turn-on Field and Ultra-high Field Emission Current Density From Pillar Array Design of Carbon Nanotubes with Optimum R/H Ratio 231
Chuan-Ping Juan and Jun-Han Lin

Trap Exploration of ZnO-Based Resistance Switching Memory Devices 234
Fu-Chien Chiu, Wen-Yuan Chang, Peng-Wei Li, Chih-Chi Chen and Wen-Ping Chiang

Superior Resistive Switching Characteristics of Cu-TiO$_2$ Based RRAM Cell 236
Yu-Chih Huang, Huan-Min Lin and Huang-Chung Cheng

Investigation into the Performance of CNT Interconnects by Spin Coating Technique 240
Wei-Chih Chiu and Bing-Yue Tsui

Promising N-Type FinFET Devices without or with Cobalt-Silicide Applied to the Gate 242
Hsin-Chia Yang, Jing-Zong Jhang, Wen-Shiang Liao, Chong-Kuan Du, Yi-Hong Lee, Sung-Ching Chi,
Quan-Hao Shen, Mu-Chun Wang and Shea-Jue Wang

The Side Effects and the Effects of Thickness of Source/Drain Fin on P-Type FinFET Devices 245
Hsin-Chia Yang, Wei-Yen Peng, Wen-Shiang Liao, Guo-Wei Wu, Cheng-Yu Tsai, Mu-Chun Wang,
Sung-Ching Chi and Shea-Jue Wang

The Adjustment of Threshold Voltage on P-Type FinFET Devices 248
Hsin-Chia Yang, Yi-Hong Lee, Wen-Shiang Liao, Chong-Kuan Du, Jing-Zong Jhang, Sung-Ching Chi,
Mu-Chun Wang and Shea-Jue Wang

The Enhancement of MOSFET Electric Performance Through Strain Engineering by Refilled SiGe as Source and Drain 251
Hsin-Chia Yang, Chao-Wang Li, Wen-Shiang Liao, Chong-Kuan Du, Mu-Chun Wang, Jie-Min Yang,
Chun-Wei Lian and Chuan-Hsi Liu

Temperature Dependence Carrier Transport Behavior of Transparent ZnO:Y Nanocrystalline Films 254
S. L. Young, C. Y. Kung, H. Z. Chen, M. C. Kao, T. T. Lin, M. C. Chang, H. H. Lin, J. H. Lin, S. H. Chin
and C. R. Ou

Multiferroic and Structural Transition Properties of $Bi_{1-x}Pr_x$ Fe 0.95 Mn 0.05 O_3 Thin Films 258
Hone-Zern Chen, Ming-Cheng Kao and San-Lin Young

High Sensing Performance of Fluorinated HfO_2 Membrane by Low Damage CF_4 Plasma Treatment for K^+ Detections 261
Chi-Hsien Huang, I-Shun Wang, Kuan-I Ho, Tzu-Wen Chiang, Chien Chou, Chu-Fa Chang and Chao-Sung Lai

Investigation of the Random Dopant Fluctuations in 20-nm Bulk MOSFETs and Silicon-On-Insulator FinFETs by Ion Implantation Monte Carlo Simulation 263
Keng-Ming Liu and Cheng-Kuei Lee

Microscopic Study of Random Dopant Fluctuation in Silicon Nanowire Transistors Using 3D Simulation 267
Chun-Yu Chen, Jyi-Tsong Lin and Meng-Hsueh Chiang

Characteristics and Hot-Carrier Effects of Strained pMOSFETs with SiGe Channel and Embedded SiGe Source/Drain Stressor 271
Min-Ru Peng, Mu-Chun Wang, Liang-Ru Ji, Heng-Sheng Huang, Shuang-Yuan Chen, Shea-Jue Wang,
Hong-Wen Hsu and Wen-Shiang Liao

Structural and Optical Properties of Cu-Doped ZnO Nanoparticles Synthesized by Co-Precipitation Method for Solar Energy Harvesting Application 274
N. Thaweesaeng, S. Suphankij, W. Pecharapa and W. Techitdheera

Self-Organized Hybrid Nanostructures Composed of the Array of Vertically Aligned Carbon Nanotubes and Planar Graphite Layer 277
V. Labunov, A. Prudnikava, B. Shulitski, B. K. Tay, X. Wang, A. Basaev, V. Galperin and Y. Shaman

Session 5A	Novel Nanofabrication Strategies II
Chair	Li Chaoyang

Synthesis of Continuous Graphene on Metal Foil for Flexible Transparent Electrode Application 281
Golap Kalita, Koichi Wakita, Masayoshi Umeno, Yasuhiko Hayashi and Masaki Tanemura

Nanoscale Mechanical Scratching of Graphene Using Scanning Probe Microscopy 285
Ryutaro Suda, Takanari Saito, Ampere A. Tseng and Jun-ichi Shirakashi

Session 5B	Carbon-related I
Chairs	Fu Yunyi and Rao VR

Retention Behavior of Graphene Oxide Resistive Switching Memory on Flexible Substrate 288
Fang Yuan, Yu-Ren Ye, Jer-Chyi Wang, Zhigang Zhang, Liyang Pan, Jun Xu and Chao-Sung Lai

A Zigbee-Based Wireless Wearable Electronic Nose Using Flexible Printed Sensor Array 291
*Panida Lorwongtragool, Reinhard R. Baumann, Enrico Sowade, Natthapol Watthanawisuth
and Teerakiat Kerdcharoen*

Session 5C	Modeling and Simulation
Chairs	Jr-Hau He and Shih-Yen Lin

Nonlinear Modeling of Compliant Mechanism 294
Raisuddin Khan, Md. Masum Billah and Mitsuru Watanabe

Doped Group-IV Semiconductor Nanocrystals 298
Latha Nataraj, Aaron Jackson, Lily Giri, Clifford Hubbard and Mark Bundy

Session 5D	Advanced Memory
Chairs	Kulkarni G. U. and Siddheswar Maikap

Robust Nitrogen Plasma Immersion Ion Implantation Treatment on Gadolinium Oxide Resistive Switching Random Access Memory 300
Yu-Ren Ye, Ying-Huei Wu, Jer-Chyi Wang and Chao-Sung Lai

Bipolar Resistive Switching Characteristics in Si_3N_4-Based RRAM with MIS (Metal-Insulator-Silicon) Structure 303
Sungjun Kim, Sunghun Jung, Jeong-Hoon Oh, Kyung-Chang Ryoo and Byung-Gook Park

Simulation Study of Dimensional Effect on Bipolar Resistive Random Access Memory (RRAM) 306
Liu Kai, Zhang Kailiang, Wang Fang, Zhao Jinshi and Wei Jun

On Pairing Bipolar RRAM Memory Element with Novel Punch-Through Diode Based Selector: Compact Modeling to Array Performance 309
R. Mandapati, A. Borkar, V. S. S. Srinivasan, P. Bafna P. Karkare, S. Lodha and U. Ganguly

Session 6A	Nanofabrication for Electronic Devices II
Chair	Heiko Wolf

NEMS Meets Bio-sensing; *There're Plenty of Things to Do in the Middle* 313
Beomjoon Kim

Novel Quantum Effect Devices Realized by Fusion of Bio-Template and Defect-Free Neutral Beam Etching 316
Seiji Samukawa

On Controlling EBL Parameters for Nanoelectromechanical Resonators Fabricated on Insulating/Semiconducting Structures 318
Ali B. Alamin Dow, H. Lin, C. Popov, U. Schmid and Nazir P. Kherani

Session 6B	Carbon-related II
Chairs	T. Masaki and Zhao Jianhua

Performance Analysis and Simulation of Two Different Architectures of (6:3) and (7:3) Compressors Based on Carbon Nano-Tube Field Effect Transistors 322
Shima Mehrabi, Keivan Navi and Omid Hashemipour

One-Step Formation of Atomic-Layered Transistor by Selective Fluorination of Graphene Film 326
Kuan-I Ho, Jia-Hong Liao, Chi-Hsien Huang, Chang-Lung Hsu, Lain-Jong Li, Chao-Sung Lai and Ching-Yuan Su

Analysis of CNT Electronics Structure to Design CNTFET 329
Soheli Farhana, Ahm Zahirul Alam, Sma Motakabber and Sheroz Khan

Session 6C	Nanophotonics IV
Chair	Alireza Nojeh

Long-Wavelength III-V Quantum-Dot Lasers Monolithically Grown on Si Substrates 333
Qi Jiang, Andrew Lee, Mingchu Tang, Alwyn Seeds and Huiyun Liu

Development of a High Sensitivity Photodetector Using Amorphous Selenium and Diamond Cold Cathode 336
K. Okano, T. Masuzawa, M. Onishi, I. Saito, A. T. T. Koh, D. H. C. Chua and T. Yamada

Fabrication and Characterization of Uni-Traveling-Carrier Photodetectors (UTC-PDs) with Dipole-Doped Structure at InGaAs/InP Interface 338
Q. Q. Meng, C. Y. Liu, H. Wang, K. S Ang, K. Manoj and T. X. Guo

6.5 nm-Thick Al_2O_3 Surface Passivated Layer Grown on Two Stacks of 10-Period InGaAs and GaAs-Capped InAs Quantum Dot Infrared Photodetector Focal Plane Arrays for High Temperature Operation 342
Shiang-Feng Tang, Tzu-Chiang Chen, Wen-Jen Lin and Shih-Yen Lin

Session 6D	MEMS and Solar Cells
Chairs	Chellappan Vijila and Teo Kie Leong

Influence of Trap Depth on Charge Transport in Inverted Bulk Heterojunction Solar Cells Employing ZnO as Electron Transport Layer 346
Naveen Kumar Elumalai Chellappan Vijila, Arthi Sridhar and Seeram Ramakrishna

A Dual-Silicon-Nanowire Based Nanoelectromechanical Switch 350
You Qian, Liang Lou, Vincent Pott, Minglin Julius Tsai and Chengkuo Lee

Device Modeling and Optimization of High-Performance Thin Film CIGS Solar Cell with $Mg_xZn_{1-x}O$ Buffer Layer 353
Saurabh Kumar Pandey and Shaibal Mukherjee

Session 7A	Terahertz and TFETs
Chairs	Sanjeev K. Manhas and Sun C. Q.

A One-Way Terahertz Plasmonic Waveguide Based on Surface Magneto Plasmons in a Metal-Dielectric-Semiconductor Structure 357
Bin Hu, Qi Jie Wang and Ying Zhang

High Frequency SAW Nanotransducer Utilizing Ultrananocrystalline Diamond/ AlN Bimorph Architecture 360
Ali B. Alamin Dow, H. Lin, C. Popov, U. Schmid and Nazir P. Kherani

Hump Phenomenon in Transfer Characteristics of Double-Gated Thin-Body Tunneling Field-Effect Transistor (TFET) with Gate/Source Overlap 364
Hyun Woo Kim, Min-Chul Sun, Sang Wan Kim and Byung-Gook Park

Epi Defined (ED) FinFET: An Alternate Device Architecture for High Mobility Ge Channel Integration in PMOSFET 367
S. Mittal, S. Gupta, A. Nainani, M. C. Abraham, K. Schuegraf, S. Lodha and U. Ganguly

FinFET Device Capacitances: Impact of Input Transition Time and Output Load 371
Archana Pandey, Swati Raycha, S. Maheshwaram, S. K. Manhas, S. Dasgupta, A. K. Saxena and Bulusu Anand

Droplet Based Lab-On-Chip Microfluidic Microsystems Integrated Nanostructured Surfaces for High Sensitive Mass Spectrometry Analysis 374
Guillaume Perry, Florian Lapierre, Yannick Coffinier, Vincent Thomy, Rabah Boukherroub, CongXiang Lu, Siu Hon Tsang, Beng Kang Tay and Philippe Coquet

Impact of Metal Contact on the Performance of Cupric Oxide Based Thin Film Solar Cells 378
S. Masudy-Panah, V. Kumar, C. C. Tan, K. Radhakrishnan, D. Z. Chi and G. K. Dalapati

Through-Silicon Via Fabrication with Pulse-Reverse Electroplating for High Density Nanoelectronics 381
Nay Lin and Jianmin Miao

Session 7B	Electron Emission
Chairs	Tseng Tseung-Yuen and Milne William

Designing a Display Unit to Drive the 8x8 LED Dot-Matrix Displays 385
Wan-Fu Huang

Physical/Process Parameter Dependence of Gate Capacitance and Ballistic Performance of $InAs_y Sb_{1-y}$ Quantum Well Field Effect Transistors 389
Iftikhar Ahmad Niaz, Md. Hasibul Alam, Imtiaz Ahmed, Zubair Al Azim, Nadim Chowdhury and Quazi Deen Mohd Khosru

Session 7C	Nanophotonic Materials and Structures II
Chair	Qi Jiang

Metal-Polymer Nanocomposite Films with Ordered Vertically-Aligned Metal Cylinders for Optical Application 393
Linda Y. L. Wu, B. Leng, W. He, A. Bisht and C. C. Wong

Ag-Doped SiO_2/TiO_2 Hybrid Optical Sensitive Thin Films with Visible Absorption Enhancement for Diffractive Optical Element Application 397
P. Junlabhut, S. Boonruang and W. Pecharapa

Study of Optical Radiation Efficiency of Nanoparticles 401
Hasan Sarwar and Md. Mydul Islam

Nano-Needle Pressure Sensor Integrated with Printed Organic Transistors 405
Jiseok Kim, Tse Nga Ng and Woo Soo Kim

Resonant Cavity Far Infrared Photo-detector based on Self-Assembled InAs/GaAs Quantum Dots 407
C. M. S. Negi, Dharmendra Kumar, Saral K. Gupta and Jitendra Kumar

Horizontally Suspended Carbon Nanotube Bundles Patterned on Silicon Trench Sidewalls 514
Jingyu Lu and Jianmin Miao

Session 7D	Spintronics
Chairs	Feng Yuanping and Chen Weimin

Poster 3	Nanophotonics (NP)

Enhanced Conversion Efficiency of Cu(In,Ga)Se$_2$ Solar Cells with Periodic Nanosphere Arrays 411
Ming-Yang Hsieh, Shou-Yi Kuo, Fang-I Lai, Hau-Vei Han, Tsung-Yeh Chuang and Hao-Chung Kuo

Sensitivity Improved Surface Plasmon Resonance Sensor Based on Graphene and Gold Nanorods 414
Shuwen Zeng, Mathieu Sylvain Bergont, Aurelien Olivier, Xuan-Quyen Dinh, Xia Yu and Ken-Tye Yong

Thickness Effect of Sputtered ZnO Seed Layer on the Electrical Properties of Li-Doped ZnO Nanorods and Application on the UV Photodetector 417
C. Y. Kung, S. L. Young, M. C. Kao, H. Z. Chen, J. H. Lin, H. H. Lin, Lance Horng and Y. T. Shih

Selective Enhancement of Red Upconvesion Luminescence of Er^{3+} by Doping with Mn^{2+} Ions 421
En-Hai Song, Fen Xiao, Shi Ye and Qin-Yuan Zhang

Indium Phosphide (InP) Colloidal Quantum Dot Based Light-Emitting Diodes Designed on Flexible PEN Substrate 425
Yohan Kim, Tonino Greco, Christian Ippen, Armin Wedel and Jiwan Kim

Modeling of the Nipip HIT Structure with the Hole Thermionic Emission Mechanism 428
H. T. Hsiao, T. Y. Kuo and C. H. Lin

Design Guidelines for (111) Si Inclined Nanohole Arrays in Thin Film Solar Cells 431
Lei Hong, Rusli, Xincai Wang, Hongyu Zheng, Hao Wang and HongYu Yu

Design Guidelines for Periodic Nanowire Arrays in Thin-Film Silicon/Organic Hybrid Solar Cell 434
Hao Wang, Lei Hong, Lining He and Rusli

Electronic Structure of Ge/Si$_x$Sn$_y$Ge$_{1-x-y}$ Quantum Dots 437
J. Chen, W. J. Fan, D. H. Zhang, Q. Xu and X. W. Zhang

Copper Oxide Based Low Cost Thin Film Solar Cells 443
Vinay. Kumar, S. Masudy-Panah, C. C. Tan, T. K. S. Wong, D. Z. Chi and G. K. Dalapati

Manipulating Surface Plasmon Polaritons on the Meta-Surface 446
Zhengji Xu, Dao Hua Zhang, Tao Li, Changchun Yan, Dongdong Li, Yueke Wang and Fei Qin

Beam Focusing by an Anisotropic Metal-Dielectric Multilayer Structure 449
Dongdong Li, Dao Hua Zhang, Yueke Wang, Zhengji Xu, Jun Wang, Fei Qin and Wenjuan Wang

Performance Improvement of Triple-Junctions GaAs-Based Solar Cell Using SiO$_2$-Nanopillars/SiO$_2$/TiO$_2$ Graded-Index Anti-Reflection Coating 452
Jheng-Jie Liu, Wen-Jeng Ho, Jhih-Kai Syu, Yi-Yu Lee, Ching-Fuh Lin and Hung-Pin Shiao

Tunable Subwavelength Terahertz Plasmonic Stub Waveguide Filters 455
Jin Tao, Qi Jie Wang, Bin Hu, Xiao Yong He and Ying Zhang

Poster 4	Nanoscience (NS)

Luminescence Properties of Cerium Doped Silicon Nitride with MgO Additive 459
Y. Y. Ma, F. Xiao, S. Ye and Q. Y. Zhang

Field Effect Transport Properties of Electrochemically Prepared Graphene Quantum Dots 463
Hemen Kalita, V. Harikrishnan and M. Aslam

In-Situ Observation of Temperature Distribution of Microheaters Using Near-Infrared CCD Imaging System 466
Takanari Saito, Weichih Lin, Ibuki Atsumo and Jun-ichi Shirakashi

Characterization of a-Se p-n Junction Fabricated Using Electrolysis in NaCl *aq* 470
M. Onishi, K. Komiyama, K. Takeno, I. Saito, W. Miyazaki, T. Masuzawa, A. T. T. Koh, D. H. C. Chua, T. Yamada, N. Sano and K. Okano

Investigation of Work Function and Surface Energy of Aluminum: An AB-Initio Study 473
Shuguang Cheng, Tianqi Deng, Feifei He, Shuai Zhang, Haibin Su and Cherming Tan

Multicolored Cell Imaging with Bioconjugated Fluorescent Quantum Dots 476
YuchengWang, Rui Hu, Guimiao Lin and Ken-Tye Yong

Nano-IGZO Layer for EGFET in pH Sensing Characteristics 480
Chia-Ming Yang, Jer-Chyi Wang, Tzu-Wen Chiang, Yi-Ting Lin, Teng-Wei Juan, Tsung-Cheng Chen, Ming-Yang Shih, Cheng-En Lue and Chao-Sung Lai

The Side Effects on N-Type FinFET Devices 483
Hsin-Chia Yang, Chong-Kuan Du, Wen-Shiang Liao, Jing-Zong Jhang, Yi-Hong Lee, Tsao-Yeh Chen, Ko-Fan Liao, Mu-Chun Wang, Sungching Chi and Shea-Jue Wang

Next Promising P-Type FinFET Devices without or with Cobalt-Silicide Applied to the Gate 486
Hsin-Chia Yang, Guo-Wei Wu, Wen-Shiang Liao, Wei-Yen Peng, Sung-Ching Chi, Mu-Chun Wang and Shea-Jue Wang

The Improvement of MOSFET Electric Characteristics Through Strain Engineering by Refilled SiGe as Source and Drain 489
Hsin-Chia Yang, Jie-Min Yang, Wen-Shiang Liao, Mu-Chun Wang, Shea-Jue Wang, Chun-Wei Lian, Chao-Wang Li and Chong-Kuan Du

Study of Surfactant Modified MWNT/Polyimide Composites by In-Situ Polymerization 492
Hung-Han Ko, Yao-Yi Cheng and Ching-Wei Wang

A Novel InGaAs Photodiode Fabrication and Its Application 495
Chii-Wen Chen, Wen-Chin Lee, Meng-Chyi Wu, Chong-Long Ho, Chia-Hao Chuang and Dong-Ying Hsieh

Study on the Characterizations and Applications of the pH-Sensor with GZO/Glass Extended-Gate FET 498
Jung-Lung Chiang and Chia-Yu Kuo

Characteristics of Al-Doped ZnO Nanorods Synthesized by the Hydrothermal Process at Low Temperature 502
Jung-Lung Chiang and Sui-Chu Tsai

Inspecting the Effects of Post-Annealing on ZnO Nanorods by Optical Second Harmonic Generation 505
Chung-Wei Liu, Shoou-Jinn Chang, Chun-Chu Liu, Ruei-Jie Huang, Yan-Shen Lin, Min-Chia Su, Peng-Han Wang and Kuang-Yao Lo

Development of Networked Electronic Nose Based on Multi-Walled Carbon Nanotubes/Polymer Composite Gas Sensor Array 508
Mario Lutz, Chatchawal Wongchoosuk, Adisorn Tuantranont, Supab Choopun, Pisith Singjai and Teerakiat Kerdcharoen

Current Matched Improving of Triple-Junctions GaAs-Based Solar Cell Using Periodic Patterns Incorporated with Indium Nanoparticle Plasmonics 511
Yi-Yu Lee, Wen-Jeng Ho, Cheng-Ming Yu, Jheng-Jie Liu, Ching-Fuh Lin and Hung-Pin Shiao

Author Index

A

Abbaspour, Elhameh
Abraham, M. C.
Ahmed, Imtiaz
Akbarzadeh, Alireza
Akimoto, Shunsuke
Alam, Ahm Zahirul
Alam, Md. Hasibul
Alam, Md. Nur Kutubul
An, Jianing
Anand, Bulusu
Ando, Masazumi
Ang, K. S
Armstrong, G. Alastair
Arscott, S.
Aslam, M.
Atsumo, Ibuki
Azim, Zubair Al

B

Bai, Ping
Baohe, Yang
Basaev, A.
Baumann, Reinhard R.
Belfiore, A.
Bergont, Mathieu Sylvain
Billah, Md. Masum
Bisht, A.
Bittencourt, C.
Bittencourt, Carla
Boonruang, S.
Borkar, A.
Boukherroub, Rabah
Bouts, N.
Brun, Christophe
Buffière, M.
Buffière, Marie
Bundy, Mark

C

Cai, Tianhao
Chang, Chi-Feng
Chang, Chia-Tsung
Chang, Chu-Fa
Chang, M. C.
Chang, Shoou-Jinn
Chang, Wen-Yuan
Chen, Chih-Chi
Chen, Chii-Wen
Chen, Chun-Yu

Chen, H. C.
Chen, H. Z.
Chen, Hone-Zern
Chen, Hsin-Chu
Chen, J.
Chen, Jian
Chen, Kuo-Ju
Chen, S. H.
Chen, S. Y.
Chen, Shuang-Yuan
Chen, T. P.
Chen, Tsao-Yeh
Chen, Tsung-Cheng
Chen, Tzu-Chiang
Chen, Z. X.
Cheng, Huang-Chung
Cheng, Shuguang
Cheng, Yao-Yi
Chi, D. Z.
Chi, Sung-Ching
Chi, Sungching
Chiang, Jung-Lung
Chiang, Meng-Hsueh
Chiang, Tzu-Wen
Chiang, Wen-Ping

Chien, Shih-Hsuan
Chin, S. H.
Chiu, Fu-Chien
Chiu, Wei-Chih
Choi, Chang-Hwan
Chongsri, K.
Choopun, Supab
Chou, Chien
Chowdhury, Nadim
Chowdhury, Zahidur R.
Chua, D. H. C.
Chuang, Chia-Hao
Chuang, Tsung-Yeh
Chung, Che-Sheng
Coffinier, Yannick
Coquet, Philippe
Czyszanowski, Toma

D

Dai, N.
Dalapati, G. K.
Dasgupta, S.
Dems, Maciej
Deng, H. Y.

Deng, Tianqi
Dinh, Xuan-Quyen
Dominique, Baillargeat
Dow, Ali B. Alamin
Du, Chong-Kuan
Du, Ke

E
Elumalai, Naveen Kumar

F
Fan, W. J.
Fan, Yu-Siang
Fang, Wang
Fang, Z.
Farhana, Soheli

G
Galperin, V.
Ganguly, U.
Gardner, J. W.
Gaudet, M.
Gautron, E.
Gebski, Marcin
Ghosh, Dipankar
Gilda, N. A.
Giri, Lily
Greco, Tonino
Guha, P. K.
Gupta, Saral K.
Guisbiers, G.
Guo, J. H.
Guo, T. X.
Guo, Yan
Gupta, S.
Guttmann, P.

H
Han, Gu-Chang
Han, H. V.
Han, Hau-Vei
Han, Kyungsup
Harikrishnan, V.
Hashem, Islam E.
Hashemipour, Omid
Hayashi, Yasuhiko
He, Feifei
He, Lining
He, W.
He, Xiao Yong
Ho, Chong-Long

Ho, Kuan-I
Ho, Wen-Jeng
Hong, Lei
Horng, Lance
Horng, Ray-Hua
Hsiao, An-Ching
Hsiao, Fu-Li
Hsiao, H. T.
Hsieh, Dong-Ying
Hsieh, Ming-Yang
Hsu, Chang-Lung
Hsu, Chia-Ming
Hsu, H. W.
Hsu, Hong-Wen
Hsu, Ying-Hui
Hu, Bin
Hu, G. J.
Hu, Rui
Hu, S. H.
Hua, Zhang Dao
Huang, Chi-Hsien
Huang, H. S.
Huang, Heng-Sheng
Huang, Hong-Huei
Huang, Po-Wei
Huang, Ruei-Jie
Huang, Sheng-Fu
Huang, Wan-Fu
Huang, Yu-Chih
Hubbard, Clifford
Hui, Lu
Hung, Yi-Ting
Huskens, Jurriaan
Hwang, Chuan-Chou

I
Ippen, Christian
Islam, Md. Mydul
Islam, Md. Rafiqul
Islam, Muhammad Shaffatul
Ito, Mitsuki

J
Jackson, Aaron
Jadhav, Sameer
Jan, Jeng-Shiung
Jang, Sheng-Lyang
Jayaraj, Vibha
Jebalin, Binola K
Jhang, Jing-Zong
Ji, Liang-Ru
Jiang, Qi
Jing, Zhang
Jinshi, Zhao

Joy, Doreen
Juan, Chuan-Ping
Juan, Teng-Wei
Jun, Wei
Jung, Sunghun
Junlabhut, P.

K
Kahattha, C.
Kai, Liu
Kailiang, Zhang
Kalita, Golap
Kalita, Hemen
Kamath, A.
Kang, Donghoon
Kao, M. C.
Kao, Ming-Cheng
Karimi, Behzad Jazizadeh
Karkare, P. Bafna P.
Kee, Jack Sheng
Kerdcharoen, Teerakiat
Khamnaulthong, Nattaporn
Khan, Raisuddin
Khan, Sheroz
Kherani, Nazir P.
Khosru, Quazi Deen Mohd
Kim, Beomjoon
Kim, Do-Bin
Kim, Garam
Kim, Hyun Woo
Kim, Hyungjin
Kim, Janghyun
Kim, Jiseok
Kim, Jiwan
Kim, Really
Kim, Sang Wan
Kim, Sung
Kumar, Dharmendra
Kumar, Jitendra
Kumar, V.
Kim, Sungjun
Kim, Wandong
Kim, Woo Soo
Kim, Yohan
Kim, Yoon
Klein, Mona J. K.
Ko, Hung-Han
Koh, A. T. T.
Komiyama, K.
Kong, J.
Konstantinidis, S.
Konstantinidis, Stephanos
Kranti, Abhinav
Krier, A.

Krishna, S. V. Hari
Kuemin, Cyrill
Kumar, N. Mohan
Kumar, P. Vijay
Kumar, Vinay.
Kung, C. Y.
Kuo, Chia-Yu
Kuo, H. C.
Kuo, Hao-Chung
Kuo, Hung-Fei
Kuo, Shou-Yi
Kuo, T. Y.
Kwong, D. L.
Kwong, Dim-Lee
Kyaw, A. K. K.

L
Labunov, V.
Lai, Chao-Sung
Lai, Fang-I
Lam, Kai-Tak
Lan, Dai
Lapierre, Florian
Lee, Andrew
Lee, Cheng-Kuei
Lee, Chengkuo
Lee, Jong-Ho
Lee, Jung Han
Lee, Kean Boon
Lee, Sang-Ho
Lee, Wen-Chin
Lee, Yi-Hong
Lee, Yi-Yu
Lei, Po-Hsun
Leng, B.
Leong, Teo Kie
Leu, Ching-Chich
Li, Chao-Wang
Li, Dongdong
Li, Er Ping
Li, Hong
Li, K. C.
Li, Lain-Jong
Li, Peng-Wei
Li, Tao
Li, Xiaohui
Lian, Chun-Wei
Liang, Gengchiau
Liao, Chin-Hsiang
Liao, Jia-Hong
Liao, Ko-Fan
Liao, Wen-Shiang
Lim, Henry

Limsuwan, Pichet
Lin, C. C.
Lin, C. H.
Lin, Chia-Te
Lin, Chien-Chung
Lin, Chih-Ting
Lin, Ching-Fuh
Lin, Guimiao
Lin, H.
Lin, H. H.
Lin, Huan-Min
Lin, Hung-I
Lin, J. H.
Lin, Jun-Han
Lin, Jyi-Tsong
Lin, K. C.
Lin, Nay
Lin, Shih-Yen
Lin, Shiuan-Huei
Lin, T. T.
Lin, Weichih
Lin, Wen-Jen
Lin, Yan-Shen
Lin, Yi-Ting
Liu, C. H.
Liu, C. Y.
Liu, Chuan-Hsi
Liu, Chun-Chu
Liu, Chung-Wei
Liu, Fu-Ken
Liu, Huiyun
Liu, Jheng-Jie
Liu, Keng-Ming
Liu, S. W.
Lo, G. Q.
Lo, Guo-Qiang
Lo, Kuang-Yao
Lo, Shih-Shou
Lodha, S.
Lorwongtragool, Panida
Lou, Liang
Lu, CongXiang
Lu, Hui
Lu, Jingyu
Lue, Cheng-En
Lun, Zhao
Lutz, Mario
Lv, Y. F.

M
Ma, Jianguo
Ma, Y. Y.
Ma, Zhenqiang
Maheshwaram, S.

Maikap, S.
Mandapati, R.
Manhas, S. K.
Manoj, K.
Masahara, Meishoku
Masudy-Panah, S.
Masuzawa, T.
Mehrabi, Shima
Mel, A. A. El
Mel, Abdel-Aziz El
Meng, Q. Q.
Miao, Jianmin
Migita, Shinji
Mittal, S.
Miyazaki, W.
Morita, Yukinori
Motakabber, Sma
Mukherjee, Shaibal

N
Naieni, Ali Kashefian
Nainani, A.
Nataraj, Latha
Nathawat, Rashi
Navi, Keivan
Negi, C. M. S.
Ng, Tse Nga
Ngo, Duc-The
Ni, Songbo
Niaz, Iftikhar Ahmad
Nirmal, D.
Nojeh, Alireza

O
Oh, Jeong-Hoon
Okano, K.
Olivier, Aurelien
Onishi, M.
Ota, Hiroyuki
Ou, C. R.

P
Palaniapan, Moorthi
Pan, Liyang
Pandey, Archana
Pandey, Saurabh Kumar
Pandey, Sushil Kumar
Parihar, Mukta Singh
Park, Byung-Gook
Park, Euyhwan
Park, Mi Kyoung
Park, Se Hwan
Park, Seulki
Patel, M.

Patil, Ganesh C.
Pecharapa, W.
Peng, Min-Ru
Peng, Wei-Yen
Perry, Guillaume
Popov, C.
Pott, Vincent
Prakash, A.
Prudnikava, A.

Q
Qian, You
Qijie, Wang
Qin, Fei
Qin, Guoxuan
Qiu, F.
Qiu, Jin-Jun
Qureshi, S.

R
Radhakrishnan, K.
Rafat, Nadia H.
Ramakrishna, Seeram
Rao, V. Ramgopal
Ray, P.
Ray, S. K.
Raycha, Swati
Rey, Antje
Rianasari, Ina
Riel, Heike
Rostami, Melika
Rusli
Ryoo, Kyung-Chang

S
Saito, I.
Saito, Takanari
Samukawa, Seiji
Sano, N.
Santra, S.
Sarvari, Reza
Sarwar, Hasan
Sasase, Masato
Saxena, A. K.
Schmid, U.
Schuegraf, K.
Seeds, Alwyn
Selvaraj, Susai Lawrence
Seo, Joo Yun
Shaman, Y.
Shen, Quan-Hao
Shiao, Hung-Pin
Shih, Min-Hsiung

Shih, Ming-Yang
Shih, Y. T.
Shin, Hyungcheol
Shirakashi, Jun-ichi
Shuhua, Wei
Shulitski, B.
Siangchaew, Krisda
Singjai, Pisith
Snyders, R.
Snyders, Rony
Soliman, Ezzeldin A.
Son, Joong-Kon
Song, En-Hai
Song, Peiyi
Sowade, Enrico
Spencer, Nicholas D.
Sridhar, Arthi
Srinivasan, V. S. S.
Su, Ching-Yuan
Su, Haibin
Su, Min-Chia
Suda, Ryutaro
Sun
Sun, Haifeng
Sun, Min-Chul
Sun, X. W.
Suphankij, S.
Syu, Jhih-Kai

T
Takeno, K.
Tan, C. C.
Tan, C. M.
Tan, Cherming
Tan, Dunlin
Tan, E.
Tanemura, Masaki
Tang, Mingchu
Tang, Shiang-Feng
Tao, Jin
Tay, Beng Kang
Techitdheera, W.
Teo, Edwin Hang Tong
Teo, K. L.
Teo, Kie-Leong
Tessier, P. Y.
Tessier, Pierre-Yves
Thaweesaeng, N.
Thomy, Vincent
Tng, Danny Jian Hang
Tong, Weihua
Tsai, Cheng-Yu
Tsai, Hsin-Han
Tsai, J. M.

Tsai, Minglin Julius
Tsai, Sui-Chu
Tsai, Y. L.
Tsang, Siu Hon
Tseng, Ampere A.
Tsui, Bing-Yue
Tuantranont, Adisorn

U
Udrea, F.
Umeno, Masayoshi

V
Vijila, Chellappan
Vinchurkar, M. S.
Vittayakorn, N.

W
Wakita, Koichi
Wang, Chao Hsun
Wang, Ching-Wei
Wang, H.
Wang, Hao
Wang, Hsun-Wen
Wang, I-Shun
Wang, Jer-Chyi
Wang, Jun
Wang, Jyh-Liang
Wang, M. C.
Wang, Mu-Chun
Wang, Nan
Wang, Peng-Han
Wang, Qi Jie
Wang, Qijie
Wang, Shea-Jue
Wang, Weizhu
Wang, Wen-Hong
Wang, Wenjuan
Wang, X. P.
Wang, Xincai
Wang, X.
Wang, Y.
Wang, Yishan
Wang, Yonggang
Wang, Yu-Pu
Wang, Yueke
Wangikar, Pramod. P.
Watanabe, Mitsuru
Watthanawisuth, Natthapol
Wedel, Armin
Weng, B. B.
Wolf, Heiko

Wong, C. C.
Wong, J. I.
Wong, T. K. S.
Wongchoosuk, Chatchawal
Wu, Cheng-Sheng
Wu, Guo-Wei
Wu, Lin
Wu, Linda Y. L.
Wu, Meng-Chyi
Wu, Ying-Huei
Wuu, Dong-Sing

X
Xiao, F.
Xiao, Fen
Xu, Jun
Xu, Naiyun
Xu, Q.
Xu, Wei
Xu, Zhengji

Y
Ya, Xia
Yaakob, Yazid
Yamada, T.
Yan, Changchun
Yang, Chia-Hao
Yang, Chia-Ming
Yang, H. Y.
Yang, Hsin-Chia
Yang, Jie-Min
Yao, Yung-Chi
Yap, Chin Chong
Yap, Qi-Jia
Ye, S.
Ye, Sheng-Jhan
Ye, Shi
Ye, Yu-Ren
Yeh, Jen-Hau
Yen, Yu-Ren
Yeo, Yee-Chia
Yin, M.
Yong, Ken-Tye
Yoo, Sungwon
Yoon, Yong-Jin
Young, S. L.
Young, San-Lin
Yu, Cheng-Ming
Yu, HongYu
Yu, P.
Yu, Pei-Chen
Yu, Xia
Yuan, Fang
Yuan, Li

YuchengWang
Yusop, Mohd Zamri

Z
Zeng, Shuwen
Zhang, D. H.
Zhang, Dao Hua
Zhang, Q. Y.
Zhang, Qin-Yuan
Zhang, Shuai
Zhang, X. W.
Zhang, Y.

Zhang, Ying
Zhang, Yongzhe
Zhang, Zhigang
Zhao, J. L.
Zhao, Wei
Zhao, Z.
Zheng, Hongyu
Zheng, K.
Zheng, Lianxi
Zhou, Xing
Zhu, Shiyang
Zhuang, Q. D.

ZnO Nanowires Lift-off From Silicon Substrate Embedded in Flexible Films

Ray-Hua Horng[1], Hung-I Lin[2], and Dong-Sing Wuu[2]

[1]Institute Graduate of Precision Engineering, National Chung Hsing University
Taichung 40227, Taiwan, Republic of China
Email: huahorng@dragon.nchu.edu.tw

[2]Department of Materials Science and Engineering, National Chung Hsing University
Taichung 40227, Taiwan, Republic of China

Abstract

A novel lifting-off method of ZnO nanowires from Si substrate and embedded in flexible films have been proposed in this study. Compared with peeling-off method by polydimethyl- siloxane (PDMS), the embedded ZnO nanowires may suffer from external forces to change the dimension. The shape of ZnO nanowires remained almost unchanged after using the novel lifting-off process. Flexible films served as the secondary and flexible substrate after ZnO nanowires transferring from the Si substrate. The lift-off fabrication is a candidate for Si substrate recycling usage and for large area fabrication.

(Keywords: ZnO, nanowires, PDMS, lift-off and flexible film)

Introduction

ZnO is a semiconductor material with large energy band-gap of 3.3 eV and exciton binding energy of 60 meV. Up to now, ZnO nanostructures have attracted much attention due to the potential application of sensors, nanogenerators, photo-detectors, actuators, solar cells, and hybrid cells... etc [1-4]. ZnO nanostructures grown on different substrates like metal [5], sapphire [6-8], glass [9], flexible films [10], and paper [11,12] have been reported. Using hydrothermal method of low growing temperature below 100 ℃ has been widely used to grow ZnO nanostructures [13]. The aqueous solution consists of an equal molar ratio of Zincnitratehexahydrate $(Zn(NO)_3 \cdot 6H_2O)$ solution mixed with hexamethylentetramine (HMTA) solution, which is used as the nutrient solution to synthesize ZnO nanowires.

Previous report of peeling-off ZnO nanowires from the substrate used a Polydimethylsiloxane (PDMS) solution spin-coating on ZnO nanowires [14]. The PDMS solution was made by mixing the base and curing agent (10:1 w/w) (Sylgard 184). After spin-coating PMDS on ZnO nanowires, the whole structure was heated to cure PDMS solution such that forming a matrix with ZnO nanowires. After that, using a mechanical force to peel-off PDMS layer embedded with ZnO nanowires.

In this study, we introduce a method to lift-off ZnO nanowires embedded in flexible films. Previous peeling-off ZnO nanowires "only" used PDMS as the flexible film because PDMS is easy to peel-off embedded ZnO nanowires from the substrate. However, in the method, many flexible films can be chosen as lifting-off method to broaden the use of ZnO nanowires applications without shape changed. We chose the epoxy resin as the flexible film owing to its flexible, lightweight, and cheap characteristics.

Epoxy is blended with the basic components and the curing agents, forming the epoxide functional group. Epoxy resin contains epoxide functional group of reactive prepolymers and polymers. Generally, epoxy resin composed of epichlorohydrin and bisphenol-A, or its equivalent derivative. Epoxy resin is a thermosetting polymer; that is , it has been widely used in industry owing to its chemical resistance, toughness, and strong adhesive properties. Epoxy resin can be used in electronics industry such as organic light-emitting-diode (OLED), solar cell, light emitting diode (LED), and printed circuit boards [15-19].

Experimental Procedure

Si (100) wafer was chosen as the ordinary substrate. A layer of polymethylglutarimide (PMGI) photoresist was spin-coated on the Si substrate. Depositing an Au/Ti (50/20 nm) layer by thermal evaporation on the surface of PMGI layer. After annealing the structure at 300 °C for 1 h, it was floated on the nutrient solution containing equal molar of 30 mM $Zn(NO)_3 \cdot 6H_2O$ and HMTA at 80 °C for 24 h by hydrothermal method. A layer of epoxy resin (Everwide™, JB646-2) was spin-coated on the ZnO nanowires at the speed of 2000 rpm for 40 s. Then, the structure was baked in the oven at 150 °C for 1 h to cure the coated epoxy resin. The structure was immersed into the developer to lift-off the ZnO nanowires from Si substrate. The morphology of ZnO nanowires embedded in the epoxy resin were characterized by using scanning electron microscope images (SEM) and X-ray diffraction (XRD).

Results and Discussion

Fig. 1 is the lifting-off process flow of separating ZnO nanowires from Si substrate and embedded in the epoxy resin. The SEM image of ZnO nanowires grown on Au/Ti layer is shown in Fig. 2. The diameter of ZnO nanowires was 100-250 nm and the height is 2 µm. The thickness of PMGI layer was 2.3 µm on the Si substrate. ZnO nanowires were grown on 50 nm Au, while 20 nm Ti was deposited on the PMGI served as an adhesion layer. PMGI is a photoresist that can endure as high temperature as 300 °C during annealing process without significant morphology change, which was purchased from MicroChem Corp. ZnO nanowires were grown dense and well-aligned. Moreover, ZnO nanowires showed hexagonal shape of cross section.

978-1-4673-4840-9/13 $31.00 © 2013 IEEE

The SEM image of cross-sectional image with 15° tilted of as-grown ZnO nanowires grown on Au surface shown in Fig. 3. XRD spectrum shows a strong peak at 34.4°(2θ) corresponding to (002) c-axis preferred orientation of the wurtzite structure of ZnO nanowires.

Fig. 5 (a) is the cross-sectional view of ZnO nanowires embedded in the epoxy resin after lifting-off process and Fig. 5 (b) is its view with a 15° tilt. PMGI photoresist was considered as a sacrificial layer. The epoxy resin was regarded as the secondary substrate and flexible film after ZnO nanowires lifting-off from the Si substrate. The thickness of the epoxy resin was 115 μm. Furthermore, the epoxy resin demonstrates excellent alkali resistance due to almost without any morphology change. After lifting-off process, ZnO nanowires embedded in the epoxy resin showed the morphology did not change. The epoxy resin forming a matrix with ZnO nanowires to prevent the physical contact.

Conclusions

We have introduced a method to lift-off ZnO nanowires and embedded in flexible films. Compared with PDMS peeling-off method, the dimension of ZnO nanowires remained almost unchanged after using the lifting-off method. The epoxy resin can be regarded as the secondary and flexible film after ZnO nanowires lifting-off from the Si substrate. The method is easy to lift-off embedded ZnO nanowires from the Si substrate. Moreover, the method is suitable for large area fabrication.

Acknowledgment

The authors would like to thank the National Science Council of the Republic of China, Taiwan for financially supporting this research under contract NSC-100-2221-E-005-092-MY3.

References

[1] Y. F. Hu, et al.,"Self-Powered System with Wireless Data Transmission," *Nano Lett.* 2011, 11, pp. 2572-2577.

[2] M. B. Lee, et al.,"Self-powered environmental sensor system driven by nanogenerators," *Energy Environ. Sci.* 2011, 4, pp. 3359-3363.

[3] Y. Zhang, et al., "Fundamental Theory of Piezotronics," *Adv. Mater.* 2011, 23, pp. 3004-3013.

[4] Y. F. Hu, et al., "High-Output Nanogenerator by Rational Unipolar Assembly of Conical Nanowires and Its Application for Driving a Small Liquid Crystal Display," *Nano Lett.* 2010, 10, pp. 5025-5031.

[5] W. T. Chang, et al., "Wind-power generators based on ZnO piezoelectric thin films on stainless steel substrates," *Curr. Appl.Phys.* 2011, 11, 1, pp. 333-338.

[6] C. C. Wu, et al., " Effects of Growth Conditions on Structural Properties of ZnO Nanostructures on Sapphire Substrate by Metal–Organic Chemical Vapor Deposition," *Nanoscale Res. Lett.* 2009, 4, pp. 377–384.

[7] C. C. Wu, et al., "Three-Step Growth of Well-Aligned ZnO Nanotube Arrays by Self-Catalyzed Metalorganic Chemical

Vapor Deposition Method," *Cryst. Growth Des.* 2009, 9 (10), pp. 4555-4561.

[8] C. C. Wu, et al., "Realization and manipulation of ZnO nanorod arrays on sapphire substrates using a catalyst-free metalorganic chemical vapor deposition technique," *J. Nanosci. Nanotechnol.* 2010, 10 (5), pp. 3001-3011.

[9] Z. R. Tian, et al., "Complex and oriented ZnO nanostructures," *Nat. Mater.* 2003, 2, pp. 821-826.

[10] C. T. Pan, et al., "Study of broad bandwidth vibrational energy harvesting system with optimum thickness of PET substrate," *Curr. Appl. Phys.* 2012, 12, pp. 684-696.

[11] A. Manekkathodi, et al., "Direct Growth of Aligned Zinc Oxide Nanorods on Paper Substrates for Low-Cost Flexible Electronics," *Adv. Mater.* 2010, 22, pp. 4059-4063.

[12] Y. Qiu, et al., "Flexible piezoelectric nanogenerators based on ZnO nanorods grown on common paper substrates," *Nanoscale,* 2012, 4, pp. 6568-6573.

[13] S. Baruah, J. Dutta, "Hydrothermal growth of ZnO nanostructures," *Sci. Technol. Adv. Mater.* 2009, 10, 013001.

[14] M. K. Kim, et al., "Tunable, Flexible Antireflection Layer of ZnO Nanowires Embedded in PDMS," *Langmuir,* 2010, 26 (10), pp. 7552–7554.

[15] T. N. Chen, et al., "Improvements of Permeation Barrier Coatings Using Encapsulated Parylene Interlayers for Flexible Electronic Applications," *Plasma Processes and Polymers,* 2007, 4, 2, pp. 180-185.

[16] R. H. Horng, et al., "Improved Conversion Efficiency of GaN InGaN Thin-Film Solar Cells," *IEEE Electron Device Letters,* 2009, 30, 7, pp. 724-726.

[17] R. H. Horng, et al., "Efficiency Improvement of GaN-Based LEDs with ITO Texturing Window Layers Using Natural Lithography," *IEEE J. Sel. Topics Quantum Electron.* 2006, 12, 6, pp. 1196 - 1201.

[18] S. C. Hsu, et al., "Power-enhanced ITO omni-directional reflective AlGaInP LEDs by two-dimensional wavelike surface texturing," *Semicond. Sci. Technol.,* 2008, 23 , 105013.

[19] K. C. Yung, et al., " Effect of AlN content on the performance of brominated epoxy resin for printed circuit board substrate," *J. Polym. Sci. Part B, 2007,* 45,13, pp. 1662–1674.

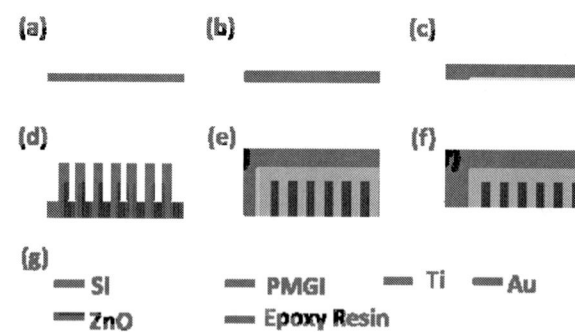

Fig. 1 Lifting-off process flow for separating ZnO nanowires from Si substrate and embedded in the epoxy resin. (a) Silicon substrate, (b) spin-coating a PMGI layer on the substrate, (c) a Au/Ti layer deposition by thermal coater, (d) ZnO NWs grown on Au surface by hydrothermal method, (e) a layer of epoxy resin spin-coated on the structure, (f) ZnO nanowires lifting-off in developer and (g) Figure legends.

Fig. 2 SEM image of as-grown ZnO nanowires with a 60° tilt.

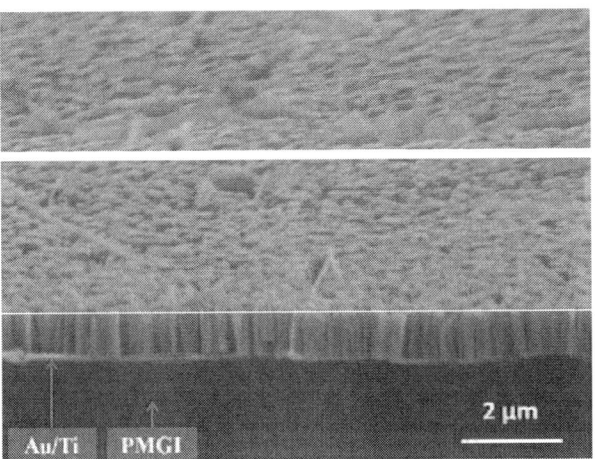

Fig. 3 Cross-sectional view with a 15° tilt of SEM image of as-grown ZnO nanowires grown on Au surface.

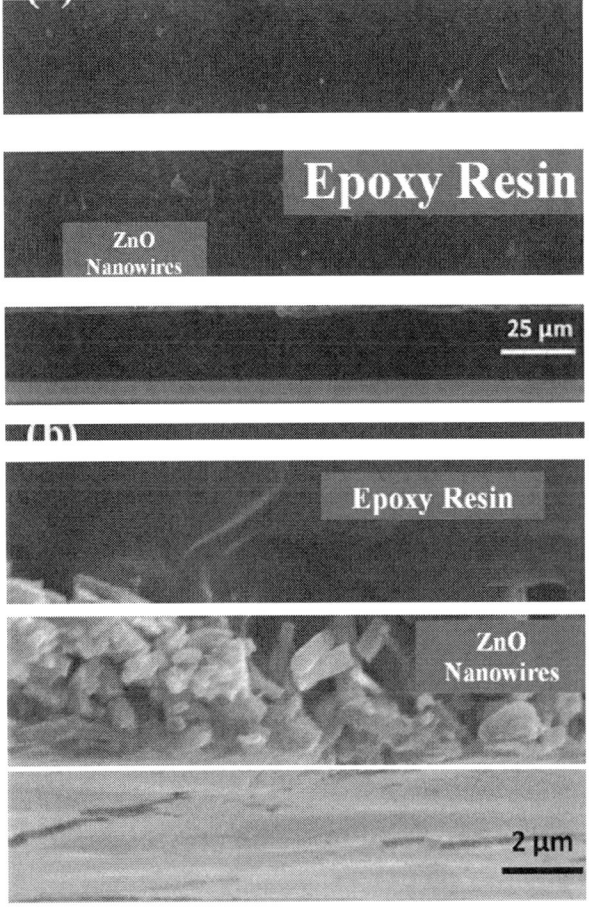

Fig. 5 (a) Cross-sectional view of SEM images of ZnO nanowires embedded in the epoxy resin after lifting-off from Si substrate. (b) Magnified view with a 15° tilt.

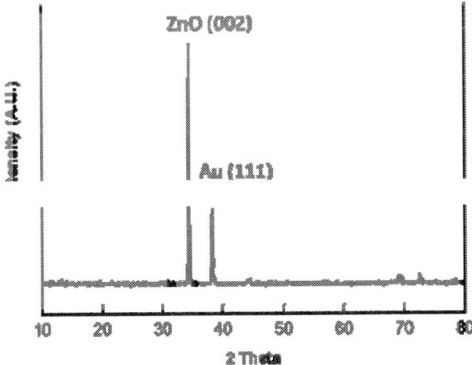

Fig. 4 XRD characterization of the as-grown ZnO nanowires. A peak at 34.4° (2θ) shows (002) c-axis preferred orientation of the wurtzite structure.

978-1-4673-4840-9/13 $31.00 © 2013 IEEE

The influence of titanium nitride barrier layer on the properties of CNT bundles

Chin Chong Yap[1,2], Dunlin Tan[1,2], Christophe Brun[2,3], Hong Li[1,2],
Edwin Hang Tong Teo[1,2,4], Baillargeat Dominique[2], and *Beng Kang Tay[1,2]

[1] School of Electrical and Electronics Engineering, Nanyang Technological University,
Block S1, 50 Nanyang Avenue, Singapore 639798

[2] CINTRA CNRS/NTU/THALES, UMI 3288, Research Techno Plaza, 50 Nanyang Drive,
Border X Block, Level 6, Singapore 637553

[3] XLIM, UMR 6172, Université de Limoges/CNRS, 123 Avenue Albert Thomas 87060 Limoges

[4] Temasek Laboratories@NTU, Research Techno Plaza , 50 Nanyang Drive,
Border X Block, Level 9, Singapore 637553

*Email: ebktay@ntu.edu.sg

Abstract

The use of carbon nanotubes (CNTs) for electrical interconnections is hinder by the possibility of growing CNT directly onto metallization. The introduction of barrier layer between catalyst and metallization is thus essential to permit the direct growth of vertically aligned CNT bundles using CVD approaches. As a result, the resultant CNT bundle resistivity is not only a function of the densities and quality of CNT growth, but is also affected by the thickness and resistivity of the barrier layer. CNT is growth on different thickness of TiN with 2 different underlying layers (Au and SiO_2). It was observed that the length of CNT grown on Au is independent of TiN thickness, whereas the height decreases for Au underlayers case. In both scenarios, both have well aligned CNTs growth when the thickness of TiN is < 90nm.

Introduction

Since the discovery of carbon nanotubes (CNTs) by Iijima and co-workers in 1991[1], many efforts have been made to study and understand the mechanism of CNTs growth. A much commonly accepted growth model was based on the vapor-liquid-solid model[2], and the development of in-situ observation of CNT growth bring the research community even closer to unlock the DNA of CNTs growth[3]. However, many of the earlier work reported growth of CNT on silicon oxide using the three most common transition metal catalysts (Fe, Ni and Co). One major challenge is that the semiconductor industry requires the CNTs to be on conductive under layers in order to maximize the use of CNTs for electrical interconnects application.

Two typical solutions to address the needs of semiconductor industry is either to grow CNTs directly on metal under layers, or transferred onto metal surface. Transferred of CNTs resulted in high contact resistance which is not favorable for electrical application[4]. While the growth of CNTs onto metal eliminate the issue of contact resistance, however differences in the growth characteristics existed between the growth of CNTs on silicon oxide and CNTs on metal due to the nature of metal-metal diffusion into each other[5]. A common solution to metal-metal diffusion is to introduce an intermediate layer, which is also known as barrier layer to limit the back diffusion of the metal catalyst [5, 6]. Since electrical application requires the CNT to be conductive, a preferred choice will be to make use of conductive barrier layers to resolve the issue of CNT growth.

An innovative and fast technique to create a combination library for investigation of catalyst with different under layers had been demonstrated by Ng et al. shows that CNT growth differ for different metal catalyst on different under layer[5]. In the same study, Ti was suggested to be the best metal barrier layer supporting CNT growth to provide the lowest contact resistance and CNTs' growth densities[5]. Besides, the differences in under layers affect growth, the quality of CNT growth was also found to be dependence on the growth temperature even though the same metal layers were used. This is due to the higher diffusion of metal catalyst at higher temperature[7]. Similarly, the grain size of the barrier layer was also crucial for the control of the diameter of CNT and the grain size are affected by the underlying metals on which the barrier layers sit on. Another important aspect to consider is the chemical formation of silicides, carbides and other compounds after the metal-metal diffusion during CNT growth. Nihei et al. used a combination of Ta and Ti barrier layer and found that ohmic contact can be readily achieved when TiC instead of TiO_2 is formed between the surface of CNT and metal [8].

With so many factors taking place at the same time, the optimum barrier layer for different metal is crucial for the success of CNT growth and practical applications. The main role of the barrier layer must be able to which the transport rate of both metal catalyst and substrate should be small, next in which the barrier layer should have thermal stability when the catalyst and substrate undergoes high temperature annealing and processing. The last condition will be that the barrier layer must adhere well to both catalyst as well as substrate and have low contact resistances[9].

In this study, Titanium nitride (TiN) is the preferred choice of barrier layer to be experimented as TiN is a triple-bond transition metal compound which has an inert nature, high mechanical hardness and high melting point[10]. It is also a commonly used as a diffusion barrier layer for Cu system. [11]. Our previous work also shows that Ni on TiN barrier layer produce CNT which have lower Id/Ig ratio which is essential for the quality of CNT growth [12]. Here, we will be comparing the growth of CNT on TiN on both Au as well as SiO_2 under layers to investigate the differences. The electrical properties are then evaluated using four point Kelvin probe method.

Experimental section

The experiments were carried out using SiO_2/Si starting wafers. First, 100 nm thin Au was deposited onto the wafers surface using e-beam evaporation to form the underlying conduction metallization layer. Standard lithography techniques were then carried out to pattern squares of 100 μm by 100 μm to create holes of barrier layer and catalyst deposition. TiN was formed by sputtering Ti in N_2 ambient at room temperature. The ratio of the N_2: Ar was 1:1. Finally, 15 nm Ni was deposited on top of the TiN layers and the wafers are subjected to lift off process to complete the patterning steps.

CNT growth was then carried out using plasma enhanced chemical vapor deposition (PECVD) mode as reported previously[13]. Briefly, the growth temperature was held at 650 °C with a gas flow rate of 1: 4 (NH_3:C_2H_2). The growth time was 10 min. Electrical characterization of CNT to CNT bumps resistivity was carried out using a kelvin four-point probe structures and the entire CNT to CNT bumps fabrication process are describe previously[13].

Results and Discussions

CNT were grown on two different sets of samples. The first was described in the fabrication methods. The second set of samples (without Au under layers) were used a control to compare the differences between the CNT growth with and without Au metallization layer. The thickness of the TiN was varied by controlled using deposition time, and measured using surface profiler. In this report, 3 min, 6 min, 10 min and 20 min deposition time was adopted yielding 30 nm, 60 nm, 90 nm and 160 nm of TiN respectively. Figure 1 shows the SEM images of CNT growth on SiO_2/Si substrate. It is observed that the growth was quite consistent in Figure 1a-c,

whereas in Figure 1d, thin strands of CNT samples can be seen crowding together.

Similarly in Figure 2, similar observations to the control samples were observed. In both figures, the occurrence of the thin CNT strands appears when the TiN thickness is ~ 160 nm. This demonstrates that the phenomenon is due to TiN thickness increased. Possible reasons include the roughing of the surface.

The length of the CNT growth was also compared for the 2 different substrate and they displayed differences in their growth height as shown in Figure 3. CNT growth on SiO_2 under layers have a constant height of ~13 um, whereas the CNT growth with Au under layers have CNT the longest CNT growth when thickness is less than 30 nm.

We also assess the DC performance of the CNT bumps by using a CNT to CNT fabrication approach. In figure 4, CNT growth on 90 nm TiN exhibits the lowest bump resistance. Possible reasons includes the CNT are shorter compare to other thickness. More work will be displayed to demonstrate the Raman signals of various samples.

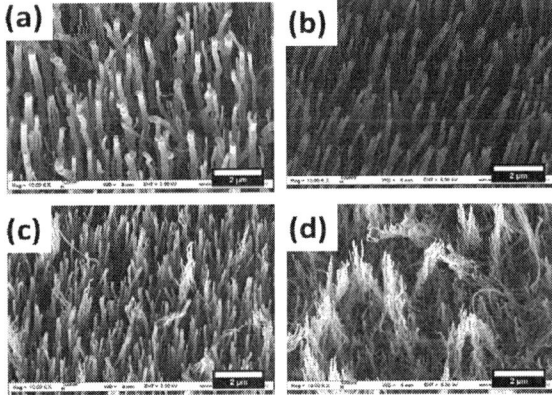

Fig. 1. SEM images of CNTs growth onto TiN/SiO_2 substrate. The TiN thickness was varied to observe the change. The thickness of TiN was (a) 30 nm, (b) 60 nm, (c) 90 nm, and (d) 160 nm TiN respectively. The white scale bar on the bottom right of each fiure represent 2 μm.

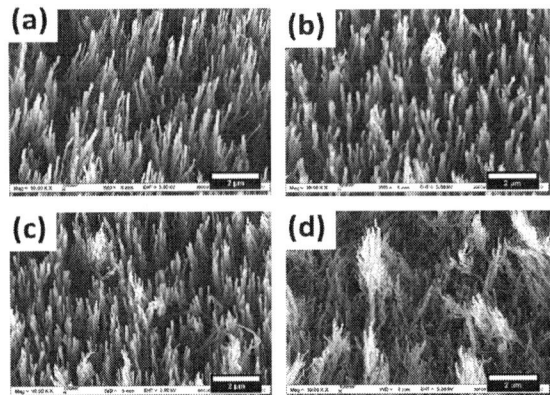

Fig. 2. SEM images of CNT growth onto TiN on 100 nm Au on SiO2 substrate. The TiN thickness was varied to observe the change. The thickness of TiN was (a) 30 nm, (b) 60 nm, (c) 90 nm, and (d) 160 nm respectively. The white scale bar on the bottom right of each fiure represent 2 μm

Conclusion

CNT can be grown directly onto metal substrate using bottom up approaches with the introduction of a thin barrier layer. However, for best electrical performance, this barrier layers need to be thin and conductivity in order not to reduce the advantage of the high conductivity of CNT. Our current results shows that TiN can be an effective barrier layer. However the resistivity we have achieved is still a few orders higher than Cu. We believe that by optimizing our barrier layers, we can achieve significant improvements in our results. Alternative approaches will be discussed in more details.

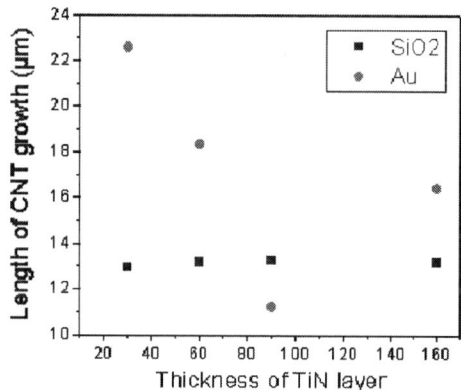

Fig. 3. Relationships between the height of CNTs growth with respect to the thickness of TiN. The black squares refer to CNTs growth on SiO2 surface, while the red circles are the CNTs on TiN/Au under layers substrates. It was observed that the height of the CNT without Au appears to be constant.

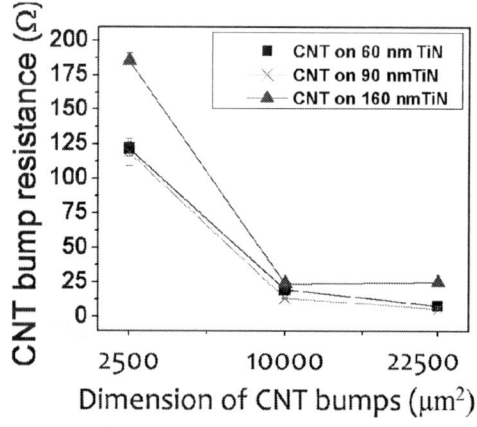

Fig. 4 kelvin probe measurements of the resistance of CNT to CNT bumps fabricated with different TiN barrier layer thickness

References

[1] Iijima, S., "Helical microtubules of graphitic carbon," Nature, Vol. 354, (1991) p. 56.

[2] Kukovitsky, E. F., et al., "VLS-growth of carbon nanotubes from the vapor," Chemical Physics Letters, Vol. 317, (2000) pp. 65-70.

[3] Hofmann, S., et al., "In situ Observations of Catalyst Dynamics during Surface-Bound Carbon Nanotube Nucleation," Nano Letters, Vol. 7, (2007) pp. 602-608.

[4] Tsai, T. Y., et al., "Transfer of patterned vertically aligned carbon nanotubes onto plastic substrates for flexible electronics and field emission devices," Applied Physics Letters, Vol. 95, (2009) pp. 013107-3.

[5] Ng, H. T., et al., "Growth of Carbon Nanotubes: A Combinatorial Method To Study the Effects of Catalysts and Underlayers," The Journal of Physical Chemistry B, Vol. 107, (2003) pp. 8484-8489.

[6] García-Céspedes, J., et al., "Efficient diffusion barrier layers for the catalytic growth of carbon nanotubes on copper substrates," Carbon, Vol. 47, (2009) pp. 613-621.

[7] Bayer, B. C., et al., "Support–Catalyst–Gas Interactions during Carbon Nanotube Growth on Metallic Ta Films," The Journal of Physical Chemistry C, Vol. 115, (2011) pp. 4359-4369.

[8] Nihei, M., et al., "Low-resistance multi-walled carbon nanotube vias with parallel channel conduction of inner shells [IC interconnect applications]," in Interconnect Technology Conference, 2005. Proceedings of the IEEE 2005 International, 2005, pp. 234-236.

[9] Nicolet, M. A. and Bartur, M., "Diffusion barriers in layered contact structures," Journal of Vacuum Science and Technology, Vol. 19, (1981) pp. 786-793.

[10] Avasarala, B. and Haldar, P., "On the stability of TiN-based electrocatalysts for fuel cell applications," International Journal of Hydrogen Energy, Vol. 36, (2011) pp. 3965-3974.

[11] Wang, Q., et al., "Plasma-assisted elaboration of macropore architectures in titanium nitride," Journal of Alloys and Compounds, Vol. 494, (2010) pp. L11-L14.

[12] Yap, C. C., et al., "Impact of the CNT growth process on gold metallization dedicated to RF interconnect applications," International Journal of Microwave and Wireless Technologies, Vol. 2, (2010) pp. 463-469.

[13] Yap, C. C., et al., "Carbon nanotubes bumps for flip chip packaging system," Nanoscale Research Letters, Vol. 7, (2012) p. 105.

Low Temperature ISSG Oxidation and Its Application in SSRW for 20nm and Below Semiconductor Devices

Weihua Tong, Really Kim, Zhao Lun, Henry Lim, Sung Kim

GLOBALFOUNDRIES
400 Stone Break Road Extension, Malta, New York. USA 12020
Phone: +1 518-305-7364
tongwh@globalfoundries.com.sg

Abstract

This paper investigates low temperature in situ steam generation (ISSG) oxidation, the correlation between the thickness growth rate and temperature, pressure, hydrogen flow and oxidation time. For low temperature ISSG oxidation, the within wafer uniformity is improved by increasing the temperature. An optimum pressure is revealed for all the possible hydrogen concentrations studied in this paper. However, the hydrogen flow itself exhibits a much more complicated relationship with the uniformity than that of temperature and pressure. Good uniformity is achieved through process optimization. It is found that the low temperature ISSG process exhibits a more robust within wafer uniformity than typical high temperature ISSG process due to the fact that the former one has better resistance to pressure and gas flow disturbance. One of the advantages that low temperature ISSG can enable SSRW application in 20nm semiconductor devices is also discussed in this paper.

Keywords- In situ steam generation; low temperature; shallow trench isolation; super steep retrograde well

Introduction

In situ steam generation (ISSG) for single wafer chamber is a widely used oxidation technology in the advanced semiconductor fabrication. Its superior shallow trench isolation (STI) corner rounding and intrinsic quality than furnace wet oxide or rapid thermal process (RTP) dry oxide enable it to dominate in the application of STI liner and thin gate oxide since 65 nm technology node [1-3]. However, traditional ISSG process employs high temperature with low hydrogen concentration which limits its application on the 20nm and beyond technology products, especially for the devices with ultra low power application.

Intra die Vt variation control is one of the key factors to determine the performance of the ultra low power application. Steep channel profiles are essential to achieve intra die variation reduction [4]. Two major approaches have been studied recently. One is to form the super steep retrograde well (SSRW) after STI formation while the other is to form the well before STI formation [5, 6]. Both approaches employ blanket selective silicon epitaxy following the well implant. However, real wafer fabrication industry is more focused on the later one

as the selectively grown epitaxial silicon has facet at the edge of active area which may generate parasitic leakage.

To realize the benefit of the SSRW, the thermal budget of the post well process should be under control. Therefore, low temperature ISSG becomes one of the key factors for the success of the SSRW. This paper investigates the ISSG correlation between the thickness growth rate and process temperature, pressure, hydrogen concentration and oxidation time. It demonstrates that excellent within wafer uniformity is achievable at low temperature through optimization of pressure and gas flow parameters. It also reveals that the optimum low temperature ISSG process exhibits more robust within wafer uniformity than typical high temperature ISSG process due to the fact that the former one has higher resistance to pressure and hydrogen flow disturbance. One of the advantages that low temperature ISSG can enable SSRW application in 20nm semiconductor devices is also discussed in this paper.

Experiments

Typical ISSG STI liner oxidation is carried out at temperature of more than 1000°C with hydrogen concentration of less than 10%. The typical thickness is around 30Å for 45nm and beyond technology products. To enable SSRW, the thermal budget requires the process temperature to be less than 850°C. Therefore, a full design of experiment for the ISSG oxidation recipe creation and optimization is planned.

The experiments are performed on 300mm p-type blanket test wafers. The wafers are cleaned with HF in a batch wet clean tool. Then the ISSG oxidation is processed in a lamp based RTP chamber. Firstly, the process pressure, oxidation time, gas and gas ratio together with all other factors are fixed but the oxidation temperature is changed from 700°C to 850°C with 25°C increment. Then, the process temperature and all other factors are fixed but the pressure is adjusted from 4 torr to 12 torr with increment of 2 torr. Thirdly, the hydrogen flow is adjusted from 1slm to 12slm with the increment of 2slm while all the other parameters are kept unchanged. Fourthly, low temperature ISSG oxidation is carried to verify the time dependence at optimized temperature, pressure and gas ratio. Finally, typical high temperature ISSG oxidations are performed at different gas ratio and pressure to compare the thickness and within wafer uniformity with that of optimized low temperature ISSG oxidation. The oxygen gas flow is set at a fixed value for all the experiments describe in this paper.

The optimized low temperature ISSG oxidation process is applied in 20nm SSRW device for the STI liner oxidation.

Results and Discussion

As shown in Figure 1, around 22Å ISSG oxide is formed successfully in the oxidation process condition of 700°C and 1slm hydrogen flow. As one can easily predict, the oxide thickness increases steadily along the increase of the oxidation temperature. The oxide thickness increases more at higher temperature than that of at lower temperature indicating that the oxidation rate shows an accelerative speed with temperature increase. At oxidation temperature of 850°C and the same hydrogen gas flow, the oxide thickness increases to around 31Å. At oxidation temperature of 700°C to 850°C, the oxide thickness increases steadily with the increase of the hydrogen gas ratio. The wafer average thickness at different hydrogen flow follows the similar trend along the oxidation temperature. These phenomena confirm ISSG wet oxide does occur at a temperature of 700°C.

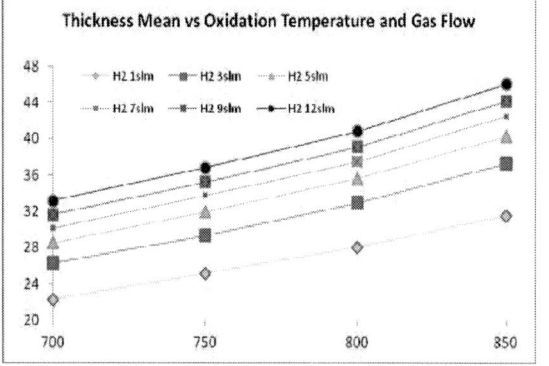

Figure 1 ISSG oxide thickness v.s. oxidation temperatrure and H_2 flow. X axis stands for temperature (°C), Y axis stands for thcikness (Å).

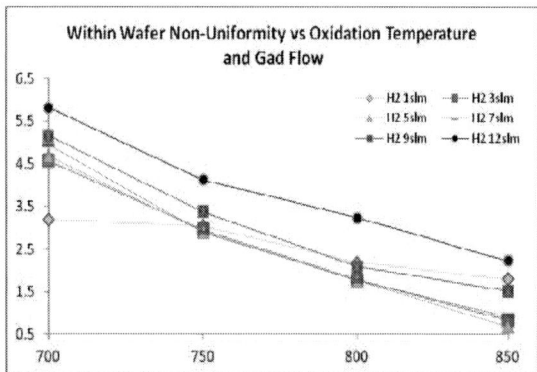

Figure 2 ISSG oxide thickness within wafer non-uniformity v.s. oxidation temperatrure and H_2 flow . X axis stands for temperature (°C), Y axis stands for thickness within wafer non-uniformity (%).

On the other hand, as shown in Figure 2, the within wafer thickness non-uniformity is more than 3% for hydrogen flow

from 1slm to 12slm and oxidation temperature of 700°C. The within wafer thickness uniformity gets worse when the hydrogen gas flow increases at oxidation temperature of 700°C. But no similar trend is observed at oxidation temperature of 750°C and above. Within wafer thickness uniformity becomes better when the temperature increases from 700°C to 850°C. These phenomena might suggest the combustion on the wafer surface is not even at temperature of less than 850°C. The best within wafer uniformity is achieved in the process condition of temperature of 850°C and hydrogen flow of 5 slm.

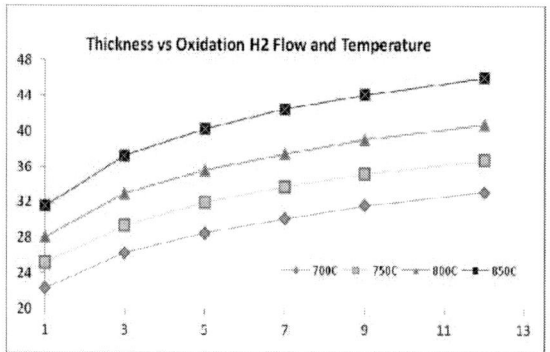

Figure 3 ISSG oxide thickness v.s. oxidation H_2 gas flow and temperature. X axis stands for H_2 gas flow (slm), Y axis stands for thcikness (Å).

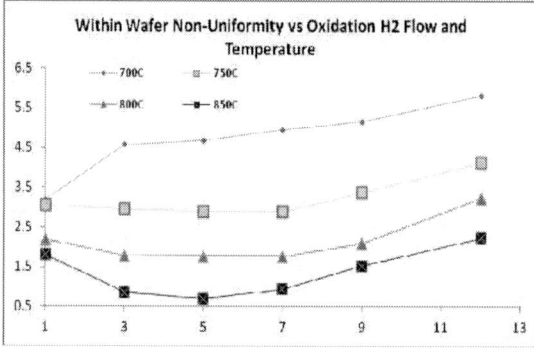

Figure 4 ISSG oxide thickness within wafer non-uniformity v.s. H_2 flow and temperature. X axis stands for H_2 flow (slm), Y axis stands for thickness within wafer non-uniformity (%).

To further understand the low temperature ISSG oxide thickness correlation with oxidation temperature and hydrogen flow, the same data in Figure 1 and 2 are reploted as X axis representing hydrogen flow and Y axis representing the oxide thickess or within wafer uniformity, as shown in Figure 3 and 4. Figure 3 further reveals that at a fixed temperature, the oxide thickness increases along the increase of hydrogen flow. At the same time, all the plots show a similar sign of thickness saturation when the hydrogen flow is increased to more than 12slm. Actually, it is not difficult to imagine that the oxidation rate will drop down once the hydrogen gas ratio exceeds the

978-1-4673-4840-9/13 $31.00 © 2013 IEEE   8

threshold. As shown in Figure 4, hydrogen flow is not only one of the major factors for the oxide thickness but also one of the key factors for the within wafer uniformity improvement. Best within wafer uniformity is achieved at hydrogen flow of 5slm to 7slm for oxidation temperature 750°C and above.

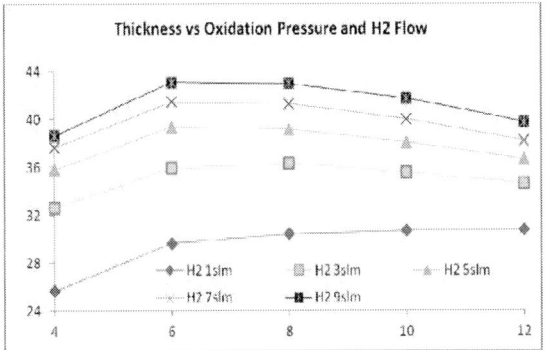

Figure 5 ISSG oxide thickness v.s. oxidation pressure and H_2 flow. X axis stands for pressure (torr), Y stands for thickness (Å)

Figure 5 and 6 show the low temperature ISSG oxide thickness as a function of oxidation pressure and hydrogen gas flow. At fixed temperature and oxidation time, the oxide thickness will increase when pressure increases from 4 torr to 6 torr, but its thickness will soon be saturated or starts to drop down when the oxidation pressure is further increased. However, similar to the temperature and hydrogen gas flow correlation, the oxide thickness increases along the increase of the hydrogen flow from 1slm to 9 slm for different oxidation pressure from 4 torr to 12 torr. The thickness shows a sign of saturation again when the hydrogen gas flow increases to beyond 9slm, as shown in Figure 6.

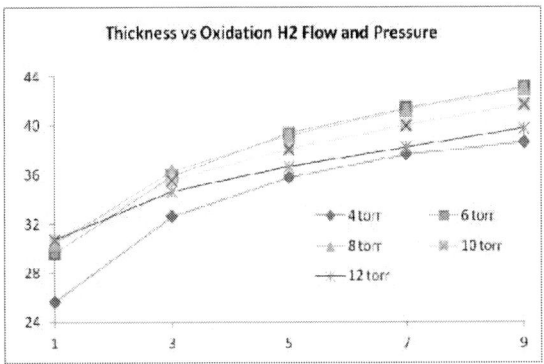

Figure 6 ISSG oxide thickness v.s. oxidation H_2 flow and pressure. X axis stands for H2 gas flow (slm), Y stands for thickness (Å)

Examination of within wafer uniformity versus oxidation pressure and hydrogen flow relveals that uniformity depends on both pressure and hydrogen flow even at fix temperature and oxidation time. For oxidation pressure increasing form 4 torr to 8 torr, the within wafer uniformrity becomes

significantly better for all investigated hydrogen flows from 1slm to 9slm. When the oxidation pressure further increases beyond 8 torr, the within wafer uniformity becomes worse than that of 8 torr, as shown in Figure 7. As illustrated in Figure 8, the within wafer uniformity is generally better at 1slm hydrogen flow for most of the tested oxidation pressure. However, the best within wafer uniformity is achieved at hydrogen flow of 5slm to 7slm in 8 torr oxidation.

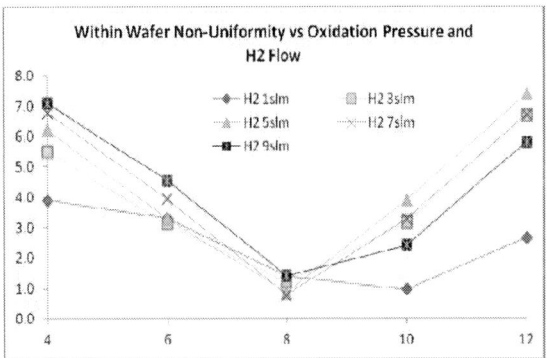

Figure 7 ISSG oxide within wafer non-uniformity v.s. oxidation pressure and H2 flow. X axis stands for pressure (torr), Y stands for within wafer unifromity (%)

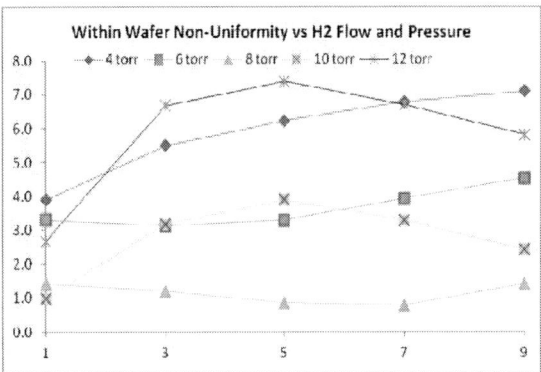

Figure 8 ISSG oxide within wafer non-uniformity v.s. oxidation H_2 flow and pressure. X axis stands for pressure (torr), Y stands for within wafer uniformity (%)

As expected, the oxide thickness increases when the oxidation temperature increases for all pressures from 3 torr to 12 torr. The oxide thickness increases rapidly when the pressure increases from 3 torr to 6 torr. But the pressure effect on thickness becomes saturated after 6 torr, the thickness variation is small even the pressure is doubled from 6 torr to 12 torr at the same oxidation temperature. However, pressure exhibits significant effect on the thickness when the oxidation pressure drops to 3 torr, as shown in Figure 9 and Figure 10. This phenomenon might suggest the wafer surface combustion is not carried out properly due to the short dwell time of the process gas at low pressure and low temperature.

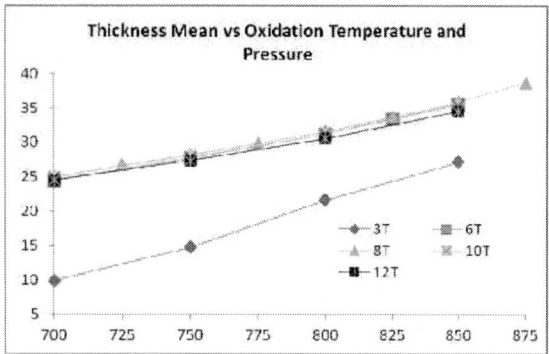

Figure 9 ISSG oxide thickness v.s. oxidation temperature and pressure. X axis stands for oxidation temperatutre (°C), Y stands for thickness (Å)

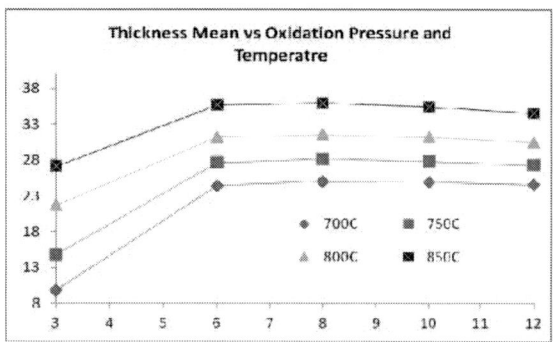

Figure 10 ISSG oxide thickness v.s. oxidation pressure and temperature. X axis stands for oxidation temperatutre (°C), Y stands for thickness (Å)

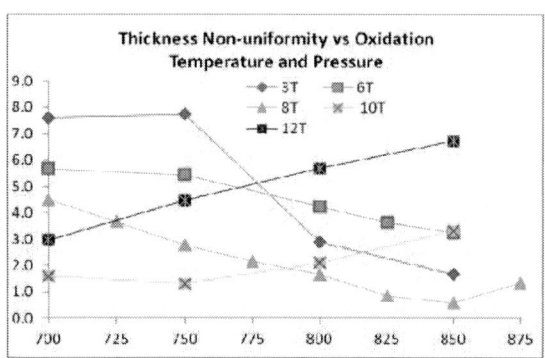

Figure 11 ISSG oxide within wafer non-uniformity v.s. oxidation temperature and pressure. X axis stands for oxidation temperature, Y stands for within wafer non-uniformity (%)

As previously discussed, oxidation pressure has significant effect on the within wafer uniformity. Here, the pressure effect on uniformity is examined again together with temperature. As shown in Figure 11, from 3 torr to 8 torr process, the thickness unifromity is generally improved by increaseing the oxidation

temperature from 700°C to 850°C. However, when temperature increases, the within wafer uniformity becomes worse when the pressure is further increased to 10 torr and 12 torr. Unlike pressure and gas flow correlation showing a single clear optimal pressure for best within wafer uniformity, pressure and temperature variation shows a much more complicated effect on uniformity, different oxidation temperatures require their own optimum pressures for uniformity.The best overall uniformity is achieved at 850°C and 8 torr process condition, as shown in Figure 12.

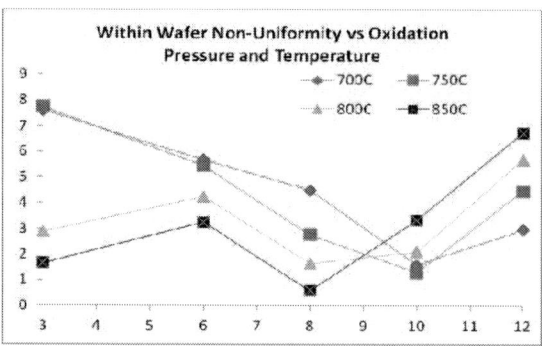

Figure 12 ISSG oxide within wafer non-uniformity v.s. oxidation pressure and temperature. X axis stands for oxidation temperature, Y stands for within wafer non-uniformity (%)

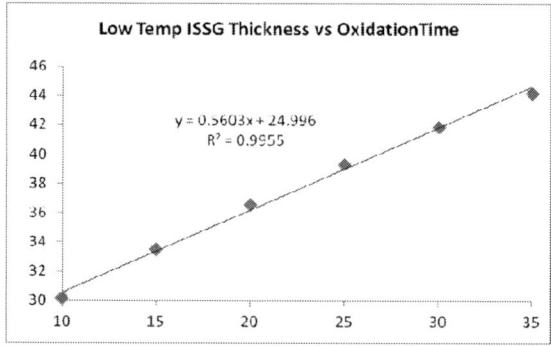

Figure 13 Low temperature ISSG oxide thickness v.s. oxidation time. X axis stands for oxidation time (seconds) Y stands for thickness (Å)

As a general verification, low temperature ISSG experiments are also performed at different process time. As well expected, the thickness increases linearly along the oxidation time for the thin oxide with short process time, as shown in Figure 13.

For comparison purpose, typical high temperature ISSG process is executed at different process pressures and hydrogen flows. As expected, the oxide thickness increases as the hydrogen flow increses from 0.2slm to 1slm. However, unlike the low temperature ISSG oxidation, process pressure

978-1-4673-4840-9/13 $31.00 © 2013 IEEE   10

shows little effect on the oxide thickness for the high temperature ISSG oxidation, as shown in Figure 14. On the other side, as shown in Figure 15, both oxidation pressure and hydrogen flow have strong effect on the within wafer uniformity. Unlike low temperature ISSG, there isn't any common optimum pressure that could lead to the best uniformity for different hydrogen flows. Small hydrogen flow variation can cause significant thickness and uniformity change which is not desireable for mass production.

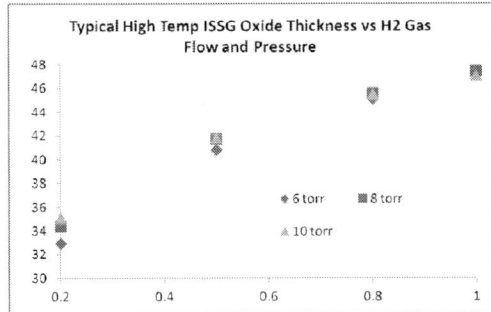

Figure 14 Typical high temperature ISSG oxide thickness v.s. H_2 flow and pressure. X axis stands for H_2 flow (slm) and Y stands for thickness (Å)

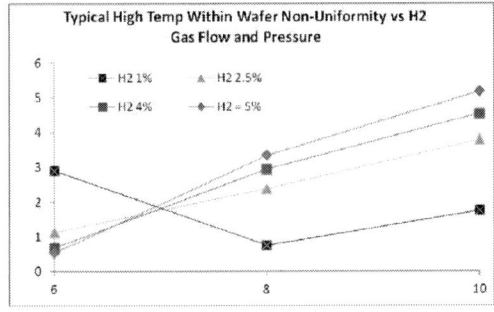

Figure 15 Typical high temperature ISSG oxide thickness within wafer non-unifromity v.s. H_2 flow and pressure. X axis stands for H_2 flow (slm), Y stands for non-uniformity (%)

I. LOW TEMPERATURE ISSG APPLICATION IN 20NM SSRW

As it is well known, SSRW requires two fundamental conditions. One is the precisely distributed dopant in the designated well area through ion implantation or in situ epitaxy process which is not the focus in this paper. The other is the thermal budget control after the dopant application to prevent excessive diffusion. Success SSRW can not only improve mobility and short channel effect but also within die Vt variation. The optimized STI low temperature ISSG process is implemented in one of the 20nm SSRW process splits. As expected, the mobility is improved by around 11% for the low temperature process compare to the normal process, as shown in Figure 16.

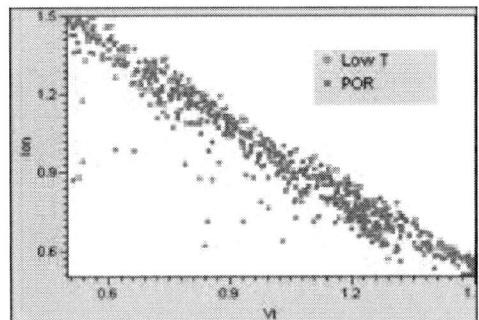

Figure 16 Mobility comparison between the low temperature and normal process

Conclusion

.ISSG oxide thickness, uniformity and their correlation to process temperature, pressure, hydrogen flow and oxidation time are investigated. It is demonstrated that low temperature ISSG with good within wafer uniformity is achievable through process parameters' optimization. The optimum low temperature ISSG is more robust than typical high temperature ISSG for small process variation. Low temperature ISSG is one of the critical processes for SSRW integration. Its application in 20nm device confirms the benefits of the low temperature process.

Acknowledgement

. The authors would like to thank Yunling Tan, GLOBALFOUNDRIES AMTD Singapore Diffusion Manager, for the fruitful discussions. Yiqun Liu, Vidas Sargunas and Bharat Krishnan would be also acknowledged for the proof reading of the final paper.

References

[1] M. Langebuch, et al., In situ steam generation (ISSG) versus standard steam technology: impact on oxide reliability, Microelectrics Reliability, pp. 875-878, 2005

[2] T.M. Pan, Electrical characterization of 13A in situ steam-generated oxynitride gate dielectrics, Electrochemical and Solid-State Letters, 2, pp. 66-68, 2006.

[3] T.Y.Luo, et al., Effect of H2 content on reliability of ultrathin in-situ steam generated (ISSG) SiO2, IEEE Electron Device Letters, 9, pp. 430-432, 2000

[4] H. Itokawa, et al., 25-nm gate length nMOSFET with steep channel profiles utilizing carbon-doped silicon layers (a P-type dopant confinement layer), Electron Devices, 5, pp. 1302-1310, 2011

[5] J. Cai, et al., Method for fabricating super-steep retrograde well Mosfet on SOI or bulk silicon substrate, and device fabricated in accordance with the method – US patent application number: 20090108350[6]

[6] J. A. Babcock, et al., Advanced CMOS using super steep retrograde wells – US patent application number 20090130805

Nanoscale Tri Gate MOSFET for Ultra Low Power Applications Using High-k Dielectrics

D Nirmal[1], , P Vijay Kumar[2], Doreen Joy[3] , Binola k Jebalin[4] N Mohan Kumar[5]

Department of Electronics and Communication Engineering[1 3 4 5]
Department of Electrical and Electronic Engineering [2]
Karunya University, Coimbatore, Tamilnadu, India[1 3]
Loyola Institute of Technology, Chennai, Tamilnadu, India[3]
Karpagam College of Engineering, Coimbatore, Tamilnadu, India[2]
SKP Engineering College, Tiruvannamalai Tamilnadu, India [5]
Phone: 9789498810, nirmal@karunya.edu[1]

Abstract

The Triple-Gate (TG) MOSFET has emerged as one of the promising devices to extend CMOS technology beyond the scaling limit of conventional CMOS technology. Triple gate MOSFET has an excellent scalability and better Short Channel Effect immunity. They are used for CMOS applications beyond the 22 nm node. In order to reduce the leakage current for device beyond the 22 nm, the gate dielectric is replaced with different High-*k* dielectric material. Triple Gate MOSFETis developed using Sentaurus simulator and its performance is analyzed for various device parameters. It is observed that the integration of high-k gate oxide dielectric material in Triple gate MOSFET significantly reduce the short channel effects and the leakage current. The parameters such as ON current, OFF current, Ion/Ioff ratio, DIBL(Drain Induced Barrier Lowering), transconductance, transconductance generation factor, output resistance, intrinsic gain and intrinsic gate capacitances are analyzed in this paper. The suitability of nanoscale Triple gate MOSFET for circuit applications is observed with the help of an inverter circuit and their gain values are calculated for VLSI low power applications.

Keywords: Triple-Gate, CMOS, VLSI, low power, high-k.

Introduction

Complementary Metal Oxide Semiconductor (CMOS) Field Effect Transistor (FET) made from silicon is the most important electronic device. The performance of CMOS devices has continued to improve over a forty year time span according to Moore's Law of scaling [1]. An implicit assumption in the Moore's law is that we can shrink the feature size of the transistor, the basic building block of an Integrated Circuit (IC) at an exponential rate. However the continuous scaling of MOS devices led to many challenges such as diminishing gate control over the channel, resulting in increased Short-Channel Effects (SCEs) and increased leakage currents(OFF current) [2].

Scaling of MOSFET lead to the invention of multiple gate devices such as double gate,Triple gate which are scalable and suppress the short channel effects by simultaneous control of the channel by more than one gate.Double gate MOSFETs are devices which control the channel using a front and a back gate. However as the double gate devices are scaled down beyond 100 nm there is considerable threshold voltage roll-off and undesirable short channel effects [3]. This undesirable problem arises in Double gate device due to the reduction in control over the channel when the gate length is reduced.The reduction of threshold voltage with decreasing channel length and increasing drain voltage is widely used as an indicator of the SCEs in evaluating CMOS cutting edge technologies [4].

Triple gate relaxes this limitation and is capable to reduce the short channel effects, threshold voltage roll-off and improve the control over the channel. Also noteworthy is the fact that at relatively large fin widths, energy quantization in TG FinFETs is higher than in DG FinFETs because of the presence of an additional gateat the top edge of silicon fin[5].

978-1-4673-4840-9/13 $31.00 © 2013 IEEE

In order to maximize ON-current per chip area, however, multi-gate structures with non-planartriple-gate transistors are favorable. Triple-gate FETs appear attractive because of the high ON current and less leakage.As the gate length decreases the gate control over the channel also decrease due to the proximity of source and drain. The gate control over the channel can be improved by geometrically placing the gate close to the channel and providing tighter gate coupling. The gate can be placed close to the channel by taking ultra-thin body known as fin type structure and tighter gate coupling can be achieved by increasing the number of gates from single gate to multi gate [6]. In triple gate MOSFET the drain current flows along the lateral channels i.e. sidewall, normally with (110) orientation as well as on the top channel with (100) orientation [7].

To maintain a proper control of the drain current flow in nanoscale CMOS devices, the thickness of silicon dioxide which has been used as the gate dielectric material for over four decades is now pushed into its technological limit of about 1 nm and theoretical limit of 0.7 nm[8].Further device downsizing would require even thinner gate dielectric films. This stringent requirement can only be achieved by using a high-dielectric constant (high-k) material.So in order to have high performance and reduced short channel effects materials with High-k dielectric material is needed in place of silicon dioxide as the gate dielectric material. High-k dielectric materials have an Equivalent Oxide Thickness (EOT) of 1nm with negligible gate oxide leakage, desirable transistor threshold voltages for n and p-channel MOSFETs, and transistor channel mobility close to those of SiO_2[9]-[10].

So in this paper High-k dielectric material is introduced which can replace SiO_2 gate dielectric and the impact of these High-k dielectrics on the performance of Triple Gate MOSFETs is analysed. Triple Gate MOSFET device is created with different device dimension using Sentaurus of ISE TCAD and the device performance of the Triple gate device is analysed for various parameters.

Device Structure and Parameters

Triple gate MOSFET is created based on the technology parameters and supply voltages according to the ITRS requirements[11]. The schematic view of the Triple gate MOSFET is shown in Figure 1.The Downscaling of MOSFETdevices are a major priority of electronic-device research. Channel lengths have alreadybeen reduced below 30 nm and according to the International Technology Roadmap for Semiconductors, this trend will continue in the coming years. In this context, the appearance of short-channel effects (SCEs) seems to be the main problem that should be overcome to continue reducing channel dimensions [12].

Silicon-On-Insulator (SOI) technology is one of the techniques that can improve the device scalability. Furthermore, SOI technology can easily be combined with new semiconductor structures such as multiple gate (MuG) devices[13]. Multiple gate devices came into existence due to the continuous scaling of MOSFETs. These scaling of MOSFETs lead to the increase of short channel effects and leakage current. Triple gate is a multiple gate device which reduces the problems of scaling to great extent due to its structure and flexible body dimensions.

Triple gate MOSFET with 22 nm technology node is created using Sentaurus structure editor. The fin thickness is chosen as 10 nm and the height of the fin as 25 nm. Poly Silicon is chosen as the gate material with a work function of 4.17 eV[14]. The thickness of the high-k gate dielectric material is chosen as 1 nm (physical) for triple gate with various high-k materials. The source and the drain regions are doped with Phosphorous active concentration of 1×10^{20} cm^{-3}[15]for creating n-channel Triple gate MOSFET. The channel region which has a thickness of 10 nm is lightly doped with boron active concentration of 1×10^{15} cm^{-3}[7]. Table 1 shows the dimensions of the Triple gate MOSFET.

Simulations of the Triple gateMOSFETare performed for a wide range of proposed gate dielectricmaterials like SiO_2, Si_3N_4, Al_2O_3, Ta_2O_5, HfO_2/ZrO_2, $LaAlO_3$, and TiO_2. The operation of the device in the subthreshold regime is the main consideration for their performance analysis.

Simulation Results and Discussions

A.Impact of high-k dielectrics on Triple gate MOSFET

Silicon dioxide has been used as a gate oxide material for decades [16]. As the thickness scales below 2 nm, leakage currents due to tunneling increase drastically, leading to unwieldy power consumption and reduced device reliability [17]. To meet the need for higher transistor speed while keeping power consumption under control, the semiconductor industry is working to introduce high-k gate dielectrics in leading-edge transistor manufacturing processes[18]. Insulators with high dielectric constants (k) play several critical roles in modern semiconductor devices[19]. So in order to construct a Triple gate device beyond 65 nm node SiO_2 is not a suitable candidate to be used as the gate oxide. Hence High dielectric constant (k) materials are an alternative to be used as the gate oxide for improved performance. Manycandidates of possible high-k gate dielectrics have been suggested to replace $SiO_2$and they include nitrided SiO_2, Hf-based oxides, and Zr-based oxides [1]. Amongthe various requirements of gate dielectric materials, the most important are thegood insulating properties and capacitance performance. Meanwhile, they must have good thermal stability,high recrystallization temperature, sound interface qualities, and so on[20]-[22].

Nanosizing of high-k materials seems to improve the electrical, mechanical andoptical properties of any compound drastically due to the increase in surface area[23]. Based on the above stated requirements and specifications of the high-kmaterials, various high-k materials are chosen for the analysis of Triple Gate MOSFET.Table 2summarizes the alternative high-k dielectric materials, its k value, energyband gap and the conduction and valence band offset. The required band gap forthe gate insulator is reported to lie in the range of 5.16 to 7.8 eV [23]. Table 2 showsthat the band gap of SiO_2 is 9 eV, which is large with sufficiently large conductionand valence band offsets. Also the band gap of HfO_2 is 5.8 which is lesser than theabove specified band gap range.

In this work the Silicon dioxide gate insulator is replaced with the alternate high-k dielectric materials and simulations are carried out for the Triple Gate MOSFET to evaluate their performance.

B.Analysis of parameters of TG MOSFET using High-k dielectrics

ON current is the drain current (I_d) due to the flow of the electrons from source to drain when the gate voltage (V_{gs}) is applied. And it can be controlled by controlling the gate voltage. The ON current of the Triple Gate MOSFET are plotted in Fig. 2 for various high-k dielectric values.The Triple Gate device with SiO_2 as gate dielectric is found tohave the highest I_{ON} of 3.02×10^{-06} A and the current decreases exponentially with increasing k values. The ON current with HfO_2 as gate dielectric material has a current of 3.771×10^{-06} A. Hence Triple gate MOSFET with high-k dielectrics shows anincrease of 18% in ON current.

Leakage current or OFF current is the drain current at a V_{ds} of 0.8V. It should be reduced for better performance. The variation of leakage current with different high-k dielectric material is shown in Fig 2. The I_{OFF} of the device with SiO_2as the dielectric is 3.65×10^{-9}A. It can be observed from the figure that as the gate dielectric value increases, I_{OFF} decreases exponentially due to the fact that the alternate dielectric materials provide higher physical thickness thereby reducing the tunnelling of carriers through the insulator.I_{OFF} for the device with HfO_2as dielectric is 4.13×10^{-13} A thereby showing a decrease of about 54%when compared to device with SiO_2 as the dielectric. Whereas double gate device has a leakage of 200 Na[24] which shows that Triple Gate MOSFET hasreduced leakage of 45% compared with Double Gate.

The ratio of ON current to OFF current is plotted against different high-k dielectric materials is shown in Fig. 3. From the figure it is inferred that I_{ON}/I_{OFF} increases exponentially withincrease in k value. This is due to a substantial decrease in the OFF currents withincrease in the k value. I_{ON}/I_{OFF} for TG MOSFET with HfO_2 as the dielectric is8900000 whereas the ratio is the lowest for the one created with SiO_2 as the gatedielectric with a value of 450000 thereby showing an increase of 82% in the ratio.This ratio should be high for the proper switching of the devices. Whereas the Double gate MOSFET has a ratio of 1900000 with high-k dielectric material as the gate oxide[4]. Hence Triple Gate MOSFET with high-k material has an increase inI_{ON}/I_{OFF}ratio of 78% compared with the double gate device. So the device switches between the off state and the on state quickly.

For logic applications, Drain Induced Barrier Lowering (DIBL) plays an importantrole as device dimensions are scaled rigorously. The DIBL is measured as thedifference

978-1-4673-4840-9/13 $31.00 © 2013 IEEE

between the linear threshold voltage and the saturation thresholdvoltage.The DIBL co-efficient is computed as in (1). The linear threshold voltage is based on the maximum transconductanceand saturation threshold voltage is based on a modified constant-current method.

$$DIBL = \frac{V_{T,LIN} - V_{T,SAT}}{V_{DD} - V_{D,LIN}} \quad (1)$$

Where, $V_{T,LIN}$ is the threshold voltage measured at linear region,

$V_{T,SAT}$ is the threshold voltage measured at saturation region,

V_{DD} is the supply voltage

The supplyvoltage of 1 V is taken as per the conventions of ITRS for 22 nm gate length inregard of logic applications. The Drain Induced Barrier Lowering as a function of various high-k materials is shown in Fig. 4. It is revealed from the plot that DIBL is the highestfor TG MOSFET having SiO2 as the dielectric material with a value of 0.691 mV/V. DIBLdecreases exponentially with increasing high-k dielectricmaterial. The device with HfO_2 as the gate dielectric has a DIBL value of 0.486mV/V.Double gate with high-k dielectric gate material has a DIBL of 4.5 mV/V –12.5mV/V [4] for different high-k materials. Thereby TG MOSFET shows a decrease of 30% with high-k dielectric gate material.Thusit is observed that the Triple Gate with High-k dielectric material provides better suppression of Short Channel Effects.

C. Analog performance

The different analog performance parameters studied in this section are thetransconductance (g_m). Transconductance Generation Factor (g_m/Id), OutputResistance (Ro) and Gate Capacitance.

Transconductance, g_m is defined as the ratio of the change in drain current to the change in gate voltage. The variation of transconductance of Triple Gate MOSFET with different gate oxidedielectric material at a drain voltage of 0.8 V is shown in Fig 5. The Triple Gate MOSFET exhibits increased transconductance. The improvement in transconductance in triple gate MOSFET can be attributed to the improved charge control by the presence of three gates surrounding the channel.As the k value is increased, the transconductance also increases almost linearly. HfO_2 has the transconductance value of 8.02 S/μA while triple gate MOSFET with SiO_2 as the gate dielectric has the lowest value of 5.78S/μA. Thereby showing an increase of 28% in transconductance obtained for TG MOSFET with high-k dielectric material. Another parameter analyzed is Transconductance Generation Factor (TGF).Transconductance Generation Factor (TGF) or g_m/I_d ratio is the available gain per unit value of power dissipation. TGF is the figure of merit of technology indicating the efficiency of devices to convert dc power into ac frequency and gain performance. In a MOS transistor, g_m/I_d is maximum when in weak inversion and degrades severely with increasing drain current in the strong inversion regime. Devices operating in the subthreshold regime are expected to provide higher gain. The comparison of TGF for the Triple Gate MOSFET with different high-k materials is shown in Fig. 6. From the figure it can be observed that the TGF increases with increase in high-k dielectric material.

The Output Resistance, R_O of a MOS transistor at any Vgs is evaluated as

$R_O = V_a/I_D (2)$

where V_a and I_D are early voltage and saturated drain current at that particular gate to source voltage. The output resistancevariation of the Triple Gate MOSFET for different k values is shown in Fig. 7. Lowest valueof $R_O$is 3.78MΩ for the device with SiO2 as the dielectric while Ro of 4.32MΩ isobtained for the device with HfO_2 as the gate dielectric. Hence Triple Gate MOSFET with high-k Dielectric material has high Output Resistance making it suitable for applications such as cascade amplifiers and feedback amplifiers.

The intrinsic gate capacitance is another important parameter in the analog design. The intrinsic gate capacitances in the Triple Gate MOSFET are gate to- source (Cgs), gate-to-drain capacitances (Cgd). The total amount of capacitance is equal to sum of all the intrinsic gate capacitances that is Cgg = Cgd + Cgs. Table 3 shows the values of Cgs, Cgd and Cgg of the Triple Gate MOSFET for various high-k dielectric materials. It can be seen from the table that the Cgs, Cgd and Cgg increases with dielectric thereby reducing the leakage current and increasing the gate control.

Conclusion

CMOS technology has seen excellent high-speed performance achieved through improved design, use of high quality materials and processing innovations over the past decade. Silicon dioxide gate dielectric is replaced with various high-k dielectric material and the simulations have been carried out. Various parameters such as OFF current, ON current, I_{ON}/I_{OFF} ratio, transconductance, DIBL,gate capacitance are observed for different dielectric constant values. It is observedthat the leakage of the device decreases by about 54% as the k value is increasedfrom $3.9(SiO_2)$ to $25(HfO_2)$ making the device applicable for low powersubthreshold analog performance. DIBL of the Triple Gate MOSFET with HfO_2 as dielectricdecreases by around 30% when compared to the conventional Triple gate device therebyproviding better suppression of Short Channel Effects. The I_{ON}/I_{OFF} ratio of the Triple Gate MOSFET increases by around 82% when dielectric material is changedfrom $SiO_2$to high-k dielectric material making the device useful for high speed switching operations.

The transconductance of the Triple Gate with high-k dielectric material is higher when compared to device with silicon dioxide ad gate dielectric. Therefore the device can be used for high gainapplications. The gate capacitance of the device increases by 71% when high-k dielectric material is used as gate dielectric. Thus the Triple Gate MOSFET with high-k dielectric as gate dielectric material can be used for low power applications and can be considered as a promising device for futureSemiconductor Industry.

References

[1] J. Robertson, "High dielectric constant oxides," The European Physical Journal Applied Physics, vol. 28, pp. 265-291, Dec 2004.

[2] E. J. Nowak, "Maintaining the benefits of CMOS scaling when scaling bogs down," Journal of Research and Development, pp. 169 – 18, March/May 2002.

[3] G. Venkateshwar Reddy, M. Jagadesh Kumar, "Investigation of the novel attributes of a single-halo double gate SOI MOSFET: 2D simulation study", Microelectronics Journal, Vol. 35, pp. 761–765, 2004.

[4] D.Nirmal, P.Vijayakumar, K Shruti, N Mohan Kumar, "Nanoscale channel engineered double gate MOSFET for mixed signal applications using high-k dielectric", International Journal of Circuit Theory and Applications,DOI: 10.1002/cta.1800, 10 Mar 2012.

[5] Raisul Islam, Md. ZunaidBaten, Emran Md. Amin, Quazi D. M. Khosru, "On the Distinction between Triple Gate (TG) and Double Gate (DG) SOI FinFETs: A Proposal ofCritical Top Oxide Thickness", 6th International Conference on Electrical and Computer Engineering (ICECE) 2010, Dhaka, Bangladesh, pp. 434-437, 18-20 December 2010.

[6] Kumar, M.P,Gupta, S.K. Paul, M., "Corner effects in SOI-Tri gate FinFET structure by using 3D Process and device simulations", Computer Science and Information Technology (ICCSIT), 2010 3rd IEEE International Conference on Vol: 9, pp. 704 – 707, 2010.

[7] J. Conde , A. Cerdeira et al, "3D simulation of triple-gate MOSFETs with different mobility regions ",Microelectronic EngineeringVolume 88 , Issue 7, pp. 1633-163, July 2011.

[8] Hei Wong, Kenji Shiraishi, KuniyukiKakushima, and Hiroshi Iwai, "High-K Gate Dielectrics", Electronic Device Architectures for the nano-CMOS era, pp. 105-140, 2009

[9]A. Bouazra, S. Abdi-Ben Nasrallah, M. Said, A. Poncet, "Current Tunnelling in MOS Devices with Al2O3/SiO2 Gate Dielectric", Research Letters in Physics, no. 6, 2008.

[10] Silicon Moore's Law, Intel® website, http://www.intel.com /research/silicon/mooreslaw.htm (Retrieved July 2004).

[11] International Technology Roadmaps for Semiconductor (ITRS); 2010.

[12] Francisco J. García Ruiz, Isabel MaríaTienda-Luna, Andrés Godoy, Member, IEEE Luca Donetti, and Francisco Gámiz, Senior Member, IEEE, "Equivalent Oxide Thickness of Trigate SOI MOSFETs With High-κ Insulators", IEEE Transactions On Electron Devices, Vol. 56, No. 11,pp. 2711-2719, November 2009.

[13] J. P. Colinge, FinFETs and Other Multigate Transistors, Berlin, Germany: Springer- Verlag, 2007.

[14]MirkoPoljak, Vladimir Jovanovic, TomislavSuligoj, "Suppression of corner effects in wide-channel triple-gate bulk FinFETs", Microelectronic Engineering, Vol.87, pp.192–199, 2010

[15] Abhisek Dixit, Anil Kottantharayil, Nadine Collaert, "Analysis of Parasitic S/D Resistance in Multiple Gate FETs", IEEE Transactions on Electron Devices, Vol.52, No.6, pp.1132–1140, 2005.

[16] D. Nirmal, Vijaya Kumar, "Gate Engineering on the Analog

978-1-4673-4840-9/13 $31.00 © 2013 IEEE

Performance of DM-DG MOSFETs with High K Dielectrics", International Journal of Advanced Science and Technology Vol. 25, December, 2010

[17] Wenjuan Zhu, Jin-Ping Han, Member, IEEE, and T. P. Ma, Fellow, IEEE "Mobility Measurement and Degradation Mechanisms of MOSFETs Made With Ultrathin High-k Dielectrics," IEEE Trans. Electron Devices, Vol.51, No.1, Jan.2004

[18] Gennadi Bersuker, Howard R. Huff, "Novel materials for future transistor generations", International SEMATECH, Austin, TX 78741, USA

[19] Charles B. Musgrave, Roy G. Gordon, "Precursors for Atomic Layer Deposition of High-k Dielectrics", section 7, Process Gases, Chemicals and Materials.

[20] A. P. Huang, Z. C. Yang, Paul K. Chu, "Hafnium-based High-k Gate Dielectrics", Advances in Solid State Circuits Technologie, pp. 333-350,April 2010.

[21] G. D. Wilk, R. M. Wallace, J. M. Anthony, "High-k gate dielectrics: Current status andmaterials properties considerations", Journal of Applied Physics, Vol. 89, No. 10, pp.5243-5275, May 2001.

[22] C.R. Manoj, V. Ramgopal Rao., "Impact of High-k gate dielectrics on the device and circuitPerformance of Nanoscale FinFETs", IEEE Electron Device Letters, Vol.28, No.4,pp. 295-297, April 2007.

[23] D.Nirmal, B.Nalini, P.Vijayakumar "Nano sized High-κ Dielectric Material for FINFET",Integrated Ferroelectrics, Vol. 121, Issue 1, pp. 31 – 35, April 2010.

[24] Jakub Kedzierski, Meikeileong, Edward Nowak, Thomas S. Kanarsky, Ying Zhang, Ronnen Roy, Diane Boyd, David Fried, and H.-S. Philip Wong, "Extension and Source/Drain Design for High-Performance FinFET Devices", IEEE Transactions on Electron Devices, Vol. 50, No. 4, pp. 952-958, April 2003.

TABLE 1

DEVICE DIMENSIONS OF TG MOSFET

Parameter	Dimension
Gate Length	22nm
Fin height	25nm
Channel thickness	10nm
Oxide thickness	1nm
Source/Drain Doping	$10^{20}/cm^3$
Channel doping	$10^{15}/cm^3$

TABLE 2

COMPARISON OF ALTERNATE HIGH-K VALUES WITH SIO_2

Dielectric Material	Dielectric Constant (k)	Band Gap (eV)	CB offset (eV)
SiO_2	3.9	9	3.2
Si_3N_4	7	5.3	2.4
Al_2O_3	9	6	2.8
Ta_2O_5	22	4.4	0.35
HfO_2/ZrO_2	25	5.8	1.4
$LaAlO_3$	30	5.6	1.8
TiO_2	40	3.2	0

TABLE 3

COMPARISON OF C_{GD},C_{GS} AND C_{GG} OF TRIPLE GATE MOSFET FOR DIFFERENT GATE DIELECTRIC MATERIALS

Dielectric Constant	Gate to Drain Capacitance, C_{gd} (fF)	Gate to Source Capacitance C_{gs} (fF)	Total Gate Capacitance, C_{gg} (fF)
k=3.9	1.25	0.64	1.89
k=7.5	1.75	0.68	2.43
k=9	2.12	0.74	2.86
k=22	2.75	0.79	3.54
k=25	4.15	0.82	4.97
k=30	4.75	0.87	5.62
k=40	5.69	0.97	6.66

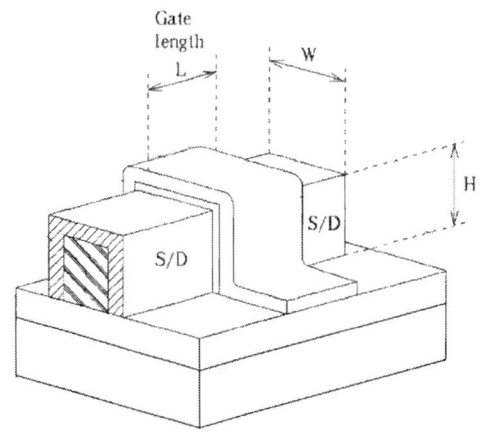

Fig. 1Schematic view of the Triple gate MOSFET

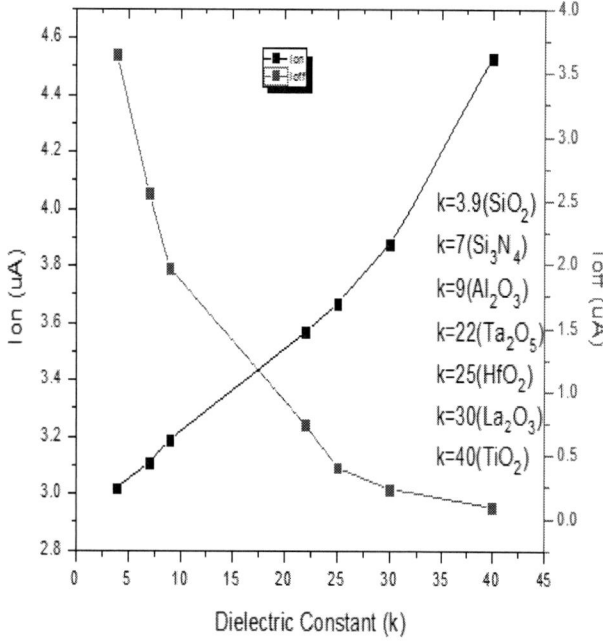

Fig. 2 Comparison of ON current and OFF current for Triple Gate with different high-k dielectric material

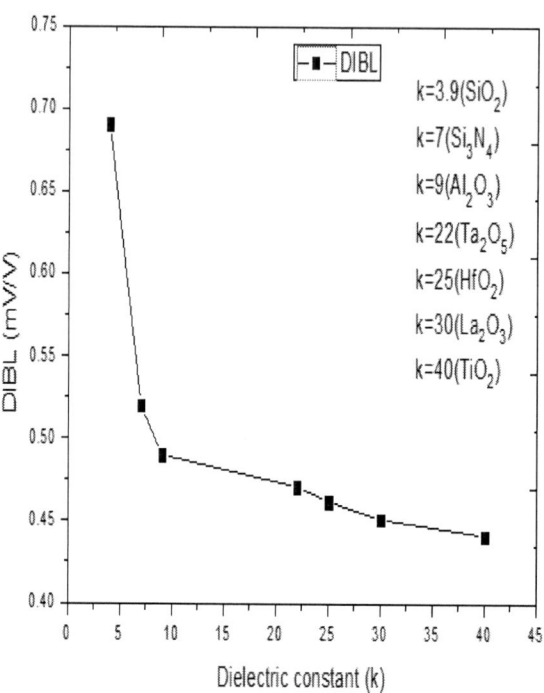

Fig. 4 Comparison of DIBL for Triple Gate with different high-kDielectric material

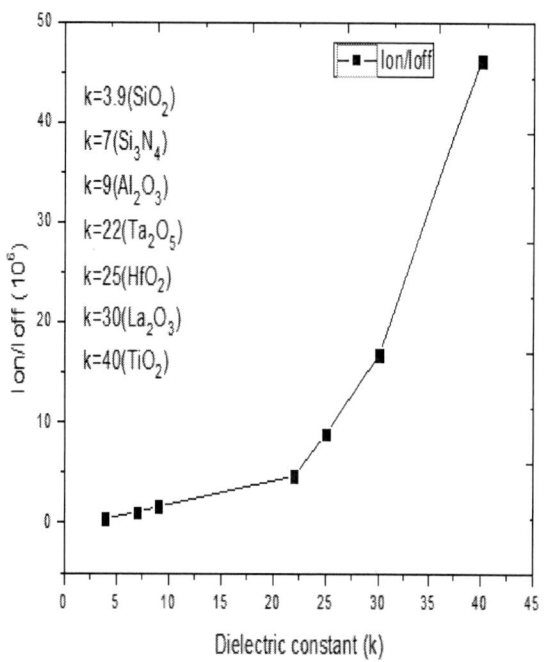

Fig. 3 Comparison of ON current to OFF current ratio for Triple Gate withDifferent high-k Dielectric material

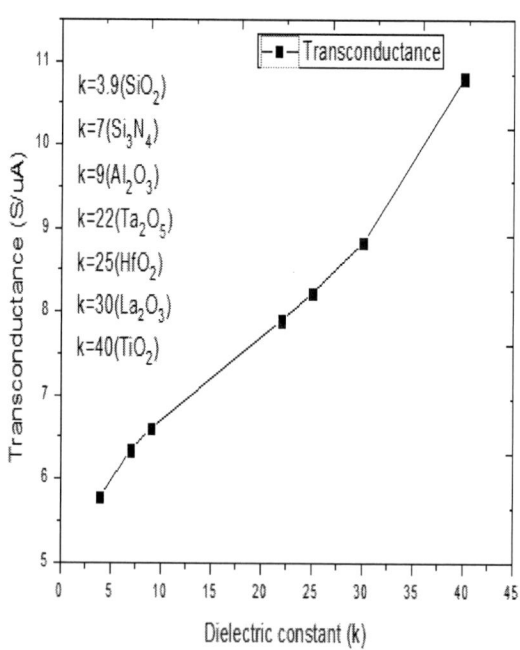

Fig. 5 Comparison of Transconductance for Triple Gate with different high-k dielectric material

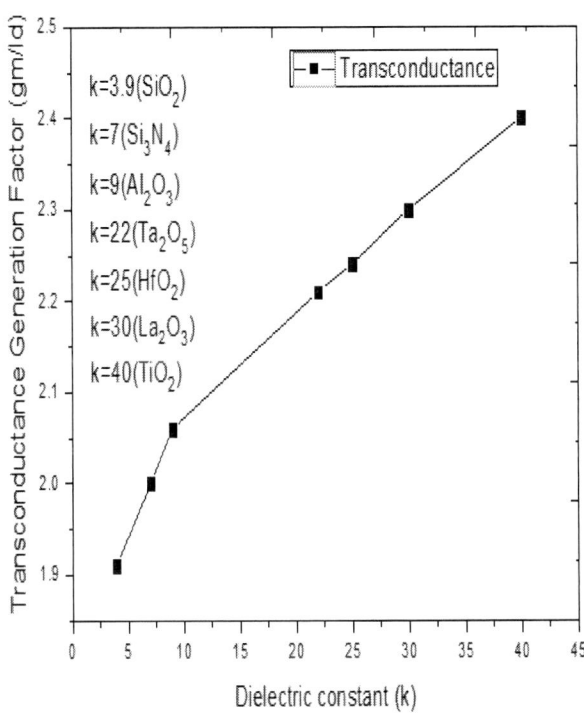

Fig. 6 Comparison of TGF for Triple Gate with different high-kDielectric material

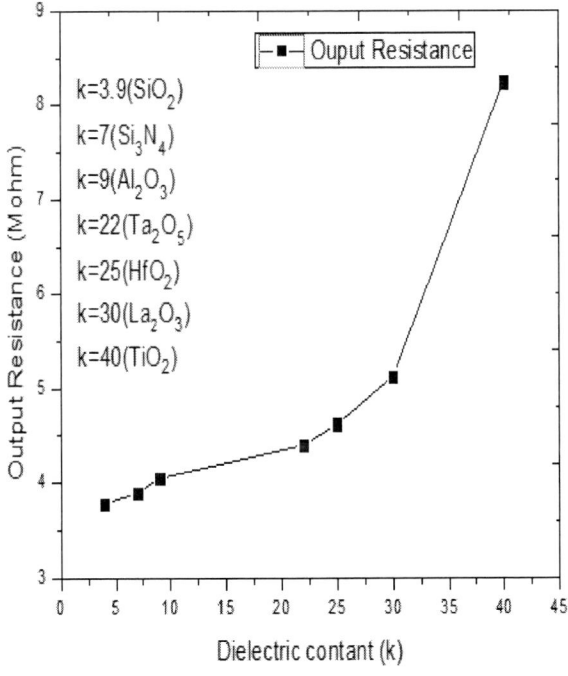

Fig. 7 Comparison of Output Resistance for Triple Gate with different high-kDielectric material

Near Flatband Mode Currents Evaluated by One-Particle Self-Consistent Calculations

Che-Sheng Chung
Department of Electronics Engineering
National Taiwan University of Science and Technology
43, Keelung Road, Section 4, Taipei, 106 Taiwan, ROC.
e-mail address: chungten@ms81.hinet.net

Sheng-Lyang Jang
Department of Electronics Engineering
National Taiwan University of Science and Technology
43, Keelung Road, Section 4, Taipei, 106 Taiwan, ROC.
e-mail address: sljjj@mail.ntust.edu.tw

Abstract—**In a compact model, an off-state current involving a recursive calculation of one-particle self-consistent (SC) potential does not vanish over a metal gate/insulators/semiconductor (MIS) structure of ultra-low temperature ($<$ 4.5°K) metal-oxide-semiconductor field-effect transistor (ULTMOSFET). When operating the ULTMOSFET in the near flatband mode, the SC potential calculation relying on SC Schrödinger-Poisson pair equations extracts both potential well width and triangular energy eigenvalues for bound states. A new calculation of one-particle SC currents is proposed to predict the current phenomena measured from the ULTMOSFET device.**

Keywords-MOSFET; off-state; one-particle; recursion; self-consistent; ultra-low temperature;

I. INTRODUCTION

Since self-consistent (SC) calculations are responsible for predicting transitions changing from quantum phenomena to classical phenomena and vice versa, as described in [1]-[4], features perturbed by small external forces are well interpreted by SC calculations in an ultra-low temperature metal-oxide-semiconductor field-effect transistor (ULTMOSFET). As described in [5], the one-particle SC calculation includes Schrödinger-Poisson pair equations. To convey information between those equations, semi-analytical approach, which consists of the relevant analytical derivations and numerical calculations, fulfills the calculations. Because measured ultra-low temperature currents showed the random patterns of plotted curves illustrated in [6]-[7], this paper proposes one-particle SC calculations characterizing near flatband mode currents also lead to similarly random characterizations produced by small external forces over the substrates of ULTMOSFET structures.

II. OPTIMIZED ALGORITHM

As compared to the multiple-path algorithm in [5], here, a new single-path symmetry algorithm, as shown in Fig. 1, optimizing the one-particle SC calculations removes the position unit's uncertainty produced by the multiple paths of recursive iterations.

III. RESULTS

Fig. 2 shows that the calculated charge sheet densities of excited carriers, as illustrated in [4], have two kinds of discrete

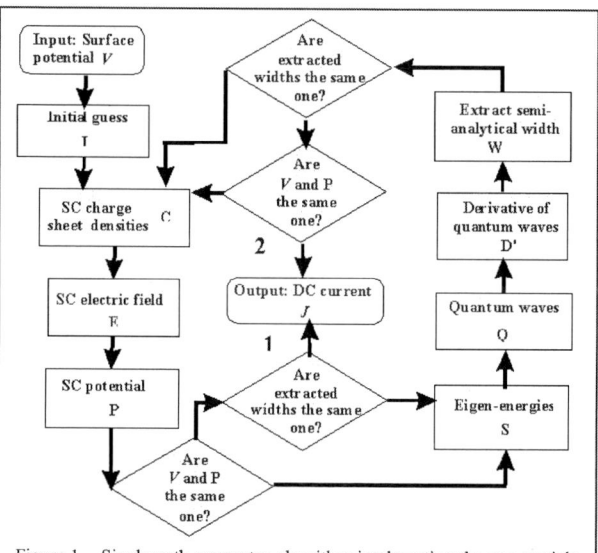

Figure 1. Single-path symmetry algorithm implmenting the one-particle SC calculation.

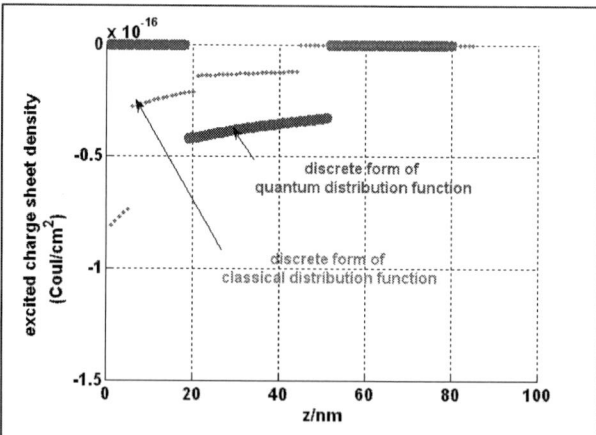

Figure 2. Electron distribution functions produced by the one-particle SC calculation with respect to different built-in surface potentials defined by inputs.

distribution functions: one is classical form; the other is

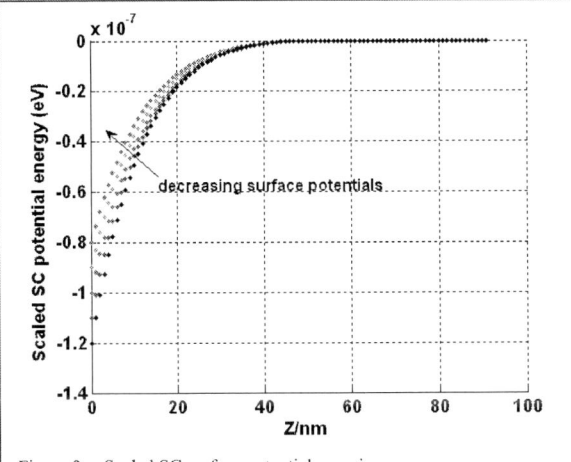

Figure 3. Scaled SC surface potential energies

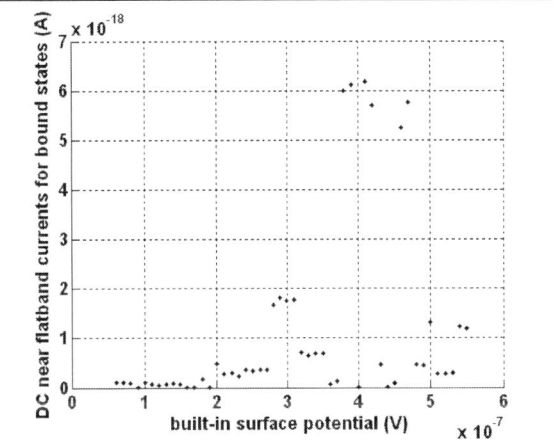

Figure 4. A random pattern of DC bound state currents over a substrate or a bulk at ultra-low temperatures.

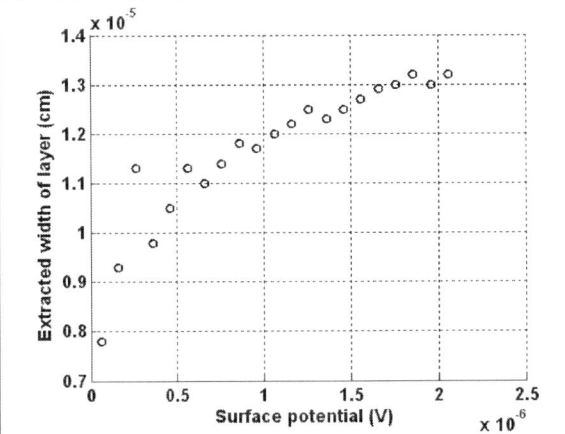

Figure 5. Extracted widths of potential well layers with respect to surface potential.

quantum form. For a quantitative calculation of the SC surface potential, a ratio scaling approach, which uses a constant ratio of calculated and defined surface potential energies, both scales up and scales down the calculated Fermi-Dirac charge sheet densities. Fig. 3 shows the scaled SC potential energies appear classical forms of potential energies although the charge sheet densities are discrete distribution functions. Comparing with the previous calculation published by JJAP in [8], this calculation offers stronger physical impacts to the relationship between the extracted width of potential well layer and surface potential. The one-particle SC calculation demonstrates quantum "bumps" shown in Fig. 4 while extracting potential well widths. DC near flatband mode currents are calculated by a modified diffusion current for bound states. Averaging the values of calculated local current densities, a modified diffusion current model works on producing near flatband mode currents for bound states. Because the measured data, as illustrated in [6] and [7], appeared the random characteristics of

current patterns, Fig. 5 produced by the one-particle SC calculation also demonstrates a similarly random pattern of DC near flatband mode currents over a channel regime, which is near the interface of the MIS of an ULTMOSFET device.

IV. CONCLUSIONS

We demonstrate the semi-analytical solution of the one-particle SC calculation and of extracting potential well widths. According to the experiences in section III, the 1D one-particle SC current is able to predict a near flatband mode current for bound states.

REFERENCES

[1] F. Stern, "Self-consistent results for n-type Si inversion layers," Phys. Rev. B, Vol. 5, pp. 4891-4899, 1972.

[2] E. Gnani, S. Reggiani, M. Rudan, and G. Baccarani, "A new approach to the self-consistent solution of the Schrodinger-Poisson equations in nanowire MOSFETs," IEEE Proc. 34th ESSDERC Conf. Belgium, Vol. 34, pp. 177, 2004..

[3] S. V. Walstra and C.-T. Sah, "Thin oxide thickness extrapolation from capacitance-voltage measurements," IEEE Trans. Electron Devices, 44, pp. 1136–1142, 1997.

[4] S. A. Hareland, S. Krishnamurthy, S. Jallepalli, C.-F. Yeap, K. Hasnat, A. F. Tasch, Jr., and C. M. Maziar, "A computationally efficient model for inversion layer quantization effects in deep submicron n-channel MOSFET's," IEEE Trans. Electron Devices, vol. 43, no. 1, pp. 90-96, Jan. 1996.

[5] C.-S. Chung and S.-L. Jang, "A numerical study of a self-consistent potential," in Proc. IEEE 4th INEC, June 2011, G12-4.

[6] F. Boeuf, T. Skotnicki, S. Monfray, C. Julien, D. Dutartre, J. Martins, P. Mazoyer, R. Palla, B. Tavel, P. Ribot, E. Sondergard, and M. Sanquer, "16 nm planar nMOSFET manufacturable within state-of-the-art CMOS process thanks to specific design and optimization," in IEDM Tech. Dig., 2001, pp. 637–640.

[7] J. P. Colinge, A. J. Quinn, L. Floyd, G. Redmond, J. C. Alderman, W. Xiong, C. R. Cleavelin, T. Schulz, K. Schruefer, G. Knoblinger, and P. Patruno, "Low-temperature electron mobility in trigate SOI MOSFETs," IEEE Electron Device Lett., Vol. 27, pp. 120-122, 2006.

[8] C.-S. Chung, "Semi-analytical depletion width evaluated by self-consistent Schrödinger-Poisson pair calculations," Jpn. J. Appl. Phys., vol. 50, no. 10, art. no. 10PF05, Oct. 2011.

Optimizing Nanoscale MOSFET Architecture for Low Power Analog/RF Applications

Dipankar Ghosh, Mukta Singh Parihar, and Abhinav Kranti

Low Power Nanoelectronics Research Group, Electrical Engineering Discipline, Indian Institute of Technology (IIT), Indore – 452 017, India. Email: akranti@iiti.ac.in

Abstract

This work reports on possible ways of improving analog/RF performance metrics, through device structure optimization, for low power applications. It is shown that underlap source/drain (S/D) design and junctionless transistor architecture can both yield improved analog/RF figures of merit in comparison to conventional abrupt source/drain MOSFETs. Junctionless devices overcome the gain–bandwidth trade–off associated with analog design. The results are significant for RFICs in emerging ultra-low-power technologies.

Keywords: junctionless, low power, analog/RF and Double Gate MOSFET.

Introduction

A fundamental challenge in analog design is the trade–off between gain and bandwidth due to the conflicting dependence of both metrics on gate length. Scaling down gate length improves the cut–off frequency whereas voltage gain is degraded [1]. In the nanoscale regime, both gain and cut–off frequency are expected to degrade due to short channel effects (SCEs) [2]. Multiple–gate architectures such as Double Gate (DG) and FinFET topologies have demonstrated excellent control of SCEs [3]. However, the fundamental gain–bandwidth trade–off and achieving enhanced metrics at lower current levels are still crucial issues.

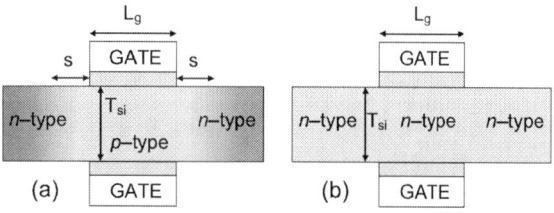

Fig.1 Schematic diagram of Double Gate (DG) (a) Underlap and (b) Junctionless (JNL) MOSFET.

Underlap source/drain (S/D) design (fig. 1*a*) has shown the potential for achieving enhanced low power analog/RF performance metrics [4]. Underlap design reduces inner fringing capacitance, suppresses SCEs thereby improving transconductance and output conductance at low current levels. Even though underlap design has shown improved performance, it requires dual spacer process and precise control of S/D profile in the extension regions [5]. An alternative to achieve improved performance metrics with

simpler fabrication process is to use Junctionless (JNL) transistor (fig. 1*b*) architecture [6-7]. In this work, we show that ultra–low–power Double Gate (DG) JNL MOSFETs perform better than conventional inversion mode MOSFETs, and achieve performance metrics that are comparable to underlap devices

Results and Discussion

JNL DG MOSFETs with gate length (L_g) of 20 nm, gate oxide thickness (T_{ox}) of 1.5 nm, and film thickness (T_{si}) of 10 nm were simulated using ATLAS simulator [8] using Lombardi mobility model [9]. The doping (N_d) of the *n*–type junctionless devices was taken to be 10^{19} cm^{-3}. A spacer width (*s*) of 18 nm and S/D doping gradient (*d*) at the gate edge was selected to be 3 nm/decade for underlap devices [4]. Drain bias (V_{ds}) was fixed at 0.5 V for all devices. As our focus is for low power applications, we will restrict the analysis up to a drain current (I_{ds}) of 40 μA/μm.

Fig. 2*a–d* shows the dependence of analog/RF performance metrics on drain current. Despite high doping concentration (10^{19} cm^{-3}), JNL devices at low current levels achieve transconductance (g_m) values that are nearly comparable to abrupt S/D devices (fig. 2*a*). This is due the formation of conducting channel through the centre of the silicon film. g_m values are nearly same for all devices up till $I_{ds} < 10$ μA/μm. JNL devices achieve higher transconductance–to–current (g_m/I_{ds}) ratio (30 V^{-1}) in comparison to abrupt S/D devices (25 V^{-1}) in the weak inversion region (fig. 2*b*). The higher peak–g_m/I_{ds} reflects on the reduction of short channel effects and a lower subthreshold slope ($S = \ln(10)/(g_m/I_{ds})_{max}$) in JNL MOSFETs. In moderate and strong inversion regions ($g_m/I_{ds} < 10$ V^{-1}), g_m/I_{ds} values degrade sharply in JNL devices due the influence of source/drain resistance and mobility.

As shown in fig. 2*c*, Early voltage ($V_{EA} = I_{ds}/g_{ds}$, where g_{ds} is the output conductance) is the highest for JNL in comparison to abrupt S/D and underlap devices. The absence of a junction results in the reduction of the peak electric field at the drain end which reduces g_{ds} (or improves output resistance, $R_{out} = 1/g_{ds}$). The combined effects of above metrics are reflected in the higher values of intrinsic voltage gain ($g_m/g_{ds} = g_m/I_{ds} \times V_{EA}$) in JNL MOSFETs. g_m/g_{ds} values in JNL devices are comparable to that of underlap MOSFETs and significantly higher ($\cong 60$ %) than that exhibited by conventional inversion mode abrupt S/D MOSFETs.

Fig. 3*a–b* shows the behavior of capacitance (C_{gg}) and cut–off frequency ($f_T = g_m/2\pi C_{gg}$) for the three different device

978-1-4673-4840-9/13 $31.00 © 2013 IEEE

architectures. JNL MOSFET exhibits the least capacitance in comparison to the other two devices. The capacitance is ~ 30 % lower than that of abrupt S/D MOSFET. The absence of *pn* junction at the gate edge allows the depletion width to extend beyond the gate (along the channel direction). This reduces the electron concentration in regions adjacent to gate edge, and lowers the inner fringing capacitance. In the subthreshold region (for low power applications), the total gate capacitance is dominated by parasitics, and the reduction in fringing capacitance is beneficial for lower capacitance in JNL devices. This reduction in fringing capacitance in JNL MOSFETs is comparable to that of underlap devices.

The dependence of f_T on I_{ds} is shown in fig. 3*b*. JNL MOSFETs achieve f_T values higher than abrupt S/D devices and comparable to that exhibited by underlap devices up to I_{ds} = 20 μA/μm. As the reduction in C_{gg} dominates over the degradation in g_m (fig. 2*a*), JNL devices achieve higher f_T values. Beyond I_{ds} = 20 μA/μm, f_T is degraded due to series resistance effect. Another interesting aspect is that for $I_{ds} < 20$ μA/μm, JNL devices exhibit an improvement in both voltage gain (g_m/g_{ds}) and cut–off frequency (f_T) without any need of channel engineering topology. Previously, the simultaneous improvement in g_m/g_{ds} and f_T was only possible either with underlap S/D design [4] or graded channel architecture [10]. As JNL devices have simpler fabrication process without the need for steep junctions, JNL devices will be most useful for ultra–low–power applications.

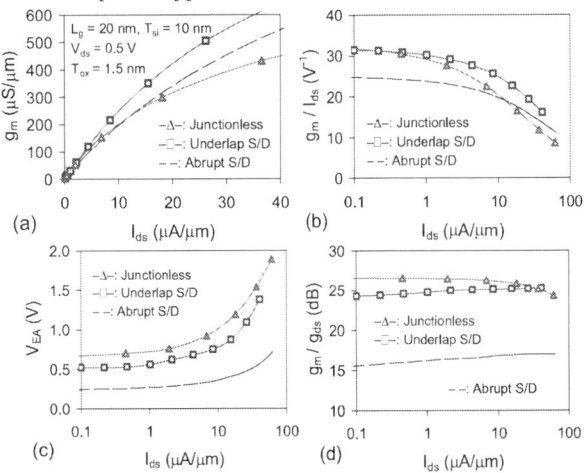

Fig.2 Dependence of (a) transconductance (g_m), (b) transconductance – to – current ratio (g_m/I_{ds}), (c) Early voltage (V_{EA}), and (d) voltage gain (g_m/g_{ds}) on drain current (I_{ds}) for junctionless, underlap and abrupt source/drain MOSFETs.

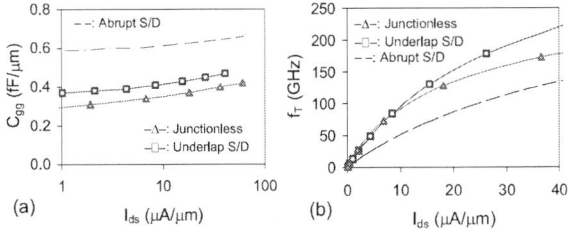

Fig.3 Dependence of (a) gate capacitance (C_{gg}), and (b) cut–off frequency (f_T) on drain current (I_{ds}) for junctionless, underlap and abrupt source/drain MOSFETs.

Conclusion

Analog/RF performance of low power Junctionless MOSFETs has been analyzed. It is shown that Junctionless transistors exhibit enhanced performance metrics in comparison to abrupt source/drain devices. Junctionless devices can overcome the gain–bandwidth trade–off at low current levels and achieve performance metrics comparable to underlap devices.

Acknowledgement

This work is supported by the Science and Engineering Research Board, Department of Science and Technology, Government of India, under Grant SR/S3/EECS/0130/2011.

References

[1] A. Kranti, and G.A. Armstrong, "Nonclassical channel design for improving OTA gain–bandwidth trade–off", *IEEE Trans. Circuits and Systems–I*, vol. 57, no. 12, pp. 3048–3054, 2010.

[2] International Technology Roadmap for Semiconductors. Available online: www.itrs.net

[3] J.-P. Colinge, Multiple-gate MOSFETs. *Solid–State Electronics*, vol. 48, no. 6, pp. 897–905, 2004.

[4] A. Kranti and G. A. Armstrong, "Design and optimization of FinFETs for ultra-low-voltage analog applications", *IEEE Trans. Electron Devices*, vol. 54, no. 12, pp. 3308–3316, 2007.

[5] B. A. Anderson, A. Bryant, W. F. Clark Jr, and E. J. Nowak, "Low capacitance FET for operation at subthreshold voltages," U.S. Patent 7009 265, 2006.

[6] J.-P. Colinge, C.-W. Lee, A. Afzalian, N. Dehdashti Akhavan, R. Yan, I. Ferain, P. Razavi, B. O'Neill, A. Blake, M. White, A.-M. Kelleher, B. McCarthy and R. Murphy, "Nanowire transistors without junctions", *Nature Nanotechnology*, vol. 5, no. 3, pp. 225-229, 2010.

[7] A. Kranti, C.-W. Lee, I. Ferain, R. Yan, N. Akhavan, P. Razavi, R. Yu, G.A. Armstrong and J.-P. Colinge, "Junctionless 6T SRAM cell", *Electronics Letters*, vol. 46, no. 22, pp. 1491 – 1493, 2010.

[8] ATLAS Users Manual, Silvaco.

[9] C. Lombardi, S. Manzini, A. Saporito, and M. Vanzi, "A physically based mobility model for numerical simulation of nonplanar devices", *IEEE Trans. Computer Aided Design of Integrated Circuits and Systems*, vol. 7, no. 11, pp. 1164-1171, 1988.

[10] A. Kranti, T. M. Chung, D. Flandre, and J.-P. Raskin, "Laterally asymmetric channel engineering in fully depleted double gate SOI MOSFETs for high performance analog applications", *Solid–State Electronics*, vol. 48, no. 6, pp. 947–959, 2004.

High Performance Gadolinium Oxide Nanocrystal Memory with Optimized Charge Storage and Blocking Dielectric Thickness

Chih-Ting Lin, Chi-Feng Chang, Yu-Ren Yen, Chin-Hsiang Liao, Po-Wei Huang, and Jer-Chyi Wang[*]

Department of Electronic Engineering, Chang Gung University

No. 259, Wen-Hwa 1st Road, Kweishan 333, Taoyuan, Taiwan

Tel: +886-3-2118800 ext.: 5784, Fax: +886-3-2118507, E-mail: jcwang@mail.cgu.edu.tw

Abstract

The gadolinium oxide nanocrystal (Gd_2O_3-NCs) memory has been demonstrated with large memory window and good reliabilities. It has been proven that the charges are stored in the crystallized Gd_2O_3-NCs which are surrounded by trapping-free amorphous phase of Gd_2O_3 film. The calculated deep trapping energy level is about 1.7eV below the conduction of the Gd_2O_3-NCs, and the energy band diagram of the Gd_2O_3-NCs memory has been proposed. In this study, we optimize the thickness of Gd_2O_3 thin film to investigate the effect on the charge storage mechanism. The thickness of Gd_2O_3 thin film and blocking oxide could simultaneously dominate the operation speed and charge loss phenomenon. Thus, the most suitable film thicknesses for the Gd_2O_3-NCs memory were proposed for future mass production nonvolatile memory application.

Keywords: Gd_2O_3, Nonvolatile memory and Nanocrystal.

Introduction

Nonvolatile memories using floating gate (FG) structure have been encountered a serious bottleneck due to the limitation of tunneling oxide thickness and the scaling induced cell-to-cell coupling effects. Many efforts have been made to dedicate the solutions on process improvement of the flash technology. One solution is to apply the isolated nanocrystals (NCs) such as Si, Ge, Au, HfO_2, RuO_2, Al_2O_3 as discrete charge storage nodes [1]. Among the metal-oxide (MO) materials, the gadolinium oxide nanocrystal (Gd_2O_3-NCs) memory has been demonstrated with large memory window and good reliabilities [2]. It has been proven that the charges are stored in the crystallized Gd_2O_3-NCs which are surrounded by the amorphous phase of Gd_2O_3 film. The calculated deep trapping energy level is about 1.7eV below the conduction of the Gd_2O_3-NCs, and the energy band diagram of the Gd_2O_3-NCs memory has been proposed [3-5]. However, to demonstrate the practicality of the Gd_2O_3-NCs on the flash memory application, it is necessary to optimize the thicknesses of the charge trapping layer and blocking oxide. In this paper, we demonstrate the physical effect of various Gd_2O_3-NCs layer and blocking oxide thicknesses. It was found that the high performance of the Gd_2O_3-NCs memory can be obtained with optimized thickness of the Gd_2O_3-NCs layer.

Experiments

The Gd_2O_3-NC memory devices were fabricated on 4 inch n-type (100) silicon wafers. After the standard RCA clean, a 3nm tunnel oxide (TO) was thermally grown at 850°C in N_2 and O_2 mixed ambience by horizontal furnace. Then, the various thicknesses of gadolinium oxide was deposited by RF sputtering with a pure gadolinium (99.9% pure) target in oxygen (O_2) and argon (Ar) mixture ambience at room temperature. The proportion of oxygen and argon ambient flow rate is 1:7 and the pressure of the chamber is 20mtorr for amorphous Gd_2O_3 formation. After the dielectric films had been formed, the samples were rapid thermal annealed (RTA) at 900°C for 30s in N_2 ambient to form the Gd_2O_3 nanocrystals. After that, the various thicknesses of blocking oxide (BO) SiO_2 using the SiH_4 and N_2O mixed gases were deposited by plasma enhanced chemical vapor deposition (PECVD) system at 300°C. The gas flow ratio of SiH_4 and N_2O gases was set to be 5:200 to lower the deposition rate for better SiO_2 quality. A 300nm Al film was deposited by thermal coater with a pure Al ingot (99.9999% pure), and a gate was defined lithographically and etched. For the electrical analysis, the capacitance-voltage (C-V) hysteresis was measured by HP4285 precision LCR meter and the P/E characteristics were measured by HP8110 to supply the gate pulse.

Results and discussion

A. The effects on the memory performance of the Gd_2O_3 thickness.

The schematic of the RTA system is shown in Fig. 1(a). The heating lamp is at the top of the system. During the heating process, the nuclei sites near the surface of the amorphous Gd_2O_3 layer will obtain the initially kinetic energy and form the crystallized phase while the heating procedure continually proceeds. In the meantime, the bottom Gd_2O_3 cannot get high enough kinetic energy and it still remains the amorphous phase. Therefore, the Gd_2O_3-NCs surrounded by the amorphous Gd_2O_3 is distributed near the surface of the film, as shown in Fig. 1(b). In order to compare the program and erase (P/E) efficiency of the Gd_2O_3-NCs memories, the electrical field (EF) of tunneling oxide and blocking oxide is fixed while examining the P/E speed for different Gd_2O_3-NCs thicknesses. Fig. 2(a) shows the program speed of the Gd_2O_3-NCs memories. The interesting phenomenon is that the program speed is reduced at higher EF. This trend is even enlarged for thicker Gd2O3 layer thickness. The extracted flat band voltage shift at 1ms pulse width is shown in Fig. 2(b). For thinner Gd_2O_3 layer thickness, the reduced flat voltage shift at higher EF is due to the electron injection through the blocking oxide to the gate terminal, resulted in the reduction of the electrons trapped in the Gd_2O_3-NCs. This phenomenon is theoretically eliminated with thicker Gd_2O_3 layer because the existence of the higher band-gap of the amorphous Gd_2O_3 layer can prevent the electron injection through the gate terminal. However, for 20nm Gd_2O_3 layer, the program speed is much lower at higher EF. This can be due to the additional hole trapping in the Gd_2O_3 layer. Fig. 3 shows the erase speed of the Gd_2O_3-NCs memory. It is observed that the erase speed is increased with

978-1-4673-4840-9/13 $31.00 © 2013 IEEE

thicker Gd_2O_3 layer, indicating that the additional hole from the substrate is injected into the Gd_2O_3 layer. The additional hole is believed to be trapped in the amorphous Gd_2O_3 region instead of the Gd_2O_3-NCs, since the additional hole trapping is only observed in the thicker Gd_2O_3 layer. Fig. 4 shows the geography location of the electron and hole trapping in the Gd_2O_3 layer.

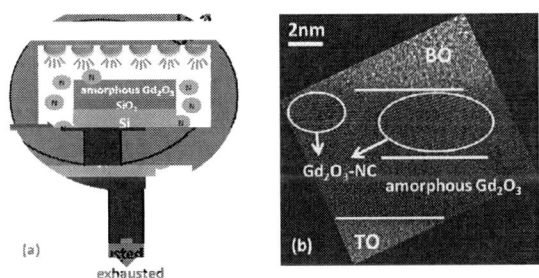

Fig. 1 (a) The schematic of the RTA system. (b) The HRTEM of the Gd_2O_3-NCs memory.

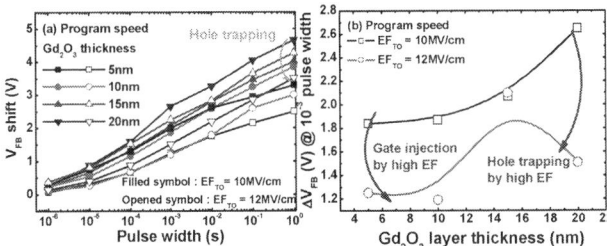

Fig. 2 (a) The program speed of the Gd_2O_3-NC memory with various Gd_2O_3 layer thicknesses. (b) The extracted flat voltage shift at 1ms pulse width.

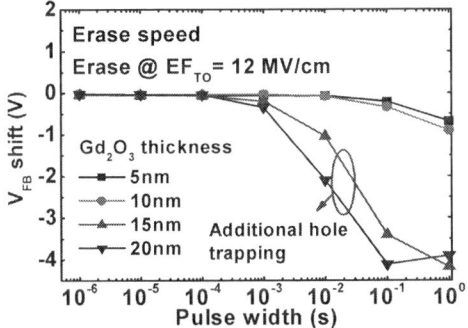

Fig. 3 The erase speed of the Gd_2O_3-NCs memory with various Gd_2O_3 layer thicknesses.

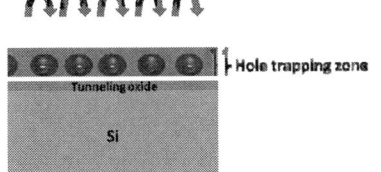

Fig. 4 The location of the electron and hole trapping zone in the Gd_2O_3 charge trapping layer

The retention characteristic of the Gd_2O_3-NCs memory for 20nm Gd_2O_3 layer thicknesses is shown in Fig. 5. With higher temperature, the charge loss is drastically enhanced due to the

hole trapping in the Gd_2O_3 layer since the initial flat voltage shift is fixed at the same for the retention measurement. The inset shows the retention characteristic at 85°C. The 10nm Gd_2O_3 layer has the lower charge loss, indicating the better performance of the Gd_2O_3-NCs memory. The endurance characteristic is shown in Fig. 6. The memory window can be remained 2.5V even after 10^4 cycles.

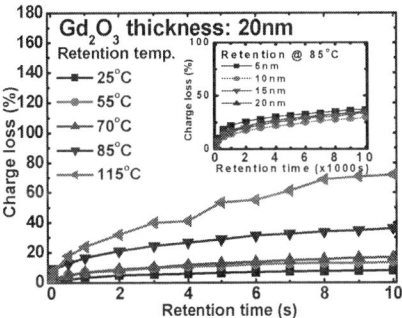

Fig. 5 The retention characteristic of the Gd_2O_3-NCs memory with 20nm Gd_2O_3 layer. Inset shows the retention characteristic at 85°C.

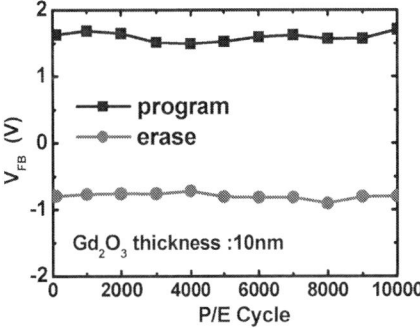

Fig. 6 The endurance characteristic of the Gd_2O_3-NCs memory with 10nm Gd_2O_3 layer.

B. The effects on the memory performance of the blocking oxide thickness.

The program speed of the Gd_2O_3-NCs memory with various blocking oxide thickness is shown in Fig. 7. The thicker blocking oxide can prevent the electron from tunneling to the gate terminal, resulting in larger amount of electrons in the Gd_2O_3-NCs. However, during the erase operation, the electron gate injection into the Gd_2O_3-NCs is observed, as shown in Fig. 8. This can be due to the Fermi-level pinning effect for different blocking oxide thickness, since the plasma damage inducing charge can be occurred during the PECVD blocking oxide deposition. In order to extract the barrier height at Al gate/blocking oxide interface, the gate current density versus gate voltage (J-V) is demonstrated. The Fowler-Nordheim tunneling mechanism is considered to extract the barrier height. The result is shown in Fig. 9 [6]. An exponential decay of the barrier height with the blocking oxide thickness is observed. This phenomenon is caused by the plasma damage during the blocking oxide deposition. With longer time deposition process, the plasma damage is enhanced and the charge or the oxygen vacancy is induced in the blocking oxide, resulted in the Fermi-Level pinning of the aluminum gate. This lower barrier height observed in the thicker blocking oxide will enhance the electron gate injection during the erase operation, and the erase speed is lower.

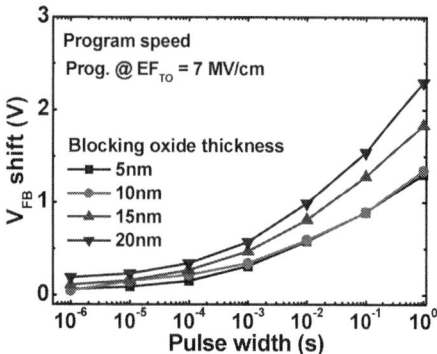

Fig. 7 The program speed of the Gd_2O_3-NCs memory with various blocking oxide thickness.

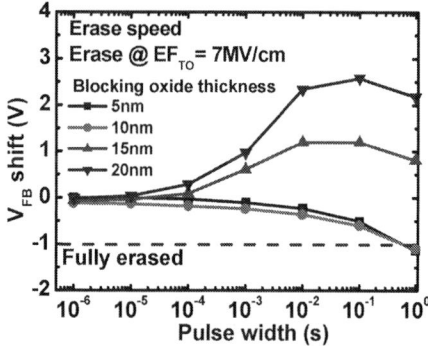

Fig. 8 The erase speed of the Gd_2O_3-NCs memory with various blocking oxide thickness.

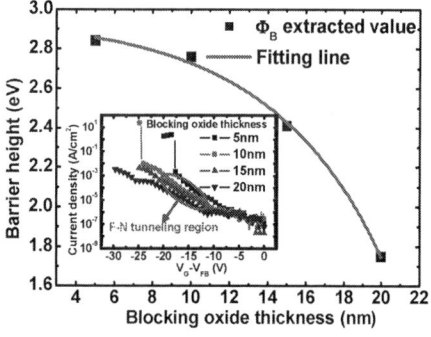

Fig. 9 The extracted barrier height with different blocking oxide thickness.

Conclusion

The physical effect of various Gd_2O_3-NCs layer and blocking oxide thicknesses applied in the memory is demonstrated in this paper. It was found that the high performance of the Gd_2O_3-NCs memory can be obtained with 11nm thickness of the Gd_2O_3-NCs layer. On the other hand, thicker PECVD blocking oxide will enhance the plasma damage, resulting in the lowering of the barrier height. This effect will affect the memory performance seriously. Thus, an alternative material for the blocking oxide is needed in the Gd_2O_3-NCs memory.

Acknowledge

We would like to express sincere thanks to the support of National Science Council, R.O.C. under the contract No. of NSC 101-2221-E-182-053.

References

[1] F. M. Yang, et al., *Appl. Phys. Lett.*, 2, pp. 212104, 2007.
[2] J. C. Wang, et al., *Electrochem. Solid State Lett.*, 12, pp. H202-H204, 2009.
[3] J. C. Wang, et al., *Microelectronic Relia.*, 52, pp. 635-641, 2012.
[4] J. C. Wang, et al., *Appl. Phys. Lett.*, 97, pp. 023513, 2010.
[5] J. C. Wang and C. T. Lin, *J. Appl. Phys.,* 109, pp. 064506, 2011.
[6] D. K. Schroder, *Semiconductor Material and Device Characterization*, 2nd ed., New York: John Wiley & Sons, 1998, pp. 392,

Threshold Voltage of Nanoscale Si Gate-all-around MOSFET: Short-channel, Quantum, and Volume Effects

Min-Chul Sun[1,2], Hyun Woo Kim[1], Sang Wan Kim[1], Jung Han Lee[1], Hyungjin Kim[1], and Byung-Gook Park[1]

[1]Inter-university Semiconductor Research Center (ISRC) and Department of Electrical Engineering and Computer Science, Seoul National University, Seoul 151-744, Republic of Korea
[2]TD Team (S.LSI), Samsung Electronics Co., Ltd., Yongin, Gyeonggi 446-711, Republic of Korea
Phone: +82-2-880-7279, Fax:+82-2-882-4658, E-mail: sunnie73@snu.ac.kr

Abstract

The threshold voltage (V_T) control of gate-all-around (GAA) MOSFETs as an extreme case of multi-gate MOSFETs is studied. After breaking down the components constituting V_T into several terms, their relative sizes are compared using TCAD technique and analytical calculation of quantum mechanical problem. As a result, the channel diameter is found to be an important knob to control V_T of a GAA MOSFET, which influences on short-channel effect, quantum confinement effect and other effects.

Keywords: gate-all-around, field-effect transistor, threshold voltage, short-channel effect, quantum effect, and volume effect

Introduction

In recent logic CMOS technology, the transition from planar bulk device to multi-gate device is rapidly occurring to achieve smaller standby power consumption and better current drivability [1]-[6]. This will make it more and more challenging for device manufactures to provide circuit designers with multi-threshold voltage options for low power design technique because the reduction of space charge region and channel doping concentration lead to a smaller contribution of depletion charge on the threshold voltage (V_T). Although the techniques of ground-plane and back-gate across thin bottom oxide are being revisited for V_T tunability of floating-body MOSFETs, these either require large swing of back-gate bias as much as 2~3 V to obtain practically useful amount of modulation or are not applicable at all as the surface-to-body volume ratio extremely increases [7]-[9]. A quite interesting trial to control V_T of a FinFET nanowire device with conventional channel doping is lately reported, but the physics behind the phenomena is not explained in detail [10].

Therefore, we investigate V_T control of gate-all-around (GAA) MOSFETs as an extreme case of multi-gate MOSFETs in this work. After breaking down the components constituting V_T into several terms, the relative sizes of the terms are quantitatively compared using TCAD simulation technique and analytical calculation of quantum-mechanical problem.

Simulation Method and Parameter Fitting with Literature Data

In order to set up a model GAA device structure and make the simulation work more reliable, the published information on Cheng's extremely-thin silicon-on-insulator (ETSOI) MOSFET is used as a reference (Fig. 1) [11]. Obtained values for the fitting parameters from the ETSOI device are used to construct the model GAA device. Electrical oxide thickness, doping concentration of the body region, series resistance, gate work function, and velocity saturation parameters are the fitting parameters to match off-current, subthreshold slope, drain-induced barrier lowering, and on-current of the ETSOI device. Considering the dimensions of the device, the energy-balance continuity equations and the density-gradient quantum-mechanical model in addition to Poisson and carrier continuity equations are used [12]. A commercial TCAD device simulator of Sentaurus Device (ver. G-2012.06) of Synopsys Inc. solves the simulation problem with all these equations coupled [13]. Fig. 2 shows the matched results of the simulated characteristics to the literature data.

Threshold Voltage of GAA device

A. Analytical Estimation

Fig. 3 and Table I summarize the model GAA device of this study. The device structure is simply obtained from cylindrical transformation of the ETSOI device in the previous section. All the device fitting parameters are used as extracted.

First, before performing TCAD simulation on this structure, we compare the relative contributions from the terms influencing the V_T of the model GAA device with simple calculations. Equation (1) shows the general terms that determine V_T of a MOSFET in a BSIM-style expression [14]. V_{FB}, $2\varphi_F$, N_A, C_{ins}, and $-Q_{dep@V_T}$ stand for flat-band voltage, surface band-bending at V_T, the average space charge density in the depletion region at V_T, the capacitance per unit area of gate insulator, and the space charge per unit area in the depleted region at V_T respectively.

$$V_T = V_{FB} + 2\varphi_F + (-Q_{dep@V_T})/C_{ins}$$

$$+ \Delta V_T^{body\_eff} - \Delta V_T^{SCE} + \Delta V_T^{QM} + \Delta V_T^{others}. \quad (1)$$

Regardless of the geometry of a device, the first two terms of (1) should remain the same by the material properties and the definition of V_T. The third term is related to the channel doping. Since $-Q_{dep@V_T}$ and C_{ins} are calculated as $qN_A D_{NW}/4$ and $\kappa_{ins}\epsilon_0/t_{ins}$ respectively, this term of the model device will be only 0.2 mV. This result implies that the influence of channel doping as a form of depletion charge effect is insignificant unless the doping concentration of the body becomes greater than a few 10^{17} cm^{-3}. A practically useful amount of V_T modulation may require a change of the doping in the range of a few 10^{18} cm^{-3}.

The quantum-mechanical term can be calculated without much complexity by solving Schrodinger equation for the confined semiconductor system to obtain the ground state energy level with an assumption of infinite potential barrier [15]. Assuming uniform distribution of properties in the azimuthal direction and the substitution of $k \cdot r$ with a new variable u, the 2-D Schrodinger equation problem is reduced to ordinary 0th-order Bessel equation of u as in (2). Here, $\hbar$, m_e, ψ, V, and E are the reduced Plank constant, electron rest mass, wave function, potential of energy barrier, and energy, respectively. Using the boundary condition of finiteness of properties at the axis of the nanowire, the ground energy level value is obtained as in (3). It is noteworthy that the contribution from the quantum-mechanical effect becomes smaller than 10 mV for the devices larger than 10 nm in the nanowire diameter.

$$-\frac{\hbar^2}{2m_e}\nabla^2\psi + V\psi = E\psi \;\rightarrow\; u^2\frac{d^2\psi}{du^2} + u\frac{d\psi}{du} + u^2\psi = 0. \quad (2)$$

, where $u = k \cdot r$ and $k^2 = 2m_e E/\hbar^2$.

$$J_0\left(\frac{kD_{NW}}{2}\right) = 0 \;\text{ or }\; E_1 = \left(\frac{\hbar^2}{2m_e}\right)\frac{(2\times2.54)^2}{D_{NW}^2} \sim \frac{1000\,\text{meV}}{D_{NW}^2}. \quad (3)$$

, where J_0 is the 0th-order Bessel function of the first kind and the unit of D_{NW} is nanometer.

The body-effect term may be ignorable since this is a case of floating-body device. The last term in (1) is to include reverse-short-channel and narrow-width effect, but these may not be physically relevant for the GAA system either. The short-channel effect term will be considered in the next section.

B. TCAD simulations

The transfer characteristics of the model GAA device are simulated with the TCAD tool to investigate the contribution of short-channel effect.

First, the change of subthreshold transfer characteristics is examined with the channel diameter varied from 30 nm to 4 nm as shown in Fig. 4. As the diameter of the nanowire shrinks, the threshold voltage of the model device increases more than 400 mV. This is a similar result that Suk et al. observed in their study on the fabricated twin silicon nanowire FETs (TSNWFETs) [16]. From the calculation in the previous section, we now understand that neither depletion charge term nor quantum-mechanical effect can explain this amount of V_T shift.

Next, the devices of 25 nm and 90 nm in gate length are compared with V_{DS} set to be 0.9 V and 0.01 V in order to see if this is due to short-channel effect or not (Fig. 5). From the result, it looks obvious that short-channel effect is the dominant factor which causes V_T to increase as the diameter becomes smaller.

Finally, the origin of the residual variation with the diameter

is studied while comparing the transfer curves with and without the quantum model as shown in Fig. 6. In order to prevent the drain-side field from influencing V_T, the drain bias is set to be 0.01 V. Even for the case of the diameter larger than 10 nm, V_T gradually increases as the diameter decreases. This must not be due to quantum-mechanical effect considering the fact that quantum effect becomes less than 10 mV for the devices of a diameter larger than 10 nm. Therefore, other physics to explain this may be needed.

Conclusions

In this work, we investigated the factors influencing V_T of a GAA MOSFET by comparing their relative contributions to V_T. The channel diameter was confirmed to be an important knob to modulate V_T of a GAA MOSFET. The dominating physics behind this modulation was the change of short-channel effect with respect to the diameter. As a secondary factor, quantum-mechanical effect became significant for the devices with the diameter smaller than 10 nm. We also observed an additional increase of ~ 60 mV from narrow channel device even without turning on the quantum model.

Acknowledgements

This work was supported by the Center for Integrated Smart Sensors funded by the Ministry of Education, Science and Technology as Global Frontier Project (CISS- 2011-0031845).

References

[1] International Technology Roadmap for Semiconductors 2011, Process Integration, Devices, and Structures, [Online]. Available: http://www.itrs.net.
[2] H. Iwai, et al, *ECS Trans.*, vol. 35, pp. 33-35, May 2011.
[3] C. Auth, et al., in *Proc. Symp. VLSI Tech.*, Honolulu, Hawaii, p. 131-132, Jun. 2012.
[4] N. Planes, et al., in *Proc. Symp. VLSI Tech.*, Honolulu, Hawaii, p. 133-134, Jun. 2012.
[5] K. Keshavarzi, et al., in *IEDM Tech. Dig.*, Washington D.C., p. 67-70, Dec. 2011.
[6] S. Bangsaruntip, et al, in *IEDM Tech. Dig.*, Washington D.C., p. 297-300, Dec. 2009.
[7] W. Xiong, et al., *Electron. Lett.* vol. 35, pp. 2059-2060, Nov. 1999.
[8] O. Weber, et al., in *IEDM Tech. Dig.*, San Francisco, CA, p.58-61. Dec. 2010.
[9] M. Saito, et al, in *Proc. Symp. VLSI Tech.*, Honolulu, Hawaii, p. 11-12, Jun. 2012.
[10] C.-H. Lin, et al., in *Proc. Symp. VLSI Tech.*, Honolulu, Hawaii, p. 15-16, Jun. 2012.
[11] K. Cheng, et al., in *Proc. Symp. VLSI Tech.*, Kyoto, Japan, p. 213-214, Jun. 2012.
[12] *Sentaurus Device User Guide*, Synopsys Inc., Mountain View, CA, ver. G-2012.06, 2012.
[13] S. Jin, et al, *J. Semicond. Tech. Sci.*, vol. 6, pp. 1-9, Mar. 2006.
[14] W. Liu, *MOSFET Models for SPICE Simulation, Including BSIM3v3 and BSIM4*, New York, John Wiley & Sons, Inc., 2001, pp. 265-268.
[15] B.-G. Park, S. W. Hwang, and Young June Park, *Nanoelectronics Devices*, Singapore, Pan Stanford Pub., 2012, pp. 238-239.
[16] S. D. Suk, et al, in *IEDM Tech. Dig.*, Washington D.C., p. 891-894, Dec. 2007.

978-1-4673-4840-9/13 $31.00 © 2013 IEEE

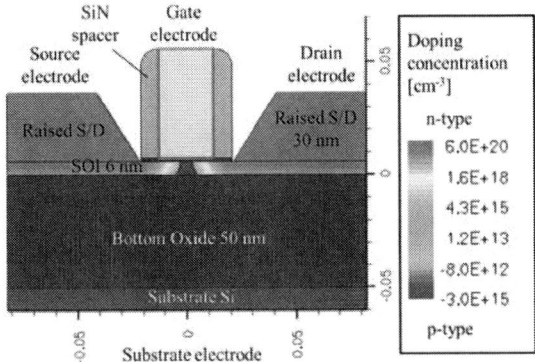

Fig. 1 ETSOI MOSFET used to extract device parameters for GAA device simulation. The structure is constructed based on the literature information [11].

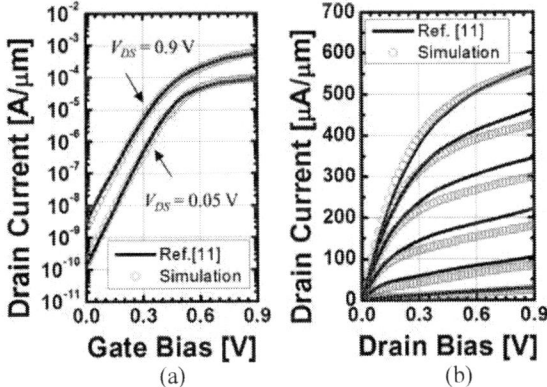

Fig. 2 Simulated and measured I-V characteristics of the ETSOI device: (a) transfer and (b) output characteristics: V_{DD} = 1.0 V and V_{GS} for (b) decreases with a step of 0.1 V.

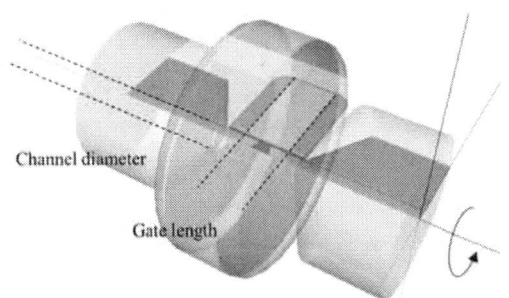

Fig. 3 Cylindrical transformation of the ETSOI device in Fig. 1 to obtain the model GAA device.

TABLE I
SUMMARY OF KEY DESIGN PARAMETERS OF THE
BASELINE GAA DEVICE

Parameters	Unit	Value
Gate length	[nm]	25
Gate work function	[eV]	4.615
Gate insulator thickness	[nm]	1.2
Raised S/D thickness	[nm]	30
S/D series resistance	[Ω]	300
Body doping	[cm⁻³]	3×10^{15}
Channel diameter	[nm]	12

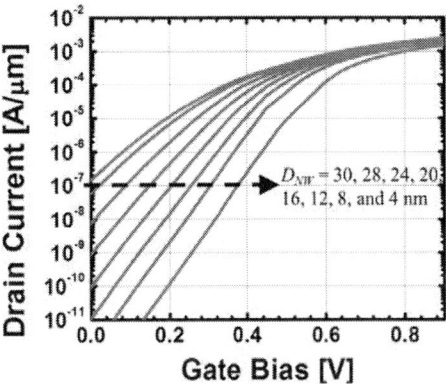

Fig. 4 Subthreshold characteristics of GAA devices of 25 nm in the channel length: The channel diameter is partitioned from 30 nm to 4 nm. Here V_{DS} is 0.9 V and the drain current is normalized with the channel diameter..

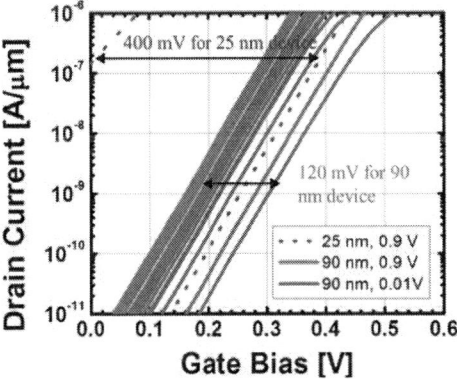

Fig. 5 Comparison of subthreshold characteristics of 25 nm and 90 nm GAA device: The left-most line shows the case of D_{NW} = 30 nm and the right-most does one of D_{NW} = 4 nm. It is noteworthy that the amounts of shift do not change much with the drain bias in the case of 90 nm devices.

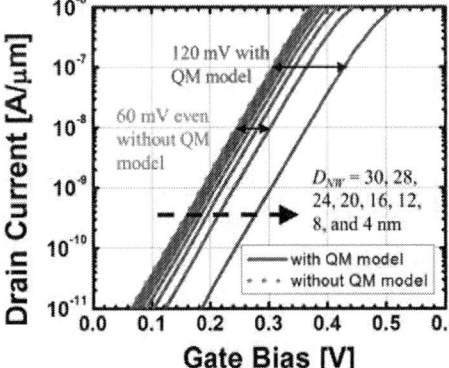

Fig. 6 Comparison of subthreshold characteristics of 90 nm GAA devices with and without the quantum model. Here V_{DS} is set to be 0.01 V to exclude the influence of short-channel effect. If the variation of I_D-V_{GS} curves in Fig. 5 were all due to the quantum effect, such variation should disappear without quantum model in Fig. 6.

978-1-4673-4840-9/13 $31.00 © 2013 IEEE

Enhancement of angular CCT by hybrid phosphor structure in White Light-emitting Diodes

Kuo-Ju Chen, Hsin-Chu Chen, Min-Hsiung Shih, Chao Hsun Wang, Shih-Hsuan Chien, Hsin-Han Tsai, Chien-Chung Lin, Hao-Chung Kuo*

Department of Photonics and Institute of Electro-Optical Engineering, National Chiao Tung University, Hsinchu 30010, Taiwan
Research Center for Applied Sciences, Academia Sinica 128 Academia Rd., Sec. 2 Nankang, Taipei 115 Taiwan
Institute of Photonics System, National Chiao Tung University, Tainan 711, Taiwan
hckuo@faculty.nctu.edu.tw
mhshih@gate.sinica.edu.tw

Abstract

The hybrid phosphor structure for white light-emitting diodes application was reported. It was demonstrated that the improvement of the luminous efficiency as well as the good uniform angular correlated color temperature (CCT) in hybrid phosphor structure compared with the conventional dispense method. The experimental results showed the CCT deviation was improved from 453K to 280K in the range of −70 to 70 degrees, which could be attributed to the large divergent angle of blue light. Moreover, the high lumen output was also achieved, as compared with dispense and conformal phosphor structures. Therefore, the experimental results indicated that the hybrid phosphor structure was feasible to use in the solid state lighting.

INTRODUCTION

In recent years, remarkable efforts have been developed to the white light-emitting diodes (LEDs), which has regarded as the promising illumination to replace the traditional lighting source such as mercury light and halogen lamp in the solid-state lighting [1-2]. Generally, the blue chip and yellow phosphor are combined to fabricate white light LEDs and exhibited as the most common approach in solid-state lighting [3]. Therefore, several articles have shown that the enhancement of internal quantum efficiency (IQE) increased by the large electron-hole wavefunction overlap quantum wells (QW) and the surface plasmon approach [4-5]. For the higher extraction efficiency, the enhanced light extraction by internal reflection (ELiXIR) is proposed from experiments [6]. In addition, the textured and air-gap structure are also demonstrated to enhance the luminous efficiency in white LEDs [7-8].

Most importantly, phosphor-dispensing method is used to fabricate white LED especially in the industry due to low cost and easy control. A disadvantage of this method is focusing on the convex surface which results in the inhomogeneous angular correlated color temperature (CCT) [9]. The non-uniform angular CCT would cause yellow ring phenomenon. To solve the problem, the conformal phosphor structure is fabricated by stacking phosphor layer to obtain the uniformity of CCT [10]. However, the influence of backscattering light from the phosphor layer would remain 60% of luminous flux [11]. Such the backscattering light not only reabsorbs by the blue chip, but also reduces the luminous efficiency. Although other studies were also demonstrated that the backscattering light could be reduced, it still hardly shows the uniform angular CCT. Therefore, how to keep the balance of the luminous efficiency and the good uniform angular CCT is the crucial point in LEDs

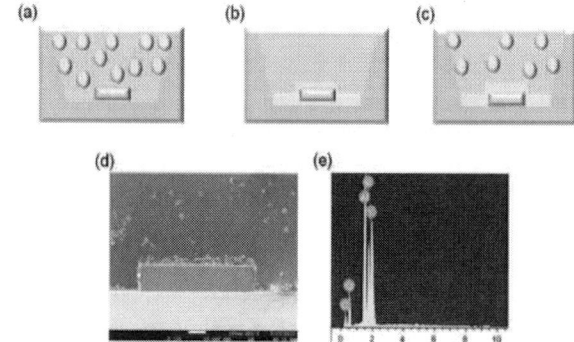

Fig. 1. Schematic diagram of three type packaging methods: (a) dispense (b) conformal (c) hybrid phosphor structure (d) SEM images of cross section of hybrid phosphor structure (e) EDS of the silicone and YAG compound.

package.

In this study, the hybrid phosphor structure is demonstrated to keep the luminous efficiency as well as uniform angular CCT by the combination of dispense and conformal phosphor structure. The CCT deviation of the hybrid phosphor structure indicated that high uniformity compared with dispense phosphor structure, which provides important application for high quality white LED.

EXPERIMENTS

In the experiment, the pulse spray coating (PSC) and the dispense method are employed to fabricate the hybrid phosphor structure. The experiment procedure shows as below: First, the phosphor slurry is sprayed to the leadframe by PSC method, which could form a uniform phosphor layer and easily control the CCT. Second, the phosphor powders are sufficiently blended with silicone and dispensed to the leadframe. The schematic diagram of three type packaging methods is illustrated in Fig.1(a-c), with the chip size of blue chip is 24mil and the wavelength is about 450 nm. The cross section of the hybrid phosphor structure is also shown in Fig.1(d). It is obvious that the clear interface between the PSC and dispense method. The thickness of the phosphor layer by PSC method is approximately 25μm and the other side is the random distribution of phosphor in dispense layer. Here, the $Y_3Al_5O_{12}$ (YAG) is used in the experiment which is analyzed by energy dispersive spectrometer (EDS) in Fig.1(e). The CCT of hybrid phosphor structure is controlled the same as the reference of dispense and conformal phosphor structure at 120mA.

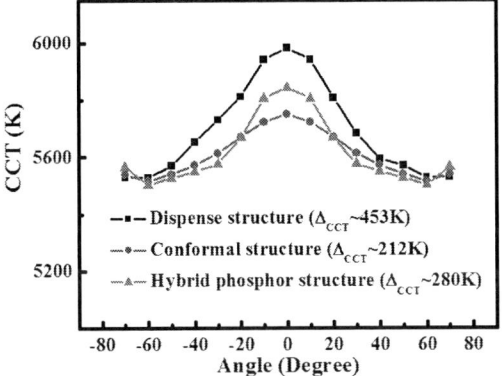

Fig. 2. (a) The angular-dependent correlated color temperature of three type packages.

RESULTS AND DISCUSSION

To understand the characteristic of CCT in LED, the CCT deviation is calculated as below:

$$\Delta CCT = CCT(Max) - CCT(Min)$$

,where CCT(Max) and CCT(Min) is represent as the CCT in the large and zero degree. The angular-dependent correlated color temperature of three type packages is showed in Fig. 2. The CCT deviations from −70 to 70 degrees of dispense, conformal and hybrid phosphor structures are 453 K, 212 K and 280 K, respectively. The yellow ring phenomenon is generated in the dispense phosphor structure due to the different ratio of blue and yellow light between the large and central degrees [12]. However, the hybrid phosphor structure demonstrates the high uniformity than dispense phosphor structure. Although the delta CCT in hybrid structure shows a little higher than the conformal structure, the performance of uniformity in angular CCT is still superior in white LEDs.

The emission spectra and lumen output of three different structures is shown in Fig. 3. These LEDs with different structures have the same CCT; ie., about 5500K. It is indicated that the hybrid structure has the higher radiation flux both in blue and yellow emission. The enhancement in yellow band could be attributed to the utilization in phosphor, resulting in the higher luminous efficiency. Furthermore, the lumen output of the hybrid structure is found to be enhanced by 4.25% and 2.57% at driving current of 120 mA, compared with dispense and conformal phosphor structures, respectively. Fig.3 (b) shows the lumen output of three kinds of structures as the function of driving current from 20 to 200 mA. With the increase of driving current, the differences of lumen output between three kinds of structures become larger. In the conformal phosphor structure, the partial of the backscattering light would loss inside the package, resulting in the reduction of the lumen flux [13]. Comparing with the conformal structure, the hybrid structure has the thin phosphor layer, which could reduce the reflection of the backscattering light, thus enhancing the lumen output. In addition, the refractive index (RI) of silicone is 1.4 and the phosphor is about 1.8. Thus, the RI of the phosphor layer with silicone is given by [14]

$$RI = V1RI1 + V2R2$$

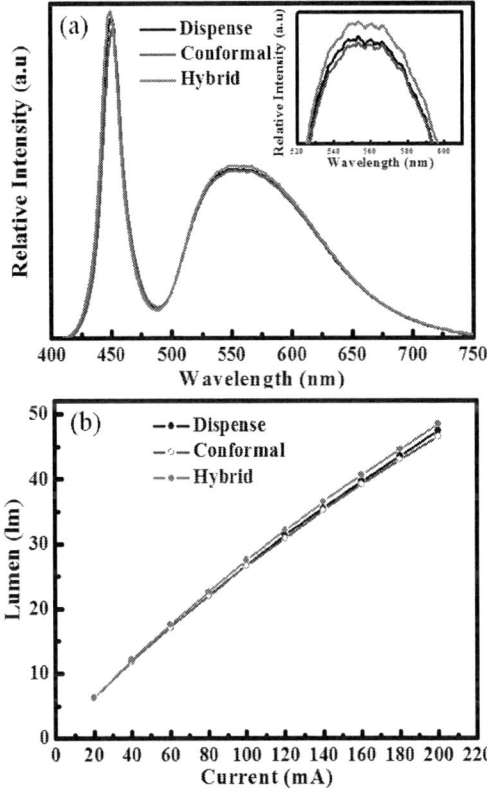

Fig. 3. (a) Emission spectra and (b) Lumen output of dispense, conformal and hybrid phosphor structure from 20mA to 200mA.

where V1 and V2 are concentration of the materials. For the hybrid structure, the mixing ratio of the phosphor layer to the silicone in the conformal coating and the dispense method are 50% and 2.5%, respectively. Therefore, RIs of the phosphor layer in each layer are 1.6 and 1.41. Comparatively, the RI of the phosphor layer is 1.42 in the dispense method. This indicates that the reason for the enhancement of the hybrid structure than dispense method is attributed to the refractive index variation between chip and phosphor layers. The RI of the hybrid structure becomes the gradual difference, which reduces the total reflect loss in the interface. Consequently, this indicates that the hybrid structure is suitable for the light output and posing better optical characteristic.

To discuss about the reason for the uniform CCT, the relative blue intensity of dispense and hybrid phosphor structure are shown in Fig. 4. It is obvious that the divergent angle of blue light in hybrid structure is larger than the dispense one. As the blue light increase in the large angle, the difference between the central and the large angle would be reduced. Therefore, the hybrid structure is verified to enhance the intensity of blue light in the large angle, which improves the light quality in white LEDs. The chromaticity coordinates shift of hybrid structure under current regulation from 20 to 200mA is shown in Fig.5. With the incensement current, the chromaticity coordinates shift keep the steady variation and the illumination characteristic in hybrid structure is almost the same. It is noted that the hybrid structure provide an appropriate solution in the white LEDs.

978-1-4673-4840-9/13 $31.00 © 2013 IEEE

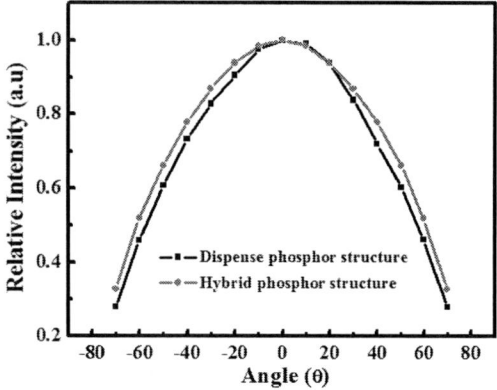

Fig. 4. Relative blue intensity of dispense and hybrid phosphor structure.

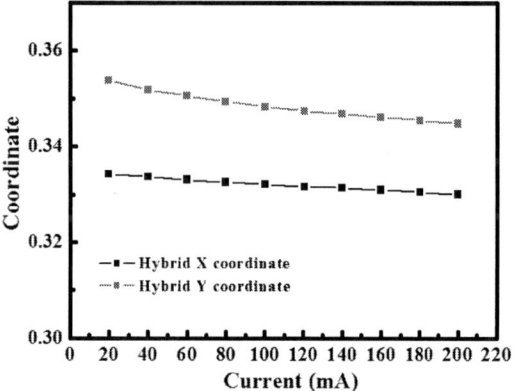

Fig. 5. Chromaticity coordinate of dispense and hybrid phosphor structure at the current from 20 to 200mA.

RESULTS AND DISCUSSION

In summary, we develop a hybrid phosphor structure to achieve high uniform white LEDs using the pulse spray and dispense method. While the CCT deviation from −70 to 70 degrees of the hybrid phosphor structures was improved to 280K from 453K in the conventional dispense method at 120 mA. This improvement could be attributed to the divergent angle of blue light in hybrid structure is larger than the dispense one. Moreover, the lumen output of the hybrid structure is found to be enhanced by 4.25% and 2.57% at driving current of 120 mA, compared with dispense and conformal phosphor structures. Finally, the high luminous efficiency as well as the good uniform angular CCT was achieved in hybrid phosphor structure, which is the suitable solution in the solid state lighting.

Acknowledgment

The authors would like to thank Helio Opto., Kismart Corporation and Wellypower Optronics for their technical support. This work was funded by the National Science Council in Taiwan under grant number, NSC 101-3113-E-009-002-CC2.

References

[1] E. F. Schubert, and J. K. Kim, "Solid-state light sources getting smart," *Science*, vol. 308, pp. 1274-1278, May 2005.

[2] Y. Renyong, J. Shangzhong, C. Songyuan, and L. Pei, "Effect of the Phosphor Geometry on the Luminous Flux of Phosphor-Converted Light-Emitting Diodes," *IEEE Photonic. Tech. L.* vol. 22, pp. 1765-7, December 2010.

[3] C. C. Lin, and R. S. Liu, "Advances in Phosphors for Light-emitting Diodes," *J. Phys. Chem. Lett.* vol. 2, pp. 1268-1277, June 2011.

[4] C. H. Lu, C. C. Lan, Y. L. Lai, Y. L. Li, and C. P. Liu, "Enhancement of Green Emission from InGaN/GaN Multiple Quantum Wells via Coupling to Surface Plasmons in a Two-Dimensional Silver Array," *Adv. Funct. Mater.* vol. 21, pp. 4719-4723, December 2011.

[5] M. Funato, T. Kondou, K. Hayashi, S. Nishiura, M. Ueda, Y. Kawakami, Y. Narukawa, and T. Mukai, "Monolithic polychromatic light-emitting diodes based on InGaN microfacet quantum wells toward tailor-made solid-state lighting," *Appl. Phys. Lett.*, vol. 1, pp.011106, January 2008.

[6] S. C. Allen, and A. J. Steckl, "ELiXIR - Solid-state luminaire with enhanced light extraction by internal reflection," *J. Disp. Technol.* vol. 3, pp. 155-159, June 2007.

[7] H. C. Chen, K. J. Chen, C. C. Lin, C. H. Wang, C. C. Yeh, H. H. Tsai, M. H. Shih, and H. C. Kuo, "Improvement of lumen efficiency in white light-emitting diodes with air-gap embedded package," Microelectron. Reliab. vol. 52, pp. 933-936, May 2012.

[8] H. C. Chen, K. J. Chen, C. H. Wang, C. C. Lin, C. C. Yeh, H. H. Tsai, M. H. Shih, H. C. Kuo, and T. C. Lu, "A novel randomly textured phosphor structure for highly efficient white light-emitting diodes," *Nanoscale Res. Lett.* vol. 7, pp. 1-5, March 2012.

[9] H. C. Kuo, C. W. Hung, H. C. Chen, K. J. Chen, C. H. Wang, C. W. Sher, C. C. Yeh, C. C. Lin, C. H. Chen and Y. J. Cheng, "Patterned structure of REMOTE PHOSPHOR for phosphor-converted white LEDs," *Opt. Express,* vol. 19, pp. A930-A936, July 2011.

[10] H. T. Huang, C. C. Tsai and Y. P. Huang, "Conformal phosphor coating using pulsed spray to reduce color deviation of white LEDs," *Opt. Express*, Vol. 18, pp. A201-A206, June 2010.

[11] N. Narendran, Y. Gu, J. P. Freyssinier-Nova, and Y. Zhu, "Extracting phosphor-scattered photons to improve white LED efficiency," Phys. Stat. Sol. (a), vol. 202, pp. 60–62, May 2005.

[12] Z. Liu, S. Liu, K. Wang, and X. Luo, "Optical Analysis of Color Distribution in White LEDs With Various Packaging Methods," *IEEE Photonic. Tech. L.* vol. 20, pp. 2027-2029, December 2008.

[13] Y. Shuai, N. T. Tran, and F. G. Shi, "Nonmonotonic Phosphor Size Dependence of Luminous Efficacy for Typical White LED Emitters," *IEEE Photonic. Tech. L.* vol. 23, pp. 552-554, May 2011.

[14] Y. H. Won, H. S. Jang, K. W. Cho, Y. S. Song, D. Y. Leon, and H. K. Kwon, "Effect of phosphor geometry on the luminous efficiency of high-power white light-emitting diodes with excellent color rendering property," *Optics Letters* vol. 34, pp. 1-3. January 2009.

Cu-Insulator-Si Hybrid Plasmonic Waveguide Based CMOS-Compatible Nanophotonic Devices

Shiyang Zhu, G. Q. Lo, and D. L. Kwong

Institute of Microelectronics, A*STAR (Agency for Science, Technology and Research)
11 Science Park Road, Singapore Science Park II, Singapore 117685
Phone: (65)67705746; Fax: (65)67731914; E-mail: zhusy@ime.a-star.edu.sg

Abstract

Vertical Cu-insulator-Si hybrid plasmonic waveguides (HPWs) along with various passive components are fabricated on a silicon-on-insulator platform using standard complementary metal-oxide-semiconductor (CMOS) technology and characterized at 1550-nm telecom wavelengths. The HPW exhibits relatively low propagation loss of ~0.12 dB/μm and high coupling efficiency of ~86% with the conventional Si strip waveguide. A plasmonic waveguide-ring resonator with 1.09-μm radius exhibits extinction ratio of ~13.7 dB, free spectral range of ~106 nm, and Q-factor of ~63. By applying a voltage between the metal cap and the Si core, the propagation property of HPW can be modulated to realize ultra-compact EO modulators.

Keywords: Nanophotonics, Nanoplasmonics, Silicon photonic integrated circuits

Introduction

In silicon photonic and electronic integrated circuits (Si EPICs), the photonic components have micrometer scale size due to the diffraction limit, much larger than the electronic components which have nanometer scale. A potential approach to scale down the photonic components to nanometer scale is to utilize surface plasmon polaritons (SPPs), which are collective oscillations of electrons at metal-dielectric interfaces [1]. Among various plasmonic waveguiding structures proposed to date, hybrid plasmonic waveguides (HPWs), in which the optical mode is confined in a low-index gap sandwiched between a metal and a high-index dielectric core, are a promising candidate because of their good tradeoff between long propagation distance and strong mode confinement [2-9]. Many kinds of HPWs have been investigated theoretically [2-5] and experimentally [6-11]. For ease of monolithic integration into a silicon-on-insulator (SOI) platform, the high-index dielectric core is naturally Si and the metal should be placed above the Si core because metal deposition is a back-end-of-line process. An example of such a vertical metal-dielectric-Si HPW named as conductor-gap-Si waveguide has been demonstrated, but it requires expensive electron beam lithography (EBL) to pattern the upper metal layer [8]. EBL can be avoided using a self-aligned approach, but it is only viable for locally oxidized Si waveguides which have blunt sidewalls [9]. An implicit assumption for the above HPWs is that the lateral light confinement should be provided by both the Si core and the patterned metal cap. However, in many other theoretical studies [2-3,5-7], the metal of HPWs has an infinite width, indicating that the lateral confinement can be provided by the Si core only. Moreover, for CMOS compatibility, the metal should be Cu or Al. At the 1550 nm

wavelength, it has been found that Cu provides a much lower metal loss than Al [12,13]. Regarding the abovementioned considerations, we develop a vertical Cu-dielectric-Si HPW structure as shown in Fig. 1. Various passive devices are experimentally demonstrated based on this plasmonic waveguide [10,11]. More importantly, the plasmonic waveguide is naturally a metal-oxide-semiconductor (MOS) capacitor, allowing a voltage to be applied between the Cu-cap and the Si core to medicate its propagation properties. An ultracompact plasmonic ring modulator is theoretically studied.

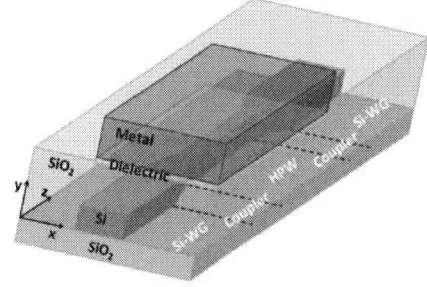

Fig. 1 The schematic metal-insulator-Si hybrid plasmonic waveguide, inserted in conventional Si channel waveguide through tapered couplers.

Experimental

The fabrication starts from SOI wafers with 0.34-μm top-Si and 2-μm buried SiO_2. After Si patterning, Si_3N_4 and SiO_2 were deposited sequentially, followed by SiO_2 chemical-mechanical polishing (CMP) to planarize the surface using Si_3N_4 as the CMP stopping layer. Then, Si_3N_4 and SiO_2 were sequentially deposited again, followed by SiO_2 window opening to expose the plasmonic area using Si_3N_4 as the etching stopping layer. After wet etching the remaining Si_3N_4 in the windows, a thin SiO_2 layer was thermally grown. A thick Cu layer was deposited, followed by Cu-CMP to remove Cu outside the windows. The thin oxide layer can prevent Cu diffuse to Si. The fabrication flow is somewhat similar to that for the horizontal $Cu-SiO_2-Si-SiO_2-Cu$ plasmonic waveguides [12] except some additional steps such as SiO_2/Si_3N_4 deposition and SiO_2-CMP, and has been reported elsewhere [10,11]. The final HPW has a ~20-nm-thick SiO_2 layer sandwiched between the ~160-nm×304-nm rectangular Si core and the wide Cu cap, as shown in Fig. 2(a). The $|E_y|$-field distribution shown in Fig. 2(b) indicates a significant $|E_y|$ enhancement in the narrow SiO_2 gap which contains ~14% power, whereas the Si core contains 42% power and the SiO_2 claddings contain 44% power. The calculated propagation loss is 0.127 dB/μm and the real part of the effective modal index (n_{eff}) is 2.08. After dicing, the chips were measured using a fiber-chip-fiber method as conventional Si photonic chips [12]. Light from a broadband (1520-1620

978-1-4673-4840-9/13 $31.00 © 2013 IEEE

nm) laser source was quasi-TM polarized by a polarizer and then coupled into the input Si waveguide through a polarization-maintaining fiber. The transmitted light from the output Si waveguide was monitored by an optical spectrum analyzer.

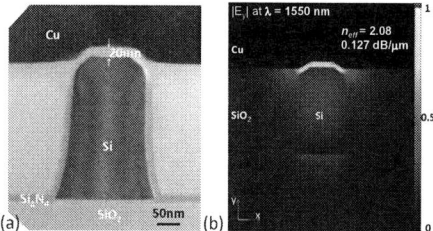

Fig. 2 (a) Cross-sectional transmission electron microscopy (XTEM) image of the fabricated HPW; (b) Electric field ($|E_y|$) distribution of the fundamental quasi-TM mode in the HPW calculated using the full-vectorial finite-difference method [3], $n_{eff} = 2.08$ and propagation loss = 0.127 dB/μm.

Experimental Results and Discussion

The straight HPWs with different lengths (L_P) are connected with the 0.5-μm-wide Si strip waveguide through 1-μm-long tapered couplers. Fig. 3 plots the transmitted power as a function of L_P, normalized by that of a reference Si waveguide without the plasmonic area.

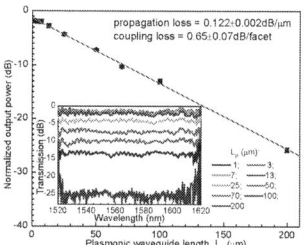

Fig. 3 Output power through the straight HPWs with different lengths (L_P), normalized by a reference Si waveguide without the plasmonic area.

From linear fitting, coupling loss of 0.65±0.07 dB/facet and propagation loss of 0.122±0.002 dB/μm are extracted, in good agreement with the theoretical prediction. Compared with the horizontal Cu-SiO₂-Si-SiO₂-Cu counterparts (which support TE-polarized light) [12], the vertical Cu-SiO₂-Si HPWs provide a ~2.5-times lower propagation loss but at the price of relatively weak power confinement. The inset of Fig. 3 plots the normalized transmission spectra of the straight HPWs. It indicates that both the propagation loss and coupling loss are almost wavelength-independent in the spectral range of 1520-1620 nm.

Fig. 4(a) illustrates the $|E_y|$-field distributions in the x-z and y-z planes of the 1-μm-long coupler, obtained from three-dimensional finite-difference time-domain (3D FDTD) simulation, showing that the TM light propagating along the 0.5-μm-wide Si waveguide can gradually launch the plasmonic mode in the HPW through the tapered coupler. The coupling loss estimated by comparing the powers monitored before and after the coupler is 0.6 dB/facet, in good agreement with the experimental value. Couplers with length (L_C) of 0.3, 0.5, 0.7, 1, and 2 μm are measured. Fig. 4(b) plots the measurement results as well as the theoretical curve. Unlike the horizontal Cu-SiO₂-Si-SiO₂-Cu counterpart [12], the coupling efficiency of the vertical HPW is not sensitive to L_C, which may be attributed to the relatively weak power confinement in the vertical HPW.

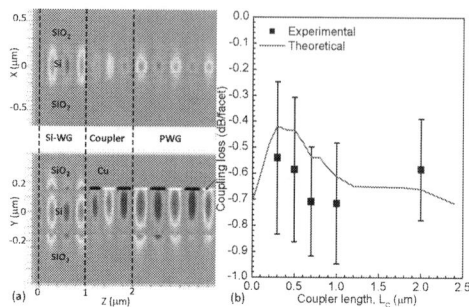

Fig. 4 (a) $|E_y|$ distribution in the x-z plane (top) and the y-z plane (bottom) of a 1-μm-long coupler, showing 1550-nm TM light in the 0.5-μm-wide Si waveguide being focused into the HPW by the coupler; (b) Experimental and theoretical coupling loss as a function of the coupler's length (L_C). Each data point is averaged from four identical HPWs and the standard deviation is indicated as the error bar. The theoretical curve is obtained from the 3D FDTD simulation.

Figs. 5(a) and (b) are microscopy pictures of ultra-compact 1×2 and 1×4 HPW-based power splitters. The insets are the scanning electron microscopy (SEM) images of their Si core patterns, whose opening angle is 90°, each branch connects with the Si strip waveguide through a 1-μm-long coupler, and the total length of each plasmonic route is 3 μm. Fig. 5(c) plots the transmission spectra measured on each output port, normalized by a 3-μm-long straight HPW. The relatively large data swings, especially for the 1×4 splitter, may arise from the weak parasitic Fabry-Perot resonances and the measurement error. Except this, the spectra depend on wavelength weakly, indicating the broadband response of the splitters.

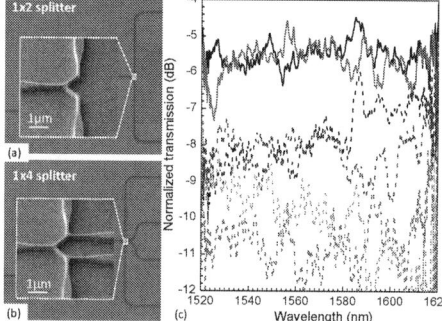

Fig. 5 Microscopy picture of a (a) 1×2 power splitter and (b) 1×4 power splitter, the inset is the SEM image of its Si core; (c) Transmission spectra measured on each output port of the 1×2 (the solid curves) and 1×4 power splitters (the dash curves), normalized by a 3-μm-long straight HPW.

We can see that the 1×2 splitter splits light equally with an excess loss of ~2.4 dB, whereas the 1×4 splitter splits light unequally: ~2.2-dB excess loss to the out-1 and out-4 and ~4.4-dB excess loss to the out-2 and out-3, respectively. They are much larger than those for horizontal Cu-SiO₂-Si-SiO₂-Cu counterparts [14], which may also be attributed to the relatively weak power confinement in the vertical HPWs. The excess loss of the HPW-based splitters may decrease significantly using a small radius S-bend [3,5].

Fig. 6 shows a HPW-based waveguide-ring resonator (WRR). Unlike the horizontal Cu-SiO₂-Si-SiO₂-Cu counterpart where an aperture is required to provide effective coupling between the ring and bus waveguides [15], the HPW WRR can provide effective evanescent coupling through a gap as conventional WRRs. Four HPW WRRs with the ring radius (R) of 1.09, 1.59, 2.09, and 2.59 μm are fabricated. The gap between the ring and bus waveguides is 0.2 μm in the layout, whereas it becomes ~0.24 μm in the final device due to the process-induced Si core shrinkage.

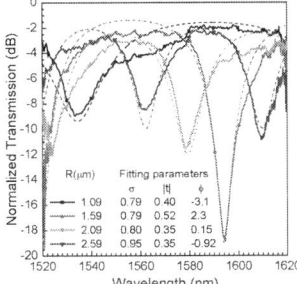

Fig. 6 Transmission spectra of HPW-based WRRs with different radius, normalized by the spectrum of a 7-μm-long straight HPW, the fitting curves are based on Eq. 1, n_g = ~3.25 for all four WRRs.

The transmission spectra in Fig. 6 are normalized by the corresponding 7-μm-long straight HPW. In general, the normalized transmission T(λ), can be expressed by [16]:

$$T(\lambda) = \frac{\alpha^2 + t^2 - 2\alpha t \cos\theta}{1 + \alpha^2 t^2 - 2\alpha t \cos\theta} \qquad (1)$$

where $\theta = (2\pi/\lambda)n_g 2\pi R$ is the phase change around the ring, $\alpha^2 = 10^{2\pi R\alpha_P/10}\sigma_b$ is the power attenuation factor per roundtrip around the ring (α_P being the propagation loss, equal to 0.122 dB/μm here, and σ_b being a parameter accounting for the pure bending loss, $\sigma_b = 1$ corresponds to no bending loss), $t = |t|\exp(i\phi)$ is the field transmission through the coupling region in the bus waveguide, n_g is the group index, and λ is the free-space wavelength. The spectra are fitted using n_g, σ_b, $|t|$, and ϕ as the fitting parameters. To reduce ambiguity, we assume that n_g is the same for four WRRs and the larger-R WRR has a larger σ_b value and a smaller $|t|$ value. The fitting results are shown in Fig. 6. The smaller-R WRR exhibits a relatively large extinction ratio (ER) because it has a larger $|t|$ value. In particular, the 1.09-μm R WRR exhibits ER of ~13.7 dB at the resonance (λ_r) of 1576 nm, free spectral range (FSR = $\lambda_r^2/(n_g 2\pi R)$) of ~106 nm, normalized insertion loss (IL) of ~2.3 dB, full width at half maximum (FWHM) of ~25 nm, and Q-factor (= $\lambda_r/FWHM$) of ~63. The Q-factor calculated by $Q = \dfrac{\pi n_g(2\pi R)\sqrt{\alpha t}}{\lambda(1-\alpha t)}$ [16] and parameters indicated in Fig. 6 is ~43 for the 1.09-μm R WRR and ~62 for 1.59-μm R WRR. For comparison, the WRRs under quasi-TE light and the WRRs without the Cu cap exhibit no clear resonance in the spectra. The n_g value extracted from WRRs (~3.25) is much larger than the n_{eff} value calculated from the straight waveguide (~2.08) due to the large dispersion effect of HPW, unlike the horizontal Cu-SiO$_2$-Si-SiO$_2$-Cu counterpart where these two values are similar [15]. By calculating n_{eff} of the straight HPW at different λ around 1550 nm, $dn_{eff}/d\lambda$ is estimated to ~ -0.0008 nm^{-1}, and n_g (= $n_{eff} - \lambda\dfrac{dn_{eff}}{d\lambda}$) is ~3.32, close to the value from WRRs.

The demonstrated performance (in terms of ER and Q-factor) of our HPW WRRs is similar to the horizontal

Cu-SiO$_2$-Si-SiO$_2$-Cu counterpart [15], but better than the dielectric-loaded plasmonic-waveguide-based WRRs if scaled to the same R [17]. As the abovementioned, the performance of our HPW WRRs can be further improved by tuning the $|t|$ value to meet the critical coupling condition of $|t| = \alpha$, which can be simply achieved by adjusting the separation gap in the layout. Numerical simulation shows that the HPW WRR may achieve a submicron R with Q up to 1000 [4]. For comparison, the aperture in the horizontal Cu-SiO$_2$-Si-SiO$_2$-Cu WRRs is difficult to be precisely controlled [13]. Although a Si strip waveguide has already achieved a 1.5-μm R WRR with Q up to 9000 [18], the HPW WRRs may provide even smaller R and some unique functionalities due to the metal cap and/or the low-index dielectric gap. For example, a heater can be placed just above the Cu layer to form an effective thermo-optic device, an effective active device such as a modulator can be realized if the SiO$_2$ gap is replaced by a dielectric with large electric-optic or thermo-optic coefficient, and the metal layer can function as an electrode, etc.

The plasmonic WRRs are measured again when the chip holder is heated to a certain temperature and keeping for a sufficiently long time before measurement to make sure that the holder and the chip have the same temperature. Fig. 7 plots the transmission spectra of the a plasmonic WRR with 2.59-μm R and 0.25-μm gap measured at 27°C, 50°C, 82°C, and 101°C, respectively. The thermo-optic (TO) coefficient (dn_g/dT) is extracted to be ~1.69×10^{-4}/°C, contributed by both the relatively large TO coefficient of Si (~1.21×10^{-4} /°C) and temperature-dependent Cu optical parameter.

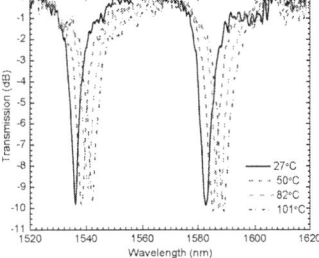

Fig. 7 Transmission spectra of the a plasmonic WRR with 2.59-μm R and 0.25-μm nominal gap measured at temperature of 27, 50, 82, and 101°C, respectively.

Theoretical Study on Electro-optic Modulators

Based on the above metal-insulator-Si hybrid plasmonic waveguide, we can design an ultra-compact electro-optic modulator as shown schematically in Fig. 8. The resonator consists of a Cu-dielectric-Si hybrid plasmonic waveguide with two electrodes located at the Cu cap and the center-donut respectively. A bus waveguide (conventional single-mode Si channel waveguide) is located near the resonator with a separation of "gap". The modulator is fabricated on SOI wafers with top Si thickness of h. Taking into the possible misalignment between the Si core layer and the SiO$_2$ window layer, we intentionally design that the window has a slightly larger size than the beneath Si core by 50 nm in each side. A dielectric layer is then deposited on the Si core through the windows, followed by Cu deposition and Cu chemical mechanical polishing (CMP) to remove the dielectric layer and the Cu layer outside the windows. We can see that the above design offers a freedom by selecting a suitable dielectric as a low-index slot between the Si core and the Cu cap. Here, we select HfO$_2$ as the gate dielectric because it has high κ value of ~25 and high reflective index of ~1.78 at 1550 nm.

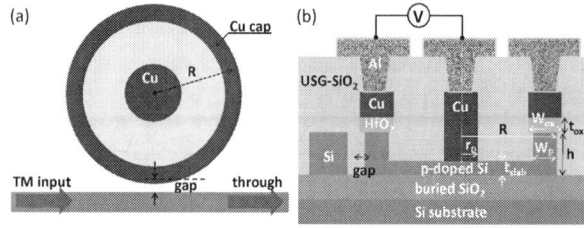

Fig. 8 (a) Top view, and (b) cross-section side view of the proposed Si plasmonic ring modulator based on Cu-insulator-Si hybrid plasmonic waveguide illustrating the ring-shaped high-κ dielectric gate, the Cu cap, and the Cu cylinder above the center-donut to contact the p-Si. A voltage is applied between the ring-shaped Cu cap and the center Cu cylinder contacted with Si.

To reach a large modulation, we design our device to work between the depletion and accumulation conditions, and the initial doping density is set to be N_A. Then, upon a voltage applied between two electrodes of our device, the beneath Si core can be in the conditions of depletion, flat-band, and accumulation, respectively. The depletion width (W_{dep}) depends on N_A as $W_{dep} \propto N_A^{-0.5}$. Here we assume $N_A = 1 \times 10^{19}$ cm^{-3}, thus W_{dep} = ~12 nm. The accumulated carriers are distributed within ~1-3 nm from the dielectric/Si interface. Here we simply assume the accumulation layer thickness (t_{AcL}) is 2 nm. The carrier density in AcL can be calculated as:

$$N_{accu} = \frac{\varepsilon_0 \cdot \varepsilon_d}{e \cdot t_{ox} \cdot t_{AcL}} (V - V_{FB}) = \frac{\varepsilon_0 \cdot \varepsilon_d}{e \cdot t_{AcL}} E_d \quad (2)$$

Fig. 9 plots the spectral transmittance of the modulator's output waveguide in the critical coupling condition in the depletion and accumulation states, respectively. The resonator has a loaded-Q value of ~400, close to that the experimental value. Δn_{eff} is calcualted to be ~0.0076, which leads to a the shift of resonant wavelength $\Delta \lambda$ of ~4.3 nm. We can see that a large extinction ratio (ER) of >6 dB is already obtained at a relatively large wavelength range. Obviously, a larger ER can be obtained simply by increasing the accumulation charge density, but it is priced by a large driving voltage as well as a large driving energy.

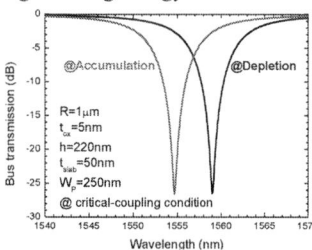

Fig. 9 calculated power transmission spectra of the optimized plasmonic donut modulator (in the critical coupling condition) in the depletion and accumulation states.

Conclusion

In conclusion, a fully CMOS-compatible vertical Cu-SiO$_2$-Si HPW structure is proposed, and various HPW components including power splitters and ring resonators are experimentally investigated. Moreover, an ultra-compact EO modulator is theoretically studied based on the developed hybrid waveguides.

References

[1] D. K. Gramotnev and S. I. Bozhevolnyi, "Plasmonics beyond the diffraction limit," *Nat. Photon.*, vol. 4, pp.83-91, 2010.

[2] R. F. Oulton, V. J. Sorger, D. A. Genov, D. F. P. Pile, and X.

Zhang, "A hybrid plasmonic waveguide for subwavelength confinement and long-range propagation", *Nat. Photon.*, vol. 2, pp. 496-500, 2008.

[3] H. S. Chu, Y. A. Akimov, P. Bai, and E. P. Li, "Hybrid dielectric-loaded plasmonic waveguide and wavelength selective components for efficiently controlling light at subwavelength scale," *J. Opt. Soc. Am. B*, vol. 28, pp.2895-2901, 2011.

[4] D. Dai, Y. Shi, S. He, L. Wosinshi, and L. Thylen, "Silicon hybrid plasmonic submicro-donut resonator with pure dielectric access waveguides," *Optics Express*, vol.19, pp.23671-23682, 2011.

[5] M. Z. Alam, J. Meier, J. S. Aitchison, and M. Mojahedi, "Propagation characteristics of hybrid modes supported by metal-low-high index waveguides and bends." *Optics Express*, vol.18, pp.12971-12979, 2010.

[6] V. J. Sorger, Z. Ye, R. F. Oulton, Y. Wang, G. Bartal, X. Yin, and X. Zhang, "Experimental demonstration of low-loss optical waveguiding at deep sub-wavelength scales," *Nat. Commun.*, vol. 2, art.331, 2011.

[7] J. Tian, Z. Ma, Q. Li, Y. Song, Z. Liu, Q. Yand, C. Zha, J. Akerman, Tong, and M. Qiu, "Nanowaveguides and couplers based on hybrid plasmonic modes," *Appl. Phys. Lett.*, vol.97, art.231121, 2010.

[8] M. Wu, Z. Han, and V. Van, "Conductor-gap-silicon plasmonic waveguides and passive components at subwavelength scale," *Opt. Express*, vol. 18, pp.11728-11736, 2010.

[9] I. Goykhman, B. Desiatov, and U. Levy, "Experimental demonstration of locally oxidized hybrid silicon-plasmonic waveguide," *Appl. Phy. Lett.*, vol.97, art.141106, 2010.

[10] S. Y. Zhu, G. Q. Lo, and D. L. Kwong, "Experimental demonstration of vertical Cu-SiO$_2$-Si hybrid plasmonic waveguide components on an SOI platform," *IEEE Photon. Techn. Lett.*, vol. 24, pp.1224-1226, 2012.

[11] S. Y. Zhu, G. Q. Lo, and D. L. Kwong, "Performance of ultracompact copper-capped silicon hybrid plasmonic waveguide-ring resonators at telecom wavelengths," *Optics Express*, vol. 20, pp.15232-15246, 2012.

[12] S. Y. Zhu, G. Q. Lo, and D. L. Kwong, "Components for silicon plasmonic nanocircuits based on horizontal Cu-SiO$_2$-Si-SiO$_2$-Cu nanoplasmonic waveguides," *Opt. Express*, vol.20, pp.5867-5881, 2012.

[13] T. K. Ng, M. Z. M. Khan, A. Al-Jabr, and B. S. Ooi, "Analysis of CMOS compatible Cu-based TM-pass optical polarizer", *IEEE Photonics Techn. Lett.*, vol. 24, pp. 724-726, 2012.

[14] S. Y. Zhu, G. Q. Lo, and D. L. Kwong, "Nanoplasmonic power splitters based on the horizontal nanoplasmonic slot waveguide," *Appl. Phys. Lett.*, vol. 99, 031112, 2011.

[15] S. Y. Zhu, G. Q. Lo, and D. L. Kwong, "Experimental demonstration of horizontal nanoplasmonic slot waveguide-ring resonators with submicrometer radius," *IEEE Photonics Techn. Lett.*, vol. 23, pp. 1896-1898, 2011.

[16] W. Bogaerts, P. D. Heyn, T. V. Vaerenhbergh, K. D. Vos, S. K. Selvaraja, T. Claes, P. Dumon, P. Bienstman, D. V. Thourhout, and R. Baets, "Silicon microring resonators," *Laser & Photonics Rev.*, vol.6, pp. 47-73 (2012).

[17] R. M. Briggs, J. Grandidier, S. P. Burgos, E. Feigenbaum, and H. A. Atwater, "Efficient coupling between dielectric-loaded plasmonic and silicon photonic waveguides," *Nano Lett.* vol. 10, pp. 4851-4857, 2010.

[18] Q. Xu, D. Fattal, and R. G. Beausoleil, "silicon microring resolators with 1.5-μm radius," *Optics Express*, vol.16, pp.4309-4315, 2008.

Electrical Probing of Multi-ions Solution by Using Graphene-based Sensor

Kuan-I Ho[1], Jia-Hong Liao[1], Chi-Hsien Huang[1], Ching-Yuan Su[1*],
Chao-Sung Lai[1,*]

Department of Electronic Engineering, Chang Gung University
259 Wen-Hwa 1st Road, Kwei-Shan, Tao-Yuan 333, Taiwan
* Corresponding author: chingyuansu@mail.cgu.edu.tw; cslai@mail.cgu.edu.tw

Abstract

The multi-ion sensing properties of graphene based field effect transistor were investigated in this paper. Hydrogen and potassium ion sensitivities were measured. The believable detection concentration region was also defined. Finally, the distribution of time-dependence drift coefficient was also discussed. (keywords: graphene, ion sensitive field effect transistor, drift coefficient, potassium, and ion)

Introduction

Graphene is a sheet of sp^2 bonded carbon atoms that are arranged in a honeycomb structure. Due to its single-atom level in thickness, the charge transport properties are particularly sensitive to the local environment around it, such as chemical doping and molecular adsorption. It have been demonstrated that filed-effect transistor based (FET-based) ionic sensor, made by graphene, shows sensitivity for probing the hydrogen ions[1]. However, for particular applications, the ionic sensor requires to operate the device in complex liquid water, where it contains multi-ions. Therefore, it is necessary to study the interactions of graphene with the multi-ions in aqueous solutions. In this paper, we investigate the multi-ions sensing properties and stability of FET-based graphene sensor, in which the high quality and large-area graphene were obtained by CVD method followed by a transferring process. Solution-gated graphene field effect transistor (SGFET) were fabricated by using Ag/AgCl as gate electrode and the sensing area is fixed to $1.0\times0.5cm^2$. The test solution is a 5-mM Tris/HCl buffer solution (pH8). The concentration of K^+ ions were ranged from $10^{-5}M$ to $10^{-1}M$ (defined as pK5 and pk1, respectively). The results show that the sensitivities for hydrogen and potassium ion are about 18.41mV/pH and 58.5mV/pH, respectively. The suitable sensing regions and sensing linearity of hydrogen and potassium ion are comprehensively studied. Moreover, to mimic the condition of ion concentration in a human blood, the SGFET were immersed in a pH6 and pK2 solution to measure the time-dependent drift coefficients. The high sensitivity of graphene-based sensor for multi-ions probing shows potential application in biosensor in the future.

Experimental

A. devices fabrication

In the first, large area single layer graphene is deposited on copper foil by chemical vapor deposition system. The following process is transfer graphene on Cu foil to Si/SiO$_2$ substrate. The thickness of silicon dioxide is 300nm. Due to graphene is easier to observe by naked eyes. To fabricate the source and drain electrode of transistor, Cr/Au (5nm/30nm) was coated by thermal evaporator and patterned by lift off technique. To avoid the variation of sensing properties between samples, the sensing area exposed to an electrolyte is kept to $1.0\times0.5cm^2$. Finally, to form a conductive line, graphene based ion sensing field effect transistors were fixed on a printed circuit board (PCB) and contacted to Cu lines by the silver gel. To package the samples, an epoxy was used to encapsulate the transistor structure and Cu line of print circuit board.

B. tested solution preparation

For the investigate of pK sensing properties, a 5 mM tris(hydroxymethyl)-aminomethane (Tris) buffer solution of pH 8 fixed with 1 N HCl was prepared. The concentration of K + ion and in a range between 10 -5 and 10 -1 M was varied by adding 0.1 M and 1 M KCl/Tris-HCl standards. In addition, For the investigation of pH-sensing properties, standard pH buffers (Merck Inc.) from pH 2 to 12 and their pH values were checked with a commercial combined pH electrode S120C (Sensorex) and pH-meter HTC-201U (HOTEC) before and after the measurements.

C. electrical measurement system

To test the current-voltage curve of graphene transistor, Keithley 4200 I-V measurement system was used and Ag/AgCl reference electrode was used to be the solution gate electrode. Fig.1 is the measurement system. Source voltage was kept at ground state and drain voltage was kept at 0.5V. Gate voltage applied on reference electrode was sweep voltage. The output sensing signal was extract from the gate voltage which is corresponding to the minimum current value of each I-V curve.

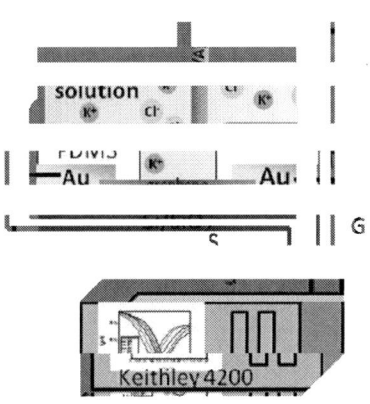

Fig. 1 Measurement system

Results and discussion

sensitivity is close to the ideal Nernst response 59.2mV/decade as well as the linearity of signals is high.

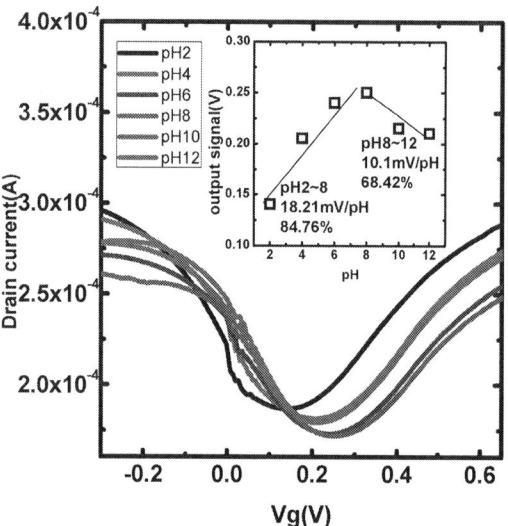

Fig. 2 I_D-V_G curves of pH sensing in pH2, 4,6,8,10 and 12

Fig.2 is the I_D-V_G curves of pH sensing. To investigate the usable sensing concentration, we test from pH2 to pH12. In pH2~pH8, pH sensitivity is about 18.21mV/pH and the linearity is 84.76%. In pH8~pH12, pH sensitivity is -10.1mV/pH and linearity is 68.42%. The insert figure in Fig. 2 is output signal against to pH value. Obviously, there are different trends in basic and acidic solution. In basic solution, dirac point did not shift with changed pH value. As a pH sensor, the usable detection region is only from pH2 to pH8.

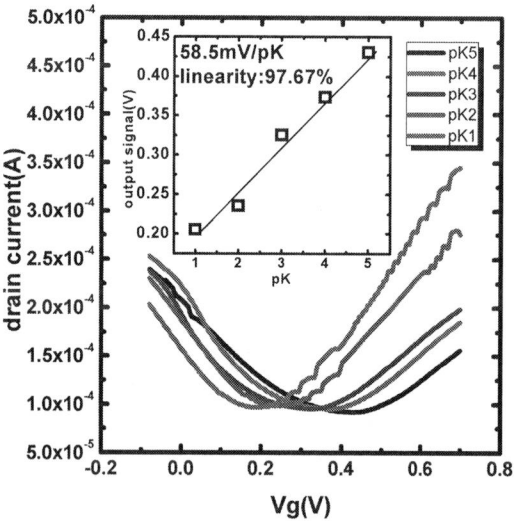

Fig. 3 I_D-V_G curves of potassium sensing in pK1 to pK5

In Fig. 3 we study the potassium ion detection properties. The pK sensitivity is about 58.5mV/pK in pK5 to pK1. Linearity is 97.67%. Graphene based sensor shows great response for potassium ion concentration. Because the

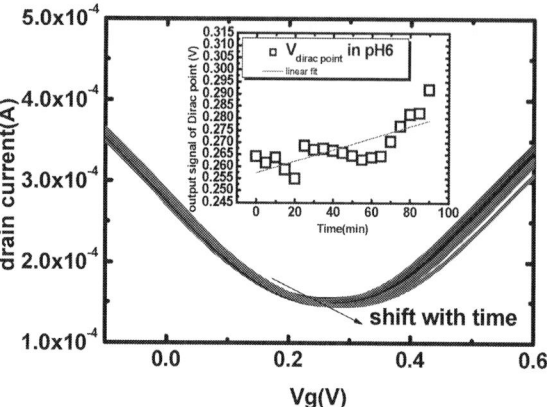

Fig. 4 Time dependence drift coefficient test in pH6

Solution at pH6 was chosen to measure the time dependence drift due to it is similar to the pH value of human blood. We record the curve each five minutes in ninety minutes. The inserted figure in Fig. 4 is the extracted dirac point voltage of each curve. Unfortunately, the signals are not stable in this 90 minutes. The drift coefficient is about 14.48mV/hr.

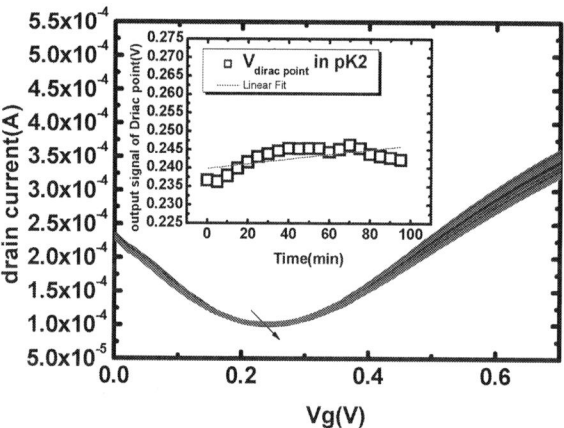

Fig. 5 Time dependence drift coefficient test in pK2

We also test the drift in pK2 solution. Graphene based sensor shows stable properties in the solution with potassium ion. The drift coefficient is 3.8mV/hr.

There is one thing which we should take care. The slopes in linear regions are different. It happened not only in the sensitivity measurement, but also in the drift research. That may cause by the ionic screening effect which may need time to approach dynamic equilibrium state [2].The screening effect can reduce the long-range scattering which made by the ion impurity charge in the interface of channel graphene and underlying silicon dioxide. That is the reason why we chosen the voltage of dirac point to be the sensing signal of ion concentration. The varying slopes may distribute the sensing accuracy if the voltage shift in the linear region is to be the sensing signal.

Table 1 comparsion of pK-sensitivity and linear range

Method	Sensi. mV/pK	Linear range (mM)
Graphene ISFET	~58.5	0.01-100
PVC matrix[3]	40~50	0.1-1000
Ion-implantation of K^+ and Al^+ into silica[4]	36~44	1-100
Ion-partitioning membrane[5]	~58	0.001-10

Conclusion

Graphene based ion sensitive field effect transistor shows detect ability to hydrogen and potassium ion. The usable sensing region is from pH2 to pH8 and pK1 to pK5. The pH and pK sensitivity is 18.41mV/pH and 58.5mV/pK. Drift coefficient of pH and pK is 14.48mV/hr for pH and 3.8mV/hr. The ionic screening effect is also observed in time-dependence test. For sensor application, the signal stability alAll half-tone illustrations (pictures/photographs) should be clear black and white prints. Do not use photocopies. These illustrations should be furnished within the copy. Make certain to include a caption in the paper for the illustration as well as to label the illustration on the back.

Acknowledgement

This work is supported by the National Science Council of the Republic of China under the contract number of NSC 101-2221-E-182-034-MY3.

References

[1] Priscilla Kailian Ang, et al., *J. AM. CHEM. SOC.*, 2008, 130, 14392–14393

[2] Fang Chen, Jilin Xia, and Nongjian Tao, *Nano Lett.*, Vol. 9, No. 4, 2009

[3] S. Wakida, M. Yamane, K. Hiiro, T. Kihara, Y. Ujihira, T. Sugano, *Analytical Sciences*, 4, 501-504 (1988); doi: 10.2116/analsci.4.501

[4] Z. M. Baccar, et al, *Materials Chemistry and Physics*, 48, 56-59 (1997); doi: 10.1016/S0254-0584(97)80077-6

[5] E. A. Moschou, N. A. Chaniotakis, *Analytica Chimica Acta*, 445, 183-190 (2001); doi: 10.1016/S00003-2670(01)01257-0

ZnO NANORODS BASED ULTRA SENSITIVE AND SELECTIVE EXPLOSIVE SENSOR

Rashi Nathawat, M Patel, P Ray, N A Gilda, M S Vinchurkar, V Ramgopal Rao

Centre of Excellence in Nanoelectronics, Department of Electrical Engineering, Indian Institute of Technology-Bombay
Mumbai-400076, Maharashtra, India
Fax: +91 22 2572 3707, Email: rrao@ee.iitb.ac.in, rashi@ee.iitb.ac.in

Abstract

A small scale (20 μm), ultra sensitive (50 ppb) and highly selective sensor based on ZnO nanostructures using Micro-electro-mechanical system (MEMS) platform has been reported here for the detection of explosive and Volatile Organic Compound (VOC) vapors. Flower and rod like architectures of nanorods were used as a sensing layer. The nanorods prepared via chemical synthesis were uniform with diameters of 50-80 nm and lengths about 3-4 μm. X-ray diffraction (XRD) and Scanning electron microscopy (SEM) reveal that the nanostructures are well oriented with the c-axis, perpendicular to the substrate. A relatively higher selectivity for 2, 4, 6-Trinitrotoluene (TNT) vapors compared to other VOCs at room temperature were observed. The intensity of deep level green emission peak associated with point defects decreases after exposure as revealed from Photoluminescence (PL) spectra.

Keywords: Nanostructure, MEMS, TNT, Sensor, XRD, PL.

Introduction

International terrorism and increased use of explosive devices in terrorist attacks is a concern world over. The need for sensors which will be able to detect hidden bombs in mails, vehicles, aircraft and so on by sensing explosive related chemical agents such as TNT, Dinitrotoluene (DNT), and RDX become vital to prevent terrorist attacks [1]. Most of explosives have a very low vapor pressure; thus very few molecules diffuse in the surroundings at room temperature, making their detection challenging. The present work is to design, realize and integrate a chemiresistor with MEMS technology based microheater to boost and elevate the selectivity and sensitivity level of the vapor monitoring system especially in the parts per billion (ppb) concentration ranges. Metal oxide semiconductors have a great potential to be used as a surface material because it possesses high chemical and thermal stability and better immunity to humidity compared to polymer based explosive sensors [2].

Device Fabrication and Experiments

A ZnO Nanorod (NR) sensor has been fabricated using a single step photolithography. The n-type Si (100) substrate is used with a 100 nm thermally grown SiO_2 insulating layer. To ease the nucleation of ZnO NRs, a 50 nm thick ZnO seed layer was deposited on the oxide layer by RF sputtering. The Au electrodes were patterned using photolithography on the ZnO thin-film. NRs were then synthesized via low temperature hydrothermal process [3] in the region excluding the Au pads.

The Pt based micro-heater is integrated along with ZnO sensor to accelerate chemical adsorption-desorption process of molecules. Surface morphology, structural and chemical characteristics were observed using Scanning electron microscopy (SEM), X-ray diffraction (XRD) and Photoluminescence (PL). The sensor response was then measured at different concentration of explosive (TNT) and at other Volatile Organic Compounds (VOCs). All the measurements were taken using Keithley 2000 multimeter with GPIB computer interface. Fig.1 represents the block diagram of the sensing setup used for the experiment. Sensor integrated with the Pt heater was assembled in the Teflon cap with inlet/outlet tubes. Vapor Generator (VG) was used to generate explosive and VOC vapor for testing. N_2 was used as carrier gas for these vapors. Mass Flow Controller (MFC) was used to maintain the flow of N_2. A 9 V power supply is used to raise the temperature of heater up to 100°C.

Results and Discussions

XRD pattern of ZnO NRs shows that they are highly crystalline, and oriented along c-axis (Fig. 2). (0001) plane confirms the wurtzite structure of ZnO at the surface. The SEM images of as grown nanorods and nanoflowers were shown in Fig. 3 (b) and (c) respectively. Fig. 4 shows the I-V characteristics of the sensor with varying temperature. The thermal stability of sensor with successive reference heating-cooling cycles represent in Fig. 5. The PL spectra reported here in Fig. 6, presents the a sharp near band-edge emission observed at 385 nm (3.2 eV), attributed to a well-known recombination of free exactions and a broad green emission at ~550 nm (2.2 eV) resulting from the recombination of photo-generated holes with a singly ionized charge state of specific defect. The green emission band intensity decreases, which confirms that the defects get reduced after TNT exposure. The free electrons of semiconductor bond with electron-acceptor nitro-aromatic explosive molecules results in an adsorption of molecules at surface. The ZnO is n-type semiconductor. The n-type semiconducting oxide represents an increase and decrease in resistance due to the presence of oxidizing and reducing gases respectively. The electrical conductivity of ZnO depends strongly on the surface states produced by molecular adsorption. This in turn will increase the resistance of device, depending on the amount of adsorption. On the other hand when a reducing gas (EtOH) comes in contact with the surface it consumes ion-adsorbed oxygen and in turn decreases the charge barriers making it easier for electrons to flow past the grain boundaries thus increasing the electrical conductance of metal oxide. Fig. 7 and 8 show the sensor response and selectivity respectively for increasing concentrations of TNT vapors and to other VOCs.

Conclusion

In summary, ZnO nanorods chemiresistor sensor shows good sensitivity towards trace TNT molecules down to few ppb concentrations. A salient feature of the sensor is its lower operating temperature (25^0C), which has been achieved by employing nanostructure material as sensing layer.

References

[1] Wang F, Gu H, Swager T M., Journal of the American Chemical Society 2008; 130 (16): 5392-5393.

[2] V Seena, A.Fernandes, P.Pant, S.Mukherji and V Ramgopal Rao, Nanotechnology 22 (2011) 295501.

[3] Min Guo, Peng Diao, Shengmin Cai, Applied Surface Science 249 (2005) 71-75.

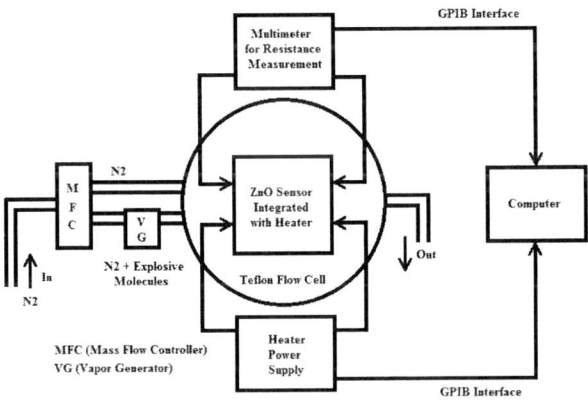

Fig. 1 Block diagram of an electrical setup used to test ZnO sensor with vapor generator.

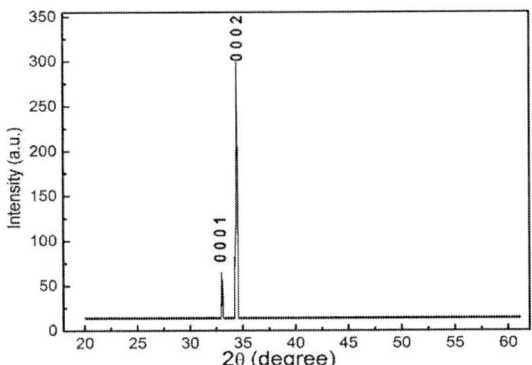

Fig. 2: X-Ray Diffraction spectra represent a sharp (0002) peak, which denotes the vertically aligned ZnO nanorods

Fig. 3: (a) SEM image of Vertical ZnO NRs rodlike, (b) ZnO NRs flowerlike.

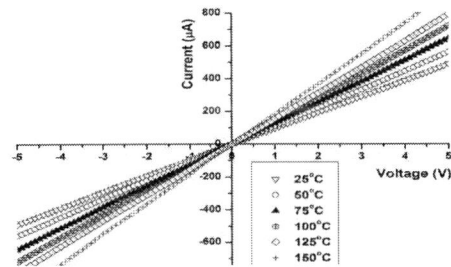

Fig. 4: I-V characteristics of device, showing linear behavior with temperature

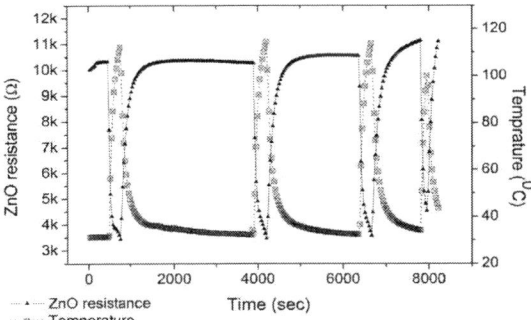

Fig. 5: Temperature stability: Base resistance of sensor remains constant after applying successive heating cycles.

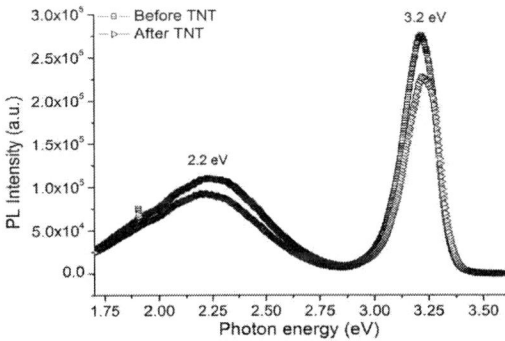

Fig. 6: Photoluminescence Spectra of ZnO surface before and after TNT exposure

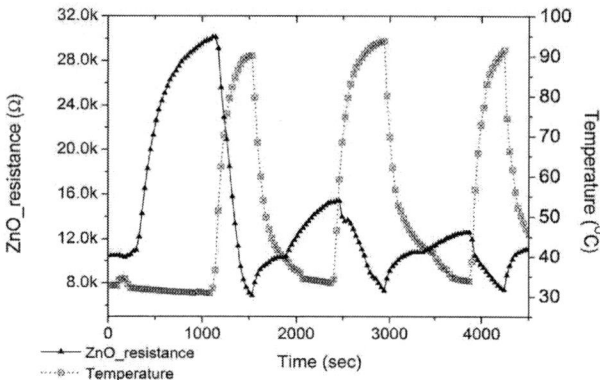

Fig. 7: Sensitivity: Response of TNT with ZnO sensor: The concentration of TNT vapors was varied gradually & response was recorded & integrated with heating

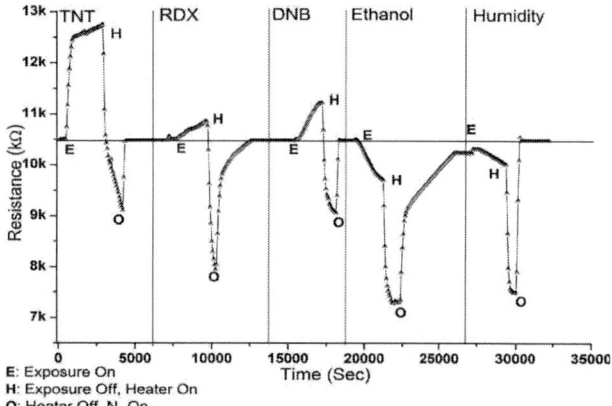

E: Exposure On
H: Exposure Off, Heater On
O: Heater Off, N₂ On

Fig. 8: Selectivity: 5 samples are tested for an identical concentration. TNT shows 3 kΩ changes in resistance as compared to others.

Patterning Metallic Films to Enhance Plasmonic Modes for mRNA detection

Ping Bai*, Lin Wu and Er Ping Li

A*STAR Institute of High Performance Computing,
1 Fusionopolis Way, #16-16 Connexis, Singapore 138632
* baiping@ihpc.a-star.edu.sg

Abstract

We pattern a metal film into different nanostructures to investigate the influence of surface plasmonic modes and the effects on the sensitivity to the change of environment's refractive indexes. We find out that 1-D nanoslit array structures and grating structures have much higher sensitivity than 2-D nanoisland array structures and nanowindow array structures when detect the molecules with moderate dimensions such as PLK1 mRNA.

Keywords: plasmonic sensors, plasmonic modes, patterned metal film, refractive index, sensitivity and mRNA

Introduction

Surface plasmon resonance (SPR) sensors are typically based on the spectral shift or intensity change caused by the molecules-binding-induced change in the refractive index at the sensor surface[1]-[4]. However, traditional SPR sensors, which require some apparatuses such as a prism to induce the plasmonic resonance, are generally very bulky and costly. Localized surface plasmon resonance (LSPR) based sensors utilize the interaction between the light and subwavelength metal nanostructures, are potential and offer promising to further develop into a low-cost point-of-care system. LSPR has similar functions as SPR, but LSPR can be generated by directly illuminating the light onto the micrometer metal nanostructures at any angle. In this paper, a systematical method is used to investigate various subwavelength metallic structures to enhance the LSPR to detect molecules with high sensitivity.

Methods

A. Methods for patterning metallic films

There are three dimensions to form patterns on a metal film. In one dimension along the thickness of the metal film, we cut through or cut halfway the metal film. In the cut-through case, we consider both periodic-finite and infinite cutting in the rest two dimensions and we obtain nanoslit array structures, nanoisland array structures and nanowindow array structures. While in the cut-halfway case, we consider only periodic-finite cutting in one dimension and infinite cutting in the other dimension, so a grating structures are obtained. The upper panel of Fig 1 shows a grating structure, nanoslit array structure, nanoisland array structure and nanowindow array structure in an Au film deposited on a glass substrate to be investigated in this paper.

B. Methods for evaluating plasmon resonance modes

We develop a comprehensive modeling method to study the plasmon resonance modes supported by various metallic structures [5]. In this method, we study the plasmonic modes by calculating the transmission spectra, reflection spectra, absorption spectra, electromagnetic field and optical-power distribution through 3-D Maxwell equations. We not only accurately reproduce the light transmission spectrum observed in experiments, but also find that the calculated light absorption spectrum can provide additional information that is missed in the transmission spectrum. Together with the studies of the electric-field distributions at resonant wavelengths, the mechanisms of the extraordinary transmission are revealed.

Results

To evaluate the sensitivity of plasmonic sensors based on different metallic structures, we use PLK1 mRNA as the test molecule in this theoretical study. The PLK1 mRNA is about 50 nm in diameter and can be bound onto the plasmonic sensor via the biomolecular recognition element (BRE). The BRE may consist of a thiol monolayer, a layer of streptavidin, and an oligonucleotide probe layer. The probe layer is the critical layer to determine the specificity and the strength of the binding, because it is defined as the single-strand DNA/RNA molecule which is complementary to part of the target analyte. Based on the binding structure, we can establish an optical model to evaluate the sensitivity (to the refractive index change) for various metal nanostructures.

The lower panel of Fig. 1 shows calculated power reflectance in the four metallic structures shown in the upper panel. There are basically three resonances α, β, or/and γ observed with water matrix (refractive index: 1.333). After 50nm PLK1 mRNA molecules (refractive index: 1.343) are bound to the metallic structures, the resonance frequencies are shifted as shown in the Fig. 1. The most shifted wavelength is observed in the slit array and grating structures, while the nanoisland array structure shows the least wavelength shift. We know that the nanowindow array or nanoisland array are two-dimensional arrays, meaning the oscillation of free electrons at metal/dielectric boundaries are restricted in the two planar directions [6]. These restrictions will mostly generate the localized surface plasmonic modes, with very limited electromagnetic field penetrations into the dielectric. On the other hand, the nanoslit array and grating structures are one-dimensional array. Propagating surface plasmonic modes can be sustained at the center of metal strips. These propagating modes will have much deeper field penetration depth. In addition, they originate from the center of the metal strips (not the edge) where most of the biomolecules are bound. As a result, the

978-1-4673-4840-9/13 $31.00 © 2013 IEEE

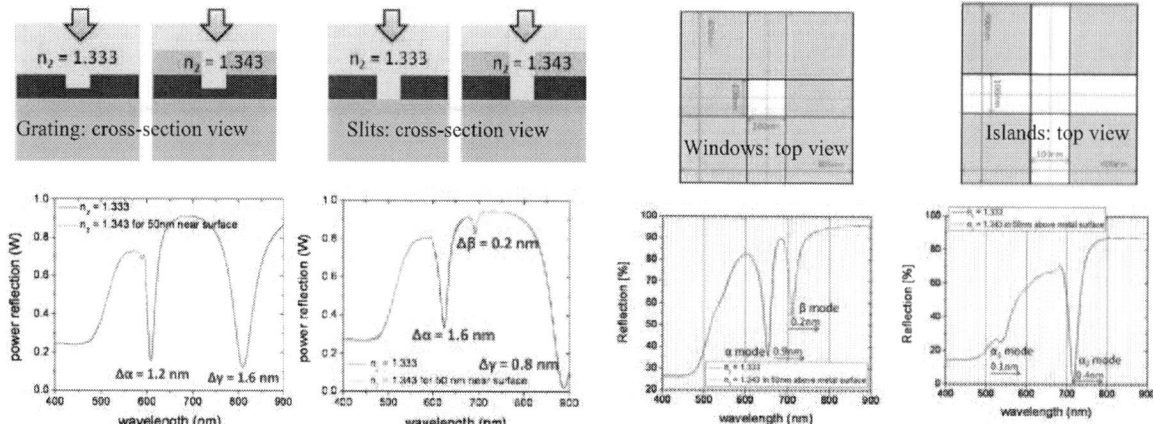

Fig. 1 Upper panel: the unit cells of grating array, nanoslit array, nanowindow array and nanoisland array structures. Lower panel: the simulated sensitivity for the grating array, nanoslit array, nanowindow array and nanoisland array structures, where the metal thickness is 100 nm and the unit cell repeat itself with the pitch of 400 nm. The grating is cut by 25nm in wide and 50nm in depth; the slit is 25nm in wide; the nanowindow is 100 nm by 100 nm; and the nanoisland is 300 nm by 300 nm. For each configuration, the sensitivity is calculated for bulk refractive of 1.333 and also the local refractive index of 1.343, only 50 nm above the metal surface.

nanoslit array and grating structures have better sensitivities than the nanowindow array or nanoisland array structures in detecting the PLK1 mRNA.

There are some difference between the grating structure and the slit array structure. The slit structure can allow light illuminate both from the matrix and substrate side, whereas the grating structure only has the conventional reflection scheme (light illuminates from the matrix side) where the scattering of the aqueous medium may cause noise. In addition, The slit structure allows the analytes to follow through the metal film to improve the binding efficiency [7] but grating does not. Slit structure may use thinner metal film in the transmission scheme to save material cost as well. As such here we focus on our study on the slit structures.

The electric-field distributions are further studied to understand why the α mode in the slit array structure has the largest shift comparing with other modes. Fig. 2 shows the electric-field distributions of the α, β and γ modes in the cross-section of the slit structure. From the field distribution, we see that the α mode is the surface mode excited by the water-metal surface; the β mode is the surface mode excited by the substrate-metal surface; and the γ mode is the edge mode excited by the substrate-metal surface close to the slit edges. The refractive index change occurs at water side of the metal. The α mode originates from the water-metal surface. There is the largest overlap between the electric-fields of the α-mode and the analyte bound on the surface where the refractive index is changed. As a result, the α mode is the most sensitive mode to the refractive index change at water side.

Furthermore, the effects of the dimensions of the nanoslit array structure on the sensitivity are studied. We find that both the plasmonic modes and the sensitivity are varied with the changes

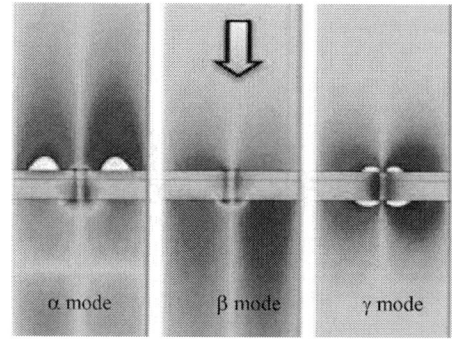

Fig. 2 The simulated electric-field distributions for the α, β and γ modes in the cross-section of the slit structure. Light is incident from the matrix (reflection scheme).

of the pitch, the width and the thickness of the slit array structures, but the effects from the change of the pitch is the most. Fig. 3 shows the power reflection changes with the pitch of the slit array. We see all three plasmonic modes are red shift when the pitch increases. The sensitivities of each mode for different pitches are also shown in the right panel of the Fig.3. The maximized sensitivity is achieved with the structure of 450-nm-pitch, 20-nm-width, and 150-nm-thickness, regardless of the reflection scheme or transmission scheme.

The effects of the dimensions and index of the analytes, such as mRNA molecule, on the sensitivity are also investigated with the slit array structure. Fig. 4 shows the resonance shift (sensitivity)

978-1-4673-4840-9/13 $31.00 © 2013 IEEE

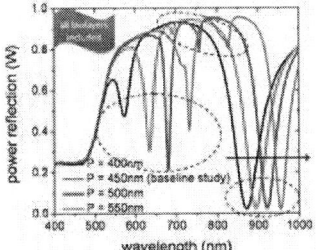

sensitivity comparison (nm)			
pitch	α	β	γ
400	6.7	-	2.4
450	6.8	0.5	2.6
500	3.0	0.5	2.7
550	1.0	0.5	2.7

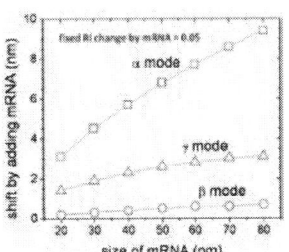 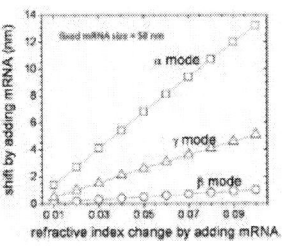

Fig. 3 The pitch effects on the plasmonic modes (left panel) of and the sensitivities (right panel) in the slit array structure, where the slit width is 37.5 nm and the metal film thickness is150 nm.

Fig. 4 The effects of the analyte sizes (left panel) and the analyte refractive index change (right panel) on the shift of resonance.

changes with the size and the index of the analytes. The shift is linearly changed with the index. This is expected as the shift should be proportionally increased with the concentration of the analyte within the sensing range. However, the shift nonlinearly changes with the variation of the size. This is due to the electromagnetic fields of the plasmonic mode can not effectively cover the whole analyte if the analyte is too large [8]. It means that the sensitivity will decrease when the size of the molecule increases.

Conclusions

We have systematically investigated the plasmonic modes of subwavelength structures patterned in a metal film deposited on a glass substrate for plasmonic sensing. Four structures including the grating, nanoslit array, nanoisland array and nanowindow array structures are studied with a comprehensive modeling method. The studies show that the nanoslit array and grating structures have much higher sensitivity than the nanoisland array and nanowindow array structures for detecting PLK1 mRNA. This is contribute to the propagating surface plasmonic modes can be supported by the 1-D nanoslit array or grating structures and the modes have a large field penetration depth to cover entire PLK1 molecules. These structures can be further developed into portable sensing systems or even a lab-on-a-chip for a wide range of applications such as diagnostics, food testing and environment monitoring.

Acknowledgment

This work was supported by Agency for Science and Technology Research (A*STAR) Singapore with Thematic Strategic Research Programme (TSRP) Grants: iNPBi – LSPR POC for Clinical Screening and Medical Diagnostics no. 102152 0014.

References

[1] S. A. Maier, *Plasmonics: Fundamentals and Applications*, Springer, 2007.

[2] J. Homola, Present and future of surface plasmon resonance biosensors, Anal. Bioanal. Chem. 377, 528-539, 2003.

[3] J. N. Anker, W. P. Hall, O. Lyanders, N. C. Shan, J. Zhao, and R. P. Van Duyne, Biosensing with plasmonic nanosensors, Nat. Mater. 7, 442-453, 2008.

[4] M. E. Stewart, C. R. Anderton, L. B. Thompsono, J. Maria, S. K. Gray, J. A. Rogers, and R. G. Nuzzo, Nanostructured plasmonic sensors, Chem. Rev. 108, 494-521, 2008.

[5] L. Wu, P. Bai, and E. P, Li, Designing surface plasmon resonance of sub-wavelength hole arrays by studying absorption, J. Opt. Soc. Am. B 29, 521-528, 2012.

[6] B. Seplveda, P. C. Angelom, L. M. Lechuga, and L. M. Liz-Marzn, LSPR-based nanobiosensors, Nano Today 4, 244-251, 2009.

[7] A. A. Yanik, M. Huang, A. Artar, T. Chang, and H. Altug, Integrated nanoplasmonic-nanofluidic biosensors with targeted delivery of analytes, Appl. Phys. Lett. 96, 021101, 2010.

[8] A. A. Yanik, M. Huang, O. Kamohara, A. Artar, T. W. Geisbert, J. H. Connor, and H. Altug, An optofluidic nanoplasmonic biosensor for direct detection of live viruses from biological media, Nano Lett. 10, 4962, 2010.

Fabrication of highly ordered hollow oxide nanostructures based on nanoscale Kirkendall effect and Ostwald ripening

Abdel-Aziz El Mel,[1] Marie Buffière,[2,3] Pierre-Yves Tessier,[4] Wei Xu,[5] Ke Du,[5] Chang-Hwan Choi,[5] Stephanos Konstantinidis,[1] Carla Bittencourt[1] and Rony Snyders[1,6]

[1]Chimie des Interactions Plasma Surface (ChIPS), CIRMAP, Research Institute for Materials Science and Engineering, University of Mons, 23 Place du Parc, B-7000 Mons, Belgium
[2]Department of Electrical Engineering, KU Leuven, KastteelparkArenberg 10, B-3001 Heverlee, Belgium
[3]IMEC, Kapeldreef 75, B-3001 Heverlee, Belgium
[4]Université de Nantes, CNRS, Institut des Matériaux Jean Rouxel, UMR 6502, 2 rue de la Houssinière B.P. 32229 - 44322 Nantes cedex 3 – France
[5]Department of Mechanical Engineering, Stevens Institute of Technology, Hoboken, NJ 07030, USA
[6]Materia Nova Research Center, Avenue Copernic 1, ParcInitialis, B-7000 Mons, Belgium

Abstract

Highly ordered oxide nanotubes are fabricated by a simple two-step method which consists of the magnetron sputtering deposition of copper nanowires on nanograted template surfaces and the thermal oxidation of the nanowires in ambient air at 300 °C. The formation of the organized copper oxide nanotubes is explained according to the nanoscale Kirkendall effect which comes into play at the metal/metal-oxide interface during the annealing process.

Introduction

Huge efforts are nowadays dedicated to the development of relevant methods to fabricate hollow nanostructures of a wide variety of materials [1-7]. Among hollow nanostructures, oxide nanotubes are considered as potential building blocks for nanoelectronics. However, the majority of synthesis methods used for producing oxide nanotubes suffers from the technical limitation such as short length and poor organization of the nanostructures. Here, we report for the first time on highly organized ultra-long metal oxide nanotube arrays (length up to several centimeters) fabricated by thermal oxidation of metal nanowire arrays. The metal nanowires were grown on nanograted surface by the plasma-assisted inverse-template method that we have recently developed [9-11]. Based on the extensive study using transmission electron microscopy, the fundamental mechanisms occurring during the formation of such oxide nanotubes are explained based on the nanoscale Kirkendall effect. We further show that such method can be extended for the fabrication of periodic zero-dimensional hollow nano-objects. This novel fabrication method of oxide nanotubes in a planar configuration opens up new strategies for advanced nanodevice fabrication and characterization.

Experimental

The nanograted substrates used as template for the selective and organized growth of the Cu nanowires were prepared by laser interference lithography coupled to deep reactive ion etching [8]. The Cu nanowires were deposited by magnetron sputtering of a copper target (99.999% pure and 30 mm in diameter) using DC argon plasma discharge. The distance between the copper target and the substrate was fixed at 75 mm. The electrical power applied to the target was 135 W and the deposition time was 2 min. The deposition rate was 1.3 nm/s. During deposition, the argon pressure was maintained at 1.34 Pa and the substrate was kept at floating potential. No intentional substrate heating was applied during the deposition. A conventional oven was use to oxidize the Cu nanowires in air at 300 °C for 1 hour.

Results and Discussion

Fig. 1A shows a SEM micrograph of the Cu nanowires grown preferentially on the top ridge surface of the silicon nanograting structures. When the sample was cleaved for the SEM examination, it was frequently observed that the nanowires were disengaged from their support (Fig. 1B). This reflects the weak adhesion between copper and silicon probably due to the presence of a silicon oxide at the interface. As shown in Fig. 2, the preferential growth of copper on the top of the silicon nanograting structures is related to the amplification of the shadowing effect due to the low directionality of the deposition process, the small width of the nanotrechnes separating the adjacent nanograting structures, and the high aspect ratio of the nanograting structures (height-to-width ratio) [9-11]. As shown in Fig. 3, after 1 hour of thermal oxidation, the copper nanowires were completely transformed into oxide nanotubes. The tubular aspect of these 1D nanostructures was checked by TEM which revealed also the columnar morphology of the tube walls (Fig. 4). The transformation of nanowires into nanotubes can be explained by the nanoscale Kirkendall effect which occurs during the thermal oxidation (Fig. 5). The Kirkendall effect describes an unbalanced mutual interdiffusion process taking place through the interface of two different materials (here: copper/copper-oxide). As a consequence of the nonequilibrium diffusion kinetics, vacancies are created at the interface and then move in the region of fast-diffusing metal (here it is copper). The coalescence of excess vacancies leads to the formation of voids located at the initial interface and within the fast-diffusing metal. These voids coalesce together and lead to the formation of hollow oxide nanostructures.

Conclusions

A simple and efficient route for the synthesis of uniform and highly ordered oxide nanotubes has been demonstrated. The fundamental mechanisms occurring during the transformation of nanowires into nanotubes have been explained to the Kirkendall effect occurring at the nanoscale. Magnetron sputtering is a powerful and scalable technique which allows synthesizing and controlling the morphology of variety of complex material at low temperature making the demonstrated nanofabrication route suitable for the synthesis of hollow nanostructures of any material types in which the Kirkendall effect can come into play. The planar architecture of nanotubes demonstrated in this study opens up the way for more advanced nanodevice construction and characterization. For example, such highly ordered and ultra-long nanotubes of a columnar morphology can be used to built-up biosensors or gas detectors since atoms and/or molecules can penetrate to the inner of the tube via the grain boundaries.

References

[1] Y. Sun et al., *Adv. Mater.*, 15, 641, 2003.
[2] Y. Sun and Y. Xia, *Adv. Mater.*, 16, 264, 2004.
[3] H. J. Fan et al., *Small*, 3, 1660, 2007.
[4] H. J. Fan et al., *Nano Lett.*, 7, 993, 2007.
[5] X. W. Lou et al., *Adv. Mater.*, 20, 3987, 2008.
[6] Y. Zhao et al., *Adv. Mater.*, 21, 3621, 2009.
[7] G. D. Moon et al., *Nano Today*, 6, 186, 2011.
[8] C. H. Choi et al., *Nanotechnology*, 17, 5326, 2006.
[9] El Mel et al. *Nanotechnology*, 23, 275603, 2012.
[10] El Mel et al., *Nanotechnology*, 21, 435603, 2010.
[11] El Mel et al. *Nanotechnology*, 22, 435302, 2011.

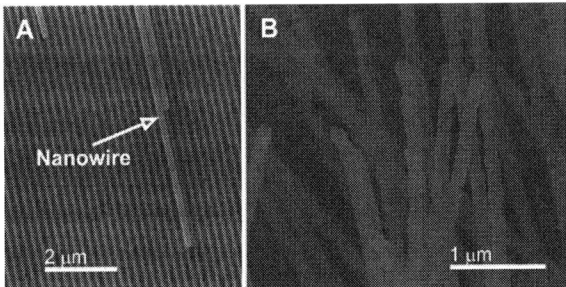

Fig. 1 SEM micrograph of: (A) copper nanowires as-grown on the nanograted substrate; and (B) copper nanowires disengaged from their silicon support (B).

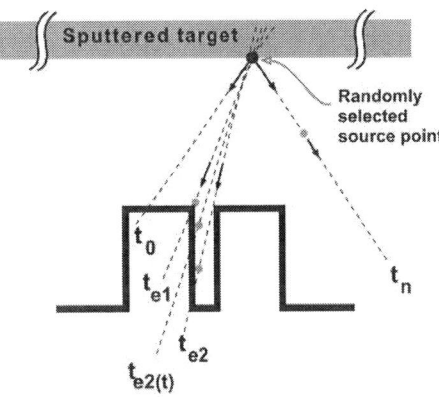

Fig. 2 Fundamental mechanisms which leads to the preferential growth of copper nanowires on the top of the silicon nanograting structures.

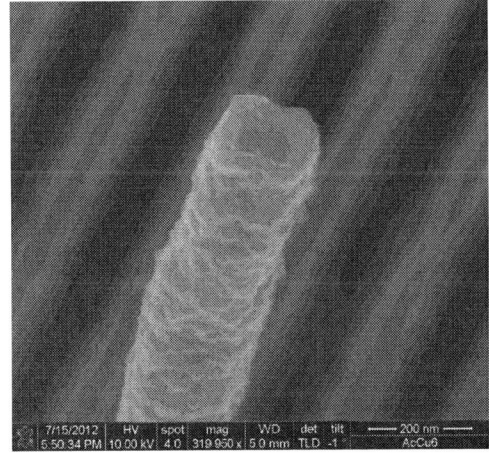

Fig. 3 SEM micrographs showing the tubular structure of the one-dimensional oxide nanostructures.

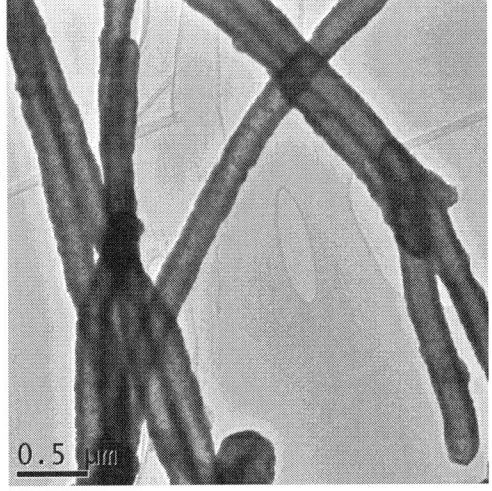

Fig. 4 TEM micrograph of copper oxide nanotubes prepared by nanoscale Kirkendall effect.

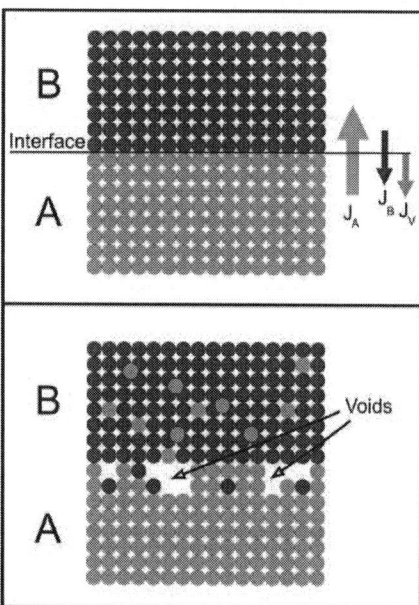

Fig. 5 Schematic illustration of the fundamental mechanisms describing the Kirkendall effect occurring upon annealing of two different stacked materials. J_A and J_B represents the fluxes of the ions diffusing from A to B and from B to A, respectively. J_V represents the flux of the vacancies generated at the interface.

Microstructure and Piezoelectric Properties of c-BN Nano-films Deposited on Si by RF Sputtering for Piezoelectric Devices

Wang Fang[1,2] Yang Baohe[1,2] Wei Jun[3] Zhang Kailiang[2*]

1. Tianjin University, Tianjin, China

2. School of Electronics Information Engineering, Tianjin Key Laboratory of Film Electronic & Communication Devices, Tianjin University of Technology, Tianjin, China

3. Singapore Institute of Manufacturing Technology, a-star 71 Nanyang Drive, Singapore

*corresponding author: kailiang_zhang@163.com; 86-22-60214196

Abstract

In this paper, boron nitride (BN) films were deposited on silicon substrates by RF magnetron sputtering under different bias voltages. Microstructure and piezoelectric properties of BN films were characterized by FTIR, SEM, AFM and PFM. The results show that surface of films are of lower roughness, while the highest cubic phase volume fraction appeared (about 95%) under bias voltage 120V. And the deposited c-BN films are of apparent piezoelectric properties both in plane and vertical direction by PFM figures.

keywords: c-BN, sputtering, bias voltage, FTIR and PFM

Introduction

Cubic boron nitride (c-BN) is of a highly desirable mechanical, thermal, and optical properties which has significant and potential technological application prospect in cutting tools, electronic and optical devices, etc[1,2]. Most of studies about c-BN films are tribology and field emission applications [3-5], while few articles on piezoelectric devices applications were reported. Recently, c-BN films attract some attention and are expected as a promising material for the fabrication of high frequency piezoelectric devices because that c-BN is of the highest acoustic velocity, highest Young's modulus and highest thermal conductivity among all piezoelectric materials. Among the high frequency piezoelectric devices, sensors based on MEMS resonators, RF microswitches, and high-frequency SAW filter for communication are focusing on the high performance c-BN films. Most studies show that the following characteristics are very important and necessary for piezoelectric thin film to piezoelectric devices: (1) Highly cubic phase volume fraction in film to provide the highest acoustic velocity, highest Young's modulus and highest thermal conductivity which could meet high frequency devices requirement; (2) Smooth and lower roughness on surface morphology for lower propagation loss and insert loss of devices; (3) Good piezoelectric properties which is the key to device performance.

In the present work, we report the fabrication and characterization of BN thin film for high frequency piezoelectric devices, and surface morphology, microstructure and piezoelectric properties of BN films were investigated. In addition, the effect of bias voltage on cubic phase volume fraction of BN films was discussed.

Experimental

BN thin films were deposited on silicon substrates by RF magnetron sputtering under different negative substrate bias voltages with a hexagonal boron nitride target. All film samples are deposited under the following condition seen from table I.

TABLE I Deposition processing conditions of samples

Parameter	Value
Substrate temperature	600 °C
RF power	300 W
Base vacuum	8×10^{-4} Pa
Sputtering pressure(Ar+N_2)	0.6Pa
Bias voltages	90V,100V,120V,140V

The surface morphologies of films are characterized by atomic force microscopy (AFM); the c-BN and h-BN phases in films are calculated based on the Fourier transform infrared (FTIR) results; micro-zone piezoelectric properties of c-BN films are investigated by piezoresponse force microscopy (PFM) in which simultaneous imaging of the morphology, as well as the amplitude and phase of the out-of-plane and in-plane piezoresponse are performed; thickness and the cross-section of the c-BN film are observed by scanning electron microscope (SEM).

Results and Discussion

Microstructure and morphology of BN films

Fig. 1 shows FTIR spectrum of c-BN films in BN/Si, and Fig. 1(a), (b), (c), (d) point at sample of bias voltage 50V, 90V, 120V, 140V, respectively. The baseline of the FTIR spectrum was founded through Origin software in Fig. 1. c-BN is with the atomic bonding of sp3 while h-BN is with that of sp2. From Fig.1, it can be seen that the c-BN film contains cubic phase at absorption peak of 1192 cm^{-1} and hexagonal phase at absorption peaks of 773cm^{-1} and 1440cm^{-1}. Relevant researches[6] indicate that infrared sensitivity factors of c-BN and h-BN are close, so volume fraction of c-BN in BN films can be presented as , and respectively represent the peak intensity of infrared absorption spectra at 1065 cm^{-1} and 1380cm^{-1}. Peaks were picked through Origin software, The c-BN contents were calculated, as shown in table II. The c-BN contents are 14.15%, 39.18%, 95.19%, 43.37% respectively with bias voltages of 90V, 100V, 120V, 140V. It was shown that, c-BN content firstly increased, and then decreased with bias voltage increase. C-BN content is the highest under 120V bias voltage which is the best conditions to form c-BN film.

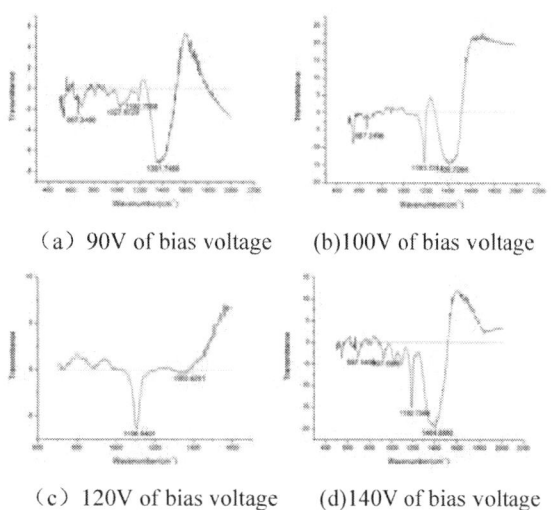

（a）90V of bias voltage   (b)100V of bias voltage

（c）120V of bias voltage   (d)140V of bias voltage

Fig. 1 FTIR spectrum of as-deposited BN films.

Among our deposited conditions, there exists an optimum bias voltage to forming highest Cubic phase content. The reason to the above results maybe as following: bombardment of plasma ions to substrate surface and grown BN thin films is key condition to the nucleation and growth of c-BN film. As we known, high energy ion to the growing surface may produce defects, form local dense zone, and disrupt the lattice structure of silicon substrate surface, which are in favor of the nucleation and forming c-BN films. The more bias voltage is, the more the energy of plasma ions is. So with the first increasing of bias voltage, the content of c-BN apparently is improved. But when bias voltage is too high, the part of growing c-BN thin films is etched, which result in the reduction of cubic content.

TABLE II   the c-BN contents of different bias voltages

Bias voltage	90V	100V	120V	140V
c-BN content	14.15%	39.18%	95.19%	43.37%

As above mentioned, surface morphology of films were important for high frequency devices, especially for propagation loss and insert loss of devices. Fig. 2 shows surface topography images by AFM of as-deposited BN films. The size of the AFM scanning images was $2 \times 2\mu m^2$. The bias voltage of Fig.2 (a), (b), (c) and (d) are 90V, 100V, 120V, 140V in which RMS values are 2.243nm, 2.256nm, 2.136nm, 3.569nm, respectively. It can be seen that the RMS values have little difference except Fig.2 (d), all samples surface are smooth and of lower roughness, and the structure compact, particle uniformity. Fig.2 (a), (b) and (c) show that the grain size in diameter is between 50-150nm. Only Fig.2 (d) with 140V bias voltages shows that there are many pits

and protrusions in the surface, the surface grain size in diameter is only about 20-50nm. High bombarding energy result in more etch of the growing BN thin films, so the grain size is small.

(a) topography of 90V   (b) topography of 100V

(c) topography of 120V   (d) topography of 140V

Fig. 2 surface topography images of BN films by AFM

The piezoelectric properties of c-BN film

From above description we can see that volume fraction of c-BN in as-deposited films are highest when the bias voltage is 120V and others technology condition are changeless. So the sample with 120V bias voltage was grown on Cu/Si film for the test of piezoelectric properties in which Cu film acting bottom electrode for testing electronic properties, and at the same time cross-sectional texture and the piezoelectric properties of the sample were researched by SEM and PFM.

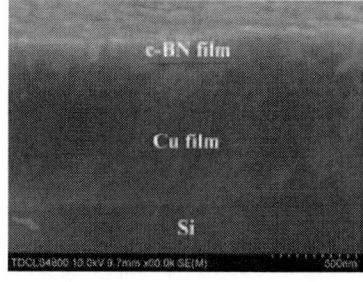

Fig. 3 Cross-sectional SEM image of c-BN/Cu/Si film

Fig. 3 is a cross-sectional SEM image of c-BN/Cu/Si film. It is that the thick of c-BN film was about of 150nm, and Cu film 1μm indicated by the SEM images. Also it can be seen that the microstructure of c-BN is the columnar structure.

978-1-4673-4840-9/13 $31.00 © 2013 IEEE

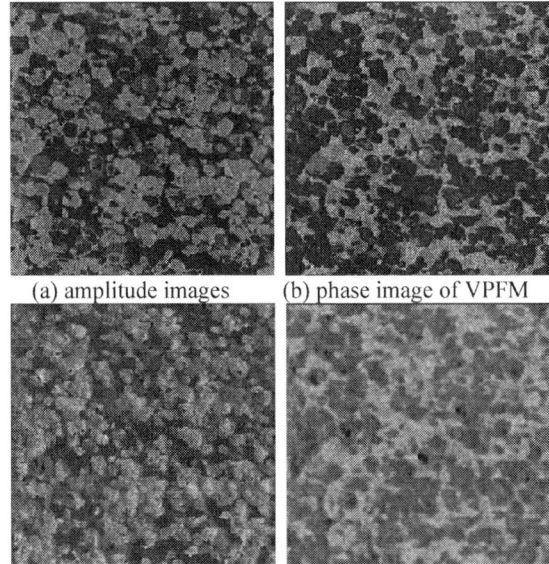

(a) amplitude images   (b) phase image of VPFM

(c) amplitude image   (d) phase image of LPFM

Fig. 4 piezoelectricity polarisation images of c-BN by PFM

To detect piezoelectric vibrations in the piezoelectric films, the AFM is operated in a contact mode with an AC modulation voltage applied between the conductive AFM tip and the bottom electrode through the piezoelectric film, which is called as the piezo-responsive mode. The signal, consisting of the DC and the first harmonic, is detected by a laser photo detector. The DC signal gives the surface topography. The first harmonic signal is amplified by a lock-in amplifier to produce the piezoelectric induced image from the piezoelectric vibration of the film arising from AC-field-induced strain.

In our measurements, we used a conductive rhodium and platinum coated silicon cantilever with a spring constant of 1.9N/m, a resonant frequency of 28 KHz, and an integrated tip of about 10nm in diameter. The size of the scanning was $2\times2\mu m^2$, the scanning velocity was 0.5 line/s, each image consisted of 256 scan lines, AC bias voltage amplitude and frequency of 10V and 25KHz. Fig. 4 is piezoelectricity respond images of c-BN film. Corresponding amplitude and phase images of VPFM are shown in Fig. 4(a) and (b) respectively, while LPFM are (c) and (d). Some of the grains have piezoresponse in the lateral direction, since these grains may be not exactly c-axis oriented, so that the polarization vectors have components in the x and z direction. In Fig. 4 (a) and (c), most of the grains with lighter contrast show greater piezoresponse, while a few grains with lower show homogeneous piezoresponse. In Fig. 4 (b) and (d), the phase image exhibits dark and light contrast. Dark and light domains correspond to grains with downward and upward spontaneous polarization orientation, respectively.

Conclusions

In this work, the microstructure and piezoelectric properties of BN nano-films are presented. The AFM results indicated that the surface morphology of all films is smoother and of lower roughness. Based on the FTIR absorption peak intensities of h-BN (I_{h-BN}) and c-BN (I_{c-BN}), the cubic phase volume fraction is obtained. The highest cubic phase volume fraction appeared (about 95.19%) at bias voltage 120V. And PFM figures and results show that the deposited c-BN films not only in plane, but also in vertical direction have apparent piezoelectric properties. At a result, higher cubic phase content, smoother surface morphology and obvious piezoelectric properties, All these properties of the sample are critical characteristics for piezoelectric materials used in piezoelectric devices, therefore the as-deposited c-BN film are compliant high frequency devices.

Acknowledgments

This work is supported by the National Natural Science Foundation of China（Grant No 61274113，11204212）, and Program for New Century Excellent Talents in University （Grant No NCET-11-1064）, and Tianjin Natural Science Foundation （ Grant No 10SYSYJC27700 and 10ZCKFGX01200 ）, and Tianjin Science and Technology Developmental Funds of Universities and Colleges（Grant No 20100703）

References

[1] Stephen J. Harris, et al., Appl. Phys. Lett., Vol. 67, pp.2314 - 2316, October 1995

[2] A. Lunk, et al., Advances in Solid State Physics, Vol. 43, pp.62-70, 2003

[3] B. Abendroth, et al., Appl. Phys. Lett. ,Vol. 85, pp. 5905- 5907, December 2004

[4] A. N. Golubenko, et al., Inorganic Materials, Vol. 39, pp.362- 365, 2003

[5] N. E. Stakhniv and L. N. Devin, Journal of Superhard Materials, 2, pp.33-36, 2011

[6] Friendmannta, etal., J Appl Phys, Vol. 76 , pp.3088- 3101, 1994

Investigation of the dynamics of carbon nanotube deposition in dielectrophoresis

Ali Kashefian Naieni, Alireza Nojeh

Department of Electrical and Computer Engineering,
The University of British Columbia
Vancouver BC, V6T 1Z4, Canada
Phone: +1 604 827 4346, Fax: +1 604 822 5949, Email: alikn@ece.ubc.ca, anojeh@ece.ubc.ca

Abstract

Dielectrophoresis is one of the popular methods in making carbon nanotube (CNT) devices using CNT solutions on pre-patterned electrodes. The interaction between the deposited and suspended nanotubes during deposition, together with the type of solution (containing surfactants or not) and electrode geometries have a significant effect in the dynamics and the results of the deposition. Here, we report the results of experiments and simulations investigating these effects. The formation of semi-periodic patterns made of stripes of nanotubes bridging the electrodes is related to the mutual effects of CNTs on each other during deposition. The periodicity of these structures depends on the geometry of the electrodes. Simulations show that the reason is changes in the electric field as a result of deposition of CNTs. These changes affect the DEP force and, therefore, the CNTs are guided toward certain spots based on their initial location.

(Keywords: Carbon nanotube, Dielectrophoresis, Deposition from solution and Finite element method)

Introduction

The promising properties of carbon nanotubes (CNTs) have motivated much research. However, fabrication challenges still stand on the way of the widespread usage of CNTs in commercial products, and more effort is needed to overcome the current obstacles.

Dielectrophoresis (DEP) is a popular method for fabricating CNT-based electronic devices [1,2]. The structures used range from single nanotubes to small mats of CNTs. Working at low temperature, deposition of nanotubes at pre-determined positions, and the simplicity of the equipment are some of the advantages of DEP.

Still, even DEP results are far from perfectly reproducible; more investigation is needed on the effect of various DEP parameters, as well as the effect that CNTs have on each other during the deposition.

We have previously reported that the solution properties and electrode shapes, among other parameters, significantly affect the DEP results [3]. Not only is the DEP force important in determining the deposition pattern, other forces such as the electrothermal force agitate the liquid medium, especially for highly conductive solutions. The DEP force is dominant in the solutions with no surfactant, and electric field lines are the main factor determining the deposition pattern. When such solutions are used to deposit CNTs on wide (a few tens of micrometers)

electrodes, the deposited CNTs show a distinct periodic pattern [4,5]. This is especially more pronounced in cases when there is a considerable gap between the electrodes [6].

Vijayaraghavan *et al.* have stated that the deposition of a CNT on sharp electrodes results in a change in the DEP force around the electrodes from attraction to repulsion during the rest of the process [7].

Two mechanisms can potentially explain these effects: the change in the DEP force because of the change in the electric field around the deposited nanotube, and/or the alteration in the movement of the solution around the electrodes and the deposited CNT due to the change in the temperature profile because of the high current passing through the deposited nanotube.

Here, the results of experiments and three-dimensional (3-D) finite element simulations to investigate the origin of these deposition patterns on wide electrodes are reported. The experiments show pattern formation at the very beginning of the DEP process. These patterns have a periodicity, which depends on the geometry of the sample. The simulations reveal that the interaction of suspended and deposited nanotubes, even if the latter do not bridge the electrodes, plays the key role in the formation of these patterns.

Methodology

A. Fabrication of the electrodes

The electrodes were fabricated using photolithography, followed by electron beam deposition of 20 nm of chromium and 50 nm of palladium on a highly p-doped [100] silicon wafer with 2 μm of oxide on the surface, and finally lift-off.

The electrodes were 50 μm wide. The gap between a pair of opposing electrodes was 4, 8 or 20 μm. These electrodes were connected to larger pads made with the same materials, which were used as contacts for micro-manipulator probes during the DEP experiments.

B. Dielectrophoresis experiments

The CNT solution was prepared by diluting a commercially available, surfactant-free aqueous CNT solution from NanoLab Inc. [8]. The final CNT concentration in the solution was 2.5 μg/milt. The suspended nanotubes had a length in the range of 1-5 μm and an average diameter of 1.5 nm.

The DEP experiments were performed by applying a potential difference with a frequency of 500 kHz to the electrodes while the sample was immersed about 3 mm deep in the solution. The applied voltages were chosen as 1 V/μm

978-1-4673-4840-9/13 $31.00 © 2013 IEEE

multiplied by the gap between the electrodes. After the deposition, the samples were rinsed with deionized water and blow-dried with nitrogen.

C. Finite element simulations

3-D finite element simulations were performed using COMSOL Multiphysics [9]. These simulations are computationally very expensive, and this puts a restriction on the size of the simulated structure. A 16 x 20 x 10 μm^3 structure (x, y, and z directions, respectively) represented the solution and was placed over a pair of 70-nm-thick electrodes, which were 4 or 8 μm apart in the y direction. At the bottom, the substrate was a 2-μm-thick silicon dioxide piece.

Quasi-static Maxwell's equations were solved to find the electric field distribution in the structure. The electrodes were set to +/- V/2 and the back gate was grounded. The DEP force was then calculated using

$$< \vec{F}_{DEP} > = \frac{\pi abc}{3} \varepsilon_m \, \mathrm{Re}\{\frac{\varepsilon_p^* - \varepsilon_m^*}{\varepsilon_m^*}\} \nabla |\vec{E}|^2$$

, in which a, b and c are half of the lengths of the major ellipsoid axes representing the CNT, E is the electric field, and ε_p^* and ε_m^* are the CNT and surrounding medium's permittivities, respectively.

To find the electrothermal motion of the solution, the temperature profile in the solution was calculated. The current passing through the solution and the deposited nanotubes (if any) act as a heat source and create a temperature gradient, which results in electrothermal movement in the solution. The electrothermal force can then be used in the Navier-Stokes equation to find the solution velocity at each point. CNTs move with the solution wherever the DEP force is negligible. On the other hand, DEP drags the CNTs proportionally to the inverse of the friction factor,

$$f = \frac{3\pi\eta l}{\ln(l/r)}$$

, where η is the solution viscosity, and l and r are the length and radius of the CNTs, respectively.

The final velocity of the CNTs in the solution, considering both the DEP force and the electrothermal motion, is calculated using

$$\vec{u}_{CNT} = \vec{u} + \frac{\vec{F}_{DEP}}{f}$$

, where $\vec{u}$ is the solution velocity, $\vec{u}_{CNT}$ is the total velocity of the nanotube, and $\vec{F}_{DEP}$ denotes the DEP force. The details of the simulations have been explained elsewhere [3].

Results and discussion

Figure 1 shows the scanning electron microscopy (SEM) images of the deposited patterns using electrodes with 8 μm of gap with various times ranging from 15 sec to 4 min. There are clearly stripes of nanotubes forming from the beginning of the process. The t = 15 sec sample shows how the stripes start to shape. Since the gap between the electrodes is almost double the maximum length of the CNTs, for these stripes to bridge the gap there must be several connected nanotubes. As the time

progresses to 30 sec, the half bridging CNTs turn into a fully bridging stripe, and from then on the stripes thicken until they cover the whole gap.

Judging from figure 1, the process seems to start by the deposition of nanotubes on the edges of one of the electrodes. Other nanotubes then deposit in alignment and connection with the half-bridging CNTs until a bridge between the electrodes is completed. Other stripes deposit parallel to each other at a certain distance and, subsequently, other nanotubes are attracted toward these stripes. The CNTs are attracted more toward the base of the bridges (nanotube-electrode contact region) than in the middle. The number of these stripes (in other words the distance between them) seem to be almost equal for all of the devices in figure 1 (note that these are different devices, and not the same device after different deposition times).

Figure 2 shows the results of similar DEP experiments with a time of 4 minutes and using electrodes with gaps of 4 and 20 μm. It is interesting that, while the overall outcome looks the same, the number of stripes seems to depend on the sample geometry. While the distance between adjacent stripes is in the range of 2.8 to 3.2 μm when the gap between the electrodes is 8 μm, it is in the range of 3.6 to 4 μm when the gap is 20 μm and less than 1.8 μm for 4 μm of gap. Also, the phenomenon is not as pronounced in 4-μm-gap devices as it is in devices with larger gaps.

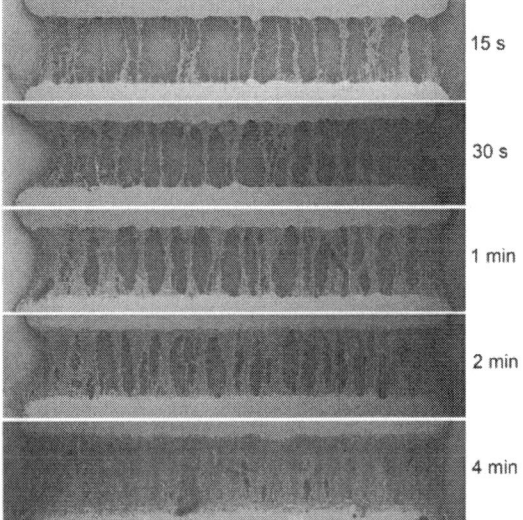

Fig. 1 SEM images of the deposition patterns for different times. The gap between the electrodes is 8 μm.

Fig. 2 SEM images of the deposition patterns using 4 minutes and devices with (a) 20 μm and (b) 4 μm of gap.

The finite element simulations were carried out for devices with 4 μm and 8 μm of gap for three different cases: no nanotube, one nanotube half bridging, and one nanotube fully bridging the gap. Figure 3 shows the velocity of the nanotubes on an xy plane (parallel to the substrate surface), 400 nm above the substrate. As can be seen, the CNTs move toward the edges of the electrodes at this stage.

A 2 nm x 2 nm x 4 μm rectangular prism was then used to emulate a CNT half bridging the 8-μm gap between the electrodes (Figure 4). The surface of the CNT was set to the electrostatic potential of the electrode it was connected to. Figure 4 shows how the velocity of the other nanotubes (in the same plane as shown in Figure 3) is affected by the presence of this deposited CNT. Because of the very high aspect ratio of the CNT, the field is considerably enhanced at the tip, and this results in a larger DEP force around it. This can explain why the half bridging nanotubes attract other nanotubes to deposit aligned and connected with them toward the opposite electrode.

For the last case, a rectangular prism with the same cross section as in the previous case, but a length of 8 μm, was placed between the two electrodes on the substrate surface. The potential drop on a metallic nanotube bridging two electrodes happens primarily within 5 nm of the nanotube-metal junction at each side [10] and, therefore, almost the entire nanotube body is at the average potential of the two electrodes, which is zero in our case.

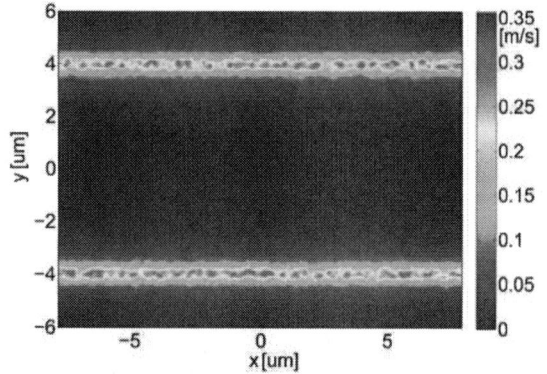

Fig. 3 Magnitude of the velocity of the nanotubes on a plane 400 nm above the substrate surface and parallel to it.

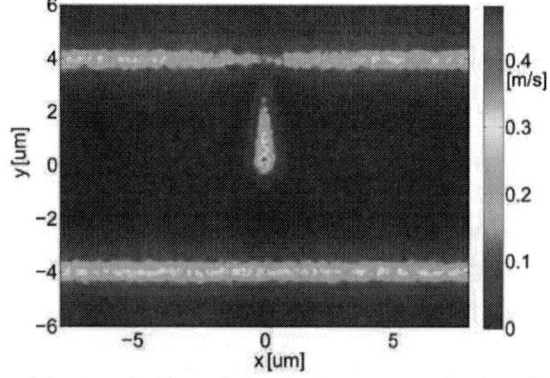

Fig. 4 Same as in Figure 3, but in the presence of a deposited nanotube half bridging the gap.

Figure 5 shows the x (parallel to the electrode edges) and z (vertical direction) components of the velocity of the nanotubes at two different heights from the surface of the substrate. At 500 nm above the surface, the x component of the velocity pushes the suspended CNTs away from the deposited nanotube. The z component is very low at this height.

Figure 5c and 5d show the velocity components 50 nm away from the substrate surface. The x component attracts the CNTs toward the nanotube especially near the edges. The z component is strongly negative (toward the surface) in the close vicinity of the deposited CNT and pins the suspended CNTs in that region to the surface. Thus, the combined effects of these two components explain the peculiar shape of the nanotube stripes and their thickening near the electrodes.

To simulate the electrothermal force and investigate if the current passing through the bridging nanotube results in a considerable increase in the agitation of the solution, the temperature of the nanotube was needed. It is noteworthy that, as we have previously shown, the electrothermal motion is negligible in surfactant-free solutions due to their low conductivity. But since the current in the CNT results in higher temperature gradients, it might agitate the solution locally.

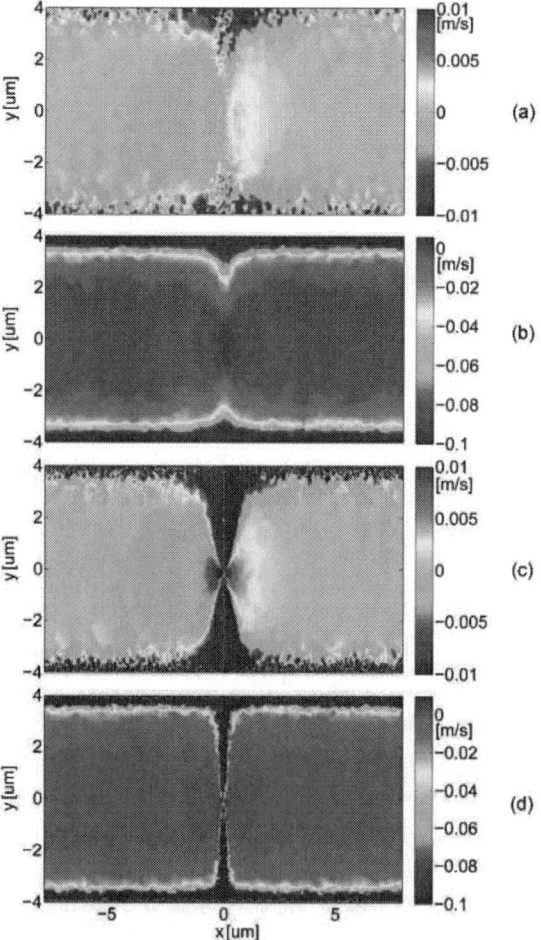

Fig. 5 Components of the velocity of the nanotubes in planes parallel to the substrate surface. (a) x, and (b) z components in the plane at a height of 500 nm from the surface. (c) x, and (d) z components in a plane 50 nm away.

The temperature of a CNT on a substrate conducting current was estimated by Pop *et al.*[11]. They showed that for most of the CNT structure on the surface, the temperature difference can be obtained from p/g, where p is the joule heating rate per unit length and g is the net heat loss rate to the substrate per unit length. They stated that this rate depends primarily on the interface and not on the thermal conductivity of the substrate. As a rough approximation, we assumed this rate to be double the case when the CNT is in air. p was found using the measured total resistance of 5 MΩ of a CNT bridging the electrodes. The average resistivity of a solution processed metallic CNT per μm of length is estimated to be 200$\pm$10kΩ [12]. The resulting temperature change was found to be around 0.18 K.

All the simulations include the electrothermal effect. The results showed that the electrothermal force is orders of magnitude smaller than DEP force in all the cases, including when a CNT connects the two electrodes. It is clear that having the deposited CNT drastically increases the electrothermal force, but not enough to make it effective compared to DEP in the vicinity of the electrodes with this type of solution.

Figure 6 shows the x velocity component of the nanotubes because of the DEP force on a line parallel to the edges of the electrodes in the middle of the device at a height of 500 nm from the surface for the 4-μm- and 8-μm-gap devices. It can be seen that the extent for which the CNTs are pushed away for the latter device is almost double that for the former. This can explain the difference in the distance between the stripes in these two devices.

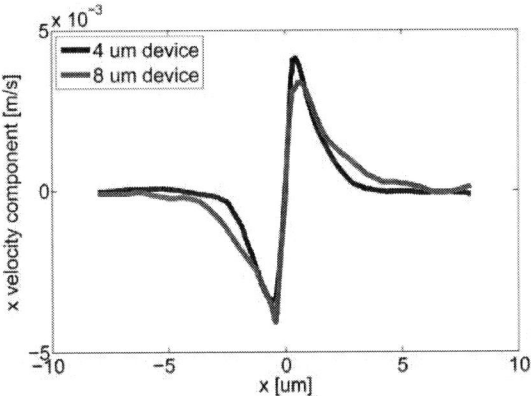

Fig. 6 The x component of the velocity for 4-μm- and 8-μm-gap devices on a line parallel to the edges of the electrodes at a height of 500 nm from the substrate surface.

In summary, the change in the electric field caused by the deposition of the first few nanotubes plays a major role in the subsequent formation of patterns during DEP. While the DEP force brings the CNTs toward the tip of a deposited CNT connected to only one electrode, it pushes most of them away laterally once a complete bridge is formed. On the other hand, a downward force pushes the CNTs toward the bridging nanotube if they are close enough to it. This effect is more pronounced for devices with larger gaps between the electrodes.

The electrothermal force increases by orders of magnitude because of changes in the temperature profile as a result of

CNT deposition, but is still negligible compared to the DEP force close to the electrodes for surfactant-free solutions.

We thank the Natural Sciences and Engineering Research Council, the BCFRST Foundation/British Columbia Innovation Council, the Canada Foundation for Innovation and the British Columbia Knowledge Development Fund for financial support.

References
[1] H. Seo, C. Han, D. Choi, K. Kim, and Y. Lee, "Controlled assembly of single SWNTs bundle using dielectrophoresis," Microelectron. Eng., vol. 81, pp. 83-89, 2005.

[2] R. Krupke, S. Linden, M. Rapp, and F. Hennrich, "Think films of metallic carbon nanotubes prepared by dielectrophoresis," Adv. Mater., vol. 18, pp. 1468-1470, 2006.

[3] A. Kashefian Naieni, and A. Nojeh, "Effect of solution conductivity and electrode shape on the deposition of carbon nanotubes from solution using dielectrophoresis," in press.

[4] S. Shekhar, P. Stokes, and S. Khondaker, " Ultrahigh density alignment of carbon nanotube arrays by dielectrophoresis," ACS Nano, vol. 5, pp. 1739-1746, 2011.

[5] P. Stokes, E. Silbar, Y. M. Zayas, and S. I. Khondaker, "Solution processed large area field effect transistors from dielectrophoreticly aligned arrays of carbon nanotubes," Appl. Phys. Lett., vol. 94, p. 113104, 2009.

[6] A. H. Monica, S. J. Papadakis, R. Osiander, and M. Paranjape, "Wafer-level assembly of carbon nanotube networks using dielectrophoresis," Nanotechnology, vol. 19, p. 085303, 2008.

[7] A. Vijayaraghavan et al., "Ultra-large-scale directed assembly of single-walled carbon nanotube devices," Nano Lett., vol. 7, pp. 1556-1560, 2007.

[8] http://www.nano-lab.com

[9] http://www.comsol.com

[10] Z. Chen, J. Appenzeller, J. Knoch, Y. Lin, and P. Avouris, "The role of metal–nanotube contact in the performance of carbon nanotube field-effect transistors," Nano Lett., vol. 5, pp. 1497-1502, 2005.

[11] E. Pop, D. A. Mann, K. E. Goodson, and H. Dai, "Electrical and thermal transport in metallic single-wall carbon nanotubes on insulating substrates," J. Appl. Phys., vol. 101, pp. 093710, 2007.

[12] Q. Cao, S. Han, G. S. Tulevski, A. D. Franklin, and W. Haensch, " Evaluation of field-effect mobility and contact resistance of transistors that use solution-processed single-walled carbon nanotubes," ACS Nano, vol. 6, pp. 6471-6477, 2012

Investigation of rock-salt CrTe thin film grown by molecular beam epitaxy toward half-metal

LU Hui* and TEO Kie Leong

Advanced Memory Laboratory, Electrical and Computer Engineering Department, National University of Singapore
4 Engineering Drive 3, Singapore 117576, Singapore
luhui@nus.edu.sg

Abstract

We report the growth of rock-salt CrTe thin film on MgO substrate by molecular-beam epitaxy (MBE). Our high-resolution transmission electron microscopy (HRTEM) results show that a metastable rock-salt (RS) CrTe film can be achieved in a crystalline phase with thickness of 50 nm. The temperature dependence of remanent magnetization shows two ferromagnetic transition temperatures at 165 K and > 400 K. The occurrence of two transition temperatures is probably due to the existence of multi-axial anisotropy in RS CrTe. A new magnetic phase of CrTe is confirmed to exist in the sample by studying the hysteresis loop at 400 K.
(keywords: CrTe, rock salt, half metal)

Introduction

Half-metallic ferromagnets (HMFs) are seen as a key ingredient in future high performance spintronics devices, because they have only one electronic spin channel at the Fermi energy and, therefore, may show nearly 100% spin polarization [1]. A lot of interests have been focused on newly created transition-metal pnictides and chalcogenides in the metastable zinc-blende (ZB) lattice structure [2-7]. These compounds are very attractive due to their structural compatibility with important III-V and II-VI semiconductors. ZB-CrSb [8, 9], CsAs [10-12] and MnAs [13, 14] pnictides have been successfully fabricated by molecular-beam epitaxy (MBE). They are ferromagnets with high Currie temperature (T_C). However, the half-metallicity (HM) could only be achieved in the expanded volume and hence the metastable energy between the ZB phase and their ground phases is rather high. Therefore, these ZB pnictides could only exist in the forms of a few monolayers or nanodots due to their large metastable energy (~1 eV/atom). On the other hand, the ZB CrTe chalcogenide is predicted to show HM in its equilibrium volume and therefore its metastable energy is lower (~0.36 eV/atom). Hence, we have previously grown ZB CrTe thin film on GaAs substrate with thickness up to ~5 nm by MBE [15]. However, the ZB CrTe has one disadvantage in that the T_C achieved in our experiment is only 100 K, which is much lower than the calculated value (>400 K) [16]. A possible explanation is that, at low value of lattice constant (< 6 Å), the dominant Cr-Cr exchange interactions in the ZB CrTe chalcogenide can be antiferromagnetic [17-19]. Meanwhile, the authors in [17, 19] has suggested that it may be possible to achieve HM in CrTe with high T_C by depositing CrTe in rock-salt (RS) phase. The dominant Cr-Cr exchange interaction in RS phase CrTe is ferromagnetic in a wide range of lattice constant and the calculated T_C is higher than that of the ZB phase. Most importantly, the RS CrTe also exhibits HM in its equilibrium volume with very small metastable energy

(~0.07 eV/f.u.). We note that the conventional semiconductor substrate, such as MgO, is also of RS phase. Hence it is of interests to grow RS CrTe on MgO substrate and investigate its magnetic properties.

Experiment

The CrTe films were directly deposited on MgO (111) substrate by MBE (an ULVAC MBC-2000 system). Elemental Cr was evaporated using conventional effusion source while an applied epivalved-cracker effusion cell was used for Te. The MgO substrate was first degassed at 250 °C for 30 minutes in preparation chamber prior to being transferred into the growth chamber. Then the MgO substrate was heated at 600 °C for 30 minutes for oxide desorption. The temperature was then lowered down at substrate temperature of 250 °C for the deposition of CrTe thin film. The flux ratio of Te/Cr was kept at ~ 2 and the beam equivalent pressure (BEP) of the Cr element remained was ~ 1.8×10^{-8} Torr. The deposition time was 60 minutes. The growth process was monitored by *in-situ* reflection high energy electron diffraction (RHEED). The RHEED patterns of CrTe film were obtained in the initial (in 10 minutes) and at the end of the growth. As shown in Fig. 1, RHEED patterns remain streaky for the whole growth progress of CrTe layer, suggesting high quality crystalline structure of the thin film.

Results and Discussions

The structural property is characterized by high resolution transmission electron microscopy (HRTEM). The cross-sectional TEM image is shown in Fig. 2. The thickness of the deposited CrTe is about 50 nm, which is much larger than that of ZB CrTe (about 5 nm). The HRTEM image in Fig. 2(b) shows that the deposited CrTe is single crystalline and its crystal structure follows the RS phase MgO substrate. Hence both the XRD results and the TEM results agree with that deposited CrTe thin film is in RS phase. From Fig. 2(b), one also finds that the lattice constant of CrTe is larger than that of MgO substrate. Furthermore, the interface is not very sharp and there is a transition zone at the interface. The thickness of this transition zone determined from the HETEM image is about 1.5 nm.

Fig. 2(c) shows the selected-area electron diffraction (SAED) patterns at the interface of MgO/CrTe. The bright dots are from the MgO substrate, while the weak dots, in yellow circles, are from CrTe thin film. Both of these diffraction patterns are very similar except that there is rotation (about 9°) in the orientations of a_1 (MgO) and a_2 (CrTe). Hence, the deposited CrTe thin film follows the RS structure of MgO substrate. As indicated by the c arrow in Fig 2(c), the out-of-plane direction of the CrTe thin film is the same as that

978-1-4673-4840-9/13 $31.00 © 2013 IEEE

of the MgO substrate, which is [111] direction. However, in the film plane, the CrTe thin film rotates about the [111] axis with 9° away from the MgO substrate.

The magnetic properties were investigated by commercial superconducting quantum interference device magnetometer (SQUID). The temperature dependence of remnant magnetization curves (M_r-T) were characterized along the out-of-plane in Fig. 3(a). The sample was cooled down from 400 K to 5 K under a magnetic field of 2 T. The magnetic field was then turned off and the remnant magnetization was recorded as the temperature is ramped up. A transition temperature (T_1) is observed between 150 K- 200 K. In order to better determine the value of T_1, derivative curve ($-dM_r/dT$) is also plotted out in Fig. 3(a). The peak of the derivative curve gives the value of T_1 ~165 K. However, the remnant magnetization does not go to zero above T_1. The inset of Fig. 3(a) is the enlarged view of the M_r-T curve above T_1. We can clearly see from the inset that there remains remnant magnetization up to 400 K. One might argue that the small magnetization detected above T_1 could be from the remnant field of the coil in SQUID. Hence, in order to further confirm the ferromagnetism in the sample above T_1, hysteresis loop was measured along out-of-plane direction at 300 K as shown in Fig. 3(b). The coercivity (H_C) is about 300 Oe and the saturation magnetization is quite low (~12 emu/cc). Hence, it is confirmed that the sample has another transition temperature (T_2), which is higher than 400 K.

A similar behavior of M_r-T curve that shows two transition temperatures has been observed in GaMnAs in [19]. The authors attributed it to the spin reorientation transition. For ferromagnetic material with multi-axial anisotropy, the spin may re-orientate from one anisotropic axis to another one with lower energy as the temperature increases and it would result in a sudden drop of magnetization. Hence, it is possible that our deposited CrTe thin film also has multi-axial anisotropy and therefore shows two transition temperatures. It is noteworthy that the T_C of all the reported CrTe compounds so far is in the range of 100 K-360 K [15]. The lowest T_C is the ZB CrTe (100 K) and the highest one is the hexagonal CrTe (360 K). Hence a new magnetic phase with T_C larger than 400 K is achieved in our sample.

Summary

In summary, thin film of RS CrTe has been successfully grown on MgO (111) substrate by MBE for the first time. Its thickness can reach 50 nm, which is much larger than any other predicted ZB HMFs. A new magnetic phase of CrTe with T_C higher than 400 K is observed in our sample. As suggested by the M_r-T curve, the RS CrTe thin film might have multi-axial anisotropy.

Acknowledge

This work is supported by Singapore Agency for Science, Technology and Research (A*STAR), under Grant No. 10210100.

References

[1] R. A. de Groot, F. M. Mueller, P. G. van Engen, and K. H. J. Buschow, Phys. Rev. Lett. **50** (1983) 2024.
[2] I. Galanakis, Phys. Rev. B **66** (2002) 012406.

[3] J. E. Pask, L. H. Yang, C. Y. Fong, W. E. Pickett, and S. Dag, Phys. Rev. B **67** (2003) 224420.
[4] J. Kübler, Phys. Rev. B **67** (2003) 220403.
[5] B. Sanyal, L. Bergqvist, and O. Eriksson, Phys. Rev. B **68** (2003) 054417.
[6] M. S. Miao and W. R. L. Lambrecht, Phys. Rev. B **71** (2005) 064407.
[7] E. Şaşioglu, I. Galanakis, L. M. Sandratskii, and P. Bruno, J. Phys.: Condens. Matter **17** (2005) 3915.
[8] J. H. Zhao, F. Matsukura, K. Takamura, E. Abe, D. Chiba, and H. Ohno, Appl. Phys. Lett. **79** (2001) 2776.
[9] J. J. Deng, J. H. Zhao, J. F. Bi, Z. C. Niu, F. H. Yang, X. G. Wu, and H. Z. Zheng, J. Appl. Phys. **99** (2006) 093902.
[10] H. Akinaga, T. Manago, and M. Shirai, Jpn. J. Appl. Phys., Part 2 **39** (2000) L1118.
[11] V. H. Etgens, P. C. de Camargo, M. Eddrief, R. Mattana, J. M. George, and Y. Garreau, Phys. Rev. Lett. **92** (2004) 167205.
[12] J. F. Bi, J. H. Zhao, J. J. Deng, Y. H. Zheng, S. S. Li, X. G. Wu, and Q. J. Jia, Appl. Phys. Lett. **88** (2006) 142509.
[13] K. Ono, J. Okabayashi, M. Mizuguchi, M. Oshima, A. Fujimori, and H. Akinaga, J. Appl. Phys. **91** (2002) 8088.
[14] T. W. Kim, H. C. Jeon, T. W. Kang, H. S. Lee, J. Y. Lee, and S. Jin, Appl. Phys. Lett. **88** (2006) 021915.
[15] M. G. Sreenivasan, J. F. Bi, K. L. Teo, and T. Liew, J. Appl. Phys. **103** (2008) 043908.
[16] Y. Liu, S. K. Bose and J. Kudrnovsky, Phys. Rev. B **82** (2010) 094435.
[17] S. K. Bose and J. Kudrnovsky, Phys. Rev. B **81** (2010) 054446.
[18] G. Y. Gao and K. L. Yao, J. Appl. Phys. **111** (2012) 113703.
[19] K. Y. Wang, K. Sawicki, K. W. Edmonds, R. P. Campion, S. Maat, C. T. Foxon, B. L. Gallagher and T. Dietl, Phys. Rev. Lett. **95** (2005) 217204.

Figures and Capture

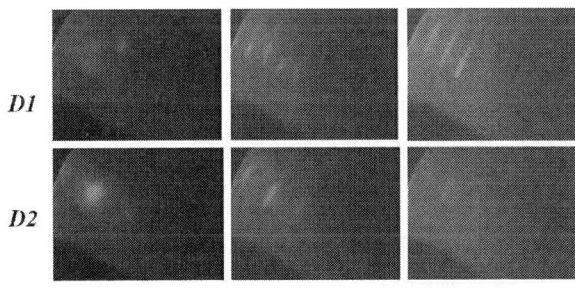

Figure 1   RHEED patterns captured during the growth along two in the film plane directions.

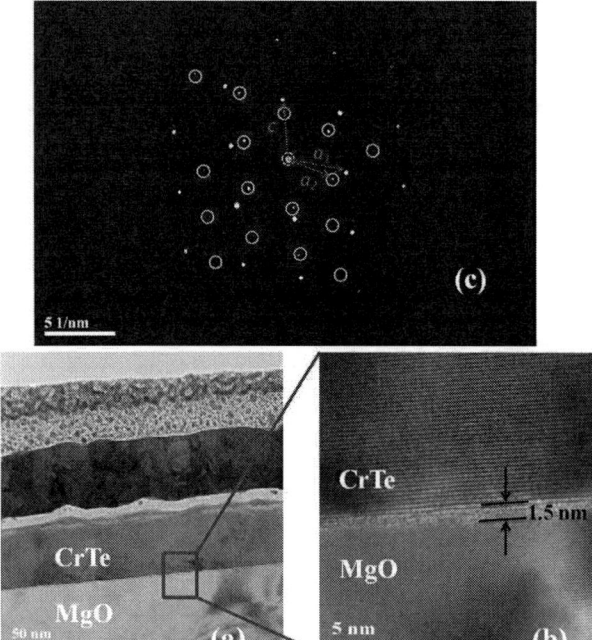

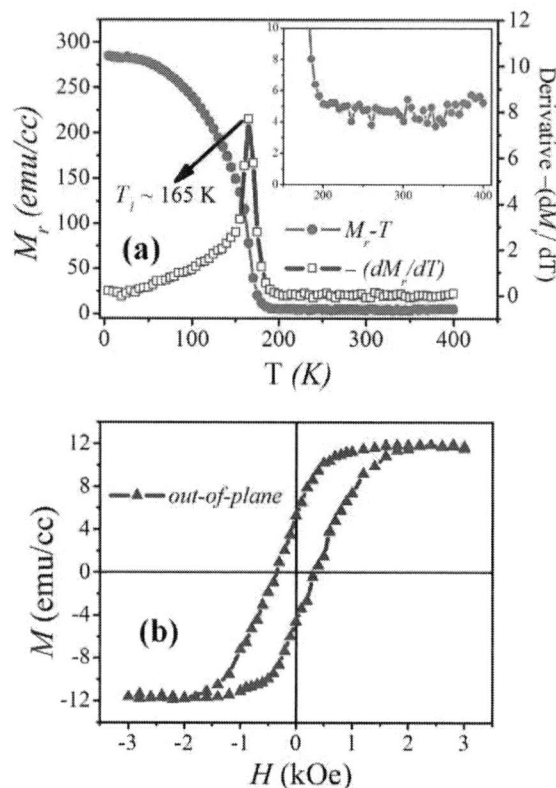

Figure 2   Cross-sectional TEM image with magnification of ×50 K (a) and ×500 K (b); SAED near the interface of MgO and CrTe layers (c). The bright dots are from MgO substrate and the weak dots (in yellow circles) are from CrTe layer. The dot in dashed yellow circle is shared by the MgO and CrTe layer.

Figure 3   (a) M_r-T curve along out-of-plane direction and its derivative. Inset of (a) is the enlarged view of M_r-T curve at high temperature. (b) Hysteresis loop measured at 300 K along out-of-plane direction.

Status Overview: Fabrication, Characterization and Modeling of Flexible RF/Microwave Nanoelectronics

Guoxuan Qin[1,2,*], Tianhao Cai[1], Zhenqiang Ma[2], and Jianguo Ma[1]

[1]School of Electronic Information Engineering, Tianjin University, Tianjin, 300072, P.R. China,
[2]Department of Electrical and Computer Engineering, University of Wisconsin, Madison, WI 53706, USA
*Tel/Fax: 86-22-27408761, Email: gqin@tju.edu.cn

Abstract — **Flexible nanoelectronics have drawn increasing attention over the past few years, for their unique properties such as bendable, rollable or foldable, can conformal attached to any shape of surfaces, light weight, etc. The device speed of flexible nanoelectronics has been increasing and reached RF/microwave regime as new fabrication/design techniques are developed. In order to provide design guidelines of employing these fast flexible nanoelectronics for flexible microwave monolithic integrated systems, characterization and modeling works have been conducted. This paper briefly reviews the recent development on fabrication, characterization and modeling for flexible RF/microwave nanoelectronics.**

Index Terms — **Flexible nanoelectronics, microwave, nanomembrane, radio frequency, single-crystal.**

Introduction

Flexible nanoelectronic devices have numerous unique advantages: bendable, rollable and foldable, can attach to any shape of surfaces, light weight, etc. These attractive characteristics make flexible nanoelectronics an increasingly hot research and application area. The widely used materials for flexible electronics are organic semiconductors, amorphous silicon and polycrystalline silicon. Universities and research institutes world wide have done a lot of work based on these materials, including material creation, device design, processing techniques and applications [1-5]. Although there's plenty of research on these kinds of flexible electronics, limitations of the materials themselves are inevitable. The carrier mobilities of these materials are significantly lower than their counterpart of inorganic single-crystal semiconductor materials. Speed is one of the most important factors for flexible electronic devices. Faster speed can greatly enhance the device performance such as data transfer, power gain and power consumption, etc. High speed flexible electronics require high carrier mobilities from the semiconductor materials. Single-crystalline silicon is the best candidate, for its advantages of low-cost, suitable for large-scale integration and compatible to current widely used industrial fabrications.

Device Fabrication

Single-crystalline Si thin film (or called nanomembrane) is first created by the research group of Prof. John A. Rogers from University of Illinois-Urbana Champaign using a novel transfer technique to release and transfer the single-crystal Si nanomembrane based on SOI (Silicon-on-Insulator) wafers [6]. With further study on flexible device design and processing, Rogers' group has successfully fabricated the first single-crystal Si thin-film transistor (TFT) and achieves electron mobility comparable to bulk Si (~500 cm^2/Vs). The single-crystal Si flexible device has higher speed than any previous works and obtains a cut-off frequency f_T of 800 MHz and maximum oscillation frequency f_{max} of 500 MHz [7].

The first microwave (or gigahertz) flexible thin-film device is fabricated by the research group of Prof. Zhenqiang Ma from University of Wisconsin-Madison, with "hot" and "cold" processes and "flip" transfer [8]. The fabrication processes enable high dosage, high energy ion implantation and high temperature furnace annealing, thus significantly reduce the contact/sheet resistance. The single-crystal Si nanomembrane TFT is thus boost up to a much higher speed level, achieving f_T of 1.9 GHz and f_{max} of 3.1 GHz [8]. With further device structure and fabrication optimizations, the flexible microwave Si nanomembrane TFT reaches f_T of 2.04 GHz and f_{max} of 7.8 GHz [9], and latest f_T of 3.8 GHz and f_{max} of 12 GHz [10], which is the fastest flexible single-crystal Si nanomembrane TFT up to date.

Based on these flexible nanoelectronic device fabrication and design techniques, a variety of flexible RF/microwave single-crystal Si nanomembrane devices/circuits have been developed, such as strained TFTs, diodes, passive components, switches, etc [11-14].

Device Characterization and Modeling

Characterization works have been conducted on the flexible microwave nanoelectronics, examples including device structure/parameter optimizations, nanomembrane transfer registration/alignment study, bending characterizations. In the work [9], the relative position between gate/dielectric layer and TFT channel is demonstrated to be a crucial factor for microwave device speed. With proper overlap of the gate/dielectric stack and source/drain doping regions, the speed of flexible microwave nanomembrane TFT can be increased dramatically. More investigations have been carried out on nanomembrane transfer reliability and local-alignment technique [10]. Further characterizations have been performed on flexible microwave nanoelectronics under mechanical bending conditions. Flexible microwave nanomembrane diodes and switches have been demonstrated to have strong performance dependence on mechanical bendings [15-17]. The high frequency responses (e.g. insertion loss) show significant and monotonic improvements with smaller bending radius (i.e. larger bending strains) (Fig. 1).

The models are also developed to better characterize the flexible microwave nanoelectronics under mechanical bending

978-1-4673-4840-9/13 $31.00 © 2013 IEEE

conditions (Fig. 2). The extracted parameters according to the models reveal the most influential factors of flexible RF/microwave diodes and switches with bending strains. The models also indicate unproportional device parameter variations with bending strains, due to the mobility enhancement under tension conditions.

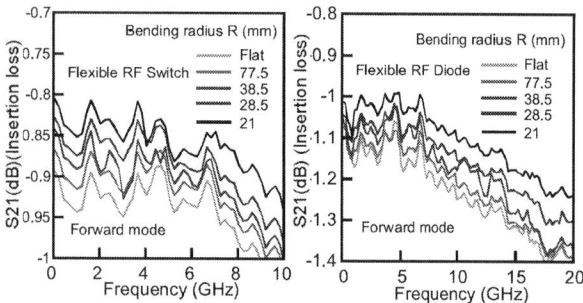

Fig. 1. Characterization of flexible microwave diodes and switches under bending conditions (data from references [15-17]).

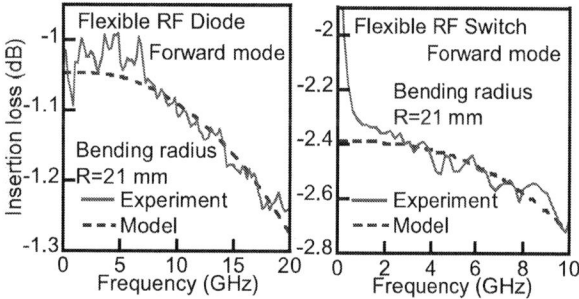

Fig. 2. Strain modeling of flexible microwave single-crystal Si diodes and switches (data from references [15-17]).

CONCLUSION

These attractive characteristics such as bendability, light weight, etc. make flexible nanoelectronics an increasingly hot research and application area. The device speed of flexible nanoelectronics has been increasing and reached RF/microwave regime as new fabrication/design techniques are developed. Characterization and modeling works have been conducted for the flexible RF/microwave nanoelectronics. The studies provide guidelines for properly designing and using these devices for realization of flexible monolithic microwave integrated systems.

ACKNOWLEDGEMENTS

This work was supported by National Natural Science Foundation of China (NSFC) under Grant No. 61006061, and Program for New Century Excellent Talents in University (NCET).

REFERENCES

[1] T. Shimoda, Y. Matsuki, M. Furusawa, T. Aoki, I. Yudasaka, H. Tanaka, H. Iwasawa, D. Wang, M. Miyasaka, Y. Takeuchi, "Solution-processed silicon films and transistors," Nature, vol. 440, pp. 783-786, 2006.

[2] R. H. Reuss, B. R. Chalamala, A. Moussessian, M. G. Kane, A. Kumar, D. C. Zhang, J. A. Rogers, M. Hatalis, D. Temple, G. Moddel, B. J. Eliasson, M. J. Estes, J. Kunze, E. S. Handy, E. S. Harmon, D. B. Salzman, J. M. Woodall, M. A. Alam, J. Y.

Murthy, S. C. Jacobsen, M. Olivier, D. Markus, P. M. Campbell, and E. Snow, "Macroelectronics: perspectives on technology and applications," P. IEEE, vol. 93, pp. 1239-1256, 2005.

[3] G. Qin, H.-C. Yuan, H. Yang, W. Zhou, Z. Ma, "Flexible thin-film transistors fabricated in polycrystalline silicon membrane transferred to a plastic substrate," Semi. Sci. &Tech., vol. 26, p. 025005, 2011.

[4] J.A. Rogers, M.G. Lagally, R.G. Nuzzo, "Review: Synthesis, assembly and applications of semiconductor nanomembranes," Nature, vol. 477, p. 45, 2011.

[5] Z. Ma, "An Electronic Second Skin", Science, vol. 333, p. 830, 2011.

[6] E. Menard, K.J. Lee, D.-Y. Khang, R.G. Nuzzo, J.A. Rogers, "A printable form of silicon for high performance thin film transistors on plastic substrates," Appl. Phys. Lett., vol. 84, pp. 5398-5400, 2004.

[7] J.H. Ahn, H.S. Kim, K.J. Lee, Z.T. Zhu, E. Menard, R.G. Nuzzo, J.A. Rogers, "High-speed mechanically flexible single-crystal silicon thin-film transistors on plastic substrates," IEEE Elect. Dev. Lett., vol. 27, pp. 460-462, 2006.

[8] H.-C. Yuan and Z. Ma, "Microwave thin-film transistors using Si nanomembranes on flexible polymer substrate", Appl. Phys. Lett., vol. 89, p. 212105, 2006.

[9] H.-C. Yuan, G.K. Celler, Z. Ma, "7.8-GHz flexible thin-film transistors on a low temperature plastic substrate," J. Appl. Phys., vol. 102, p. 034501, 2007.

[10] L. Sun, G. Qin, J.H. Seo, G.K. Celler, W. Zhou, Z. Ma, "12-GHz thin-film transistors on transferable silicon nanomembranes for high-performance flexible electronics," Small, vol. 6, pp. 2553-2557, 2010.

[11] H.-C. Yuan, M. M. Roberts, D. E. Savage, M. G. Lagally, Z. Ma, "High-speed strained-single-crystal-silicon thin-film transistors on flexible polymers", J. Appl. Phys., vol. 100, p. 013708, 2006.

[12] G. Qin, H.-C. Yuan, G.K. Celler, W. Zhou, Z. Ma, "Flexible microwave PIN diodes and switches employing transferrable single-crystal Si nanomembranes on plastic substrates," J. Phys. D: Appl. Phys., vol. 42, p. 234006, 2009.

[13] H.-C. Yuan, G. Qin, G. K. Celler, Z. Ma, "Bendable high-frequency microwave switches formed with single-crystal silicon nanomembranes on plastic substrates," Appl. Phys. Lett., vol. 95, p. 043109, 2009.

[14] L. Sun, G. Qin, H. Huang, H. Zhou, N. Bahdad, W. Zhou, Z. Ma, "Flexible high-frequency microwave inductors and capacitors integrated on a polyethylene terephthalate substrate," Appl. Phys. Lett., vol. 96, p. 013509, 2009.

[15] G. Qin, H.-C. Yuan, G.K. Celler, J. Ma, Z. Ma, "Impact of strain on radio frequency characteristics of flexible microwave single-crystalline silicon nanomembrane p-intrinsic-n diodes on plastic substrates," Appl. Phys. Lett., vol. 97, p. 233110, 2010.

[16] G. Qin, H.-C. Yuan, G. K. Celler, J. Ma, Z. Ma, "Influence of bending strains on radio frequency characteristics of flexible microwave switches using single-crystal silicon nano membranes on plastic substrate," Appl. Phys. Lett., vol. 99, p. 153106, 2011.

[17] G. Qin, L. Yang, J.-H. Seo, H.-C. Yuan, G. K. Celle, J. Ma, Z. Ma, "Experimental characterization and modeling of the bending strain effect on flexible microwave diodes and switches on plastic substrate", Appl. Phys. Lett., vol. 99, p. 243104, 2011.

[18] G. Qin, H.-C. Yuan, G.K. Celler, W. Zhou, J. Ma, Z. Ma, "RF model of flexible microwave single-crystalline silicon nanomembrane PIN diodes on plastic substrate," Microelectron. J., vol. 42, pp. 509-514, 2011.

[19] G. Qin, H.-C. Yuan, G. K. Celler, J. Ma, Z. Ma, "RF model of flexible microwave switches employing single-crystal silicon nanomembranes on a plastic substrate", Microelectronic Engineering, vol. 95, p. 21, 2012.

Characterization of MIM Diodes based on Nb/ Nb$_2$O$_5$

Islam E. Hashem[1,2,*], Nadia H. Rafat[1], Ezzeldin A. Soliman[3]

[1] Department of Engineering Math. and Physics, Faculty of Engineering, Cairo University, Giza 12613, Egypt
[2] Youssef Jameel Science and Technology Research Center, American University in Cairo, Egypt
[3] Department of Physics, School of Science and Engineering, American University in Cairo, New Cairo 11835, Egypt
[*] E-mail: eemm_hashem@ieee.org

Abstract—**MIM diodes based on Nb/Nb$_2$O$_5$ are analyzed. The tunneling probability calculation is based on the transfer matrix method and the non equilibrium green's function. Contour plots are presented showing the effect of the oxide thickness and the work function difference between the left metal electrode, Nb and the right metal electrode on the diode resistance, responsivity, and non linearity. Also, the total rectenna efficiency is analyzed for various MIM structures.**

Keywords-MIM; Rectenna; Tunneling

I. INTRODUCTION

Active research on the solar energy harvesting is now directed towards the use of antenna coupled Metal Insulator Metal (MIM) diodes [1-4]. The nano-antenna, confines a highly localized field in the gap between the two metals, and the MIM diode, rectify the incident electromagnetic waves [4, 5]. In order to rectify this fast THz signal, the MIM diode should maintain an insulator thickness less than 4 nm, to keep the tunneling current as the main current transport in this device. This tunneling current is modeled using different analytical models, these models are all based on the WKB approximation, however, the WKB does not take into consideration the reflections at the interface between the Left-Metal/Insulator interface and the Insulator/Right-Metal interface. In this paper, the tunneling probability calculation through a trapezoidal barrier, is calculated using the transfer matrix method based on the Airy function. The model gains its importance for its fast simulation time, in comparison with other numerical techniques, like the Non Equilibrium Green's Function (NEGF), which has a large computing time. In this paper, the simulation methodology for modeling the tunneling current using the transfer matrix method is presented in Section II. The performance of MIM diodes at zero bias is investigated in Section III. Next, simulating devices based on Nb/Nb$_2$O$_5$ takes place in Section IV taking into consideration the whole system efficiency calculation. Finally, the conclusion is presented in Section V.

II. SIMULATION METHODOLOGY

The tunneling current throughout the MIM structure is modeled using the Transfer Matrix Method (TMM) and the Non Equilibrium Green's Function (NEGF) [6]. In the following, the tunneling current throughout the MIM structure is analyzed using the Transfer function method based on the Airy function.

The 1D schematic band diagram of the device under applied voltage bias, V_b, is shown in Fig. 1. Where, $V(x)$ is the potential difference across the insulator layer at a certain position x, and is given by:

$$V(x) = (\phi_L - \chi) - (\phi_L - \phi_R - V_b)\frac{x}{d}$$ (1)

where $q\phi_L$ and $q\phi_R$ are the work functions of the left and right metal electrodes respectively, χ is the insulator electron affinity, and d is the insulator thickness.

Assuming the electron tunneling energy and the momentum transverse component are conserved, the tunnel current density is given by [7]:

$$J = J_{L-R} - J_{R-L}$$ (2)

where

$$J_{L-R} = \frac{4\pi m_L^2 q}{h^3}\int_0^{E_m} T(E_x)\int_0^{\infty} f_L(E)\,dEdE_x$$

$$J_{R-L} = \frac{4\pi m_R^2 q}{h^3}\int_0^{E_m} T(E_x)\int_0^{\infty} f_R(E+qV_b)\,dEdE_x$$ (3)

where E, E_x and $T(E_x)$ are the total energy, the transverse energy of a tunneling electron, and the tunneling probability respectively. m_L and m_R are the transverse electron effective masses in the left and right electrodes, respectively. f_L and f_R are the Fermi-Dirac distributions of the left and right metal electrodes.

The 1D time-independent single-particle Schrödinger equation is given by:

$$-\frac{\hbar^-}{2}\frac{d}{dx}\left[\frac{1}{m}\frac{d\psi(x)}{dx}\right] + U(x)\psi(x) = E_x\psi(x)$$ (4)

where $\hbar$ is the reduced Plank constant, $\psi(x)$ is the electron wave-function and m is the effective mass.

The solutions of the Schrödinger equation within the left and right metal regions are:

$$\psi_L(x) = A_L e^{ik_L x} + B_L e^{-ik_L x}, \quad x < X_1$$ (5a)

$$\psi_R(x) = A_R e^{ik_R x} + B_R e^{-ik_R x}, \quad x > X_2$$ (5b)

where ψ_L, ψ_R are the wave functions in the left and right electrodes respectively, $k_{L,R}$ are the electron wave vectors of the plane wave functions in the left and right electrodes, respectively. By applying the normalization factor at each interface as follows:

$$u_1 = s\left(\frac{2m}{\hbar^-}|F|\right)^{\frac{1}{3}}\left(\frac{U^l - E_x}{F}\right) \text{ and}$$ (6a)

$$u_2 = s \left(\frac{2m}{\hbar} |F| \right)^{\frac{1}{3}} \left(\frac{U^r - E_x}{F} + d \right) \tag{6b}$$

where $F = \dfrac{U^l - U^r}{X_2 - X_1}$, and $s = \mathrm{sgn}(F)$ (7)

Hence, Schrödinger equation turns to $\dfrac{d^2\psi}{du^2} - u\psi = 0$

The electron wave function through the insulator can be expressed as a linear combination of Airy functions, Ai and Bi [8]:

$$\psi(x) = A * Ai(u) + B * Bi(u), \quad X_1 < x < X_2 \tag{8}$$

The wave function $\psi(x)$ and its spatial derivative (i.e. $\frac{1}{m} \frac{d\psi(x)}{dx}$) should be continuous at the interface between the left metal electrode and the insulator, and the insulator and the right metal electrode. Hence, the wave functions are related to each other at the interfaces as:

$$\begin{pmatrix} A_L \\ B_L \end{pmatrix} = \begin{pmatrix} T_{11} & T_{12} \\ T_{21} & T_{22} \end{pmatrix} \begin{pmatrix} A_R \\ B_R \end{pmatrix} = T \begin{pmatrix} A_R \\ B_R \end{pmatrix} \tag{9}$$

where

$$T = \frac{1}{2}\pi \begin{pmatrix} 1 & \dfrac{m_L}{ik_L} \\ 1 & \dfrac{m_L}{ik_L} \end{pmatrix} \begin{pmatrix} Ai(u_1) & Bi(u_1) \\ \dfrac{u'}{m} Ai'(u_1) & \dfrac{u'}{m} Bi'(u_1) \end{pmatrix}$$

$$\begin{pmatrix} Bi'(u_2) & -\dfrac{m}{u'} Bi(u_2) \\ -Ai'(u_2) & \dfrac{m}{u'} Ai(u_2) \end{pmatrix} \begin{pmatrix} 1 & 1 \\ \dfrac{ik_R}{m_R} & -\dfrac{ik_R}{m_R} \end{pmatrix} \tag{10}$$

where $u' = s \left(\dfrac{2m}{\hbar} |F| \right)^{\frac{1}{3}}$.

In order to find simple expressions for the probability amplitude, initial conditions are applied. Suppose the electron is incident from the left with $A_L = 1$, and there is no reflected traveling wave in metal 2, i.e. $B_R = 0$. It follows that $A_R = 1/T_{11}$. So, the tunneling transmission probability $T(E_x)$ can be computed as follows:

$$T(E_x) = \frac{m_L k_R}{m_R k_L} \frac{1}{|T_{11}|^2} \tag{11}$$

As shown in Fig. 2, complete agreement is achieved between the presented TMM and the NEGF for the tunneling probability calculation through Nb/Nb$_2$O$_5$/Nb structure with 1.5 nm insulator thickness simulated at zero bias. Comparing with the WKB, the tunneling probability calculated using the TMM and NEGF, takes into consideration the reflection at the interface between the left metal and the dielectric, and the dielectric and the right metal electrode, while the WKB gives a tunneling

probability equals to one for all energy values greater than the potential barrier. In addition, before proceeding with investigating the effect of various parameters on the MIM diode performance, we carry out a comparison with a previously published experiment. The TMM and NEGF are compared with the fabricated MIM presented in ref. [9]. As it is clear from Fig. 3, a good agreement between the used theoretical models and the experiment is shown. This agreement validates the simulations we carry in the next section.

III. MIM Performance

In order to maximize the rectenna efficiency, active research is now directed towards operating the MIM diode at zero bias. In this section, Nb is used as the left metal electrode, Nb$_2$O$_5$ is used as an insulator, while the work function of the right metal is changed from 4.3 eV to 5.65 eV; metals of work function covering that range are: Nb (4.3 eV), Cr (4.5 eV), Cu (4.65 eV), NbN (4.7 eV), Au (5.1 eV), and Pt (5.65 eV), where the bracketed quantities represent the work function of these metals. Each of these devices is simulated for different insulator thickness. Contour plots are presented in Fig. 4. These contour plots show the effect of thickness and barrier asymmetry (difference between the work function of the right and left metal electrodes) on the resistance, responsivity, and non linearity. As depicted in Fig. 4, the responsivity and resistance increases as the oxide thickness and barrier asymmetry increases, the variations of the resistance and responsivity is very sensitive for thicker oxides. Also, it is clear that non-linearity increases as the thickness increases and barrier asymmetry increase. For a certain oxide thickness, as the barrier asymmetry increases the shape of the tunneling barrier for the positive and negative bias vary a lot, resulting in high non linear current voltage (J/V) characteristics. In these contours, it is clear that low resistance diode may result in a low responsivity with a linear J/V relation which is inefficient for rectenna application. Also, high non-linear diode may have high resistance, which may degrade the matching of the diode with the antenna, and low responsivity, resulting in low output DC current. Hence, optimization maybe required here on the choice of the right metal electrode and insulator thickness.

IV. RECTENNA MODELING

The equivalent circuit of the nano-rectenna is described by an antenna and a diode connected in series as shown in Fig. 5. From the circuit analysis, it is clear that the coupling efficiency, which is the ratio between the ac power delivered to the diode resistance, R_D, to the power of AC voltage source, is given by [10, 11]:

$$\eta_c = \frac{4(R_A R_D / (R_A + R_D)^2)}{1 + (\omega C_D (R_A // R_D))^2} \tag{12}$$

The coupling efficiency has two main parameters that highly affect it [11]: (1) The ratio of the diode to the antenna resistance (R_D/R_A), and (2) The frequency dependence term $(\omega C_D(R_D//R_A))$. Both the unity of the first term and the zero of the second term lead to a unity coupling efficiency. In the following we show how each term affect the diode overall efficiency. In this section, the left-metal/insulator is Nb/Nb$_2$O$_5$, while the right metal is changed; Nb, Cr, and Cu are simulated. The wavelength of the incident electromagnetic radiation is swept from 0.5 um to 15 um. While, the thickness of the insulator layer Nb$_2$O$_5$ is fixed at 1.5 nm and the effect of these parameters on the total rectenna efficiency $(\eta_c B)$ [12], is shown in Fig. 6, where B is the diode responsivity. The antenna

978-1-4673-4840-9/13 $31.00 © 2013 IEEE

resistance, R_A is set to 1000 Ohm at zero bias. In addition, the diode area is assumed as A= 100* 100 nm².

It is clear that increasing the barrier asymmetry leads to the increase of the diode resistance independent of the frequency of operation. At small wavelengths, the responsivity of the three simulated MIM diodes are almost the same, however increasing the range of wavelength simulation shows the variation of the three MIM diodes responsivity. For the diode coupling efficiency, the behavior of the three MIM diodes shows a certain peak. This peak occurs when the diode resistance is equal to the antenna resistance (effect of the first term). Afterwards the coupling efficiency decreases due to the increase of the diode resistance. Then, the coupling efficiency starts to increase again. This increase is due to the decrease of the frequency of the diode operation, where the second term effect starts to glow up. Also, it is worth noting here that the total efficiency follows the same behavior of the coupling efficiency.

V. CONCLUSION

The performance of MIM diodes using Nb/Nb₂O₅ is analyzed. Different materials are simulated and the effect of the insulator thickness and barrier asymmetry on the structure performance is showed. The effect of the coupling between the antenna and diode resistance, and frequency of operation on the rectenna efficiency is analyzed.

REFERENCES

[1] M. Bareiss, A. Hochmeister, G. Jegert, G. Koblmuller, U. Zschieschang, H. Klauk, B. Fabel, G. Scarpa, W. Porod, and P. Lugli, "Energy harvesting using nano antenna array," in *Nanotechnology (IEEE-NANO), 2011 11th IEEE Conference on*, 2011, pp. 218-221.

[2] B. R. K. Richard M. Osgood III, and Joel Carlson, "Nanoantenna-coupled MIM nanodiodes for efficient vis/nir energy conversion " in *Proceeding SPIE*, San Diego, CA, USA, 2007.

[3] A. M. A. Sabaawi, C. C. Tsimenidis, and B. S. Sharif, "Bow-tie nano-array rectenna: Design and optimization," in *Antennas and Propagation (EUCAP), 2012 6th European Conference on*, 2012, pp. 1975-1978.

[4] S. G. Richard Osgood III, Carlson Gustavo, E. Fernandes, Jin Ho Kim, Jimmy Xu Matthew Chin, Barbara Nichols, Madan Dubey Philip Parilla, Joseph Berry, David Ginley Prakash Periasamy, Harvey Guthrey, Ryan O'Hayre, and Walter Buchwald, "Diode-coupled Ag nanoantennas for nanorectenna energy conversion," in *Proc. SPIE* San Diego, California, USA.

[5] S. D. Novack, D. K. Kotter, D. Slafer, and P. Pinhero, *SOLAR NANTENNA ELECTROMAGNETIC COLLECTORS*, 2008.

[6] S. Datta, "Nanoscale device modeling: the Green's function method," *Superlattices and Microstructures*, vol. 28, pp. 253-278, 2000.

[7] J. G. Simmons, "Potential Barriers and Emission-Limited Current Flow Between Closely Spaced Parallel Metal Electrodes," *Journal of Applied Physics*, vol. 35, pp. 2472-2481, 1964.

[8] M. Abramowitz and I. A. Stegun, *Handbook of Mathematical Functions*. New York: Dover, 1965.

[9] S. Grover and G. Moddel, "Engineering the current–voltage characteristics of metal–insulator–metal diodes using double-insulator tunnel barriers," *Solid-State Electronics*, vol. 67, pp. 94-99, 2012.

[10] J. A. Bean, A. Weeks, and G. D. Boreman, "Performance Optimization of Antenna-Coupled Tunnel Diode Infrared Detectors," *IEEE Journal of Quantum Electronics*, vol. 47, pp. 126-135, 2011.

[11] S. Grover and G. Moddel, "Applicability of Metal/Insulator/Metal (MIM) Diodes to Solar Rectennas," *IEEE Journal of Photovoltaics*, vol. 1, pp. 78-83, 2011.

[12] S. Grover, O. Dmitriyeva, M. J. Estes, and G. Moddel, "Traveling-Wave Metal/Insulator/Metal Diodes for Improved Infrared Bandwidth and Efficiency of Antenna-Coupled Rectifiers," *Nanotechnology, IEEE Transactions on*, vol. 9, pp. 716-722, 2010.

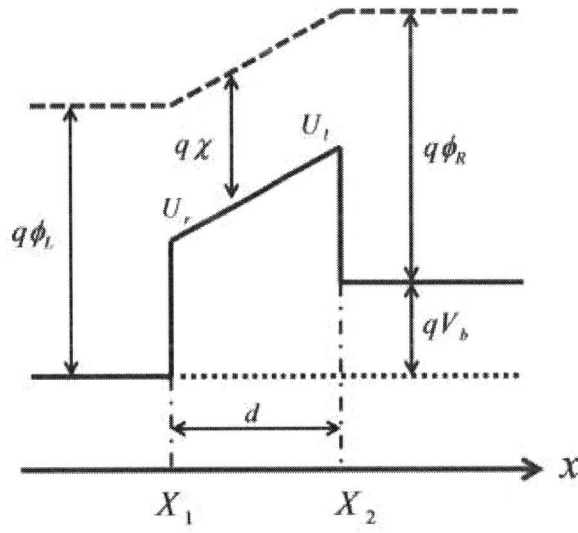

Figure 1 Potential barrier of MIM diode

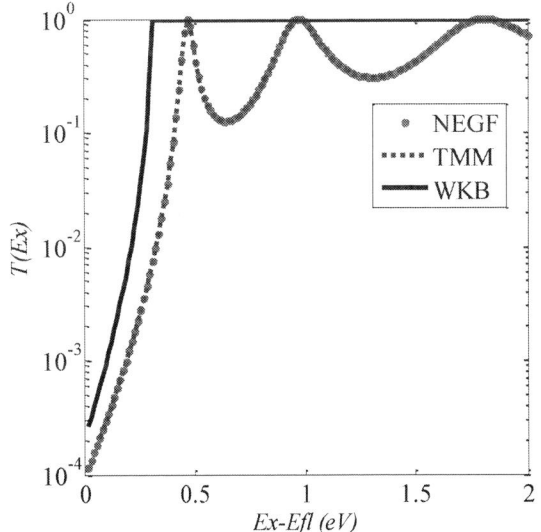

Figure 2 Tunneling probability versus the transmission energy for Nb/Nb₂O₅/Nb.

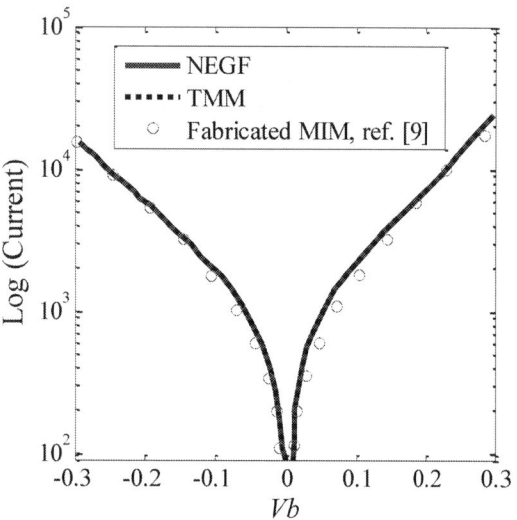

Figure 3 The current density of the fabricated Nb/Nb$_2$O$_5$/Nb presented in ref. [9] and comparison with the results obtained using the TMM, and NEGF.

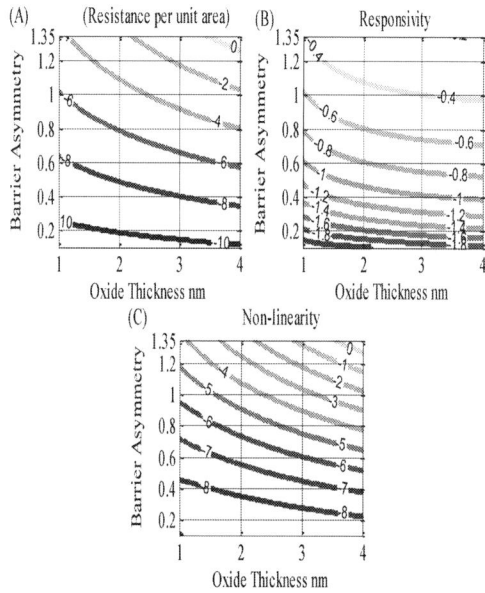

Figure 4 Contour plots showing the effect of the insulator thickness and barrier asymmetry on the MIM diode resistance (log-scale), responsivity(log-scale), and non-linearity (log-scale).

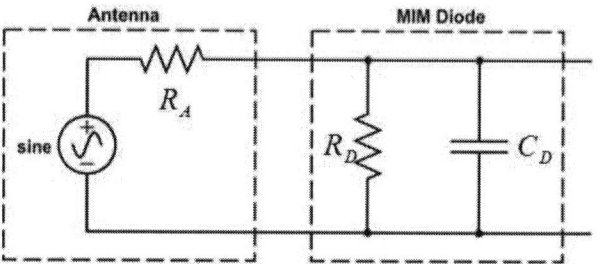

Figure 5 Equivalent circuit of rectenna.

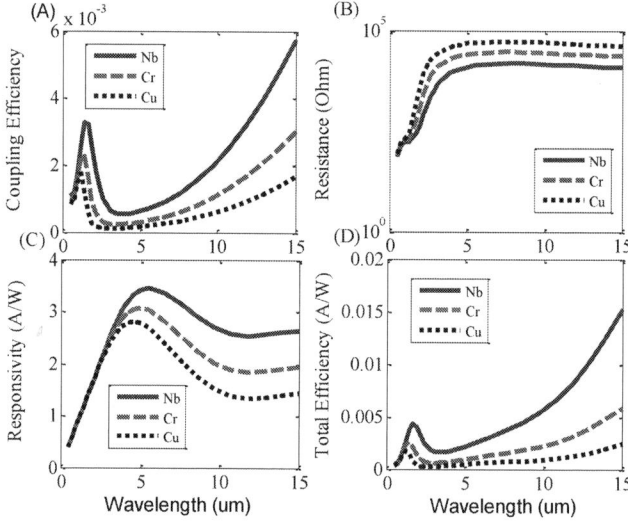

Figure 6 Rectenna coupling efficiency, MIM diode resistance, MIM responsivity, and total rectenna efficiency for Nb/Nb$_2$O$_5$/Nb, Nb/Nb$_2$O$_5$/Cr, and Nb/Nb$_2$O$_5$/Cu.

UTBB with Ground-Plane Dopant-Segregated Schottky Barrier SOI MOSFET for Thermally Efficient Low-Variability Nanoscale CMOS Circuits

S. Qureshi and Ganesh C. Patil

Department of Electrical Engineering,
Indian Institute of Technology Kanpur
Kanpur, Uttarpradesh, India
E-mail: qureshi@iitk.ac.in, pganesh@iitk.ac.in

Abstract

In this paper, a comparative study on self-heating effect, scalability, threshold voltage variability and CMOS logic performance of dopant-segregated Schottky barrier (DSSB) ultrathin body (UTB) thick buried oxide (BOX) and DSSB UTB thin BOX (UTBB) silicon-on-insulator (SOI) MOSFETs has been carried out by using the two dimensional MEDICI simulator. A novel DSSB UTBB-ground plane (GP) SOI MOSFET has also been proposed to improve the scalability and CMOS logic performance of DSSB SOI MOSFET. It has been found that, the presence of GP in DSSB UTBB-GP device not only improves the on-state drive current but also reduces the off-state leakage current of the device. Further, since the presence of GP cuts-off the path of the fringing field lines arising from the drain and the screening effect due to GP suppresses the random dopant fluctuations, both drain induced barrier lowering and threshold voltage variability in the proposed device are also low. In addition to this, the intrinsic gate delay and the static power dissipation in the case of DSSB UTBB-GP MOSFET are also reduced by ~80% and ~40% respectively over the DSSB thick BOX and DSSB UTBB SOI MOSFETs. Thus, significant reduction in self-heating effect, threshold voltage variability and the significant improvement in CMOS logic performance make the proposed device suitable for nanoscale CMOS logic circuits.
Keywords: ultrathin BOX, dopant segregation, SOI, variability, Schottky barrier, ground plane, self-heating

Introduction

Recently, due to superior scaling properties and availability of thin buried oxide (BOX) wafers, ultrathin-body on thin BOX (UTBB) SOI MOSFET has become an emerging device for future technology nodes [1]-[3]. However, the increased source/drain (S/D) series resistance (R_{SD}) at ultrathin Si film and the depletion fields at thin-BOX limits the use of this device for CMOS logic circuits. Employing ground-plane (GP) under the thin-BOX alleviates the problem of depletion fields due to S/D-to-substrate coupling and replacement of doped S/D by metal silicides can alleviate the problem of increased R_{SD} [3]-[4]. However, the presence of Schottky barrier (SB) at the metal-semiconductor (M-S) junction formed between S/D and the channel reduces the on-state drive current (I_{ON}) of the device [5]. The use of low-SB silicides such as erbium silicide (ErSi$_{1.7}$) for n-channel MOSFET (NMOS) and platinum

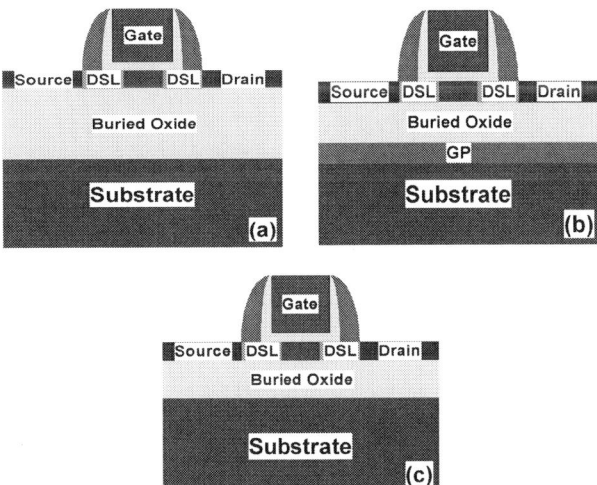

Fig. 1. (a) DSSB thick BOX, (b) proposed DSSB UTBB-GP and (c) DSSB UTBB SOI MOSFETs used in the MEDICI simulations. Here, DSL indicates the dopant-segregation layer.

silicide (PtSi) for p-channel MOSFET (PMOS) and the extra barrier lowering and thinning induced dopant-segregation layer (DSL) can improve the I_{ON} [5]-[9]. However, employing DSL during silicidation leads to an increase in the process induced device parameter variability and the inhomogeneity in SB height affects the saturation threshold voltage (V_{TSAT}) of the device [10]-[11]. To alleviate these challenges, in this work we have shown that, employing thin-BOX and GP in dopant-segregated SB (DSSB) SOI MOSFET not only improves the I_{ON} but also suppresses the short-channel effects (SCEs), self-heating (SH) effect and the threshold voltage variability induced by the process fluctuations in SOI film thickness (T_{Si}) of the device.

The rest of the paper is organized in three sections. Section II presents the device structures and the simulation setup used whereas the detailed discussion on the simulation results obtained in this work is presented in Section III. Finally, the conclusion is given in Section IV.

Device Structures and Simulation Setup

The device structures and the parameters used in the MEDICI simulator [12] are shown in Fig. 1(a)-(c) and TABLE I respectively. The range of physical gate length (L_G) has been considered as per the high performance logic technology nodes

978-1-4673-4840-9/13 $31.00 © 2013 IEEE

TABLE I
DEVICE PARAMETERS

Parameter	Unit	DSSB Thick BOX	DSSB UTBB-GP	DSSB UTBB
Physical gate length	nm	variable	variable	variable
Gate oxide thickness	nm	1	1	1
Nominal Silicon film thickness	nm	8	8	8
BOX thickness	nm	100	20	20
Substrate doping	cm^{-3}	1×10^{15}	1×10^{15}	1×10^{15}
Spacer thickness	nm	10	10	10
Doping in DSL	cm^{-3}	1×10^{20}	1×10^{20}	1×10^{20}
Channel doping	cm^{-3}	1×10^{15}	1×10^{15}	1×10^{15}
Overlap channel length	nm	1	1	1
Length of DSL	nm	11	11	11
Electron SB for NMOS	eV	0.3	0.3	0.3
Hole SB for NMOS	eV	0.82	0.82	0.82
Electron SB for PMOS	eV	0.82	0.82	0.82
Hole SB for PMOS	eV	0.3	0.3	0.3
Supply voltage	V	1	1	1
Nominal saturation threshold voltage	V	0.2	0.2	0.2
Doping in GP*	cm^{-3}	#	3×10^{18}	#
Thickness of GP	nm	#	5	#

* p-type doping for NMOS and n-type doping for PMOS

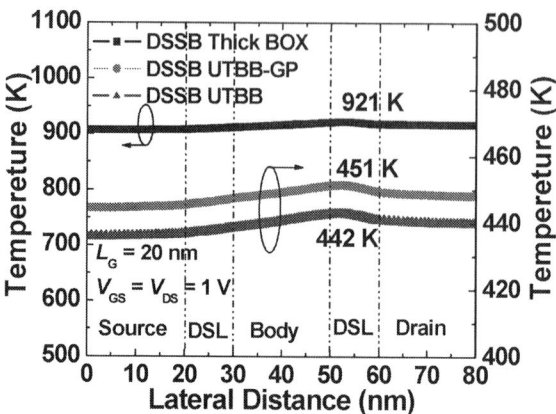

Fig. 2. Temperature profile of DSSB thick BOX, DSSB UTBB-GP and DSSB UTBB n-channel SOI MOSFETs.

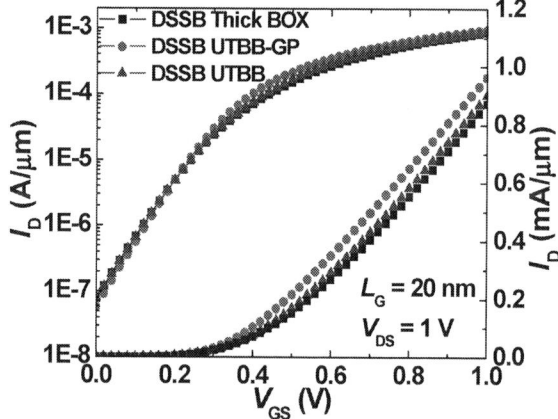

Fig. 3. I_D vs V_{GS} characteristics of DSSB thick BOX, DSSB UTBB-GP and DSSB UTBB n-channel SOI MOSFETs.

specified in the international technology roadmap for semiconductors (ITRS)-2009 [13]. The drift-diffusion (DD) simulations coupled with lattice heat equation have been carried out to capture the thermal effects. The SB tunneling, image force SB lowering and the band-to-band tunneling models have been incorporated along with the mobility models. To study the impact of process fluctuations on T_{Si}, different sets of NMOS/PMOS devices with varying T_{Si} have been simulated. The standard deviation (σ) obtained in each set of 25 devices for T_{Si} is 1.3 nm.

Results and Discussion

A. Self-heating effect

Fig. 2 shows the temperature profile of DSSB thick BOX, DSSB UTBB-GP and DSSB UTBB n-channel SOI MOSFETs. From the figure it can be seen that at $V_{DS} = V_{GS} = 1$ V the peak channel temperature in DSSB thick BOX MOSFET is 921 K whereas at the same V_{DS} and V_{GS} values the peak channel temperature in DSSB UTBB and DSSB UTBB-GP devices is 442 K and 451 K respectively. Thus, employing thin BOX in DSSB SOI MOSFET reduces SH effect of the device.

B. DC Performance and Scalability

Fig. 3 shows the I_D vs V_{GS} characteristics of DSSB thick BOX, DSSB UTBB-GP and DSSB UTBB NMOS devices for $L_G = 20$ nm. From the figure it can be seen that both I_{ON} and off-state leakage (I_{OFF}) in DSSB thick BOX and DSSB UTBB devices are almost equal. On the other hand, I_{ON} in the case proposed DSSB UTBB-GP device improves whereas I_{OFF} reduces. The main reason for this improvement in I_{ON} and I_{OFF} is the presence of GP under the thin BOX of the device.

The improvement in I_{ON} and I_{OFF} can also be observed from the I_{ON}-I_{OFF} characteristics shown in Fig. 4. In this figure I_{OFF} is extracted at $V_{DS} = V_{DD} = 1$ V and $V_{GS} = V_{GOFF} = 0$ V whereas I_{ON} is extracted at $V_{DS} = V_{DD}$ and $V_{GS} = V_{GOFF} + V_{DD}$. From Fig. 4 it

can also be seen that, at lower L_G value i.e at $L_G = 14$ nm, I_{OFF} in the case of DSSB thick BOX and DSSB UTBB SOI devices is almost equal whereas I_{ON} in DSSB UTBB device improves over the DSSB thick BOX device. This difference in I_{ON} is mainly due to different BOX thicknesses of the devices. Although due to presence of thin BOX I_{ON} in the case of DSSB UTBB device improves over the DSSB thick BOX device, in comparison to proposed DSSB UTBB-GP device this improvement in I_{ON} is significantly low. This is mainly because of the reduction in substrate potential coupling due to GP under the thin BOX of the proposed device. Thus, employing GP in DSSB UTBB MOSFET improves the I_{ON} of the device.

Fig. 5 shows the variation of drain induced barrier lowering (DIBL) as a function of L_G for DSSB thick BOX, DSSB UTBB-GP and DSSB UTBB NMOS and PMOS devices. Here, the DIBL is defined as $(V_{TLIN} - V_{TSAT})/0.95$ where V_{TLIN} and V_{TSAT} are the threshold voltages at $V_{DS} = 50$ mV and 1 V respectively. The V_{TLIN} and V_{TSAT} of all the NMOS devices have been extracted by varying the gate bias to achieve $I_D = 1\times10^{-7}(W/L_G)$ A where $W = 1$ μm is the width of the transistor. Further, since the electron mobility is twice as that of the hole mobility and the width of both NMOS and PMOS transistors is considered to be equal (i.e. $W = 1$ μm), the threshold voltage of PMOS transistors has been defined at $I_D = -0.5\times10^{-7}(W/L_G)$ A. For better comparison V_{TSAT} of all the devices has also been adjusted to 0.2 V by tuning the gate work

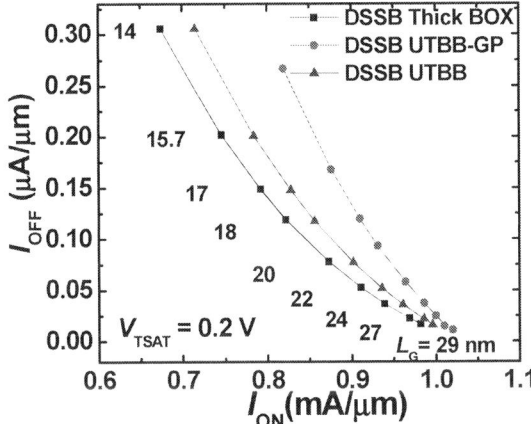

Fig. 4. I_{ON}-I_{OFF} characteristics of DSSB thick BOX, DSSB UTBB-GP and DSSB UTBB n-channel SOI MOSFETs at various L_G values. The L_G values are considered as per the high performance logic technology nodes specified in ITRS-2009.

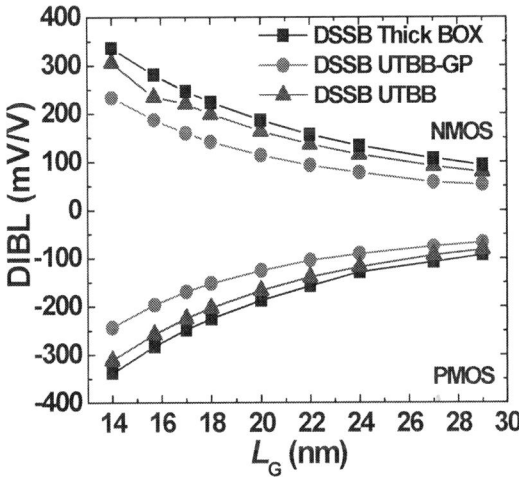

Fig. 5. Variation of DIBL as a function of L_G for DSSB thick BOX, DSSB UTBB-GP and DSSB UTBB n-channel and p-channel SOI MOSFETs.

function. From the figure it can be seen that, in comparison to DSSB thick BOX and DSSB UTBB NMOS/PMOS devices, the DIBL in the case of proposed DSSB UTBB-GP NMOS and PMOS devices is significantly low. This is mainly because of the presence of GP under the thin BOX of the proposed device at which the fringing field lines arising from the drain region are terminated.

C. Threshold Voltage Variability

From Fig. 6 it can also be seen that, in comparison to DSSB thick BOX and DSSB UTBB NMOS/PMOS devices, the V_{TSAT} variability induced by the fluctuations in T_{Si} of DSSB UTBB-GP NMOS and PMOS devices is significantly low. The main reason for the reduction in V_{TSAT} variability in the proposed DSSB UTBB-GP device is the presence of GP under the thin BOX of the device. Since the screening effect due to GP under the thin BOX suppresses the threshold voltage variations due to random dopant fluctuations, the V_{TSAT} variability induced by the fluctuations in T_{Si} of this device reduces.

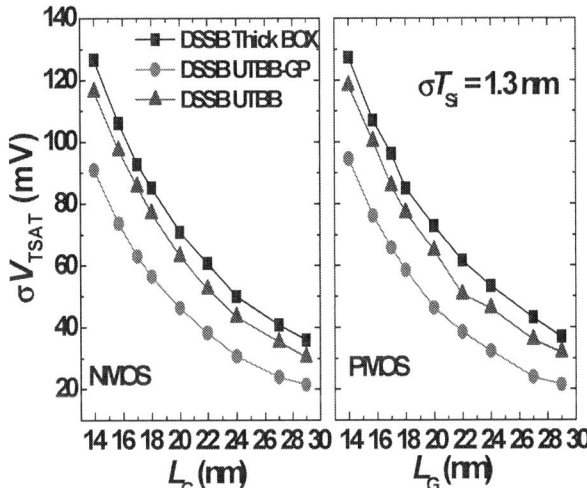

Fig. 6. Variation of standard deviation in threshold voltage with L_G for DSSB thick BOX, DSSB UTBB-GP and DSSB UTBB n-channel and p-channel SOI MOSFETs.

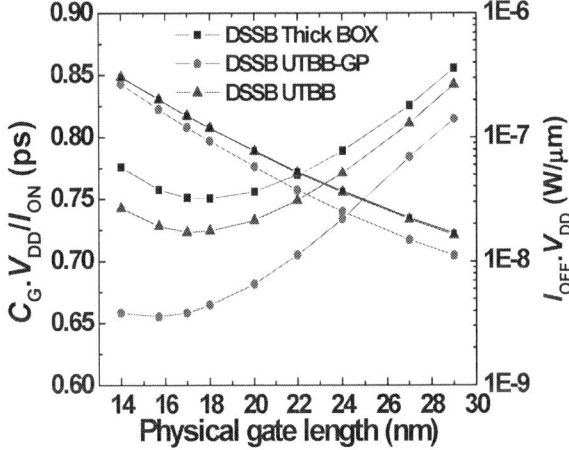

Fig. 7. Variation of intrinsic gate delay and static power dissipation as a function of L_G for DSSB thick BOX, DSSB UTBB-GP and DSSB UTBB n-channel SOI MOSFETs.

D. Intrinsic Device Performance

Fig. 7 shows the combined effect of gate capacitance (C_G), I_{ON} and I_{OFF} on the intrinsic device performance of DSSB thick BOX, DSSB UTBB and the proposed DSSB UTBB-GP n-channel SOI MOSFETs. From this figure it can be seen that, the static power dissipation ($I_{OFF}.V_{DD}$) in the case of DSSB thick BOX and DSSB UTBB devices is almost equal whereas due to significant reduction in I_{OFF} of the proposed DSSB UTBB-GP device $I_{OFF}.V_{DD}$ of this device is reduced by ~40% and ~35 % respectively over the DSSB thick BOX and DSSB UTBB devices. Further, due to significant improvement in I_{ON} of the proposed DSSB UTBB-GP device, the intrinsic gate delay ($C_G.V_{DD}/I_{ON}$) of this device is also reduced by ~80% and ~70% respectively over the DSSB thick BOX and DSSB UTBB devices. This clearly indicates that the proposed DSSB UTBB-GP device is a most suitable candidate for low-power and high performance CMOS logic circuits.

978-1-4673-4840-9/13 $31.00 © 2013 IEEE

Conclusion

In this work, a novel DSSB UTBB-GP Schottky barrier SOI MOSFET has been proposed for thermally efficient low-variability nanoscale CMOS logic circuits. The presence of ultrathin BOX and the GP not only reduces the self-heating effect, off-state leakage current, short-channel effects and the process induced threshold voltage variability but also improves the on-state drive current of the device. It has also been found that, the intrisic gate dealy and the static power dissipation in the case of proposed device are also reduced by ~80% and ~40% respectively over the DSSB thick BOX and DSSB UTBB SOI MOSFETs. Thus, employing GP in UTBB DSSB SOI MOSFET makes it a better choice for future nanoscale CMOS logic circuits.

Acknowledgement

The authors would like to thank The Ministry of Communication and Information Technology, Government of India, for their valuable support, without which this work could not have been conducted.

References

[1] A. Ohata, Y. Bae, C. F.-Beranger and S. Cristloveanu, "Mobility enhancement by back-gate biasing in ultrathin SOI MOSFETs with thin BOX", *IEEE Electron Device Lett.*, vol. 33, no. 3, pp. 348-350, Mar. 2012.

[2] S. Chouksey, J. G. Fossum and S. Agrawal, "Insights on design and scalability of thin-BOX FD/SOI CMOS", *IEEE Trans. Electron Devices*, vol. 57, no. 9, pp. 2073-2079, Sep. 2010.

[3] C. F.-Beranger et al., "FDSOI devices with thin BOX and ground plane integration for 32 nm node and below", *Solid-State Electronics*, vol. 53, no. 7, pp. 730-734, Jul. 2009.

[4] G. C. Patil and S. Qureshi, "A novel δ-doped partially insulated dopant-segregated Schottky barrier SOI MOSFET for analog/RF applications", *Semiconductor Science and Technology*, vol. 26, no. 8, pp. 085002, Aug. 2011.

[5] S.-J. Choi, C.-J. Choi, J.-Y. Kim, M. Jang and Y.-K. Choi, "Analysis of transconductance (g_m) in Schottky-barrier MOSFETs", *IEEE Trans. Electron Devices*, vol. 58, no. 2, pp. 427-432, Feb. 2011.

[6] J. Knoch, M. Zhang, Q. T. Zhao, S Lenk and S. Mantl, "Effective Schottky barrier lowering in silicon-on-insulator Schottky-barrier metal-oxide-semiconductor field-effect transistors using dopant segregation", *Applied Physics Lett.*, vol. 87, no.26, pp. 263505-3, Dec. 2005.

[7] G. Larrieu, D. A. Yarekha, E. Dubois, N. Breil, and O. Faynot, "Arsenic-segregated rare-earth silicides junctions: Reduction of Schottky barrier and integration in metallic n-MOSFETs on SOI", *IEEE Electron Device Lett.*, vol. 30, no.12, pp. 1266-1268, Dec. 2009.

[8] G. C. Patil and S. Qureshi, "Underlap channel metal source/drain SOI MOSFET for thermally efficient low-power mixed-signal circuits", *Microelectronics Journal*, vol. 43, no. 5, pp. 321-328, May. 2012.

[9] G. C. Patil and S. Qureshi, "Engineering spacers in dopant-segregated Schottky barrier SOI MOSFET for nanoscale CMOS logic circuits", *Semiconductor Science and Technology*, vol. 27, no. 4, pp. 045004, Apr. 2012.

[10] M. Zhang, J. Knoch, S.-L. Zhang, S. Feste, M. Schroter and S. Mantl, "Threshold voltage variations in SOI Schottky-barrier MOSFETs", *IEEE Trans. Electron Devices*, vol. 55, no. 3, pp. 858-865, Mar. 2008.

[11] S. F. Feste, M. Zhang, J. Knoch and S. Mantl, "Impact of variability on the performance of SOI Schottky barrier MOSFETs", *Solid-State Electronics*, vol. 53, no. 4, pp. 418-423, Apr. 2009.

[12] MEDICI User Manual, Ver. Y-2006.06, TMA, Palo Alto, CA, USA (2006)

[13] International Technology Roadmap for semiconductors (ITRS), 2009, [online] http://public.itrs.net

A Study on Gate-All-Around (GAA) Polycrystalline Silicon Channel SONOS Flash Memory

Joo Yun Seo, Sang-Ho Lee, Yoon Kim, Se Hwan Park, Wandong Kim, Do-Bin Kim, and Byung-Gook Park

Inter-university Semiconductor Research Center (ISRC) and
School of Electrical Engineering and Computer Science, Seoul National University
San 56-1, Sillim-dong, Gwanak-gu, Seoul 151-742, Republic of Korea
Tel.: +82-2-880-7279, Fax: +82-2-882-4658, E-mail address: jooyun@snu.ac.kr

Abstract

In this study, the gate-all-around (GAA) poly-Si channel flash memories with charge trap layer (Si_3N_4) have been successfully fabricated. Electric characteristics of fabricated devices including threshold voltage shift with program/erase operation have been investigated. Gate configurations were structured differently according to each defined channel width. Results show that devices with gate-all-around structure have superior program efficiency. To investigate the effect of gate configuration on the program efficiency, TCAD simulation was carried out. (Keywords: SONOS flash memory, gate-all-around (GAA), charge trap memory)

Introduction

As the quest for higher storage capability continues, NAND flash memory has become the most attractive storage device. For ultra high density applications, silicon-oxide-nitride-silicon (SONOS) nonvolatile semiconductor memories have advantages over the conventional floating-gate flash memories, such as low program/erase voltage, easiness of fabrication, and stronger immunity to coupling between cells [1–3].

Also, SONOS memory devices are desirable to adopt multiple gate structure. Previous literatures demonstrate that SONOS device with a cylindrical nanowire (NW) channel and gate-all-around (GAA) structure shows superior program and erase efficiency by introducing a field enhancement effect. [4], [5]

In this work, we successfully fabricated poly-Si channel SONOS flash memories with gate-all-around structure. With a simple fabrication process, Si-nanowire can be separated from the bulk wafer. Results demonstrate that devices having narrow channels show superior program characteristics because they are desirable to get the field enhancement effect of GAA structure.

Experimental

Flash memory cells were fabricated by initially depositing a 400-nm-thick tetra-ethyl-ortho-silicate (TEOS) oxide on a 6-inch silicon wafer. A 40-nm-thick undoped amorphous-silicon (a-Si) layer was deposited by low-pressure chemical vapor deposition (LPCVD) at 550 ℃. Then, channels were defined with a mix-and-match process consisting of e-beam and photolithography, followed by dry etch using HBr and O_2 gas plasma (HBr : 40 sccm, O_2 : 2 sccm) (Fig. 1(a)). To separate the Si channels from the bulk, the buried oxide was removed by wet etch with HF solution (Fig. 1(b)). After SC1 cleaning (NH_4OH : H_2O_2:D.I. water=1:8:64 solution at 65℃), medium temperature oxide (MTO) was formed as a tunneling oxide layer, 3-nm-thick. A 6-nm-thick Si_3N_4 layer and a 8-nm-thick TEOS layer were sequentially deposited by LPCVD as a charge trap layer and a blocking oxide, respectively. To form gate electrodes, in situ n+ doped poly-Si layer was deposited, followed by mix-and-match process and dry etch process (Fig. 1(c)). After source/drain ion implantation, the back end process including contact hole etch, metal deposition, and metal line patterning was carried out. Fig. 2 shows SEM images of nanowire after active patterning. Nanowires were successfully formed, with widths varying from 40 nm to 100 nm. Undergoing a SC1 cleaning process, the width of Si nanowire was slightly decreased.

Results and Discussion

Fig. 3 shows electrical results from the fabricated devices (W_{ch} = 50 nm, L_g = 100 nm). 1.9V of ΔV_T from program operation could be achieved, but the amount of threshold voltage shift in erase operation was saturated due to the quality of the blocking oxide. The injected carriers from the gate to the nitride layer are the main cause of erase saturation. To alleviate this issue, O/N/O stack engineering is required [2], [6].

Differently defined gate lengths (L_g) and channel widths (W_{ch}) resulted in different electrical characteristics. Fig. 4 presents the amount of V_T shift when programming with 12 V for 1 ms as a function of gate length. As gate length increased, wider program windows were achieved. Fig. 5 shows the program speed with different channel widths. In this device, channel widths affected the gate configuration. Devices with narrow channels show improved cell characteristics due to stronger field concentration effect. Fig. 6 illustrates the different gate configurations with different channel widths. As etch time to remove buried oxide layer (Fig. 1 (b)) is the same, the amount of remained buried oxide below the channel is varied among channels with different widths. In other words, wider channels are not thoroughly surrounded by O/N/O stack and gate, which is closer to an omega-gate structure. To verify the field concentration effect, simulation was carried out. When gate is all around the channel, electric field applied to the tunneling oxide is much larger compared to the case of a triple-gate. Stronger electric field enables carriers to be injected from the channel to the nitride layer effectively. Besides the gate configurations, narrower nanowires show stronger field enhancement as shown in fig. 7.

978-1-4673-4840-9/13 $31.00 © 2013 IEEE   69

Conclusion

Charge trap flash memory devices with poly-Si nanowire channel and multiple gate structure were fabricated with a simple fabrication process. Measurement results show that devices with narrower channels have efficient program characteristics.

TCAD simulation was carried out to verify the experimental results. Variation in channel widths resulted in different structures of gate configurations, and the applied field to the tunneling oxide is much higher for the GAA devices compared to the triple-gate devices. Thanks to the field enhancement effect of GAA structure and narrow nanowires, fabricated devices with narrow nanowires show superior program characteristics.

Acknowledgment

This work was supported by the IT R&D program of MKE/KEIT (10035320, Development of novel 3D stacked devices and core materials for the next generation flash memory).

References

[1] M. H. White, D. A. Adams, and J. Bu, "On the go with SONOS," *Circuits and Devices Magazine, IEEE DOI - 10.1109/101.857747*, vol. 16, no. 4, pp. 22–31, 2000.

[2] Hang-Ting Lue, Szu-Yu Wang, Erh-Kun Lai, Yen-Hao Shih, Sheng-Chih Lai, Ling-Wu Yang, Kuang-Chao Chen, J. Ku, Kuang-Yeu Hsieh, Rich Liu, and Chih-Yuan Lu, "BE-SONOS: A bandgap engineered SONOS with excellent performance and reliability," in *Electron Devices Meeting, 2005. IEDM Technical Digest. IEEE International*, 2005, pp. 547–550.

[3] Chang Hyun Lee, Kyung In Choi, Myoung Kwan Cho, Yun Heub Song, Kyu Charn Park, and Kinam Kim, "A novel SONOS structure of SiO2/SiN/Al2O3 with TaN metal gate for multi-giga bit flash memories," in *Electron Devices Meeting, 2003. IEDM '03 Technical Digest. IEEE International*, 2003, pp. 26.5.1–26.5.4.

[4] Hung-Bin Chen, Yung-Chun Wu, Chao-Kan Yang, Lun-Chun Chen, Ji-Hong Chiang, and Chun-Yen Chang, "Impacts of Poly-Si Nanowire Shape on Gate-All-Around Flash Memory With Hybrid Trap Layer," *Electron Device Letters, IEEE*, vol. 32, no. 10, pp. 1382–1384, 2011.

[5] J. Fu, N. Singh, K. D. Buddharaju, S. H. G. Teo, C. Shen, Y. Jiang, C. Zhu, M. B. Yu, G. Q. Lo, N. Balasubramanian, D. L. Kwong, E. Gnani, and G. Baccarani, "Si-Nanowire Based Gate-All-Around Nonvolatile SONOS Memory Cell," *Electron Device Letters, IEEE*, vol. 29, no. 5, pp. 518–521, May 2008.

[6] Hang-Ting Lue, Sheng-Chih Lai, Tzu-Hsuan Hsu, Pei-Ying Du, Szu-Yu Wang, Kuang-Yeu Hsieh, R. Liu, and Chih-Yuan Lu, "Understanding barrier engineered charge-trapping NAND flash devices with and without high-K dielectric," *Reliability Physics Symposium, 2009 IEEE International*, pp. 874–882, 26.

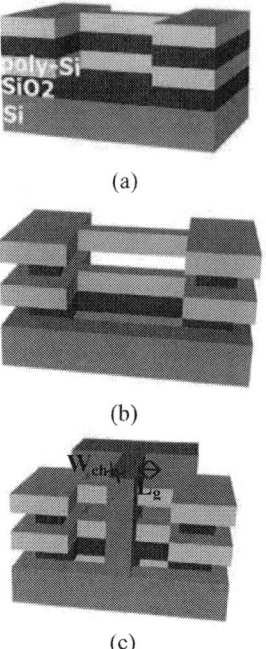

(a)

(b)

(c)

Fig. 1 Fabrication process (a) Active patterning (b) Buried oxide removal (c) Gate patterning (Gate length : L_g, Channel width : W_{ch}).

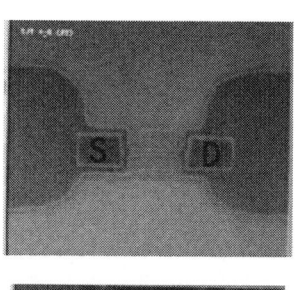

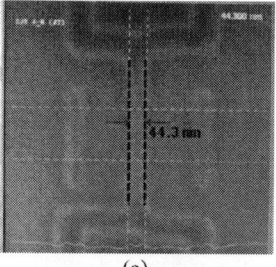

(a)

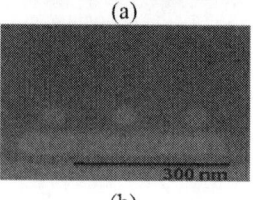

(b)

Fig. 2 SEM images (a) Top images of NW channel after active patterning (b) vertical image of NW

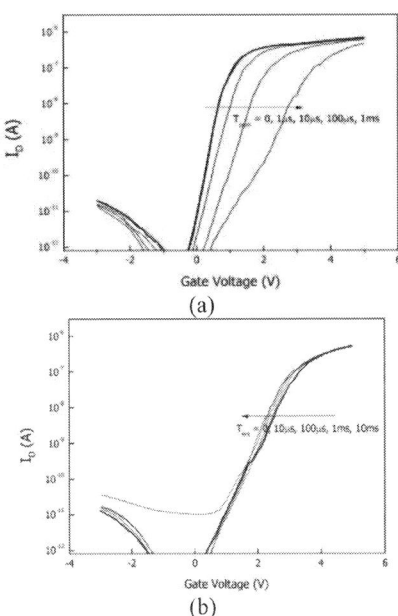

(a)

(b)

Fig. 3 (a) Program speed (V_{pgm} = 12V) (b) Erase speed (V_{ers} = -10V), (L_g=100 nm, W_{ch}=50 nm)

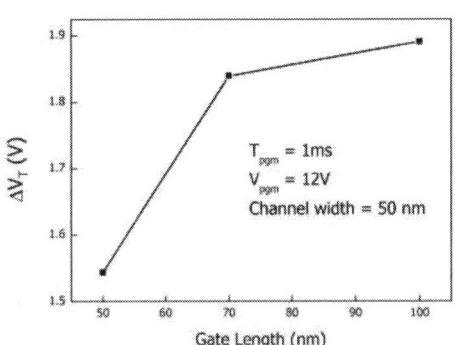

Fig. 4 V_T shift as a function of gate length

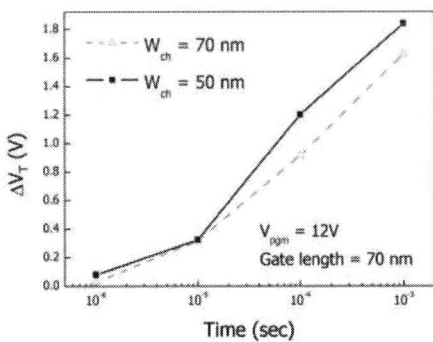

Fig. 5 Program speed with different channel widths

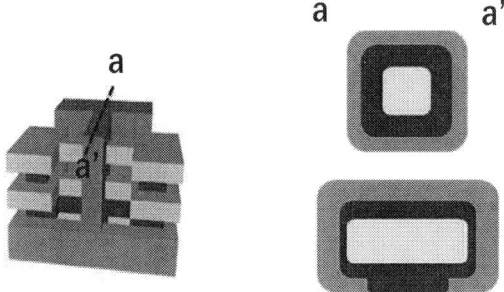

Fig. 6 Different gate configurations with different channel widths

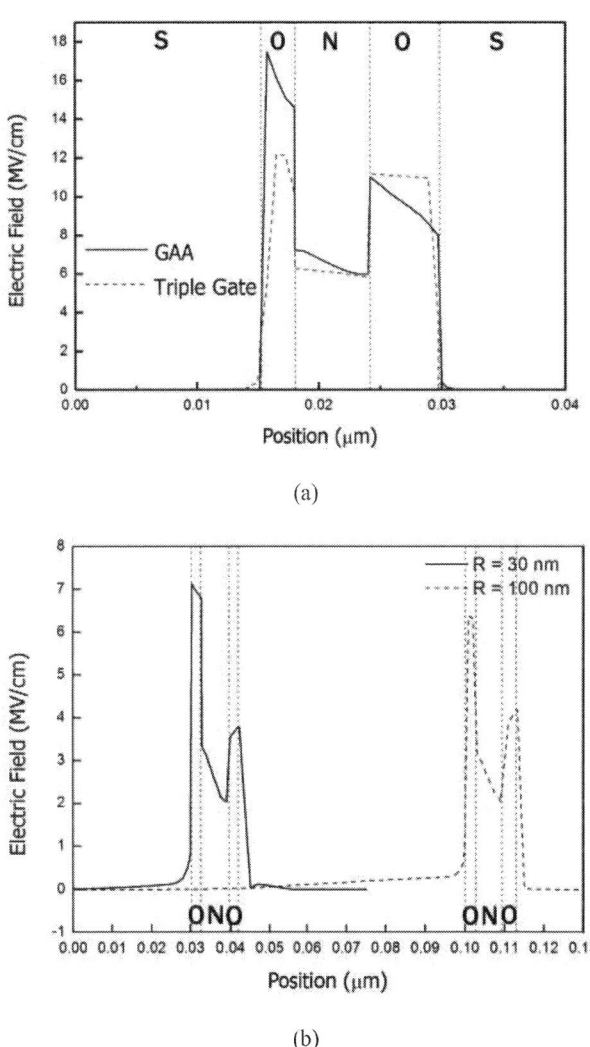

(a)

(b)

Fig. 7 (a) Simulated electric field distribution along the position from the channel to the gate when program voltage (15 V) is applied with different gate configurations (b) Simulated electric field distribution with different radius of nanowires

978-1-4673-4840-9/13 $31.00 © 2013 IEEE   71

Single Transistor Latch Phenomena in Junctionless Nanotransistors

Mukta Singh Parihar[1], Dipankar Ghosh[1], G. Alastair Armstrong[2], and Abhinav Kranti[1]

[1]Low Power Nanoelectronics Research Group, Electrical Engineering Discipline, Indian Institute of Technology (IIT),
Indore – 452 017, India. Email: akranti@iiti.ac.in
[2]School of Electronics, Electrical Engineering and Computer Science, Queen's University Belfast, UK

Abstract

In this work, we analyze the dependence of steep subthreshold (S–slope) on device and bias parameters of Junctionless (JL) MOSFETs. It is observed that for certain parameters and bias conditions, the JL transistor cannot be turned OFF resulting in a single transistor latch. This phenomenon is an extreme case of impact ionization in JL MOSFETs. It is shown that thicker values of silicon film thickness and gate oxide along with higher drain bias can drive the JL MOSFET in to the latch state.

Keywords: junctionless, MOSFET, impact ionization and Steep S–slope.

Introduction

Junctionless MOSFETs in Silicon–on–Insulator (SOI) technology have emerged as serious contenders for device scaling at the end of ITRS projections [1-4]. Apart from the simpler fabrication process without doping gradients, JL transistors offer improved immunity from short channel effects (SCEs) in comparison to conventional inversion mode MOSFETs [1]. JL MOSFETs have also shown to exhibit subthreshold slope (S–slope) less than the theoretical minimum (< 60mV/decade) at room temperature due to impact ionization process [5-6]. As S–slope represents the switching capability of a device, lower values are highly desirable. It has been demonstrated that higher degree and wider region of impact ionization is achieved in JL MOSFETs in comparison to inversion mode devices [5].

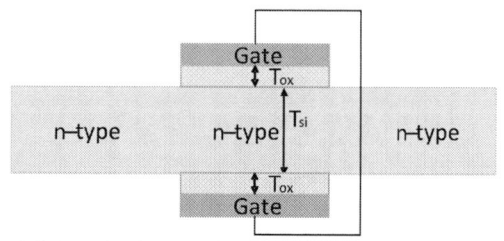

Fig. 1 Schematic diagram of Double Gate (DG) Junctionless (JL) MOSFET.

In this work, we analyze the dependence of S–slope on device parameters (gate oxide thickness and film thickness), and drain bias. It is shown that proper selection of the device parameters and voltage is crucial for achieving steep S–slope and avoiding the latch phenomenon.

Single Transistor Latch

In order to analyze critical parameters for achieving steep S–slope, JL MOSFETs (fig. 1) with gate length (L_g) of 60 nm and doping (N_d) of 10^{19} cm^{-3} were simulated with ATLAS software [7] using modules for doping, bipolar and impact ionization effects. Fig. 2a shows drain bias (V_{ds}) induced latching of drain current (I_{ds}). At lower V_{ds} values (~ 1 V), JL transistor exhibits the classical S–slope of 60 mV/decade. Increasing the drain bias raises the impact of lateral field and carriers gain sufficient energy to start impact ionization process. Generated electrons take part in the conduction, increasing the current level while holes accumulate at the minimum potential region (at the surface). The successive accumulation of holes raises the potential of gate–film surface which forward biases source–channel junction. The induced forward bias further increases I_{ds} due to the positive feedback loop. To overcome the positive feedback loop and the switch–off the device, a negative gate bias is required. The steepness of S–slope increases with drain bias and is in good qualitative agreement with results of Lee et al., [5]. Interestingly, beyond a particular V_{ds} value (> 2.5 V in fig. 2a), the JL MOSFET fails to turn OFF and latches to the ON state. This phenomenon has been previously reported for inversion mode SOI MOSFETs [8-9]. The gate voltage at which the device latches to the ON state is known as the holding voltage [8].

Apart from the drain voltage, we report on the latching effect in JL transistors as a function of film thickness (T_{si}) and gate oxide thickness (T_{ox}). As shown in fig. 2b, an increase in T_{si} from 8 to 9 nm, results in the very high values of drain current and device latches to the ON state. Please note that S–slope reduces with increase in film thickness only up to T_{si} = 8 nm. Similarly, an increase in gate oxide thickness from 1 nm to 3 nm results in a decrease in S–slope (fig. 2c). Very steep values of 1 mV/decade are obtained at T_{ox} = 3 nm with drain current increasing by nearly 5 decades at the transition gate voltage. Any increase in T_{ox} beyond 3 nm results in the onset of latching state. Though, gate oxide facilitates a larger range of values (1 nm to 3 nm) in comparison to T_{si} (7 nm to 8 nm) to achieve steep S–slope and avoid the latching condition, yet along with T_{si} and V_{ds}, it imposes limits on the selection of JL device parameters to avoid the latch state.

The latch phenomenon observed in fig. 2a–c, is explained in terms of Generation–Recombination (G–R) rates in fig. 3a. A drain bias of 2.5 V produces a steep S–slope and JL MOSFET can be turned OFF, whereas at V_{ds} = 3 V the transistor latches to the ON state. In the reverse gate bias sweep, the substantial difference between G–R rates is obtained in the ON state, whereas in the OFF state, G–R rates decrease and are broadly comparable in magnitude in the subthreshold region. However, at $V_{ds} = 3$ V, the much higher G–R rates maintain an approximately constant difference over the entire operating voltage range, and cannot be lowered by reducing the gate voltage. Fig. 3b shows the variation of centre (φ_c) and surface (φ_s) mid-gate potentials. The condition $\varphi_c > \varphi_s$ is indicative of the conduction channel at the centre of the silicon layer. A sharp increase in the values of φ_c and φ_s occurs at the gate bias corresponding to steep increment in I_{ds}. Although the difference between φ_c and φ_s remains constant for the gate bias range in fig 3b, the absolute values of potential change at the onset of steep I_{ds} transition at $V_{ds} = 2.5$ V. However at $V_{ds} = 3$ V, φ_c and φ_s both attain high values over the V_{gs} range. This is indicative of the high current and higher G–R rates ($> 10^{29}$ cm^{-3} sec^{-1}), which lead to the device latching to the ON state and not being able to turn OFF. Results (not shown here) similar to fig. 3a–b, were obtained for T_{si} and T_{ox} values shown in fig. 2b–c. In order to benefit from the steep S–slope in JL MOSFETs, device and bias parameters should be carefully selected to avoid the latching condition.

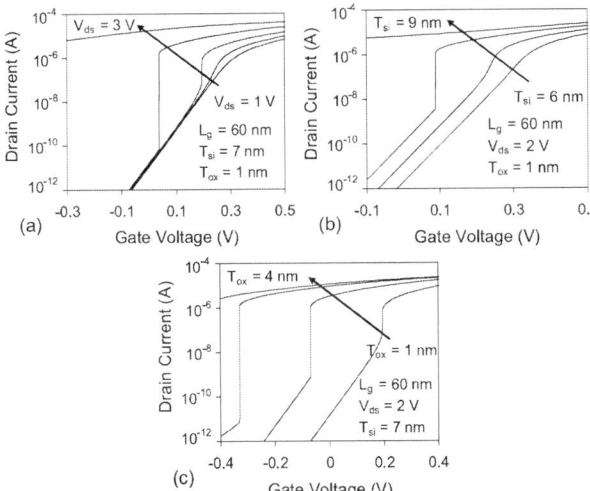

Fig. 2 Transfer characteristics (I_{ds}–V_{gs}) characteristics exhibiting latch phenomena as a function of (a) drain bias (V_{ds}), (b) silicon film thickness (T_{si}), and (c) gate oxide thickness (T_{ox}).

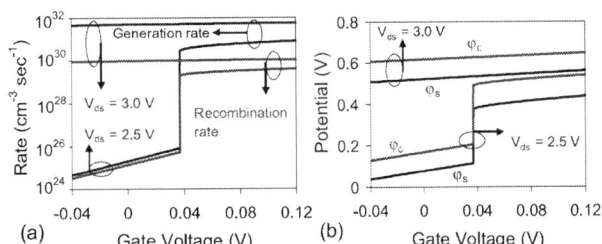

Fig. 3 (a) Generation–Recombination (G–R) rates as a function of gate voltage for two different drain voltages. (b) Variation of surface (φ_s) and centre (φ_c) potentials with gate bias.

Conclusion

The single transistor latch effect in Junctionless MOSFETs has been analyzed. It is shown that JL nanotransistors exhibit steep S–slopes (1 mV/decade) due to impact ionization effect. The single transistor latch effect limits the maximum operating drain bias voltage and maximum film thickness and gate oxide thickness in order to obtain steep S–slope and avoid the latching regime.

Acknowledgement

This work is supported by the Science and Engineering Research Board, Department of Science and Technology, Government of India, under Grant SR/S3/EECS/0130/2011.

References

[1] C.-W. Lee, A. Afzalian, N.D. Akhavan, R. Yan, I. Ferain, and J.-P. Colinge, Junctionless multigate field-effect transistor, *Applied Physics Letters*, vol. 94, no. 5, article 053511, 2009.

[2] J.-P. Colinge, C.-W. Lee, A. Afzalian, N. Dehdashti Akhavan, R. Yan, I. Ferain, P. Razavi, B. O'Neill, A. Blake, M. White, A.-M. Kelleher, B. McCarthy and R. Murphy, "Nanowire transistors without junctions", *Nature Nanotechnology*, vol. 5 no. 3, pp. 225-229, 2010.

[3] A. Kranti, R. Yan, C.-W. Lee, I. Ferain, R. Yu, N.D. Akhavan, P. Razavi and J.-P. Colinge, "Junctionless nanowire transistor (JNT): properties and design guidelines", *In Proc. European Solid State Device Research Conference (ESSDERC)*, Seville, Spain, pp. 357–360, 2010.

[4] A. Kranti, C.-W. Lee, I. Ferain, R. Yan, N. Akhavan, P. Razavi, R. Yu, G.A. Armstrong and J.-P. Colinge, "Junctionless 6T SRAM cell", *Electronics Letters*, vol. 46, no. 22, pp. 1491 – 1493, 2010.

[5] C.-W. Lee, A.N. Nazarov, I. Ferain, N. Akhavan, R. Yan, P. Razavi, R. Yu, R.T. Doria and J.-P. Colinge, "Low subthreshold slope in junctionless multigate transistors", *Applied Physics Letters*, vol. 96, no. 10, article no. 102106, 2010.

[6] R. Yu, S. Das, I. Ferain, P. Razavi, N. Akhavan, C.A. Colinge and J.-P. Colinge, "Simulation of impact ionization effect in short channel junctionless transistors", *In Proc. Workshop of the Thematic Network on Silicon on Insulator technology, devices and circuits (EuroSOI)*, pp. 37-38, 2012.

[7] ATLAS Users Manual, Silvaco.

[8] G.A. Armstrong and W.D. French, "Improved physical modeling for accurate simulation of bipolar effects thin film SOI transistors", *In Proc. IEEE SOI Conference*, pp. 46-47, 1991.

[9] G.A. Armstrong and W.D. French, "Hysteresis effects in thin film SOI transistors", *Microelectronic Engineering*, vol. 22, no. 1-4, pp. 375-378, 1993.

Modeling and analysis of 1.3 μm InAs/GaAs self-assembled quantum dot lasers with rate equation

C. Y. Liu[1,*] H. Wang[1, 2], and Q. Q. Meng[1]

1. TemasekLaboratories@NTU (TL@NTU), Nanyang Technological University
50 Nanyang Drive, Singapore 637553
2. School of Electrical and Electronic Engineering, Nanyang Technological University, Nanyang avenue, Singapore, 639798
*: Corresponding author. E-mail: liucy@ntu.edu.sg,

Abstract

Temperature (20-100 °C) and excitation power (10-700 mW)-dependent photoluminescence (PL) measurements have been carried out on the 1.3 μm GaAs-based InAs quantum dot (QD) laser structures. Rate equation was used to interpret the PL behavior of the QD. The exciton behavior of the carriers in InAs QD has been verified. The excitonic modal gain has been calculated (under exciton picture) and compared with the experimentally obtained modal gain value from the QD laser.

Keywords: quantum dot laser, rate equation, carrier dynamics, modal gain.

Introduction

Since its first demonstration [1], self-assembled quantum dot (QD) lasers have attracted great research interest due to their unique physical properties. QD structures have promised and demonstrated many intrinsic advantages compared to other semiconductor gain materials, such as bulk structure, quantum well (QW) structure, for photonic devices applications [2-13]. In spite of these numerous achievements, further intense investigations are highly desirable to fully realize the advantages of QDs in order to reveal the ultimate limits of QD devices [10]. For instance, the carrier dynamics in the QD structure are still under debate. For example, it is believed that the electrons and holes form excitons due to the strong confinement of QDs [2, 12], while some other researchers [3, 4, 6, 8] (and references therein) believe that electrons and holes are existing in the form of free carriers in QD structure. Furthermore, based on the theoretical work, Dikshit and Pikal believed that both excitons and free carriers co-existing in the QD structures, while free carriers are predominant [14].

In this paper, we experimentally verify that the carriers will form excitons in the QD structures, even at elevated temperature 373 K (100 °C), using temperature and excitation-dependent photoluminescence (PL) study. The modal gain of the QD laser structure has also been calculated in the exciton picture and compared with the measured value.

Experimental details

The self-assembled InAs/InGaAs QD structures have been grown using molecular beam epitaxy (MBE) on Si-doped GaAs (100) substrate. The InAs QD active region consists of ten-layer InAs (2.32 monolayer)/In$_{0.15}$Ga$_{0.85}$As (5 nm) QDs separated by a 33-nm-thick GaAs spacer. The detailed laser structure has been reported in [9]. The PL measurements were performed under excitation of the 514.5-nm line of an Ar$^+$ laser, a closed temperature controller, a spectrometer, a liquid nitrogen cooled Ge detector, and a lock-in amplifier. Before the PL measurement, the top contact layer and most part of the top cladding layer of QD structures were etched away for about 1 μm. The temperature has been varied from room temperature (RT) to 373 K. At each temperature, the excitation power has been varied from 12 mW to 700 mW. Using the same structure, ridge waveguide lasers were also fabricated with a contact ridge width of 4 μm and isolated by pulsed anodic oxidation from the same structure. The fabricated QD lasers were tested under continuous wave (CW) condition.

Results and discussion

Fig. 1 shows the PL spectra from the InAs QD laser structure at 303 K. The excitation power was varied from 12 mW to 700 mW. The 1st emission peak located at 1298 nm and the 2nd emission peak located at 1218.4 nm. With increasing the excitation power, the 2nd peak is more and more distinct with respect to the 1st peak, which is a typical feature of the excited state (ES) transitions in QD. We therefor attributed the two peaks to QD ground state (GS) and 1stES transitions, respectively. The inset of Fig. 1 showed the logarithm plot of integrated PL (I_{PL}) intensity as a function of excitation power, from which we know that the slope is ~0.978, close to unity.

Fig. 2 also shows the PL spectra at 373 K (100 °C) under varied excitation power from 12 to 700 mW. Strong PL signals at both GS and 1st ES emission are still observed, and GS emission is still dominant, which implies the excellent QD structure quality. Similarly, the inset of Fig. 2 shows the logarithm plot of I_{PL} vs. excitation power, from which the slope is also ~1.

In order to interpret the PL behavior of the photo-generated carriers in GaAs-based InAs QDs, a set of steady state rate equations are used, which are shown below:

$$\frac{dn}{dt} = a\frac{P_0}{E_0} - \frac{n}{\tau_n} - g_e np + d_e n_e \tag{1}$$

$$\frac{dp}{dt} = a\frac{P_0}{E_0} - \frac{p}{\tau_n} - g_e np + d_e n_e \tag{2}$$

$$\frac{dn_e}{dt} = -c_e n_e + g_e np - d_e n_e \tag{3}$$

Where n, p, and n_e are the densities of laser excitation generated electrons, holes and excitons, respectively. P_0 is the excitation laser power, a is the absorption coefficient of InAs QDs, and E_0 is the photon energy of the excitation laser. τ_n and τ_p are the lifetimes of electrons and holes associated with the nonradiative recombination processes, respectively. g_e, c_e, and d_e are the generation rate, recombination rate, and dissociation rate of excitons, respectively. Assuming carrier neutrality (hole=electron), under stable conditions, we can obtain (4) from re-arrangement of (1) – (3).

$$a\frac{P_0}{E_0} = [g_e + \frac{g_e d_e}{c_e + d_e}]np + \frac{n}{\tau_n} \quad (4)$$

If we assume n=p, therefore, the integrated QD PL intensity (I_{PL}) of the exciton recombination signal can be written as (5):

$$I_{PL} = c_e n_e = \frac{c_e}{c_e + d_e} g_e np \propto np \quad (5)$$

From (4), we know that if radiative recombination dominates, then, n/τ_n is negligible. Therefore, we have $P_0 \propto np$, while from (5), $I_{PL} \propto np$, so $I_{PL} \propto P_0$. Namely, exciton recombination has the first order power law in the excitation-power dependent PL.

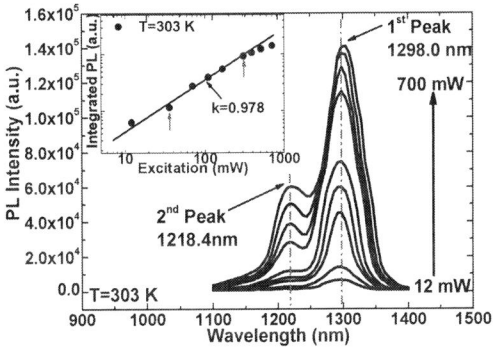

Fig. 1: PL as a function of excitation power at 303 K, inset: logarithm plot of I_{PL} vs. power, k=~0.978.

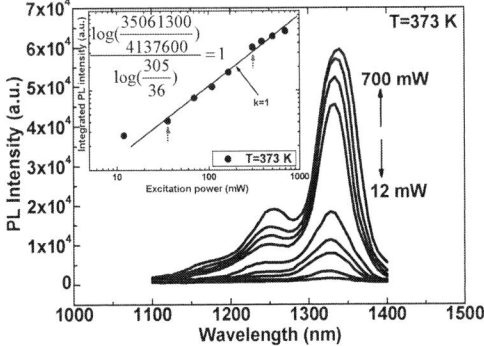

Fig. 2: PL as a function of excitation power at 373 K, inset: logarithm plot of I_{PL} vs power, k=~1.0.

On the other hand, when the nonradiative recombination dominated the photo-generated carriers, then from (4) we know that $P_0 \propto n/\tau_n$ (or $P_0 \propto p/\tau_p$, which is the same), while $I_{PL} \propto np$ (as in (5)), so $I_{PL} \propto P_0 P_0 \propto P_0^2$. Namely, the nonradiative recombination has the second order power law with excitation power. The above analysis is in accordance with other power-dependent PL study [15].

Based on above analysis and experimental results presented in Figs. 1 and 2, we know that the carriers form excitons in the QD structures even at high temperature. These experimental results support the exciton picture described in [2]. This is due to the exciton Bohr radius for InAs is much larger that the dot radius, therefore these excitons are in the strong confinement regime [2]. The authors would like to highlight that, to the best of our knowledge, this is the first time report on the excitation-dependent PL study on the QD laser structure at high temperature up to 100 °C.

We have previously built a comprehensive rate equation model [8], considering all the possible carriers relaxation paths and carrier transport, to describe the carrier dynamics in the QD structure. For better explanation, we list the major rate equations [8].

$$\frac{dN_B}{dt} = \frac{J}{qb} - \frac{N_B}{t_{bw}} + \frac{N_w}{t_{ebw}}\frac{V_w}{V_B}(1 - f_B) - \frac{N_B}{t_{rB}} \quad (6.1)$$

$$\frac{df_w}{dt} = \frac{1 - f_w}{g_w}(\frac{N_B b}{t_{bw}} + \frac{N_d d}{t_{bd}}) - \frac{f_w}{t_{ewb}} +$$

$$\sum_{i=0}^{2}[\frac{2p_i \rho}{g_w}Es_{rw}f_i(1 - f_w) - R_{wi}f_w(1 - f_i)] - \frac{f_w}{t_{rw}} \quad (6.2)$$

$$\sum_{i=0}^{2}\left[\begin{array}{l}\frac{g_w}{2\rho}R_{wi}\left(f_w(1 - f_i) - \exp(\frac{-E_{iw}}{k_B T})f_i(1 - f_w)\right) \\ -\frac{p_i f_i}{t_{ri}} - p_i PG_i\end{array}\right] = 0 \quad (6.3)$$

Where all the symbols have the same meaning as those reported in [8].

However, the electrons and holes were regarded as free carriers (isolated carriers independent of each other) in above model in [8]. In order to accurately describe the carrier's behavior in QDs, based on our experimental results, we need to modify the rate equation analysis under exciton picture. Since the carrier population inversion in free carriers case is $f_e + f_h$ -1, while in exciton picture, the holes follow the electrons. The population inversion will be $2f_e$-1. Together with the modified rate equation, the material gain of the QD structure has been calculated using the following equation [2], which considered both the inhomogeneous broadening $g_{inh}(E)$ and the homogeneous broadening $g_{hom}(E)$.

$$g_{material}(E) = C_g \frac{1}{E}\int_{-\infty}^{\infty}|M_b|^2|M_{env}|^2 g_{inh}(E)\times$$

$$[2f(E, E_{fc}) - 1]g_{hom}(E)dE \quad (7)$$

$$C_g = \frac{\pi e^2 \hbar}{m_0^2 \varepsilon_0 c_0 n_r}$$

Where E is photon energy, M_b is the Bloch matrix element and M_{env} is the QD electron hole wave function overlap integral, which is ~ 1 in QD for excitons. The Fermi function $f(E)$ determines the occupation probability of the single dot states by electrons. m_0 is the electron mass, n_r is the refractive index, ε_0 is permittivity of free space.

In self-assembled QD structures, the inhomogeneous broadening is much larger than that of homogeneous broadening. For the inhomogeneous broadening, which is due to the different dot has different size, and it can be described using the following Gaussian function [2]:

$$g_{inh}(E_i) = \frac{1}{\sqrt{2\pi}\sigma} \exp(-\frac{(E_i - E)^2}{2\sigma^2}) \qquad (8)$$

Where σ is related to the QD size fluctuations. $\sigma = 2.35$ is adopted in this work.

For the homogeneous broadening, we used the secant function. This expression for the gain does not cause spurious absorption below the band gap, as does broadening with a Lorentzian curve [16]. The broadening parameter γ_{hom} is given by carrier scattering and can be used as fitting parameter.

$$g_{hom}(E) = \frac{1}{\pi\gamma_{hom}} \sec h(\frac{E}{\gamma_{hom}}) \qquad (9)$$

From (7) to (9), we can calculate the modal gain from QD lasers if optical confinement factor (Γ) is known. We compare the calculated modal gain and compared it with the measured value from the fabricated QD laser.

Fig. 3 shows the measured net modal gain ($G_{net} = \Gamma \cdot g_{material} - \alpha_i$) spectra from the fabricated QD laser with a cavity length of 1500 μm at 25 °C under several bias currents. G_{net} is computed from multiple amplified spontaneous emission (ASE) spectra obtained at different injection currents using Hakki-Paoli method (10) mentioned in [17]. The ASE spectra were recorded with an optical spectral analyzer (OSA, AQ6317C) with the spectral resolution of 0.01 nm.

$$G_{net} = \Gamma g_{material} - \alpha_i = \frac{1}{L} \ln\frac{\sqrt{S}-1}{\sqrt{S}+1} + \frac{1}{2L} \ln\left(\frac{1}{R_1 R_2}\right) (10)$$

Where α_i is the internal loss, S is the ratio of intensity maximum and minimum in the FP resonances, L is the cavity length, and R_1 and R_2 are facet power reflectivity.

In calculating the G_{net}, we used the known laser cavity length L. The laser facet reflectivity R is determined by $[(n-1)/(n+1)]^2$, where n is the group refractive index, obtained from ASE spectrum. When the current is 45 mA, just below the threshold, the maximum gain reached. From (10), we know that, at lasing or approaching lasing condition, the first part approaches to zero, since S in (10) approaches to infinity. As expected, the maximum modal gain is limited by the second part in (10), i.e., the mirror loss, $1/L(\ln(1/R))$, which is around 7.6 cm^{-1}.

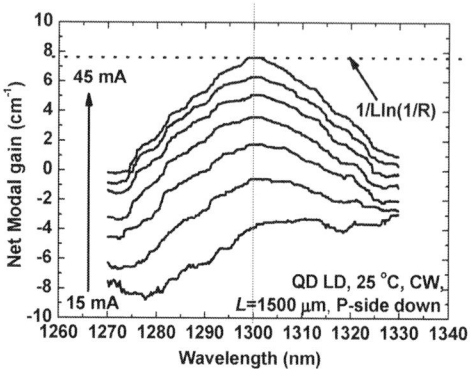

Fig. 3: Net modal gain vs. injection current at 25 °C from a QD laser.

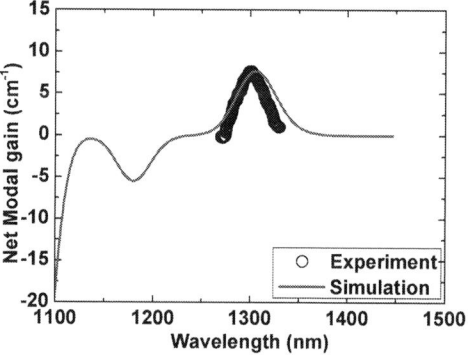

Fig. 4: Comparison of net modal gain by experiment (dots) and simulation (line).

Fig. 4 shows the comparison of the measured modal gain from Hakki-Paoli method with the theoretically calculated one. The dots are the experimental results, and the solid line is from theoretical calculation. For the calculation, the QD density is assumed to be 5×10^{10}/cm^2/layer with 10 QD layers, the inhomogeneous broadening of 30 meV for a 1500×4 μm^2 QD laser. It should note that the simulation has included both the 1st ES transitions and 2nd ES transitions, and therefore covers longer wavelength range. However, from the modal gain measurement with Hakki-Paoli method, we can only get the modal gain data from GS transition (1270 nm to 1330 nm). We can see that, at higher energy side (shorter wavelength region), the simulation agrees pretty well with the experimental data. At longer wavelength (lower energy region) side, the simulation results deviate slightly with the experimental data. But the general trend is the same. The deviation may possible be due to the simulation is too idealistic. Furthermore, the QD size, QD density, inhomogeneous broadening, homogeneous broadening values as well as the optical confinement factor cannot be accurately easily. Further investigation and improvement on the simulation is on the way. Nevertheless, the important feature in our model is that, with exciton picture, the ground state

978-1-4673-4840-9/13 $31.00 © 2013 IEEE

lasing can be achieved first, prior to the lasing of the excited state as reported in [9].

Conclusion

In this work, temperature-dependent and excitation power-dependent PL measurement has been carried out on the QD laser structure. From the rate equation analysis, the 1st power law was obtained in integrated PL vs. excitation power the excitation-dependent PL. The dominating exciton recombination behavior of the QD structure, up to 100 °C, has been verified. Based on this finding, we modified the previously built rate equations (based on the free carriers' picture) with exciton picture. We calculated the modal gain under exciton picture and compared it with the experimentally measured modal gain value from Hakki-Paoli method with reasonable good agreement.

References

[1]   N. Kirstaedter, N. N. Ledentsov, M. Grundmann, D. Bimberg, V. M. Ustinov, S. S. Ruvimov, M. V. Maximov, P. S. Kopev, Z. I. Alferov, U. Richter, P. Werner, U. Gosele, and J. Heydenreich, "LOW-THRESHOLD, LARGE T-O INJECTION-LASER EMISSION FROM (INGA)AS QUANTUM DOTS," *Electronics Letters,* vol. 30, pp. 1416-1417, Aug 1994.

[2]   D. Bimberg, N. Kirstaedter, N. N. Ledentsov, Z. I. Alferov, P. S. Kop'ev, and V. M. Ustinov, "InGaAs-GaAs quantum-dot lasers," *Selected Topics in Quantum Electronics, IEEE Journal of,* vol. 3, pp. 196-205, 1997.

[3]   L. V. Asryan and R. A. Suris, "Temperature dependence of the threshold current density of a quantum dot laser," *Ieee Journal of Quantum Electronics,* vol. 34, pp. 841-850, May 1998.

[4]   G. Park, O. B. Shchekin, and D. G. Deppe, "Temperature dependence of gain saturation in multilevel quantum dot lasers," *Ieee Journal of Quantum Electronics,* vol. 36, pp. 1065-1071, Sep 2000.

[5]   A. Markus, J. X. Chen, O. Gauthier-Lafaye, J. G. Provost, C. Paranthoen, and A. Fiore, "Impact of intraband relaxation on the performance of a quantum-dot laser," *Selected Topics in Quantum Electronics, IEEE Journal of,* vol. 9, pp. 1308-1314, 2003.

[6]   M. Sugawara, N. Hatori, H. Ebe, M. Ishida, Y. Arakawa, T. Akiyama, K. Otsubo, and Y. Nakata, "Modeling room-temperature lasing spectra of 1.3-μm self-assembled InAs/GaAs quantum-dot lasers: Homogeneous broadening of optical gain under current injection," *Journal of Applied Physics,* vol. 97, Feb 2005.

[7]   J. Kim and S. L. Chuang, "Theoretical and experimental study of optical gain, refractive index change, and linewidth enhancement factor of p-doped quantum-dot lasers," *Ieee Journal of Quantum Electronics,* vol. 42, pp. 942-952, Sep-Oct 2006.

[8]   C. Z. Tong, S. F. Yoon, C. Y. Ngo, C. Y. Liu, and W. K. Loke, "Rate equations for 1.3-μm dots-under-a-well and dots-in-a-well self-assembled InAs-GaAs quantum-dot lasers," *Ieee Journal of Quantum Electronics,* vol. 42, pp. 1175-1183, Nov-Dec 2006.

[9]   C. Y. Liu, S. F. Yoon, Q. Cao, C. Z. Tong, and H. F. Li, "Low transparency current density and high temperature operation from ten-layer p-doped 1.3 μm

InAs/InGaAs/GaAs quantum dot lasers," *Applied Physics Letters,* vol. 90, Jan 2007.

[10]   N. N. Ledentsov, D. Bimberg, and Z. I. Alferov, "Progress in epitaxial growth and performance of quantum dot and quantum wire lasers," *Journal of Lightwave Technology,* vol. 26, pp. 1540-1555, May-Jun 2008.

[11]   D. W. Xu, C. Z. Tong, S. F. Yoon, W. Fan, D. H. Zhang, M. Wasiak, L. Piskorski, K. Gutowski, R. P. Sarzala, and W. Nakwaski, "Room-temperature continuous-wave operation of the In(Ga)As/GaAs quantum-dot VCSELs for the 1.3 μm optical-fibre communication," *Semiconductor Science and Technology,* vol. 24, May 2009.

[12]   D. Bimberg and U. W. Pohl, "Quantum dots: promises and accomplishments," *Materials Today,* vol. 14, pp. 388-397, Sep 2011.

[13]   C. Y. Liu, M. Stubenrauch, and D. Bimberg, "Spontaneous emission study on 1.3 μm InAs/InGaAs/GaAs quantum dot lasers," *Nanotechnology,* vol. 22, Jun 2011.

[14]   A. A. Dikshit and J. M. Pikal, "Carrier distribution, gain, and lasing in 1.3-mu m InAs-InGaAs quantum-dot lasers," *Ieee Journal of Quantum Electronics,* vol. 40, pp. 105-112, Feb 2004.

[15]   T. Schmidt, K. Lischka, and W. Zulehner, "Excitation-power dependence of the near-band-edge photoluminescence of semiconductors," *Physical Review B,* vol. 45, pp. 8989-8994, 1992.

[16]   U. T. Schwarz, H. Braun, K. Kojima, M. Funato, Y. Kawakami, S. Nagahama, and T. Mukai, "Investigation and comparison of optical gain spectra of (Al,In)GaN laser diodes emitting in the 375 nm to 470 nm spectral range - art. no. 648506," in *Novel In - Plane Semiconductor Lasers IV.* vol. 6485, C. Mermelstein and D. P. Bour, Eds., ed, 2007, pp. 48506-48506.

[17]   B. W. Hakki and T. L. Paoli, "GAIN SPECTRA IN GAAS DOUBLE-HETEROSTRUCTURE INJECTION LASERS," *Journal of Applied Physics,* vol. 46, pp. 1299-1306, 1975.

Formation of Thick Textured Carbon Film using Filtered Cathodic Vacuum Arc Technique

Naiyun Xu[1], Siu Hon Tsang[1], Edwin Hang Tong Teo[2], Beng Kang Tay[1]

[1]School of Electrical and Electronic Engineering, Nanyang Technological University, 50 Nanyang Avenue, Singapore 639798, Singapore

[2]Temasek Laboratories@NTU, 50 Nanyang Avenue, Singapore 639798, Singapore

Corresponding author: Edwin Hang Tong Teo: HTTEO@ntu.edu.sg
Tel: +65 6790 4390/6593; Fax: +65 6896 7448

Abstract

In this work, thick textured carbon film (800nm) was grown on Si substrate using filtered cathode vacuum arc technique. Multi-cycle of deposition was performed to avoid the high film stress issue hence improve the film adhesion. Optical microscope was used for delamination check, the bonding composition was investigated using X-Ray photoelectron spectroscopy, and the microstructure of the film was studied using visible Raman spectroscopy. The electrical property was also studied by measuring the resistivity of the film.

Keywords: FCVA, Textured carbon, XPS, Raman and Electrical

Introduction

Amorphous carbon (a-C) films attract many attentions because of their interesting properties. The two main hybridizations—sp2 and sp3 bonding are the key factor to determine the film properties [1]. The high sp3 content a-C film, known as diamond-like carbon (DLC) is reported to have superior mechanical hardness, chemical inertness, smooth surface, wide band gap which can be used in applications such as protective coatings [2-3], RF devices [4]. In contrast of DLC, another type of a-C contains high fraction of sp2 bonding, which is graphite-like. Recently a new kind of graphite like carbon—textured carbon was developed [5-7], which contains large number of sp2 clusters perpendicular to the growth substrate. Textured carbon has been reported to have excellent electrical and thermal conductivity along the plane [5-9], hence it is considered as a promising material for future electrical and thermal applications.

Fabrication of textured carbon film using FCVA technique has been demonstrated in our previous work [6, 10-11]. One of the most direct methods is in-situ thermal growth, which is realized by applying suitable substrate heating during deposition. However, the film thickness is limited by the nature of the carbon film. Due to its high intrinsic compressive stress, film delamination issue limits the growth of thick carbon film [12]. The other reason is the caused by overheating of the source of FCVA system, which restricts the deposition time of FCVA process. Multilayering has been proved to be an effective way to improve the adhesion of the as-deposited film [13-14]. However, the feasibility of multilayering for textured carbon fabrication has not yet been examined. In this work, to develop the method to fabricating thick textured carbon film, FCVA deposition process with both single and multi-cycle was

carried out. Delamination of films was examined by optical microscope, the bonding composition of the thick textured was investigated using X-Ray photoelectron spectroscopy (XPS), the microstructure of films was studied using visible Raman spectroscopy and the electrical performance between thin and thick textured carbon film was carried out by measuring their electrical resistivity.

Experimental Details

The carbon films were fabricated by FCVA on (100) n-type Si wafer. The Si wafer was firstly cleaned using acetone, followed by ultrasonic isopropanol (IPA) before deposition. To achieve the texture structure within the carbon films, the growth temperature was set to 600°C through the substrate heating. The substrate bias was set to zero, and all depositions were carried out at a base pressure of 10^{-6}Torr. In this work, three different carbon films were fabricated, the deposition conditions is listed in table 1. The as-deposited films were firstly check by Olympus BX 60 optical microscope, the delamination was only found on film 2. And the film thickness was measured using a Tencor P10 surface profiler meter, the thickness of film 1 and film 3 was measured to be about 100nm and 800nm. The bonding composition of film 1 and film 2 was obtained from a VG ESCA 220i-XL Monochromatic Al Kα X-ray (hν=1486.6eV) Imaging XPS, the spectra were collected by scanning the film with photoelectron take-off angle of 90° respect to surface plane. Carbon 1s peaks were fitted based on Lorentzians broadened by a Gaussion using the manufacturer's standard software, with an estimated ±0.2eV error of binding energy. The microstructure of films was studied using WITEC CRM200 system using excitation energy of 532nm, the obtained spectra were then fitted using a Breit-Wigner-Fano (BWF) line shape for the G band, and a Lorentzian line shape for the D band. The two point probing method was employed to measure the electrical resistivity, with a cascade 200-nm probe station with a Hewllet Packard 4265A precision semiconductor parameter analyzer.

TABLE I
DEPOSITON CONDITIONS FOR C FILMS

	Cycle No.	Deposition Time per Cycle (min)	Deposition Temperature (°C)
Film 1	1	3	600
Film 2	1	25	600
Film 3	5	5	600

Results and Discussion

The deposition rate and deposition time are factors which determine the film thickness. For the FCVA system used in this work, the deposition rate is approximately 35nm/min. Film 1 is considered as to be a thin film with 3mins deposition and the stress of the film with this thickness does not give any negative impact for the film adhesion. To grow thick textured carbon film, film 2 was deposited for 25mins. However, the film is delaminated, which is shown in Fig. 1(a). This is due to the stress of the film increases with the increasing of film thickness, which degrades the film adhesion [13]. Film 3 was deposited using 5 cycles, with 5mins deposition each cycle, the substrate heating was maintained at 600°C throughout the deposition process. 10mins relaxing time was set between two deposition cycles. The film surface shows no delamination in Fig. 1(b). The reason is suggested to be the stress relaxation during the period between two cycles [15], hence the film stress caused by the increasing of film thickness has less impact on the film adhesion.

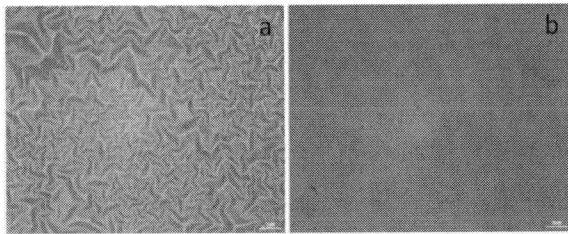

Fig. 1 Optical microscope image of the carbon films (a) film deposited using single cycle (25mins), (b) film deposited using multi cycle (5cycle 5mins deposition)

Fig. 2 shows the carbon 1s core level XPS spectra of film 1 and film 3. The two major peaks for the spectra include sp2 bonded carbon at 284.5eV, and sp3 bonded carbon at 285.2eV. The fitting of the spectra provides the detailed bonding composition of these two films. For film 1, the sp2/sp3 ratio is about 4:1, while for film 3, the ratio is about 6:1. It indicates the sp2 content of film 3 is about 85.7%, which is higher compared to 80% for film 1. The reason is because of the relaxing period between the two deposition cycles is considered as an annealing step, where some sp3 bonding transforms into sp2 bonding [15].

The Raman spectra of film 1 and film 3 are shown in Fig. 3. There are two main Raman peaks in these spectra, the D peak and G peak. D peak is only caused by the ring-like sp2 bonding, which indicates the appearance of graphitic clusters. While G peak indicates all sp2 bonding, which includes the chain-like sp2 bonding [16]. The fitting results provide two important features—G peak position and I(D)/I(G) ratio. G peak position reflects the amount of graphitic sp2 bonding, and I(D)/I(G) ratio indicates the size of the graphitic clusters [17-18]. In this work, the G peak position of film 1 and film 3 are measured to be 1560cm^{-1} and 1563cm^{-1} respectively, with its corresponding I(D)/I(G) ratio of 0.71 and 0.74. In our previous work, Film 1 had been shown to be textured carbon [6]. The Raman results of film 3 shows similar microstructure as Film 1, which suggests the texture structure exists in the film.

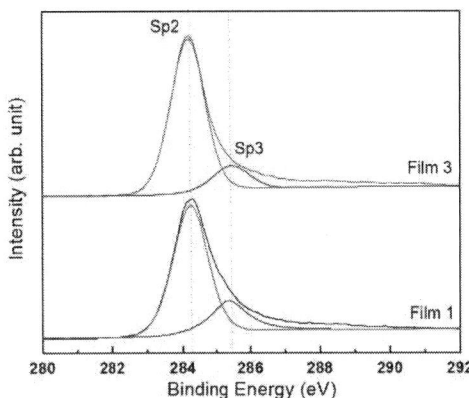

Fig. 2 XPS core-level spectra of thin and thick textured C films

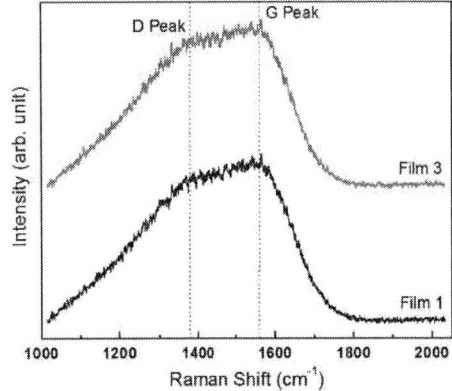

Fig. 3 Raman Spectra of thin and thick textured C films

One of the properties of textured carbon film is known for its excellent electrical conductivity. The electrical performance has been shown in previous work [5-6, 19]. Figure 2 shows the I-V characteristics of the thin and thick textured carbon films. Both the I-V curves show ohmic behavior for film 1 and film 3. The resistance ratio between film 1 and film 3—R_1/R_3 is estimated to be 1:3.6, based on the I-V curves, and the film thickness ratio—T_1/T_3 is 1:8, hence the ratio of resistivity—ρ_1/ρ_3 is about 20:9. The difference in the resistivity is mainly due to the higher amount sp2 bonding in the thick textured carbon film. However the electrical performance is still maintained and even better for thick textured carbon film fabricated using multi-cycle deposition.

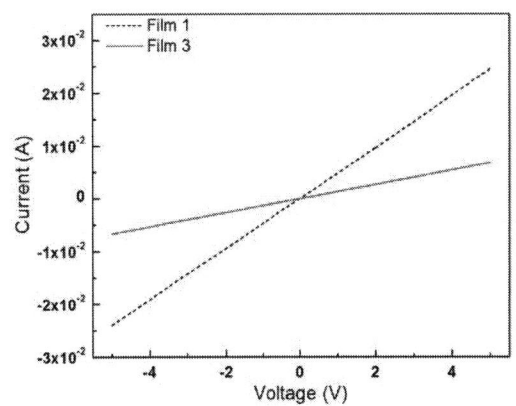

Fig. 4 I-V characteristics of thin and thick textured C films

Conclusions

Thick textured carbon film with 800nm thickness has been successfully fabricated on Si substrate, without any delimitation. XPS and Raman results show the similar binding composition and structure for thick carbon film fabricated using multi-cycle deposition, comparing with the thin textured carbon film. The electrical measurement confirms the thick textured carbon film fabricated using this method has the unique property of textured carbon.

References

[1] J. Robertson, "Diamond-like amorphous carbon," *Materials Science & Engineering R-Reports*, vol. 37, pp. 129-281, May 2002.

[2] T. Nakatani, *et al.*, "Application of diamond-like carbon to a rotary engine," *New Diamond and Frontier Carbon Technology*, vol. 16, pp. 187-200, 2006.

[3] J. K. Luo, *et al.*, "Diamond and diamond-like carbon MEMS," *Journal of Micromechanics and Microengineering*, vol. 17, pp. 147-63, 2007.

[4] W. Yanqing, *et al.*, "High-frequency, scaled graphene transistors on diamond-like carbon," *Nature*, vol. 472, pp. 74-8, 2011.

[5] N. Xu, *et al.*, "Electrical properties of textured carbon film formed by pulsed laser annealing," *Diamond and Related Materials*, vol. 23, pp. 135-139, 2012.

[6] C. W. Tan, *et al.*, "Microstructure and through-film electrical characteristics of vertically aligned amorphous carbon films," *Diamond and Related Materials*, vol. 20, pp. 290-293, 2011.

[7] M. Shakerzadeh, *et al.*, "Field emission enhancement and microstructural changes of carbon films by single pulse laser irradiation," *Carbon*, vol. 49, pp. 1018-1024, 2011.

[8] N. Xu, *et al.*, "Effect of initial sp3 content on bonding structure evolution of amorphous carbon upon pulsed laser annealing," *Diamond and Related Materials*, vol. 30, pp. 48-52, 2012.

[9] M. Shakerzadeh, *et al.*, "Thermal conductivity of nanocrystalline carbon films studied by pulsed photothermal reflectance," *Carbon*, vol. 50, pp. 1428-1431, 2012.

[10] M. Shakerzadeh, *et al.*, "Plasma density induced formation of nanocrystals in physical vapor deposited carbon films," *Carbon*, vol. 49, pp. 1733-1744, 2011.

[11] D. W. M. Lau, *et al.*, "Abrupt Stress Induced Transformation in Amorphous Carbon Films with a Highly Conductive Transition Phase," *Physical Review Letters*, vol. 100, p. 176101, 2008.

[12] D. Sheeja, *et al.*, "Effect of film thickness on the stress and adhesion of diamond-like carbon coatings," *Diamond and Related Materials*, vol. 11, pp. 1643-1647, 2002.

[13] E. H. T. Teo, *et al.*, "Mechanical properties of alternating high-low sp3 content thick non-hydrogenated diamond-like amorphous carbon films," *Diamond and Related Materials*, vol. 16, pp. 1882-1886, 2007.

[14] E. H. T. Teo, *et al.*, "Fabrication and characterization of multilayer amorphous carbon films for microcantilever devices," *IEEE Sensors Journal*, vol. 8, pp. 616-620, 2008.

[15] D. G. McCulloch, *et al.*, "Mechanisms for the behavior of carbon films during annealing," *Physical Review B*, vol. 70, Aug 2004.

[16] A. C. Ferrari and J. Robertson, "Interpretation of Raman spectra of disordered and amorphous carbon," *Physical Review B*, vol. 61, p. 14095, 2000.

[17] S. Reich and C. Thomsen, "Raman spectroscopy of graphite," *Philosophical Transactions of the Royal Society a-Mathematical Physical and Engineering Sciences*, vol. 362, pp. 2271-2288, Nov 2004.

[18] G. Abrasonis, *et al.*, "Growth regimes and metal enhanced 6-fold ring clustering of carbon in carbon-nickel composite thin films," *Carbon*, vol. 45, pp. 2995-3006, 2007.

[19] M. Shakerzadeh, *et al.*, "Microstructure and electrical properties of in-situ annealed carbon films," in *2010 IEEE 3rd International Nanoelectronics Conference (INEC 2010), 3-8 Jan. 2010*, Piscataway, NJ, USA, 2010, pp. 230-1.

Effects of carbon loading on the performance of functionalized carbon nanotube polymer heat sink for ultra high power light-emitting diode

S H Chen[2*], C M Tan[†1], E Tan[†1], J Kong[†1]

*Corresponding author

Email: shchen@SIMTech.a-star.edu.sg

1School of Electrical and Electronic Engineering, Nanyang Technological University,
Nanyang 639798, Singapore

2Singapore Institute of Manufacturing Technology, A*STAR, Nanyang 638075,
Singapore

†Contributed equally

Abstract

This research focuses on the effect of carbon loading, in 0.2wt% increment, on the performance improvements of the elastomer heat sinks. The results shows an interesting change in heat transfer mechanism of surface phonon transfer at higher carbon loading and a bulk phonon transfer at lower carbon loading. Finally, the results were compared to an aluminum heat sink and were found that the heat dissipation for the heat sink with higher carbon loading performed better than the aluminum heat sink.

Introduction

Heat sinks are used for thermal management in many industries especially for application such as automotive solid state lighting where its performance very much depends on thermal management [1]. Conventional heat sinks are made of metal, such as aluminum, copper etc [2]. However, some corrosion prone [3] and physical weight calls for radical improvements.

Although polymer appears to solve the limitations of metal heat sink, poor thermal conductivity hinders the practicality [4, 5]. The research frontier for increasing heat transfer in polymers revolves around filler addition to achieve a hybrid composite plastic material with high thermal conductivity [6]. Some common high thermally conductive filler used are carbon nano-tubes (CNTs) [7-12], metal filling [13-15] and carbon black [9, 16].

In our previous study [11], three-dimensional functionalization of multi-walled carbon nano-tubes (MWCNTs) fillers was done to improve the heat transfer performance of silicone elastomer. It shows a significant improvement in heat transfer over pure silicone. However, the heat transfer performance still lags that of metal heat sink. This work aims to study the effects of carbon loading on the heat transfer performance of the heat sink. In this study, silicone elastomer will be loaded with different weight percentages of the MWCNTs fillers functionalized using the reported method [11] and heat sinks will be fabricated. Infra-red scans will be conducted to study the heat dissipation performance of the plastic heat sinks. Evolution of heat over time will be used as a qualitative measure of the heat spreading mechanism. All heat sinks will then be subjected to thermal resistance and capacitance characterization with a 1W LED mounted on each of the plastic heat sinks. LED is used as the heat source in this work and the cumulative structure function of the LED and heat sink system is used as a quantitative indicator of performance.

The purpose of using LED is because they depend very much on their thermal management [17, 18]. As LEDs produce great amount of heat flux which causes lumens degradation and change in colour of LED, it is important to have an efficient heat sink to dissipate the heat [19]. Since the LEDs are usually incorporated into automotive systems, due to the need of lighter weight, they would be the best experimental choice for the testing of polymer heat sink.

Methodology

A. Functionalization

In order for heat transfer to take place, the MWCNT and host polymer have to be covalently bonded through the process called functionalization. Before the MWCNT can be covalently bonded, it is first oxidized to form defects on its wall. 3-methacryloxypropytrimethoxysilane (3-MPTS) is added to hydrolyze the MWCNT [20]. After which, the MWCNT is attached with the covalent trisilanol group. The trisilanol group forms covalent bonds with the silicone elastomer to allow a miscible blend and strong chemical bond between the elastomer molecules and the carbons in CNT [20]. These processes will produce the functionalized MWCNT.

Dow Corning® Sylgard® 184 Silicone Elastomer part A (Dow Corning Corporation, Midland MI, USA) is then blended together with the functionalized MWCNT for 1 hour duration through ultrasound, using Elmasonic S15/H ultrasonic bath (Elma Hans Schmidbauer GmbH & Co. KG, Singen, Germany). This method will create a miscible MWCNT and silicone matrix. Next, Dow Corning® Sylgard® 184 Silicone Elastomer part B (Dow Corning Corporation, Midland MI, USA) is added and stirred until no bubbles were observed. Finally, the MWCNT and silicone compound is placed into a mold to cure in an oven at 70°C for 24 hours.

B. Thermal imaging with heat signature analysis

A total of four plastic heat sinks comprising of 0.2wt%, 0.4wt%, 0.6wt% and 0.8wt% of functionalized MWCNT(Fs-MWCNT) composites and silicone elastomer blend were fabricated. Fig. 1 shows the heat sink that we fabricated.

978-1-4673-4840-9/13 $31.00 © 2013 IEEE

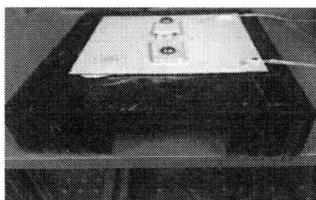

Fig. 1 Fs-MWCNT-silicone elastomer nano-composite heat sink

Thermal heat profiling was done to characterize the heat spreading abilities of the composite heat sinks. The heat source is chosen to be a high power LED driven at 1 W power. The surface of the heat sink with the LED mounted is scanned in 1mm step size for infra-red emission in 10 min repetitions. The thermal scans were repeated for 3 times in order to capture the evolution of heat spreading until a steady state is reached. Fig. 2 shows the experimental set-up for this work.

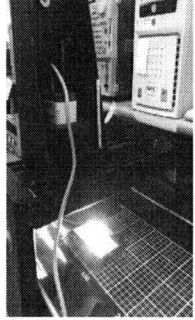

Fig. 2 Thermal heat profiling characterization setup

The heat transfer signatures of heat sinks with different weight percentage of Fs-CNTs were compared. This heat signature could be derived from the cross sectional plot of sample temperature points of the thermal images, shown in Fig. 5 to 8. A change in temperature (ΔT) from transient to steady state over "points on scale" analysis was plotted. The cross sectional plot of ΔT of the heat sinks over 30mins with respect to "points on the scale" from the center of the LED is presented in Fig. 3. A total of ten points (Point A to point J) were chosen for the analysis.

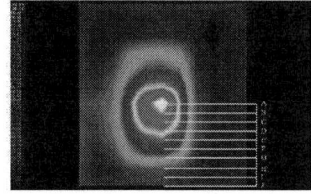

Fig. 3 Cross sectional analysis of thermal scans

C Cumulative Structure Function

In order to compare quantitatively the significance of the heat spreading signatures, the thermal resistances and capacitances of the heat sinks were investigated using the cooling curve of the high power LEDs mounted on each heat sink.

First, each LED is calibrated for voltage response against temperature to find the temperature sensitive parameter (TSP). Then, a heating power of 0.89W, (i.e. $P_{electrical} - P_{optical}$) was supplied to the heat sink. At steady state, the power was dropped instantly to a sensing power and the voltage evolution against time was captured at 60s interval.

From the voltage information, we can obtain the thermal resistance (K/W) and thermal capacitance (W.s/K) of the setup. Using the Cauer model of single thermal resistance and capacitance as shown in Fig. 4 [21], we can plot the cumulative structure function of each LED heat sink system.

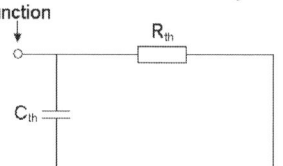

Fig. 4 Cauer Model of single thermal resistance and capacitance [21]

Results and discussion

A. Thermal Images and Heat Spreading Signatures

Figs. 5-8 show the results from the thermal profiling of each heat sink are shown below.

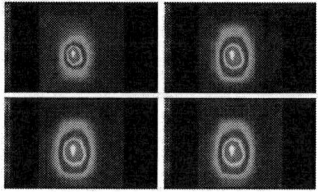

Fig. 5 Heat Sink 0.2wt% IR scan results

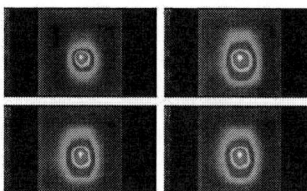

Fig. 6 Heat Sink 0.4wt% IR scan results

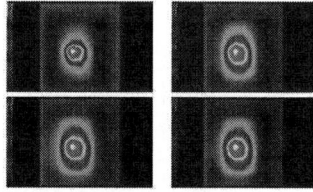

Fig. 7 Heat Sink 0.6wt% IR scan results

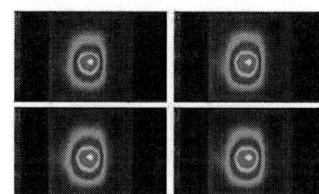

Fig. 8 Heat Sink 0.8wt% IR scan results

As confirmed by our previous studies [11], the heat is removed from the LED by the composite as shown by the spread of the thermal contours in all the thermal images. This proves the potentials of the composite heat sink for high power LED applications. However, the thermal images provide only a qualitative comparison between the differently carbon loaded heat sinks. Thus heat signature has to be done to quantitatively compare the performances of the composite heat sinks at different weight percentages Fs-MWCNT. The heat spreading signature is shown in Fig. 9.

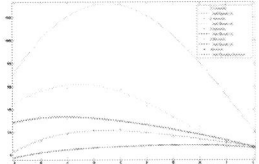

Fig. 9 Average ΔT over 30mins with respect to units away from center of LED on respective heat sinks

It can be observed that the heat sink comprising of 0.2wt% and 0.4wt% of Fs-MWCNT exhibit a similar heat spreading signature as the aluminium heat sink. It could be described as the reservoir effect as heat was drawn away from the LED and gathered at the area next to the heat source and was later dissipated to the fins of the heat sink. This effect is largely due to the high thermal capacitance of the matrix material which draws and stores high amount of heat [22].

For the case of the 0.8wt% heat sink, this reservoir effect is absent from the heat spreading signatures. Instead, the heat spreading signature follows that of a material with very low thermal resistance that is a smooth curve with very low ΔT value across the whole cross section. A negative change in temperature could be observed at the origin which symbolizes the heat source is dropping in temperature rather than heating up. This heat spreading signature is desirable as such steady state behaviour allows the heat sink to dissipate heat constantly without the need to form a reservoir where heat is gathered at a point.

It is suspected that this behaviour was a result of curing at room temperature instead of high temperature. This finding was backed by Martin's group [23] which proved that the heat dispersion of nanotubes is better in nano composites cured at low temperature. High concentration of Fs-MWCNT saturates and aligns to the top surface of the heat sink due to gravitational force and mold design, causing a surface of thermally conductive Fs-MWCNT to be formed. This surface of concentrated CNTs thus dissipates heat more rapidly and reduces the reservoir effect as compared to the 0.2wt% and 0.4wt% where heat dissipation was mainly done by the Fs-MWCNT in the bulk with high Kapitza thermal resistance [24-26] between silicone chains and the Fs-MWCNT.

The heat spreading signature of 0.6wt% heat sink was observed to be between the reservoir effect of that of the 0.2wt% and 0.4wt% and the surface phonon transport effect of the 0.8wt% heat sink. Therefore it is suspected that 0.6wt% heat sink is the point which the heat spreading mechanism changes. The concentration of the 0.6wt% carbon loading was not enough to form a complete and aligned network of CNTs with high thermal conductivity and therefore unable to spread heat in a consistent manner as compared to that of the 0.8wt%. At the same time, the bulk of CNTs and silicone mixture in the 0.6wt% heat sink is conducting substantial amount of heat from the LED, forming reservoir effect. The combination of both effects thus results in a heat spreading trend as observed. This is in turn detrimental to the degradation of the LED as the

stored heat cancels out the surface phonon transport effect which caused more heat trapping than dissipation.

A. Cumulative Structure Function

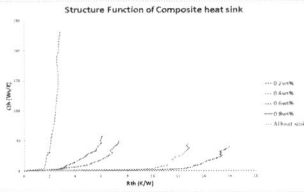

Fig. 10 Cumulative structure function of various weight percentages of Fs-MWCNT and Aluminum heat sinks

The cumulative structure function is presented in Fig. 8. As seen from the structure functions plot, the thermal resistances decrease as the weight percentages of the Fs-MWCNT increases. This confirms that the heat signatures obtained by the thermal scans accurately represent the heat spreading capabilities of the composite heat sinks. It can be seen from the cumulative structure functions of the composite heat sinks that a sharp decrease in the thermal resistance occur at the transition between 0.4 to 0.6 wt% which coincides with our findings from the heat signatures of the thermal scans. This confirms our suspicion in which the surface phonon transport mechanism starts to occur at this transition point.

When compared to the metal heat sink, we can see that a relatively small weight percentage of Fs-MWCNT results in a rivaling performance to aluminum heat sink of the same geometry. However it could be seen that the increase in thermal capacitance is slow compared to that of the aluminum heat sink while the heat signatures suggests a quick dissipation of heat. This phenomenon results from the competing nature of the bulk, Fs-MWCNT and the covalent interfaces of the composite.

Conclusion

In this research, the effects of carbon loading on MWCNT-silicone nano-composite heat dissipation performance were studied. The performance indicators used are the heat transfer signatures as well as the thermal resistance and capacitance of the heat sinks with LED as the heat source. The novelty of this research lies in finding the ideal weight percentage of MWCNTs to be added to silicone matrix in order to allow a significant increase in thermal performance with relatively low weight percentages of MWCNT fillers by three-dimensional covalent functionalization. A surface phonon transfer mechanism has been presented and proven with thermal analysis.

The following concludes the findings of the research:

1. Curing the plastic heat sinks slower and at a lower temperature allows better dispersion of the CNT composites.

2. The model of thermal scan of heat sinks proved the performance of thermal dissipation of aluminium heat sink falls in between 0.6wt% and 0.8wt% of plastic heat sink.

3. The model of thermal scan of heat sinks proved that 0.8wt% of filler in silicone exhibits a different trend of heat dispersion.

4. The model of thermal scan of heat sinks proved that 0.8wt% of filler in silicone has better heat dispersion then conventional metal heat sinks.

References

[1] M. Arik, et al., "Thermal challenges in the future generation solid state lighting applications: light emitting diodes," pp. 113-120, 2002.

[2] S. Lee, "Optimum design and selection of heat sinks," Components, Packaging, and Manufacturing Technology, Part A, IEEE Transactions on, vol. 18, pp. 812-817, 1995.

[3] L. H. Hihara and R. M. Latanision, "Corrosion of metal matric composites," International Materials Reviews, vol. 39, pp. 245-263, 1994.

[4] C. P. Wong and R. S. Bollampally, "Thermal conductivity, elastic modulus, and coefficient of thermal expansion of polymer composites filled with ceramic particles for electronic packaging," Journal of Applied Polymer Science, vol. 74, pp. 3396-3403, 1999.

[5] J. Liu and R. Yang, "Tuning the thermal conductivity of polymers with mechanical strains," Physical Review B, vol. 81, pp. 174122, 2010.

[6] H. K. Lyeo and D. G. Cahill, "Thermal conductance of interfaces between highly dissimilar materials," Physical Review B, vol. 73, pp. 144301, 2006.

[7] T. Zhou, et al., "Improved thermal conductivity of epoxy composites using a hybrid multi-walled carbon nanotube/micro-SiC filler," Carbon, vol. 48, pp. 1171-1176, 2010.

[8] A. Yu, et al., "Enhanced thermal conductivity in a hybrid graphite nanoplatelet – Carbon nanotube filler for epoxy composites," Advanced Materials, vol. 20, pp. 4740-4744, 2008.

[9] F. H. Gojny, et al., "Evaluation and identification of electrical and thermal conduction mechanisms in carbon nanotube/epoxy composites," Polymer, vol. 47, pp. 2036-2045, 2006.

[10] A. Bagchi and S. Nomura, "On the effective thermal conductivity of carbon nanotube reinforced polymer composites," Composites Sciences and Technology, vol. 66, pp. 1703-1712, 2006.

[11] S. H. Chen, C. M. Tan, M. H. Tan, B. K. Chen, "Performance Evaluation of Covalently Functionalized Carbon Nano-tube Polymer Heat Sink for Ultra High Power LED", Proceedings - International NanoElectronics Conference, INEC, 2011, 4th IEEE International NanoElectronics Conference, INEC 2011.

[12] M. Moniruzzaman and K. I. Winey, "Polymer nanocomposites containing carbon nanotubes," Macromolecules, vol. 39, pp. 5194-5205, 2006.

[13] F. Danes, et al., "Predicting, measuring, and tailoring the transverse thermal conductivity of composites from polymer matrix and metal filler," International Journal of Thermophysics, vol. 24, pp. 771-784, 2003.

[14] Y. Xu, et al., "Thermally conducting aluminium nitride polymer-matrix composites," Composites Part A: Applied Science and Manufacturing, vol. 32, pp. 1749-1757, 2001.

[15] Y. P. Mamunya, et al., "Electrical and thermal conductivity of polymers filled with metal powders," European Polymer Journal, vol. 38, pp. 1887-1897, 2002.

[16] Y. Agari and T. Uno, "Thermal conductivity of polymer filled with carbon materials: Effect of conductive particle chains on thermal conductivity," Journal of Applied Polymer Science, vol. 30, pp. 2225-2235, 1985.

[17] D. G. Todorov and L. G. Kapisazov, "LED Thermal Management," ELECTRONICS, pp. 139-144, 2008.

[18] M. Arik, et al., "Thermal Management of LEDs: Package to System," pp. 64-75, 2004.

[19] M. Arik, et al., "Development of a High-Lumen Solid State Down Light Application," IEEE TRANSACTIONS ON COMPONENTS AND PACKAGING TECHNOLOGIES, vol.

33, pp. 668-679, 2010.

[20] D. S. Bag, et al., "Chemical functionalization of carbon nanotubes with 3-methacryloxypropyltrimethoxysilane (3-MPTS)," Smart Mater. Struct, vol. 13, pp. 1263-1267, 2004.

[21] Yang, L., Hu, J., & Shin, M. W. (2008). Dynamic thermal analysis of high-power LEDs at pulse conditions. Electron Device Letters, IEEE, 29(8), 863-866.

[22] S. Y. Yang, et al., "Effect of functionalized carbon nanotubes on the thermal conductivity of epoxy composites," CARBON, vol. 48, pp. 592-603, 2010.

[23] Martin CA, et al., "Formation of percolating networks in multi-wall carbon nanotube-epoxy composites. Compos Sci Technol (2004) in press.

[24] T. C. Clancy and T. S. Gates, "Modeling of interfacial modification effects on thermal conductivity of carbon nanotube composites," Polymer, vol. 47, pp. 5990-5996, 2006.

[25] K. Yang, et al., "Effects of carbon nanotube functionalization on the mechanical and thermal properties of epoxy composites," CARBON, vol. 47, pp. 1723-1737, 2009.

[26] H. M. Duong, "Computational modelling of the thermal conductivity of single-walled carbon nanotube-polymer composites," Nanotechnology, vol. 19, 8pp, 2008.

A Neural Network Approach to Classify Inversion Regions of High Mobility Ultralong Channel Single Walled Carbon Nanotube Field-Effect Transistors for Sensing Applications

Hari Krishna S.V[1*], Jianing An[2] and Lianxi Zheng[3]

[1,2,3] Nanyang Technological University
School of Mechanical and Aerospace Engineering
50 Nanyang Avenue 639798, Singapore
*Phone: 65-83589019; fax: 65-67906995; e-mail: hari1@ e.ntu.edu.sg.

Abstract

Millimetre long individual single walled carbon nanotubes (SWCNTs) were consistently grown and fabricated into carbon nanotube field effect transistors (CNTFETs). In this work, we extracted the effective mobilities in the strong inversion region, near-threshold region and subthreshold region respectively for these long-channel CNTFETs. Using the mobility data as an input parameter, an artificial neural network (ANN) employing multi-layer perceptron (MLP) architecture was used to classify the different inversion regions of the mobility curves with an accuracy of 90%.

Keywords: artificial neural network, carbon nanotube, field-effect transistor, mobility, multi-layer perceptron.

Introduction

Semiconducting single walled carbon nanotubes (SWCNTs) have tremendous potential in the field of chemical and biosensing.[1, 2] On account of their tubular structure, CNTs contain a large number of surface atoms, which gives them excellent electronic properties that modulate upon the slightest electrical disturbance in the vicinity. Previously, many investigations using field-effect transistors (FETs) bio/chemical applications were done using short CNTs. The use of short-channel CNTs in FETs suffered from multiple competitive mechanisms like Schottky barrier and charge transfer effects making the true detection mechanism debatable. [3] On the other hand, the use of long CNT helps to truly harness the large number of surface mobile charges. Long CNT based FETs have minimum interference from the contact electrodes and can pave way for sensing based upon intrinsic property modulation of CNT upon interaction with target of interest. Their large surface coverage area along with dimensions comparable to biomolecules makes them worthy contenders for biosensing. Mobility is one such intrinsic attribute that determines the sensitivity of the CNTFET. Long-channel CNTFETs have much higher mobilities than Si MOSFETS or short channel CNTFETs. [4]

Many research groups have used classification strategies for data obtained from sensory arrays for bio/chemical sensing.[5] Herein, we fabricate FETs using ultralong SWCNTs and extract their mobility. Following this, we utilize an artificial neural network (ANN) with a multi-layer perceptron (MLP) architecture to classify the regions of inversion using the mobility data obtained from a number of long-channel CNTFETs. Also, the potential use of this approach in future automated nanosensing has been discussed in brief.

Device Fabrication

Ultralong SWCNTs as seen in Fig. 1 were grown using ethanol chemical vapour deposition (CVD) on n+-doped Si wafers capped by 1μm thermally grown SiO_2 using 0.01 M $FeCl_3$ catalyst solution. The catalyst solution is applied with a dig-pen onto one end of Si/SiO2 substrate followed by placing the sample into a horizontal 1-inch quartz tube furnace at 950 °C under a flow of 120 sccm argon (Ar) and 30 sccm hydrogen (H_2). After 10 minutes, Ar was bubbled through ethanol (20 °C) at a controlled rate of 20 sccm to initiate the CNT growth. After a 40 minute synthesis period, the ethanol supply was terminated and the furnace was cooled to room temperature under ~120 sccm Ar. CNTFETs were fabricated by shadow mask facilitated E-beam evaporation of Ti/Au (5nm/50nm) onto these substrates. The average diameter of the SWNT found using atomic force microscope (AFM) was found to be ~2nm.

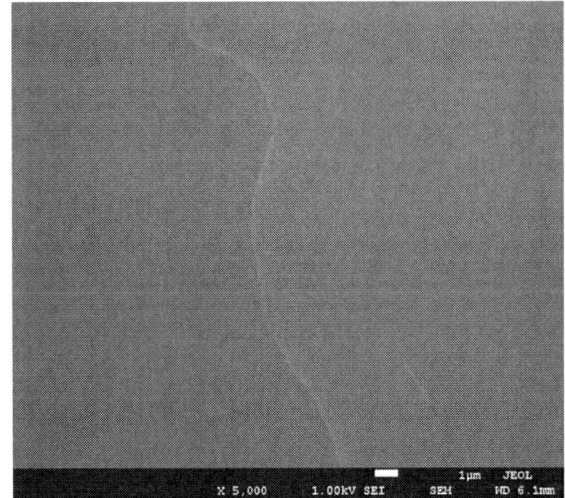

Fig. 1. FE-SEM image of 1mm long SWCNT grown by ethanol based thermal CVD.

Electrical Characterization

Electrical characterization was done using Agilent 4156B semiconductor parameter analyzer for about 10 ultralong channel CNTFETs with 1mm long channel. The output and transfer curves are given in Figs. 2a and 2b respectively for a particular device. The as-grown CNTs were observed to be p-type due to oxygen adsorption during growth. A positive gate voltage causes formation of a depletion zone devoid of holes while a negative gate voltage results in attraction of the holes towards the CNT-SiO₂ interface. The characteristic of a CNTFET is very similar to that of a Si-channel MOSFET. [6]

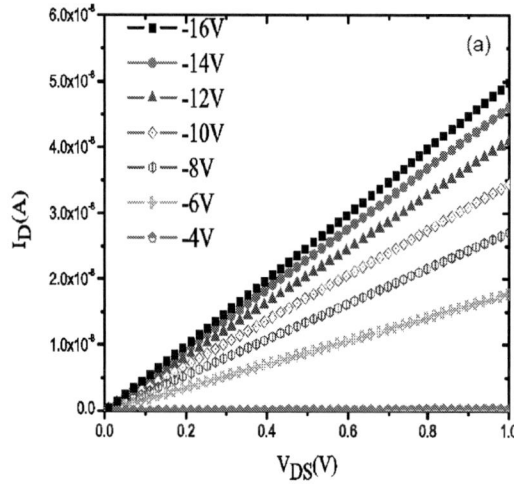

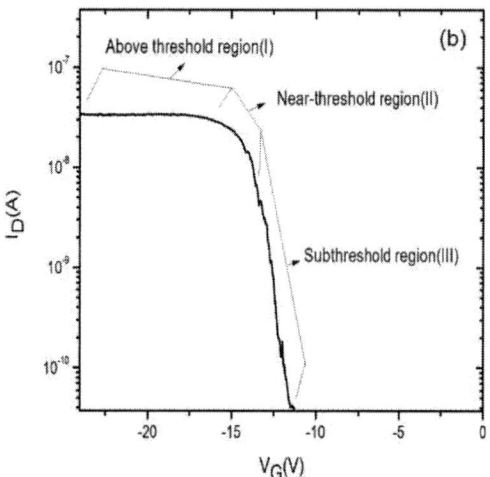

Fig. 2. (a) I_{DS}-V_{DS} curves for V_{GS} -16V to 0V in steps of 2V and (b) the I_{DS}-V_{GS} curves in semi–log scale at a particular V_{DS}.

In the moderate and weak inversion regime, the behaviour of the charge carriers differs from the strong inversion region on account of the comparable magnitude of the transverse gate electric field (E_{GS}) with the longitudinal drain-source electric field (E_{DS}). The gate capacitance per unit length of the CNT modelled as a cylinder on a planar substrate is given by

$$C/L = 2\pi\varepsilon_0\varepsilon_r/ln~(2t_{ox}/r) \qquad (1)$$

where ε_0 is the permittivity of free space (8.8542×10^{-12} Fm⁻¹), ε_r is the dielectric constant of SiO₂ gate insulator (~3.9), t_{ox} is the SiO₂ thickness, r is the radius of the nanotube and L the length of the CNT ~1mm.[5] Mobility is a measure of how fast

the charge carriers respond to an external electric field. It is defined as

$$\mu = v/E = e\tau/m* = \sigma n/e \qquad (2)$$

where v is the drift velocity of charge carriers, E is the applied electric field, σ is the conductivity, and ne is the charge density. Mobility measures the momentum scattering rate of carriers, and therefore is of fundamental interest in understanding the scattering processes in a given system. The effective mobility is closest to the true physical mobility of the CNT without including the contribution of device attributes and is given by,

$$\mu_{eff} = L^2G/C(V_T - V_G) \qquad (3)$$

where $G = \partial I_D/\partial V_{DS}$ at constant V_{DS}. G and V_T represent the channel conductance at a particular gate voltage and the threshold voltage respectively.[5,7] The transfer characteristics and the effective mobility versus gate voltage curves for one particular device are shown in Fig 3 with regions I, II and III indicated.

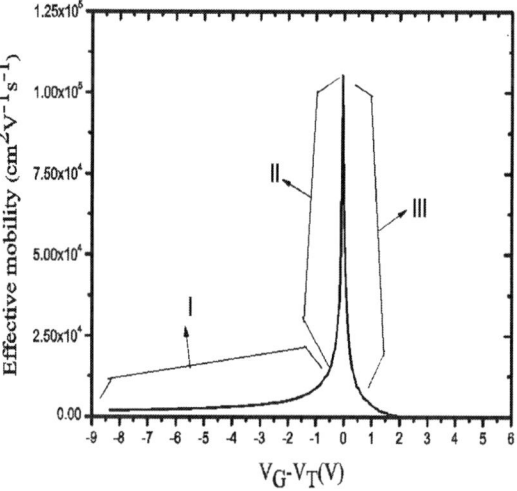

Fig. 3. μ_{eff} vs V_{GS}-V_T for 1mm long CNTFET with the three inversion regions indicated as I, II and III.

Artificial Neural Network (ANN)

Artificial neural networks (ANNs) are networks with nodes analogous to the biological neurons, which are interconnected to weighted links.[8] By using a learning process for training, the weights can be adjusted. ANNs usually have the input layers-where the input features are given, the hidden layers-where the data is processed by the weight functions and the output layer- where the classification result appears.

Fig 4. Multi-layer perceptron architecture.

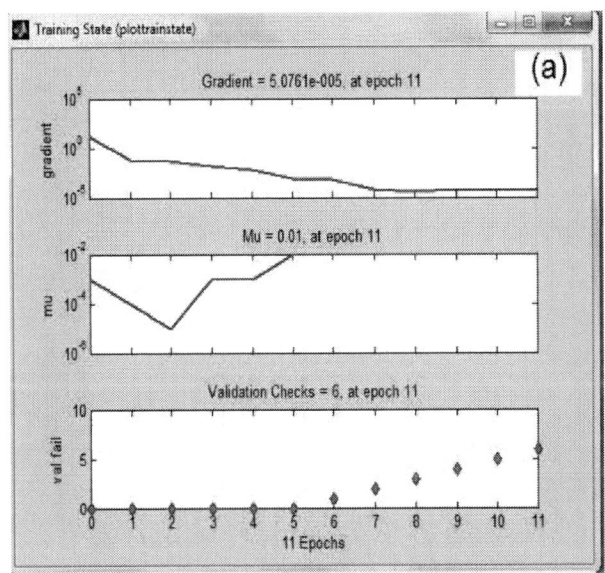

The number of hidden units determines the complexity of neural network and hence an optimum number of the hidden layers and neurons should be determined and applied before using the predictions of the model. In this study, the feed-forward multilayer perceptron (MLP) as shown in Fig. 4 was used and trained with back propagation algorithm. [9]

TABLE I
SUMMARY OF TWO REGION AND THREE-REGION CLASSIFICATION USING MLP ANN.

Classificati-on regions	Total sample points	Correct classificati-on (%)	Sensitivi-ty (%)	Specificity (%)
I and II	10008	97.39	98.25	95.55
I and III	14712	93.65	96.31	97.36
II and III	12144	99.28	99.41	98.71
I, II and III	22600	90.41	84.99	91.61

Results

MATLAB 2008 was used to perform neural network classification for the mobility data obtained from 11 ultralong CNTFETs for classifying the three regions. The effective mobilities of the three regions given as input to the MLP neural network taken two at a time produced classification accuracies of 97.39% (region I and II), 93.65% (region I and III) and 99.28% (region I and II) respectively as indicated in Table 1. The train to test data was taken in the ratio 3:2 and 10 validation checks were performed for the given input samples (Figs. 5(a) and 5(b)). The mobility of the three regions given simultaneously as inputs gave a classification accuracy of 90%. Such a classification strategy bodes well for mobility based sensing, where a target upon interaction with a long CNT introduces a scattering center, which leads to reduction in mobility.

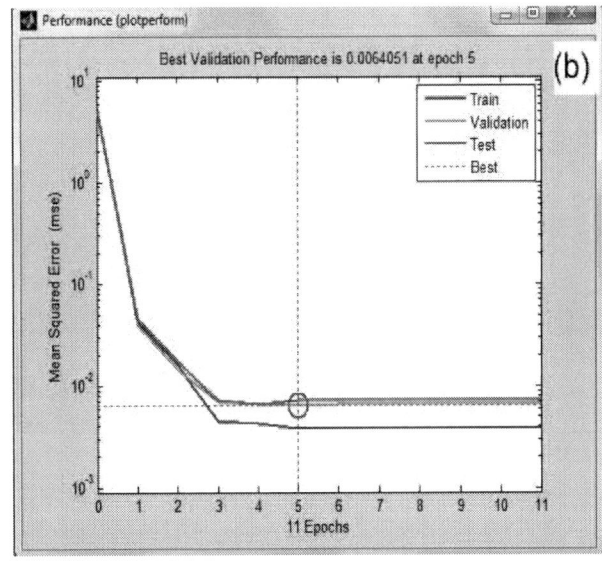

Fig 5. (a) Training state and (b) performance validation graphs using MLP neural network in MATLAB.

This mobility data obtained from CNTFETs with and without attachment of a target can be given as input to the MLP neural network. This can classify the mobility changes in regions I, II and III into two classes-detected and undetected as shown in Fig. 6.

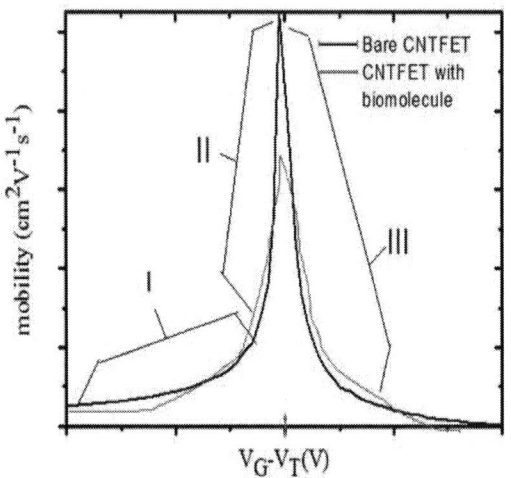

Fig 6. Change in mobility upon interaction with target of interest.

Conclusions and Future Work

In conclusion, ultralong CNTFETs were successfully fabricated on a large scale and their effective mobilities were found to be consistently greater than $10^4 cm^2 V^{-1} s^{-1}$, which are quite high for semiconducting CNTs at room temperature. The long-channel CNTFETs could pave way for solely mobility based sensing based on interaction between the CNT and the target of interest due to freedom from hindrance of metal contacts. The mobility data obtained here were used for classifying the subthreshold region, near-threshold region and strong inversion region using multi-layer perceptron neural network. This sort of classification promises the possibility of an expert system that can automatically detect changes in mobility of a sensor based on its ability to classify them according to the operation regions of the CNTFET. Here, the mobilities obtained in the strong inversion, near-threshold region and the subthreshold region were fed as inputs to an artificial neural network, which efficiently classified the regions with an accuracy, sensitivity and specificity of about 90%, 85% and 92% respectively.

The future work would include exposing these high mobility CNTFETs to certain biomolecules like deoxy ribo nucleic acid (DNA) to observe the change in mobility and classify the different regions of inversion of the CNTFETs with and without CNT-biomolecule interaction.

Acknowledgments

We gratefully acknowledge the financial support provided by Nanyang Technological University.

References

[1] B.L. Allen, P.D. Kichambare, and A. Star, "Carbon Nanotube Field-Effect-Transistor-Based Biosensors," *Advanced Materials*, vol. 19, pp. 1439-1451, 2007.

[2] P. Hu, J. Zhang, L. Li, Z. Wang, W. O'Neill, and P. Estrela, "Carbon Nanostructure-Based Field-Effect Transistors for Label-Free Chemical/Biological Sensors," *Sensors*, vol. 10, pp. 5133-5159, 2010.

[3] S. Heinze, J. Tersoff, R. Martel, V. Derycke, J. Appenzeller, and P. Avouris, "Carbon Nanotubes as Schottky Barrier Transistors," *Physical Review Letters*, vol. 89, p. 106801, 2002.

[4] A.B. Artyukhin, M. Stadermann, R.W. Friddle, P. Stroeve, O. Bakajin, and A. Noy, "Controlled Electrostatic Gating of Carbon Nanotube FET Devices," *Nano Letters*, vol. 6, pp. 2080-2085, 2006.

[5] S. Li, Z. Yu, C. Rutherglen, and P.J. Burke, "Electrical Properties of 0.4 cm Long Single-Walled Carbon Nanotubes," *Nano Letters*, vol. 4, pp. 2003-2007, 2004.

[6] N. Arora, *MOSFET Modeling for VLSI Simulation*, 2006.

[7] T. Dürkop, S.A. Getty, E. Cobas, and M.S. Fuhrer, "Extraordinary Mobility in Semiconducting Carbon Nanotubes," *Nano Letters*, vol. 4, pp. 35-39, 2003.

[8] M. Hayati, A. Rezaei, and M. Seifi, "CNT-MOSFET modeling based on artificial neural network: Application to simulation of nanoscale circuits," *Solid-State Electronics*, vol. 54, pp. 52-57, 2010.

[9] M.W. Gardner and S.R. Dorling, "Artificial neural networks (the multilayer perceptron)—a review of applications in the atmospheric sciences," *Atmospheric Environment*, vol. 32, pp. 2627-2636, 1998.

Impact of Stress Induced by Stressors on Hot Carrier Reliability of Strained nMOSFETs

H.W. Hsu[1], H. S. Huang[1], S. Y. Chen[1], M. C. Wang[2], K.C. Li[1], K. C. Lin[3], C. H. Liu[4, *]

[1]Institute of Mechatronic Engineering, National Taipei University of Technology (NTUT), Taiwan, ROC
[2]Dept. of Electronic Engineering, Minghsin University of Science and Technology (MUST), Taiwan, ROC
[3]Dept. of Electronic Engineering, Ming Chuan University (MCU), Taiwan, ROC
[*, 4] Dept. of Mechatronic Technology, National Taiwan Normal University (NTNU), Taiwan, ROC
[*]Tel: +886-2-77343515, Fax: +886-2-23583074, E-mail: liuch@ntnu.edu.tw

Abstract

In this study, the nMOSFETs with contact-etch-stop-layer (CESL) stressor and SiGe channel have been fabricated with a modified 90-nm technology. The performance of nMOSFETs and stress distribution in the channel region have been investigated. The hot carrier reliability of the SiGe-channeled nMOSFETs with various CESL nitride layers has also been extensively studied. In addition, the impact of stress induced by CESL stressor and SiGe-channel on hot-carrier reliability of the strained nMOSFETs has been analyzed through experimental measurements and stress simulation results.

KEYWORD: Contact-etch-stop-layer (CESL), SiGe channel, hot carrier reliability

Introduction

In recent years, various CMOS fabrication technologies using the process induced stress to enhance the carrier mobility have been successfully demonstrated. The strained engineering is basically divided into biaxial and uniaxial strain approaches. The strain induced by process is called local strain technology in which the stress in the channel depends on the stressor location in the device. This approach, including shallow trench isolation (STI), nitride contact-etch-stop-layer (CESL), embedded SiGe in source/drain and so on [1-3], has three main advantages: (i) strain can be more effectively to optimize performance enhancement of CMOSFETs, (ii) threshold voltage shift is less in strained MOSFETs, and (iii) local strain techniques are cheaper and compatible with standard CMOS process [4]. In CESL stressor devices, depending on the deposition conditions, the nitride etch-stop layer can result in different internal strains and hence the strain transmitted into the channel region can also be different. Among all the feasible strain approaches, SiN capping technique receives much attention because it is easy to execute in modern ultra-large scale integration technology. In addition, strain level ranging from highly tensile to highly compressive is adjustable [5]. Although the mobility enhancement in strained-Si has been extensively studied, the reliability issue such as hot-carrier effect has seldom been investigated. In this work, the performance and hot-carrier reliability of the devices with CESL stressor and SiGe channel is investigated.

The device structure in this study, consisted of SiN CESL (1300 Å), SiGe channel (22.5% Ge concentration), and Si-Cap (24 Å), and was fabricated with 90-nm process. This research first analyzed the device characteristics for different CESL stressors, including highly compressive (-2.0 GPa) and highly tensile (+1.1 GPa) stressing layers. The devices were then stressed under CHC (channel hot carrier) test conditions. Device parameters, such as threshold voltage (Vth), drain current (Id), and transconductance (Gm), during CHC testing were measured.

Results and Discussion

A. Basic Characteristics

From Fig. 1, we can see that the performance of nMOSFETs with either compressive or tensile CESL stressor is improved. However, the improvement of devices with compressive CESL has been ascribed to the SiGe channel because it is known that compressive CESL is not so effective to enhance the electron mobility for short channel nMOSFETs, as indicated by Fig. 2. The curve of on-current (Ion) versus channel length (L) was shown in Fig. 3, from which it is clear that the combination of CESL and SiGe channel is more effective as the channel length decreases more.

B. Hot-Carrier Reliability

As shown in Fig. 4, the interface states (Nit) and oxide trapped charges (Not) increased after hot-carrier stress. The Vth shift versus stress time under stress temperatures of 25, 85, and 125 ℃ were shown in Fig. 5, 6, and 7, respectively. For all three stress temperatures, the degradation of nMOSFETs with compressive CESL stressors is more serious than those with tensile CESL stressors. In summary, a careful examination among Figs. 4–7 suggests that the hot-carrier degradation be caused by the damages that are induced by the CESL stressor.

Conclusions

The nMOSFETs with SiGe channel and CESL stressor have been fabricated. Tensile CESL results in a better performance. Hot-carrier reliability of SiGe-channeled nMOSFETs with various CESL nitride layers was studied. The devices with compressive CESL have the worst degradation. Moreover, the experimental results suggest that the damages, induced by the CESL, located at the interface between gate oxide and underlying channel be responsible for the hot-carrier degradation, implying that the device reliability may be a potential concern in integrating the tensile or compressive SiN CESL stressors into SiGe-channel devices.

978-1-4673-4840-9/13 $31.00 © 2013 IEEE

References

[1] G. Scott et al., "NMOS Drive Current Reduction Caused by Transistor Layout and Trench Isolation Induced Stress" *IEDM Tech. Dig.*, 1999, pp. 827-830.

[2] S. Ito et al., "Mechanical Stress Effect of Etch-stop Nitride and Its Impact on Deep Submicron Transistor Design" *IEDM Tech. Dig.*, 2000, pp. 247-250.

[3] K. J. Chui et al., "Source-Drain Germanium Condensation for P-Channel Strained Ultra-Thin Body Transistors" *IEDM Tech. Dig.*, 2005, pp. 493-496.

[4] E. Ungersboeck et al., "Strain Engineering for CMOS Devices" *ICSICT.*, 2006, pp. 124-127.

[5] A. Shimizu et al., "Local Mechanical-Stress Control (LMC): A New Technique for CMOS-performance Enhancement" *IEDM Tech. Dig.*, 2001, pp. 433-436.

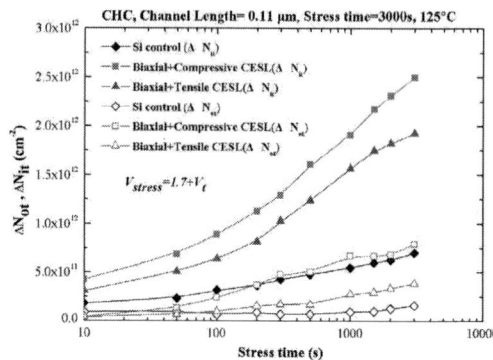

Fig. 4 Nit and Not versus stress time after 3000 s at 125℃.

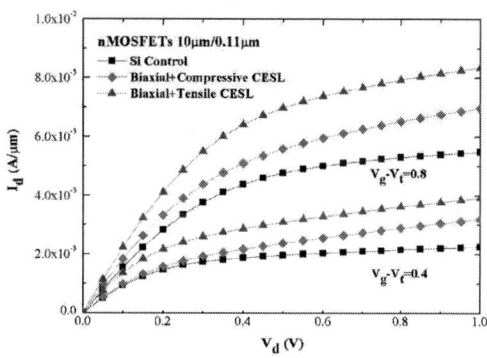

Fig. 1 Id-Vd characteristics of devices with tensile, compressive, or without CESL.

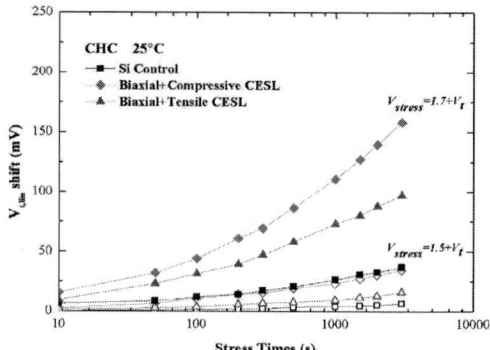

Fig. 5 Vth shift under 25 ℃ CHC stress.

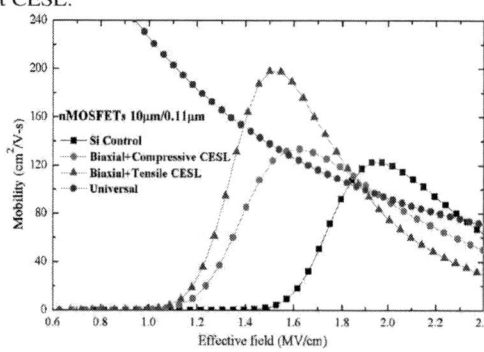

Fig. 2 The effective mobility of Si control, compressive, and tensile CESL.

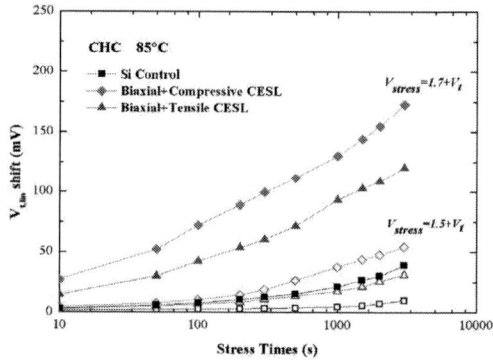

Fig. 6 Vth shift under 85℃ CHC stress.

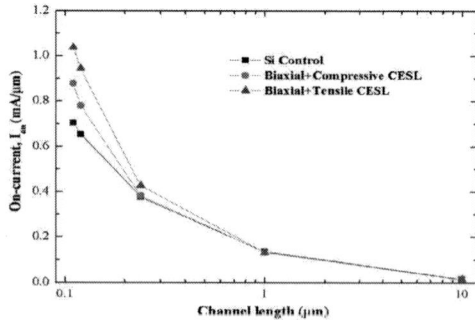

Fig. 3 On-current versus channel length in three types of devices.

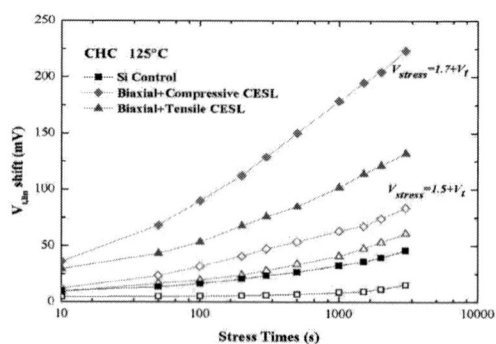

Fig. 7 Vth shift under 125℃ CHC stress.

978-1-4673-4840-9/13 $31.00 © 2013 IEEE

Capture Cross Section Analysis of Four-level Random-Telegraph-Noise in Gate-Induced Drain Leakage Current

Seulki Park, Sungwon Yoo, and Hyungcheol Shin

Inter-University Semiconductor Research Center (ISRC) and School of Electrical Eng. and Computer Sci., Seoul Nat. Univ
Gwanak 599, Gwanak-gu
Seoul 151-742, Republic of Korea
Phone: +82-2-880-7279, Fax: +82-2-882-4658, E-mail: horhor8166@snu.ac.kr

Abstract

In this paper, we measured four-level Random Telegraph Noise (RTN) in Gate Induced Drain Leakage (GIDL) current of Metal Oxide Semiconductor Field Effect Transistor (MOSFET). Using RTN measurement data, we extracted fundamental parameters of each trap, such as the trap depth (x_T) and energy level (E_{Cox}-E_T). To correctly interpret capture and emission process, capture cross section (σ_c) of the traps was extracted by applying more accurate capture cross section model and bias dependence of σ_c was analyzed using potential barrier lowering.

Keywords: RTN, GIDL, MOSFET, trap depth, energy level, capture cross section, MPE, lattice coordinate reconfiguration and barrier lowering

Introduction

As devices are scaling down, the effect of an individual defect on device performance and reliability becomes important issue. In Metal Oxide Semiconductor Field Effect Transistor (MOSFET), Random Telegraph Noise (RTN) occurs due to random trapping and de-trapping of charges into the trap located in the gate oxide or at Si/SiO$_2$ interfaces [1]. The previous researches have mainly focused on RTN in channel and gate leakage current in MOSFET [2]-[4]. However, Gate Induced Drain Leakage (GIDL) current, which is composed of band-to-band tunneling or trap-assisted tunneling in a gate/drain overlapped region, is a significant component of leakage current in modern MOSFETs [5], [6]. In this paper, we observed four-level RTN in GIDL [7]. Using measurement data, we extracted fundamental parameters such as trap depth and energy level [8]. Also, we extracted capture cross section (σ_c) and performed analysis by applying new capture cross section model [9]-[11]. Therefore, we could analyze the charge trapping mechanism and trap characteristics well.

Result and discussion

A. Background and GIDL RTN measurement

GIDL current is mainly affected by electric field at the gate/drain overlapped region. If an electron is captured to an oxide trap, the electric field at the surface is enhanced as shown in Fig. 1. As tunneling can take place more easily due to this increased electric field, I_{GIDL} is increased. Therefore, the time duration of low state of the I_{GIDL} is the capture time (τ_c) and the time duration of high state of the I_{GIDL} is the emission time (τ_e). Each time constants are determined by the difference between the energy level of a trap and the Fermi level (E_T-E_F).

The origin of four-level RTN showing different amplitude of RTN is distinct two traps. We observed amplitude changes when another trap is empty and filled are almost same. Therefore, each switching events was considered as independent and traps leading to four-level RTN could be characterized individually. Fig. 2 shows definition of time constants in four-level RTN. τ_{cs} and τ_{es} are capture and emission time in slow trap, whereas τ_{cf} and τ_{ef} are capture and emission time in fast trap.

The currents of all terminals were monitored in both the time and frequency domains. The experimental set-up and method for measuring RTN has been described in detail elsewhere, being consisted of an SR570 low noise amplifier (LNA), B1500 semiconductor parameter analyzer, and a HP35670A dynamic signal analyzer. SR570 LNA amplifies the drain current and converts it to voltage, and then we observed low frequency noise behavior both in time and frequency domains through HP35670A. B1500 is used to bias DC voltage and observe time domain signal.

The device measured in this experiment was n-MOSFET with W/L = 2μm/0.24μm and T_{ox} is 3.7nm. We observed changes of four-level GIDL RTN as a function of drain to gate bias voltage (V_{DG}) as shown in Fig. 3. Gate, source and body were biased to 0V for neglecting other leakage current except GIDL current during measurement. From RTN measurement, we extracted capture, emission time constants and time constant ratio ($\ln(\tau_c/\tau_e)$) as function of V_{DG} as shown in Fig. 4. As shown in Fig. 4(a), emission time changes more rapidly than capture time with V_{DG} in slow and fast trap.

As emission time changes more rapidly than capture time with V_{DG} in slow and fast trap, the traps could be considered as donor-like traps. The x_T of oxide trap can be obtained from previous researches as follows [8].

$$x_T = \frac{T_{ox}\left[\dfrac{k_B T}{q}\dfrac{d\ln(\overline{\tau}_c/\overline{\tau}_e)}{dV_{DG}} - \dfrac{d|\psi_s|}{dV_{DG}}\right]}{1 - \dfrac{d|\psi_s|}{dV_{DG}}} \quad (1)$$

Using (1), x_T of slow trap is 1.03nm, whereas x_T of fast trap is 0.44nm. (E_{Cox}-E_T) of traps are 4.24eV and 3.69eV, respectively as shown in Fig. 5. From this result, the fast trap is located more closely from Si/SiO$_2$ interface than the slow trap.

B. Capture cross section analysis

Using extracted parameters, the σ_c of trap can be extracted by below equation [10].

978-1-4673-4840-9/13 $31.00 © 2013 IEEE

$$\frac{1}{\tau_c} = \sigma_c v_{th} n_s \times \exp\left(-\frac{\Delta E}{k_B T}\right) \quad (2)$$

Where v_{th} is average thermal velocity of electrons given by $(8k_B T/\pi m^*)^{1/2}$ with Boltzmann constant k_B, electron effective mass m^*, and electron concentration in conduction band surface, n_s. Fig. 6(a) shows calculated σ_c with respect to V_{DG} in both slow and fast trap. For both slow and fast trap σ_c increase with respect to V_{DG}. Also, σ_c of slow trap is much larger than that of fast trap. In this analysis, we apply capture cross section model which considers Multi-Phonon Emission (MPE) theory and trap energy, which can be written as [11]-[13]

$$\sigma_c = \sigma_{oc} \exp\left(-\frac{|E_{oc}|}{k_B T}\right) \quad (3)$$

Where E_{oc} is a function of the activation energy, E_B, and is given by $E_{oc} = E_B - k_B T$. The σ_{oc} is given by

$$\sigma_{oc} = \sqrt{\frac{E_1}{S\hbar w}}\left(\frac{\pi^2}{2e}\right)\left(\frac{\hbar^2}{2m^* k_B T}\right) \quad (4)$$

Where $\hbar$ is reduced Planck constant and e is Euler's number. The lattice relaxation energy, E_{relax}, are given by $S = E_{relax}/\hbar w$. E_{relax} is the energy difference between ground state of the trap and the oxide conduction band edge. Therefore, E_{relax} is equal to $(E_{Cox} - E_T)$. We explained electron trapping process by using lattice coordinate reconfiguration as shown in Fig. 7(a) [12]. On electron capture process, electrons with large energy to overcome barrier energy can cross the conduction band and move to trap site [14]. Fig. 6 shows calculated σ_c and E_{oc} using (3) and (4) with V_{DG} in slow and fast trap. As shown in Fig. 6(b), E_{oc} of fast trap is much larger than that of slow trap and had decreasing bias dependency as expected.

Bias dependence of σ_c and E_{oc} in both traps can be explained by using potential barrier lowering as illustrated in Fig. 7(b) [15]. As V_{DG} increases, magnitude of electric field in gate oxide and at drain surface also increases, which results in potential barrier lowering of donor-like trap. Therefore, trapping of electron into trap site occurs more easily. Moreover, trapped electron loses more kinetic energy since oxide field is applied toward the direction that disturbed the electron trapping. Thus, the probability for reemitting reduces, while capturing probability increases. Consequently, σ_c increases with larger V_{DG}.

From above analysis, we concluded that although two traps exist in gate oxide at the same time, two traps show different values and dependency in σ_c and E_{oc} according to each trap characteristics.

Conclusion

We observed four-level RTN in GIDL current that stems from independent two traps. Based on measurement data and equations, we extracted various parameters such as trap depth and energy level respectively. By applying coulomb energy theory, we also extracted σ_c. Furthermore, we used MPE model to explain σ_c and explained bias dependency of σ_c and E_{oc} by

using potential barrier lowering theory.

Acknowledgements

This work is supported by Inter-university Semiconductor Research Center (ISRC) and Samsung Electronics, Inc.

References

[1] M. J. Kirton and M. J. Uren, "Noise in solid-state microstructures: A new perspective on individual defects, interface states and low-frequency (1/f) noise," *Adv. Phys.*, vol. 38, no. 4, pp. 367-468, 1989.

[2] Zhongming Shi, Jean-Paul Mieville and Michel Dutoit, "Random telegraph signals in deep submicron n-MOSFET's," *IEEE Transactions on Electron Devices*, vol. 41, no. 7, pp. 1161-1168, Jul. 1994.

[3] Ma Zhong-Fa, et al., "Accurate extraction of trap depth responsible for RTS noise in nano-MOSFETs", *Chin. Phys. B.*, vol. 19, no. 3, pp. 0372011-0372014, 2010.

[4] Heung-Jae Cho, et al., "Extraction of trap energy and location from random telegraph noise in gate leakage current (I_g RTN) of metal-oxide-semiconductor field effect transistor (MOSFET)," *Solid-State Electronics*, vol. 54, no. 4, pp. 362-367, Apr. 2010.

[5] H. Kim, et al., "RTS-like fluctuation in gate induced drain leakage current of saddle-fin type DRAM cell transistor," *IEDM Tech. Dig.*, pp. 271-274, 2009.

[6] Ju-Wan Lee, et al., "Investigation of random telegraph noise in gate-induced drain leakage and gate edge direct tunneling currents of high-k MOSFETs," *IEEE Transactions on Electron Devices*, vol.57, no. 4, pp. 913-918, Apr. 2010.

[7] N. V. Amarasinghe, Z. C. Butler, "Complex random telegraph signals in 0.06 um2 MDD n-MOSFETs," *Solid-State Electronics*, Vol. 44, pp. 1013-1019, 2010.

[8] Byoungchan Oh, et al., "Characterization of an oxide trap leading to random telegraph noise in gate-induced drain leakage current of DRAM cell transistors," *IEEE Transactions on Electron Devices*, vol. 58, no. 6, pp. 1741-1747, Jun. 2011.

[9] M. Schulz, "Coulomb energy of traps in semiconductor space-charge regions," *J. Appl. Phys.*, vol. 74, no. 4, pp. 2649-2657, Aug. 1993.

[10] M.-P. Lu and M.-J. Chen, "Oxide-trap-enhanced coulomb energy in a metal-oxide-semiconductor system," *Phys. Rev. B, Condens. Matter*, vol. 72, no. 23, pp. 2354171-2354175, Dec. 2005.

[11] M. Schulz and N. M. Johnson, "Evidence for multiphonon emission from interface states in MOS structures," *Solid State Commun.*, vol. 25, pp. 481-484, Feb. 1978.

[12] C. H. Henry and D. V. Lang, "Nonradiative capture and recombination by multiphonon emission in GaAs and GaP," *Phys. Rev. B*, vol. 15, no. 2, pp. 989-1016, Jan. 1977.

[13] Younghwan Son, et al., "A simple model for capture and emission time constants of random telegraph signal noise," *IEEE Transactions on Nanotechnology*, vol. 10, no. 6, pp. 1352-1356, Nov. 2011.

[14] Nuditha V. Amarasinghe, Zeynep Celik-Butler and Abdol Keshavarz, "Extraction of oxide trap properties using temperature dependence of random telegraph signals in submicron metal-oxide-semiconductor field-effect transistors," *J. Appl. Phys.*, vol. 74, no. 4, pp. 2649-2657, Aug. 2001.

[15] T. H. Ning, "Hot-electron emission from silicon into silicon dioxide," *Solid-State Electronics*, vol. 21, pp. 273-282, Oct. 1978.

978-1-4673-4840-9/13 $31.00 © 2013 IEEE

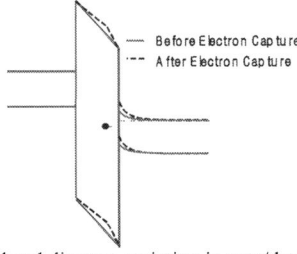

Fig. 1 Energy band diagram variation in gate/drain overlap region as electron is captured (dashed line) and emitted (solid line).

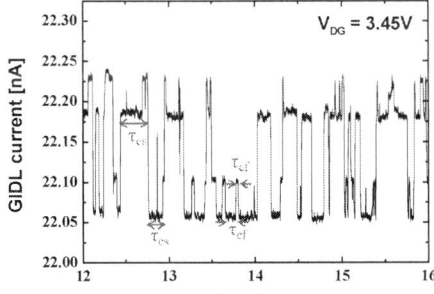

Fig. 2 Definition of time constants in four-level RTN.

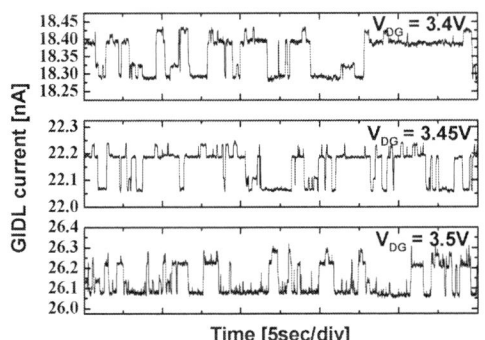

Fig. 3 Change of four-level RTN with respect to V_{DG}.

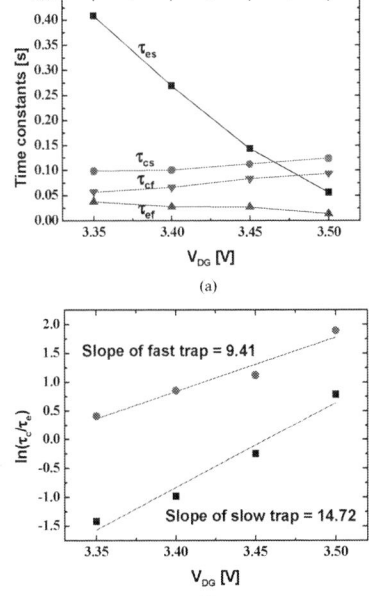

Fig. 4 (a) Time constants and (b) Time constants ratio with V_{DG}.

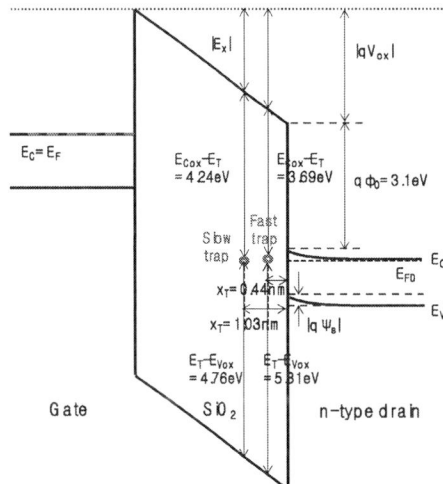

Fig. 5 Energy band diagram representing the trap depth (x_T) and energy level ($E_{Cox}-E_T$) of two traps.

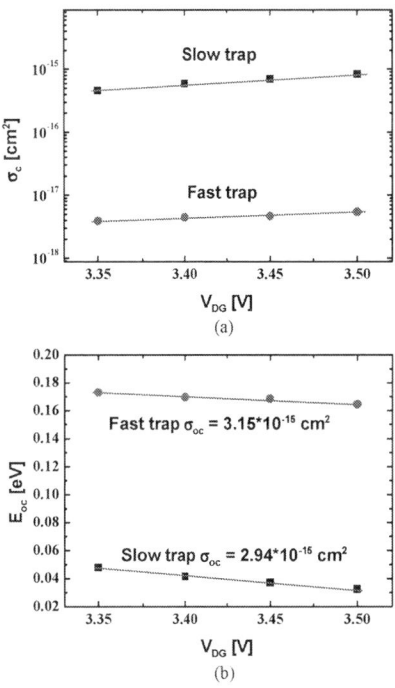

Fig. 6 (a) σ_c of GIDL RTN with V_{DG} in both slow and fast trap (b) the σ_{oc} and E_{oc} with V_{DG} in both slow and fast trap.

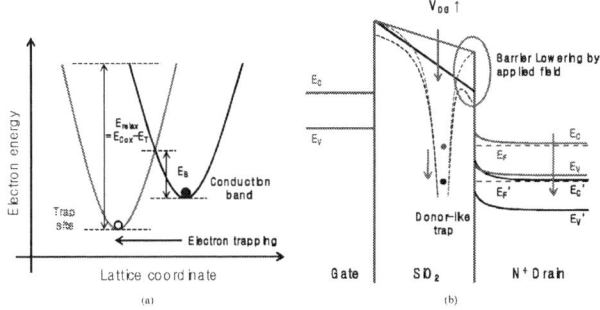

Fig. 7 (a) Lattice coordinate reconfiguration diagram (b) Schematic of potential barrier lowering by applied field in donor-like trap.

Self-Consistent Quasi Static CV Characterization of In$_x$Ga$_{1-x}$Sb Buried Channel n-MOSFET

Muhammad Shaffatul Islam[1*], Md. Nur Kutubul Alam[1] and Md. Rafiqul Islam[1]

Department of Electrical and Electronic Engineering
Khulna University of Engineering & Technology
Khulna-9203, Bangladesh
*Corresponding author : mashru.islam@yahoo.com

Abstract

The quasi-static capacitance-voltage (CV) characteristics of buried channel n-InGaSb MOSFET is investigated using SILVACO's ATLAS device simulation package. Self-consistent method is applied to solve the coupled one dimensional Schrödinger-Poisson equation taking into account of wave function penetration and strain effect. It is found that the CV profiles and threshold voltage are strongly depended on some important process parameters like oxide thickness, channel thickness, channel composition and temperature for buried channel InGaSb n-MOSFET.

INTRODUCTION

Silicon is the most widely used and established semiconductors, which dominates the electronic industry for the last 25 years. The mainstream digital electronics fabricated by CMOS approach is a method for minimizing standby power [1]. When the dimension of the silicon-based MOSFETs is scaling down to their ultimate limits, many adverse effects arise that degrade the device performance. Parasitic resistance as well as the scattering at the oxide-silicon interface results the degradation of mobility that lowers the device speed. As a result, it is impossible to operate it in the ballistic limit(corresponding to channel length approaching zero) [1]. For the limitations of Si and the quest for ever better performance as envisioned by the ITRS (International Technology Roadmap for Semiconductors) has had researchers looking into the possibilities of CMOS circuits that utilize III-V materials either alone or in hybrid form with germanium [2].

III-V materials have emerged as a promising candidate for channel material of MOSFETs. Due to excellent transport and electronic properties, III-V materials can be used in the channel of MOSFETs for future logic applications, where high performance and low power are required [3]. It is well known that InSb shows highest electron mobility and using InSb the low power and high speed quantum well devices were obtained in previous studies [4,5]. Recently, InGaSb-based surface channel MOSFET has been reported in [6] where mobility of the device was experimentally demonstrated. Self-consistent capacitance-voltage (CV) characterization including direct tunneling gate leakage current was reported for surface channel n-MOSFET in [7]. It is well established that the buried channel MOSFET shows better performance than surface channel MOSFET. To best of our knowledge there is no work

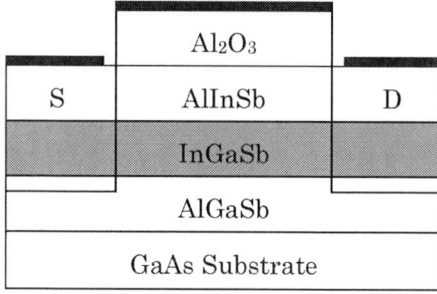

Fig 1. Proposed In$_x$Ga$_{1-x}$Sb buried channel n-MOSFET structure.

on CV characterization of buried channel MOSFET. To understand the process parameters dependence of CV characteristics and threshold voltage of buried channel n-MOSFET, the present study is performed.

PROPOSED DEVICE STRUCTURE

The buried channel structure shown in Fig. 1 consists of Al gate contact and 10nm Al$_2$O$_3$ gate dielectric. 7nm In$_x$Ga$_{1-x}$Sb channel is buried below the 3nm Al$_x$In$_{1-x}$Sb spacer layer which leads to isolate the channel from the direct contact of the oxide layer. 1µm Al$_x$Ga$_{1-x}$Sb buffer layer is used on the top of the GaAs substrate to minimize the lattice mismatch between the substrate and the channel material. In this study, quasi static CV characteristics are investigated using SILVACO ATLAS device simulation package as a function of different process parameters. For the proposed structure, the self-consistent Schrodinger-Poisson equation is solved [8] including the wave function penetration and strain effects. Doping concentration of 10^{17} cm^{-3} is used in both channel and buffer regions. To have a high quality spacer-channel interface, undoped spacer layer is used.

RESULTS AND DISCUSSION

In the present study, we have simulated gate CV characteristics of the proposed buried channel In$_x$Ga$_{1-x}$Sb n-MOSFET as a function of different process parameters like oxide thickness, channel thickness, channel composition and temperature solving coupled one dimensional Schrödinger-Poisson equation using SILVACO's ATLAS device simulation package.

978-1-4673-4840-9/13 $31.00 © 2013 IEEE

A. Effect of oxide thickness

Figure 2 shows the room temperature CV characteristics as a function of gate voltage with different oxide thicknesses. The minimum capacitance is found for all oxide thickness near the threshold voltage due to the series combination of oxide and depletion layer capacitances. Small change in gate capacitance is observed in the accumulation, depletion, and strong inversion regions with the variation of gate voltage when oxide thickness changes from 6 nm to 10 nm. The change of capacitance is found to significant in the accumulation and depletion regions and more significant in the inversion region for the smaller oxide thickness (2 nm) and inversion layer charge. To understand the oxide thickness dependent shift in threshold voltage, Fig. 3 is plotted. It is found that the threshold voltage decreases with decreasing oxide thickness. This is due to the fact that the threshold voltage varies inversely with oxide capacitance, which increases with decreasing oxide thickness.

B. Effect of strain

The effect of strain on the gate capacitance is illustrated in Fig. 4. Inversion layer charge density increases with increasing compressive strain that leads to have a strong dependence of the gate capacitance in the depletion, weak and moderate inversion regions. However, in the strong and accumulation regions, the capacitance is the same for all the compositions of the channel material. This is because, in these regions, the gate capacitance is obtained either from the contribution of accumulation region charge or inversion region charge. Further, the shift in threshold voltage is also observed with compressive strain in the channel material due to strain dependent change in band gap. Figure 5 shows the shift in threshold voltage with respect to In composition of the channel material. It is found that the threshold voltage changes from 0.01V to -0.012V for the variation of In composition from 0.2 to 0.45. It is evident from Fig. 5 that the polarity of threshold voltage depends on the indium composition due to composition dependent change in bandgap energy of the channel material.

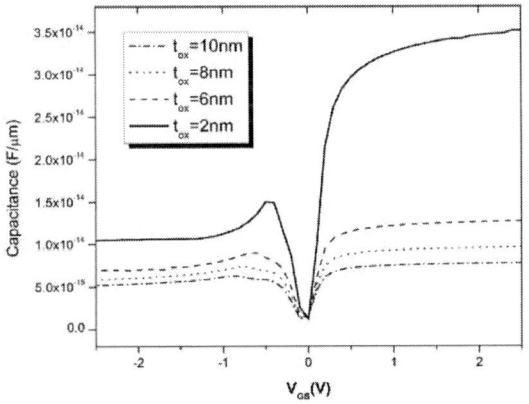

Fig. 2 Room temperature gate capacitance as a function of gate voltage for different oxide thickness and channel material $In_{0.2}Ga_{0.8}Sb$.

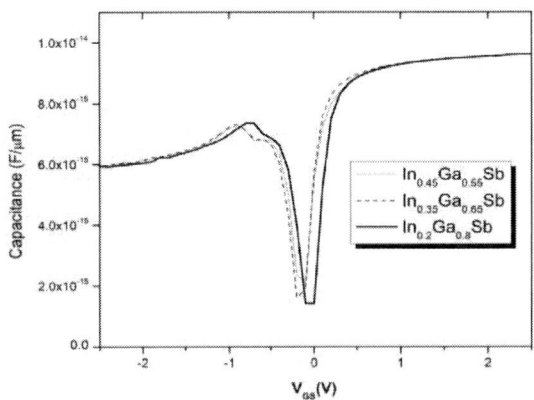

Fig. 4 Room temperature gate capacitance as a function of gate voltage for different compositions of the channel material.

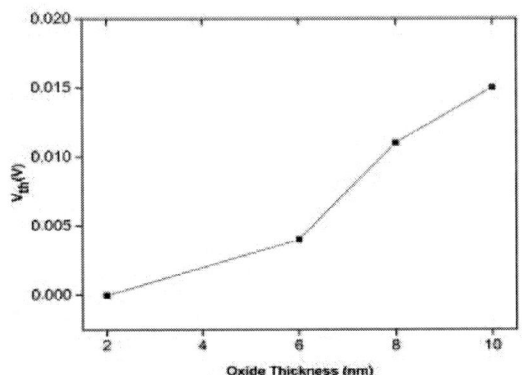

Fig. 3 Variation of threshold voltage as a function of oxide thickness at room temperature.

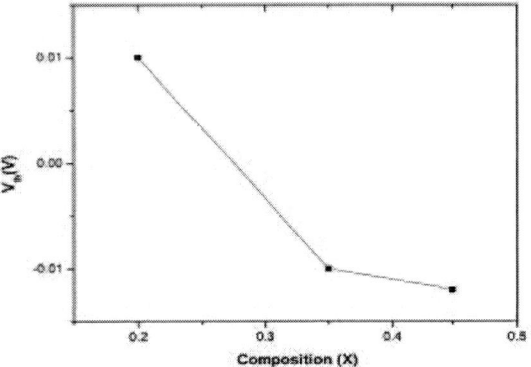

Fig. 5 Variation of threshold voltage as a function of Indium composition at room temperature.

C. Effect of Temperature

It is observed in Fig. 6 that the capacitance decreases significantly with increasing temperature in the accumulation and deletion regions, which may be due to the temperature dependent bandgap energy of the channel material. In contrast, almost constant capacitance is obtained in the strong inversion region for the temperatures 100K and 200K, although gate capacitance decreases when temperature rises to 300K. This may be due to considerable decreasing bandgap energy at this temperature compared to100K and 200K. It is also found in Fig. 6 that the capacitance changes significantly with temperature in the weak and moderate inversion regions. The shift in threshold voltage is plotted as a function of temperature in Fig. 7. It is found that the threshold voltage decreases with increasing temperature.

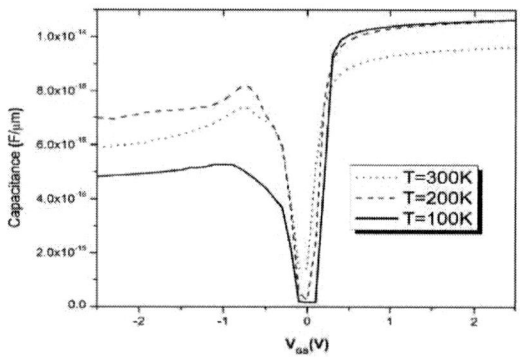

Fig. 6 Gate capacitance as a function of gate voltage for different temperatures when channel composition $x = 0.2$.

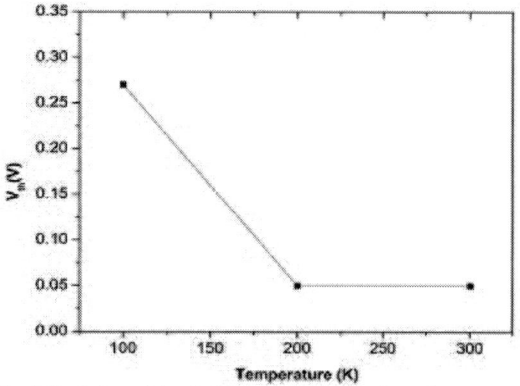

Fig. 7 Variation of threshold voltage as a function of temperature.

D. Effect of channel thickness

It is found in Fig. 8 that the channel thickness have almost no impact on CV profiles over the entire range of gate voltages. Because, the gate capacitance has no relationship with channel thickness, rather, it mainly depends on the dynamic change of charges in the accumulation, depletion and inversion regions with gate voltage. Further, in Fig. 8, the minimum capacitance is found near the threshold voltage and the maximum in the strong inversion region.

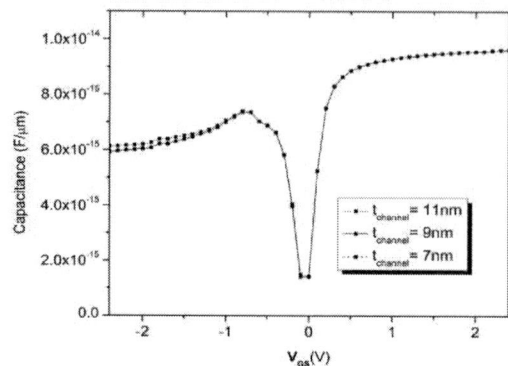

Fig. 8 Gate capacitance as a function of gate voltage for different channel thicknesses with channel composition $x = 0.2$

CONCLUSION

Finally, the influence of process parameters on the CV characteristics are simulated for InGaSb buried channel n-MOSFET. It is concluded that the CV profiles and the shift in threshold voltage strongly depend on the process parameters except the channel thickness of the device. The results obtained in the present study suggest that careful selection of the process parameters lead to have high performance MOS devices.

REFERENCES

[1] Farzin Assad, Zhibin Ren, Dragcia Vasileska, Supriyo Datta, Mark Lundstrom "On the Performance Limits for Si MOSFETs : A Theoretical Study" *IEEE Transactions On Electron Devices* ,vol. 47,No 1, pp.232-240 January 2000.

[2] See,e.g, Heyns M,T sai w,editors.MRS Bulletin entiteled: "Ultimate Scaling of CMOS logic devices with Ge and III-V materials:" July 2009.

[3] Annesh Nainani et al "In$_x$Ga$_{1-x}$Sb channel p-metal–oxide semiconductor filed effect transistors: Effect of strain and heterostructure design*" Journal of Applied Physics* 110,014503, (2011).

[4] M.Radosavljevic et al "High performance 40 nm gate length InSb p-channel compressively strained quantum well field effect transistors for low power (Vcc=0.5v) logic applications" . *International Electron Devices Meeting*, pp.1-4,(2008).

[5] S.Datta et al " 85 nm gate length enhancement and depletion mode InSb quantum well transistor for ultra high speed and very low power logic applications". *Technical Digest. International Electron Devices Meeting*, pp.763 - 766 (2005).

[6] Ze Yuan, Nainani, A. ; Bennett, B.R. ; Boos, J.B. ; Ancona, M.G.;Saraswat, K.C. "Heterostructure design and demonstration of InGaSb channel III-V CMOS transistors". *International Semiconductor Device Research Symposium* ,pp.1-2,(2011).

[7] Md. Hasibul alam et al , "In$_x$Ga$_{1-x}$Sb MOSFET: Performance Analysis by Self Consistent CV Characterization and Direct Tunneling Gate Leakage Current" *IEEE International Conference on Electro / Information Technology,*pp.1-6,(2012).

[8] F. Stern, "Self-Consistent results for n-type si inversion layers," *Phy. Rev. B,* vol. 5, pp. 3175-3183,1986.

Investigation of Conduction Mechanism in Ti/Si$_3$N$_4$/p-Si stacked RRAM

Sunghun Jung[1], Sungjun Kim[1], Jeong-Hoon Oh[2], Kyung-Chang Ryoo[2], Jong-Ho Lee[1], Hyungcheol Shin[1], and Byung-Gook Park[1]

[1]Inter-university Semiconductor Research Center (ISRC) and

School of Electrical Engineering and Computer Science, Seoul National University,

San 56-1, Sillim-dong, Gwanak-gu, Seoul 151-742, Republic of Korea.

[2]DRAM Process Architecture Team, Memory Division, Semiconductor Business, Samsung Electronics Co., Ltd.,

Nongseo-dong, Giheung-gu, Yongin-si, Gyeonggi-Do, Korea, 445-701

Tel.: +82-2-880-7279, Fax: +82-2-882-4658, E-mail address: jsh1127@snu.ac.kr

ABSTRACT

The conduction mechanism in Ti/Si$_3$N$_4$/p-Si memory stack is described. In order to analyze the conduction mechanism, we measured the *I-V* characteristics in voltage sweep mode and conducted *I-V* curve fitting. And the temperature dependence in Ti/Si$_3$N$_4$/p-Si stacked cell is also investigated because we cannot claim the conduction mechanism just based on the *I-V* curve fitting. From *I-V* curve fitting and temperature measurement data, we found that space charge limited conduction (SCLC) model is well fitted in both high resistance state (HRS) and low resistance state (LRS).

(Keywords: RRAM, conduction mechanism and Si$_3$N$_4$)

I. INTRODUCTION

As the demand for high-density memory device increases and charge-based memory technology such as NAND Flash memory suffers from scalability problems, resistive random access memory (RRAM) has attracted much attention as one of the most promising candidates replacing the conventional memory due to its simple structure and good scalability property. Although transition metal oxides such as TaO$_2$, HfO$_2$, etc. are widely used as resistive switching material of RRAM stack, optimized RRAM stack is still debatable due to uniformity problem. It is because conduction and resistive switching mechanisms is unclear [1, 2].

In this paper, we investigate *I-V* characteristics and conduction mechanism of the silicon nitride (Si$_3$N$_4$)-based RRAM device with Ti/Si$_3$N$_4$/p-Si stack. The resistive switching of Si$_3$N$_4$-based RRAM is related to not oxygen vacancy but nitride-related traps [3]. The analysis of trap-related conduction mechanism is important to improve the memory performance in Si$_3$N$_4$-based RRAM.

II. FABRICATION

Fig. 1 shows the schematic diagram of fabricated Ti/Si$_3$N$_4$/p-Si RRAM cell. Fabrication process is as follows. First, sacrificial oxide was grown by 10 nm using dry oxidation. Then, in order to form the heavily doped p-Si bottom electrode, BF$_2$ was implanted into bulk Si wafer at a dose of 10^{15}/cm^2 and at implantation energy of 40 keV and rapid thermal annealing was performed at 1000 ℃ for 10 sec. After sacrificial oxide strip, Si$_3$N$_4$ film with 10 nm thickness was deposited as resistive switching material using low pressure chemical vapor deposition. In order to form the top electrode, a 100 nm thick titanium layer was deposited by sputter and patterned by using a photolithography and etching. A diameter of top electrode is 100 μm. Our device has simple metal – insulator - semiconductor (MIS) structure with titanium (Ti) top electrode and p-type doped silicon (p-Si) bottom electrode. CMOS-friendly metal is adopted without noble metals such as platinum and ruthenium. Memory stack materials can be easily adapted to 3D RRAM structure for high density memory application. For measurement, we applied positive or negative bias to the top electrode in the *I-V* sweep mode while the bottom electrode was grounded as common.

III. RESULTS AND DISCUSSION

Fig. 2 shows the typical bipolar resistive switching characteristics of Ti/Si$_3$N$_4$/p-Si RRAM. The set and reset processes occur under positive and negative biases, respectively. And forming-free behavior can be seen.

At first, in order to investigate the conduction mechanism, *I-V* fitting is performed. Figure 3 shows the fitting results about Schottky emission and Poole-Frenkel (PF) models. In low resistance state (LRS), *I-V* curves do not fit with both of two models. In high resistance state (HRS), however, it seems to follow the both models.

For more detail analysis, temperature dependence on current

in HRS is investigated by varying from 25 ℃ to 150 ℃. From temperature measurement, we found that the HRS current has a weak dependence on the temperature. Therefore, we can rule out the possibility of Schottky emission. As shown in Fig. 4 (a), we plot ln(I/V) vs. 1/kT curve to extract activation energy (E_a) for PF model fitting [4]. The activation energy can be extracted from the slope at each bias. From the activation energies, trap energy level can be extracted by extrapolating from the E_a vs. sqrt(V) plot in Fig. 4 (b). The values of trap energy (E_t) in positive and negative biases are 0.14 and 0.16 eV, respectively. The slope of Fig. 4 (b) indicates dielectric constant of Si_3N_4. The value of the extracted dielectric constant is about 100, which is much higher than the known dielectric constant about 7 for Si_3N_4. This discrepancy explains that PF model does not meet the conduction mechanism in HRS [4].

Fig. 5 shows the fitting results about space charge limited conduction (SCLC) in HRS. In low voltage regime, the slope retains 1. As the voltage increases, the slope changes from 1 to 2 in SCL region. *I-V* fitting results and a weak dependence of current on temperature support that SCLC is dominant conduction mechanism in HRS [5]. Fig. 6 shows the fitting results about SCLC in LRS. In SCL region, *I-V* relationship is well fitted with exponential function. It reveals that traps are generated during set process and distributed uniformly in energy below the conduction band [6].

IV. CONCLUSION

The conduction mechanism in $Ti/Si_3N_4/p$-Si memory stack is investigated from *I-V* fitting and temperature measurement. We found that trap controlled SCLC is the most promising conduction mechanism in $Ti/Si_3N_4/p$-Si RRAM cell.

ACKNOWLEDGEMENT

This work was supported in part by the Smart IT Convergence System Research Center funded by the Ministry of Education, Science and Technology as Global Frontier Project (CISS-2011-0031845) and in part by the IT R&D program of MKE/KEIT. (10035320, Development of novel 3D stacked devices and core materials for the next generation flash memory)

REFERENCES

[1] Lin Yang, *et al., Appl. Phys. Lett.* **95**, 013109 (2009).

[2] B. Gao, *et al., IEDM Tech. Dig.*, 2011, pp. 417-420.

[3] Hee-Dong Kim, *et al., Semicond. Sci. and Technol.* **25**, 065002 (2010).

[4] Shimeng Yu, *et al., Appl. Phys. Lett.* **99**, 063507 (2011).

[5] Kyung Min Kim, *et al., Nanotechnology*, **22**, 254010 (2011).

[6] A. Rose, *Phys. Rev.*, Vol. 97, No. 6, Mar., 1955

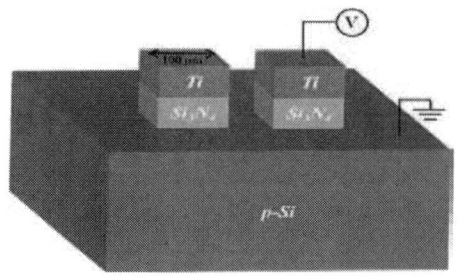

Fig. 1. A schematic diagram of $Ti/Si_3N_4/p$-Si stacked RRAM cells.

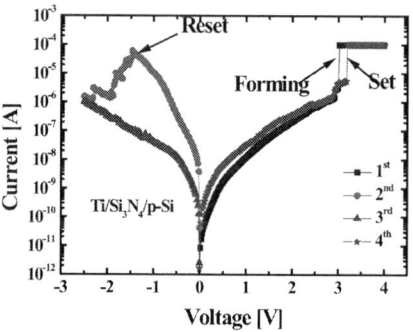

Fig. 2. Typical *I-V* curves of $Ti/Si_3N_4/p$-Si stacked RRAM device showing bipolar resistive switching behavior.

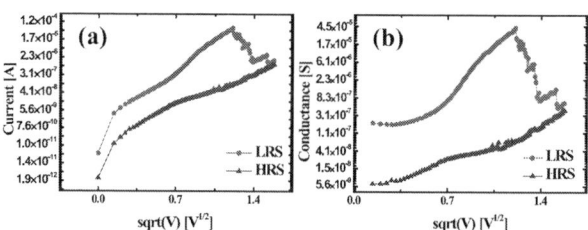

Fig. 3. Fitting results about (a) Schottky emission model and (b) Poole-Frenkel model.

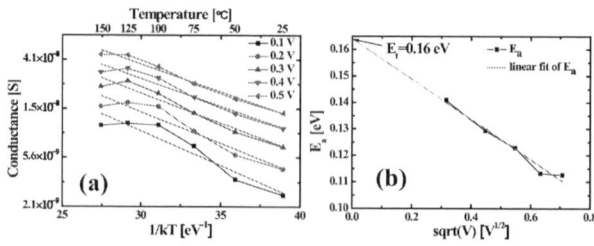

Fig. 4. The plots of (a) conductance as a function of temperature to extract activation energy and (b) activation energy as a function of read voltage to extract the trap energy level

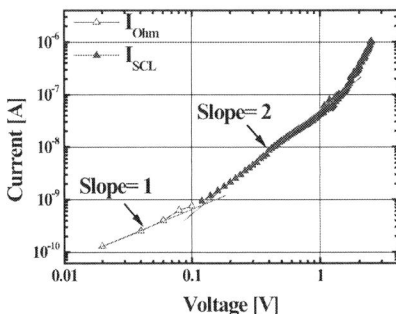

Fig. 5. SCLC fitting result in HRS.

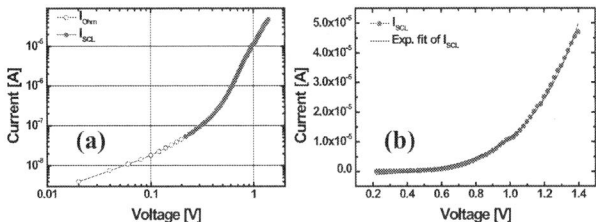

Fig. 6. (a) SCLC fitting result in LRS and (b) exponential function fitting in SCL region.

Experimental Demonstration of Fano Resonance in Microfabricated Phononic Crystal Resonators Based on Two-dimensional Silicon Slab

Nan Wang[1,2], Fu-Li Hsiao[1,3], J.M. Tsai[2], Moorthi Palaniapan[1], Dim-Lee Kwong[2], and Chengkuo Lee[1]

[1] Department of Electrical and Computer Engineering, National University of Singapore, Singapore 117576
[2] Institute of Microelectronics, A*STAR (Agency for Science, Technology and Research), Singapore 117685
[3] Graduate Institute of Photonics, National Changhua University of Education, Taiwan, R.O.C.
Phone: +65 6516-5865; Fax: +65-677911034; E-mail:elelc@nus.edu.sg

Abstract

In this paper, we demonstrate the experimental study of Fano resonance in a microfabricated phononic crystal (PnC) resonator by creating defects on a two-dimensional (2-D) silicon PnC slab. Our experimental results show that for this type of PnC resonator, the local resonance mechanism plays a dominant role in the acoustic wave transmission, rather than the non-resonant propagating mechanism.

Keywords: Phononic crystal, resonator, and Fano resonance

Introduction

Phononic crystals (PnCs), also known as Phononic Band Gap Materials (PnBGs), have received much attention recently due to its renewed ability to control and manipulate acoustic waves. PnCs are the acoustic wave equivalent of photonic crystals, where scattering inclusions arranged periodically in a homogeneous host material enable certain frequencies to be completely reflected by the structure [1-3]. Various PnC band gap structures, such as cylindrical pillars inside air holes [4, 5], cylindrical rods imbedded inside the membrane [6], as well as inverse acoustic band gap (IABG) structure [7] have been studied because the 2-D nature of PnC slabs guide the acoustic wave within the slab with better confinement of elastic energy. When defects are created by removing some of the scattering inclusions, resonant cavities can be formed and resonant peaks can be observed.

The Fano resonance, after its discovery in 1961[8], has been widely studied in various types of metamaterials[9]. Due to its excellent sensitivity to changes in geometry or local environment, i.e., significant changes in resonance or line shape for small amount of disturbance to the local environment, it has been applied in the development of chemical and biological sensors. So far, the Fano resonance in microfabricated PnCs has not been experimentally reported. In this paper, we experimentally report the Fano resonance in 2-D silicon PnC micromechanical resonators which operate in the radio frequency range. The PnC structure consists of air holes arranged in a square lattice on a 2-D silicon slab. The PnC resonator is formed by removing three lines of air holes at the centre of the PnC region, forming a Fabry-Perot resonant cavity.

Design and Modeling

The inset of Fig. 1 sketches the unit cell of our phononic crystal structure where the lattice constant (a) is 18.18μm, the thickness of the silicon slab (d) is 10μm and the radius of the air holes (r) is 8.18μm. The above geometric parameters are optimized in order to obtain the widest band gap in the frequency domain [2] under the limitation of our fabrication capability. The band structure shown in Fig. 1 reveals that a band gap exists between the two red horizontal lines, extending from 143.3MHz to 186.3MHz, which renders the gap-to-midgap frequency ratio to be 26.1%.

When defects are created on an otherwise perfect PnC, a resonant peak appears within the originally forbidden band gap. As such, a mechanical resonator can be formed. The PnC resonator that we designed here is a cavity-mode Fabry-Perot resonator formed by removing three rows of air holes at the center of the active 2-D PnC region. When air holes are removed, a Fabry-Perot resonant cavity can be formed, thus various resonant modes can be supported within the PnC structure.

Fabrication

The device is fabricated on a silicon-on-insulator (SOI) substrate. The thickness of device layer and the buried oxide (BOX) layer is 10μm and 1μm, respectively. 1μm of aluminum nitride (AlN) is deposited followed by metallization of 0.5μm aluminum (Al) as the top electrode. The deposited AlN layer and top Al electrode were patterned subsequently using dry etching with a combinational gas of $Cl_2/BCl_3/Ar$ in the reactive ion etching (RIE) tool. This will form the interdigital transducers (IDT) which act as the input and output of the acoustic waves on the configured AlN layer. After that, a two-step deep reactive ion etching (DRIE) is performed, one on the front side and the other from the rear side of the wafer, forming cylindrical air holes of the square lattice of PnCs in the silicon device layer. The latter will release the structure from the bulk substrate, resulting in a free-standing 2-D silicon PnC slab. Lastly, the device is fully released by hydrofluoric (HF) vapour that etches away the BOX underneath the PnC slab.

Fig. 2(a) shows a microfabricated PnC structure of 10 periods (rows) of air holes, with IDT electrodes formed by Al on the two sides of the PnC structure. In Fig. 2(b), the resonators are formed by removing 3 periods of air holes from the centre, i.e., four periods (rows) of air holes are patterned at each side of the Fabry-Perot resonant cavity.

Experimental Characterization and Discussion

A typical experimentally measured transmission spectrum of the pure Fabry-Perot resonator (Fig. 2b) is shown in Fig. 3 by red circles. The vertical axis indicates the normalized

978-1-4673-4840-9/13 $31.00 © 2013 IEEE

transmission by percentage. A symmetric resonant peak is observed. The resonant frequency is 152.75MHz which is in good agreement with the FEM simulated results. This symmetric profile of Fig. 3 can be explained by Fano quantum interference effect[8]. Similar effect has been observed in locally resonant material for elastic wave [10]. The effect arises when the elastic wave propagate through a structure via both resonant and non-resonant ways. The interference between these two pathways leads to the symmetric profile. The detailed physical picture can be explained by the mode profiles which we will discuss in a later paragraph. The measured data can be fitted with the Fano line shape:

$$T(f) = A + B\frac{[q + 2(f - f_0)/\Gamma]^2}{1 + [2(f - f_0)/\Gamma]^2} \quad (1)$$

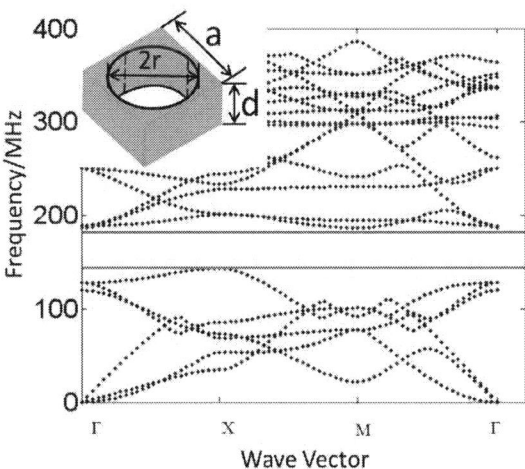

Fig. 1: The band structure of the phononic crystal structure. A unit cell for numerical calculation of the band structure is shown in the inset. a is the lattice constant, d is the thickness of the silicon slab and r is the radius of the air holes. In our study, we use d=10μm, r=8.18μm and a=18.18μm, which gives r/a=0.45 and d/a=0.55. A band gap from 143.3MHz to 186.3MHz can be clearly seen from the band structure, which renders the gap-to-midgap frequency ratio to be 26.1%.

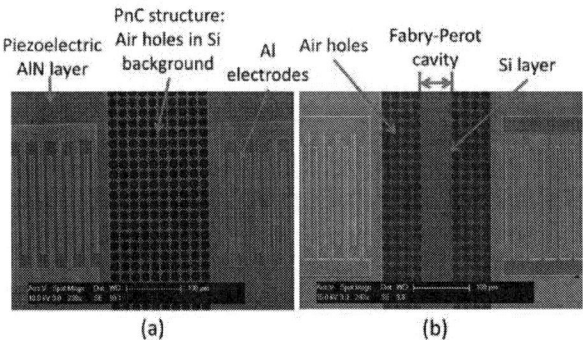

Fig. 2: SEM images of the microfabricated devices. (a) A PnC structure of 10 periods (rows) of air holes, with IDT formed by Al on the two sides of the PnC structure. (b) PnC micromechanical resonator formed by removing 3 lines of air holes from the centre.

where A and B are constants, f_0 is the resonant frequency and Γ is the resonant line width. The dimensionless parameter q,

which may be positive or negative, indicates the amplitude ratio between the resonant and the non-resonant transmission. When the amplitudes of the resonance and the non-resonance are comparable, i.e. the value of q is about unity, the resonant profile is strongly asymmetric. In Fano quantum interference effect, $q \gg 1$ means that the amplitude of resonant scattering is much stronger than that of non-resonant ones. On the other hand, $q \ll 1$ means the non-resonant scattering dominates in the transmission. In the case of Fig. 3, the observed transmission peak for the pure Fabry-Perot resonator is almost symmetric. By fitting the experimental data to equation (1), q = -0.063 is obtained, i.e., only one resonance dominates the transmission. The estimated quality factor is 693.

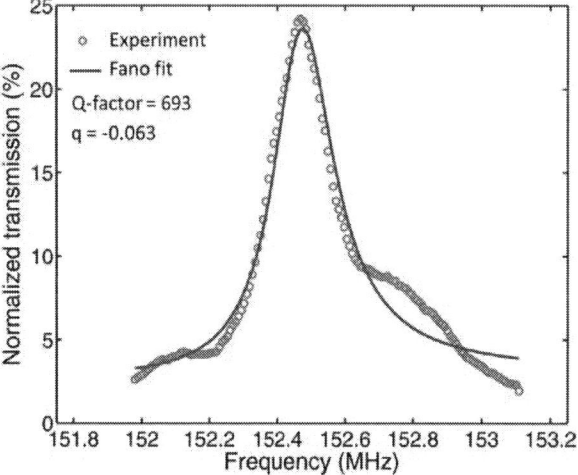

Fig. 3: The transmission spectrum of the pure Fabry-Perot resonator shown in Fig. 2b. The vertical axis indicates the normalized transmission by percentage. Experimental data are shown in red circles while the blue line indicates the Fano fitting curve using equation (1) to the experimental data. The dimensionless parameter q, which may be positive or negative, indicates the amplitude ratio between the resonant and the non-resonant transmission.

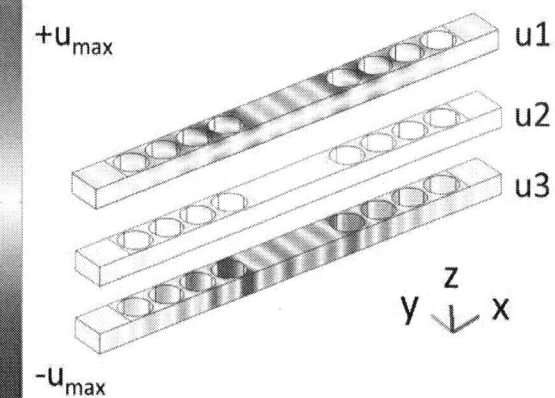

Fig. 4: Simulated mode profiles of displacement of the Fabry-Perot PnC resonator. The u1, u2 and u3 represent the displacement vector components in x, y, and z directions, respectively. The colour bar indicates the amplitude of displacements in an arbitrary unit.

In order to clarify the physical picture for the Fabry-Perot resonator, the displacements of steady resonant modes for Fig.

3 are shown in Fig. 4. FEM is adopted to calculate the modes in each unit cell in Fig. 4 with periodic boundary along y direction. As the elastic waves in the silicon plate propagate by the interactions among the silicon atoms when they are displaced from their equilibrium positions, the energy stored in any solid structure is then associated with the displacements of the silicon atoms within the silicon plate. Thus, by analyzing the displacements of all the silicon atoms within the silicon plate (mode profiles of displacement), we can get information about the energy distribution along the structure. In Fig. 4, the u1, u2 and u3 represent the displacement vector components in x, y, and z directions, respectively. The colour bar indicates the amplitude of displacements in an arbitrary unit. We observe that for our designed PnC resonator, the displacement vector components in x and z directions are concentrated at the central defect region. With the advantage of good confinement of the elastic energy by the phononic structure surrounding the central Fabry-Perot resonant cavity in the proposed design, a high Q factor is expected to be achieved. From the aspect of the mode type, a typical Fabry-Perot resonant mode is shown. With the elastic energy concentrating at the central defect region uniformly and the polarizations are in X and Z direction.

Conclusions

In conclusion, we investigated theoretically and experimentally the performance of the micromechanical resonators based on two-dimensional silicon PnC slab. The PnC resonator is formed by creating a Fabry-Perot resonant cavity on an otherwise perfect PnC structure. The Fabry-Perot resonant cavity is formed by removing three rows of air holes on the PnC. Our experimental result reveals a symmetric resonant profile for this type of resonator, with a very small q value obtained after the Fano fitting, indicating that the local resonance mechanism plays a dominant role in the acoustic wave transmission, rather than the non-resonant propagating mechanism. The displacements of steady resonant modes are also analyzed to explain the modes which are supported by this type of resonator from a physical perspective.

Acknowledgements

This work was partially supported in research grants of SERC 1021650084 (R-263-000-643-305), SERC 1021010022 (R-263-000-612-305) at the National University of Singapore.

References

[1] S. Mohammadi, A. A. Eftekhar, A. Khelif *et al.*, "Evidence of large high frequency complete phononic band gaps in silicon phononic crystal plates," *Applied Physics Letters*, vol. 92, no. 22, Jun, 2008.

[2] S. Mohammadi, A. A. Eftekhar, A. Khelif *et al.*, "Complete phononic bandgaps and bandgap maps in two-dimensional silicon phononic crystal plates," *Electronics Letters*, vol. 43, no. 16, pp. 898-899, August, 2007.

[3] T. T. Wu, L. C. Wu, and Z. G. Huang, "Frequency band-gap measurement of two-dimensional air/silicon phononic crystals using layered slanted finger interdigital transducers," *Journal of Applied Physics*, vol. 97, no. 9, May, 2005.

[4] I. El-Kady, R. H. Olsson, and J. G. Fleming, "Phononic band-gap crystals for radio frequency communications," *Applied Physics Letters*, vol. 92, no. 23, Jun, 2008.

[5] F.-L. Hsiao, A. Khelif, H. Moubchir *et al.*, "Waveguiding inside the complete band gap of a phononic crystal slab," *Physical Review E*, vol. 76, no. 5, pp. 056601, Nov, 2007.

[6] R. H. Olsson, I. F. El-Kady, M. F. Su *et al.*, "Microfabricated VHF acoustic crystals and waveguides," *Sensors and Actuators a-Physical*, vol. 145, pp. 87-93, Jul-Aug, 2008.

[7] K. Nai-Kuei, Z. Chengjie, and G. Piazza, "Demonstration of inverse acoustic band gap structures in AlN and integration with piezoelectric contour mode wideband transducers." pp. 10-13.

[8] U. Fano, "EFFECTS OF CONFIGURATION INTERACTION ON INTENSITIES AND PHASE SHIFTS," *Physical Review*, vol. 124, no. 6, pp. 1866-&, 1961.

[9] B. Luk'yanchuk, N. I. Zheludev, S. A. Maier *et al.*, "The Fano resonance in plasmonic nanostructures and metamaterials," *Nature Materials*, vol. 9, no. 9, pp. 707-715, Sep, 2010.

[10] C. Goffaux, J. Sanchez-Dehesa, A. L. Yeyati *et al.*, "Evidence of Fano-like interference phenomena in locally resonant materials," *Physical Review Letters*, vol. 88, no. 22, Jun, 2002.

ZnS/ZnO/ZnS Core/Shell/Shell Hollow Spheres and Its Optical Properties

Yi-Ting Hung* and Shih-Shou Lo

Department of Photonics, Feng Chia University

No.100, Wenhwa Road

Taichung ,Taiwan ,R.O.C.

e-mail:M0020038@fcu.edu.tw

Abstract

ZnS/ZnO/ZnS core/shell/shell (CSS) hollow spheres have been successfully developed. The ZnO CSS hollow structures were prepared by immerging ZnO hollow spheres in Na_2S aqueous solution.Photoluminescence spectroscopy of the as-synthesized ZnO is found that the UV emission energy of free exciton recombination evidently shifts to the high energy with the longer reaction times. The effect of the ZnS on the optical properties of as-synthesized nanostructure was presented.

Key words: ZnO, ZnS, core/shell/shell

Introduction

ZnO has some very attractive properties, including high transparency in the visible wavelength, a high piezoelectric constant, and a large electro-optic coefficient [1]. Several recent studies have examined the fabrication of ZnO films, which have great potential and have attracted a considerable amount of interest due to their potential application in solar cells, gas sensors, piezoelectric transducers and varistors [2–7].

ZnS is also an important member of group II-VI semiconductor phosphor with wide band gap energy of 3.7 eV, have also attracted much research interest due to its potential applications in optoelectronic device such as solar cells and infrared windows. ZnS has application in the fields of flat display, sensors and lasers [8-10].

Recently, much effort has been devoted to designing and controlling fabrication of nanostructured materials with optical properties. In some reports, the hybrid or core/shell composite materials demonstrate novel properties. Therefore, surface modification of nanometer sized inorganic core with different inorganic shell to form core-shell type nanostructures has become an important route to functional nanomaterials. Such modification has brought about interesting physical and chemical properties of the nanostructured materials that have shown important technological application. Li et al. reported increased ultraviolet (UV) emission from ZnS-coated ZnO nanowires fabricated by a self-assembling method [11]. Zhu et al. prepared ZnO–ZnS core-shell microspheres by a solution method through the chemical conversion of spherical templates and reported an obvious enhancement of the UV emission [12].

In this study, we report a new ZnS/ZnO/ZnS core/shell/shell (CSS) hollow nanostructure. The structural properties were characterized by Energy Dispersive Spectrometer (EDS), X-ray Diffraction (XRD), high-resolution transmission electron microscopy (HRTEM) and the related luminescent properties were analyzed by room- temperature photoluminescence (PL) and Raman spectra.

Results and Discussions

Fig.1 (A) shows the field emission SEM images of ZnO hollow structures, the diameter of the hollow sphere is around 500 nm. Fig.1 (B) shows the ZnS/ZnO/ZnS (CSS) hollow structures. The chemical composition of the nanostructures is investigated by EDS, as shown in Fig.1 (C) and (D). In the EDS spectrum of the nanostructure, it reveals the composite of O, Zn, S elements, providing powerful evidence for successful incorporation of S element into ZnO spheres.

A typical XRD pattern is shown in Fig.2, in which the diffraction peaks corresponding to both ZnO and ZnS can be clearly seen. From Fig.2, deduce the CSS hollow structured composites are composed of crystalline ZnS and ZnO.

To detail investigate the structure characterizations of the as-synthesized sample, the high resolution transmission electron micro-spectroscopy are used. Fig. 3(A)-(B) shows a representative HR-TEM image of the CSS nanostructures, and its corresponding SAED pattern (Fig.3 (B) inset). The materials of ZnS nanoparticles coating on the surface of ZnO hollow sphere are clearly displayed in Fig. 3(A) and (B). The apparent contrast between the inner and the outside layer suggests the existence of a coaxial structure, while the interface is fairly sharp and there appears to be no intermediate layer in the HR-TEM image of ZnS/ZnO/ZnS core/multi-shells. It can be seen that the ZnS nanoparticles grows on ZnO hollow sphere, and this result is also consistent with XRD. From the HRTEM image, we can see that ZnS nanoparticles (about 50nm) are well distributed on the surface of ZnO. The SAED pattern (Fig. 3(B) inset) demonstrated the details of the nanostructure, and the concentric rings could be assigned as diffractions from the (100), (002) and (103) planes of ZnS, (100) plane of ZnO.

Fig.4 presents the room-temperature photoluminescence spectra (PL) of the as-synthesized nanostructures at treatment times 12 hours. In comparison of the PL intensity, the bare ZnO hollow spheres are also displayed in this figure. The PL spectrum of ZnO hollow spheres shows a strong ultraviolet (UV) emission peak at 383 nm and a broad green emission band centered at 530 nm. The origin of 383 nm is attributed to the free exciton recombination of ZnO. The green emission originates from the defects of the ZnO. In Fig.4, the PL spectrum of as-synthesized sample reveals an UV emission peak with small blue shift about ~6 nm. A weaker side band 420 nm was observed. The blue emission 420 nm can be explained by the surface state of ZnS.

The Fig. 5 shows the Raman spectra of the ZnO hollow and CSS hollow structures. Spectra regard to the main peak at 435 cm^{-1} in the bulk ZnO. CSS hollow structures have the same peak at 254cm^{-1} and 343^{-1} in different sulfurization times. We find the main peak of CSS structure has small red shift that change with sulfurization times.

References

[1] Polla D L, Muller R S and White R M 1996 IEEE *Electron Device Lett.***7** 254

[2] Beek W J E, Wienk M M and Janssen R A J 2006 Adv. Funct. 16 1112

[3] Wang Z L and Song J 2006 Science 312 242.

[4] Wang J X, Sun X W, Yang Y, Huang H, Lee Y C, Tan O K and Vayssieres L 2006 Nanotech. 17 4995

[5] Liu H Y, Kong H, Ma X M and Shi W Z 2007 J. Mater. Sci. 42 2637

[6] Gupta V and Mansingh A 1996 *J. Appl. Phys.* **80** 1063

[7] Bertolotti M, Laschena M V, Rossi M, Ferrari A, Qian L S,Quaranta F and Valentini A 1990 J. Mater. Res. 5 1929

[8] W. Chen, Z. Wang, Z. Lin, L. Lin, 1997, Appl. Phys. Lett. 70, 1465.

[9] C. Falcony, M. Gareia, A. Ortiz, J. C. Alonso, J. Appl, Phys. (1992), 72, 1525

[10] M. Bredol, J. Merikhi, 1998, J. Mater. Sci. 471.

[11] Li JH, Zhao DX, Meng XQ, Zhang ZZ, Zhang JY, Shen DZ, Lu YM, Fan XW (2006) Enhanced ultraviolet emission from ZnS-coated ZnO nanowires fabricated by selfassembling method. J Phys Chem B 110:14685–14687

[12] Zhu YF, Fan AH, Shen WZ (2008) A general chemical conversion route to synthesize various ZnO-based core/shell structures. J Phys Chem C 112:10402–10406

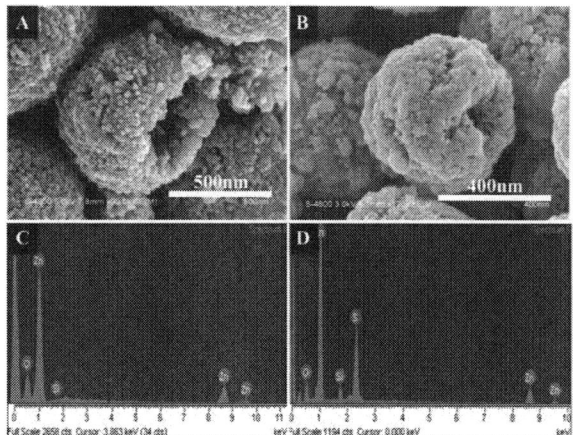

Figure1. (A) FE-SEM images of ZnO hollow spheres. (B)High magnification FE-SEM of image of ZnO/ZnS/ZnS core/multi-shells structure with sulfurization 12 hrs. (C) EDS spectrum of ZnO hollow structure and (D) for various sulfurization time.

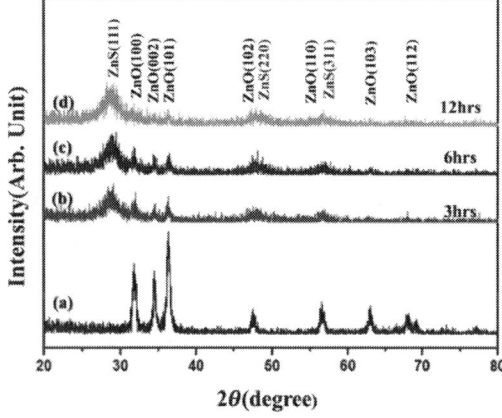

Figure 2. XRD patterns of ZnO hollow spheres with various sulfurization times.

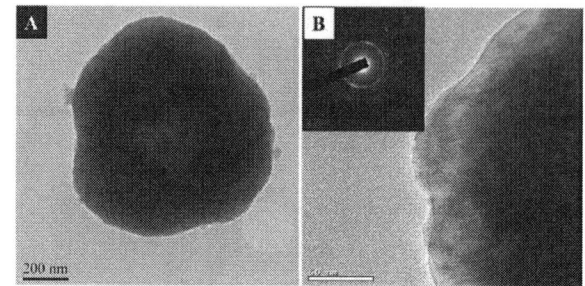

Figure 3.(A) Low magnification HR-TEM images of as synthesized nanostructure (B) High magnification HR-TEM images of ZnO hollow nanostructure with sulfurization time 12 hrs. The select area electron diffraction pattern was shown in the inset of Figure (B).

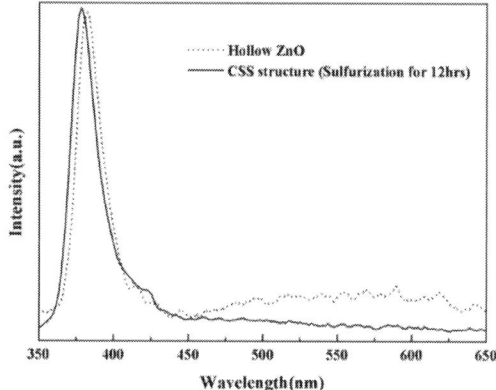

Figure 4. Room temperature PL spectra of ZnO hollow sphere and CSS hollow structure (sulfurization for 12hrs).

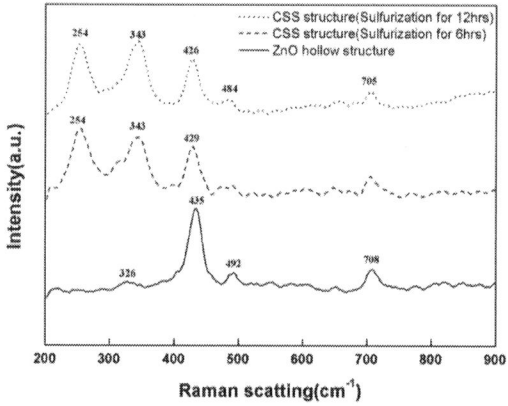

Figure 5.Raman scatting spectra of ZnO nanostructure for different sulfurization times.

Selenium Surface Energy Determination from Size-Dependent Considerations

G. Guisbiers[1,*], S. Arscott[2], M. Gaudet[2], A. Belfiore[3], R. Snyders[1,3]

[1] University of Mons
Avenue Nicolas Copernic, 1
7000 Mons, Belgium
[2] Institut d'électronique, de microélectronique et de nanotechnologies (CNRS UMR-8520)
Avenue Poincaré
Villeneuve d'Ascq, France
[3] Materia Nova, R&D research center
Avenue Nicolas Copernic, 1
7000 Mons, Belgium
*gregory.guisbiers@physics.org

Abstract

Nanothermodynamics is used successfully to calculate the size and shape dependencies of several material properties. The goal of this paper is to prove that the proposed nanothermodynamics model can also be used to predict bulk surface energies. To check this prediction, the bulk solid surface energy of selenium has been measured using the sessile drop technique. The experimental value, 0.291±0.025 J/m², is found to be in excellent agreement with the theoretical prediction, 0.285±0.022 J/m².

Keywords : Size-dependency, Shape-dependency, Surface tension, Nanothermodynamics, Selenium.

Introduction

Understanding how materials behave at tiny length scales is crucial for developing future nanotechnologies. The advances in nanomaterials modeling coupled with new characterization tools are the key to study new properties and capabilities and then to design devices with improved performance [1]. This study of size and shape effects on material properties has attracted enormous attention due to their scientific and industrial importance [2-5]. Nanomaterials have different properties from the bulk due to their high surface area to volume ratio and possible appearance of quantum effects at the nano-scale [5-7]. The determination of nanomaterials properties is still in its infancy and many materials properties are unknown or ill-characterized at the nano-scale [8, 9]. Therefore, nanothermodynamics can be particularly helpful since traditional methods as molecular dynamics simulations and density functional theory (DFT) techniques are very computationally demanding and thus are size limited. Indeed, molecular dynamics generally consider less than 10^5 atoms in order to keep calculations time within reasonable values [10]. Predictions using DFT is generally limited to a few nanometers [11]. Therefore, nanothermodynamics and computational techniques are not competing but complementary.

Nanothermodynamics

The roots of nanothermodynamics go back to the original work of Gibbs [12] and Kelvin [13] in the nineteenth century when the importance of surface contributions to the thermodynamic potentials of small systems was realized. The interest on this topic came back on the front of the stage due to the recent development of nanosciences. Indeed, when decreasing the size of one material, its surface area to volume ratio increases dramatically. As the total number of atoms, N, scales linearly with the volume, the fraction of atoms at the surface scales with $N^{-1/3}$. Nanothermodynamics can be defined as the study of small systems using the methods of statistical thermodynamics.

Thermodynamics implies that we are dealing with a large number of particles and therefore we require a size limit on the applicability of thermodynamics at the nano-scale. A first answer can be given by the fact that thermodynamics describes a material in thermodynamic equilibrium, defined as a volume where thermal fluctuations are small i.e. fluctuate by less than 1% and this occurs for sizes higher than ~4 nm [14, 15]. A second answer can be given by the appearance of quantum effects at the nano-scale which signifies that classical thermodynamics is no more applicable when the discrete character of the energy levels appears. It occurs when the energy bandgap between two successive levels becomes larger than the thermal energy. Approximately, energy level spacings of about 1 K can be found in particles with sizes around ~10 nm. This size depends on the material and varies only within a range of about 50% [16].

To describe a system thermodynamically, we have to use a thermodynamic potential, depending on the constraints imposed on the system. Let us note, as the number of atoms present in nanostructures is limited, we have to use a thermodynamic potential developed for a closed system i.e. fixed particle number but varying energy. The Gibbs free energy corresponds to closed systems, coupled thermally and mechanically to the outside world and is therefore well suited to describe nanostructures [17]. The statistical mechanical ensemble linked to the Gibbs free energy is the constant-pressure ensemble. It has been shown by one of the pioneers studying thermodynamics at the nano-scale, Terrell Hill that the rules describing stable equilibrium states remain the same for nanostructures [18, 19]. When the size of a material is reduced, the excess free energy induced by the surface has to be considered [20]. The Gibbs free energy of a nanostructure can then be expressed as a sum of the bulk Gibbs free energy,

G_∞, with another term considering the effect of the surface at the nano-scale, G_Σ.

$$G = G_\infty + G_\Sigma. \qquad (1)$$

Assuming that the surface may be characterized by a single value of surface energy, γ, the surface term can be expressed as $G_\Sigma = \gamma(A/V)$ where A and V are the surface area and volume of the nanostructure respectively.

To describe the solid-liquid phase transition at the nano-scale, the Gibbs free energy difference between the liquid and the solid phases of a nanostructure, which is given by (2), should be equal to zero.

$$G_l - G_s = G_{l,\infty} - G_{s,\infty} + (A/V)(\gamma_{lv} - \gamma_{sv}). \qquad (2)$$

Where γ_{lv} and γ_{sv} are the surface energy in the liquid and solid phases (J/m²), respectively. γ_{lv} and γ_{sv} are considered size-independent. This is justified by the fact that the size effect on the surface energies is less than 4% for sizes higher than 4 nm [21, 22]. Indeed, below this size, edges and corners of the structures begin to play a significant role in the surface energy [23].

Then, at the temperature, T, corresponding to melting at the nano-scale i.e. $T = T_m$, (2) becomes $0 = \Delta H_{m,\infty} - T_m \Delta S_{m,\infty} + (A/V)(\gamma_{lv} - \gamma_{sv})$. Furthermore, the bulk melting temperature, $T_{m,\infty}$, can be written as the following ratio, $T_{m,\infty} = \Delta H_{m,\infty}/\Delta S_{m,\infty}$, where $\Delta H_{m,\infty}$ and $\Delta S_{m,\infty}$ are the bulk melting enthalpy and bulk melting entropy, respectively. Therefore, the melting temperature at the nano-scale, T_m, for free-standing nanostructures can be expressed as function the size of the structure, D, and one shape parameter, α_{shape} [24].

$$T_m/T_{m,\infty} = 1 - \alpha_{shape}/D, \qquad (3)$$

where the shape parameter, α_{shape}, is defined as $\alpha_{shape} = AD(\gamma_{sv} - \gamma_{lv})/(V\Delta H_{m,\infty})$. To be complete, a lot of α_{shape} values are given in Ref. [25].

According to the Lindemann's work [26], the melting temperature is inversely proportional to the thermal expansion coefficient :

$$\alpha_{therm} T_m = C, \qquad (4)$$

where C is a constant. Equation (4) is known as the Lindemann's melting rule [27]. Therefore, the size-dependency of the thermal expansion coefficient can be written as:

$$\alpha_{therm}/\alpha_{therm,\infty} = (1 - \alpha_{shape}/D)^{-1}. \qquad (5)$$

TABLE I MATERIAL PROPERTIES OF SELENIUM

$T_{m,\infty}$ (K) [28]	494
$\Delta H_{m,\infty}$ (J/m³) [28]	4.06 10⁸
γ_l (J/m²) [28]	0.106
γ_s (J/m²) [29]	0.175
$\alpha_{therm,\infty}$ (K⁻¹) [30]	9.45 10⁻⁵

As mentioned by (5), the thermal expansion coefficient of nanoparticles increases when the size is reduced. The theory is compared with selenium experimental data in Figure 1, where divergence appears for sizes below ~25 nm between theory and experiment. One possible reason is that the solid surface energy of selenium is not around 0.175 J/m² as announced by Ref. [29]. To fit the experimental data, we proposed the following solid surface energy for selenium: 0.285±0.022 J/m².

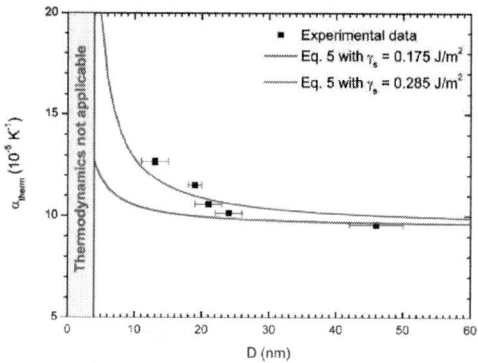

Fig. 1. Size dependency of the thermal expansion coefficient of selenium nanoparticles. The points indicate the experimental values and the lines correspond to the theoretical predictions.

The reasons why we directly suspected the accuracy of the selenium surface energy are firstly that the theory applies successfully to describe the thermal expansion coefficient size-dependency of other materials [30, 31] and more generally this theory can be applied to deduce the size-dependency of other material properties [14, 15]. Secondly, among the material properties involved in the definition of α_{shape}, this is the solid surface energy which is the most sensible to determine. Indeed, the liquid surface energy can be accessed via contact angle measurements on different substrates, and the bulk melting enthalpy can be determined by differential scanning calorimetry.

Selenium surface energy measurement

Most of the experimental solid surface energy data stems from surface tension measurements in the liquid phase extrapolated to zero temperature. In 1971, a lower limit value of the solid surface energy of selenium (γ_{Se}) was firstly determined around ~0.175 J/m², by extrapolating the surface energy of the melt of selenium at 20°C. An attempt to determine γ_{Se}, by using contact angle measurements involving various liquids presenting a surface energy ranging from 0.028 J/m² to 0.052 J/m², failed to be more precise than the extrapolated value due to the adsorption of water or other

978-1-4673-4840-9/13 $31.00 © 2013 IEEE

contaminants on the selenium surface [29].

In this paper, five different probe liquids have been used to measure the surface energy of selenium by contact angle measurements (Figure 2). The selected probe liquids should be liquid below the bulk melting temperature of selenium (217°C), and should have liquid surface energy higher than the expected γ_{Se} (0.285 J/m²). For organic and aqueous solutions, the surface energy of liquids is often less than 0.1 J/m². Therefore, mercury and gallium, which are liquid at room temperature and at 30°C, respectively, and which have a liquid surface energy around 0.476 and 0.718 J/m², respectively, were selected. Gallium was heated to 40°C prior to deposition onto the Se surface – which was also heated to 40°C. To complete the study, de-ionized water, glycerol and ethylene glycol, three common liquids used in surface energy determination, were also selected. Therefore, the wide range of liquid surface energies (0.047-0.718 J/m²) used in this study encompasses γ_{Se}. For each probe liquid, five drops were deposited onto the Se surface and the contact angle was measured for each drop. The contact angle data was gathered using a commercial Contact Angle Meter (GBX Scientific Instruments, France) - θ values were extracted using an interpolation model.

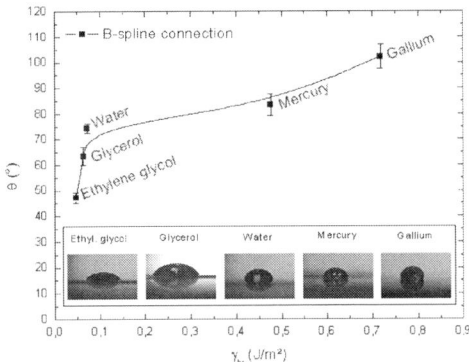

Fig. 2. Contact angle versus the liquid surface energy of the liquid probe. A B-spline connection is inserted to guide the eye. Inset: The pictures of the contact angles formed by the different probe liquids on the selenium surface are illustrated.

The θ measurement has been carried out in a class ISO 5/7 cleanroom ($T = 20°C\pm0.5°C$; $RH = 45\%\pm2\%$). To avoid any contamination, the Se surface was cleaned using VLSI grade solvents, then rinsed with de-ionized water and finally dried with nitrogen gas. After this cleaning procedure, a droplet of probe liquid was deposited by the sessile drop technique onto the film and a high resolution camera captured the image (Inset of figure 2). The diameter of each droplet was less than its capillary length to ensure that the effects of gravity can be ignored.

The possibility of estimating the surface energy from contact angle measurements relies on a relation which has been proposed by Young in 1805 [32]. From the Young's equation, we can access to the solid surface energy by evaluating for which value of γ_{lv}. $d\left(\gamma_{lv}\cos\theta\right)/d\gamma_{lv} = 0$. It occurs at 0.291±0.025 J/m² which is in excellent agreement with the theoretical prediction, 0.285±0.022 J/m². For historical reasons, a plot of $\gamma_{lv}\cos\theta$ versus γ_{lv} has been used in conjunction with Young's equation to argue the fact that $\gamma_{lv}\cos\theta$ increases and approaches γ_{sv}.

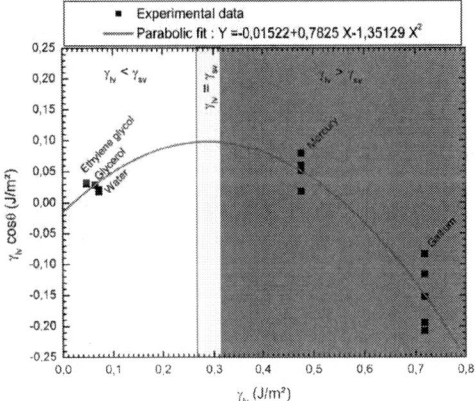

Fig. 3. $\gamma_{lv}\cos\theta$ versus γ_{lv} for a selenium thin film. The red line indicates the evolution of the contact angle with the liquid surface energy of the probe liquid. Three zones are represented : $\gamma_{lv} < \gamma_{sv}$, $\gamma_{lv} = \gamma_{sv}$ and $\gamma_{lv} > \gamma_{sv}$. The zone where $\gamma_{lv} = \gamma_{sv}$ indicates the possible values for the solid surface energy of selenium.

Selenium surface characterization

The selenium thin film has been deposited from high purity selenium pellets (purity >99.999%, Sigma-Aldrich) by thermal evaporation on a <100> silicon substrate. The thickness of the film has been measured by a profilometer Dektak from Veeco to be around 190 nm.

The selenium surface has been observed with a scanning electronic microscope (Ultra 55, Zeiss). The grain boundaries are clearly visible in Figure 4. The grain size varies between 40 nm to 170 nm.

Fig. 4. SEM image of the selenium surface.

The roughness of the surface has been determined by Atomic Force Microscopy (AFM) using a Dimension D3100, Bruker - Veeco. The RMS roughness is around 1.04 nm. The fractal dimension of the surface is measured to be around 2.28±0.04.

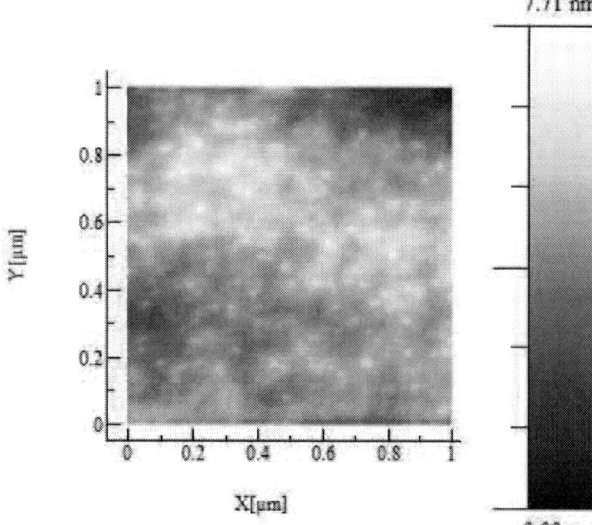

Fig. 5. AFM image of the selenium surface (1 μm x 1 μm).

The chemical composition of the selenium surface was studied by X-ray photoelectron spectroscopy (XPS) using a PHI 5000 VersaProbe apparatus. The XPS spectrum is obtained by irradiating a material with a beam of X-rays while simultaneously measuring the kinetic energy and number of electrons that escape from the top 1 to 10 nm of the material being analyzed. The XPS spectrum delivers information exclusively on the surface. Figure 6 indicates that the selenium film is pure and not contaminated with carbon. The inset indicates that selenium at the surface is only bonded to selenium and that it is not oxidized. Therefore, the surface state is ideal for contact angle measurements.

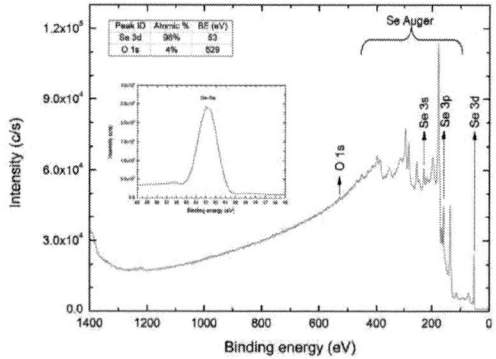

Fig. 6. XPS spectrum.

The orientation of the selenium film has been determined by X-ray Diffraction (XRD) using a D5000 from Siemens. The selenium film is oriented along the <101> direction. With XRD, the surface and bulk of our selenium/silicon system are both analyzed. Here, we can see that at the interface thin film-substrate, we have formation of an alloy, $SiSe_2$ (Figure 7).

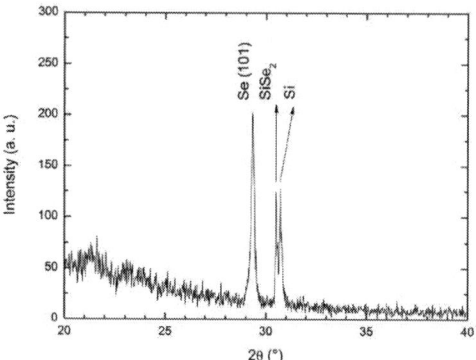

Fig. 7. XRD spectrum.

Conclusions

Small systems of decreasing dimensions to the nanoscopic level are well described by the Gibbs potential. Physical properties change with size according to relative simple scaling equations involving a power law dependence on the system size due to the energy contribution of the surface on the Gibbs free energy of the system. From these scaling equations, it has been possible to predict the bulk solid surface energy of selenium. Finally, the surface energy for a <101> facet of selenium has been measured using the sessile drop technique to be around 0.291±0.025 J/m^2 and it is found to be in excellent agreement with the theoretical prediction.

References

[1] M. J. Pitkethly, "Nanomaterials-the driving force," *Materials Today*, vol. 7 (Supplement 1), pp. 20-29, 2004.

[2] H. Gleiter, "Nanostructured materials: basic concepts and microstructure," *Acta Materialia*, vol. 48, pp. 1-29, 2000.

[3] R. Ferrando, *et al.*, "Nanoalloys: From theory to applications of alloy clusters and nanoparticles," *Chemical Reviews*, vol. 108, pp. 845-910, 2008.

[4] F. Baletto and R. Ferrando, "Structural properties of nanoclusters: Energetic, thermodynamic, and kinetic effects," *Review of Modern Physics*, vol. 77, pp. 371-423, 2005.

[5] E. Roduner, "Size matters: why nanomaterials are different," *Chemical Society Reviews*, vol. 35, pp. 583-592, 2006.

[6] M. J. Yacaman, *et al.*, "Structure shape and stability of nanometric sized particles," *Journal of Vacuum Science & Technology B*, vol. 19, pp. 1091-1103, Jul-Aug 2001.

[7] C. Q. Sun, "Size dependence of nanostructures: Impact of bond order deficiency," *Progress in Solid State Chemistry*, vol. 35, pp. 1-159, 2007.

[8] T. A. Campbell, "Measuring the nanoworld," *Nano Today*, vol. 4, pp. 380-381, 2009.

[9] E. K. Richman and J. E. Hutchison, "The

978-1-4673-4840-9/13 $31.00 © 2013 IEEE

nanomaterial characterization bottleneck," *ACS Nano*, vol. 3, pp. 2441-2446, 2009.

[10] F. Delogu, "Melting behaviour of a pentagonal Au nanotube," *Nanotechnology*, vol. 18, p. 325706, 2007.

[11] A. S. Barnard, "Modelling of nanoparticles: approaches to morphology and evolution," *Reports on Progress in Physics*, vol. 73, p. 086502, 2010.

[12] J. W. Gibbs, *The Collected Works of J. Willard Gibbs* New Haven: Yale University Press, 1948.

[13] W. Thomson, "On the equilibrium of vapour at a curved surface of liquid," *Philosophical magazine*, vol. 42, pp. 448-452, 1871.

[14] G. Guisbiers and L. Buchaillot, "Universal size/shape-dependent law for characteristic temperatures," *Physics Letters A*, vol. 374, pp. 305-308, Dec 28 2009.

[15] G. Guisbiers, "Size-Dependent Materials Properties Toward a Universal Equation," *Nanoscale Research Letters*, vol. 5, pp. 1132-1136, Jul 2010.

[16] W. P. Halperin, "Quantum size effects in metal particles," *Review of Modern Physics*, vol. 58, p. 533, 1986.

[17] L. E. Reichl, *A Modern Course in Statistical Physics*. New York: Wiley-Interscience, 1998.

[18] T. L. Hill, "Thermodynamics of Small Systems " *Journal of Chemical Physics*, vol. 36, pp. 3182-3197, 1962.

[19] T. L. Hill, "A different approach to nanothermodynamics," *Nano Letters*, vol. 1, pp. 273-275, 2001.

[20] J. Weissmüller, *et al.*, "Two-phase equilibrium in small alloy particles," *Scripta Materialia*, vol. 51, pp. 813-818, 2004.

[21] L. H. Liang, *et al.*, "Melting enthalpy depression of nanocrystals based on surface effect," *Journal of Materials Science Letters*, vol. 21, pp. 1843-1845, Dec 1 2002.

[22] H. M. Lu and Q. Jiang, "Size-dependent surface energies of nanocrystals," *Journal of Physical Chemistry B*, vol. 108, pp. 5617-5619, 2004.

[23] A. S. Barnard and P. Zapol, "A model for the phase stability of arbitrary nanoparticles as a function of size and shape," *Journal of Chemical Physics*, vol. 121, p. 4276, 2004.

[24] G. Guisbiers and L. Buchaillot, "Modeling the Melting Enthalpy of Nanomaterials," *Journal of Physical Chemistry C*, vol. 113, pp. 3566-3568, Mar 5 2009.

[25] G. Guisbiers, *et al.*, "Mechanical and thermal properties of metallic and semiconductives nanostructures," *Journal of Physical Chemistry C*, vol. 112, pp. 4097-4103, 2008.

[26] F. A. Lindemann, "The calculation of molecular vibration frequencies," *Zeitschrift für Physik*, vol. 11, pp. 609-612, 1910.

[27] A. V. Granato, *et al.*, "Melting, thermal expansion, and th eLindemann rule for elemental substances," *Applied Physics Letters*, vol. 97, p. 171911, 2010.

[28] W. Martienssen and H. Warlimont, *Springer Handbook of Condensed Matter and Materials Data*. Berlin: Springer, 2005.

[29] L.-H. Lee, "Solid surface tensions of amorphous and crystalline selenium," *Journal of Non-Crystalline Solids*, vol. 6, pp. 213-220, 1971.

[30] C. C. Yang, *et al.*, "Size effects on Debye temperature, Einstein temperature, and volume thermal expansion coefficient of nanocrystals," *Solid State Communications*, vol. 139, pp. 148-152, 2006.

[31] G. Guisbiers, "Size and shape dependencies of nanomaterial properties: Thermodynamic considerations," *MRS Proceedings*, vol. 1371, pp. imrc11-1371-s1-03 2012.

[32] T. Young, "An essay on the cohesion of fluids," *Philosophical Transactions of the Royal Society of London*, vol. 95, pp. 65-87, 1805.

Facile Nanometer Thick Native Oxide Based Passivation of Silicon for High Efficiency Photovoltaics

Nazir P. Kherani* and Zahidur R. Chowdhury

Department of Electrical and Computer Engineering
University of Toronto, 10 King's College Road
Toronto, Ontario, Canada
*Email: kherani@ecf.utoronto.ca

Abstract

Fabrication of low cost solar cells using ultra-thin (approximately 20 μm) silicon wafers is a viable route given the significant potential of reduced material cost and its versatility to a range of portable and terrestrial applications. Low temperature processing is a compelling opportunity for the synthesis of high-efficiency ultra-thin silicon wafers. Further, excellent surface passivation attainable through facile low temperature processing techniques is an essential enabler for effective manufacturing of ultra-thin silicon solar cells, and thus paving the way for high-efficiency low-cost silicon foil photovoltaics. This article presents a novel low temperature passivation scheme using approximately 1 nm thick facile native oxide and 75 nm PECVD SiN_X. A maximum lifetime of 1.7 ms has been obtained for the passivation scheme. Moreover, the passivated wafers were also used to fabricate Back Amorphous-Crystalline Silicon Heterojunction (BACH) cells using double side polished n-type FZ wafers. A maximum cell efficiency of 16.7% is obtained for facile native oxide - PECVD SiN_x bilayer passivated cells having V_{OC} of 641 mV, J_{SC} of 33.7 mA/cm^2 and fill-factor of 0.77 for a 1 cm^2 untextured cell (all measurements having been performed under AM 1.5 global spectrum illumination).

Introduction

Over the last three decades crystalline silicon (cSi) based solar cells have undergone remarkable technological advances, experienced a significant drop in silicon feedstock cost, and today are at or approaching grid parity in a number of regions of the world [1].

The next step in the evolution of silicon photovoltaics is the realization of high efficiency ultra-thin silicon photovoltaics (PV), rendering silicon PV an inevitable economically competitive reality. In this framework the advent of amorphous-crystalline silicon heterojunction PV coupled with compatible low temperature passivation schemes [2] provide a path for the production of high efficiency silicon foil solar cells.

The interface quality of the absorbing material of a solar cell, especially as the absorber thickness (silicon foil) becomes smaller, plays a great role in determining the cell performance. The interface quality at the native silicon oxide - crystalline silicon junction has traditionally been deemed to be of poor and non-uniform quality *vis-à-vis* its interfacial defect density. Contrary to conventional wisdom, we have recently demonstrated that the controlled growth of native oxide on crystalline silicon followed by an over layer of silicon nitride yields a high quality silicon oxide - silicon interface with reduced defect density [3]. This paper details the growth and characterization of facile native oxide-silicon nitride (f-SiO_X-SiN_X) passivation layer, showing the attainment of low surface recombination velocity, and illustrating its application for high-efficiency silicon photovoltaics.

The f-SiO_X-SiN_X passivation scheme is a low temperature scheme and is suitable for ultra-thin silicon wafers. The excellent surface passivation quality is achieved not only by reducing the dangling bond density at the interface but also through suitable charge density at the interface that results in field passivation. Consequently, the positive charge density of the f-SiO_X-SiN_X passivation layer also serves as the front surface field for n type cSi wafers. Moreover, wide band-gap materials used in the passivation scheme results in reduced parasitic absorption at the front surface. Also, the use of ultra-thin facile silicon oxide, placed between SiN_X and cSi, makes the passivation scheme more suitable from a light trapping perspective in comparison with passivation schemes where thicker (greater than or equal to 10 nm) thermal oxide and PECVD SiN_X are used [4-7]. Excellent passivation qualities in terms of surface recombination velocities (SRV) have been reported for alternative passivation schemes. Specifically, SRVs of 10 cm/s [4-7] and 6 cm/s [4] were reported for the thermal oxide (10 nm thick)-SiN_X and PECVD oxide (50 nm thick)-SiN_X passivation schemes, respectively.

This article presents a novel passivation scheme with an ultra-thin layer of native oxide (approximately 1 nm) grown at room temperature followed by the deposition of PECVD SiN_X layer at 400°C on cSi surfaces. The passivation scheme was then integrated within the Back Amorphous-Crystalline silicon Heterojunction (BACH) photovoltaic cell [8], substituting for hydrogenated amorphous silicon or alternative passivation schemes, in order to demonstrate the potential of the novel passivation scheme for high efficiency silicon solar cells.

Experimental details

n-type double-side polished 300 $μm$ thick (100) FZ crystalline silicon wafers with 1-5 Ω resistivity were used for this study. The wafer was passivated using f-SiO_x and PECVD SiN_x. The f-SiO_x was grown in a controlled clean room ambient at room temperature for different time period to determine the suitable growth time or/and the suitable native oxide thickness. At the conclusion of the facile native oxide growth period, 75 nm

978-1-4673-4840-9/13 $31.00 © 2013 IEEE

thick SiN_X layers were deposited using Oxford PlasmaLab 100 PECVD tool. Once the surfaces were passivated, subsequent processing steps were carried out to fabricate the BACH cell by selective etching of SiN_X layer from the back surface and depositing approximately 20 *nm* thick *n* and *p* doped hydrogenated amorphous silicon (aSi:H) layers. Chrome-silver metal layers were deposited on the back surface using *e*-beam evaporation.

The oxide thicknesses were measured using parallel angle resolved x-ray photoelectron spectroscopy (PARXPS). Theta Probe from Thermo Scientific[TM] was used for the PARXPS measurements. The excess carrier density (ECD) dependent effective minority carrier lifetime, τ_{eff}, was measured using a Sinton Silicon Lifetime Tester WCT-120 system. Transient and Quasi-Steady-State Photo-Conductance (QSSPC) methods were used to measure injection-dependent τ_{eff} of the sample. Spatial distributions of the lifetime of the sample were measured using Semilab's WT-2000 PVN Microwave Photo Conductance Decay (μ-PCD) instrument.

The *I–V* characteristics were measured using the neonsee's IV AM 1.5 Solar Simulator. The light intensity was adjusted using a calibrated cell. The aperture area of 1×1 cm^2 was defined by a shadow mask on the cell front side for cells having approximately the same cell area.

Results and Discussion

Oxide Growth Time (OGT) on silicon surface in cleanroom ambient was varied from 20 min to 40K min for the f-SiO_X-SiN_X passivation of crystalline silicon. PARXPS shows that the oxide growth saturates to a thickness of approximately 1nm in 40K minutes. Moreover, the passivation quality also depends on the thickness of the facile oxide layer. Accordingly, a small enhancement in passivation quality with the increased OGT was observed for the passivation scheme where OGTs were below 20K minutes. Excellent passivation quality, having SRV values lower than 20 cm/sec with increasing OGTs, were realized for OGTs above 20K minutes. A maximum minority excess carrier lifetime of approximately 1.7 ms was obtained for this passivation scheme as measured using the Sinton Lifetime Testing Instrument for a facile oxide layer grown for four months. Fig. 1 shows the spatial profile of the passivation quality achieved using f-SiO_X-SiN_X as measured by the μ-PCD instrument.

Multiple cells were fabricated in a wafer where the cells had different doped-region widths and varying inter-digital gaps (IDGs) between doped-regions. The width of the *n*-doped hydrogenated amorphous silicon (a-Si:H) region varied from 180 μm to 410 μm while the ratio of *p*-doped region width to *n*-doped region width (*p/n* ratio) was varied from 2.1 to 2.67. The *n*-width to the IDG ratio (*n*/IDG ratio) was also varied from 1.5 to 4. The maximum cell efficiency obtained for f-SiO_X-SiN_X passivation was 16.7% with V_{OC} of 641 mV, J_{SC} of 33.7 mA/cm^2 and fill-factor of 77% under AM1.5 solar spectrum; the corresponding IV curve is shown in Fig. 2.

Conclusion

The novel facile native oxide and PECVD SiN_X bilayer passivation provides excellent surface passivation for crystalline silicon surfaces. A maximum lifetime of 1.7 ms was

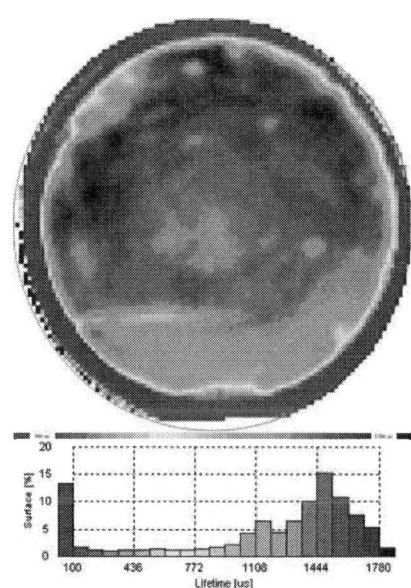

Fig. 1 Passivation quality achieved by f-SiO_x - SiN_x passivation measured using μ-PCD.

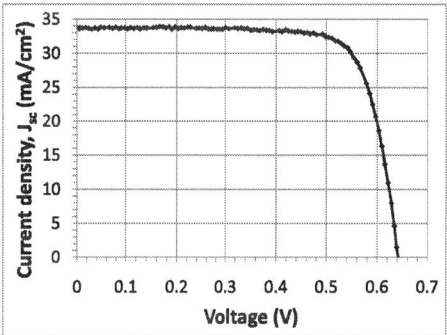

Fig. 2 IV curve for the BACH cell with maximum cell efficiency of 16.7% with V_{OC} of 641 mV, J_{SC} of 33.7 mA/cm^2 and fill-factor of 77% under AM1.5 solar spectrum.

obtained for this low temperature passivation scheme. BACH cells with different design conditions were successfully fabricated using the passivation scheme on a double side polished crystalline silicon wafer. The maximum cell efficiency of 16.7% was obtained for f-SiO_X-SiN_X passivation under AM1.5 solar spectrum.

Acknowledgement: Ontario Research Fund – Research Excellence program, Natural Sciences and Engineering Research Council of Canada, and University of Toronto.

References

[1] Applied Materials, "International Solar Energy Survey", 2012.
[2] M. Taguchi et al., *Proceedings of the 24th European Photovoltaic Solar Energy Conference*, 1690–1693, 2009.
[3] Z Chowdhury et al., *Appl. Phys. Lett.*, 101, 021601, 2012.
[4] G. Dingemans et al., *Appl. Phys. Lett.* 98, 222102, 2011.
[5] J. Schmidt et al., *Semicond. Sci. Technol.* 16, 164, 2001.
[6] S. Narasimha and A. Rohatgi, *Appl. Phys. Lett.* 72, 1872. 1998.
[7] Y. Larionova et al., *Appl. Phys. Lett.* 96, 032105, 2010.
[8] A. Hertanto et al., *34th IEEE PVSC*, 2009, pp. 1767-1770.

Various size images mapping technique to analyze trap-assisted non-radiative recombination mechanism using Cathodo- and Electro- Luminescences measurement in GaN-based LEDs

Euyhwan Park[1], Garam Kim[1], Janghyun Kim[1], Donghoon Kang[2], Joong-Kon Son[2], Byung-Gook Park[1]

[1] Inter-university Semiconductor Research Center (ISRC) and
School of Electrical Engineering and Computer Science, Seoul National University
San 56-1, Sillim-dong, Gwanak-gu, Seoul 151-742, Republic of Korea
[2] Samsung Electronics Co., Ltd., Republic of Korea
Tel.: +82-2-880-7279, Fax: +82-2-882-4658 Email: park77@snu.ac.kr

Abstract

We have investigated the non-radiative recombination of GaN-based light-emitting diodes (LEDs) using electroluminescence (EL) and cathdoluminescence (CL) techniques. Comparing between EL and CL measurements, we can map the defects causing non-radiative recombination centers. At first, same size mapping techniques of these two measurements show defect-assisted non-radiative recombination. And enlarged CL images give more detail contents about this mechanism.

It is concluded that the image mapping of EL and CL in GaN have a strong correlation between the two techniques, and this result shows that non-radiative recombinations are mainly affected by defects of GaN LEDs.
(Keywords: Electroluminescense, Cathodoluminescense, GaN LED, image mapping, non-radiative recombination)

Introduction

Gallium nitride (GaN) based semiconductors are promising materials to make blue light emitting diodes (LEDs) and GaN-based LEDs are playing an important role as energy efficient light sources in solid state lighting.

An important subject in developing GaN technology is to reliably and efficiently determine the defect density in GaN. [1, 2] Reducing defect density in nitride active layers is critical for improving device performance and lifetime. The recent progress made in terms of increasing the availability of low dislocation density templates should facilitate the improvement of reliability of nitride devices.

These devices work in spite of high density of threading dislocations affecting the lifetime and efficiency of the devices. So far, many authors have studied the electrical and optical properties of dislocations and defects in GaN epitaxial layer using cathodoluminescence (CL) imaging technique combined with a scanning electron microscope (SEM-CL). [3-10]

In this report, comparison between electroluminescence and cathodoluminescence techniques is made by characterizing low-defect density GaN LEDs, indicating that defects causing each dark spots are strong non-radiative recombination centers. Moreover, it is also shown that CL mapping can give additional information about internal defect structures in GaN.

Experimental

For this study, in all the experiments, we have used InGaN / GaN blue LEDs. The experimental device structure is similar to a typical commerical InGaN / GaN LED. It consists of undoped-GaN layer on the sapphire substrate, and InGaN / GaN MQW active layer sandwiched between p-type and n-type GaN cladding layers. Indium Tin Oxide (ITO) is used for current spreading layer to the bottom of the p-metal contact.

The fabricated LEDs had the InGaN / GaN layers grown by metal organic chemical vapor deposition (MOCVD) on the sapphire substrate. The device chip size was 500 $\mu m \times$ 1000 μm. Fig. 1 shows schematic cross-section of GaN-based LED and it represents mesa type lateral mode commerical LED structure.

Because we want to only check the defects-assisted non-radiative recombination and investigate accurate CL measurement, we eliminate Induim-Tin-Oxide (ITO) electrode just above p-GaN when we check CL measurement.

Results and Discussion

At first, we examine the emission property of GaN-based LED by electroluminescence through light-emitting pattern. Using IR Confocal Emission Microscope System equipment (PHEMOS-1000), we measure the relative emission by light-emitting pattern of LED chips which were driven to 1uA constant current direct probing to p-type and n-type contacts. Figure 2 is illustrated by the density of radiative-recombination of electron-hole pairs of GaN LED partial surface. In other words, the rest of black regions are not self-emissive. There exist differences between emissive and not emissive region of same LED chip in constant operating current. Therefore, we examine the phenomenon in detail which causes differences between the emissive regions using other luminescence method.

Cathodoluminescence measurement method is widely used to check and analysis the defects and dislocation distribution on compound semiconductor epitaxial layer, so we focus on defect-assisted effect of non-emissive region in GaN layer using CL measurement.

To accurate confirm, we examine exact identical region image mapping between EL and CL techniques. Figure 2 and 3 show the results of same size 1:1 mapping and we verify the matching differences in same size region. For more accurate

978-1-4673-4840-9/13 $31.00 © 2013 IEEE   112

comparing, four regions are selected in EL image and we move these squares and check the differences in same regions of CL image. As shown in table I, More emissive region in EL regions is matched to brighter image of CL measurement. That is, brighter region of CL image means that electron-hole pairs recombine more radiatively. But darker regions are less radiatively recombined, in other words non-radiative recombination by defects.

And it is necessary to check defect-assisted effect of CL image, so we select more emissive region (Region (a)) and less emissive region (Region (b)) in EL image. Thus, we acquire enlarged CL images of these selected two regions, and it is obvious that emissive mechanism is mainly affected by defects. Figure 5 and 6 show the difference of dark contrast of CL image through same incident electron beam voltage. Like same size mapping technique, enlarged CL images have identical results, so these two techniques must be strongly correlated.

For more analysis, Fig. 7 shows CL spectrum of certain areas in region (a) and (b). The more emissive region (a) has higher main peak in blue wavelength (450nm), and the spectrum is concentrated at the main peak. In general, CL intensity depends on the number of electron-hole pairs excited by the incident beam. Thus, more emissive region has lower defect and dislocation density, so more radiative reocombinations are occurred by recombining e-h pairs and this fact give better efficiency in GaN LED chips

Conclusion

In summary, we have reported EL and CL measurement in GaN-based LEDs. To establish the relation between EL and CL measurements, same size mapping technique of two measurement images was performed.

Given the advantages of EL and CL image mapping of defects in GaN, establishing a strong correlation between the two techniques should be very beneficial for further validating the two techniques for GaN LEDs.

In other words, non-radiative recombinations are mainly affected by defects. From all evidences of our measured data, it is concluded that non-radiative recombinations are affected by defect-assisted processes.

Acknowledgement

We would like to acknowledge the technical supports provided by Samsung Electronics Co., Ltd and Inter-university Semiconductor Research Center (ISRC) in Seoul National University.

This work was supported by the Smart IT Convergence System Research Center funded by the Ministry of Education, Science and Technology as Global Frontier Project (SIRC-2011-0031845).

References

[1] M. Meneghini, A. Tazzoli, G. Mura, G. meneghesso, and E. Zanoni, "A Review on the Physical Mechanisms That Limit the Reliability of GaN-Based LEDs", IEEE Transactions on Electron Devices,Vol. 57, No.1, pp. 108-118, 2010.

[2] M. Meneghini, L.Trevisanello, G. Meneghesso, and E. Zanoni, "A Review on the Reliability of GaN-Based LEDs", IEEE Transactions on Device and Materials Reliability, Vol.8, No.2, pp.323-331, 2008

[3] Hai Lu, X.A.Cao, S.F.LeBoeuf, H.C.Hong, E.B.Kaminsky, S.D.Arthur, "Cathodoluminescence mapping and selective etching of defects in bulk GaN", Journal of Crystal Growth, Vol.291, pp.82-85, 2006.

[4] Yong Xia, et.al., "Low-Temperature Cathodoluminescence Mapping of Green, Blue, and UV GaInN / GaN LED Dies", Mater. Res. Soc. Symp. Proc., Vol.955, 2007.

[5] A.Cremades, et.al., "Cathodoluminescence study of GaN epitaxial layers", Materials Science and Engineering, Vol.42, pp.230-234, 1996.

[6] Jun Xu, et.al., "Cathodoluminescence Study of InGaN/GaN Quantum-Well LED Structures Grown on a Si Substrate", Journal of Electronic materials, Vol.36, pp.1144-1148, 2007.

[7] N.Yamamoto, et.al., "Cathodoluminescence characterization of dislocations in gallium nitride using a transmission electron microscope", Journal of Applied Physics, Vol.94, No.7, pp.4315-4319, 2003.

[8] F.Ishikawa, H.Hasegawa, "Characterization of recombination process in GaN by Cathodoluminescence in-depth spectroscopy", Phys.stat.sol. No.7, pp.2707-2711, 2003

[9] M.G. Cheong, et.al., "Properties of InGaN/GaN quantum wells and blue light emitting diodes", Journal of Luminescence, Vol.99, pp. 265-272, 2002

[10] Jun Xu, et.al., "Cathodoluminescence Study of InGaN/ GaN Quantum-Well LED Structures Grown on a Si Substrate", Journal of Electronic Materials, Vol. 36, No. 9, pp. 1144-1148, 2007

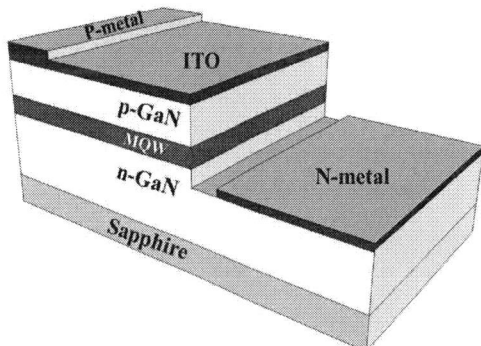

Fig 1. Schematic cross-section of GaN-based LEDs.

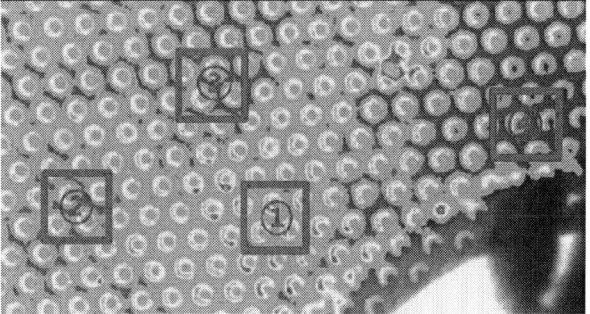

Fig 2. Electroluminescence image of GaN LED

Fig 3. Cathodoluminescence image of GaN LED.

TABLE I
Comparison of selected regions between EL and CL image

	Region ①	Region ②	Region ③	Region ④
EL				
CL				

Fig 4. Two selected regions of electroluminescence image of GaN LED (1μA was driven to the LED chip).

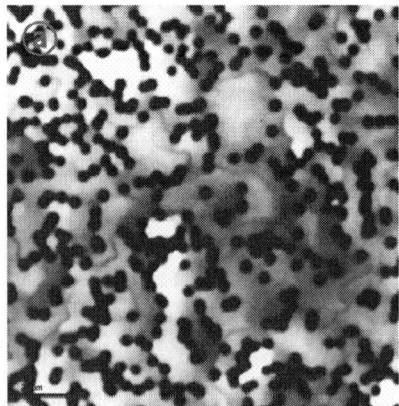

Fig 5. Cathodoluminescence panchromatic image of more emissive region (Region (a)).

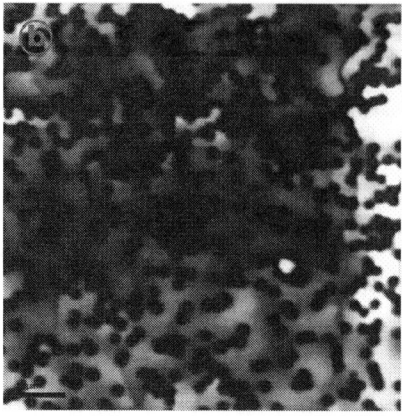

Fig 6. Cathodoluminescence panchromatic image of less emissive region (Region (b)).

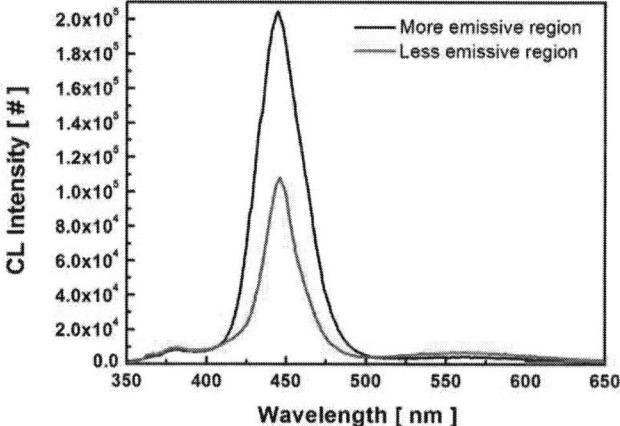

Fig 7. Comparison of two regions (selected in Fig 4) in CL spectrum between 350-650nm.

The Temperature Dependent TCAD and SPICE Modeling of AlGaN/GaN HEMTs

Li Yuan[1], Weizhu Wang[1], Kean Boon Lee[1], Haifeng Sun[1], Susai Lawrence Selvaraj[1], Xing Zhou[2] and
Guo-Qiang Lo[1]

[1]Institute of Microelectronics, A*STAR (Agency for Science, Technology and Research), 11 Science Park Road, Singapore
Science Park II, Singapore 117685
[2]School of Electrical and Electronic Engineering, Nanyang Technological University, 50 Nanyang Avenue, Singapore 639798
Phone: +65-6770 5603, FAX: +65-6773 1914, Email: yuanl@ime.a-star.edu.sg

Abstract

In this work, the temperature dependent TCAD and SPICE modeling platform of AlGaN/GaN HEMTs has been established by using Sentaurus TCAD and Silvaco UTMOST IV. Typically, the temperature co-efficient of series resistance, trans-conductance, sub-threshold swing and gate/buffer leakage has been extracted from the TCAD simulation. Based on that, the compact SPICE device model of HEMTs has been built up, which can be used for DC/transient SPICE simulation and power electronics circuit demonstration. The SPICE modeling results agree well with our TCAD simulation, indicating good revealing of the physical mechanisms of the AlGaN/GaN HEMTs' operation.

Introduction

AlGaN/GaN high electron mobility transistors (HEMTs) device technology, especially grown on large size Si substrate recently attracts a lot of attentions owing to its superior advantages for next generation cost effective, high speed, high efficiency and high operating temperature power electronics applications [1-3]. In this work, for the first time we have demonstrated a complete TCAD and SPICE device modeling platform for AlGaN/GaN HEMTs to predict the device performance with considering of the influence of environment temperature which is important for the demonstration of the power electronics system's temperature tolerance [4-7]. Physical models, such as spontaneous and piezoelectric polarization, unintentional background doping, surface states trapping/de-trapping, and high electrical field impact

ionization have also been taken into account in our modeling platform.

TCAD simulation of HEMTs

In this project, a physical based device simulation of AlGaN/GaN HEMTs has been carried out by using Sentaurus TCAD from Synopsys [8]. Pivotal physical mechanisms, such as gate Schottky tunneling, surface cap layer leakage, avalanche breakdown and GaN buffer leakage have been considered.

For the device simulation, the standard AlGaN/GaN HEMTs on high resistivity Si wafer with GaN cap layer and AlN spacer have been modeled as shown in Fig. 1. From top to bottom, the hetero-structure consists of a GaN cap layer (2 nm), an $Al_{0.24}Ga_{0.76}N$ barrier layer (18 nm), an AlN spacer (1 nm), a GaN buffer layer (3 μm), an AlGaN/AlN transition/nucleation

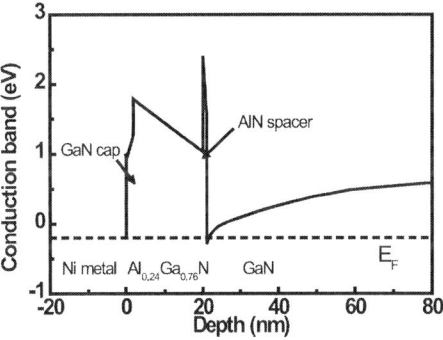

Fig. 2 The TCAD simulated conduction band profile of AlGaN/GaN HEMTs.

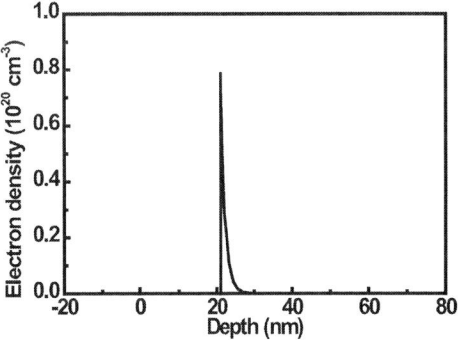

Fig. 3 The TCAD simulated electron carrier distribution profile of AlGaN/GaN HEMTs.

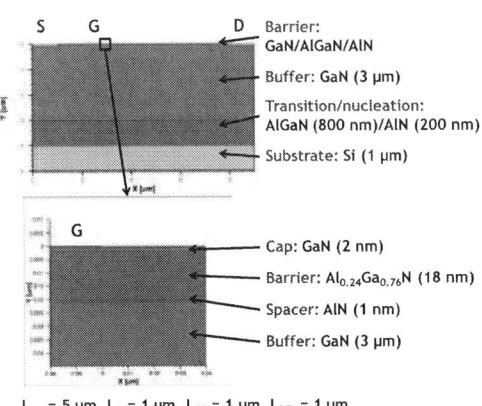

L_{GD} = 5 μm, L_G = 1 μm, L_{GS} = 1 μm, $L_{S/D}$ = 1 μm

Fig. 1 The TCAD device model structure of AlGaN/GaN HEMTs.

978-1-4673-4840-9/13 $31.00 © 2013 IEEE   115

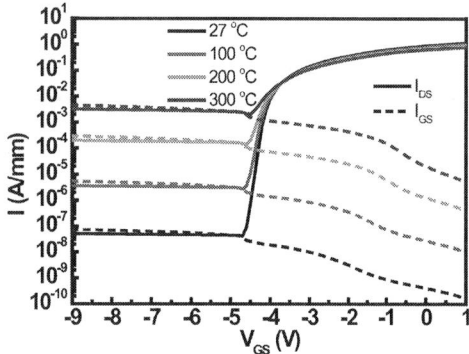

Fig. 4 The TCAD simulated transfer I_{DS}-V_{GS} and I_{GS}-V_{GS} characteristics of AlGaN/GaN HEMTs in log scale from 27 °C to 300 °C.

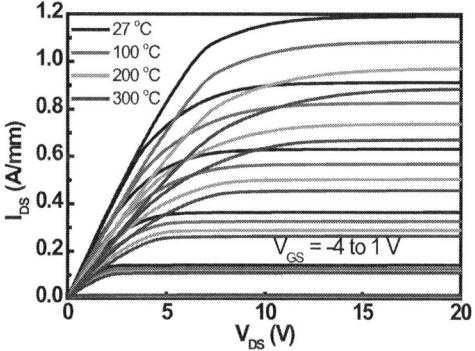

Fig. 5 The TCAD simulated I_{DS}-V_{DS} characteristics of AlGaN/GaN HEMTs from 27 °C to 300 °C.

layer (1 μm) and the Si substrate (1 μm). The unintentional background n-type doping in AlGaN and GaN layer is set to be 10^{16} cm^{-3} [9], and the Fe compensation doping (2×10^{17} cm^{-3}) in GaN buffer [10] is also considered. High density donor-like traps are added in the AlGaN/AlN transition/nucleation layer to model the lattice dislocations. The Si layer here is un-doped to model the high-resistivity substrate.

The simulated conduction band diagram of AlGaN/GaN hetero-junction is shown in Fig. 2. With the lattice stress relax of 10%, the spontaneous and piezoelectric polarization of the AlGaN/GaN hetero-junction are 0.78×10^{13} cm-2 and 0.51×10^{13} cm-2, respectively. On the surface of AlGaN barrier,

there are donor-like surface states of $\sim 10^{13}$ cm^{-2} arising from the Ga adatom dangling bonds [11]. In our simulation, the AlGaN surface trap state is set to be located at 3.2 eV above the valance band. The distribution of electron carriers within the AlGaN/GaN hetero-junction is shown in Fig. 3.

The transfer and output I-V characteristics of AlGaN/GaN HEMTs are shown in Figs. 4 to 5. As we can see, when temperature goes up, the HEMTs suffer higher gate leakage from the stronger electron tunneling and thermionic emission at the gate Schottky contact, which also dominates the off-state drain current behavior. At 300 °C, the gate leakage of HEMTs can reach more than 4 mA/mm, which dramatically limits the application feasibility of the Schottky gate based GaN HEMT in power switches at this high temperature. In addition from Fig. 5 we can found at 300 °C the maximum on-state current density degrades from 1.2 A/mm to 0.89 A/mm ($V_{GS} = 1$ V) due to the stronger lattice scattering and lower mobility in the 2DEG channel, indicating the current driving capability can be >25 % worse at higher temperature (300 °C), which should be noted in the power electronics design.

The high off-state leakage at 300 °C will also affect the parasitic capacitance of the AlGaN/GaN HEMTs. From the C-V simulation results shown in Fig. 6, it can be seen that at 300 °C, because of the high leakage current shown in Fig. 4, in the off-state the gated channel of HEMTs cannot be fully pinched-off, thus the extracted capacitances are still very high, similar to the values in the on-state, indicating a loss of effective gate control. While in the on-state, the capacitances can keep the same as at the room temperature.

SPICE modeling

The compact SPICE model of the transistor has been established to simulate the AlGaN/GaN HEMTs' characteristics, as demonstrated in Fig. 7. The C_{GD}, C_{GS} and C_{DS} are parasitic capacitances between gate, source and drain electrodes. The R_G, R_S and R_D stand for the contact resistances from electrodes to the gated channel, including the series gate to source or gate to drain channel resistance. GM reproduces the intrinsic gated channel conductance of the AlGaN/GaN HEMT. The detailed model descriptions are shown in equations (1-6).

At higher environment temperature, the trans-conductance degrades due to the lower electron carrier mobility; the

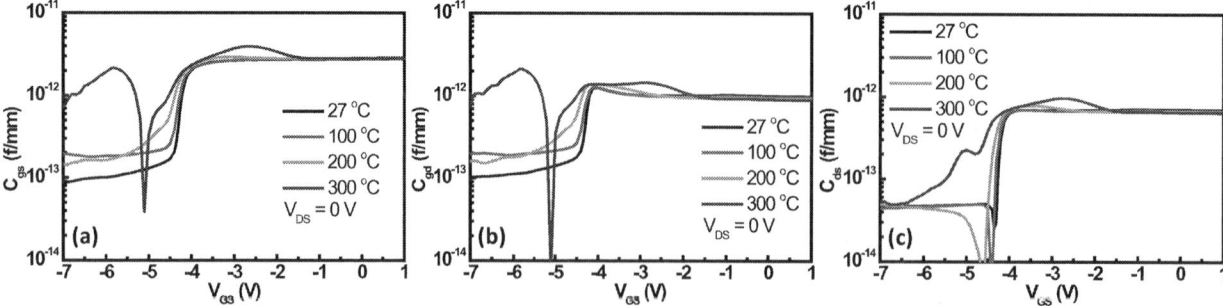

Fig. 6 The TCAD simulated C-V_{GS} characteristics of AlGaN/GaN HEMTs from 27 °C to 300 °C.

978-1-4673-4840-9/13 $31.00 © 2013 IEEE

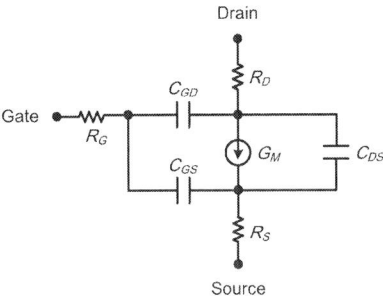

Fig. 7 The compact SPICE model of AlGaN/GaN HEMTs.

series-resistance increases owing to the stronger lattice scattering; the off-state leakage increases because of the higher carrier density, stronger thermionic emission and triangle-barrier tunneling and the sub-threshold swing increases as a result of the higher thermal voltage of the gate Schottky contact.

$$I_{DS} = G_m * W_G * (SS/\ln 10) * \ln\left\{1 + \exp\left(\frac{V_{GS} - V_{th}}{SS/\ln 10}\right)\right\} * \{1 - \exp(-V_{DS})\} \tag{1}$$
$$+ G_{leak\_DS} * W_G * V_{DS}$$

$$I_G = G_{leak\_GS} * W_G * V_{GS} + G_{leak\_GD} * W_G * V_{GD} \tag{2}$$

$$G_m = G_{m0} * [1 - agTc * (T - 27)] \tag{3}$$

$$R_{D/S} = R_{D0/S0} * [1 + acTc * (T - 27)] \tag{4}$$

$$G_{leak} = G_{leak0} * \exp\left(\frac{T - 27}{akTc}\right) \tag{5}$$

$$SS = SS_0 * [1 + asTc * (T - 27)] \tag{6}$$

$$C_{xx} = C_0 * [1 + cxxTc * (T - 27)] \tag{6}$$

where agTc, acTc, akTc and asTc are temperature co-efficient.

By fitting to the TCAD simulation results, the model parameter used in the equations (1-6) can be extracted as shown in table 1. The comparisons between TCAD and SPICE simulations of AlGaN/GaN HEMTs are shown in Figs. 8 to 11. As we can see our SPICE model can successful represent the operation behavior of TCAD simulated AlGaN/GaN HEMTs,

TABLE I
The data sheet of the AlGaN/GaN compact SPICE model.

	D-HEMT GaN		D-HEMT
G_m	389.55 mS/mm	agTc	3.175×10^{-5} K^{-1}
SS	100 mV/dec	asTc	4.762×10^{-3} K^{-1}
V_{th}	-4.07 V		
R_D	3.2525 Ω·mm	acTc	3.095×10^{-3} K^{-1}
R_S	1.7669 Ω·mm	acTc	3.095×10^{-3} K^{-1}
R_G	0.1 Ω·mm		
C_{gs}	2.73 pf/mm	cgsTc	1.23×10^{-4} K^{-1}
C_{gd}	1.03 pf/mm	cgdTc	-3.67×10^{-4} K^{-1}
C_{ds}	0.713 pf/mm	cdsTc	-3.03×10^{-4} K^{-1}
G_{leak_GD}	2.38 nA·mm^{-1}V^{-1}	akgTc	24.80 K
G_{leak_GS}	3.17 nA·mm^{-1}V^{-1}	akgTc	24.80 K
G_{leak_DS}	557.0 pA·mm^{-1}V^{-1}	akdTc	25.63 K

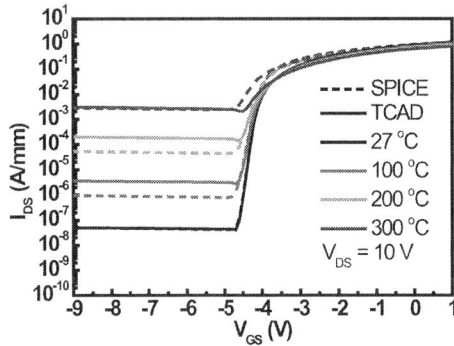

Fig. 8 The TCAD and SPICE simulated transfer I_{DS}-V_{GS} characteristics of AlGaN/GaN HEMTs from 27 °C to 300 °C.

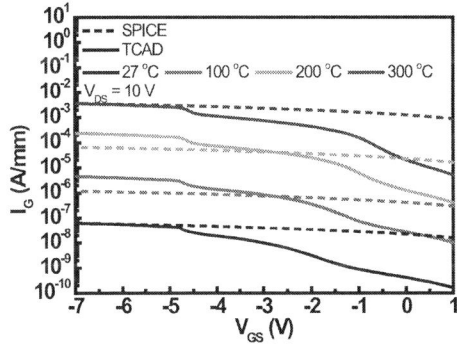

Fig. 9 The TCAD and SPICE simulated transfer I_G-V_{GS} characteristics of AlGaN/GaN HEMTs from 27 °C to 300 °C.

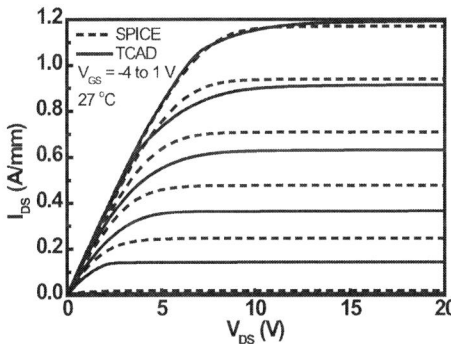

Fig. 10 The TCAD and SPICE simulated output I_{DS}-V_{DS} characteristics of AlGaN/GaN HEMTs at 27 °C.

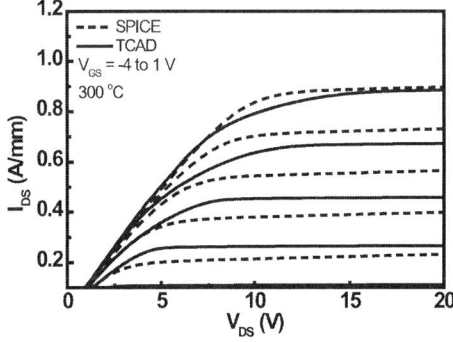

Fig. 11 The TCAD and SPICE simulated output I_{DS}-V_{DS} characteristics of AlGaN/GaN HEMTs at 27 °C.

978-1-4673-4840-9/13 $31.00 © 2013 IEEE

even at high temperature by considering the temperature coefficient in the SPICE model. It should be noted that here we assume the leakage current is as a linear function of the bias voltage to simplify our SPICE model setup, which will not be suitable for Schottky gate forward turn on or breakdown.

Conclusion

The temperature dependent behaviors of AlGaN/GaN HEMTs have been investigated and revealed by using TCAD and SPICE modeling platform. The trans-conductance, off-state leakage, series-resistance and sub-threshold swing have been found a strong degradation at high temperature, due to the physical mechanisms such as lattice scattering, tunneling, thermionic emission, etc. during the device operation. For power electronics applications, e.g. power switches, it is suggested to keep the active transistors under 300 °C to avoid any potential large off-state leakage and loss of the effective gate control.

References

[1] U. K. Mishra, "AlGaN/GaN transistors for power electronics," *IEDM Tech. Dig.*, pp. 312, 2010.

[2] M. Kanechika, T. Uesugi and Tetsu Kachi, "Advanced SiC and GaN Power Electronics for Automotive Systems," *IEDM Tech. Dig.*, pp. 324, 2010.

[3] K. J. Chen, L. Yuan, M. J. Wang, H. Chen, S. Huang, Q. Zhou, C. Zhou, B. K. Li and J. N. Wang, "Physics of Fluorine Plasma Ion Implantation for GaN Normally-off HEMT Technology," *IEDM Tech. Dig.*, pp. 465, 2011.

[4] K.Y. Wong, W. Chen and K.J. Chen, "Integrated Voltage Reference and Comparator Circuits for GaN Smart Power Chip Technology," *ISPSD 2009*, pp. 57, 2009.

[5] N. Zhang, V. Mehrotra, S. Chandrasekaran, B. Moran, L. Shen, U. Mishra, E. Etzkorn and D. Clarke, "Large Area GaN HEMT Power Devices for Power Electronic Applications: Switching and Temperature Characteristics," *PESC 2003*, pp. 233, 2003.

[6] J.A. Charter, J. Acord, D. Hoffmann, A. Trageser and C. Pagel, "Thermal Factors Influencing the Reliability of GaN HEMTs," *IEEE SEMI-THERM Symposium 2012*, pp. 182, 2012.

[7] A.M. Darwish, A.J. Bayba and H.A. Hung, "Utilizing Diode Characteristics for GaN HEMT Channel Temperature Prediction," *IEEE Trans. Microwave theory and techniques*, vol. 56, pp. 3188, 2008.

[8] L. Yuan, W. Wang, K.B. Lee, H. Sun, S.L. Selvaraj, S. Todd and G. Lo, "On the operation mechanism and device modeling of AlGaN/GaN high electron mobility transistors (HEMTs)," *ICECECE 2012*, pp. 165-169, 2012.

[9] A. Armstrong, et al., "Impact of Growth Pressure on Defects in GaN Grown - by Metalorganic Chemical Vapor Deposition," *Compound Semiconductors: Post-Conference Proceedings*, 2003 International Symposium on, pp. 42-48, 2003.

[10] Y. C. Choi, et al., "The Effect of an Fe-doped GaN Buffer on OFF-State Breakdown Characteristics in AlGaN/GaN HEMTs on Si Substrate," *IEEE Trans. Electron Devices*, vol. 53, pp. 2926-2931, Dec. 2006.

[11] A. Rizzi, et al., "Surface and interface electronic properties of AlGaN(0001) epitaxial layers," *Appl. Phys. A*, vol.87, pp. 505-509, Mar. 2007.

SOI CMOS Integrated Zinc Oxide Nanowire for Toluene Detection

S. Santra[1*], P. K. Guha[2], S. K. Ray[1], F. Udrea[3] and J. W. Gardner[4]

[1] Department of Physics & Meteorology, Indian Institute of Technology, Kharagpur, India – 721302
[2] Department of Electronics & Electrical Communication Engineering, Indian Institute of Technology, Kharagpur, India – 721302
[3] Department of Electrical Engineering, CAPE Building, University of Cambridge, UK – CB3 0FA
[4] School of Engineering, University of Warwick, UK – CV4 7AL
*Phone: +91-3222-2882285, Fax: +91-3222-255303, e-mail: ss778@cam.ac.uk

Abstract

The paper reports on the in-situ growth of zinc oxide nanowires (ZnONWs) on a complementary metal oxide semiconductor (CMOS) substrate, and their performance as a sensing element for ppm (parts per million) levels of toluene vapour in 3000 ppm humid air. Zinc oxide NWs were grown using a low temperature (only 90°C) hydrothermal method. The ZnONWs were first characterised both electrically and through scanning electron microscopy. Then the response of the on-chip ZnONWs to different concentrations of toluene (400 – 2600ppm) was observed in air at 300°C. Finally, their gas sensitivity was determined and found to lie between 0.1% and 0.3% per ppm.

Keyword: Gas sensor, SOI CMOS, MEMS, zinc oxide nanowires, toluene sensor

Introduction

Nanomaterials are promising candidates in the gas sensors market because of their high surface to volume ratio, which will lead to much higher sensitivity. Hence gas sensors based on nanomaterials are expected to play a crucial role for environmental monitoring, automobiles, health care system and even in military scenario. Gas sensors available today in the market are generally discrete, bulky (1 cm^3), expensive ($20) and typically power hungry (100 mW). But integration with SOI (silicon on insulator) CMOS MEMS (Micro Electro Mechanical System) not only allows sensing devices with a smaller size (1 mm^2), but also offers low power (10 mW) consumption, option for batch fabrication and most importantly the possibility for on-chip circuitry for sensor interfacing. But conventional bulk materials are not suitable for this type of micro-sensor. Nanomaterials are a strong contender for miniaturized sensors. So such a single chip solution (sensor-on-a-chip) is the key for the development of a new generation low-cost, low-power sensors in high volume.

Among the different metal oxide nanostructures ZnO is a promising material for nano gas sensors. They have been tested in the presence of different gases and reported previously [1-4].

ZnONWs are n-type semiconductors with high thermal stability. The high temperature (more than 500°C) growth of ZnONWs (e.g., growth using chemical vapour deposition (CVD)) is not suitable for a CMOS substrate. In this paper the integration of ZnONWs (grown at 90°C using CMOS friendly hydrothermal technique) with the CMOS MEMS substrate and their response to toluene vapour in air is reported.

Micro-hotplate fabrication

The basic gas sensing device is a micro-hotplate structure. It mainly consists of a micro-heater and interdigitated electrodes (IDEs). The micro-hotplates were designed in Cadence and fabricated at the wafer level in a commercial foundry using 1.0 µm SOI CMOS process. The micro-heater was made of p^+ single crystal silicon of radius 75 µm. The IDEs were formed from the top metal layer of the CMOS process. They were exposed by removing silicon oxide/nitride passivation layer at the same step of bond-pad formation. The IDEs are used to measure the change in resistance of sensing material in the presence of a gas. The membrane structure (which was made of silicon nitride/oxide of radius 282 µm) was formed using Deep Reactive Ion Etching technique after the CMOS process at a separate commercial MEMS foundry to reduce the power consumption. Detailed descriptions of these devices may be found in [4-7]. An optical microscope picture of the fabricated device is shown in Fig. 1. Power versus temperature plot is shown in Fig. 2. The micro-heaters were characterised on several devices across the wafer and also from wafer to wafer. It was found that an operating temperature of 500°C can be reached at expense power consumption of just 30±2 mW.

Zinc oxide nanowires fabrication and characterization

The Zinc oxide nanowires were synthesized on-chip using a low temperature (~90°C) hydrothermal method - rather than a conventional high temperature CVD technique. In our method a very thin zinc oxide seed layer was deposited on the interdigitated electrode (using a metal shadow mask) using a radio frequency sputtering system.

Then an equimolar (25 mM) aqueous solution of zinc nitrate hexahydrate ($Zn(NO_3)_2 \cdot 6H2O$, Sigma–Aldrich) and hexamethylenetetramine (HMTA, Sigma Aldrich) were made and heated to 90°C. The sputter deposited chips were dipped into the solution and held there for a period of 2 hours. After growth, the chips were cleaned with the deionised water and dried under nitrogen flow.

The current versus voltage plot is shown in Fig. 3 and was taken at room temperature. The almost linear plot confirms that there is a good Ohmic contact between the ZnONWs and the

978-1-4673-4840-9/13 $31.00 © 2013 IEEE   119

sensing electrodes. A field emission scanning electron microscope picture showed that the electrodes were covered by ZnONWs and that they were nearly aligned (see Fig. 4).

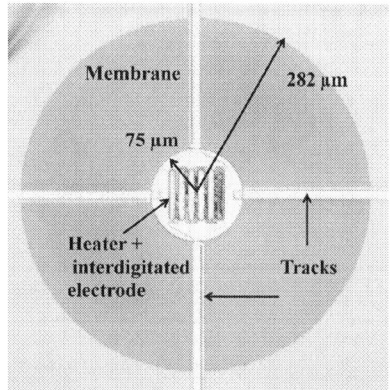

Fig. 1 Top view of fabricated SOI device with IDEs

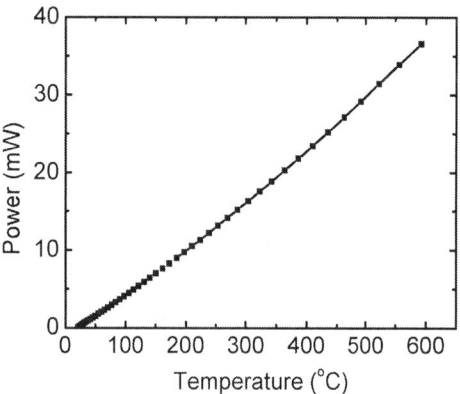

Fig. 2 Power versus temperature plot of the silicon micro-heater on an SOI CMOS membrane

Toluene test and discussion

The gas sensing measurements was performed at the Microsensors & Bioelectronics Laboratory, Warwick University (UK) using a custom test facility. The gas sensor device was bonded onto a 16 pin DIL (duel in line) package and then kept in a stainless steel chamber. The sensor was connected to the test and measurement system which was interfaced through a National Instruments DAQ card within a PC. The sensor was given a constant current and the voltage across the sensor was measured. The silicon micro-heater was used to heat the sensing material locally and kept it at a constant temperature of 300°C. The ZnONWs response to six different toluene concentrations (400 – 2600 ppm) was measured in 3000 ppm humid air as shown in Fig. 5. The sensor was exposed to air for 25 minutes and then toluene vapour for 25 minutes.

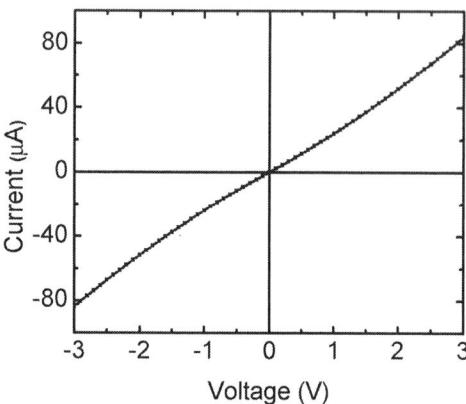

Fig.3 Current vs voltage plot of the ZnONWs measured in laboratory air

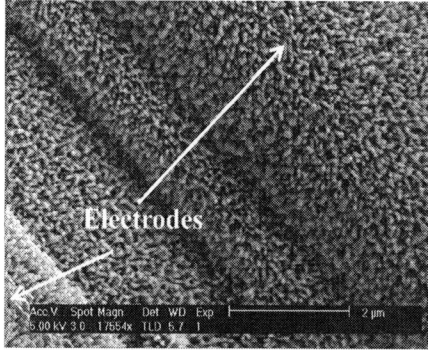

Fig. 4 SEM picture of the ZnONWs on interdigitated electrode

The working principle of the metal oxide resistive gas sensor is that the conductivity changes of the semiconducting materials upon the interaction with the target gas [8]. When the semiconductor materials are exposed to air an oxygen molecule adsorbs on the surface of the ZnONWs. As a result, an oxygen ion (O_{ads}^-, O_{2ads}^- and O_{ads}^{2-}) is formed by capturing an electron from the conduction band which in turn produces an electron-depleted space-charge layer in the surface region of the nanowires and results in a higher resistance [9]. The reaction kinetics can be described as follows:

$$O_{2gas} \leftrightarrow O_{2ads} \qquad (1)$$

$$O_{2ads} + e^- \leftrightarrow O_{2ads}^- \qquad (2)$$

$$O_{2ads} \leftrightarrow 2O_{ads} \qquad (3)$$

$$O_{ads} + e^- \leftrightarrow O_{ads}^- \qquad (4)$$

$$O_{ads}^- + e^- \leftrightarrow O_{ads}^{2-} \qquad (5)$$

978-1-4673-4840-9/13 $31.00 © 2013 IEEE   120

When these sensors are exposed to reducing gases (like toluene) the gas reacts with these surface adsorbed oxygen species and release the trapped electrons back to the conduction band increasing the electron concentration. The reaction between the toluene and ionic oxygen species is described by

$$C_7H_8 + 2O^- \leftrightarrow H_2O + C_7H_{6-}O + 2e^- \quad (6)$$

The ZnONWs response was found to increase with increasing concentration of toluene vapour as shown in Fig. 6. A straight line fitted well through the experimental points. The sensor sensitivity S (i.e. response per ppm, where response is the ratio of resistance of ZnONWs in the presence of air (R_a) to the resistance of ZnONWs in the presence of toluene (R_g)) was found to be 0.3%/ppm and 0.1%/ppm at 400 and 2600 ppm, respectively.

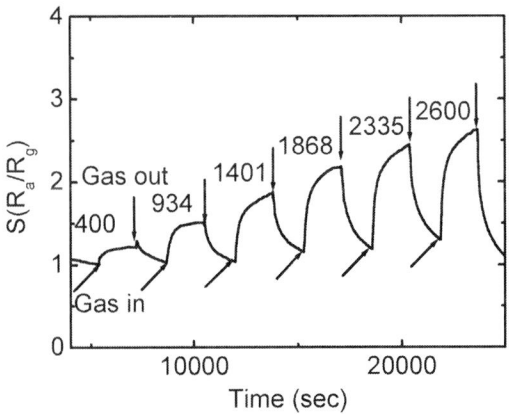

Fig. 5 Change in resistance of the ZnONWs in presence of different ppm concentrations of toluene vapour in air

Fig. 6 Response (linear) of the ZnONWs as a function of toluene concentration in air.

Conclusions

This paper describes the growth of ZnONWs on an SOI CMOS MEMS substrate using a low temperature (90°C) hydrothermal method. Electrical characterisation showed good Ohmic contacts between the NWs and sensing electrodes. SEM showed that the NWs were nearly aligned. The sensor was tested at six different concentration of toluene vapour at humid air (3000 ppm) and sensitivities were found to be 0.3%/ppm and 0.1%/ppm at 400 and 2600 ppm, respectively. We believe that ZnONWs CMOS based sensors could be important in the future development of low-cost (<$5), low-power (< 10 mW) smart sensor systems for the environmental monitoring of toxic volatile organic compounds (such as toluene).

Acknowledgment

This work has been supported in part by department of science and technology (DST), India under the project number SR/S2/RJN-104/2011.

References

[1] Y. Zeng, T. Zhang, L. Wang, R. Wang, W. Fu, H. Yang, *J. Physical Chemistry C*, 113 pp. 3442–3448, 2009.

[2] R. Ferro, J. A. Rodriguez, P. Bertrand, *Thin Solid Films*, 516, pp. 2225–2230, 2008

[3] N. Hongsith, C. Viriyaworasakul, P. Mangkorntong, N. Mangkorntong, S. Choopun, *Ceramics International*, 34, pp. 823–826, 2008

[4] S. Santra, P. K. Guha, S. Z. Ali, P. Hiralal, H. E. Unalan, J. A. Covington, G. A. J. Amaratunga, W. I. Milne, J. W. Gardner, F. Udrea, *Sensors and Actuators B*, 146, pp. 559–565, 2010.

[5] S. Santra, S. Z. Ali, P. K. Guha, J. A. Covington, W. I. Milne, J. W. Gardner, F. Udrea, *Nanotechnology* vol 21 no 48 7, 2010.

[6] P. K. Guha S Z Ali, CCC Lee, F Udrea, W I Milne, T Iwaki, J A Covington, J W Gardner, *Sensors and Actuators B*, 127, pp. 260-266, 2007.

[7] S. Z. Ali, F. Udrea, W. I. Milne, J. W. Gardner *Journal of Microelectromechanical Systems*, 17, No. 6, pp. 1408-1417, 2008.

[8] J. Cerda Belmonte, J. Manzano, J. Arbiol, A. Cirera, J. Puigcorbe, A. Vila, N. Sabate, I. Gracia, C. Can'e, J.R. Morante, *Sens. Actuators B* 114, pp. 881–892, 2006.

[9] N.J. Dayan, S.R. Sainkar, R.N. Karekar, R.C. Aiyer, *Thin Solid Films*, 325, pp. 254–258, 1998.

Critical Thickness of Diamond-like Carbon Study using X-ray Photoelectron Spectroscopy Depth Profiling

Nattaporn Khamnaulthong[1, 2], Krisda Siangchaew[2], and Pichet Limsuwan[1, 3]

[1]Department of Physics, Faculty of Science, King Mongkut's University of Technology Thonburi, Bangkok, 10140, Thailand
[2]Western Digital (Thailand) Co.Ltd, Ayutthaya, 13160, Thailand
[3]Thailand Center of Excellence in Physics, CHE, Ministry of Education, Bangkok 10400, Thailand

nattapornkh@gmail.com, krisda.siangchaew@wdc.com, opticslaser@yahoo.com

Abstract

A bi-layer stack of ion-beam sputtered Si-Si_3N_4 and tetrahedral amorphous diamond-like carbon made by filtered cathodic arc thin films were grown and heated to 200 °C at varying time to study the protective capability of diamond-like carbon film in preventing oxidation of underlying material. Depth profiling of such film by X-ray photoelectron spectroscopy showed that tetrahedral amorphous diamond-like carbon can slow down oxidation and with 1.0 nm thickness film can prevent Si-Si_3N_4 film from oxidizing when heated up to 200 °C.

Introduction

Tetrahedral amorphous diamond-like carbon (ta-DLC) films made by Filtered Cathodic Arc (FCA) are widely used in many applications such as for protecting corrosion-sensitive magnetic materials in the magnetic recording head and magnetic disk. [1,2] In such application, the ta-DLC film is expected to have thickness of few nanometers and yet still provide a good protection against environmental corrosion and tribomechanical wear. ta-DLC film can be thinned down to few nanometers and yet still has superior mechanical and chemical properties compared to many other films. [3,4] High fraction of carbon in ta-DLC film are bonded with sp^3-bonded structure. Carbon with high fraction of sp^3 bonded structure can be packed tightly to produce a high density and low permeability film. [1,2] It is because of this high density and hardness that fundamentally allow ta-DLC film to be thinned down. Therefore, it is the purpose of this study to determine the ta-DLC critical film thickness below which the film loses its protective ability against thermal oxidation.

Experiment Setup

A. Process condition

Samples were prepared using Veeco's DC-FCA tool. Pre-deposition surface etching, adhesion material film deposition and ta-DLC film deposition were made within a single chamber which equipped with two RF ion beam sources and FCA source with 90° curvilinear magnetic solenoid filter. In-situ process control is being performed by two multi-wavelength ellipsomters for separate monitoring for Si-Si3N4 sputtering and ta-DLC deposition [5]. Ion beam etching was performed with a 120V Ar+ ion to clean the sample surface before adhesion and ta-DLC deposition. 1.0 nm

of Si-Si_3N_4 was then deposited to act as an adhesion layer prior to the DLC coating. DLC films were then deposited from a FCA to a thickness 0, 0.5, 1.0, and 1.5 nm on NiFe coupons.

As-grown ta-DLC/ Si-Si_3N_4 films were then isothermally stress in oven at 200 °C for 30, 60, 90, 120, and 150 minutes. Samples were loaded into the oven at room temperature and thermally ramped to set annealing temperature at 20 °C/min. Heating was done in an air atmosphere.

B. Film characterization

PHI Quantera SXM X-ray photoelectron spectroscopy (XPS) was used to spatially-resolved the presence of O1s and Si2p in this study. The sputtering gun setting was set at 1 kV at 2x2 mm²; and the spectroscopy data were collected using an Al Kα X-ray source with 20 μm spot size, and a take-off angle of 45 degree. XPS depth profiling was collected from as-deposited and heated samples. Sputtering rate of 0.5 nm per cycle was used in depth profiling scanning.

The oxidation rate and initial oxide of each ta-DLC thickness that calculated from slope and intercept of the plot of maximum oxide detected in depth profiling as a function of heating time.

Result and Discussion

Fig 1 and 2 show the example of depth profiling of Si-Si_3N_4 adhesion layer oxide (Si-O) as-deposited and after heating 150 min of 1.0 nm adhesion layer thickness and ta-DLC 0 nm, 0.5 nm, 1.0 nm, and 1.5 nm. The percent atomic concentration of Si-O will be considered the maximum peak of each profile.

Table I shows the percent atomic concentration of the Si-O obtained from samples with varying DLC thicknesses as-is and after they have been heated to various times. The results show that the Si-O percentage increases as the heating time increases; and it significant decreases as ta-DLC thickness increases. The results also indicated that with 0.5 nm ta-DLC, the amount of Si-O found can be reduced by half when compared to non-DLC coated parts. At 1.0 and 1.5 nm DLC thicknesses, there are no Si-O detected from the as-deposited condition; and the amount of Si-O detected after these samples have been heated are comparable.

Fig 3 shows the calculated oxidation rate and the initial (adhesion layer) Si-O thickness from each sample. The result shows that ta-DLC film can significant slow down the thermal oxidation of the Si-Si_3N_4 adhesion layer under specified thermal stressing condition. ta-DLC thickness of 1.0 nm is sufficient in reducing the Si-Si_3N_4 adhesion layer oxidation

978-1-4673-4840-9/13 $31.00 © 2013 IEEE

rate; and there was no further oxidation rate reduction when the ta-DLC thickness was thicker.

Summary

XPS reveals that ta-DLC film can significantly reduce the thermal oxidation of underlying material. At 0.5 nm, ta-DLC can already reduce the oxidation rate by half; and an optimal ta-DLC thickness for protecting against thermal oxidation of the Si-Si_3N_4 adhesion layer is within 1.0 to 1.5 nm thickness range when the heating condition is in the range of 200 °C.

Acknowledgements

This work was supported by Magnetic Head Operations at Western Digital (Thailand) company limited and we would also like to thank DSTAR, KMITL and NSTDA, NECTEC for Ph.D. scholarship in data storage research.

References

[1] J.Robertson, Material Science and Engineering, R37, pp. 129-281, 2002

[2] A. Anders, et al., IEEE Trans. Plasma Sci., Vol. 29, No.5, pp.768-775, Oct. 2001

[3] H. Tsai, et al., J.Vac. Sci. Technol. A, Vol.5, pp. 3287-3312

[4] N. Yasui, et al., IEEE Trans. Magn., Vol. 45, No.2, pp. 805-809, Feb. 2009

[5] B. Druz, et al., Diamond Relat.Mater., Vo. 14, pp. 1508-1516, 2005

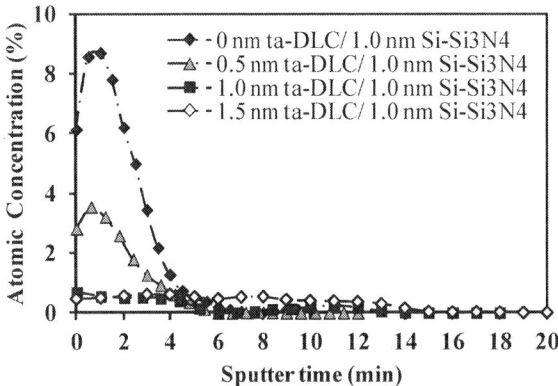

Fig. 1 Adhesion layer oxide (Si-O) depth profiling as-deposited of 1.0 nm adhesion layer thickness and ta-DLC 0 nm, 0.5 nm, 1.0 nm, and 1.5 nm

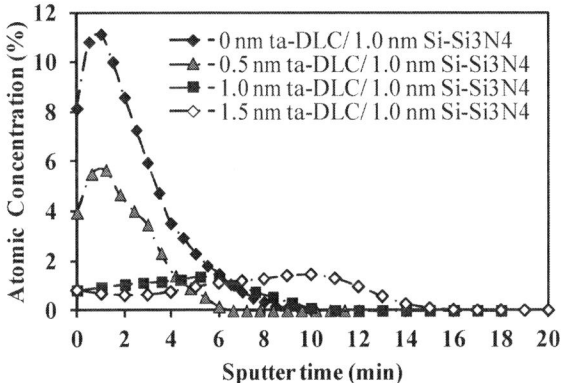

Fig. 2 Adhesion layer oxide (Si-O) depth profiling after heating 150 min of 1.0 nm adhesion layer thickness and ta-DLC 0 nm, 0.5 nm, 1.0 nm, and 1.5 nm

TABLE I
%ATOMIC CONCENTRATION OF Si-Si_3N_4 ADHESION LAYER OXIDE (Si-O) DETECTED ON PART WITH DIFFERENT ta-DLC THICKNESS AS-DEPOSITED AND AFTER HEATING AT VARIOUS TIMES

Heating time (min)	ta-DLC thickness			
	0 nm	0.5 nm	1.0nm	1.5 nm
0	8.70	3.55	0.00	0.00
30	9.04	4.29	0.80	0.71
60	9.04	4.79	0.94	1.07
90	9.73	5.39	1.37	1.07
120	11.51	5.72	1.30	1.45
150	11.3	5.67	1.58	1.45

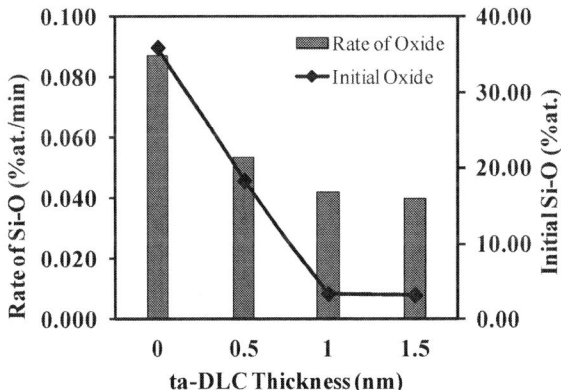

Fig. 3 Si-O rate and initial Si-O of ta-DLC 0 nm, 0.5 nm, 1.0 nm, and 1.5 nm

Impact of source pupil shapes on process windows in EUV lithography

Hung-Fei Kuo

Graduate Institute of Automation and Control, National Taiwan University of Science and Technology
#43, Sec.4, Keelung Road
Taipei, 106, Taiwan, R.O.C.
Phone: +886-2-27303695, Fax: +886-2-27301265, E-mail: hungfeikuo@mail.ntust.edu.tw

Abstract

International Technology Roadmap for Semiconductors (ITRS) report proposes extreme ultraviolet (EUV) lithography to be the key candidate of lithography tools to manufacture devices at the 22nm node and beyond. The image effects on wafer critical dimensions (CDs) in the EUV lithography are different from the effects in the conventional lithography caused by the off-axis illumination and refelective optics design. This research investigates process windows of line/space (L/S) with the target CD 22nm and contact hole (CH) features with the target CD 35nm illuminated by the conventional, annular, dipole, and quasor source shapes. The diffraction amplitudes by the EUV mask are summarized. The research suggests that the dipole is the better illumination source shape to print L/S features and the quasor is the better one for CH features. In addition, the research reports the best dipole illumination setting for the L/S features and the best quasor illumination setting for the CH features. The exposure latitude and depth of focus (DOF) for L/S feature illuminated by the dipole to print the target CD 22nm are 4% and 100nm respectively. The exposure latitude and DOF for CH feature illuminated by the quasor to print the target CD 35nm are 24% and 300nm respectively. Full field analysis of CH CDs displays the minimized CD error through slit due to the quasor illumination in the EUV lithography.

Keywords: source pupils, EUV lithography, process windows, critical dimension, exposure latitude, and depth of focus.

Introduction

A trend of device down-scaling continues to achieve the dense device integration and low power consumption over the constant chip area [1]-[3]. The half pitch for a memory cell in the CMOS process is toward the 22nm node and beyond according to the International Technology Roadmap for Semiconductor (ITRS) report [4]. One of the key challenges proposed by ITRS is to enable the new lithography tool to process devices at the new technology node. Extreme ultraviolet (EUV) lithography at the wavelength 13.6nm is the leading candidate among other lithography tools due to its high-volume manufacturability [5]-[7]. Previous research works focused on imaging effects on wafer caused by off-axis illumination and reflective optics design in the EUV different from the conventional immersion lithography tools [8]-[11]. However, one of the unresolved issues for the EUV tool is to scale the source power efficiency with the desired throughput rate [12]-[14]. The siginificance of the research is to investigate the effects of various source pupils related with exposure latitudes on process stacks in the EUV lithography.

This paper explores the response of process windows to the illumination sources using full field analysis. In addition, this study identifies the contribution of illumination source parameters to critical dimension (CD) variations. The studies of process windows are carried out for lithography features including line/space (L/S) and contact hole (CH) patterns on the EUV mask. The key benefit for the study is to breakdown the souce pupil impacts to achieve a better lithographic control of EUV tools implemented in production for the semiconductor industry.

Lithography Process Modeling

Fig 1 sketches unit cells of L/S and CH features on two EUV masks for the research study. The white and the black areas indicate 100% and 0% transmittance, respectively. The line feature width is 88nm and the pitch is 332nm in Fig. 1(a). The rectangular hole feature width is 140nm and the pitch is 560nm on the mask in Fig. 1(b). The EUV light at the wavelength 13.6nm illuminates the mask by the oblique angle 6° and azimuthal angle φ from 66° to 114°. The EUV mask uses a sequence of 20 alternating layers of molybdenum and silicon to reflect the light. The multilayer stack is protected by a ruthenium capping layer. An absorber layer on the top of the capping layer provides spatially modulated reflectivity of the mask. An objective lens with the reduction ratio 4 and numerical aperture 0.25 collects the backward diffracted beams and focuses the beams without aberration to print the L/S and CH patterns on the wafer process stack. The process stack consists of the resist layer and silicon substrate. The thickness of the resist layer is 80nm. The position of the mask focus is placed at 80nm above the mask. The position of the wafer focus is placed in the middle of the resist layer. The illumination source shapes are summarized in Table I for the study including the conventional with σ=0.8, annular, dipole, and quasor. The σ is defined in (1).

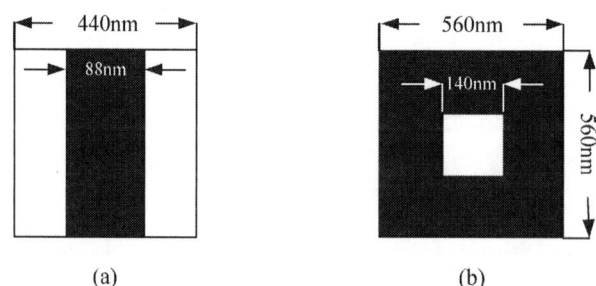

(a)   (b)

Fig. 1 A unit cell on the EUV mask for (a)L/S; and (b) CH features.

Table I			
Various illumination source shapes in the EUV lithography.			
Conventional	Annular	Dipole	Quasor

$$\frac{n\sin(\theta_{max})}{NA} = \frac{source\ diameter}{lens\ diameter} \tag{1}$$

,where NA is the numerical aperture and n is the refractive index of the medium , and θ_{max} is the maximum half-angle of the light that can enter the objective lens.

Numerical Analysis of Process Windows

The complex amplitude of the diffraction by the EUV mask with L/S or CH features is calculated based on the Rayleigh-Sommerfeld diffraction formula in (2) [15] [16].

$$u(p_0) = \frac{1}{j\lambda}\iint_A u(p_1)\frac{\exp(jkr_{01})}{r_{01}}\cos(\hat{n},\hat{r}_{01})dA \tag{2}$$

,where u is the complex amplitude in space, λ is the wavelength, j is the imaginary number $\sqrt{-1}$, r_{01} is the measurement distance, and $\cos(\hat{n},\hat{r}_{01})$ represents the cosine of the angle between the vector $\hat{r}_{01}$ and the outward normal $\hat{n}$. Table II and Table III summarize the diffraction amplitudes by EUV mask with L/S and CH features, respectively. Table II reports the amplitudes of the -1st, 0th, $+1^{st}$ diffraction orders by the EUV mask with the L/S features in Fig. 1(a). The maximum diffraction amplitude is 0.8 occured at the 0th order. Table III reports amplitudes of the -1st, 0th, and $+1^{st}$ diffraction orders by the EUV mask with the CH features in Fig 1(b). The maximum diffraction amplitude is 0.063 occured at the (0,0) order.

Table II
The amplitude for the -1st, 0th, $+1^{st}$ diffraction orders by the EUV mask with L/S features

Diffraction Order	Diffraction Amplitude
-1	0.187
0	0.8
+1	0.187

Table III
The amplitude for the -1st, 0th, $+1^{st}$ diffraction orders by the EUV mask with CH features

Diffraction Order in (X, Y)	Diffraction Amplitude
(-1,-1)	0.05
(-1,0)	0.056
(-1,+1)	0.05
(0,-1)	0.056
(0,0)	0.063
(0,+1)	0.056
(+1,-1)	0.05
(+1,0)	0.056
(+1,+1)	0.05

The analysis of process windows in the research is carried out on L/S features with the target CD 22nm and CH features with the target CD 35nm. The tolerence CD specification is within the range of target CD ±10% and the tolerence side wall angle is between 70° and 90°. Fig. 2 shows the focus exposure matrix of the L/S patterns using different illuminations. The quasor indicated by the pink line and annular illumination indicated by the green line display the better performance in the depth of focus (DOF) than the dipole indicated by the red line and conventional indicated by the blue line. However, the dipole has the better performance in the exposure latitude with 4% than the other illuminations. The dipole illumination on L/S features is further analyzed to locate the proper dipole radius sigma and dipole center sigma settings to meet target wafer CD 22nm of L/S features as shown in Fig. 3. The dipole center sigma and dipole radius sigma specify the distance of each pole from the center of the illumination source and the radius of each pole respectively. Fig. 3 illustrates the calculated wafer CDs by different colors in Fig. 3. The CDs within the range of the target CD±10% are labelled by the yellow color.

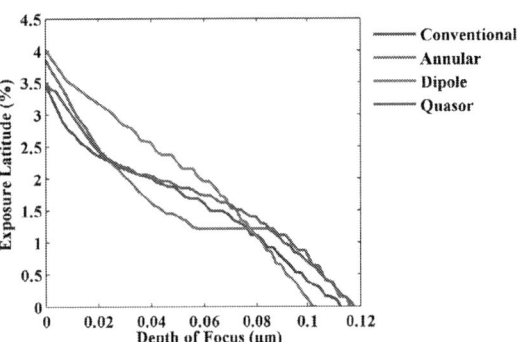

Fig. 2 Focus exposure matrix for L/S features using the conventional, annular, dipole, and quasor illuminations.

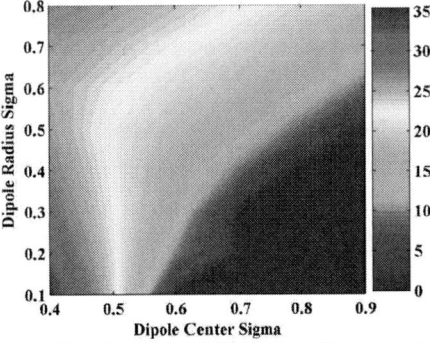

Fig. 3 The CD map of L/S features illuminated by the dipole as a function of dipole radius and dipole center sigmas.

Fig. 4 shows the focus exposure matrix for the CH features illuminated by different sources. The quasor illumination indicated by the pink line has the maximum exposure latitude 24% and the maximum DOF 300nm. The quasor illumination displays the better performance in the DOF and the exposure latitude than the annular indicated by the green color, dipole indicated by the red color, and conventional indicated by the blue color. This research further explores the quasor parameter to meet the target CD of the CH features on wafer as shown in Fig. 5. The quasor outer sigma and inner sigam specify the outer radius and inner radius, respectively. The result indicates the significant outer sigma selections are from 0.6 to 0.8 in order to print wafer CDs within the range of target CD±10% for the blade angle sector opened from 25° to 60°. When selecting the outer sigma 0.8, the difference of wafer CD is 1nm between the quasor blade angle 25° and 60°. The wafer CD is 36.5nm for blade angles from 25° to 60° at outer sigma 0.6. Fig. 6 shows the full field analysis of CDs for CH features illuminated by different source shapes. The azimuthal angles of the ring slit in EUV lithography are from 66° to 114°. The results indicate that the errors of CH CDs are varied through the slit positions. The dipole illumination displays larger drift from the target CD 35nm than the other illuminations. The quasor illumination shows the minimized CD error through slit positions.

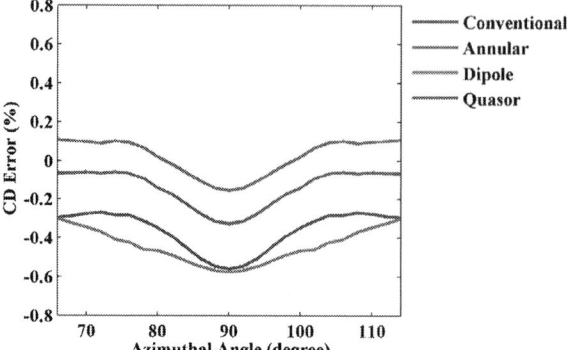

Fig 6 The CD errors as a function of the azimuthal angle of the ring slit in the EUV lithography.

Conclusion

We have presented the effects of source shapes on the process windows. The numerical investigation shows that the dipole and quasor source shapes are better illuminators to exposure L/S and CH features, respectively. The CD map of L/S features exposed by the dipole source reports the optimum range of the dipole center sigma and dipole radius sigma. The curves of CDs as a function of quasor outer sigma setting suggest the range of optimum outer sigma settings from 0.6 to 0.8. The research also discusses CD errors as a function of slit angles in the full field analysis of CH features. The result reports the quasor illumination shows the minimized CD drift from the target CD than the other illuminations.

Acknowledgments

The author would like to thank the funding support from National Science Council of Taiwan under project number NSC 100-2221-E-011-019 and also thank KLA-Tencor for supporting the Prolith simulator required to complete the numerical work.

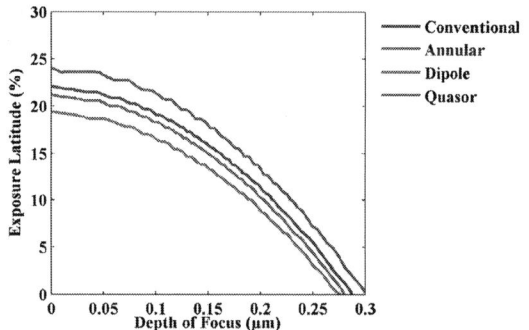

Fig. 4 Focus exposure matrix for CH features using the conventional, annular, dipole, and quasor illuminations.

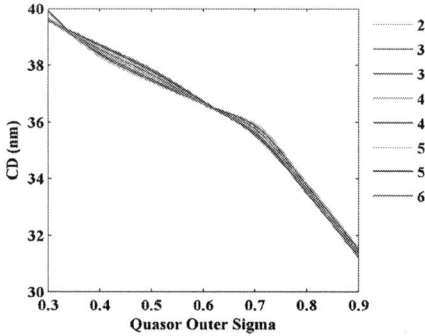

Fig. 5 The CH CDs as a function of quasor outer sigma using the inner sigma 0.3 with respect to various blade angles.

References

[1] H. Kawasaki, V. S. Basker, T. Yamashita, C. H. Lin, Y. Zhu, J. Faltermeier, *et al.*, "Challenges and solutions of FinFET integration in an SRAM cell and a logic circuit for 22 nm node and beyond," in *Electron Devices Meeting (IEDM), 2009 IEEE International*, 2009, pp. 1-4.

[2] K. J. Kuhn, "CMOS scaling for the 22nm node and beyond: Device physics and technology," in *VLSI Technology, Systems and Applications (VLSI-TSA), 2011 International Symposium on*, 2011, pp. 1-2.

[3] K. Ronse, P. Jansen, R. Gronheid, E. Hendrickx, M. Maenhoudt, V. Wiaux, *et al.*, "Lithography options for the 32 nm half pitch node and beyond," *Trans. Cir. Sys. Part I*, vol. 56, pp. 1884-1891, 2009.

[4] *International Technology Roadmap*, Executive Summary ed., 2011.

[5] A. M. Goethals, R. Jonckheere, G. F. Lorusso, J. Hermans, F. Van Roey, A. Myers, *et al.*, "EUV lithography program at IMEC," in *Emerging Lithographic Technologies XI*, *Proc. SPIE*, 2007, pp.

978-1-4673-4840-9/13 $31.00 © 2013 IEEE

651709-1-651709-12.

[6] O. Wood, C.-S. Koay, K. Petrillo, H. Mizuno, S. Raghunathan, J. Arnold, *et al.*, "EUV lithography at the 22nm technology node," in *Extreme Ultraviolet (EUV) Lithography, Proc. SPIE*, 2010, pp. 76361M-1-76361M-8.

[7] N. Harned, M. Goethals, R. Groeneveld, P. Kuerz, M. Lowisch, H. Meijer, *et al.*, "EUV lithography with the Alpha Demo Tools: status and challenges," in *Emerging Lithographic Technologies XI, Proc. SPIE*, 2007, pp. 651706-1-651706-12.

[8] H. Kang, S. Hansen, J. van Schoot, and K. van Ingen Schenau, "EUV simulation extension study for mask shadowing effect and its correction," in *Emerging Lithographic Technologies XII, Proc. SPIE*, 2008, pp. 69213I-1-69213I-12.

[9] G. McIntyre, C.-s. Koay, M. Burkhardt, H. Mizuno, and O. Wood, "Modeling and experiments of non-telecentric thick mask effects for EUV lithography," in *Alternative Lithographic Technologies, Proc. SPIE*, 2009, pp. 72711C-1-72711C-12.

[10] H. Song, L. Zavyalova, I. Su, J. Shiely, and T. Schmoeller, "Shadowing effect modeling and compensation for EUV lithography," in *Extreme Ultraviolet (EUV) Lithography II, Proc. SPIE*, 2011, pp. 79691O-1-79691O-9.

[11] M. C. Smayling, T. H. Coskun, and V. Kamat, "Custom source and mask optimization for 20nm SRAM and logic," in *Optical Microlithography XXIV, Proc. SPIE*, 2011, pp. 79731W-1-79731W-8.

[12] T. Yezheng, M. S. Tillack, S. Yuspeh, R. A. Burdt, N. M. Shaikh, N. Amin, *et al.*, "Interaction of a CO_2 laser pulse with Tin-Based plasma for an extreme ultraviolet lithography source," *Plasma Science, IEEE Transactions on,* vol. 38, pp. 714-718, 2010.

[13] E. Wagenaars, A. Mader, K. Bergmann, J. Jonkers, and W. Neff, "Extreme ultraviolet plasma source for future lithography," *Plasma Science, IEEE Transactions on,* vol. 36, pp. 1280-1281, 2008.

[14] A. Hassanein, V. Sizyuk, T. Sizyuk, and S. Harilal, "Effects of plasma spatial profile on conversion efficiency of laser-produced plasma sources for EUV lithography," *Journal of Micro/Nanolithography, MEMS, and MOEMS,* vol. 8, pp. 041503-1-041503-6, 2009.

[15] C. Mack, *Fundamental Principles of Optical Lithography: The Science of Microfabrication*: John Wiley & Sons, 2011.

[16] J. W. Goodman, *Introduction to Fourier Optics*: McGraw-Hill, 1996.

Technology Options for Reducing Contact Resistances in Nanoscale Metal-Oxide-Semiconductor Field-Effect Transistors

Yee-Chia Yeo

Department of Electrical and Computer Engineering, National University of Singapore.
4 Engineering Drive 3, Singapore 117576.
Tel: +65 6516 2298, Fax: +65 67791103, Email: yeo@ieee.org

Abstract

Technology options for reducing contact resistances in advanced transistors will be discussed. With scaling down of the size of metal-oxide-semiconductor field-effect transistors (MOSFETs) and with channel strain engineering to increase carrier mobility or reduce channel resistance, the external series resistance R_{ext} has become a more significant component of the total resistance between the source and the drain regions. An important component of R_{ext} is the contact resistance R_C between contact metallization and the heavily doped source or drain (S/D) region. Solutions for reducing R_C or the Schottky barrier height between the contact metal and the S/D regions will be reviewed.

Keywords: contact resistance, series resistance, Schottky barrier height, interface.

Introduction

For scaling Complementary Metal-Oxide-Semiconductor (CMOS) field-effect transistors (FETs) well into the sub-20 nm regime, advanced device architectures such as the multi-gate field-effect transistors (MuGFETs or FinFETs) and the ultra-thin body (UTB) structure are required for suppression of short-channel effects. Strain engineering of the silicon channel has been used to increase carrier mobility and drive current I_{Dsat}. Strain engineering would also be adopted in FinFETs or UTB FETs to enable a low channel resistance R_{ch} to be achieved. High mobility channel materials may also be used.

As R_{ch} is reduced, parasitic series resistance or external resistance R_{ext} becomes an increasingly important limiting factor for I_{Dsat}. A large portion of R_{ext} is contributed by contact resistance R_C, which is found between the source/drain (S/D) region and the metal contact material. There is a strong motivation to reduce R_C to boost I_{Dsat} performance.

In this paper, technology solutions for reducing contact resistance for nanoscale transistors will be discussed. R_C is strongly dependent on the effective Schottky barrier height between the metal contact material and the S/D regions of transistors. Novel techniques for decreasing the electron and hole barrier heights between the metallic contact and the S/D regions in n-FET and p-FET, respectively, will be examined. Technology options for integrating these techniques in advanced device architectures will be discussed.

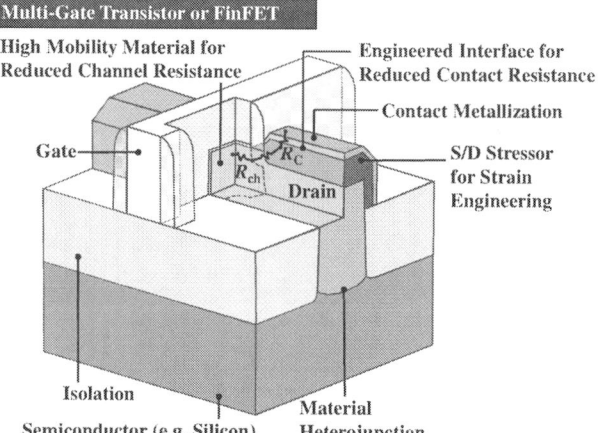

Fig. 1. Three-dimensional schematic of a multi-gate field-effect transistor or FinFET (top) and a cross-sectional schematic of an ultra-thin-body transistor (bottom). The channel material may be strained Si, Ge, GeSn, InGaAs, or other high mobility materials. The carrier mobility in the channel may be further enhanced by strain engineering techniques commonly deployed in advanced Si CMOS technology. In addition to a low channel resistance, a low contact resistance is required for achieving high drive current.

Approaches for Reduction of Contact Resistance, R_C

R_C is dependent on contact resistivity ρ_c, sheet resistance of the S/D region R_{sd}, width W_C and length L_C of the contact hole and the transfer length L_T, as given by [1]

$$R_C = \sqrt{\rho_c R_{sd}} \big/ [W_C \times \tanh(L_C/L_T)]. \qquad (1)$$

In the S/D regions of a MOSFET, the contact resistivity ρ_c is given by

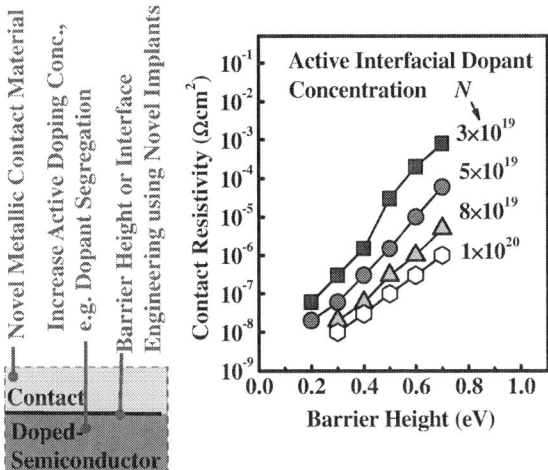

Fig. 2. Various approaches that may be adopted to reduce contact resistivity ρ_c and contact resistance R_C (left). ρ_c is reduced with higher interfacial doping concentration N and lower Schottky barrier height Φ_B (right).

$$\rho_c \propto e^{\left(\frac{4\pi\sqrt{\varepsilon_s m^*}}{h} \left[\frac{\Phi_B}{\sqrt{N}} \right] \right)}, \qquad (2)$$

where ε_s, m^*, and N are the permittivity, effective mass, and active dopant concentration of the semiconductor, respectively, h is Planck's constant, and Φ_B is the Schottky barrier height (SBH) between the the metal contact and the doped semiconductor S/D. ρ_c is an exponential function of Φ_B, and a small reduction of Φ_B can significantly reduce ρ_c (Fig. 2) and R_C. Φ_B well below 0.5 eV would be needed to realize contact resistivities in the order of 10^{-8} $\Omega \cdot cm^2$ needed for UTB FET and FinFET in advanced technology nodes.

R_C can be reduced the following approaches (Fig. 2): (1) Reducing Φ_B by choice of contact material [2]-[4], (2) Increasing N through dopant segregation [5]-[8], and (3) Reducing Φ_B by interface engineering techniques [9]-[17].

Choice of Contact Materials for Barrier Height Tuning

A first approach to reduce R_C is to reduce Φ_B through appropriate selection of the work function of the contact material. This essentially involves material selection and searching for an integration method. For n-FETs, the work function of the contact material should be small, so that the SBH for electrons $\Phi_B{}^n$ is low. For Si p-FETs, the work function of the contact material should be large, so that the SBH for holes $\Phi_B{}^p$ is low.

A. Contacts for n-FETs

Rare earth metal (e.g. Er, Yb) silicides may be used as contact materials for Si n-FETs with Si or Si:C S/D [18]-[20]. However, there are difficulties in integrating rare earth silicide modules in transistor fabrication. Rare earth metals are very reactive and tend to react with SiO_2 in the isolation regions

during the silicidation anneal, leading to inter-well leakage and loss of electrical isolation.

An alternative method of tuning metal silicide work function through formation of Ni-alloy silicide is found to be more integration friendly [21]. This involves alloying a metal M in Ni, before the silicidation to form an alloy silicide [Ni(M)Si], and is shown to be effective for Φ_B tuning. Here, M may be a metal such as Yb or Al.

Ni-Al alloy silicide (NiAlSi) has been investigated [22]-[24] for achieving low $\Phi_B{}^n$. $\Phi_B{}^n$ depends on the Al concentration in NiAlSi, and decreases from 0.65 eV for NiSi to about 0.41 eV for NiAlSi with ~20% Al. The sheet resistance for NiAlSi increases with higher Al concentration. Therefore, an optimum trade-off between $\Phi_B{}^n$ reduction and sheet resistance degradation must be carefully considered for device integration.

B. Contacts for p-FETs

For the p-FETs, metal silicides with large work functions would be needed. As SiGe S/D has been adopted for inducing compressive strain in the Si channel, research efforts on $\Phi_B{}^p$ reduction are focused on germanosilicide formation on SiGe [4],[25],[26]. One attractive approach is to use NiPt or even Pt [27] for germanosilicide contact formation. Increasing the Pt content in $Ni_{1-y}Pt_ySiGe$ contacts on SiGe S/D reduces $\Phi_B{}^p$.

Reduction of R_C through Dopant Segregation

Another approach to reduce the contact resistance is to exploit dopant segregation effects [8],[28]-[31] to boost the active dopant concentration in the semiconductor region right beneath the contact.

In this approach, traditional dopants such as arsenic ions (As^+) are implanted at a very shallow depth prior to silicidation. For arsenic segregation, doses of As^+ implanted may range from 1×10^{14} cm^{-2} to 1×10^{15} cm^{-2}. Arsenic segregates at the NiSi/Si interface during the nickel silicidation process. The heavily doped region of segregated As^+ at the contact interface effectively reduces R_C. The dopant segregation effect has been demonstrated in Ni silicidation on Si and Ni germanidation on Ge [31].

Furthermore, it has been shown that the dopant segregation method can be combined with nickel-alloy silicidation or work function tuning method to yield additive effects. NiAl silicidation, when combined with As^+ segregation, gives greater reduction in $\Phi_B{}^n$ than with either one of the methods alone [29].

Interface Engineering

It was found that with the introduction of certain atomic species such as Sulfur (S) [10]-[13] and Selenium (Se) [9]-[10] at the interface between NiSi and Si, $\Phi_B{}^n$ is reduced. The method of introduction is similar to dopant segregaton, in that S or Se ions were implanted prior to silicidation, and the silicidation process resulted in the segregation effects at the contact interface. As such atomic species are not usual dopant species in Si, it was thought that the SBH tuning effect could be due to interface effects.

978-1-4673-4840-9/13 $31.00 © 2013 IEEE

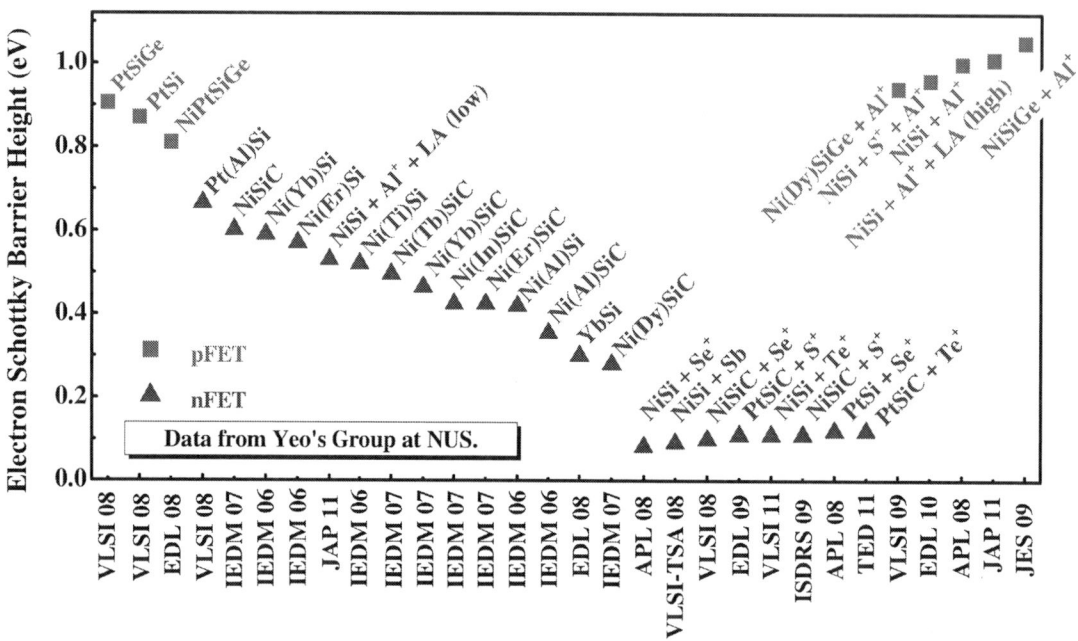

Fig. 3. Electron Schottky barrier heights obtained for various metal silicides on Si or Si:C. Barrier height tuning can be achieved by pre-silicide S^+, Se^+, Te^+, or Al^+ implants, as well as with alloy silicides [Ni(M)Si, where M is the metal alloyed with Ni prior to silicidation] or germano-silicides.

Sulfur segregation at interface between NiSi:C and Si:C was used for $\Phi_B{}^n$ and R_C reduction in FinFETs with Si:C S/D [32]. As compared with control FinFETs without introduction of S, S segregation leads to reduction in series resistance and increase in I_{Dsat} [32].

It was found that S segregation at the interface between silicon and a contact material with high work function such as Pt could also attain low $\Phi_B{}^n$ [13]. This is very interesting because a PtSi contact material which gives low $\Phi_B{}^p$ for p-FET could chosen, while on the n-FET side, its $\Phi_B{}^n$ could be optimized using a S^+ implant. Sulfur incorporated PtSi:C/Si:C contacts were also demonstrated in strained FinFETs with Si:C source/drain stressors, providing reduction in external resistance and enhancement in I_{Dsat} as compared to devices without S segregation [13].

The reduction in $\Phi_B{}^n$ due to S was explained using charge transfer and dipole formation at the silicide/semiconductor interface. However, the amount of S present at the interface may not be sufficient to quantitatively account for the magnitude of $\Phi_B{}^n$ reduction observed. Ab-initio studies revealed that the lowest energy configuration for a S atom in crystalline Si is the substitutional site. It is likely that S atoms also reside in the substitutional sites, enabling them to behave as donor-like traps. This gives rise to positive charges in the vicinity of the contact interface, and induces a high electric field near the contact interface for enhanced trap-assisted tunneling.

On the other hand, for p-type contacts, it was reported that Al implant and segregation at the NiSi/Si interface leads to reduction of $\Phi_B{}^p$ [15]. In fact, $\Phi_B{}^p$ decreases with increasing Al concentration at the NiSi/p-Si interface. This technique of pre-silicide Al implant followed by segregation during silicidation was integrated in p-FinFETs for R_C reduction [15],[33],[34]. However, when Al is incorporated into the NiSi contact material, the effect is similar to Ni(Al)Si alloy silicide, where R_C reduction on n^+ S/D is observed. Recently, we reported a way to engineer Al profiles separately for n- and p-FETs such that R_C for both n^+ and p^+ S/D are reduced [33],[34].

Summary

Various approaches for Schottky-barrier height engineering have been extensively explored for the reduction of R_C in advanced MOSFETs. Modulation of Schottky barrier height with metal alloying in NiSi and/or NiSiGe shows effective electron and hole barrier reduction for application in n-FET and p-FET, respectively. For further reduction of R_C, a combination of dopant segregation and novel metal contact materials such as NiSiAl may be used. An interfacial segregation technique with atomic species such as S or Se was also developed for R_C reduction in n-FETs. For p-FETs, Al implant and segregation was observed to reduce R_C. The approaches presented in this work may be promising for adoption in sub-20 nm technology nodes.

978-1-4673-4840-9/13 $31.00 © 2013 IEEE

Acknowledgement

The author acknowledges a research grant from National Research Foundation, Singapore (NRF-RF2008-09).

References

[1] S. D. Kim *et al.*, *IEEE Trans. Elect. Dev.* 49, pp. 467, 2002.

[2] R. T.-P. Lee *et al.*, *IEEE Elect. Dev. Lett.* 28, pp. 164, 2007.

[3] R. T.-P. Lee *et al.*, *IEEE Elect. Dev. Lett.* 29, pp. 89, 2008.

[4] R. T.-P. Lee *et al.*, *IEEE Elect. Dev. Lett.* 29, pp. 438, 2008.

[5] Y.-C. Yeo, *Semi. Sci. Tech.* vol. 22, pp. S177, 2007.

[6] K. W. Ang *et al.*, *IEDM 2004*, pp. 1069.

[7] H.-S. Wong *et al.*, *IEEE Elect. Dev. Lett.* 28, pp. 703, 2007.

[8] A. Kinoshita *et al.*, *Symp. VLSI Tech.*, pp. 158, 2005.

[9] H.-S. Wong *et al.*, *IEEE Elect. Dev. Lett.* 29, pp. 841, 2008.

[10] H.-S. Wong *et al.*, *IEEE Elect. Dev. Lett.* 28, pp. 1102, 2007.

[11] Q. T. Zhao *et al.*, *Appl. Phys. Lett.* 86, 062108, 2005.

[12] S.-M. Koh *et al.*, *ISDRS*, 2009.

[13] R. T.-P. Lee *et al.*, *IEEE Elect. Dev. Lett.* 30, pp. 472, 2009.

[14] E. Alptekin *et al.*, *IEEE Elect. Dev. Lett.* 30, pp. 331, 2009

[15] M. Sinha *et al.*, *Appl. Phys. Lett.* 92, 222114, 2008.

[16] M. Sinha *et al.*, *IEEE Elect. Dev. Lett.* 30, pp. 85, 2009.

[17] M. Sinha *et al.*, *J. Electrochem. Soc.* 156, pp. H233, 2009.

[18] S. Y. Zhu *et al.*, *IEEE Elect. Dev. Lett.* 25, pp. 565, 2004.

[19] E. Alptekin *et al.*, *IEEE Elect. Dev. Lett.* 30, pp. 949, 2009.

[20] E. Alptekin *et al.*, *J. Electrochem. Soc.*, 156, pp. H378, 2009

[21] R. T.-P. Lee *et al.*, *IEDM Tech. Dig.*, pp. 851, 2006.

[22] A. T.-Y. Koh *et al.*, *J. Electrochem. Soc.*, 155, pp. H151, 2008.

[23] R. T.-P. Lee *et al.*, *IEEE Elect. Dev. Lett.* 29, pp. 382, 2008.

[24] R. T.-P. Lee *et al.*, *IEDM Tech. Dig.*, pp. 685, 2007.

[25] R. T.-P. Lee *et al.*, *IEEE Trans. Elect. Dev.* 56, pp. 1458, 2009.

[26] E. Alptekin *et al.*, *IEEE Trans. Elect. Dev.* 56, pp. 1220, 2009.

[27] R. T.-P. Lee *et al.*, *Symp. VLSI Tech.*, pp. 28, 2008.

[28] A. Kinoshita *et al.*, in *IEDM Tech. Dig.*, pp. 79 , 2006.

[29] R. T.-P. Lee *et al.*, *Symp. VLSI Tech.*, pp. 108, 2007.

[30] C.-H. Ko *et al.*, *Symp. VLSI Tech.*, pp. 98, 2006.

[31] P. S.-Y. Lim *et al.*, *216th Electrochem. Soc. Meeting*, 2009.

[32] S.-M. Koh *et al.*, *IEEE Trans. Elect. Dev.* 59, pp. 1046, 2012.

[33] S.-M. Koh *et al.*, *IEDM Tech. Dig.*, p. 845, 2011.

[34] S.-M. Koh *et al.*, *J. Appl. Phys.* 110, 073703, 2011.

Impact of Atomic-Scale Structural Design on Ultra-Short Channel (3 nm) MOSFETs

Shinji Migita, Yukinori Morita, Meishoku Masahara, and Hiroyuki Ota

Collaborative Research Team Green Nanoelectronics Center,
National Institute of Advanced Industrial Science and Technology
AIST Tsukuba West 7A, 16-1 Onogawa
Tsukuba, Ibaraki 305-8569, Japan
E-mail: s-migita@aist.go.jp

Abstract

Electrical performances of ultra-short channel MOSFETs were investigated on SOI substrates. The channel length was scaled to 3 nm using the anisotropic wet etching technique. A difficulty of junction technology was solved by fabrication of Junctionless-FET, which consists of uniform high concentration dopants through the body of device. Superior Junctionless-FET performance was confirmed when the channel thickness and gate dielectric film thickness were scaled close to 1 nm. Experimental and simulation studies suggest that variation of performance originates from atomic-scale fluctuation in device structures.

Keywords: MOSFET, Junctionless-FET, V-groove, Short Channel, variation.

Introduction

Scaling of MOSFETs is the fundamental booster for the development of VLSI circuits, and the critical dimension of advanced MOSFETs are approaching to 10 nanometers presently. For the sustainable progress of future VLSI, fabrication and investigation of nanometer-scale MOSFETs is indispensable.

One of the big challenges in ultra-short channel MOSFETs is the integration of junction technology. PN junctions at source and drain edges have been utilized to operate inversion-mode FETs (Fig. 1(a)). The strict positioning of PN junctions, however, will fail when the channel lengths of MOSFETs are scaled to several nanometers. Junctionless-FET that consists of uniform doping concentration and has no PN junction is expected to solve this issue [1]. Drain current of Junctionless-FET flows through the body of channel smoothly, and is cut off by the extension of depletion layer (Fig. 1(b)).

In this paper, we present experimental demonstration of 3 nm-channel length MOSFETs that is operate in Junctionless-FET mode [2, 3]. In combination with simulation study, the impacts of structural design in atomic-scale in future nano-devices are discussed.

Experimental

Junctionless-FETs were fabricated on SOI wafers (Fig. 2). Firstly, V-grooves were formed on SOI using the anisotropic wet etching nature of alkaline solution [4-6]. We used Tetramethylammonium hydroxide (TMAH) solution. Etching rates of undoped Si (100) and (111) surfaces at room temperature were 27 nm/min and 2.2 nm/min, respectively (Fig. 3(a)). Owing to the large difference of etching rates, the etched holes forms V-shape grooves that consist of Si (111) facets (Fig. 3(b)). The depths of V-grooves are proportional to

the aperture sizes. Thus ultra-thin channel can be formed on SOI substrates by precisely controlling the aperture size of oxide mask (Fig. 2(a)). Following to the deposition and patterning of poly-Si/metal/high-k gate stack structure, a heavy dose of ion implantation was executed at the thick source and drain region (Fig. 2(b)). P^+ and BF_2^+ ion sources were used for n-type and p-type doping, respectively. Junctionless-FET was completed by activation annealing at 1000°C for 10min that promotes lateral diffusion of dopants into channel region (Fig 2(c)). Simulation study of dopant diffusion behavior show that 10 min is long enough to form a highly doped channel region in our device structure (Fig. 4). Devices were finished by metallization and forming gas anneal in H_2 gas at 400°C.

Electrical performances were measured using the semiconductor parameter analyzer (Agilent 4156C) combined with a semi-auto prober system at room temperature. Device simulation was performed using SILVACO simulator in which quantum module is added.

Results and Discussion

TEM image of a Junctionless-FET is shown in Fig. 5. Ultra-thin channel is formed at the edge of V-groove while the source and the drain region consist of thick SOI layers. This device structure contributes to both the superior controllability of channel potential by the gate bias and the reduction of parasitic resistance at the source and drain region. A magnified TEM image of the V-groove shows that the sidewalls consist of atomically flat Si (111) facets. It is interesting that the edge of V-groove is not sharp-pointed in atomic-scale but has a rounded shape with small curvature. The shortest channel length thus determined is 3 nm.

Electrical performances of Junctionless-FETs with 3.2 nm equivalent oxide thicknesses (EOT) are shown in Fig. 6. Both n-type and p-type FETs show clear cut-off performances. However these FETs require a gate bias swing as large as 5 V, and the operations are in "normally-on" style. This is because the EOT is too thick to control the channel potential effectively. Improved performances are obtained by scaling the EOT from 3.2 nm to 1.5 nm (Fig. 7). Superior cut-off performances are obtained with a small gate bias swing less than 2 V. The difference of drain currents between n-type and p-type Junctionless-FETs is proportional to the difference of electron and hole mobilities. The mobility of carries still has a strong influence on the performance of nanometer-scale MOSFETs.

EOT scaling clearly contributes to the reduction of both the subthreshold swing (SS) and the drain-induce barrier lowering (Fig. 8(a)). Furthermore, the threshold voltages (V_{TH}) were adjusted to optimal values both for n-FET and p-FET cases (Fig. 8(b)). The effective workfunction of the gate stack structure in this work was evaluated to be between 4.4 and 4.5

978-1-4673-4840-9/13 $31.00 © 2013 IEEE

eV that corresponds to the midgap of Si band gap. This experimental result suggests that CMOS devices with ultra-short channel Junctionless-FETs can be simply integrated by a midgap workfunction metal gate. It is a great advantage against the development of independent metal gate technologies for n-type and p-type FETs, respectively.

Many Junctionless-FETs were fabricated on a single SOI wafer using the same process condition. Electrical performances show variations both in p-type and n-type FETs (Fig. 9). The relationship of I_{ON} current ($V_G=1$ V for n-FET and -1 V for p-FET) and I_{OFF} current ($V_G=0$ V) are summarized in Fig. 14. They show that the variation is not random, but the I_{ON} and the I_{OFF} changes proportionally. Major origins of these variations are considered to be EOT, SOI, and channel concentration (N_{CH}). Thus impacts of these structural parameters on variations were examined by simulation. Starting form the standard structure (Fig. 10), EOT, SOI, and N_{CH} were changed individually.

Impact of EOT variation is shown in Fig. 11. Increment of EOT by 20% causes 240 mV reduction of V_{TH}. As a result both I_{ON} and I_{OFF} increase proportionally. Increment of SOI thickness by 20% also has similar effect, but the reduction of V_{TH} is 600 mV (Fig. 12). Reduction of N_{CH} from 10^{20}/cm^3 to 10^{19}/cm^3 causes 440 mV increment of V_{TH} (Fig. 13). Influence of these parameters on SS variation was negligibly small within the estimated range. Important knowledge of the simulation result is that variation of structural parameters in atomic-scale induces a large variation in electrical performances.

Another origin of variation, which is not discussed here, is the effective workfunction of gate stack caused by the orientation of metal crystals [7, 8]. Simulation of workfunction variation indicated that this factor shows similar trend with the variation of N_{CH}, and the change of V_{TH} is smaller than the case of N_{CH}. Thus the effect of workfunction is not included in this discussion although it is a significant subject of MOSFET variation that came to notice from the several 10 nanometers technology nodes.

Simulation results plotted on Fig. 14 help to consider the origin of variation in ultra-short channel Junctionless-FETs. In our experiment, EOT variation within the wafer was evaluated to be less than 0.1 nm. Therefore, the influence of EOT variation is small. In contrast, variations of SOI thickness and N_{CH} seem to affect largely. The variation of SOI thickness has two causes; one is the precision of V-groove etching and another is the uniformity of initial SOI substrate. We think that both of these causes are responsible to the variation of experimental results. Variation of N_{CH} is also noteworthy. In general, dopant diffusion in Si crystal accompanies out diffusion of dopants at the surface and interface. In case of the dopant diffusion into V-groove channel, dopants are possibly lost by out diffusion into gate dielectric and BOX layers. This effect is not negligible at the channel region where the thickness is scaled to 1 nm. A loss of a small amount of dopants at the silicon/oxide interface results in a large reduction of dopant concentration in the ultra-thin channel region. Experimental variations might be generated by the combination of SOI variation and N_{CH} reduction.

It is interesting that the trends of n-type and p-type FETs show different slope. Furthermore, the variation range of n-type FET is larger than that of p-type. This difference cannot be explained by the variation of EOT and SOI. It suggests that the nature of dopant species in a small volume of Si might be different. Therefore, individual technique might be required to control the N_{CH} for n-FET and p-FET, respectively.

Conclusion

Experimental demonstration of 3 nm channel length MOSFETs with Junctionless-FET concept encourages the scaling of MOSFETs beyond 10 nanometers technology node. Optimal threshold voltages for n-FET and p-FET are obtained using a midgap metal gate technology. Variations of electrical performances in these FETs are largely influenced by a sub-nanometer fluctuation of device structures and dopant concentration. Thus atomic-scale design of device structures and control of high concentration dopants in the channel through the understanding of dopants behavior in a small volume are indispensable for the construction of practical nano-devices in VLSI industry.

Acknowledgements

This research was granted by the Japan Society for the Promotion of Science through the First Program initiated by the Council for Science and Technology Policy. Device fabrication was supported by ICAN in AIST.

References

[1] J.P. Colinge, C.W. Lee, A. Afzalian, N. D. Akhavan, R. Yan, I. Ferain, P. Razavi, B. O'Neill, A. Blake, M. White, A.M. Kelleher, B. McCarthy, and R. Murphy: Nature Nanotech. **5**, (2010) 225.

[2] S. Migita, Y. Morita, M. Masahara, and H. Ota: Ext. Abstract of SSDM 2012, p. 725.

[3] S. Migita, Y. Morita, M. Masahara, and H. Ota: Tech, Dig. IEDM 2012, 8.6.

[4] H. Seidel, L. Csepregi, A. Heuberger, and H. Baumgartel: J. Electrochem. Soc. **137** (1990) 3612 and 3626.

[5] T.D. Mok and C.A.T. Salma: Solid-State Electron. **19** (1976) 159.

[6] J. Appenzeller, R. Martel, P. Solomon, K. Chan, Ph. Avouris, J. Knoch, J. Benedict, M. Tanner, S. Thimas, K.L. Wang, and J.A. del Alamo: Microele. Eng. **56**, (2001) p.213.

[7] K. Ohmori, T. matsuki, D. Ishikawa, T. Morooka, T. Aminaka, Y. Sugita, T. Chikyow, K. Shiraishi, Y. Nara, and K. Yamada: Tech. Dig. IEDM 2008, p.409.

[8] K. Endo, S.-I. O'uchi, Y. Ishikawa, Y. Liu, T. Matsukawa, K. Sakamoto, J. Tsukada, H. Yamauchi, and M. Masahara: IEEE Electron Device Lett. 31 (2010) 546.

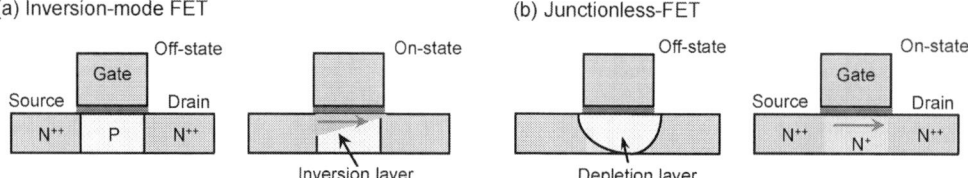

Fig. 1 Structures and operation principles of two types of MOSFETs. (a) Inversion-mode FET has PN junctions at source and drain edges. Drain current flow through the inversion layer which is formed by the gate bias. (b) Junctionless-FET has no PN junction at source and drain edges. Drain current flows through the body of channel, and is cut off by the depletion layer formed by the gate bias.

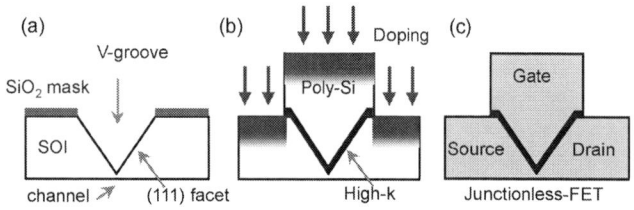

Fig. 2 Process flow of ultra-short channel Junctionless-FET in this work. (a) V-groove formation on undoped SOI using TMAH wet etching. (b) Gate stack deposition and patterning followed by heavy dose ion implantation. (c) Activation and diffusion anneal at 1000°C for 10 min, and salicidation process at S/D region optionally.

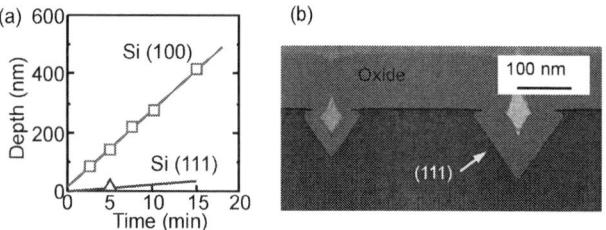

Fig. 3 (a) Etching rate examinations of Si (100) and Si (111) surfaces by TMAH solution at room temperature. Etching depths were measured by DEKTAK. (b) TEM image of anisotropic wet etching of undoped Si (100) substrate by TMAH solution. Grooves form identical V-shape, and the depths of grooves are corresponding with the sizes of apertures.

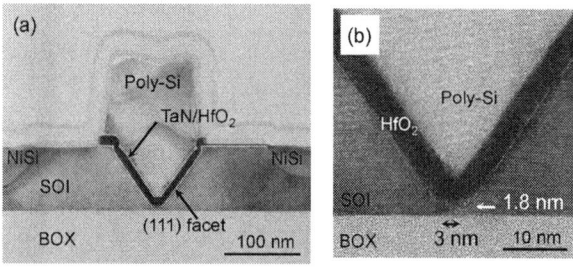

Fig. 5 (a) Whole TEM image of a Junctionless-FET integrated with poly-Si/TaN/HfO$_2$ gate stack and NiSi at S/D region. (b) An enlarged TEM image at the edge region of V-groove. Note that the channel length and the thickness are scaled to 3 nm and 1.8 nm, respectively. Many TEM observations suggest that the edge of V-groove has a rounded shape with a small curvature.

(a) BF$_2$ implanted

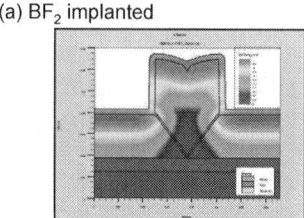

(b) 1000°C 10s

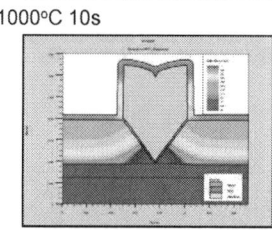

(c) 1000°C 5min

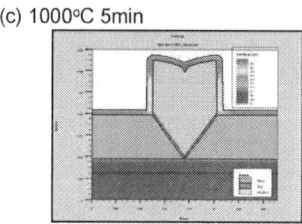

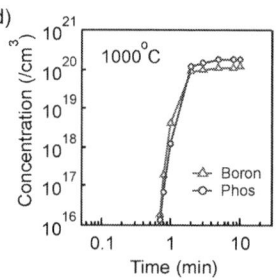

Fig. 4 Simulation of dopant diffusion into channel region with time. (a) BF$_2$ ion implanted into 80 nm SOI with 25 keV energy. Activation anneal at (b) 1000°C for 10s and (c) 1000°C for 5 min. (d) Progress of dopant concentration at the edge of V-groove with time.

978-1-4673-4840-9/13 $31.00 © 2013 IEEE

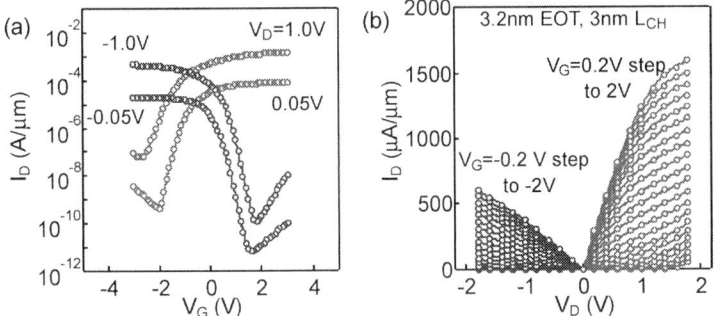

Fig. 6 Characteristics of N-type and P-type ultra-short channel Junctinless-FETs integrated with poly-Si/TaN/HfO$_2$ gate stack (3.2 nm EOT) and without NiSi process. (a) I$_D$-V$_G$ and (b) I$_D$-V$_D$ curves.

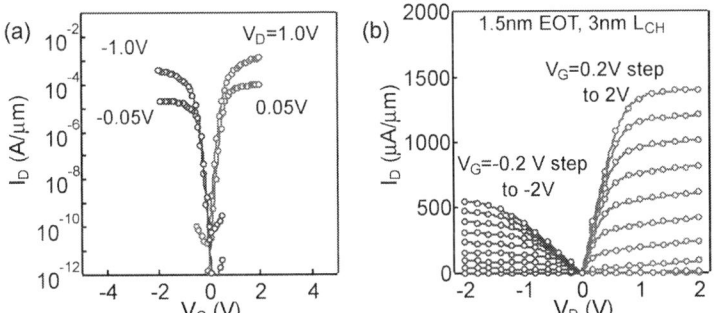

Fig. 7 Characteristics of N-type and P-type ultra-short channel Junctionless-FETs integrated with poly-Si/TaN/HfO$_2$ gate stack (1.5 nm EOT) and with NiSi process. (a) I$_D$-V$_G$ and (b) I$_D$-V$_D$ curves.

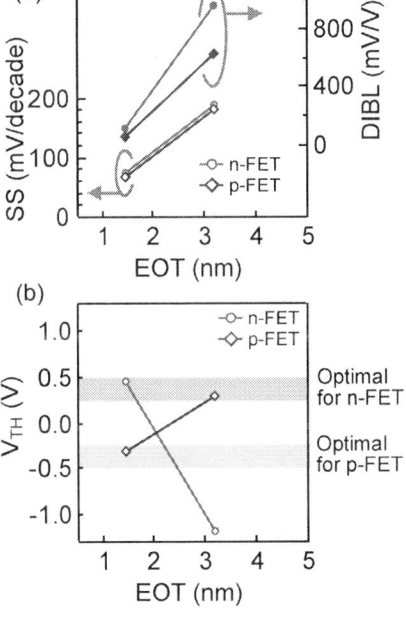

Fig. 8 EOT dependences of (a) subthreshold swing (SS) and drain-induced barrier lowering (DIBL), and (b) threshold voltages (V$_{TH}$).

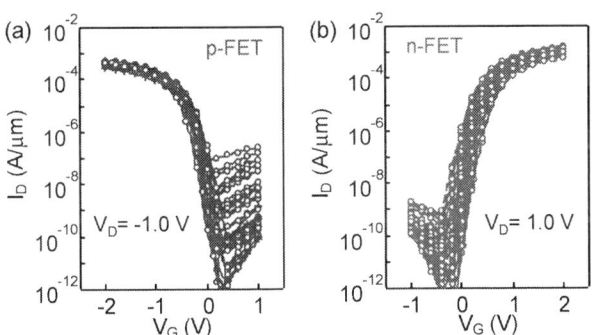

Fig. 9 I$_D$-V$_G$ curves of (a) 35 number of p-type Junctionless-FETs and (b) 41 number of n-type Junctionless-FETs on an SOI wafer.

Single gate JL-FET	
W.F.	4.5 eV
SOI	1 nm
L$_{CH}$	3 nm
N$_{CH}$	1 x 10^{20}/cm^3
S/D	1 x10^{20}/cm^3

Fig. 10 Structure and standard parameters for the simulation of ultra-short channel JL-FETs.

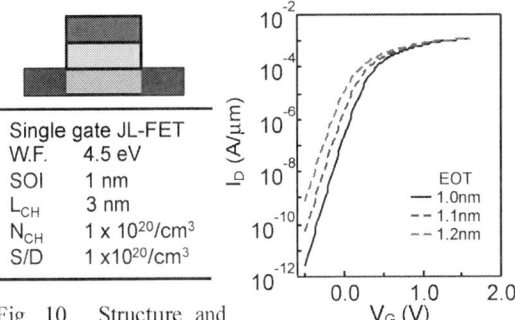

Fig. 11 Dependence of JL-FET performance on EOT variation.

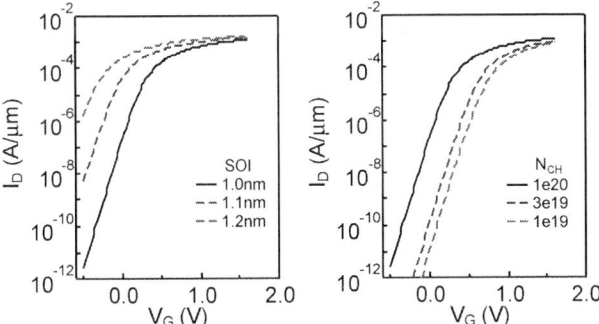

Fig. 12 Dependence of JL-FET performance on SOI thickness variation.

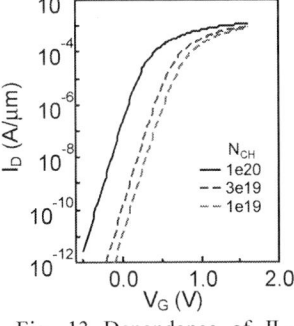

Fig. 13 Dependence of JL-FET performance on channel concentration.

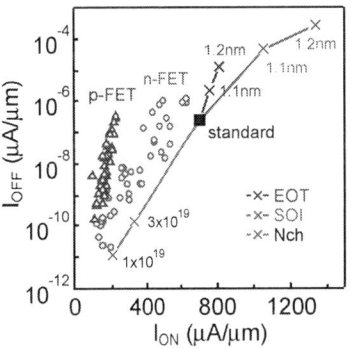

Fig. 14 Experimental ON-OFF relationship for n-type and p-type JL-FETs plotted with the simulated impacts of structural variations.

Improved Resistance Memory Characteristics and Switching Mechanism Using TiN Electrode on TaO$_x$/W Structure

A. Prakash and S. Maikap*

Thin Film Nano Tech. Lab., Department of Electronic Engineering, Chang Gung University,
259 Wen-Hwa 1st Road, Kwei-Shan,
Tao-Yuan, 333, Taiwan

*Corresponding author: Tel: 886-3-2118800 ext. 5785 Fax: 886-3-2118507 E-mail: sidhu@mail.cgu.edu.tw

Abstract

Improvement in the resistive switching characteristics using TiN bottom electrode on TaO$_x$/W structure as compared to W electrode is reported. The thickness of TaO$_x$ layer is confirmed by high-resolution TEM image. The memory device with TiN electrode has shown formation-free repeatable bipolar resistive switching with resistance ratio of >30. Besides, improved resistive switching performance in terms of lower operation current (50 vs. 300 μA), longer read endurance (>10^5 cycles), 85 °C data retention of >10^4 seconds and device yield of >90% is also achieved for TiN electrode. The TiN/TaO$_x$ interface is playing a key role in controlling the oxygen vacancy defects which in turn control the formation/dissolution of conducting filament.

Keywords: RRAM, high-κ tantalum oxide, interfacial layer, oxygen ion migration, filament, and bipolar switching.

Introduction

Resistance switching phenomena, observed in simple metal-insulator-metal stacks so-called resistance random access memory (RRAM) is attracting everyone's attention in the field of non-volatile memory technology due to possible alternative candidate for next generation memory technology [1-2]. Although various materials have been reported for resistive switching, binary oxides have advantage over other resistive switching materials in terms of simple composition and easy fabrication process. Among other binary oxides such as TiO$_x$, [3-5] NiO, [6-7] HfO$_x$, [8-10] ZrO$_x$,[11-12] AlO$_x$, [13-14] etc., tantalum oxide [15-18] is one of the promising choices due to its thermal stability and complementary metal oxide semiconductor (CMOS) compatibility as it is being used in the dynamic random access memories in the current semiconductor industry. Moreover, TaO$_x$ based resistive memory devices have shown good memory performance reported in various studies [19-20].However, formation-free operation, resistance ratio (HRS/LRS), low power consumption, switching uniformity and device yield are important parameters. In this study, formation-free resistive switching properties of W/TaO$_x$/TiO$_x$N/TiN memory device fully fabricated at room temperature has been investigated and compared with W/TaO$_x$/W structure. The memory device with TiN bottom electrode (BE) has shown improved resistive switching with small set/reset voltage of <1.5 V and resistance ratio of >30 at a small current compliance of 50 μA as compared to W BE device. The improvement is due to the formation of an oxygen deficient interfacial layer of TiO$_x$N

between TaO$_x$ and TiN BE which controls oxygen vacancy defects. The switching mechanism is attributed to the formation/dissolution of oxygen vacancy filament due to the migration of oxygen ions when voltage is applied on TE. Good read endurance of >10^5 times and data retention at 85 °C are also obtained.

Experiment

At first, 100nm thick titanium nitride (TiN) layer and tungsten (W) layer as bottom electrodes (BE) were deposited separately by RF sputter system on 8" p-type SiO$_2$/Si substrates. In a next step, a low temperature silicon dioxide (SiO$_2$) of 150nm was deposited. Different via sizes of 0.2-8μm and BE contacts were then defined by lithography and etching process followed by another lithography step to pattern the devices for lift-off. High-κ TaO$_x$ layer with a thickness of ~7nm was deposited by electron-gun system on TiN as well as W BEs. Then W (approximately 100nm) layer as a top electrode (TE) was deposited by rf magnetron sputter using W target. Final devices were obtained after lift-off process. Electrical characterizations were performed using HP4156C precision semiconductor parameter analyzer in voltage sweep mode with two-probe DC measurement method. All voltages were applied on the TE and BE was electrically grounded during all measurements. All electrical measurements were done in the air at room temperature until otherwise stated.

Results and discussion

Figure 1 shows the cross-sectional high resolution transmission electron microscopy (HRTEM) image of fabricated resistive memory device in W/TaO$_x$/W structure. The thickness of tantalum oxide (TaO$_x$) layer is approximately 7 nm. The TaO$_x$ film is amorphous as no crystalline region is visible while W is partially crystalline. Figure 2 (a) and (b) show the current-voltage (I-V) curves of resistive memory devices in W/TaO$_x$/TiO$_x$N/TiN and W/TaO$_x$/W structures respectively. The voltage sweep direction used for switching cycle measurement is indicated by the arrows (1→4). As can be seen, both devices exhibit bipolar resistive switching behavior and can be set/reset by applying positive/negative voltage on TE respectively. However, memory device with TiN BE shows improved switching with smaller set/rese t voltage of less than 1.5 V and large resistance ratio of more than 10 at a much lower current compliance (CC) only 50 μA (Fig. 2(a)) as compared to 300 μA for W BE devices (Fig. 2(b)). Moreover, formation voltage of about +3.8 V (not shown here)

978-1-4673-4840-9/13 $31.00 © 2013 IEEE

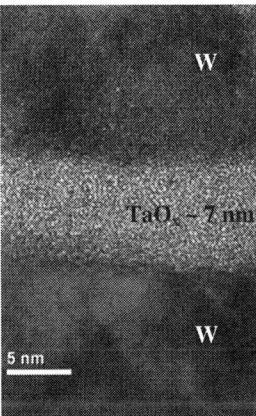

Fig. 1 Cross-sectional HRTEM image of W/TaO$_x$/W resistive memory stack. The thickness of amorphous TaO$_x$ layer is approximately 7 nm.

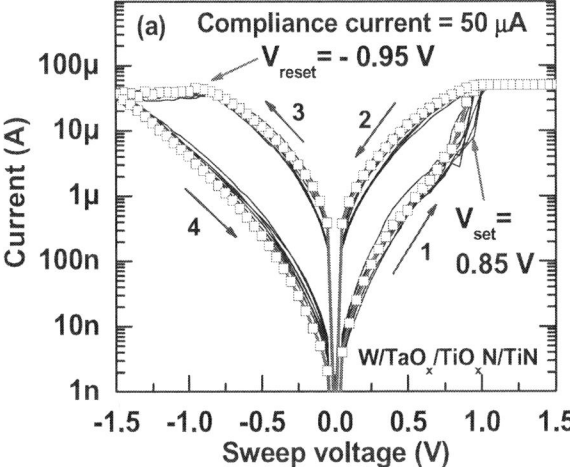

Fig. 2(a) Current-voltage curves of ten consecutive switching cycles of fabricated resistive memory device in W/TaO$_x$/TiO$_x$N/TiN structure.

was needed to initiate the resistive switching in W BE device whereas device with TiN BE is formation-free which is required for practical realization of RRAM device. The improvement in the resistive switching device is attributed to the formation of an interfacial layer of TiO$_x$N at TaO$_x$/TiN interface. This oxygen deficient interfacial layer is formed due to the oxygen getter property of Ti which can take oxygen from TaO$_x$ layer to form partially oxidized TiO$_x$N layer. This will create more oxygen vacancies in the TaO$_x$ layer owing to which TiN BE device is formation-free whereas preexisting defects are not sufficient in the TaO$_x$ layer in W BE device to show resistive switching and hence extra formation step is needed to make the device switchable. Moreover, the formation/dissolution of oxygen vacancy filament is more uniform due to the oxygen vacancy modulation controlled by oxygen getter property of Ti in TiN electrode device. The switching mechanism can be attributed to the

formation/dissolution of oxygen vacancy filament due to the application sufficient voltage. When positive voltage is applied

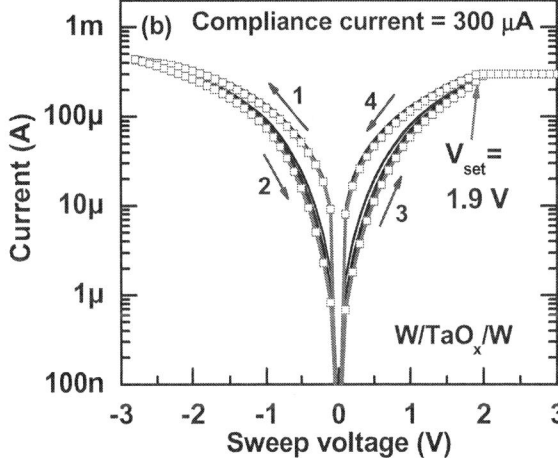

Fig. 2(b) Current-voltage curves of ten consecutive switching cycles of resistive memory device in W/TaO$_x$/W structure.

on the W TE, oxygen ions (O^{2-}) migrate towards TE and form thin O$_2$-rich layer. As a result oxygen vacancy filament can be formed and the device comes to low resistance state (LRS).When negative voltage is applied on the TE, O^{2-} repelled from TE and annihilate the oxygen vacancies. In a consequence filament breaks bringing the device into high resistance state (HRS). In case of TiN BE devices, filament formation is better controlled due to the presence of interfacial layer of TiO$_x$N leading to improved resistive switching performance as compared to W BE devices.

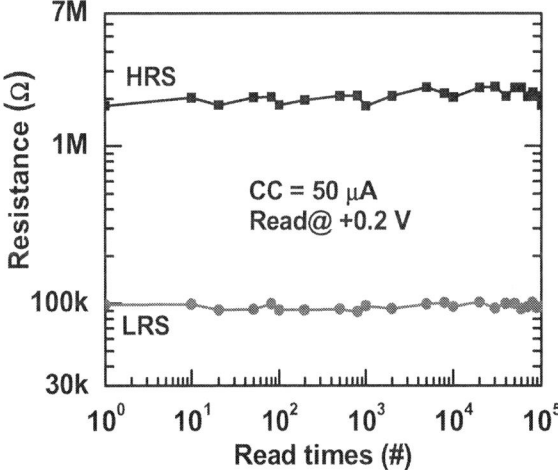

Fig. 3 Read endurance of resistive memory device with TiN BE at a read voltage of 0.2 V. Both the memory states can be read for more than 10^5 times without any disturbance.

To test the non-destructive read out the read endurance of the device with TiN BE is carried out and the result is presented in

Figure 3. The device has been read successfully for more than 10^5 cycles at a read voltage of 0.2 V with stable HRS and LRS. Figure 4 shows the data retention characteristics of TiN BE resistive memory device at 85 °C. The current compliance is 50 µA. As can be seen both low and high resistance states remain stable without any noticeable degradation after 3 hours of testing which confirms the nonvolatility of memory device.

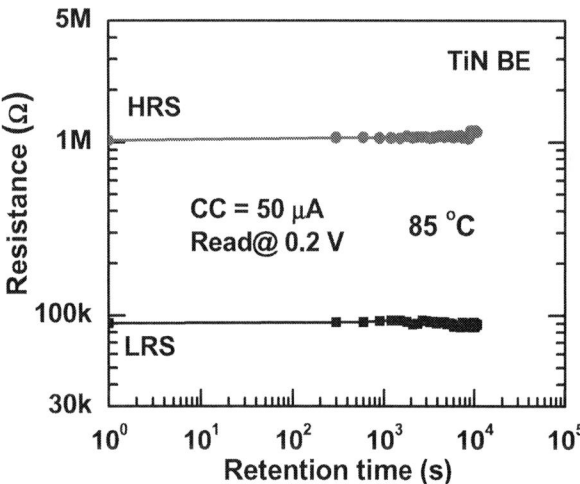

Fig. 4 Data retention characteristics of more than 3 hours of fabricated resistive memory device with TiN BE at 85 °C.

Conclusion

Resistive switching characteristics using TiN and W as bottom electrodes on TaO_x/W structure have been investigated. Cross-sectional TEM image confirms the thickness of amorphous TaO_x switching layer to be approximately 7 nm. The memory device with TiN bottom electrode has shown repeatable bipolar resistive switching with small set/reset power of less than 50 µW and resistance ratio of more than 30 as compared to only 3 in W BE device. The TiN/TaO_x interface is playing a key role in controlling the oxygen vacancy defects which in turn control the formation/rupture of conducting filaments. The switching mechanism is attributed to the formation/dissolution of oxygen vacancy filament due to the oxygen ion migration on the voltage application. Moreover, improved resistive switching parameters in terms of lower operation current (50 vs. 300 µA), longer read endurance ($>10^5$ cycles), 85 °C data retention of $>10^4$ seconds and device yield of $>90\%$ is also achieved. It is expected that the device can be useful in future low power non-volatile memory applications.

Acknowledgement

This work was supported by National Science Council (NSC), Taiwan under contract no. NSC-101-2221-E-182-061. The Chang Gung University group is grateful to EOL/ITRI, Hsinchu, Taiwan, to support this work.

References

[1] A. Sawa, *Mater. Today*, 11 [6], p. 28, 2008.
[2] R. Waser and M. Aono, *Nat. Mater.*, 6, p. 833, 2007.
[3] K. M. Kim, et al., *Appl. Phys. Lett.*, 91, p. 012907, 2007.
[4] Y. L. Chung, et al., *ACS Appl. Mater. Interfaces*, 3, p. 1918, 2011.
[5] B. J. Choi, et al., *J. Appl. Phys.*, 98, p. 033715, 2005.
[6] D. C. Kim, et al., *Appl. Phys. Lett.*, 88, p. 232106, 2006.
[7] F. Nardi, et al., *Solid-State Electron*, 58, p. 42, 2011.
[8] H. Y. Lee, et al., *Jpn. J. Appl. Phys., Part 1*, 46, pp. 2175, 2007.
[9] H. Y. Lee, et al., *IEEE Electron Device Lett.*, 30, 703, 2009.
[10] J. Lee, et al., et al., *Appl. Phys. Lett.*, 97, p. 172105, 2010.
[11] Y. Li, et al., *IEEE Electron Device Lett.*, 31, p. 117, 2010.
[12] S.-Y. Wang, et al., *Appl. Phys. Lett.*, 95, p. 112904, 2009.
[13] C.-Y. Lin, et al., *J. Electrochem. Soc.*, 154, pp. G189-G192, 2007.
[14] W. Banerjee, et al., *J. Electrochem. Soc.*, 159, pp. H177, 2012.
[15] S. Z. Rahaman, et al., *J. Appl. Phys.*, 111, p. 063710, 2012.
[16] S. Z. Rahaman, et al., *Nanoscale Res. Lett.*, 7, p. 345, 2012.
[17] A. Prakash, et al., *Solid-State Electron*, 77, pp. 35-40, 2012.
[18] A. Prakash, et al., *Jpn. J. Appl. Phys.*, 51, p. 04DD06, 2012.
[19] Wei, Z., et al., *Highly reliable TaO_x ReRAM and direct evidence of redox reaction mechanism*, Technical Digest - International Electron Devices Meeting, IEDM, p. 293, 2008.
[20] J. J. Yang, et al., *Appl. Phys. Lett.*, 97, p. 232102, 2010.

A single-walled carbon nanotube wall paper as an absorber for simultaneously achieving passively mode-locked and Q-switched Yb-doped fiber lasers

Xiaohui Li[1,2], Yonggang Wang[3], Yishan Wang[2*], Qijie Wang[1*], Wei Zhao[2], Yongzhe Zhang[1], Xia Ya[4], Ying Zhang[4]

1. Division of Microelectronics, School of Electrical and Electronic Engineering, Nanyang Technological University, 50 Nanyang Ave., 639798, Singapore.
2. State Key Laboratory of Transient Optics and Photonics, Xi'an Institute of Optics and Precision Mechanics, Chinese Academy of Sciences, Xi'an 710119, China.
3. Department of Applied Physics and Materials Research Centre, the Hong Kong Polytechnic University, Hong Kong, China.
4. Singapore Institute of Manufacturing Technology, 71 Nanyang Drive, 638075 Singapore
Corresponding author: yshwang@opt.ac.cn,
qjwang@ntu.edu.sg

Abstract

We demonstrate a fiber ring laser based on a single-walled carbon nanotube (SWCNT) wall paper absorber. It is found that the proposed Yb-doped fiber laser can either be operated in the mode-locked states or Q-switched states. First, SWCNT wall paper acts as a mode locker when the pump power is below 80 mW. Self-started mode locking can be obtained when the pump power is about 47 mW. The proposed Yb-doped mode-locked fiber lasers can be operated in the dissipative soliton regime that the spectra have a narrow peak. Second, SWCNT wall paper acts as a Q switcher when the pump power is above 80 mW. The Yb-doped fiber laser can work in the Q-switched states at higher pump power which is quite different from the conventional pulse fiber lasers. The repetition rate increases from 30 kHz to 50 kHz and the pulse duration decreases from 2.7 µs to 1 µs with the increase of pump power. This is due to the SWCNT wall paper induced loss in the cavity which leads to Q-switched state at higher pump powers. The combination of the Q-switching and mode-locking in one fiber laser have potential application in the fields that require different pulse fiber lasers.

Introduction

Passively mode-locked and Q-switched fiber lasers have wide applications in materials processing, biomedicine, and telecommunications. Various kinds of saturable absorbers (SAs) have been utilized to achieve mode-locking in fiber lasers such as semiconductor saturable absorber mirrors (SESAM), single-walled carbon nanotube (SWCNT), or few layers graphene saturable absorber [1-12]. However, these approaches have several drawbacks. SESAM has a low damaged threshold, a complex semiconductor structure, a narrow turning range, and a limited recovery time. Few layers grapheme absorbers, which have been successfully utilized to achieve mode-locking as a new materials, still have the disadvantages of complex fabrication processing and difficulty of realizing mode locking in 1-µm region [9, 10]. Single-walled carbon nanotubes (SWCNTs) have attracted a lot of interests recently due to their broad operation range, low cost, high optical damage threshold and short ultrafast recovery times (~1

ps) in the near-infrared region [3-8].

Generally, SAs could act as mode lockers or Q switchers in a fiber laser under different conditions. Sun *et al.* demonstrated an ultrafast fiber laser mode locked by carbon nanotubes [3]. S. Kobtsev *et al.* reported a passively mode-locked Yb-doped fiber laser based on carbon nanotubes [5]. Recently, the carbon nanotube as a Q-switcher has also been intensively studied. B. Dong et al. reported a tunable passively Q-switched fiber laser with different cavity configurations [13-14]. To the best of knowledge, most of the reported CNT-SA based fiber lasers were solely focused on either the mode-locked operation or the Q-switched operation [3-6, and 13-16].

In this paper, we demonstrate a Yb-doped fiber lasesr incorporating a SWCNT wall paper. The proposed fiber laser can operate in passively mode-locked states or Q-switched states depending on operation conditions. Dissipative solitons with a center wavelength of 1061 nm can be obtained in mode-locked states at a low pump power (from 47 mW to 80 mW). When the pump power is high, the proposed fiber laser could work in Q-switched states. When the pump power increases from 80 mW to 120 mW, the repetition rate increases from 30 kHz to 50 kHz and the pulse duration decreases from 2.7 µs to 1 µs. To the best of our knowledge, it's the first report to experimentally demonstrate that first the mode-locked states and then the Q-switched states in the same fiber laser just by increasing the pump power.

Expeimental setup

The SWCNTs were grown by the electric arc discharge technique. The mean diameter of the SWCNTs is about 1.5 nm. First, several milligrams of SWCNT powder were poured into 10 ml 0.1% SDS (sodium dodecyl sulfate) aqueous solution. Here SDS was used as a surfactant. In order to obtain SWCNT aqueous dispersion with high absorption, SWCNT aqueous solution was ultrasonically agitated for 10 hours. After the ultrasonic process, the dispersed solution of SWCNT was centrifuged to induce sedimentation of large SWCNT bundles. The upper portion of the centrifuged solution was diluted and decanted into a bottle. Some polyvinyl alcohol (PVA) powder was poured into the bottle and dissolved at 90 ℃ with ultrasonic agitation for 3 hours. Then, the SWCNT/PVA

978-1-4673-4840-9/13 $31.00 © 2013 IEEE

dispersion was poured into a polystyrene cell. Finally we put these cells in an oven for evaporation. The temperature of the oven was kept at 40 °C. It took about one or two days for completing the evaporation. After the evaporation was finished, the wall and the bottom of the cell were coated with a thin plastic film. The PVA aqueous solution has strong viscosity to the polystyrene cell so that it adheres to the wall of the cell. When the cell was dry, the PVA film lost the viscosity to the cell, thus we can strip the PVA film cell off the polystyrene cell by a tweezers easily. The SWCNTs were carried by PVA to the surface of the wall and the bottom of the polystyrene cell during the evaporation process. The SWCNT/PVA film on the wall of the cell had higher quality than that on the bottom so that we can use the former as absorber for mode locking and hereafter we call it as "SWCNT wallpaper absorber". In practice, we cut off the wallpaper absorber cell into many pieces for use in mode locking.

Figure 1 shows the schematic diagram of the YDFL which incorporates SWCNT wallpaper absorber in the cavity. A 0.8-m-long Yb-doped fiber (YDF) with absorption coefficients of 500 dB/m at 976 nm is used as the gain medium. The other fiber in the cavity together with the pigtail of the passive components is HI-1060 fiber. In order to achieve mode-locking solely by SWNTs and make the light propagating unidirectionally, a polarization-independent isolator (PI-ISO) is inserted in the cavity. A polarization controller (PC) is mounted on the passive fiber to achieve certain polarization state in the cavity which may effect on the absorption characteristic of the SWNTs. A 90:10 output coupler is used to output the signal. Since the fiber cavity in this case is all normal dispersion, a spectral filter effect is essential. As a result, a fiber filter at 1064 nm with bandwidth of about 10 nm is inserted in the cavity to make sure mode-locking at certain wavelength. The films are cut into 1 x 1 mm pieces and sandwiched between the FC/PC fiber connectors. The fiber laser is pumped by a 976-nm laser diode (LD). The total length of the fiber cavity is about 9.56 m, which corresponds to the repetition rate of 21 MHz. The output spectra are detected by the optical spectral analyzer (OSA, AQ-6315A). And the pulse trains are monitored by an oscilloscope with bandwidth of 6 GHz (LeCroy SDA) together with a high speed photodetector with bandwidth of 10 GHz (Kangguan).

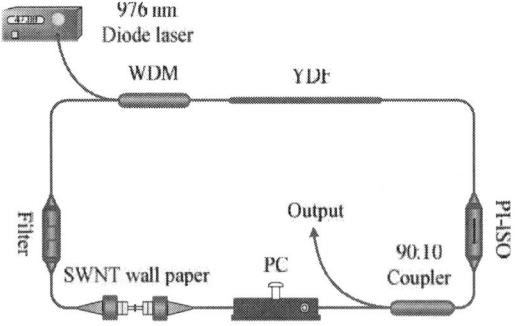

Fig. 1. The schematic diagram of the proposed fiber laser based on a carbon nanotube absorber.

Experimental results and analysis

Self-started mode-locking can be achieved by increasing the pump power to 47 mW. A typical spectrum in the log scale is shown in Fig. 2(a) and the spectrum in the linear scale is shown in Fig. 2(b) with a step-edge spectrum. The center of the spectrum shows a peak which we can call it the peak spectral profile. The spectral width is about 0.17 nm which is smaller than the step edge-to-edge width of 0.6 nm. The corresponding oscilloscope trace is shown in Fig. 2(c), which indicates the proposed fiber laser operates in the repetition rate of 21.5 MHz corresponding to the cavity length of 9.56 m. And the corresponding autocorrelation trace is shown in Fig. 2(d) with the pulse duration of about 310 ps, if a Gaussian profile is assumed. Since the proposed fiber laser operates in the all normal dispersion state, spectral filtering effect is needed on the pulse shaping. The output power is about 1.5 mW.

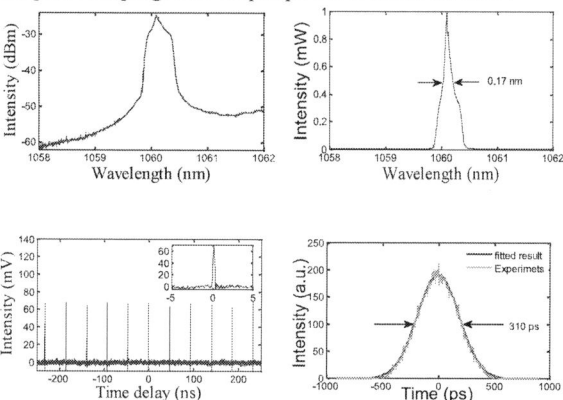

Fig. 2. Typical mode locked state of the proposed fiber laser when the pump power is about 47 mW, (a) the spectrum in log scale and (b) the linear scale, (c) the corresponding oscilloscope trace with the single pulse, (d) corresponding autocorrelation trace with experimental and fitted results.

By adjusting the polarization controllers, different types of spectra can be obtained. This effect changes not only the characteristics of saturable absorption of the SWCNT, but also the phase delay of the cavity, which leads to different mode-locked states.

In addition, Q-switched states can be obtained, when the pump power increases above 80 mW. The repetition rate and the pulse duration can be varied by increasing the pump power. Figures 4(a) and (b) show the typical optical spectrum and the corresponding oscilloscope trace, respectively, when the pump power is about 80 mW. The spectral width becomes narrower than the one in mode-locked states, and the interpulse-interval is 26.62 µs that is far larger than the cavity round trip time of 46.5 ns, which indicates that the proposed fiber laser operate in the Q-switched states. Figure 5(a) shows the pulse-train evolution by increasing pump power. Figure 5(b) shows the pulse duration and the repetition rate as a function of the pump power, respectively. Larger pump intensity could lead to smaller interpulse-interval (or larger repetition rate) and smaller pulse width. As the pump power increases from 80 mW to 120 mW, the repetition rate varies from 30 kHz to 50 kHz, and the corresponding pulse width decreases from 2.7 µs to 1 µs. Further increasing the pump power would lead to optical power induced thermal damage of the SWCNT wallpaper.

When the pump power is 120 mW the output power is about 988 µW corresponding to the pulse energy of 18.38 nJ.

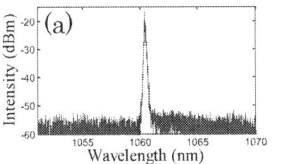 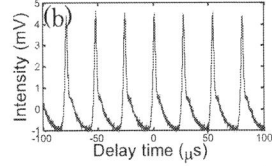

Fig. 4. Typical spectrum of Q-switched operation when the pump power is about 80 mW.

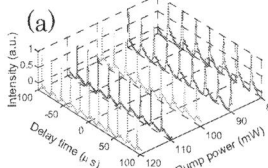 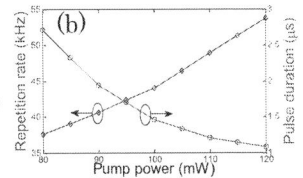

Fig. 5. (a) The pulse train evolution versus pump power. The repetition rate becomes larger and the pulse duration decreases with the increase of the pump power. (b) The repetition rate and pulse duration versus the pump power, respectively. The green curve shows the pulse duration versus the pump powers, and the blue curve shows the repetition rates versus the pump powers.

Conventionally, a fiber laser operates in the Q-switched mode-locked states at the lower pump power, while it operates in the mode-locked states when the pump power is higher. However, due to some other effects, such as the characteristics of the SAs, the proposed fiber laser can operate in the mode-locked states when the pump power is lower, while it operates in the Q-switched states when the pump power is higher. This is quite different from the former experiments observed. We believed that the SWCNT wall paper performed like a mode locker when the pump power is low, and it perform as a Q switcher when the pump power is high enough. In higher pump power, the wall paper induced cavity loss is increased. So there exists a Q-switched effect formed in the cavity. Further experiments about the absorber of the SWNT should be done in order to make clear of the mechanism of the SA.

Conclusion

In conclusion, we demonstrate a special and simple technique for the fabrication of single-walled carbon nanotube wallpaper absorbers. By using this absorber, Yb-doped fiber laser can work at mode-locked and Q-switched states in different pump power levels. Dissipative solitons with peak spectral profile generate in the cavity when the pump power is lower. Mode-locked state occurs at a lower pump power and Q-switched state occurs at a higher pump power. The wall paper induced cavity loss which plays an important role in the formation of Q-switched states at the higher pump powers.

Acknowledgments

This work was supported by the CAS Special Grant for Postgraduate Research, Innovation and Practice, the CAS/SAFEA International Partnership Program for Creative Research Teams, and the National Natural Science Foundation of China under Grant 61107034. It is also partially funded by A*STAR SERC grant (Grant Number 1122904018).

Reference

[1] H. Zhang, D. Y. Tang; X. Wu, L. M. Zhao, "Multi-wavelength dissipative soliton operation of an erbium-doped fiber laser," Opt. Express, 17(15), 12692-12697 (2009).

[2] C. M. Ouyang, P. Shum, K. Wu, J. H. Wong, H. Q. Lam, and S. Aditya, "Bidirectional passively mode-locked soliton fiber laser with a four-port circulator," Opt. Lett. 36(11), 2089-2091 (2011)

[3] Z. P. Sun, A. G. Rozhin, F. Wang, V. Scardaci, W. I. Milne, I. H. White, F. Hennrich, A. C. Ferrari, "L-band ultrafast fiber laser mode locked by carbon nanotubes." Appl. Phys. Lett. 93, 061114 (2008).

[4] F. Wang, A. G. Rozhin, V. Scardaci, Z. Sun, F. Hennrich, I. H. White, W. I. Milne, A. C. Ferrari, "Wideband- tuneable, nanotube mode-locked, fibre laser." Nat. Nanotechnol., 3, 738–742 (2008).

[5] S. M. Kobtsev, S. V. Kukarin, Y. S. Fedotov. "Mode-locked Yb fiber laser with saturable absorber based on carbon nanotubes," Laser Phys., 21(2), 283-286 (2011).

[6] F. Shohda, M. Nakazawa, J. Mata, and J. Tsukamoto "A 113 fs fiber laser operating at 1.56 µm using a cascadable film-type saturable absorber with P3HT-incorporated single-wall carbon nanotubes coated on polyamide," Opt. Express, 18(9). 9712-9721 (2010).

[7] C. M. Ouyang, P. Shum, R. M. Li, H. H. Wang, E. J. R. Kelleher, J. H. Wong, A. I. Chernov, K. Wu, and E. D. Obraztsova, S. N. Fu, "Observation of timing jitter reduction induced by spectral filtering in a fiber laser mode locked with a carbon nanotube-based saturable absorber," Opt. Lett. 35(14), 2320-2322 (2010).

[8] X. H. Li, Y. G. Wang, Y. S. Wang, X. H. Hu, W. Zhao, X. L. Liu, J. Yu, C. X. Gao, W. Zhang, Z. Yang, C. Li, and D. Y. Shen, "Wavelength-Switchable and Wavelength-Tunable All-Normal-Dispersion Mode-Locked Yb-Doped Fiber Laser Based on Single-Walled Carbon Nanotube Wall Paper Absorber," IEEE Photon. J., 4(1), 234-241 (2012).

[9] H. Zhang, D. Y. Tang, R. J. Knize, L. M. Zhao, Q. L. Bao, K. P. Loh, "Graphene mode locked, wavelength-tunable, dissipative soliton fiber laser," Appl. Phys. Lett., 96(11), 111112 (2010).

[10] Z. Q. Luo, M. Zhou, Z.P. Cai, C.C. Ye, J. Weng, G. M. Huang, H. Y. Xu, "Graphene-Assisted Multiwavelength Erbium-Doped Fiber Ring Laser," IEEE Photon. Tech. Lett. 23(8), 501-503 (2011).

[11] E. J. R. Kelleher, J. C. Travers, Z. Sun, A. G. Rozhin, A. C. Ferrari, S. V. Popov, and J. R. Taylor, "Nanosecond-pulse fiber lasers mode-locked with nanotubes," Appl. Phys. Lett. 95 111108 (2009).

[12] E. J. R. Kelleher, J. C. Travers, E. P. Ippen, Z. Sun, A. C. Ferrari, S. V. Popov, and J. R. Taylor, "Generation and direct measurement of giant chirpin a passively mode-locked laser," Opt. Lett. 34(22), 3526-3528 (2009).

[13] B. Dong , J. Z. Hao, J. H. Hu, C. Y. Liaw, "Short linear-cavity Q-switched fiber laser with a compact short carbon nsanotube based saturable absorber" Opt. Fiber Technol., 15, 105-107, (2011).

[14] B. Dong, J. H. Hu, C. Y. Liaw, J. Z. Hao, and C. Y. Yu, "Wideband-tunable nanotube Q-switched low threshold erbium doped fiber laser," Appl. Opt. 50, 1442-1445 (2011).

[15] D. P. Zhou, L. Wei, B. Dong, and W. K. Liu, "Tunable Passively-switched Erbium-Doped Fiber Laser with Carbon Nanotubes as a Saturable Absorber," IEEE Photon. Tech. Lett., 22(1), 9-11, (2010).

[16] L. Zhang, Y. G. Wang, H. J. Yu, L. Sun, L. Guo, W. Hou, J. Tang, X. C. Lin, and J. M. Li, "880 nm LD pumped passive Q-switched and mode-locked Nd:YVO4 laser using a single-walled carbon nanotube saturable absorber," Laser Phys., 21(3), 454–458, (2011).

Simulation of Grading Double Hetero-junction non-polar InGaN Solar cell

Hsun-Wen Wang[2], Pei-Chen Yu[1], Hau-Vei Han[1], Chien-Chung Lin[3], Hao-Chung Kuo[1*], Shiuan-Huei Lin[2]

[1] Department of Photonics & Institute of Electro-Optical Engineering, National Chiao-Tung University, Hsinchu, 300, Taiwan.

[2] Department of Electrophysics, National Chiao Tung University, Hsinchu, Taiwan.

[3] Institute of Photonics System, National Chiao Tung University, Tainan, Taiwan

* Corresponding author. Tel.: +886-3-571-2121 Ext. 31986; fax: +886-3-571-6631.
E-mail address: hckuo@faculty.nctu.edu.tw

Abstract

In this study, the characteristics of non-polar double heterojunction GaN/ $In_xGa_{1-x}N$ solar cells with various indium contents are numerically investigated under AM 1.5 global spectrum using finite element analysis. By smoothing the interface band edge offset with graded junction, we see the enhancement on short circuit current and power conversion efficiency. The maximum efficiency of the simulation results reached 24.32 % when the major absorption region contains 65% of indium composition..

Introduction

InGaN alloys are promising for achieving high reliability and high output characteristics in light-emitting diodes (LEDs) and laser diodes. Recently, InGaN solar cells are attracting much attention for the terrestrial and space solar cell application. One of the great features about this GaN based material is the direct bandgap energy for the entire alloy range, which varies from 0.7 eV for InN to 3.4 eV for GaN, and thus provides wide range absorption of solar spectrum from ultraviolet to visible and infrared. [1,2] Another important physical properties are high carrier mobility and drift velocity, high radiation resistance and optical absorption of $\sim 10^5$ cm^{-1} near the band edge of III-V nitride-based materials and good thermal stabilities.[3-5]

In general, GaN/$In_xGa_{1-x}N$ hetero-junction p-i-n solar cells had In contents lower than 15%, grown on c-facet sapphire substrate by Metalorganic vapour phase epitaxy (MOCVD). [6-8] The i-InGaN layer as active region is usually grown 60~300nm thick which sandwiches in between n-GaN and p-GaN layer.[6-8] InGaN epitaxial layer in their structure possesses large polarization charges include spontaneous electrical polarization and strain-induced piezoelectric polarization.[9, 10] However, the experimental efficiencies of InGaN solar cells are inadequate. The most serious problem is the strong built-in electric field cause by polarization effect.

In this work, we calculate the theoretical efficiencies of $In_xGa_{1-x}N$ double heterojunction solar cells in non-polar orientation with various indium components and evaluate the effect of sandwiching a graded region between InGaN/GaN interfaces. The optimized structure is important for the development of the full-solar-spectrum solar cells, and it can

be a guideline for the experimental work on the fabrication of high efficiency $In_xGa_{1-x}N$ solar cells.

Simulation and method

Our calculations use advanced physical models of semiconductor devices (APSYS) software to analyze InGaN solar cells. The two-dimensional flow of electrons and holes is simulated based on the generic semiconductor transport equations, e.g. Poisson equation, drift-diffusion model, and carrier-transport equation.

The p-i-n structure consisted of a 2 m-thick Si-doped GaN bottom layer (n-doping = 5×10^{18} cm^{-3}), a thin 300 nm intrinsic $In_xGa_{1-x}N$ absorption layer, and a 100 nm-thick Mg-doped GaN layer (p-doping = 5×10^{17} cm^{-3}) on top of the absorption region. The schematic diagram of InGaN double heterojunction solar cell is shown in Fig. 1 (a). The calculation assumed a perfect anti-reflection coating for zero reflection losses and ignored series resistance of ohmic contact in the cell. In the unit cell, the mesh points are discretized into 55130 points in a two-dimensional compatibly to approximate the boundary and minimize error. The solar cell devices are simulated under air mass 1.5 global solar conditions.

First we need to adapt the theoretical effects of polarization charges on IV curve of non-polar and polar $In_{0.12}Ga_{0.88}N$ double hetero-junction solar cell. The spontaneous (P^{sp}) and piezoelectric polarization (P^{pz}) of InxGa1-xN alloy, in C/m2, are defined as [9, 10]

$$P^{sp}_{In_xGa_{1-x}N} = -0.042x - 0.034(1-x) + 0.038x(1-x) \ , \ (1)$$

$$P^{pz}_{InN} = -1.373\varepsilon + 7.559\varepsilon^2 \ , (2)$$

$$P^{pz}_{GaN} = -0.918\varepsilon + 9.541\varepsilon^2 \ , (3)$$

The strain is described by the lattice constant of substrate (a_{subs}) and material ($a(x)$) as

$$\varepsilon(x) = [a_{subs} - a(x)]/a(x) \ , (4)$$

And the strain induced piezoelectric polarization can be expressed as vegard's law :

$$P^{sp}_{In_xGa_{1-x}N} = xP^{pz}_{InN}[\varepsilon(x)] + (1-x)P^{pz}_{GaN}[\varepsilon(x)] \ , (5)$$

The polarization charges, which includes spontaneous and piezoelectric polarization effect at the interfaces from p-GaN to n-GaN, are -2.12×10^{13}, 1.10×10^{13}, -1.10×10^{13}, and

2.12×10^{13} /cm^2, respectively.

Results and discussion

Fig. 1(b) shows the typical current-voltage characteristics of non-polar and polar $In_{0.12}Ga_{0.88}N$ solar cell. The polar cell had the open circuit voltage (V_{oc}) of 2.673 V, the short circit current (J_{sc}) of 0.331 mA/cm2, and fill-factor (F.F.) of 92.9 % and conversion efficiency (η) of 0.82 %. We notice that the non-polar cell showed the better performance, which had a V_{oc} of 2.203 V, J_{sc} of 2.021 mA/cm^2, and F.F. of 89.3 % and η of 3.976 %. We attribute this to the improvement of polarization effect. In the traditional polar InGaN solar cell, even before illumination, the interface-charge-infected conduction band and valence band are bending shown in Fig. 2(a). However, the non-polar device presents the normal p-i-n band structure. Under solar radiation, the corresponding built-in field for non-polar condition will be much higher than the polar case, as shown in Fig. 2(b), and therefore it will be much easier to collect the minority carriers in non-polar case.

Fig. 3 shows the results of calculation of non-polar $In_xGa_{1-x}N$ double hetero-junction solar cells with indium content x varying from 0.05 to 0.5. We could see the increasing indium content leads to a lower Voc, higher Jsc and higher efficiency when x is smaller than 0.3. The maximal short circuit current is 7.94 mA/cm^2 and the efficiency is 10.85 % at $In_{0.3}Ga_{0.7}N$. After this critical point, when indium content is more then 0.3, the short circuit current and the efficiency decrease sharply due to large discontinuity of band gap at interface.

One solution to alleviate the discontinuity issue is to insert graded $In_xGa_{1-x}N$ layer between both sides of GaN/ InGaN interface which can remove the discontinuity of interface. Fig. 4 shows the optimized structure and calculation results of the graded InGaN heterojunction solar cell with i-$In_xGa_{1-x}N$ as a function from 0.05 to 0.95. The short circuit current of graded-$In_{0.35}$ is enhanced to 10.1 mA/cm^2 and the efficiency is improved to 12.77 %. The garded layer structure smooth the abrupt interface band structure which the discontinuities of the conduction band and valence band have been removed between GaN/InGaN interface. The short circuit current increases mainly due to graded layer which improves the minority carrier transmission. The maximum value of the short circuit current density with graded structure is 52.54 mA/cm^2 at $In_{0.85}Ga_{0.15}N$.

Conclusion

In conclusion, the non-polar InGaN heterojunction solar cells can improve the band-bending due to polarization effect. The optimal structure of non-polar graded InGaN solar cell can remove the band discontinuity at interface and also improve the minority carrier collection. The maximal value with $In_{0.65}Ga_{0.35}N$ of conversion efficiency is predicted to display the non-polar cells with indium content equal to 65%.

References

[1] V. Yu. Davydov, et al., Phys. Status Solidi B, vol. 229, pp.r1-r3 , 2002.

[2] T. Matsuoka, et al., Appl. Phys. Lett., vol.81, pp.1246-1248, 2002.

[3] O. Jani, et al., Appl. Phys. Lett., vol. 91, pp.132117-132119, 2007.

[4] M. Vazquez, et al., Progr. Photovoltaics, vol. 15, pp.477-491, 2007.

[5] Y. Nanishi, et al., Jpn. J. Appl. Phys., vol. 42, pp.2549-2559, 2003.

[6] O. Jani, et al., Appl. Phys. Lett., vol. 91, pp. 132117-1-132117-3, 2007.

[7] C.J. Neufeld, et al., Appl. Phys. Lett., vol. 93, pp. 143502-1143502-3, 2008.

[8] E. Matioli, et al., Appl. Phys. Lett., vol. 98, pp. 021102-1021102-3, 2011.

[9] V. Fiorentini, et al., Appl. Phys. Lett., vol. 80, pp.1204-1206, 2002.

[10] O. Ambacher, et al., J. Phys. Condens. Matter, vol. 14, pp.3399-3434, 2002.

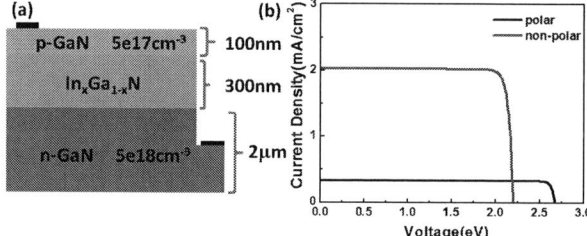

Fig. 1. (a) Schematic of $In_xGa_{1-x}N$ heterojunction solar cell. (b) IV-characteristic of non-polar and polar $In_{0.12}Ga_{0.88}N$ solar cell.

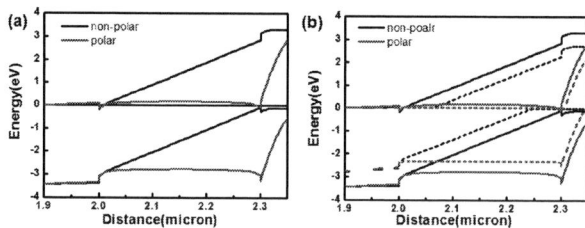

Fig. 2. (a) The band structure of non-polar and polar $In_{0.12}Ga_{0.88}N$ without solar irradiation. (b) The band structure of non-polar and polar $In_{0.12}Ga_{0.88}N$ with solar irradiation.

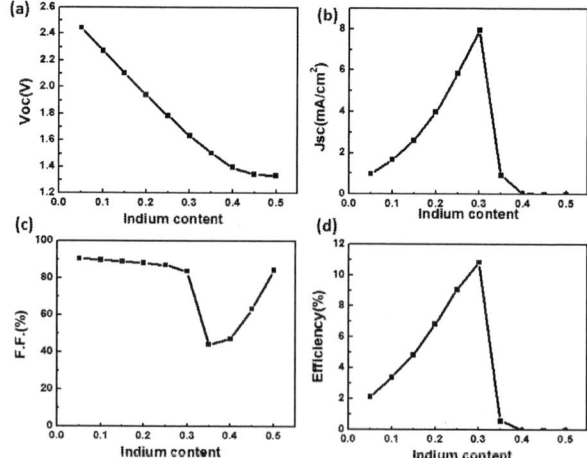

Fig. 3. The (a) V_{oc}, (b) J_{sc}, (c) Fill factor, and (d) efficiency of InGaN heterojunction solar cell were calculated with various indium contacts.

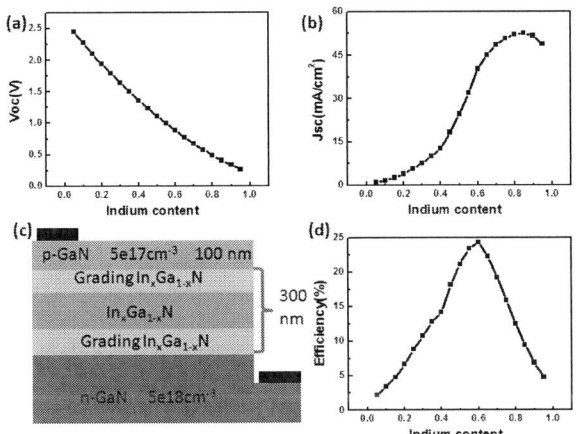

Fig. 4(a) the schematic of the graded InGaN heterojunction solar cell. The graded InGaN heterojunction solar cell and the (b) V_{oc}, (c) J_{sc} and (d) efficiency were calculated with various indium.

Optical properties of Si/SiO2 and GaAs/AlOx sub-wavelength HCG mirrors on GaAs substrate and an impact of structural imperfections on their performance

Marcin Gebski[1], Maciej Dems[1], Jian Chen[2], Wang Qijie [2,3], Zhang Dao Hua[2], and Tomasz Czyszanowski[1]

[1]Institute of Physics, Lodz University of Technology
219 Wolczanska str., 90-924 Lodz, Poland
marcin.gebski@p.lodz.pl, maciej.dems@p.lodz.pl, tomasz.czyszanowski@p.lodz.pl
[2]Division of Microelectronics, School of electrical and electronic engineering, Nanyang Technological University,50 Nanyang Ave., 639798, Singapore
[3]Division of Physics and Applied Physics, School of Physical and Mathematical Sciences, Nanyang Technological University, 637371, Singapore
chen0498@e.ntu.edu.sg, edhzhang@ntu.edu.sg, qjwang@ntu.edu.sg

Abstract

In order to obtain low threshold VCSELs high reflectivity mirrors are crucial in their designs. Typically used Distribute Bragg Reflectors (DBR) provide reflectivity over 99.8 % in the spectral range of 70 nm. To achieve polarization control numerous of shallow etchings approaches has been proposed but already lastly developed High Contrast Grating (HCG) has an ability to discard one of the mode polarization completely and additionally to provide twice broader spectrum of large reflectivity. We present simulation results of various HCG structures based on Si and GaAs high index materials combined with SiO2 and AlOx cladding low index material layers respectively. Reflectivity simulations at 980 nm wavelength show that reflectivity exceeding 99.8% can be obtained in a broad range of grating parameters. Additionally the same parameters assure very strong discrimination between TE and TM modes. Manufacturing process of a grating might involve undesirable imperfections leading to deterioration of optical properties of HCG. Hence we consider error analysis assuming random nonperiodicity of a grating structure based on Gaussian distribution of manufacture inaccuracy with standard deviation as inaccuracy measure varying from 0 nm to 50 nm. We found that inaccuracy in the range 0 nm – 25 nm does not deteriorates significantly the reflection of HCG.

Keywords: HCG, manufacturing imperfections, numerical methods.

Introduction

Recent studies show that high contrast gratings (HCG) are efficient mirrors for various applications including integrated optoelectronic devices and focusing reflectors and lenses [1]. Properly designed and manufactured structure may provide broad reflectivity spectrum [2]. Moreover, HCG structures are characterized by very strong polarization discrimination thanks to their inherent anisotropy. Those features makes HCG superior to distributed Bragg reflectors (DBRs), commonly used in vertical cavity surface emitting lasers (VCSELs). Promising experimental investigations on HCG VCSELs [3,4] are a motivation for detailed study of optical properties of HCGs.

Due to high order diffraction underlying large reflectivity

of discussed reflectors, scalar numerical models can not be utilized in the simulations. In order to perform detailed numerical modeling, state of art plane-wave reflection transformation (PWRT) method, a modification of plane wave admittance method (PWAM), was used. In general, PWRT solves Maxwell equations in frequency domain using transformation of electro-magnetic field into diagonal coordinates and plane-wave expansion within each layer of analyzed structure [5]. This approach changes problem of solving differential equation to algebraic problem of finding eigenvalues of matrices in the basis of plane-wave functions. Those in general infinite matrices are truncated by choice of number of plane waves.

Optical properties of arsenide and dielectric HCGs are discussed in the following parts of the article. First, simulation overview together with investigated structures' parameters are presented. Next chapter comprise simulation results for ideal structures. Dependencies of reflectors' parameters on its reflectivity and polarization discrimination are shown and discussed. The last chapter deals with impact of manufacturing imperfections on mirrors' optical quality.

Investigated structures

The analysis were performed for structures presented in the Fig. 1. All layers were considered infinite in xy plane with leading z dimension (thickness). The bottom layer was GaAs substrate which thickness was considered infinite. Since reflectivity of separated mirror was to be determined, zero absorption in this layer was considered. Second layer was a low

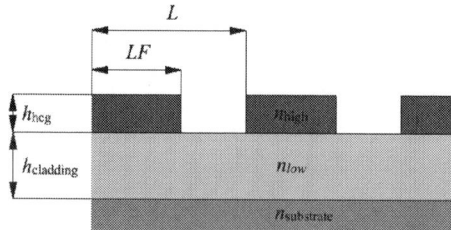

Fig. 1. Intersection of investigated structure in xz plane. L is HCG period, F is fill factor, hhcg is high refractive index material layer thickness, hcladding is low refractive index material layer thickness, nhigh, ncladding and ncladding refers to high refractive index, low refractive index and GaAs substrate refractive index respectively.

978-1-4673-4840-9/13 $31.00 © 2013 IEEE   146

refractive index cladding layer of thickness varying from 0 µm to 0.5 µm. Grating layer comprised of high refractive index material bars separated by air gaps was embedded on the cladding. It's thickness was varied from 0 µm to 0.3 µm. The top layer was an air layer which thickness was considered infinite.

Optimal optical parameters of mirrors were investigated with respect to four grating parameters: cladding layer thickness ($h$cladding), high index material thickness ($h$hcg), HCG period (L) and fill factor (F) defined as ratio between bar width and HCG period. Investigated gratings were illuminated from

the substrate. Incident light was propagating perpendicularly to layers surfaces. In order to calculate reflectivity of ideal structures, 7 plane waves were used.

Two kinds of structures were simulated distinguished by material systems used. In the first one, GaAs was used as a high refractive index material with typical parameters $n_{Re} = 3.52$ and $\alpha = 9.4$ cm^{-1}. For the low refractive index material AlO$_x$ was used. In [6] it is shown that wet oxidatation of AlGaAs layers, typically used in such structures, does not produce homogenous, fully oxidized Al$_2$O$_3$ layers, but mixture of a few types of aluminium oxides characterized by different oxidation level as well as different optical parameters. For that reason we assumed optical parameters of such AlO$_x$ layer to be $n_{Re} = 1.55$ and $\alpha = 0$ cm^{-1}. Moreover, strain effects increasing with increase of AlO$_x$ layer thickness may produce cracks deteriorating optical properties of the structure for layers of thickness exceeding 0.4 µm. Future development in oxidation techniques may enable producing thicker cracks-free AlO$_x$ layers. This possibility motivated us to theoretically consider AlO$_x$ layers as thick as 0.5 µm.

TABLE I
OPTICAL PARAMETERS OF MATERIALS USED

Material	n_{Re}	α [cm^{-1}]
GaAs	3.52	9.4
AlO$_x$	1.55	0
Si	3.60	69.73
SiO2	1.54	0
substrate GaAs	3.52	0

In the second material system high refractive index bars, made of crystalline Si, were coupled with crystalline SiO2 cladding layers. Optical parameters were considered to be $n_{Re} = 3.60$ and $\alpha = 69.73$ cm^{-1} for Si and $n_{Re} = 1.54$ and $\alpha = 0$ cm^{-1} for SiO$_2$. Optical parameters of materials used in simulation are summarized in Table 1.

Simulation results

A. HCG composed of GaAs on AlOx

Fig. 2 shows map of reflectivity of light polarized perpendicularly to HCG bars with respect to HCG period L and fill factor F for arbitrary chosen layers thicknesses - $h$cladding = 0.25 µm and $h$hcg = 0.21 µm. One can see that broad large reflectivity region (LRR) for HCG period between 0.4 µm and 0.6 µm is formed. However, only high end of fill factor range is feasible due to its parallelism to fill factor axis. This feature results in need of a careful control of only one mirror parameter which is HCG period. On the other hand, LRR in low end of fill factor range is much narrower and tilted

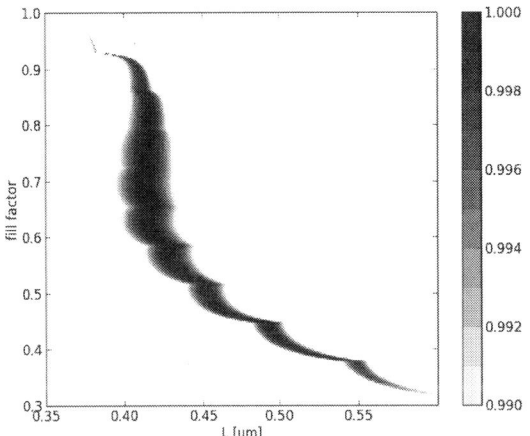

Fig. 2. Map of reflectivity of light polarized perpendicularly to HCG bars.with respect to HCG period L and and fill factor F. Layers thicknesses were fixed on $h$hcg = 0.21 µm and $h$cladding = 0.25 µm.

to both axes what additionally forces precise control of fill factor for a given HCG period.

Fig. 3 presents reflectivity as a function of cladding layer (AlOx) thickness for light polarized both perpendicularly and in parallel direction to HCG bars. Other grating parameters were $h$hcg = 0.21 µm, $F = 0.75$ and $L = 0.42$ µm. Low reflectivity for parallel polarization for cladding layer thinner than 0.47 µm together with high reflectivity for perpendicular polarization for cladding as thin as 0.1 µm makes investigated structure highly polarization discriminatory in favour of perpendicular polarization. Reflectivity of 0.998, crucial for high efficiency VCSEL devices, is exceeded for perpendicularly polarized light for cladding thicknesses between 0.14 µm and 0.27 µm and higher than 0.37 µm. In contrary, reflectivity for light polarized in parallel never exceeds 0.992.

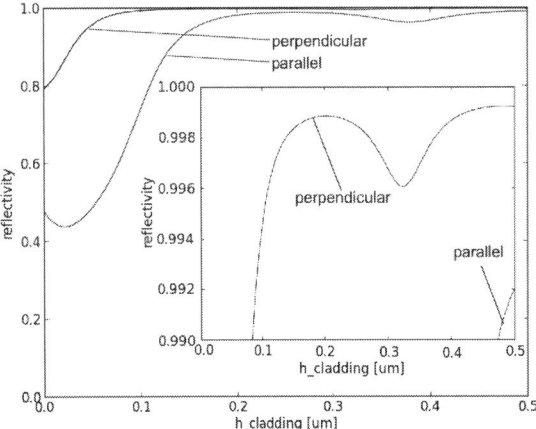

Fig. 3. Reflectivity as a function of cladding layer (AlOx) thickness for light polarized perpendicularly to HCG bars. Structure parameters were fixed on $h$hcg = 0.21 µm, $F = 0.75$ and $L = 0.42$ µm

B. HCG composed of Si on SiO2

Fig. 4 shows map of reflectivity of light polarized in parallel to HCG bars with respect to HCG period L and fill factor F for arbitrary chosen layer (SiO2) thicknesses - $h$cladding = 0.45 μm and $h$hcg = 0.1 μm. Broad, parallel to fill factor axis LRR is formed for HCG period between 0.57 μm and 0.58 μm with reflectivity exceeding 0.998 in the fill factor range between 0.3 and 0.4.

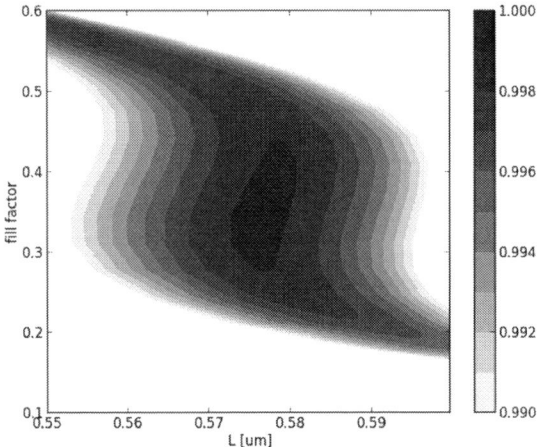

Fig. 4. Map of reflectivity of light polarized in parallel to HCG bars.with respect to HCG period L and and fill factor F. Layers thicknesses were fixed on $h$hcg = 0.1 μm and $h$cladding = 0.45 μm.

Reflectivity as a function of cladding layer (SiO2) thickness is shown in fig. 5. Other fixed grating parameters were HCG period L = 0.575 μm, fill factor F = 0.35 and $h$hcg = 0.1 μm. It is evident that cladding layer thickness has a strong impact on reflectivity of light polarized in parallel to HCG bars. Increase of reflectivity with increase of cladding layer thickness observed for low thickness cladding layers reaches two maxima for $h$hcg = 0.18 μm where reflectivity exceeds 0.967 and $h$hcg = 0.47 μm where reflectivity exceeds 0.998. Corresponding reflectivity of light polarized perpendicularly to HCG bars does not exceed 0.3 what proves that

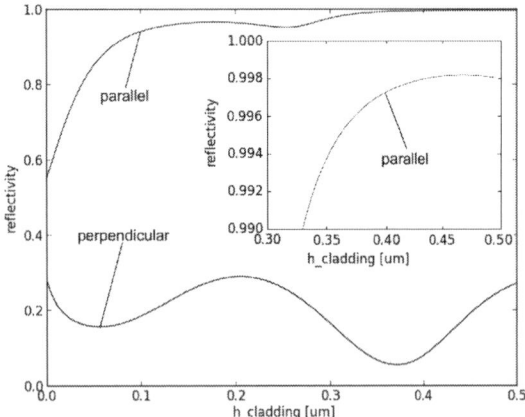

Fig. 5. Reflectivity as a function of cladding layer (SiO2) thickness for light polarized perpendicularly to HCG bars. Structure parameters were fixed on $h$hcg = 0.1 μm, F = 0.35 and L = 0.575 μm.

structure produces very strong polarization discrimination for the whole considered cladding layer thickness range.

Properties of imperfect HCG

In order to properly estimate manufacturing imperfections, every bar edge position was chosen randomly and independently from other draws with mean value of Gaussian distribution equal to position of the edge in the ideal structure. Standard deviation of this distribution was considered to be a measure of an imperfection and was chosen arbitrary between 0 nm and 50 nm. Periodic numerical cell, one HCG period wide, containing one HCG bar and one air gap used in ideal structure calculations, was substituted by supercell containing five randomly imperfect standard cells. 50 plane waves were used to simulate this highly nonperiodic structure.

Functions of reflectivity with respect to fill factor for GaAs on AlOx reflector structure for 5 nm (solid line), 15 nm (dashed line), 25 nm (dotted line), 35 nm (long dashed line) and 45 nm (thin solid line) standard deviation are presented in the Fig. 6. Optimum structure for high reflectivity of in parallel polarized light characterized by HCG period L = 0.58 μm, cladding layer thickness $h$cladding = 0.45 μm and $h$hcg = 0.1 μm was analyzed.

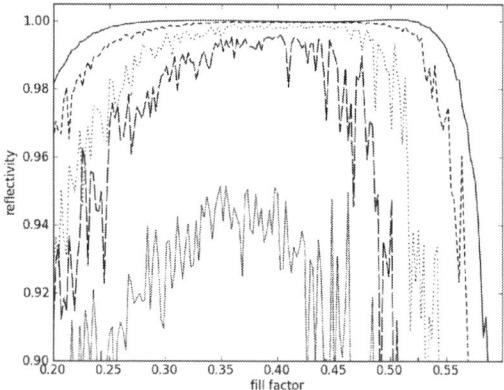

Fig. 6. Reflectivity of light polarized in parallel to HCG bars.as a function of fill factor F for 5 nm (solid line), 15 nm (dashed line), 25 nm (dotted line), 35 nm (long dashed line) and 45 nm (thin solid line) standard deviation based on 25 probes. GaAs on AlOx mirror structure parameters were fixed on L = 0.58 μm $h$hcg = 0.1 μm and $h$cladding = 0.45 μm.

Obtained results show that LRR region for fill factor between 0.3 and 0.5 is constricted with increase of standard deviation value. On the other hand only slight deterioration of reflectivity values for fill factor between 0.35 and 0.45 occurred for standard deviation lower or equal 25 nm. One can see, that significant deterioration of reflectivity values and constriction of LRR fill factor range is observed for standard deviation greater than 25 nm. It's worth pointing out that large oscillations of reflectivity values in high end and low end of fill factor range for in every case occur. It shows that the reflectivity is more susceptible to deterioration resulting from imperfections for fill factor values smaller and greater than optimal value, which in this case is F = 0.4.

Conclusions

Performed analysis shows that high reflectivity HCG on GaAs/AlO$_x$ and Si/SiO$_2$ are feasible for light polarized in both directions – parallel and perpendicular to HCG bars. In both cases, for arbitrary chosen h_{hcg} parameters large freedom in choice of cladding layer thickness occurs. If layers thicknesses are fixed, broad, parallel to fill factor axis LRR emerges for both structures what results in broad fill factor values range. Critical cladding layer thickness for high reflectivity of perpendicular modes in analyzed GaAs/AlO$_x$ structure is 0.14 µm, while for high reflectivity of parallel modes in analyzed Si/SiO$_2$ structure is 0.45 µm. Simulations of imperfect structure optimized for high reflectivity of parallel mode show that such structure is characterized by slightly deteriorated and constrained LRR for standard deviation of manufacturing imperfections not greater than 25 nm. High oscillations in both high end and low end of fill factor LRR region occur what translates into higher fragility of LRR there.

Acknowledgements

Authors would like to acknowledge Polish-Singapore grant no. 1/3/POL-SIN/2012 "A Novel Photonic Crystal Surface Emitting Lasers Incorporating a High-Index-Contrast Grating". M. Dems acknowledges support of Polish National Centre for Research and Development within project LIDER.

References

[1] C. J. Chang-Hasnain and W. Yang, *AOP*, 4, pp. 379-440, 2012.

[2] M. C. Y. Huang, Y. Zhou and C. J Chang-Hasnain, *Nature Photonics*, Vol. 2 , 2008, pp. 180-184.

[3] M. C. Y. Huang, Y. Zhou and C. J Chang-Hasnain, *Nature Photonics*, 1, 2007, pp. 119-122.

[4] M. Ortsiefer et al., *IEEE Phot. Tech. Lett.,* Vol. 22, No. 1, 1 Jan 2010, pp. 15-17 .

[5] M. Dems, "Plane-wave admittance method and its applications to modeling semiconductor lasers and planar photonic-crystal structures" *PhD Dissertation*, Lodz University of Technology, 2007,

[6] K. J. Knopp, R. P. Mirin, D. H. Christensen, K. A. Bertness, A. Roshko et al., Appl. Phys. Lett. 73, 3512 (1998); doi: 10.1063/1.122821

Enhance current density and light trapping effect in a-Si thin film solar cells by flexible textured PDMS film

H. V. Han[1], H. C. Chen[1], Y. L. Tsai[1], C. C. Lin[2, *], H. C. Kuo[1], and P. Yu[1]

[1]Department of Photonics and Institute of Electro-Optical Engineering,
National Chiao Tung University, Hsinchu, Taiwan
[2] Institute of Photonic System, National Chiao Tung University, Tainan 711, Taiwan
*Email: chienchunglin@faculty.nctu.edu.tw

Abstract

The light trapping effect of a textured polydimethylsiloxane (PDMS) film on the suppression of surface reflection in a-Si:H solar cell is investigated. The angular dependence of scattered light from this PDMS film was measured and a symmetric scattering intensity peak was found at +/- 20^0. The enhancement of power conversion efficiency in the a-Si:H solar cell with textured PDMS film can reach 16.37% compared to the a-Si:H solar cells without textured PDMS film. This improvement is mainly attributed to the enhanced efficiencies of light trapping and scattering. Such a design concept and fabrication process can be widely adapted to a variety of thin film solar cells.

Introduction

Because of the global warming effect and energy crisis, the usage of solar energy gets more and more important in recent years. Hydrogenated amorphous silicon (a-Si:H) based thin film solar cells are attractive candidates for large-scale photovoltaics due to their materials being highly abundant on earth and their compatibility for roll-to-roll processing [1-3]. A typical film thickness of approximately 1 μm is required for effective light absorption in the a-Si:H active layers. However, the minority carrier diffusion length is typically limited to 300 nm [4,5]. The mismatch of light absorption depth and minority diffusion length can cause insufficient absorption or carrier collection losses. As a result, efficient light management by reducing the surface reflection as well as increasing the optical path for low energy photons is important for efficiency improvements and cost reduction. In this work, we demonstrate a platform to combine the flexible textured PDMS film with a-Si:H solar cell. The advantages of using PDMS film are the low-cost, non-vacuum system and simple process (only spin coating and imprinting needed). Other than ease of fabrication, PDMS film provides a refractive index gradient to serve as an anti-reflection layer. Extra benefits of light trapping and scattering can be added when the film is stamped with high texture pattern. As the results, the measured photovoltaic current density-voltage (I–V) and the external quantum efficiency (EQE) characteristic demonstrate great enhancement.

Fabrication

Fig. 1 shows the schematic plot of a single-junction a-Si:H solar cell with textured PDMS film. A typical single junction a-Si:H solar cell was deposited on the Asahi-U type substrate, which consists of an fluorine-doped SnO_2 layer as both top electrode, a 292-nm-thick a-Si:H active layer (p-i-n, 12/260/20 nm), an 80-nm-thick ITO layer as the back electrode, and a 500-nm-thick Al as a back reflector. To prepare a flexible textured PDMS film, a randomly textured crystalline silicon (c-Si) mold was prepared by wet-etching in potassium hydroxide (KOH) [6, 7]. This mold was used to shape PDMS film, and the cross-sectional scanning electron microscope (SEM) image of the randomly textured silicon substrate with a height of about 6 μm was shown Fig. 2(a). Second, the PDMS pre-polymer solution (in the form of a viscous liquid) was dispensed on the textured c-Si mold-pattern surface. Then, the spin coating method was employed to form a uniform PDMS layer which fill the c-Si mold-pattern, and then the substrate was baked at 100 ℃ for one hour. After detaching from the c-Si mold, a flexible textured PDMS film was successfully obtained and the height of the textured structure was about 5 μm, as shown in Fig. 2(b). Finally, the flexible textured PDMS film was pasted up on the surface of a-Si solar cell.

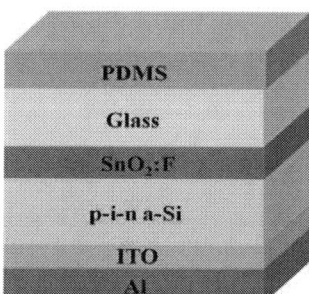

Fig. 1 A schematic plot of the fabricated single-junction a-Si solar cell with flexible textured PDMS.

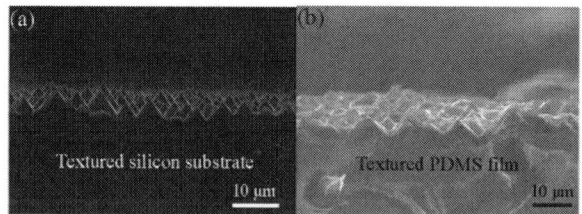

Fig. 2 (a) SEM image of randomly textured c-Si mold-pattern and (b) flexible textured PDMS film.

Results and discussions

For understand the material properties of PDMS, we measure the optical characteristic of flat and textured PDMS films. Fig. 2 shows the transmittance and reflectance of the two

978-1-4673-4840-9/13 $31.00 © 2013 IEEE

type PDMS. The transmittance is above 90% in almost all region of a-Si:H absorption spectrum, but the reflectance is lower than 5%. Although the transmittance of textured PDMS is a little lower than the transmittance of flat PDMS film, the haze of the textured PDMS is tremendous higher than the flat one, as shown in Fig. 4. The cell with high haze means the most of transmitted light is not zero order pass, and this characteristic can lead to increasing the optical path for low energy photons. Therefore, the PDMS have so suitable characteristic to use on the a-Si solar cell.

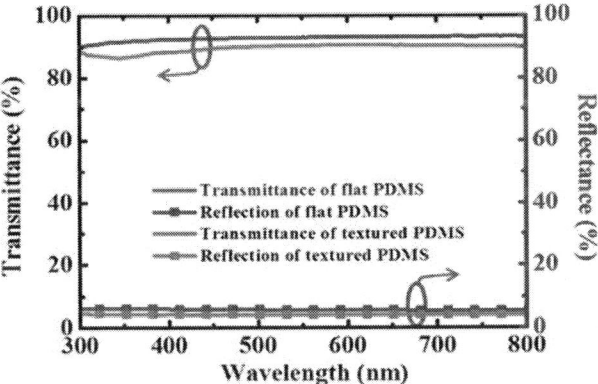

Fig. 3 The measured transmittance and reflectance of the flat and textured PDMS films.

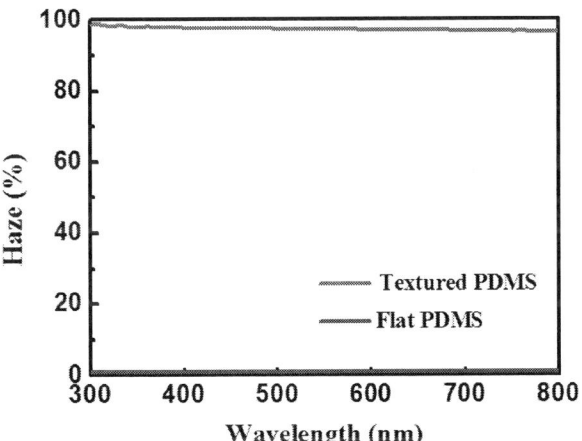

Fig. 4 The haze of the flat and textured films.

To understand how the scattering light capability of the flexible textured and flat PDMS films varies in the far-field pattern, we measured the angle-dependent intensity of transmittance by bidirectional transmittance distribution function (BTDF) system with an incident light of 380 nm and ultraviolet-visible spectrophotometer. The view angle of the flexible textured PDMS film at the full-width at half-maximum was enlarged from 12° to 50° compared to the flat PDMS film, which could be attributed to the increased light scattering by the textured structure as shown in Fig. 5. The maxima peaks of scattering curve of BDTF measurement were found at +/-20 degree of scattering angle. As the results, the introduction of the flexible textured PDMS film can deflect the photons in much wider angle and lengthen the traveling distance in the cell, which implies higher possibility of getting absorbed.

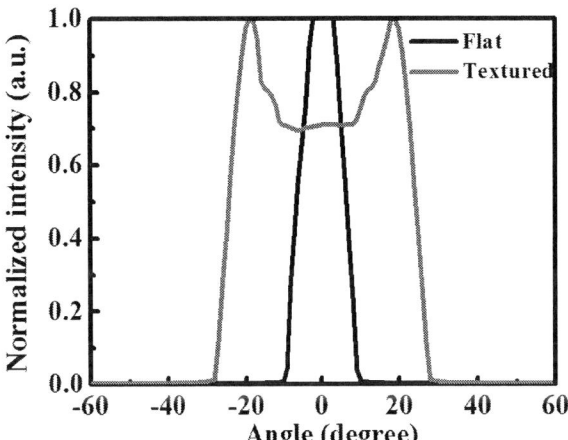

Fig. 5 The measured angular-dependent intensity of transmittance in flat/textured PDMS structures.

The external quantum efficiency (EQE) spectra were measured for the reference and textured PDMS solar cell devices, as shown in Fig. 6. The EQE enhancement in the whole spectral range confirms the optical absorption enhancement as seen in Fig. 4. The open-circuit voltage (V_{oc}) and short-circuit current density (J_{sc}) are measured under a simulated AM1.5G illumination condition at room temperature, and the current density-voltage curves of reference and textured PDMS solar cells are plotted in Fig. 7. The textured PDMS solar cell shows a power conversion efficiency of 6.48% ($V_{oc} = 0.81$ V; $J_{sc} = 20.55$ mA/cm^2; fill factor = 39.15%), while the reference cell shows an efficiency of 5.57% ($V_{oc} = 0.80$ V; $J_{sc} = 17.81$ mA/cm^2; fill factor = 39.28%). The efficiency enhancement (~16%) is mostly contributed by the photon-current enhancement, demonstrating the sufficient level of a-Si:H absorption improved by the scattering of textured PDMS. The V_{oc} and fill factor remain approximately the same, which shows the electrical property does not deteriorate as a result of the use of PDMS film on the cell.

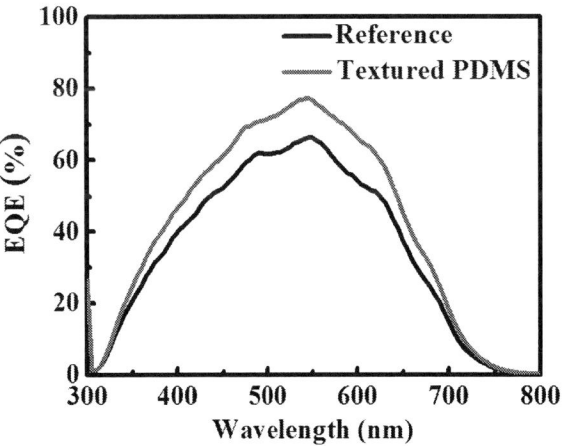

Fig. 6 Measurement of External quantum efficiency of a-Si solar cells with bare and textured PDMS film.

978-1-4673-4840-9/13 $31.00 © 2013 IEEE

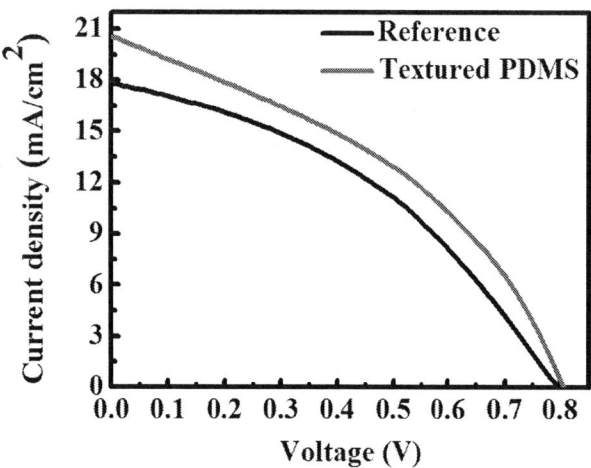

Fig. 7 Photovoltaic J-V characteristics of a-Si solar cells with bare and textured PDMS film.

Conclusion

In conclusion, we successfully improve power conversion efficiency by combining a flexible textured PDMS film with a-Si:H solar cell. Especially, the textured PDMS film can significantly enhance short-circuit current density under air mass 1.5 global illuminations without hurting the open circuit voltage and fill factor. The main mechanism of the enhancement can be attributed to anti-reflection and light scattering. Consequently, the overall power conversion efficiency is enhanced by 16.37%, when compared to the cell without textured PDMS film. Furthermore, the angle-dependent intensity of transmittance in flexible flat PDMS and textured PDMS films by BTDF system was measured to confirm that the view angle of the full-width at half-maximum was enlarged from 12° to 50°. Finally, we believe this technology shall be a great candidate for next generation of highly efficient and low-cost photovoltaic devices.

References

[1] R. E. I. Schropp, and M. Zeman, Amorphous and microcrystalline silicon solar cells: modeling, materials, and device technology, (Kluwer Academic Publishers, Norwell, Mass., 1998).

[2] D. E. Carlson, C. R. Wronski, "Amorphous silicon solar cell,"Appl. Phys. Lett. **28** (11), 671-673 (1976).

[3] R. C. Chittick, J. H. Alexande, H. F. Sterling, "The preparation and properties of amorphous silicon," Journal of the Electrochemical Society, **116**, 77- (1969).

[4] A. V. Shah, H. Schade, M. Vanecek, J. Meier, E. Vallat-Sauvain, N. Wyrsch, U. Kroll, C. Droz, J. Bailat, "Thin-film Silicon Solar Cell Technology," Prog. Photovolt: Res. Appl. **12**, 113 (2004).

[5] R. E. I. Schropp, M. Zeman, "Amorphous and Microcrystalline Silicon Solar Cells: Modeling, Materials, and Device Technology," Kluwer Academic Publishers: Boston, MA, (1998).

[6] K. Sato, M. Shikida, T. Yamashiro, M. Tsunekawa, and S. Ito, Sensors Actuat. A-Phys. **73**, 122 (1999).

[7] E. D. Palik, O. J. Glembocki, I. Heard, P. S. Burno, and L. Tenerz, J. Appl. Phys. **70**, 3291 (1991).

A Sustainable Approach to Individualized Disease Treatment: the Engineering of a Multiple Use MEMS Drug Delivery Device

Danny Jian Hang Tng, Peiyi Song, Rui Hu, Guimiao Lin and Ken-Tye Yong[*]

School of Electrical and Electronic Engineering, Nanyang Technological University, Singapore 639798, Singapore
*Corresponding author: Ken-Tye Yong, PhD, School of Electrical and Electronic Engineering, Nanyang Technological University, Singapore 639798, Singapore Tel: +65-6790-5444, email: ktyong@ntu.edu.sg

Abstract

Individualized disease diagnosis and therapy has emerged as a new direction in the research of future medication. Over the past several years, innovative approaches based on microelectromechanical system (MEMS) technology have demonstrated promising potential in individualized therapy. In this contribution, a sustainable approach for the individualized treatment of chronic disease is presented using a compact, implantable and refillable MEMS drug delivery device with an electrolysis based actuator. As a demonstration, we utilized the device for programmable delivery of a chemotherapy drug for the treatment of pancreatic cancer with an in vitro configuration based on cancer cell colonies. After the delivery of drug using the device, the growth of the colonies has been greatly inhibited as compared with the control samples. These results show that our new approach has a great potential for future in vivo studies and opens up promising opportunities for future medication.

Keywords: MEMS, drug delivery, theranostics, implantable, individualized treatment.

Introduction

Over the years, the advancements in conventional drug delivery using oral, topical and injectable drugs have allowed the treatment of a wide range of diseases. However, despite these improvements, many limitations still need to be addressed [1, 2]. Conventional drug delivery is ineffective at treating chronic illnesses which are dose dependent and require frequent, precise and targeted drug delivery [3]. Oral and topical drugs have to be administered in large doses as only a small amount can cross the physiological barriers of the body [4], raising concerns of overdose [5, 6]. Injectable drugs reduce this problem by offering direct delivery using an injection needle or catheter [7]. Frequent injections at the same site are required to maintain the therapeutic drug dose for effective treatment of the chronic disease [8], causing serious physiological trauma and risk of infection at the site of injection due to the damage inflicted by the needles used [5, 9]. Implantable MEMS drug delivery devices have shown great promise in overcoming the problems associated with conventional drug delivery [10]. These devices are micrometer scale, drug delivery devices made using MEMS microfabrication technology [9, 11, 12] incorporating: i) microactuators [13], ii) integration of multiple functional components [14] and iii) biocompatible polymers such as polydimethylsiloxane (PDMS) [15], Parylene or Polyimide [16]. Currently, major improvements have been made in their sustainability, featuring refillable reservoirs allowing long term usage without re-implantation [17, 18], more efficient and programmable actuators [5, 9, 12]. Additionally, delivery mechanisms such as cannulas [19] assist by providing more localized drug delivery. Through these MEMS drug delivery devices, the limitations of conventional drug delivery can be overcome and sustainable long term drug delivery can be carried out from the device, unlocking the potential of individualized disease treatment. This enables a programmed, long term, individualized drug delivery scheme for every patient [20]. For example, a continuous pulsatile drug release profile can be used to achieve long term therapeutic effects [6, 9, 12]. This reduces the side effects associated with overdose encountered in conventional drug treatments which utilize systemic drug delivery [9]. In this contribution, using an in vitro scheme, we present a sustainable approach to individualized drug delivery using an implantable MEMS drug delivery device compatible with a wide range of drugs from nanoparticle drugs to conventional medicines. The device utilizes an efficient electrochemical actuator with a long cannula for targeted delivery and a large refillable reservoir for long term use and sustainability [5, 9, 21]. It uses a two-chamber design which separates the drugs from the electrolysis reaction, thereby avoiding drug oxidization and increasing its compatibility with a wide range of drugs [21]. Individualized drug delivery was demonstrated through the programmed delivery of the chemotherapy drug, doxorubicin to pancreatic cancer cell cultures and monitored for changes to determine the effect of the drug delivery.

Fabrication and Experimental Methods

The interdigitated electrodes were fabricated using photolithography and lift-off process. On a Silicon wafer, AZ5214 (MicroChemicals GmbH, Germany) photoresist was spin coated at 4000 rpm for 45 s and soft baked at 105 °C for 4 min. Lithography using hard contact for 8 s with a 420W UV lamp was used to transfer the positive pattern of the electrodes (MJB4, SUSS MicroTec, Germany). Image reversal was performed by a post exposure bake at 120 °C for 4 min and a flood exposure for 30 s. The wafer was then developed and Ti/Pt (50/50 nm) was deposited using an e-beam evaporator. The wafer was dipped into acetone, to perform the lift-off process and the electrodes are formed. The reservoir with attached cannula and pump chamber with flexible membrane were both fabricated individually using micromachined Plexiglas molds. Degassed PDMS was poured into the molds and cured at 120 °C for 20 min. The individual components, including the electrode, were assembled using PDMS as the bonding agent under a vaccum at 120 °C for 30 min.

To show the concept of individualized drug delivery, the chemotherapy drug, doxorubicin was delivered in vitro to

Panc-1 cell colonies. The experiment aims to: i) illustrate the use of the presented MEMS drug delivery device in individualized drug administration and ii) to show that individualized drug administration can produce similar end results seen in conventional drug administration, potentially without the associated side effects. Three petri dishes of Panc-1 cell colonies (suspended in agar medium) corresponding to three test groups were used. The drug was administered to each test group to simulate the individualized treatment of cancer: i) the control group (Control), with no drug delivered for the whole experiment, ii) the single dose group (Direct), simulating a dose from an injection, where 18 µg of drug was added on day 0 and iii) a programmed delivery group (Program), where 6 µg of drug was delivered daily on days 0 - 2. Drug administration was performed at the same hour daily from the same batch of doxorubicin (1 mg/ml) for consistency. Once the drug delivery is performed, each group is returned to the incubator (37 °C, 5% CO_2). To maintain the hydration level, temperature and pH of the cell cultures, 100 µl of PBS solution was added to each group every 3 days. The cell colonies in each group were examined daily for 7 days at the same time each day using an inverted microscope (Nikon Eclipse Ti-U). Bright field and fluorescence pictures were taken at 4x magnification using constant exposure settings. To monitor the changes in the size of the cell colonies in each group, the bright field images were observed and the colonies were labeled. The colony sizes were determined based on its area in the acquired bright field pictures using an image processing software (NIS Elements BR V4.0). The uptake of the delivered drugs into the cell colonies was determined from the acquired fluorescence pictures, using a 500 nm wavelength light for illumination.

Results and Discussion

A. Working Principle of the Device

The presented device (Figure 1a-b) consists of a refillable drug reservoir and pump chamber (Figure 1b) which can be refilled using a 30 gauge non-coring needle and syringe, allowing multiple usages for enhanced sustainability. Additionally, during the inactive time of the device, Platinum (Pt) electrodes (Figure 1d) in the pump chamber act as a catalyst for the recombination of H_2 and O_2 gas into H_2O [17], providing a more sustainable solution to individualized drug delivery, reducing the number of elements needed to be refilled. As the pump chamber and the drug reservoir are separated by a flexible membrane, oxidation damage to the drugs in the drug reservoir due to the electrochemical reactions within the pump chamber are avoided. Thus, a wider range of drugs are compatible with the device, no special modification for resistance against the oxidation is needed. Electrochemical actuation takes place within the pump chamber which contains the electrodes on a Si substrate. When a current is supplied to the electrodes, the electrolysis of water begins, generating H_2 and O_2 gas. Pressure in the pump chamber increases, causing the flexible membrane to expand and deflect upwards, displacing the contents of the drug reservoir outwards through the cannula. The rate of gas formation is proportional to the current supplied. Therefore, the rate and of drug delivered to the patient is controllable, allowing an individualized drug delivery profile for treatment. The cannula carries the displaced drug and releases the drug at the target site providing localized delivery. A simple flap check valve at the end of the cannula prevents any backflow of fluids from the target to avoid contamination and accidental drug release (Figure 1e).

B. Study of Individualized Treatment of Cancer

The presented device was used to study its role in the treatment of a chronic disease, cancer. Doxorubicin, a common fluorescent chemotherapy drug was used. It inhibits tumor cell growth and division as it damages cellular DNA by chelation, limiting cell nucleus function [22]. The first step in individualized disease treatment is the targeted delivery of the drug. In this case, the delivery involves a small amount of drug solution (6 µl – 18 µl), targeted delivery is therefore crucial to enhance the drug's uptake into the target cell colonies. Fig. 2 shows fluorescent and bright field microscope images of Panc-1 colonies at different stages of treatment of the direct and programmed drug delivery. To study the uptake of doxorubicin into the target cells, fluorescent images of the cell colonies during the treatment were compared (Fig. 2f-j, Fig. 2p-t). Prior to the administration of doxorubicin (Fig. 2f, Fig. 2p), no florescence could be seen from the cell colonies. After 1 day of doxorubicin administration (Fig. 2g, 2q), the cell colonies have become fluorescent under green excitation, suggesting that doxorubicin was taken in. The fluorescence intensity in both the programmed and direct schemes can be seen to increase as time progresses, indicating that more doxorubicin entered the cell colonies over time (Fig. 2g-j, 2q-t).

Once the targeted delivery of the drug is achieved, the next step in individualized disease treatment would be to administer the drug in a programmed manner, to maximize the therapeutic effects and minimize the dose dependent side effects of the drug. Fig. 2a and Fig. 2k shows the initial state of the cell colonies before treatment. For the programmed delivery (Fig. 2a-e), through the course of 7 days, the colonies showed increasing darkening and a reduction of their overall size, indicating an inhibition of the disease. Additionally, as more and more doxorubicin was added, the boundaries of the cell colonies become increasingly indistinct. For the direct delivery (Fig. 2k-o), since 18 µl of doxorubicin was administered on day 1, which was significantly higher than that of the programmed delivery, the effects of doxorubicin on the cell colonies were more rapid. Besides showing darkening and a reduction of overall size, the cellular boundaries within the colonies were more indistinct and there was evidence of cellular disintegration after 7 days of treatment (Fig. 2o). The disintegrated components diffuse into the surroundings, giving the illusion of an increase in area. Therefore, from the comparison of the bright field images, it can be seen that the programmed drug delivery exhibits a similar inhibition in the growth of the cancer cells, albeit it is much slower.

From the analysis of how the size of the cell colonies respond to individualized treatment, the potential of customized treatment can be seen. The therapeutic needs of an individual may vary depending on the stage of one's disease. In the case of cancer, cancer colonies are large in the advanced stages, and smaller in the initial stages. Thus, the analysis of individualized treatment is performed for 2 groups, one group consisting of smaller sized colonies (1K – 6K µm^2) representing the initial stages and another group of larger

978-1-4673-4840-9/13 $31.00 © 2013 IEEE

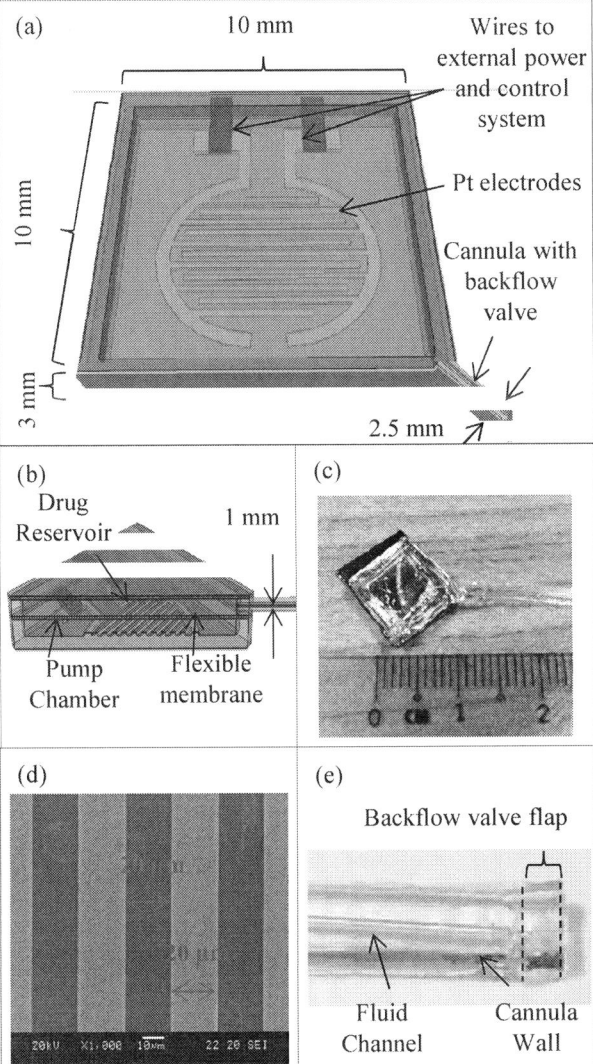

Fig. 1 Implantable MEMS electrochemical drug delivery device: (a) overview and dimensions of the device, (b) cross sectional view (bold lines are in the same plane), (c) fabricated device showing the electrode housed within the pump chamber and the drug reservoir lying on top, connected to the cannula, (d) SEM image of the electrode fingers showing a finger width and spacing of 20 μm, (e) the flap backflow valve connected to the end of the cannula and (f) diagram determining the resistance coefficient of the device.

sized colonies (6K – 12K μm²) representing the advanced stages of the disease. The same dosage effective for an initial stage may be ineffective for treating the advanced stages. Individualized delivery allows the user to employ a drug delivery scheme which can change on demand, based on the progression of the disease. The direct delivery group signifies two situations: i) the simulation of the drug delivery from an injection, where a single large dose is administered and ii) the ability of the drug delivery device to release large doses for the treatment of advanced stage diseases. Compared to the control groups which were given no medication throughout the test, the direct delivery groups which were treated with 18 μg of drug on day 0, showed a significant decrease in cell colony size throughout the 7 days (Fig. 3). This suggests that the treatment

using a large dose was effective. However, this raises issues about the potential side effects of such large doses. To address this concern, the same amount of drug was delivered within a time span of 3 days using programmed delivery. For cell colonies of smaller sizes, the programmed delivery exhibited a size reduction that was similar to the direct delivery, showing a significant decrease in size after 7 days. This suggests that for the small sized colonies, the programmed delivery was able to achieve the same effect as conventional drug delivery through injection, as simulated by the direct delivery. More importantly, by using smaller pulsed doses each time, the risk of overdose caused by side effects is reduced. For cell colonies with larger sizes however, the same programmed delivery was not as effective as that of the direct group. However, compared with the control groups, it can be seen that for the first 3 days, the growth of the colonies increased at a slower rate, indicating that the treatment showed some growth inhibition. A sharp drop in the size of the colonies was observed from day 3 to day 4, suggesting that the therapeutic dose for of the larger colonies was in between 12 μl to 18 μl. This highlights the need for individualized drug delivery, as it shows that for different individuals with different stages of the same disease, different drug delivery profiles are required.

Conclusion

With the advancements made in conventional drug delivery, the concept of individualized disease treatment can be applied to optimize the therapeutic effects while minimizing the potential side effects. In this contribution, a sustainable approach to individualized disease treatment was demonstrated in vitro with a refillable and multiple use implantable MEMS drug delivery device. The device features PDMS construction for high biocompatibility, a controllable electrochemical actuator for precise drug administration and a long cannula for targeted drug delivery. Individualized disease treatment was illustrated with the programmed administration of doxorubicin in vitro to cell colonies of pancreatic cancer cells, showing that similar cancer cell growth inhibition and regression as conventional drug delivery was possible. The programmed administration allows the regulation the treatment of the cancer cell colonies, thus enabling therapeutic strategies to be tailored to the needs of the patient. This underscores the importance of individualized disease treatment, as excessive dosages leading to overdose are minimized, whilst still maintaining therapeutic treatment. We anticipate that further development on this device will allow it to leverage on the strengths of the many advanced drugs made for conventional drug delivery methods such as nanoparticle drugs and provide even more effective individualized treatment.

Acknowledgement

This study was supported by the Start-up grant (M4080141.040) of Nanyang Technological University and partially from the Singapore Ministry of Education under Tier 2 Research Grant MOE2010-T2-2-010 (4020020.040 ARC2/11) and Tier 1 Academic Research Funds (M4010360.040 RG29/10).

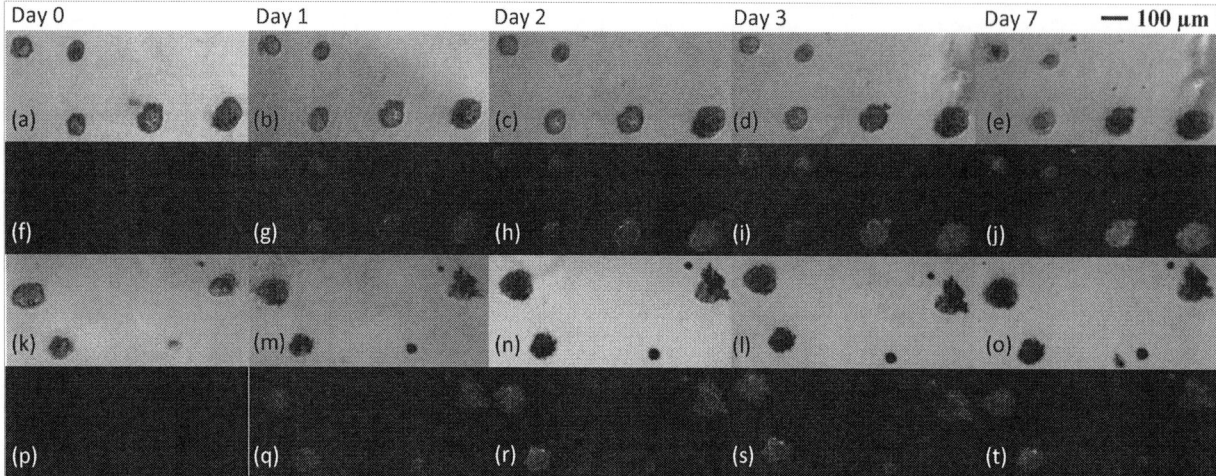

Fig. 2 Comparison of the cancer cell colonies from the direct and programmed groups after doxorubicin delivery using the MEMS drug delivery device. All pictures were taken at 4x magnification: (a)-(e) brightfield images of programmed group, T = 0 days, T = 1 day, T = 2 days, T = 3 days and T = 7 days (With a total of 18 µl of 1 mg/ml doxorubicin delivered, 6µl daily on T = 1 day, T = 2 days and T = 3 days), (f)-(j) fluorescence images of programmed group, T = 0 days, T = 1 day, T = 2 days, T = 3 days and T = 7 days, (k)-(o) brightfield images of direct group, T = 0 days, T = 1 day, T = 2 days, T = 3 days and T = 7 days (With 18 µl of 1 mg/ml doxorubicin delivered on day 1) and (p)-(t) fluorescence images of direct group, T = 0 days, T = 1 day, T = 2 days, T = 3 days and T = 7 days.

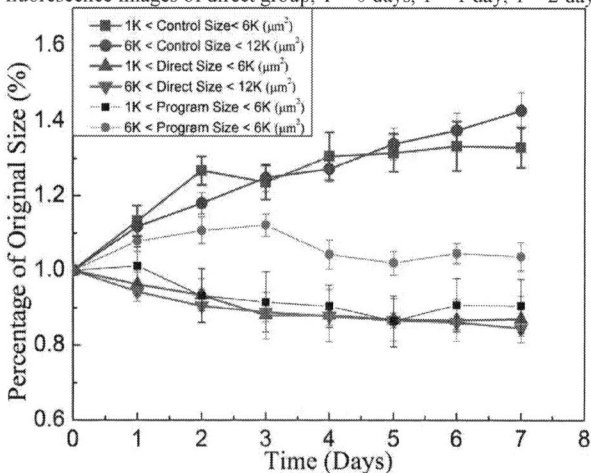

Fig. 3 In vitro test results expressed as % of original size over time. The results from each group are categorized according to their size into 2 groups, one group consisting of smaller colonies of sizes between 1K µm² - 6K µm² and larger colonies of sizes between 6K µm² - 12K µm². Control populations with no drug delivered ($13 \leq n \leq 20$, mean ± SE); Direct delivery with 18 µg of doxorubicin added on T = 0 ($17 \leq n \leq 20$, mean ± SE); Program delivery with 6 µg of doxorubicin added on T = 0, T = 1 and T = 2 ($13 \leq n \leq 16$, mean ± SE).

References

[1] S. V. Sastry, J. R. Nyshadham, and J. A. Fix, *Pharmaceutical Science & Technology Today*, vol. 3, pp. 138-145, 2000.

[2] M. Staples, K. Daniel, M. Cima, *et al.*, *Pharmaceutical Research*, vol. 23, pp. 847-863, 2006.

[3] R. Lo, P. Y. Li, S. Saati, *et al.*, *Lab on a Chip*, vol. 8, pp. 1027-1030, 2008.

[4] R. K. Jain, *Clinical Cancer Research*, vol. 5, pp. 1605-1606, July 1, 1999 1999.

[5] P. Y. Li, J. Shih, R. Lo, *et al.*, *Sensors and Actuators a-Physical*, vol. 143, pp. 41-48, May 2008.

[6] N. M. Elman and U. M. Upadhyay, *Current Pharmaceutical*

Biotechnology, vol. 11, pp. 398-403, 2010.

[7] R. Gref, A. Domb, P. Quellec, *et al.*, *Advanced Drug Delivery Reviews*, vol. 16, pp. 215-233, 1995.

[8] M. E. Davis, Z. Chen, and D. M. Shin, *Nat Rev Drug Discov*, vol. 7, pp. 771-782, 2008.

[9] H. Gensler, R. Sheybani, P. Y. Li, *et al.*, *Biomedical Microdevices*, vol. 14, pp. 483-496, Jun 2012.

[10] D. A. LaVan, T. McGuire, and R. Langer, *Nature Biotechnology*, vol. 21, pp. 1184-1191, Oct 2003.

[11] B. Ziaie, A. Baldi, M. Lei, *et al.*, *Advanced Drug Delivery Reviews*, vol. 56, pp. 145-172, Feb 2004.

[12] N. C. Tsai and C. Y. Sue, *Sensors and Actuators a-Physical*, vol. 134, pp. 555-564, Mar 2007.

[13] N.-C. Tsai and C.-Y. Sue, *Sensors and Actuators A: Physical*, vol. 134, pp. 555-564, 2007.

[14] S. Zafar Razzacki, P. K. Thwar, M. Yang, *et al.*, *Advanced Drug Delivery Reviews*, vol. 56, pp. 185-198, 2004.

[15] A. Nisar, N. Afzulpurkar, B. Mahaisavariya, *et al.*, *Sensors and Actuators B: Chemical*, vol. 130, pp. 917-942, 2008.

[16] G. Voskerician, M. S. Shive, R. S. Shawgo, *et al.*, *Biomaterials*, vol. 24, pp. 1959-1967, 2003.

[17] P.-Y. Li, J. Shih, R. Lo, *et al.*, *Sensors and Actuators A: Physical*, vol. 143, pp. 41-48, 2008.

[18] H. Gensler, R. Sheybani, L. Po-Ying, *et al.*, in *Micro Electro Mechanical Systems (MEMS), 2010 IEEE 23rd International Conference on*, 2010, pp. 23-26.

[19] R. Lo and E. Meng, in *Micro Electro Mechanical Systems, 2009. MEMS 2009. IEEE 22nd International Conference on*, 2009, pp. 236-239.

[20] J. K. Nicholson, *Mol Syst Biol*, vol. 2, 2006.

[21] P. Y. Li, R. Sheybani, C. A. Gutierrez, *et al.*, *Journal of Microelectromechanical Systems*, vol. 19, pp. 215-228, Feb 2010.

[22] F. A. Fornari, J. K. Randolph, J. C. Yalowich, *et al.*, *Molecular Pharmacology*, vol. 45, pp. 649-656, Apr 1994.

Microfluidic device optimization for cell growth

Vibha Jayaraj , Pramod. P.Wangikar and Sameer Jadhav*

Department of Chemical Engineering
Indian Institute of Technology Bombay
Mumbai, India
+91 22 2576424, srjadhav@che.iitb.ac.in

Abstract

Intensive biological research has been directed towards why and how cells respond to a dynamic environment. Microfabrication has emerged as crucial tool to capture dynamic response. Our current study describes the approaches taken to fabricate, optimize, and validate a biocompatible device. Our experimental observations revealed that attempt to increase the channel depth induced surface roughness, which affects the device characteristics. Rough surface may not facilitate smooth pattern transfer to PDMS and more or less affects opacity of device. Thus Multiobjective optimization (MOO) helps us to choose the right combination of parameters to develop a device suitable for cell growth.

Keywords- Microfluidics, Reactive ion etching (RIE) and Multiobjective optimization (MOO).

Introduction

Over the last decade, incentives offered by microfabrication have driven the fusion of microfluidics and biological sciences. Fabrication of low cost microfluidic channels which can mimic in vivo conditions is a highly sought objective. In this study, we attempted to fabricate a biocompatible microfluidic device. A permanent master mould for PDMS device was fabricated combining photolithography and reactive ion etching (RIE). Etching experiments have been reported to induce roughness on silicon surface. Literature reports roughness in silicon surface was reduced by using scavenging gases CHF_3 or Ar [1]. Mixture of gases are generally used for chemical etching of silicon in first place, secondly species for inhibitor film formation and thirdly species to suppress the inhibitor film at horizontal surface [2]. Objective of our study is to use single flow gas SF6, and study the effect of process flow chamber pressure, RF power and etch time. The novelty of the work is to design microfluidic channel with increased depth and lowered surface roughness. Since both the objective functions are competing, we considered the optimization to be a MOO problem. Three fold approach was considered, which involves experimentation, model optimization and validation.

Experimental Method

The patterned silicon wafer was etched using 20sccm Sulphur hexa fluoride SF_6 in STS RIE320. Channel depth and roughness were measured by surface profilometer Dektak. Surface reflectance was measured using PerkinElmer lambda-950 spectrophotometer equipped with integrated sphere. Based on the experimental data obtained, response surface method (RSM), was used to track the functional relationship between response and process variable. MOO was carried out based on ε-constraint approach. The quasi Newton and line search methods as available in the fmincon subroutine in MATLAB (Mathworks, Natick, NA) was used to solve quadratic problem with quadratic constraints [3]. PDMS based device was fabricated to track *Bacillus pumilus* cell growth. 0.1 OD innoculum was used for on-chip experiments. Every 2hrs interval, images were captured to track cell event.

Results and Discussion

Device was optimized for channel depth and surface roughness. To identify an optimum operating condition, each process variables were varied at three levels, hence 27 different combinations were studied. Channel depth continued to increase with time while increment in pressure had marginal effects on the same as shown in Fig.1a and b. At power 150 and 200 w, channel depth increased with time but when power was increased to 400 w, channel depth increase between 10 and 20 min's was insignificant (Fig. 1b). Etching experiments have been reported to induce roughness on silicon surface. Surface roughness showed a drastic change on increasing chamber pressure from 20 to 80mtorr as shown in Fig. 1c & d. RF power 150 and 200w showed surface roughness range of $9-20$ μm for 5 and 10 min's etched sample respectively. The highest tested power 400w increased the surface roughness up to 60 μm as plotted in Fig 1b. Our experimental results revealed that the response was influenced by process variables like pressure, RF power and etch time.

Based on the above observation, quadratic model was built using D-optimal design in the design experiment software. Model data indicated that increase in pressure showed a cyclic response of increased and decreased channel depth as shown in Fig 2. Beyond 84mtorr pressure, the channel depth declined at all time points. The plausible reason may be that increase in pressure leads to more collisions of the reactive ions before they reach the silicon substrate [4]. Apart from the influence of pressure, power and time also play a role. This is possibly due to increased disassociation of SF_6 molecules [5], which results in high density plasma, which in turn increases the ion bombardment on the substrate and an increased etch rate. However when chamber power was reduced, increased collision of ions affects the reactive species mean free path which reduces the ion energy and directionality leading to a reduction in etch rate. Our results were in agreement with experiments conducted by Legtenberg [6]. Surface roughness declined when chamber pressure was increased. On the contrary, increment in power increased surface roughness up to 74μm as shown in Fig. 1d. Experimental results indicate that increased mean roughness (Fig. 3b) significantly reduced the

978-1-4673-4840-9/13 $31.00 © 2013 IEEE   157

reflectance of silicon surface (Fig. 5). Our objective of parameter optimization was to obtain a device of increased channel depth and lowered surface roughness. In order to identify an optimal solution satisfying both objectives, pareto front was obtained by MOO as shown in Fig. 4. Pareto front indicated the maximized channel depth of 32μm with a surface roughness of 20μm. Generation of such pareto front gives flexibility to choose the most appropriate operating recipe from several choices corresponding to different tradeoffs [2].

Conclusion

The novelty of this work lies in the device fabrication by using a combination of parameter optimization with a single flow gas SF_6, unlike the earlier work where mixtures of gases were adopted. Our experiments conclude that desired channel depth and surface roughness can be obtained by choosing the right combination of pressure, power and time. This study permitted us to resolve a region in which both the objectives were satisfied. The fabricated device was tested for cell growth and it was found that bacterial cells continued to grow and divide for more than 24hrs (Fig 6). Generation of concentration gradients for cell migration and nanopatterning in microfludic device is suggested future scope of the work.

Acknowledgment-The authors are thankful to Centre for Excellence in Nanoelectronics (CEN), IITBombay for the microfabrication facillities.

References

[1] Siti Azlina Rosli, A. A. A. a. H. A. H. (2006). ICSE2006 Proc., Kuala Lumpur, Malaysia, IEEE

[2] Henri, J., G. Han, et al. (1996). Journal of Micromechanics and Microengineering 6(1): 14

[3] Maiti, S. K., A. E. Lantz, et al.,2011 Bioresource Technology 102(13): 6951-6958

[4] Chang, Y. C., G. H. Mei, et al. (2007). Nanotechnology 18: 285-303.

[5] KnizikeviÄ ius, R. and V. Kopustinskas (2004). Vacuum 77(1): 1-4.

[6] Legtenberg, R., H. Jansen, et al. (1995).Journal of The Electrochemical Society 142(6): 2020-2028.

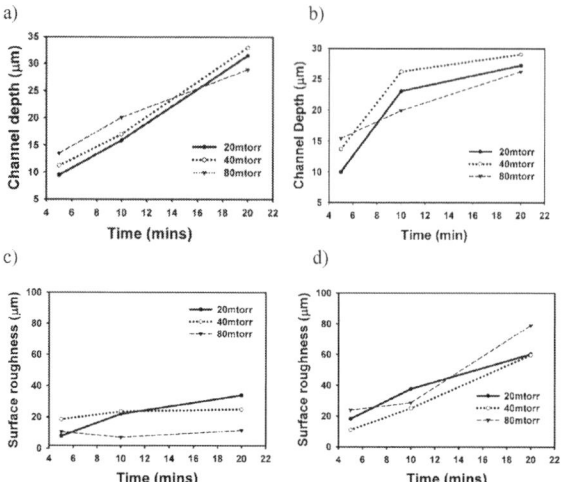

Fig. 1 Channel depth and surface roughness of etched silicon sample. At constant power (a, c) at 150w and (b, d) 400w, pressure and time was varied.

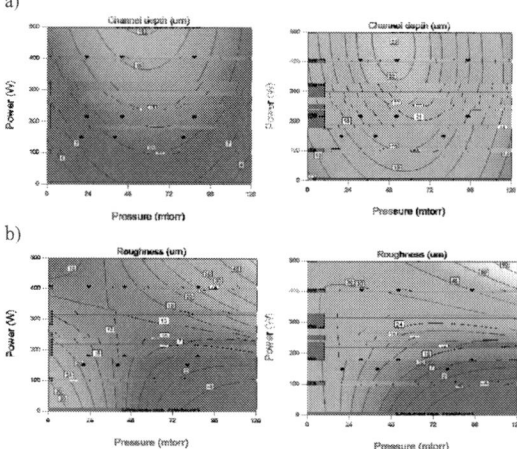

Fig. 2 Channel depth and surface roughness as a function of power, pressure and time are captured. Response obtained are shown as function of power and pressure (a) 5 min (b) 10min

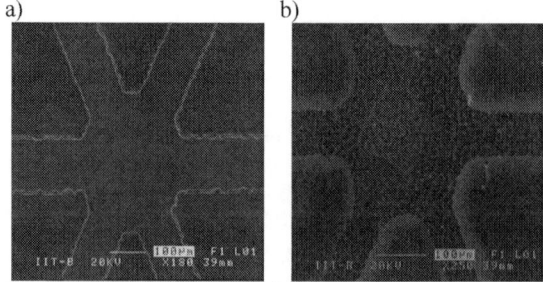

Fig. 3 Scanning electron micrograph showing etched silicon surface characteristics. a) 80mtorr 200w 5min   b) 40mtorr 150w 20min.

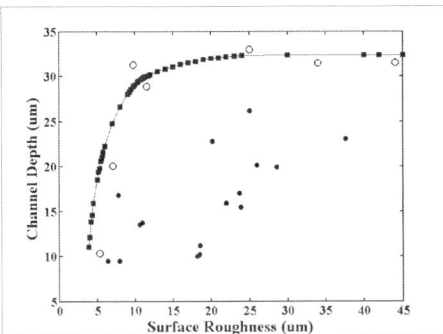

Fig. 4 Pareto front generated from MOO. Solid circle indicates non-optimal solutions obtained, square indicates simulated pareto front, empty circle indicates experimental data that are close to simulated pareto front

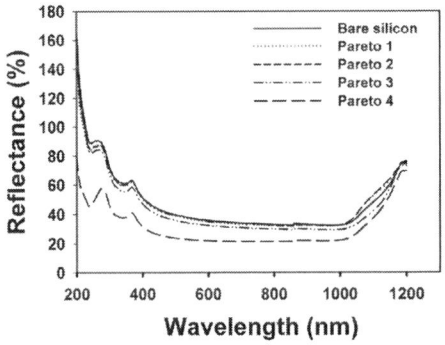

Fig. 5. Total reflectance measurement of bare and etched (100) silicon wafer using Integrated sphere.

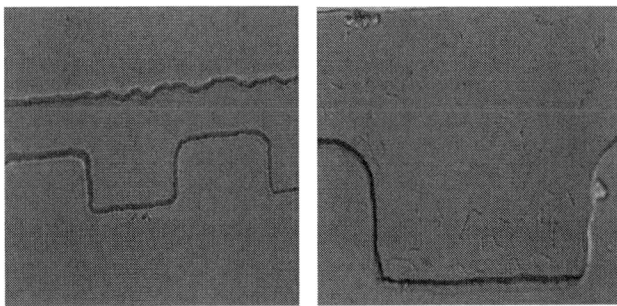

Figure 6. Optical micrograph of PDMS device (a) Replica of smooth silicon surface (b) Bacterial cells in the device

Stoichiometric Amorphous Hydrogenated Silicon Carbide Thin Film Synthesis Using DC-Saddle Plasma Enhanced Chemical Vapour Deposition

Behzad Jazizadeh Karimi, Ali B. Alamin Dow, Nazir P. Kherani*

Department of Electrical and Computer Engineering
University of Toronto, 10 King's College Road
Toronto, Ontario, Canada
*Email: kherani@ecf.utoronto.ca

Abstract

Silicon carbide is a versatile material amenable to a variety of applications ranging from electrical insulation, surface passivation and diffusion barrier to optical devices. The DC saddle-field plasma enhanced chemical vapour technique is an alternative large area deposition technique. Here we report on the synthesis of stoichiometric hydrogenated amorphous silicon using the dc saddle-field PECVD technique. We also report on the attainment of very smooth surface morphology for the stoichiometric a-SiC:H films in contrast to low carbon content films. Surface roughness of 1 nm *rms* was demonstrated for films grown at a temperature as low as 225°C.

Keywords: DC saddle-field PECVD, SiC, XPS, AFM, low temperature

Introduction

Hydrogenated amorphous silicon carbide (a-Si$_{1-x}$C$_x$:H) is an attractive alloy for a variety of applications due to its manifold of excellent properties. These include mechanical (hardness) [1], chemical (inertness, corrosion resistance), thermal (stability, conductivity), electrical (resistivity, passivation) [2] and optical (transmittance and wide tuneable band-gap) properties [3]. Further, hydrogenated amorphous silicon carbide is a tuneable alloy wherein its elemental composition can be readily modulated to meet specific performance requirements and an appealing material given its simplicity of synthesis using large area plasma enhanced chemical vapour deposition (PECVD) techniques .

Hydrogenated amorphous silicon carbide films have been prepared using a variety of techniques including rf PECVD [4 – 9], electron cyclotron resonance [10], rf sputtering [11], and high temperature chemical vapour deposition [12]. In the context of rf PECVD, a-Si$_{1-x}$C$_x$:H films have been studied extensively, leading to the successful preparation of stoichiometric (x = 0.5) amorphous silicon carbide.

Recently, the dc saddle-field PECVD technique has been used for the preparation of hydrogenated amorphous silicon carbide as a wide bandgap passivation layer on crystalline silicon surface [13] albeit the carbon content is relatively dilute and far from stoichiometry.

The objective of the present study is to investigate the preparation of stoichiometric hydrogenated amorphous silicon carbide using the dc saddle-field PECVD technique. Further, it is also the objective of this work to carry out the synthesis at a relatively low temperature of 225-240°C.

DC Saddle Field PECVD

The dc saddle-field plasma enhanced chemical vapour deposition technique is a novel glow discharge method that has been reported to produce device quality hydrogenated amorphous silicon [14], high quality interfacial passivation at the amorphous-crystalline silicon heterojunction [15], high sp^3 content hydrogenated amorphous carbon [16], and diamond-like carbon-based low-emissive coatings [17].

The dc saddle-field method utilizes a semitransparent mesh anode symmetrically positioned between two cathodes to create a dc saddle field that effectively serves to extend the path length of the electron and thus result in a stable plasma attainable at lower pressures than the dc diode PECVD [18].

Experiment

Hydrogenated amorphous silicon carbide films were prepared in the dc saddle field PECVD deposition chamber which is illustrated in Fig. 1. The deposition system consists of a deposition chamber equipped with a substrate heater, precursor gas mixing and delivery system, chamber pressure control valve and an ultrahigh vacuum system. The semitransparent electrodes consist of semitransparent stainless steel mesh.

Typically there are two different electrode configurations that are used inside the chamber, namely the pentode and triode arrangements. The triode configuration was employed for the purpose of the experiments in this study. The motivation for the triode choice, as shown in Figure 1, was to mitigate surface roughness by ensuring that the plasma region was sufficiently remote from the growth surface. The triode choice also features uniform deposition over a larger area.

978-1-4673-4840-9/13 $31.00 © 2013 IEEE

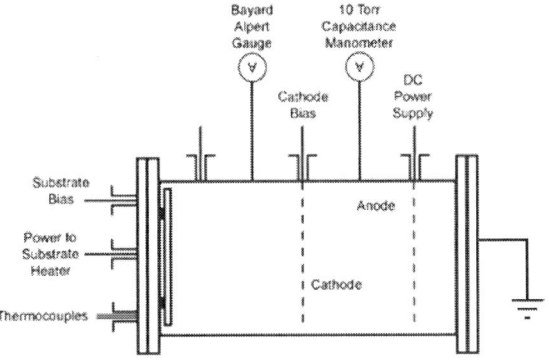

Fig. 1 Schematic of the triode configuration of the DC-SF PECVD system where the cathode, substrate and chamber are at ground potential. The remote plasma provides for a low ion energy film growth environment at the substrate.

The deposition parameters are given in Table I. All depositions were carried out at a substrate temperature which ranged from 225°C to 240°C, corresponding to a control heater temperature of 400°C. Deposition times were varied from 40 to 300 minutes, which principally affected the film thickness. While keeping all the deposition parameters fixed, the precursor gas mixture - which consisted of silane (SiH_4) and methane (CH_4) gases – was varied systematically to obtain a-$Si_{1-x}C_x$:H films. A series of hydrogenated amorphous silicon carbide films were synthesized as a function of the methane mole fraction χ where $\chi \equiv [CH_4]/([CH_4]+[SiH_4])$. The mole fraction χ was varied from 0.5 to 0.93 in small increments as illustrated by the abscissa of Figure 2.

Table I
Fixed Deposition Parameters for the Synthesis of Hydrogenated Amorphous Silicon Carbide Films

Parameter	Value (Unit)
Heater Temperature	400 (°C)
Substrate Temperature	225-240 (°C)
Anode Current	17.5 (mA)
Anode Voltage	550 (V)
Total Mixing Gas Pressure	1000 (Torr)
Deposition Pressure	160 (mTorr)
Gas Flow	30 (sccm)

The hydrogenated amorphous silicon carbide films were deposited on 25 mm square crystalline silicon samples. The samples were cut from single side polished *n*-type (100), 1 – 5 Ω.cm, 100 mm diameter, 525 μm thick float zone wafers. All samples were cleaned prior to the deposition using successive ultrasonic cleaning in acetone, isopropanol alcohol (IPA) and deionized (DI) water for 10 minutes each and subsequently dried with nitrogen. The samples were then loaded onto the stainless steel substrate holder and held in place with sample clips.

The deposited films were characterized for thickness using the Tencor Alphastep 200 Automatic Step Profiler. X-ray Photoelectron Spectroscopy (XPS) was used to probe the elemental makeup of the film and thus determine its composition. XPS measurements were performed using the Thermo Scientific K-Alpha instrument. The morphology of the surface of the deposited films was characterized using Atomic Force Microscopy (AFM) by Digital Instruments' Nanoscope Dimension 3100.

Results and Discussion

The hydrogenated amorphous silicon carbide film growth rate as a function of the methane mole fraction χ is shown in Figure 2. The growth/deposition rate is observed to decrease from 1.9 nm/min to 1.5 nm/min with increasing mole fraction χ, indicating an inversely proportional dependence on the methane mole fraction χ. Close observation of the data in Figure 2 shows an inflection point in the growth rate at a methane mole fraction of approximately 0.78, suggesting that beyond this point the characteristics of a methane plasma dominate the growth rate and composition of the films; this is better understood when examining the composition data presented in Figure 3 below. It is worth highlighting that the growth rate levels off at a practically viable deposition rate of 1.5 nm/min as the films approach stoichiometry.

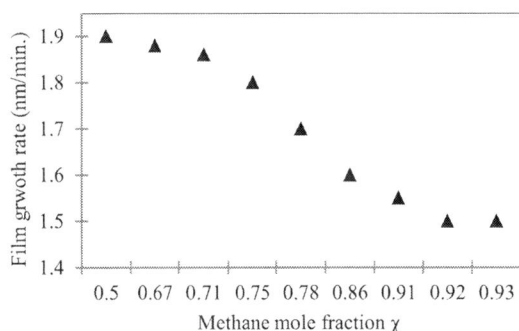

Fig. 2 Evolution of the a-$Si_{1-x}C_x$:H film growth/deposition rate as a function of the increasing methane mole fraction χ in the precursor gas mixture.

The carbon mole fraction x and the silicon mole fraction $1-x$ in the a-$Si_{1-x}C_x$:H films as a function of the methane mole fraction χ are shown in Figure 3 (a). The dashed lines represent least squares fitting to a parabolic curve with correlation coefficient R^2 for the two curves approaching unity. It is interesting to observe that in order to obtain stoichiometric a-SiC:H requires a methane mole fraction of approximately 93%, or equivalently a silane mole fraction of about 7%. This indicates that the occlusion of C is dominated by the methane plasma and corresponding growth processes. To illustrate this further, the dependence of the rate of increase in the carbon mole fraction x with respect to the methane mole fraction as a function of x is shown in Figure 3(b). The rate of C occlusion drops off rapidly with increase methane mole fraction as the films approach stoichiometry. It is noteworthy that these films are essentially at stoichiometric parity in C and Si albeit there is oxygen in the films in the range of 1 to 4 atomic percent as determined by XPS. The source of the oxygen is attributed to the grid and chamber surfaces owing to the fact that the system was not baked prior to the depositions.

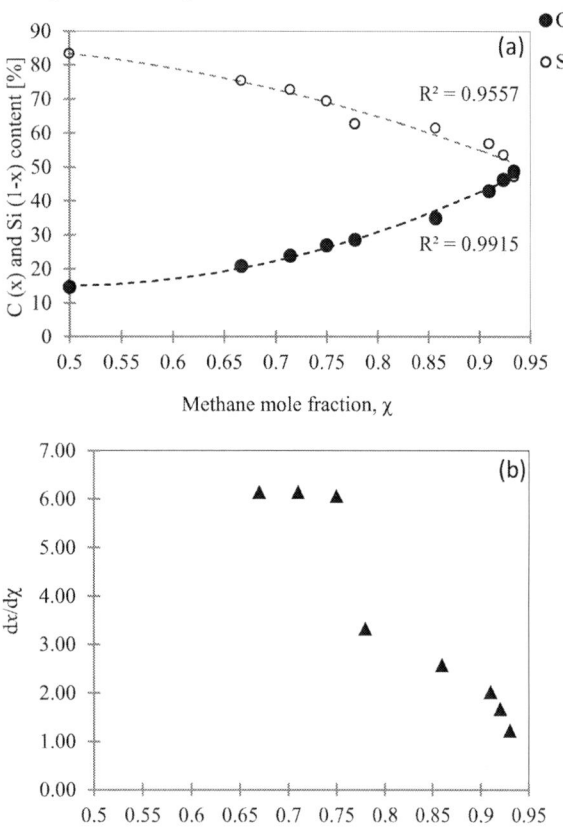

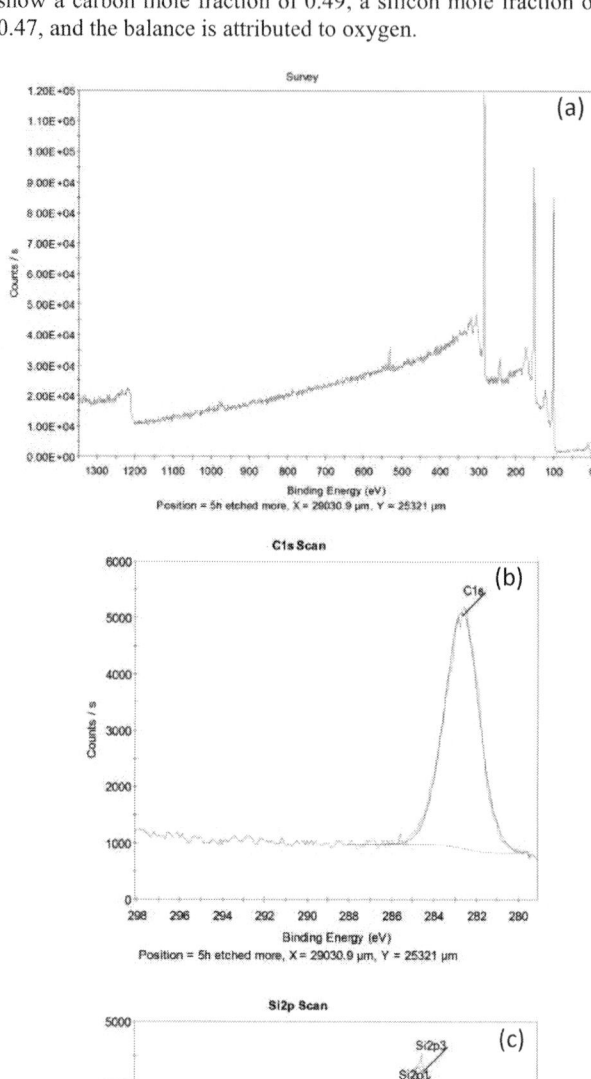

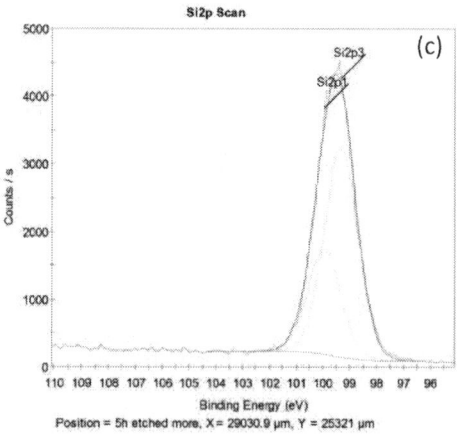

show a carbon mole fraction of 0.49, a silicon mole fraction of 0.47, and the balance is attributed to oxygen.

Fig. 3 (a) Carbon mole fraction x and silicon mole fraction $1-x$ in the a-Si$_{1-x}$C$_x$:H films as a function of the methane mole fraction χ in the precursor gas mixture. (b) The rate of increase in the carbon mole fraction x with respect to the methane mole fraction χ as a function of the methane mole fraction χ.

The XPS spectra for the stoichiometric a-Si$_{1-x}$C$_x$:H, prepared at χ of 0.93, are shown in Figure 4. Figure 4 (a) shows the full XPS spectrum which clearly indicate the predominant absence of oxygen given that the samples were argon etched prior to XPS analysis; specifically, the O1s peaks at 531 eV and 980 eV are absent in relation to the unetched sample (XPS spectrum not shown). The peaks around 150 eV correspond to the silicon substrate. Figure 4(b) shows the single C1s peak at 282.6 eV which indicates the C – Si chemical bond. Figure 4 (c) show the two deconvoluted peaks representing Si2p3 and Si2p1 at 99.4 and 99.9 eV, respectively, and which collectively indicate the Si – C chemical bonding. Detailed analysis of the XPS spectra

Fig. 4 XPS spectra of stoichiometric a-SiC:H deposited at $\chi = 0.93$. (a) The full XPS scan. (b) The XPS carbon scan. (c) The XPS silicon scan with two deconvoluted peaks.

978-1-4673-4840-9/13 $31.00 © 2013 IEEE

The surface roughness of the a-Si$_{1-x}$C$_x$:H films as a function of the methane mole fraction χ are shown in Figure 5(a). The rms roughness is observed to decrease almost exponentially, from approximately 11 nm to 1 nm *rms*, as the films approach stoichiometry. It is also interesting to observe that both the decrease in film roughness and decrease in film growth rate correlated, which is generally valid for thin film growth. The surface morphology of the stoichiometric a-SiC:H film as measured using AFM is shown in Figure 5(b).

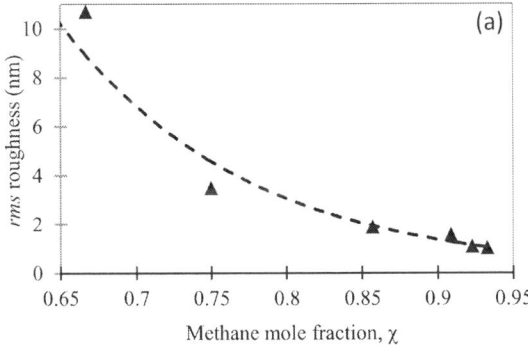

Fig. 5 (a) The *rms* surface roughness as a function of increasing methane mole fraction. (b) The surface morphology of the stoichiometric a-SiC:H film obtained at $\chi = 0.93$. The *rms* surface roughness of this sample was measured to be 1.014nm.

Conclusion

In this study we presented the synthesis of hydrogenated amorphous silicon carbide using DC Saddle-Field PECVD. Silane and methane were used as precursor gases at a deposition temperature of approximately 225°C. Stoichiometric a-SiC:H was acheived at a methane mole fraction of 0.93 with a corresponding very low surface roughness of 1.014 nm *rms*. These results pave the way for using DCSF-PECVD technique for the synthesis of stoichiometric amorphous silicon carbide films at low temperatures.

References

[1] H. Abderrazak and E. S. B. Hadj Hmida, *Silicon Carbide: Synthesis and Properties*, Properties and Applications 388 of Silicon Carbide, intechopen, pp. 361 – 388, April 2011.

[2] I. Martín et. al., *J. Appl. Phys.* 98, p. 114912, December 2005.

[3] D. Brassard and M. A. El Khakani, *J. Appl. Phys.*, Vol. 93, Issue 7, p. 4066, 2003.

[4] I. Martín et. al., *Appl. Phys. Letters*, 79, 14, p. 2199, October 2001.

[5] E. Pascual et. al., *Diam. Rel. Mat.*, 4, 5, pp. 702 – 705, 1995.

[6] D. Wuu et. al., *Appl. Surf. Sci.*, 144, 1-4, pp. 708 – 712, 1999.

[7] Y. H. Wang et. al., *Mat. Sci. Forum*, 338-342, pp. 325 – 328, 2000.

[8] W. K. Choi and S. Gangadharan, *Mat. Sci. Eng. B*, Vol. 75, Issue 2, pp. 174 – 176, 2000.

[9] T. Kaneko et. al., *Thin Sol. Films*, 409, 1, pp. 74 – 77, 2002.

[10] J. Cui et. al., *J. Appl. Phys.*, 89, 11, p. 6153, June 2001.

[11] G. De Cesare et. al., *Surf. Coat. Tech.*, 80, 1, pp. 237 – 241, 1996.

[12] D. Kuhman et. al., *Thin Sol. Films*, 177, 1, pp. 253 – 262, 1989.

[13] C. C. Yang, *Hydrogenated Amorphous Silicon Carbide Prepared using DC Saddle Field PECVD for Photovoltaic Applications*, MASc Thesis, University of Toronto (2011).

[14] P. Mahtani et. al., *J. Non-Cryst. Sol.* 358 pp. 3396 – 3402 , 2012.

[15] B. Bahardoust et. al., *Phys. Status Solidi A*, 207, 3, pp. 539–543, 2010.

[16] S. Zukotynski, F. Gaspari, D. Manage, V. Pletnev and E. Sagnes, *Mat. Res. Soc. Symp. Proc.* 595, pp 239-248 (2000).

[17] P. Mahtani et. al., *Solar Energy Mat. & Sol. Cells*, 95 pp. 1630 – 1637, 2011.

[18] R.V. Kruzelecky and S. Zukotynski, *DC Saddle-Field Plasma-Enhanced Vapour Deposition*, Mat. Sci. Forum, Vol. 140-142, pp. 89 – 106, 1993.

Acknowledgements

This work was supported by the Self Powered Sensor Networks program under the aegis of the Ontario Research Fund – Research Excellence, and by grants from the Natural Sciences and Engineering Research Council of Canada and the University of Toronto.

Realization and Application of Nanometer E-beam Lithography System

Wei Shuhua, Dai Lan, Zhang Jing

Microelectronic Center, College of Information Engineering, North China University of Technology
No.5 Jingyuanzhuang Road, Shijingshan District
Beijing, China
Phone: 86-010-88803508 E-mail: jslwsh@hotmail.com

Abstract

The electron beam lithography system is a kind of important nanofabrication equipment with high resolution and excellent flexibility. In this paper, nanometer electron beam lithography (EBL) system based on scanning electron microscope is introduced. Its main components include a modified SEM, a laser interferometer controlled stage, a versatile high speed pattern generator, and a fully functional and easy-operational software system. In order to explain this EBL system design principle, realization method, this paper mainly introduces each component's design basis, main structures and functions. Stitching experiments and overlay experiments have been done on this EBL system based on JSM-35CF SEM. The lithography results show that stitching and overlay error is less than 100 nm. This kind of EBL system based on SEM can meet the need of micro-nanofabrication research and design activities at flexibility and low price.

Keywords: Micro-nanofabrication, Electron Beam Lithography System, SEM, Pattern Generator, Stitching, Overlay

Introduction

Lithography is an essential step for the fabrication of passive/active devices in electronic and electrical manufacturing industries. With the rapid development of nanotechnology, nanolithography, which is a kind of important fabrication of nanostructures and nanodevices, has attracted more and more attention. Electron beam lithography (EBL) especially plays an irreplaceable role in nanolithography with high resolution and excellent flexibility. The spot size of electron beam can be focused to less than one nanometer and ultra high resolution patterns can be generated. So the EBL has a great potential to be used in the nanoelectronics, nano-optics and most other nanofabrication fields [1-3].

The EBL system is the most important nanofabrication equipment, which combines the electrical, mechanical, vacuum and computer technologies. However, the commercial EBL systems are considerably more expensive for many educational or research laboratories which are just interested in the development of technologies for innovative devices. So a high performance, low cost and flexible operation EBL system is a good solution. Ref. [4] proposed a simple and general-purpose EBL system based on scanning electron microscopy (SEM). In this paper, a new EBL system is introduced, which is composed of a modified SEM to allow external signals to control beam position, a laser interferometer controlled stage, a versatile high speed pattern generator, and a fully functional and easy-operational software system [5-7]. This EBL system based on SEM is flexible and low cost. It has a great potential to be used in the micro-electronics, micro-optics, micro-mechanics and most other micro-nanofabrication fields.

EBL System Main Components

EBL technique was developed from the SEM, whose work principle is familiar with the EBL. So the suitable SEM can be chosen and assembled with beam blanker, nanometer pattern generator, embedded precision stage and EBL control software to form the EBL system. The main components are shown in Fig.1. This kind of EBL system based on SEM is inexpensive, easy to operate, and has a good prospect in micro-nanofabrication application.

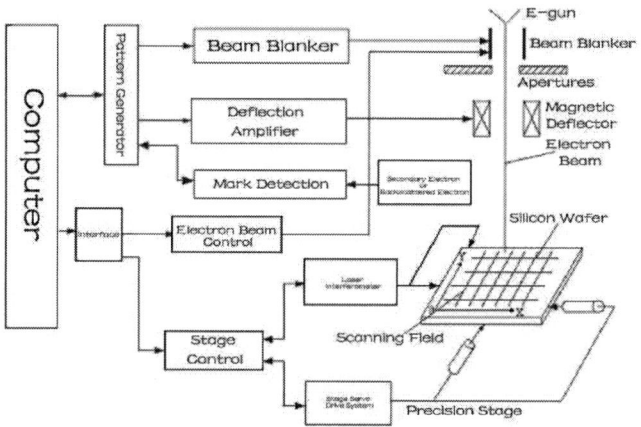

Fig.1   EBL System based on SEM

A. Scanning Electron Microscopy

The SEM is the core component, which provides the electron-optical system of the EBL system. The electron-optical performance has a direct effect on the resolution and stability of the EBL system, so the suitable SEM must be chosen. After analysis and comparison, we found that the thermal field emission SEMs were better than cold field emission for stabilizing total electron beam current, maximizing probe current and reducing electron beam noise, as well as in sensitivity to the environment[7]. The main function of SEM is to produce electron beam, focus on electron beam and control electron beam on and off to realize electron beam scanning.

B. Precision Stage

The positioning accuracy of SEM stage is usually in range of 1-5 μm, and mobile range is limited. So it can't meet the requirements of EBL scanning field stitching [8]. In order to achieve high accuracy of field stitching, the precision laser interferometer controlled stage is needed. It is composed of work stage machinery structure, laser interferometer measurement system, XY positioning control system, CCD alignment system and automatic transmission tablet control system. The laser interferometer measurement system and XY positioning control system constitute a closed loop measurement control system, which

978-1-4673-4840-9/13 $31.00 © 2013 IEEE

can locate the stage in the target location. CCD alignment system is used to make the silicon wafer in the depth of focus of electronic optical system, to get the best exposure effect.

The main functions of precision stage are reflected in two aspects. The first one is to solve the contradiction between high resolution and large area exposure, and to realize the EBL scanning field stitching. Because of electronic optical aberration and distortion restriction, electron beam scanning field scale is restricted, so in order to guarantee the resolution and to realize large area exposure, the precision stage must be assembled. The second function is to realize accurate positioning, guarantee alignment accuracy between layer and layer.

C. Pattern Generator

The pattern generator is the key component of making use of SEM to assemble the EBL system. The main functions of pattern generator are to interpret data produced by a software package and control beam deflection and beam blanking coils of SEM for high resolution lithography. Fig. 2 shows a block diagram of the hardware structure of the pattern generator. It consists of operation control unit, scanning unit, image acquisition unit, and some others.

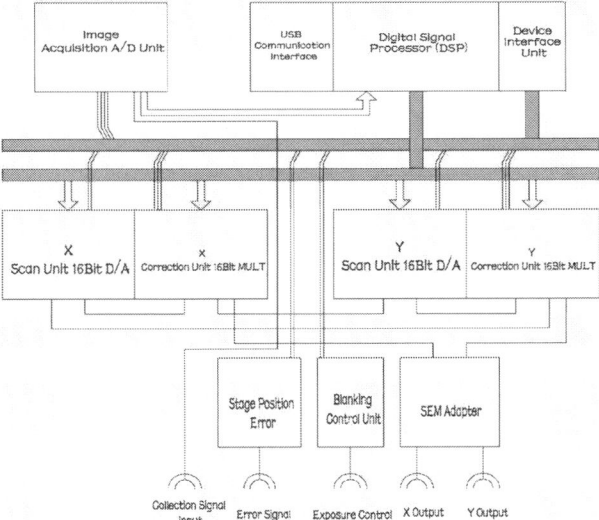

Fig. 2 Block diagram of the pattern generator hardware structure.

The pattern generator requires high speed and high accuracy in the translation of pattern data to shot data. So the digital signal processor (DSP) is employed in the operation control unit. The DSP has powerful computing capability, which can complete 32 float of multiplication and division operations during 80 clock cycles. Therefore, circles, rings and other complex curved shapes can be interpreted with a very high speed.

The scanning unit is controlled by two set of 16-bit digital-to-analog converter (DAC) models. Either set of DACs includes one main DAC and three multiplicative DACs. The main DAC receives pattern coordinates, and three multiplicative DACs receive the gain, offset, rotation and stage position corrections. The scanning unit can also generate blanking signals to control beam blanking coils.

In order to correct scanning field distortion, standard image must firstly be acquired. The function of image acquisition unit is to scan marks and standard chess graphics to acquire image information. The main component is DAC.

These data are converted from analog signal of image information collected by sensor to data signal by DAC. They are transferred to computer by USB2.0 interface and displayed on the screen.

D. Software System

The EBL system is so complicated and sophisticated, which needs a fully functional and easy-operational software system to ensure it run correctly. The main functions of the software system include initializing the system, generating exposure data, detecting the status of system components, correcting the scanning field, transferring exposure data and parameters, and controlling the exposure process [9]. According to these functional requirements, the software system has been designed three modules: exposure layout processing functional module, alignment control functional module, and exposure control functional module. The software system is developed based on Visual C++6.0 development environment [6].

The main purpose of exposure layout processing module is to generate exposure data format (EDF) files. This is accomplished by two processes, one of which is exposure layout design, and another is format conversion. Various layouts can be designed directly by drawing and editing figures. Another way of creating exposure layout is to import common industrial layout such as Caltech Intermediate Format (CIF) and Graphic Design System II (GDSII) format file, which can be edited conveniently. The file format is parsed by recursive descent parsing method on basis of BNF (Backus-Naur Form) rule [10]. Both layouts designed directly and common industrial layouts imported can be transferred to EDF file.

The alignment control module is to implement scanning field alignment and coordinates alignment. This can be implemented by scanning and acquiring standard chessboard image, adjusting marks positions, calculating correction parameters and then transmitting them to pattern generator. Then pattern generator controls beam deflection according to these correction parameters scanning again to accomplish scanning field and coordinates alignment.

The exposure control module is to control the whole process of exposure, which is the final procedure and also integrated operation of many processed. The exposure parameters are important to determine the dose of exposure, which is the description of resist absorbing the electronic energy when expose layouts. Various graphics have different exposure dose. The formula is as follows:

Area exposure dose: $D_A = 0.1 \cdot \dfrac{A \cdot T_A}{S_A^2}(\mu As/cm^2)$ (1)

Line exposure dose: $D_L = 10 \cdot \dfrac{A \cdot T_L}{S_L}(pAs/cm)$ (2)

Dot exposure dose: $D_D = A \cdot T_D (fAs)$ (3)

Where, A is the size of electron beam current, whose unit is pA. T_A, T_L and T_D are respectively the exposure dwell time of area, line and dot, whose unit is ms. S_A and S_L are respectively the step size of area and line, whose unit is μm. Then the EDF file acquired from the exposure layout processing module and these exposure parameters can be transferred to pattern generator, which will control beam deflection according to layouts information stored in EDF file to exposure layouts.

978-1-4673-4840-9/13 $31.00 © 2013 IEEE

Exposure Experiments

Exposure experiments have been done on the electron beam lithography system based on JSM-35CF SEM. Exposure experiments include stitching experiments, overlay experiments and patterns exposure. Stitching and overlay accuracy is an important evaluation indicator of EBL equipment performance.

A. Exposure stitching experiments

As mentioned earlier, because of electron optical design limitations，the EBL single exposure scanning field scale is restricted when a nanometer structure pattern is etched. So in order to realize large area exposure, the EBL system must be capable of field stitching. However, electronic equipment drift, various magnification of electronic optical cylinder and other factors lead to the scanning field distortion. So in order to ensure stitching accuracy, it is necessary to calibrate the scanning field.

The calibration can be realized by use of coordinate system linear transformation, and its mathematical expression is as follows:

$$dx = A + B x + Cy$$
$$dy = E + Fx + Gy$$

Where, dx and dy are deviation of the actual position and the ideal position; x and y are the sample stage position of the mark; A, E represent shift parameters; B, F represent gain parameters; C, G represent rotation parameters. For solving the six coefficients, three marks in a scanning field are needed as shown in Fig. 3. The pattern generator controls the SEM to scan the three marks and gain the actual position coordinates. The software system acquires these coordinates and calculates the six equations to get calibration coefficients, then send these parameters to pattern generator. The scanning unit of pattern generator control beam deflection coils according to these parameters. This process will be executed several times until the precise scanning field can be obtained.

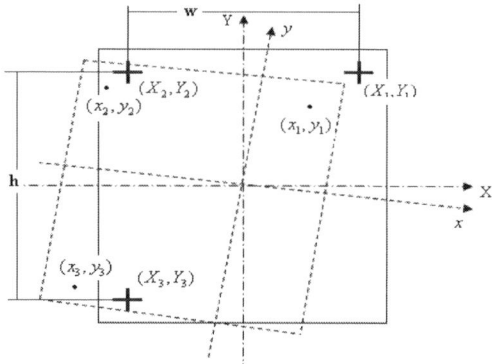

Fig.3 Principle diagram of scanning field calibration.

The stitching test pattern is designed directly by the software. It is a 6×6 array of $100\ \mu m$ size vernier cursor fields. Each field pattern is shown in Fig.4. The four "L" graphics is to observe the stitching condition roughly. The top left corners of each field are XY main verniers, each cursor with an interval of $2\ \mu m$, and the bottom right corners are XY deputy verniers, each cursor with an interval of $1.98\ \mu m$. The measurement resolution is $20\ nm$ and the measurement range is $-200\ nm \sim +200\ nm$.

Fig. 5 shows the SEM micrograph of this experimental result. According to the error calculation formula:

$$\sigma_n = \sqrt{\dfrac{\sum\limits_{i=1}^{n}\left(\bar{x} - x_i\right)^2}{n}} \qquad (4)$$

Where, n is sample size, $\bar{x}$ is sample average and σ_n is sample mean-square deviation. Statistical results show that this exposure test error σ_x is $31.19\ nm$ and σ_y is $26.53\ nm$.

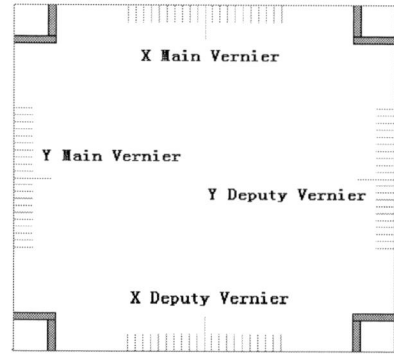

Fig.4 Stitching test pattern of each field.

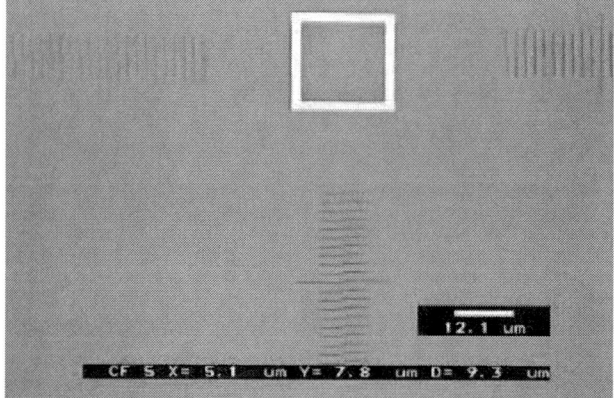

Fig.5 The SEM micrograph of stitching experimental result.

B. Exposure overlay experiments

Multilayer lithography is needed for some MEMS structure and semiconductor fabrication. In this process, each layer pattern is exposed and then removed out to do post-treatment. When this silicon chip is back into the work stage, its relative position of the work stage is changed. So in order to guarantee overlay accuracy, it is needed to alignment marks of chip, determine the position and azimuth of chip.

The test patterns are shown in Fig.6. The red XY main verniers are the first layer, each cursor with an interval of $2\ \mu m$. The blue XY deputy verniers are the second layer, each cursor with an interval of $1.98\ \mu m$. The actual exposure pattern are two 6×6 array of $100\ \mu m$ size fields by these two layers verniers respectively. Each layer pattern is saved as an exposure data format (EDF) file. Experiment operation steps are as follows: a. Put the sample with marks into the stage, and implement the coordinate system correction to make the stage coordinate and silicon wafer coordinate consistently. b. Control the stage to exposure area, and implement the scanning field

978-1-4673-4840-9/13 $31.00 © 2013 IEEE   166

calibration, then the EDF file of main verniers is exposure. c. After the first layer exposure, take the silicon wafer out. d. Put the silicon wafer into the stage again, and then implement the coordinate system correction once again. e. Control the stage to exposure area, and implement the scanning field calibration, then the EDF file of deputy verniers is exposure. The SEM micrograph of this experimental result is shown in Fig.7. The error calculation formula is the same with the stitching experiments.

Statistical results show that this exposure test error σ_x is 31.95 nm and σ_y is 33.38 nm.

All of these experiments were done at the 30 kV acceleration voltage, monolayer PMMA resist and 5pA beam current.

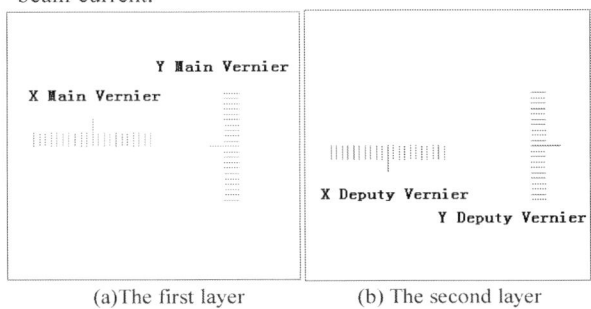

(a)The first layer　　(b) The second layer

Fig.6 Overlay test pattern.

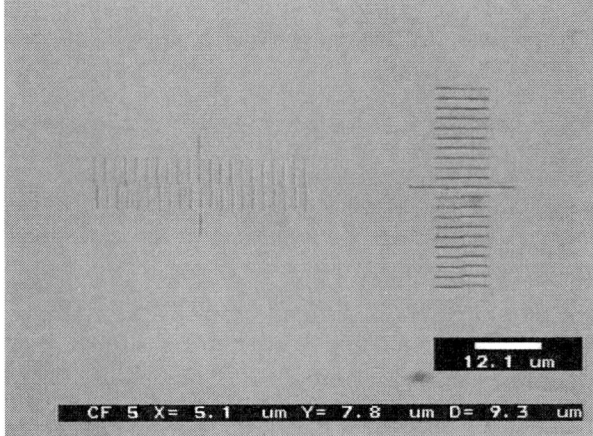

Fig.7 The SEM micrograph of overlay experimental result.

Conclusion

The above experiments testify the feasibility of this EBL system based on a modified SEM. This EBL system can be used for large-scale micro-nanofabrication and functional MEMS or micro electrical parts. This EBL fabrication system can meet most lithography applications in university laboratories with its powerful functions, friendly manipulation and low cost. It has made very important contributions in quantum effect devices, integrated optical device manufacturing and nanostructure manufacturing.

Acknowledgments

This work was supported in part by the National Natural Science Foundation of China (No.61001052) and Beijing Natural Science Foundation （4123096）.

REFERENCES

[1] C.Vieu, F.Carc enac, A.Pepin, et al., "Electron beam lithography: resolution limits and applications", Applied Surface Science, Vol.164, pp.111-117, 2000.

[2] A.Ampere, Tseng, et al., "Electron Beam Lithography in Nanoscale Fabrication: Recent Development", Electronics Packaging Manufacturing, IEEE Transactions, Vol.26, pp. 141-149, 2003.

[3] Peter H, Olaf F. "50 years of electron beam lithography: Contributions from Jena", Microelectronic Engineering, Vol.86, pp.438-441, 2009.

[4] K Molhave, DN Madsen, P Boggil, "A simple electron-beam lithography system", Ultramicroscopy, Vol.102(3), pp.215-219, 2005.

[5] S.H.Wei, W.Liu, L.HAN, "A new versatile high speed pattern generator for nanolithography", Proceedings of the IEEE, pp.824-828, 2008.

[6] S.H.Wei, J. Z. Zhang, L.HAN. "Design and Implementation of Software System of E-beam Lithography Based on SEM", Proceedings of the IEEE, pp.547-550, 2009.

[7] B.H.Yin, G.R. Fang, J.B. Liu, et al., "Miniature Electron Beam Lithography System for Micro /Nanometer Pattern Fabrication", Nanotechnology and Precision Engineering, Vol.8(4), pp. 290-294, 2010.

[8] Lü S L, Song Z T, Feng S L. "Fabrication of arrays of line with nanoscale width and large length by electron beam lithography with high-precision stage", Microelectronics Journal, Vol. 39(9), pp.1126-1129, 2008.

[9] A.D.Wilson, "Electron-beam systems for precision micron and submicron lithography", Proceedings of the IEEE, Vol.71, pp.575-584, 1983.

[10] J.H. Chen, H, Xue, "Parse the layout description language using recursive descent parsing method", Microcomputer and its applications, Vol.21, pp.52-54, 2002.

Oxide thin film transistors with ink-jet printed In-Ga-Zn oxide channel layer and ITO/IZO source/drain contacts

Y. Wang, X. W. Sun, S. W. Liu and A. K. K. Kyaw
School of EEE, Nanyang Technological University,
Nanyang Avenue, Singapore
exwsun@ntu.edu.sg

J. L. Zhao
College of Science, Tianjin University, Tianjin,
China

Abstract

In-Ga-Zn oxide (IGZO) TFTs was fabricated by ink-jet printing technology on a silicon substrate with SiO_2 on top. The device fabrication process includes printing ITO electrodes and IGZO semiconductor layer. A typical printed TFT shows a mobility of 0.32 cm^2/V s and a contact resistance of ~1 MΩ. Device performance was further improved by inserting an IZO layer between the source/drain electrode and IGZO channel, which results in a contact resistance of 0.05 MΩ, and an enhanced mobility of 0.82 cm^2/V s.

Keywords: ink-jet printing, In-Ga-Zn oxide, contact resistance.

Introduction

Recently, all printed electronic device technology has received more and more attentions because it offers the possibility of printing functional device by roll-to-roll manufacturing in ultra-large area under ambient condition with low cost [1]. Over the last decade, research on printable semiconductor materials is mainly focused on organics, especially for semiconductor channel materials. Sirringhaus et al. have fabricated all polymer poly(9, 9-dioctylfluoreneco-bithiophene) (F8T2) thin film transistors (TFTs) by ink-jet printing method and achieved a mobility of 0.02 cm^2/V s [2]. Afterward, pentacene [3] and poly (3-hexylthiophene) (P3HT) etc. [4-6] were used as channel layer of transistors fabricated by printing method. Up to now, only a few inorganic materials have been ink-jet printed due to the limited solubility of inorganic in printable solvent. Ridley et al. reported cadmium selenide (CdSe) TFTs using printing method and achieved a mobility of 1 cm^2/V s, and an on/off ratio of 3.1 × 10^4 [7]. Recently, Shimoda et al. have successfully printed polycrystalline silicon TFTs with a mobility of 6.5 cm^2/V s and an on/off ratio is 1 × 10^3 [8]. However, previous organic and inorganic TFTs have been processed in a glove box to control moisture and oxygen content to avoid film decay and oxidization which in turn increasing manufacturing cost.

In contrast to previous materials, metal oxide semiconductor is advantageous in terms of mobility, stability in the air and low cost, and some groups have been fabricated metal oxide semiconductor TFTs by ink-jet printing technology [9-12]. In-Ga-Zn oxide (IGZO) is a good potential alternative channel layer material for TFTs [13, 14]. It is transparent in the visible region due to the large band gap, and has a high mobility even for an amorphous structure due to s-electron conduction [13].

Recently, we have fabricated IGZO TFTs by ink-jet printing method with a coplanar architecture, and the device shows a typical n-type transistor property with a controllable threshold voltage by controlling the thickness of the channel layer [15]. However, electrodes of those devices are fabricated by magnetron sputtering and lift-off, which results in high cost and is not suitable for ultra-large device fabrication. In this chapter, we fabricated all printed IGZO TFTs including printed ITO source/drain electrodes and IGZO semiconductor layer on silicon substrate with SiO_2 on top. This study will significantly promote the progress of all printed oxide TFTs applications in large area with low cost.

Experimental details

The device fabrication process consists of printing ITO electrodes and IGZO semiconductor layer. The IGZO ink formulation and printing can be found elsewhere [15, 16]. In brief, the IGZO ink was prepared by dissolving 0.1 M of zinc acetate dihydrate [Zn(OAc)$_2$·2H$_2$O], 0.1 M indium chloride, and 0.025 M gallium chloride, in 2-methoxyethanol with 0.2 M MEA. was then added in the precursor solution as a sol-gel stabilizer. The solution was stirred at 50 °C for 2 hours and then aged for 24 hours. A heavily doped p^+-Si wafer (carrier concentration ~10^{19} cm^{-3}) was employed as the bottom gate of TFT. A 150 nm thick SiO_2 film as gate dielectric layer was thermally grown on top of the silicon wafer. Before printing, silicon substrate was cleaned by acetone, isopropanol and de-ionized water sequentially, followed by nitrogen blow dry. ITO ink used was commercial ink bought from ULVAC Technologies, Inc. 100 ~ 150 nm thick ITO electrodes were deposited by a DMP 2831 ink-jet printer and heated at 400 °C for 30 mins. IGZO ink was printed by the same printer twice on the prepared substrate and heated at 300 °C in air for 10 mins after each printing. Post-annealing was performed at 500 °C for 1 hour in air to remove residual chemicals and improve the quality of IGZO films. The width (W) of the channel was fixed at 500 μm, and the length (L) was varied from 50 μm to 200 μm with a 50 μm step. Since carrier concentration of IGZO films is increased by reducing Gallium mole ratio, device performance was further improved by introducing an IZO interlayer (Gallium mole ratio is 0) between IGZO channel layer and ITO electrodes.

The electrical characteristics of IGZO and IZO films were carried out by a Hall effect system (BIORAD HL5500) with a buffer amplifier (HL5580 buffer amplifier) with a magnetic field of 0.32 T by a four-point probe van der Pauw method at room temperature. The film morphology was characterized by atomic force microscopy (AFM). The transistor performance

978-1-4673-4840-9/13 $31.00 © 2013 IEEE

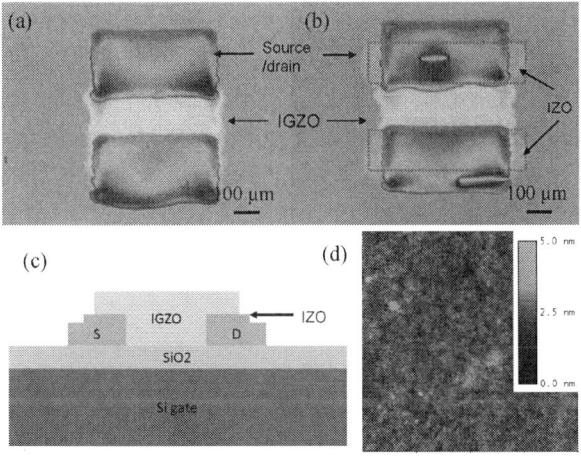

Fig. 1. Optical pictures of all printed IGZO TFTs with printed ITO electrodes, (a) without and (b) with IZO interlayer, respectively. The IZO interlayer is indicated by the red line in (b). (c) Schematic diagram of the device architecture and (d) AFM image of surface morphology of printed IGZO thin films, the size of the image is 1 μm × 1 μm.

of IGZO TFTs was measured with an HP 4156A semiconductor parameter analyzer. The device fabrication and characterization were all conducted under ambient conditions without taking precautionary measures to avoid ambient lights, moisture and oxygen.

Results and discussions

Fig. 1(a) and (b) show typical optical pictures of all printed IGZO TFTs with printed ITO electrodes, without and with IZO interlayer, respectively. The device architecture is bottom gate bottom contact which is suitable for printed process, as shown in Fig. 1(c). Two parallel ITO pads were printed firstly as source and drain electrodes, followed by printing IGZO or IZO/IGZO layer. As we can see, ITO and IGZO ink were only dropped on source & drain (S&D) electrode area and between S&D electrodes without wasting ink, respectively. Since printed films were composed of lateral overlapped printed dots, small zigzag shape can be seen at the edge of ITO and IGZO films. The film formed by lateral overlapped dots avoids coffee ring formed by single drop which is commonly observed during printing process [10, 17]. The surface morphology of IGZO thin films is presented in Fig. 1(d). The root mean square (RMS) roughness of IGZO film is only 0.358 nm, which is consistent with our previous studies [15].

Fig. 2(a) shows typical output characteristics of printed IGZO TFT at various gate voltages V_{GS}, exhibiting a typical n-channel enhancement field effect behavior. The width and length are 500 and 150 μm, respectively. Fig. 2(b) shows the transfer characteristics of IGZO TFT at a drain-source voltage V_{DS} of 20 V, and corresponding $I_{DS}^{1/2} \sim V_{GS}$ curve. Extra plotting linear portion of $I_{DS}^{1/2} \sim V_{GS}$ curve, threshold voltage (V_{th}) of the TFT can be obtained as 8 V, as indicated in Fig. 2(b). From the slope of the $I_{DS}^{1/2} \sim V_{GS}$ curve, the field effect mobility can also be obtained as 0.32 cm²/V s. The current on/off ratio is 2 × 10⁵, seen from the $I_{DS} \sim V_{GS}$ curve in Fig. 2

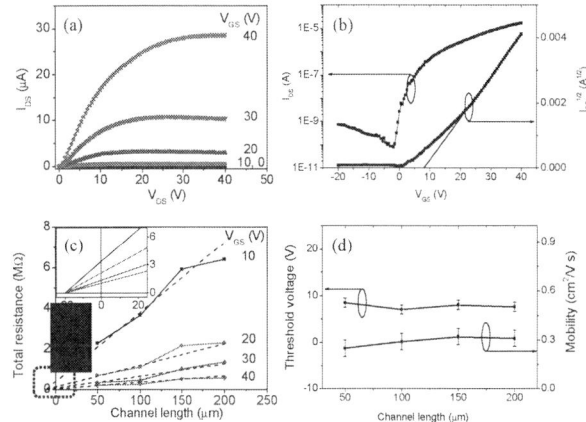

Fig. 2. Typical output characteristics and (b) transfer characteristics of the printed IGZO TFTs (W/L = 500/150 μm) with printed ITO source/drain electrodes. (c) Contact resistances estimated at various V_{GS} (10 ~ 40 V) and (d) threshold voltage and mobility as a function of the channel length of ink-jet printed IGZO TFTs.

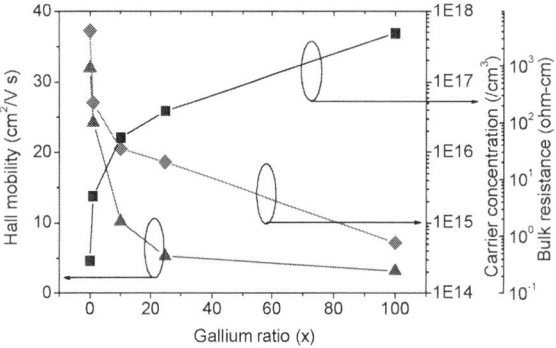

Fig. 3. Carrier concentration, bulk resistance and Hall mobility of IGZO thin films as a function of the Gallium content x, (the mole ratio of Ga: In: Zn = x: 100: 100, x change from 0 ~ 100).

(b). Source/drain contact resistance is one of the key factors to determine the TFT performance. Previous study proved that the drain on current level of a-Si:H TFTs can be increased by one order (from ~40 to ~400 nA) by reducing the contact resistance [18]. According to the gradual channel approximation [19], contact resistance under different gate voltage of the ink-jet printed IGZO TFTs is shown in Fig. 2(c). It can be seen that the contact resistance is estimated to be about 1 MΩ under a 40 V gate voltage. The width-normalized R_cW value is 50 kΩcm (channel width is 500 μm), which is higher than IGZO TFTs fabricated by magnetron sputtering (1.25 kΩcm) [20]. The high contact resistance may originate from the barrier at the ITO/IGZO interface due to the work function difference (ITO: 4.5-4.75 eV [21], IGZO: 4.25-4.45 eV [22]) and the interface defect states.

As well known, one of commonly used methods to reduce the contact resistance is to introduce a heavily doped layer (with high carrier concentration) into the contact area between the source/drain and channel layer. For a-Si:H TFT, previous

978-1-4673-4840-9/13 $31.00 © 2013 IEEE   169

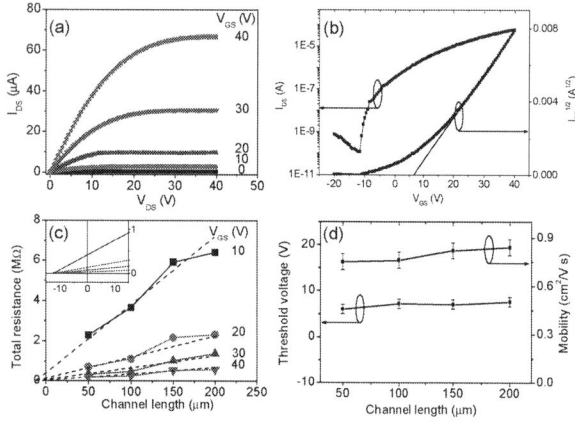

Fig. 4. (a) Output characteristics and (b) transfer characteristics of the printed IGZO TFTs with an IZO layer inserting between IGZO and source/drain electrodes. (c) Contact resistances estimated at various V_{GS} (10 ~ 40 V) and (d) threshold voltage and mobility as the function of the channel length of ink-jet printed IGZO TFTs.

studies found that Ohmic contact can be easily to be formed when the surface region of a-Si is heavily doped, which can significantly improve the TFT performance [23]. Different from a-Si and crystalline semiconductors where substitutional doping is effective, an effective doping of amorphous oxide semiconductors is achieved by altering stoichiometry of oxygen ion [20, 24], which can be realized by controlling gallium mole ratio in the precursor solution preparation [25]. Because Ga ions form stronger chemical bonds with oxygen than Zn and In ions, more Ga ions in IGZO films suppress oxygen vacancy formation resulting in carrier concentration reduction [13]. Therefore, electron concentration decreases with the increase of the gallium mole ratio. And the threshold voltage of IGZO TFTs shifts from positive to negative with the increase of the gallium ratio as discussed by Lim et al. [26]. Hall measurement results of IGZO films with different gallium ratio (the atom ratio of Ga: In: Zn = x: 100: 100, x ranges from 0 to 100) are shown in Fig. 3. It can be clearly seen that the carrier concentration of IZO films (gallium mole ratio x is 0) is 2 orders higher than that of IGZO film (gallium mole ratio x is 25, which was used in this report), and resistance is 430 times lower than that of IGZO.

Fig. 4(a) and (b) show output and transfer characteristics of TFTs with IZO interlayer. The output characteristic exhibits a good Ohmic contact property. The improved device shows a field effect mobility 0.82 cm^2/V s, the current on/off ratio is 6 × 10^5, and the threshold voltage is 7 V. Compared with the device without IZO layer, although threshold voltage and on/off ratio change slightly, the mobility is improved from 0.32 to 0.82 cm^2/V s. At the same time, contact resistance is reduced remarkably from 1 to 0.05 MΩ as indicated in Fig. 4(c). With the improvement of ITO/IGZO contact via IZO interlayer, higher performance all printed IGZO TFT is demonstrated, which provides a good start point for the development of low cost oxide circuits by ink-jet printing. Na et al. reduced the contact resistance of the IGZO TFTs by introducing a high electron concentration low resistivity IGZO layer by sputtering [20]. Du Ahn et al. employed Ar plasma

treatment to increase the carrier concentration in the contact area from 10^{14} cm^{-3} (pristine) to ~10^{20} cm^{-3}, and the contact resistance reduced from 1550 to 330 Ω cm, leading to ~3 fold increase of field effect mobility [27, 28]. Because they fabricated IGZO TFTs by sputtering, therefore, it is feasible to introduce the low resistivity layer by PVD equipment. However, in our process, it is more compatible with the roll-to-roll process by just printing a IZO layer into IGZO TFTs with low cost.

Conclusion

In conclusion, all printed IGZO TFTs including ITO source/drain electrodes and IGZO channel layer are fabricated. The device shows a threshold voltage of 8 V, a field effect mobility of 0.32 cm^2/V s and a drain current on/off ratio of 2 × 10^5. The field effect mobility is improved remarkably from 0.32 to 0.82 cm^2/V s by introducing a high electron concentration, low contact resistance IZO layer between source/drain electrode and IGZO channel layer. The contact resistance was significantly reduced from 1 to 0.05 MΩ. Our result shows that whole IGZO TFTs fabrication process can be accomplished by the low-cost printing in the air in the future.

Acknowledgement

The work is partially supported by the National Natural Science Foundation of China (NSFC) (Project Nos. 61006037, 61177014 and 61076015), and Tianjin Natural Science foundation (Project Nos. 11JCZDJC21900 and 11JCYDJC25800).

References

[1] B. Ong, "Semiconductor ink advances flexible displays," *Laser Focus World*, vol. 40, no. 6, pp. 85-88, Jun. 2004.

[2] H. Sirringhaus, T. Kawase, R. H. Friend, *et al.*, "High-resolution inkjet printing of all-polymer transistor circuits," *Science*, vol. 290, no. 5499, pp. 2123-2126, Dec 2000.

[3] S. K. Volkman, S. Molesa, B. Mattis, *et al.*, "Inkjetted organic transistors using a novel pentacene precursor," *Flex. Electron. Mater. Device Technol.*, vol. 769, p. 369-374, 2003.

[4] S. K. Park, Y. H. Kim, J. I. Han, *et al.*, "High-performance polymer TFTs printed on a plastic substrate," *IEEE Trans. Electron Devices*, vol. 49, no. 11, pp. 2008-2015, Nov 2002.

[5] G. Lu, H. Usta, C. Risko, *et al.*, "Synthesis, characterization, and transistor response of semiconducting silole polymers with substantial hole mobility and air stability. Experiment and theory," *J. Amer. Chem. Soc.*, vol. 130, no. 24, pp. 7670-7685, Jun 2008.

[6] S. K. Park, Y. H. Kim, J. I. Han, *et al.*, "Electrical characteristics of poly (3-hexylthiophene) thin film transistors printed and spin-coated on plastic substrates," *Synthetic Metals*, vol. 139, no. 2, p p. 377-384, Sep 2003.

[7] B. A. Ridley, B. Nivi, and J. M. Jacobson, "All-inorganic field effect transistors fabricated by printing," *Science*, vol. 286, no. 5440, pp. 746-749, Oct 1999.

[8] T. Shimoda, Y. Matsuki, M. Furusawa, *et al.*, "Solution-processed silicon films and transistors," *Nature*, vol. 440, no. 7085, pp. 783-786, Apr. 2006.

978-1-4673-4840-9/13 $31.00 © 2013 IEEE

[9] D. H. Lee, Y. J. Chang, G. S. Herman, *et al.*, "A general route to printable high-mobility transparent amorphous oxide semiconductors," *Adv. Mater.*, vol. 19, no. 6, pp. 843-847, Mar 2007.

[10] Y. H. Kim, K. H. Kim, M. S. Oh, *et al.*, "Ink-Jet-Printed Zinc-Tin-Oxide Thin-Film Transistors and Circuits With Rapid Thermal Annealing Process," *IEEE Electron Device Lett.*, vol. 31, no. 8, pp. 836-838, Aug 2010.

[11] D. H. Lee, S. Y. Han, G. S. Herman, *et al.*, "Inkjet printed high-mobility indium zinc tin oxide thin film transistors," *J. Mater. Chem.*, vol. 19, no. 20, pp. 3135-3137, 2009.

[12] S. K. Park, Y. H. Kim, and J. I. Han, "All solution-processed high-resolution bottom-contact transparent metal-oxide thin film transistors," *J. Phys. D: Appl. Phys.*, vol. 42, no. 12, p. 125102, Jun 2009.

[13] K. Nomura, H. Ohta, A. Takagi, *et al.*, "Room-temperature fabrication of transparent flexible thin-film transistors using amorphous oxide semiconductors," *Nature*, vol. 432, no. 7016, pp. 488-492, Nov 2004.

[14] K. Nomura, H. Ohta, K. Ueda, *et al.*, "Thin-film transistor fabricated in single-crystalline transparent oxide semiconductor," *Science*, vol. 300, no. 5623, pp. 1269-1272, May 2003.

[15] Y. Wang, X. W. Sun, G. K. L. Goh, *et al.*, "Influence of Channel Layer Thickness on the Electrical Performances of Inkjet-Printed In-Ga-Zn Oxide Thin-Film Transistors," *IEEE Trans. Electron Devices*, vol 58, no. 2, pp. 480-485, Feb 2011.

[16] Y. Wang, S. W. Liu, X. W. Sun, *et al.*, "Highly transparent solution processed In-Ga-Zn oxide thin films and thin film transistors," *J. Sol-Gel Sci. Technol.*, vol 55, no. 3, pp. 322-327, Sep 2010.

[17] M. Singh, H. M. Haverinen, P. Dhagat, *et al.*, "Inkjet Printing-Process and Its Applications," *Adv. Mater.*, vol. 22, no.6, pp. 673-685, Feb 2010.

[18] C. S. Chiang, S. Martin, J. Kanicki, *et al.*, "Top-gate staggered amorphous silicon thin-film transistors: Series resistance and nitride thickness effects," *Jpn. J. Appl. Phys.*, vol. 37, no. 11, pp. 5914-5920, Nov 1998.

[19] S. Martin, C. S. Chiang, J. Y. Nahm, *et al.*, "Influence of the amorphous silicon thickness on top gate thin-film transistor electrical performances," *Jpn. J. Appl. Phys.*, vol. 40, no. 2A, pp. 530-537, Feb 2001.

[20] J. H. Na, M. Kitamura, and Y. Arakawa, "High field-effect mobility amorphous InGaZnO transistors with aluminum electrodes," *Appl. Phys. Lett.*, vol. 93, no. 6, p. 063501, Aug 2008.

[21] K. Sugiyama, H. Ishii, Y. Ouchi, *et al.*, "Dependence of indium-tin-oxide work function on surface cleaning method as studied by ultraviolet and x-ray photoemission spectroscopies," *J. Appl. Phys.*, vol. 87, no. 1, pp. 295-298, Jan 2000.

[22] T. C. Fung, C. S. Chuang, C. Chen, *et al.*, "Two-dimensional numerical simulation of radio frequency sputter amorphous In-Ga-Zn-O thin-film transistors," *J. Appl. Phys.*, vol. 106, no. 8, p. 084511, Oct 2009.

[23] K. D. Mackenzie, A. J. Snell, I. French, *et al.*, "The characteristics and properties of optimized amorphous-silicon field-effect transistors," *Appl. Phys. A-Mater.*, vol. 31, no. 2, pp. 87-92, 1983.

[24] H. Hosono, "Ionic amorphous oxide semiconductors: Material design, carrier transport, and device application," *J. Non-Cryst. Solids*, vol. 352, no. 9-20, pp. 851-858, Jun 2006.

[25] D. Kim, C. Y. Koo, K. Song, *et al.*, "Compositional influence on sol-gel-derived amorphous oxide semiconductor thin film transistors," *Appl. Phys. Lett.*, vol. 95, no. 10, p. 103501, Sep 2009.

[26] J. H. Lim, J. H. Shim, J. H. Choi, *et al.*, "Solution-processed InGaZnO-based thin film transistors for printed electronics applications," *Appl. Phys. Lett.*, vol. 95, no. 1, p. 012108, Jul 2009.

[27] B. Du Ahn, H. S. Shin, H. J. Kim, *et al.*, "Comparison of the effects of Ar and H2 plasmas on the performance of homojunctioned amorphous indium gallium zinc oxide thin film transistors," *Appl. Phys. Lett.*, vol. 93, no. 20, p. 203506, Nov 2008.

[28] J. K. Jeong, "The status and perspectives of metal oxide thin-film transistors for active matrix flexible displays," *Semicond. Sci. Technol.*, vol. 26, no. 3, p. 034008, Mar 2011.

Enhancement of Nanoelectronic Sensor Performance with Microfluidic Device

Kyungsup Han[1,2], Yong-Jin Yoon[2], Jack Sheng Kee[1], Mi Kyoung Park[1,*]

[1]Institute of Microelectronics, A*STAR (Agency for Science, Technology and Research),
11 Science Park Road, Singapore Science Park II,
Singapore 117685.
[2]School of Mechanical and Aerospace Engineering,
50 Nanyang Avenue, Nanyang Technological University
Singapore 639798.
*Email: parkmk@ime.a-star.edu.sg, +65 6464 0517

Abstract

In this paper, we suggest a simple and inexpensive method for the enhancement of binding performance with low pressure drop using density difference between core fluid and sheath fluid. Simulations and parametric studies of the method have been performed using computational fluid dynamics (CFD) simulation program and the results were validated by fluorescence experiments using fluorescent polystyrene beads and streptavidin labeled with fluorescent dye.

Keywords: Microfluidic device, Nanoelectronic sensor, Self-rotation and Hydrodynamic vertical focusing

Introduction

Nanoelectronic devices are widely used in the fields of food safety medical diagnostics and nanomedicine. Especially, the nanoelectronic devices have been actively studied to detect biomolecules including viruses, proteins, cells and DNAs for use as direct, real-time, and multiplexing biosensors in the field of medical diagnostics [1]. To establish the real-time biosensors as one platform system, the nanoelectronic devices are often positively integrated with microfluidic devices to transport the biomolecules to near the sensor surface [2].

Recently, with development of the microfluidic device for the integrated system, investigations to enhance performance of the biosensors are being carried out. Most of devices and methods to enhance the sensor performance, however, are complex and expensive due to the necessity of additional devices or systems and has a limitation of biomolecule selection because of particular target molecules [3, 4].

A. Sadana and D. Sii (1992) have analyzed relationship between surface concentration and diffusional limitations using Damkohler number (Da) to enhance the binding kinetics [5]. In this paper, the surface concentration is enhanced with low Da and a characteristic length of microfluidic channel (L). In general, the easiest way to reduce the L is decrease of a channel height in microfluidic devices for enhancement of the binding probability. The height-reduced microfluidic device, however, has a critical drawback due to augmented pressure drop as much as square of height-reduced ratio. For that reason, development of a method enabling not only to enhance the binding probability, but also to maintain bonding of the microfluidic device is urgent.

In this paper, we suggest a simple and inexpensive method for enhancement of the binding probabilities of biomolecules with low pressure drop and validate the method through simulation and experimental demonstrations.

Concept of method

A. Concept of Self Rotational Hydrodynamic Vertical Focusing

For enhancement of the binding kinetics, reducing a characteristic length of a microfluidic channel is needed. The Hagen-Poiseuille equation in rectangular channels is represented in (1),

$$\Delta P = \frac{12 \mu L}{H^2} V \approx \frac{V}{H^2} \tag{1}$$

where ΔP is the pressure drop, μ is the dynamic viscosity, L is the characteristic length, V is the velocity of fluid and H is the channel height [6]. In case of a height-reduced channel, high pressure drop is accompanied due to term of H square as in (1).

Fig. 1 represents a schematic diagram of concept of a self-rotational hydrodynamic vertical focusing method. Once two fluids which have different densities are injected horizontally, a target fluid with biomolecules is rotated to the bottom side in a microfluidic channel by gravitational force. In order to achieve highly probability of binding between target molecules and immobilized prove molecules, a less dense fluid is injected faster than the target fluid. As a result, by changing a volumetric fraction ratio between the target fluid and the less dense fluid, the target fluid with biomolecules is hydrodynamic focused near surface of sensors resulting in the enhancement of the binding probability as shown to the enlarged view of a cross section A in Fig. 1. By comparison with a height-reduced channel, a self-rotational hydrodynamic vertical focusing method provides relatively lower pressure drop which is a few increase by injection velocity of a less dense fluid, V, without changing H ($V << 1/H^2$).

Results and Discussions

A. Simulation Result

As we mentioned earlier, a method of self-rotational hydrodynamic vertical focusing is enable to enhance the binding probability. To investigate a vertical focusing performance of the method, simulations regarding a behavior of biomolecules in the method have been carried out using CFD program. Widths and lengths were changed through parametric studies at 200 μm which is the maximum height of

978-1-4673-4840-9/13 $31.00 © 2013 IEEE

a device fabricated of height of the device to find an optimum design for the generation of a self-rotational vertical focusing by UV-lithography. As a result, a dimension in Fig. 2(A) (width = 300 μm, height = 200 μm and length = 6000 μm) was decided to generate the self-rotational vertical focusing flow with biomolecules. Fig. 2(B) shows the rear view of CFD simulation result for behavior observation of biomolecules in the device. Biomolecules, which are a colored ball type, are injected from the left inlet with a target fluid device and rotated to the bottom side in the device toward outlet of the device. The self-rotational hydrodynamic vertical focusing of target fluid with biomolecules was successfully achieved with 0.25 of volumetric fraction ratio between target fluid and less dense fluid to enhance the binding probability effectively.

B. Experimental Result

To validate simulation results, experiments using fluorescent polystyrene beads and ethanol have been done in the same device dimension and flow condition with simulations. The injected fluorescent polystyrene beads were rotated to bottom side along the microfluidic channel by density difference with ethanol as shown in Fig. 3(A) and a self-rotational hydrodynamic vertical focusing phenomenon is demonstrated by flow patterns of the target fluid in experimental and simulation results (Fig.3(B)~(D)).

Fig. 4 represents comparison data between non-treated, general, self-rotational vertical focusing and self-rotational hydrodynamic vertical focusing methods. To compare binding performances of target molecule in various methods, three of microfluidic channels were first treated with 3-aminopropyltriethoxysilane followed by n-succinimidyl ester activated biotin. After treatment of biotins, a 10 ng/ml of fluorescent strepavidin was injected to be bound to the immobilized biotin using different methods as described in Table 1. As a result, the binding performance of self-rotational hydrodynamic vertical focusing was demonstrated by comparing with other methods in Fig. 4.

Conclusion

In this paper, we suggested a simple and inexpensive method for the enhancement of binding performance with low pressure drop using density difference between target fluid with biomolecules and less dense fluid. Simulations and parametric studies have been performed to find optimal condition for the self-rotational hydrodynamic vertical focusing method using CFD simulation program. To validate simulation results, experiments using fluorescent polystyrene beads and streptavidin have been carried out in same device dimension and flow condition with simulations. Finally, by comparing with other methods, the binding performance of self-rotational hydrodynamic vertical focusing was successfully demonstrated.

References

[1] S. M. Stavis, et al., *Lab Chip*, 12, pp. 1174-1182, 2010
[2] C. F. Carlborg, et al., *Lab Chip*, 10, pp. 281-290, 2010.
[3] Y. Liu, et al., *Comput. Methods Appl. Mech. Engrg.*, 197, pp. 2156–2172, 2008.
[4] O. Frey, et al., *Lab Chip*, 10, pp.2226-2234, 2010.
[5] A. Sadana and D. Sii, *Biosensors and Bioelectronics*, 7, pp.

559-568, 1992.
[6] B.R. Munson, D.F. Young, T.H. Okiishi, Fundamentals of Fluid Mechanics, Wiley, New York, 1998.

Acknowledgement

This work was supported by the Agency for Science Technology and Research (A*STAR) Joint Council Office (JCO) grant (1234e00018), Singapore.

TABLE I
Flow conditions in various injection method

Injection methods	Left inlet	Right inlet
Non-treated method	0	0
General method	3 μl/m of streptavidin	3 μl/m of streptavidin
Self-rotational vertical focusing	3 μl/m of ethanol	3 μl/m of streptavidin
Self-rotational hydrodynamic vertical focusing	4.5 μl/m of ethanol	1.5 μl/m of streptavidin

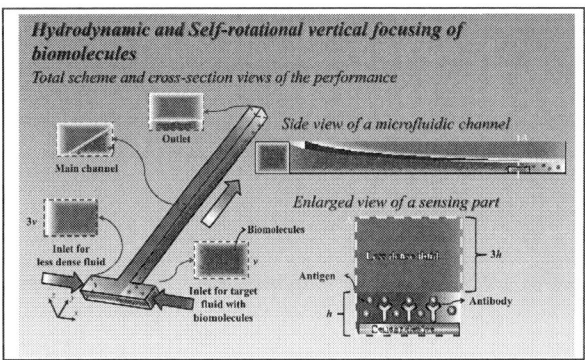

Fig. 1 Schematic diagram of a method of self-rotational hydrodynamic vertical focusing of biomolecules.

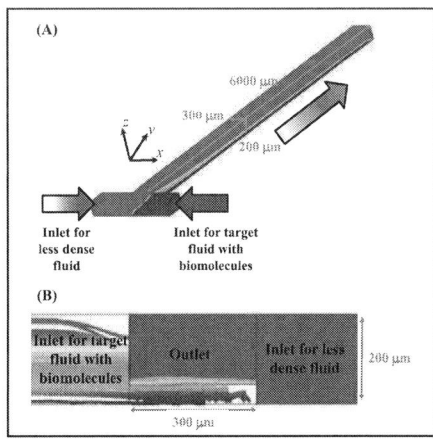

Fig. 2 Simulation results of the self-rotational vertical focusing of a target fluid with biomolecules in different viewpoints; (a) tilted view and (b) rear view.

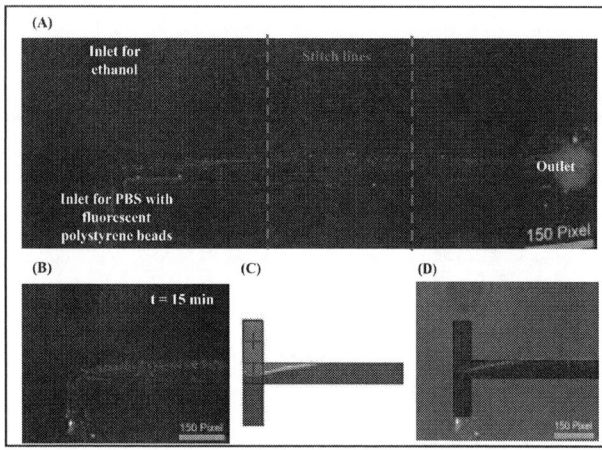

Fig. 3 (A) Stitched image of interval inverted microscope taken at the beginning of experiment, (B) image of experimental result taken at inlet part after 15 minute, (C) CFD simulation result in same flow condition as experimental condition at the bottom part of the device and (D) overlap view between (B) and (C).

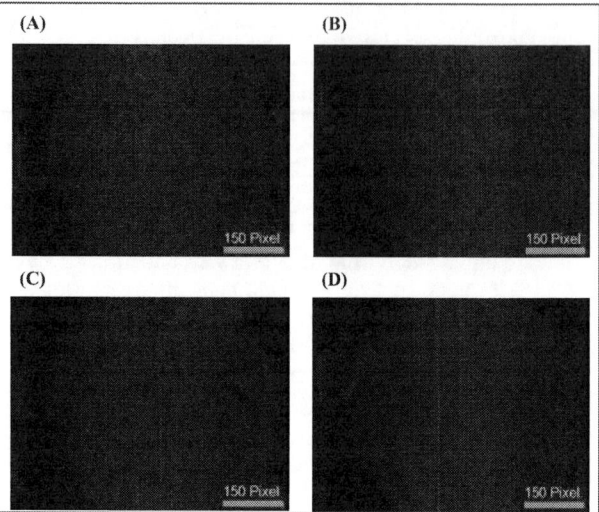

Fig. 4 Fluorescence microscope observation of four kinds of microfluidic channels immobilized fluorescent streptavidin by various injection methods; (A) non-treated, (B) general, (C) self-rotational vertical focusing and (D) self-rotational hydrodynamic vertical focusing.

A Novel Density Control of Carbon Nanotubes by Partial Oxidation of Catalyst Metal and its Field Emission Enhancement

Chuan-Ping Juan[1,2], Chia-Tsung Chang[2], Jyh-Liang Wang[3], Chuan-Chou Hwang[3]

1. Department of Electronic Engineering, St., John's University, Taipei, Taiwan R.O.C
2. Nano Electronics and Display Technology Lab.Department of Electronics Engineering and Institute of Electronics,National Chiao Tung University Hsinchu, Taiwan, R.O.C.
3. Department of Electronic Engineering, Ming Chi University of Technology, Taipei, Taiwan R.O.C
* E-mail address: cpjuan @mail.sju.edu.tw

Abstract

A novel density control of carbon nanotubes is fabricated using partial oxidation of catalyst metal prior to the CNTs growth. The results show that CNTs are aligned, closely spaced, and divided into two groups with some long nanotubes protruding among short ones. Field emission improvement is achieved due to those long nanotubes are subjected less field screening effect from the surrounding nanotubes. The obtained results show that a low turn-on field (1.9 V/μm) and an ultra high field emission current density (160mA/cm^2 at 6 V/μm) can be achieved through this novel morphology of CNTs.

Keywords: density control, carbon nanotubes, field screening, partial oxidation, turn-on field

Introduction

Due to CNTs' high aspect ratio, well chemical stability, high mechanical strength and small radii of curvature, Carbon nanotubes (CNTs) have attracted considerable interest as electron field emission application [1-3] since the first discovery of CNTs by Iijima in 1991[4].However, the density of CNTs grown by chemical vapor deposition (CVD) is always very high (~10^{10}/cm^2) and suffered serious screening effect. Therefore, the morphology control of CNTs plays a significant role in the field emission characteristics, and the optimal CNT density was theoretically calculated to be 2.5x10^7 emitters/cm^2 [5]. There are several methods for controlling CNT density, such as the use of different aspect ratios of anodic aluminum oxide (AAO) nanochannels [6] the NH$_3$ plasma treatment of a metal catalyst [7] oxide capping layer on metal catalyst [8] and the plasma post-treatment of CNTs film [9]. However, CNT density is still very high and cannot be controlled effectively. In this paper, an effective control of CNTs density is first proposed using partial oxidation of catalyst metal prior to the growth of CNTs. The results show that two groups of CNTs with very different lengths and effective density reduction of CNTs (~ 10^8 emitters/cm^2) is achieved.

Experimental Procedures

In this study, Fe (15nm)/Ti (50nm) multilayer catalyst are subsequently deposited on the photoresist patterned Si substrate. CNTs are grown selectively on the iron layers by the atmospheric pressure thermal chemical vapor deposition.

Samples are placed horizontal up into the reactor at about 400°C for 2 minutes with furnace open about 1cm to proceed partial oxidation of catalytic film, and then be heated to 700°C in pure N$_2$ with the flow rate at 1000 sccm. Then, the partial oxidation of catalytic film is pre-treated in pure H$_2$ for 10 minutes with the gas flow rate at 300sccm. Finally, CNTs are grown in C$_2$H$_4$ (20sccm)/ N$_2$ (500) mixture gases for 10min. Scanning electron microscopy (SEM) is used to examine the morphology of carbon nanotubes. The crystallinity of CNTs was analyzed by Raman spectrum. The field emission tests are measured in a parallel plate diode configuration at room temperature in vacuum of ~ 5•10^{-6} Torr. The spacer between the anode and CNTs cathode is approximately 100μm and the emitting area is 25x10^{-4} cm^2. The field emission current is measured as a function of the anode voltage.

Results and Discussions

The SEM images of CNTs grown with and without partial oxidation of catalytic film are shown in Fig. 1. The SEM results of CNTs grown with partial oxidation in Fig.1 (a) show that CNTs are divided into long and short groups and the length of long group is more than twofold of the length of short group. Especially, the long CNTs are protruding outside the short group and its density is obviously decreased to about 10^8 emitters/cm^2. Besides, some α-carbon covering on the shot CNTs and spaghetti-like carbon nanotubes are also found. Figure 1 (b) shows no obvious two groups' phenomenon of CNTs and no α-carbon are covering on the CNTs. Schematic illustration of CNTs' growth with partial oxidation of catalytic film is proposed in Fig. 2.The surface of catalytic film become metal oxide under oxidation at about 400°C for 2 minutes. The catalytic film turns into nanoparticles during the pre-treatment process in H$_2$ gas. It is probable that some metal oxide nanoparticles are reduced to metal nanoparticles or even some metal nanoparticles are directly formed from metal oxide film in the pre-treatment process. Therefore CNTs can grow first and long CNTs are produced. On the other hand, most catalytic nanoparticles' top surfaces are still in metal oxide state, amorphous carbon will cover on their surface. However, most parts of each nanoparticle are still catalytic to carbon atoms which can diffuse into the space between nanoparticles, thus CNTs can be grown but with a low growth rate because of the diffusion of carbon atoms are blocked by amorphous carbon. Therefore, short CNTs group with amorphous carbon covered on their surface will be grown which can be confirmed in Fig.1. Besides, a lot of spaghetti-like CNTs are also found intermixing with α-carbon. Fig. 3 shows the Raman spectrum of the CNTs grown with partial oxidation of catalytic film. The higher intensity of D peak (1350 cm⁻¹) in the spectrum demonstrates that amorphous carbon exists on top of the short CNTs group. Fig. 4 shows the field emission properties of CNTs. A low turn-on field (1.9 V/μm) and an ultra high field emission current density (160mA/cm² at 6 V/μm) can be achieved through this novel morphology of CNTs. The linearity of the F-N plot in Fig. 4 (b) confirms the field emission phenomenon.

Conclusions

Contrary to the other methods that need another expensive and complicated processes, a simple and novel method of morphology control of carbon nanotubes fabricated by partial oxidation of catalyst metal just prior to CNTs' growth is first proposed. The results show that CNTs' morphology is divided into long and short groups. Long CNTs group suffers less field screening effect due to the obvious decrease of density The obtained results show that a low turn-on field (1.9 V/μm) and an ultra high field emission current density (160mA/cm² at 6 V/μm) can be achieved through this novel morphology control of CNTs

Acknowledgements

This research was supported in part by the National Science Council in Taiwan under contracts NSC99-2221-E-129-019 Technical support from the Nano Facility Center of National Chiao Tung University is also acknowledged.

References

[1] M. Chhowalla, C. Ducati, N. L. Rupesinghe, K. B. K. Teo, and G. A. J. Amaratunga, Appl. Phys. Lett. **79**, pp.2079, 2001.

[2] J. Haruyama, Y. Sato, Appl. Phys. Lett. **77**, pp.2891, 2000.

[3] A. Modi, and P. M. Ajayan, Nature (London) **424**, pp.171, 2003

[4] S. Iijima, Nature (London) **354**, **pp.**56, 1991.

[5] L. Nilsson, O. Groening, C. Emmenegger, O. Kuettel, E. Schaller, L.Schlapbach, H. Kind, J. M. Bonard, and K.Kern, Appl. Phys. Lett. 76, pp. 2071, 2000.

[6] S. H. Jeong, O. J. Lee, K. H. Lee, S. H. Oh, and C. G. Park, Chem.Mater. 14, pp. 4004, 2002.

[7] J. H. Choi, T. Y. Lee, S. H. Choi, J. H. Han, J. B. Yoo, C. Y. Park, T.Jung, S. Yu, W. Yu, I. T. Han, and J. M. Kin, Thin Solid Films 435, pp. 318, 2003.

[8] C.P Juan, K.J Chen, C.C Tsai, K.C Lin, W.K Hong, C.Y Hsieh, W.P Wang, R.L Lai, K.H Chen, L.C Chen and H.C Cheng, Jpn. J. Appl. Phys. 44, pp.365, 2005.

[9] C.P. Juan, C.C Tsai, K.H Chen, L.C Chen and H.C Cheng, Jpn. J. Appl. Phys. 44, pp8231,2005.

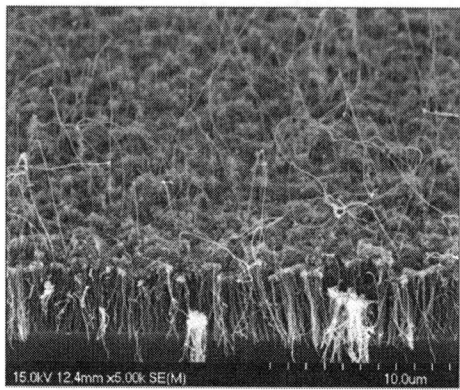

(a)

(b)

Fig. 1 (a) SEM images of CNTs grown with partial oxidation of catalytic film (b) cross section view of CNTs grown without partial oxidation of catalytic film.

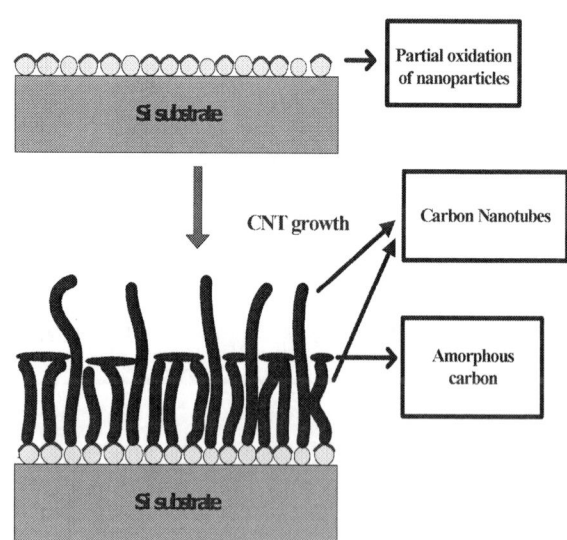

Fig. 2 Schematic diagram of CNTs growth with partial oxidation of catalytic film

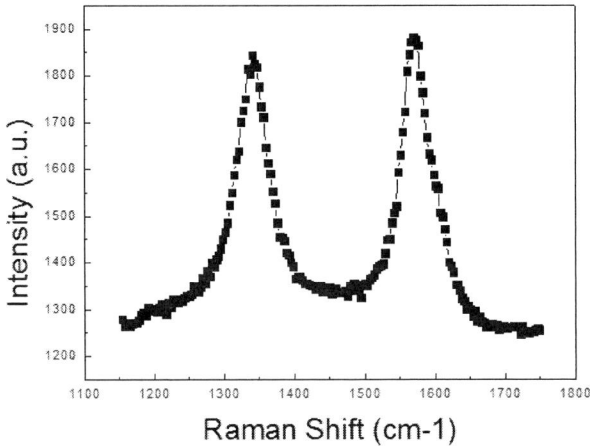

Fig. 3 Raman spectra of CNTs grown with and without partial oxidation of catalytic film

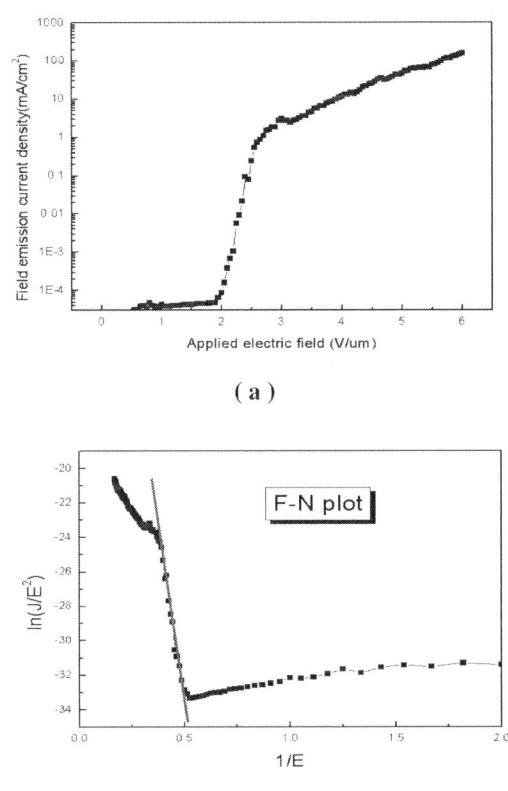

Fig. 4. Field emission properties of CNTs (a) Field-emission current density vs applied electric field. (b) Corresponding F–N plot.

Magnetic and Leakage Current Properties of $Bi_{1-x}Gd_xFeO_3$ Thin Films

Ming-Cheng Kao[*], Hone-Zern Chen[*], San-Lin Young

Department of Electronic Engineering, Hsiuping University of Science and Technology, Taichung, Taiwan.
No.11, Gongye Rd., Dali Dist. Taichung City 412, Taiwan
Tel:+886-4-24961100#1500, E-mail: kmc@mail.hust.edu.tw and hzc@mail.hust.edu.tw

Abstract

$Bi_{0.9}Gd_{0.1}FeO_3$ (BGFO) thin films were deposited on Pt(111)/Ti/SiO$_2$/Si(100) substrates using the sol-gel method. The thin films showed ferromagnetic properties with remnant magnetization (Mr) of 1.2 emu/g and saturation magnetization (Ms) of 5.3 emu/g. It was found that more than one conduction mechanism is involved in the electric field range used in these experiments. In the low electric field region, the leakage current was controlled by Poole-Frenkel emission. On the other hand, the mechanism can be explained by Schottky emission from the Pt electrode in the high field region.

Keywords: Leakage current, Bismuth ferrite, Magnetic.

Introduction

Bismuth ferrite BiFeO$_3$ (BFO) is a well-known multiferroic material, has been given considerable attention recently, owing to its high phase transition temperature, good multiferroic properties, and G-type antiferromagnetic behavior to make it appealing for applications in ferroelectric nonvolatile memories and high temperature electronic devices [1–7]. Recently, it was reported that lanthanide-substituted BiFeO$_3$, $(Bi_{1-x}Ln_x)FeO_3$ (Ln = La, Nd and Eu) could improve its ferroelectric and electrical properties [8-15]. The reason is that the enhanced ferroelectric properties of these films can be attributed to the suppressed formation of impure phases and oxygen vacancies, which is caused by the substitution of stable rare-earth ions for the volatile Bi ions. In addition, the studies of the leakage currents in multiferroic thin films are also important for most electrical applications. The leakage current in thin films often become nonohmic under applied voltage. The reason for this is complex and depends on the nature of the contact, the film composition and its internal structure. In this study, the characteristics of leakage currents in the $Bi_{0.9}Gd_{0.1}FeO_3$ (BGFO) thin films were investigated. Various conduction models, including the Poole-Frenkel (PF) emission and schottky emission, have been proposed.

Experimental

The general chemical formulas of the precursor solutions of $Bi_{0.9}Gd_{0.1}FeO_3$ prepared by the sol-gel method. Bismuth acetate, $Bi(OOCCH_3)_3$ (Alfa, 99.99%+ purity), gadolinium isopropoxide, $Gd[OCH(CH_3)_2]_5$ (Alfa, 99.9%+ purity) and iron acetate, $Fe(CO_2CH_3)_2$ (Alfa, 99.9%+ purity) were used as source materials and 2-methoxyethanol was used as solvent. The gravimetrically assayed bismuth acetate and gadolinium isopropoxide reagents were dissolved in the 2-methoxyethanol, to obtain sol compositions. The stock solutions were spin-coated onto Pt/Ti/SiO$_2$/Si(100) substrates using a commercial photoresist spinner. BGFO thin films were annealed at 600 $^{\circ}$C for 2 min by the rapid thermal annealing in an oxygen atmosphere. The crystallization and microstructures of thin films were analysed by X-ray diffraction (XRD) and atomic force microscopy (AFM), respectively. Measurements of the leakage current were carried out using a metal-ferroelectric-metal (MFM) configuration. The magnetic hysteresis loops of the samples were analyzed at room temperature by a vibrating sample magnetometer (VSM). The leakage current was measured using an HP-4156A semiconductor parameter analyzer in a MFM configuration.

Results and discussion

The XRD patterns of the BGFO thin films annealed at temperatures ranging from 400 °C to 700 °C are shown in Fig. 1. The XRD peaks are quite similar to those of the standard diffraction pattern data of BiFeO$_3$ in the JCPDS card. The results (see Fig. 1) show a randomly oriented perovskite phase structure, but there are also evidence of a minor presence of $Bi_2Fe_4O_9$ phase structure in the film at temperatures lower than 700 °C. In addition, it can be seen that the intensities of (010), (012) and (020) peaks are relatively broad and weak at high annealing temperatures (700 °C). The thin films annealed at 700 $^{\circ}$C shows the better (110) orientation compared with those annealed at other temperatures.

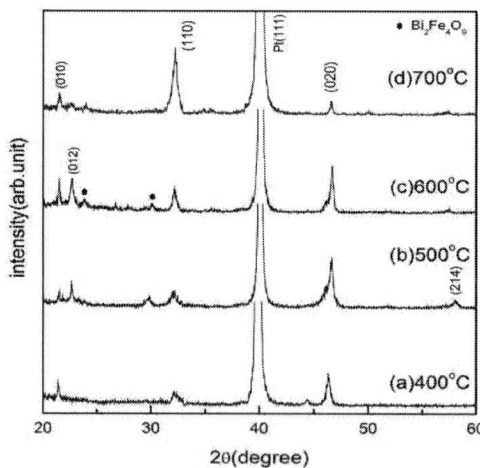

Fig. 1. XRD patterns of the BGFO thin films with various annealing temperatures.

The AFM images of the BTO and BGFO thin films deposited on the Pt coated silicon substrates are shown in Fig. 2. It shows that the Gd ions in the BTO thin film will affect the grain growth mechanism. The BGFO thin films have no platelike crystal owing to that the BGFO thin film with the $(Bi_2O_2)^{2+}$ layers in the *ab*-plane, which result the BGFO thin film with the larger grain size than the BTO thin film.

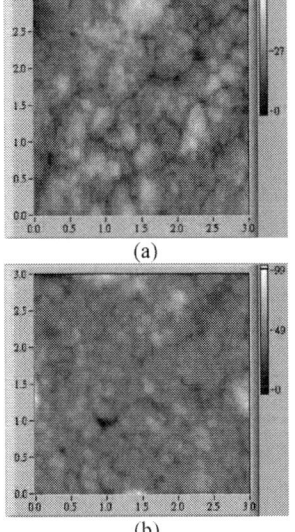

(a)

(b)

Fig. 2. The AFM images of the BTO thin film (a), and BGFO thin film (b).

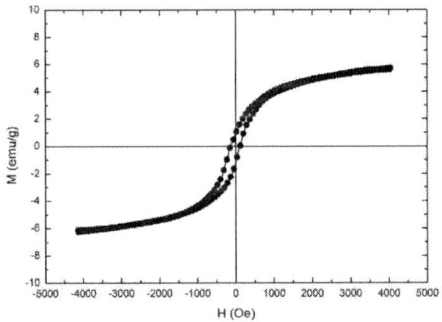

Fig. 3. Magnetic hysteresis loops of the BGFO thin films annealed at 600 °C.

Figure 3 presents the magnetic hysteresis loops of the BGFO thin films annealed at 600 °C. Magnetic hysteresis loops of the BGFO thin films were measured at room temperature with applied magnetic field parallel to the plane of the samples. From the magnetic hysteresis loops, a well-developed M-H loops can be observed in the BGFO thin films, indicating the presence of an ordered magnetic structure. In addition, the BGFO thin film exhibits a remnant magnetization (Mr) of 1.2 emu/g and a saturation magnetization (Ms) of 5.3 emu/g, respectively.

The leakage current behaviors of the BGFO thin films crystallized at 600 °C can be investigated further on the leakage current (*J*)-electric field (*E*) characteristics such as

the *ln J* vs. $E^{1/2}$ plots, as shown in Fig. 4. The Schottky-Richardson emission generated by the thermionic effect is caused by the electron transport across the potential energy barrier via field-assisted lowering at a metal-insulator interface. The current density (*J*) in the Schottky emission can be quantified by the following equation [7]:

$$J = A^* T^2 \exp\left(\frac{\beta_s E^{1/2} - \phi_s}{k_B T}\right), \qquad (1)$$

where $\beta_s = (e^3 / 4\pi\varepsilon_0\varepsilon)^{1/2}$, e is the electron charge, ε_0 is the dielectric constant of free space, A^* is the effective Richardson constant, T is the absolute temperature, E is the applied electric field, φ_s is the contact potential barrier, and k_B is the Boltzmann constant. The Poole-Frenkel emission is owing to field-enhanced thermal excitation of trapped electrons from the valence band into the conduction band. The current density is given by [8]

$$J = J_0 \exp\left(\frac{\beta_{PF} E^{1/2} - \phi_{PF}}{k_B T}\right), \qquad (2)$$

where $J_0 = \sigma_0 E$ is the low-field current density, $\beta_{PF} = (e^3 / \pi\varepsilon_0\varepsilon)^{1/2}$, σ_0 is the low-field conductivity, φ_{PF} is the height of trap potential well.

The relative dielectric constant of the BGFO thin films annealed at 600 °C was 250. The various values of β_{PF} and β_s can be calculated from the equation (1) and (2), respectively. The experimental β_{exp} values can be calculated from the slope ($=\beta/k_B T$) of the curve lnJ-$E^{1/2}$ straight line portion in Fig. 4. In the low electric field region (0~100 kV/cm), the β_{exp} is closer to β_{PF} than β_s. By increasing the external field above 100 kV/cm, the β_{exp} is closer to β_s than β_{PF} which implies that the current conduction is governed by another mechanism. These results reveal that carriers are transported through the BGFO thin films of Pt/Ti/SiO$_2$/Si by the field-enhanced Poole-Frenkel mechanism in the low electric field region. However, the mechanism can be explained by Schottky emission from the Pt electrode in the high electric field region.

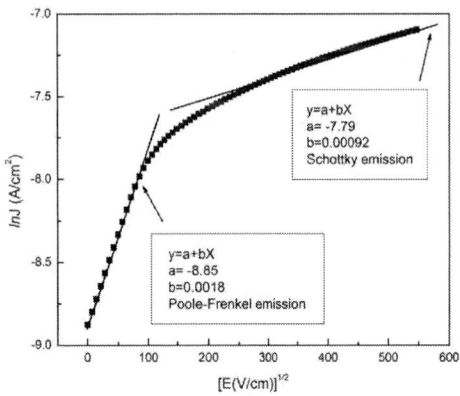

Fig. 4. The leakage current-electric field characteristics of the

$Bi_{0.9}Gd_{0.1}FeO_3$ thin films annealed at 600 °C.

Conclusion

Ferroelectric BGFO thin films were deposited onto Pt(111)/Ti/SiO$_2$/Si(100) substrates, using a sol-gel process and rapid thermal processing. The BGFO thin films showed ferromagnetic properties with remnant magnetization (Mr) of 1.2 emu/g and saturation magnetization (Ms) of 5.3 emu/g. The leakage current characteristics of the BGFO thin films varied with the external electric field. In the low electric field region, the leakage current revealed a Poole-Frenkel emission. In addition, the mechanism could be explained by Schottky emission in the high electric field region, implying thermionic emission.

Acknowledgment

This study was supported by the National Science Council, R. O. C., under contrast No. NSC 100-2112-M-164-001 and NSC 101-2112-M-164-001.

References

[1] D. Lebeugle, A. Mougin, M. Viret, D. Colson, and L. Ranno, "Electric field switching of the magnetic anisotropy of a ferromagnetic layer exchange coupled to the multiferroic compound BiFeO3," Phys. Rev. Lett., vol. 103, art. no. 257601, 2009.

[2] W. Eerenstein, N. D. Mathur, and J. F. Scott, "Multiferroic and magnetoelectric materials," Nature, vol. 442, pp. 759–765, Aug. 2006

[3] R. Ramesh and N. A. Spaldin, "Multiferroics: Progress and prospects in thin films," Nature Mater., vol. 6, pp. 21–29, Jan. 2007.

[4] J. Wang et al., "Epitaxial BiFeO multiferroic thin film heterostructures," Science, vol. 299, pp. 1719–1722, Mar. 2003.

[5] R. J. Zeches et al., "A strain-driven morphotropic phase boundary in BiFeO ," Science, vol. 326, pp. 977–980, Nov. 2009.

[6] C. F. Chung, J. P. Lin, and J. M. Wu, "Influence of Mn and Nb dopants on electric properties of chemical-solution-deposited BiFeO films," Appl. Phys. Lett., vol. 88, p. 242909, Jun. 2006.

[7] J. G. Wu, and J. Wang, "BiFeO$_3$ thin films of (111)-orientation deposited on SrRuO3 buffered Pt/TiO$_2$/SiO$_2$/Si(100) substrates, " Acta Mater., vol. 58, pp. 1688-1697, Mar. 2010.

[8] H. R. Liu, and X. Z. Wang, "Rectangular and saturated hysteresis loops of BiFeO$_3$ film on LaNiO$_3$ bottom electrode," J. Alloy. Compd., vol. 485, pp. 769-772, Oct. 2009.

[9] J. G. Wu, and J. Wang, "Effects of SrRuO$_3$ buffer layer thickness on multiferroic $(Bi_{0.90}La_{0.10})(Fe_{0.95}Mn_{0.05})O_3$ thin films, " J. Appl. Phys., vol. 106, pp.054115-054119, Sep. 2009.

[10] F. Z. Huang, X. M. Lu, W. W. Lin, X. Cai, X. M. Wu, Y. Kan, H. Sang, and J. S. Zhu, "Multiferroic properties and dielectric relaxation of $BiFeO_3/Bi_{3.25}La_{0.75}Ti_3O_{12}$ double-layered thin films, " Appl. Phys. Lett., vol. 90, pp. 252903-252905, Jun. 2007.

[11] G. Catalan, and J. F. Scott, "Physics and Applications of Bismuth Ferrite, " Adv. Mater., vol. 21, pp. 2463-2485, Jun. 2009.

[12] X. Zheng, Q. Xu, Z. Wen, X. Lang, D. Wu, T. Qiu, and M.X. Xu, "The magnetic properties of La doped and codoped BiFeO$_3$, " J. Alloy. Compd., vol. 499, pp. 108-112, Jun. 2010.

[13] S. Karimi, I. M. Reaney, I. Levin, I. Sterianou, "Nd-doped BiFeO3 ceramics with antipolar order, " Appl. Phys. Lett., vol. 94, pp. 112903-112905, Mar. 2009.

[14] V. A. Khomchenko, D. A. Kiselev, I. K. Bdikin, V. V. Shvartsman, P. Borisov, W. Kleemann, J. M. Vieira, and A. L. Kholkin, "Crystal Structure and Multiferroic Properties of Gd-Substituted BiFeO$_3$, " Appl. Phys. Lett., vol. 93, pp. 262905-262907, Dec. 2008.

[15] H. Katsura, N. Nagaosa, and A. V. Balatsky, "Electric-field control of local ferromagnetism using a magnetoelectric multiferroic," Phys. Rev. Lett., vol. 95, art. no. 057205, 2005.

A novel synthesis approach of gold nanoparticles by amino acids

Cheng-Sheng Wu[1], Hong-Huei Huang[1], Fu-Ken Liu[2], and Ching-Chich Leu[1,*]

[1]Department of Chemical and Materials Engineering, National University of Kaohsiung
[2]Department of Applied Chemistry, National University of Kaohsiung
700, Kaohsiung University Rd., Nanzih District, 811.
Kaohsiung, Taiwan, R.O.C.
Phone: 886-7-5919456 E-mail: ccleu@nuk.edu.tw

Abstract

Monodisperse colloidal particles have attracted much attention not only for their scientific interest but also for many technological applications. However, many kinds of toxic chemicals or solvents are usually involved in the particle preparation process. Recently, anionic surfactants derived from diverse amino acids have been applied to the synthesis of nanosized materials. Amino acid is non-toxic and harmless to the environment or human. These accomplishments have inspired us to use amino acids in preparing materials unique in structure.

In this work, we used Lysine to assist the synthesis of gold nanoparticles. Lysine is an essential amino acid in the body. The experimental process of low complexity, and in pure water phase with no organic solvents participate, is consistent with the currently popular green chemistry. The resulting gold nanoparticles were characterized via UV-Vis Spectrometer, and Scanning Electron Microscopy. The critical roles of lysine in the formation of Au NPs were investigated based on the experimental results. The influences of the lysine concentration, solution pH value, temperature, and stirring speed on particle size and size distribution are discussed. By employing such an appropriate reduction method and process condition, we can make gold nanoparticles process simplification, non-polluting and has a good size distribution and stability, with the value on the practical application.

Introduction

In recent years, the nonvolatile flash memory and SONOS memory devices utilizing floating gate and silicon nitrides as charge storage nodes had met the leakage and reliability challenges involved with scaling down the devices size. Therefore, in order to overcome these problems, nanocrystal memory is proposed.

In the applications of materials of NPs, silicon, gallium [1,2] or metal NPs (ex. Au, Pt, W…) [3,4] are main choices. The particle size, uniformity and coverage density are so important that the choices of materials and the way of synthesis greatly influence the properties of cells. There are so many methods of NPs fabricating; according to the synthesis processes, they can be classified into physical way and chemical way. The physical way is called Top-Down, for instance, breaking bulk or alloy into nano scale by heat or milling etc. In contrast, the chemical way is called Bottom-Up, the nano-materials are fabricated by aggregation or reduction of precursors.

Recently, the concept of green chemistry gradually arose, therefore, those harmless materials and simple preparation techniques are preferred. Lysine is an essential in the body so non-toxic and harmless to human, which is completely soluble in pure water and in line with the norms of green chemistry. In this experiment, lysine-assisted nanoparticle synthesis, an environmental friendly process, was employed.

Illustrations

The Au NPs synthesis process is illustrated in Figure 1. Since the size of Au NPs is much smaller than the wavelength of visible light, Surface Plasmon Resonance (SPR) of the Au NPs will lead to a high absorption coefficient and show the color of pink (the inset in Figure 2-(a)). The optical spectra of the Au NPs solution (Figure 2-(a)) showed a characteristic absorption peak at 518nm indicating that there does some Au NPs formed. The SEM image (Figure 3) shows that the particle size of Au NPs is ~8nm with good uniformity, but the absorbed amount of particles on substrate is really low. To understand the influences of the lysine concentration on the synthesis of Au NPs, four solution concentrations of lysine were prepared at 60°C, during a stirring speed of 550rpm. As shown in Figure 3 and Table 1, we found that the particle size decreased with the increase of lysine concentrations. But the lysine molecule seems to get cross-linked into sheet structure at high concentration, which is unfavorable to applications. Then, we would like to understand how the pH of the solution affects the synthesis of NPs, since adjustment of concentrations of lysine also lead to a variations of pH values in solutions. In Fig. 4, HCl or NaOH were added to adjust the pH value, and the particle size becomes smaller when the pH was higher. The influences of solution temperatures on the synthesized particle size were also investigated. As shown in Fig. 5, the particle size decreased as the temperature rose except at 100°C. According to the literature [5], the stirring speed has significant influence on NPs; therefore the effects of stirring speed was also considered. In Fig. 6, the SEM images display the Au NPs on Si/5nm SiO$_2$/APTMS substrate deposited from the solutions with 150rpm and 950rpm stirring speed, respectively. The result indicated that the stirring speed did little influence on the size of Au particle here. Although the dispersion and yield of the Au NPs have to be further improved, it is a simple and environmental protection process.

Table 1
Au NPs sizes as functions of the concentrations of lysine solution.

Concentration (mole)	25	2.5	0.25	0.02
Particle size (nm)	7.6±2	8.4±2	39.4±18	≒500

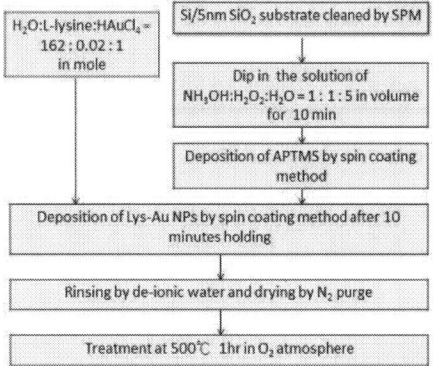

Fig.1. The synthesis procedure of Au NPs on Si/5nm SiO$_2$/APTMS substrate.

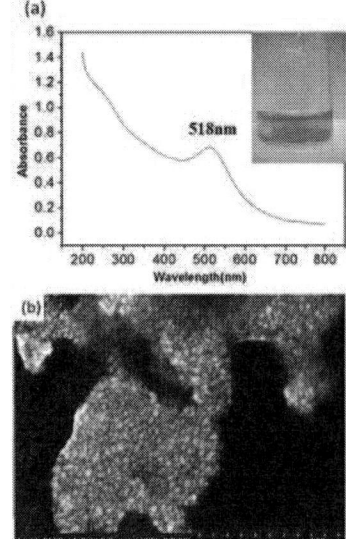

Fig.2. (a) UV–vis absorption spectra of Lys-Au NPs. The inset displays the Lys-Au solution, showing a color of pink. (b) FE-SEM image of the Lys-Au NPs .

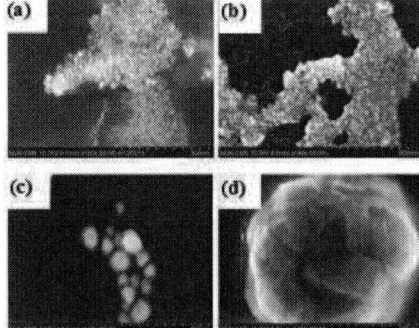

Fig.3. FE-SEM images of the Lys-Au NPs synthesized with the molar composition of 1:X:162 HAuCl$_4$:Lysine: H$_2$O.(a)X=25, (b)X=2.5,(c)X=0.25 and (d)X=0.02.

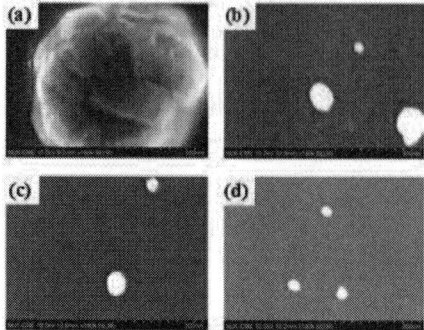

Fig.4. FE-SEM images of the Lys-Au NPs formed in different pH of solutions with the molar composition of 1:0.02:162 HAuCl$_4$:Lysine:H$_2$O. (a)pH=4, (b) pH=6,(c) pH=8 and (d) pH=10.

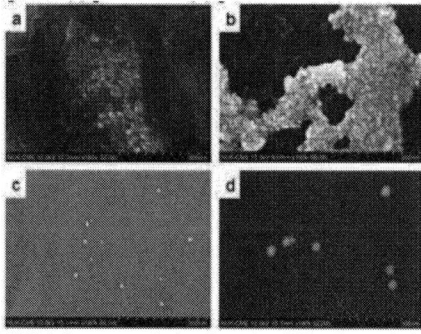

Fig.5. FE-SEM images of the Lys-Au NPs formed in different temperature of solutions. (a) 25,(b)60,(c) 80 and (d)100℃.

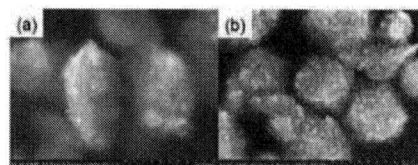

Fig.6. FE-SEM images of the Lys-Au NPs synthesized with various stirring rates: (a) 150 and (b) 950 rpm.

References

[1] W. Guan, S. Long, M. Liu, Q. Liu, Y. Hu, Z. Li, R. Jia, "Modeling of retention characteristics for metal and semiconductor nanocrystal memories", Solid-State Electronics 51, 806-811(2007)

[2] W. Guan, S. Long, M.Liu, Z. Li, Y. Hu, and Q. Liu, "Fabrication and charging characteristics of MOS capacitor structure with metal nanocrystals embedded in gate oxide", J. Phys. D: Appl. Phys. 40, 2754-2758(2007)

[3] J. Y. Yang, J. H. Kim, J. S. Lee, S. K. Min , H. J. Kim, K. L. Wang ,J. P. Hong, "Electrostatic force microscopy measurements of charge trapping behavior of Au nanoparticles embedded in metal-insulator-semiconductor structure", Ultramicroscopy, 108, 1215-1219(2008)

[4] J. Dufourcq, S. Bodnar, "High density platinum nanocrystals for non-volatile memory applications", Appl. Phys. Lett. 92, 073102 (2008)

[5] Toshiyuki Yokoi, Junji Wakabayashi, Yuki Otsuka, Wei Fan, Marie Iwama, Ryota Watanabe, Kenji Aramaki, Atsushi Shimojima, Takashi Tatsumi, Tatsuya Okubo, "Mechanism of Formation of Uniform-Sized Silica NanospheresCatalyzed by Basic Amino Acids", Chem. Mater, 21 , 3719–3729 3719(2009)

The studies of poly (amino acid) assisted synthesis of gold nanoparticles

Ying-Hui Hsu[a], Jen-Hau Yeh[a], Jeng-Shiung Jan[b], Ching-Chich Leu[a] *

[a]National University of Kaohsiung, Kaohsiung Department of Chemical and Materials Engineering
[b]National Cheng Kung University Department of Chemical Engineering
700, Kaohsiung University Rd., Nanzih District, 811.
Kaohsiung, Taiwan, R.O.C.
Tel：886-7-591-9456　E-Mail：ccleu@nuk.edu.tw

Abstract

In the past decade, bio-inspired chemistry using organized biomolecules, such as peptides, proteins, viruses, and enzymes, has demonstrated the power for assembling and structure-directing small species into unique materials due to their highly elaborate, self-assembling capability in nature, leading to the emergence of unique physiological properties. In this work, we used Poly-L-Lysine (PLL) or Poly-L-Glutamic acid (PLGA) to assist the synthesis of gold nanoparticles. Both PLL and PLGA are poly-peptides with non-toxic, and harmless for the environment and human. Therefore, our nanoparticle synthesis process is consistent with the recent trend of green chemistry.

The PLL or PLGA solutions ware prepared respectively by dissolving PLL or PLGA in D.I. water. After adding a freshly prepared $HAuCl_4$ solution to the poly-peptide solutions and stirring, the solution color slowly changed with time. The variation of solution color was attributed to the surface plasmon resonance (SPR) of gold nanoparticles, which was demonstrated by the SEM investigation. The experimental results showed that both the PLL and PLGA successfully assisted the synthesis of gold nanoparticles. The particle size increased with the reaction time but with the reduction of pH value. The influences of reaction time and solution pH value on the precipitated Au nanoparticles were discussed. By employing an appropriate solution conditions (e.g. reduction time and solution pH value), we can obtain the Au nanoparticle with an adequate size for application.

(Keywords: bio-inspired chemistry, poly-peptide, gold nanoparticles)

Introduction

During the past years, there have been growing interests in designing novel materials to deposit thin films or prepare nanoparticles (NPs). Comparing with traditional chemical methods, the advantages of bioassisted synthesis of inorganic materials include ambient conditions, biocompatible templates and nontoxic solvent. To date, a wide range of inorganic nanomaterials have been synthesized by using biomolecules as both mediating agents and templates. For example, peptides and proteins have recently been found as molecular templates for the synthesis of metal nanoparticles. Among all of the metal nanoparticles, gold nanoparticles have attracted rapidly growing interests due to their catalytic and chemical

properties, which are industrially and environmentally important to many reactions. Synthesis of gold nanoparticles by poly (amino acid) is a desirable solution process, which includes only three starting materials, namely, poly(amino acid), $HAuCl_4$, and water. In this study, we report the synthesis- of inorganic gold nanoparticles by using poly(amino acid),poly-glutamic acid or poly-lysine. Preparation of the gold nanoparticles was carried out in a simple one-pot process and the nano-sized particles were obtained in ambient temperature.

CURRENT RESULTS

The PLL or PLGA solutions ware prepared respectively by dissolving PLL or PLGA in D.I. water. PLGA with different molecular numbers used in this study were PLGA1 (Mn=5200), PLGA2 (Mn=15000), PLGA3 (Mn=60000), respectively. After adding a freshly prepared $HAuCl_4$ solution to the poly-peptide solutions and stirring, the solution color slowly changed with time, as displayed in Fig. 1a. The variation of solution color was attributed to the response of SPR as a function of sizes of the gold nanoparticles. As shown in Fig. 1a, the color firstly changed in PLGA1 within 10 mins, which showed a quick appearance of light pink color in the solution. Since the appearance of pink solution color suggested the formation of gold nanoparticles, the reduction rate was considered to be faster by the poly-glutamic with a less molecular number. The UV-vis spectrum (Fig. 1b) shows the evolution of gold nanoparticle with time in PLGA1. The red shift of absorption peak, as well as the solution color, indicated the growth of nanoparticles. The reduction experiment was also performed by using PLL (Mn:62500) as the reduction agent. When the $HAuCl_4$ solution was added into the PLL solutions, the mixture displayed a color of light yellow then gradually turned to wine red, as shown in Fig. 2a. Fig. 2b shows the SPR response of the PLL solution, and the reduction rate of PLL seems lower than that of PLGA. The influence of pH values on the reduction conditions of gold nanoparticles was studied. Fig. 3a shows the solution image of the reaction mixtures for four different pH values, which were adjusted by dilute HCl or NaOH solution. The solution colors are quite different suggesting the pH value indeed do some effects on the reduction of nanoparticles. At pH~3, the solution was colorless and some precipitates occurred at the bottom. Notably, the UV-vis absorption spectra (Fig. 3b) showed a signfifcant red-shift of the solution at pH 6, indicating a fast growth of nanoparticles in size. But the growth rate of the solutions at pH9 and pH11 were slower. On the other hand, it takes longer times for PLL to change the colors of solutions, as shown in Fig. 4. The pH values for the solutions were adjusted to be 2.88, 6.03, 9.4 and 11.0, respectivelya. Interesting, the solution color immediately turned to yellow and kept unchanged after the pH value had been adjusted to be 2.88 . . The colors of the solutions at relatively high pH values (pH 9~11) changed very slowly. On the contrary, the color of the solution at pH ~6 quickly changed to pink and the plasmon absorption spectrum showed a significant red shift (λ = 572 nm) as displayed in Fig.4b. To summarize, both the PLGA and the PLL solutions show highest reduction abilities at pH ~6. When the pH value of solution was lowered, the reaction seems to freeze. It is interesting to note that the variation in reduction of gold nanoparticles obtained by tuning the reaction pH was more significant than that obtained by varying the polymer solution concentration. The experimental results suggested that the pH-dependent reactivity of the amino functional group in the poly-peptides may play a determining role in the current system.

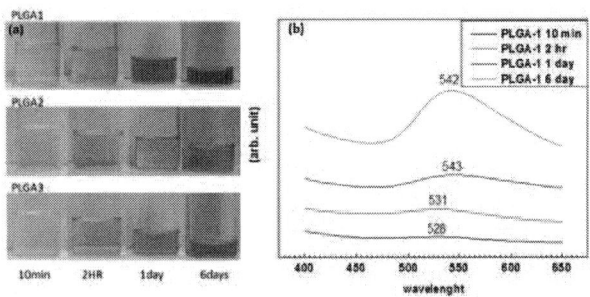

Figure1. (a)The images of PLGA solution as functions of molecular numbers and reaction times; (b) UV-vis spectrum of the PLGA1 solution with reaction time in 10min, 2hr, 1day, and 6days.

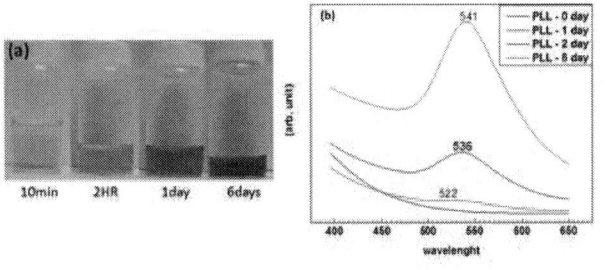

Figure2. (a)The images of PLL solution as a function of reaction times; (b) UV-vis spectrum of the PLL solution with reaction time in 10min, 2hr, 1day, and 6days.

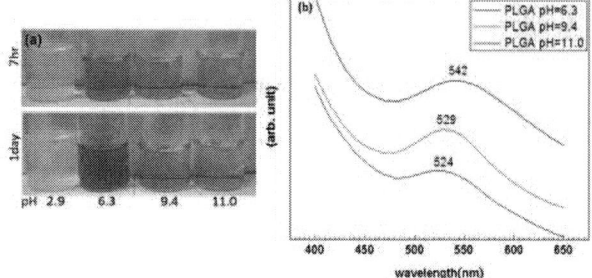

Figure3. (a) The images of PLGA solution as a function of pH values and reaction times; (b)UV-vis spectrum of PLGA solution as a function of pH values in 2hr.

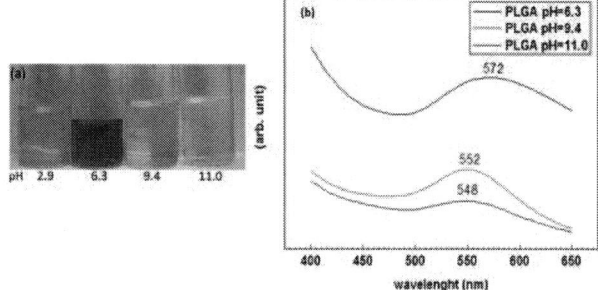

Figure4. (a)The image of PLL solution as a function of pH values in a reaction time of 7days ; (b) UV-vis spectrum of PLL solution as a function of pH values in 1 day.

Reference

[1] C. M. Pradier, V. Humblot, L. Stievano, C. Methivier, J. F. Lambert, "Salt Concentration and pH-Dependent Adsorption of Two Polypeptides on Planar and Divided Alumina Surfaces. In Situ IR Investigations", Langmuir, 23, 2463-2471(2007)

[2] Michael Faraday, "The Bakerian Lecture: Experimental Relations of Gold (and Other Metals) to Light", Trans. R. Soc. Lond. 147, 145-181 (1857)

[3] Tom C. Wang, Michael F. Rubner, Robert E. Cohen, "Polyelectrolyte Multilayer Nanoreactors for Preparing Silver Nanoparticle Composites: Controlling Metal Concentration and Nanoparticle Size", Langmuir, 18, 3370-3375 (2002)

[4] Jinhua Dai, Merlin L. Bruening, "Catalytic Nanoparticles Formed by Reduction of Metal Ions in Multilayered Polyelectrolyte Films", Nano Lett., Vol. 2, No. 5 (2002)

[5] Suresh K. Bhargava, Jamie M. Booth, Sourabh Agrawal, Peter Coloe, Gopa Kar, "Gold Nanoparticle Formation during Bromoaurate Reduction by Amino Acids", Langmuir, 21, 5949-5956 (2005)

Memory property of APTMS-mediated Au-SiO₂ core-shell nanocrystal memory

Sheng-Fu Huang[1], An-Ching Hsiao[1], Fu-Ken Liu[2] and Ching-Chich Leu[1,*]

[1]Department of Chemical and Materials Engineering, National University of Kaohsiung
[2]Department of Applied Chemistry, National University of Kaohsiung
700, Kaohsiung University Rd., Nanzih District, 811.
Kaohsiung, Taiwan, R.O.C.
Phone: 886-7-5919456 E-mail: ccleu@nuk.edu.tw

Abstract

Conventional flash memory, utilizing floating gate as charge storage nodes, had met the leakage challenges as the devices tend to be scaling down and have high densities. Nanocrystal (NC) floating gate memory devices, containing nanocrystals as the discrete charge traps, have attracted attention as strong candidates to act as nonvolatile memories devices due to their scalability, electrical isolation and low charge leakage through the tunneling oxide. In this work, an all-solution processed MOS memory structure containing colloidal Au NPs within a sol-gel derived HfO_2 oxide layer was fabricated. We use chemical reduction method to synthesize Au nanoparticles, and use the 3-aminopropyltrimethoxysilane (APTMS) to make the nanoparticles being self-assembled on SiO_2 oxide layer. Finally, it is covered by sol-gel derived HfO_2 as a control oxide to construct a NC memory structure.

To improve the memory device property, we use a simple but effective SAM method to construct an Au-SiO₂ core-shell NC capacitor by means of APTMS as a mediator. The Au-SiO₂ core-shell structure was constructed by a two-run APTMS SAM process which was conducted before and after the SAM of the colloidal Au NPs. The first-run APTMS formed a well-organized monolayer on substrate which was responsible for capturing the high-density of the Au NPs. Next, the second-run APTMS formed an APTMS bilayer around the Au NPs. A following 500 °C-annealing completed the Au-SiO₂ core-shell structure within the HfO_2 layer to form a $Si/SiO_2/Au-SiO_2$ core-shell/HfO_2 floating structure. The effects of APTMS-mediated SiO_2 shell on the property of memory device were investigated.

Introduction

In past years, the memory device based on nanocrystals (NCs) have attracted attentions as one of the strong candidates for nonvolatile memories due to scaling limitation of the traditional floating gate memory, because of its scalability, electrical isolation, and low charge leakage. The NC memories have been invested into research from materials and structures. Different kinds of nanocrystals, such as Au [1], Pt [2], Ag, W, Ni, Al [3], Ru, Zn, Si [4], Ge [5], SrTiO₃, and Al₂O₃ have been prepared and their electrons storage abilities are studies. Recently, the core-shell NC memory devices have attracted interest because the shells have the ability to resist the thermal process, and effectively improvise the retention of NC memory due to the additional barrier [6]. The shell layer has been prepared by a variety of methods, such as the treatment of core NPs with ambient O₂, sputtering, atomic layer deposition [6] and self

assembly (SAM) process [7] etc. In this work, the Au NPs were prepared by chemical reduction method and self-assembly method (SAM) on silica substrate by spin coating. Then the APTMS were covered over Au NPs by spin coating or dipping method. The aminated surfaces of the APTMS assembled layer are positively charged to capture the negatively Au NPs which are protected by the citrate ion. Finally, the sol-gel derived HfO_2 control layer was spin coating to fabricate the MOS structure. The whole device structure was constructed by an all-chemical solution method. This method have several advantages such as low temperature process, high flexibilityand high uniformity [8].

Illustrations

In this work, two kinds of MOS structure, Si/ SiO₂/ one-layer spin coated APTMS/ Au NPs/ four-layers dip coated APTMS (Sample 1) and Si/ SiO₂/ one-layer spin coated APTMS / Au NPs/ four-layers spin coated APTMS (Sample 2), were established. The experiment procedure of the MOS structure is illustrated in Fig.1. The property of these two devices was investigated. Firstly, the silica surface was cleaned by Sulfuric Peroxide Mixture (SPM) method, then the Au NPs were SAM on silica substrates by APTMS. The Au NPs was prepared by NaBH₄ reduction of HAuCl₄ in an ice bath condition, and the color of solution changed from pale yellow to wine red, as shown in Fig.2 (a). Fig.2 (b) shows that the surface plasma resonance (SPR) absorption peak of aqueous Au NPs is located at 512nm illustrating the particle size of ~ 3 nm. Following, the Au NPs were capped by dipping or spin coating of APTMS for four times, and they were baked at 100 °C for 20 min after each APTMS coating. The FESEM images, as shown in Fig.3(a) and (b) demonstrate that after APTMS dip coating or spin coating, the APTMS-capped Au NPs on Silica substrates also exhibited high coverage densities of $(1.07\pm0.05)\times10^{12}$ cm⁻² and $(1.12\pm0.19)\times10^{12}$ cm⁻², respectively. Finally, the samples were covered by spin coating of sol-gel-synthesized HfO_2 control oxide layer (~13 nm) to construct a Si/ SiO₂/ Au-SiO₂ core-shell NPs/ HfO_2 memory structure. The memory device was further annealed at 500 °C in O₂ atmosphere for 10 min to improve oxide qualities [9]. Top and bottom Aluminum electrodes were deposited by electron beam thermal evaporation, followed by post-metal annealing at 300 °C for 30 min to improve interface property between substrate and electrode. The schematic diagram of Au-SiO₂ core-shell NC devices is illustrated in Fig.4. The normalized high frequency (100 kHz) capacitance-voltage (C-V) measurement were conducted using a HP 4294A capacimeter at room temperature to investigate the electrical properties of memory structure. As shown in Fig.5,

978-1-4673-4840-9/13 $31.00 © 2013 IEEE

we observed a clear counterclockwise C-V hysteresis curves, indicating the electron charging and discharging effects from Au NPs at a swapping voltage between ±5. However, such a C-V curves exhibits a slightly deformation of a hump in the depletion region. We attribute the deformation of MOS capacitors to the defects produced by the residual APTMS carbon chains, which was not completely removed during 500℃ post-annealing . The observed shifts of V_{FB} at both bias polarities are due to the successive charging of the Au NPs by electrons or holes; but the V_{FB} of Sample 2 is larger than that of Sample 1. The better property could be ascribed to the relatively thicker SiO_2 shell around Au NPs in sample 2.

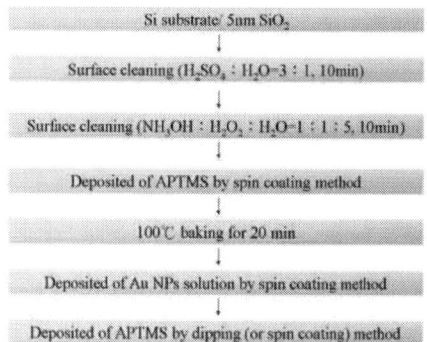

Fig.1. The experimental procedure of the Au-SiO_2 core-shell NPs prepared on 5nm Si/SiO_2 substrate.

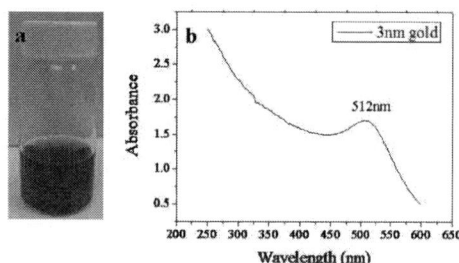

Fig.2. (a) The image of the Au NPs solution (b) The UV-visible adsorption spectrum obtained for the Au NPs solution

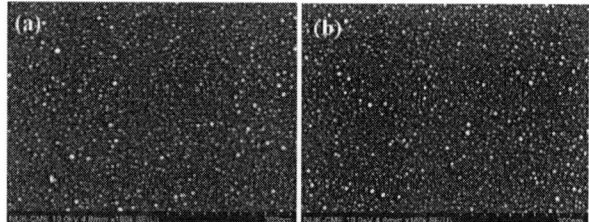

Fig.3. FESEM image of (a) Sample 1 and (b) Sample 2

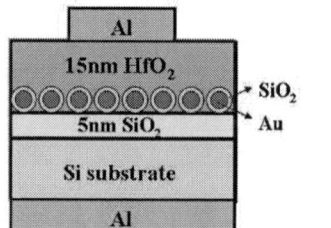

Fig.4. A schematic illustration of a fabricated memory device structure embedded with Au-SiO_2 core-shell NPs

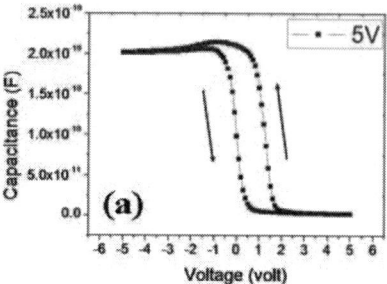

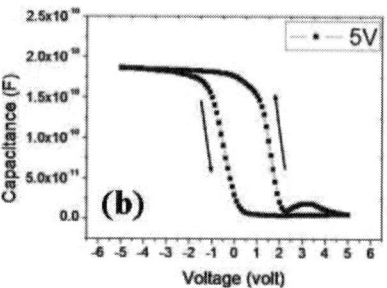

Fig.5. The high frequency C-V curves of (a) Sample 1 and (b) Sample 2

Reference

[1] V. Mikhelashvili, B. Meyler, S. Yoffis, J. Salzman, M. Garbrecht, T. Cohen-Hyams, W. D. Kaplan, and G. Eisenstein, "Nonvolatile low-voltage memory transistor based on SiO_2 tunneling and HfO_2 blocking layers with charge storage in Au nanocrystals", Appl. Phys. Lett. 95, 023104 (2009).

[2] R. C. Jeff, Jr., M. Yun, B. Ramalingam, B. Lee, V. Misra, G. Triplett, and S. Gangopadhyay, "Charge storage characteristics of ultra-small Pt nanoparticle embedded GaAs based non-volatile memory", Appl. Phys. Lett. 99, 072104 (2011).

[3] Z. G. Xu, Z. L. Huo, C. X. Zhu, Y. X. Cui, M. Wang, Z. W. Zheng, J. Liu, Y. M. Wang, F. H. Li, and M. Liu, J, "Improved performance of non-volatile memory with Au-Al2O3 core-shell nanocrystals embedded in HfO2 matrix", Appl. Phys. 110, 104514 (2011).

[4] S. Tiwari, F. Rana, K. Chan, L. Shi, and H. Hanafi, "Charge storage and photoluminescence characteristics of silicon oxide embedded Ge nanocrystal trilayer structures", Appl Phys. Lett. 69, 1232 (1996).

[5] Z. Liu, C. Lee, V. Narayanan, G. Pei, and E. C. Kan, "Study of charge distribution and charge loss in dual-layer metal nanocrystal embedded high-κ/SiO2 gate stack", IEEE Trans. Electron Devices 49, 1614 (2002).

[6] Z. G. Xu, Z. L. Huo, C. X. Zhu, Y. X. Cui, Y. M. Wang, "Improved performance of non-volatile memory with Au-Al_2O_3 core-shell nanocrystals embedded in HfO_2 matrix", Appl. Phys. Lett. 100, 203509 (2012)

[7] R. K. Gupta, D. Y. Kusuma, P. S. Lee, and M. P, "Srinivasan, Covalent assembly of gold nanoparticles for nonvolatile memory applications", ACS, 3, 4619-4625 (2011)

[8] F. K. Liu, Y. C. Chang, F. H. Ko, T. C. Chu, B. T. Dai, "Rapid fabrication of high quality self-assembled nanometer gold particles by spin coating method", Microelectronic Engineering 67-68, 702-709 (2003).

978-1-4673-4840-9/13 $31.00 © 2013 IEEE

Optimization of Al-doped Zinc Oxide Grown on Sapphire Using Dual-Plasma-Enhanced Metal Organic Chemical Vapor Deposition for InGaN/GaN Light-emitting Diodes

Po-Hsun Lei*, Chia-Ming Hsu, Chia-Te Lin, Yu-Siang Fan, and Sheng-Jhan Ye

Institute of Electro-Optical and Material Science, National Formosa
University, No. 64, Wunhua Rd., Huwei, Yunlin County 632, Taiwan
Phone: 886-5-6315668, *E-mail: pohsunlei@gmail.com

Abstract

We prpose Al-doped zinc oxide (AZO) grown on sapphire substrate using dual-plasma-enhanced metal-organic chemical vapor deposition (DPEMOCVD). The crystalline quality, optical properties, and electrical characteristics of AZO depend on the deposition temperature. The AZO thin film grown at 185 ^{o}C shows the highest intensity of (002) preferent orientation for X-ray diffraction (XRD) pattern, highest transmittance of 89 %, highest photoluminence (PL) intensity and lowest full-width at half-maximum (FWHM), and highest carrier concentration and mobility of 3.66×10^{21} cm^{-3} and 10.08 V/cms, which results a very low resistivity of 1.86×10^{-4} Ωcm. The PL peak shows a blue-shift for AZO grown at 185 ^{o}C as compared with ZnO bexause of the Burstein-Moss (BM) effect. In addition, the experimental results represent that the optimized Al content for Al-doped ZnO is 2.88 at % under the deposition temperature of 185 ^{o}C. Finally, the AZO was deposited on InGaN/GaN light-emitting deodes (LEDs) as transparent conductive layer (TCL). InGaN/GaN LEDs using DPEMOCVD-deposited AZO TCL show a lowest forward resist and highest light output intensity as compared to the those without and with TCL composed of commercial indium-tin-oxide (ITO).

Introduction

Gallium nitride (GaN) based blue and green light-emitting diodes (LEDs) have been achieved a tremendous progress and extensively used in many applications including large size full-color displays, short-haul optical communication, traffic signal lights, and backlight for the color liquid-crystal displays [1-3]. Among these applications, high-output-intensity LEDs are very important. The high-output intensity is determined by the external quantum efficiency, which is the product of internal quantum efficiency (radiative recombination in active region) and extraction efficiency (escape of photons from active region) [3, 4]. Because of the high refractive index contrast between air and semiconductor, the extraction efficiency can be reduced by the total internal reflection and Fresnel losses. Even with very high internal quantum efficiency, the conventional GaN-based LEDs show low external quantum efficiency due to the low extraction efficiency. In addition, the total internal reflection also makes a narrow escape cone and parasitic nonradiative losses during photon recycling. The ratio of escaped photons and photons generated in active region for GaN-based LEDs can be approximately $1 / (4n^2)$ where n is the effective index of GaN [3, 5]. An anti-reflection layer with the refractive index about

1.7-1.9 is necessary. Furthermore, GaN-based LEDs with mesa-structure can cause current crowding due to lateral current-injection. As a result, the current tends to crowd at the edge of the mesa contact adjoining the n-type contact. To avoid these problems, the transparent conductive oxides (TCOs) have been widely used in GaN-based LEDs. The most popular TCO is indium-tin-oxide (ITO) for GaN-based LEDs [6-8]. However, the indium is rare on the Earth`s crust and it is urgent to find the substitute for indium.

In this study, the recorded low resistive and highly transparent Al-doped ZnO (AZO) thin films grown by dual-plasma-enhanced metal-organic chemical vapor deposition (DPEMOCVD) were used as the transparent conductive layer (TCL) for InGaN/GaN LEDs. The turn-on voltage, light output intensity, and emission spectrum for the InGaN/GaN LEDs with DPEMOCVD-deposited AZO TCL shows superior characteristics as compared to those without TCL.

Experimental details

A. fabrication of InGaN/GaN LED

The GaN epi-wafers were grown on a c-face (0001) sapphire substrate by metal-organic chemical vapor deposition (MOCVD) system. The device structure consists of a low temperature grown GaN buffer layer, a highly Si-doped n-type GaN layer, an InGaN/GaN multiple-quantum-wells (MQWs) active region, and an Mg-doped p-type GaN layer. The wafer was then patterned using the standard photolithographic process to define the square mesas as emitting regions by partially etching the exposure of p-GaN/n-GaN, ZnO/p-GaN/n-GaN, and ITO/p-GaN/n-GaN. Cr-Au alloy was used as ohmic contact metal on p- and n-GaN contact region, and the wafer was then alloyed in N$_2$ atmosphere for 5 minutes under 450°C. The size of the emission window for the InGaN/GaN LEDs without TCL and with ZnO and ITO was 300×300 μm^2.

B. Deposition of AZO

A high ionization rate for oxygen gas is very important to react to the DEZn completely, allowing a high-quality AZO thin film with well-controlled stoichiometry to grow on the sapphire substrate. However, the ionization rate in conventional PECVD systems is very low, especially O$_2$ gas with strong bond energy. To enhance the dissociation of O$_2$ gas and increase the concentration of oxygen free radicals (O*), the dual-plasma-enhanced metal organic chemical vapor deposition (DPEMOCVD) system, which the oxygen free radicals ionized by DC voltage driven plasma system and then

978-1-4673-4840-9/13 $31.00 © 2013 IEEE

diffuse to the substrate surrounding by a RF plasma system to reduce the recombination of oxygen free radicals during the growth process was used. The experiments in this study used a working pressure of 2.25×10^{-4} Pa. The substrate temperature, which was controlled by a PID-controlled resistive heater mounted on the back-side of the substrate holder, changed from 170 to 190 °C. The flow rates of DEZn TMAl, and O_2 gas were 13, 18 and 100 sccm, respectively.

Results and discussion

Fig. 1 shows the X-ray diffraction patterns for AZO grown at varied substrate temperature. The peak intensity of (002) preferential orientation increases with rising substrate temperature and achieves to the maximum value at 185 °C. This is attributed to the increase in thermal energy to enhance the crystalline quality. However, the (002) peak intensity decreases at 190 °C due to the low sticking coefficient of the reactive elements.

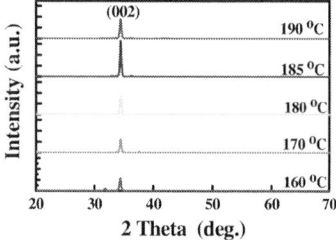

Fig. 1 The X-ray diffraction (XRD) patterns for AZO thin films grown at varied substrate temperature

Fig. 2 shows the transmittance of AZO grown at varied substrate temperature over the visual range. Due to the high crystalline quality, the AZO grown at 185 °C show the highest transmittance of 89 % over the visual range.

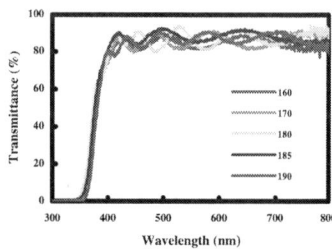

Fig. 2 The transmittance of the AZO thin film grown at varied substrate temperature in the wavelength range from 300 to 800 nm

Fig. 3 shows the electrical resistivity, carrier mobility, and carrier concentration for the AZO thin film grown at varied substrate temperature. The carrier concentration increases with the substrate temperature due to the increasing thermal energy and then decreases at 190 °C because of the low sticking coefficient of the reactive elements. The carrier mobility increases with the raising substrate temperature, which is attributed to the scattering of grain boundary. AZO grown at 185 °C using DPEMOCVD shows a very low resistivity of 1.85×10^{-4} Ωcm.

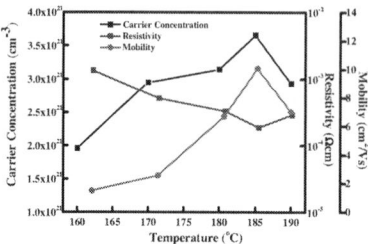

Fig. 3 The dependence of the electrical resistivity, carrier mobility, and carrier concentration for the AZO thin film grown at varied substrate temperature

Finally, the optimized AZO thin film was used as TCO for InGaN/GaN LED to enhance the emitting intensity and to reduce the current crowding effect. Fig. 4 shows the L-I-V characteristics for InGaN/GaN LED without TCO (LED I), InGaN/GaN LED with AZO (LED II) and InGaN/GaN LED with commercial ITO (LED III). The forward resistance for LED II and III is lower than that of LED I, which is attributed to the reduction of metal/p-GaN contact resistance. In addition, LED II shows the highest light output intensity due to the low resistivity and reduction of current crowding effect.

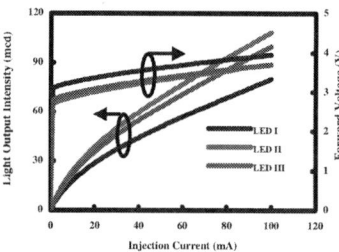

Fig. 4 The L-I-V characteristics for InGaN/GaN LEDs without TCO (LED I) and with AZO (LED II) and commercial ITO (LED III)

Acknowledgements

The authors gratefully acknowledge the financial support of the National Science Council (NSC 100-2221-E-150-059-).

References

[1] M. Koike, N. Shibata, H. Kato and Y. Takahashi, *IEEE J. Sel. Topics Quantum Electron.*, vol. 8, No. 2, pp. 271-277, 2002.

[2] E. F. Schubert and J. K. Kim, *Science*, vol. 308, no. 5726, pp. 1247-1278, 2005.

[3] E. F. Schubert, *Light –Emitting Diodes*, Cambridge, U. K.: Cambridge University Press, 2003.

[4] T. Mukai, *IEEE J. Sel. Topics Quantum Electron.*, vol. 8, No. 2, pp. 264-270, 2002.

[5] M. Broditsky and E. Yablonovitch, " *Light-emitting-diode extraction efficiency,*" in Proc. SPIE, vol. 3002, pp. 119-122, 1997.

[6] T. K. Kim, S. H. Kim, S. S. Yang, J. K. Son, K. H. Lee, Y. G. Hong, K. H. Shim, J. W. Yang, K. Y. Lim, S. J. Bae, and G. M. Yang, *Appl. Phys. Lett.*, vol. 94, pp. 161107-1, 2009.

[7] R. H. Horng, Y. L. Tsai, T. M. Wu, D. S. Wuu, and C. H. Chao, *IEEE J. Sel. Topics Quantum Electron.*, vol. 15, pp. 1327, 2009.

[8] C. C. Lin, and C. T. Lee, *IEEE Photon. Technol. Lett.*, vol. 22, pp. 1132, 2010.

Characterization and photocurrent study of Al, Mg-doped ZnO transparent thin films prepared by sol-gel route

W. Techitdheera[a*], K. Chongsri[b] W. Pecharapa[c,d]

[a] School of Applied Physics, King Mongkut's Institute of Technology Ladkrabang,

Ladkrabang, Bangkok 10520, Thailand

[b]Department of Applied Physics, Faculty of Science and Technology, Rajabhat Rajanagarindra University,

Chachoengsao 24000, Thailand

[c]College of Nanotechnology, King Mongkut's Institute of Technology Ladkrabang,

Ladkrabang, Bangkok 10520, Thailand

[d]ThEP Center, CHE,328 Siayuthtaya Rd., Bangkok 10400,Thailand

*wdheera@gmail.com

Abstract

In this work, ZnO, Al-doped ZnO (AZO) and Mg-doped ZnO (MZO) thin films were deposited onto glass substrates by sol-gel spin-coating method. The doping composition was varied from 0-10 at.%. Structural properties and surface morphologies of as-prepared films were investigated by X–ray diffraction (XRD) and scanning electron microscope (SEM). Optical properties of the films were interpreted from their transmission spectra using UV-VIS spectrophotometer. The photocurrent spectroscopy was employed to study their optical response and sensitivity of the films. The XRD and SEM results disclosed that the crystallinity and grain size of as-prepared films were strongly influenced by Al and Mg doping. UV-VIS spectrophotometer results indicated that Mg and Al additives could not only efficiently improve the optical transparency but also obviously induce the blue-shift in optical band gap of ZnO films. The photocurrents of as-prepared films disclosed that the enhancement in photosensitivity of ZnO films can be achieved by specific content of either Al or Mg additives

Keywords: ZnO thin films, Al doping, Mg doping, Sol-gel route and photocurrent

Introduction

Recently, considerable attention has been paid to ZnO material for realizing room-temperature UV-Visible optoelectronic application, which is enabled by the wide direct band gap (E_g = 3.34 eV) and large exciton binding energy (60 meV) at room temperature [1]. Moreover, ZnO-based materials are being applied in various electronic and optical devices such as light emitting diodes [2], laser diodes [3] and ultraviolet photodetector. Many attempts such as doping, alloying, and heterostructures [4] have been purposed to adjust its electronic and optical properties to meet required operating wavelength. To obtain high quality devices, it is essential to confine carriers and photons in optoelectronic devices. The major problem in this endeavor is to adjust the band gap while keeping the lattice constants matching between the materials constructing the heterojunction. One of the most practical approaches to adjust its optical band gap can be carried out by alloying. Magnesium oxide (MgO) has a band gap of 7.8 eV and a cubic crystal structure. The ionic radius of Mg^{2+} (0.57 Å) is close to that of Zn^{2+} (0.6 Å) [1], replacement of Zn by Mg consequently should not cause significant change in lattice constant. By suitably doping Mg into ZnO, it may be possible to obtain a ternary $Mg_xZn_{1-x}O$ alloy with a wider band gap than pure ZnO, yet still having a lattice constant similar to that of pure ZnO. These properties may enable $Mg_xZn_{1-x}O$ to form heterostructures with ZnO that can be used in UV detectors [5]. Furthermore, Al is one of the most proper element widely used as a dopant for ZnO. Al-doped ZnO is reported to be a useful material for a variety of technological applications [6]. Metal-doped ZnO can be obtained by various growth techniques such as radio frequency magnetron sputtering (RFMS), pulsed laser deposition (PLD), laser molecular beam epitaxy (P-MBE), spray pyrolysis , metal organic vapour-phase epitaxy (MOVPE), metal organic chemical vapour deposition (MOCVD) [7], and sol-gel spin coating method [8]. For a decade, a number of researches have reported on the development of ZnO-based and many methods have dedicated to enhance its properties [9]. However, for our best knowledge, few recent works on photocurrent study of sol-gel derived ZnO-based thin films doped with Al and Mg have been yet reported.

In this work, we report on the preparation, characterization, optical properties and optical response of spin-coated Al and Mg-doped ZnO films deposited on borosilicate glass substrate using spin coating method.

Materials and Methods

The AZO and MZO films were prepared by sol-gel spin-coating method using zinc acetate dehydrate $(CH_3COO)_2Zn \cdot 2H_2O$, magnesium acetate tetrahydrate $(CH_3COO)_2Mg \cdot 4H_2O$, aluminum acetate $(C_4H_7AlO_5)$ as starting precursors for Zn, Mg and Al sources, respectively. The precursors with concentration of 0.5 M were dissolved and stabilized by absolute ethanol (C_2H_5OH) and

978-1-4673-4840-9/13 $31.00 © 2013 IEEE

diethanolamine ((HOCH$_2$CH$_2$)$_2$NH, DEA). The solutions were prepared with various Mg and Al compositions (0% ≤ x ≤ 10%). The mixed solution was stirred at 100 °C for 6 hr and cooled to room temperature for 24 hr. All films were annealed at 400-550 °C for 4 hr in air. The films crystal structure and morphology was determined by X-ray diffraction ((XRD, Panalytical x'Pert Pro MPD) using Cu-K_α radiation over a 2θ in the range of 20-60°) and scanning electron microscope (SEM: JSM-6510), respectively. The optical properties of the films were executed from their transmission spectra carried out by a UV-VIS spectrophotometer (Thermo Electron Corporation Heliosα). The photodetectors were fabricated using as-prepared films by depositing silver contacts in form of planar MSM structure. The photoresponse using Xenon arc lamp as light source dispersed by a monochromator to cover the range of 250-500 nm.

RESULTS AND DISCUSSIONS

The crystalline structure of ZnO films annealed at 400 °C and 550 °C, AZO and MZO films are shown in Fig. 1. The noticeable diffraction peaks positioned at 2θ =31.8°, 34.5°, 36.2° and 47.5° are assigned to (100), (002), (101) and (102) orientation planes of ZnO with hexagonal wurtzite structure, respectively. Upon increasing annealing temperature, crystallinity of the films improves significantly, accompanying intense and sharp reflection of XRD peaks. Referring the XRD patterns of AZO and MZO films, diffracted graphs exhibit the decrease in their peak intensities, suggesting the deterioration in their crystallinity and grain size reduction caused by Mg and Al doping. These features were able to observed in previous literatures [10]. Moreover, for MZO, the board XRD pattern appeared at 43.0° is attributable to (200) plane of MgO, implying the formation of secondary phase of MgO due to relatively high Mg doing content.

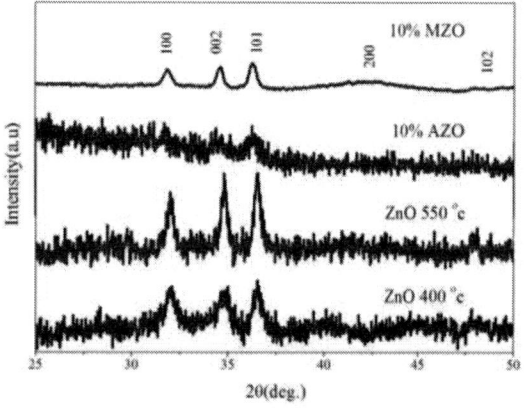

Fig.1 XRD patterns of ZnO films with various annealing temperature ranging from 400 and 550 °C, 10% AZO film and 10% MZO film.

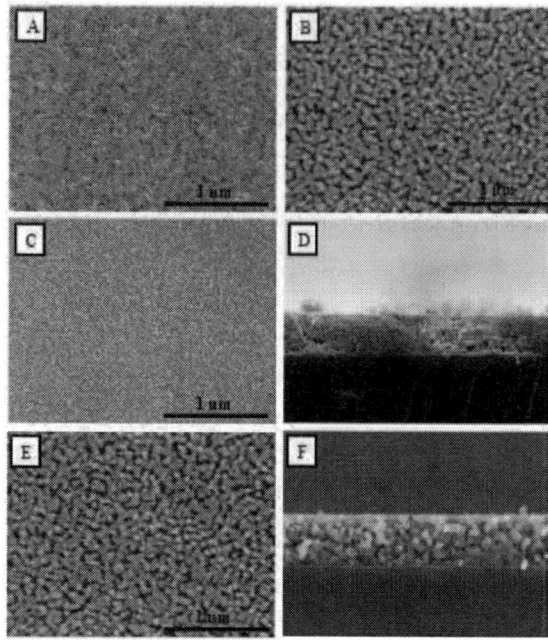

Fig.2 SEM micrographs of ZnO films annealed at (A) 400 °C and (B) 550 °C, (C) 10 at. % Al-doped ZnO film,(D) cross-sectional image of 10 at. % Al doped ZnO film, (E) 10 at. % Mg doped ZnO film and (F) cross-sectional image of 10 at. % Mg doped ZnO film.

The surface morphologies of all samples are shown in Fig. 2. Fig.2. (A) and (B) illustrate surface morphologies of ZnO films annealing at 400 °C and 550 °C, respectively. It is clearly observed that all films have smooth surface comprising uniform grain size. As the annealing temperature increases, the crystallite structure and grain size of the film improves. Fig. 2 (C) and (E) display surface morphologies of the 10 at.% Al-doped ZnO film and the 10 at. %-Mg doped ZnO film annealed at 550 °C, respectively. As seen in 10% AZO film, the crystallinity of the film significantly decreases, accompanying decreasing grain size. Moreover, The morphology of the 10% MZO as shown in Fig.2 (E) displays polycrystalline of MZO which consists of grains with large number of grain boundary. The average thickness of the AZO and MZO films are approximately 400 nm and 450 nm measured by cross-sectional SEM as shown in Fig. 2 (D) and (F), respectively. The optical transmission spectra of all samples at room temperature are illustrated in Fig. 3(A) and (B). All transmission spectra indicate sharp absorption edge. Moreover, as Mg composition increases, the transmission spectra exhibits the obvious blue shift of absorption edge of optical band gap of the film and highly transparent within visible region with increasing Mg composition. As observed in the Fig.5 (A) and (B) for AZO samples, when Al composition increases, the transmission spectrum exhibits obvious blue shift in absorption edge of optical band gap of the film and better transparency within visible region with more than 90-100% in its transparency. The band gap of the MZO can be varied from 3.29 to 3.42 eV by varying Mg content meanwhile band gap of the AZO can be increased to 3.29 eV by varying Al content.

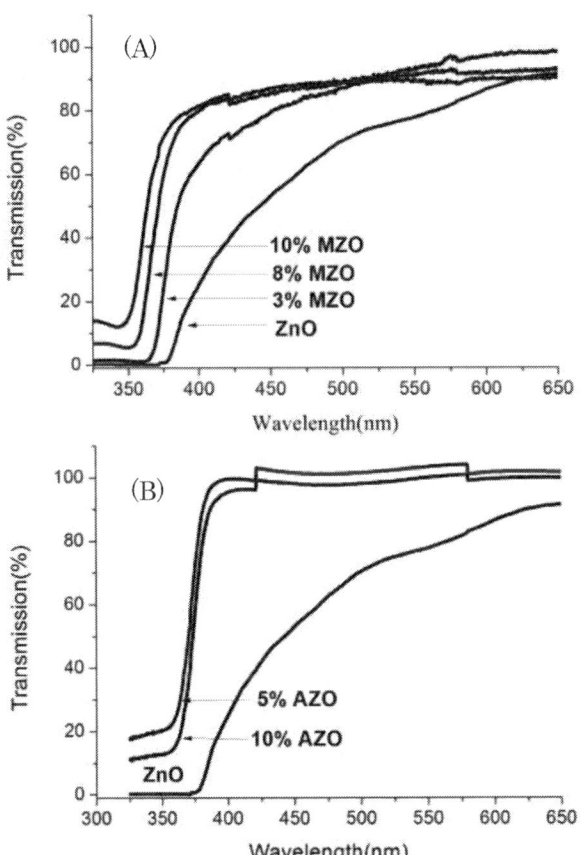

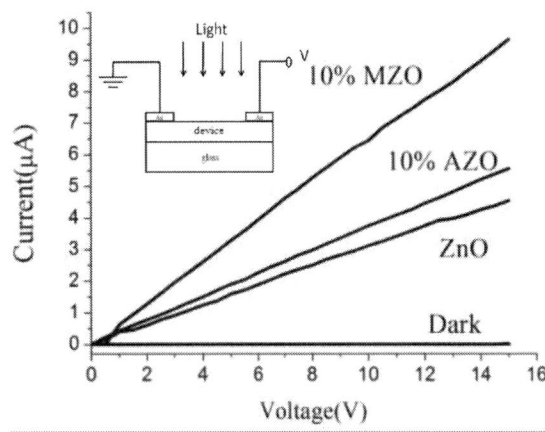

Fig. 4. Dark and photoilluminated (λ= 360 nm) current of ZnO film and MZO, AZO films with different Mg and Al doping contents. Inset shows schematic structure of the detectors.

Fig. 3 Optical transmittance spectra of (A) MZO films with different Mg doping contents and (B) AZO films with different Al doping contents.

Fig. 4 shows the photoilluminated current and dark current of ZnO, MZO and AZO - MSM photodetector (schematically shown in the inset of Fig. 4) with different Mg and Al doping composition. As seen in the figure, the illuminated current linearly increases with increasing bias voltage, indicating the good Ohmic contact of the device. Moreover, as Mg and Al composition increases, the illuminated current exhibits the obvious increase in current of the film. At a bias voltage of 10 V, the dark and photoilluminated current of ZnO, AZO and MZO photodetector are approximately 1.05, 2.08, 3.19 and 6.44 μA, respectively. Fig. 5(A) and (B) show the photoresponse spectra of MZO and AZO detectors at bias voltage of 7.3 V, respectively. The well-defined wavelength response of all detectors covers the UV range from 300- 400 nm (A) and 275-380nm (B). The peak wavelength of the detector based on pure ZnO is at 377 nm with shaft cutoff at around 450 nm. The photoresponse profile of the detector can be narrowed and wavelength can be tuned to shorter wavelength by introducing more Mg and Al composition into films as seen in Fig.5 (A) and (B).

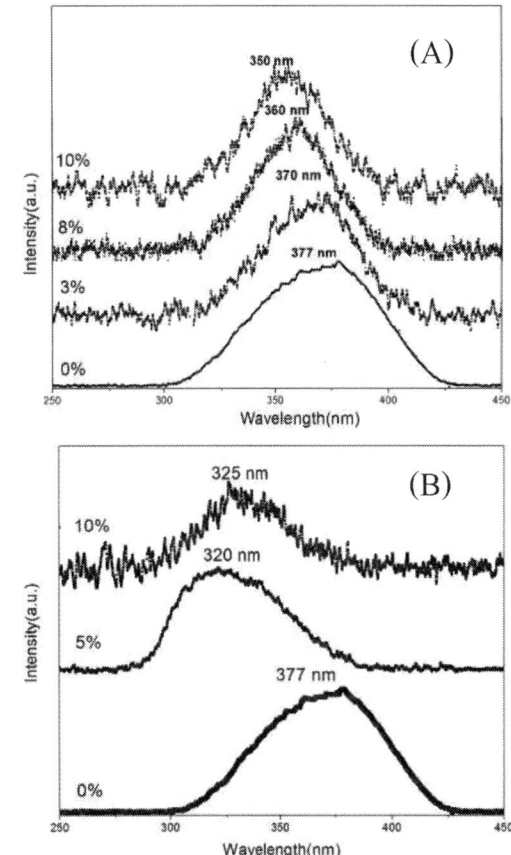

Fig. 5 Photoresponse of (A) MZO films with different Mg doping contents and (B) AZO films with different Al doping contents as a function of wavelength.

978-1-4673-4840-9/13 $31.00 © 2013 IEEE

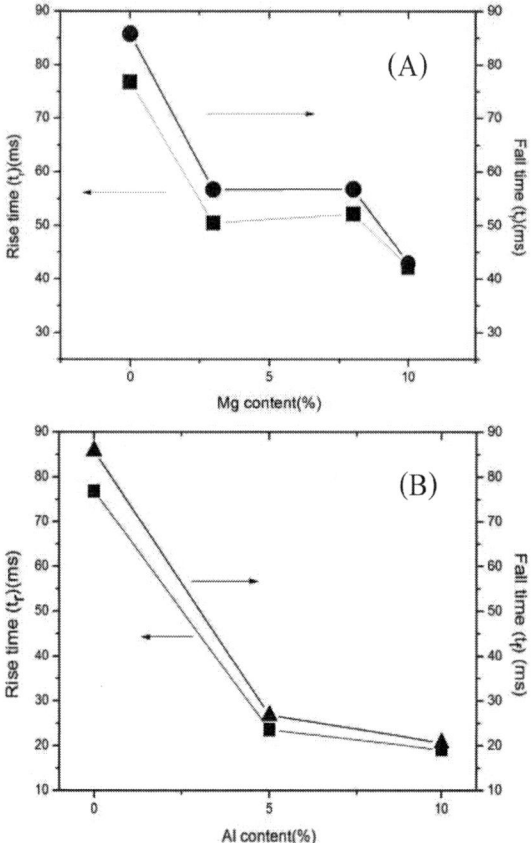

Fig. 6 UV photoresponse of (A) MZO films with different Mg doping contents and (B) AZO films with different Al doping contents with property under illumination by a 360 nm UV source and 5 V bias.

Photoresponse properties of the MZO and AZO films with different doping concentrations measured under the excitation of 365 nm UV light are shown in Fig. 6 (A) and (B), respectively. Rise time (t_r) of MZO films with various Mg contents is in range of 36 to 75 ms whereas fall time (t_f) of these films are in range of 28 to 85 ms depending on Mg content. Meanwhile, rise time (t_r) of AZO films with various Mg contents is in range of 36 to 75 ms whereas fall time (t_f) of these films are in range of 28 to 85 ms. At this stage, it is suggested that the optical response of sol-gel derived ZnO optical thin films including wavelength response extension to UV region and shorter response time can be effectively ameliorated by incorporation of either Mg or Zn additive.

CONCLUSION

In summary, MZO and AZO fims were deposited on glass substrate by spin coating technique. Effects of Mg and Al doping on the structural, optical and photoresponse properties of ZnO films have been investigated and discussed. The relevant results disclosed that the crystallization quality and grain size of as-prepared films are highly influenced by Mg and Al doping. All the films exhibit high transmittance in visible region. Optical band gad energy is found to increase with Mg and Al doping concentration. The photoconductive

measurement revealed that the incorporation of either Mg or Al into ZnO films results to the wavelength response extension to shorter wavelength and the reduction in response time, that is one of crucial requirement for photo detection applications.

ACKNOWLEDGMENTS

This work has partially been supported by the National Nanotechnology Center (NANOTEC), NSTDA, Ministry of Science and Technology, Thailand, through its program of Center of Excellence Network. Authors would like to thank Rajamangala University of Technology Thanyaburi (RMUTT) for XRD and SEM measurement, King Mongkut's Institute of Technology Ladkrabang Research Fund (KMITL Research Fund) and Department of Applied Physics, Faculty of Science and Technology, Rajabhat Rajanagarindra University Chachoengsao for research fund support.

References

[1] J. Zeng, S. Wang, P. Tao and J. Xu, J Alloys Compd. vol 476, pp.60-63, 2009.

[2] W. Mingsong, J-K. Eui, K. Sunwook, S-C. Jin, Y. Ik-Keun, W-S. Eun., H-H. Sung and P. Chinho, Thin Solid Films. vol 516,pp. 1124-1129, 2008,.

[3] L. Weiwei, Y. Bin,L. Yongfeng, L. Binghui, Z. Changji, Z. Bingye, S. Chongxin, Z. Zhenzhong, Z. Jiying, S. Dezhen, Appl Surf Sci. vol 255, pp.6745-6749, 2009.

[4] W. Pecharapa ,C. Kahattha and W. Techitdeera, "Optical properties of MgZnO alloyed films characterized by transision spectroscopy", Proc. International workshop and symposium on science and technology 2008 December 15 – 16, 2008 Nongkhai, Thailand.

[5] J.H. Li, Y.C. Liu, C.L. Shao, X.T. Zhang, D.Z. Shen, Y.M. LU, J.Y. Zhang, X.W. Fan, J Colloid Interf Sci. vol 283, pp. 513-517, 2005.

[6] M.J. Alamand and D.C Cameron, J. Vac. Sci. Technol. Vol. A19, pp. 1642-1646, 2001.

[7] R.N. Gayen, S.N. Das, S. Dalui, R. Bhar and A.K. Pal, J. Cryst Growth. vol 310, pp. 4073-4080, 2008.

[8] H. Wang, K.P. Yan, J. Xie and M. Duan, Mat.Sci. Semicon. Proc. vol 11, pp. 44-47, 2008.

[9] L.K.Wang, Z.G.Ju, C.X.Shan, J.Zheng, D.Z.Shen, B.Yao, D.X.Zhno, Z.Z.Zhang, B.H.Li, J.Y.Zhang, Solid. State. Commun, vol. 149, pp. 2021-2025, 2009.

[10] M.H. Mamat, M.Z. Sahdan, Z. Khusaimi, A. Zain Ahmed, S. Abdullah, M. Rusop, Opt Mater, vol.32, pp. 696–699, 2010.

Titanium dioxide/Vanadium oxide nanocomposites synthesized via sonochemical and hydrothermal process for energy storage application

C. Kahattha[1,2*], W. Techitdheera[3], N. Vittayakorn[1,4] and W. Pecharapa[1,2]

[1]*College of Nanotechnology, King Mongkut's Institute of Technology Ladkrabang, Chalongkrung Rd.,*

Ladkrabang Bangkok 10520, Thailand

[2]*ThEP Center, CHE, 328 Siayuthtaya Rd., Bangkok 10400, Thailand*

[3]*School of Applied Physics, Faculty of Science,*

King Mongkut's Institute of Technology Ladkrabang, Bangkok 10520, Thailand

[4]*Advanced Materials Science Research Unit, Department of Chemistry, Faculty of Science,*

King Mongkut's Institute of Technology Ladkrabang, Bangkok 10520, Thailand

*kahattha@gmail.com

Abstract

V_2O_5/TiO_2 nanocomposites were synthesized via sonochemical and hydrothermal process. At first, titanium dioxide (TiO_2) nanopowders were synthesized by sonochemical process using titanium isopropoxide as a titanium source. Meanwhile, hydrothermal process was carried out to modify the structure of commercial V_2O_5 powder to be nanorod-like structure V_2O_5 with increasing specific surface area. Structural and morphological properties of the composites were characterized by XRD and SEM, respectively. The XRD results indicated that the crystalline of the composites corresponds to anatase and orthorhombic structures of TiO_2 and V_2O_5, respectively. The significant variation of charge storage properties of the nanocomposites under ultraviolet irradiation were obtained by varying V_2O_5 content in the composite. Results suggest that V_2O_5 loaded into the nanocomposite play a key role as a storage material of photoelectron generated by TiO_2 illuminated under ultraviolet.

Introduction

Vanadium pentoxide (V_2O_5) has been intensively studied in the field of nanomaterial due to its excellent and suitable properties for various applications such as sensors, catalysts, cathode materials for batteries and electrochemical applications [1-2]. Previously, many techniques such as sputtering, atmospheric chemical vapour deposition, co-precipitation and hydrothermal method [3-4] have been employed to synthesize functional V_2O_5 nanostructures. M.B. Sahana *et. al* [5] prepared V_2O_5 by spin coating and observed electrochemical properties of the thin films. Metalorganic, organic and inorganic were selected as the sol-gel precursors. The results indicated that the Li^+ intercalation capacity and Li^+ diffusion coefficient was increased by an order of magnitude in the non-stoichiometric films. A. Dhayal *et. al.* [6] reported on gas sensing properties of V_2O_5 hollow spheres made up of self-assembled nanorods synthesized by solvothermal method and the corresponding results indicated that V_2O_5 nanorods had superior sensing response for ethanol when compared to that of ammonia. Keng-Che *et. al.* [7] gave a report on

electrochromic properties of V_2O_5 nanowires using commercial V_2O_5 powder. The deposition of V_2O_5 nanowires were carried out by thermal evaporation onto ITO substrate and the results indicated that V_2O_5 nanowires were obtained after kept in a pressure of 8 x 10^{-4} Torr and 650 °C. The transmittance spectrum change of V_2O_5 nanowires is 37.4% at 415 nm. For enhancement the properties of V_2O_5, the composites with functional materials have been dramatically attended. Among many metal oxide compound, TiO_2 has been attracted numerous attention due to its wide band gap and excellent response in ultraviolet region. From this ability, TiO_2 can be generated photoelectron under UV irradiation, which can effectively assist the optical performance of V_2O_5.

In this work, we reported on the improvement of electrochemical properties of V_2O_5 using TiO_2 nanoparticles in from of functional composite. The effect of weight ratio of V_2O_5:TiO_2 on photoelectrochemical properties of the products were studied and discussed.

Materials and Method

A. Synthesis of TiO_2 by sonochemical process

TiO_2 powders were synthesized by sonochemical-assisted process. In the synthesis process, certain titanium isopropoxide was dissolved into designated solution of absolute ethanol and acetylacetone and then stirred at room temperature for 24 h until transparent pale yellow solution was obtained. 10 mL of the stocked solution and 50 mL of deionized water was filled into the chamber and then the mixed liquid was irradiated with high intensity ultrasound (650 W 20 kHz) by a Sonics Model VCX 750 at room temperature in ambient air for 30 min until the completely precipitated product was reached. After cooled down to room temperature, the resulting precipitates were washed with deionized water and ethanol. After that the cleaned precipitates were calcined at 500 °C for 4 h.

B. Synthesis of V_2O_5 nanorods by hydrothermal method

The V_2O_5 nanorods were synthesized by hydrothermal method using commercials V_2O_5 powder as the source of vanadium and *n*-butanol, acetylacetone as the reducing agents. In a typical process, 3.62 g of commercial V_2O_5 powder, 10 mL of *n*-butanol, 10 mL of acetylacetone and 100 mL of deionized water were vigorously magnetically stirred at room temperature for 1 h, the suspension was transferred into a 250 mL Teflon-lined stainless autoclave, which was then filled with deionized water up to 200 mL of total volume. The autoclave was sealed and kept at 120 °C for 24 h and then cooled down to room temperature for 24 h. The obtained dark blue precipitate was filtered and washed for several times with deionized water, acetone and absolute ethanol and dried in air at 80 °C for several time, followed by calcinations at 500 °C for 4 h.

C. Fabrication of V_2O_5/TiO_2 nanocomposite films

In this process, V_2O_5/TiO_2 nanocomposites with different ratio of V_2O_5:TiO_2 were dissolved in the solution of nitric acid, DI water, absolute ethanol and terpineol. After that, the mixed solution was stirred at room temperature for 30 min and assigned as solution A. The solution B was prepared using ethylcellulose dissolved in absolute ethanol and sonicated until the opaque solution was obtained. After that, the final mixed solution between solution A and B was homogenized at 6000 rpm for 30 min and stirred at 120 °C until the viscous yellow suspension was obtained. Secondly, the well mixed suspension was slowly dropped and spread onto the FTO glass substrate followed by drying in air at 80 °C for 10 min to yield the as-prepared thin film. Finally, the as-prepared composites film was further proceeded residual removal by heating at 500 °C in air for 2 h to obtain the V_2O_5/TiO_2 nanocomposite film. The schematic of the nanocomposite films is depicted in Fig. 1.

D. Film characterization

The crystal structures of the samples were investigated by X' Pert PRO X-ray diffraction with a monochromatic source of Cu K_α (λ=0.15405 nm). Their morphologies were monitored with JEOl JSM-6510 scanning electron microscope with an accelerating voltage of 5.0 kV. Its optical absorption was investigated by Heλios γ UV-Vis spectrophotometer.

E. Photoelectrochemical measurement

Photoelectrochemical measurement was conducted using Autolab PGSTAT302 with V_2O_5/TiO_2 nanocomposite film as the working electrode and Pt films as the counter electrode. Photocurrent measurement was carried out in the electrolyte of 0.1 M $LiClO_4$ at room temperature. The current-potential curves were measured at a potential sweep rate of 20 mV/s in dark and under sun light irradiated by solar simulator lamp.

Glass slide
FTO
Pt film
Electrolyte
Nanocomposites film
FTO
Glass slide

Fig. 1 The schematic of V_2O_5/TiO_2 nanocomposite films.

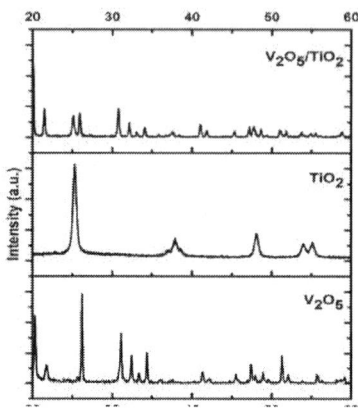

Fig. 2 XRD patterns of V_2O_5 nanocomposite films.

Results and discussion

The crystalline structure of the V_2O_5/TiO_2 nanocomposite films were investigated by XRD and their patterns are shown in Fig. 1. The noticeable diffraction peaks, which appeared in diffraction spectra of V_2O_5 synthesized by hydrothermal method and TiO_2 synthesized via the sonochemical process attributed to orientation plane of orthorhombic structure of V_2O_5 corresponded to JCPDS file No. 89-061 and anatase phase of TiO_2 corresponded to JCPDS file No. 89-4921. These results indicated that V_2O_5 orthorhombic structure and TiO_2 with pure anatase phase were obtained with calcinations route at 500 °C. Meanwhile, dual spectra were appeared in the diffraction peaks of V_2O_5/TiO_2 nanocomposite and disappearance of unusual diffraction peaks in the spectra. These results imply that the V_2O_5/TiO_2 nanocomposite can be obtained.

The SEM image of the nanorod-like V_2O_5 is shown in Fig. 3(a). Meanwhile the SEM image of TiO_2 nanopowders synthesized by sonochemical-assisted process is exhibited in Fig. 3(b). Fig. 3(c) illustrates the morphology of V_2O_5/TiO_2 nanocomposite with weight ratio 0.9:0.1. It is clearly seen that the nanorod-like V_2O_5 with diameter about 170 nm and 900 μm in length can be synthesized by hydrothermal method. From Fig. 3(c), it is observed that TiO_2 nanoparticles are uniformly dispersed in V_2O_5 nanorod matrix. These results implied that V_2O_5/TiO_2 nanocomposite films can be prepared by mixing oxide materials using facile technique.

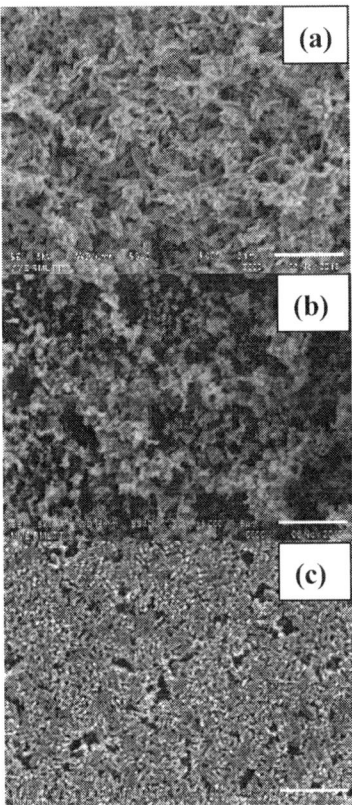

Fig. 3 SEM images of (a) V_2O_5 nanorods, (b) TiO_2 nanoparticles and (c) V_2O_5/TiO_2 nanocomposites (the scale bar = 5 μm)

The optical absorption of TiO_2 was investigated by UV-Vis spectrum as shown in Fig. 4. It clearly seen that the UV light in range 300-400 nm was absorbed by TiO_2 nanoparticle due to its typical optical band gap of TiO_2 [8]. From the result, the absorption in UV region of TiO_2 nanoparticles can initiate the generation of photoelectron under illumination, which supported the photoelectrochemical process.

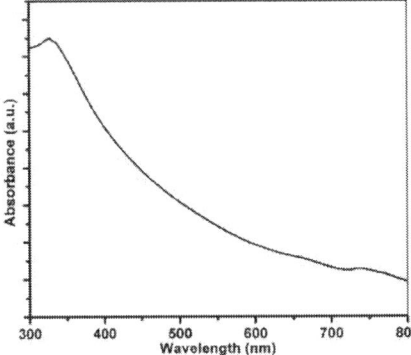

Fig. 4 Optical absorption spectrum of TiO_2 nanoparticles synthesized by sonochemical method.

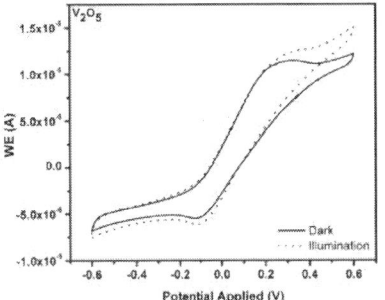

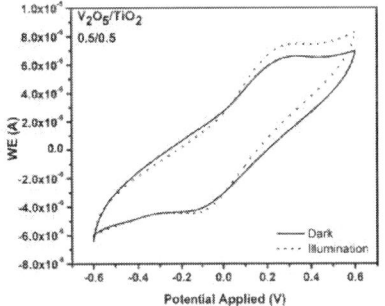

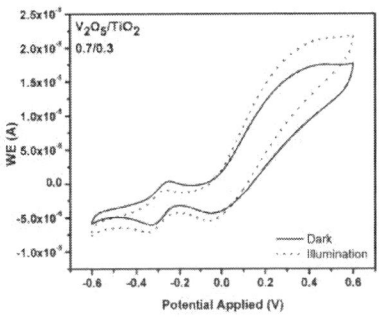

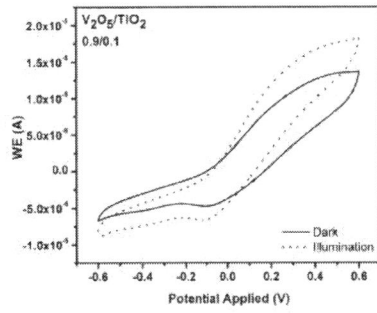

Fig. 5 Cyclic voltammograms of V_2O_5/TiO_2 nanocomposites with different weight ratio at scan rate 20 mV/s. (Electrode area was 0.5x0.5 cm^2)

The photoelectrochemical properties of V_2O_5/TiO_2 nanocomposites with different weight ratio are shown in Fig. 5. For V_2O_5 case, it is found that the current density of the film is increased when illuminated by solar simulator light due to its absorption properties in the visible region of this material [9]. Moreover, the significantly increasing current density in the nanocomposites films can be observed. This phenomenon implies that the enhancement in light-induced photoelectron generation can be achieved by the incorporation of TiO_2 into V_2O_5 due to the photoelectron generated by TiO_2 during illumination. The optimized weight ratio of V_2O_5/TiO_2 for superiority in current density is found to be 0.7:0.3, which provides the current density of 22 μA per 0.25 cm^2.

Conclusion

In summary, V_2O_5/TiO_2 nanocomposites were successfully prepared by screen print technique. The XRD results revealed that the diffraction spectrum of nanocomposites film consist of dual diffraction peaks between orthorhombicV_2O_5 and anatase TiO_2 structures. SEM results indicated that the TiO_2 nanoparticles are well-dispersed in V_2O_5 host matrix. The photoelectrochemical of nanocomposites films were enhanced by loading TiO_2 in V_2O_5 nanorods and the maximum current is about 22 μA at the weight ratio is 0.7:0.3.

Acknowledgment

This work has partially been supported by the National Nanotechnology Center (NANOTEC), NSTDA, Ministry of Science and Technology, Thailand, through its program of Center of Excellence Network and was financially supported by KMITL research fund. Authors would like to thank Energy Policy and Planning Office, Ministry of Energy, Thailand, for research funding support. Authors would like to thank Rajamangala University of Technology Thanyaburi (RMUTT) for XRD and SEM measurement.

References

[1] L.M. Chen, et al., *J. Alloy. Compd*, 467, pp. 465–471, December 2009.
[2] T. Ivanava, et al., *Mat. Res. B*ull, 40, pp. 411–419, December 2005.
[3] N. Krishnan, et al., *Thin Solid Films*, 519, pp. 3663 – 3668, March 2011.
[4] C. Cai, et al., *J. Alloy. Compd*, 509, pp. 909–915, September 2011.
[5] M.B. Sahana, et al., *Mat. Sci. Eng. B-Solid*, 143, pp. 42–50 August 2007.
[6] A. D. Raj, et al., *Curr. Appl. Phys*, 10, pp. 531–537, July 2010.
[7] K.C. Cheng, et al., *Sol. Energ. Mat. Sol. C*, 90, pp. 1156 – 1165 December 2006.
[8] J. Liu, et al., *Mat. Sci. Eng. B-Solid*, 172, pp. 142–145 April 2010.
[9] L. Boudaoud, et al., *Catal. Today*, 113, pp. 230–234, January 2006.

In$_x$Ga$_{1-x}$Sb n-channel MOSFET: Effect of Interface States on C-V Characteristics

Muhammad Shaffatul Islam [1*], Md. Nur Kutubul Alam[1]. Md. Rafiqul Islam[1]

[1] Department of Electrical and Electronic Engineering
Khulna University of Engineering & Technology
Khulna-9203, Bangladesh
*Corresponding author : mashru.islam@yahoo.com

ABSTRACT

The Capacitance-Voltage (CV) characteristics of InGaSb based n-MOSFET are investigated by quantum mechanical calculation solving 1D self-consistent Schrodinger-Poisson equation using Silvaco's ATLAS device simulation package. The charge density profile is determined by wave function penetration effect within the oxide layer and Neuman boundary condition. Low and high frequency CV characteristics are studied both for the positive and negative interface charge densities. The results obtained from the simulation demonstrate that the CV characteristics are not so sensitive in the shift of threshold voltage for high frequency operation. On the other hand, a significant shift in threshold voltage is noticed for the low frequency operation and the shift is entirely depends on the density of interface charge density and its type.

INTRODUCTION

Silicon is the most widely used semiconductor in digital logic applications. As silicon based devices are scaling down to their ultimate limits , some adverse effects arise which degrade the device performance. For the limitation of the silicon and quest for better performing material, researchers are looking into the possibilities of CMOS circuits that utilize III-V material either alone or in hybrid form with Si or Ge [1].

III-V materials have emerged as a promising candidate for channel material of MOSFET to be used in future logic applications due to excellent electron and transport properties [2]. Since the modern complementary metal oxide semiconductor technology is aggressively scaling down, the device dimension and the thickness of the material has also been scaled [3]. For compensating gate leakage and good control over electrostatic potential, High-k oxides are suitable. In scaled down III-V devices High-K gate oxide integration on compound semiconductor is a challenging issue. The low quality interface can generate interface states in the oxide-semiconductor interface. The generation of interface trap states at oxide/semiconductor interface and oxide trapped charge of MOS structures has long been demonstrated as the prime factors behind the degradation of MOS device performances [4,5]. When bias is applied on the gate at elevated temperature or for long time, an increase in oxide trap charge density causes the shift in threshold voltage and decrease in carrier mobility. However, charge trapping induced through bias is extremely high in high-k gate stacks because of high densities of intrinsic defect in material [6.7]. The bias-induced charging and discharging greatly influence the device performance by reducing the drive current due to electrostatic interaction with trapped charges [8].This kind of degradation on device parameters can led to circuit failures in analog and digital applications [5].

Among all the III-V semiconductors, antimonide based materials have higher electron mobility and can be used as a potential material in the channel of MOSFET. Recently, InGaSb-based surface channel p-MOSFET [2] and n-MOSFET [9] was experimentally demonstrated. In this study logic figure of merits (e.g. SS,DIBL,I$_{ON}$/I$_{OFF}$ ratio) and mobility for p-MOSFET and channel mobility behavior for n-MOSFET has been investigated. Self-Consistent capacitance-voltage (CV) characterization was reported considering direct tunneling gate leakage current for InGaSb n-channel MOSFET in[10]. To best of our knowledge there is no study on effect of interface states on InGaSb n-channel MOSFETs including the CV characteristics.

In this work coupled Schrodinger and Poisson equations have been solved self consistently [11] in order to calculate the quantum mechanical charge distribution in MOS devices incorporating wave function penetration effect within the oxide layer using SILVACO's ATLAS device simulation package. The solver employs Finite Element Method. The CV characteristics are calculated using the electrostatics of the device revealed by the solver. The results obtained from the simulation indicate that the density of interface states and its type have significant influence on the shift in threshold voltage for low frequency (LF) operation. However, at the high frequency (HF) operation, the shift in threshold is found to be almost negligible with interface states.

SELF CONSISTENT MODELING

A. Device Structure

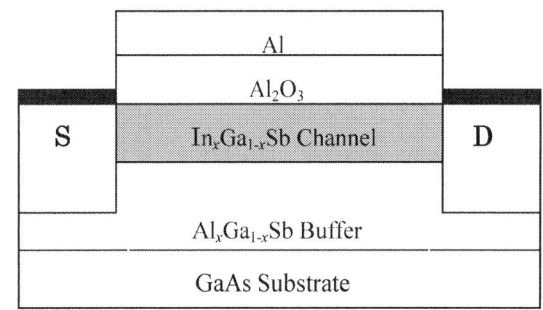

Fig. 1 Proposed In$_x$Ga$_{1-x}$Sb surface channel MOSFET structure.

978-1-4673-4840-9/13 $31.00 © 2013 IEEE

The structure of the device under study shown in Fig.1 consists of Al as gate contact and Al_2O_3 as gate dielectric with 10nm dimension, The $In_xGa_{1-x}Sb$ is used as channel material with very thin dimension of 7nm, and 1µm $Al_yGa_{1-x}Sb$ buffer layer is used on the top of the GaAs substrate to minimize the lattice mismatch between the substrate and the channel material. In this study, we have investigated HFCV and LFCV characteristics of InGaSb n-MOSFET for both positive and negative interface charge densities at room temperature (300K).

B. *Self- Consistent Theory*

The quantum mechanical solver used in SILVACO's Atlas device simulation package uses classical PDE to solve the Schrodinger and Poisson equation. The Poisson equation in coefficient form is given as.

$$-\nabla.(c\nabla u) = f \qquad (1)$$

The equation for MOS electrostatic potential is

$$\varepsilon_0\varepsilon\frac{d^2v(y)}{dy^2} = q[p(y) - n(y) + N_D - N_A] \qquad (2)$$

where $n(y)$ and $p(y)$ are electron and hole concentration and N_D and N_A are ionized donor and acceptor concentration, respectively. ε is the relative dielectric constant and ε_0 is the free space permittivity. The electron concentration $n(y)$ is obtained for the n-MOS structure [11]by

$$n(y) = \sum_{ij} N_{ij} |\psi_{ij}(y)|^2 \qquad (3)$$

$$N_{ij} = \frac{n_{vi}\,m_{di}\,KT}{\pi\hslash^2}\ln[1+\exp(\frac{E_F-E_{ij}}{KT})] \qquad (4)$$

Here N_{ij} is the carrier concentration in the j th subband of the I th valley, n_{vi} and m_{di} are the i th valley degeneracy and the i th density of effective mass of the channel material. N_{inv} is the total inversion layer carrier concentration and E_F is the Fermi energy. E_{ij} and ψ_{ij} are the eigenvalue and eigenfunction of the j th energy level of the i th valley, which are obtained from the one dimensional solution of the Schrodinger equation.

The equation of one dimensional Schrodinger equation with effective mass approximation can be defined as

$$\left[-\frac{\hslash^2}{2m^*}\frac{d^2}{dy^2} + V(y)\right]\psi_{ij}(y) = E_{ij}\psi_{ij}(y) \qquad (5)$$

At first the Poisson equation has been solved (1) using the semi-classical approximation assuming zero charge density. Then the charge density profile $n(y)$ is determined from (3) and (4) by solving (5) using SILVCO's quantum effect module with Neuman boundary condition. The charge density profile is added to the Poisson equation, and (1) and (5) are solved iteratively until the Fermi level. After this processes, the CV characteristics can be found from the device electrostatics.

C. *Effect of interface states on CV profile*

In III-V devices, interface states are originated from native interface defects in oxide-semiconductor interface [12]. The interface states are located at or very close to the semiconductor/oxide interface with energy distributed along the bandgap of the semiconductor , designated as D_{it} ($cm^{-2}ev^{-1}$), Q_{it} (C cm^{-2}). Electron and holes get trapped in these states and act like generation/recombination center and threshold voltage shift [13] which is given by

$$\Delta V_T = -\frac{\Delta Q_{it}(\Phi_S)}{C_{ox}} \qquad (6)$$

where Φ_S is the surface potential, C_{ox}.is the capacitance in oxide layer. The position of the donor and acceptor like interface states with respect to charge neutrality level is shown in Fig. 2. This often expressed by an equivalent distribution with a charge neutrality level E_{CNL} above which the states are acceptor type and below which they are donor type. The energy levels marked by E_C, E_V, and E_F represent conduction, valance, and Fermi energies, respectively.

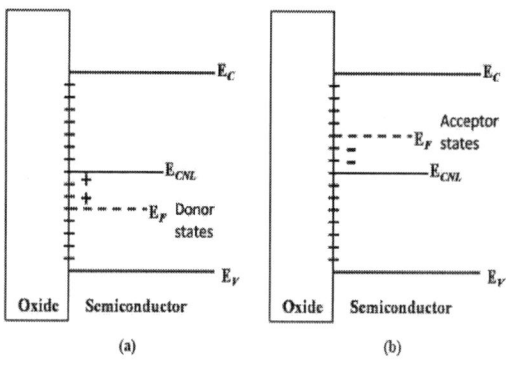

Fig. 2 Typical interface system consisting of (a) donor states and (b) acceptor states.

In HFCV measurement, HF AC is superimposed on a DC bias that is changed from accumulation to inversion. Interface trapped charges cannot respond to HF AC signal but, they can respond to the DC bias. As a result, the measured capacitance does not depend on the amount of interface traps present. But the threshold voltage is modified. In LFCV measurement, the interface traps respond to the LF AC signal along with DC bias. Thus the capacitance changes with the alteration of the threshold voltage is given by

$$C_g = \frac{\Delta Q_{total}}{\Delta V_G} \qquad (7)$$

where ΔQ_{total} is the change in total charge due to change in gate voltage ΔV_G. The total charge can be calculated by

$$Q_{total} = Q_{depletion} + Q_{inversion} + Q_{it} \qquad (8)$$

RESULTS AND DISCUSSIONS

Room temperature HFCV and LFCV characteristics are studied considering positive and negative interface charge densities at oxide semiconductor interface. It is seen from Figs. 3 (a) and (b) that the shift in threshold voltage is less significant both for positive and negative interface charge density of states in HF operation. With increasing positive interface charge density, the CV profiles are shifted from more negative gate voltage side to less negative gate voltage side. The opposite results are found with increasing the amount of negative interface charge density. In case of positive interface charge, the position of the CV profile is found to shift in opposite direction with respect to negative interface charge. The shift in CV profile with the amount of interface charge density, as well as, with its polarity, is occurred to oppose the band banding at oxide/semiconductor interface in order to have flat band condition. It is also found in Figs 3 (a) and (b) that the capacitance changes slightly in accumulation and depletion regions both for negative and positive interface charge densities. However, almost negligible change in capacitance is found in weak, moderate and strong inversion regions with both types of interface charge densities. Figs. 4 (a) and (b) show the LFCV characteristics obtained for different interface charge densities. It is evident that the shift in threshold voltage is more significant in LF operation and found in opposite direction for positive and negative interface charge densities. In contrast to HF operation, interface charges respond to low LF operation with DC bias. Consequently, the capacitance at the oxide/semiconductor interface is changed and, thereby the threshold voltage is also changed. To understand the shift in threshold voltage with the amount of interface charge density, as well as with its polarity, the threshold voltage is plotted as a function of interface charge density in Fig. 5. The maximum shift is found from 0 V to -0.47V for the variation of positive interface charge density and 0V to 0.35V for the negative interface charge density.

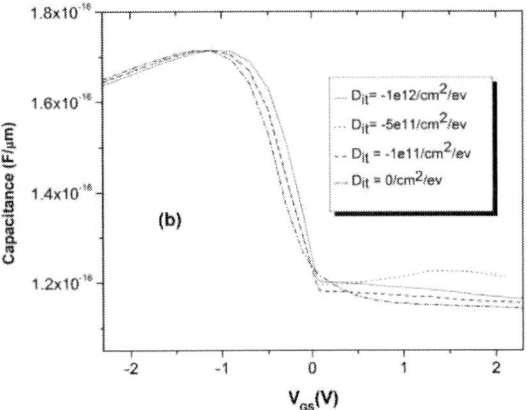

Fig. 3 Room temperature HFCV characteristics (a) positive interface charge density and (b) negative interface charge density.

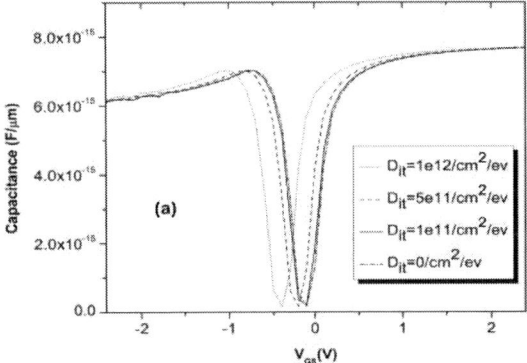

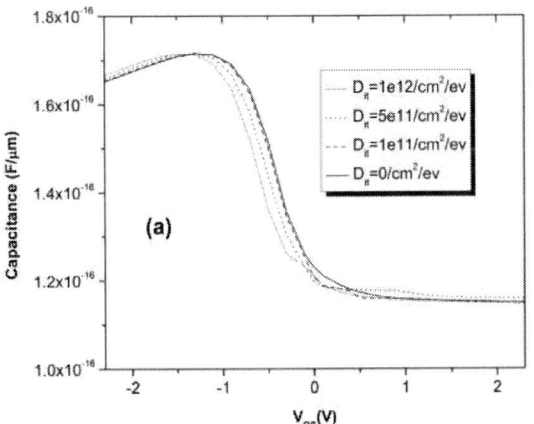

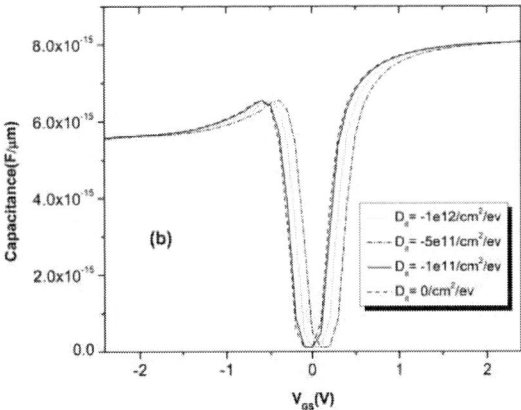

Fig. 4 Room temperature LFCV characteristics (a) positive interface charge density and (b) negative interface charge density.

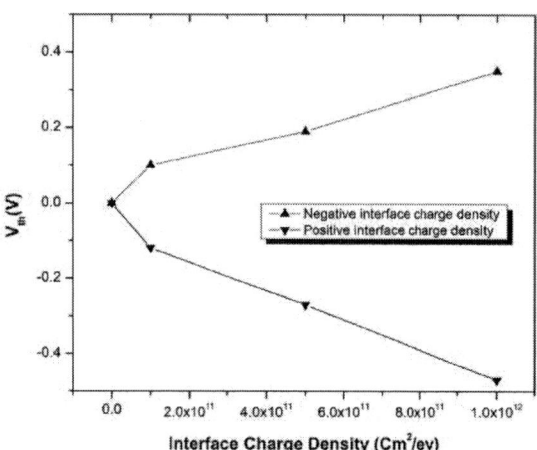

Fig. 5 Threshold voltage as a function of interface charge density

CONCLUSION

In this study, the HFCV and LFCV characteristics are simulated for InGaSb-based surface channel MOS structure incorporating the effect of interface states using Silvaco's ATLAS device simulation package. The study is carried out for different positive and negative interface states. The results obtained from the simulation indicate that, at LF operation, the effect of interface states on CV characteristics is more prominent than at HF operation. The position of the CV profile depends of the amount and polarity of the interface states. It is also found that the shift in threshold voltage have strong dependence on the positive and negative interface charge densities. Thus the operation of InGaSb-based surface channel MOSFET's in VLSI circuits to be modeled considering the effect of interface states and its polarity.

REFERENCES

[1] See,e.g,HeynsM,Tsai w,editors. MRS Bulletin entiteled:"Ultimate Scaling of CMOS logic devices with Ge and III-V materials:" July 2009.

[2] Annesh Nainani et al "InxGa1-xSb channel p-metal–oxide semiconductor filed effect transistors: Effect of strain and heterostructure design" *Journal of Applied Physics* 110,014503, (2011)..

[3] S. Mahapatra, P. B. Kumar, and M. A. Alam, "Investigation and modeling of interface and bulk trap generation during negative bias temperature instability of p-MOSFET", *IEEE Transactions on Electron Devices*, vol. 51,no. 9, pp.1371-1379, Sep. 2004.

[4] C. E. Blat, E. H. Nicollian, E. H. Poindexter, "Mechanism of negative-bias-temperature instability", *Applied Physics Letters* , vol. 69, no. 3, pp.1712-1720, Feb. 1991.

[5] J. F. Zhang and W. Eccleston, "Positive bias temperature instability in MOSFET", *IEEE Transactions on Electron Devices*, vol. 45, no.1, pp.116-124, Jan. 1998.

[6] J. A. Felix, M. R. Shaneyfelt, D. M. Fleetwood, T. L. Meisenheimer, J.R. Schwank, R. D. Schrimpf, P. E. Dodd, E. P. Gusev, and C. D. Emic, "Radiation-induced charge trapping in thin Al2O3 /SiO x N y/Si(100) gate dielectric stacks" , *IEEE Transactions on Nuclear Science*, vol. 50, no. 6,pp. 1910-1918, Dec. 2003.

[7] E. P. Gusev et al , "Advanced gate stacks with fully silicided (FUSI) gates and high-κ dielectrics: enhanced performance at reduced gate leakage" ,*IEEE International Electron Devices Meeting (IEDM) Technical Digest* ,pp. 79-82, Dec. 2004..

[8] V. Fomenko, E. P. Gusev, E. Borguet, Optical second harmonic gen-eration studies of ultrathin high- κ dielectric stacks , *Journal of Applied Physics*, vol. 97, article no. 083711, 2005.

[9] Ze Yuan, Nainani, A. ; Bennett, B.R. ; Boos, J.B. ; Ancona, M.G.;Saraswat, K.C. "Heterostructure design and demonstration of InGaSb channel III-V CMOS transistors". *International Semiconductor Device Research Symposium* ,pp.1-2,(2011).

[10] Md. Hasibul alam et al , "In$_x$Ga$_{1-x}$Sb MOSFET: Performance Analysis by Self Consistent CV Characterization and Direct Tunneling Gate Leakage Current" *IEEE International Conference on Electro / Information Technology*,pp.1-6,(2012).

[11] Stern, F., and W. E. Howard, "Properties of Semiconductor Surface Inversion Layers in the Electric Quantum Limit", *Phys. Rev.* , 163 (1967): 816

[12] J. Robertson: "Model of interface states at III-V oxide interfaces." *Appl. Phys .Lett.* **94**, 152104, 2009.

[13] F. Stern, "Self-consistent results for n-type Si inversion layers," *Phys. Rev. B*, vol. 5, pp. 4891– 4899, 1972.

Growth control of CuO nanowires on copper thin films: Toward the development of pn nanojunction arrays

A. A. El Mel,[1] M. Buffière,[2,3] N. Bouts,[4] E. Gautron,[4] C. Bittencourt,[1] P. Guttmann,[5] P. Y. Tessier,[4] S. Konstantinidis,[1] and R. Snyders[1,6]

[1]Chimie des Interactions Plasma Surface (ChIPS), CIRMAP, Research Institute for Materials Science and Engineering, University of Mons, 23 Place du Parc, B-7000 Mons, Belgium
[2]Department of Electrical Engineering, KU Leuven, KastteelparkArenberg 10, B-3001 Heverlee, Belgium
[3]IMEC, Kapeldreef 75, B-3001 Heverlee, Belgium
[4]Université de Nantes, CNRS, Institut des Matériaux Jean Rouxel, UMR 6502, 2 rue de la Houssinière B.P. 32229 - 44322 Nantes cedex 3 – France
[5]Helmholtz-Zentrum Berlin für Materialien und Energie GmbH, Institute for Soft Matter and Functional Materials, D-12489 Berlin, Germany
[6]Materia Nova Research Center, Avenue Nicolas Copernic 1, Parc Initialis, B-7000 Mons, Belgium

Abstract

The growth of single crystal CuO nanowires by thermal annealing of copper thin films deposited by DC magnetron sputtering is studied. We show that by tuning the morphology and structure of the sputtered copper thin films, the density, length, and diameter of the CuO nanowires can be controlled accurately.

Keywords: Nanowires, Nanotechnology, Photovoltaic

Introduction

The thermal annealing of copper thin films at ambient air has been proposed as an outstanding route for the growth of CuO nanowires [1]. Contrary to other synthesis strategies, thermal annealing is a simple, low-cost, and catalyst-free method. The growth of oxide nanowires can be explained as a result of the unbalanced outward and inward diffusion of metal and oxygen ions, respectively [2-4]. The preferential segregation of Cu ions at the top of the copper oxide grains leads to the formation of aligned CuO nanowires. Therefore, the synthesis conditions are expected to have a strong impact on the characteristics of the grown nanowires. The impact of the annealing temperature, oxygen partial pressure, and annealing time on the morphology of nanowires has been widely investigated using bulk copper specimens [1-5], whereas very little is known about the growth control of CuO nanowires using copper thin films.

In this work copper thin films of different morphologies and structures were used to investigate the growth of CuO nanowires. Magnetron sputtering was selected as a powerful and scalable technique allowing to growth copper thin films over a large surface area.

Experimental

The copper thin films used as a support for the growth of CuO nanowires by thermal annealing were deposited by magnetron sputtering on thermally oxidized silicon substrates. A copper target of 60 mm in diameter located at 131 mm from the substrate was used. The electrical power applied to the target was 200 W. The base pressure and the argon deposition pressure were 10^{-4} and 0.64 Pa, respectively. No substrate bias was applied in these conditions. For the growth of the nanowires, the films were annealed at ambient air at 500 °C for 4 hours. The heating rate was 8 °C/min, whereas the cooling rate was 2 °C/min.

Scanning electron microscopy (SEM) imaging was performed at 5 kV on a JEOL JSM 7600 F microscope. Transmission electron microscopy (TEM) imaging was performed on a Philips CM200 microscope (tungsten filament, 200 kV, Scherzer resolution: 0.4 nm). High-resolution TEM imaging and selected area electron diffraction (SAED) were performed on a Hitachi H9000NAR microscope (LaB$_6$ filament, 300 kV, Scherzer resolution: 0.18 nm).

Results and discussion

Fig. 1A shows a top-view SEM micrograph of the copper thin film used for the growth of the CuO nanowires. The film is formed of grains with a mean diameter of about 126 nm; the inter-grain distance is about 12 nm. When cutting the films for cross-sectional SEM observation (Fig.1B), we remarked that they are highly ductile and therefore cannot be cut properly. However, according to the structure zone model of metal thin films grown by physical vapor deposition, in such deposition conditions the films may exhibit a columnar morphology (Fig. 1C) [6]. After annealing in ambient air at 500 °C during 4 hours, highly aligned CuO nanowires were formed on the top surface of the films (Fig. 2A). The density of the wires (number of wires/μm^2) is ~ 3. The mean length and diameter are, respectively, ~ 1 μm and ~ 150 nm (Fig. 2B). Fig. 3A shows a TEM micrograph of a CuO nanowire on a carbon-coated copper grid. The high-resolution TEM micrograph (Fig. 3B), on which the lattice fringes can be seen, is a direct proof of the single crystal structure of the nanowire. This is further supported by the selected area electron diffraction (SAED) which also reveals the monoclinic structure of CuO (Fig. 3C).

References

[1] X. Jiang et al., *Nano Lett.*, 2, 1333, 2002.
[2] L. Yuan et al., *Acta Materialia*, 59, 2491, 2011.
[3] A. M. B. Gonçalves et al., *J. Appl. Phys.*, 106, 034303, 2009.
[4] Y. Yue et al., *Scripta Materialia*, 66, 81, 2012.
[5] K. Zhang et al. *Nanotechnology*, 21, 235602, 2010.
[6] I. Petrov et al. *J. Vac. Sci. Technol. A*, 21, 117, 2003.

Acknowledgment

This work was funded in part by the Directorate of Research in Wallonia, under the scope of the ERA-NET MATERA programme and by the COST Action MP0901, the European Commission (contracts RII3-CT 2004-506008 (IASFS). The French Community of Belgium is acknowledged through the "Cold Plasma" project. S. K. is a research associate of the National Funds for Scientific Research (FNRS, Belgium).

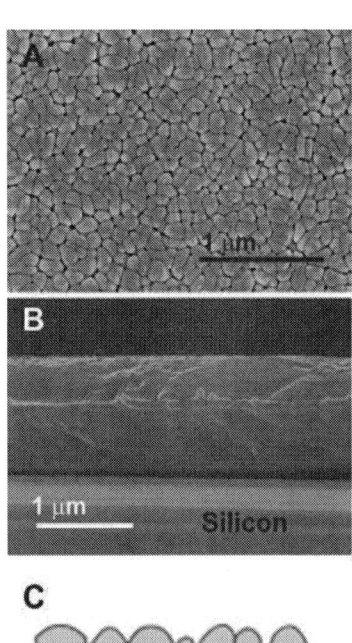

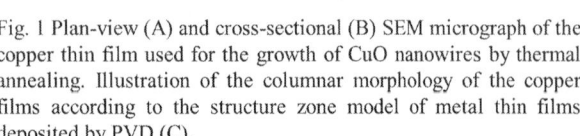

Fig. 1 Plan-view (A) and cross-sectional (B) SEM micrograph of the copper thin film used for the growth of CuO nanowires by thermal annealing. Illustration of the columnar morphology of the copper films according to the structure zone model of metal thin films deposited by PVD (C).

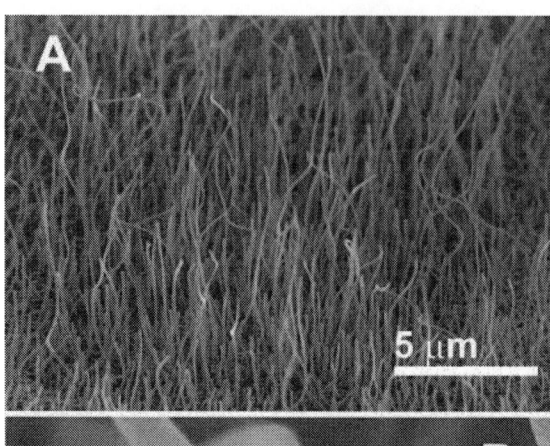

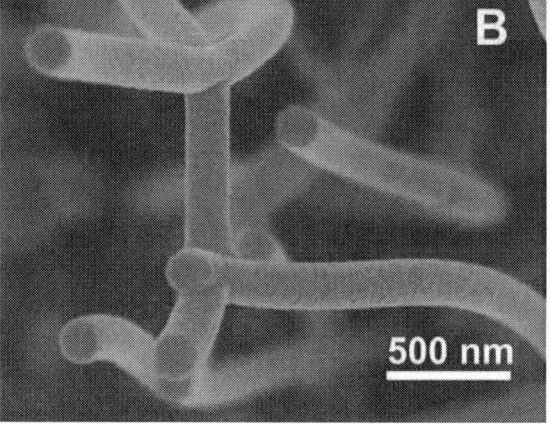

Fig. 2 Low (A) and high (B) magnification SEM micrograph of CuO nanowires grown by thermal annealing in air.

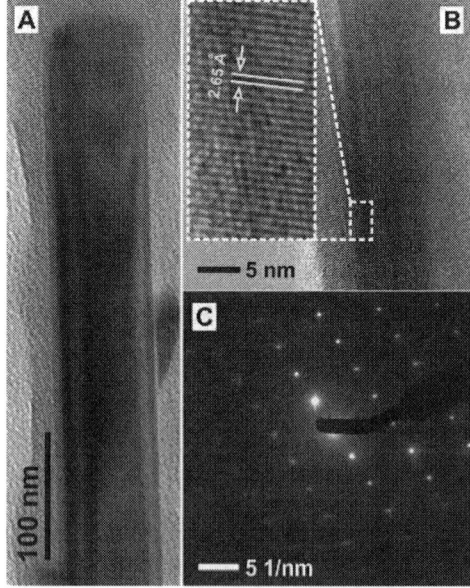

Fig. 3 Low magnification TEM (A) and high-resolution TEM (B) micrographs and the associated SAED pattern (C).

Growth, structure, and optical properties of GaSb quantum dot by LPE technique

F. Qiu, Y. Zhang, Y. F. Lv, J. H. Guo, G. J. Hu, Sun, H. Y. Deng, S. H. Hu* and N. Dai*

National Laboratory for Infrared Physics, Shanghai Institute of Technical Physics,
Chinese Academy of Sciences, Shanghai 200083, PR China

Q.D. Zhuang, M.Yin, A.Krier

Physics Department, Lancaster University, Lancaster LA1 4YB, United Kingdom

Z. Zhao

Hefei National Laboratory for Physical Sciences at the Microscale, University of Science
and Technology of China, Hefei 230026, PR China

* E-mail for Corresponding author: hush@mail.sitp.ac.cn (S. H. Hu)
ndai@mail.sitp.ac.cn (N. Dai)

Abstract

In this paper, we have grown self-assembled GaSb quantum dots (QDs) on GaAs (100) substrate by liquid phase epitaxy (LPE) technique. The surface morphology, density and size distribution of GaSb QDs are investigated by High-Resolution Scanning Electron Microscope and Atomic Force Microscopy, respectively. Cross-sectional transmission electron microscopy is employed to obtain a cross sectional image of single quantum dot and to present the composition of QDs by focused energy dispersive X-ray (EDX). Feature of QDs of room-temperature photoluminescence (PL) spectroscopy is obvious, and the peak of the QDs at ~1.1eV is well separated from wetting layer (WL) at ~ 1.34 eV.

Keywords: GaSb QDs, X-TEM, PL spectrum and liquid phase epitaxy

Introduction

In recent years, great interests have been aroused by GaSb/GaAs QDs structures due to its type-II band alignment. The spatial separation of electrons and holes of this system is physically different from behavior comparing to type- I heterostructure, Holes in this band structure have strong confinement, while electrons are very weakly localized by coulomb binding. The GaSb/GaAs QDs, thereby, are promising for both fundamental physical effects and novel devices.

Although the majority of GaSb/GaAs QDs structures have been grown by MBE and MOCVD, to our knowledge there are no reports of GaSb QDs grown on GaAs substrate using liquid phase epitaxy (LPE) technique. The epitaxial layer with high quality, however, can be achieved by the near-equilibrium LPE method. In this paper, we have successfully grown the GaSb quantum dots on semi-insulating (100) GaAs substrate for the first time by rapid sliding LPE technique to decrease the contact time of melt-substrate in the magnitude of milliseconds. The surface density, shapes, and dimensions of QDs depend on the growth conditions (growth temperature, degree of supercooling, ramp cooling rate, and melt–substrate contact time) have been characterized by HRSEM and AFM. The cross-sectional of single GaSb quantum dot is presented by cross-sectional transmission electron microscopy (X-TEM), and composition of QDs is confirmed by EDX of TEM. The emission band of QDs is observed in room-temperature PL spectroscopy by 808 nm line of a diode laser.

II. Experiments

Sample labeled 61# of unencapsulated GaSb QDs and sample labeled 58# of GaAs-encapsulated were grown for structure and PL investigation, respectively. The sample labeled 61# is grown by rapid sliding LPE, which is combined with thin layer of melt solution and contact time of a few milliseconds with rapid slider driven by Copley linear motor through LabVIEW program. The growth process was initially grown GaAs buffer at 580°C with contact times of 180s, then, short contacted with melt solution of GaSb QDs of 4 milliseconds on semi-insulating GaAs (100) substrate with ramp cooling rate of 0.2°C/min and 20 degrees of supercooling by super cooling version. This sample is to confirm the topography, size distribution and cross sectional of QDs. PL sample labeled 58# is completed by buffer layer and QDs layer under the same growth condition as 61#, and then is capped with GaAs layer at 510°C with ramp cooling rate of 0.2°C/min and 20 degrees of supercooling for 30 seconds, and the thickness of cap layer is about few tens of nanometers.

III. Results and discussion

Figure.1 shows the clear surface morphology of GaSb QDs by HRSEM. The round shape confirm the Stranski–Krastanov (SK) growth mode with the lattice mismatch of GaSb/GaAs up to 7.8%.This large misfit is benefit for the self-organized GaSb island formation. The width or diameter distribution of QDs from Fig.2 is allowed to fit them with Gauss function to define their width and central position, and present narrow width distribution. The width of majority QDs range from 30 to 40 nanometers.

Figure.3 gives the obvious topography of GaSb QDs with density up to $2.7 \times 10^9 \text{cm}^{-2}$ by AFM. The QDs distribution is not order but statistic uniformity. The statistical distribution of height is rather broad, and mainly ranges from 8 to 12 nanometers in Fig.4 and could be fitted with Gaussian fit. The growth dynamics and Raman analysis of GaSb QDs on GaAs matrix by LPE have already been carefully researched and been reported by our group [1].

Figure.5 presents the cross sectional image of single GaSb QDs on the surface with 9 nm in height and 38 nm in base width by X-TEM. But these sizes depend on the position of single QDs when it comes to be naturally cleaved on (110) surface. From the two sides of the individual dot we can see the crystallization orientation of GaSb QDs in accordance with the GaAs matrix, but the interface under the dot is misty. This phenomenon indicates that these interface exist a gradient distribution of Sb element. The composition of the single dot is not easy to determine, but here that is first identified through focusing a region contained the dot with the atomic ratio of Sb of 7.5%.

Figure.6 plots the emission wavelength dependent of the PL intensity at room temperature.PL spectrum is obtained for excitation power density of 2 W/cm². The band of the QDs at ~1.1eV is well separated from wetting layer (WL) at ~ 1.34 eV. The position of emission spectrum of GaSb QDs is consistent with that of sample grown by MBE[2] or MOCVD[3]. The full width at half maximum (FWHM) of QDs peak and wetting layer (WL) peak is 42.4 meV and 28 meV, respectively. The interesting thing of our sample made by LPE, however, is the strong PL signal of QDs at room temperature in comparison with other conventional semiconductor growth techniques, such as MBE or MOCVD.

IV. Conclusion

The medium dots density of $2.7 \times 10^{9} cm^{-2}$ was obtained by LPE technique. Round quantum dot indicate that the growth process of strain-induced structure is S-K growth mode. The width and height distribution are determined by HRSEM, AFM, respectively, and size distribution of statistical dots was fitted well by Gauss function. The cross sectional image of surface single QDs is showed by XTEM and the composition information of GaSb QDs on GaAs matrix is first presented by EDX in this paper. At room temperature we give the rather strong PL band of GaSb QDs with narrow emission spectrum of 42.4 meV. Narrow band of QDs indicate that range of the size statistical distribution is narrow, which would be confirmed by Gaussian fit of size statistical distribution.

Acknowledgment

The authors would like to thank the support of the Innovation Fund Program of Shanghai (No. 11DZ1140500), the National Science Foundation in China (No. 61274139), the Innovation Program of Shanghai Institute of Technical Physics of the Chinese Academy of Sciences (No.Q-ZY-78), and the Shanghai City of Committee for Science and Technology (No.12ZR1435500).

References

[1] F. Qiu, *et al.*, "Growth and Raman spectra of GaSb quantum dots in GaAs matrices by liquid phase epitaxy," *Chinese Optics Letters*, in press.

[2] F. Hatami, *et al.*, "Radiative Recombination in Type-Ii Gasb/Gaas Quantum Dots," *Applied Physics Letters*, vol. 67, pp. 656-658, Jul 31 1995.

[3] L. Muller-Kirsch, *et al.*, "Temporal evolution of GaSb/GaAs quantum dot formation," *Applied Physics Letters*, vol. 79, pp. 1027-1029, Aug 13 2001.

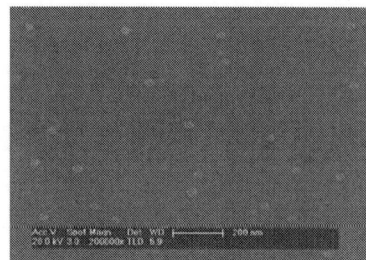

Fig.1. HRSEM profile of GaSb Quantum dots

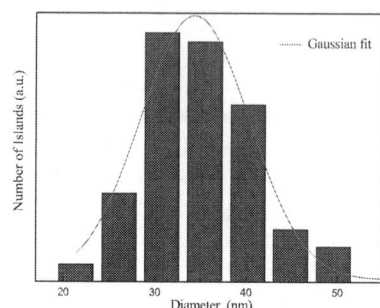

Fig.2. Diameter distribution of GaSb Quantum dots

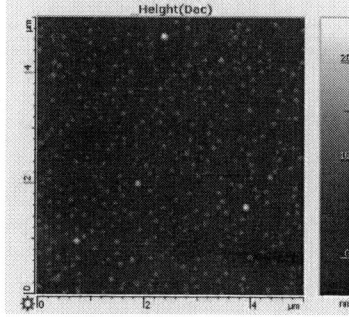

Fig.3. AFM profile of GaSb Quantum dots

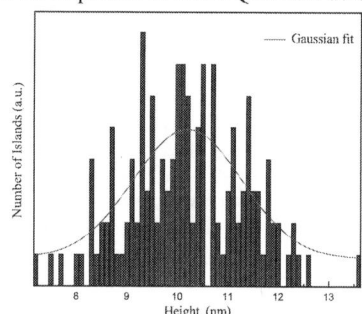

Fig.4. Height distribution of GaSb Quantum dots

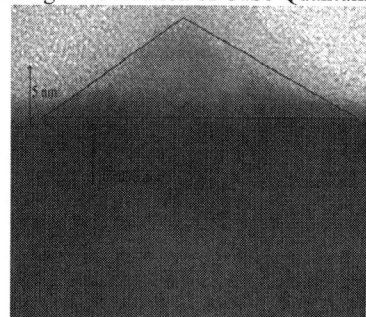

Fig.5. XTEM profile of GaSb Quantum dots

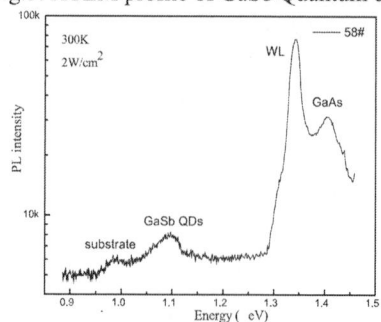

Fig.6. PL spectrum at RT from GaSb Quantum dots

Design and growth optimization by dual ion beam sputtering of ZnO-based high-efficiency multiple quantum well green light emitting diode

Sushil Kumar Pandey, Saurabh Kumar Pandey, and Shaibal Mukherjee

Hybrid Nanodevice Research Group (HNRG), Discipline of Electrical Engineering, Indian Institute of Technology, Indore-453441, India

Phone: +91-732-4240715, Fax: +91-731-2361482, +1-484-2317876

Email: shaibal@iiti.ac.in

Abstract

This paper presents an in-depth analysis of $Cd_{0.4}Zn_{0.6}O/ZnO$ multiple quantum well light emitting diode (LED) using commercial simulation software and experimentally optimized growth conditions of n-type ZnO on Si (001) substrate by dual ion beam sputtering deposition (DIBSD) system. Theoretical study reveals an internal quantum efficiency ~93.5% is achieved at room temperature from the device, emitting at 510 nm with a turn-on voltage of 3 V. The effect of substrate temperature and gas composition on ZnO growth has been investigated. Growth parameters optimization is performed using structural, electrical, and optical characterizations. ZnO grown at 600 °C shows a strong ZnO (002) X-ray diffraction (XRD) peak at 34.6°, indicating the realization of high-quality c-axis orientation of ZnO layer. Four probe Hall measurements demonstrate achievements of a maximum carrier mobility of ~500 $cm^2/V.s$ with a low electrical resistivity of ~10^{-3} Ω.cm and a carrier concentration of ~10^{18} cm^{-3} from the grown ZnO samples at room temperature. Results from atomic force microscope (AFM) measurements depict that RMS roughness of ZnO (10 μm × 10 μm) reduces from 44 Å to 10 Å when the substrate temperature is increased from 100 °C to 400 °C and then increased to 22 Å as the substrate temperature is increased to 600 °C. Photoluminescence (PL) studies conducted at room temperature describe a strong band-edge emission at 380 nm from ZnO samples. Prominent PL shoulder peaks are observed at ~485 nm and 618 nm from ZnO grown at 400 °C.

Keywords: DIBSD, green LED, multiple quantum well, XRD, ZnO.

Introduction

These instructions give you basic guidelines for p   Zinc Oxide (ZnO) is a II-VI semiconductor material with wide direct energy band gap of 3.37 eV at room temperature [1]. The binding energy of ZnO material is around 60 meV for free exciton which is larger than the thermal energy at room temperature, so efficient excitonic transitions are possible above the room temperature [1]. In comparison to other semiconductor materials such as GaN, SiC, and AlN, ZnO offers several fundamental advantages such as high temperature operation, high breakdown strength and transparency in the visible region [1-5]. Recent years have witnessed increased interest in ZnO for various applications such as light emitting diodes (LED), laser diodes (LD), dye sensitized solar cells (DSSC), transparent thin film transistors (TTFT) [6, 7]. As-grown ZnO always shows n-type conductivity [8-10]. Usually n-type conductivity in unintentionally doped ZnO is due to Zinc interstitials (Zn_i) and oxygen vacancies (O_v) in the films [8, 9]. By optical and electrical excitation, ZnO predominantly emits light in the UV region centred at around 380 nm due to near band edge excitonic transitions. Visible emission has also been observed with UV emission due to various intrinsic and extrinsic defects [5, 11-14]. Therefore it is essential to know the relation between the native point defects with optical and electrical properties of ZnO for its applications in high quality electronic and optoelectronic devices. We have grown the ZnO thin film on p-Si (100) substrate because the study of ZnO film growth on Si is of high significance for large-scale optoelectronic device fabrication in the future.

In order to realize ZnO-based LEDs, first it is necessary to perform an indepth theoretical analysis to optimize the LED structure and performance before the actual fabrication of the device. The optical band gap of ZnO can be tuned by alloying with CdO (band gap 2.5 eV at room temperature) to have luminescence from ultraviolet to visible light spectra. A number of researches are being carried out to optimize the performance of CdZnO/ZnO multiple quantum-well LEDs. In this paper, an attempt has been made to optimize growth temperature, gas composition for ZnO thin films grown on Si (100) substrate by DIBSD system for high electrical, optical and structural performance. It is also mentioned that the LED active region is formed by three pairs of $Cd_{0.4}Zn_{0.6}O/ZnO$ quantum-well/barrier regions. In the following section a detailed description of $Cd_{0.4}Zn_{0.6}O/ZnO$ multiple quantum well LED is given at room temperature.

Device structure

The green LED active region is formed by three pairs of $Cd_{0.4}Zn_{0.6}O/ZnO$ quantum-well/barrier regions. The cross section of the device structure, designed for a 516 nm wavelength emission, is shown in Fig. 1. The device is grown on ZnO substrate. It is assumed that the layer stack is grown along the (0001) wurtzite direction. Initially doping concentration in p-ZnO (cladding layer), p-MgZnO (EBL), n-MgZnO [(hole barrier layer (HBL)], n-ZnO (cladding layer) and n-ZnO substrate is 1×10^{18} cm^{-3}. Barrier layer ZnO is undoped. Well layers have $Cd_{0.4}Zn_{0.6}O$ material with doping concentration 1×10^{17} cm^{-3}. EBL and HBL have material p-$Mg_{0.1}Zn_{0.9}O$ and n-$Mg_{0.1}Zn_{0.9}O$ respectively. All variations have been done considering this structure as base.

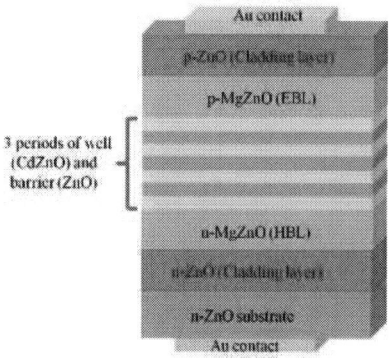

Fig. 1 Schematic cross section of the considered quantum-well LED structure.

Results and Discussion

Section A illustrates the effects of thickness variation of the barrier layers of active region on EL intensity and current-voltage characteristics. Section B illustrates the effects on EL intensity of doping and Mg composition in HBL. Section C describes the design methodology for the electron barrier layers (EBL). Section D describes the study of photoluminescence (PL), electrical properties, surface morphology, and crystalline properties of ZnO thin films, with growth temperature.

A. Barrier thickness variation

Fig. 2 shows luminous power variation anode voltage characteristic for different barrier thicknesses. The biasing voltage range is 0-5 V. The luminous power curve shows a clear rectifying behavior and can be linked to the p-n junction with a turn-on voltage around 3 V. Thicker barrier layer produce high series resistance of device. So at 5V biasing voltage, current will be less for thicker barrier layers. Thus the luminous power is increasing with decreasing barrier layers thickness. Another reason is, reducing the barrier thickness to a level where the wells are coupled, enhances the hole transport through the barrier layers and increases hole recombination with injected electrons, therefore reducing the electron leakage at bias voltage and improving the luminous power [15].

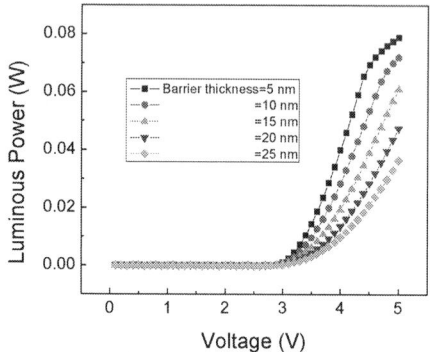

Fig. 2 LED Luminous Power variation with anode at different barrier thicknesses.

B. HBL

Introducing HBL in the device structure produces conduction band discontinuity, which may introduce a barrier for electron injection in the active region. The Mg molar fraction in HBL should be chosen to produce just enough valance band discontinuity to block holes while not preventing electron injection. Fig. 3 shows that EL intensity is almost same for Mg molar fraction 10% and 12%, but it is decreasing rapidly for Mg molar fraction 14% and 16% due to blocking of electron in the active region. From fig. 3 it is also clear that the EL peak intensity is centered around 510 nm wavelength.

Fig. 4 shows that EL intensity increases with increasing the doping in the HBL ($Mg_xZn_{1-x}O$) due to the higher doping makes more electrons available for recombining with holes. EL intensity is almost constant for doping concentration 10^{18} cm^{-3} and 10^{19} cm^{-3} possibly due to the fact that electron quasi-Fermi level is almost constant for these doping concentrations.

A. EBL

Fig. 5 shows the calculated IQE as a function of the $Mg_xZn_{1-x}O$ EBL thickness variation for an Mg molar fraction x=10%. For very thin EBL, the electron blocking is hampered by tunneling of electron. When the EBL thickness is effective enough to reduce tunneling, i.e., above 50 nm, the IQE increases with increasing EBL thickness. It is observed that thicknesses that are larger than 150 nm provide marginal reduction in IQE. Previously it is reported that an EBL improves the IQE only for operation at low current density, while for high injection levels, its role would be less noticeable [16].

Fig. 6 shows EL intensity verses wavelength characteristics for different Mg molar fraction in $Mg_xZn_{1-x}O$ EBL at 4V biasing. The EL intensity is increased when Mg molar fraction increased from 10% to 12% because increasing the Mg molar fraction increases barrier for electron to cross EBL so, electrons are confined in well-barrier region resulting in more radiative recombination in the well. But at higher Mg molar fraction (as 14%, 16%), the valance band offset effectively limits the supply of holes from the p-contact [17, 18] so EL intensity peak drop very rapidly.

978-1-4673-4840-9/13 $31.00 © 2013 IEEE

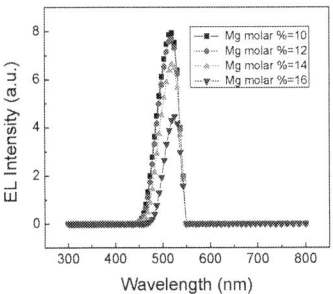

Fig. 3 LED EL intensity variation with Wavelength for different Mg molar fraction in (n-Mg$_x$Zn$_{1-x}$O) HBL.

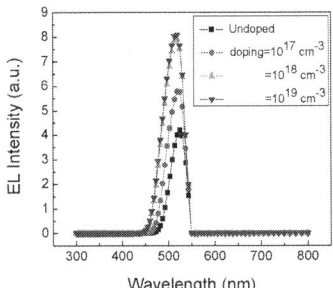

Fig. 4 LED EL intensity variation with Wavelength for different doping concentration in (n-Mg$_x$Zn$_{1-x}$O) HBL.

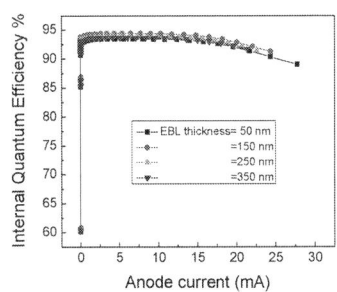

Fig. 5 LED IQE variation with anode current different (p-Mg$_x$Zn$_{1-x}$O) EBL thickness.

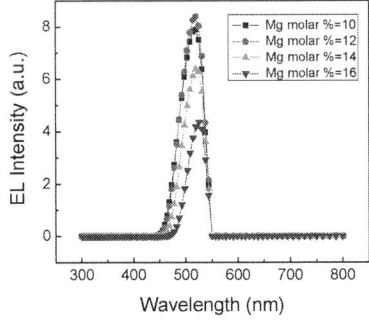

Fig. 6 LED EL intensity variation with wavelength for different Mg molar fraction in (p-Mg$_x$Zn$_{1-x}$O) EBL at 4V bias.

B. ZnO thin film characterization

Fig. 7 shows an XRD spectra of the (002) plane for ZnO thin films deposited on p-type Si(100) substrate as a function of growth temperature with constant RF power 40 W and O$_2$/(O$_2$+Ar)% of 66.667%. The crystal structure of ZnO films deposited at substrate temperatures 600 °C was identified to be *c*-axis oriented. The position of (002) peak was at 34.60° for the film deposited at 600 °C substrate temperature which was closer to the angular peak position of ZnO powder at 34.42° [19]..

Fig. 8 shows room-temperature PL spectra of ZnO thin film grown at substrate temperature 400 °C at O$_2$/(O$_2$+Ar)% of 66.67% with RF power 40 W .The optical properties of this ZnO film were analyzed by photoluminescence measurement set-up affixed with a 20 mW continuous wave He-Cd laser having an excitation wavelength of 325 nm. Excitonic near band-edge emission (NBE) was observed at ¯380 nm, corresponding to band gap of ZnO (¯3.3 eV) at room temperature for ZnO thin film grown at substrate temperature of 400 °C. Apart from the excitonic emission in the ultraviolet region, a broad PL intensity in visible region due to the deep level emission (DLE) was also observed. The DLE peak was observed to be centered on spectral wavelengths of 485 nm (green spectrum) and 618 nm (orange spectrum). Visible photo-emission from ZnO originates from defects present in the film [20-22].

Fig. 9 shows surface morphology of ZnO films, as characterized by AFM measurements. Root-mean-square (RMS) roughness of ZnO (10 µm × 10 µm) was observed decreasing constantly from 44 Å at 100 °C to 10 Å at 400 °C, and then increased continuously to 22 Å when growth temperature was further enhanced to 600 °C.

Four probe Hall measurements considering Van der Pauw geometry were carried out with a magnetic field of 0.50 Tesla at room temperature. ZnO film, grown at 100 °C, was found to have the highest carrier mobility of ~530 cm^2/V.sec at room temperature with resistivity of ~10^{-3} ohm-cm and electron concentration of ~10^{18} cm^{-3}.

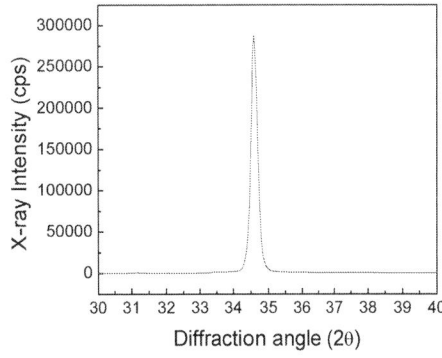

Fig. 7 XRD spectra of ZnO thin film grown at 600 °C with O$_2$/(O$_2$+Ar)% of 66.67% and 40 W RF power.

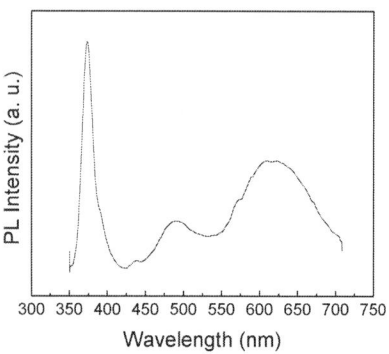

Fig. 8 Room-temperature PL spectra of ZnO film grown at substrate temperature 400 °C at $O_2/(O_2+Ar)$% of 66.67% with RF power 40 W.

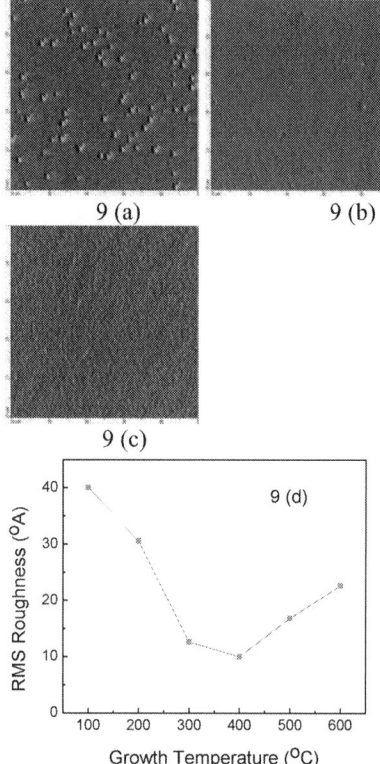

9 (a) 9 (b)

9 (c)

9 (d)

Fig. 9 AFM images of ZnO films deposited at (a) 100 °C (b) 300 °C (c) 500 °C (d) Surface RMS roughness of ZnO films deposited at substrate temperature range of 100-600 °C, with $O_2 /(O_2+Ar)$% of 66.67% and 40 W RF power.

Conclusion

In conclusion, we have theoretically demonstrated near band edge green electroluminescence around 510 nm with an IQE value of around 93.5% from the ZnO material system based multiple quantum well LED. Factors largely affecting the device internal quantum efficiency and EL intensity; including doping concentration, composition, thickness of well and barrier layers are numerically examined in detail. Also ZnO thin film deposition using DIBSD system was characterized and, AFM, XRD, Hall measurement and PL of the same have been reported.

Acknowledgment

This work is partially supported by Department of Science and Technology (DST) Fast Track Scheme for Young Scientist No. SR/FTP/ETA-101/2010. This work is also supported by DST Science and Engineering Research Board (SERB) project number SR/S3/EECE/0142/2011. We are also grateful to the Atomic Force Microscopy (AFM) Facility equipped at Sophisticated Instrument Centre (SIC), IIT Indore.

References

[1] D. C. Look, *Mater. Sci. Eng.*, B 80, 383 2001
[2] D. C. Reynolds, D. C. Look, and B. Jogai, *Solid State Commun.* 99, 873 1996.
[3] S. J. Pearton, D. P. Norton, K. Ip, Y. W. Heo, and T. Steiner, *Prog. Mater. Sci.* 50, 293 2005.
[4] D. C. Look, G. M. Renlund, R. H. Burgeber II, and J. R. Sizelove, *Appl. Phys. Lett.* 85, 5269 2004.
[5] T. Makino, Y. Segawa, A. Tsukazaki, A. Ohtomo, and M. Kawasaki, *Appl. Phys. Lett.* 87, 022101 2005.
[6] M. Chen, Z. L. Pei, X. Wang, C. Sun, L. S. Wang, *J. Vac. Sci. Technol. A.* 19, 963 2001.
[7] R. G. Gordon, MRS Bull. 25, 52 2000.
[8] Pearton S J, Abernathy C R, Overberg M E, Thaler G T and Norton D P, *J. Appl. Phys.* 93 (2003) 1.
[9] D. C. Look, D. C. Reynolds, J. R. Sizelove, R. L. Jones, C. W. Litton, G. Cantwell, and W. C. Harsch, *Solid State Commun.* 105 (1998) 39.
[10] C. G. Van de Walle, Phys. Rev. Lett. 85 (2000) 1012
[11] E. G. Bylander, *J. Appl. Phys.* 49, 1188 ~1978.
[12] K. Vanheusden, C. H. Seager, W. L. Warren, D. R. Tallant, and J. A. Voiget, *Appl. Phys. Lett.* 68, 403 ~1996.
[13] M. Liu, A. H. Kitai and P. Mascher, *J. Lumin.* 54 (1992) 3.
[14] B. Lin, Z. Fu, and Y. Zia, *Appl. Phys. Lett.* 79, 943 ~2001.
[15] J. Xie, X. Ni, Q. Fan, R. Shimada, Ü. Özgür, and H. Morkoç, *Appl. Phys. Lett.* 93, 121107 ı 2008.
[16] S.-H. Han, D.-Y. Lee, S-J. Lee, C.-Y. Cho, M.-K. Kwon, S. P. Lee, D. Y. Noh, D.-J. Kim, Y. C. Kim, and S.-J. Park, "Effect of electron blocking layer on efficiency droop in InGaN/GaN multiple quantum well light-emitting diodes," *Appl. Phys. Lett.*, vol. 94, no. 23, p. 231 123, Jun. 2009.
[17] M. Schubert, M. H. Kim, J. K. Kim, and E. F. Schubert, Uncovering the LED's darkest secret. London, U.K.: Compound Semiconductor, 2008, pp. 19–21.
[18] E. Bellotti, N. Sucena Almeida, A. Moldawer, T. D. Moustakas, S. Chiaria, F. Bertazzi, E. Furno, M. Goano, and G. Ghione, "Physics-based design of III–nitride and ZnO LEDs: From material properties to device optimization," in Proc. 17th HETECH, Venezia, Italy, Nov. 2008, pp. 21–26.
[19] American Standard for Testing of Materials ~ASTM 36-1451.
[20] J. P. Zhang,L. D. Zhang, L. Q. Zhu, Y. Zhang, M. Liu, and X. J. Wang, *J. of Appl. Phys.* 102, 114903 2007.
[21] Vanheusden K, Seager C H, Warren W L, Trallant D R, Caruso J, Hampden-Smith M J and Kodas T T, *J. Lumin.* 75 11 1997.
[22] Sekiguchi T, Ohashi N and Terada Y, Japan. *J. Appl. Phys.* 36 L289 1997.

Accurate Numerical Model for Surface Scattering, Grain Boundary Scattering, and Anomalous Skin Effect of Copper Wires

Elhameh Abbaspour, Reza Sarvari, Alireza Akbarzadeh, Melika Rostami
Sharif University of Technology
Tehran, Iran
abbaspourelhame@gmail.com

Abstract

In this paper we have studied both DC size effect and anomalous skin effect caused by surface and grain boundary scattering on the resistivity of Cu thin films by a Monte Carlo method. Contribution of each scattering mechanism and the interaction between them are analyzed separately. A simple and fast numerical recursive method is also introduced to guess the structure of electric field and distribution of current inside the thin film to evaluate the surface resistance instead of complicated analytical formulas.

Introduction

Shrinking the dimension of a metal film to less than the mean free path of electrons will cause a huge increase in the resistivity that will affect performance of integrated circuit negatively. According to the last prediction of Roadmap international technology, metal thickness will be 12 nm in 2026, which is less than electron mean free path at room temperature [1]. In order to design future interconnects efficiently we have to investigate the behavior of electrons inside thin films of these dimensions. For preparing suitable small conductors, the relative contribution of each scattering mechanism should be investigated [2]. When the main factor of resistivity increase is determined, then we can focus on minimizing it more than the other factors [3]. Studying the interaction between different scattering mechanisms is also helpful to figure out when we can simply neglect their correlation and when we need carefully consider it. In this paper we have done a thorough studying of both DC and AC size effects at very high frequencies and nanometer thin films at room temperature.

Methodology and Simulation

A. Resistivity Calculation

Transition of N_1 electrons inside the thin film and N_2 collisions to phonons of network for each electron are going to be studied. Under DC applied electric field E_0, electron's displacement caused by drift velocity will be,

$$x = \frac{e \tau^2 E_0}{2m} + v_f \tau \quad (1)$$

In which τ is the average time between two collisions to the network. The average velocity of electrons movement inside the thin film will be,

$$v = \frac{\sum_{i=1}^{N_1} \sum_{j=1}^{N_2} x_{ij}}{\sum_{i=1}^{N_1} \sum_{j=1}^{N_2} \tau_{ij}} \quad (2)$$

Now we can use $J = n_e e v$ or $J = \sigma E_0$ to obtain conductivity. We have used random generators to create time between two collisions and initial angle of movement of electrons after colliding to the network and the initial position of each electron inside the film.

B. Simulated Structure in AC Case

In order to meet quasi-TEM assumption for simulating the thin film, we should have the below condition:

$$D \omega \ll \frac{1}{\sqrt{\mu\varepsilon}} \quad (3)$$

Where D is the thickness of the conductor, ω the signal frequency, μ the magnetic permeability and ε the dielectric permittivity [4].

In AC case we have used a numerical recursive method to calculate electric field and current distribution in both normal and anomalous skin effect. The steps are as follow,

Step1: considering an arbitrary value for electrical field.
Step2: calculating current density according to electric field.
Step3: If the solution has converged, quit.
Step4: using Maxwell equations to calculate electric field by using current density from part2 and go to step2.

Results and Discussions

In order to visualize how reducing the size of a thin film increases size effect, trajectory of an electron with 10 collisions to phonons is shown for two thin films with thickness of 30 and 60 nm in Fig. 1 (a) and (b) respectively. 16 and 2 collisions occur for $D = 20$ and 100 nm respectively which obviously shows the impact of surface scattering by shrinking the dimensions.

Fig. 2 shows normalized resistivity versus thickness of the thin film obtained by Monte Carlo simulations and analytical formulas which totally fit together, and size effect on interaction between different scattering mechanisms. It can be seen that we have more interaction between surface and phonon scattering rather than grain boundary and phonon scattering by reducing dimensions. Fig. 3 illustrates interaction between surface and grain boundary scattering which increases by decreasing dimensions. This figure also shows that grain boundary scattering serves higher contribution to resistivity than surface scattering for the same value of R and p (based on modeled introduced in [5]).

For studying anomalous skin effect we have assumed the normal skin effect solution for electric field as initial value for the method presented in part B. By this assumption after 3 iterations the solution converges. At room temperature transition frequency to anomalous skin effect at which skin depth is five times of electron mean free path is $f'_c = 150\text{GHz}$ [6]. Hence, for a thin film with $D = 40\text{nm}$ and $f = 5\text{THz}$ we will have anomalous skin effect. Fig. 4 shows this fact. Because of departure of Ohm's law there is no agreement between real part of current density and σE_r. But the area under both curves are the same, which shows that the total amount of current flow through any cross section of the film is the same as normal case, while the distribution of current changes. Fig. 5 depicts magnitude of electric field inside this film. Solid and dash lines show magnitude of electric field calculated by simulation and Maxwell equations, respectively. There are two main differences between normal and anomalous states according to the magnitude of electric field. First, there is a local maximum at the center of film in anomalous case while normal case has a minimum. Second, anomalous field dose not decrease as much as normal field by getting farther from external surfaces of the film [7].

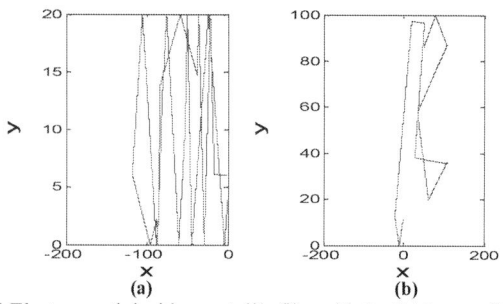

Fig. 1 Electron path inside a metallic film with the thickness of (a) 20 nm and (b) 100 nm while collision to the surfaces occurs

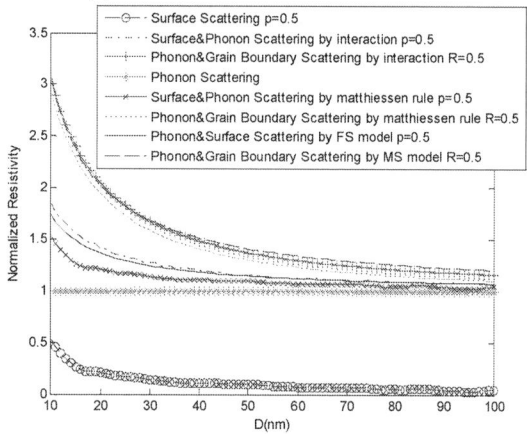

Fig. 2 Interaction between phonon and surface scattering and phonon and grain boundary scattering at room temperature

Conclusion

We developed an accurate and efficient Monte Carlo method to study DC size effect on electrical properties of Cu thin films. Contribution of scattering mechanisms was studied independently and by considering correlation between them. We introduced a recursive Monte Carlo method to study anomalous skin effect suitable for thin films operating at nanometer dimensions and THz frequencies.

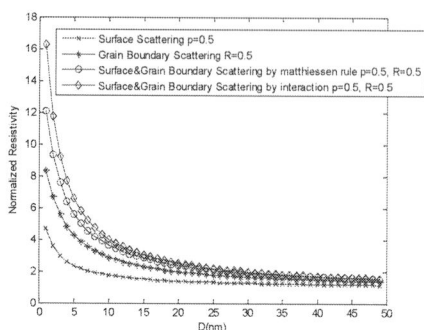

Fig. 3 Resistivity contribution of surface and grain boundary scatterings

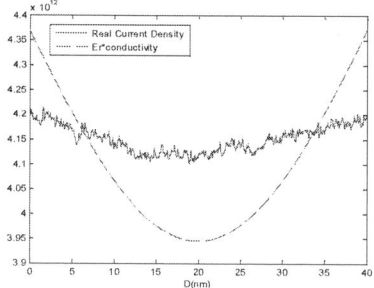

Fig. 4 Real part of current distribution in a film with D = 40nm, f=5THz and p=1

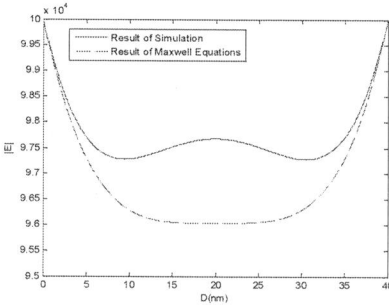

Fig. 5 Magnitude of electric field in film with D = 40nm, $f = 5\text{THz}$ and $p = 1$

References

[1] "International technology Roadmap for Semiconductors, 2012 Edition, Executive Summary," http://public.itrs.net/.

[2] Tik Sun, " Classical Size Effect in Copper Thin Films: Impact of Surface and Grain Boundary Scattering on Resistivity", PhD. Thesis, B.A.New College of Florida, 2005. Supervisor: Prof. Kevin R. Coffey

[3] Hao Zhang, LilinTian, Wenjian Yu, and Zhiping Yu, "Monte Carlo Simulation of Cu-Resistivity Considering Size-Effects", *proceeding of ASIC*, 2007

[4] Francesc Moll, Miquel Roca, "Interconnection Noise in VLSI Circuits", Boston: Kluwer Academic Publishers, pp.20–23, 2004.

[5] A. F. Mayadas and M. Shatzkes, "Electrical Resistivity Model for Polycrystalline Films", *Phys Rev B*, Vol 1, 1970.

[6] Reza Sarvari, "Impact of size effects and anomalous skin effect on metallic wires as GSI interconnects", PhD. Thesis, Georgia Institute of Technology, Aug 2008. Supervisor: Prof. Meindl.

[7] A. F. Mayadas, G. R. Henry, and M. Shatzkes, "Surface Impedance of Thin Metal Films under Anomalous Skin Effect Conditions", *J. Appl. Phys. Lett.*, Vol. 20, pp. 417–419, 1972

Conduction Mechanism of Single-Electron Transistors Fabricated by Field-Emission-Induced Electromigration

Shunsuke Akimoto, Ryutaro Suda, Mitsuki Ito, Masazumi Ando, and Jun-ichi Shirakashi[*]

Department of Electrical and Electronic Engineering, Tokyo University of Agriculture & Technology,
2-24-16 Nakacho, Koganei, Tokyo 184-8588, JAPAN.
[*]Phone/Fax: +81-042-388-7919, E-mail: shrakash@cc.tuat.ac.jp

Abstract

We report the conduction mechanism of the single-electron transistors (SETs) fabricated by field-emission-induced electromigration, which is so-called "activation". By applying the activation to Ni nanogaps, we were easily able to fabricate the SETs operating at room temperature. Additionally, strong Coulomb staircases were clearly obtained, and the quasi-periodic current oscillations were also observed at room temperature. These results indicate that the higher charging energy associated with a smaller Ni island structure within the multiple islands causes a bottleneck mechanism in conduction, improving the Coulomb staircase structures.
Keywords: electromigration, field emission current, nanogap, Coulomb blockade and single-electron transistor

Introduction

Single-electron transistors (SETs) based on Coulomb blockade effects are promising candidates for the basic devices of future nanoelectronics and can be scaled down to atomic dimensions, enabling high-density integration, along with low-power dissipation. Room-temperature operation of single-electron devices requires a very small structure of less than 10 nm, which is much smaller than present lithography resolutions. Therefore, some attempts have been made to advance nanofabrication technologies [1-4]. However, it has been known that these approaches are complicated and require special techniques and systems.

So far, we have reported that planar-type nanogap-based SETs were easily fabricated by field-emission-induced electromigration. We call this "activation" method. The activation method is based on moving the atoms induced by the Fowler-Nordheim (F-N) field emission current passing through the nanogaps between source and drain electrodes [5-18]. In the activation procedures, it is expected that the moved atoms accumulate within the gaps and play the dual role of reducing the gap width and forming SETs islands [15-18]. In fact, we were easily able to fabricate the SETs operating at room temperature [15, 18]. Furthermore, the charging energy of the SETs and the number of islands were precisely controlled and adjusted by both the field emission current through a nanogap and the initial gap separation of the nanogaps [15-18]. Therefore, by precisely tuning the activation parameters, it is suggested that we can easily and simply fabricate the SETs with single and multiple islands.

In this report, we investigate the conduction mechanism of the SETs with multiple islands fabricated by the activation. Here, current-voltage characteristics of the SETs with multiple islands exhibited the strong Coulomb staircases at room temperature. We discuss this point by considering the bottleneck mechanism in conduction.

Experimental Details

The fabrication of the SETs using the activation procedure is representatively illustrated in Figs. 1(a)-1(d). First, metallic contact pads were processed by electron-beam evaporation of 5 nm Ti and 25 nm Au on top of an oxidized Si substrate. Then, nanogaps composed of source and drain electrodes with separations of a few tens of nanometers, having side-gate electrodes, were formed by electron-beam evaporation of 30 nm Ni within the gaps between the Ti/Au metallic contact pads (Fig. 1(a)). Next, the F-N field emission current is applied to the nanogap, as shown in Fig. 1(b). The Ni atoms at the tip of the source electrode (cathode) are activated by the field emission current. Then, Fig. 1(c) shows the migration of activated atoms from the source (cathode) to drain (anode) electrode, which is caused along the direction of electron flow. By precisely adjusting the field emission current in combination with initial gap separation of the nanogaps, SET islands were successfully formed by accumulation of Ni atoms within the gaps, resulting in the formation of the SETs with multiple islands (Fig. 1(d)).

The activation procedure was carried out in a vacuum chamber at room temperature. The electrical properties of the nanogaps during and after the activation were controlled and measured with a semiconductor parameter analyzer using high resolution source-monitor units in a shielded room. The details of the fabrication method of the SET using the activation are representatively outlined below. First, we determine the magnitude of preset current I_S. Then, we apply the voltage V to

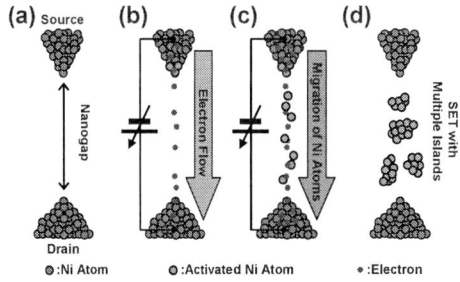

Fig. 1 Schematic of activation procedure. (a) Initial nanogap before performing the activation. (b) Field emission current passes through the nanogap. (c) The Ni atoms at the tip of the source electrodes (cathode) are migrated from the source (cathode) to the drain (anode). (d) SETs with multiple islands are formed by the accumulation of Ni atoms within the gap.

978-1-4673-4840-9/13 $31.00 © 2013 IEEE

the initial nanogap while monitoring the current *I* passing through the nanogap. Finally, we stop the voltage *V* when the current *I* reaches a preset current *Is*. Following the activation, electrical properties of the activated nanogaps are obtained to confirm SET characteristics. This procedure was continuously repeated to the devices with increasing the preset current. Here, SETs with multiple islands are fabricated by the activation procedure.

Results and Discussion

A. Typical Nanogaps Before and After Performing Activation

Figs. 2(a) and 2(b) show a scanning electron microscope (SEM) and an atomic force microscope (AFM) images of a typical initial Ni nanogap before the activation. In the SEM image, initial gap separation is estimated to be approximately 48 nm. On the other hand, SEM and AFM images of the nanogap after performing the final activation step with the preset current *Is* of 1 μA are exhibited in Figs. 2(c) and 2(d), respectively. As shown in the figures, Ni atoms are accumulated in the gap between source and drain electrodes. Furthermore, as seen in inset of Fig. 2(c), accumulated Ni atoms within the gap can be also observed. Hence, the result clearly indicates that Ni atoms controlled by the activation can form island structures within the gap.

B. Current-Voltage Characteristics of Sample A with Initial Gap Separation of 41 nm

Fig. 3(a) shows the drain current-drain voltage (*Id-Vd*) characteristics of the sample A, having initial gap separation of 41 nm, activated with the preset current *Is* from 1 nA (1st) to 1 μA (6th). When the preset current *Is* in 1st activation step was set to 1 nA, the *Id-Vd* characteristics of the activated nanogap showed the insulating-like *Id-Vd* properties. On the other hand, the *Id-Vd* characteristics controlled with the preset current *Is* = 100 nA in 2nd activation step displayed the suppression of electrical currents at low-bias voltages, known as Coulomb blockade, at room temperature. Moreover, the

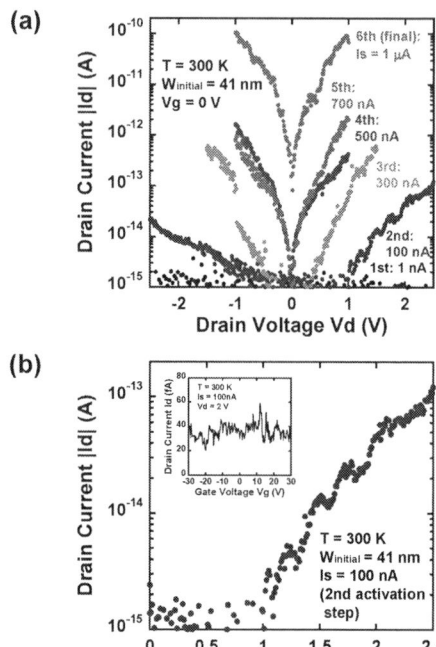

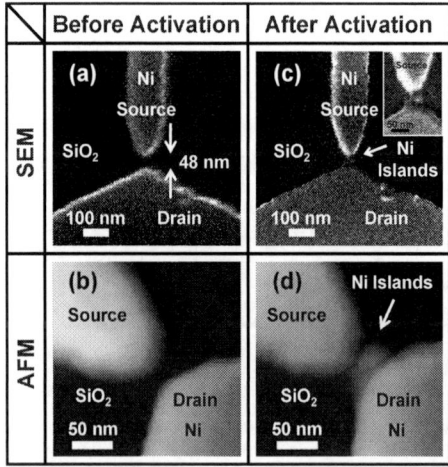

Fig. 2 SEM and AFM images of typical nanogap before and after performing the final activation step with the preset current *Is* = 1 μA. The inset shows an enlarged SEM image of the nanogap after performing activation.

Fig. 3 (a) *Id-Vd* characteristics of the sample A after performing the activation with the preset current *Is* = 1 nA (1st), 100 nA (2nd), 300 nA (3rd), 500 nA (4th), 700 nA (5th) and 1 μA (6th). (b) *Id-Vd* characteristics of the sample A activated with the preset current *Is* = 100 nA, measured at 300 K. The inset shows current oscillation in the *Id-Vg* characteristics of the sample A activated with the preset current *Is* = 100 nA at room temperature and *Vd* = 2 V.

Coulomb blockade voltage was decreased with increasing the preset current *Is*, as shown in Fig. 3(a). Finally, in the sample A activated with the final (6th) preset current *Is* of 1 μA, the *Id-Vd* characteristics show the strong nonlinear properties without Coulomb blockade. Therefore, it is suggested that the wide range of control of SETs properties can be achieved by activation method.

Strong Coulomb staircase can be seen at room temperature in sample A activated with the preset current *Is* = 100 nA (2nd activation step), as shown in Fig. 3(b). Three/four current steps can be identified in the characteristics, at approximately 1, 1.3, 1.6, and 1.9 V. The inset of Fig. 3(b) also shows current oscillations in the drain current-gate voltage (*Id-Vg*) characteristics at room temperature, at *Vd* = 2 V. The current oscillations are weak and do not show a single clear period at room temperature, resulting in the formation of multiple islands (multiple tunnel junctions) in the SETs. These characteristics indicate that formation of multiple islands is achieved by the activation method and the SET operates well at room temperature. Hence, it is considered that the reduction in capacitance at an island within the multiple tunnel junctions causes a bottleneck mechanism in conduction, reducing the drain current at low drain voltages and enhancing the drain current steps in Coulomb staircase, as seen in Si-nanochain SETs [19]. Therefore, it is expected that drain current flows through multiple tunnel junction arrays of planar-type nanogap-based Ni SETs formed by activation method.

978-1-4673-4840-9/13 $31.00 © 2013 IEEE

(a)

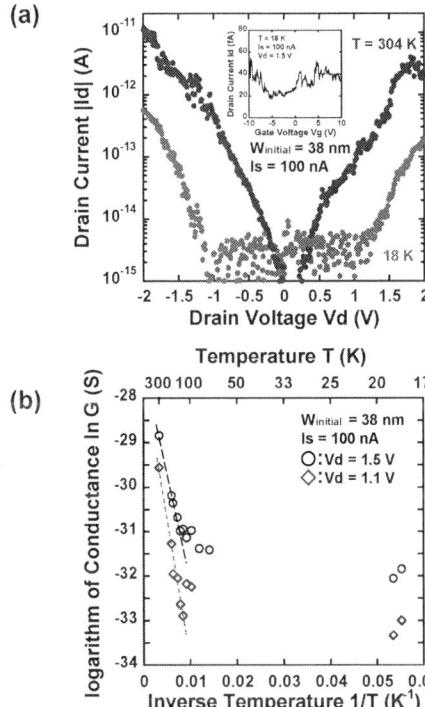

(b)

Fig. 4 (a) *Id-Vd* characteristics of the sample B measured at 304 K and at 18 K. The inset shows drain current as a function of gate voltage at a temperature of 18 K. (b) Arrhenius plot of the logarithm of conductance as a function of inverse temperature.

C. Temperature Dependence of Electrical Characteristics of Sample B with Initial Gap Separation of 38 nm

We performed the activation using the preset current *Is* of 100 nA (5th activation step) to the sample B with initial gap separation of 38 nm. Fig. 4 (a) shows the *Id-Vd* characteristics at 304 K and at 18 K. The Coulomb blockade region is seen to increase greatly, from ~0.25 V at 304 K to ~2 V at 18 K. At 18 K, complex current oscillations are also seen in the inset of Fig. 4(a). A few irregular peaks are expected in the current oscillation for multiple islands. Furthermore, Fig. 4(b) exhibits an Arrhenius plot of the conductance of sample B as a function of inverse temperature. The conductance was measured at drain voltage *Vd* = 1.5 V and 1.1 V. For temperatures from 304 K to 100 K, the conductance decreases linearly as the thermally activated conductance diminishes. This corresponds to an activation energy E_A of 40-60 meV. These tendencies are similar to those observed in Si-nanochain SETs [19], and may be attributed to the highest charging energy in the planar-type nanogap-based Ni SETs with multiple islands.

D. Fabrication Yield of Single-Electron Transistors Formed by Activation

In the present study, we have measured 12 samples (N = 12) after completing the activation procedures. Fig. 5 summarizes the percentage of statistical distribution for which we observe each of the *Id-Vd* characteristics at room temperature. In particular, 11 samples (92 %) show Coulomb blockade (CB) and 5 samples (42 %) represent strong Coulomb staircases (SCS) at room temperature. The *Id-Vg* characteristics of the

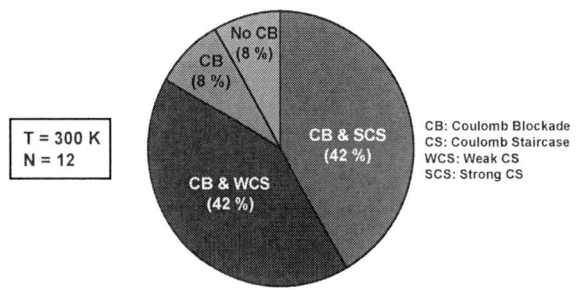

Fig. 5 Pie chart summarizing the statistics of *Id-Vd* characteristics observed in the present study.

5 samples suggested the presence of multiple islands. Therefore, the SCS can be explained by the bottleneck mechanism in conduction. These results clearly indicate that the activation is suitable for the fabrication of SETs operating at room temperature. Moreover, it is considered that Ni atoms are accumulated within the gap by activation and act as multiple islands, associated with inhomogeneity in the tunnel junctions.

Conclusions

We have investigated the conduction mechanism of the SETs fabricated by activation method. When the activation was applied to the nanogaps, strong Coulomb staircases were observed at room temperature. In addition, the current oscillations in *Id-Vg* characteristics do not show a single clear period at room temperature, resulting in the formation of multiple islands in the SETs. Therefore, it is considered that the effects of the bottleneck mechanism in conduction along the multiple tunnel junctions are very strong. Furthermore, an activation energy E_A of 40-60 meV was obtained from the temperature dependence of the conductance. These results imply that drain current flows through inhomogeneously formed multiple tunnel junction arrays of planar-type nanogap-based Ni SETs, leading to the strong Coulomb blockade effects at room temperature.

References

[1] J. Shirakashi, et al., *Appl. Phys. Lett.*, 72, pp. 1893-1895, April 1998.

[2] J. Shirakashi, et al., *Jpn. J. Appl. Phys.*, 37, pp. 1594-1598, March 1998.

[3] Yu.A. Pashkin, et al., *Appl. Phys. Lett.*, 76, pp. 2256-2258, April 2000.

[4] Y. Nakamura, et al., *Jpn. J. Appl. Phys.*, 35 pp. L1465-L1467, November 1996.

[5] S. Kayashima, et al., *Jpn. J. Appl. Phys.*, 46, pp. L907-L909, September 2007.

[6] S. Kayashima, et al., *J. Phys. Conf. Ser.*, 100, pp. 052022-1-4, March 2008.

[7] Y. Tomoda, et al., *J. Vac. Sci. Technol. B*, 27, pp. 813-816, March 2009.

[8] Y. Tomoda, et al., *IEEE Trans. Mag.*, 45, pp. 3480-3483, October 2009.

[9] Y. Tomoda, et al., *J. Phys. Conf. Ser.*, 200, pp. 062035-1-4, January 2010

[10] T. Watanabe, et al., *J. Appl. Phys.*, 109, pp. 07C919-1-3, April 2011.

[11] K. Takiya, et al., *J. Nanosci. Nanotechnol.*, 11, pp. 6266-6270, July 2011.

[12] K. Takiya, et al., *Appl. Surf. Sci.*, 258, pp. 2029-2033, January 2012.

[13] M. Ito, et al., *IEEE-NANO 2011*, pp. 260-263, 15-19 August 2011.

[14] R. Suda, et al., "Formation Scheme of Quantum Point Contacts Based on Nanogaps Using Field-Emission-Induced Electromigration," *J. Nanosci. Nanotechnol.*, in press.

[15] W. Kume, et al., *J. Nanosci. Nanotechnol.*, 10, pp. 7239-7243, November 2010.

[16] S. Ueno, et al., *J. Nanosci. Nanotechnol.*, 11, pp. 6258-6261, July 2011.

[17] S. Ueno, et al., *Appl. Surf. Sci.*, 258, pp. 2153-2156, January 2012.

[18] S. Akimoto, et al., "Control Parameters for Fabrication of Single-Electron Transistors Using Field-Emission-Induced Electromigration," *J. Nanosci. Nanotechnol.*, in press.

[19] M. A. Rafiq, et al., *Jpn. J. Appl. Phys.*, 51, pp. 025202-1-6, February 2012.

Structural and Magnetic Properties of Fe$_2$CrSi Heusler Alloy and Tunneling MagnetoResistance of its Magnetic Tunneling Junctions

Yu-Pu Wang[1,2], Jin-Jun Qiu[2], Hui Lu[1], Qi-Jia Yap[2], Wen-Hong Wang[2], Gu-Chang Han[2], Duc-The Ngo[1], and Kie-Leong Teo[1]

[1]Department of Electrical and Computer Engineering, Advanced Memory Laboratory, National University of Singapore, 4 Engineering Drive 3, Singapore 117583

[2]Data Storage Institute, Agency for Science, Technology and Research (A*STAR), 5 Engineering Drive 1, Singapore 117608

wangyupu@nus.edu.sg

Abstract

We report the magnetic properties, microstructure and surface morphology of epitaxially grown Fe$_2$CrSi films. Highly ordered B2 films were obtained by deposition at room temperature followed by annealing at 400°C. Magnetic tunnel junctions using Fe$_2$CrSi show a tunnelling magnetoresistance (TMR) of 2.5%. The low TMR is ascribed to the oxidation of Fe$_2$CrSi at the interface with MgO. An enhancement of TMR to 8.1% was achieved by inserting a 0.3nm Mg between Fe$_2$CrSi and MgO to prevent the oxidation of Fe$_2$CrSi.

Keywords: Heusler alloy, half metal, MTJ, TMR, Fe$_2$CrSi

Introduction

The emerging field of spintronics exploits both the charge and spin properties of electrons to make devices more compact and faster. An important application in spintronics involves the injection of spin polarized current through a non-magnetic barrier layer, such as MgO barrier in magnetic tunnelling junctions (MTJs). The spin injection is low in the metal based spintronics device due to the mismatch of conductivity at the interface [1]. One possible solution to overcome the conductivity mismatch is to use half metallic ferromagnets (HMFs) with 100% spin polarization [2]. Categories of promising HMFs include Heusler alloys, oxide-based HMFs, perovskite manganites and zinc-blende HMFs. Among them, Heusler alloys have the greatest potential to realize half metallicity at room temperature due to their lattice constant matching with the III-V semiconductors, high Curie temperature (>400 K) and large bandgap at Fermi level [3]. Many research groups have focused on investigating Co-based Heusler alloys, such as Co$_2$MnSi, Co$_2$MnAl, and Co$_2$CrAl [4-8]. An alternative candidate for HMF is Fe-based Heusler alloy, such as Fe$_2$CrSi (FCS). FCS is predicted to be a half metal in the L2$_1$ structure [9]. Compared with other Heusler alloys, FCS possesses the following advantages [9,10]: Firstly, the calculated majority DOS of FCS at Fermi level is as high as 7.0 states/eV, which is about 5 times higher than that of Co$_2$MnSi, leading to a high spin polarization; Secondly, the magnetization of FCS is found to be small (~2 μ_B/f.u.), in contrast to that of the Co-based Heusler alloy (4-6 μ_B/f.u.) [4-9], allowing for a low critical current for spin transfer torque switching in spin torque devices [11]; Thirdly, FCS has a suitable Curie temperature of 630K, which is low enough for thermally assisted recording, yet high enough

compared to room temperature.

The FCS film was reported to be obtained in B2 structure instead of L2$_1$ structure due to its structural instability [9]. However, according to theoretical studies, FCS in B2 structure is also expected to show high spin-polarization [12-14]. The same group also attempted to utilize FCS films in a magnetic tunnelling junction (MTJ) structure [9], but no TMR was reported, probably due to a rough surface of the FCS layer and a resultant strong coupling between the free layer and the pinned layer. Hence, a high-quality MTJ involving FCS has yet to be achieved.

In this paper, we focus on the implementation of B2 phase FCS thin film in MTJ structure. Highly-ordered FCS in B2 structure has been successfully grown by magnetron sputtering, and exhibits excellent magnetic properties and surface roughness (RMS~0.18 nm) for application in MTJ structures. The tunnel magnetoresistance (TMR) ratio and the resistance-area product (RA) of the barrier MgO layer were measured by the current-in-plane-technique (CIPT) on un-patterned MTJ samples. TMR ratio and RA were studied as a function of post-annealing temperature. It is found that a well-defined exchange bias is established for our MTJ samples and a TMR ratio of 2.5% is obtained at room temperature in the MTJ sample with a 1.5 nm MgO barrier. This TMR ratio is greatly enhanced to 8.1% with an insertion of Mg layer between FCS and MgO barrier that could prevent the oxidation of the bottom electrode FCS layer.

Experimental Procedure

All samples were prepared by ultra high vacuum magnetron sputtering system with a base pressure of 5×10^{-10} Torr. The FCS thin films were grown by the sputtering of a stoichiometric Fe$_{46}$Cr$_{30}$Si$_{24}$ target and were deposited at room temperature on MgO (100) substrates with Cr and Ag buffer layers. Prior to any deposition, the MgO substrate was optimized by pre-heating at 600°C for 1 hour. In-situ thermal treatments for the FCS layer were performed at various temperatures of 300°C, 400°C, and 500°C for 15 minutes to form ordered structure. After optimizing the FCS films, MTJs with a stacking structure of MgO(100)/Cr 40nm/FCS 30nm/MgO t nm/CoFe 5nm/IrMn 12nm/Ru 8nm were fabricated subsequently. The MgO barriers were grown by RF sputtering in a plasma oxidation chamber. A barrier MgO thickness of 1.5 nm was used to obtain a reasonable RA for CIPT measurements. Furthermore, an Mg layer was inserted between FCS and MgO barrier layer which is expected to enhance the TMR ratio. During the deposition of the stack, a

978-1-4673-4840-9/13 $31.00 © 2013 IEEE

series of *in-situ* thermal treatments were done as follows: 700°C for Cr buffer layer for 30 minutes; 400°C for FCS thin film for 15 minutes. All these samples were naturally cooled to room temperature in vacuum. The whole junctions were then *ex-situ* post-annealed at various temperatures from 200°C to 500°C in high vacuum in the presence of an in-plane magnetic field of 1 Tesla.

Vibrating sample magnetometer (VSM) and alternating gradient magnetometer (AGM) were used to characterize the magnetic properties. X-ray diffraction (XRD) and transmission electron microscopy (TEM) were employed to characterize the crystalline structure. The surface roughness was investigated using atomic force microscope (AFM). TMR ratio and RA of the MTJs were measured by CIPT on un-patterned samples.

Results and Discussions

Fig. 1(a) shows the XRD (θ-2θ) patterns for MgO (100)/Cr/Ag/FCS samples under various thermal treatment conditions. When the *in-situ* annealing temperature is 400°C and above, FCS (200) and (400) peaks are clearly observed, indicating the formation of highly ordered B2 structure. However, the intensity of these peaks becomes weaker when the annealing temperature increases to 500°C. AFM measurements indicate that smooth films were obtained when the annealing temperature is below 500°C. The best surface roughness of 0.18 nm is obtained at an annealed temperature of 400°C as shown in Fig. 1(b), whereas the surface of the sample annealed at 500°C becomes rough with particles protruding from the surface clearly observed, as shown in Fig. 1(c). These artifacts are somehow similar to that reported previously [9] in which the film was deposited at high temperatures that led to a rough surface. As shown in Fig. 1(d), the rocking curve of FCS(400) of the film *in-situ* annealed at 400°C shows a narrow peak with the full width at half maximum (FWHM) as small as 0.53°, in contrast to 1.3° reported in the literature [9]. These results indicate that the FCS film was grown in the highly ordered B2 structure with the c-axis perpendicular to the plane. The optimized condition for the FCS films in terms of surface roughness and crystalline structure is obtained when the *in-situ* annealing temperature is 400°C. Cross-sectional TEM image for the film is illustrated in Fig. 1(e). The image clearly shows that the smooth and abrupt interfaces are formed between the layers in the structure of MgO substrate/Cr/Ag/FCS. The high resolution TEM image in Fig. 1(f) as well as the fast Fourier transform (FFT) image [inset of Fig. 1(f)] confirms that the deposited FCS layer is single crystalline.

Fig. 2 depicts *M-H* loop for the FCS thin film with *in-situ* annealing at 400°C. The field was applied along the in-plane [011] direction, which is parallel to the easy axis of the crystalline anisotropy of FCS. Sharp magnetization reversal is observed and the ratio of remnant magnetization (M_r) over saturation magnetization (M_S) is about 1, reflecting the easy-axis is along [011] direction. The M_S and coercivity (H_C) of FCS are 322 emu/cc and 2.2 Oe, respectively. The M_S of FCS is relatively smaller than that of Co_2MnSi, Co_2FeSi, and Co_2FeAl bulk (1000-1200 emu/cc), which is favourable for the low switching current density required for spin-transfer torque devices, such as STT-MRAM [11]. FCS with small H_C is also a suitable ferromagnetic material for soft magnetic electrodes in MTJ applications.

MTJs with stacking structure of MgO substrate/Cr 40nm/FCS 30nm/MgO 1.5nm/CoFe 5nm/IrMn 8nm/Ru 3nm were fabricated. We examined the post-annealing effect with temperatures ranging from 200°C to 500°C. Fig. 3(a) shows the XRD (θ-2θ) patterns for MTJs fabricated under different post-annealing temperatures. The FCS(200), Cr(200), and FCS(400) peaks are clearly observed, indicating that our samples are (100) orientation of epitaxial growth. Even without any post-annealing for the MTJ, clear FCS(200) and FCS(400) peaks are observed, indicating that (100)-orientated highly ordered B2 structure of FCS was already formed during *in-situ* annealing. When the post-annealing temperature increases to 200°C, the intensity of FCS(400) peak remains the same but that of FCS(200) peak becomes weaker. When the post-annealing temperature raises to 300°C, the FCS(400) peak is still quite strong but the FCS(200) peak disappears, which indicates poorer crystalline structure of FCS. When the post-annealing temperature further increases to 350°C, the FCS(400) peak becomes weaker in addition to the disappearance of the FCS(200) peak, showing that the crystalline structure of FCS becomes even more deteriorated. We ascribed the deterioration of the crystalline structure after post-annealing to possible oxidation of FCS during annealing. Fig. 3(b) presents the M_S and H_C of FCS as a function of post-annealing temperature. The M_S obtained is as high as 360 emu/cc under 200°C post-annealing, which is close to the calculated M_S for FCS with an $L2_1$ ordered structure (398 emu/cc) [15]. Band calculation predicted that the half metallicity of FCS is robust against most defects (e.g., antisites, swaps, and vacancies) [16]. Hence, the presented FCS with a B2 order and 360 emu/cc M_S is expected to have a high spin polarization. The increase in M_S can be attributed to the improvement of the chemical order and local stoichiometry after thermal treatment. The decrease in H_C with annealing temperature reveals possible reduction in defect density accompanied with an enhancement of crystalline quality. However, the annealing at higher temperatures results in lower M_S of FCS, which is suspected to be related to a migration of oxygen from the MgO barrier layer into FCS layer. In other words, oxidation of the underlying FCS ferromagnetic layer suppresses the total magnetic moment of the FCS film.

Fig. 4(a) shows the room temperature *M-H* loop of the MTJs post-annealed at different temperatures measured by AGM. An exchange bias is observed with 200°C post-annealing. However, the H_C of the top CoFe electrode is quite large, which indicates CoFe is not well pinned by the antiferromagnetic IrMn layer due to low post-annealing temperature. When the post-annealing temperature increases to 350°C, as shown from the *M-H* loop in Fig. 4(a), a well-defined exchange bias between the top CoFe and IrMn is established and H_C of the CoFe layer reduces significantly. The M_S of the FCS free layer and the CoFe pinned layer are 322 emu/cc and 1102 emu/cc, respectively. This indicates that FCS layer and CoFe layer with a highly ordered structure are obtained in the MTJ structures. It is worthwhile to note that the *M-H* loop of the free layer and pinned layer in our MTJ samples post-annealed at 350°C are well-separated with a shift of 220 Oe. The exchange bias in *M-H* loop of MTJ structure we obtained is better-defined than that reported previously [9]. However, when the post-annealing temperature is above 500°C, the exchange bias is destroyed and the magnetization of two magnetic layers switches simultaneously, resulting in no TMR effect. The vanishing of the exchange bias in this case could be attributed to the diffusion of the Mn from IrMn to the other layers. The post-annealing temperature dependence of RA and TMR ratio of these samples are shown in Fig. 4(b). The highest TMR ratio (2.5 %) measured by CIPT on un-patterned samples is achieved for the sample annealed at 200°C. The annealing of the MTJs at an appropriate temperature improves not only the structural properties of the upper electrode but also the interfacial structural properties, leading to an enhancement of the TMR ratios [17-20]. As seen in Fig. 4(b), TMR ratio decreases with the post-annealing temperature above 200°C, although the exchange bias is well developed. This decrease in TMR ratio is highly likely due to the oxidation of FCS layer [21, 22] at the interface with the MgO barrier. The oxidized FCS surface may act as additional scattering centres for spin polarized tunnelling electrons, which would suppress TMR. The oxidation of FCS at the interface is further confirmed from the monotonically increasing RA with increase in post-annealing temperature. The TMR ratio and RA relationship is illustrated in the inset of Fig. 4(a). The highest TMR ratio is obtained with RA of 2000 ohm/μm^2. The decrease in TMR ratio at higher RA values is probably because of the polarization reduction of the FCS layer due to the oxidation. We propose that the FCS surface is considered oxidized because it was exposed to oxygen during MgO deposition, and more serious oxidation of FCS occurs as the post-annealing temperature is increased. In order to avoid the FCS oxidation, a 0.3 nm Mg layer was inserted between the FCS layer and the barrier MgO layer to enhance the TMR effect. Since the Mg layer will be oxidized eventually forming MgO, the thickness of the MgO layer is now reduced to 1.2 nm to retain the total thickness of the tunnelling barrier (Mg+MgO) as 1.5 nm. An enhancement of the TMR ratio up to ~8.1% is obtained apparently when the sample is annealed at 250°C. However, this TMR ratio is still lower than the expected from the first-principle calculations [23]. As shown in Ref. [9], DOS profile of the majority spin of FCS at Fermi level is very sharp and close to the conduction band edge. Hence, the DOS profile would be easily affected by thermal fluctuations and imperfections of FCS crystalline structure. As a result, a little shift of the Fermi level may cause a large reduction of the DOS of the majority spin, resulting in low spin polarization and TMR ratio. Therefore, the tuning of the Femi level would be critical to obtain high TMR ratio when using FCS as an electrode in MTJ structure.

Conclusion

In summary, this study investigated magnetic tunnel junctions with an (100)-epitaxial FCS bottom electrode and MgO tunnel barrier. The epitaxial FCS showed very low surface-roughness (RMS~0.18nm) and highly ordered B2 structure. The plasma oxidation process of the MgO tunnel barrier may have caused oxidation of the bottom FCS surface. Oxidation of the bottom electrode interface increased the RA value and reduced TMR ratio. With 1.5nm MgO tunnel barrier, a TMR ratio of 2.5% was obtained. This TMR ratio is suppressed by bottom electrode oxidation. Insertion of Mg layer between bottom electrode and MgO tunnel barrier helped to enhance the TMR ratio to 8.1% by protecting the FCS layer from the oxidation. Optimization of FCS electrode (e.g.,

adjustment of Fermi level of FCS, interfacial studies) is expected to engender higher and more stable spin polarization and enhanced TMR ratios.

References

[1] Mark Johnson, and R. H. Silsbee, Phys. Rev. Lett. 55, 17, pp. 1790-1793 (1985).

[2] R. A. de Groot and F. M. Mueller, Phys. Rev. Lett. 50, 25, pp. 2024-2027 (1983).

[3] J. Dubowik, I. Goscianska, A. Szlaferek, Y. V. Kudryavtsev, Mater. Sci. Poland 25, 2 (2007).

[4] K. Yakushiji, K. Saito, S. Mitani, K. Takanashi, Y. K. Takahashi, and K. Miyazaki, and H. Kubota, Appl. Phys. Lett. 88, 192508 (2006).

[5] Y. Sakuraba, M. Hattori, M. Oogane, Y. Ando, H. Kato, A. Sakuma, T. Hono, Appl. Phys. Lett. 88, 222504 (2006).

[6] S. Ishida, S. Fujii, S. Kashiwagi, and S. Asano, J. Phys. Soc. Jpn. 64, 2152 (1995).

[7] I. Galanakis, P. H. Dederiches, and N. Papanikolaou, Phys. Rev. B 66,174429 (2002).

[8] S. Picozzi, A. Continenza, and A. J. Freeman, Phys. Rev. B 66, 094421 (2002).

[9] S. Yoshimura, H. Asano, Y. Nakamura, K. Yamaji, Y. Takeda, M. Matsui,S. Ishida, Y. Nozaki, and K. Matsuyama, J. Appl. Phys. 103, 07D716 (2008).

[10] V. Ko, J. Qiu, P. Luo, G.C. Han, and Y.P. Feng, J. Appl. Phys. 109, 07B103 (2011).

[11] S. Mangin, Y. Henry, D. Ravelosona, J. A. Katine, and Eric E. Fullerton, Appl. Phys. Lett. 94, 012502 (2009).

[12] S. Ishida, S. Mizutani, S. Fujii, and S. Asano, Mater. Trans. 47, 31 (2006).

[13] S. Ishida, S. Mizutani, S. Fujii, and S. Asano, Mater. Trans. 47, 464 (2006).

[14] S. Mizutani, S. Ishida, S. Fujii, and S. Asano: Mater. Trans. 47, 25 (2006).

[15] H. Sukegawa, H. Xiu, O. Ohkubo, T. Furubayasi, T. Niizeki, W. Wang, S. Kasai, S. Mitani, K. Inomata, and K. Hono, Appl. Phys. Lett. 96, 212505 (2010).

[16] B. A. Hamad, Eur. Phys. J. B 80, 11-18 (2011).

[17] T. Ishikawa, S. Hakamata, K.-i. Matsuda, T. Uemura, and M. Yamamoto:J. Appl. Phys. 103, 07A919 (2008).

[18] T. Ishikawa, H. Liu, T. Taira, K.-i. Matsuda, T. Uemura, and M.Yamamoto: Appl. Phys. Lett. 95, 232512 (2009).

[19] M. Yamamoto, T. Ishikawa, T. Taira, G. Li, K.-i. Matsuda, and T. Uemura:J. Phys.: Condens. Matter 22 , 164212 (2010).

[20] S. Tsunegi, Y. Sakuraba, M. Oogane, K. Takanashi, and Y. Ando: Appl.Phys. Lett. 93, 112506 (2008).

[21] Y. Sakuraba, J. Nakata, M. Oogane, Y. Ando, H. Kato et al., Appl. Phys. Lett. 88,022503 (2006).

[22] Sungjun Joo, K. Y. Jung, B. C. Lee, Tae-Suk, Kim, K. H. Shin et al., Appl. Phys. Lett. 100,172406 (2012).

[23] Shinpei Fujii, Shoji Ishida, and Setsuro Asano, J. Phys. Soc. Jap 81, 034716 (2012).

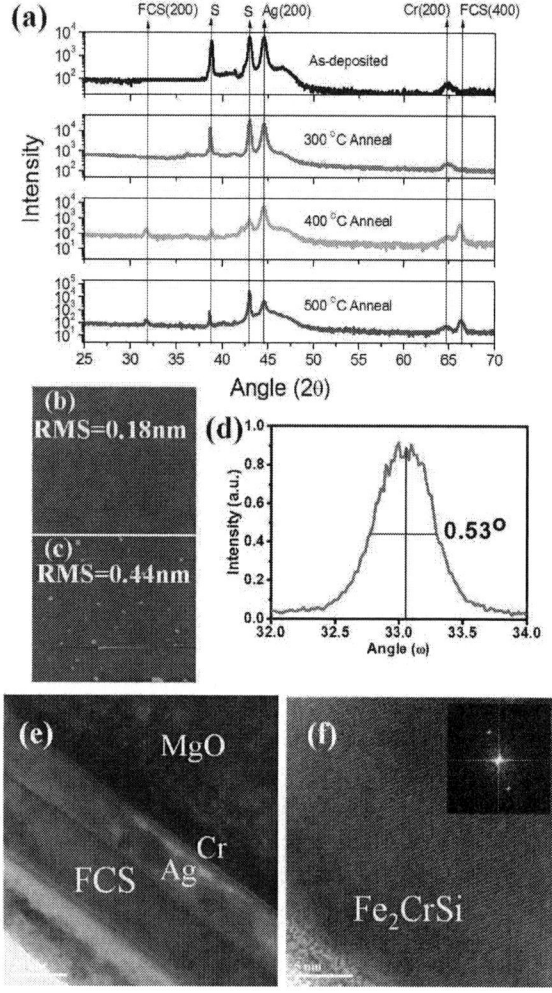

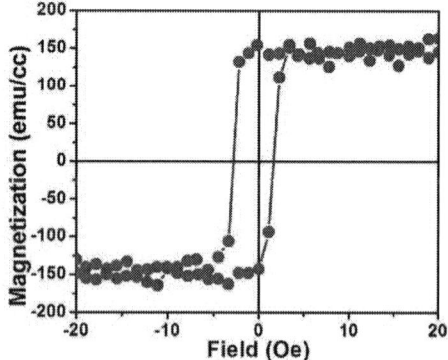

Fig. 1 (a) X-ray Diffraction patterns for Fe₂CrSi thin film. From top to bottom, *in-situ* annealing are as deposited, 300°C, 400°C, and 500°C, respectively. S refers to substrate peak from K_α and K_β, respectively. (b) AFM image of Fe₂CrSi thin film with 400°C, and (c) 500°C *in-situ* annealing. (d) Rocking curve measurement of Fe₂CrSi(400) peak. (e) TEM image of MgO/Cr/Ag/ Fe₂CrSi sample. (f) High resolution TEM of Fe₂CrSi thin film. The inset shows FFT image of the Fe₂CrSi layer.

Fig. 2 *M-H* loop of Fe₂CrSi film with 400°C *in-situ* annealing along [011] direction using VSM.

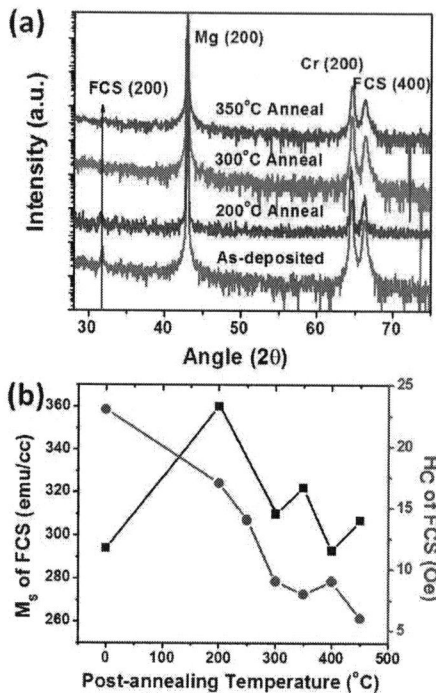

Fig. 3 (a) X-ray Diffraction patterns for MTJs. From bottom to top, post-annealing are as deposited, 200°C, 300°C, and 350°C, respectively. (b) M_S and H_C of the bottom electrode Fe₂CrSi for MTJs, as a function of post-annealing temperature.

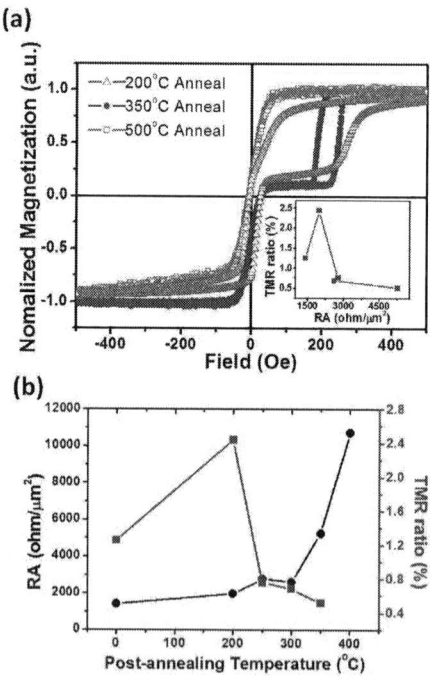

Fig. 4 (a) *M-H* loop for MTJs with 200°C, 350°C, and 500°C post-annealing, respectively. The inset figure is TMR ratio as a function of Resistance-Area. (b) Resistance-Area production and TMR ratio as a function of post-annealing temperature.

978-1-4673-4840-9/13 $31.00 © 2013 IEEE

The effect of intrinsic defects on resistive switching based on p-n heterojunction

K. Zheng[1], X. W. Sun[1], and K. L. Teo[2]

1. School of Electrical & Electronic Engineering, Nanyang Technological University, Nanyang Avenue, Singapore
2. Department of Electrical and Computer Engineering, National University of Singapore, 4 Engineering Drive 3, Singapore 117608

Abstract

We report a unidirectional bipolar resistive switching in an n-type GaO_x/p-type NiO_x heterojunction fabricated by magnetron sputtering at room temperature. The resistive switching (RS) of the heterojunction directly relate with the concentration of intrinsic defects in oxide, such as oxygen vacancies and oxygen ions. Under external electric field, these electromigrated defects accumulate at the pn junction interface and modify the interface barrier, forming or rupturing the filamentary paths between n-GaO_x and p-NiO_x, leading to the switching between Ohmic and diode characteristics of the device.

Introduction

Resistive switching random access memory (RRAM) has attracted much interest with great potential in non-volatile data storage. To avoid the thermal disadvantages of unipolar RS (usually related with a filamentary path), recently preferable attention have focused on bipolar RS, which mainly attribute to the interface engineering in basic metal-insulator-metal (MIM) structure. It is generally accepted that the RS occurs mainly at interface between oxide and electrode rather than in the bulk, and such switching originates from the redox process based on migration of oxygen vacancies (or oxygen ions) and change in Schottky barrier height or width by trapping/detrapping effects near interface region [1]. Thus, RS phenomenon could also be expected by adjusting the interface between p-type and n-type oxide films in metal-insulator-insulator-metal (MIIM) structure. Different from some early works observed the resistance hysteresis in some perovskite oxide heterojunction, recently we reported the transition from highly stable rectifying behavior to RS behavior in $In/GaO_x/NiO_x/ITO$ structure [2]. According to the initial rectifying characteristic of p-n junction (HRS under reverse bias, LRS under forward bias), the reproducible resistance change could be obtained by an opposite operation; the proper soft breakdown under reverse bias could turn the device from HRS to LRS, whereas the forward bias will recover the device from LRS back to HRS (preferably in low voltage region). In this paper, we shall discuss the effect of intrinsic defects on the RS in pn heterojunction RRAM.

Experimental Detail

Commercial ITO glass was employed as the substrate with ITO as the bottom electrode in our device. A NiO_x and GaO_x film with a thickness of 90 nm and 70 nm respectively was deposited successively on the ITO substrate by RF magnetron sputtering. 28 sccm Ar was input into the sputtering chamber to maintain the pressure of 6×10^{-3} Torr at room temperature. Oxygen was also introduced to fabricate some samples for comparison. Finally, indium shots were employed as simple top electrodes for test. The schematic structure of device is shown in the inset of Fig. 1. The I-V characteristics were measured by a semiconductor parameter analyzer in a diode I-V sweeping mode at room temperature. During the test, the bottom ITO electrode was always grounded while different biases were applied on top electrode (In). Moreover, the XPS was also adopted to analyze the intrinsic defects in NiO_x and GaO_x film.

Result and Discussion

Fig. 1 shows the current-voltage (IV) characteristics of the stack of $In/GaO_x/NiO_x/ITO$ without oxygen in film depositing. We can see a typical intrinsic rectifying behavior of diode with a turn-on voltage of about -2 V for as-prepared stack as the red curve indicates. As reported previously, non-stoichiometric NiO_x (x > 1) is a p-type semiconductor due to the excess oxygen ions (Ni-deficient) as acceptors for generating holes [3], and non-stoichiometric GaO_x (x < 1.5) is an n-type semiconductor due to the oxygen vacancies (oxygen-deficient) as donors for generating electrons [4]. With Hall Effect measurement, we confirmed the carrier type for GaO_x and NiO_x and the sheet carrier concentration of GaO_x and NiO_x is $-8.603 \times 10^{11}/cm^2$ and $+2.012 \times 10^{14}/cm^2$, respectively. According to the polarity of the forward and reverse bias, the rectifying behavior comes from the p-NiO_x/n-GaO_x barrier instead of the possible Schottky barrier at In/GaO_x or NiO_x/ITO (actually these possibilities was excluded [2]). Therefore, under external electric field, the alteration of the potential barrier at this junction should be responsible.

The cluster of black curves in Fig. 1 demonstrates the typical resistive switching cycles of the $In/GaO_x/NiO_x/ITO$ stack. When a positive bias on In sweeps to a critical value, the current suddenly increases due to a restorable breakdown and the device is set from HRS to LRS. An Ohmic behavior is detected as the bias sweeps back to lower negative region and this is usually related with filamentary conduction. When the negative voltage increases further, an abrupt decrease of current occurs and the device is reset from LRS to HRS. In fact, after reset process, the device returns to rectifying behavior again and the HRS state just exists in a narrow negative voltage range which caused by the interface barrier. It is worth mentioning that, the device could only be unidirectionally set by applying positive voltage and reset by applying negative voltage on In, which also provides the evidence that RS behavior originates from the n-p interface. As is widely accepted, the forming process and compliance current (CC) are

978-1-4673-4840-9/13 $31.00 © 2013 IEEE

two important factors for RS operation. In our measurement, there exists no apparent forming process, while an optimized CC of 5mA was applied to the set operation to prevent the device from permanent breakdown.

As illustrated in our previous work, the RS process in the p-n heterojunction could be explained by the electromigration of intrinsic defects [2]. When the p-n junction is reversely biased, external electric field will enlarge the depletion region and elevate the interface barrier, which suppresses the drift current passing through the junction as a usual diode behaves. The conductive paths at both sides were blocked by the interface barrier and the device shows HRS. Further increasing the reverse bias leads to accumulation of positive oxygen vacancies (from inside GaO_x) and negative oxygen ions (from inside NiO_x) at the interface, bringing a significant reduction of effective barrier height and depletion region width until the electron tunneling the barrier with filamentary paths penetration, which bring the device to LRS. Due to the high density of defects near the interface, the junction behaves like a resistor with high current and non-volatile LRS. On the contrary, a high enough forward bias will repels these defects away from interface region, then the barrier recovers back, cutting off the connected filaments and recovering the device to HRS. With further increase of negative bias, the height of barrier reduces and an increasing current run through the interface.

To study the defects role in RS behavior of our device, we introduced 1 sccm O_2 with 27 sccm Ar during depositing the NiO_x and GaO_x. More oxygen will increase the self-doping of Ni vacancies in p-type NiO_x, while reduce the self-doping of oxygen vacancies in n-type GaO_x. Therefore we sign the + to those samples with higher self-doping concentration. The intrinsic defects in as prepared oxide film were analyzed by XPS measurement. Fig. 2a shows the spectrum of p NiO_x and p+ NiO_x for Ni 2p core level. The peak at 853.1 eV corresponds to the metallic Ni (Ni^0) state while the 857.5 eV and 858.1 peaks refer to the Ni^{2+} and Ni^{3+} bonding states respectively [5]. The decrease peak of Ni^0 in p+ NiO_x manifests that there exist more Ni vacancies as predicted. Fig. 2b gives the O 1s core level spectrum of n GaO_x and n+ GaO_x. The 528.1 peak in both curves is attributed to the O-Ga bond while the 529 peak in n+ GaO_x is usually associated with the O^{2-} ions in the oxygen deficient regions [6], which proves the more oxygen vacancies in the n+ GaO_x.

The oxide films with different self-doping concentration were combined for devices and the difference is apparent as shown in Fig 3. The device with p+n+ stack has large resistance in HRS and wider resistance window than previous pn stack. The difference may mainly be attributed to the higher barrier formed in p+n+ heterojunction than pn one, and further prevent more drift current with a larger resistance.

The distribution of operating voltages in 150 cycles was also investigated as shown in Fig. 4. We can see that, the reset voltage (V_{reset}) distributes randomly in a smaller range (-0.7V ~ -3.5V) compared to the set voltage (V_{set}) (0.9V ~ 7.8V), which can be observed in Fig. 1 as well. To reset the device, a relatively fixed power is required to drive these oxygen vacancies away from interface, resulting in a relatively small-range distribution of V_{reset}. However, each reset sweeping will exceed V_{reset}, which would redistribute these oxygen vacancies into the semiconductor film randomly [7]. As a result, the power needed for attracting the same amount of

oxygen vacancies to the interface region in the next set cycle will be much different, resulting in a much more random distribution of V_{set} than V_{reset}.

Therefore, to summarize, the alteration of junction barrier due to the electric field and the migration of intrinsic defects (oxygen vacancies in GaO_x and oxygen ions in NiO_x) control the filamentary paths passing through the interface, which accounts for the RS behavior observed.

This work is supported by advanced memory research program, Grant No. 092 151 0088 from Singapore Agency for Science, Technology and Research (A*STAR) SERC.

References

[1] A. Sawa, *Materials Today*, vol.11, no.6, pp. 28-36 , 2008

[2] K. Zheng, et al, *Appl. Phys. Lett.*, vol. 101, no. 14, p. 143110, 2012

[3] K. Kinoshita, et al, *Appl. Phys. Lett.*, vol. 96, no. 14, p. 143505, 2010

[4] L. Nagarajan, et al, *Nat. Mater.*, vol. 7, no. 5, pp. 391-398, 2008

[5] R. Jung, et al, *Appl. Phys. Lett.*, vol. 91, no. 2, p. 022112, 2007

[6] S. M. Park, et al, *Thin Solid Films*, vol. 513, no. 1-2, pp. 90-94, 2006

[7] H. Stocker, et al, *Appl. Phys. A*, vol. 100, no. 2, pp. 437-445, 2010

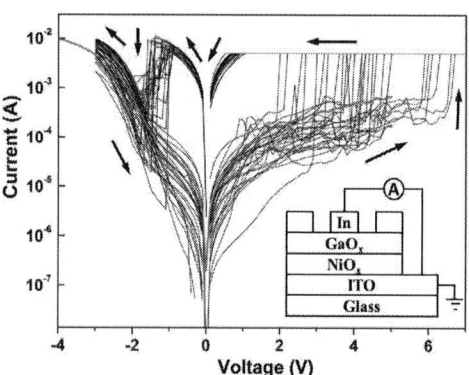

Fig. 1 The bipolar RS cycles with the device structure

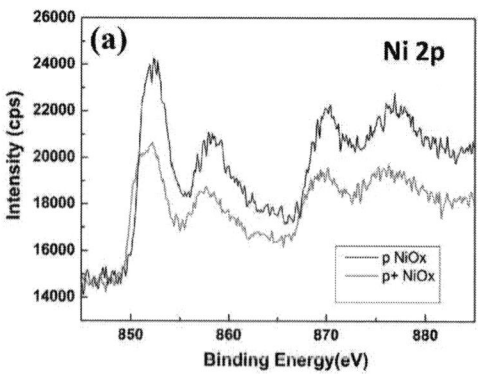

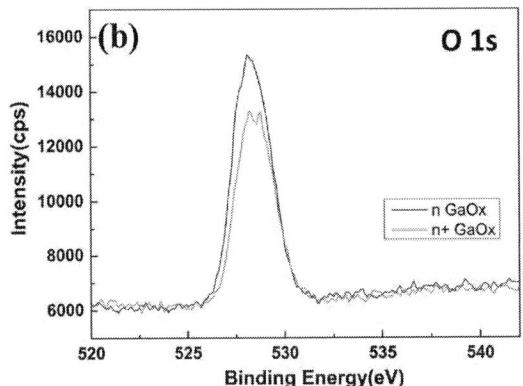

Fig. 2 The XPS spectrum of (a) Ni 2p for NiO_x (b) O 1s for GaO_x

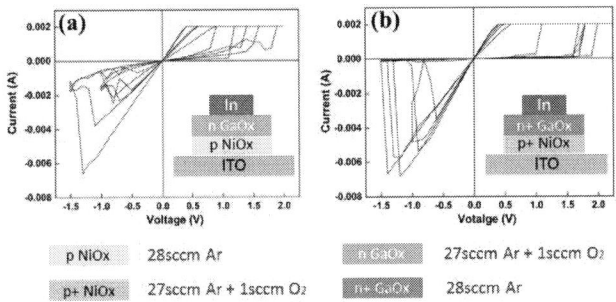

Fig. 3 The RS of devices based on pn and p+n+ heterojunction

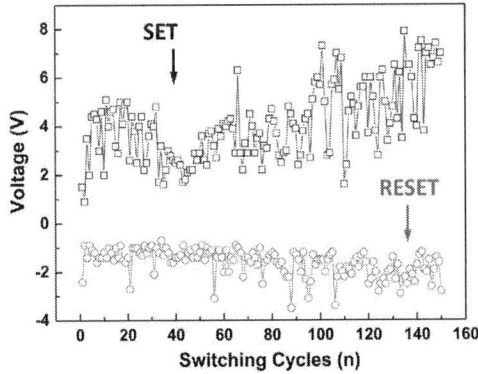

Fig. 4 Distribution of set and reset voltage in 150 repeating cycles

Ink-jet printed In-Ga-Zn oxide nonvolatile TFT memory utilizing silicon nanocrystals embedded in SiO₂ gate dielectric

Y. Wang, T. P. Chen and X. W. Sun
School of EEE, Nanyang
Technological University, Nanyang
Avenue, Singapore
exwsun@ntu.edu.sg

J. I. Wong and H. Y. Yang
Pillar of Engineering Product
Development, Singapore University
of Technology and Design,
Singapore

J. L. Zhao
College of Science,
Tianjin University,
Tianjin, China

Abstract

A nonvolatile memory based on ink-jet printed In-Ga-Zn oxide (IGZO) thin film transistor with bottom gate bottom contact architecture is reported. The memory device contains SiO₂ gate dielectric layer embedded with silicon nanocrystals, which act as charge trapping sites. Memory effects were observed by a clockwise loop in V_{gs}-I_d curves, which is attributed to the charging and discharging of the silicon nanocrystals. The printed IGZO memory exhibits a high (about 1×10^3) on/off ratio.

Keywords: ink-jet printing, In-Ga-Zn oxide, nonvolatile memory.

Introduction

Found by Hanafi et al. [1], nanocrystal memory has recently attracted much attention as one of the promising solutions to the scaling issues of conventional floating gate non-volatile memory (NVM) devices. This kind of memory device utilizes nanocrystals located in close proximity of several nanometers to the channel of transistor as the localized discrete charge storage nodes to store charge embedded in gate oxide of a field-effect transistor (FET) [1-3]. Compared with FET, thin film transistor (TFT) structure allows the device to be fabricated on glass with large area and low cost.

Metal-oxide semiconductor is advantageous in terms of transparency, mobility and stability. In-Ga-Zn oxide (IGZO) is a potential alternative channel layer material for TFTs, compared to conventional a-Si and poly-Si [4-7]. It is transparent in the visible region due to the large band gap and has a high mobility, even for amorphous structure due to s-electron conduction [4]. Though IGZO is mainly studied for TFT applications, there are some reports of IGZO memories [8-12]. However, all previously reported IGZO memory devices are fabricated by conventional physical vapor deposition (PVD), such as sputtering or pulse laser deposition (PLD). Ink-jet printing technology offers the possibility of fabricating functional device by roll-to-roll manufacturing in ultra-large scale under ambient condition with low cost [13]. As a low waste and maskless process, ink-jet printing reduces material usage and process complexity in comparison with the PVD processes [14]. Therefore, this promising technology has received substantial interests over the last decade. In this paper, we shall report a nonvolatile memory device fabricated by ink-jet printing IGZO TFT with nanocrystal-Si (nc-Si) embedded SiO₂ gate dielectric, where the nc-Si acts as the charge trapping centre and the printed IGZO film serves as the active layer [15, 16].

Experimental details

A 15 nm SiO₂ thin film was thermally grown in dry oxygen at 950 °C on a highly doped p-type Si (100) wafer. Si ions with a dose of 2×10^{16} cm^{-2} were then implanted into the SiO₂ thin film with an acceleration voltage of 2 kV. Thermal annealing was carried out in N₂ ambient at 1000 °C for 1 hour. A 100 nm thick ITO was deposited by DC sputtering on SiO₂ and was then patterned for source and drain electrodes by lift-off. The IGZO ink formulation and printing can be found elsewhere [17], [18]. In brief, IGZO ink was prepared by dissolving 0.1 M of zinc acetate dehydrate [Zn(OAc)₂·2H₂O], 0.1 M indium chloride and 0.0025 M gallium chloride (the atom ratio of Ga: In: Zn = 25: 100: 100) in 2-methoxyethanol. A 0.2 M monoethanolamine (MEA) was then added in the precursor solution as a sol-gel stabilizer. After thoroughly mixing all components, the solution was stirred at 50 °C for 2 hours and then aged for 24 hours. IGZO thin film was deposited by a DMP 2831 ink-jet printer on the prepared substrate and heated at 300 °C in air for 10 mins after each printing. Printing and heating cycle was repeated for 3 times to achieve desired film thickness. Post-annealing was performed at 500 °C for 1 hour in air to remove residual chemicals and improve the quality of IGZO film. After that, 100 nm SiO₂ film was deposited by PECVD as the passivation layer of IGZO TFT. This sample is named as sample A. For comparison purpose, another IGZO TFT was fabricated on a pure 15 nm thick SiO₂ without embedded nc-Si using the same procedure, and it is named as sample B. The film morphology was characterized by atomic force microscopy (AFM). The transistor and memory performance was measured with a Keithley 4200 semiconductor analyzer.

Results and discussions

The basic device architecture is bottom gate bottom contact as shown in Fig. 1(a). The highly doped Si substrate is served as gate electrode, SiO₂ is employed as dielectric layer, nc-Si inside is acted as charge trap center, and printed IGZO is used as channel layer of TFTs. As simulated by the freeware called stopping and range of ion in matter (SRIM) [3], the implanted Si distributes from the oxide surface with a center depth of ~5 nm. The formation of nc-Si was confirmed by cross-sectional transmission electron microscopy (TEM) measurement [3]. The optical image of printed IGZO TFT memory and surface morphology of IGZO thin films are presented in Fig. 1(c) and

978-1-4673-4840-9/13 $31.00 © 2013 IEEE

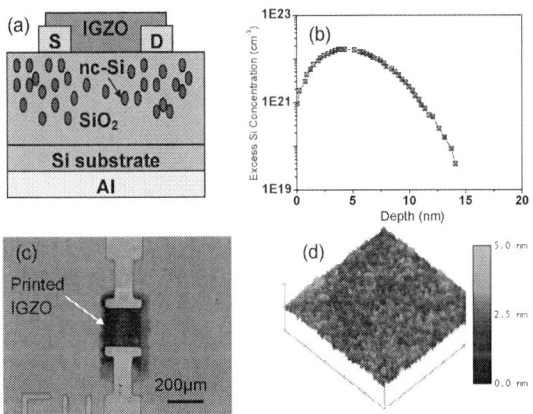

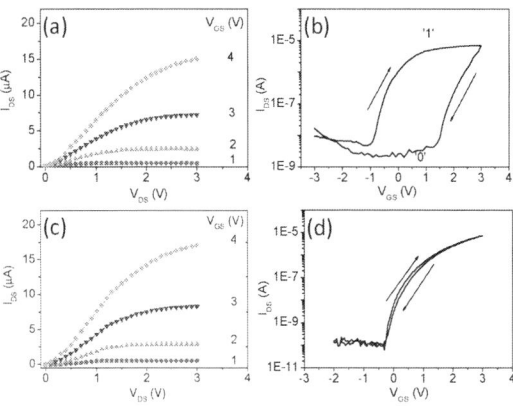

Fig. 1. (a) Schematics of ink-jet printed IGZO memory device architecture based on silicon nanocrystals embedded in SiO₂ gate dielectric. (b) Distributions of nc-Si in 15 nm SiO₂ obtained from SRIM simulation. (c) Optical image of the printed IGZO memory and (d) AFM image of surface morphology of the printed IGZO thin films.

Fig. 2. Output characteristics and (b) transfer characteristics of sample A, which is printed IGZO TFT on top of SiO₂ with nc-Si embedded inside of SiO₂. (c) Output characteristics and (d) transfer characteristics of sample B, which is the conventional printed IGZO TFT on SiO₂/Si substrate.

(d). Both the channel width and length of IGZO TFT are 200 μm. The root mean square (RMS) roughness of IGZO film is about 0.362 nm.

The transport properties of the sample A (the printed IGZO memory TFT on top of SiO₂ with nc-Si embedded inside) are shown in Fig. 2(a) and (b). Fig. 2(b) shows the transfer characteristic at $V_{DS} = 2$ V and a much larger hysteresis is observed as compared with sample B (Fig. 2(d)), due to the charging effect of Si nano-crystal in SiO₂ as proven in previous studies [1, 22]. The small clockwise hysteresis loop is observed in the transfer characteristics in sample B. Such kind of hysteresis is also observed in other FETs, commonly ascribed to the SiO₂/channel interface charge traps, defects in SiO₂ or impurities attracted to channel layer [19-21].

The threshold voltage of the sample A (defined as the gate voltage when the drain current reaches 4 nA in this paper) varies from -0.8 to 1.3 V as the gate voltage sweeps forward and backward, respectively. Supposing an electron is extracted per nano-crystal, the magnitude of the threshold voltage shift can be approximated by [1]:

$$\Delta V_T = \frac{qn_{well}}{\varepsilon_{ox}} \left(t_{cntl} + \frac{1}{2} \frac{\varepsilon_{ox}}{\varepsilon_{Si}} t_{well} \right) \quad (1)$$

where, ΔV_T is the threshold voltage shift, t_{cntl} is the thickness of the control of oxide under the gate, t_{well} is the linear dimension of the nano-crystal well, ε is the permittivity, q is the magnitude of electronic charge, and n_{well} is the density of nano-crystals. For $t_{well} = 4$ nm, $n_{well} = 4 \times 10^{12}$ cm⁻³, and $t_{cntl} = 10$ nm, the threshold voltage shift is 1.98 V, which is consistent with the experiment result of 2.1 V (shifted from -0.8 to 1.3 V).

Schematic diagrams of the write and erase process are shown in Fig. 3 [1]. When a positive bias voltage is applied on gate, negative charges are induced in IGZO channel layer and injected into nc-Si, as shown in Fig. 3(a). Positive charges are induced by injected negative charges in nc-Si after removing gate voltages, leading to the threshold voltage shifted to be more positive. Therefore, a small current can be obtained at a certain V_{GS}, 0.5 V for example, this state is '0', as shown in Fig.

2(b). This is the write '0' process. When a high enough negative bias voltage is applied on the gate electrode, electrons in the nc-Si are rejected into the TFT channel, leading to threshold voltage shifted to be more negative. Therefore, a large current can be obtained at a certain V_{GS}, 0.5 V for example, this state is '1'.

The current retention was evaluated by continuously monitoring the on and off state currents as a function of time after gate was programmed (Fig. 4). The drain current I_{ds} was continuously monitored at a fixed $V_{ds} = 2$ V after a gate voltage of +5 V and -5 V lasting for 10 s for off and on state respectively. The on/off currents are very stable in 1000 s and the window keeps at more than 10³, which enables reliable storage. In contrast, memory window of sample B is very small and gradually disappears in the period tested (Fig. 4). Therefore, sample B which is lack of nanocrystal embedded in dielectric layer is not suitable for memory application. Both on and off currents of sample B slightly increase with the increase of testing time after 200 s, which is caused by the increase in carriers in the channel due to Joule heating [21].

The endurance of the device is carried out by measuring the drain current stability as a function of program and erase cycles (i.e. switching between '1' and '0' state). Both on and off state currents were measured at a fixed $V_{ds} = 2$ V in between the interval (10 s) of gate voltage pulses ($V_{gs} = +5$ V (off state) and -5 V (on state) for 10 s). Each complete cycle lasts for 40 s. From the results shown in Fig. 5(a), the on state drain current maintains at about 2 μA after 100 cycles. The off state drain current is about 0.6 nA at the beginning and gradually increases to 2 nA after 100 cycles. The possible reasons of the increased off current include: (1) nc-Si not fully charged, (2) thermal effect introduced by multi-measuring, and (3) TFT fatigue. It has been demonstrated that nc-Si can work well for thousands of program/erase cycles [22]. If the increased off current is due to the thermal effect, the on current should be also increased which is not the case here (Fig. 5(a)). Therefore, the most possible reason is the fatigue problem of IGZO TFT, which has been also observed in solution processed ZnO TFTs, where the electrodes were eroded by the Joule heating [23]. We think that coupled with Joule heating, the inner micropores

978-1-4673-4840-9/13 $31.00 © 2013 IEEE

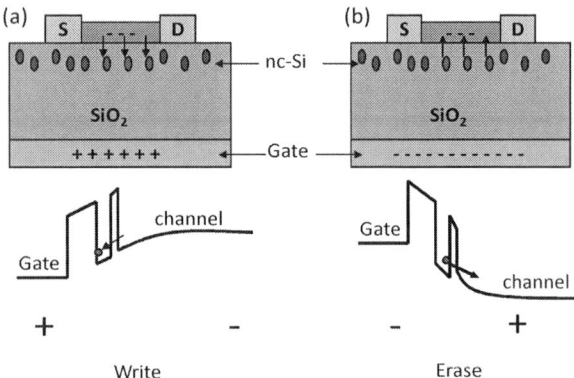

Fig. 3. A schematic diagram and related energy band diagram during (a) write and (b) erase '0' process of memory.

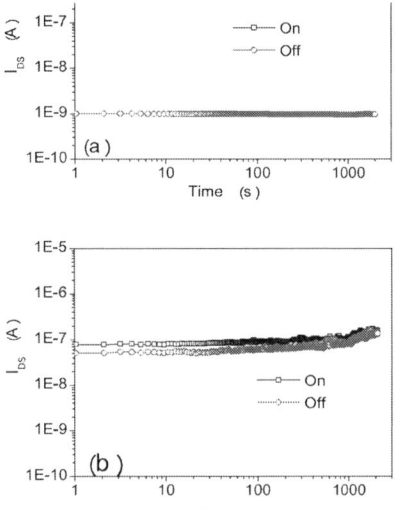

Fig. 4. Time evolution of the drain currents of ink-jet printed TFT devices with a fixed $V_{ds} = 2$ V. (a) Memory. (b) Conventional TFT on SiO$_2$/Si substrate. The gate pulse was ±5 V lasted for 10 s for (a) and (b).

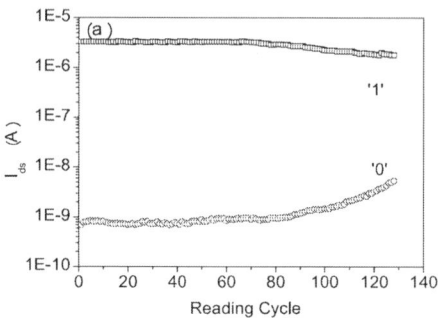

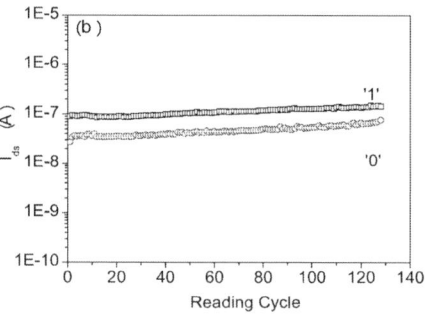

Fig. 5. (a) Endurance tests by measuring the off and on state drain current at a fixed $V_{ds} = 2$ V as a function of programming cycles of the printed IGZO devices. (a) Memory. (b) Conventional TFT on SiO$_2$/Si substrate.

memory device contains SiO$_2$ gate dielectric layer embedded with silicon nanocrystals, which act as charge trapping centers for the memory effect. The memory effect was evidenced by the clockwise loop of V_{gs}-I_d curve. The printed IGZO memory exhibits a high (up to 1×10^3) on/off ratio. Our results show that it is potential to fabricate low-cost transparent nonvolatile memory devices in large area by ink-jet printing technology.

Acknowledgement

The work is partially supported by the National Natural Science Foundation of China (NSFC) (Project Nos. 61006037, 61177014 and 61076015), and Tianjin Natural Science foundation (Project Nos. 11JCZDJC21900 and 11JCYDJC25800).

generated in the printed IGZO film during the process of vaporization, decomposition and crystallization is the main reason. The micropores are not stable after repeated current stress leading to IGZO fatigue. However, the exact reason is still under further investigation. The current on/off ratio remains more than 1×10^3 after 100 program and erase cycles. For comparison, the memory endurance of the IGZO on SiO$_2$/Si is shown in Fig. 5(b). The measurement condition is the same as that of sample A. In contrast to sample A, the current on/off ratio of sample B is less than 2, which is too small to be distinguished between on and off state.

Conclusion

In conclusion, a nonvolatile memory based on ink-jet printed In-Ga-Zn oxide thin film transistor was fabricated. The

References

[1] H. I. Hanafi, S. Tiwari, and I. Khan, "Fast and long retention-time nano-crystal memory," *IEEE Trans. Electron Devices*, vol. 43, no. 9, pp. 1553-1558, Sep 1996.

[2] E. Kapetanakis, P. Normand, D. Tsoukalas, et al.,"Charge storage and interface states effects in Si-nanocrystal memory obtained using low-energy Si$^+$ implantation and annealing," *Appl. Phys. Lett.*, vol. 77, no. 21, pp. 3450-3452, Nov 2000.

[3] J. I. Wong, T. P. Chen, M. Yang, et al., "Current conduction in Al/Si nanocrystal embedded SiO$_2$/p-Si diodes with various distributions of Si nanocrystals in the oxide," *J. Appl. Phys.*, vol. 106, no. 1, p. 013718, Jul 2009.

[4] K. Nomura, H. Ohta, A. Takagi, et al., "Room-temperature fabrication of transparent flexible thin-film transistors using amorphous oxide semiconductors," *Nature*, vol. 432, no. 7016,

pp. 488-492, Nov 2004.

[5] K. Nomura, H. Ohta, K. Ueda, et al., "Thin-film transistor fabricated in single-crystalline transparent oxide semiconductor," *Science,* vol. 300, no. 5623, pp. 1269-1272, May 2003.

[6] J. K. Jeong, J. H. Jeong, H. W. Yang, et al., "High performance thin film transistors with cosputtered amorphous indium gallium zinc oxide channel," *Appl. Phys. Lett.,* vol. 91, no. 11, p. 113505, Sep 2007.

[7] M. Kim, J. H. Jeong, H. J. Lee, et al., "High mobility bottom gate InGaZnO thin film transistors with SiO_x etch stopper," *Appl. Phys. Lett.,* vol. 90, no. 21, p. 212114, May 2007.

[8] Y. S. Park, S. Y. Lee, and J. S. Lee, "Nanofloating Gate Memory Devices Based on Controlled Metallic Nanoparticle-Embedded InGaZnO TFTs," *IEEE Electron Device Lett.,* vol. 31, no. 10, pp.1134-1136, Oct 2010.

[9] N. C. Su, S. J. Wang, and A. Chin, "A Nonvolatile InGaZnO Charge-Trapping-Engineered Flash Memory With Good Retention Characteristics," *IEEE Electron Device Lett.,* vol. 31, no. 3, pp. 201-203, Mar 2010.

[10] A. Suresh, S. Novak, P. Wellenius, et al., "Transparent indium gallium zinc oxide transistor based floating gate memory with platinum nanoparticles in the gate dielectric," *Appl. Phys. Lett.,* vol. 94, no. 12, p.123501, Mar 2009.

[11] H. X. Yin, S. Kim, H. Lim, et al., "Program/erase characteristics of amorphous gallium indium zinc oxide nonvolatile memory," *IEEE Trans. Electron Devices,* vol. 55, no. 8, pp. 2071-2077, Aug 2008.

[12] M. C. Chen, T. C. Chang, C. T. Tsai, et al., "Influence of electrode material on the resistive memory switching property of indium gallium zinc oxide thin films," *Appl. Phys. Lett.,* vol. 96, no. 26, p. 262110, Jun 2010.

[13] B. Ong, "Semiconductor ink advances flexible displays," *Laser Focus World,* vol. 40, no. 6, pp. 85-88, Jun 2004.

[14] B. J. de Gans, P. C. Duineveld, and U. S. Schubert, "Inkjet printing of polymers: State of the art and future developments," *Adv. Mater.,* vol. 16, no. 3, pp. 203-213, Feb 2004.

[15] C. Y. Ng, T. P. Chen, L. Ding, et al., "Memory characteristics of MOSFETs with densely stacked silicon nanocrystal layers in the gate oxide synthesized by low-energy ion beam," *IEEE Electron Device Lett.,* vol. 27, no. 4, pp. 231-233, Apr 2006.

[16] L. Ding, T. P. Chen, M. Yang, et al., "Photon-induced conduction modulation in SiO2 thin films embedded with Ge nanocrystals," *Appl. Phys. Lett.,* vol. 90, no. 10, p. 103102, Mar 2007.

[17] Y. Wang, S. W. Liu, X. W. Sun, et al., "Highly transparent solution processed In-Ga-Zn oxide thin films and thin film transistors," *J. Sol-Gel Sci. Technol.,* vol 55, no. 3, pp. 322-327, Sep 2010.

[18] Y. Wang, X. W. Sun, G. K. L. Goh, et al., "Influence of Channel Layer Thickness on the Electrical Performances of Inkjet-Printed In-Ga-Zn Oxide Thin-Film Transistors," *IEEE Trans. Electron Devices,* vol 58, no. 2, pp. 480-485, Feb 2011.

[19] L. Liao, H. J. Fan, B. Yan, et al., "Ferroelectric Transistors with Nanowire Channel: Toward Nonvolatile Memory Applications," *Acs Nano,* vol. 3, no. 3, pp. 700-706, Mar 2009.

[20] O. Hayden, M. T. Bjork, H. Schmid, et al., "Fully depleted nanowire field-effect transistor in inversion mode," *Small,* vol. 3, no. 2, pp. 230-234, Feb 2007.

[21] W. Kim, A. Javey, O. Vermesh, et al., "Hysteresis caused by water molecules in carbon nanotube field-effect transistors," *Nano Lett.,* vol.3, no. 2, pp. 193-198, Feb 2003.

[22] J. I. Wong, T. P. Chen, M. Yang, et al., "Influence of excess Si distribution in the gate oxide on the memory characteristics of MOSFETs," *Appl. Phys. A-Mater.,* vol. 91, no. 3, pp. 411-413, Jun 2008.

[23] C. S. Li, Y. N. Li, Y. L. Wu, et al., "Fabrication conditions for solution-processed high-mobility ZnO thin-film transistors," *J. Mater. Chem.,* vol. 19, no. 11, pp. 1626-1634, 2009.

Ultimate Performance Projection of Ballistic III-V Ultra-Thin-Body MOSFET

Yan Guo[†], Kai-Tak Lam, Yee-Chia Yeo, *and* Gengchiau Liang*

Department of Electrical and Computer Engineering
National University of Singapore (NUS)
4 Engineering Drive 3, Singapore 117576, Republic of Singapore
†guoy@nus.edu.sg, *elelg@nus.edu.sg

Abstract—we investigate the device performance of III-V ultra-thin-body field-effect transistors with the consideration of the effects of materials, body thickness and dielectric effect based on the top-of-barrier model. These three major factors predominate the transport performance in double-gate ultra-thin-body MOSFET for the same transport and surface orientation. Firstly, we observe that among these selected III-Vs for n-type FETs, GaSb has the largest ON-state current due to large charge density caused by subband degeneracy at high bias. Moreover, the effect of the number of layer on FET performance is investigated. The current increases as the number of layers increases from 12 layers to 24 layers but degrades as it keep increasing. Lastly, the advantage of increased dielectric constant on ballistic transport is reduced for material having less density of states such as InSb due to its smaller quantum capacitance.

Keywords- III-V, ultra-thin-body, ballistic transport, atomistic simulation.

I. INTRODUCTION

Research on III-V ultra-thin-body (UTB) transistors draws much attention because of their near perfect control of the OFF-state current due to the ultra-thin thickness, and high ON-state current with a high injection velocity [1][2], indicating that the III-Vs are potential candidates of the novel channel materials for next generation FET devices [3]. In this work, therefore, we investigate a variety of potential III-V UTB transistors and evaluate their ultimate performance based on ballistic transport model [4]. Our results show that GaSb and its ternary have the largest ON-state current. Furthermore, by varying the body thickness from 12 atomic layers (AL, 1.83nm) to 48 AL (7.32nm) of GaSb, we observed that 24 AL DG-UTB transistor show the largest ON-state currents due to its largest electron density and high injection velocity. In addition, the increase of ON-state current (V_G=0.8V) for GaSb is four times larger by replacing SiO_2 with HfO_3 while the increase for InSb is less than twice because of its relatively small quantum capacitance [5][6].

II. SIMULATION SETUP

Fig. 1 shows the side view of the simulated UTB structure and the atomistic arrangement of the III-V channel materials along the confinement direction. The transport is along [100]/(001). In this paper, we firstly study electronic structures of III-V ultra-thin-body (UTB) MOSFETs based on $sp^3d^5s^*$ tight-binding model with solving the Poisson's equation self-consistently. The developed self-consistent mechanism could accurately model the electrostatic potential at each atomic site, according to previously reported on Si nanowire [7]. The top-of-barrier model is implemented for the ballistic transport characteristics of the 2-D devices [4]. Source and drain effect on the electrostatics of the channel is neglected to focus on the band structure effect of the material.

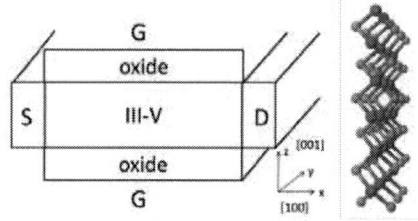

Figure 1. (Left) The DG-UTB structure simulated in this paper. (Right) Atomic representations of ultra-thin-body MOSFET structure. The insets show the transport (x-axis, [100]) and confinement (z-axis) directions. The atomic arrangement is along the confinement. The brown and purple balls represent the anion and cation respectively.

III. RESULTS AND DISCUSSION

Fig. 2 shows the drain current for I_D-V_G (V_D=0.8V) for the DG-UTB device with EOT=1nm and body thickness of 12 AL. The OFF current is manipulated by adjusting the source Fermi level position to maintain at the 0.1 μA /μm. Among these various III-V channel materials, GaSb shows the highest ON-state current for n-type, which agrees with the previous report [8]. As shown in the table 1, the charge density contributing to the ballistic current and injection velocity at Vg=0.8V are calculated. The product of electron density and injection velocity could well reflect the ON-state current. According to Table 1, we found that GaSb has extremely high carrier density, more than twice of the second highest material ($In_{0.3}Ga_{0.7}Sb$). It can be attributed to that GaSb UTB offers multiple conduction subbands projected from L-valley, capable of carrying high-velocity electrons.

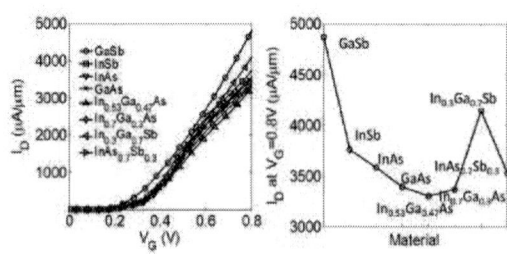

Figure 2. (a) The I_D-V_G (V_D=0.8V) curve for the eight III-V materials investigated. I_{off} is adjusted to 0.1 μA/μm by tuning the source Fermi level position. (b) Comparison of Ion for different III-V materials at Vg=0.8 μA/μm.

978-1-4673-4840-9/13 $31.00 © 2013 IEEE

Material	Electron Density $(\times 10^{15}/cm^2)$	Average Injection Velocity $(\times 10^7 cm/s)$	Product $(\times 10^{22}/(cm\ast s))$
GaSb	**1.835**	1.653	3.033
InSb	0.483	4.909	2.371
InAs	0.530	4.200	2.226
GaAs	0.901	2.346	2.114
$In_{0.53}Ga_{0.47}As$	0.691	2.997	2.071
$In_{0.7}Ga_{0.3}As$	0.609	3.508	2.136
$In_{0.3}Ga_{0.7}Sb$	0.970	2.692	2.611
$InAs_{0.7}Sb_{0.3}$	0.504	4.324	2.179

Table 1. Electron density, average injection velocity at ON-state and the product of the two for eight different III-Vs.

Next, the transport characteristics of GaSb with different body thickness, i.e., 12 AL (1.83 nm), 24 AL (3.66 nm), 36 AL (5.49 nm) and 48 AL (7.32 nm), are investigated. It can be found that from 12 to 24 AL, ON-state current increases simply due to the size effect. On the other hand, for 36 AL and 48 AL, the current drops even less than the 12 AL, and it can be attributed to its largest electron densities compared to 36 and 48 AL (Fig. 3(c)) caused by the stronger body inversion of the thinner body thickness. As shown in Fig. 3(b), it can be clearly found that electron density in the centre of the channel becomes flatter as the thickness increases, in which case electrical field penetrates harder into the centre of the body, and it tends to become the surface inversion as the body becomes thicker. Furthermore, we investigate the thickness effects of UTB on the average electron injection velocity, and observe that 24, 36 and 48 AL have significant improvement of injection velocity over 12 AL. This confirms that as the thickness increases, relieved confinement helps reduce the effective mass resulting in increasing the electron velocity of the UTB structures.

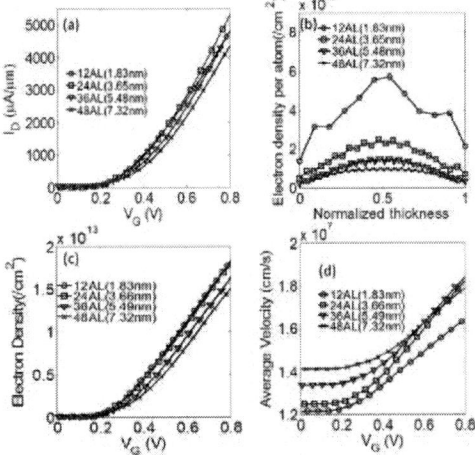

Figure 3. (a) The I_D-V_G (V_D=0.8V) curve for GaSb with 12, 24, 36 and 48 atomic layers. I_{off} is adjusted to 0.1 µA/µm. (b) Charge density distribution along the channel for the four different layers. (c) Injection velocity for four different layers with V_G varying from 0V to 0.8V. (d) Electron density for four different layers with V_G varying from 0V to 0.8V.

In addition to material and body thickness, dielectric effect is equally important for device performance of FETs because the increase of oxide capacitance improves the overall gate control, in principle. This effect is more significant for materials with the large quantum capacitance, such as GaSb, because the large insulator capacitance makes its effect less significant for the overall gate capacitance, resulting in the larger weights of the quantum capacitance in the channel to determine the gate capacitance. As shown in Fig. 4, we compare the I_D-V_G curves of GaSb (large C_Q) and InSb (small C_Q) by using SiO_2 (ε=4) and HfO_3 (ε=25). GaSb current is 4 times larger by switching from SiO_2 to HfO_3 while for InSb only twice larger. Therefore, the material with larger density-of-state has larger improvement of its ON-state current by increasing the insulator dielectric constant. It also demonstrated that the low density-of-state materials will suffer the bottleneck of improving device performance with shrinking down the size of the devices although their injection velocity is much high.

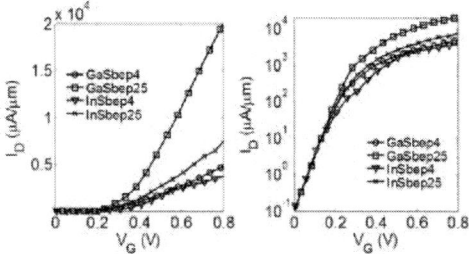

Figure 4. (a) The ID-VG (VD=0.8V) curve for GaSb and InSb with SiO_2 (ε=4) and HfO_3 (ε=25) as the insulator layer. Ioff is adjusted to 0.1 µA/µm. (b) log scale.

IV. CONCLUSION

We present a study of material property, body thickness and insulator dielectric effect on the transport performance of III-V UTB transistor. The GaSb is found to have the largest ON-state current among eight different III-Vs. By increasing the body layer thickness of GaSb transistor, the ON-state current increases from 12 AL to 24 AL but drops to the lowest of 36 AL due to the combinational effects of electron density and injection velocity. The improvement of the insulator capacitance plays a significant role in material with a large density-of-state/quantum capacitance.

ACKNOWLEDGMENT

This work was supported by the National Research Foundation (NRF) of Singapore under project grant NRF-CRP6-2010-04.

REFERENCES

[1] M. Luisier, "Performance comparison of GaSb, strained-Si, and InGaAs double-gate ultrathin-body n-FETs", IEEE Electron Device Lett, vol. 32, no. 12, December 2011.

[2] J.A del Alamo, "Nanometre-scale electronics with III-V compound semiconductors," Nature, vol. 479, pp. 317-323, November 2011.

[3] Y. Liu, N. Neophytos, G. Klimeck and M.S. Lundstrom, "Band-structure effects on the performance of III-V ultrathin-body SOI MOSFETs", IEEE Trans. on Electron Devices, vol. 55, no. 5, May 2008.

[4] A. Rahman, J. Guo, S. Datta and M.S. Lundstrom, "Theory of ballistic nanotransistors", IEEE Trans. on Electron Devices, vol. 50, no. 9, September 2003.

[5] T. Krishnamohan, D. Kim and K.C. Saraswat, "Chapter 2 Properties and trade-offs of compound semiconductor MOSFETs", pp 9-30, "Fundamentals of III-V semiconductor MOSFETs", Springer, 2010.

[6] N. Takiguchi, S. Koba, H. Tsuchiya and M. Ogawa, "Comparisons of performance potentials of Si and InAs nanowire MOSFETs under ballistic transport", IEEE Trans. on Electron Devices, vol. 59, no. 1, January 2012.

[7] N. neophytou, A. Paul, M.S. Lundstrom and G. Klimeck, "Bandstructure effects in silicon nanowire electron transport", IEEE Trans. on Electron Devices, vol. 55, no. 6, June 2008.

[8] R. Kim, T. Rakshit, R. Kotlyar, S. Hasan and C.E. Webber, "Effects of surface orientation on the performance of idealized III-V thin-body ballistic n-MOSFETs", IEEE Electron Device Lett, vol. 32, no. 6, June 2011.

Fully CMOS Compatible 1T1R Integration of Vertical Nanopillar GAA Transistor and Oxide Based RRAM Cell for High Density Nonvolatile Memory Application

Z. Fang[*], X. P. Wang, B. B. Weng, Z. X. Chen, A. Kamath, G. Q. Lo and D. L. Kwong

Institute of Microelectronics
A*STAR (Agency for Science, Technology and Research)
Singapore 117685
[*]e-mail: FANGZ@ime.a-star.edu.sg

Abstract

Fully CMOS compatible vertical nanopillar GAA transistor integrated with Oxide based RRAM cell to realize $4F^2$ footprint has been demonstrated and systematically characterized. Nanopillar transistor exhibits excellent transfer characteristics with diameter down to a few tens nanometer. Three type of resistive switching behavior have been found in the fabricated 1T1R cell, namely pre-forming ultralow current switching, unipolar switching and bipolar switching after forming process. Reset current of only 200pA has been observed in pre-forming ultralow current switching; while for unipolar and bipolar switching after forming process, good memory performance and operation parameter uniformity is demonstrated. Furthermore, reset current is found to decrease with reducing nanopillar transistor design diameter, which is beneficial for circuit power consumption concern.

Introduction

Resistive random access memory (RRAM) has attracted significant considerations for potentially becoming the next generation nonvolatile memory [1-5]. Compared to conventional Flash memory, RRAM device exhibits lower operation voltage and higher access speed, in additional to low power consumption, superior data retention, high-density capacity and CMOS compatibility [6-8]. Among the various transition metal oxides being explored, HfO_x has shown outstanding resistive switching performance [6-9].

In order to achieve high density memory array, crosstalk between adjacent cells must be avoided by adapting a one selection device one resistor or similar structures. Due to large feature size in planar transistor [10] and high leakage current in bipolar junction transistor (BJT) [11], a more direct method to achieve $4F^2$ density is to integrate RRAM cell with vertical nanopillar gate-all-around (GAA) transistor which has been demonstrated feasibility as an option for 15nm and beyond technology nodes with sub-10nm channel length devices through both simulations and experiments [13-14].

In this work, we demonstrate the integration of vertical nanopillar GAA transistor with Oxide based RRAM cell to achieve $4F^2$ footprint and systematically investigate 1T1R architecture in nanometer scale for high density nonvolatile memory application.

Fabrication Process

Device fabrication was done on standard 8 inch CMOS platform, the architecture of 1T1R memory cell is schematically shown in Fig. 1. A more detailed process flow of similar vertical transistor can be found in the reference [12].

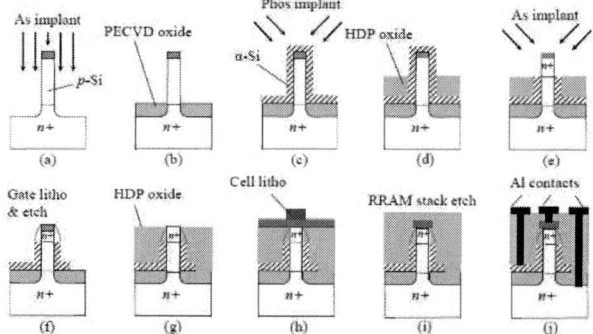

Figure 1. Fabrication process flow of the vertical nanopillar 1T1R memory cell: (a) Si nanopillar formation with nitride hard mask etching, trimming with oxidation for a smaller diameter and source implantation doping; **(b)** plasma enhanced chemical vapor deposition (PECVD) isolation oxide; **(c)** gate oxide growth and amorphous Si (α-Si) gate deposition and implantation; **(d)** HDP isolation oxide deposition and etch back to expose Si nanopillar; **(e)** excess gate α-Si removal and As drain implantation; **(f)** spacer formation and oxide strip, followed by α-Si gate patterning; **(g)** HDP oxide deposition and CMP to expose and remove hard mask; **(h)** RRAM memory cell deposition and patterning after second nitride spacer formed on nanopillar tip sidewall; **(i)** pre-metal dielectric (PMD) oxide deposition for passivation; **(j)** metallization with Al metal pad.

Results and Discussion

The fabricated device exhibits three modes of resistive switching operations, namely pre-forming ultralow current switching, unipolar switching and bipolar switching. The different resistive switching characteristics are presented as followed.

A. Ultralow current switching

A forming process is usually necessary to initialize the resistive switching in RRAM [6], however, in the 1T1R device fabricated in this work, a pre-forming resistive switching with ultralow current is observed. An external 20nA current compliance is set to prevent unexpected current jump during voltage sweeping and reset current of below 200pA is demonstrated as shown in Fig. 2.

978-1-4673-4840-9/13 $31.00 © 2013 IEEE

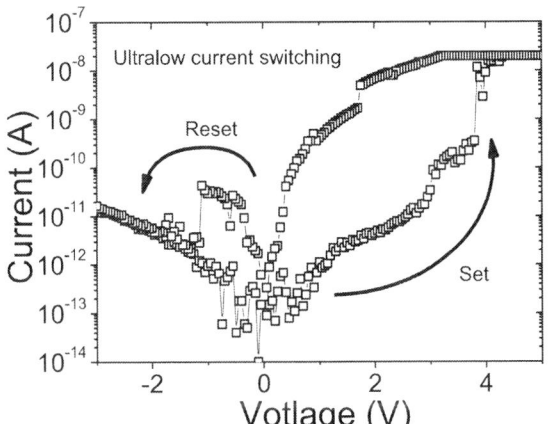

Figure 2. Typical pre-forming ultralow current switching IV characteristics with reset current of below 200pA.

Such low current bipolar switching is possibly due to oxygen ion movement in HfOx layer. Without high voltage forming process, a thin layer of SiO2 exists between nanopillar transistor tip and HfOx switching layer, which limits the current levels. When positive voltage is applied on the transistor drain (Ni/TiN electrode), negatively charged oxygen ions in HfOx tends to be attracted to Ni layer, which results in a more oxygen deficient oxide and higher conductance of the stack. While for reset process, reverse movement of oxygen ions recovers the resistance of the memory cell.

B. Unipolar switching

Besides pre-forming ultralow current switching, 1T1R cell in this work also demonstrates unipolar and bipolar switching behaviors after proper forming process around 5-6V at positive direction.

Typical unipolar switching curve of the memory device is shown in Fig. 3. Gate controlled transistor current is used to prevent device from hard breakdown. Unipolar switching behavior of memory stack n+-Si/HfOx/Ni/TiN is very similar to that reported by Tran, et al. [16]. It is commonly accepted that the origin of unipolar switching is the formation and thermal dissolution of conduction filament in NiO layer, in this case NiO layer at HfOx/Ni interface.

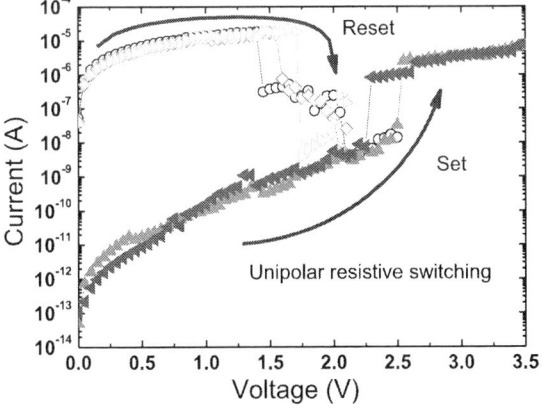

Figure 3. Typical resistive switching IV characteristics of unipolar switching in n+-Si/HfOx/Ni/TiN stack.

It is also observed in Fig. 4 that reset current decreases as critical dimension (CD) of the nanopillar transistor decreases. This may be due to the current limiting effect of the transistor on resistive switching.

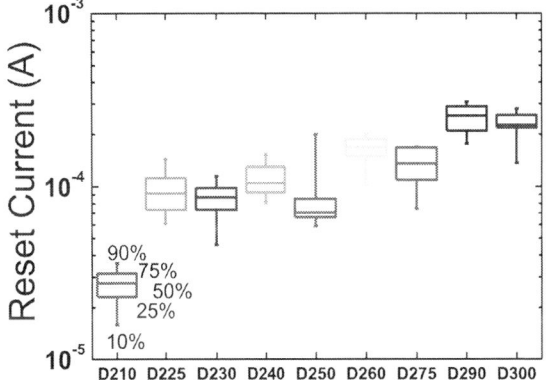

Figure 4. Reset current in unipolar switching plots versus nanopillar transistor diameter. In this box chart plot, upper and lower whisker indicate 90% and 10%, upper and lower of rectangular box indicate 75% and 25% and the middle line indicate 50% percentage distribution respectively.

C. Bipolar switching

Bias voltage polarity dependent bipolar switching is also observed in the fabricated memory cells. Fig. 5 plots the typical bipolar switching IV characteristics. Set process occurs at positive bias with nanopillar transistor limited current without system current compliance; while reset occurs at negative bias side without any current compliance.

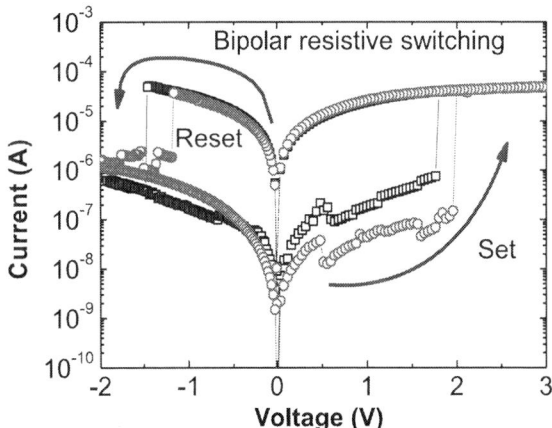

Figure 5. Bipolar resistive switching IV characteristics from 1T1R configuration after forming process. Device set in positive bias with current limited by nanopillar transistor while reset in negative direction without current compliance.

Similar to unipolar switching mode, reset current in bipolar mode is also found to reduce with shrinking CD of nanopillar transistor as shown in Fig. 6. Median reset current value reduces from more than 200uA in 295nm device to about 30uA in the 220nm device. With further device scaling, this value is expected to lower further, which is beneficial for lowering of power consumption in high density memory integration.

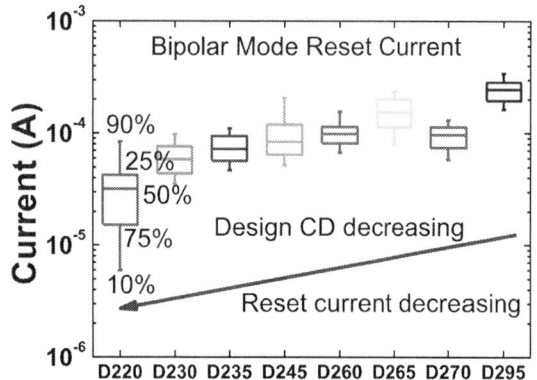

Figure 6. Reset current statistic in bipolar mode versus different nanopillar transistor design CD. A similar decreasing trend versus transistor diameter can be seen as in unipolar switching mode.

Conclusion

In this work, vertical nanopillar GAA transistor integrated with Oxide based RRAM cell has been demonstrated and systematically investigated. It is the first time to realize 1T1R configuration with $4F^2$ footprint. Three type of resistive switching behavior have been observed in the fabricated cell. For pre-forming ultralow current switching, reset current of only 200pA has been achieved. While for unipolar and bipolar switching after forming process, good operation parameter uniformity as well as device uniformity is demonstrated. More than that, reset current is found to decrease with reducing nanopillar transistor design diameter, which is beneficial for circuit power consumption concern.

Acknowledgement

This work is supported by Future Data Center Technologies Thematic Strategic Research Programme: NVM based on integration of PCRAM and RRAM cells with ultra scaled vertical Si nanowire devices (SERC Grant No: 1121720016).

References

[1] A. Asamitsu, Y. Tomioka, H. Kuwahara, and Y. Tokura, "Current switching of resistive states in magnetoresistive manganites," *Nature,* vol. 388, pp. 50-52, 1997.

[2] A. Beck, J. G. Bednorz, C. Gerber, C. Rossel, and D. Widmer, "Reproducible switching effect in thin oxide films for memory applications," *Applied Physics Letters,* vol. 77, pp. 139-141, 2000.

[3] I. G. Baek, M. S. Lee, S. Seo, M. J. Lee, D. H. Seo, D. S. Suh, J. C. Park, S. O. Park, H. S. Kim, I. K. Yoo, U. I. Chung, and J. T. Moon, "Highly scalable nonvolatile resistive memory using simple binary oxide driven by asymmetric unipolar voltage pulses," in *IEDM Technical Digest.* 2004, pp. 587-590.

[4] I. G. Baek, D. C. Kim, M. J. Lee, H. J. Kim, E. K. Yim, M. S. Lee, J. E. Lee, S. E. Ahn, S. Seo, J. H. Lee, J. C. Park, Y. K. Cha, S. O. Park, H. S. Kim, I. K. Yoo, U. I. Chung, J. T. Moon, and B. I. Ryu, "Multi-layer cross-point binary oxide resistive memory (OxRRAM) for post-NAND storage application," in *IEDM Technical Digest.* 2005, pp. 750-753.

[5] C. H. Ho, E. K. Lai, M. D. Lee, C. L. Pan, Y. D. Yao, K. Y. Hsieh, R. Liu, and C. Y. Lu, "A Highly Reliable Self-Aligned Graded Oxide WOx Resistance Memory: Conduction Mechanisms and Re-

liability," in *VLSI Technology Symp.,* 2007, pp. 228-229.

[6] R. Waser and M. Aono, "Nanoionics-based resistive switching memories," *Nature Materials,* vol. 6, pp. 833-840, 2007.

[7] A. Sawa, "Resistive switching in transition metal oxides," *Materials Today,* vol. 11, pp. 28-36, 2008.

[8] H. Akinaga and H. Shima, "Resistive Random Access Memory (ReRAM) Based on Metal Oxides," *Proceedings of the IEEE,* vol. 98, pp. 2237-2251, 2010.

[9] Z. Fang, H. Y. Yu, X. Li, N. Singh, G. Q. Lo, and D. L. Kwong, "HfOx/TiOx/HfOx/TiOx Multilayer-Based Forming-Free RRAM Devices With Excellent Uniformity," *IEEE Electron Device Letters,* vol. 32, pp. 566-568, 2011.

[10] T. Yuan Heng, H. Chia-En, C. H. Kuo, Y. D. Chih, and L. Chrong Jung, "High density and ultra small cell size of Contact ReRAM (CR-RAM) in 90nm CMOS logic technology and circuits," in *IEDM Technical Digest.* 2009, pp. 5.6.1-5.6.4.

[11] W. Ching-Hua, T. Yi-Hung, L. Kai-Chun, C. Meng-Fan, K. Ya-Chin, L. Chrong-Jung, S. Shyh-Shyuan, C. Yu-Sheng, L. Heng-Yuan, F. T. Chen, and T. Ming-Jinn, "Three-dimensional 4F2 ReRAM cell with CMOS logic compatible process," in *IEDM Technical Digest.* 2010, pp. 29.6.1-29.6.4.

[12] D.-L. Kwong, X. Li, Y. Sun, G. Ramanathan, Z. X. Chen, S. M. Wong, Y. Li, N. S. Shen, K. Buddharaju, H. Y. Yu, S. J. Lee, N. Singh, and G. Q. Lo, "Vertical Silicon Nanowire Platform for Low Power Electronics and Clean Energy Applications," *Journal of Nanotechnology,* vol. 2012, p. 492121, 2012.

[13] E. Gnani, S. Reggiani, M. Rudan, and G. Baccarani, "Design Considerations and Comparative Investigation of Ultra-Thin SOI, Double-Gate and Cylindrical Nanowire FETs," in *European Solid-State Device Research Conference,* 2006, pp. 371-374.

[14] Y. Jiang, T. Y. Liow, N. Singh, L. H. Tan, G. Q. Lo, D. S. H. Chan, and D. L. Kwong, "Performance breakthrough in 8 nm gate length Gate-All-Around nanowire transistors using metallic nanowire contacts," in *VLSI Technology Symp.,* 2008, pp. 34-35.

[15] J. H. Stathis, "Percolation models for gate oxide breakdown," *Journal of Applied Physics,* vol. 86, pp. 5757-5766, 1999.

[16] X. A. Tran, B. Gao, J. F. Kang, X. Wu, L. Wu, Z. Fang, Z. R. Wang, K. L. Pey, Y. C. Yeo, A. Y. Du, M. Liu, B. Y. Nguyen, M. F. Li, and H. Y. Yu, "Self-rectifying and forming-free unipolar HfOx based-high performance RRAM built by fab-avaialbe materials," in *IEDM Technical Digest.* 2011, pp. 31.2.1-31.2.4.

Ultra-low Turn-on Field and Ultra-high Field Emission Current Density from Pillar Array Design of Carbon Nanotubes with Optimum R/H Ratio

Chuan-Ping Juan[1], Jun-Han Lin[2]

1. Department of Electronic Engineering, St., John's University, Taipei, Taiwan R.O.C
2. Nano Electronics and Display Technology Lab. Department of Electronics Engineering and Institute of Electronics, National Chiao Tung University Hsinchu, Taiwan, R.O.C.
* E-mail address: cpjuan @mail.sju.edu.tw

Abstract

It is reported that field emission can effectively enhanced for aligned carbon nanotubes (CNTs) as field emitters when the ratio of distance between neighboring nanotubes to the height of each individual CNT is about 2. In this paper, we proposed pillar array design of CNTs to fulfill the optimum R/H ratio (the ratio between inter-pillar distance R and pillar height H) with high density of aligned CNTs by changing H while maintaining R at 100μm, respectively. From the obtained experimental results, ultra-low turn-on field at 0.81V/μm and the ultra-high field emission current density (1A/cm^2 at 2.8 V/μm) were achieved at the optimum R/H ratio of 2 with the inter-pillar distance of 100μm.

Introduction

Since the first discovery of CNTs by Iijima in 1991[1], carbon nanotubes (CNTs) have drawn considerable interest as cold cathodes for field-emission displays[2-5], tips for scanning probe microscopes or biological probes,[6] sensors,[7] and quantum wires for nanoscale electronics[8-9], because of their unique properties such as high aspect ratio, well chemical stability, high mechanical strength, small radii of curvature at the tip apex, extremely low applied threshold field, and easily deliver a relatively high emission current density [5,9-10].However, serious screening effect caused by the high density of CNTs (~10^{10}/cm^2) grown by chemical vapor deposition (CVD) deteriorate their field emission characteristics. According to the calculations performed by Nilson et al., the field emission can effectively enhanced for aligned carbon nanotubes(CNTs) as field emitters when the ratio of distance between neighboring nanotubes to the height of each individual CNT is about 2. [11].Therefore, the density control of CNTs plays a significant role to improve the field emission properties. There are several methods for controlling CNT density, such as the use of electron beam lithography [12], self-assembly nanosphere lithography [13] oxide capping layer on metal catalyst [14], and the plasma post-treatment of CNTs film [15]. However, CNT array is either difficult in manipulating individual CNTs over a large area or the density is still very high and cannot be controlled effectively. In this paper, we proposed arrays of CNTs' bundles with uniform length, small diameter and easily controlled length by changing the

process parameters and succeeded in fabricating different ratios of inter-pillar distance (R) to pillar height (H). Ultra-low turn-on field at 0.81V/μm and the ultra-high field emission current density (1A/cm^2 at 2.8 V/μm) were achieved at the optimum R/H ratio of 2 with the inter-pillar distance of 100μm. Besides, well graphite structure of CNTs was confirmed by Raman analyses and well field emission image uniformity was also achieved at the optimum R/H ratio of 2 with the inter-pillar distance of 100μm.

Experimental Procedure

In this study, an array of 50×50 square patterns with 100μm inter-pillar distance and each square is 20×20 μm^2 were used to grow CNTs pillar arrays. Co (5nm)/Ti (50nm) multilayer catalyst were deposited on the photoresist patterned Si substrate. CNTs were grown selectively on the cobalt layers by the atmospheric pressure thermal chemical vapor deposition. Co/Ti bi-layered catalysts were pre-treated in N$_2$ (500sccm)/H$_2$ (300sccm) for 15min at 700°C. Length control of CNTs' pillar arrays to obtained different R/H ratios were grown at 700°C for 15min, 30min, 45min and 60 minutes by adding C$_2$H$_4$ at the flow rate of 20sccm to the N$_2$/H$_2$ mixture gases. The field emission tests were measured in a parallel plate diode configuration at room temperature in vacuum of ~ 3•10^{-6} Torr. The spacer between the anode and CNTs cathode is approximately 150μm and the emitting area is 10^{-2} cm^2. The field emission current is measured as a function of the anode voltage. Scanning electron microscopy (SEM) is used to examine the morphology of carbon nanotubes. The crystallinity of CNTs was analyzed by Raman spectrum.

Results and Discussion

Figure 1 showed the pixel design of CNTs' pillar arrays. There were 2500 square patterns with 100μm inter-pillar distance and each square is 20×20 μm^2. The SEM results of CNTs showed in Fig.2 demonstrated that CNTs were aligned, closely spaced. The lengths of CNTs' pillar were 10.1um, 21.6 um, 31.9 um and 49 um, respectively. The R/H ratios were 9.9, 4.63, 3.15 and 2.04, respectively. Figure 3 shows the Raman spectrum of the CNTs' pillar array grown under 15 min pre-treatment in N$_2$/H$_2$ gas mixtures

978-1-4673-4840-9/13 $31.00 © 2013 IEEE

(500 sccm/300 sccm) and growth in $C_2H_4/N_2/H_2$ gas mixtures (20sccm/500 sccm/300 sccm) for 30 min. The higher intensity of D peak (1350 cm^{-1}) clearly demonstrates that the CNTs have a highly crystalline graphite structure. Carbon nanotubes on the edge of each pillar can be envisioned as a major emission sites because of a dominant electric field concentration on their tops. Thus, pillar arrays of CNTs can be regarded as individual field emission sites. In this paper, we replace single CNT bundle as an individual field emission site and fulfill the calculations performed by Nilson. Figure 4 shows the field emission properties of CNTs. The ultra-low turn-on field at 0.81V/μm and the ultra-high field emission current density (1A/cm^2 at 2.8 V/μm) were achieved at the optimum R/H ratio of 2 with the inter-pillar distance of 100μm. The linearity of the F-N plot confirmed the field emission phenomenon. Well field emission uniformity of CNTs' pillar array with electric field at 2.3V/um was showed in Fig.5.

Conclusions

In this study, high quality of CNTs' bundle arrays with different length was used to demonstrate different R/H ratios. The turn-on field decreased from 2.27 V/um to 0.81 V/um when the R/H ratio decreased from 9.9 to 2. We have successfully fulfilled the calculations performed by Nilson. Ultra-low turn-on field at 0.81V/μm and the ultra-high field emission current density (1A/cm^2 at 2.8 V/μm) were achieved at the optimum R/H ratio of 2 with the inter-pillar distance of 100μm. Therefore, design of CNT pillar array is one of the best way for achieving higher field emission current density and lower operating voltage.

Figure 1 Pixel design of CNTs' pillar arrays

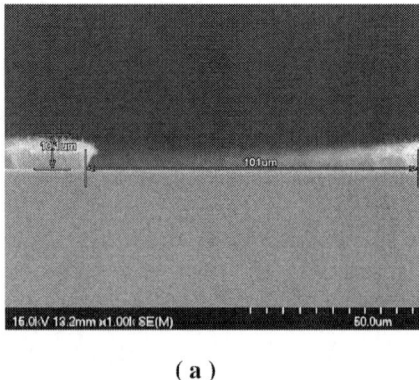

(a)

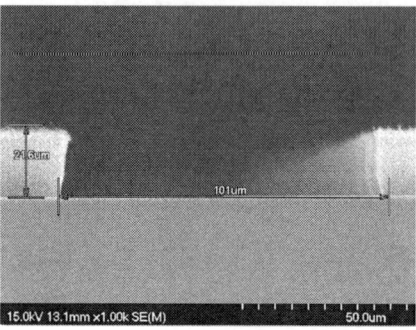

(b)

(c)

(d)

Figure 2 SEM images of CNTs pillar arrays for different R/H ratios with 100 μm inter pillar distance. (a) R/H ratio at 9.9 (b) R/H ratio at 4.63 (c) R/H ratio at 3.15(d) R/H ratio at 2.04.

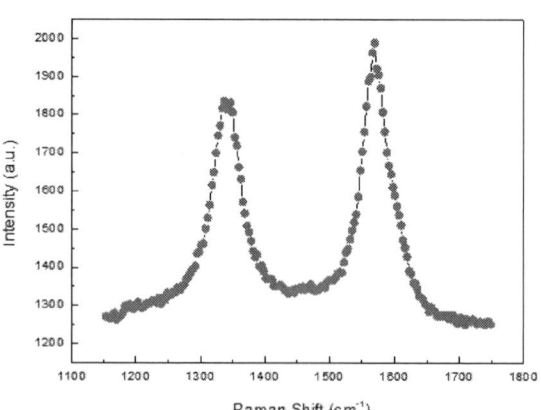

Figure 3 Raman spectra of CNTs' pillar array grown under 15 min pre-treatment in N_2/H_2 gas mixtures (500 sccm/300 sccm) and growth in $C_2H_4/N_2/H_2$ gas mixtures (20sccm/500 sccm/300 sccm) for 30 min.

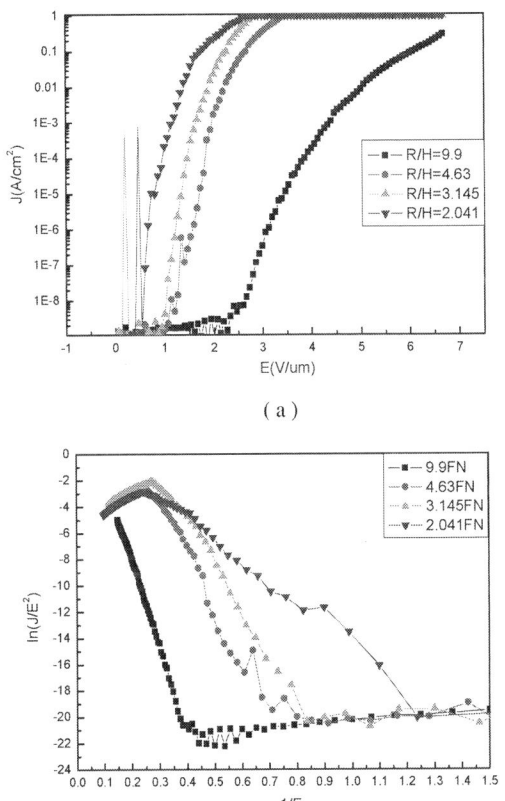

(a)

(b)

Figure 4 Field emission characteristic at different R/H ratios (a) Field-emission current density vs applied electric field. (b) Corresponding F–N plot.

Figure 5.Field emission image of CNTs' pillar array with electric field at 2.3V/um

Acknowledgements

This research was supported in part by the National Science Council in Taiwan under contracts NSC99-2221-E-129-019 Technical support from the Nano Facility Center of National Chiao Tung University is also acknowledged.

References

[1] S. Iijima, Nature (London) 354, PP.56, 1991.
[2] W.A. de Heer, A. Chatelain, D. Urgate, Science 270, PP.1197 ,1995.
[3] S. Fan, M.G Chapline, N.R. Franklin, T.W. Tombler, A.M. Cassell, H.Dai, Science 283,PP. 512,1999.
[4] W.B. Choi, D.S. Chung, J.H. Kang, H.Y. Kim, Y.W. Jin, I.T. Han, Y.H.Lee, J.E. Jung, N.S. Lee, G.S. Park, J.M. Kim, Appl. Phys. Lett. 75, PP.3129, 1999.
[5] M. Chhowalla, C. Ducati, N. L. Rupesinghe, K. B. K. Teo, and G. A. J. Amaratunga, Appl. Phys. Lett. 79, PP. 2079, 2001.
[6] S. S. Wong, J. D. Harper, J. P. T. Lansbury, and C. M. Lieber: J. Am.Chem. Soc. 120, PP. 603, 1998.
[7] A. Modi, N. Koratkar, E. Lass, B. Wei, and P. M. Ajayan: Nature (London) 424, PP.171, 2003.
[8] W. B. Choi, J. U. Chu, K. S. Jeong, E. J. Bae, J. W. Lee, J. J. Kin, and J. O. Lee: Appl. Phys. Lett. 79, PP.3696, 2001
[9] J. Haruyama, I. Takesue, and Y. Sato: Appl. Phys. Lett.77, PP.2891, 2000.
[10] A. Modi, and P. M. Ajayan, Nature (London) **424**, 171 (2003)
[11] L. Nilsson, O. Groening, C. Emmenegger, O. Kuettel, E. Schaller, L.Schlapbach, H. Kind, J. M. Bonard, and K.Kern: Appl. Phys. Lett. 76(2000) 2071.
[12] K.B. K. Teo, M. Chhowalla, G. A. J. Amartunga, W. I. Milne, P. Legagneux,G. Pirio, L. Ganglogg, D. Pribat, V. Semet, V. T. Binh, W. H.Bruenger, J. Eichholz, H. Hanssen, D. Friedrich, S. B. Lee, D. G. Hasko, and H. Ahmed, J. Vac. Sci. Technol. B 21, PP.693, 2003.
[13] Kempa, B. Kimball, J. Rybczynski, Z. P. Huang, P. F. Wu, D. Steeves,M. Sennett, M. Giersig, D. V. G. L. N. Rao, D. L. Carnahan, D. Z. Wang,J. Y. Lao, W. Z. Li, and Z. F. Ren, Nano Lett. **3**, PP.13, 2003.
[14] C.P Juan, K.J Chen, C.C Tsai, K.C Lin, W.K HONG, C.Y Hsieh, W.P Wang, R.L Lai, K.H Chen, L.C CHEN and H.C Cheng, Jpn. J. Appl. Phys. 44, PP.365, 2005.
[15] C.P Juan, C.C Tsai, K.H Chen, L.C CHEN and H.C Cheng, Jpn. J. Appl. Phys. 44, PP.8231, 2005.

Trap exploration of ZnO-based resistance switching memory devices

Fu-Chien Chiu [1,*], Wen-Yuan Chang [2], Peng-Wei Li [1], Chih-Chi Chen [1], and Wen-Ping Chiang [1]

[1]Department of Electronic Engineering, Ming-Chuan University, Gui-Shan, Taoyuan 333, Taiwan ROC
[2]Department of Materials Science and Engineering, National Tsing-Hua University, Hsinchu 300, Taiwan ROC
Tel: 886-3-350-7001 ext.3753; Fax: 886-3-359-3877; E-mail: fcchiu@mail.mcu.edu.tw

Abstract

The trap exploration of Pt/ZnO/Pt memory cells was made. Based on the temperature dependence of *I-V* characteristics, the conduction mechanisms in ZnO films are dominated by the hopping conduction and ohmic conduction in high-resistance state (HRS) and low-resistance state (LRS), respectively. Simulation results show that the trap spacing and trap energy level in HRS are around 2 nm and 0.46 eV, respectively. Also, the Fermi level, the electron mobility, and the effective density of states in conduction band in LRS in ZnO films are extracted. Keywords: ZnO, resistive switching, trap spacing.

Introduction

Resistance random access memory (RRAM) has attracted a great deal of attention because of its good compatibility with the CMOS process, nonvolatility, low power consumption, low cost price, high switching speed, high durability, small cell size, simple cell structure, and multistate switching [1-4]. There are several types of materials used in RRAM, such as perovskite-type oxides [1, 3], binary metal oxides [2-4], solid-state electrolytes [4], organic compounds [5], and amorphous Si [6]. Among the RRAM materials being studied, binary metal oxides are most favorable because of their simple constituents, compatible with CMOS processes, and resistive to thermal/chemical damages [2, 4, 7].

Experiment

The metal-insulator-metal (MIM) diodes using ZnO were fabricated. The ZnO films of 25 nm were deposited by *rf* magnetron sputtering. Pt top electrodes of 400 um in diameter were defined by *rf* magnetron sputtering using a metal shadow mask. Fig. 1 shows the structure of Pt/ZnO/Pt MIM capacitors.

Results and Discussion

A typical *I-V* switching characteristic in Pt/ZnO/Pt structure is shown in Fig. 2. An initial forming process is required to achieve the bipolar resistive switching behavior. To investigate the trap characteristics in ZnO thin film, the temperature dependence of *I-V* characteristics were studied. Results show that the current density (*J*) increases with increasing temperature both in HRS and LRS, as shown in Fig. 3. The calculated data match the theory of hopping conduction very well in HRS in the electric field (*E*) between 1.5×10^5 V/cm and 2.5×10^5 V/cm, as shown in Fig. 4. The hopping conduction can be expressed as [8]:

$$J = qanv \exp\left[\frac{qaE}{kT} - \frac{\Phi_t}{kT}\right] \quad (1)$$

where q is the electronic charge, a is the mean spacing between trap sites (i.e., the hopping distance), n is the electron concentration in the conduction band of the dielectric, v is the frequency of thermal vibration of electrons at trap sites, T is the absolute temperature, k is Boltzmann's constant, and Φ_t is the

energy level from the trap states to the bottom of conduction band (E_C) in ZnO. Therefore, the trap spacing in ZnO is determined to be about 2.0 nm according to Fig. 4. Besides, the trap energy level is determined to be about 0.46 eV according to the temperature dependence of current density, as shown in Fig. 5. The trap energy level of 0.46 eV in HRS may come from the defect state of interstitial zinc [9] which may be produced during the initial forming process.

In LRS, the current density increases with increasing temperature. The *J-E* curves are shown in Fig. 6 in a double-logarithmic plot. The linear relation between current density and electric field is observed, which matches the ohmic conduction very well because the slopes are very close to 1. The ohmic conduction can be expressed as [8]

$$J = \sigma E = q\mu N_C E \exp\left[\frac{-(E_C - E_F)}{kT}\right] \quad (2)$$

where σ is the electrical conductivity, μ is the electron mobility, N_C is the effective density of states of the conduction band, and E_F is the Fermi energy level; the other terms are as defined above. Figure 7 shows the linear relation between electrical conductivity and inverse temperature in LRS. According to the Arrhenius plot, the Fermi level (E_F) of ZnO in LRS is determined to be about 0.4 eV below the E_C edge of ZnO as shown in the inset of Fig. 7. Hence, the product of electron mobility (μ) and effective density of states of the conduction band (N_C) at each temperature can be extracted by the combination of E_F and electrical conductivity (σ). In addition, N_C is a function of temperature, which is proportional to $\beta T^{3/2}$, where β is a constant [10]. The N_C in ZnO at room temperature is 4.8×10^{18} cm^{-3} [11]. Therefore, the temperature-dependent μ and N_C in ZnO can be extracted, as shown in Fig. 8. At room temperature, the electron mobility is about 4.6 cm^2/V·s.

Conclusion

The trap characteristics in ZnO thin film were explored by the temperature dependence of I-V curves. Simulation results indicate that the trap spacing and the trap energy level in HRS are around 2 nm and 0.46 eV, respectively. In addition, the Fermi level, the electron mobility, and the effective density of states in conduction band in LRS in ZnO are also obtained.

Acknowledgment

The authors would like to thank the National Science Council, Taiwan, under Contract No. NSC 101-2221-E-130-006.

References

[1] W. W. Zhuang et al., in *IEDM Tech. Dig.*, 193 (2002).
[2] H. Akinaga and H. Shima, *Proc. IEEE*, **98**, 2237 (2010).
[3] T. Zhang et al., *Nanoscale Res. Lett.*, **4**, 1309 (2009).
[4] R. Waser et al., *Adv. Mate.*, **21**, 2632 (2009).
[5] M. Colle et al., *Organic Electronics*, **7**, 305 (2006).
[6] S. H. Jo, K. H. Kim, and W. Lu, *Nano Lett.*, **9**, 496 (2009).
[7] I. G. Baek et al., in *IEDM Tech. Dig.*, 587 (2004).

978-1-4673-4840-9/13 $31.00 © 2013 IEEE

[8] N. F. Mott and E. A. Davis, *Electronic Processes in Non-crystalline Materials*, Oxford University Press, 1979.

[9] A. B. Djurisic and Y. H. Leung, *Small*, **2**, 944 (2006).

[10] S. M. Sze, *Physics of Semiconductor Devices*, 2nd ed., New York, Wiley, 1981.

[11] M. Nakano et al., *Appl. Phys. Lett.*, **91**, 142113 (2007).

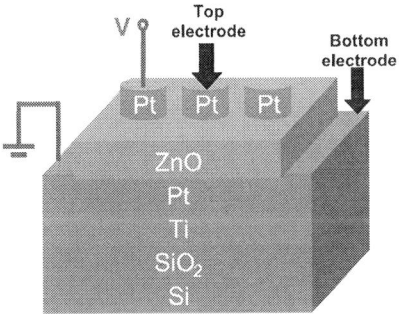

Fig. 1 Schematic of Pt/ZnO/Pt MIM memory cell.

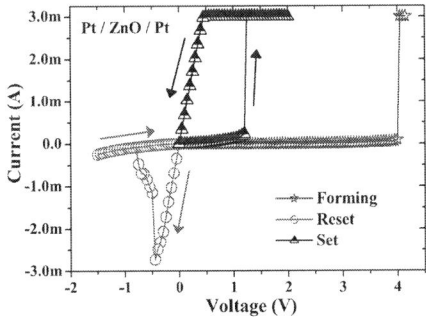

Fig. 2. Typical bipolar I-V switching characteristics in Pt/ZnO/Pt.

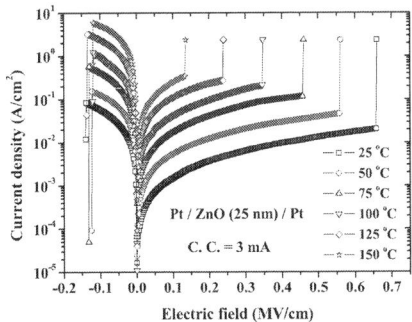

Fig. 3. Temperature dependence of J-E characteristics in Pt/ZnO/Pt.

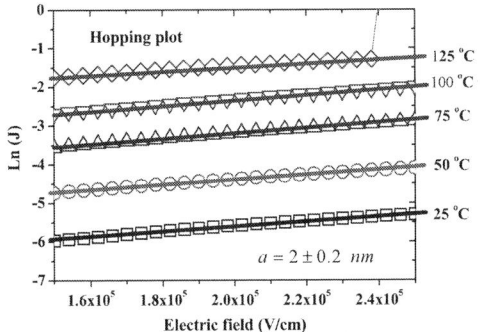

Fig. 4. Characteristics of Ln (current density) versus electric field for the extraction of trap spacing in the hopping conduction in HRS.

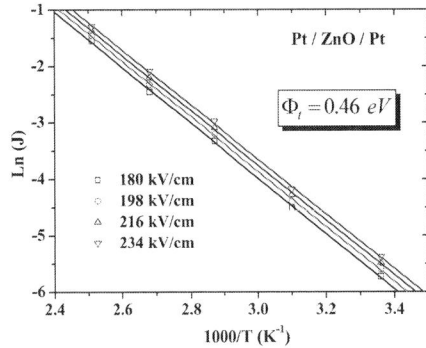

Fig. 5. Arrhenius plot for the extraction of trap energy level.

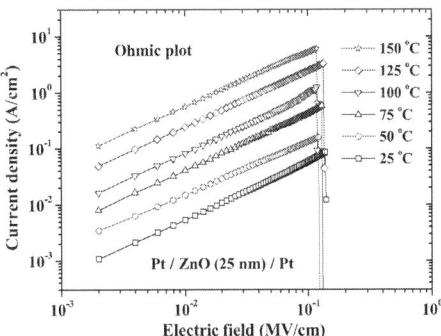

Fig. 6. Linear J-E curves at 25-150 °C in LRS.

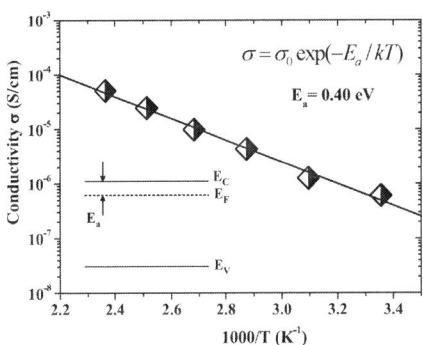

Fig. 7. Temperature-dependent electrical conductivity in LRS.

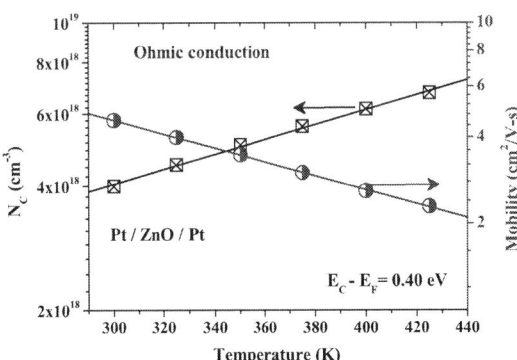

Fig. 8. Temperature dependence of the electron mobility and the effective density of states of the conduction band in LRS in ZnO film.

978-1-4673-4840-9/13 $31.00 © 2013 IEEE

Superior resistive switching characteristics of Cu-TiO2 based RRAM cell

Yu-Chih Huang, Huan-Min Lin and Huang-Chung Cheng

Department of Electronics Engineering and the Institute of Electronics, National Chiao Tung University
Room 309b, Engineering Building 4, 1001 University Road. 30010
Hsinchu City, Taiwan, R.O.C.
Phone:+88635712121#54218, E-mail: orionmasterkimo@gmail.com

Abstract

The random access memory (RRAM) cells consisted with the Cu doped TiO_2 film and both Pt inert electrodes has been purposed. Compared with the common conductive-bridging mode sample structure which have a Cu active electrode, a non-doped TiO_2 film and a Pt inert electrode. The Cu doped film with both inert electrodes sample exhibited the endurance of 1000 cycles more than the common ones of 400 cycles under the same operating conditions. Furthermore, the doped sample required lower setting voltage of -0.7V than the common ones of -1.5V. The differences in resistance resistive switching characteristics were possible caused by the amounts of Cu ions distribution and the forming of the conductive filaments in the TiO_2 layers, according the results of the electrical characteristics measurements and the material analysis.

Introduction

In recent years, resistive random access memory (RRAM) has attracted great attention owing to the simple fabrication process, highly stackable, high operation speed, low power consume and low cost for next-generation nonvolatile memory (NVM) technology [1]. The basic cell structure of RRAM is an insulator sandwiched between two metal electrodes, called the metal–insulator–metal (MIM) structure [2].

The widely acceptable conduction mechanisms were base on reduction-oxidation (redox) reactions on the nanoscale and nanoionic transport which have also been studied by the electrical stimulation [3-4]. Furthermore, the conduction mechanisms could be classified into three types; Electrochemical (ECM), Valence-change (VCM) and Thermo-chemical (TCM). The ECM type RRAM also called conductive-bridging RAM (CBRAM) or programmable metallization cell (PMC), which were fabrication of the electrochemically active metal electrode such as Ag or Cu. The resistive switch of the ECM RRAM cell were involving the electrode metal ions (cations) transport through the insulator and corresponding redox processes for the formation and dissolution of metallic filaments [3-5].

A novel conductive-bridging type RRAM cell fabricated by the copper doped titanium dioxide (TiO_2:Cu) films as the insulator was proposed in this study. The amounts of metal ions source could be limited and the growth of the filaments could be controlled , via adjustment the distribution of copper ions sources in the TiO_2:Cu films. We also compared the resistive switching characteristic of the doped (TiO_2:Cu) samples to the non-doped (TiO_2) samples with both active copper (Cu) electrode and inert platinum (Pt) electrode

samples. Those samples could be denote as Pt\$TiO_2$:Cu\Pt, Cu\$TiO_2$:Cu\Pt, Pt\$TiO_2$\Pt and Cu\$TiO_2$\Pt. The Pt\$TiO_2$:Cu\Pt sample exhibits the best characteristic of all the samples and shown the good endurance and low programming power consume.

Experiment

The 300nm thick Cu and Pt layers were deposited on the silicon dioxide substrates by DC sputtering system as the bottom active Cu electrodes and inert Pt electrodes respectively. Then, the 200nm thick non-doped TiO_2 layers and 200nm thick Cu doped TiO_2 (TiO_2:Cu) layers and non-doped TiO_2 layers were deposited on the top of both Cu and Pt electrodes respectively.

The TiO_2 layers were deposited by the DC reactive sputtering system with titanium target in the power of 190W. The TiO_2:Cu layers were also deposited by the DC reactive sputtering system with titanium target and copper target in the power of 190W and 10W respectively. Both the TiO_2 and TiO_2:Cu layers were deposited at room temperature in the atmosphere of the argon-oxygen mixed sputtering gases. The Ar/O_2 was 7/3. Finally, the 50nm thick Pt layers were deposited as the top inert electrode for all the samples.

The chemical bonding states, composition elements and crystal structures of prepared TiO_2 layers were examined by the x-ray photoelectron spectroscopy (XPS) and the x-ray diffraction (XRD). Furthermore, the resistive switching characteristics were measured at room temperature in the dark with an Agilent 4156C precision semiconductor parameter analyzer.

Results and Discussion

Fig.1 show the resistive switch characteristic after the first bipolar sweep, begin in the reverse of 0V to -5V and forward of 0V to 5V. the schematic diagrams of the samples also shown inset.

As shown in Fig.1, the sample of Pt\$TiO_2$\Pt exhibits no resistive switch characteristic and the largest resistance of all the samples. Thus the non-doped TiO_2 films could be used as the insulator for the metal ions (Cu cations) transport through it. On the other hand, the sample of Cu\$TiO_2$:Cu\Pt also exhibits no resistive switch characteristic too, but the lowest resistance of all the samples. According to the measurement results of the XPS. We speculated that the electrical characteristics of Cu\$TiO_2$:Cu\Pt might due to the excess amounts of released Cu and dissolved in the TiO_2:Cu film. The excess amounts of Cu atoms caused the sample exhibits the metal-like characteristics (The detail will be reported later in the this article).

978-1-4673-4840-9/13 $31.00 © 2013 IEEE   236

There have two samples have the resistive switch characteristics, one was the Pt\TiO$_2$:Cu\Pt and the another was Cu\TiO$_2$\Pt. It can be shown in Fig.1 The Pt\TiO$_2$:Cu\Pt has an lower set voltage, but larger off-state current then the Cu\TiO$_2$\Pt.

In order to compare the two samples under the same conditions. We choosing the same set/reset pulse voltages amplitude of -5V/3V, pulse width of 1 µs, and compliance currents of 0.01A when setting for both samples.

As shown in Fig.2, the Pt\TiO$_2$:Cu\Pt sample exhibits better endurance than the Cu\TiO$_2$\Pt sample. The off-state current of Cu\TiO$_2$\Pt sample were gradual increase with the switching times, and break down (resetting failure) when it over 400 switch cycles. In contrast, the switch cycles of Cu-doped TiO2 film with Pt electrode sample could exhibits over 1000 times. Moreover, the increase trend in the off-state currents of Cu\TiO$_2$\Pt sample was much larger than the Pt\TiO$_2$:Cu\Pt ones.

The chemical states and atomic compositions of both the Copper-doped (TiO$_2$:Cu) and non-doped TiO$_2$ films have also been analyzed by XPS system. The atomic concentration ratios of O 1s, Ti 2p and Cu 2p3 can be calculated by the areas of the binding energy peaks for all the samples, given in Table I.

As shown in Table I, the atomic compositions ratio of oxygen to titanium was close to 2:1 of the TiO$_2$ film. On the other hand, the copper component concentration of the TiO$_2$:Cu film was about the 12.9%. The ratio of oxygen to titanium was over 2:1 for TiO$_2$:Cu film. Thus it can be expected that the Cu atoms might not only in the form of the metallic Cu but also in the form of the copper oxides and exist in the TiO$_2$:Cu film. The Cu atoms chemical bonding states of TiO$_2$:Cu film could be measured by the XPS as shown in Fig.3.

As shown in Fig.3, the XPS spectra of Cu 2p$_3$ core level could be divided into three peaks approximately centered at 532.2eV, 533.8eV and 532.7eV by Gaussian fitting. The sub-peaks which centered at 532.2eV are respectively the copper atoms with the metallic bond in the TiO$_2$:Cu film denoted by Metallic Cu. The centered at 533.8eV ones are respectively the copper ions (cations) with the oxidation state of +2 called Cu(II) or in the form of cupric (CuO), denoted by Cu^{2+}. The centered at 532.7 eV ones are respectively the copper ions (cations) with the oxidation state of +1 called Cu(I) or in the form of cuprous (Cu$_2$O), denoted by Cu$^+$. The ratio of the three sub-peaks components to the total Cu 2p$_3$ core level could be analyzed by the method of peak areas [6–8], and the results given in Table II.

As shown in Table II, the most components of the doped Cu atoms were in the form of CuO. The second multiple were in the form of metallic Cu and the least of Cu$_2$O. The CuO is the higher oxide of copper, that the Cu atoms were have stronger bonding with the oxygen atoms and more difficult to be released in the form of CuO than the Cu$_2$O [9-11]. Thus, it might the reason of the Pt\TiO$_2$:Cu\Pt sample has the larger OFF-state current than the Cu\TiO$_2$\Pt ones due to the narrower band gap of CuO than the TiO$_2$. Furthermore, besides the metallic copper, the redox reaction might also occurred on the Cu$_2$O in the TiO$_2$:Cu film.

The XRD pattern was measured and shown in Fig.4, both the non-doped and Cu-doped TiO2 films were polycrystalline contain the TiO$_2$ of anatase and rutile phases. There have no crystalline signals of copper oxide were detected. Thus, the copper oxide seem dissolved in the Cu-doped TiO2 films in the form of amorphous, clusters or substitial.

The Pt\TiO$_2$\Pt sample exhibits no resistive switch characteristic and the large resistance showed in Fig.1. It might caused by there have no transportable Cu cations whether in the electrode or the non-doped TiO2 film. On the other hand, the Cu\TiO$_2$:Cu\Pt ones also exhibits no resistive switch characteristic, but lowest resistance. It might due to the excess amounts of Cu cations released from the Cu sources (metallic Cu and Cu$_2$O) due to the oxidation process in the TiO$_2$:Cu and the Cu electrode during the first voltage sweep. Then dissolved into the TiO$_2$:Cu film. The oxidation process of the Cu sources driven by the electronic current can be describe as the following equation:

$$Cu \rightarrow Cu^+ + e^- \qquad (1)$$
$$Cu_2O \rightarrow CuO + Cu^+ + e^- \qquad (2)$$

The metallic conductive paths (filaments) were speculated formation from the opposite end of Pt electrode when the drifting Cu cation (Cu$^+$) reduction by the electronics, as the following equation:

$$Cu^+ + e^- \rightarrow Cu \qquad (3)$$

Thus the Cu\TiO$_2$:Cu\Pt sample exhibits very low resistance and without resistive switch characteristic might caused by the overfull huge conductive paths formation between the electrodes. Led to the resistive characteristic of the sample just like a metal.

According the results of the electrical characteristics measurements and the material analysis. We speculated that, the first voltage reverse sweep of 0V to -5V was the key process to determine the growth conditions of filaments for both the Pt\TiO$_2$:Cu\Pt and Cu\TiO$_2$\Pt samples. Due to the different distribution of Cu sources, the growth of the filaments would also different. The concept diagrams are shown in Fig.5.

As shown in Fig.5 (a), before the formation of the early several filaments. The Cu sources were oxidized to the Cu cations, then drift toward the cathode and reduction to the Cu during the first reverse voltage sweep. As shown in Fig.5 (b), when the filaments grown longer, the filaments of the Pt\TiO$_2$:Cu\Pt sample would obtain less Cu cations for reduction. Because the highly conductivity of filaments, the majority voltage would drop across the anode to the front-end of the filaments. Thus the majority oxidation process caused by the electronics current would majority occurred in the same region of the voltage drop. Furthermore, the region contained oxidation-able Cu sources of the Pt\TiO$_2$:Cu\Pt sample would narrowing with the growth of the filaments length. In comparison, the Cu\TiO$_2$\Pt sample would not occurred the same situation. Because the oxidation-able Cu sources of the Cu\TiO$_2$\Pt sample were always in the front of the filaments growth directs (in the Cu electrode), and the amounts didn't significant changes with the growth of the filaments due to the thick pure copper buck electrode.

As shown in Fig.5 (c), the filaments of the Pt\TiO$_2$:Cu\Pt sample would difficult to grown thicker because of the narrowing oxidation-able Cu sources region and the Cu sources consume of this region near the anode, but might grown faster due to the nearer Cu sources with the more concentrated oxidation current caused by the concentrated local field. Thus, It could be expected, the filaments of the

Pt\TiO$_2$:Cu\Pt sample would become thinner and thinner with the filaments growth longer than the Cu\TiO$_2$\Pt sample. Finally, when the filaments growth after the first voltage sweep. The Cu\TiO$_2$\Pt sample was expected have the thicker and the relatively wider front-end of filaments than the Pt\TiO$_2$:Cu\Pt ones, as shown in Fig.5 (d).

Because of the thinner front-end of the filaments, the local electric fields should larger than the thicker ones. On the other hand, the thinner front-end of the filaments should have the more concentrated current path near the anode than the thicker ones, that led to the larger current density than the thicker ones and contribute to the migration of ions [12]. The Pt\TiO$_2$:Cu\Pt sample could have the advantages of the lower set voltage than the Cu\TiO$_2$\Pt ones due to the larger local electric field on the front-end of filaments in the reduction process of Cu cations, as the result of Fig.1. After the first forward voltage sweep of 0V to 5V, the filaments would redox and transport to the cathode as the first resetting process.

The resistive switch process were recycle able as the concept diagrams shown in Fig.6. The filaments would growth when the set voltage applied, then switching to the ON-state from the OFF-state.

With the resistive switch, the residual Cu atoms were possible exist in the TiO$_2$ layer as the not completely dissolved incomplete filaments [13-14]. Furthermore, the amounts of the residual Cu atoms might increase with the increase of switch cycles and the probability were positively correlated with the amounts of the Cu sources. Thus the gradually increased off currents with the switch times of the Cu\TiO$_2$\Pt sample was speculated cause by this reason. The Cu sources of Cu\TiO$_2$\Pt sample were almost unlimited supplied from the Cu electrode. But the reduced Cu sources close to the anode of the Pt\TiO$_2$:Cu\Pt sample when setting led to the limited amounts Cu atom from the redox reactive of filaments. Therefore, compared to the Cu\TiO$_2$\Pt sample, the Pt\TiO$_2$:Cu\Pt ones exhibits better endurance and less off current increase trend, might caused by the limited redox Cu sources and the thinner filaments due to the growth conditions of the filaments controlled by the Cu sources distribution TiO$_2$:Cu film.

Conclusions

The Pt\TiO$_2$:Cu\Pt sample exhibits the better endurance of 1000 cycles and requested lower setting voltage of -0.7V then the Cu\TiO$_2$\Pt ones of 400 times switch cycles and -1.5V required setting voltage. According to the measurement results of electrical characteristics, XPS and XRD. Those were the possible reasons that the amounts and the distribution of redox-able Cu sources dominant the resistive switch of the samples. Thus the Pt\TiO$_2$\Pt and the Cu\TiO$_2$:Cu\Pt samples exhibits no resistive switch characteristics, might due to the insufficient and the excess amounts of the Cu sources respectively. Moreover, we speculated that the distribution of Cu might the important factor of the filaments growth. Compared to the Cu sources supplied by the Cu electrode of the Cu\TiO$_2$\Pt sample, the Cu sources supplied by the TiO$_2$:Cu film of the Pt\TiO$_2$:Cu\Pt sample should be able to grown the thinner filaments sharper and faster. On the other hand, the thinner filaments would formed the limited oxidation-able Cu sources close to the inert electrode. That led to the lower set voltage caused by the more concentrated current. And the less off current increase trend with the resistive switch times.

References

[1] Shyh-Shyuan. S., Kuo-Hsing. C. and Meng-Fan. C., "Fast-Write Resistive RAM (RRAM) for Embedded Applications," *IEEE Design & Test of Computers*, Vol. 28, Issue. 1 (2011), pp. 64-71.

[2] H.-S. P. Wong., H. Y. Lee., S. Yu., Y. S. Chen., Y. Wu., P. S. Chen., B. Lee., F. T. Chen., and M. J. Tsai., "Metal–Oxide RRAM," *Proceedings of the IEEE.*, Vol. 100, No. 6 (2012), pp. 1951-1970.

[3] Rainer. W., Regina. D., Georgi. S. and Kristof. S., "Redox-Based Resistive Switching Memories –Nanoionic Mechanisms, Prospects, and Challenges," *Advanced Materials*, Vol. 21, Issue. 25-26 (2009), pp. 2632-2663.

[4] Rainer W., Stephan M. and Vikas. R., "Recent Progress in Redox-Based Resistive Switching," *Circuits and Systems IEEE International Symposium on.*, ISCAS (2012), pp.1596-1599.

[5] J. Billen., S. Steudel., R. Müller., J. Genoe. and P. Heremansa., "A comprehensive model for bipolar electrical switching of CuTCNQ memories," *Appl. Phys. Lett.*, Vol. 91, No. 263507 (2007), pp. 1-3.

[6] S.W. Tsao., T.C. Chang., S.Y. Huang., M.C. Chen., S.C. Chen., C.T. Tsai., Y.J. Kuo., Y.C. Chen. and W.C. Wub., "Hydrogen-induced improvements in electrical characteristics of a-IGZO thin-film transistors," Solid-State Electronics, Vol. 54, Issue. 12 (2010), pp. 1497-1499.

[7] G. H. Kim., H. S. Kim., H. S. Shin., B. D. A., K. H. Kim. and H. J. Kim., "Inkjet-printed InGaZnO thin film transistor," *Thin Solid Films*, Vol. 517, Issue. 14, pp. 4007–4010.

[8] G. H. Kim., W. H. Jeong. and H. J. Kim., "Electrical characteristics of solutionprocessed InGaZnO thin film transistors depending on Ga concentration," *Phys. Status Solidi A*, Vol. 207, No. 7 (2010), pp. 1677–1679.

[9] K. Nagase., Y. Zheng., Y. Kodama. and J. Kakuta., "Dynamic Study of the Oxidation State of Copper in the Course of Carbon Monoxide Oxidation over Powdered CuO and Cu2O," *Journal of Catalysis*, Vol. 187, Issue. 1 (1999), pp. 123–130.

[10] A.O. Musa., T. Akomolafe. and M.J. Carter. "Production of cuprous oxide, a solar cell material, by thermal oxidation and a study of its physical and electrical properties," *Solar Energy Materials and Solar Cells*, Vol. 51, Issues. 3–4 (1998), pp. 305–316.

[11] A. Soon., M. Todorova., B. Delley. and C. Stampfl., "Surface oxides of the oxygen–copper system: Precursors to the bulk oxide phase?," *Surface Science*, Vol. 601, Issue. 24 (2007), pp. 5809–5813.

[12] DiGiacomo. and Giulio., "Metal Migration (Ag, Cu, Pb) in Encapsulated Modules and Time-to-Fail Model as a Function of the Environment and Package Properties," *Reliability Physics Symposium*, 20th Annual (1982), pp. 27-33.

[13] N. Banno., T. Sakamoto., S. Fujieda. and M. Aono., "On-state reliability of solid-electrolyte switch," *Reliability Physics Symposium*, IRPS. (2008), pp. 707-708.

[14] A. Vena., E. Perret., S. Tedjini., C. Vallée., P. Gonon. and C. Mannequin., "A Fully Passive RF Switch Based on Nanometric Conductive Bridge," *Microwave Symposium Digest*, IEEE MTT-S International. (2012), pp. 1-3.

TABLE I

Atomic concentration of the Cu-doped (TiO_2:Cu) and non-doped TiO_2 films were determined by x-ray photoelectron spectrum (XPS).

XPS Peaks Areas	Atomic concentration (%)		
	O 1s	Ti 2p	Cu $2p_3$
Non-doped TiO_2	67.5	32.5	0
Cu-doped TiO_2	60.5	26.6	12.9

TABLE II

The XPS peaks area ratio of Metallic, Cu^{2+} and Cu^+ to total Cu $2p_3$, corresponding to the copper atoms form of the TiO_2:Cu film.

The chemical bonding states	Ratio of the different peak area (%)		
	Metallic Cu	Cu^{2+} (CuO)	Cu^+ (Cu_2O)
Cu-doped TiO_2	11.12	88.1	0.78

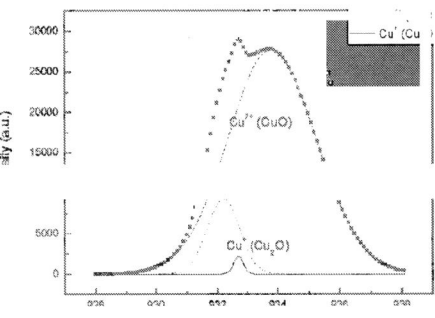

Fig. 3 XPS data of the Cu 2p3 core level of the TiO_2:Cu film. Three peaks (Metallic Cu, Cu2+ and Cu+) were distributed across the original Cu 2p3 core level peak by Gaussian fitting.

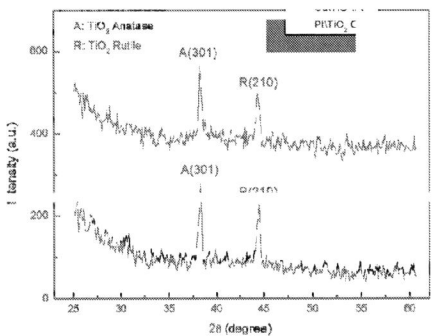

Fig. 4 XRD peaks of the of the non-doped TiO2 and Cu-doped TiO2 film.

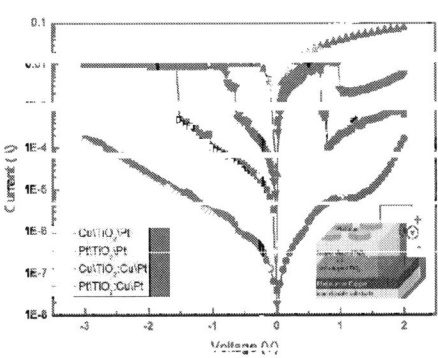

Fig. 1 The I-V curves and bipolar resistive switching characteristic of the Cu\$TiO_2$\Pt (black-line), Pt\$TiO_2$\Pt (red-line), Cu\$TiO_2$:Cu\Pt (blue-line) and Pt\$TiO_2$:Cu\Pt (green-line). The structures and the applied voltage diagrams of samples (inset).

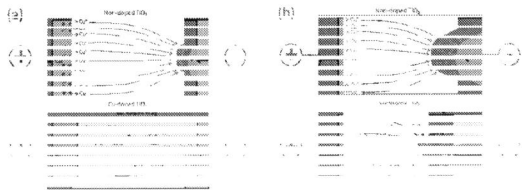

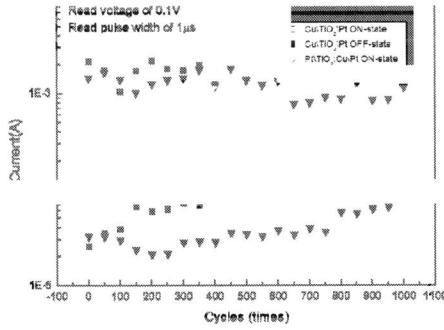

Fig. 2 The endurance of Pt\$TiO_2$:Cu\Pt and Cu\$TiO_2$\Pt samples.

Fig. 5 Schematic diagrams of (a) both the Pt\$TiO_2$:Cu\Pt and Cu\$TiO_2$\Pt fresh samples starting first voltage sweep. (b) The filaments of both samples growth during the first reverse sweep of 0V to -5V. (c) The different growth filaments due to the different Cu source distribution of the two samples. (d) The different shape of the filaments after the first reverse sweep finished.

Investigation into the Performance of CNT-Interconnects by Spin Coating Technique

Wei-Chih Chiu and Bing-Yue Tsui

Department of Electronics Engineering and Institute of Electronics
National Chiao-Tung University, Hsinchu, Taiwan, R.O.C.
scsa.ee94g@nctu.edu.tw

Abstract

In this work, we proposed a simple fabrication process, spin coating and dry etching, to construct CNT-interconnects. The CNT-interconnects formed by the slow rate spin coating method have conductivity of more than 10^2 (S/cm) and the CNT-interconnects with 10^2 square numbers possess about 30% conductive probability. In addition, we inserted metal bridges into 1000-μm-long CNT-interconnects to effectively facilitate the performance.

Introduction

Since the late 1990s, copper has been adopted to be the main material for interconnect in VLSI technology due to the exceptional conductivity, 5.88×10^5 (S/cm). However, with the scaling of integrated circuit, the occurrence of size effect and electromigration has been deeply influencing the performance of Cu-interconnect [1-2]. Besides, the fabrication difficulty of Cu-interconnects constructed by electroplating becomes more and more challenging [3]. New interconnect technology should be developed.

Since 1991, the discovery of carbon nanotubes revolutionizes the post-Si technology. Carbon nanotubes have been adopted for various applications due to the excellent electrical properties. In theory, the current capability of a single metallic CNT with diameter 1nm, 2.4×10^8 (A/cm^2) of current density, is one thousand times greater than that of copper without any adverse effects [4]. In this study, we used spin coating and dry etching to fabricate various sizes of CNT-interconnects and investigated the characteristics of the CNT-interconnects.

Fabrication

Two sets of CNT-interconnects were designed for investigation. The width varying set has the width ranging from 5μm to 500μm with fixed length at 100μm, and the length varying set has the length ranging from 5μm to 1000μm with fixed width at 5μm as illustrated in Fig.1. For better performance of the 1000-μm-long CNT-interconnects, Pd/Ti metal bridges were inserted at every specified distance. The conductance of each CNT-interconnect is measured by the four-point bridge resistors as shown in Fig.2. The starting material, silicon wafer, was first capped by a 200-nm-thick wet oxide grown in high temperature furnace and by a 1-nm-thick Al$_2$O$_3$ by an atomic-layer deposition system (ALD). The CNT solution was prepared by dissolving 2mg arc-discharge grown CNTs provided by Carbolex in dimethylformamide (DMF) and then uniformly dispersing by 24hr sonication. The first type of CNT film is formed by total 400 drops of the CNT solution with spin speed 500 rpm for 30 sec per cycle. This sample is named the normal rate (NR) sample. The other type is formed by 100 rpm for 10 sec per cycle and is named the slow rate (SR) sample. A 120°C baking after each cycle was adopted for the entire evaporation of the DMF solution. Next, the Pd/Ti metal was deposited by a sputtering system with the ratio 8/1 and patterned by a lift off process. Finally, the CNT-interconnects were patterned by O$_2$ plasma etching.

Results and Discussions

Figure 3 shows the conductance distribution of the width varying CNT-interconnect set. From the statistical results, the NR samples with less than 6 square numbers are probable to be conductive while every CNT-interconnect of the SR sample has at least 70% conductive probability. As for the CNT-interconnect conductance distribution of the length varying set shown in Fig.4, the CNT- interconnects of the SR samples also demonstrate higher conductivity and conductive probability than the NR sample. The results interpret that slow rate spin coating and heating process could effectively keep most of the well suspended CNTs deposited on the wafer and possibly form more conductive paths within the interconnect regime. In addition, by such fabrication process, the size of 5μmx500μm CNT -interconnects has about 30% conductive probability.

According to percolation theory, the conductivity of randomly distributed CNT sticks, as shown in Fig.5, has a power law dependent relation with the density. For a two dimensional conductive plane, the exponent of the power function can be theoretically calculated as 1.33 [5]. In this study, we interpret the size variation of CNT- interconnects under the same CNT density as CNT density varying at a designated interconnect regime. Figures 6, 7 show the characteristics and fitting curves of the average conductance as a function of the square numbers of the CNT-interconnects. The high exponents of the power fits for both NR samples, about 1.5, means that there are only few CNT connected path within the interconnect regime and these interconnects are sorted into percolation region in this study. The SR samples in width varying set have the exponent 1.45 close to 1.33 implies that part of CNT-interconnects had two dimensional CNT connected conductive planes, while in the length varying set, all of the CNT-interconnects are classified into power region based on 1.34 of the exponent. The size effect occurred in CNT-interconnect conductance shown in Fig.8 arises from the bundling effect between CNT sticks during baking process.

Figure 9 shows the conductance distribution of the 5μmx1000μm CNT-interconnects with metal bridges crossed at every specified distance. The result shows that inserting metal bridges within the CNT-interconnects is indeed able to improve the conductive probability of long interconnect. However, too many bridge metals may conversely exacerbate the conductivity owing to the contact resistance from Pd/Ti to CNTs.

Summary

In summary, the investigation into performance of CNT-interconnects by spin coating technique has been completed. Through slow rate spin coating method, the conductivity as well as conductive probability obtained great enhancement. Next, the size-dependently percolative characteristics of the CNT-interconnect conductance were demonstrated. At final, the conduction of 1000-μm-long CNT-interconnects with the metal bridges inserted approves the potential for future VLSI interconnects.

Acknowledgement

This work was supported in part by the Ministry of Education in Taiwan under ATU Program, and was supported in part by the National Science Council, Taiwan, R.O.C. under the contract No. ： NSC 100-2221-E-009- 010-MY2.

References

[1]. S. M. Rossnagel, et al., J.Vac.Sci.Technol., **B22** (2004) 240.
[2]. Y. Chai, et al.,Electro.Dev.Lett., **29** (2008) 1001
[3]. P. C. Andricacos, et al., Soc.Interface, (1999) 32
[4]. B. Q. Wei, et al., Appl.Phys.Lett., **79** (2001) 1172
[5]. L. Hu, et al., Nano Lett., **4** (2004) 2513.

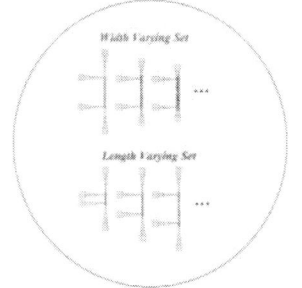

Fig.1 Schematic width and length varying sets of CNT-interconnects with four-probe bridge resistors designed for this work.

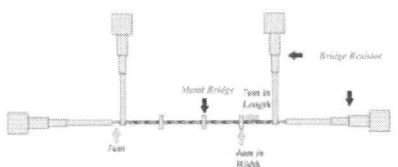

Fig.2 Schematic layout of the four-probe bridge resistor with metal bridges. The length and width of the metal bridge is 7μm and 4μm, respectively. The distance between metal bridges ranges from 25μm to 500μm. The 5μm space between two probes is for precise measurement of CNT-interconnect conductance.

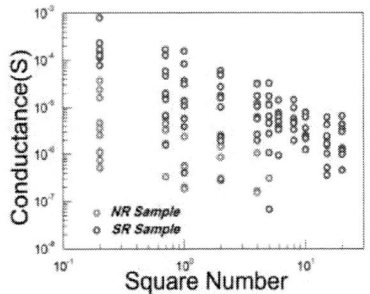

Fig.3 Conductance distribution of the width varying set of CNT-interconnects (100μm in length). The SR samples have higher conductivity and conductive probability.

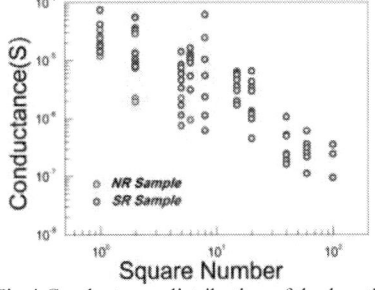

Fig.4 Conductance distribution of the length varying set of CNT-interconnects (5μm in width). The SR samples have higher conductivity and conductive probability.

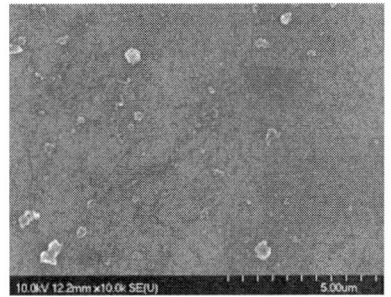

Fig.5 Top-view SEM image of CNT networks formed by 200 drops CNT solution and spin coating at 500rpm for 30sec. The 1-nm-thick ALD Al_2O_3 layer is used to improve the uniformity.

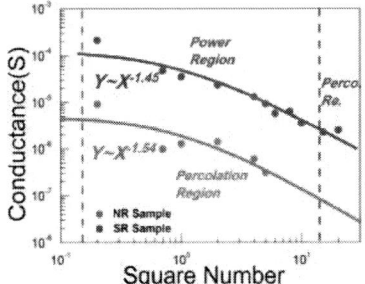

Fig.6 Power fit for average conductance and square number of the width varying set of CNT-interconnects (100μm in length). The entire NR samples belong to percolation region, while the SR samples with less than 15 square numbers are around power region.

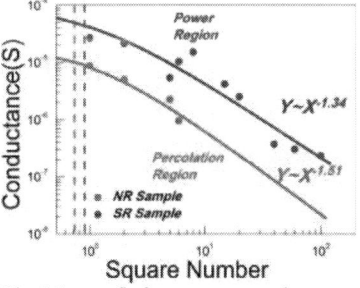

Fig.7 Power fit for average conductance and square number of the length varying set of CNT-interconnects (5μm in width) All NR samples locate in percolation region, while all SR samples are in power region.

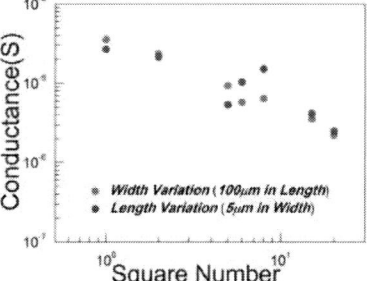

Fig.8 The distribution of the average conductance of width and length varying sets of the SR samples within the same range of square number. The difference of conductance results from the size effect caused by bundling phenomena.

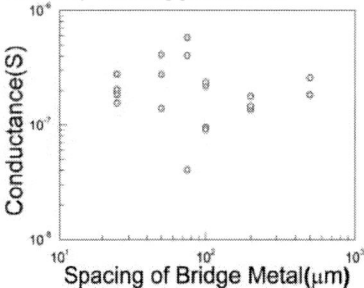

Fig.9 Conductance distribution of CNT-interconnects (1000μm in length) made by slow rate spin coating with various square numbers between Pd/Ti metal bridges.

Promising N-Type FinFET Devices without or with Cobalt-Silicide Applied to the Gate

Hsin-Chia Yang*[1], Jing-Zong Jhang[1], Wen-Shiang Liao[1], Chong-Kuan Du[1], Yi-Hong Lee[1],
Sung-Ching Chi[1], Quan-Hao Shen[1], Mu-Chun Wang[1], Shea-Jue Wang[2]

[1]Dept. of Electronic Engineering, Minghsin University of Science and Technology, Taiwan

[2]Dept. of Materials and Resources Engineering, National Taipei University of Technology, Taipei, Taiwan

*Corresponding email: hsinchia_yang@yahoo.com.tw

Abstract

Channels of FinFET are 3 dimensional fin-like structures which are thin enough to be fully depleted as the gate is appropriately biased leaving no leaky neutrally-charged leaky body. Different widths (fin thickness), different gate materials, and the Vt-adjustment using different implant energies are taken into account in this paper. It is then found that different fin thicknesses do affect the electrical performance. But cobalt silicide replacing the gate poly-silicon does not show apparent benefits on n-channel FinFET. (Keywords: fully silicide, $CoSi_2$ silicide, work function, threshold voltage).

Introduction

The fin-like channel devices are feasibly fabricated using the silicon on oxide insulator (SOI). The thickness of the channels is termed as channel width is thin enough so that the channels can be fully depleted as the gates crossing over them are biased. The two opposite sides of a device channel with the gate unto it are called Source and Drain. As a bias, V_{DS}, is applied to Source and Drain without Gate charged, the leakage current is outrageously out of control at short channel length. But this leaky body becomes much less leakier as Gate is biased because the whole body is fully depleted. The structure is quite amazing compared to the one of traditional CMOSFET devices. One thus expects that the FinFET [1] devices may possess higher on-current $I_{DS(on)}$ through the high aspect-ratio standing channels. Nevertheless, to achieve the structure requires selective etching techniques and sustainable materials strength followed by Somehow, the good step coverage of poly-silicon deposition. Somehow, it is thought to be worth paying the price for the goal coming true.

In this study, NMOS devices using Fin-FET structure are taken into account to understand the enhancement because of the fully carrier-depleted region without leaky body. Ones therefore expect a better threshold voltage (Vt) comparatively, and further consider the effects between with and without cobalt-silicide [2]. The P-well Vt adjustment is completed using the same boron dosage at different energies, 10KeV and 6KeV.

Experiments and Preparation of FinFET Devices

Sacrificing oxide is grown by Furnace on the SOI wafers followed by dispensing positive photo resist (PR). The photo mask with letters "I" patterns is applied to protect PR underneath the letters from exposure. After development, sacrificing oxide and the silicon beneath are plasma dry etched [3] leaving the high aspect-ratio fin-like body. P-well Vt adjustment implants at two different energies (10KeV and 6KeV) mainly turn the channel into P-type silicon where the strong inversion is to happen for n-channels. After HF dip, 14 angstrom silicon dioxide adulterated with nitrogen is grown. Two different materials, heavy-doped poly-silicon and fully cobalt silicide, are deposited to form gates as shown in Fig.1. Either end of "I" is called Source or Drain.

Results and Discussions

The I_D-V_D and I_D-V_G characteristic curves in Fig.2 to Fig.5 demonstrate different electric performances corresponding to different channel widths and lengths, in which w7 and w8 signify the Vt implant energies 10KeV and 6KeV, respectively, while w9 signifies the device with poly-silicon Gate replaced with fully cobalt silicide [4]. It is quite amazing that three conditions all enjoy their own superiorities at different channel widths/lengths. This gives the clues that the variations may only be due to the boundary conditions. For example, (W/L=0.11μm/0.10μm) in Fig.2 (A) prefers w8, (W/L=0.12μm/0.10μm) in Fig.2 (B) prefers w9, and (W/L=0.12μm/0.12μm) in Fig.3 (B) prefers w7.

What is more, take a closer look at the I_D-V_G curves near V_G =0.where one can find the Vt_linear. A lot of n-channel devices having negative threshold voltages really surprise ones even though only few devices have positive threshold voltages. One shall recall the model about parameters describing I-V as follows:

$$V_{th} = [-0.56 - \frac{k_B T}{e}\ln(\frac{p}{n_i})] + 2\frac{kT}{e}\ln(\frac{p}{n_i}) + \frac{Q_{dep}}{C_{ox}^{(1)}} \quad (1),$$

In equation (1), k_B is Boltzmann constant (1.38×10^{-23} J/K), e is carrier charge (1.6×10^{-19}C), p is p-type dopant concentration, n_i is the intrinsic dopant concentration, and Q_{dep} is the depleted space charge that causes the burden to strongly inverse p-type into n-type for n-channel MOSFET devices. For FinFET devices, the depleted region is thinned too much so that the last term in Eq.1 is not large enough to compensate the first negative term which is called flat-band voltage.

Furthermore, the performances of devices can be also

978-1-4673-4840-9/13 $31.00 © 2013 IEEE

checked on I_D-V_G in the sub-threshold current region in which Swing (mV / decade) can be determined according to the slope formula as follows:

$$S \equiv \frac{dV_G}{d \log I_D}\bigg|_{V < V_{th}} \quad (2)$$

Swing [5] for poly-silicon gate with channel width/length = 0.10μm/0.10μm in Fig.4 (A) is 85mV/decade while it is 115mV/decade corresponding to the poly-silicon gate with channel width/length = 0.12μm/0.1μm in Fig.4 (B). In addition, I_{off} (V_D=0.05V, w9)= 7.24×10^{-8}A in Fig.2 (B) and I_{off} (V_D=0.05V, w8)= 2.28×10^{-6}A in Fig.2 (A) both give satisfactory leakage currents at V_G=0.0V

Conclusion

FinFET devices perform much better at channel width = 0.11μm (fin thickness) than 0.12μm for either channel length L=0.11μm or L=0.12μm. But FinFET devices with fully cobalt silicide gate do show the apparent benefits from thicker fin thickness. It may help give us roadmap in the near future. It is found that the enhancement is to be observed as the cobalt silicide is formed at the gate for adjusting the work function. As Fin-FET devices of ten channels in parallel compared to the ones of single channel, I_D gears up from 0.01mA to over 0.1mA at V_D=1.2 V with the same applied Gate voltages. The swing correlated with the switching capability and the mobility associated with the trans-conductance g_m are to be determined and discussed

Acknowledgement

All the authors sincerely thank National Nano Device Laboratory in Taiwan for being so fully supportive.

Reference

[1] Su Hsin-Wen, *et al.*, "Drain-induced-barrier lowering and subthreshold swing fluctuations in 16-nm-gate bulk FinFET devices induced by random discrete dopants," *2012 70th Annual Device Research Conference (DRC)*, pp.109-110,18-20 Jun. 2012.
[2] H.A. Elgomati, *et al.*, "Cobalt silicide and titanium silicide effects on nano devices," *2011 IEEE Regional Symposium on Micro and Nanoelectronics (RSM)*, pp.282-285, 28-30 Sept. 2011.
[3] Michael A. Marrs, *et al.*, "Comparison of wet and dry etching of zinc indium oxide for thin film transistors with an inverted gate structure," *Journal of Vacuum Science & Technology A: Vacuum, Surfaces, and Films*, pp.011505-011505-6, Jan. 2012.
[4] T. Schulz. *et al.*, "Fin thickness asymmetry effects in multiple-gate SOI FETs (MuGFETs) ," *2005. Proceedings IEEE International SOI Conference*, pp. 154- 156, 3-6 Oct. 2005.
[5] P. Chotimanus, *et al.*, "Real swing extraction for video indexing in golf practice video," *2012 Computing, Communications and Applications Conference (ComComAp)* , pp.420-425, 11-13 Jan. 2012.

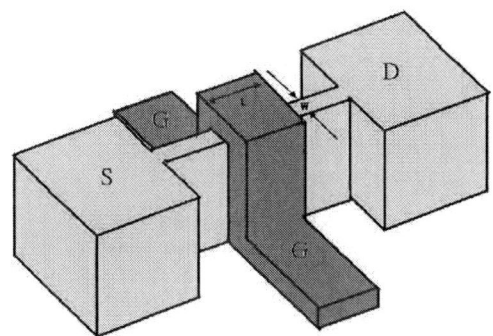

Fig. 1 The top view of the 3-D FinFET device

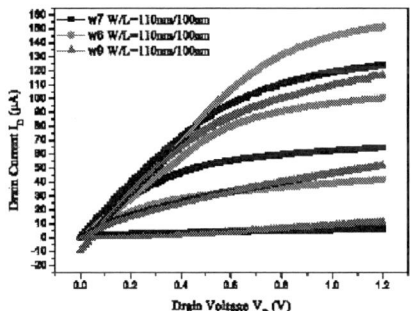

(A)

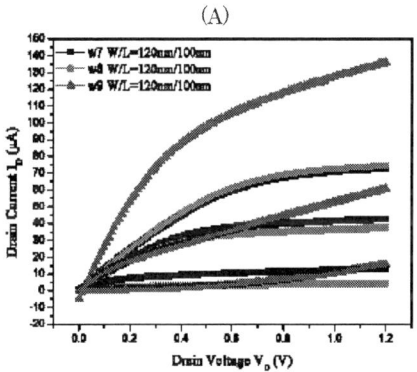

(B)

Fig.2 I_D-V_D characteristic curves with channel width/length (A) (W/L = 0.11μm/0.10μm) and (B) (W/L = 0.12μm/0.10μm)

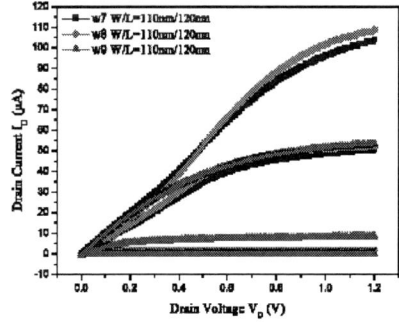

(A)

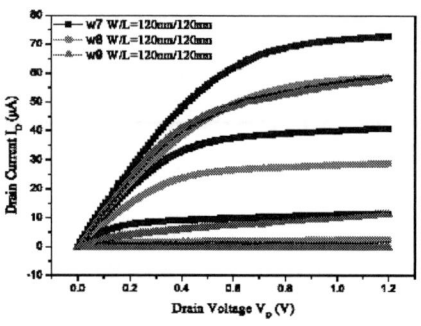

(B)

Fig.3 I_D-V_D characteristic curves with channel width/length (A) (W/L = 0.11μm/0.12μm) and (B) (W/L = 0.12μm/0.12μm)

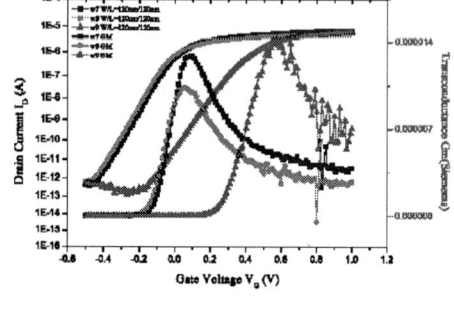

(B)

Fig.5 I_D-V_G characteristic curves with channel width/length (A) (W/L = 0.11μm/0.12μm) and (B) (W/L = 0.12μm/0.12μm)

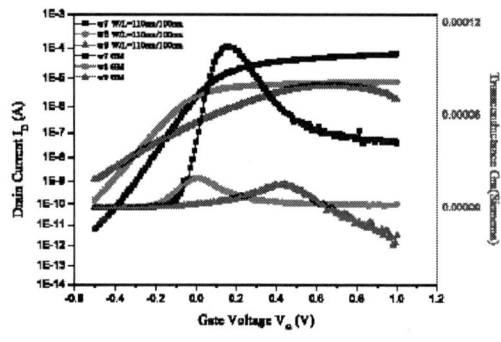

(A)

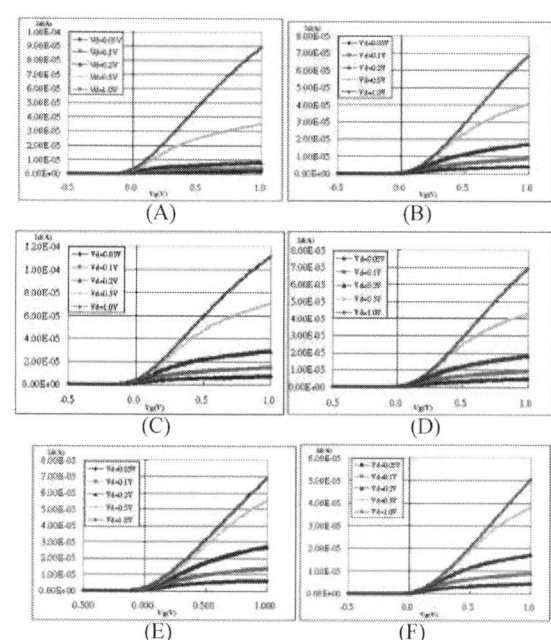

Fig.6 I_D-V_G curves at P-Well Vt implant energy 10KeV. (A) W/L = 0.11μm/0.1μm (B) W/L =0.11μm/0.24μm. (C) W/L =0.115μm/0.1μm (D) W/L = 0.115μm/0.24μm. (E) W/L = 0.12μm/0.1μm (F) W/L = 0.12μm/0.24μm.

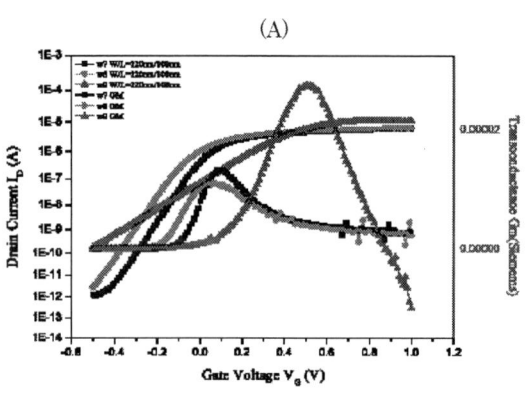

(B)

Fig.4 I_D-V_G characteristic curves with channel width/length (A) (W/L = 0.11μm/0.10μm) and (B) (W/L = 0.12μm/0.10μm)

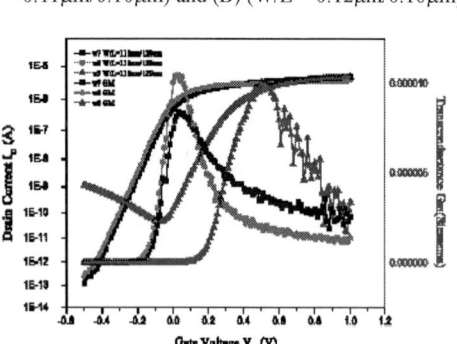

978-1-4673-4840-9/13 $31.00 © 2013 IEEE   244

The Side Effects and the Effects of Thickness of Source/Drain Fin on P-Type FinFET Devices

Hsin-Chia Yang[*,1], Wei-Yen Peng[1], Wen-Shiang Liao[1], Guo-Wei Wu[1], Cheng-Yu Tsai, Mu-Chun Wang[1], Sung-Ching Chi[1], Shea-Jue Wang[2]

[1]Dept. of Electronic Engineering, Minghsin University of Science & Technology, Taiwan

[2]Dept. of Materials and Resources Engineering, National Taipei University of Technology, Taipei, Taiwan
[*]Corresponding email: hsinchia_yang@yahoo.com.tw

Abstract

The 3-D structural fin-like channels of FinFET suppress the leakage current as the sizes of devices get substantially shrunk. In this study, the fin-thickness effects on the electrical performances are mainly observed. Two different kinds of thickness (namely, 110nm, and 120nm) with the same channel length (0.1 micron) are put into comparison. The phosphorus implants of the same dose with different energies for N-well threshold voltage adjustment are also taken into account. (Keywords: FinFET Devices, 3-D structure, Leakage current, Over Exposure, Over Etching)

Introduction

The so-called fin-like 3-D structure, Fin-FET structure [1, 2], is formed by depositing poly-silicon, as Gate, unto a standing less than a micron thick plate with two ends of Source and Drain. Instead of the traditional MOSFET devices suffering high leakage current [3] and thus Gate uncontrollability, Fin-FET devices are to be noticeably promoted as the Gate length is deliberately shrunk down to tens of nanometer. One possibly develops the channel length less than 90nm by using 90nm photo mask followed by over-exposure [4] and over-etching [5]. Fin-FET devices are so expected to provide lower I_{off} current as the channel length gets shorten continuously. They may thus be the candidates to keep the threshold voltage and the swing from rolling-up. Those electrical characteristics may be demonstrated by the fully depleted region and the lack of leaky body as the gate gets biased.

In this study, the side effects of single-channel and eleven-channel PMOS devices using Fin-FET structure are to be addressed. The discount of multiplication of I_D is to be found, and the side effects are subjected to discussion. One would also determine the correlated threshold voltage (Vt) and the resulted swing. The phosphorus implants for N-well Vt adjustment at two different energies, 20KeV and 15KeV, are put into comparison, resulting to the variation of threshold voltages. Of course, the mobility accompanied with the trans-conductance g_m shall be analyzed as the thickness of the fin varies.

Experiment and Preparation of FinFET Devices

A layer of <100> silicon on the oxide dielectric as an insulator (SOI) is dispensed with photo resist (PR) and gets exposed [4] for defining the active area of devices, which is a 3-D fin-like structure. After the development and the anisotropic plasma dry-etching [5], the channels of devices are high aspect-ratio formed in the shape of "I", followed by phosphorous N-Well Vt implants at different energies and 14 angstrom grown nitrided oxide as gate dielectric. The two ends of "I" are deliberately set to Source and Drain. Then a layer of 150nm heavy-doped CVD poly-silicon [3] is chemically deposited onto the grown gate oxide of the thin high wall as shown in Fig.1.

Results and Discussions

PFinFET devices with two different channel width/length sizes, (W/L=0.11μm/0.1μm, 0.12μm/0.1 μm), are to be evaluated. As shown in Fig. 2 and Fig.3, the N-Well Vt implant energy is set to 20KeV and 15KeV respectively. As the maximum gate bias is controlled under 1V, I_D-V_D data in Fig.2 demonstrate that devices of width 0.11μm are generally superior to ones of width 0.12μm by over 10% to 15% addressing the influences of depletion region in the width (the thickness of the fin). Normally, the whole fin is intentionally designed to be fully depleted for eliminating the leakage current as the channel length is short enough. As plotted in Fig.4, the superiority of electric performance of devices with W=0.11μm can be verified as the leakage currents of the devices with W=0.12μm are higher than the ones with W=0.11μm for both single and eleven channel devices. In addition, it is so amazing that the drain currents of the devices with eleven channels are over twenty times higher than the ones with single channel in Fig.2 and Fig.3. That is so encouraging and promising that FinFET devices do demonstrate the potential dominating the future development.

In Fig. 6 and Fig.7, all the I_D-V_G characteristics curves tend to be linear or two slopes linear. The two curves always merge as V_D is set to 0.5V or 1.0V simply because both characterize the saturating behaviors. It is worth pointing out that the width of channel, the thickness of the fin, does impact the electrical performances. For example, 0.115 microns wide fins always enhance the drain current for both 0.1 micron long channel and 0.24 micron long channel at different Vt adjustment implant energies. In addition, the linear threshold voltages approach to zero because of the narrower depletion region as the gate bias is to be applied. If one subtracts the un-expected leakage current at V_G=0 from all the curves at V_D=0.05V or 0.1V, the linear threshold voltages are to be as listed in Table.1.

In the end, one is interested in Swing (mV / decade) in the sub-threshold current according to the slope formula as follows:

$$S \equiv \left. \frac{dV_G}{d \log I_D} \right|_{V < V_{th}} \qquad (1)$$

It may further help check with the performances of promising devices in Fig.4 (A), e.g., $S_{0.11}$=85mV/decade. In addition, I_{off} (V_D=0.05V)= 1.88×10^{-8}A in Fig.4 (A) and I_{off} (V_D=0.05V)= 1.01×10^{-9}A in Fig.5 (A) both give satisfactory leakage currents at V_G=0.0V

Conclusion

The FinFET devices are expected to well control the leakage current, especially for the shorter channel length. In this paper, the channel length is fixed to 0.1μm and many other concerning characteristics are shown to be promising.

Acknowledgement

All the authors sincerely thank National Nano Device Laboratory in Taiwan for being so fully supportive.

Reference

[1] F Jafari, et al., "Designing Robust Asynchronous Circuits Based on FinFET Technology," 2011 14th Euromicro Conference on Digital System Design (DSD), pp.401-408, Aug. 31 2011-Sep. 2 2011.

[2] Mingu Kang, et al., "FinFET SRAM Optimization With Fin Thickness and Surface Orientation," IEEE Transactions on, Electron Devices, vol.57, no.11, pp.2785-2793, Nov. 2010.

[3] Byoungseon Choi, et al., "Improvement of drain leakage current characteristics in metal-oxide-semiconductor-field-effect-transistor by asymmetric source-drain structure," 2012 IEEE International Meeting for Future of Electron Devices, Kansai (IMFEDK), pp.1-2, 9-11 May. 2012.

[4] Young-Su Moon, et al., "A simple ghost-free exposure fusion for embedded HDR imaging," 2012 IEEE International Conference on Consumer Electronics (ICCE), pp.9-10, 13-16 Jan. 2012.

[5] Tao Zhang, *et al.*, "Surface uniform wet etching of ZnO films and influence of oxygen annealing on etching properties," *2011 IEEE International Conference on Nano/Micro Engineered and Molecular Systems (NEMS)*, vol., pp.626-629, 20-23 Feb. 2011.

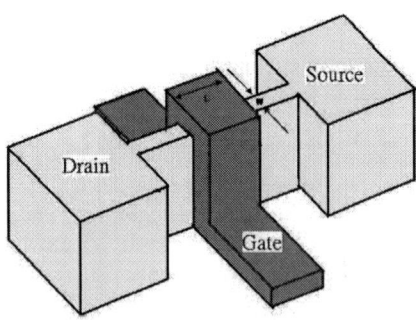

Fig. 1 The top view of the 3-D FinFET device

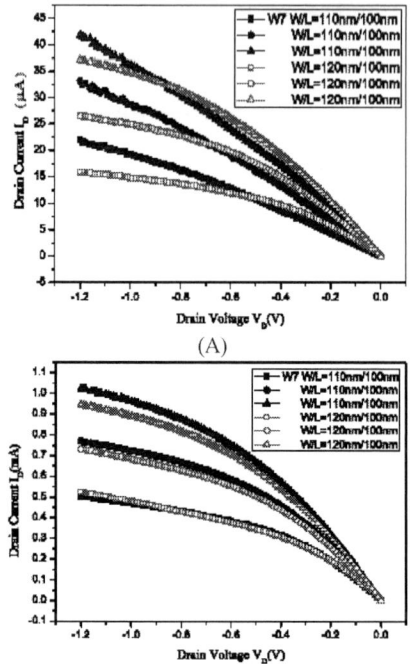

Fig.2 I_D-V_D curves at N-Well Vt implant energy 20KeV. (A) Single channel; (W/L = 0.11μm/0.1μm) and (W/L = 0.12μm/0.1μm). (B) Eleven channels; (W/L = 0.11μm/0.1μm) and (W/L = 0.12μm/0.1μm)

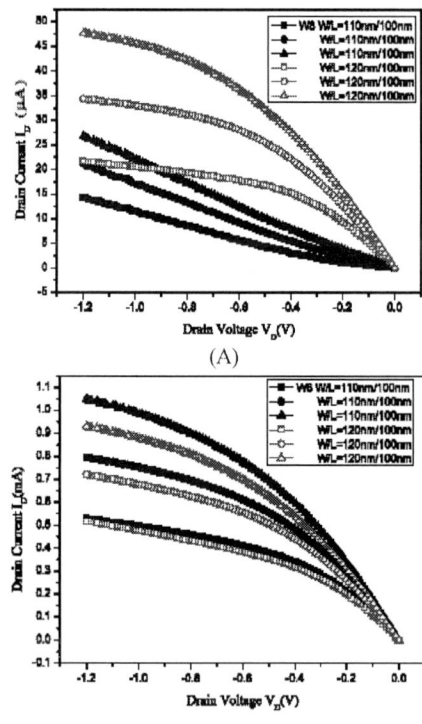

Fig.3 I_D-V_D curves at N-Well Vt implant energy 15KeV. (A) Single channel; (W/L = 0.11μm/0.1μm) and (W/L = 0.12μm/0.1μm) (B) Eleven channels; (W/L = 0.11μm/0.1μm) and (W/L = 0.12μm/0.1μm)

Fig.5 I_D-V_G and trans-conductance curves at N-Well Vt implant energy 15KeV. (A) Single channel; (W/L = 0.11μm/0.1μm) and (W/L = 0.12μm/0.1μm) (B) Eleven channels; (W/L = 0.11μm/0.1μm) and (W/L = 0.12μm/0.1μm)

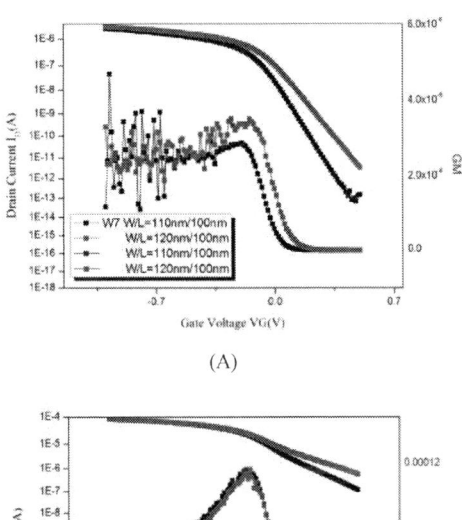

(A)

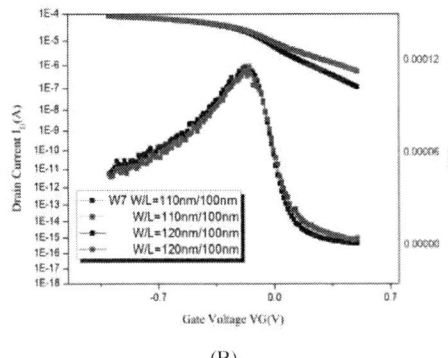

(B)

Fig.4 I_D-V_G and trans-conductance curves at N-Well Vt implant energy 20KeV. (A) Single channel; (W/L = 0.11μm/0.1μm) and (W/L = 0.12μm/0.1μm) corresponding to (V$_{th\_0.11}$=0.047V, V$_{th\_0.12}$=0.0074V) and (S$_{0.11}$~85mV/decade, S$_{0.12}$~110mV/decade) (B) Eleven channels; (W/L = 0.11μm/0.1μm) and (W/L = 0.12μm/0.1μm)

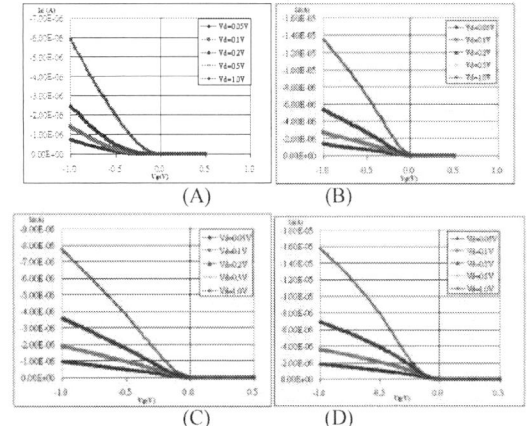

Fig.6 I_D-V_G curves at N-Well Vt implant energy 20KeV. (A) W/L = 0.11μm/0.1μm (B) W/L = 0.115μm/0.1μm. (C) W/L = 0.11μm/0.24μm (D) W/L = 0.115μm/0.24μm.

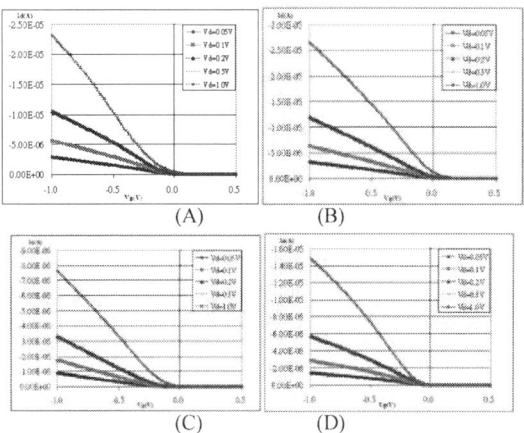

Fig.7 I_D-V_G curves at N-Well Vt implant energy 15KeV. (A) W/L = 0.11μm/0.1μm (B) W/L = 0.115μm/0.1μm. (C) W/L = 0.11μm/0.24μm (D) W/L = 0.115μm/0.24μm.

Tabel.1 Modified threshold voltages at V$_D$=0.05V or 0.1V

Size(w/L)	Vt-lin(0.05V)	Vt-lin(0.1V)
0.110/0.100	1.59E-06	3.07E-06
0.115/0.100	1.64E-06	2.87E-06
0.120/0.100	1.07E-06	1.37E-06
0.110/0.240	1.09E-06	1.86E-06
0.115/0.240	8.86E-07	1.64E-06
0.120/0.240	8.74E-07	1.29E-06

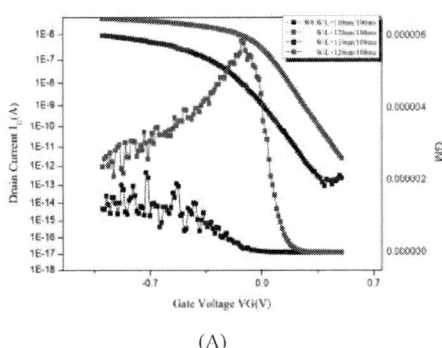

(A)

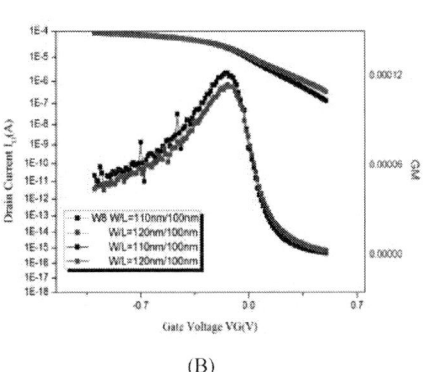

(B)

978-1-4673-4840-9/13 $31.00 © 2013 IEEE

The Adjustment of Threshold Voltage on P-Type FinFET Devices

Hsin-Chia Yang[*1], Yi-Hong Lee[1], Wen-Shiang Liao[1], Chong-Kuan Du[1], Jing-Zong Jhang[1], Sung-Ching Chi[1], Mu-Chun Wang[1], Shea-Jue Wang[2]

[1]Dept. of Electronic Engineering, Minghsin University of Science and Technology, Taiwan

[2]Dept. of Materials and Resources Engineering, National Taipei University of Technology, Taipei, Taiwan
*Corresponding email: hsinchia_yang@yahoo.com.tw

Abstract

FinFET devices have the structure of 3-D fins as channels, which are capable of being fully depleted as the gate is biased and potentially suppress the leakage currents. In this paper, one compares poly-silicon gates with fully cobalt silicide gate to see how much they are affected by the fin widths. Two channel lengths (0.1 micron and 0.12 micron) at two different fin widths (namely, 110nm, and 120nm) are taken into account. (Keywords: fully silicided gate, mobility, work function, threshold voltage)

Introduction

The traditional MOSFET devices have come to the bottleneck as the channel length gets shrunk to below a hundred nanometers. One of the reasons is due to the leakage current[1,2] even as the devices are switched off. The 3-D Fin-FET [3,4] structure is ever proposed and proved to be impressively resolving on the issues. The fin is formed by depositing poly-silicon[5,6] as gate electrode, fully-metal-silicided or partly-metal-silicided, unto a fin-like with both sides called Source and Drain. Fin-FET is then feasible and promising by combining lithography and selective processing. For one thing, Fin-FET devices are capable of lowering I_{off} current as the channel length gets substantially shorten. They can also keep the swing[7,8] from rolling-up because of well-controllability of Gate.

In this paper, p-channel FinFET devices are fabricated on SOI wafers (silicon on insulator). And N-well is defined by N-well threshold voltage [9,10] implant after the fins are prepared. Fin thickness (channel width) effects on the FinFET performances are observed with poly-silicon or fully cobalt silicide[11,12] as Gate.

Experiments and Preparation of FinFET Device

A layer of sacrificing oxide is grown by Furnace on the silicon, which is deliberately fabricated on insulator (SOI), followed by the dispensing of positive photo-resist (PR) covering the oxide. After the exposure and development, there forms an "I". The sacrificing oxide and silicon underneath "I" get protected from plasma ion dry etching and become a high-aspect 3-D fin-like channel. The channel turns into N-type body as the N-well Vt adjustment implant is applied. After HF dip, 14 angstrom gate oxide dielectric is grown and adulterated with nitrogen by remote plasma annealing for raising the dielectric constant. Two different gate materials, heavy-doped poly-silicon and fully cobalt silicide, are separately deposited unto the gate oxide for splitting as shown in Fig.1. The two ends of "I" are called Source and Drain.

Results and Discussions

The I_D-V_D characteristic curves of p-channel FinFET devices with two different gate materials are shown in Fig.2 (0.1 micron channel length) and Fig.3 (0.12 micron channel length). It is interesting that, for both channel lengths, the 0.11 μm wide devices with poly-silicon gates always perform much better and even better for over 80% enhancement. As for fully cobalt silicide gate devices, channel width 0.11μm does not apparently show superiority to channel width 0.12 μm. Nevertheless, it gives us some ideas that p-channel FinFET device with fully cobalt silicide gates, corresponding to I_{off}=1.46×10^{-8}A, is quite insensitive for either channel width. But it is for sure that the work function of cobalt silicide does promote the I_D-V_D performance in Fig.2 (B) as compared to the one with poly-silicon gate.

Moreover, the performances of devices can be also checked on the sub-threshold current where Swing can be determined as follows:

$$S \equiv \frac{dV_G}{d \log I_D}\bigg|_{V<V_{th}} \qquad (1)$$

It is found that the swing for cobalt silicide gate with channel width/length = 0.1μm/0.1μm in Fig.2 (A) is 84mV/decade while it is 110mV/decade corresponding to the poly-silicon gate with channel width/length = 0.12μm/0.1μm in Fig.3 (B). The threshold voltage in the linear region in Fig.2 (A) can be obtained through the following equation by extrapolation,

$$I_D = \frac{\mu_N C_{ox}^{(1)} W}{2L}(V_G - V_{th})V_D \cdot \qquad (2)$$

Therefore, Vt_linear can be found by examining I_D-V_G curves near V_G =0. And it was determined to be V_{th}=-0.47V at W/L = (0.11μm / 0.1μm) with poly-silicon as gate materials. And it is reasonable that many of the p-channel devices have negative threshold voltages. Even though few p-channel devices have positive threshold voltages, the values are very close to zero. It reminds us that linear extrapolated values for determining threshold voltage no longer fit the current cases. One should always get back to the trivial model about parameters describing V_{th} as follows:

$$V_{th} - [\Phi_{work} + \frac{k_B T}{e}\ln(\frac{n}{n_i})] - 2\frac{kT}{e}\ln(\frac{n}{n_i}) - \frac{Q_{dep}}{C_{ox}^{(1)}} \qquad (3)$$

978-1-4673-4840-9/13 $31.00 © 2013 IEEE

where Φ_{work} is work-function related value that is thought to be negative, k_B is Boltzmann constant (1.38×10^{-23} J/K), e is carrier charge and equal to (1.6×10^{-19}C), n is n-type dopant concentration, n_i is the intrinsic dopant concentration, and Q_{dep} is the depleted space charge. As the gate is biased crossing over the threshold voltage, n-type channels thus get strongly inversed into p-type channels resulting in routes between conductive Sources and Drains. For PFinFET devices, the depleted region is too thin to make the threshold voltage distinguishably negative, and sometimes close to Ground. Somehow, the threshold voltages may be adjusted by imposing the various gate materials like fully-covered cobalt silicide, whose work function may play crucial roles on controlling the values of threshold voltages for integration issues. As shown in Fig.4, there emerges a competitive current that happens at $V_G = 0.0$ V and demonstrates the benefits of a device with a positive threshold voltage.

As for n-channel devices, the threshold voltages, instead, are to be expressed a little differently as follows:

$$V_{th} = [\Phi_{work} - \frac{k_B T}{e}\ln(\frac{n}{n_i})] + 2\frac{kT}{e}\ln(\frac{n}{n_i}) + \frac{Q_{dep}}{C_{ox}^{(1)}} \qquad (4),$$

If Φ_{work} is negative again, V_{th} may turn out to be negative if the width of the channel is too thin making Q_{dep} less just as predicted.

Conclusion

At channel length 0.10μm and 0.12μm with poly-silicon gate, FinFET devices perform much better at channel width = 0.11μm (fin thickness) than 0.12μm. As for FinFET devices with fully cobalt silicide gate, the enhancement is not apparently seen because the use of different gate materials is for modifying the threshold voltages. Swing is determined to be 84mV/decade, which is quite reasonable.

Acknowledgement

All the authors sincerely thank National Nano Device Laboratory in Taiwan for being so fully supportive.

References

[1] Ramirez I , et al., "Measurement of leakage current for monitoring the performance of outdoor insulators in polluted environments," *IEEE Electrical Insulation Magazine*, vol.28, no.4, pp.29-34, July-Aug. 2012.
[2] M.A Alam, et al., "Influence of Molding Compound on Leakage Current in MOS Transistors," *IEEE Transactions on Components, Packaging and Manufacturing Technology*, vol.1, no.7, pp.1054-1063, Jul. 2011.
[3] Hsin-Wen Su, "Drain-induced-barrier lowering and subthreshold swing fluctuations in 16-nm-gate bulk FinFET devices induced by random discrete dopants," *2012 70th Annual Device Research Conference (DRC)*, pp.109-110, 18-20 Jun. 2012.
[4] S.H Rasouli, et al., "Design Optimization of FinFET Domino Logic Considering the Width Quantization Property," *IEEE Transactions on Electron Devices*, vol.57, no.11, pp.2934-2943, Nov. 2010.

[5] Hyejeong, et al., "Properties of poly silicon films deposited on silicon seed-layers prepared by AIC process," *2011 37th IEEE Photovoltaic Specialists Conference (PVSC)* pp.003005-003007, 19-24 Jun. 2011.
[6] T. Inoue, et al., "Life cycle costing analysis for poly-silicon photovoltaics production processes in Japan," *2010 IEEE International Conference on Industrial Engineering and Engineering Management (IEEM)*, pp.1990-1993, 7-10 Dec. 2010.
[7] P. Chotimanus, et al., "Real swing extraction for video indexing in golf practice video," *2012 Computing, Communications and Applications Conference (ComComAp)*, pp.420-425, 11-13 Jan. 2012.
[8] M. Afzali, et al., "A novel algorithm to identify power swing based on superimposed measurements," *2012 11th International Conference on Environment and Electrical Engineering (EEEIC)*, pp.1109-1113, 18-25 May 2012
[9] Xingsheng Wang, et al., "Statistical Threshold-Voltage Variability in Scaled Decananometer Bulk HKMG MOSFETs: A Full-Scale 3-D Simulation Scaling Study," *IEEE Transactions on Electron Devices*, vol.58, no.8, pp.2293-2301, Aug. 2011.
[10] Jia Sun, et al., "Low-Voltage Oxide-Based TFTs Self-Assembled on Paper Substrates With Tunable Threshold Voltage," *IEEE Transactions on Electron Devices*, vol.59, no.2, pp.380-384, Feb. 2012.
[11] H.A. Elgomati, et al., "Cobalt silicide and titanium silicide effects on nano devices," *2011 IEEE Regional Symposium on Micro and Nanoelectronics (RSM)*, pp.282-285, 28-30 Sept. 2011.
[12] Liang. Yu-Hsin, et al., "Growth of single-crystalline cobalt silicide nanowires with excellent physical properties," *Journal of Applied Physics*, vol.110, no.7, pp.074302-074302-4, Oct 2011

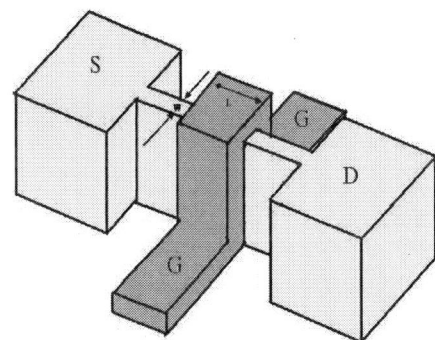

Fig. 1 The top view of the 3-D FinFET device

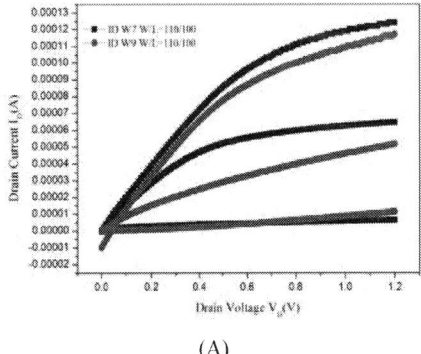

(A)

978-1-4673-4840-9/13 $31.00 © 2013 IEEE

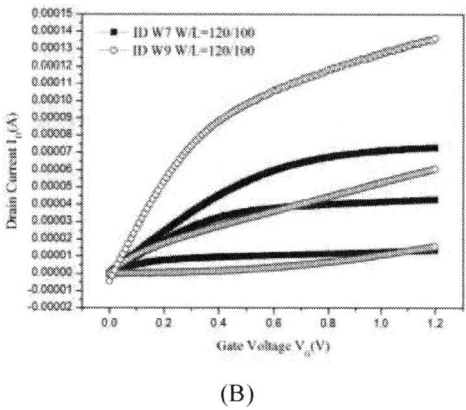

(B)

Fig.2 The I_D-V_D characteristic curves with channel width/length (A) (W/L = 0.11 μm /0.1 μm) and (B) (W/L = 0.12 μm /0.1 μm)

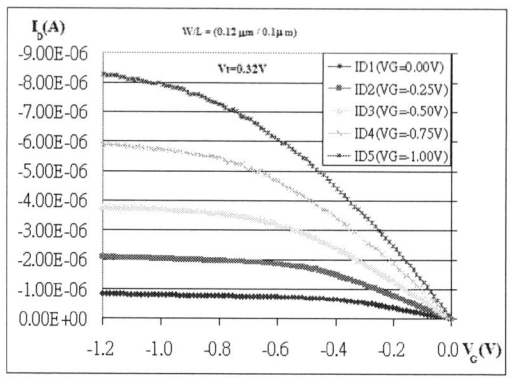

Fig.4 The I_D-V_D characteristic curves with channel width/length (W/L = 0.12 μm /0.1 μm)

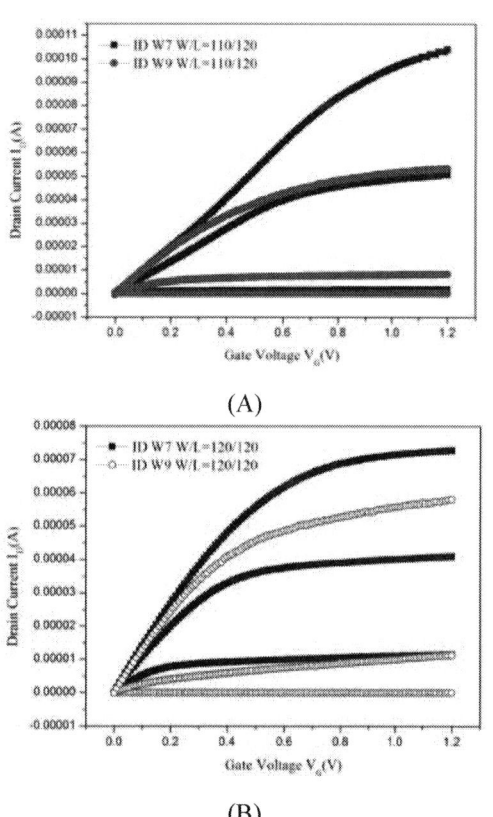

(A)

(B)

Fig.3 The I_D-V_D characteristic curves with channel width/length (A) (W/L = 0.11 μm /0.12 μm) and (B) (W/L=0.12 μm /0.12 μm)

The Enhancement of MOSFET Electric Performance through Strain Engineering by Refilled SiGe as Source and Drain

Hsin-Chia Yang[1]*,Chao-Wang Li[1], Wen-Shiang Liao[1], Chong-Kuan Du[1], Mu-Chun Wang[1]*, Jie-Min Yang1,Chun-Wei Lian[1], Chuan-Hsi Liu[3]

[1]Dept. of Electronic Engineering, Minghsin University of Science & Technology, Taiwan
[2]Dept. of Materials and Resources Engineering, National Taipei University of Technology, Taipei, Taiwan
[3]Dept. of Mechatronic Technology, National Taiwan Normal University, Taipei, Taiwan
*Correspondence email: mucwang@must.edu.tw

Abstract

Mismatched lattice constants between SiGe and silicon can cause the strain making the mobility improved. SiGe are grown underneath the channel apparently to form global strain over the whole devices, while Source/Drain refilled with SiGe would squeeze or pull up the devices uni-axially. The I_D-V_G characteristics curves and the maximum trans-conductance (g_m) using strain engineering are observed to be superior to the baseline. Nevertheless, the breakdown voltages with strain engineering no longer enjoy as robustly as ones without.
Keywords: Strained Engineering, SiGe-Refilled Source/Drain

Introduction

Strained Engineering [1] has been applied to enhance the electrical performances of nano-process MOSFET devices. Mismatched lattice constants between SiGe and silicon can cause the strain making the mobility improved. SiGe are grown underneath the channel apparently to form global strain[2] over the whole devices, while Source/Drain refilled with SiGe, on the other hand, would squeeze or pull up the devices uni-axially [3] (including "compressive" or "tensile"). Both kinds of strains do provide beneficial circumstances for carriers to travel in, and hence the mobility gets enhanced.

In this paper, PMOSFET devices are fabricated on the <110> substrate. The set of 90nm photo masks with over exposure and over etching technique can generate even channel lengths less than 90nm line width. One thus examines the junction breakdown voltage [6] [7] and the punch-through [8] voltage, and check if the enhancement of electrical performances is impressive..

Experiments and Preparation of PMOSFET Devices

A 5nm buffer layer of epi-silicon is grown onto <110> substrate for the adjustment of concentration profile during subsequent thermal budget, followed by 10nm epitaxial Silicone Germanium (SiGe) layer. A 3nm epi-silicon cap layer is then grown on SiGe partly for growing 1.4nm nitrided oxide as gate dielectric. 150nm thick heavy-doped poly-silicon is chemically deposited on the gate dielectric and dry etched to form the gate, followed by TEOS process to form side wall. As seen in Fig.1, the areas of source and drain are dug up for 50nm in depth [4] and refilled with 90nm silicon germanium [5] in which there exists 30% germanium. An Agilent Instruments 4156C is interfaced through GPIB card to a personal computer having ICS software for the measurements. The chuck temperatures can be kept at $25^{\circ}C \pm 1\,^{\circ}C$.

Results and Discussions

PMOSFET devices with three different channel width/length sizes, (W/L=10μm/10μm, 10μm/1μm, 10μm/0.24μm) are to be evaluated. As shown in Fig.3, the strained devices show superior I_D over 30% to the non-strained ones. This improvement assures ones that p-type carriers move faster in the constructed environments.

The limit of 1 μA is set up to test the junction breakdown voltage as the drain is biased, as shown in Fig.4 (A). For determining junction breakdown voltages, the junction doped concentration shall be taken into account. It is reasonable that the refilled and thus strained devices, of course, possess higher breakdown voltage with the help of the buffer layer.

As for the leakage current for determining punch-through voltage, 1 μA is also taken to be the limit. As found in Fig.4 (B), it is so encouraging that the refilled and thus strain devices enjoy the higher punch through voltage. Punch-through voltages involve the dosage concentrations of n-type substrate. Due to the re-grown SiGe, the conditioning processing is good enough to eliminate the dangling bonds and the reduced nearby concentrations therefore strengthen the capability of facing the stresses as the biased is applied. After the testing, the test results are data-plotted in Fig.4 (A) and (B). Both demonstrate superior sustainability as compared with non-strained ones.

After testing breakdown and punch through voltages, I_D-V_D performances are verified again and proved to be non-deteriorated as referred to Fig.5. As in Fig.6 (A), I_D-V_G curves of refilled NMOS devices show slightly better swing than unstrained ones even though the leakage currents at V_G=0V of them get higher. Also as demonstrated in Fig.6 (B), the g_m performances are upgraded after the strain engineering is applied. One can easily see the improved mobility correlated with the swing (~70mV/decade) and lower threshold voltages owing to the strain engineering

Conclusion

Promising features of the electrical performances, such as better threshold voltages rolling trend, higher breakdown voltages, higher punch-through voltages, and somewhat good swing, are so impressive that it is worthwhile to promote the strain engineering techniques for the future uses.

Acknowledgement

Many thanks are due to National Nano Device Laboratory in Taiwan for being so fully supportive.

Reference

[1] E. Ungersboeck, V. Sverdlov, H. Kosina, S. Selberherr, "Strain engineering for CMOS devices," Solid-State and Integrated Circuit Technology, 2006. ICSICT '06. 8th International Conference on , pp.124-127, Oct. 2006.

[2] Takagi, S, Tezuka, T, Irisawa, T, Nakaharai, S, Numata, T, Usuda, K, Maeda, T, Sugiyama, N, , "Mobility-Enhanced CMOS Technologies Using Strained Si/SiGe/Ge Channels," Integrated Circuit Design and Technology, 2006. ICICDT '06. 2006 IEEE International Conference on , pp.1-2, 0-0 0.

[3] Woo Sik Yoo; Ueda, T, Kang, K , "Characterization of uni-axially stressed Si and Ge concentration in Si1-xGex using polychromator-based multi-wavelength Raman spectroscopy," Junction Technology, 2009. IWJT 2009. International Workshop on , pp.79-83, 11-12 June 2009.

[4] F.K .Moghadam, X.-C. Mu, "A study of contamination and damage on Si surfaces induced by dry etching, " *IEEE Transactions on Electron Devices*, vol.36, no.9, pp.1602-1609, Sep. 1989.

[5] K.Shenai, "A novel trench planarization technique using polysilicon refill, polysilicon oxidation, and oxide etchback, " IEEE Transactions on Electron Devices, vol.40, no.2, pp.459-463, Feb. 1993.

[6] V.V.Obreja, "Leakage current voltage dependence and performance of power semiconductor devices in the breakdown (avalanche) region, " *Power Electronics Specialists Conference, 2008. PESC 2008. IEEE* , pp.1777-1782, 15-19 Jun. 2008.

[7] Obreja, V.V , "Leakage current voltage dependence and performance of power semiconductor devices in the breakdown (avalanche) region," Power Electronics Specialists Conference, 2008. PESC 2008. IEEE , pp.1777-1782, 15-19 June 2008.

[8] C. Betancourt, J. Wright, A. Bielecki, Z. Butko, C. Parker, N. Ptak, V .Fadeyev ,H.F.-W. Sadrozinski, "Punch-through effect and collapse of the electric field in silicon strip detectors, " *Nuclear Science Symposium Conference Record (NSS/MIC), 2010 IEEE* , pp.388-391, Oct. 30 2010-Nov. 6 2010.

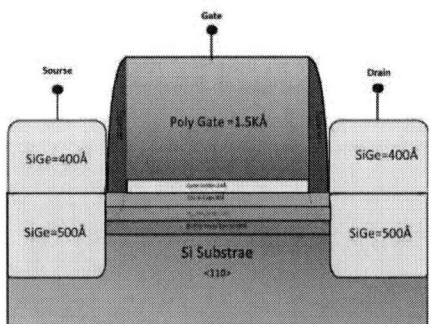

Fig. 1 Cross section of the device with refilled SiGe as Source and

Drain

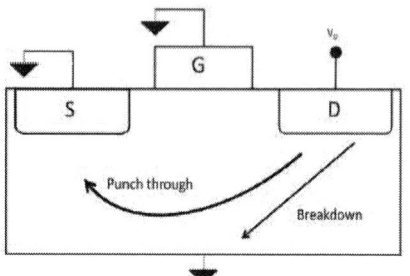

Fig. 2 Punch through and junction breakdown tests

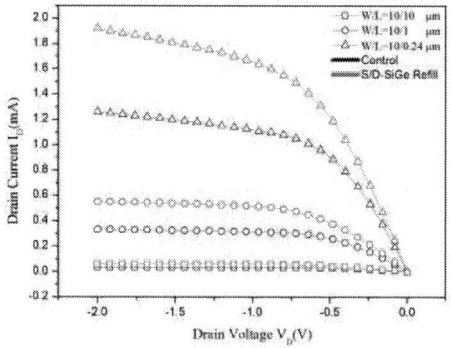

Fig. 3 The comparison between non-strained PMOS and

strained PMOS

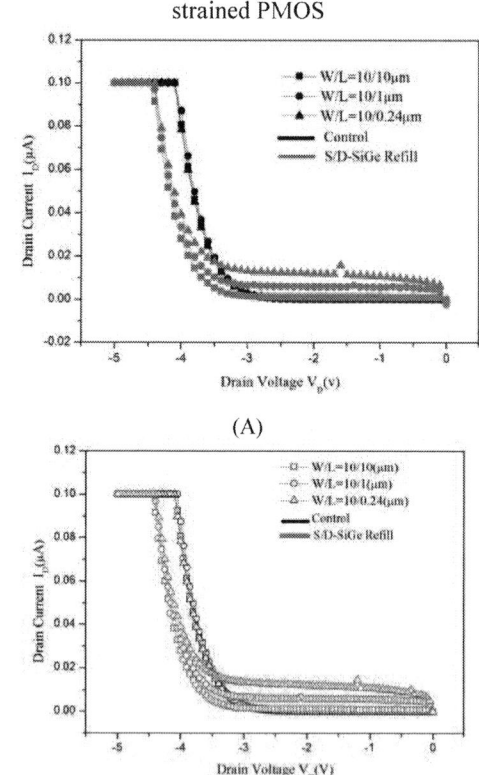

(A)

(B)

Fig. 4 (A) Breakdown tests (B) Punch through tests

978-1-4673-4840-9/13 $31.00 © 2013 IEEE   252

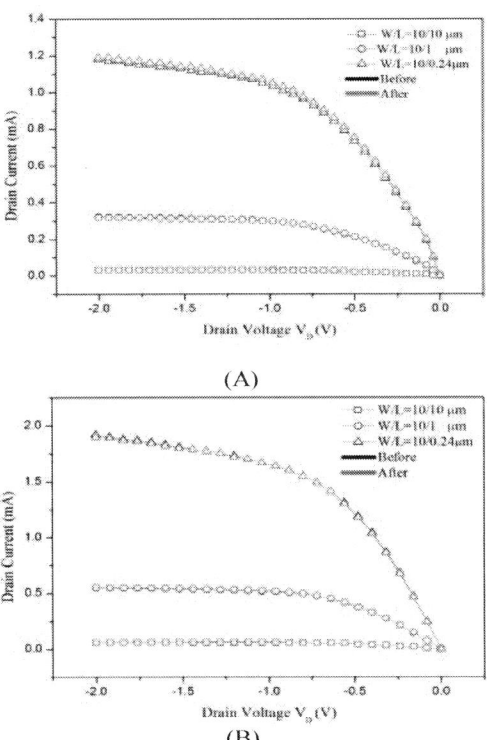

Fig. 5 I_D-V_D curves before and after punch-through and breakdown voltage tests (A) Un-strained (B) Strained

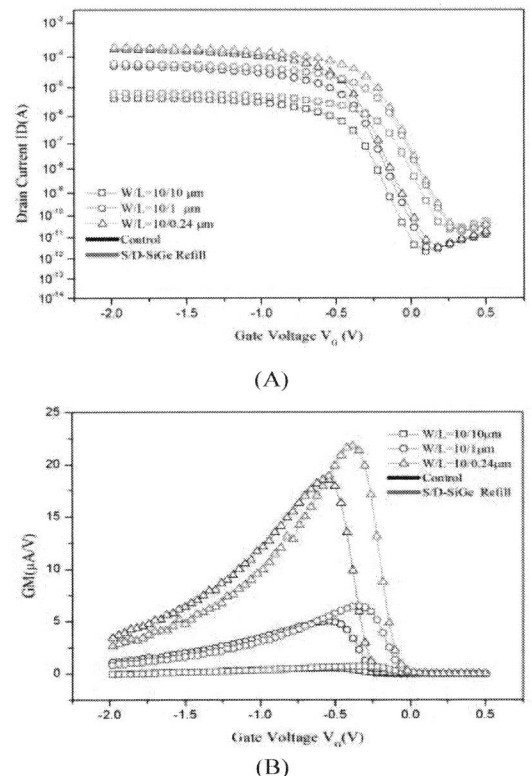

Fig. 6 (A) I_D-V_G curves (B) Trans-conductance g_m

Temperature Dependence Carrier Transport Behavior of Transparent ZnO:Y Nanocrystalline Films

S. L. Young[a,*], C. Y. Kung[b,*], H. Z. Chen[a], M. C. Kao[a], T. T. Lin[b], M. C. Chang[b], H. H. Lin[c], J. H. Lin[c], S. H. Chin[b], C. R. Ou[c]

[a]Department of Electronic Engineering, Hsiuping University of Technology, Taichung, Taiwan
[b]Department of Electrical Engineering, National Chung Hsing University, Taichung, Taiwan
[c]Department of Electrical Engineering, Hsiuping University of Technology, Taichung, Taiwan
Phone: (+886) 4-24961100ext1206, E-mail: slyoung@mail.hust.edu.tw and cykung@nchu.edu.tw

Abstract

The Y-doped ZnO nanocrystalline films were deposited on the glass substrates by sol-gel method. X-ray diffraction measurements of the films showed the same wurtzite hexagonal structure and preferential orientation along the c-axis. The grain size of the ZnO films was decreased by the doping of Y. Temperature dependence resistivity showed a semiconductor transport behavior for the nanocrystalline films. At low temperature region, the resistivity can be fitted well with the behavior of Mott variable range hopping, $\sigma(T)=\sigma_{h0} \exp[-(T_0/T)^n]$ with n=1/4. On the contrary, at high temperature region, the transport mechanism can be fitted with semiconductor behavior by Arrhenius equation, $\sigma(T)=\sigma_0 \exp[-(E_a/kT)^m]$ with m=1. The activation energy E_a is increased from 0.47 meV for nondoped ZnO film to 0.83 meV for $Zn_{0.98}Y_{0.02}O$ film obtained from equation. The results demonstrate that the crystallization and the corresponding carrier transport behavior of the Y-doped ZnO films are affected by the doping of Y.

Keywords: ZnO film, sol-gel method, resistivity, Arrhenius equation and variable range hopping.

Introduction

ZnO has recently drawn much interest for the possible application in optoelectronics devices [1-3] due to the large direct bandgap of 3.37 eV and exciton energy of 60 meV. These properties are important for application to commercial electronic products, such as photoconductors for electrophotography [4], varistors for electrical circuits [5], sensors for gas detection [6], and active layer for thin film transistors [7]. There have been also reported highly conductive and optical transparent ZnO films in the visible range suitable for transparent electrodes in solar cell and liquid crystal display applications [8, 9].

In addition, ZnO doped with transitional magnetic metal, such as $Zn_{1-x}TM_xO$ (TM = Mn, Fe, Co, Ni, and Cr) [10-14] were found to be diluted magnetic semiconductors with high Curie temperature. Magnetic doped ZnO has attracted broad interests for their possible use as spintronic materials [15, 16]. Some previous researches reported that the inhomogeneous distribution of transitional metal might greatly influence the magnetic and electronic properties of the ZnO-based materials [17-20].

The ZnO-based thin films have been investigated through various kinds of fabrication methods, such as sputtering, pulsed laser deposition, chemical vapor deposition, spray pyrolysis, sol-gel method, etc. However, compared with the high vacuum processes to fabricate the ZnO thin films, solution-based processes of spray pyrolysis and sol-gel methods offer more merits, such as easier control of chemical composition and much simpler method for large area coating at a low cost. Meanwhile, ZnO thin films prepared by spray pyrolysis and their corresponding electrical characterization had been well studied [21]. Moreover, the optical properties of doped ZnO thin films deposited by sol-gel method had also been reported [22, 23]. Nevertheless, the carrier transport behavior of the ZnO-based films deserves further discussion.

The temperature dependence of resistivity for ZnO films can be controlled by the fabrication process. The conductivity is mainly due to zinc excess at interstitial position. In the present work, we concentrate on an attempt to investigate the variation of the structural and electrical characteristics caused by Y doping in the transparent ZnO nanocrystalline films. At the same time, the report also discusses on the temperature dependence of carrier transport mechanisms in varying temperature ranges, as a guiding rule for possible application.

Experimental Procedure

The $Zn_{1-x}Y_xO$ (x = 0, 0.02) films were deposited on the glass substrate by sol-gel method. The source solutions were prepared by zinc acetate dehydrate $Zn(C_2H_3O_2)_2 \cdot 2H_2O$, iron (II) acetate tetrahydrate $Y(C_2H_3O_2)_3 \cdot 4H_2O$, 2-methoxyethanol $C_3H_8O_2$, and 2-aminoethanol (ethanolamine) C_2H_7NO. Zinc acetate dehydrate and iron (II) acetate tetrahydrate were firstly dissolved in 2-methoxyethanol in stoichiometric proportions. The concentration of metal ions was kept at 0.5 M. Then, 2-methoxyethanol was added into the solutions to form a stable sols. After stirring at 150°C for 1 h on a hotplate, transparent solutions were obtained. The $Zn_{1-x}Y_xO$ thin films deposited on glass substrate were prepared by spin coating technique. The precursor sols were dropped on the glass substrate and spun at 2500 rpm for 30 sec. Then, the samples were annealed by rapid thermal annealing treatment in air at the temperature of 800°C for 2 min with a heating rate of 600°C/min.

978-1-4673-4840-9/13 $31.00 © 2013 IEEE

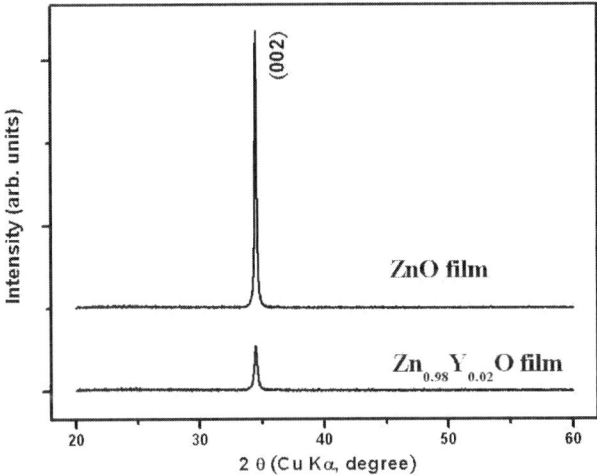

Fig. 1 X-ray diffraction patterns of ZnO and $Zn_{0.98}Y_{0.02}O$ films grown on glass substrate at the sintering temperature of 800°C.

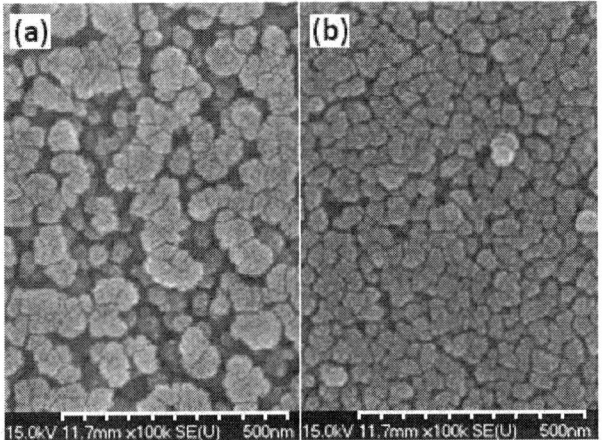

Fig. 2 SEM surface images of (a) ZnO film and (b) $Zn_{0.98}Y_{0.02}O$ film grown on glass substrate.

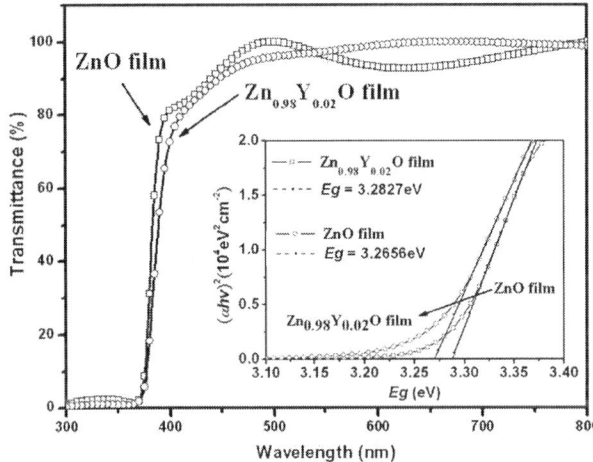

Fig. 3 Optical transmittance of ZnO and $Zn_{0.98}Y_{0.02}O$ films. Inset is the plot of $(\alpha h v)^2$ versus E_g of ZnO and $Zn_{0.98}Y_{0.02}O$ films.

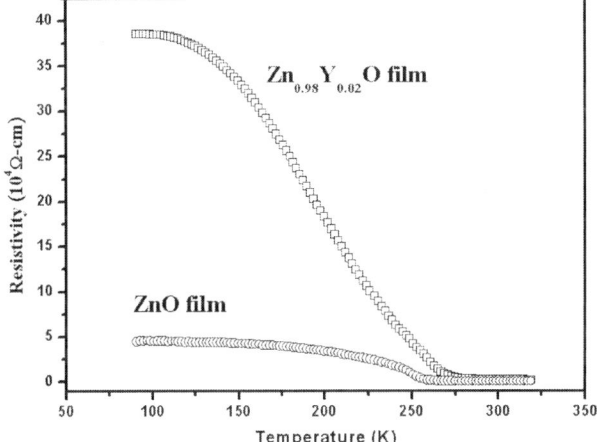

Fig. 4 Temperature dependence of resistance for ZnO and $Zn_{0.98}Y_{0.02}O$ films.

The crystal structure and grain orientation of ZnO films were determined by the x-ray diffraction (XRD) patterns using a Rigaku D/max 2200 x-ray diffractometer with Cu-Kα radiation. The XRD data were recorded at room temperature under the 2θ range from 20° to 60° with a step width of 0.01° and a scan speed of 0.5°/min. Morphological characterization was observed using a field emission scanning electron microscopy (FE-SEM, JEOL JSM-6700F) at 3.0 kV. Finally, the temperature dependence of resistivity was measured using the four-probe method in an Oxford gas helium cooled cryostat system.

Results and Dissusion

Fig. 1 shows the XRD patterns of pure ZnO and $Zn_{0.98}Y_{0.02}O$ samples. As shown from the XRD patterns, the $Zn_{0.98}Y_{0.02}O$ sample is found to have single polycrystalline phase with the same wurtzite hexagonal structure of space group $P6_3/mc$ (JCPDS 36-1451) as pure ZnO film, and no secondary phase is found. Stick and ball representation of the ZnO crystal structure is shown as the inset in Fig. 1. Both compositions

exhibit an (002) diffraction peak at about $2\theta = 34.5°$ indicating that they are c-axis preferred orientation which are consistent with the main diffraction planes as shown in the inset of Fig. 1. However, the increase of full wavelength half maximum of the XRD peak for the $Zn_{0.98}Y_{0.02}O$ sample is due to the decrease of grain size. The average grain sizes of the samples calculated by the classical Scherrer formula are 39.27 nm for ZnO film and 24.9 nm for $Zn_{0.98}Y_{0.02}O$ film. The decrease of grain size by the doping of Y in the ZnO film reveals the increase of boundary to volume ratio and, hence, the increase of resistivity which will be examined by the following measurement of the temperature dependence of resistivity. In addition, the decrease of the (002) peak intensity also shows that the crystallization is restrained by the doping of Y in the ZnO film.

Fig. 2 (a) and (b) show the FE-SEM surface images of ZnO and $Zn_{0.98}Y_{0.02}O$ films that reveal porous and granular morphology for both samples. It is noted that the grain size of the $Zn_{0.98}Y_{0.02}O$ film is obviously smaller than that of the ZnO film, which implies the crystallization is restrained by Y doping. This is consistent with the result indicated in Fig. 1. The insets in Fig. 2 (a) and (b) show the cross-section views of ZnO and

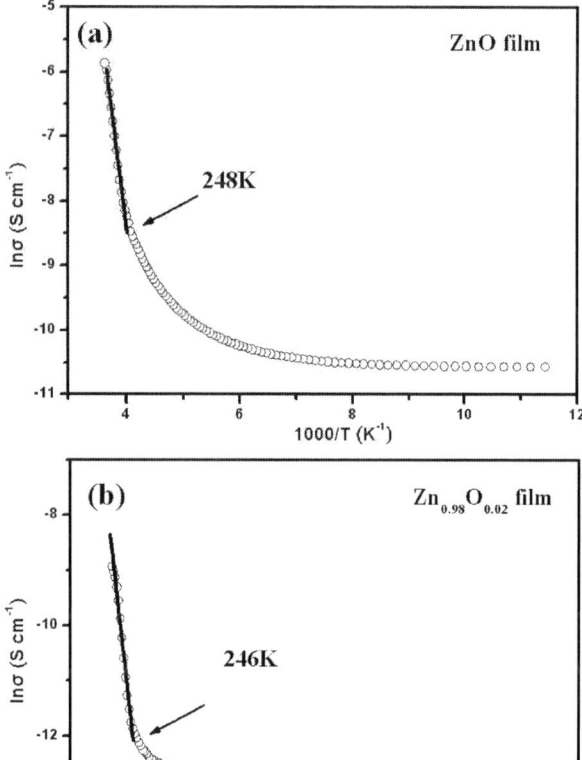

Fig. 5 Plots of $ln\sigma$ versus $1000/T$ for (a) ZnO film and (b) $Zn_{0.98}Y_{0.02}O$ film at high temperature range.

$Zn_{0.98}Y_{0.02}O$ films, in which the average thickness of the films are about 345 nm and 330 nm, respectively.

Fig. 3 gives the transmittance spectra of ZnO and $Zn_{0.98}Y_{0.02}O$ films measured at room temperature in the wavelength range from 300 to 800 nm. Optical spectra are the most direct and perhaps simplest method for probing the band structure of semiconductors and the optical band gap E_g can be calculated by the following relation [24]

$$\alpha = A(h\nu - E_g)^2/(h\nu), \qquad (1)$$

where A is a constant. The details of the mathematical determination of the absorption coefficient α could be found in literature [25]. The estimated band gaps from the plots of $(\alpha h\nu)^2$ versus $h\nu$ as shown in the inset of Fig. 3 are 3.2827 eV for ZnO film and 3.2656 eV for $Zn_{0.98}Y_{0.02}O$ film. As shown in Fig. 3, the $Zn_{0.98}Y_{0.02}O$ film reveals a decrease of transmission and an increase of absorbance after the doping of Y into the ZnO film. This is speculated due to the decrease of the grain size, resulting in the increase of grain boundary to volume ratio and possible defects. Clearly, from the optical spectra shown in Fig. 3, the transmission edge slightly shifting towards a longer wavelength for the $Zn_{0.98}Y_{0.02}O$ film indicates the decrease of the optical band gap E_g.

Fig. 4 shows the temperature dependence of resistance of the $Zn_{1-x}Y_xO$ films (x=0.00 and 0.02). For both of the pure ZnO

and $Zn_{0.98}Y_{0.02}O$ films, the resistance decreases with the increase of temperature measured from 75K to 320K indicating the semiconductor behavior. The resistance for the $Zn_{0.98}Y_{0.02}O$ film is obviously greater than that of pure ZnO film which can be explained by the decrease of grain size and corresponding increase of grain boundary resistance due to the Y doping in the film.

Fig. 5 shows approaching plots of the pure ZnO and $Zn_{0.98}Y_{0.02}O$ films fitted by Arrhenius equation at high temperature region. The relationship of conductivity σ versus temperature T can be expressed as

$$\sigma(T) = \sigma_o \, exp[-(E_a/kT)^m], \qquad (2)$$

where $m=1$, σ_0 is a constant, E_a is the activation energy and k is the Boltzmann's constant. As shown in Fig. 5, the experimental data can be fitted well by equation (2) above 248 K for pure ZnO film and above 246 K for that of $Zn_{0.98}Y_{0.02}O$ film, respectively. The activation energy is 0.47 meV for pure ZnO film and 0.83 meV for $Zn_{0.98}Y_{0.02}O$ film obtained from equation (2). However, the conductivity deviates from the fitting line at lower temperature for both samples. Hence, the deviation from the fitted line with experimental data for both samples can evidence the existence of different conduction mechanism at lower temperature.

At low temperature range, there are two different conduction mechanisms, nearest neighbor hopping (NNH) and variable range hopping (VRH), had been reported for carrier transport behavior of semiconductors. If the plot of $ln\sigma$ vs. $1/T$ can be fitted at low temperatures, it indicates the presence of NNH conduction [26]. However, data from the ZnO and $Zn_{0.98}Y_{0.02}O$ samples cannot be fitted well in the low temperature region. Hence, NNH is not the dominant conduction mechanism of $Zn_{1-x}Y_xO$ films in the present cases. For the VRH conduction mechanism, the conductivity σ can be expressed as

$$\sigma = \sigma_{h0} \exp[-(T_0/T)^{1/4}], \qquad (3)$$

where σ_{h0} and T_0 are given by the following expressions [26, 27],

$$\sigma_{h0} = \frac{3e^2 v_{ph}}{(8\pi)^{1/2}} \left[\frac{N(E_F)}{\alpha kT} \right]^{1/2} \qquad (4)$$

and

$$T_0 = \left[\frac{16\alpha^3}{kN(E_F)} \right], \qquad (5)$$

in which v_{ph} (about 10^{13} s^{-1}) is te phonon frequency at Debye temperature, k is the Boltzmann's constant, $N(E_F)$ is the density of state at Fermi level and α is the inverse localization length at the localized state .

From equations (3) and (4), the following relationship,

$$\ln\left(\sigma T^{1/2}\right) \alpha T^{-1/4}, \qquad (6)$$

can be obtained. Then, we plot the $ln(\sigma T^{1/2})$ with $T^{-1/4}$ of $Zn_{1-x}Y_xO$ films as shown in Fig. 6 to examine the possibility of VRH conduction mechanism. The result illustrates that the experimental data can be fitted well at the low temperature range from 100 K to 140 K for pure ZnO as shown in Fig. 6(a) and from 100 K to 167 K for $Zn_{0.98}Y_{0.02}O$ film as shown in Fig. 6(b). The results demonstrate that the VRH for both samples dominates the conduction mechanism at low temperature, and

978-1-4673-4840-9/13 $31.00 © 2013 IEEE

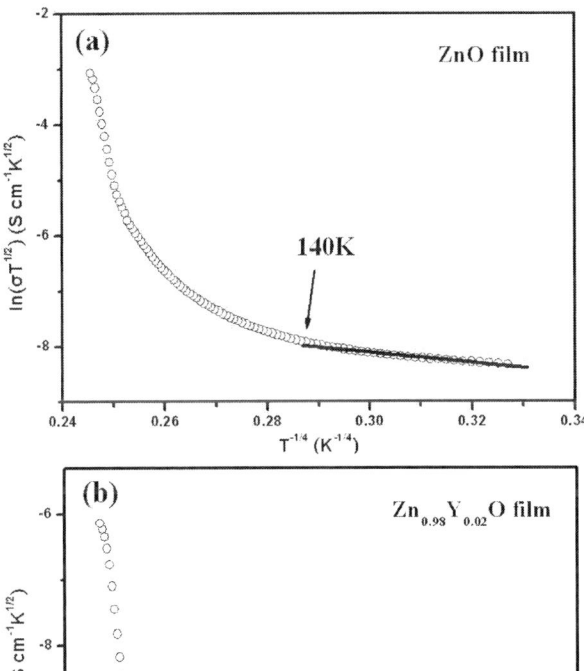

Fig. 6 Plots of $ln(\sigma T^{1/2})$ with $T^{-1/4}$ for (a) ZnO film and (b) $Zn_{0.98}Y_{0.02}O$ film at low temperature range.

the introduction of Y into ZnO film increases the temperature range toward higher temperature. As a consequence, the temperature dependence of resistivity ($\rho=\sigma^{-1}$) at low temperature range decreases sharply as temperature increases with the relationship, $ln(\sigma T^{1/2}) \propto T^{-1/4}$ from equation (4), while the resistivity at high temperature range with the relationship, $ln\sigma \propto T^{-1}$ from equation (2).

Conclusions

The transparent nanocrystalline ZnO and $Zn_{0.98}Y_{0.02}O$ films were deposited on the glass substrate by sol-gel method for microstructure and carrier transport behavior comparison. XRD patterns showed that both samples are found to exhibit the single polycrystalline phase with the same wurtzite hexagonal structure of group space $P6_3/mc$. FE-SEM Surface images show the grain size of the granular $Zn_{1-x}Y_xO$ films decreases with the doping of Y. The result of temperature dependence resistivity indicates that the resistivity is increased with the doping of Y in the ZnO film. At high temperature region, the conductivity vs. temperature demonstrates that the activation type dominates the transport behavior indicated by the Arrhenius plot for both composition of films. On the contrary, the VRH is the dominant conduction mechanism at low temperature region.

Acknowledgment

This work was sponsored by the National Science Council of the Republic of China under the grants No. NSC 101-2221-E-164-004 and NSC 101-2221-E-005-014.

References

[1] P.I. Reyes, C.J. Ku, Z. Duan, Y. Lu, A. Solanki, K.B. Lee, Appl. Phys. Lett. 98 (2011) 173702.
[2] V. Chivukula, D. Ciplys, M. Shur, P. Dutta, Appl. Phys. Lett. 96 (2010) 233512.
[3] M. Ahmad, J. Zhu, J. Mater. Chem. 21 (2011) 599.
[4] K.K. Kim, S. Niki, J.Y. Oh, J.O. Song, T.Y. Seong, S.J. Park, J. Appl. Phys. 97 (2005) 066103.
[5] A. Sedky, T.A. El-Brolossy, S.B. Mohamed, J. Phys. Chem. Solids 73 (2012) 505.
[6] S. Öztürk, N. Kılınç, N. Taşaltin, Z.Z. Öztürk, Thin Solid Films 520 (2011) 932.
[7] A. Alias, K. Hazawa, N. Kawashima, H. Fukuda, K. Uesugi, Jpn. J. Appl. Phys. 50 (2011) 01BG05.
[8] N. Hirahara, B. Onwona-Agyeman, M. Nakao, Thin Solid Films 520 (2012) 2123.
[9] N. Yamamoto, T. Yamada, H. Makino, T. Yamamoto, J. Electrochem. Soc. 157 (2010) J13.
[10] W. Yu, L.H. Yang, X.Y. Teng, J.C. Zhang, Z.C. Zhang, L. Zhang, G.S. Fu, J. Appl. Phys. 103 (2008) 093901.
[11] A.P. Palomino, O. P. Perez, R. Singhal, M. Tomar, J. Hwang, P. M. Voyles, J. Appl. Phys. 103 (2008) 07D121.
[12] Z. Yang, J.L. Liu, M. Biasini, W.P. Beyermann, Appl. Phys. Lett. 92 (2008) 042111.
[13] Y.F. Tian, S.S. Yan, Y.P. Zhang, H.Q. Song, G.L. Liu, Y.X. Chen, L.M. Mei, J. Appl. Phys. 100 (2006) 103901.
[14] E.J. Kan, L.F. Yuan, and J.L. Yang, J. Appl. Phys. 102 (2007) 033915.
[15] Y. Fukuma, K. Goto, S. Senba, S. Miyawaki, H. Asada, T. Koyanagi, and H. Sato, J. Appl. Phys. 103 (2008) 053904.
[16] Y. Niwayama, H. Kura, T. Sato, M. Takahashi, T. Ogawa, Appl. Phys. Lett. 92 (2008) 202502.
[17] K. Ueda, H. Tabata, T. Kawai, Appl. Phys. Lett. 79 (2001) 988.
[18] T.S. Herng, S.P. Lau, L. Wang, B.C. Zhao, S.F. Yu, M. Tanemura, A. Akaike, K.S. Teng, Appl. Phys. Lett. 95 (2009) 012505.
[19] X.X. Wei, C. Song, K.W. Geng, F. Zeng, B. He, F. Pan, J. Phys.: Condens. Matter 18 (2006) 7471.
[20] S.S. Lee, G. Kim, S.C. Wi, and J.S. Kang, S.W. Han, Y.K. Lee and K.S. An, S.J. Kwon, M. H. Jung, H. J. Shin, J. Appl. Phys. 99 (2006) 08M103.
[21] A. Goyal, S. Kachhwaha, Mater. Lett. 68 (2012) 354.
[22] C.-Y. Tsay, H.-C. Cheng, Y.-T. Tung, W.-H. Tuan, C.-K. Lin, Thin Solid Films 517 (2008) 1032.
[23] K.J. Chen, F.Y. Hung, S.J. Chang, Z.S. Huc, Appl. Surf. Sci. 255 (2009) 6308.
[24] J. Robertson, Phil. Mag. B 63 (1994) 307.
[25] N.M. Ahmed, Z. Sauli, U. Hashim, Y. Al-Douri, Int. J. Nanoelectron. Mater. 2 (2009) 189.
[26] R. Kumar, N. Khare, Thin Solid Films 516 (2008) 1302.
[27] G. Paasch, T. Lindner, S. Scheinert, Synth. Met. 132 (2002) 97.

Multiferroic and Structural Transition Properties of $Bi_{1-x}Pr_xFe_{0.95}Mn_{0.05}O_3$ Thin Films

Hone-Zern Chen[*], Ming-Cheng Kao[*], San-Lin Young

Department of Electronic Engineering, Hsiuping University of Science and Technology, Taichung, Taiwan.
No.11, Gongye Rd., Dali Dist. Taichung City 412, Taiwan
Tel:+886-4-24961100#1500, E-mail: hzc@mail.hust.edu.tw and kmc@mail.hust.edu.tw

Abstract

$Bi_{1-x}Pr_xFe_{0.95}Mn_{0.05}O_3$ (BPFMO) thin films were successfully deposited on Pt(111)/Ti/SiO$_2$/Si(100) substrates by spin coating with a sol-gel technology and rapid thermal annealing. The effects of Pr content on the microstructure, magnetic and multiferroic properties of thin films were investigated. The result of X-ray diffraction analysis shows that the BPFMO thin films have rhombohedral-to-tetragonal $R3c \rightarrow P4mm$ phase transition at $x = 0.15$. The Pr doping on the A-site of BiFeO$_3$ could induce the appearance of the spontaneous magnetization and polarization by the phase transition of rhombohedral-to-tetragonal. The BPFMO thin films with $x = 0.2$ exhibits the maximum remanent magnetization (2Mr) of 0.44 emu/g. Keywords: Sol-gel process, Bismuth ferrite, Phase transition, Magnetic.

Introduction

Bismuth ferrite BiFeO$_3$ (BFO) thin films with simultaneous ferroelectric and magnetic properties have been extensively studied in the recent years due to their potential applications in information storage, spintronics, magnetoelectrical/optical devices, sensors, actuators, and transducers[1-5]. It has been recently reported that lanthanide-substituted BiFeO$_3$, $(Bi_{1-x}Ln_x)FeO_3$ (Ln = La, Nd, Gd and Pr) could improve its ferroelectric and electrical properties. The reason is that the enhanced ferroelectric properties of these films can be attributed to the suppressed formation of impure phases and oxygen vacancies, which is caused by the substitution of stable rare-earth ions for the volatile Bi ions. In addition, some A-site or B-site substitution in the BiFeO$_3$ thin films could improve its ferroelectric and fatigue resistance properties [6-15]. The reason was that the improvement of ferroelectric and ferromagnetic in these films can be attributed to the suppression of the spatially modulated spin structure and release the locked magnetization, which was caused by the substitution of the ions with the different ionic radii and valence states. In this paper, we deposited the $Bi_{1-x}Pr_xFe_{0.95}Mn_{0.05}O_3$ (BPFMO) thin film onto Pt(111)/Ti/SiO$_2$/Si(100) substrates using the sol-gel process and annealed by the rapid thermal annealing (RTA). The crystallization and microstructures of thin films were analysed by X-ray diffraction (XRD) and atomic force microscopy (AFM), respectively. In addition, effects of the Pr-substitution on the magnetic properties of the BPFMO thin films were also investigated.

Experimental

The general chemical formulas of the precursor solutions of BPFMO prepared by the sol-gel method. Bismuth acetate, $Bi(OOCCH_3)_3$ (Alfa, 99.99%+ purity), praseodymium(III) acetate, $Pr(OOCCH_3)_3 \cdot xH_2O$, iron acetate, $Fe(CO_2CH_3)_2$ (Alfa, 99.9%+ purity) manganese(II) acetate, Mn $(C_2H_3O_2)_2 \cdot 4H_2O$, were used as source materials and 2-methoxyethanol was used as solvent. The gravimetrically assayed bismuth acetate and praseodymium acetate reagents were dissolved in the 2-methoxyethanol, to obtain sol compositions. The solutions were refluxed at 60 °C for 0.5 h under one atmosphere pressure. After the addition of iron acetate and manganese acetate, the solutions were further refluxed at 80 °C for 2 h to promote solution homogeneity; a stock solution of ~1M concentration was obtained. The obtained solution had a golden color. Inductively coupled plasma mass spectrometry was used to confirm that the deviation from stoichiometry was within approximately ±1 %. The stock solutions were spin-coated onto Pt/Ti/SiO$_2$/Si(100) substrates at a spin rate of 2500 rpm for 30 s using a commercial photoresist spinner. After each coating step, the gel films were pyrolyzed at 300 °C for 2 min by the hot plate before final annealing. After multi-coating, the BPFMO thin films were annealed at 600 °C for 2 min by the rapid thermal annealing (RTA) in an oxygen atmosphere.

The crystallization and microstructures of thin films were analysed by X-ray diffraction (XRD) and atomic force microscopy (AFM), respectively. A top electrode was prepared by DC sputtering platinum through a mask onto the thin film surface for the electrical measurements. Measurements of the leakage current was carried out using the metal-ferroelectric-metal (MFM) configuration. The magnetic hysteresis loops of the samples were analyzed at room temperature by a vibrating sample magnetometer (EV7, ADE, USA). The leakage current was measured using an HP-4156A semiconductor parameter analyzer in a MFM configuration.

Results and discussion

XRD patterns of the BPFMO thin films are shown in Fig. 1. The XRD peaks are quite similar to those of standard diffraction patterns of BiFeO$_3$ on the joint committee on powder diffraction standards (JCPDS #71-2494) card. The pure BiFeO$_3$ thin film shows a polycrystalline rhombohedral perovskite structure with the space group R3c. No obvious second phase can be obtained for all the samples, indicating that the Pr-substitution can stabilize the perovskite structure. In addition, the (110) and (104) diffraction peaks in the 2θ ranges of 31°–33° for the pure BFO thin films are changed

978-1-4673-4840-9/13 $31.00 © 2013 IEEE

to a single peak by substituting Bi ions with Pr doping. It reveals that a phase structure distortion can be obtained by Pr doping from rhombohedral (R3c) at x=0 to tetragonal (P4mm) at x= 0.15 of the BPFMO thin films.

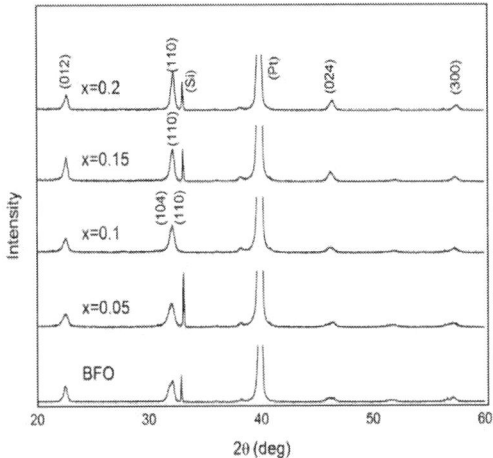

Fig. 1. XRD patterns of the BPFMO thin films.

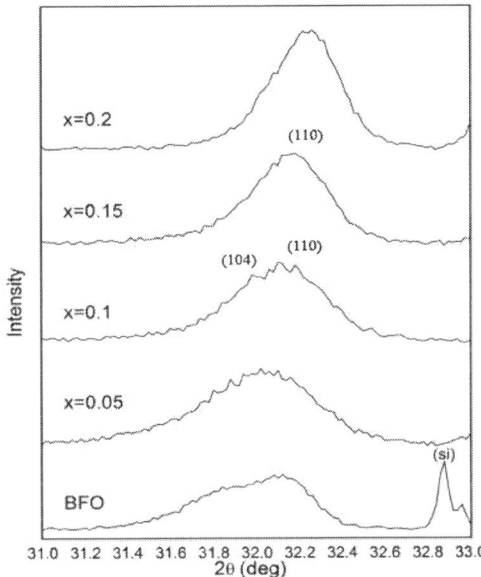

Fig. 2. Enlargement of Figure 1 in the range 31-33 degrees.

The AFM images of the BPFMO thin films deposited on the Pt coated silicon substrates are shown in Fig. 3. When the Pr and Mn ions exist in the BTO thin film, the grain size and porosity of thin films increases as the Pr content increased. It shows that the Pr and Mn ions in the BFO thin film will affect the grain growth mechanism. In addition, the BPFMO thin films with x=0.15 have no platelike crystal owing to that a phase structure distortion of rhombohedral (R3c) at x=0 to tetragonal (P4mm) at x= 0.15.

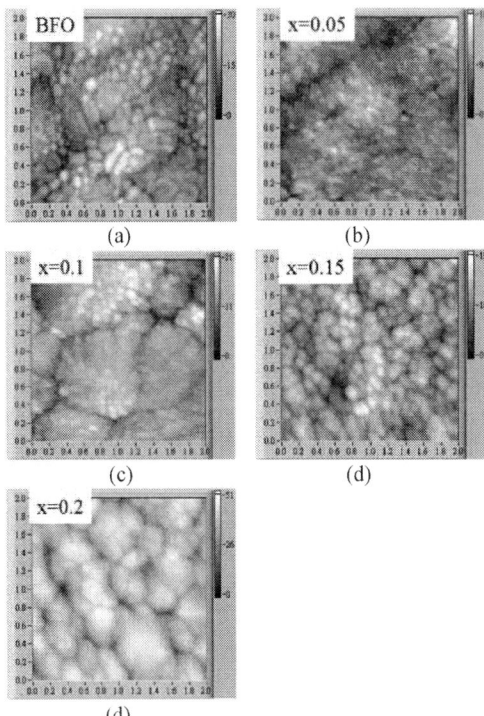

Fig. 3. The AFM images of BPFMO thin films with (a)x=0, (b)x=0.05, (c)x=0.1, (d) x=0.15 and (e)x=0.2.

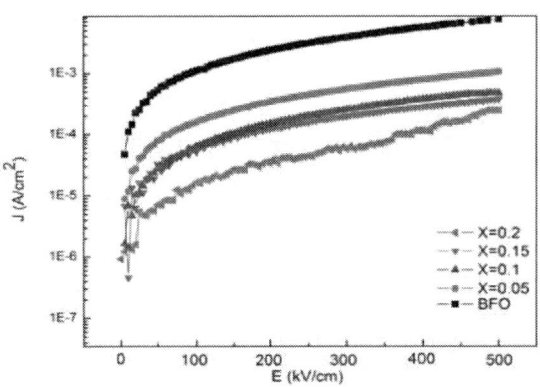

Fig. 4. Leakage current density (J)-Electric field (E) plot of the of the BPFMO thin films.

Figure 4 is a plot of leakage current density (J) as a function of the electric field (E) for the BPFMO thin films. The result indicates that the lower leakage current density can be obtained for the BPFMO thin film. This phenomenon is due to the substitution with Pr^{3+} for Bi^{3+} in the BFO thin film assisted elimination of phase structure distortion. The result indicates that the lowest leakage current density can be obtained for BPFMO thin films with x=0.2. This phenomenon is attributed to the phase structure distortion of rhombohedral (R3c) to tetragonal (P4mm), as shown in Fig. 2.

978-1-4673-4840-9/13 $31.00 © 2013 IEEE   259

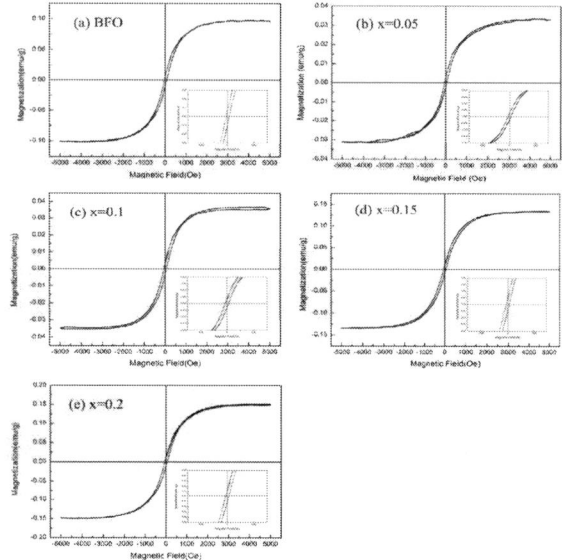

Fig. 5. Magnetic hysteresis loops of the BPFMO thin films.

Figure 5 shows the magnetic hysteresis loops of the BPFMO thin film annealed at 600°C. Magnetic hysteresis loops of the thin films were measured at room temperature with applied magnetic field parallel to the plane of the samples. The results show typical curve of magnetic hysteresis loops, indicating the presence of an ordered magnetic structure. From the results, the BPFMO thin films with x=0.2 exhibited a better remnant magnetization (2Mr) of 0.025 emu/g. It is obvious that the measured 2Mr of the BFO thin films was larger than those of the pure BFO thin film (2Mr =0.008 emu/g). The improved magnetic properties can be attributed to the substitution with Pr^{3+} for Bi^{3+} and Mn^{4+} for Fe^{3+} in BFO thin film assisted the structural transition of rhombohedral (R3c) to tetragonal (P4mm), respectively.

Conclusion

The multiferroic BPFMO thin films with high quality have been fabricated on Pt(111)/Ti/SiO₂/Si(100) substrates. In this study, we proposed a more convenient process to fabricating crack-free BPFMO thin films. From the XRD pattern analysis, a phase structural transition from rhombohedral (R3c) to tetragonal (P4mm) occurred in the BPFMO thin films. The $Bi_{1-x}Pr_xFe_{0.95}Mn_{0.05}O_3$ thin films with x=0.2 showed better magnetic properties with a remnant magnetization (2Mr) of 0.025 emu/g, which are better than those of the pure $BiFeO_3$ thin films.

Acknowledgment

This study was supported by the National Science Council, R. O. C., under contrast No. NSC 100-2112-M-164-001 and NSC 101-2112-M-164-001.

References

[1] W. Eerenstein, N. D. Mathur, and J. F. Scott, "Multiferroic and magnetoelectric materials," Nature, vol. 442, pp. 759–765, Aug. 2006

[2] R. Ramesh and N. A. Spaldin, "Multiferroics: Progress and prospects in thin films," Nature Mater., vol. 6, pp. 21–29, Jan. 2007 .

[3] J. Wang et al., "Epitaxial BiFeO multiferroic thin film heterostructures," Science, vol. 299, pp. 1719–1722, Mar. 2003.

[4] R. J. Zeches et al., "A strain-driven morphotropic phase boundary in BiFeO ," Science, vol. 326, pp. 977–980, Nov. 2009.

[5] C. F. Chung, J. P. Lin, and J. M. Wu, "Influence of Mn and Nb dopants on electric properties of chemical-solution-deposited BiFeO films," Appl. Phys. Lett., vol. 88, p. 242909, Jun. 2006.

[6] J. G. Wu, and J. Wang, "$BiFeO_3$ thin films of (111)-orientation deposited on SrRuO3 buffered Pt/TiO₂/SiO₂/Si(100) substrates, " Acta Mater., vol. 58, pp. 1688-1697, Mar. 2010.

[7] H. R. Liu, and X. Z. Wang, "Rectangular and saturated hysteresis loops of $BiFeO_3$ film on LaNiO₃ bottom electrode," J. Alloy. Compd., vol. 485, pp. 769-772, Oct. 2009.

[8] J. G. Wu, and J. Wang, "Effects of SrRuO₃ buffer layer thickness on multiferroic $(Bi_{0.90}La_{0.10})(Fe_{0.95}Mn_{0.05})O_3$ thin films, " J. Appl. Phys., vol. 106, pp.054115-054119, Sep. 2009.

[9] F. Z. Huang, X. M. Lu, W. W. Lin, W. Cai, X. M. Wu, Y. Kan, H. Sang, and J. S. Zhu, "Multiferroic properties and dielectric relaxation of $BiFeO_3/Bi_{3.25}La_{0.75}Ti_3O_{12}$ double-layered thin films, " Appl. Phys. Lett., vol. 90, pp. 252903-252905, Jun. 2007.

[10] G. Catalan, and J. F. Scott, "Physics and Applications of Bismuth Ferrite, " Adv. Mater., vol. 21, pp. 2463-2485, Jun. 2009.

[11] S. Vijayanand, M. B. Mahajan, H. S. Potdar, and P.A. Joy, "Magnetic characteristics of nanocrystalline multiferroic $BiFeO_3$ at low temperatures, " Phys. Rev. B, vol. 80, pp. 064423-064427, Aug. 2009.

[12] X. Zheng, Q. Xu, Z. Wen, X. Lang, D. Wu, T. Qiu, and M.X. Xu, "The magnetic properties of La doped and codoped $BiFeO_3$, " J. Alloy. Compd., vol. 499, pp. 108-112, Jun. 2010.

[13] S. Karimi, I. M. Reaney, I. Levin, I. Sterianou, "Nd-doped BiFeO3 ceramics with antipolar order, " Appl. Phys. Lett., vol. 94, pp. 112903-112905, Mar. 2009.

[14] V. A. Khomchenko, D. A. Kiselev, I. K. Bdikin, V. V. Shvartsman, P. Borisov, W. Kleemann, J. M. Vieira, and A. L. Kholkin, "Crystal Structure and Multiferroic Properties of Gd-Substituted $BiFeO_3$, " Appl. Phys. Lett., vol. 93, pp. 262905-262907, Dec. 2008.

[15] B. F. Yu, M. Y. Li, Z. Q. Hu, L. Pei, D. Y. Guo, X. Z. Zhao, and S. X. Dong, "Enhanced Multiferroic Properties of the High-Valence Pr Doped $BiFeO_3$ Thin Film, " Appl. Phys. Lett., vol. 93, pp. 182909-182911, Nov. 2008.

High sensing performance of fluorinated HfO₂ membrane by low damage CF₄ plasma treatment for K⁺ detections

Chi-Hsien Huang, I-Shun Wang, Kuan-I Ho, Tzu-Wen Chiang, Chien Chou,
Chu-Fa Chang and Chao-Sung Lai

Chang Gung University
259 Wen-Hwa 1ˢᵗ Road, Kwei-Shan
Tao-Yuan, Taiwan
Phone: +886-3-2118800 Ext. 5786 Email: cslai@mail.cgu.edu.tw

Abstract

A low damage CF₄ plasma with a filter in a PECVD system was proposed to incorporate fluorine atoms into an HfO₂ sensing membrane in an EIS structure for K⁺ ion sensor application. The highest sensitivities of 82mV/pK was obtained when the low damage plasma treatment with RF power of 100W was used for 30min. Compared with the conventional CF₄ plasma, the sensitivity significantly improved. The reasons were the high polarization induced by fluorine atoms and elimination of plasma damage. Keywords, low damage, CF₄ plasma, ion sensor, HfO₂, electrolyte-insulator-semiconductor (EIS)

Introduction

Potentiometric sensors, such as the ISFET[1] and electrolyte-insulator-semiconductor (EIS)[2] using dielectric layer as sensing membrane, have been used extensively for measuring concentrations of biologically relevant ions. Recently, in order to comply with the tendency of downscaling the gate dielectric thickness for next generation complementary metal-oxide semiconductors, many high-k materials were used as sensing membranes in ISFETs and EISs. From amongst many high-k materials, hafnium oxide (HfO₂) is identified as a promising ion-sensitive membrane due to high hydrogen ion sensitivity, low drift, and low body effect. On the other hand, the detection of potassium (K⁺) in the biomedical field is also very important because variations in the K⁺ ion concentration in human serum raise the risk of acute cardiac arrhythmia.[3] However, pristine HfO₂ exhibits a low sensitivity to K⁺ in ISFETs and EISs.[4] In our previous study[5], we proposed carbon tetrafluoride (CF₄) plasma treatment to improve the K⁺ sensitivity owing to the polarization at the sensing membrane surface using CF₄ plasma resulting from fluorine atoms. In addition, the sensing characteristics remained stable for more than 15 months. However, the conventional plasma process always causes damages to the dielectric characteristics of high-k materials and the semiconductor/insulator interface[6,7], which results from energetic electrons and ions, and ultraviolet (UV) irradiation, and may lead to a loss of any sensitivity improvement that might be gained by the incorporation of fluorine using CF₄ plasma. In this study, we therefore proposed the concept of a filter inserted in a plasma chamber to block those species that damage the high-k material and the semiconductor/insulator interface to achieve a low damage plasma treatment.

Experiment

A p-doped silicon wafer with resistivity of 8–12 Ωcm was used as a starting substrate to fabricate the EIS structure, as shown in Fig. 1. A 15-nm thick HfO₂ layer was deposited on the silicon substrate by sputtering as a sensing membrane after standard RCA cleaning. After deposition, the samples were treated by low damage CF₄ plasma for 30 mins in a PECVD system with a quartz filter, as schematically shown in Fig. 2. The filter includes upper and lower parts, which are separated by a quartz spacer, and each has many stripes and slits with the same width. The slits of the upper part are aligned with the stripes of the lower part. One side of the upper part of the quartz filter is coated with a thin metal thin and the quartz filter is placed onto a supporter so that the energetic electrons and ions can be confined to the plasma area between the upper electrode and the filter[21]. Around 95% of the anisotropic UV photons can be eliminated by this quartz filter (data are not shown here). Therefore, the filter can efficiently block the plasma damage while allowing only highly reactive radicals to reach the heated substrate (300 °C). After CF₄ plasma treatment, a 300-nm thick aluminum film contact layer was evaporated on the backside of the p-type wafer after removing native oxide. Finally, EIS structures were assembled with silver gel on a copper line of printed circuit board. To package the samples, an epoxy was used to encapsulate the EIS structures.

To investigate the K⁺ sensing characteristic, a 5 mM Tris/HCl buffer solution of pH 8.0 was prepared. The concentrations of K⁺ from 10^{-1} to 10^{-4} M were prepared. The sensitivity of the EIS samples were determined by capacitance-voltage (C-V) measurements using a HP4284A high precision LCR meter at a frequency of 100 Hz. A commercial Ag/AgCl electrode was used as a reference electrode. To obtain a stable ionic response, all EIS samples were immersed in the 5 mM Tris-HCl buffer solution for 12 h before measurement. For all EIS samples, the responsive voltages were defined as the voltage at 80% of the maximum capacitance in the linear region of the measured C-V curves.

Results and discussion

XPS measurements were performed to examine the incorporation of fluorine atoms into the HfO₂ sensing membrane. Figure 3 shows the XPS spectra of sample surfaces with and without low damage CF₄ plasma treatment. As shown in Fig. 3, a distinct F 1s peak at around 685.5 eV, which indicates Hf-F bonding[23], can be observed in all samples except the untreated sample. The result shows that the low damage CF₄ plasma treatment performed with a filter introduces fluorine atoms into the HfO₂ sensing membrane. Furthermore, the F 1s peak of the sample with a longer plasma treatment time displays a higher intensity, which indicates that more fluorine atoms were introduced into the HfO₂ sensing membrane and it became a fluorinated HfO₂ thin film. Even though the filter is inserted in the PECVD system to block many species in the CF₄ plasma, the XPS results suggested that the fluorine atoms were successfully incorporated into the HfO₂ thin film.

Figure 4 (a) and (b) show the representative C-V curves and responsive voltages for various concentrations of KCl, respectively, of the HfO₂ EIS sample treated by low-damage CF₄ plasma at 100W for 30 min. A clear flat band voltage shift was observed due to potential change at various K⁺ concentrations on the surface of HfO₂ sensing membrane resulting from K⁺ ions adsorption. Figure 5 shows the sensitivities of the HfO₂ EIS structure and the fluorinated-HfO₂ EIS structure treated by low-damage CF₄ plasma at various RF powers (30, 60, 100W). The sensitivity increased with an increase in the RF power. All linearities are more than 98%. These high sensitivities show a significant improvement compared with our previously optimized results (50 mV/pK).[5] Furthermore, in our previous study, rapid thermal annealing was performed after CF₄ plasma treatment without a filter to obtain a good quality of fluorinated-HfO₂ sensing membrane. From these results, we suppose that the filter inserted into the PECVD system effectively eliminated the plasma damage to realize a low damage plasma treatment. The main reason for the further improvement in the K⁺ sensitivities can be explained by the followings:

1. High polarization induced by fluorine atom

   After low-damage CF₄ plasma treatment, the Hf-F bonds form in HfO₂ thin film, as shown in Fig. 2, that results in higher dipole property between Hf and F atoms than that between Hf and O atoms. As a result, attraction of positive K⁺ ions in the electrolyte becomes easier resulting in higher sensitivity.

2. Reduction of plasma damage

   Plasma process indispensably causes damage leading to accumulation of positive charges on the HfO₂ thin film surface. The more positive charges accumulated on the HfO₂ thin film surface, the more repulsion of positive ions is. Namely, the lower damage resulting from plasma, the more positive ions are on the HfO₂ thin film surface leading to higher sensitivity.

Conclusions

A low damage CF₄ plasma with a filter in a PECVD system was proposed to incorporate fluorine atoms into an HfO₂ sensing membrane in an EIS structure for K⁺ ion sensor application. The chemical compositions at the F 1s of the HfO₂ sensing membrane were examined by XPS to confirm the incorporation of fluorine atoms after low damage CF₄ plasma treatment. As the plasma RF power was increased, the F 1s intensity increased. The corresponding K⁺ sensitivities also increased with increasing RF power. The highest sensitivities of 82 mV/pK was obtained when the low damage CF₄ plasma treatment with an RF power of 100 W was used for 30 min. Compared with the conventional CF₄ plasma process, the sensitivity to K⁺ significantly improved. The reasons for this significant improvement were the high polarization induced by fluorine atoms and elimination of plasma damage.

References

[1] C. S. Lai., et al., *Sens. Actuator B* **143** (2010) 494.

[2] M. J., et al., *J. Solid State Electrochem.* **13** (2009) 115.

[3] A. Errachid, et al., *Mater. Sci. Eng.* C **21** (2002) 9.

[4] K. I. Ho, et al., *Proc. IEEE Sensors* (2011) 1128.

[5] T. F. Lu, et al., *J. Electrochem. Soc.* **158** (2011) J91.

[6] M. Okigawa, et al., *J. Vac. Sci. Technol. B* **22** (2004) 2818.

[7] A. Uedono, et al., *J. Appl. Phys.* **100** (2006) 064501

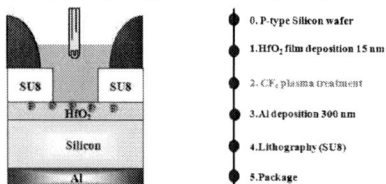

Figure 1 Fabrication process of EIS device

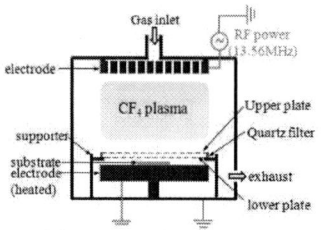

Figure 2 Schematic of PECVD system with filter.

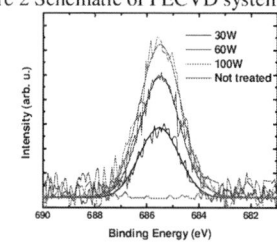

Figure 3 XPS spectra of sample surfaces without and with low damage CF₄ plasma treatment at various RF powers for 30 mins.

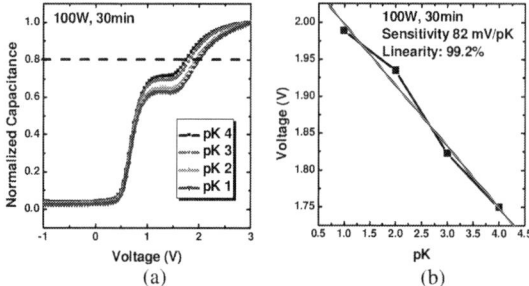

Figure 4 Representative (a) C-V curves and (b) responsive voltages for various concentrations of KCl of the HfO₂ EIS sample treated by low-damage CF₄ plasma at 100W for 30 mins.

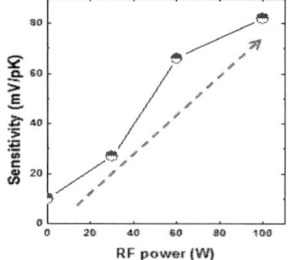

Figure 5 Sensitivities of the HfO₂ EIS structure treated by low-damage CF₄ plasma at various RF powers (0, 30, 60, 100W).

Investigation of the Random Dopant Fluctuations in 20-nm Bulk MOSFETs and Silicon-on-Insulator FinFETs by Ion Implantation Monte Carlo Simulation

Keng-Ming Liu, and Cheng-Kuei Lee

Department of Electrical Engineering, National Dong Hwa University
Hualien 974-01, Taiwan
Phone: +886-3-863-4085; Fax: +886-3-863-4060; E-mail: kmliu@mail.ndhu.edu.tw

Abstract

In this paper, we proposed a simulation approach for studying the random dopant fluctuation (RDF) effects in the nanoscale MOSFETs using only the TCAD tools. We use this approach to simulate the RDF effects in the 20-nm gate-length bulk MOSFETs, silicon-on-insulator (SOI) single-gate (SG) and triple-gate (TG) FinFETs for demonstration. This approach utilizes the stochastic nature of the Monte Carlo (MC) simulation of ion implantation to capture the RDF phenomena in the devices. The simulation results show that the standard deviation of the threshold voltage (σV_T) is approximately proportional to the cube root of the channel doping concentration for the conventional bulk MOSFETs. For the SOI SG and TG FinFETs, the σV_T increases much less than that of the conventional bulk MOSFETs as the channel doping concentration increases. Besides, the σV_T of the SOI TG FinFETs show about 30% to 40% reductions comparing to those of the SOI SG FinFETs. The average of the MC simulation results agrees with the implant table simulation results. The reasonable simulation results verify the validity of this TCAD simulation scheme for the RDF study.

Keywords: Random dopant fluctuation (RDF), multiple-gate SOI MOSFET, FinFET, TCAD simulation, Monte Carlo simulation, ion implantation, 3D

Introduction

As MOSFETs keep scaling down in the nanometer regime, the variability among devices is becoming serious. The sources of variability are classified into two groups. One is called random or intrinsic variation, which is caused by the inherent stochastic phenomena associated with each device, like random dopant fluctuation (RDF) and line edge roughness (LER). The other is called systematic or extrinsic variation, which is caused by the process variations, like the variations on the gate oxide thickness (t_{ox}) and the gate length (L_g). Among all the variations, RDF has been identified as one of the major sources of variability [1], [2] and, moreover, RDF becomes worse as device area decreases. Over the past two decades, RDF effects were simulated and modeled extensively [1], [3]–[10]. Most of the RDF simulation studies are based on the atomistic simulation approach which involves generating a certain number of dopants in the channel or other region and then determining the position of each individual dopant [1], [3], [6]–[10]. This approach will require the extra efforts to develop the program for producing the random discrete dopants (RDDs) and their positions. Besides, incorporating the information of generated RDDs into the device simulator also needs to be implemented. Furthermore, if the nonuniform doping profiles are also taken into account in the RDF simulation, then even more complicated techniques need to be used in the simulation [1], [7], [8]. In this paper, we utilize the Monte Carlo (MC) simulation of ion implantation in the TCAD process simulator (Sentaurus Process 3D, one of the Synopsys TCAD products, is used here) to simulate the RDF effects. By this approach, no extra programs are required for simulating the RDF effects, and the whole RDF simulation can be done simply through the commercial TCAD tools. In addition, the RDF effects of the devices with the multiple ion implantation processes can be easily simulated through this approach although in this work we only demonstrated the RDF effects resulted from the single ion implantation process. We use this approach to examine the RDF effects in the 20-nm conventional bulk MOSFETs and multiple-gate (MG) silicon-on-insulator (SOI) FinFETs. The validity of this approach is then verified according to the simulation results. The details of this simulation approach and its simulation results are presented and discussed in the following sections.

Simulation Approach

A. Simulated Device Structures

Figs. 1(a)–(c) show the three simulated device structures which are the conventional bulk MOSFET structure, the SOI single-gate (SG) FinFET structure, and the SOI triple-gate (TG) FinFET structure, respectively. The gate lengths of all the simulated device structures are 20 nm and the gate oxide (SiO$_2$) has a thickness of 10 Å. The gate material is n$^+$ polysilicon with the doping concentration of 10^{20} cm^{-3}. For the conventional bulk MOSFET structure, the gate width (W) is 10 nm. For the SOI SG and TG FinFET structures, the height and width of the silicon fin are both 10 nm. In this paper, for simplicity, we examine the RDF effects of the channel region only. Therefore, the source and drain (S/D) regions are assumed to be uniformly doped with an n-type doping concentration of 10^{20} cm^{-3} and the S/D junctions are abrupt. The lengths of the S/D regions are both 10 nm. For the conventional bulk MOSFET structure, the junction depths of the S/D regions are 10 nm. For the SOI SG and TG FinFET structures, the S/D junction depths are just equal to the height of the silicon film which is also 10 nm. The doping profile of the channel region is produced by the single ion implantation process (channel implant) in our simulation. The implanted impurity for the channel implant is boron or indium and the implant energy is 10 keV. The implant dose of the channel implant varies in our simulation study. No tilt and rotation are used for the channel implant.

B. RDF Sample Preparation

Fig. 2 illustrates how the samples for RDF simulation study are prepared. Here we use the SOI TG FinFET as an example.

978-1-4673-4840-9/13 $31.00 © 2013 IEEE

Fig. 2(a) shows the SOI structure with an area of 40 nm × 50 nm viewing from the top after the channel implant process. The thicknesses of the top silicon film and the buried oxide are 10 nm and 30 nm, respectively. The doping profile shown in Fig. 2(a) is the result of the MC simulation of the channel implant. When the ion implantation process is simulated by the MC method, we set the number of sampling particles equal to the actual implantation dosage. For example, if the dose of channel implant is 5×10^{12} cm^{-2}, then theoretically 100 impurities or dopants will be implanted into the area of 40 nm × 50 nm. In the MC simulation of the channel implant, the number of sampling particles is then set to be 100. Under the same number of sampling particles, the MC simulation result of each ion implantation will be different if the random number seed chosen for each ion implantation MC simulation is different. By changing the random number seed, we can generate the devices with various doping profiles as the samples for RDF study. In this work, the number of samples for RDF study is 100. After the MC simulation of channel implant process, the silicon film on the buried oxide is etched to form a silicon fin with the fin width of 10 nm located at the center as shown in Fig. 2(b). By this way, we can make the number and the positions of the dopants in the channel region are different for each device and hence the RDF effects can be captured. Sequentially, the gate oxide and the n$^+$ polysilicon film are deposited and etched to form the final SOI TG FinFET as shown in Fig. 2(c). Fig. 3 shows the comparison between the channel doping profiles produced by the conventional implant table simulation and the MC simulation. For this particular sample generated by the MC simulation (Fig. 3(b)), its average channel doping concentration is higher than that of the device generated by the conventional implant table simulation (Fig. 3(a)).

After the device samples which are produced by Sentaurus Process 3D are ready, we then perform the device simulation for each device sample by Sentaurus Device 3D. We use the hydrodynamic (HD) model combined with the density gradient (DG) model for quantum corrections [11] to simulate the device characteristics. The simulation results will be presented and discussed in the next section.

Simulation Results and Discussion

Figs. 4(a) and (b) show the simulated I_D–V_G curves of the conventional bulk MOSFET samples in linear and logarithm scales, respectively. The implanted impurity for the channel implant is indium. Two doses 5×10^{12} cm^{-2} and 5×10^{13} cm^{-2} for the channel implant are simulated. The dosages used in this work are chosen to make the average channel doping concentrations in the order of 10^{18} cm^{-3}. The simulated I_D–V_G curves based on the channel doping profiles generated by the implant table simulation are also shown in Figs. 4(a) and (b). Table I shows the statistics of the threshold voltage fluctuation induced by the channel random dopants of the conventional bulk MOSFET. The threshold voltage (V_T) is determined by the constant current method under the drain bias of 50 mV with the current criterion of (500 nA) × (W/ L_g). In Table I, the standard deviation of the threshold voltage (σV_T) of the channel implant dose 5×10^{13} cm^{-2} is 2.3 times that of the dose 5×10^{12} cm^{-2}. According to the analytical model of the σV_T in

[1], [5], [6] which is derived under the assumption of the uniform channel doping, the σV_T is proportional to the fourth root of the channel doping concentration. Thus, the analytical model predicts 1.8 times increase as the channel implant dose increases from 5×10^{12} cm^{-2} to 5×10^{13} cm^{-2}. Therefore, for the channel doping dependence of the σV_T, our simulation results agree with the analytical model in trend but show the approximately cube-root dependence instead of the fourth-root dependence [1], [5], [6]. The significant negative V_T-shift for the channel implant dose 5×10^{13} cm^{-2} in Table I is possibly due to the inhomogeneity of the channel potential resulted from the random dopants and the formation of early turn-on paths in the channel.

Figs. 5 and 6 show the simulated I_D–V_G curves of the SOI SG and TG FinFET samples, respectively. The implanted impurity for the channel implant is boron. Two doses 10^{13} cm^{-2} and 10^{14} cm^{-2} for the channel implant are simulated. The simulated I_D–V_G curves based on the channel doping profiles generated by the implant table simulation are also shown in Figs. 5 and 6. Table II and III show the statistics of the threshold voltage fluctuations induced by the channel random dopants of the SOI SG and TG FinFET, respectively. Comparing Table II with Table III, the σV_T of the SOI TG FinFETs are 57% and 67% of those of the SOI SG FinFETs under the channel implant doses 10^{13} cm^{-2} and 10^{14} cm^{-2}, respectively. The results indicate the TG structure can effectively suppress the σV_T caused by the channel random dopants. Besides, the TG structure has the smaller subthreshold swing and the larger drive current comparing to the SG structure. We also found that the σV_T of the SOI SG and TG FinFETs increase much slower than that of the conventional bulk MOSFETs as the channel implant dose increases. That means the deeper (> 10 nm) channel random dopants which are not supposed to affect the device characteristics of the SOI FinFETs still play a role in the threshold voltage fluctuations of the conventional bulk MOSFETs. Also, the analytical models of the σV_T appeared in [1], [5] and [6] which are derived for the conventional bulk MOSFETs are not suitable for predicting the σV_T of the SOI SG and TG MOSFETs. The V_T-shifts in Table II and III are quite small which implies that the MC simulation of ion implantation is very consistent with the implant table simulation of ion implantation for the SOI SG and TG FinFETs.

Conclusion

In this paper, we proposed a simulation scheme for investigating the RDF effects in the nanoscale planar and non-planar MOSFETs using only the TCAD tools. This scheme utilizes the stochastic nature of the MC simulation of ion implantation to capture the RDF phenomena in the devices. This TCAD simulation approach is used to simulate the RDF effects of the 20-nm gate-length conventional bulk MOSFETs, SOI SG and TG FinFETs for demonstration. The simulation results show that the σV_T is approximately proportional to the cube root of the channel doping concentration for the conventional bulk MOSFETs. For the SOI SG and TG FinFETs, the σV_T increases much less than that of the conventional bulk MOSFETs as the channel doping concentration increases. Besides, the σV_T of the SOI TG FinFETs show about 30% to 40% reductions comparing to those of the SOI SG FinFETs. The average of the MC simulation results agrees with the implant table simulation results. The reasonable simulation results verify the validity of this TCAD simulation scheme for the RDF study. In the future, Pelgrom-plot analysis for the RDF study will be made based on this approach. We believe this approach can provide a quick and accurate RDF analysis after the calibrations of both the device characteristics and the doping profiles resulted from the ion implantation are carefully done.

Acknowledgment

This work was supported by the Taiwan National Science Council (NSC) under Contract NSC-101-2221-E-259-022. We are grateful to the National Center for High-performance Computing for computer time and facilities.

References

[1] P. A. Stolk, F. P. Widdershoven, and D. B. M. Klaassen, "Modeling Statistical Dopant Fluctuations in MOS Transistors," *IEEE Trans. Electron Devices*, vol. 45, no. 9, pp. 1960–1971, 1998.

[2] K. J. Kuhn, "Reducing Variation in Advanced Logic Technologies: Approaches to Process and Design for Manufacturability of Nanoscale CMOS," in *IEDM Tech. Dig.*, 2007, pp. 471–474.

[3] H. S. Wong, and Y. Taur, "Three-Dimensional "Atomistic" Simulation of Discrete Random Dopant Distribution Effects in Sub-0.1□m MOSFET's," in *IEDM Tech. Dig.*, 1993, pp. 705–708.

[4] X. Tang, V. K. De, and J. D. Meindl, "Intrinsic MOSFET Parameter Fluctuations Due to Random Dopant Placement," *IEEE Trans. VLSI Systems*, vol. 5, no. 4, pp. 369–376, 1997.

[5] K. Takeuchi, T. Tatsumi, and A. Furukawa, "Channel Engineering for the Reduction of Random-Dopant-Placement-Induced Threshold Voltage Fluctuation," in *IEDM Tech. Dig.*, 1997, pp. 841–844.

[6] A. Asenov, "Random Dopant Induced Threshold Voltage Lowering and Fluctuations in Sub-0.1 □m MOSFET's: A 3-D "Atomistic" Simulation Study," *IEEE Trans. Electron Devices*, vol. 45, no. 12, pp. 2505–2513, 1998.

[7] A. Asenov, G. Slavcheva, A. R. Brown, J. H. Davies, and S. Saini, "Increase in the Random Dopant Induced Threshold Fluctuations and Lowering in Sub-100 nm MOSFETs Due to Quantum Effects: A 3-D Density-Gradient Simulation Study," *IEEE Trans. Electron Devices*, vol. 48, no. 4, pp. 722–729, 2001.

[8] T. Ezaki, T. Ikezawa, and M. Hane, "Investigation of Realistic Dopant Fluctuation Induced Device Characteristics Variation for Sub-100nm CMOS by Using Atomistic 3D Process/Device Simulator," in *IEDM Tech. Dig.*, 2002, pp. 311–314.

[9] D. Reid, C. Millar, G. Roy, S. Roy, R. Sinnott, G. Stewart, G. Stewart, and A. Asenov, "Prediction of Random Dopant Induced Threshold Voltage Fluctuations in NanoCMOS Transistors," in *Proc. SISPAD Conf.*, 2008, pp. 21–25.

[10] C. Shin, X. Sun, and T. J. K. Liu, "Study of Random-Dopant-Fluctuation (RDF) Effects for the Trigate Bulk MOSFET," *IEEE Trans. Electron Devices*, vol. 56, no. 7, pp. 1538–1542, 2009.

[11] *Sentaurus Device User Guide*, Version C-2009.06, Synopsys.

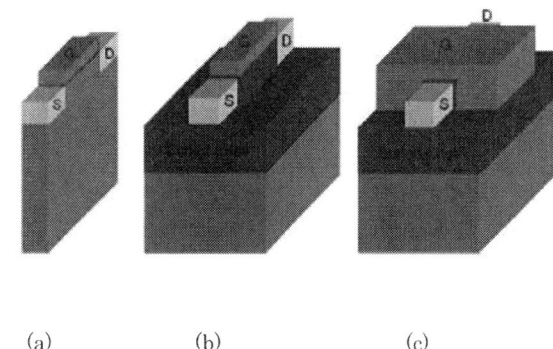

(a)　　　　(b)　　　　(c)

Fig. 1. The three simulated device structures: (a) the conventional bulk MOSFET structure, (b) the SOI single-gate (SG) FinFET structure, and (c) the SOI triple-gate (TG) FinFET structure.

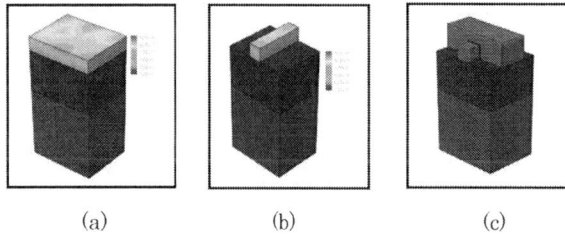

(a)　　　　(b)　　　　(c)

Fig. 2. The flow of how the samples for RDF simulation study are prepared. Here the SOI TG FinFET is used as an example.

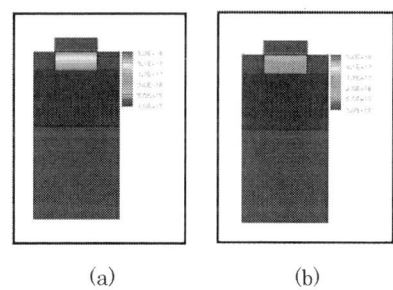

(a)　　　　(b)

Fig. 3. The comparison between the channel doping profiles produced by (a) the conventional implant table simulation and (b) the MC simulation.

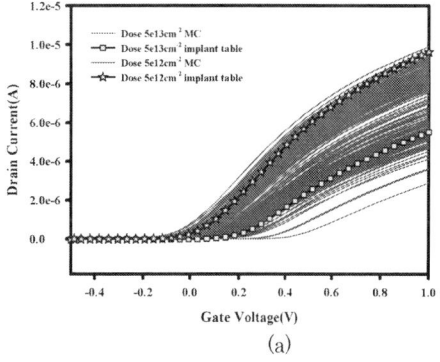

(a)

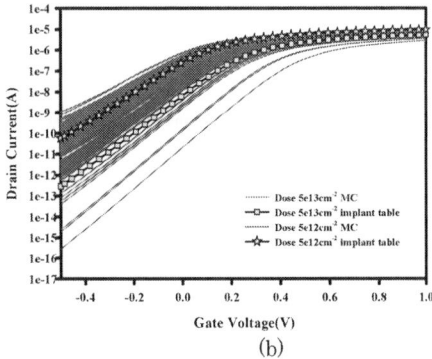

(b)

Fig. 4. The simulated I_D–V_G curves of the conventional bulk MOSFET samples in (a) linear and (b) logarithm scales under the drain bias of 50 mV. The implanted impurity for the channel implant is indium. The simulated I_D–V_G curves based on the channel doping profiles generated by the implant table simulation are also shown.

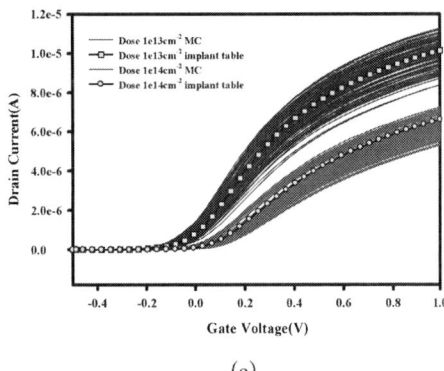

(a)

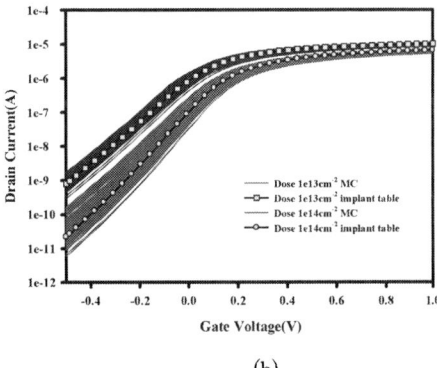

(b)

Fig. 5. The simulated I_D–V_G curves of the SOI SG FinFET samples in (a) linear and (b) logarithm scales under the drain bias of 50 mV. The implanted impurity for the channel implant is boron. The simulated I_D–V_G curves based on the channel doping profiles generated by the implant table simulation are also shown.

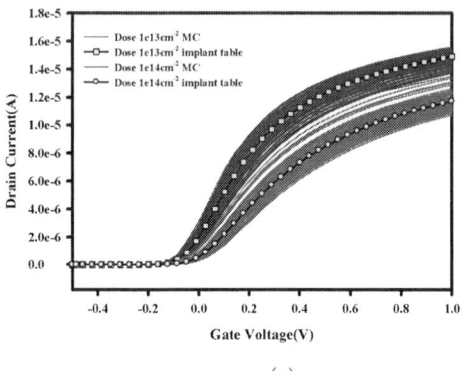

(a)

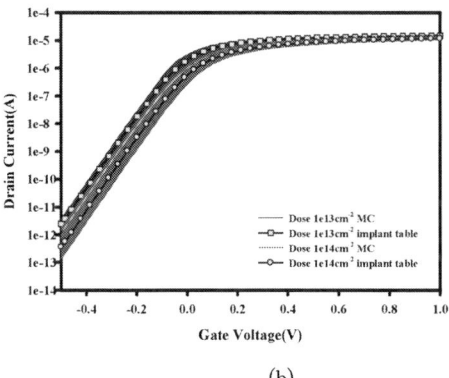

(b)

Fig. 6. The simulated I_D–V_G curves of the SOI TG FinFET samples in (a) linear and (b) logarithm scales under the drain bias of 50 mV. The implanted impurity for the channel implant is boron. The simulated I_D–V_G curves based on the channel doping profiles generated by the implant table simulation are also shown.

TABLE I
THE STATISTICS OF THE THRESHOLD VOLTAGE FLUCTUATION INDUCED BY THE CHANNEL RANDOM DOPANTS OF THE BULK MOSFET

parameter Dose (cm^{-2})	$\overline{V_T}^a$ (V)	σV_T^a (mV)	$V_{T,imp}^b$ (V)	$V_T - shift^c$ (mV)
5e12	0.0423	26.18	0.0171	25.2
5e13	0.1294	60.58	0.3114	−181.7

[a] $\overline{V_T}$ and σV_T represent the mean threshold voltage and the standard deviation of the 100 samples, respectively.

[b] $V_{T,imp}$ represents the threshold voltage of the device in which the channel doping profile is generated by the implant table simulation.

[c] $V_T - shift = \overline{V_T} - V_{T,imp}$

TABLE II
THE STATISTICS OF THE THRESHOLD VOLTAGE FLUCTUATION INDUCED BY THE CHANNEL RANDOM DOPANTS OF THE SOI SG FINFET

parameter Dose (cm^{-2})	$\overline{V_T}$ (V)	σV_T (mV)	$V_{T,imp}$ (V)	$V_T - shift$ (mV)
1e13	−0.1033	23.75	−0.0965	−6.8
1e14	0.0528	25.78	0.0512	1.6

TABLE III
THE STATISTICS OF THE THRESHOLD VOLTAGE FLUCTUATION INDUCED BY THE CHANNEL RANDOM DOPANTS OF THE SOI TG FINFET

parameter Dose (cm^{-2})	$\overline{V_T}$ (V)	σV_T (mV)	$V_{T,imp}$ (V)	$V_T - shift$ (mV)
1e13	−0.0625	13.69	−0.0585	−4
1e14	0.0123	17.38	0.0149	−2.6

Microscopic Study of Random Dopant Fluctuation in Silicon Nanowire Transistors Using 3D Simulation

Chun-Yu Chen[1], Jyi-Tsong Lin[1] and Meng-Hsueh Chiang[2]

[1]Department of Electrical Engineering, National Sun Yat-Sen University,
Kaohsiung 804, Taiwan.
[2]Department of Electronic Engineering, National Ilan University,
I-Lan 260, Taiwan.
Phone: +886-7-525-2000 ext. 4122, E-mail: jtlin@ee.nsysu.edu.tw

Abstract

Variability impact of random dopant fluctuation in nanometer-scale silicon nanowire MOSFET is assessed via TCAD numerical simulations. We have simulated ensembles of 629 devices, which differ from each other due to the physical manifestation of the dopant variability in the channel location, including the detailed microscopic pattern of a discrete sphere dopant from drain to source. Based on our study, the variations of leakage and drive currents are not great, but not negligible. Implications from random dopant fluctuation for design are also discussed

Introduction

Work towards continuous device scaling is becoming increasingly difficult due to complicate channel doping profile required for threshold voltage control and short-channel effect (SCE) suppression. Due to random nature of impurity dopants, variability issue is inevitable and must be taken care of. Random dopant fluctuation (RDF) and random discrete dopant (RDD) have been investigated theoretically and statistically for long years [1]-[10]. New transistor structures have been sought for ultimate device scaling. Among emerging devices, nanowire transistors, as illustrated in Fig. 1, which could be done with the undoped or doped body, have drawn much attention for good scaling capability and technology compatibility [11]. However, within such an advanced structure, unintended impurity dopant in the "undoped" channel will greatly influence the effective doping density in terms of number of impurity atoms divided by the volume, thus affecting the device characteristics. Whether the novel nanowire structure is immune to RDF is thus questionable. In this work, we evaluate the impact of RDF with assumption of one impurity atom in the channel. For our study, the nominal transistor dimension is defined as follows: gate length (L) of 17 nm, nominal wire diameter (D) of 8 nm, buried oxide of 200 nm, and truncated substrate of 30 nm following ITRS [12]. The channel of the ideal nanowire is assumed undoped with source/drain doping of 2×10^{20} cm^{-3}, as shown in Fig. 1. Fermi-Dirac statistics and drift-diffusion transport with density gradient of quantum model for carrier confinement were included in simulation [13].

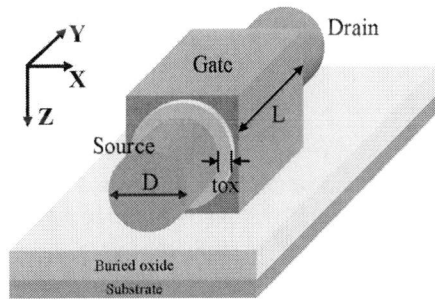

Fig. 1. 3-D view of nanowire transistor device where y axis is in the direction along the channel and x-z surface corresponds to the cross section of the wire (not to scale).

Simulation Methodology

To understand and further resolve the issues induced by RDF and RDD efficiently, a macroscopic simulation methodology has been used in conventional as well as non-planar MOSFETs [14]-[16]. Such methodology is applicable to intrinsic body of non-bulk transistors. In order to simulate the discrete distribution of impurity atoms, we divided the nanowire channel into many spheres of equal volume. If the impurity dopant is located in one of the spheres, one can use following equation [14]:

$$N = \frac{4}{3}\pi \times l^3 \times N_A \qquad (1)$$

to estimate the doping density (N_A). As shown in Fig. 2(a), the sphere radius of l is set to 0.5 nm and then N_A is found following (1). In our simulation, a single number ($N = 1$) of dopant with diameter of 1 nm and the corresponding N_A of 1.9×10^{21} cm^{-3} are assumed. In addition, the channel diameter and channel length are in multiple of l's (e.g., L = 17 nm and D = 8 nm), and x-z plane (cross section) of the silicon wire is divided into 37 regions for allocation of the impurity atom, as shown in Fig. 2(b). For presentation of the dopant location, we label 17 regions from left to right in y-direction along the channel. After the device structure is defined, the physical insight into RDF can now be analyzed via 3D numerical device simulation.

978-1-4673-4840-9/13 $31.00 © 2013 IEEE   267

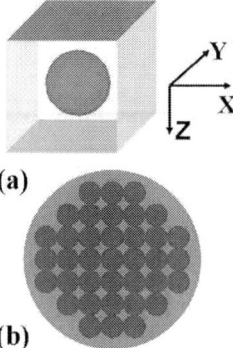

(a)

(b)

Fig. 2. (a) Schematic diagram of a random dopant in a sphere with 1 nm in diameter. (b) The cross section of the channel is divided into 37 regions for macroscopic modeling.

Variability Study

Based on the 3D macroscopic study, we assess RDF impact from design perspective. Due to its random nature, the implication from RDF for design is best understood via statistics using the simulation data for all 629 devices, each of which represents the case of a single acceptor impurity dopant located at each respective location in y direction, as demonstrated in Fig. 2(b), plus uniformly doped counterpart for comparison. Before RDF study, we first evaluate the threshold voltage dependence on uniform doping level. Fig. 3 shows the simulated V_T's versus N_A for the nanowire transistor in uniform doping, where V_T is extracted by maximum transconductance at $V_{DS} = 0.05$ V. As expected, their V_T's are virtually flat for low doping level ($< 10^{18}$ cm^{-3}) and increase substantially for high doping ($> 10^{19}$ cm^{-3}). Fig. 4 shows predicted V_T's and mean for a single dopant located in the channel of the wire at respective y's from source to drain. The predicted V_T variation is most significant in the center of the channel as shown in Fig. 5. Fig. 6 and Fig. 7 show SS and DIBL characteristics and their mean values and standard deviations in the channel of the wire at respective y's. SS is nearly independent of the doping location in y. Besides the random dopant in y, we also assess the variability in x. Fig. 8 and Fig. 9 show V_T's characteristics and mean and standard deviation in the channel of the wire diameter at respective x's. Fig. 10 and Fig. 11 show SS and DIBL characteristics and their means and standard deviations in the channel of the wire at respective x's. When the dopant is located near the center of the wire, higher variation is predicted. Fig. 12 shows simulated I_{OFF} versus I_{ON} from 629 cases. Noticeable difference between the undoped case and the discrete dopant is shown. For scaled devices with undoped channel, preventing contamination from impurity atoms is still important. Fig. 13 shows the occurrences of cumulative percentage versus threshold voltage variation due to random dopant fluctuation in the channel. The percentage of threshold voltage increases rapidly near lower value of 0.2 V as the threshold voltage increases quickly when the dopant impurity moves towards the center

of the channel.

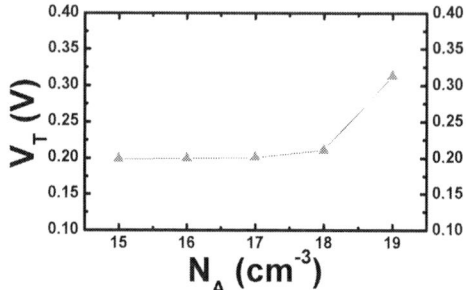

Fig. 3. Simulated V_T's versus N_A at $V_{DS} = 0.05$ V.

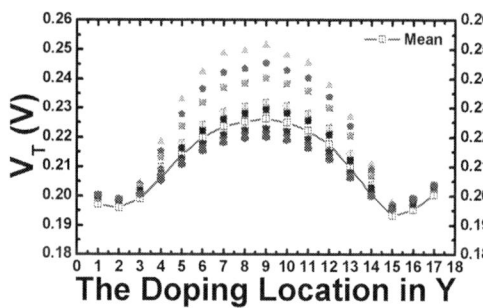

Fig. 4. Predicted V_T's and mean for a single impurity atom located in the channel of the wire at respective y's.

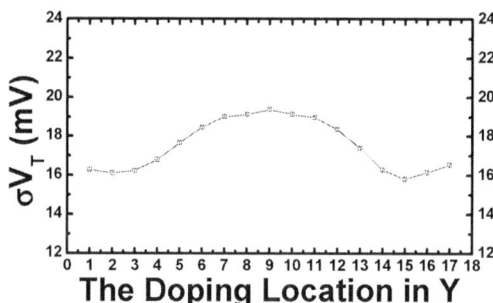

Fig. 5. Predicted standard deviation of V_T for a single impurity atom located in the channel of the wire at respective y's.

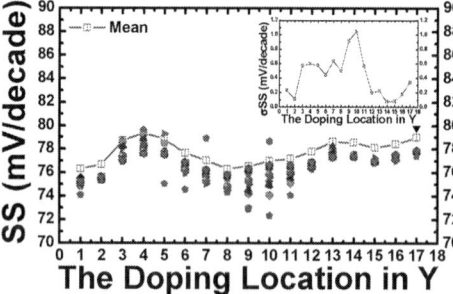

Fig. 6. Predicted SS, mean and standard deviation for a single impurity atom located in the channel of the wire at respective y's.

978-1-4673-4840-9/13 $31.00 © 2013 IEEE

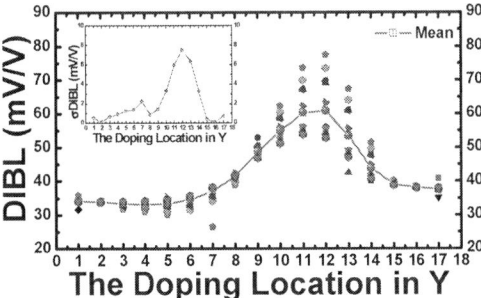

Fig. 7. Predicted DIBL, mean and standard deviation for a single impurity atom located in the channel of the wire at respective y's.

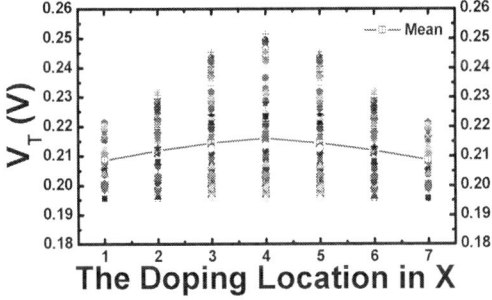

Fig. 8. Predicted V_T's and mean for a single impurity atom located in the channel of the wire at respective x's.

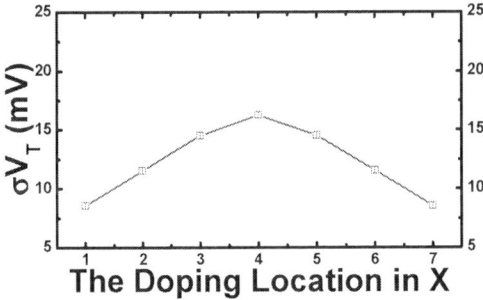

Fig. 9. Predicted standard deviation of V_T for a single impurity atom located in the channel of the wire at respective x's.

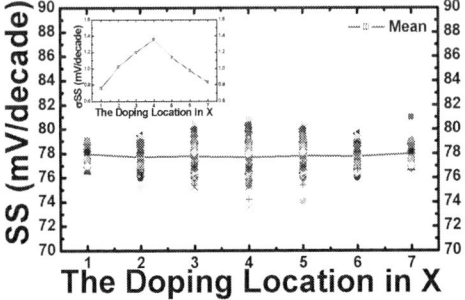

Fig. 10. Predicted SS, mean and standard deviation for a single impurity atom located in the channel of the wire at respective x's.

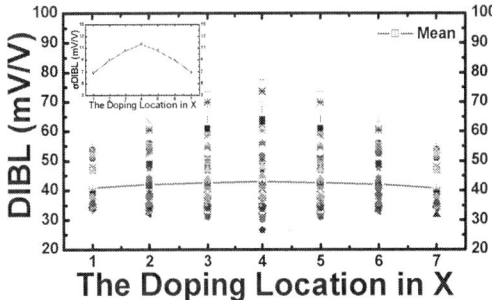

Fig. 11. Predicted DIBL, mean and standard deviation for a single impurity atom located in the channel of the wire at respective y's.

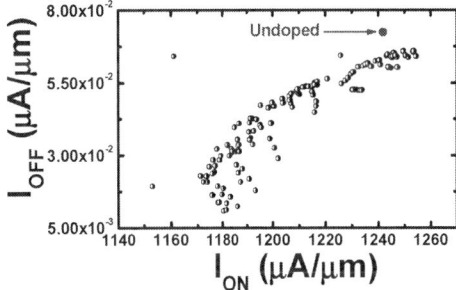

Fig. 12 Simulated I_{ON} versus I_{OFF} for a single discrete dopant at 629 respective locations in channel region.

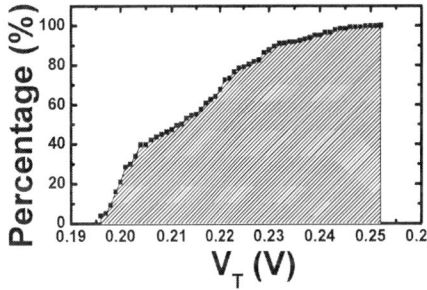

Fig. 13. Cumulative percentage versus random dopant for threshold voltage of 629 devices.

Conclusion

We have presented a detailed assessment of random dopant fluctuation for nanowire transistors. The cases with uniform dopant and discrete dopant show very different characteristics. The discrete dopant located in the center of the channel moving along the diameter between the surrounding gate has the most significant impact.

Acknowledgments

The authors were partially supported by the National Science Council of Taiwan under NSC-99-2221-E-110-078 and NSC-100-2628-E-197-003-MY2. We are grateful to the National Center for High-Performance Computing for support of computational facilities.

References

[1] D. J. Frank, Y. Taur, M. Ieong, and H. P. Wong, "Monte Carlo modeling of threshold variation due to dopant fluctuation," *in VLSI Symp. Tech. Dig.*, 1999, p. 169..

[2] P. A. Stolk, F. P. Widdershoven, and D. B. M. Klaassen, "Modeling statistical dopant fluctuation in MOS transistors," *IEEE Trans. Electron Devices*, vol. 45, no. 9, pp. 1960–1971, Sep. 1998.

[3] H. S. Wong and Y. Taur, "Three-dimensional "atomic" simulation of discrete random dopant distribution effects in sub-0.1 μm MOSFETs," *in IEDM Tech. Dig.*, 1993, pp. 705–708.

[4] A. Asenov, "Random dopant induced threshold voltage lowering and fluctuations in sub-0.1 μm MOSFET's: A 3-D 'atomistic' simulation study," *IEEE Trans. Electron Devices*, vol. 45, no. 12, pp. 2505–2513, Dec. 1998.

[5] A. Asenov and S. Saini, "Suppression of random dopant-induced threshold voltage fluctuations in sub-0.1-μm MOSFET's with epitaxial and δ-doped channels," *IEEE Trans. Electron Devices*, vol. 46, no. 8, pp. 1718–1724, Aug. 1999.

[6] A. Asenov, G. Slavcheva, A. R. Brown, J. H. Davies, and S. Saini, "Increase in the random dopant induced threshold fluctuations and lowering in sub-100 nm MOSFETs due to quantum effects: A 3-D density-gradient simulation study," *IEEE Trans. Electron Devices*, vol. 48, no. 4, pp. 722–729, Apr. 2001.

[7] I. D. Mayergoyz and P. Andrei, "Statistical analysis of semiconductor devices," *J. Appl. Phys.*, vol. 90, no. 6, pp. 3019–3029, Sep. 2001.

[8] P. Andrei and I. Mayergoyz, "Quantum mechanical effects on random oxide thickness and doping fluctuations in ultrasmall semiconductor devices," *J. Appl. Phys.*, vol. 94, pp. 7163–7172, Dec. 2003.

[9] S. Roy and A. Asenov, "Where do the dopants go?" *Sci. Mag.*, vol. 309, no. 5733, pp. 388–390, Jul. 2005.

[10] A. R. Brown, A. Asenov and J. R. Watling, "Increase in the Random Dopant Induced Threshold Fluctuations and Lowing in Sub-100 nm MOSFETs Due to Quantum Effects: A 3-D Density-Gradient Simulation Study," *IEEE Trans. Electron Devices*, vol. 48, no. 4, pp. 722–729, Apr. 2001.

[11] C.-Y. Chen, Y. B. Liao, and M.-H. Chiang, "Scaling study of nanowire and multi-gate MOSFETs," *in Proc. The 9th Internat. Conf. on Solid-State and Integrated-Circuit Tech.*, Oct. 2008, pp. 57-60.

[12] International Technology Roadmap for Semiconductors, 2007 Edition. [Online]. Available:http://public.itrs.net/.

[13] Taurus-Device, User Guide, Synopsis Inc., ver. X-2005.10, Oct. 2005.

[14] X. Tang, V. K. De, and J. D. Meindl, "Intrinsic MOSFET parameter fluctuations due to random dopant placement," *IEEE Trans. Very Large Scale Integr. Syst.*, vol. 5, no. 4,pp. 369–376, Dec. 1997.

[15] S. Mudanai, W. K. Shih, R. Rios, X. Xi, J. H. Rhew, K. Kuhn and P. Packan, "Analytical Modeling of Output Conductance in Long-Channel Halo-Doped MOSFETs," *IEEE Trans. Electron Devices*, vol. 53, no. 9, pp. 2091–2097, Sep. 2006.

[16] M. H. Chiang, J. N. Lin, K. Kim and C. T. Chuang, "Random Dopant Fluctuation in Limited-Width FinFET Technologies," *IEEE Trans. Electron Devices*, vol. 54, no. 8, pp. 2055–2059, Aug. 2007.

Characteristics and Hot-Carrier Effects of Strained pMOSFETs with SiGe Channel and Embedded SiGe Source/Drain Stressor

Min-Ru Peng[1], Mu-Chun Wang[1,3,*], Liang-Ru Ji[1], Heng-Sheng Huang[1], Shuang-Yuan Chen[1], Shea-Jue Wang[2], Hong-Wen Hsu[1], Wen-Shiang Liao[3]

[1]Graduate Institute of Mechatronic Engineering, National Taipei University of Technology, Taipei, Taiwan
[2]Dept. of Materials and Resources Engineering, National Taipei University of Technology, Taipei, Taiwan
[3]Dept. of Electronic Engineering, Minghsin University of Science and Technology, Hsinchu, Taiwan
* Corresponding email: mucwang@must.edu.tw

Abstract

The embedded SiGe source/drain stressor helpful to promote the drive current involves etching out the source/drain silicon and replacing it with SiGe filler. This process uses the lattice mismatch between silicon and germanium atoms making the silicon channel compressive. This compressive stress enhances hole mobility, and the pMOSFET performance can be enhanced.

In this study, the characteristics of devices contained biaxial strain in channel and embedded SiGe source/drain stressor with different channel lengths and the channel hot carrier (CHC) in short channel pMOSFETs was explored, too.

Introduction

The performance of MOSFETs has been improved by the aggressive scaling of the device feature sizes. However, there are some problems as the device dimensions continue to scale down, such as short channel effect, gate oxide thickness limitation and shallow junction depth limitation. The short channel effect causes the modification of the threshold voltage due to the shortening channel length. The gate leakage current increases rapidly when the gate oxide thickness is too thin. The reduction of parasitic resistance is due to reducing the depth of shallow junction. It is challenging to continue scaling down even if advanced anneal technologies. The improvement of MOSFETs performance becomes difficult by these limitations. In order to continue to improve MOSFETs performance, the strain silicon techniques have become the best choice to improve MOSFETs performance. The strained-Si technology can be used to increase carrier mobility, including applying stressors from various regions, such as mechanical stress, contact etch stop layer (CESL), shallow trench isolation (STI), embedded SiGe source/drain, and metal-induced strain.

Among above techniques, the embedded SiGe S/D is one of the most common strain methods. In this work, the characteristics and hot-carrier reliability of strained-Si pMOSFETs with SiGe channel as biaxial strain and embedded SiGe source/drain stressors playing uniaxial strain are investigated.

Experiments

The strain pMOSFET devices were fabricated on (110) wafer surface. The SiGe channel pMOSFETs were fabricated by using an advanced 90nm generation technology, shrunk down to 45nm gate length with over photo exposure and etching technology. Figure 1 and 2 show the cross sections of the device structure. The thickness of buffer-Si and $Si_{1-x}Ge_x$ (x=22.5%) were 50 Å and 100 Å, respectively. And the Si-cap layer was fabricated with 30 Å thicknesses, but consumed with 6 Å in oxidation process.

The Si-cap layer is beneficial to reduce the interface-trap states between the gate dielectric and the SiGe channel. Subsequently, a 1500-Å thick undoped poly-silicon gate electrode was deposited [1-3].

To fabricate uniaxial strain, the source/drain was etched by 500 Å. Then a nickel-salicidation process was used, following by a recipe-varied plasma-enhanced chemical-vapor-deposition (PECVD) ILD-SiN$_x$ CESL layers. After ILD-oxide deposition, chemical-mechanical planarization, contact patterning, etching, and W-plug formation, a standard copper-metal interconnection process was processed. Finally, a passivation layer was followed by the formation of aluminum-bonding pads.

Results and Discussions

Figure 3 depicts that threshold voltages roll-off happened with various channel lengths on different strained-Si pMOSFETs, but the pMOSFETs' threshold voltage should be reverse roll-off with channel length scaling down. For the above phenomenon, it may use halo implant processes to reduce threshold voltage and decrease short channel effect.

Figure 4 to 6 show in the initial pMOSFETs I_d-V_d curves which are measured at V_g-V_t = -0.4, -0.7 and -1V in different channel length. The drain current enhancement of embedded SiGe stressor is more obvious in short channel. Maybe, the phenomenon is that local strained induces more compressive stress in short channel.

For the previous phenomenon, it may use halo implant processes to reduce threshold voltage and decrease short channel effect. Figure 7 shows hole mobility with various strained-Si devices in different channel length. The values of mobility are extracted by transconductance method. The tested pMOSFETs have L_g= 0.11 μm and the channel width is 10 μm. The stress conditions are shown as Table I.

Figure 8 under a high voltage stress demonstrates the combination of biaxial and embedded SiGe devices propose a higher $V_{t,lin}$ shifts than Si -control and biaxial devices. In Figure 9 under a high voltage stress, the combination of biaxial and embedded SiGe devices show a higher $I_{d,sat}$ degradation than Si control and biaxial devices. According to the experimental results at room temperature, the saturation current of the embedded SiGe source/drain pMOSFETs degraded 3.3% as the W/L= 10/0.11 (μm/μm) and stress voltage= -1.7V+V_t, which is more serious than Si-control devices with 2.1% degradation. It is presumable that embedded SiGe source/drain induces more traps in the interface between S/D sites and channel or gate fringe. [4-6]

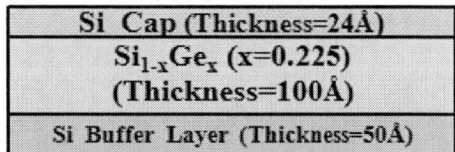

Fig. 1. Schematic showing the thickness of buffer- Si/Si$_{1-x}$Ge$_x$ (x=22.5%)/Si-cap in channel.

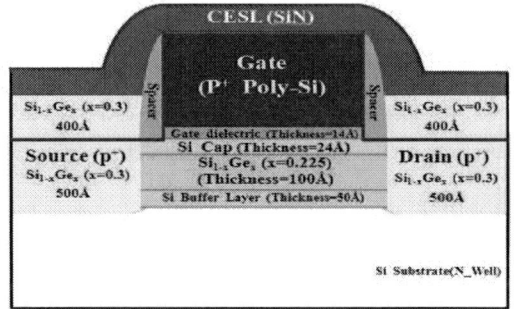

Fig. 2. Schematic diagram of a proposed MOSFET structure with embedded SiGe in the S/D region.

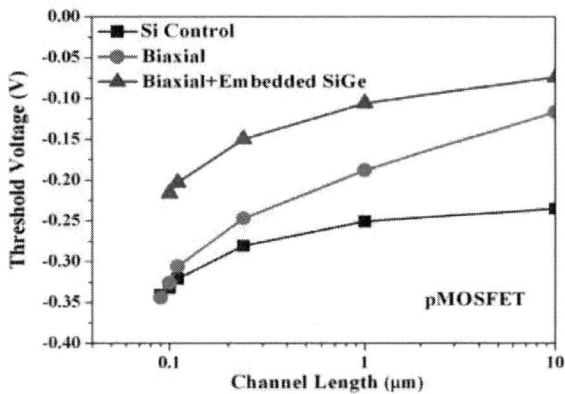

Fig. 3. Linear threshold voltage versus channel length for 90-nm pMOSFEs with Si-control, biaxial and the combination of biaxial and embedded SiGe stressors.

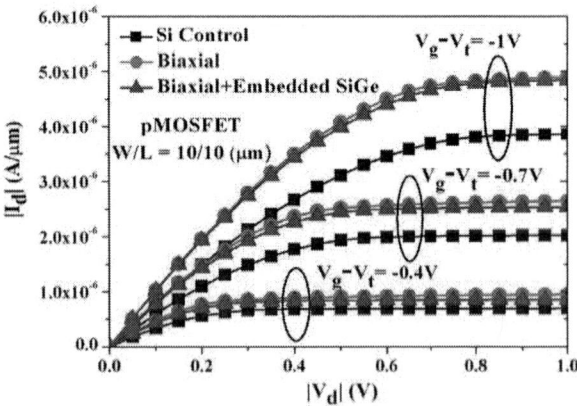

Fig. 4. Drain current versus drain voltage (I$_d$-V$_d$) for pMOSFETs with Si-control, biaxial and the combination of biaxial and embedded SiGe stressors. The devices width/length are 10μm/10μm and were measured at V$_d$ = -1 V.

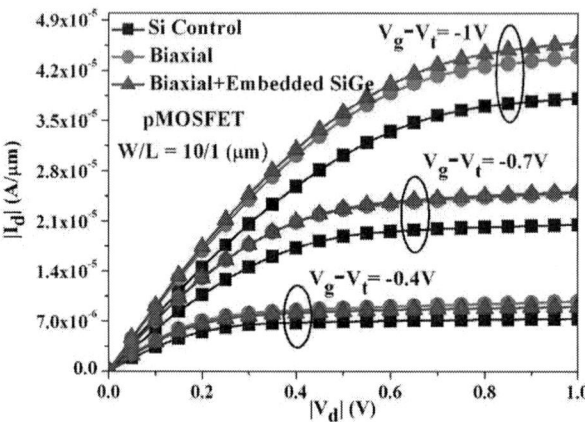

Fig. 5. Drain current versus drain voltage (I$_d$-V$_d$) for pMOSFETs with Si-control, biaxial and the combination of biaxial and embedded SiGe stressors. The devices width/length are 10μm/1μm and were measured at V$_d$ = -1 V.

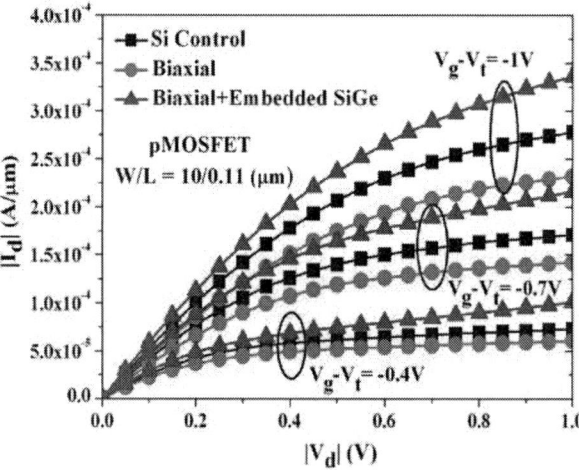

Fig. 6. Drain current versus drain voltage (I$_d$-V$_d$) for pMOSFETs with Si-control, biaxial and the combination of biaxial and embedded SiGe stressors. The devices width/length are 10μm/0.11μm and were measured at V$_d$ = -1 V.

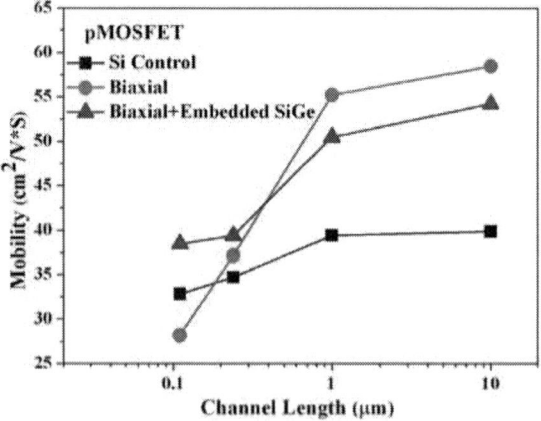

Fig. 7 Hole mobility of pMOSFETs with Si-control, biaxial and the combination of biaxial and embedded SiGe stressors in different channel length.

978-1-4673-4840-9/13 $31.00 © 2013 IEEE   272

TABLE I
STRESS CONDITIONS FOR CHC STRESS

Dimension (μm/μm)	W/L=10/0.11		
V_{cc} (V)	-1		
V_{stress} (V)	1.5V$_{cc}$	1.6V$_{cc}$	1.7V$_{cc}$
Stress voltage (V)	$V_g = V_{stress} + V_t$ $V_d = V_{stress}$		
Temperature (°C)	25		
Stress time (sec)	0 to 3000		

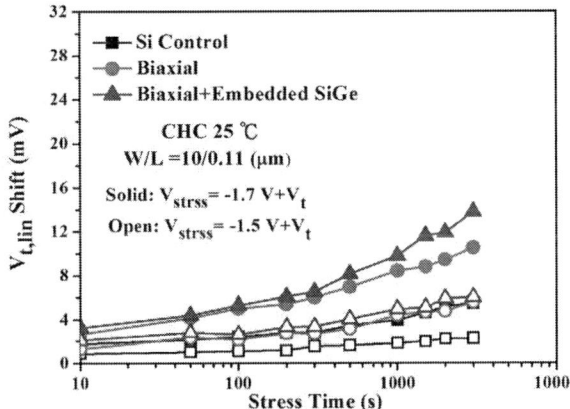

Fig. 8. Comparison of $V_{t,lin}$ shifts versus stress time at 25℃ under hot-carrier stress on pMOSFETs.

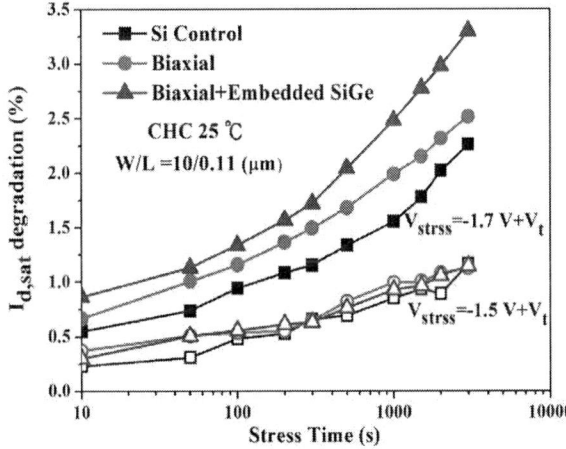

Fig. 9. Comparison of $I_{d,sat}$ degradation versus stress time at 25℃ under hot-carrier stress on pMOSFETs.

Conclusions

The impact of biaxial and embedded SiGe source/drain stressors for pMOSFETs performance and reliability has been investigated. The biaxial stressor shows higher mobility in the long channel, but the mobility degrades rapidly with channel length scaling down. Therefore, the biaxial stressor is suitable for long channel of pMOSFETs. In contrast, the experiment

result represents that the embedded SiGe source/drain can enhance pMOSFETs performance in short channel. It may be that the embedded SiGe increase more compressive stress in short channel.

CHC stress reliability issues of Si-control, biaxial and combination of biaxial and embedded SiGe devices are also studied in this work. The devices combining biaxial and embedded SiGe show the highest degradation under CHC stress. It is believed that the embedded SiGe source/drain inducing defects are mostly located at the surface of extension of source/drain. For our experiment results, it suggests that the device reliability could be a major concern when the biaxial and embedded SiGe source/drain stressors are combined in fabrication.

References

[1] S. H. Olsen, et al., "Study of Single- and Dual-Channel Designs for High-Performance Strained-Si-SiGe n-MOSFETs," *IEEE Transactions on Electron Devices*, Vol. 51, No. 7, pp. 1245-1253, July 2004.

[2] K. D. Goutam, et al., "Impact of Strained-Si Thickness and Ge Out-Diffusion on Gate Oxide Quality for Strained-Si Surface Channel n-MOSFETs," *IEEE Transactions on Electron Devices*, Vol. 53, No. 5, pp. 1142-1152, May 2006.

[3] T. Krishnamohan, et al., "High-Mobility Ultra-Thin Strained Ge MOSFETs on Bulk and SOI with Low Band-to-Band Tunneling Leakage: Experiments," *IEEE Transactions on Electron Devices*, Vol. 53, No. 5, pp. 990–999, May 2006.

[4] E.Amat, et al., "Channel hot-carrier degradation on strained MOSFETs with embedded SiGe or SiC Source/Drain" *IEEE ICSICT*, pp.1648-1650, November 2010.

[5] W. Yeh, et al., "The Improvement of High-*k*/Metal Gate pMOSFET Performance and Reliability Using Optimized Si Cap/SiGe Channel Structure," *IEEE Transactions on Devices and materials reliability*, Vol. 11, No. 1, pp. 7-12, March 2011.

[6] H. Okamoto, et al., "In situ Doped Embedded-SiGe Source/Drain Technique for 32nm Node p-Channel Metal–Oxide–Semiconductor Field-Effect Transistor," *Japan Society of Applied Physics*, Vol. 47, No. 4, pp. 2564-2568, 2008.

Structural and optical properties of Cu-doped ZnO nanoparticles synthesized by co-precipitation method for solar energy harvesting application

N. Thaweesaeng[1]*, S. Suphankij[2], W. Pecharapa[2,3] and W. Techitdheera[1]

[1] School of Applied Physics, Faculty of Science

King Mongkut's Institute of Technology Ladkrabang, Bangkok 10520, Thailand

[2] College of Nanotechnology, King Mongkut's Institute of Technology Ladkrabang, Chalongkrung Rd.,

Ladkrabang Bangkok 10520, Thailand

[3] ThEP Center,CHE,328 Siayuthtaya Rd., Bangkok 10400,Thailand

*nthaweesaeng@gmail.com

Abstract

Cu-doped ZnO nanoparticles were synthesized by co-precipitation method using zinc nitrate ($Zn(NO_3)_2 \cdot 6H_2O$), copper(II) nitrate trihydrate ($Cu(NO_3)_2 \cdot 3 H_2O$) as starting precursors for Zn and Cu sources, respectively. The structural properties of powders were characterized by X-ray diffraction (XRD), field-emission scanning electron microscopy (FESEM) and X-ray Photoelectron Spectroscopy (XPS). The XRD results disclose that Cu-doped ZnO nanopowders are in hexagonal wurtzite structure and their crystallinities are deteriorated with increasing Cu doping content. Meanwhile, XPS results indicate the existence of Cu ion with relevant electronic state in ZnO. Optical properties of the samples were investigated by mean of their absorption and the results suggest that Cu additive has significant effects on their crucial optical properties that can be tuned to meet the requirement for practical solar energy harvesting applications such as sun-light photocatalyst and photovoltaic devices.

KEYWORDS : Cu-doping, ZnO, co-precipitation

Introduction

Recently, research and development of new material that can perform multi-functionality in the form of alloys, composites or compound have been proposed. Due to its has direct wide band gap (3.37 eV) and large exciton binding energy (60 mV) at room temperature, ZnO has attracted much attention because of particular optical, chemical, physical, and electrical properties. ZnO has been synthesized in form of various types of attractive nanostructures such as nanoparticle, nanorod, nanoneedle, nanosphere and so on. Because of its versatility, ZnO as one of multipurpose semiconductor has widely been utilized for various applications such as blend solar cell [1], gas sensor [2-3], photodetector [4-5], and photocatalytic activity [6]. Doping ZnO with transition metal element such as Ni [7], Co [8], Al [9],Cu [10-13] has been verified as an effective method to adjust its functionality including electrical and optical properties. More recently, Cu-doped ZnO has shown singnificant improvement in relevant properties such as electrical, magnetic, photocatalytic performance and gas sensing properties [10-13].

In this work, The synthesis of high-quality Cu-doped ZnO nanoparticles via co-precipitation method using zinc nitrate ($Zn(NO_3)_2.6H_2O$), copper(II) nitrate trihydrate ($Cu(NO_3)_2.3 H_2O$) and sodium hydroxide as starting precursors. The properties of annealed Cu-doped ZnO nanoparticles were investigated by XRD, FESEM ,XPS and UV-Vis spectroscopy.

Experiment

Nanoparticles of Cu-doped ZnO with various doping contents; x (x=0-0.10) have been synthesized via co-precipitation method using zinc nitrate ($Zn(NO_3)_2.6H_2O$) and copper(II) nitrate trihydrate ($Cu(NO_3)_2.3 H_2O$) as starting precursor sources of Zn and Cu, respectively. Preparation of Cu-doped ZnO nanopowders is briefly described. Zinc nitrate was dissolved in 100 ml deionized water to obtain 0.5 M precursor and certain amount copper (II) nitrate trihydrate was added according to doping condition in solution. NaOH solution was slowly added into the precursor under vigorous stirring until pH of the solution reached to 14, leading to the precipitated product. After that, as-precipitated products were washed several times with deionized water and ethanol in the last step until its pH became to 7. The products were dried in oven at 80°C for 6 hr. At last, the products were annealed at 550 °C for 2 hr in an ambient air to finally obtain the ZnO and Cu-doped ZnO nanoparticles.

Result and discussion

Field-emission scanning electron microscopy (FESEM) was used to investigate the morphorogy, shape and size of Cu-doped ZnO nanoparticles. Fig.1 shows a image of $Zn_{0.94}Cu_{0.06}O$ nanoparticles that possess homogeneity and uniform distribution. The morphology of Cu-doped ZnO nanopowder is found to be in spheroid-like with average grain size of 30 nm. The crystalline structure and phase purity of Cu-doped ZnO nanopowders with x=0, 0.02, 0.06, 0.08 and 0.10 were examined by the X-ray diffraction (XRD) and the corresponding results are illustrated in Fig. 2. All diffraction peaks position are nicely indexed to (100), (002), (101), (102) and (110) orientation planes, confirming the formation of the typical hexagonal wurtzite structure of ZnO according to the

978-1-4673-4840-9/13 $31.00 © 2013 IEEE

standard JCPDS card (No.76-0704). The XRD patterns of the samples with x=0.06, 0.08 and 0.10 additionally exhibit the split of CuO phase [14]. This appearance indicates that with low doping content, Cu ions may properly incorporate to Zn site without significant effect of the crystal structure of ZnO due to relatively close radius of Cu^{2+} (0.73Å) to that value of Zn^{2+}(0.74 Å). However increasing doping content beyond critical value may initiate the phase separation of CuO and ZnO. The average crystallite size of Cu-doped ZnO nanopowders can be calculated from well-known Debye-Scherer equation;

$$D = \frac{\kappa \lambda}{\beta \cos \theta} \qquad (1)$$

where D is the average crystallite size, κ is the shape factor , λ is the wavelength of incident x-ray beam , β is the full - width at the half maximum (FWHM) in radius on 2θ scale and θ is the Bragg's diffraction angle . The average crystallite sizes was found to be about 17-40 nm, which is in good accordance with the value determined by SEM image. In addition, the lattice constant for hexagonal wurtzite ZnO nanopowders are estimated from equation ;

$$\frac{1}{d^2} = \frac{4}{3\left(h^2 + hk + k^2 / a^2\right)} + \frac{l^2}{c^2} \qquad (2)$$

where a and b are the lattice constant, h, k and l are the miller indices and d is the interplanar spacing which can be calculated from Bragg's low ;

$$2d \sin \theta = n\lambda . \qquad (3)$$

The crystalline size and lattice parameter are presented in Table 1.

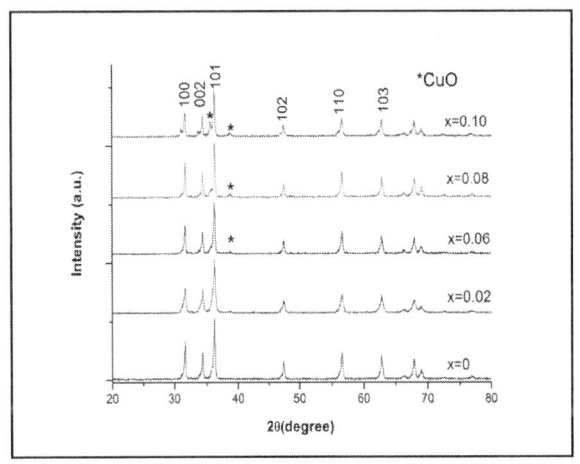

Fig. 2 X-ray diffraction patterns Cu-doped ZnO nanoparticles annealed at 550 °C (x=0, 0.02, 0.06 , 0.08 and 0.10)

Table 1 The crystalline size and lattice parameter of Cu-doped ZnO nanoparticles annealed at 550 °C

Copper content (wt%)	Crystalline size(nm)	Lattice constant	
		a=b (Å)	c (Å)
0	28.83	2.8197	5.2152
2	18.91	2.8201	5.5139
6	31.89	2.8165	5.2099
8	40.03	2.8129	5.2035
10	36.10	2.8147	5.2061

Fig. 1 Field-emission scanning electron microscopy (FESEM) images of Cu-doped ZnO nanoparticle (x= 0.06) annealed at 550 °C.

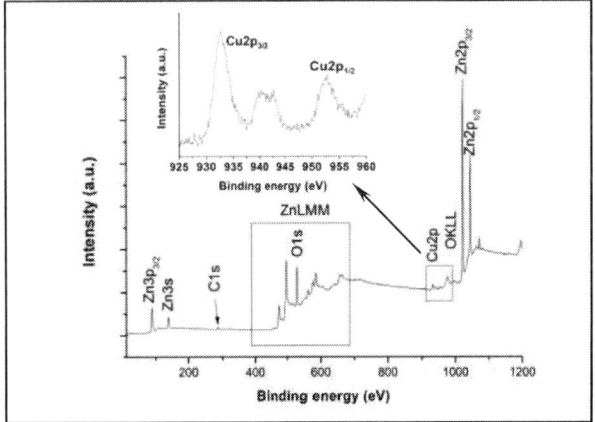

Fig. 3 The survey XPS spectrum of Cu-doped ZnO nanoparticle annealed at 550 °C (x= 0.08). Inset: Cu 2p spectra.

978-1-4673-4840-9/13 $31.00 © 2013 IEEE   275

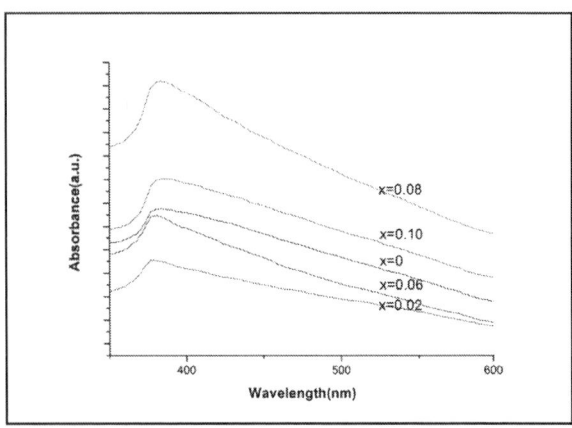

Fig. 4 UV-Visible absorption spectra of Cu-doped ZnO (x=0, 0.02, 0.06 , 0.08 ,0.10) nanoparticle annealed at 550 ºC.

X-ray Photoelectron Spectroscopy (XPS) spectrum was employed to characterize the oxidation state of Cu-doped ZnO. The survey XPS spectrum of the nanoparticle with x=0.08 is displayed in Fig. 3. All indexed peaks can be specified to C,O, Cu and Zn .The two main peaks detected at the binding energy position at 1201.1 eV and 1044.2 eV correspond to the Zn $2p_{3/2}$ and Zn $2p_{1/2}$ respectively . Fig.3 (inset) shows the Cu 2p core-level XPS spectrum of Cu-doped ZnO nanoparticle. The Cu $2p_{3/2}$ and Cu $2p_{1/2}$ spectra are observed at binding energy of 932.1 eV and 951.9 eV , respectively, confirming the existence of Cu in the ZnO nanoparticle. The peaks at binding energy of 932.1 is typically assigned to Cu^{2+} and the $2p_{3/2}$ peak of Cu^{2+} in Cu_2O appeared at 932.7 eV [15].

Fig 4. illustrates UV-visible absorption spectra of Cu-doped ZnO with various Cu doping compositions in range of 300-600 nm. The absorption peak position of the samples for x= 0, 0.02, 0.06, 0.08 and 0.10 were found to be at 381, 369, 381, 383 and 387 nm, respectively. ZnO is generally an n-type semiconductor with wide band gap of 3.3 eV corresponding to wavelength of about 370 nm in UV region. The absorption peak position of sample is noticeably increased as Cu doping concentration increases. The feature implies the decrease in optical band gap of ZnO with increasing in Cu doping content. Furthurmore, the absorption spectra of doped sample exhibits the increase in its absorption intensity and extension into longer wavelength. It is suggested that, by doping ZnO with Cu, its optical absorption can be enhanced, that is the requirement for solar energy harvesting applications.

Conclusions

In summary, Cu-doped ZnO nanoparicles (x=0, 0.02, 0.06, 0.08, 0.10) nanopowders have been successfully prepared using co-precipitation method in combination with annealing process. The XRD results confirmed that the crystal structure of Cu-doped ZnO nanoparicles is hexagonal wurtzite structure. The appearance of additional CuO peak is observed in the doped samples with high doping content. The morphology of the sample is found to be in spheroid-like with average grain size about 30 nm. XPS results indicate the existence and chemical state of C, O, Cu and Zn. The absorption peak position of sample is noticeably increased as Cu doping concentration increases. The feature implies the decrease in optical band gap of ZnO with increasing in Cu doping content. These alternation in its optical properties, especially the optical absorption enhancement suggests the potential utilization of this material in solar energy harvesting applications.

Acknowledgments

Corresponding author (N. Thaweesaeng) would like to thank Nakhon Ratchasima Rajabhat University for scholarship funding support. This work has partially been supported by the National Nanotechnology Center (NANOTEC), NSTDA, Ministry of Science and Technology Thailand, through its program of Center of Excellence Network.

References

[1] S. R. Ferreira, el al, *Organic Electronics*, pp.1258–1263, April 2011.

[2] N. Han, el al, *Sensors and Actuators B: Chemical*, pp. 230–238, July 2010 .

[3] J, Hyung Juna, el al, *Sensors and Actuators B: Chemical*, pp. 412–417, May 2009.

[4] J, Hyung Juna, el al, *Ceramics International* ,pp. 2797–2801, April 2009.

[5] Y. Li, el al, *Physica B*, pp. 4282– 428, August 2009.

[6] J. Xie, el al , *Power Technology* , pp 140-144, November 2010.

[7] M. El-Hilo, A.A.Dakhel, A.Y.Ali-Mohamed, *Journal of Magnetism and Magnetic Materials* , pp. 2279–2283, July 2009.

[8] S. A. Ansari, el al , *Materials Science and Engineering* , pp. 428-435, March 2012.

[9] M.H. Mamat, el al, *Optical Materials*, pp. 696-699, April 2010.

[10] S. Singhal, el al , *Physica B*, pp. 1223-1226. February 2012.

[11] H. Liu, el al , *Applied Surface Science*, pp. 4162-4165, April 2010.

[12] S. Muthukumaaran , R Gopalakrishnan, *Optical Materials*, September 2012.

[13] Y.S. Sonowane, el al , *Materials Research Bulletin*, pp. 2719-2726. October 2006.

[14] C.C. Vidyasagar, el al, *Powder technology*, pp.337-343 ,Semtember 2011.

[15] P. jongavakit ,et al , *Applied Surface Science*, pp. 8192-8198. May 2012.

Self-Organized Hybrid Nanostructures Composed of the Array of Vertically Aligned Carbon Nanotubes and Planar Graphite Layer.

V. Labunov[1], A. Prudnikava[1], B. Shulitski[1], B.K. Tay[2],
X. Wang [2], A. Basaev[3], V. Galperin[3], Y. Shaman[3]

[1]Belarusian State University of Informatics & Radioelectronics, Brovka St.6,
220013 Minsk, Belarus,
Tel./Fax: +375 17 2021005, e-mail: labunov@bsuir.by
[2]Nanyang Technological University, 50 Nanyang Avenue,
639798, Singapore
Tel.: +65 6790 4533, Fax: +65 6790 9313, e-mail: ebktay@ntu.edu.sg
[3]SMC Technological Center, K-498,
Moscow 103498, Russia
Tel.: +7 (095) 532 8906, Fax: +7 (095) 913 2192, e-mail: as@tcen.ru

Abstract

The hybrid carbon nanostructures composed of an array of vertically aligned carbon nanotubes (CNTs) and a self-organized planar graphite layer (PGL) located on the top of the array (CNT-PGL nanostructures) have been obtained by the CVD method with the volatile catalyst. The fundamental characteristics (morphology, elemental composition and structure) of these nanostructures were characterized by Scanning and Transmission Electron Microscopy, Energy-dispersive X-ray (EDX), Raman and Auger spectroscopy. It has established that CNTs grow simultaneously both via root - growth mechanism (attached to the substrate) and tip-growth mechanism (attached to the planar layer). This planar layer is a layered - graphitic structure and connected with CNTs through the catalyst nanoparticles. It consists of disordered graphite flakes with the sizes of some tens of nanometers randomly oriented and overlapping with one another. These graphitic flakes have a crystallite of ~5.2 nm in size. （Keywords: self-organization and self-assembling processes, allotropic forms of carbon, carbon nanotubes, graphite, grapheme, carbon hybrids, hybrid CNT-Planar graphite layer nanostructure, CNT-PGL nanostacks ）

Introduction

Among various processes of functional nanomaterials synthesis and nanodevices fabrications, the self-organization and self-assembling processes are playing a fundamental role. One of the examples is the synthesis of hybrid nanostructures on the basis of different allotropic forms of carbon, such as carbon nanotubes (CNTs) filled with fullerenes, named as "peapods" [1].

In the nanoscale regime, mainly two carbon allotropies - carbon nanotubes and graphene - have attracted enormous interest of researchers since their discoveries. While both of these materials possess unique electronic, mechanical and thermal properties but different dimensionality, there are prospects of deriving benefits from CNT-graphene hybrids. The latest obtained carbon-carbon hybrid nanostructures are arrays of vertically aligned CNTs with a planar graphite layer (PGL) located at the top (CNT-PGL nanostructure). Single-layer [2,3], multi-layer nanostructures [4], as well as

nanostacks [5] were created. These hybrid nanostructures possess a number of new specific properties and open a way to the creation of essentially new functional devices in the area of nanoelectronics, nanophotonics and nanoelectromechanics. As predicted theoretically, the CNT-PGL nanostacks when doped with lithium cations can be efficient structures for hydrogen storage [6] and moreover these nanostructures are considered as a novel material with tailored multidimensional thermal transport characteristics [7]. First experimental realization of CNT-PGL nanostacks was reported in reference [4] as super efficient electrodes of supercapacitors. Despite of its attractiveness for supercapacitors application, the proposed fabrication method is extremely complicated. Herein, we present a simple and low cost method to fabricate the CNT-PGL nanostructure. The fundamental characteristics (morphology, elemental composition and structure) of the CNT-PGL nanostructures will also be discussed. The simplest case, one layer CNT-PGL nanostructure, is considered here.

Experimental

A feeding solution of ferrocene in xylene was used in the CVD process and it was injected into the reaction zone of the tubular silica reactor. Argon was used as a gas-carrier with a constant flow of 100 $cm^3 \cdot min^{-1}$ through the reactor. The synthesis process was carried out at the temperature of 850 °C for 60 min under atmospheric pressure with 1.0 wt.% feeding solution. Wafers (5×5 mm^2) of n-type silicon coated with a layer of SiO_2 were used as substrates.

The obtained nanostructures were characterized by Scanning Electron Microscopy (SEM, Hitachi S4800), Transmission Electron Microscopy (TEM, Jeol JEM-2010). For Raman measurements, Renishaw micro-Raman Spectrometer (Series1000) with laser beam of 1.5 mW incident power and 514 nm wavelength was used. Auger Analysis was performed using Perkin Elmer, PHI–660 spectrometer equipped with Auger micro probe. Energy-dispersive X-ray Analysis (EDX) was performed on SEM SUPRA–55WDS (Zeiss).

Results and Discussion

Figure 1 shows the SEM image and the Auger spectrum of the obtained hybrid nanostructure, consisting of an array of vertically aligned CNTs and a self-organized planar layer located at the top of the array.

SEM image of the obtained nanostructure is presented in fig. 1a. The height of nanostructure is ~2.3μm. Despite the long synthesis process (T=60 min), the height of the CNTs are relatively short because a small injection rate of the reaction

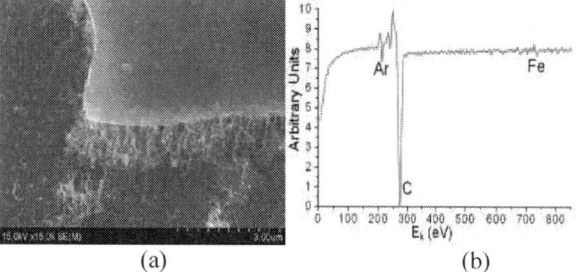

(a)   (b)

Fig. 1 Hybrid nanostructure composed of an array of vertically aligned carbon nanotubes and planar layer on the top of the array: (a) SEM image; (b) Auger spectrum of the planar layer.

mixture was used, which is crucial for the formation of the planar layers. [3]

From the Auger spectroscopy at figure 1b, it is shown that planar layer consists of carbon with negligible amount of Fe and Ar. Fe represents catalyst and Ar is used in the synthesis process.

In figure 2, the EDX analysis results are shown. Fig. 2a demonstrates the EDX mapping of elements over the SEM image. It is seen that the whole nanostructure (both CNT array and planar layer) has a homogenous color (green), which corresponds to the same element. Si/SiO$_2$ substrate is in a darker green color. EDX spectrum (fig. 2b) showing strong C signal testifies the fact that the nanostructure represents all-carbon structure. Strong Si peak belongs to the Si/SiO$_2$ substrate. O peak (red droplets in fig. 2a) belongs to the silicon

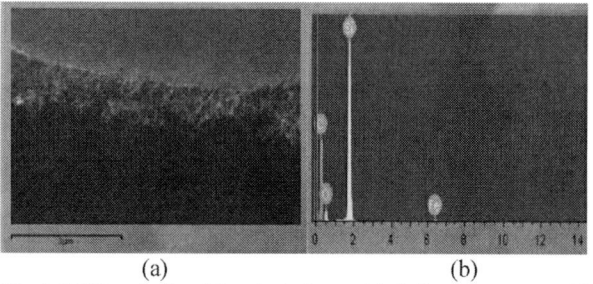

(a)   (b)

Fig. 2 EDX analysis of the single layered hybrid nanostructure: (a) EDX image (the map of the elements), (b) EDX spectra (unit for the transverse axis: KeV)

dioxide substrate and traces of Fe (blue droplets in fig. 2a) to the catalyst used for growing the nanostructure.

The SEM images of nanostructure at different aspect angles and magnifications are presented in figure 3. In fig. 3a, it is seen that the planar layer is composite of layer-built structure, sometimes referred to as "graphene multi-layer" [2]. For the more general case, we would consider this layer as Planar Graphite Layer (PGL) and hybrid nanostructure with CNTs as CNTs - PGL nanostructure. It should be noted that the surface of this layer is rough in the nano-scale. From figure 3b, it can be seen that CNTs are connected to PGL through the catalyst nanoparticles (dark spots) and testifies the fact that the growth

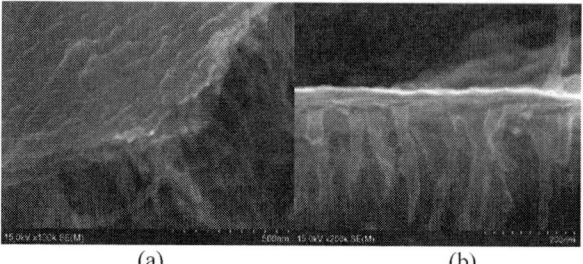

(a)   (b)

Fig. 3 SEM images of one-layer CNTs - PGL nanostructure: (a) ×100 000; (b) ×200 000.

of CNTs is by tip–growth mechanism. The diameter of CNTs varies from 10 to 25 nm.

From the SEM image in figure 4a, the CNT-PGL nanostructure presented a uniform height and a plane-parallel surface of PGL, which can be useful for numerous applications in microelectronics and MEMS. From figure 4b, the thickness of the planar layer is of ~35 nm. Using high resolution TEM and high speed Furrier method, the distance between layers

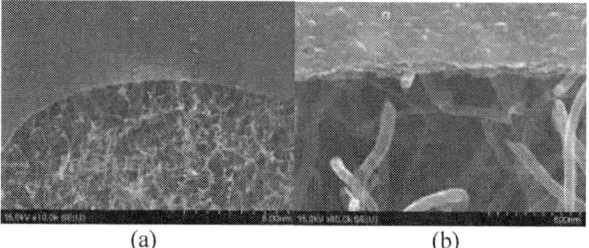

(a)   (b)

Fig. 4 SEM images of the CNT-PGL nanostructure at different magnifications: (a)×10000; (b)×80000.

was calculated to be 0.37 nm. As a result, the number of graphene planes in the ~35 nm layer was about 95.

(a)   (b)

Fig. 5 Formation of plane graphite layers via PGL detachment from CNT-PGL nanostructure (SEM): (a) strips of PGL; (b) magnified image of an opposite side of a single strip.

Figure 5 shows a portion of PGL, mechanically detached from CNT array and broken into the strips. Figure 5a shows strips of PGLs and the opposite side with CNT attached is presented in fig. 5b. From the attachment of the CNTs, it can be confirmed that CNTs grow simultaneously both via root - growth mechanism (attached to the substrate) and tip-growth mechanism (attached to the planar layer).

The morphology and structure of the CNT-PGL junction, which can give information on the growth mechanism of the nanostructure, attract particular interest. In Figure 6, some fragments of the junction are presented. In Figure 6a, one can recognize that CNTs growth with open ends (root - growth mechanism). While from Figure 6b, it is seen that catalyst nanoparticles are embedded within the channels of CNTs and

978-1-4673-4840-9/13 $31.00 © 2013 IEEE

(a)   (b)   (c)

Fig. 6 TEM images of the fragments of the CNT-PGL nanostructures: (a) CNTs grown from the planar layer with open ands, (b) CNTs with catalyst nanoparticles located at the tips of, (c) catalyst nanoparticles on/in the surface of the planar layer and some disordered CNTs.

most of them are found at the tip of CNTs (tip - growth mechanism). Figure 6c represents a spot at the junction containing mostly catalyst nanoparticles at the surface of the planar layer and some disordered CNTs. The diameters of the nanoparticles are about 50 nm.

The thorough investigation of the CNT-PGL junction had been performed by the HRTEM. In particular, main components of the junction have been investigated separately – the catalyst nanoparticles, the CNTs on the surface of planar layer and the planar layer itself.

In figure 7, the HRTEM images of the catalyst nanoparticles are presented. From figure 7a, one can observe that the catalyst nanoparticles with different shapes and sizes ranging from 20

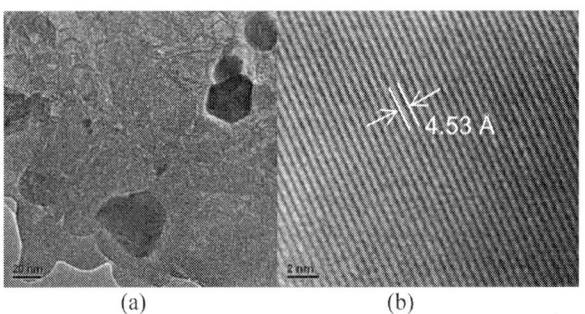

(a)   (b)

Fig. 7 HRTEM images of the catalyst nanoparticles: (a) catalyst nanoparticles of different shapes and sizes, (b) the crystalline structure of one of the catalyst nanoparticles at the surface of the planar layer.

to 40 nm in diameter. Also, it is clearly visible that the planar layer is composite of disordered flakes overlapping with each other in the sizes of tens of nanometers.

From figure 7b, the crystalline structure of the catalyst nanoparticle at the surface of the planar layer is presented. It demonstrates a highly crystalline structure of the catalyst nanoparticle. The accurate measurement of the crystal interplane distance of the nanoparticle performed using

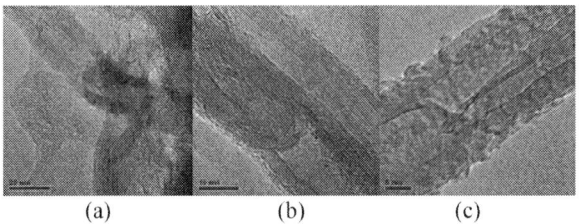

(a)   (b)   (c)

Fig. 8 HRTEM images of the CNTs: (a) CNTs with planar layer, (b) and (c) different structures of CNTs.

Fourier transformation gives a value of approximately 4.53Å, which is very close to the carbide Fe_3C interplane distance (4.524 Å).

Various structures of CNTs are presented in figure 8. There

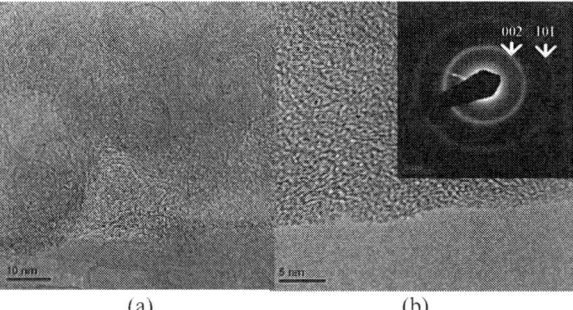

(a)   (b)

Fig.9 HRTEM images of the planar layer: (a) the image of planar layer made by the electron beam perpendicular to the layer, (b) another spot of the PGL (insert - electron diffraction diagram of that spot).

are a variety of CNTs combinations with planar layers (figure 8a) and CNTs structures (figure 8 b and 8c). Obviously, all the CNTs are multiwall.

HRTEM images of the planar layer are presented in figure9. In figure 9a, the image of planar layer made by the electron beam perpendicular to the layer is presented. It is clearly visible that the disordered flakes of graphite with the sizes of some tens of nanometers overlapping each other and figure 9b shows another magnification spot. The electron diffraction pattern is shown in the insert of figure 8b. The solid rings indicate the polycrystalline structure of graphite with their small grain size.

Raman spectroscopy is performed on the CNT-PGL presented in figure 10(a). In this specimen, the height of

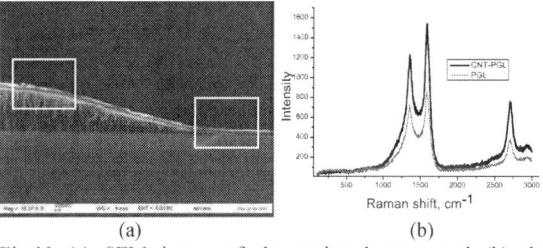

(a)   (b)

Fig.10 (a) SEM image of the analyzed areas and (b) the corresponding Raman spectra of CNT-PGL (black) and PGL (red)

CNT-PGL nanostructure varies along the sample and decreases gradually from 7.5 μm to flat where no CNT is left. This would allow us to observe the Raman spectra of CNT-PGL nanostructure in a whole and as well as the PGL alone.

Raman spectra of two analyzed areas indicated in figure 10(a) (white enclosed areas) are presented in figure 10(b). The intensity is generally higher for the area containing both CNTs and PGL. The ratio of I_G/I_D for CNT-PGL is found to be 1.25 and PGL alone is 1.18. This means that the structure of PGLs for both is relatively similar, despite one was synthesized on top of CNT arrays instead of SiO_2 surface. The I_G/I_D ratio can be related to the in-plane crystallite size L_a, which, according to the relationship L_a (nm)$=4.4$ (I_G/I_D) [7], while single layered PGL is equal to ~5.2 nm.

Conclusion

The carbon nanostructures composed of an array of vertically aligned CNTs and a self-organized planar graphite layer (PGL) located at the top of the array (CNT-PGL nanostructures) have been obtained by the CVD method with the volatile catalyst.

The catalyst nanoparticles, the CNTs and the planar layer of CNT-PGL junctionthe have been investigated separately. The catalyst nanoparticles have different shapes and sizes ranging from 20 to 40 nm in diameter and highly crystalline structure with the interplanar distance indicating Fe carbide (Fe_3C). A planar layer represents the disordered flakes of graphite with the sizes of some tens of nanometers randomly oriented under different angles to the substrate and overlapping each other. The CNT-PGL nanostructures are expected to have unique electro-physical properties and therefore are likely to find many applications in nanoelectronics, nanophotonics and nanoelectromechanics.

References

[1]   B.Smith, M.Monthioux, D.Luzzi. Nature 396, 323 (1998)

[2]   D.Kondo, S. Sato, Y.Awano. Applied Physics Express, 1, 074003 (2008).

[3]   Labunov, V. A.; Shulitski, B. G.; A.L. Prudnikava; Y.P. Shaman; Basaev, A. S., Semiconductor Physics, Quantum Electronics & Optoelectronics, 2 (2010) 137-141.

[4]   Z. Fan, J. Yan, L. Zhi, Q. Zhang, T. Wei, J. Feng, M. Zhang, W. Qian, and F. Wei, Advanced Materials 22, 3723 (2010).

[5]   Labunov, V.; Shulitski, B.; Prudnikava, A.; Basaev, A., Physica Status Solidi (a) (2010) 1-6.

[6]   Dimitrakakis, G.; Tylianakis, E.; Froudakis, G., Nano Letters, 10 (2008) 3166-3170.

[7]   Varshney, V.; Patnaik, S.; Roy, A.; Froudakis, G.; Farmer, B., ACS nano, 2 (2010) 1153-1161.

Synthesis of Continuous Graphene on Metal Foil for Flexible Transparent Electrode Application

Golap Kalita*[1,3], Koichi Wakita[2], Masayoshi Umeno[2], Yasuhiko Hayashi[3] and
Masaki Tanemura[3]

[1]Center for Fostering Young and Innovative Researchers, Nagoya Institute of Technology, Gokiso-cho, Nagoya, 466-8555, Japan
[2]Department of Electronics and Information Engineering, Chubu University, 1200 Matsumoto-cho, Kasugai 487-8501, Japan
[3]Department of Frontier Materials, Nagoya Institute of Technology, Gokiso-cho, Nagoya 466-8555, Japan
*Corresponding author: kalita.golap@nitech.ac.jp; Phone/Fax: +81-52-735-5216

Abstract

We demonstrate synthesis of large area graphene on a metal (Ni and Cu) foil by the thermal chemical vapor deposition (CVD) process using the solid camphor ($C_{10}H_{16}O$) as a carbon source. The graphene growth process on a polycrystalline metal foil significantly influence by the gas composition and quantity of solid precursor. Synthesis of high quality continuous graphene film is achieved in the developed technique. Fully flexible transparent conductor is fabricated by transferring the graphene film on a plastic substrate.

Keywords: Graphene, camphor, chemical vapor deposition, metal foil and transparent electrode.

Introduction

Graphene and related carbon materials have attracted significant attention in the last few years; due to their fascinating electrical, optical, thermal, chemical and mechanical properties for electronics application [1-2]. There have been efforts to utilize the outstanding properties of graphene for transparent electrode fabrication [3-6]. In the present scenario, indium tin oxide (ITO) and related materials have been widely used as electrode material in electronic devices as it poses high conductivity and transmittance. The use of ITO, however, is appearing increasingly problematic due to limited availability of Indium (In), intrinsic chemical and electrical drawbacks.

Synthesis or deriving process of graphene is one of the most important aspects and several routes have been explored. Graphene synthesized by chemical exfoliation and CVD process has been recognized as most suitable for fabrication of flexible transparent electrode. However, solution processable chemically exfoliated graphene shows much higher sheet resistance due to smaller in size and higher defects. Recent studies shows possibility of synthesizing continuous large-area graphene by the CVD approach [3-10]. In the CVD process the hydrocarbons are dehydorgentated at an elevated temperature and followed by the precipitation of atomic carbon on the top surface of metal surface during the cool-down step as the solid solubility limit is reached [11-12]. Graphene synthesized by CVD process have much higher potential as an alternative of metal oxide based transparent electrode. In this paper, we tackle synthesis of large area graphene film from the solid camphor with a simplified CVD process.

Experimental

Deposition of graphene films on metal foil (Ni and Cu) were achieved by thermal CVD technique using the botanical hydrocarbon camphor ($C_{10}H_{16}O$) as carbon source. For the experiments a quartz tube with length around 90 cm and diameter 50 mm is taken as a CVD reactor and kept horizontally inside two furnaces. For different experiment camphor quantities is varied from 3 mg to 7.5 gm and kept in the first furnace zone. Camphor is evaporated at 200 °C and pyrolyzed in the second furnace at 800 °C with Ar/H$_2$ as a carrier gas. The schematic diagram of the CVD technique is shown in fig. 1.

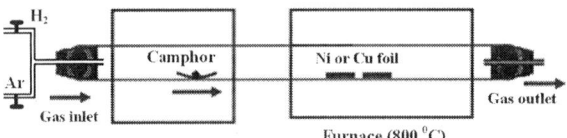

Fig. 1 CVD process using solid camphor as carbon precursor for graphene synthesis on metal foil.

Large-area transparent electrodes are fabricated from synthesized graphene films by transferring to an arbitrary substrate. The synthesized graphene film on metal foil was coated with the Poly(methyl methacrylate) (PMMA) solution and dried. The underneath Cu foil of the PMMA coated graphene film was chemically etched in the Fe(NO$_3$)$_3$ solution with a concentration of 50 mg/ml. Completing the etching of Cu foil, the PMMA/Graphene stack was put on the plastic substrate and PMMA layer was dissolved in acetone. Finally, the graphene film was treated with a diluted nitric acid solution to remove the residual Fe(NO$_3$)$_3$.

The graphene sheets were characterized with Raman spectroscopy having laser excitation energy of 532.08 nm in NRS 300 laser Raman spectrometer, scanning electron microscopy (SEM) with in FESEM, Hitachi S-4300 and high-resolution transmission electron microscope (HRTEM) with JEOL:JEM-2100F operated at 200 kV. Optical transmittance characterization of the graphene film on a plastic was carried out with the UV-Vis-NIR on JASCO-V570 spectrophotometer.

Result and discussions

Camphor is a natural solid botanical hydrocarbon source consisting of hexagonal, pentagonal carbon rings along with methyl carbon. Our studies shows that the molecular structure of the camphor have significant influence in formation of high quality graphene structure. The formation of graphene domain and number of layers greatly influence by the synthesis process along with the composition of gas mixture, temperature and quantity of camphor molecules. Graphene films were deposited on Ni foil at 800 °C by pyrolyzing the camphor molecules in Ar atmosphere. Raman studies show formation of highly ordered graphene sheet with minimum defects.

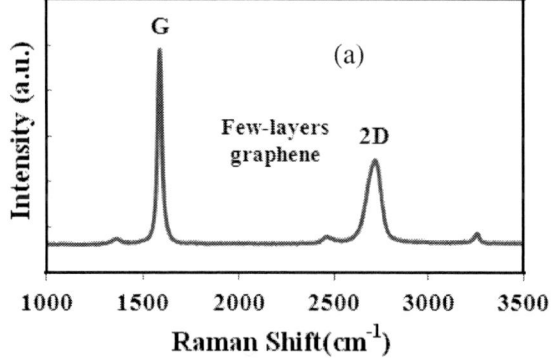

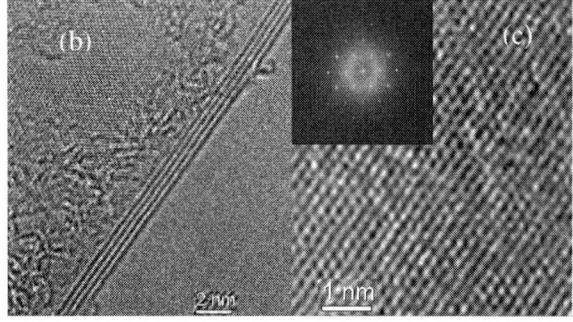

Fig. 2 (a) Raman spectra of a few-layers graphene sheet synthesized on Ni foil (b) TEM image of a 4 layers graphene (b) high resolution TEM image of lattice pattern of the graphene sheet.

Fig. 2(a) shows the Raman spectra of the few-layers graphene with a sharp sp^2 carbon peak at 1583 cm^{-1}, having a very weak defect related D peak at 1331 cm^{-1}. A 2D peak is observed at around 2702 cm^{-1} with less intensity than the G peak, presenting a few-layers graphene. The FWHM of G and 2D peaks for the few-layers graphene is found to be 30.3 and 78.8 cm^{-1}, respectively. Formation of few-layers and single-layer graphene was observed depending on the metal catalytic layer and deposition conditions. Fig. 2(b) shows a TEM image for 4 layers graphene sheet deposited using the solid camphor precursor. The edges of the suspended film can fold back, allowing for a cross-sectional view. The observation of these edges by TEM provides an accurate way to measure the number of layers. Formation of graphene sheets with

different number of layers were observed in the TEM studies. The inter-planer spacing of the graphene is found to be around 0.345 nm. Fig. 2(c) shows high resolution TEM (HRTEM) image of highly ordered lattice pattern of the synthesized graphene. HRTEM studies give a very good insight of the atomic packing structure in the graphene sheet by directly imaging the atomic structure. Inset of the figure shows a FFT pattern, presenting a set of hexagonal pattern. The Raman and TEM studies showed high quality graphene formation on the Ni foil. However, it is observed that there is formation of graphene sheets with different number of layers on polycrystalline Ni foil. The CVD growth process of graphene can be explained from the absorption and successive surface segregation of reactive carbon atoms on metal surface. The solubility of carbon on metal surface increases with temperature and the carbon atoms dissolved in a metal at high temperature can precipitate during cooling process to form graphene. It is observed that the grain boundaries in the polycrystalline Ni foil can serve as nucleation sites for growth of few-layers graphene [13,14]. Again, it is quite difficult to control the number of graphene layers on Ni foil due to high carbon solubility (0.1 at % at around 1000 °C).

Graphene films were also synthesized on low carbon soluble Cu foil (<0.001 at % at around 1000 °C) to have a precise control in number of layers. Fig. 3(a) shows Raman spectra of a single layer graphene detected at various points on the Cu foil. The Raman spectra for the graphene show a sharp sp^2 carbon peak at 1581 cm^{-1}, having a very weak defect related D peak at 1340 cm^{-1}. A strong 2D peak is observed at around 2698 cm^{-1}, corresponding second order Raman spectra. The 2D peak intensity of the single layer graphene is much higher than that of the G peak; whereas for the few-layers graphene,

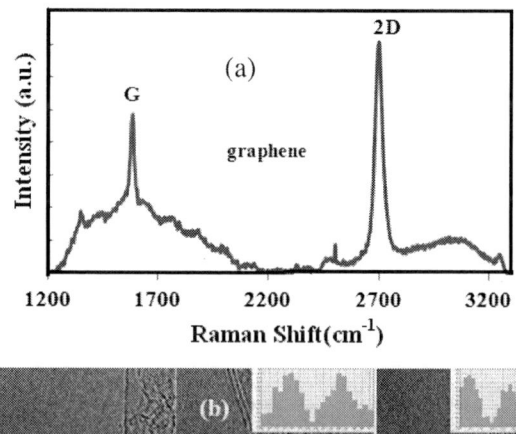

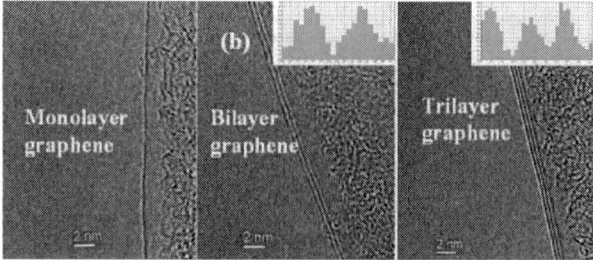

Fig. 3 (a) Raman spectra of a single layers graphene sheet synthesized on Cu foil (b) TEM image of the single-layer, bi-layers and tri-layers graphene synthesized on Cu foil.

978-1-4673-4840-9/13 $31.00 © 2013 IEEE

the 2D peak intensity is lower than that of G peak. The FWHM of G and 2D peaks for the single layer graphene is found to be 18.5 and 38.7 cm^{-1}, respectively [14-15]. HRTEM images of pristine graphene sheets derived from camphor are shown in fig. 2(b). The edges of the suspended film always fold back, allowing for a cross-sectional view of the film. The observation of these edges by TEM provides an accurate way to measure the number of layers. Figure 2(a), (b) and (c) present single, bi layers and tri-layers graphene sheet. The graphene film seems to be predominately single and bi-layers when analyzed using Raman spectra. However, as shown in the TEM study presence of tri-layers and few-layers graphene sheets is also observed. In the inset of the figure an inter-planer spacing of about 0.34 nm for the bi-layers and tri-layers graphene is estimated. It is observed that the presence of 6 member carbon ring in camphor molecules assist in growth of high quality graphene film in a CVD process. Again, camphor pyrolysis on Ni substrates showed formation of few-layers graphene sheets rather than monolayer graphene, this is due to fact that Ni has high carbon solubility. Recently, we demonstrate that with control in deposition condition monolayer or bi-layers graphene along with few-layers graphene can be obtained on Ni foil.[14] The graphene synthesized on Cu foil have much higher percentage of monolayer and bi-layers graphene. The control in the number of graphene layers is an important factor for device applications. These results indicate that the suppression of the carbon dissolution and substrate preferential diffusion or absorption is a key factor for single-layer graphene growth.

Synthesized graphene film on metal foil was transferred to a PET substrate using pre-coated PMMA layer, such that large area thin graphene film does not collapse during the transfer process. The etching process is presented in the schematic diagram as shown in fig. 4. The photograph of a transparent graphene on to the plastic substrate is also presented. In our study, the transferred graphene film on plastic substrate is continuous and optically uniform.

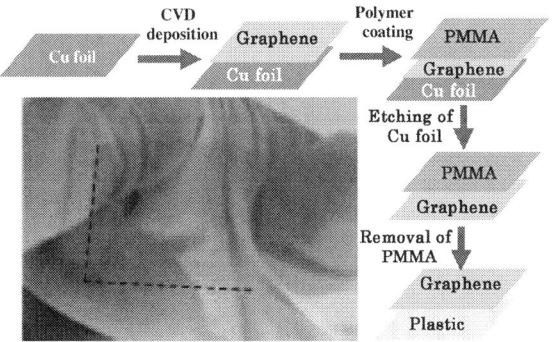

Fig.4 schematic diagram of the transfer process and a photograph of the transparent graphene on plastic substrate.

Fig. 5(a) presents an optical microscope image of the transferred grphene film on plastic substrate. The graphene film shows very good adhesion to the plastic substrate. The fully flexible graphene film on plastic can be excellent material as a smart optical window for different electronic devices.

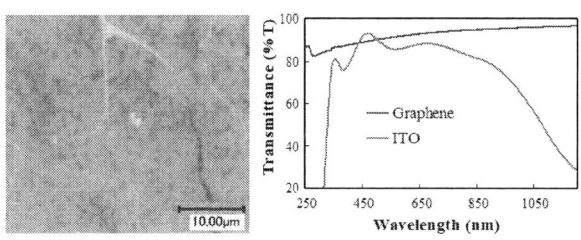

Fig. 5(a) a optical microscope image of the graphene film on the plastic substrate (b) transparency of the graphene film in comparison of a ITO electrode.

Fig. 5(b) shows the transmittance of the graphene film after transferring to a plastic substrate. Transferred graphene sheets showed very good transmittance for a wide wavelength range (0.3-2 μm). In contrast to ITO, which show strong absorptions in the region of near and short-wavelength infrared (0.8-2 μm), the graphene remain transparent in these regions. Transmittance of the graphene sheet with less thickness was achieved as high as 91 % at 550 nm wavelength (in the visible range) as shown in figure 1(b). The sheet resistance of the graphene film with a transmittance of 91.6 % at 550 nm wavelength is obtained as 860 Ω/sq. The sheet resistance of the graphene/PET flexible transparent electrode was measured after bending several times and shows almost similar sheet resistance. Our findings show that excellent large area graphene can be synthesized with a simplified CVD process using solid camphor molecules. The synthesized graphene film can be excellent material for application as a smart optical window and conducting electrode.

Conclusions

We demonstrated synthesis of large area continuous graphene film on Ni and Cu foil using the solid camphor as the carbon source materials. It is observed that the presence of 6 member carbon ring in camphor molecules assist in growth of high quality graphene film in a CVD process. Raman and HRTEM studies show high quality few-layers graphene structure on Ni foil. Single-layer or bi-layers graphene were synthesized on Cu foil. The solubility of carbon on the Ni and Cu foil and growth conditions plays important role in controlling the number of graphene layers. Flexible transparent electrodes were fabricated transferring the graphene film by wet etching process. Continuous and optically uniform graphene film was obtained on plastic substrate by the surface to surface transfer process. Fabricated trnaprent electrode shows sheet resistance of 860 Ω/sq with a transmittance of 91.6 % at 550 nm wavelength. Graphene based transparent electrode can be suitable alternative to ITO or other conventional transparent electrode and optical window based materials.

Acknowledgments

The Authors are grateful to human resource development, Japan for the financial supports as grants-in-aid for science and technology.

References

[1] K.S. Novoselov, et al., *Science*, Vol. 306, pp. 666-669, (2004).

[2] A.K. Geim et al., *Nature mater.*, Vol. 6, pp. 183-191, (2009).

[3] K. S. Kim, et al., *Nature*, Vol. 457, pp. 706-710, (2009).

[4] X. S. Li et al, Science, Vo. 324, pp. 1312-1314. (2009).

[5] G. Kalita, et al., *Mater. Lett.*, Vol. 64, pp. 2180-2184, (2010).

[6] G. Kalita, et al., *J. Mater. Chem.*, Vol. 20, pp. 9713-9717, (2010).

[7] S. P. Somani, *Chem. Phys.Letts.*, Vol. 430, pp. 56-59, (2006).

[8] G. Kalita, et al., *J. Mater. Chem.*, Vol 21, pp. 15209, (2011).

[9] G. Kalita et al., *Physica E*, Vol. 43, pp. 1490-1493, (2011).

[10] G. Kalita, et al. *Mater. Lett.*, Vol 65, pp. 1569-1572, (2011).

[11] R. Hirano et al., *Nanoscale*, DOI:10.1039/C2NR31723K, (2012).

[12] T. Wu, *Adv. Funct. Mater.*, DOI: 10.1002/adfm.201201577, (2012).

[13] Y. Zhang et al., *J. Phys. Chem. Lett.*, Vol. 1, pp. 3101–3107, (2010).

[14] S. Sharma et al., *Mater. Lett.*, *in press*, (2012).

[15] A. C. Ferrari et al., *Phys. Rev. Lett.*, Vol. 97, pp. 187401, (2006).

Nanoscale Mechanical Scratching of Graphene Using Scanning Probe Microscopy

Ryutaro Suda[1], Takanari Saito[1], Ampere A. Tseng[2], and Jun-ichi Shirakashi[1*]

[1]Department of Electrical and Electronic Engineering, Tokyo University of Agriculture & Technology,
2-24-16 Nakacho, Koganei, Tokyo 184-8588, JAPAN.
[2]Department of Mechanical and Aerospace Engineering, Arizona State University, Tempe, Arizona 85287-6106, USA.
*Phone/Fax: +81-042-388-7919, E-mail: shrakash@cc.tuat.ac.jp

Abstract

Nanolithography of graphene surfaces using scanning probe microscopy (SPM) scratching with a diamond-coated tip was systematically investigated. The graphene films were obtained by mechanical exfoliation of pyrolytic graphite sheet (PGS). The groove size increased linearly with the applied force. Furthermore, there were no effects of scan speed and scratch angle on the groove size. These results imply that SPM scratch nanolithography is promising for the fabrication of nanoscale graphene devices.
Keywords: scanning probe microscope, scratch, diamond coated tip, nanolithography and graphene.

Introduction

Graphene has attracted intense interest for a wide range of nanoelectronics applications. This material is made of a single layer of carbon atoms assembled in a hexagonal structure, providing extraordinary electrical and mechanical properties [1, 2]. Especially, the energy gap structures in patterned graphene can be tuned during fabrication with the appropriate choices of geometry [3]. Hence, sophisticated nanolithography techniques are required because of the nanometer-scale dimension of the graphene devices with well-controlled energy gap structures.

Recently, scanning probe microscopes (SPMs) have been known as not only observation tools but also nanolithography ones [4-7]. In particular, the mechanical scratching using SPM is one of powerful methods for the modification of surfaces on the nanometer scale [8-11]. We have already reported that it is possible to perform sub-20-nm scratch nanolithography on Si with a diamond-coated tip by precisely tuning the scan parameters [12, 13]. Furthermore, we fabricated Au quantum point contacts (QPCs) using SPM scratching and controlled quantum states of the QPCs in ambient conditions at room temperature [14]. Hence, it is expected that one can easily fabricate graphene nanostructures by SPM scratching. In this study, we explore the possibility of performing the nanolithography of exfoliated multilayer graphene surfaces using SPM scratching.

Experimental Details

SPM scratching nanolithography was applied to graphene surfaces, as shown in Fig. 1(a). First, graphene films were prepared by using mechanical exfoliation of pyrolytic graphite sheet (PGS), which is commercially available from an industrial materials company [15], and were deposited on Si/SiO2 substrate with approximately 760 nm thermally grown oxide. Fig. 1(b) shows an optical microscopy image of relatively large multilayer graphenes on top of an oxidized Si

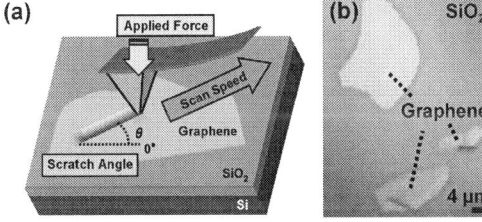

Fig. 1 (a) Schematic of scratching of graphene surface using an SPM. The influence of various scan parameters (applied force, scan speed, and scratch angle) on the groove size was investigated. (b) Optical microscopy image of relatively large graphene films on top of oxidized Si wafer.

substrate. The thickness of the multilayer graphenes was 30-100 nm. As shown in Fig. 2(a), the atomic force microscopy (AFM) image of the graphene film on Si/SiO2 substrate shows that moderately long linear steps (>100 nm) are available at the surfaces. In Fig. 2(b), the step height estimated from the cross-sectional profile, taken along the dashed line in Fig. 2(a), is approximately 0.33 nm. Moreover, the height histogram obtained from the cross-sectional profile yields conspicuous peaks in the histogram. In the figure, the peak-to-peak differences are 0.34 nm and 0.31 nm, which correspond to the layer spacing of graphite (0.335 nm) [17].

The scratch experiments were carried out in ambient conditions, with a commercially available diamond-coated tip having a nominal radius of curvature of 100-200 nm and a cantilever spring constant of 46 N/m. Grooves were patterned by scratching the samples at various applied forces, scan speeds, and scratch angles. The imaging of the surfaces after scratching was typically performed using contact-mode

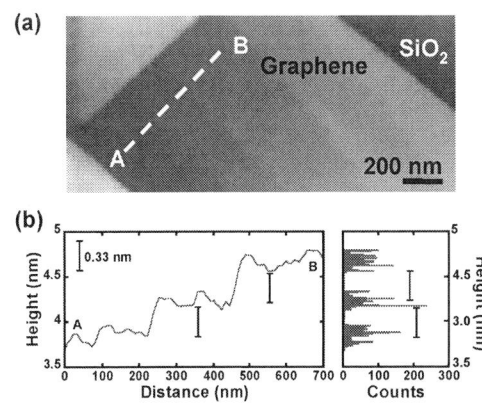

Fig. 2 (a) AFM image of exfoliated multilayer graphene on Si/SiO2 substrate. (b) Cross-sectional profile along the dashed line in the AFM image and the height histogram (right).

978-1-4673-4840-9/13 $31.00 © 2013 IEEE

operation with a Si₃N₄ tip to prevent additional damage to the surfaces. The width and depth of the grooves were measured from the groove profiles over ten points per groove and were then averaged. The line width of the grooves is defined by the full width at half maximum (FWHM).

Results & Discussion

A. Typical Images of Scratched Graphene Surfaces

Fig. 3(a) shows an AFM image of a typical groove on a graphene surface fabricated using SPM scratching. The scratch was performed with the applied force of 13 μN and the scan speed of 10^{-1} μm/s. The groove with the width of 133 nm and the depth of 7.4 nm was obtained. It can be seen that residual pieces of graphene clearly tend to accumulate along the grooves. Then, the same patterned grooves were observed using scanning electron microscopy (SEM), and similar results were obtained, as shown in Fig. 3(b). It is found that the formation of wear debris around/along the groove differs from that of the SPM-scratched materials ever reported [13, 16]. The result may be due to the strong sp² bonds of graphene [1, 18].

B. Influence of Applied Force on Groove Size

AFM image of five scratches generated at various forces (from left to right: 11, 13, 15, 17, and 19 μN) and a scan speed of 10^{-1} μm/s on a graphene surface is shown in Fig. 4(a). Fig. 4(b) shows cross-sectional profiles of the grooves, marked in Fig. 4(a) with white boxes. As expected, the groove size increases with the normal applied force. In Fig. 4(c), the width and the depth of the grooves are plotted as a function of the applied force. The results show that the size of the grooves increases linearly with the applied force and is well controlled. Moreover, grooves with the width of 20-120 nm and the depth of 1-25 nm were obtained with the normal applied forces above 10 μN. However, no grooves were formed with the forces blow 10 μN. The results indicate that the threshold force, which is the minimum applied normal load needed for scratching the graphene surfaces, is about 8-10 μN. In the case of Si, the threshold force for SPM scratching is equal to or slightly larger than 1 μN [16]. Furthermore, nano-indentation study for graphene confirmed the elastic modulus ranging from 0.1 to 0.9 TPa [2], suggesting the strong material like diamond. Thus, as expected, it can be seen that the graphene is a hard material, rather than Si [16]. The result might be explained as a consequence of strong chemical bonds within the layer of graphene [1, 18].

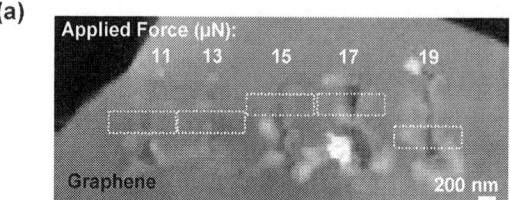

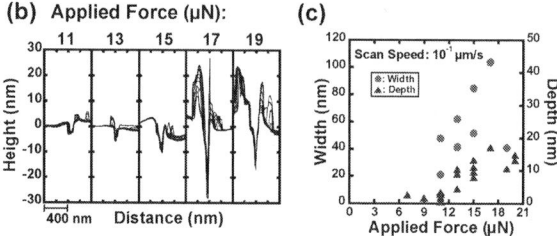

Fig. 4 (a) AFM image of scratched graphene surface. The normal applied force varies from left to right as follows: 11, 13, 15, 17, and 19 μN. (b) Cross-sectional profiles of the grooves, taken from the AFM image indicated by the boxes. (c) Dependence of groove size on applied force.

C. Effect of Scan Speed on Scratched Geometry

To determine the influence of scan speed on the groove size, grooves were generated at scan speeds ranging from 10^{-2} to 10^1 μm/s and an applied force of 15 μN, as shown in Fig. 5(a). From the cross-sectional profiles of the grooves (Fig. 5(b)) and relation between scan speed and groove size (Fig. 5(c)), the width and depth of the grooves remain almost constant as the scan speed is varied. Previously, similar results were also observed in other materials, such as Al [9], Si [13], and NiFe [16]. Therefore, it is concluded that high-speed nanolithography can be achieved without degradation of the graphene patterns by SPM scratching.

D. Dependence of Groove Size on Scratch Angle

Fig. 6(a) shows the grooves scratched by an applied force of

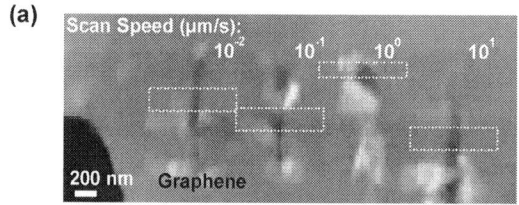

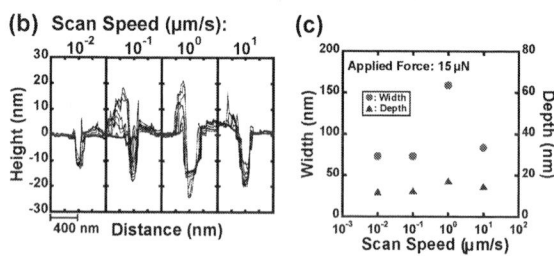

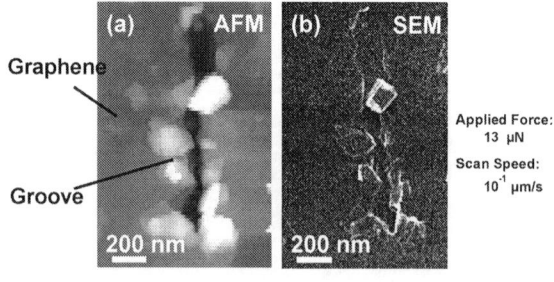

Fig. 3 (a) AFM and (b) SEM images of a groove fabricated by SPM scratching on graphene surface.

Fig. 5 (a) AFM image of scratched graphene surface. The scan speed varies from left to right as follows: 10^{-2}, 10^{-1}, 10^0, and 10^1 μm/s. (b) Cross-sectional profiles of the grooves, taken from the AFM image indicated by the boxes. (c) Dependence of groove size on scan speed.

978-1-4673-4840-9/13 $31.00 © 2013 IEEE

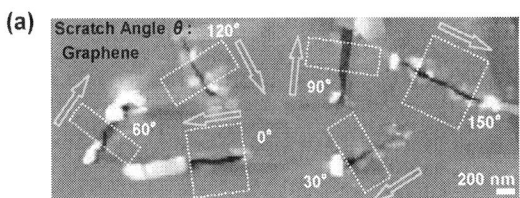

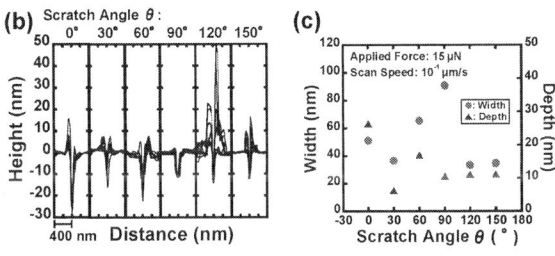

Fig. 6 (a) AFM image of graphene scratched in different directions. (b) Cross-sectional profiles of the grooves, taken from the AFM image indicated by the boxes. (c) Dependence of groove size on scratch angle.

15 μN in the different directions: 0°, 30°, 60°, 90°, 120°, and 150°. The scratch angle is defined with respect to the horizontal line. As shown in Figs. 6(b) and 6(c), no clear relationship is found between the groove size and the scratch angle. Since the SPM scratching is based on the directly controlled mechanical modification, honeycomb crystal structure of graphene might not affect the scratch properties, such as the width and depth of the fabricated grooves. Hence, it is indicated that more complex graphene nanostructures could be easily obtained by the mechanical patterning using SPM.

Conclusions

SPM scratching with a diamond-coated tip on a graphene surface was performed. The dimensional characteristics of the grooves formed were studied as functions of the applied force, scan speed, and scratch angle. Then, groove size was well controlled by the applied normal force. Moreover, based on the experimental results, the threshold force of graphene should be about 8-10 μN. In addition, the size of the grooves no longer depends on the scan speed and the scratch angle. It is suggested that one can easily and quickly obtain more complicated graphene nanostructures using the SPM scratching. These results imply that the SPM scratching could be a key technique for the fabrication of nanoscale graphene devices.

References

[1] A. K. Geim and K. S. Novoselov, *Nat. Mater.*, 6, pp. 183-191, March 2007.

[2] Y. Zhang and C. Pan, *Diamond Relat. Mater.*, 24, pp. 1-5, April 2012.

[3] Z. Moktadir et al., *Electron. Lett.*, 47, pp. 199-200, February 2011.

[4] A. A. Tseng, *J. Vac. Sci. Technol. B*, 23, pp. 877-894, May 2005.

[5] J. Shirakashi, *J. Nanosci. Nanotechnol.*, 10, pp. 4486-4494, July 2010.

[6] S. Nishimura, et al., *Jpn. J. Appl. Phys.*, 47, pp.718-720, January 2008.

[7] T. Toyofuku, et al., *J. Nanosci. Nanotechnol.*, 10, pp. 4543-4547, July 2010.

[8] Y. Kim and C. M. Lieber, *Science*, 257, pp. 375-377, July 1992.

[9] T. H. Fang, et al., *Nanotechnology*, 11, pp. 181-187, June 2000.

[10] C. K. Hyon, et al., *Jpn. J. Appl. Phys.*, 38, pp. 7257-7259 , December 1999.

[11] L. Santinacci, et al., *Appl. Phys. Lett.*, 79, pp. 1882-1884, July 2001.

[12] T. Ogino, et al., *Jpn. J. Appl. Phys.*, 46, pp. 6908-6910, October 2007.

[13] T. Ogino et al., *Jpn. J. Appl. Phys.*, 47, pp. 712-714, January 2008.

[14] R. Suda, et al., *IEEE-NANO 2012*, pp. 1-5, 20-23 August 2012.

[15] Industrial Devices Company, Panasonic Corporation. (2012) http://panasonic.net/id/jp/ Cited 2012

[16] A. A. Tseng, et al., *J. Appl. Phys.*, 106, pp.044314-1-8, August 2009.

[17] Y. Niimi et al., *Phys. Rev. B*, 73, pp. 085421-1-8, February 2006.

[18] S. Marchini, et al., *Phys. Rev. B*, 76, pp. 075429-1-9, August 2007.

Retention Behavior of Graphene Oxide Resistive Switching Memory on Flexible Substrate

Fang Yuan[1,2], Yu-Ren Ye[2], Jer-Chyi Wang[2], Zhigang Zhang[1], Liyang Pan[1], Jun Xu[1], Chao-Sung Lai[2]

1. Institute of Microelectronics, Tsinghua University, Beijing 100084, China
2. Department of Electronic Engineering, Chang Gung University, Kweishan, 333 Taoyuan, Taiwan
Tel: +86-10-62789252 Fax: +86-10-62771130 E-mail: yuanf09@mails.tsinghua.edu.cn

Abstract

This work presents a flexible carbon based memory with the Al/graphene oxide (GO)/ITO structure fabricated at room temperature. The Al/GO/ITO devices show the unipolar resistive switching behavior with the resistance ratio to over 30, and sustain over 250 cycling without any resistance window closure. However, the retention fails due to the resistance increase of low resistance state (LRS). The mechanisms of switching and retention failure are studied.

Keywords: graphene oxide, resistive switching, RRAM, unipolar, retention

Introduction

A variety of metal oxide thin films including the transition metal oxide and rare earth materials have been discovered for the resistive switching memories (ReRAM) applications. The metal oxide based ReRAM can be operated with low power consumption, high speed and good reliability. Recently, the graphene oxide (GO) has been proposed as potential resistive switching material on flexible substrate [1-3]. Nevertheless, the reliabilities such as the retention and endurance of the GO ReRAM were not yet thoroughly discussed. In this work, the unipolar resistive switching of GO ReRAM on flexible substrate was reported and the retention properties as well as the failure mechanisms were studied.

Device Fabrication

The Al/graphene oxide (GO)/ Indium Tin Oxides (ITO) MIM (metal-insulator-metal) ReRAM devices were fabricated at room temperature on the flexible substrate of polyethylene terephthalate (PET). After O_2 plasma treatment to modify the surface, the GO dispersion which made by modified Hummer's method was drop-cast onto the ITO bottom electrode and followed by 12 hours vacuum evaporation. The GO film can be viewed as stacks of single-layered graphene oxide flakes. Finally, the Al film as top electrode was deposited and patterned by circular shadow mask with diameter of 200 μm. The schematic device structure was shown in inset of Fig. 1.

Measurements

The current-voltage (I-V) and resistive switching characteristics of the fabricated ReRAM devices were obtained using a Agilent B1500A and Keithley 4200 semiconductor parameter analyzer with biased top and grounded bottom electrodes at room temperature.

Results and Discussion

Typical unipolar I-V characteristics of the GO ReRAM are shown in Fig. 1. Form this figure, the SET and RESET processes can be performed at around +5V and +3V respectively. Fig. 2 shows the I-V curves of the device in a double logarithmic scale in SET and RESET memory operations respectively. It has been widely reported and well-studied that the ohmic conduction relationship is described by $I(V) = \alpha V$ and the space charge limited current (SCLC) is by $I(V) = \alpha V + \beta V^2$ for the MIM ReRAM devices [4]. The values shown in Fig. 2 display the slope of the $log\,I - log\,V$ curve. For the low resistance state (LRS) and low bias of the high resistance state (HRS), the slope is almost 1, indicating the ohmic conduction mechanism. When the applied voltage increases, the SCLC plays a more important role. As shown in Fig. 2, for the high bias of HRS, the slope is 1.63, meaning that both ohmic conduction and SCLC will contribute to the carrier transportation.

Fig. 3 shows the endurance reliability of the GO devices which can be repeatedly operated over 250 cycles. No or little deterioration of the HRS/LRS resistance ratio window is found during the switching cycles, which performs better than most of the other groups. Fig. 4 shows the retention properties. However, the retention fails at 5000s due to the sharp resistance increase of LRS, which is firstly observed and reported in the GO based resistive memory. To further investigate the retention behavior of the GO ReRAM, various read interval times (read at 0.1V at every 100s, 250s, 500s and 1000s) and read voltages (read at 0.1V, 0.2V and 0.5V) were conducted and the results are presented in Fig. 5(a) and (b) respectively. It can be observed that the LRS of GO ReRAM can preserve for longer time when the read interval time decreases and voltage increases.

Y. Li and K. S. Vasu et al. have proposed that the unipolar switching of GO film is resulted from the NEM (Nano Electro-Mechanical) effect [5-6]. The switching mechanism of the Al/GO/ITO ReRAM device is schematically illustrated in Fig. 6. Initially, the GO flakes are stacked but not joined all together, therefore, the resistance is high. After the forming process, the voltage will induce some Al ions into the films bulk, and the GO flakes join at the edge by Al ions and also by electrostatic attraction, which turns the ReRAM into LRS. The RESET process can break the inter-flaked joints by voltage-induced damage [5], while the SET process can enable flakes rejoin through electrostatic attraction because the broken sheets remain very close to each other. Unfortunately, in the retention measurements of LRS, the flakes would separate and lead the GO film into HRS. The increased read duration and voltage can keep the GO flakes joint by Al ions and the retention characteristics are significantly improved.

Conclusion

The Al/GO/ITO ReRAM device has stable unipolar switching properties and the switching mechanism is dominated by NEM effect. The superior endurance was obtained and the retention was failed owing to the resistance increase of LRS. The increased read duration and voltage can significantly improve the retention which can be ascribed to the GO flakes joint by Al ions.

Acknowledgements

This work was supported (in part) by the State Key Development Program for Basic Research of China (No. 2011CBA00602), and the National Natural Science Foundation of China (No. 20111300789).

References

[1] C. L. He, et al., *Appl. Phys. Lett.*, 95, 232101 (2009).
[2] S. K. Hong, et al., IEEE Nanoelectronics Conf. (INEC), 2011.
[3] Z. Wang, et al., *Journal of The Electrochemical Society*, 159 (6) K177-K182 (2012).
[4] A. Rose, *Phys. Rev.*, 97(6), 1538 (1955).
[5] Y. Li, et al., *Nat. Mater.*, 7, 966 (2008).
[6] K. S. Vasu, et al., *Solid State Communications*, 151, 1084–1087 (2011).

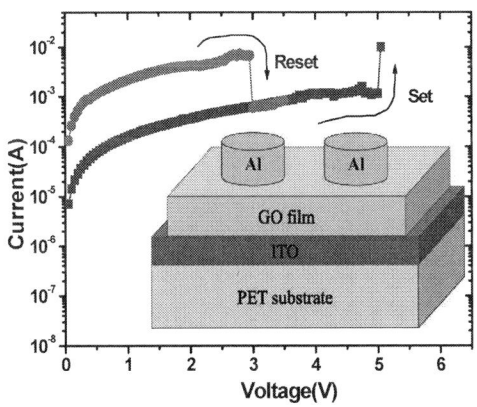

Fig. 1 Schematic of Al/GO/ITO ReRAM device on PET substrate and the typical current versus voltage (I-V) characteristics.

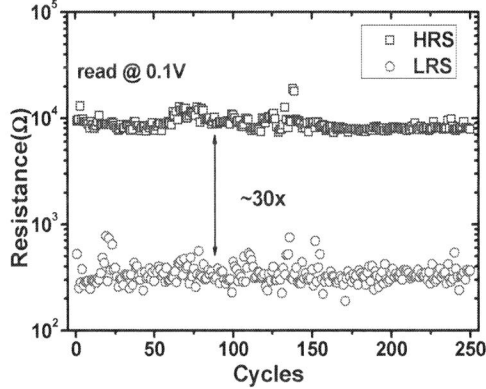

Fig. 3 The endurance properties of the device.

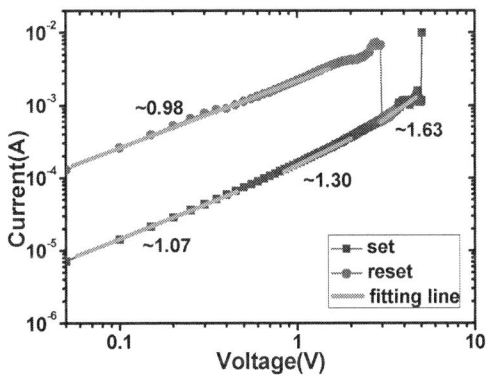

Fig. 2 The I-V curves in a double logarithmic scale. The indicated numbers are the slope value of *log I - log V* curve.

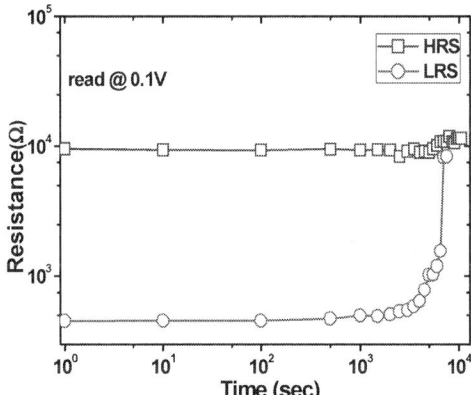

Fig. 4 The retention properties of the device.

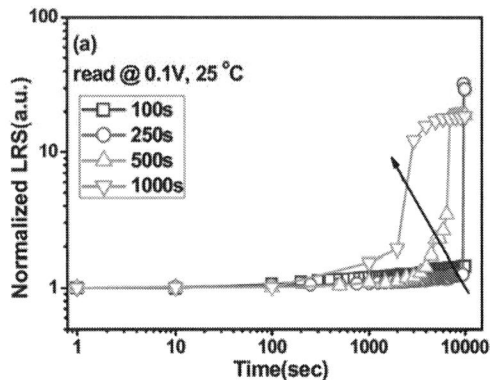

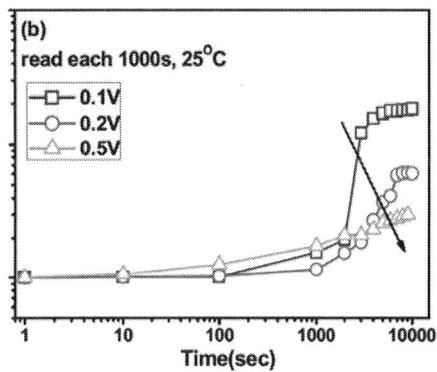

Fig. 5 The retention behaviors of the device: (a) read at shorter interval times and (b) with larger read voltages lead LRS maintain longer time.

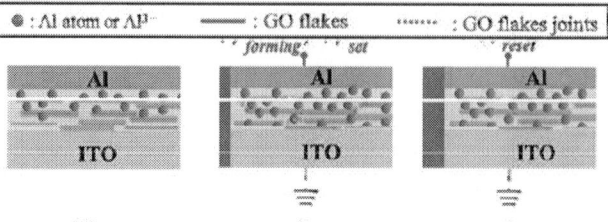

Fig. 6 Schematic of the proposed mechanisms for Al/GO/ITO device. (a) As fabricated. (b) Forming process and SET process. (c) RESET process.

A Zigbee-Based Wireless Wearable Electronic Nose Using Flexible Printed Sensor Array

Panida Lorwongtragool[1], Reinhard R. Baumann[2,3,*], Enrico Sowade[2], Natthapol Watthanawisuth[4], Teerakiat Kerdcharoen[5,*]

[1] Faculty of Science and Technology, Rajamangala University of Technology Suvarnabhumi, Nonthaburi, Thailand
[2] Department of Digital Printing and Imaging Technology, Institute for Print and Media Technology, Chemnitz University of Technology, Chemnitz, Germany
[3] Printed Functionalities, Fraunhofer Institute for Electronic Nano Systems (ENAS), Chemnitz, Germany
[4] Nanoelectronic and MEMS Lab National Electronic and Computer Technology Center Pathumthani, Thailand
[5] Department of Physics and NANOTEC's Center of Excellence, Faculty of science, Mahidol University, Bangkok, Thailand
*Corresponding author: teerakiat.ker@mahidol.ac.th

Abstract

A wearable electronic nose (e-nose) has been developed by integrating a low cost chemical sensor array with a wireless communication for applications in healthcare. Its sensing unit was fabricated by a fully inkjet-printing technique, comprising eight different sensor elements manufactured by varying printing patterns and sensing materials. These sensors have shown response to a wide variety of complex odors. A wearable e-nose prototype using Zigbee wireless technology was designed as a compact armband for monitoring the axillary odor released from human body. Preliminary results based on principal component analysis (PCA) could classify different odors released from the human body upon various activities.

Keywords: electronic nose, wearable device, inkjet printing, Zigbee, body odor, healthcare monitoring, flexible sensor

Introduction

Nowadays, the development of wearable systems has advanced rapidly due to their promising applications in the fields of sports, security and healthcare. The key requirements to design the hardware components in wearable devices are usually addressed in terms of compact and flexible, low-weight, low-power and low-cost systems that continuously monitor environmental conditions and communicate with wireless technology [1-3]. Specifically, wearable e-nose can directly collect the data of complex VOCs as emitted from the human body. The approach thus allows monitoring of disorder conditions or health status as well as body hygiene of individuals [4-5]. Moreover, the device offers not only routine monitoring of health status but it is also suitable for home-based point-of-care diagnostics.

One significant factor concerning reliability of a wearable e-nose is the design of the sensing unit. In this work, we have manufactured a chemical sensor array based on polymer/CNT nanocomposites using inkjet printing technology. Inkjet printing as direct-writing technology enables the precise deposition of the designed patterns on a desired area. . With the same materials, distinct sensors can be produced by printing in different patterns to achieve the different responses [6-7]. The fabricated chemical sensor array has been installed in the sensing unit of the wearable e-nose and also applied to detect a primary odor source from human body, i.e. the odor around the axillary skin. A Zigbee module based on wireless technology was used as an appropriate platform to transfer the data from the wearable e-nose to a computer due to a compact and low-cost component.

This work proposes a novel system of wearable e-nose for healthcare application using a sensing unit based on a fully inkjet-printing technique. Preliminary results have showed the success of classify different odors released from the body of volunteer upon various activities.

Wearable Electronic Nose System

A. Flexible Printed Sensor Array

The sensor array comprises eight different elements produced by varying printed patterns and sensing materials is shown in detail in Table 1. These printed chemical gas sensors are based on polymer/multi-walled carbon nanotubes (MWCNTs) nanocomposites. Polyethylene naphthalate (PEN film, Dupont Teonex Q56FA) film was chosen as substrate material due to its flexible property, smooth surface and heat stability. In this work, the ink depositions were performed by the printing systems Dimatix Materials Printer 2831 (DMP) and the Autodrop micro dispensing system (Microdrop Technologies). Silver ink was inkjet-printed as an array of silver interdigitated electrodes (SIDEs) on top of the substrate then followed by printing the sensing layers. We designed two different patterns that are double printed layers (DPLs) and blended single layer (BSL) for obtaining different response layers [6-7]. The structures of the printed thin films are shown in Fig.1. Five repeating printings of the both DPLs and BSL were performed to increase the amount of sensing layers and therefore to improve the efficiency of the sensing responses.

TABLE I
DETAILS OF AN INDIVIDUAL ELEMENT IN A PRINTED SENSOR ARRAY

Sensor No.	Sensing Material	Printing Pattern
1	PVC/MWCNTs	DPLs
2	Cumene-PSMA/MWCNTs	DPLs
3	PSE/MWCNTs	DPLs
4	PVP/MWCNTs	DPLs
5	PVC/MWCNTs	BSL
6	Cumene-PSMA/MWCNTs	BSL
7	PSE/MWCNTs	BSL
8	PVP/MWCNTs	BSL

* PVC: Polyvinyl chloride
Cumene-PSMA: Cumene terminated polystyrene-co-maleic anhydride
PSE: Poly(styrene-co-maleic acid) partial isobutyl/methyl mixed ester
PVP: Polyvinylpyrrolidon

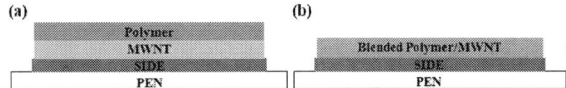

Fig.1 Two different printed architectures for sensing elements: (a) double printed layers and (b) blended single layer

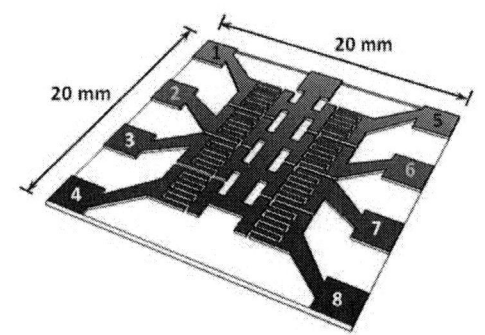

Fig.2 Three dimensional structure of the designed SIDE array; sensor numbers refer to the sensing material and printed pattern as shown in table 1

Fig.2 shows the three dimensional structure of the designed SIDE array. It consists of eight SIDEs for deposition of eight different thin films as shown in Table 1. The fingers of each SIDE were defined for the width and separation distance of 200 μm. The size of the array is about 20 mm x 20 mm. The sensing materials were inkjet-printed on top of each SIDE as square in an area of 3 mm x 3 mm. Silver ink dispensing parameters and further information about the ink formulations and printing systems can be found in our previous work [6-7].

B. Sensor Testing

We have tested the fabricated sensor array in a static chamber to observe the electrical responses when exposed to individual volatile organic compounds (VOCs). A measurement circuit was performed using voltage divider method to obtain the resistances of each sensor. The output voltages of eight elements were acquired through an 8-channel analog multiplexer connected to a USB DAQ device. The Ohm's law was employed to calculate the individual sensor resistances. The period of reference baseline was defined during the first 2 min without exposure to any VOCs (ambient condition). In the second period, the analyte was injected into the static chamber and the output signals were recorded for 7 min to obtain a steady-state condition. The fractional method was applied on the determined sensor resistances to obtain the sensor responses [8].

In this work, the printed gas sensor array was exposed to VOCs of ammonia, acetic acid, acetone and ethanol.

C. Monitoring of Axillary Odor

The flexible printed sensor array was integrated into the wearable e-nose prototype, which was designed as a compact armband appropriate for monitoring the axillary odor as shown in Fig. 3. Therefore, body odor could be directly collected from the armpit region. The odor data was recorded in terms of sensor resistances as function of the presented VOCs and their concentrations.

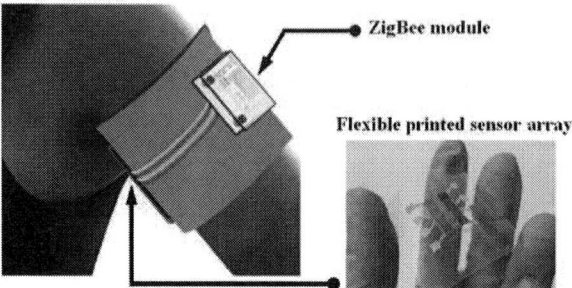

Fig. 3 Prototype of a wearable e-nose, the sensors is based on inkjet printing and the communication is based on Zigbee technology

Based on a preliminary study, we have collected odor data from the armpits (left (L) and right (R) sides) of a volunteer (33 years-old healthy subject) using the constructed wearable device. Body odors were monitored based on the following activities of the volunteer as simulated in three stages: (i) before exercise (normal stage and used as reference), (ii) after exercise and (iii) after exercise and relaxing for a long time.

Before the experiment was performed, the reference baseline was obtained by exposing the sensor to the ambient air. Then the volunteer wore the device and the data was collected for 5 min. The experiments in each stage were carried out for three times. The sensors were recovered to the reference baseline after each experiment before the next experiment started (by exposure of the sensor to the ambient air). We have used principal component analysis (PCA) based on the unsupervised learning method to observe the odor progression in every minute. For one state, we obtained 15 data points (5 points x 3 repeating times) on the space of the first two principal components.

Results and Discussion

According to different printed sensing layers as shown in Table 1, we have found that the sensor responses of the fabricated chemical sensors yield distinguishable patterns when exposed to the VOCs. Fig. 4 shows the electrical responses of the sensors towards ammonia, acetic acid, acetone and ethanol that are mainly presented in the complex volatiles released from axillary skin [9]. Therefore, based on these characteristics the designed wearable e-nose is considered as suitable system to investigate the VOC fingerprints of human odors.

The results also demonstrate an advantage of the inkjet-printing technology in order to fabricate the alternative chemical sensors, which provide a unique pattern for each odorant based on using the same sensing materials. They support our previous work about inkjet-printed chemical sensors based on CNT/PSE nanocomposites [6]. The methods to produce the different sensing elements were proposed in the same way, i.e. DPL and BSL. However, in this work we have proven the operation for other polymers and used these elements in the sensing unit of the e-nose for real applications. In addition, we also improved the sensor responses by printing multi-layers to increase the amount of sensing material as suggested in the previous work.

978-1-4673-4840-9/13 $31.00 © 2013 IEEE

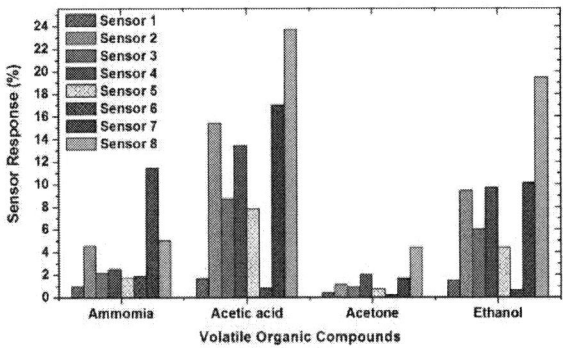

Fig. 4 Percent sensor response of eight sensor elements in a static system when exposed to volatiles of ammonia, acetic acid, acetone and ethanol with a concentration of 200 ppm

The sensing unit of this prototype is based on a flexible substrate, thereby posing an advantage for wearable device. Moreover, this feature could be considered as an important step to enable roll-to-roll processing in order to scale up the productivity in the future.

Fig. 5 shows the classification of the armpit odors under the simulated activities using PCA technique. The results can be observed by the variations of principal component scores of all three stages. Odor progression after exercise was monitored and it was found that the data groups were moved out from a normal stage (reference). In addition, we can observe recovering of underarm odor after relaxing of the volunteer. This result provides a straightforward understanding about the relationship between the observed signals and changing of the human odors related to different activities. In other words, the wearable e-nose could be used to evaluate unusual odors of body resulted of health status, i.e., the level of skin hygiene and unusual odors from ailments or stress.

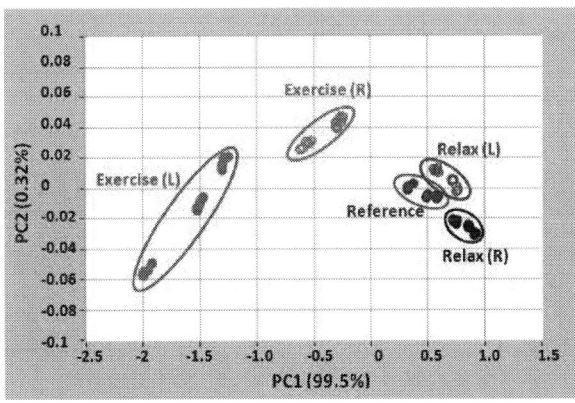

Fig. 5 PCA scores plot of underarm odor in different activities

Conclusions

The Zigbee-based wireless wearable e-nose was developed using a sensing unit based on a flexible, inkjet-printed sensor array. Eight different sensor elements can respond and generate fingerprints of various VOCs due to different printed patterns and various sensing materials. Classifications based on PCA of armpit odors of the volunteer using the wearable

e-nose can provide a qualitative correlation of the amount of generated VOCs due to the activities in comparison to the reference stage.

Acknowledgments
This work was supported by Rajamangala University of Technology Suvarnabhumi and Mahidol University. A research career development grant from National Nanotechnology Center to TK is acknowledged (Project No. P-12-01157).

References
[1] T. Yilmaz, R. Foster, Y. Hao, "Detecting Vital Signs with WearableWireless Sensors" *Sensors* 2010, 10, pp.10837-10862.
[2] A. D. Wilson and M. Baietto, "Advances in Electronic-Nose Technologies Developed for Biomedical Applications", *Sensors* 2011, 11, pp.1105-1176.
[3] Winston H. Wu, A. A.T. Bui, M. A. Batalin, L. K. Au, J. D. Binney, W. J. Kaiser, "MEDIC: Medical embedded device for individualized care", *Artificial Intelligence in Medicine* 2008 42, pp.137—152.
[4] A. D. Wilson (2011). Future Applications of Electronic-Nose Technologies in Healthcare and Biomedicine, Wide Spectra of Quality Control, Isin Akyar (Ed.), ISBN: 978-953-307-683-6, InTech.
[5] C. Wongchoosuk, M. Lutz, T. Kerdcharoen, "Detection and Classification of Human Body Odor Using an Electronic Nose", *Sensors. 9,* 2009, pp.7234-7249.
[6] P. Lorwongtragool, T. Kerdcharoen, R. R. Baumann, " All Inkjet-Printed Chemical Gas Sensors Based on CNT/Polymer Nanocomposites: Comparison between Double Printed Layers and Blended Single Layer" in Proceedings of 9th ECTI-CON2012, May 2012, Hua Hin, THAILAND.
[7] P. Lorwongtragool, E.Sowade, T. Kerdcharoen, R. R. Baumann, "Inkjet printing of chemiresistive sensors for the detection of volatile organic compounds" in Proceedings of NIP28 – 28th International Conference on Digital Printing Technologies / Digital Fabrication September 9-13, 2012, Quebec, Canada.
[8] T.C. Pearce, S.S. Schiffman, H.T. Nagle, and J.W. Gardner, *Handbook of Machine Olfaction; Electronic Nose Technology*, Wiley-VCH, 2002.
[9] S. K. Pandey, K.-H. Kim, "Human body-odor components and their determination", *Trends in Analytical Chemistry*, 30 (5), 2011, p.784-796.

Nonlinear Modeling of Compliant Mechanism

Raisuddin Khan, Md. Masum Billah and Mitsuru Watanabe

Department of Mechatronics Engineering, Faculty of Engineering
International Islamic University Malaysia
53100 Kuala Lumpur, Malaysia
Phone: +603 6196 4575 and Fax:+603 6196 3344 and E-mail: masum.uia@gmail.com

Abstract

In this paper, feasibility of using pseudo-rigid-body modeling technique in examined on a simple modular compliant mechanism. A mathematical model is constructed to determine the non-linear behavior of large-deflections. The problem is solved using an iterative method. The solution to the non-linear model is then compared to empirical results collected from a mock-up of the compliant device. The pseudo-rigid-body model is then combined with a linear model to develop the complete solution, which shows behavior fairly close to the actual behavior of the mock-up.

Keywords: Compliant, nonlinear, modular.

Introduction

A compliant mechanism can be defined as single piece flexible structure, which uses elastic deformation to achieve force and motion transmission [1-2]. In recent years compliant mechanisms have been exhibiting great promise, as simpler and often better solutions to designs of mechanical systems are achieved by flexible bodies. It gains some or all of its motion from the relative flexibility of its members rather than from rigid body joints alone [3]. Such mechanism, with built-in flexible segments, is simpler and replaces multiple rigid parts, pin joints and add-on springs. Hence, it can often save space and reduce costs of parts, materials and assembly labor. Other possible benefits of designing compliance into devices may be reductions in weight, friction, noise, wear, backlash and importantly, maintenance. Much research has been done in the field of compliant mechanism, as its application has already grown substantially, particularly in the area of microelectromechanical systems (MEMS). However, compliant mechanisms are not limited only to MEMS applications, novel uses of compliant mechanism in the macroscopic world is also gaining ground. Especially in the area of robotics, compliant mechanisms are increasingly incorporated. In robotic applications the issue of kinematics coupled with elastic behavior presents an interesting problem. With roots in the concept of flexible linkages [4], first approach to compliant mechanism design was based on kinematic synthesis converting analogous rigid-body mechanisms into compliant mechanisms [5]. This approach, however, is limited in its capability to produce only lumped compliance such as flexible hinges [6]. A second major approach is a continuum synthesis using design techniques originally developed for structural optimization. A favorite avenue is to use the topology optimization methods rooted with the homogenization theory of mechanics and flourished in the past decade with various variations and improvements [7].

Due to the vast use of computers, the finite element method is perhaps the most common approach to investigating compliant systems, but Howell [8] examines other current techniques used in compliant structure investigations. One of the useful methods outlined by Howell has been used in this research. The propose design is simple, back-to-basics design for a modular large-deflection mechanism and analyze it using the pseudo-rigid-body modeling technique. Howell also used the very pseudo-rigid-body model to examine compliant MEMS structures, but tackled the problem from a mathematical approach through the use of both open-loop and closed-loop equations. Our method, however, utilizes a numerical approach where the problem is solved through an iterative process in the environment of MATLAB. Using this method, the achieved convergence over a very short number of iterations, and the complete model is able to accurately model the true behavior of compliant structures.

Proposed Structural Model

The model itself that has designed is indeed an extremely simple one. The design process was begun with several variations and proposals. Most of the designs we initially developed seemed too specific; that is, only applicable for the particular task in question. Hence, the developing modular design would provide the building blocks for more complicated compliant structures. In fact, a simpler design is easy to fabricate and mass produce, and hence a more attractive alternative. The design is based on the cross-section of an I-beam as shown in Fig. 1, where the upper and lower portions of the "I" would be relatively thick to maintain rigidity while the central vertical column would be much thinner to enable compliance. From the upper platform to the lower, an actuator would be added. In the MEMS application this could possibly be replaced by an electrostatic comb actuator. Furthermore, the single unit of the mechanism can be linked to form a chain to achieve a wider range of motion, especially for manipulator application. The modular units can also be linked in an orthogonal configuration such that two or more units can be put together to traverse a surface in Cartesian space.

978-1-4673-4840-9/13 $31.00 © 2013 IEEE

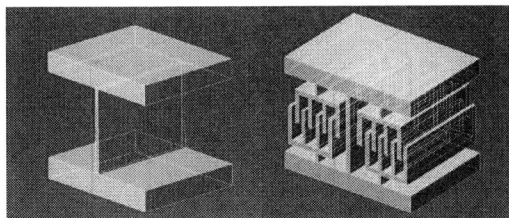

Fig. 1 The "I" shaped design and a possible MEMS application with comb actuators

In the macro model mock-up, shown in Fig. 2, a cable actuation has used to examine the behavior of the structure. The mock-up was constructed using wood and a steel blade. The steel blade normally used in a saw acted as the central compliant member, while a wooden structure stood in for the platforms. A copper wire was used to pull the upper platform down to the lower, mimicking the action of an actuator.

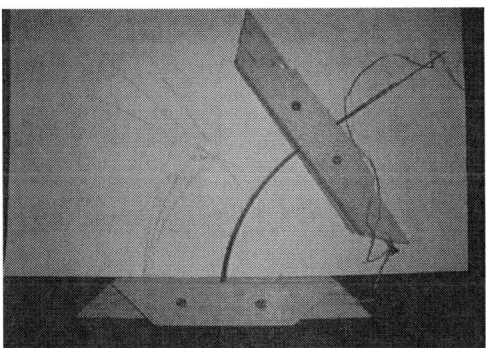

Fig. 2 Mock-up structure used for comparing results

Mathematical Model

The first behavior about the compliant mechanism is that the bending configuration is symmetrical both at small and large deflections. For small deflections of the beam, often a linear relationship between angle, deflection, and force is enough to model the behavior, but with large deflections the problem clearly becomes nonlinear. For such nonlinear systems, finite element analysis is normally conducted. However, in our model we have opted to use the pseudo-rigid-body modeling technique as outlined by Howell [8].

To begin, the free body diagram set up and to determine what loads and moments are acting upon the beam. The free body diagram of the model is shown in Fig. 3. Due to its symmetric nature it will orient the beam sideways so the left side would be symmetric to the right.

Fig. 3 Free body diagram

The beam, therefore, is undergoing both compressive force as well as a moment. The forces on both sides will be the same, which will be exerted by the tension of the spring, while the moment will also be equivalent on both sides. This moment is determined by multiplying the tension of the cable by the perpendicular distance from the line of force to the end of the beam. The pseudo-rigid-body equivalent model for this configuration therefore is created by attaching three rigid beams with two torsional springs and is shown in Fig. 4.

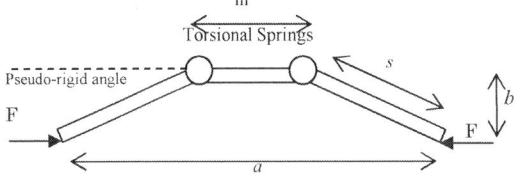

Fig. 4 Pseudo-rigid-body model

In the resulting scheme a few key dimensions must be noted. The variable a denotes the distance from end to end while b denotes the deflection from the horizontal. The pseudo-rigid angle is that formed between the end links and the horizontal. This angle is related to the true angular deflection of the ends by a factor of 1.26 [9].

The motivation then of using this pseudo-rigid-body is to find the horizontal and vertical deflections of the beam given a desired angle between the two platforms. That is, going back to the original upright diagram, the model should be able to determine the x-y deflection the center of the upper platform that will result in a desired angular deflection of that same platform. Likewise, the model should also determine the amount of force that shall achieve such a configuration.

To start the analysis, the contributions of moment and force must be separated. Moments cause a uniform circular curvature when applied to a beam, and hence we know the moment component will create a particular curvature in the beam. This curvature is then referred to as the initial curvature, and the system is treated as if a force is acting upon an initially curved beam [8]. However, since the breakdown of the moment and force aspects are not known, an initial guess is made regarding the initial curve. From this initial guess of the curve, κ_o, both guesses for the force and initial positions can be made.

$$F_o = \frac{2\kappa_o EI}{wl\sin\frac{\theta}{2}} \qquad (1)$$

where l = original length of compliant beam, w = length of the platform, and θ = desired angular deflection of the platform.

$$a_i = \frac{2l}{\kappa_o}\sin\frac{\kappa_o}{2} \qquad (2)$$

$$b_i = \frac{l}{\kappa_o}\left(1-\cos\frac{\kappa_o}{2}\right) \qquad (3)$$

The pseudo-rigid angle associated with the initial curve is then given by

$$\Theta_i = \tan^{-1}\left(\frac{2b_i}{a_i - l(1-\gamma)}\right) \qquad (4)$$

The parameter γ is determined by the value of the initial curvature [8], and is used also to determine the characteristic length m.

$$m = (1 - \gamma)l \qquad (5)$$

The second parameter, ρ, is determined by the initial positions and parameter γ, which in turn determines characteristic length s.

$$\rho = \left\{ \left[\frac{a_i}{l} - (1 - \gamma) \right]^2 + \left(\frac{2b_i}{l} \right)^2 \right\}^{\frac{1}{2}} \qquad (6)$$

$$s = \rho l \qquad (7)$$

Using these values, the final pseudo-positions are determined.

$$a = 2l(1 - \gamma + \rho \cos \Theta) \qquad (8)$$
$$b = \rho l \sin \Theta \qquad (9)$$

As mentioned earlier the pseudo-rigid angle is related to the true angle by a parameter know as c, the parametric angle coefficient, which in this case is 1.26 [8].

$$\theta = c\Theta \qquad (10)$$

From here we continue to solve the problem using the stiffness coefficient, which is in fact a behavior dependent upon the initial curve [8].

$$K_\Theta = 2.568 - 0.028\kappa_o + 0.137\kappa_o^2 \qquad (11)$$

The effective spring constants of the two springs are therefore determined from the stiffness coefficient.

$$K = 2\rho K_\Theta \frac{EI}{l} \qquad (12)$$

Finally, from this, the effective force and the maximum stress are determined.

$$F = \frac{K(\Theta - \Theta_i)}{b} \qquad (13)$$

$$\sigma_{max} = \pm \frac{Fbc}{I} - \frac{F}{A} \qquad (14)$$

In this particular problem with a beam of rectangular cross-section, A would be the cross-sectional area, while c would be half of the cross-sectional beam height.

One will note, however, that we have determined two values for the same force, Fo and F, and herein lies the solution. As above, the first value is the initial guess regarding the force that ought to cause the moment responsible for initial curvature, and from this initial force to calculate the force taking into account all components. In other words, an initial force that will equal the force determined by the full equation, then that would be the solution. That is, for a given angle of deflection then, it should be able to find the horizontal and vertical deflections and the required force. By simply inputting the angle of deflection, the x-y position of the center of the upper platform was found.

Solution of the Model

The iterative technique was used for this solution purpose. Implementing the calculations in MATLAB, started with an initial curvature guess and then compared the two force outputs. Another guess is then made and the two are compared. The guess is made on a binary system. That is, firstly the range of solutions is identified. Then the range is divided into two, and the median values from both halves are compared. The half with the lower error difference will then be taken to be the whole and all the steps are again repeated. Then depending on the error values, the next guess for initial curvature would be closer to the one with less error.

Desired angular deflection (in deg.) = 58
Fo = 1.109657e+001, F = 1.012283e+001
Fo = 1.294599e+001, F = 9.787877e+000
Fo = 1.017185e+001, F = 1.029012e+001
Fo = 1.063421e+001, F = 1.020645e+001
Fo = 1.086539e+001, F = 1.016463e+001
Fo = 1.051862e+001, F = 1.022736e+001
Fo = 1.034524e+001, F = 1.025873e+001
Fo = 1.025854e+001, F = 1.027443e+001
Fo = 1.027299e+001, F = 1.027181e+001
linear x = 4.504419e-002 and y = 8.126188e-002
nonlinear x = 4.249552e-002 and y = 7.666395e-002
Maximum Stress = 6.093707e+008, -6.123055e+008

In the MATLAB environment, to achieve an error rate of 0.1% only nine iterations were performed to find the solution to the problem. This same algorithm was then used to find the full range of motion of the compliant beam. We based our calculations on the mock-up structure described earlier, where the length of the steel blade is 9.7cm and the length of the wooden platform 17.2cm. The maximum stress caused in the beam in the above example would be 6.12 X 108 Pa, which is clearly within the yield strength of 1.64 X 109 Pa for steel.

Though the iterative method for solving the non-linear behavior at high-deflections gave positive results, the results were off for smaller deflections. In Fig. 5 is a comparison of the linear and non-linear solutions for the range of motion.

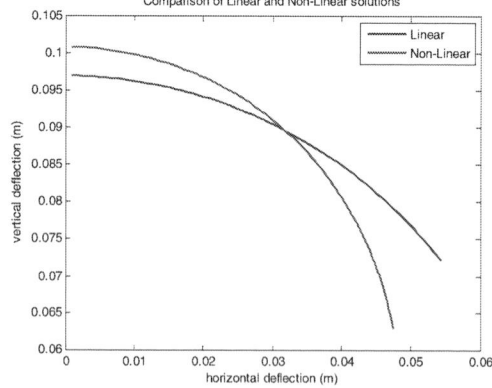

Fig. 5 Comparison of Linear and nonlinear solutions

Clearly the non-linear solution breaks down at lower deflections, since at 0° the beam cannot reach beyond it original height of 9.7cm, while the linear solution differs greatly from the non-linear at higher deflections. Upon examination we found that the graphs that gave lower values along the whole range gave the best approximations. The intersection of the two graphs was found to be at 39°, and hence a new solution was created by utilizing the linear solution prior to 39°, and the non-linear afterwards. The graph was then smoothed out using weighted averages around the point of 39° as shown in Fig. 6.

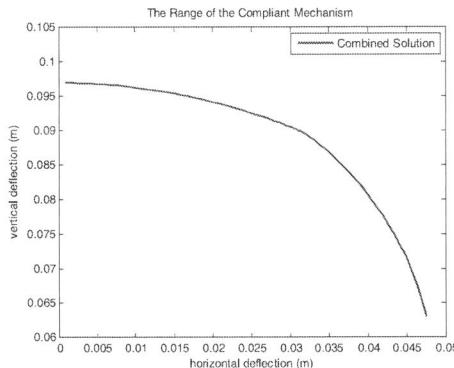

Fig. 6: Combined solution of linear and nonlinear models

The pseudo-rigid-body solution was then compared to actual mock-up and the results are tabulated in Table I.

TABLE I
COMPARISON OF EXPERIMENT AND ANALYTICAL RESULTS

Angle	Coordinates	Mock-Up/ True (cm)	Model/ Calculated (cm)	Error
9°	x	0.7	0.76	7.8%
	y	9.6	9.66	0.62%
15°	x	1.3	1.26	3.1%
	y	9.5	9.58	0.83%
27°	x	2.0	2.24	11%
	y	9.1	9.34	2.6%
45°	x	3.5	3.56	1.7%
	y	8.5	8.61	1.3%
57°	x	4.3	4.24	1.4%
	y	7.9	7.67	2.9%

The model contains the error rate may reach up to 11%, however, there are also areas where the error may be as low as 0.62%. The average error from the above would be put at around 3.3%. One must keep in mind that even when measuring the mock-up model inherent uncertainties crept into the problem as the instrument used was only able to measure to the nearest millimeter.

Conclusions

The investigation demonstrated the feasibility of the pseudo-rigid-body model in modeling compliant mechanisms, and illustrated how a simple modular compliant structure could be developed at least at the macro level. Though similar techniques and designs may be applied to the micro level of MEMS, further studies will need to be conducted to examine the maximum stress for the silicon based materials used in MEMS fabrication. Furthermore, when it comes to the specific MEMS design presented in this paper, further studies need to be conducted regarding the structure. Such studies would only truly be conclusive if such mechanisms could actually be fabricated in robotics labs. Some of the characteristics that would have to be determined would be the maximum force delivered by comb actuators so as to determine the maximum achievable deflection. In terms of MEMS application, such devices could be used to actuate optical modulators and micro-mirrors, or even linked together to form micro-manipulators.

References

[1] M.B. Sanders, "Novel compliant mechanisms could simplify," *improve aircraft*, http://www.afosr.af.mil/pdfs, 2003.

[2] L.J. Mary, "On campus: Compliant mechanisms make braking news", http://magazine.byu.edu/bym/2000sum, 2003.

[3] M.I. Frecker, and S. Canfield, "Methodology for systematic design of compliant mechanism with integrated smart materials," Forthcoming.

[4] Y.L. Mei, X.M. Wang, "A level set method for structural topology optimization and its applications," *Adv Eng Softw*, 35:415–41, 2004..

[5] Z. Luo, L.Y. Tong, "A level set method for shape and topology optimization of large-displacementcompliant mechanisms," *Int J Numer Methods Eng*. doi:10.1002/nme.2352, 2008.

[6] G.W. Jang, K.J. Kim, Y.Y. Kim, "Integrated topology and shape optimization software for compliant MEMS mechanism design," *Adv Eng Software*, 39:1–14.

[7] M. Moghadasian and B. Moghadassian, "3D inverse Boundary Design Problem of Conduction-radiation Heat Transfer," *Journal of Applied Sciences*, 12(3), pp. 233-243, 2012, DOI: 10.3923/jas 2012.233.243.

[8] L.L. Howell, "Compliant Mechanisms," *John Wiley & Sons, Inc.*, New York, 2001.

[9] C. Haug, S. Regnier and P. Bidaud, "Compliant Beam Networks Optimization for Microsystems," *Proceedings of the 2000 /EEE/RSJ International Conference on Intelligent Robots and Systems*, 2000.

Doped Group-IV Semiconductor Nanocrystals

Latha Nataraj*, Aaron Jackson, Lily Giri, Clifford Hubbard, and Mark Bundy

U.S. Army Research Laboratory, Aberdeen Proving Ground, MD

*email : latha.nataraj.ctr@mail.mil

Abstract

Doped semiconductor nanocrystals offer great potential for microelectronics and integrated optoelectronics. Nanocrystal based technology is promising for several fields of technology such as thin conducting films, light emitting devices, tunable lasers, transistors, photovoltaics, and less harmful alternatives to toxic fluorescent dyes for application in bio-imaging, to name a few. While properties of nanocrystals are tunable through their size, considerable research is underway to explore the influence of dopants on the properties of semiconductor nanocrystals. The introduction of dopants has been elusive for these confined structures due to their nanoscale sizes as well as the possibility of making them degenerate even with the addition of very small quantities of the dopant. Here, we present a facile, low-cost procedure that we have developed for the synthesis of non-degenerate doped semiconductor nanocrystals and study their properties.

Background

Semiconductor nanocrystals have demonstrated enhanced optical properties [1, 2, 3, 4]. Nanocrystals synthesized from several materials are being actively investigated for various applications. Undoped germanium nanocrystals have demonstrated direct band gap behavior [4] providing great enhancement in optical properties for applications in optoelectronics and silicon nanocrystals have been demonstrated to be attractive for photovoltaics [5].

Dopants serve to significantly alter the electronic, optical, and magnetic properties of semiconductors [6]. The addition of dopants influences the characteristic electronic band structure of a material. With the addition of dopant atoms into the host material, more and more dopant states are introduced into the system. Several nanocrystals have been demonstrated to be good emitters. The fundamentals of quantum systems have established that the wavelength of emission from the nanocrystals is size-dependent. However, many times, the emission efficiency is not sufficient for application in lasers. Among various approaches that help to achieve an improvement in this area [7], the addition of extra electrons into the system is an effective mechanism [8], which may be achieved through n-type doping of the nanocrystals. For bio labeling, such highly luminescent doped nanocrystals provide less toxic alternatives to the fluorescent and radioactive dyes currently being used [9, 10]. In case of photovoltaic devices, doping serves to efficiently absorb the energy from the incoming photons, localizing the excitation and minimizing surface reactions that lead to nanocrystal degradation from thermal and environmental effects [9]. In addition such stability to heat and environment is advantageous to lasers and LEDs, making them more durable [9]. Carrier confinement enhances interactions of magnetic dopants with other carriers or quantum mechanical spins for application in spintronics [11, 12].

The doping of semiconductor nanocrystals is extremely challenging due to their small size [13, 14, 15] and the fact that the introduction of even a few atoms of the dopant into the nanocrystal that contains just a few hundred atoms could result in a degenerate material or compromise the crystal structure. Some progress has been made with doping in nanocrystals through strategies such as remote doping, substitutional doping, and electrochemical carrier injection to yield heavily doped nanocrystals. Most involve complex chemical synthesis processes and do not provide consistently successful results; many questions are still unanswered. Here we present a simple, cost-effective mechanism for consistently obtaining doped semiconductor nanocrystals through top-down mechanical milling techniques and report on their properties.

Fabrication process

Highly pure silicon and germanium nanocrystals have been obtained using a simple, cost-effective, high-yielding mechanical milling process. Fine nanocrystalline powders have been obtained using pristine n-type Si and Ge wafers (100), doped with Sb, using high energy ball-mill. It should be noted that this approach provides very little control over the size of the NCs, resulting in a wide size distribution.

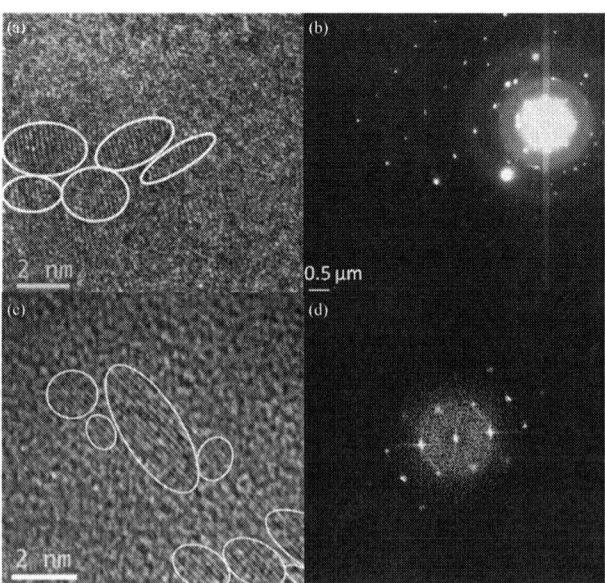

Fig 1. (a) TEM image of doped silicon nanocrystals (b) Selected area electron diffraction pattern from a doped silicon nanocrystal (c) TEM image of doped germanium nanocrystals (d) Fast Fourier transform pattern from a doped germanium nanocrystals

However, this does not adversely affect, and in fact, would be advantageous, for the application of these NCs in photovoltaics, as this would allow for absorption of a wide range of energies from the electromagnetic spectrum.

Figures 2(a) and 2(c) are TEM images from Si and Ge nanocrystals indicated by the white ovals. As expected, nanocrystals synthesized using mechanical grinding techniques as these [4, 16] have a wide size distribution, from 2 nm to a few hundred nanometers. The selected area electron diffraction (SAED) pattern from the Si nanocrystals illustrated in figure 2(c) and the Fast Fourier transform (FFT) pattern of Ge nanocrystals figure 2(d) clearly indicate that the nanocrystals are indeed crystalline.

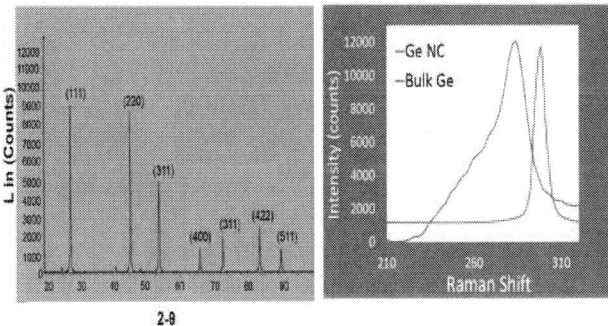

2-8

Fig 2. (a) XRD pattern from n-ytpe germanium nanocrystals (b) Raman spectroscopy results comparing n-type nanocrystals to the bulk wafer they were synthesized from

The XRD pattern from the Ge nanocrystals in figure 2(a) confirms the crystalline nature of the doped nanocrystals. Raman spectroscopy measurements shown in figure 2(b) indicates a shift to lower wave numbers and an inhomogenous broadening of the nanocrystal peak in comparison with the peak from bulk Ge. This confirms that the nanocrystals synthesized by mechanical milling of crystalline bulk Ge is highly tensile-strained, as expected [4, 16].

It is this strain in the nanocrystals that modifies the near-direct band structure of Ge. This direct band gap behavior, provides greatly enhanced optical properties, namely absorption of a wider range of energies from the electromagnetic spectrum as well as dramatically increased photoluminescence, demonstrated in our earlier work [4]. Further work is underway to determine the effects of combining the advantage of low-dimensional structures with those derived from the addition of dopants for the application of these doped nanocrystals in the area of photovoltaics.

In summary, we demonstrate a facile process to synthesize tensile-strained semiconductor nanocrystals using mechanical grinding techniques. HRTEM, SAED, FFT data indicate the crystalline nature of the nanocrystals which is confirmed by the XRD measurements. Raman spectroscopy measurements point to high tensile-strains in the nanocrystals. Such tensile-strained nanocrystals synthesized using mechanical grinding techniques modify the material to demonstrate direct band gap behavior resulting in greater absorption and superb room-temperature light-emission, indicating highly improved optical properties. We believe that such band engineering

techniques, combined with the traditional advantages offered by semiconductor doping offer great possibilities for photovoltaics and monolithic integration with silicon-photonics.

References

[1] V. G. Prakash, M. Cazzanelli, Z. Gaburro, L. Pavesi, F. Iacona and F. Priolo, *Materials Research Society Symposium*, 2002.

[2] O. Bisi, S. Ossicini and L. Pavesi, *Surface Science Reports,* vol. 38, no. 1-3, p. 1, 2000.

[3] J. Linnros, O. Edited by Bisi, S. Campisano, L. Pavesi and F. Priolo, Silicon based Microphotonics: from basics to applications, Netherlands: IOS, 1999, p. 47.

[4] L. Nataraj, F. Xu and S. G. Cloutier, *Optics Express,* vol. 18, no. 7, p. 7085, 2010.

[5] , D. Mariotti, T. Nagai, Y. Shibata, I. Turkevych and M. Kondo, *The Journal of Physical Chemistry,* vol. 115, no. 12, p. 5084, 2011.

[6] S. M. Sze, Physics of Semiconductor Devices, New York: Wiley-Interscience, 1981.

[7] V. I. Klimov, S. A. Ivanov, J. Nanda, M. B. I. Achermann, J. A. McGuire and A. Pirtinski, *Nature,* vol. 447, p. 441, 2007.

[8] C. Wang, B. L. Wehrenberg, C. Y. Woo and P. Guyot-Sionnest, *Journal of Chemistry B,* vol. 108, p. 9027, 2004.

[9] N. Pradhan, D. Goorskey, J. Thessing and X. Peng, *Journal of American Chemical Society,* vol. 127, p. 17586, 2005.

[10] W. C. W. Chan and S. Nie, *Science,* vol. 281, p. 2016, 1998.

[11] J. D. Bryan and D. R. Gamelin, *Progress in Inorganic Chemistry,* vol. 54, p. 47, 2005.

[12] A. L. Efros, E. I. Rashba and M. Rosen, *Physical Review Letters,* vol. 87, p. 206601, 2001.

[13] T. L. Chan, M. L. Tiago, E. Kaxiras and J. R. Chelikowski, *Nano Letters,* vol. 8, p. 596, 2008.

[14] G. M. Dalpian and J. R. Chelikowsky, *Physical Review Letters,* vol. 96, p. 226802, 2006.

[15] D. Turnbull, *Journal of Applied Physics,* vol. 21, p. 1022, 1950.

[16] P. K. Giri, *Journal of Physcs D: Applied Physics,* vol. 42, p. 245402, 2009.

978-1-4673-4840-9/13 $31.00 © 2013 IEEE

Robust Nitrogen Plasma Immersion Ion Implantation Treatment on Gadolinium Oxide Resistive Switching Random Access Memory

Yu-Ren Ye, Ying-Huei Wu, Jer-Chyi Wang*, and Chao-Sung Lai

Department of Electronic Engineering, Chang Gung University,
No. 259, Wen-Hwa 1st Road, Kweishan 333,
Taoyuan, Taiwan
Tel: +886-3-2118800 ext.: 5784, Email: jcwang@mail.cgu.edu.tw

Abstract

In this paper, we demonstrate gadolinium oxide RRAM with nitrogen plasma immersion ion implantation (PIII) treatment technique at first time. For nitrogen plasma treatment, the nitrogen ions were incorporated with gadolinium oxide. We controlled the implantation voltage that the nitrogen ions exist near the surface of gadolinium oxide and it was forming a $Gd_xO_yN_z$ layer. This can reduce the leakage current to reach a low current and power consumption operation. In addition, the retention and endurance characteristics were also improved.

Keyword: RRAM, plasma immersion ion implantation, gadolinium, nitrogen plasma, and oxynitride.

Introduction

Resistive random access memory (RRAM) is one of the candidates for future nonvolatile memories, because there are many advantages of RRAM such as the fast operation speed, excellent scalability, simple structure, easy fabrication, multi-level potential, and compatible with CMOS process. Many researches proposed that the simple metal-insulator-metal (MIM) structure with a transition metal oxide (TMO) material like Al_2O_3, NiO, HfO_2, TiO_2, ZrO_2, and Gd_2O_3[1-6], can exhibit excellent bi-stable resistive switching behavior. The switching phenomenon depending on the measurement of polarity can be defined as unipolar, bipolar. The set (high change to low resistance state) and reset (low to high resistance state) operation process are the same polarity. On the other hand, bipolar are operated on difference polarities. The different resistive switching behavior is associated with various mechanisms. For unipolar resistive switching, "filament" theory is widely recognized. The metal ions or oxygen vacancies are the main composition for the filament conducting path. Then it was broken by "Joule heat effect", which is called reset operation process. The bipolar resistive switching mechanism was attributed to the variation of the Schottky barrier height or the width caused by trapped charge carriers (oxygen ions or vacancies) near the metal-insulator interface. Difference MIM material were shown to have various mechanisms and properties. In addition, the electrodes are also important in RRAM performance. Platinum electrodes have been found to have superior properties, except for the retention and stability, owing to the ease of movement of oxygen ions or atoms along grain boundaries into the atmosphere.

In our previous work, gadolinium oxide RRAM with low operation voltage, high resistance ratio (high resistance/low resistance), and stable bipolar switching have been demonstrated. The bipolar resistive switching can be attributed to the change of Schottky barrier height by the oxygen vacancies near the metal-insulator interface. Thus, the interface between the top electrode and insulator plays an important role for the gadolinium oxide RRAM. To modify the interface, we introduce the nitrogen atoms at the gadolinium oxide surface by using the plasma immersion ion implantation (PIII) technique. The PIII technique is an abundant and efficient implantation treatment. We can control the implantation power and time to change the position and concentration of implanted nitrogen ions. Then, annealing was used to passivate the implantation damage of gadolinium oxide and form a $Gd_xO_yN_z$ layer. The layer can reduce the leakage current and increased the high resistance to reach a low power consumption operation. In addition, the retention and endurance characteristics were also improved. According to this study, the plasma immersion ion implantation treatment can be used for future resistive switching memory application.

Experiment

A gadolinium oxide (Gd_xO_y) resistive switching memory was fabricated on 4-inch Si wafers. All wafers were first cleaned by a standard RCA cleaning method. A 600nm thickness oxide was thermally grown at 900°C in a pyrogenic hydrogen/oxygen mixture gas as an isolation layer. Next step, a 100 nm tungsten bottom electrode was deposited using a RF sputtering method. Then, a 20 nm-thick amorphous gadolinium oxide layer was deposited by RF sputtering a 99.9% pure Gd target in an ambient of oxygen and argon mixture (O_2: Ar = 10: 10). Subsequently, the key process, the key process of nitrogen plasma was treated on Gd_xO_y with 1kV implantation voltage and 300 w plasma power for 1, 5 and 10 min by plasma immersed ion implantation (PIII) system. The schematic PIII system was shown in the Fig. 1. Then the samples denoted as 1 min, 5 min, and 10min, respectively. Besides, the sample without treatment was denoted as w/o. Then a 50-nm-thick platinum top electrode was deposited by an electron beam evaporator and then a pattern of the diameter of 20μm was formed by using a lift-off process. Finally, a post-metallization annealing (PMA) process at 300°C was conducted for 10 min by furnace system. The schematic device structure and process flow of this study is shown in Fig. 2. The electrical properties of the samples were measured by using an Agilent 4156C analyzer.

Results and Discussion

Fig. 3(a) shows the bipolar resistance switching behavior of all samples. The set and reset processes are operated at negative and positive voltages, respectively. In the set process, the resistance was switching from high to low. The second step was reset process that the resistance was switching from low to high. Forming process is unnecessary for all samples. Fig. 4(a) and (b) demonstrated the set/reset voltage, high/low resistance state (HRS/LRS) and ratio characteristics, respectively. Because the defects were passivated in the $Gd_xO_yN_z$ layer and the leakage current was reduced, the sample of 1 min treatment with higher ratio characteristic was the optimized condition. During 10 min N_2 PIII treatment, the plasma damage was increased resulting of switching property degrades. A sample with N_2 PIII treatment could be able to operate stably for over 600 cycles, as shown in Fig. 5. Fig. 6 shows the high resistance state retention properties of Gd_xO_y RRAM with N_2 PIII treatment for 1, 10 min, and w/o. The stable retention was obtained for all samples. Fig. 7 showed the yield for all samples. The highest yield for 1 and 5 min N_2 PIII treatment was much larger than the other samples. This was due to the $Gd_xO_yN_z$ layer passivated defects and reduced leakage current.

Conclusion

In this work, the defect was reduced by nitrogen ion plasma immersed ion implantation on Pt/Gd_xO_y/W RRAM. The high resistance state and yield were improved at 1min treatment. Furthermore, the endurance and retention properties became stable after N_2 PIII treatment. Too long treated time will induce damage. Thus, the optimized treatment time is 1min for low power consumption memory application.

Acknowledgement

This work was supported by the National Science Council under the contract of NSC 100-2221-E-182-012.

References

[1] C.-Y. Lin, et al., Surface and Coatings Technology, vol. 203, 2008, pp. 628-631.
[2] J. Y. Son, et al., ACS nano, vol. 4, 2010, pp. 2655-2658.
[3] C.-S. Peng, et al., Electrochemical and Solid-State Letters, vol. 15, 2012, pp. H88-H90.
[4] W.-T. Wu, J.-J. Wu, and J.-S. Chen, ACS Applied Materials and Interfaces, vol. 3, 2011, pp. 2616-2621.
[5] H. Zhang, et al., Applied Physics Letters, vol. 96, 2010, art. no. 123502.
[6] K.-C. Liu, et al., Microelectronic Engineering, vol. 88, 2011, pp. 1586-1589.

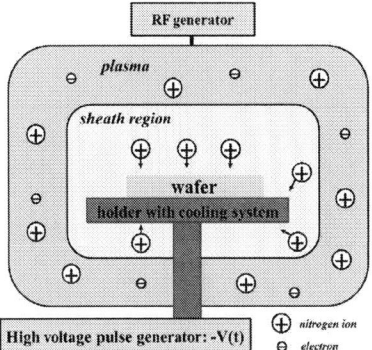

Fig. 1. The plasma ion immersion implantation system schematic.

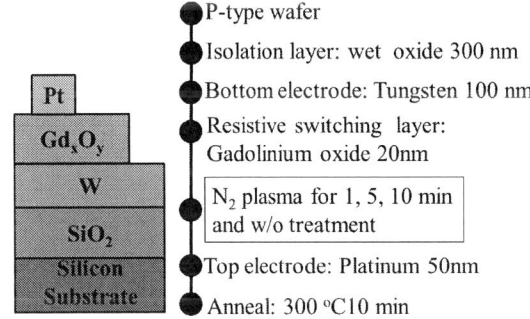

Fig. 2. The fabrication procedure of Pt/Gd_xO_y/W RRAM with and without N_2 PIII treatment.

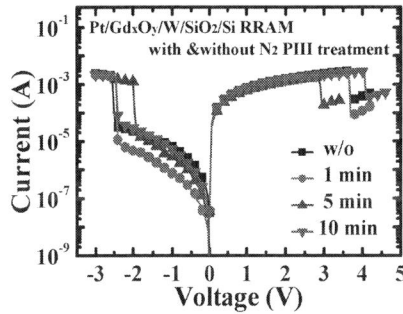

Fig. 3. Bipolar current-voltage characteristics of Gd_xO_y RRAM without and with N_2 PIII treatment.

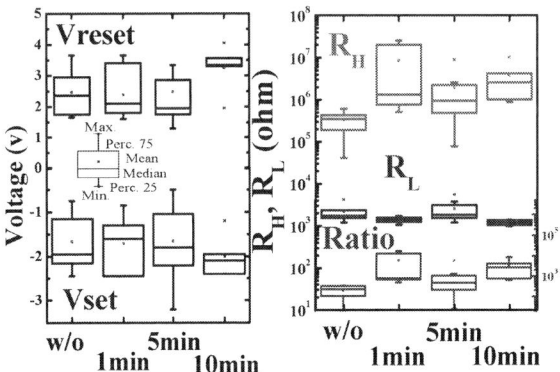

Fig. 4. The statistical distributions of the set/reset voltages, the high/low resistances, and the resistance ratio of these samples are shown in (a) and (b). The samples treated with N_2 PIII treatment demonstrated high R_H and rario for low-voltage operation.

978-1-4673-4840-9/13 $31.00 © 2013 IEEE

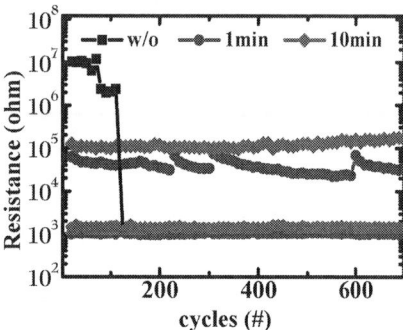

Fig. 5 The cycling endurance properties with DC sweeping mode of Gd_xO_y RRAM with N_2 PIII treatment.

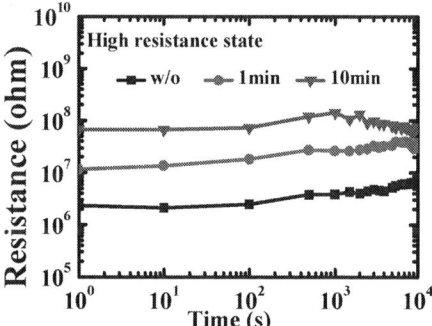

Fig. 6. The retention properties of Gd_xO_y resistive switching memories with and without N_2 PIII treatment.

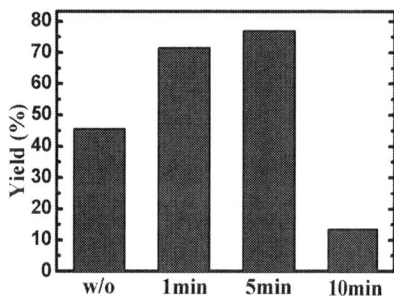

Fig. 7. The yield statistics of all Samples with and without N_2 PIII treatment.

Bipolar Resistive Switching Characteristics in Si₃N₄-based RRAM with MIS (Metal-Insulator-Silicon) Structure

Sungjun Kim[1], Sunghun Jung[1], Jeong-Hoon Oh[2], Kyung-Chang Ryoo[2], and Byung-Gook Park[1]

School of Electrical Engineering and Computer Science, Seoul National University,

San 56-1, Sillim-dong, Gwanak-gu, Seoul 151-742, Republic of Korea.

[2]DRAM Process Architecture Team, Memory Division, Semiconductor Business, Samsung Electronics Co., Ltd.,

Nongseo-dong, Giheung-gu, Yongin-si, Gyeonggi-Do, Korea, 445-701

Tel.: +82-2-880-7279, Fax: +82-2-882-4658, E-mail address:lizzym19@gmail.com

ABSTRACT

In this paper, bipolar resistive switching was investigated in our fabricated Ti/Si₃N₄/p⁺-Si resistive random access memory (RRAM) devices. Heavily doped p-type Si was used instead of a conventional bottom electrode (BE) using metal such as Pt. We found that forming-free process, self-compliance and gradual reset were shown in this device. The operation voltage was with 1.8~3.5 V during set process due to forming-free process. And self-compliance was observed by restriction of parasitic resistance without external current limiter. Finally, multi-level cell (MLC) feasibility was achieved using voltage stop during gradual reset.
(Keywords: Si₃N₄ based RRAM, MIS structure)

I. INTRODUCTION

Extensive research about resistive switching characteristics including a variety of resistive materials is ongoing to replace NAND flash since resistive switching device using transition metal oxide (TMO) was reported in 2004 [1]. Recently, Nitride-based resistive random access memory (RRAM) has been reported as one of the candidates of next generation nonvolatile memory due to fast programming speed [2], superior endurance and retention [3]. In addition, nitride-based RRAM has a lot of traps which is used as defects related to resistive switching. In order to compete with NAND flash, other requirements of RRAM as storage memory are MLC and 3D structure. In this paper, silicon nitride (Si₃N₄)-based RRAM with MIS (metal-insulator-silicon) structure was fabricated. Silicon used as the BE is CMOS friendly material in backend of line (BEOL). And this structure can avoid noble metals such as Pt which must follow deposition and patterning when they are used as the BE [4]. Si₃N₄ as switching layer (SL) was deposited by LPCVD which is suitable to 3D RRAM due to good step coverage [5]. Bipolar resistive switching including forming-less process, self-compliance and gradual reset were observed in our devices. Self-compliance effect is helpful

with respect to memory circuit design and gradual reset can be used for realization of MLC [6].

II. FABRICATION

The schematic drawing and process flow of our fabricated Ti/Si₃N₄/p-Si RRAM device are shown in Fig. 1 and cross-sectional TEM image in Fig. 2. 10-nm-thick SiO₂ was by dry oxidation for protection of damage cluster of during ion implantation. Subsequently, Si was doped with BF₂ (dose: 1×10^{15}, energy: 40 keV) by ion implanter in order to form p⁺-Si on Si substrate. Next, 5nm-thick Si₃N₄ was deposited by low pressure chemical vapor deposition (LPCVD) at 785 °C after HF cleaning to remove SiO₂. Finally, Ti top electrode (TE) was formed using an RF sputtering system and patterned by a conventional photolithography. TE size and thickness are 100×100 μm² and 100 nm, respectively. All electrical characterizations were performed by DC voltage sweep mode using Agilent 4156C semiconductor parameter analyzer. And the bottom electrode (p⁺-Si) was grounded and the positive voltage bias was applied to the TE (Ti).

III. RESULTS AND DISCUSSION

Figure 3 shows that typical bipolar resistive switching is observed in our fabricated Ti/Si₃N₄/p-Si RRAM. Device is set under a current compliance of 100 μA to prevent permanent dielectric breakdown. Figure 4 shows that forming voltage ($V_{FORMING}$) is comparable to set voltage (V_{SET}). First set process (Forming) is within dispersion of subsequent set process in all samples. This is attributed to an initially high leakage from nitride based trap which helps to lower $V_{FORMING}$.

Measured on/off resistance, set/reset voltage, and reset current distributions are shown with 40 cycles in fig. 5-7. Overlap between on and off resistance is found in some cycles. Thus, it is needed to more accurate control during reset process to make sure a reasonable on/off resistance ratio and avoid overlap between on and off resistance.

978-1-4673-4840-9/13 $31.00 © 2013 IEEE

For better resistive characteristics, device is set a current compliance of 1 mA in fig. 8. The current is limited by self-compliance effect before 1 mA. It is originated from the restriction of parasitic resistance in device cell. The permanently breakdown can be stopped by self-compliance during set process without external current limiter such as transistor. The gradual reset is also observed by controlling stop voltage (from -2 V to -2.4 V) during the reset process in fig. 5. The more reduction of reset current transition for a margin among the reset states is produced due to high compliance current (1 mA) than that of I-V curve in Fig. 3 (I_{COMP} = 100 µA). Thus, 2-bit/cell MLC states are possible in our device.

IV. CONCLUSION

This work investigated the resistive characteristics of Ti/Si$_3$N$_4$/p$^+$-Si. Silicon nitride as switching layer is deposited using LPCVD on p$^+$-Si. Our fabricated MIS structure shows reproducible bipolar switching under an operation voltage (<3V) because forming-less process was observed. In addition, self-compliance is found less than 1 mA. Finally, 2 bit/ cell is achieved using voltage stop during gradual reset process.

ACKNOWLEDGEMENT

This work was supported in part by the Smart IT Convergence System Research Center funded by the Ministry of Education, Science and Technology as Global Frontier Project (CISS-2011-0031845) and in part by the IT R&D program of MKE/KEIT. (10035320, Development of novel 3D stacked devices and core materials for the next generation flash memory)

REFERENCES

[1] I. G. Baek, *et al., IEDM Tech. Dig.* 2004, pp. 587-590.
[2] H.-D. Kim, *et al., IEEE TED.* VOL. 59, NO. 9, 2012.
[3] H.-D. Kim, *et al., IEEE SNW.* p.99, 2012.
[4] Y.-H. Wu, *et al., IEEE TED.* VOL. 33, NO. 3, 2012.
[5] A. Stoffel, *et al., JMM.* VOL. 6, NO. 1, 1996.
[6] J.-H. Oh, *et al., IEEE SSDM.* 2011, pp. 154-155.

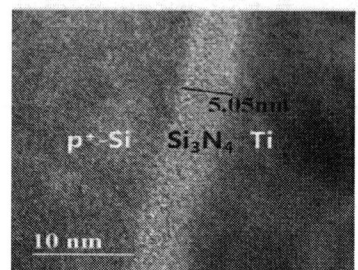

Fig. 2. Cross-sectional TEM image of our fabricated Ti/Si$_3$N$_4$/p$^+$-Si RRAM device.

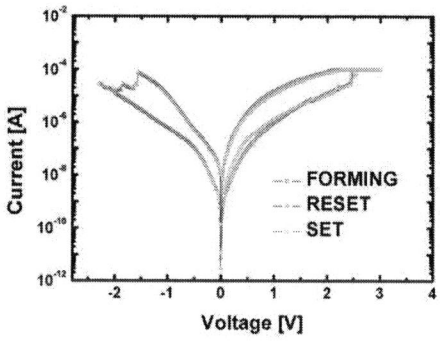

Fig. 3. Typical *I-V* curves of fabricated Ti/Si$_3$N$_4$/p-Si RRAM device showing bipolar resistive switching behavior

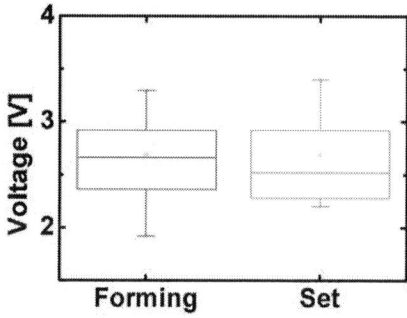

Fig. 4. Statistic distribution of forming voltage and set voltage of fabricated Ti/Si$_3$N$_4$/p-Si RRAM device

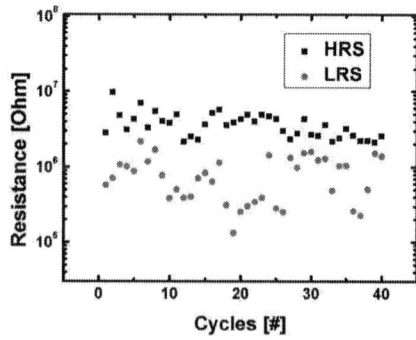

Fig. 5. On/off resistance cycling of fabricated Ti/Si$_3$N$_4$/p$^+$-Si RRAM device.

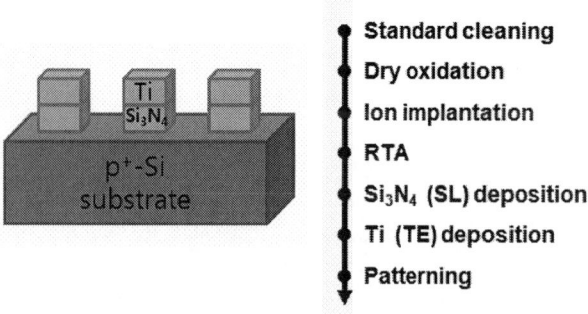

Fig. 1. Schematic drawing and process flow of our fabricated Ti/Si$_3$N$_4$/p$^+$-Si RRAM device.

978-1-4673-4840-9/13 $31.00 © 2013 IEEE

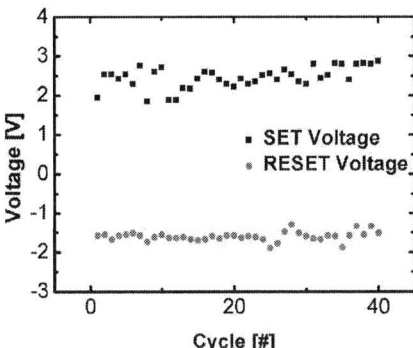

Fig. 6. Set/reset voltage cycling of fabricated Ti/Si$_3$N$_4$/p$^+$-Si RRAM device.

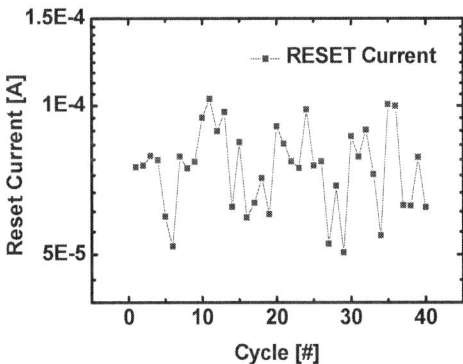

Fig. 7. Reset current cycling of fabricated Ti/Si$_3$N$_4$/p$^+$-Si RRAM device.

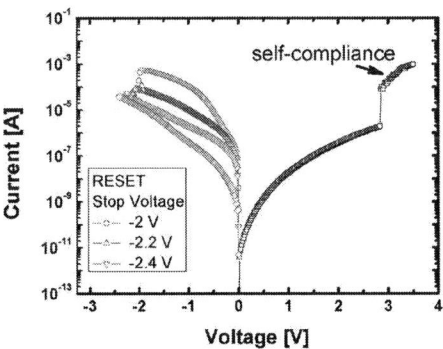

Fig. 8. I-V curves of Ti/Si$_3$N$_4$/p$^+$-Si RRAM device showing self-compliance and the possibility of 2-bit/cell MLC.

978-1-4673-4840-9/13 $31.00 © 2013 IEEE

Simulation Study of dimensional effect on Bipolar Resistive Random Access Memory (RRAM)

Liu Kai[1], Zhang Kailiang[1]*, Wang Fang[1,2], Zhao Jinshi[1], and Wei Jun[3]

1. School of Electronics Information Engineering, Tianjin Key Laboratory of Film Electronic & Communication Devices, Tianjin University of Technology, Tianjin, China
2. Tianjin University, Tianjin, China
3. Singapore Institute of Manufacturing Technology, a-star 71 Nanyang Drive, Singapore

*corresponding author: kailiang_zhang@163.com，86-22-60214196

Abstract

The dependency of the RRAM device electrical parameters such as set voltage, reset current and resistance on the RRAM cell dimensional scalability is investigated with Monte Carlo simulation to optimize the power consumption of bipolar RRAM. It is found in the simulation that the switching process in bipolar RRAM is related to the cell dimension in the sub-nm region in terms of its horizontal length. The suppressing effect of existing conducting filament is also discussed. With optimal cell size sufficient initial resistance and a low forming voltage will be achieved, accelerating the feasibility of the high-density low-power RRAM.

Keywords: RRAM, power consumption, electric parameter and computational simulation

Introduction

To replace conventional flash memory, various new memory devices have been proposed. Among them, resistive switching random access memory (RRAM) devices are one of the possible candidates due to its simple structure, fast switching speed and high scalability [1]. The RRAM device can be fabricated with various materials such as binary oxides or perovskite oxides [2]. However, the current switching uniformity and reliability of current RRAM devices are not sufficient for high density memory applications. In addition, a detailed clarification on the switching mechanism has not yet been fully developed [3].

In many literatures of RRAM, the researcher studied resistive switching behavior of large area devices, which might not be applicable for future nano-scale high density memory application. So far, very limited researchers have focused on the scaling effect of RRAM devices. In this paper, we investigated the scaling issues of RRAM device with computational simulation. Our Monte Carlo simulation results show that the charged ions can be reasonably described as charged point particles which play an important role in the switching process of RRAM device. The present approach also provides a new way to probe the inner process of dielectric breakdown.

Computational Methods for RRAM Structure Simulation

We employ the kinetic Monte Carlo (KMC) method by considering the ion dynamics in a simplified way. It is assumed that the oxidation, transportation and reduction of electrode atoms in switching layers are simulated in a two-dimensional matrix. The electrochemical transitions, switching time and final state of the device are determined by the random-test method. The hopping rate P between jumps is obtained from the following Boltzmann relationship:

$$P = \nu \ e^{\wedge}(-E_a/kT) \qquad (1)$$

where ν, k and T are attempting frequency of atoms, Boltzmann constant and local temperature, respectively. E_a is the migration barrier along the jump direction. The magnitude of E_a and its variation in different situation are based on our previous work [4], where more detailed simulation descriptions were also provided.

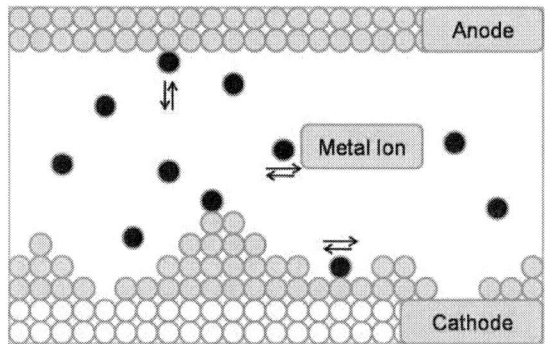

Fig.1 Schematic diagram of simulation box in KMC simulation for RRAM cell

Simulation box of different length in x dimensional are studied in the modeling process and periodic boundary conditions are also used. The intrinsic defect is introduced and randomly generated when initializing simulation, the concentration of which is assumed to be around 1% in the simulation box. In the KMC simulation process we adopt a simple residence-time algorithm. For the sake of simplicity, we consider only the movement to the nearest neighbor positions, and the interaction between charged ions is not considered.

Results and Discussion

A. Dimensional scaling effect I :Degradation of the electric characteristics

Previous studies have revealed that the switching mechanism of RRAM stack structure was attributed to conductive filament formation/rupture by oxygen vacancy or metal ions [5]. Although RRAM device with oxygen vacancy (VOs) conducting path and metal conducting bridge have different switching mechanism, similar conducting filaments (CFs) dynamic behavior was observed [6]. In our research the simulation model is mainly based on metallic conducting bridge RRAM (also called CBRAM) device, but it should be pointed out that the simulation results also apply VO based RRAM.

These improvements can be attributed to the difference in the amount of uncontrollable defects in the active memory areas of the devices. It is well known that vacancies and grain boundaries are major causes of intrinsic defects in metal oxide dielectric films. These intrinsic defects may be responsible for the formation of filaments comprising metallic atoms during resistive switching, thus leading to instable resistive switching. If the deviation from the defects in the nano-scale device can be minimized with device area shrinking, resistive switching uniformity can be improved.

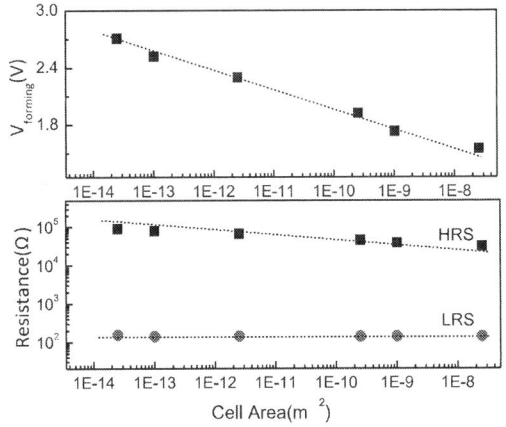

Fig. 2 Resistance and forming voltage as a function of cell area

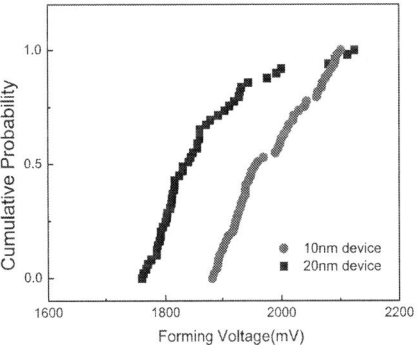

Fig.3 Distribution of switching voltages for different nano- scale RRAM devices

As shown in Fig. 2, $V_{forming}$ and R_{HRS} of the device are strongly related to active device area. With device area scaling down, $V_{forming}$ and R_{HRS} increase. The increase in voltage with scaling area can be attributed to the reduced number of defects. Generally, the formation of a conducting filament can be induced by connection of defects in the active memory area, where metal ions could be more easily accumulated. In large area device, a relatively small voltage is sufficient to induce the formation of a conducting filament in the dielectric layer. Because of reduction in the defects, a higher voltage is required for filament formation in nano devices. Forming voltage also increase with cell area scaling due to defect reduction in nano device.

In the low resistance state (LRS) case, however, same resistances were observed regardless of the device area. In this situation, the resistance is mainly affected by the formed conducting filaments. According to the suppressing effect of existing conducting filament to its neighboring area, the CFs dynamics don't show much difference with device area scaling down.

B. Dimensional scaling effect II: Uniformity improvement

By simulating the forming operation of a large number of devices, it is possible to extract the statistical distribution of the forming voltage. Fig. 3 shows the statistical charts of the resistive switching parameters for different devices with linearly scaled active memory area. 100 randomly initialized simulation boxes are selected to evaluate the switching uniformity of devices. Compared with large area devices, the nano devices show more uniform and sharp distribution of switching voltage.

C. Dimensional scaling effect III: Local heat effect

We also investigated the effect of area scaling on the local heating of filament. It is well known that charged ion movement is closely related to local temperature. To confirm the temperature effect on the resistive switching, we simulated the response time under different background temperature.

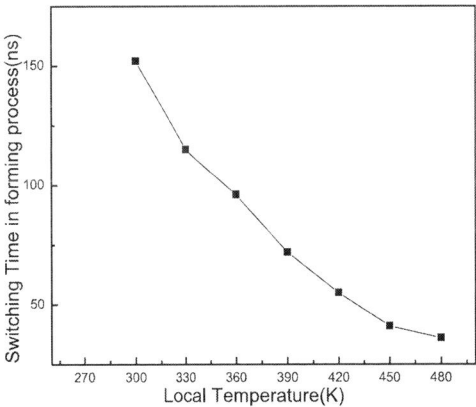

Fig.4 Trend of switching time as a function of local temperature

As shown in Fig. 4, with increasing temperature, the response speed of the device increases significantly, this confirmed the acceleration of metal ions movement during switching. In other words, if additional heat is generated in active area of the device during switching, significant speed improvement is possible to be obtained. Compared with large

978-1-4673-4840-9/13 $31.00 © 2013 IEEE

area device, current density of small area device increases significantly in the switching process, which induced the high temperature under same bias, thus the enhanced switching speed which can be explained by local heat effect.

Based on empirical estimates from literature, we expect the effective temperature of nano devices to be approximately 100°C higher than that of large area device. However, in order to obtain more precise simulation results, further study for thermal barrier calculation is required, and other simulation method should be introduced into the calculation process.

Conclusions

This paper has clarified some properties of RRAM devices resulting from area scaling down for clarification of the switching mechanism. When adopting device area scaling, three phenomena can be confirmed: 1) variation of electric characteristics; 2) improvement of uniformity; 3) local heating effect. It is shown that the process of formation of conducting filaments is greatly affected by the scaling of device area. The present work can provide a feasible way for simulating RRAM operations and obtaining reliable and uniform device for future high-density memory application.

Acknowledgments

This work is supported by the National Natural Science Foundation of China（Grant No 61274113，11204212），and Program for New Century Excellent Talents in University （Grant No NCET-11-1064），and Tianjin Natural Science Foundation （ Grant No 10SYSYJC27700 and 10ZCKFGX01200），and Tianjin Science and Technology Developmental Funds of Universities and Colleges（Grant No 20100703）．

References

[1] R. Waser, et al., Adv. Mater. 21, pp.2632-2663, July 2009.

[2] N. Banno, et al., IEEE Transactions on Electron Devices, 55, pp. 3283-3287, December 2008.

[3] Y. Li, et al., IEEE Electron Device Letters, 31, pp. 117-120, February 2010.

[4] K. Liu, et al., ECS Transactions, 44(1), pp. 93-98, 2012.

[5] G. Bersuker, et al., Journal of Applied Physics, 110, pp. 124518 , 2011.

[6] D. Ielmini, IEEE Electron Device Letters, 58, pp. 4309-4313, December 2011.

On Pairing Bipolar RRAM memory element with novel punch-through diode based selector: Compact modeling to array performance

R. Mandapati, A. Borkar, V.S. S. Srinivasan, P. Bafna P. Karkare, S. Lodha, U. Ganguly*

Dept of Electrical Engineering, IIT Bombay, Mumbai, India 400076 email: *udayan@ee.iitb.ac.in phone: +912225767698

Abstract

To reduce sneak currents in high density non-volatile Bipolar RRAM technology the bipolar selector diode with high on-current density and larger on-off current ratio is required. Recently, we have experimentally demonstrated an n+/p/n+ stack based epitaxial Si punch-through diode for selector application with excellent TCAD matching. This selector technology provides flexibility in on-voltage (V_{on}), on-current (I_{on}) and on-off current ratio. Here we present a performance evaluation of bipolar RRAM array using NPN selector. First we develop a compact circuit model of the novel punch-through diode. Second we develop a methodology of pairing a specific NPN selector with a bipolar RRAM memory based on cross-point requirements. SPICE implementation based array performance analysis shows an optimal cross-point on-off current ratio (e.g. $3x10^4$ for 1M array) for minimum array power.

Keywords: *Bipolar RRAM, Cross-point memory array, Punch-through selector, Resistance ratio.*

I. Introduction

Large bipolar RRAM cross-point arrays are enabled by a memory element (1M) in series with a selector (1S) to reduce the sneak currents [1]. The promising bipolar RRAM memory elements span a space of symmetrical set/reset voltages ($V_{set}=V_{reset}$) with lower set currents (1μA-10μA) [2]-[7]. Recently, we have experimentally demonstrated a punch-through diode with epitaxial-Si n+pn+ stack for Bipolar RRAM selector application [8]-[9]. The performance of the punch-through selector compares favorably to other selector devices [1], [10]-[12]. The punch-through selector device can be designed to a range of on-voltage (V_{on}) for required on-current (I_{on}) and on-off current ratio. As the designability in performance of the punch-through diode is established, the methodology of pairing a specific selector with a chosen memory element based on array performance needs to be developed. In this paper, we developed a compact circuit model of novel punch-through selector and present a methodology of *selector pairing* with chosen memory element based on array performance. The existence of an optimum on-off current ratio (K_I) of a cross point for minimum array power is demonstrated using array performance based pairing.

II. Punch-through diode performance

We have recently proposed punch-through diode for bipolar RRAM selector application with excellent experimental and TCAD matching [8]. Experimentally validated TCAD simulation bench is used to generate punch-through (NPN) selector I-V characteristics at 20nm node [9]. As I-V is symmetric only positive I-V is shown in Fig. 1a. Selector on-off current ratio (K_{IS}- selector non-linearity) is extracted for a required on-current (I_{on}) at corresponding on-voltage (V_{on}) and off-current (I_{off-s}) at $V_{on}/2$ using V/2 biasing scheme [13]. Fig. 1a shows that a higher p-region length (e.g. L_P=52nm) device gives a higher K_{IS} value with higher on-voltage (V_{on}) for a required on-current (I_{on}). The selector device in series with memory element should able to pass require I_{set} for successful set operation. The selectors that meet this criterion ($I_{set}=I_{on}$) can be characterized by K_{IS} vs. V_{on} plot as shown in Fig. 1b which can be extracted from Fig. 1a at a given $I_{on}(=I_{set})$. This signifies the extent of non-linearity in current available versus the operating voltage penalty of the selector.

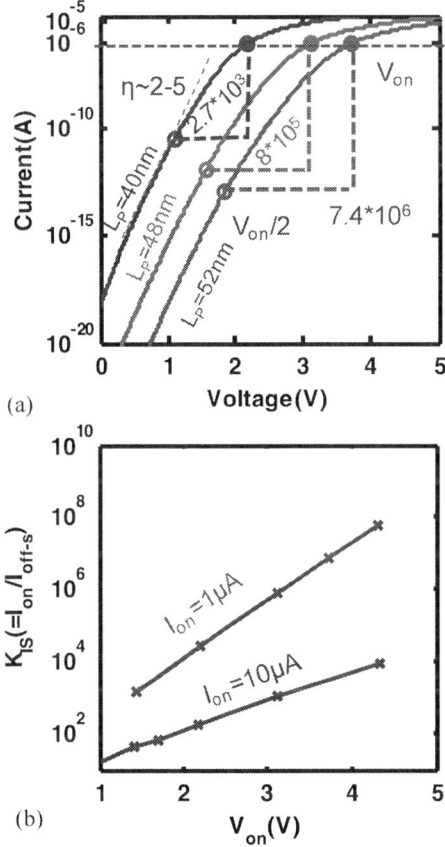

(a)

(b)

Fig. 1 (a) I-V characteristics of a punch-through (NPN) selector with N_A of 4×10^{18} cm^{-3} and different p-region lengths(L_P). Shows different $K_{IS}(=I_{on}/I_{off-s})$ values with V_{on}(closed circles) and $V_{on}/2$(open circles) voltages for a fixed I_{on}. (b) Selector performance in terms of K_{IS}(non-linearity of selector) versus V_{on}(required on-voltage) for different on-currents (I_{on}).

III. Circuit Compact models and Pairing

A single cross-point element (1S1M) composed of memory device (1M) in series with selector device (1S).

978-1-4673-4840-9/13 $31.00 © 2013 IEEE   309

A. Selector device

For cross point and array simulation, we modeled selector device (1S) as two diodes in parallel with a series resistance (R_s) for symmetrical I-V characteristics as shown in Fig. 2a. For single polarity it is simplified as a simple diode and a series resistor (R_s) as shown in Fig. 2b. For set current (I_{set}) the voltage across the selector during set operation i.e. $V_{set-s} = V_{on}$. The set current (I_{set}) through 1S circuit at V_{set-s} can be modeled as (1), similarly selector off current (I_{off-s}) at $V_{set-s}/2$ can be modeled as (2).

$$I_{set} = I_o e^{\frac{(V_{set-s}-I_{set}R_s)}{\eta k_B T}} \tag{1}$$

$$I_{off-s} = I_o e^{\frac{(V_{set-s}/2-I_{off-s}R_s)}{\eta k_B T}} \tag{2}$$

By dividing (1) and (2) and taking log on both sides, we get a simple relationship between selector non-linearity K_{IS} ($=I_{set}/I_{off-s}$), V_{set-s} and I_{set}.

$$\eta k_B T ln(1/K_{IS}) - I_{set}R_s(1-(1/K_{IS})) + V_{set-s}/2 = 0 \tag{3}$$

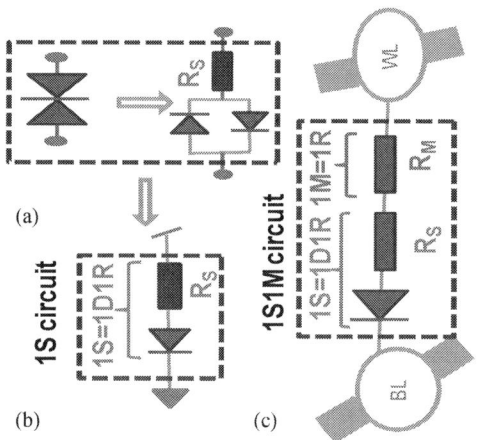

(a)

(b)   (c)

Fig. 2 Circuit schematic of (a) Bipolar selector as two diodes in parallel with series resistance (b) Selector (1S) as diode and series resistor for single polarity (c) Cross-point(1S1M) as memory element(1M) in series with selector(1S).

B. Memory Element

The resistive memory device can be defined as low resistance state (LRS) and high resistance state (HRS) with set and reset switching threshold voltages. For selector pairing and array performance, bipolar RRAM element (1M) is modeled as memory resistor (R_M) with V_{set}, I_{set} ($R_{LRS}=V_{set}/I_{set}$) and R_{HRS}. The memory element non-linearity is negligible as compared to selector non-linearity (K_{IS}).

C. Cross-point element

The circuit schematic of the cross-point is shown in Fig. 2c, where memory element (1M) is added in series with selector (1S). For set current (I_{set}) the voltage across the cross-point during set operation i.e. V_{set-c} ($=V_{set}+V_{set-s}$). Similar to (1) in selector device, the set current (I_{set}) through 1S1M circuit at

V_{set-c} can be modeled as

$$I_{set} = I_o e^{\frac{(V_{set-c}-I_{set}R)}{\eta k_B T}} \tag{4}$$

where $R=R_S+R_M$.

At cross-point off voltage i.e $V_{off-c}=V_{set-c}/2$, $I=I_{off-c}$ can be extracted from (4). The cross-point on-off ratio (K_I) can be calculated in a similar way as (3), which is given by

$$\frac{(V_{set-c}-I_{set}R)-(V_{set-c}/2-(1/K_I)I_{set}R)}{\eta k_B T} = -\ln(1/K_I) \tag{5}$$

$$K_V = (V_{set-c} - V_{set})/V_{set} = V_{set-s}/V_{set} \tag{6}$$

$$K_R = R_S/R_{LRS} \tag{7}$$

By substituting the definitions of K_V which is the ratio of voltage drop across selector and memory during set process (6) and K_R which is the ratio of the series resistance of selector and memory in low resistance state (LRS) (7), into (5), we get the linear relationship between K_V and η (diode ideality factor) for different K_I and K_R values.

$$K_V + \eta \frac{2k_B T ln(1/K_I)}{V_{set}} + 1 - \frac{2I_{set}R_{LRS}(K_R+1)(1-(1//K_I))}{V_{set}} = 0 \tag{8}$$

Equation (8) defines the linear relationship between K_V and eta (η-ideality factor of a diode) for different K_I and K_R values as plotted in Fig. 3. By using (1) and (8), for a chosen memory (V_{set}, I_{set}), we can get relationship between K_{IS}, K_V for a known K_R and given cross-point non-linearity (K_I) values as plotted in Fig. 4a. For required cross-point non-linearity (K_I), lower K_V (lower voltage drop across the selector) requires higher selector non-linearity (K_{IS}). To identify the choice of selector for desired cross-point non-linearity (K_I), we propose the following. We translate V_{set-s} to K_V ($=V_{set-s}/V_{set}$) in Fig1b which defines punch-through selector performance. We superimpose translated plot on selector performance requirement Fig. 4a as shown in Fig. 4b. The intersection point identifies the choice of punch-through selector corresponding to the memory element (V_{set}, I_{set}), for a given circuit requirement of cross-point non-linearity (K_I) to reduce sneak path currents.

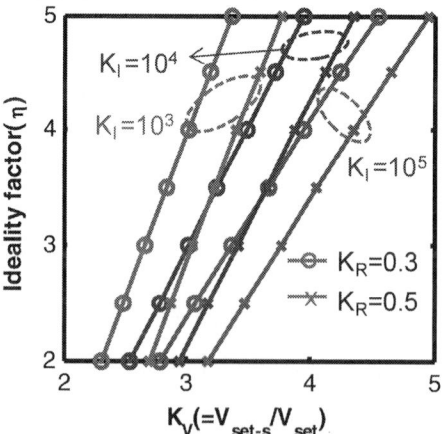

Fig. 3. Diode ideality factor (η) and K_V (defines voltage drop across selector for required I_{set}) are linearly related for a given K_R and K_I values.

978-1-4673-4840-9/13 $31.00 © 2013 IEEE   310

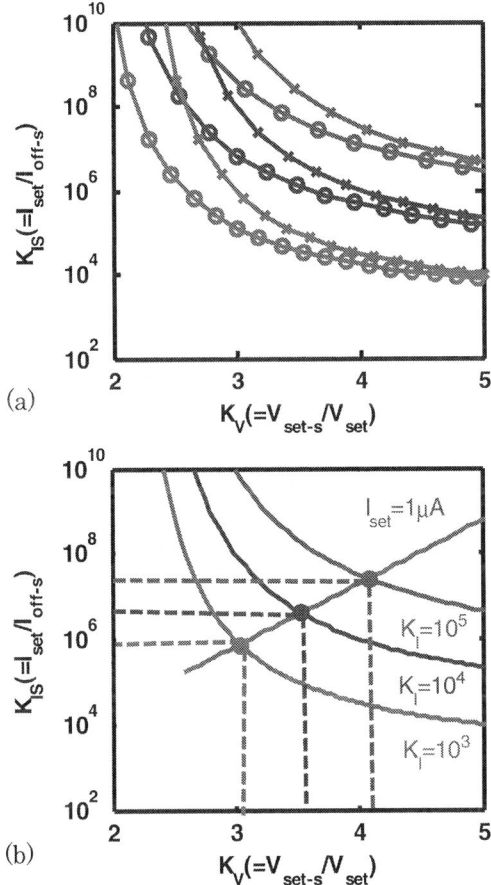

(a)

(b)

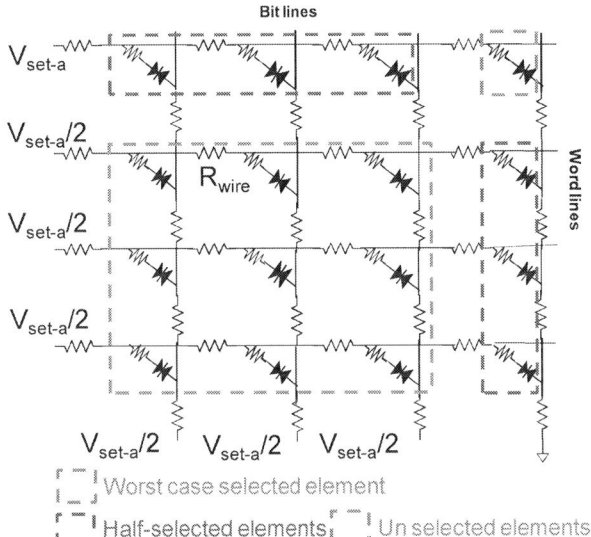

Fig. 5 Cross-point memory array with wire resistance (R_{wire}) using V/2 biasing scheme [13] for set operation. Worst case selected element (green) farthest from bias source with bias of V_{set-a}, half-selected elements (red) with bias of $V_{set-a}/2$, unselected elements (blue) with zero bias.

A. Power degradation in set operation

The V/2 biasing scheme is used for array set operation where the selected word line is biased at V_{set-a} and selected bit line at ground which could ideally allows voltage (V_{set-a}) across the selected element [13] as shown in Fig. 5. All other word lines and bit lines are biased at $V_{set-a}/2$ which creates some half-selected elements leads to excess power dissipation by sneak currents. The worst-case selected element is the farthest element from word-line/bit-line bias sources as shown in Fig. 5. The analysis is done during set operation of worst case selected element with all memory elements are in LRS state.

$$P_{set-a} = V_{set-c}I_{set} + \left(\frac{V_{set-c}}{2}\frac{I_{set}}{K_I}2(n-1) \right) \quad (8)$$

By replacing $V_{set-c} = V_{set}(1+K_V)$

$$P_{set-a} = V_{set}I_{set}(1 + K_V)\left(1 + \frac{(n-1)}{K_I}\right) \quad (9)$$

The total power dissipation of an n x n memory array for single bit set operation is given by (8). The first term is set power of single cross-point element and second term is excess power due to sneak currents through half-selected elements. By replacing V_{set-c}, we get (9), which shows 2 factors of power dissipation. The first factor represents excess power due to sneak currents through half-selected elements, scales as n/K_I (as n>>1). The second factor of excess power is by the factor K_V, which represents voltage drop across the selector to pass set-current (I_{set}). The power dissipation of different array sizes with cross-point non-linearity (K_I) is shown in Fig. 6a. It shows two regimes of K_I dependence. For lower K_I array power decreases as K_I increases (K_I dependent regime) and for higher K_I power dissipation increases independent of array

Fig. 4 a. Selector performance requirement for cross–point non-linearity (K_I) with known memory (V_{set}=1V, I_{set}=1μA). For specified K_I lower K_V (lower drop across the selector) requires higher K_{IS} value. (b) Intersection points are selector performance requirement and punch-through selector performance based on Fig. 1b by converting V_{set-s} to K_V (=V_{set-s}/V_{set}) for known memory (V_{set}, I_{set}).

Voltage penalty required across the selector for given cross-point non-linearity (K_I) is shown in Fig. 4b. This establishes a facile methodology for memory/selector pairing.

IV. Array Performance

In cross-point memory array, the cross-point element is placed at intersection of word and bit lines [13]. The read operation of the cross-point will be at lower voltage than set operation [13]. So the power dissipation of the cross-point memory array is defined by set operation. A representative memory element (arbitrarily chosen as V_{set}=1V, I_{set}= 1μA typically in literature [2]-[7]) is paired with a selector to generate various cross-point on/off current ratio (K_I) and consequent excess voltage across the selector(K_V) by the methodology developed in compact model and pairing section. The effect of K_I on array power dissipation metric is presented. All simulation are done using H-SPICE [14] with an interconnect resistance of 2.5Ω between adjacent cross-point elements, for a 4F² cross-point structure according to the ITRS road map for 22nm-techonology node [15].

978-1-4673-4840-9/13 $31.00 © 2013 IEEE   311

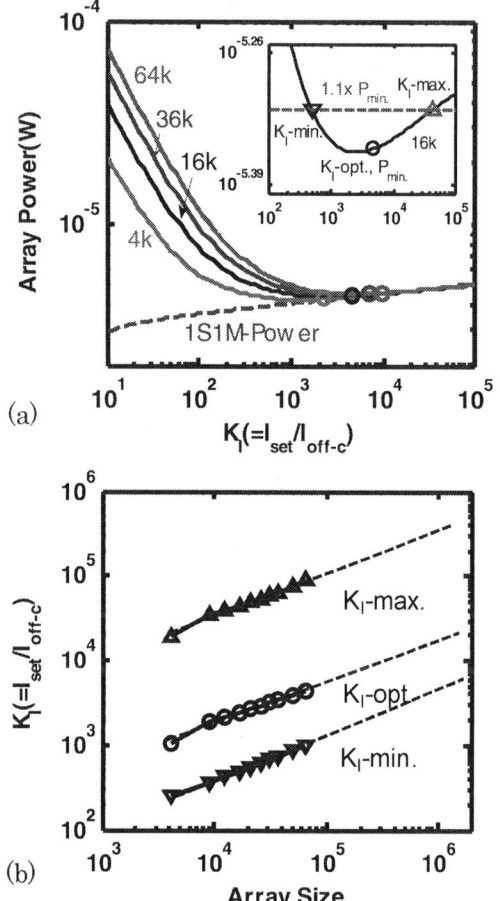

(a)

(b)

Fig. 6 a. Array and single cross-point power vs K_I for different array sizes, open circles are minimum power (K_I-opt.). K_I-min. and K_I-max. are extracted for 10% of P_{min} (see inset). (b) The various K_I's are plotted with array size shows that K_I of 10^4 to 10^5 is optimum for 1M array size.

size (single cross-point (1S1M) power) due to higher V_{on} for higher K_{IS} (Fig. 1b) or due to the factor K_V from (9). So the optimal K_I (K_I-opt.) exist for minimum array power. We define K_I-min. and K_I-max. with 10% of minimum power. Fig. 6b shows extraction of K_I-min., K_I-max. and K_I-opt. for different array sizes and it also shows optimum K_I of 3×10^4 for 1M array size.

V. Conclusion

In summary, we developed a methodology of pairing the punch-through diode with chosen memory element based on array performance. Higher cross-point non-linearity can be achieved with voltage penalty across the selector. The compact circuit model of a punch-through diode is developed and implement in HSPICE for array performance analysis with cross-point non-linearity (K_I) as a metric. The optimal K_I (e.g. $\sim 10^4$ for 1M array size) exists after which power performance actually degrades with increasing K_I as opposed to simply diminishing return with higher K_I.

References

[1] A. Kawahara, R. Azuma, Y. Ikeda, et al., ISSCC Dig. Tech. papers, pp. 432, Feb. 2012.

[2] Y. Y. Chen, L. Goux, J. Swerts, et al., IEEE Electron Device Lett., vol. 33, No. 4, pp. 483-485, 2012.

[3] Y–H. Wu, J–R. Wu, C –Y. Hou, et al., IEEE Electron Device Lett., vol. 33, No. 3, pp. 435-437, 2012.

[4] H. Heng, P. C. Chen, Y. H. Wu, et al., Solid-state electronics, vol. 73, pp. 60-63, 2012.

[5] H. Ho, C–L. Hsu, C-C. Chen, et al., IEDM Tech. dig., pp 436-439, 2010.

[6] X. Liu, K. P. Biju, S. Park, et al., Current Applied Physics 11, e58-e61, 2011.

[7] X. Liu, I. Kim, M. Siddik, et al., Journal of Korean physic society, vol. 59, No. 2, pp. 497-500, 2011.

[8] V. S. S. Srinivasan, S. Chopra, P. Karkare, et. al, IEEE Electron Device Lett., vol. 33, no. 10, pp. 1396, 2012.

[9] P. Bafna, P. Karkare, S. Srinivasan, et al., Device Research conference, pp. 115, 2012.

[10] J. J. Huang, Y. M. Tseng, C. W. Hsu, et al., IEEE Elect. Dev. Letters, vol. 32, no. 10, pp. 1427, 2011.

[11] K. Gopalakrishnan, R. S. Shenoy, C. T. Rettner, et al., Symposium on VLSI technology, pp. 205, 2010.

[12] M. Son, J. Lee, J. Park, et al., IEEE Electron Device Lett.,vol. 32, no. 11, pp. 1579, 2011.

[13] J. Mustafa, Ph. D dissertation, Dept. Elect. Eng., RWTH Aachen Univ., Aachen, Germany, 2006.

[14] "http://www.synopsys.com/tools/verification/amsverification/circuitsimulation/hspice/Pages/default.aspx".

[15] ITRS, "International Technology Roadmap for semiconductors", 2007.

NEMS meets Bio-sensing;
There're plenty of things to do in the middle

Beomjoon Kim

CIRMM, Institute of Industrial Science, The University of Tokyo
4-6-1, Komaba, Meguro-ku
Tokyo 153-8505, Japan
Tel:+81-3-5452-6224, fax:6225 email: bjoonkim@iis.u-tokyo.ac.jp

Abstract

We propose a new fabrication method for nanofluidic device with novel shapes using thermal shadow evaporation and wet anisotropic etching in aqueous KOH solution. The shadow evaporation of metal onto the sidewalls of topographic features, followed by the selective etching of the substrate, generates narrow channel structures with nanometer scales. And the wet anisotropic etching of silicon substrate in KOH solution provides the well-defined fluidic channels with triangular cross-sections as well as chamber structures with micrometer scales. We also demonstrate that a single DNA molecule could be driven in the nanofluidic device by electrophoresis, in which the linear stretching and electrophoretic migration of a DNA polymer molecule could be observed at a single molecule level. Finally, we investigate length-dependent mobility of chromosome-sized DNA in nanofluidic channel and successfully separate different lengths of DNA under simultaneous action of electric field as well as pressure gradient.

Introduction

Our research goals are to build nanosystems and fabricate nanoscale devices, in particular for bio-sensing in singular level, through both bottom-up and top-down approaches.

We focus on interdisciplinary research about local "bottom-up" surface modification using functional self-assembled monolayers and "top-down" approaches for micro/nano patterning technologies. Key technologies concentrate on high-resolution surface patterning with simple, low-cost techniques such as micro-contact printing (µCP), flexible polymer based soft lithography, and micro even nano shadow-masks patterning. In spite of its great versatility, µCP has still many difficulties in the application as a practical patterning technology for a large area. Therefore, we developed the optimized µCP methods in liquid environment and designed hybrid stamp and stamping device to increase the uniformity on the pattern and decrease the deformation of a stamp. Moreover, liquid-µCP technique reduced the collapse of PDMS tips and the diffusion of SAM molecules so that it showed the possibility for practical application to the nano-patterning process in a large area.

Based on these studies on nano/micro components systems for the fabrication of novel nano devices, we investigate to develop various micro sensors for biological applications, such as i) CMOS compatible fabrication of top-gated FET silicon nanowires for detection of proteins, pH level, even metastatic related cancer makers and label-free biosensor components, ii) temperature measurement on resistively heated nanowires for the study on single molecules, and iii) arbitrary-shaped nanochannels fabrication to achieve single DNA stretching, etc.

A single biomolecule, DNA now draws much attention, since relevant dimension of nanometer level chips are possible to be made by nano-fabrication techniques. Among many DNA analysis devices, recently nanochannel is highlighted as it provides a proper platform based on DNA stretch phenomenon inside nanochannel. We will continue on the development of complete fabrication of these nanochannels and deep investigation with various DNA or enzyme, bio molecules. Finally, we now aim to realize a tool for the study of temperature dependent phenomena of biomolecules, e.g. DNA and proteins, at a single molecule level.

On the other hands, thermal conductivity in nanoscale, specially affected by contribution of surface phonon-polaritons (SPPs), will be investigated with micro/nano heaters. We aim at the modeling, the fabrication and the characterization of micro/nanostructures (glass tubes) exhibiting anomalous thermal conductivity due to the contribution of SPPs. The dependence to temperature and to nanostructure sizes will be explored in order to possibly reveal several SPPs features such as attenuation length and predominant wavelength.

Here, three topics about NEMS for bio/molecular engineering are introduced in brief.

Nanofluidic device to achieve length dependent mobility of long DNA molecules & separation

The uniformly charged polymer such as DNA molecule moves with length-independent mobility in the electric field because the friction force is proportional to DNA contour length as well as the electrostatic force. This size-independent migration prevents separation in free buffer solution, and thus the sieving matrix such as agarose gels should be used. However, the DNAs above a critical length (typically ~20,000 basepairs) show the length-independent electrophoretical mobility even in sieving matrix, because the long DNA molecule becomes highly oriented along the direction of electric field in the gels. As a result, the pulsed field gel electrophoresis (PFGC) is generally used for the long DNA separation, which is typically one-day process.

To achieve the size-dependent behavior of the long DNAs, a novel concept, the electrophoresis under pressure gradient, is proposed. A fluidic device of nanoslit style is fabricated on silicon wafer with microfabrication technique. Then, the electrode for electrophoresis was patterned on the fluid access holes and the PEEK® tubes for hydrodynamic pressure was installed and connected with a high performance liquid chromatography (HPLC) pump. The electric potential and the hydrodynamic pressure were applied simultaneously, but with opposite direction. As a result, the different two kinds of DNA show the length-dependent behavior, where YOYO-I stained λ-DNA (48.5 kbp) and T4-DNA (166kbp) were used as the standard of long DNA molecules. The figure 1 shows fluorescent sequential movie frames which two different lengths of long DNA electrophoresis under pressure gradient in nanoslit [1-2].

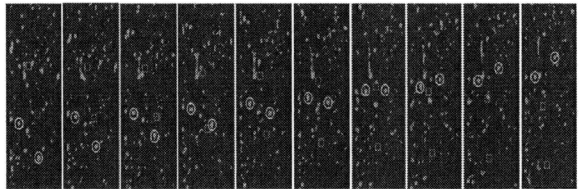

Fig. 1 DNA electrophoresis under pressure gradient in nanoslit.
(The time gap between frames is ~ 300 ms. The height of the slit is ~ 200 nm, fabricated on silicon wafer. The bigger DNAs (T4 DNA; the white circles) move along the direction of the electric field, while the smaller DNAs (λ-DNA; the red rectangles) move oppositely, along the direction of the pressure gradient. The width of the frame is 50 μm, which is also the width of the nanoslit device.)

Supramolecular interaction between biomolecules and calixerenes capped silver nanoparticles
(Joint research with Prof. A. W. Coleman from Univ. of Lyon and D. Collard, H. Fujia from IIS)

The calix[n]arenes are one of the most widely studied organic host classes. In recent years their interactions with biological molecules, and indeed their biochemistry have been widely studied. In contrast to the large body of work on their complexation with amino acids, peptides or proteins the study of their interactions with nucleosides, nucleotides and RNA or DNA is sparse.

Very recently we have shown that nucleotides, nucleosides and deoxynucleosides bind differentially to para-sulphonato-calix[4]arene capped silver nanoparticles again inducing aggregation. A possible mechanism at the molecular level can be derived from the solid state structure of the cytosine para-sulphonato-calix[4]arene complex which shows cytosine hydrogen bonded bridging of the calix-arenes [3].

To futher investigate the change of plasmonic properties caused by the complexation between para-sulphonato-calix[4]arene capped silver nanoparticles and biomolecules, we propose to develop a new nanoheaters microsystem. Furthermore, a micro electro-mechanical system (MEMS) -based nano-tweezer is used to measure the effects of the interaction of calix[n]arene capped silver nanoparticles on the mechanical properties of DNA molecules [4].

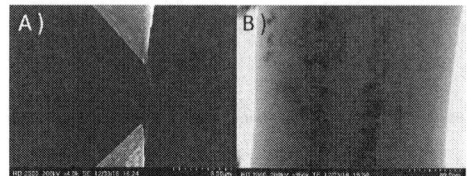

Fig. 2 TEM picture of tweezer + DNA immersed inside nanoparticles solution (A) shows the bundle of DNA (B) shows silver nanoparticles along the DNA.

Heat Transfer in Micro/nano glass tubes mediated by Surface Phonon Polariton
(Joint research with Prof. S. Volz from ECP, Paris)

We investigate thermal nanostructures that allow the Surface Phonon Polariton guiding. The glass tubes were designed to guide monochromatic thermal energy to improve heat conduction or serve as radiative source. Surface Phonon Polaritons (SPPs) are surface waves produced by the coupling of atomic vibrations with the electromagnetic field. Because they are monochromatic waves in the Infra-Red, SPP might able to provide monochromatic sources just based on simple heating. At ambient temperatures, they also might produce abnormal heat conduction phenomena related to ballistic regimes of transport. Considering heat transfer in the propagative direction of SPPs, quite recent works have tried to show heat conduction improvement in nanofilms.

However, the coupling is not favored in this configuration because the film is dissipating the electromagnetic energy and the coupling distance is too small. We aim at designing glass microtubes in order to measure the SPP heat flux and its abnormal behavior. The modelling of the tube consists in solving the Maxwell equations in the approximation of small wall thickness and small outer diameter compared to the polariton wavelengths. The experimental set-up for proving abnormal heat conduction consists in sticking a glass capillary sample on an AC heating Pt wire on one end and measuring the temperature on the other end by using a deposited thermocouple. The phase between heating and temperature signals directly provides thermal diffusivity.

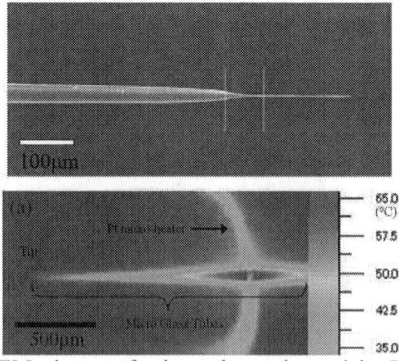

Fig. 3 SEM picture of micro glass tube and its IR-thermo image with Pt micro heater in vaccum chamber.

The first theoretical results reveal a very strong enhancement of the Free Path appearing in a narrow range of tube radius. Experimental works have proven that the signal to noise ratio

978-1-4673-4840-9/13 $31.00 © 2013 IEEE

was satisfying and the spectrum of the temperature response was obtained. The temperature profile by using IR microscopy should yield the proof of the SPP contribution in the heat transfer along the tube [5].

Micro/nano patterning on Nonplanar Substrates and Roll Micro contact Printing

Micro structuring on nonplanar substrates has not been fully established yet, although several fabrication methods, such as laser direct writing and modified photolithography, were proposed. Moreover, those techniques require still expensive and complex processes. To overcome those drawbacks, optical softlithography using flexible photomasks was developed. Firstly, SU-8 micro structures were fabricated on concave substrates by optical softlithography using PDMS flexible photomasks. As a result, SU-8 structures with 2.5 μm line width and high aspect ratio over 7.9 were fabricated on a concave substrate. Also, experimental parameters for optical softlithography were investigated and established for further fabrications and applications. In addition, the tilting structures were confirmed due to the vertical UV exposure method.

Next, Based on these novel 3D patterning technologies, PDMS roller stamps were fabricated and roll microcontact printing was performed. The roll microcontact printing was investigated by using the customized roll microcontact printing apparatus and the flexible pressure sensor system. We expect this technique can provide various sized roller stamps with various micro patterns for μCP process as well as roll microcontact printing process.

Micron Scale Wire Electrode Heating for Single Cell Level Analysis

The multiple biochemical reactions that occur in individual cells are extremely sensitive to temperature variations. Among the many reactions inherent to the cell, an abundant literature addresses those related to stresses where heat shock proteins (HSPs) are involved. The current temperature range observed for triggering heat shock is obtained under thermostat conditions with isothermal heating of the cell culture. However, understanding single cell intracellular mechanisms requires local and short time analysis. The research project aims to demonstrate for the first time the heat shock conditions of a single cell. Metal electrodes with micron scale width allows for heating less than a dozen of cells in a confluent layer at predictable temperatures up to 85°C with accuracy lower than 2°C.

Those performances were obtained by a preliminary robust temperature calibration based on Biotin-Rhodamin fluorescence and by designing relevant rules for device fabrication as well as for heating protocol. Inducing the expression of heat proteins in four transfected NIH-3T3 cells through a predicted confined and precise temperature rise proved the temperature accuracy. Electrode deposition being one of the most common processes, the fabrication of electrodes array with a simple control circuit appears as a simple step. This configuration allows for probing several key mechanisms such as, for instance, cell shock, thermoporation, death and signaling.

Figure 4 is a picture of the heated cells after 12 hours and shows that some of the cells around the micro-heater were induced GFPs. It indicates that the area on which GFP expressed cells were placed was approximately 40 degrees because the GFPs were expressed when the temperature was around 40 degrees by an hour heating.

We have proven that the temperature range for heat shock conditions was the same as the ones applied with thermal bath techniques. We also showed that the over heated cells were not dead, that there was no HSPs signaling between cells and finally that the cell were unaffected by extremely high temperature gradients[6]. Perspectives concern cell thermoporation based on ultra-thin hot wires. We also foresee Ca2+ signaling during heat shock and intend to reveal it.

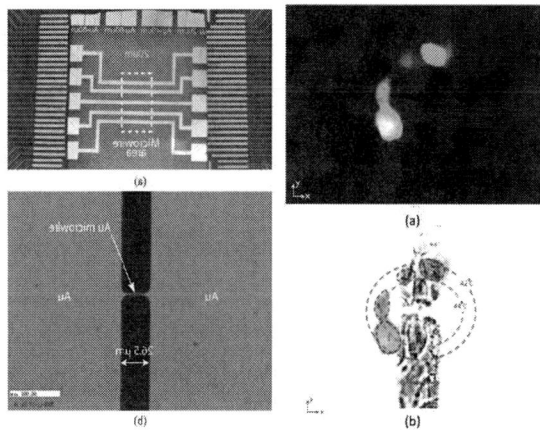

Fig. 4 (left a) Image of a device consisting of five gold electrodes deposited on a glass substrate. (left b) Optical view of a gold microwire 3 microns in width deposited on a glass substrate with connections to two electrodes. (right a) Provides the raw fluorescence images revealing the GFP (green color) after the heating period. (right b) reports the superposition of the fluorescence and the optical signals.

References

[1] K. Park, B.J. Kim, *2012 MRS Spring Meeting & Exhibit*, April 9-13, California, 2012 (10.April, oral presentation/1269295, NN3.4).

[2] B.J. Kim and K. Park, *The 16th International Conference on Miniaturized Systems for Chemistry and Life Sciences (MicroTAS 2012)*, Proceedings of MicroTAS 2012, pp. 1750-1752, 2012.

[3] Y. Tauran, A. Brioude, P. Shahgaldian, A. Cumbo, B.J. Kim, F. Perret, A. W. Coleman and I. Montasser, *Chemical Communications (Chem. Commun.)* Vol. 48, Issue 76, pp. 9483-9485, 2012.

[4] Y. Tauran, M. Kumemura, N. Lafitte, R. Ueno, L. Jalabert, Y. Takayama, D. Collard, H. Fujita, A. W. Coleman, and B.J. Kim, *The 16th International Conference on Miniaturized Systems for Chemistry and Life Sciences (MicroTAS 2012)*, Proceedings of MicroTAS 2012, pp. 1882-1884, 2012.

[5] T. Tokunaga, L. Tranchant, N. Takama, S. Volz and B. Kim, *Journal of Physics*, conference series, 2012 in press.

[6] P. Ginet, K. Montagne, S. Akiyama, A. Rajabpour, A. Taniguchi, T. Fujii, Y. Sakai, B. Kim, D. Fourmy and S. Volz, *Lab Chip*, Vol.11, Issue 8, pp.1513-1520, 2011.

Novel Quantum Effect Devices realized by Fusion of Bio-template and Defect-Free Neutral Beam Etching

Seiji Samukawa[1,2,*]

[1]Institute of Fluid Science, Tohoku University, Japan,
[2]Japan Science and Technology Agency (JST), CREST, Japan
* TEL/FAX:+81-22-217-5240, E-mail: samukawa@ifs.tohoku.ac.jp

Abstract

An original top-down process involving a bio-template and damage-free neutral beam etching (NBE) has been developed to fabricate a high-quality nanodisk superlattice. The self-assemble ferritin 2D array (iron core diameter: 4.5 nm) acts as uniform etching mask and our developed neutral beam etching eliminates UV photons and high-energy charged particles to achieve a damage-free etching. As a result, we have fabricated a high-quality nanodisk superlattice with a high density (1.4×10^{12} cm^{-2}), uniform size (Si nanodisk diameter: 6.4 nm), and well-ordered arrangement (hexagonal close packing). The Si nanodisk 2D array with SiC interlayer had an extremely high optical absorption coefficient and high carrier transport due to the formation of a wide miniband. This advanced nano-process brings high-efficiency quantum dot (QD) solar cell (SC) .

Introduction

Recently, an all-silicon tandem solar cell comprising a quantum dot superlattice (QDSL) has attracted much attention due to its potential to breakthrough the Shockley-Queisser limit.[1,2] One of the advantages of the QDSL is that the required energy band gap for each cell can be engineered by changing the quantum dot size.[3] Reportedly, the maximum conversion efficiency can be improved up to 47.5% for three-cell tandem stacks.[4] However, not only the uniformity and control of QD size but also of the spacing between QDs are equivalently essential to generate the miniband in the QDSL for carrier transport.[5] The ideal spacing between QDs is approximately 2 nm or less in the SiO_2 matrix.[6] The technique widely used to fabricate the Si quantum dot superlattice is depositing alternately multiple layers of amorphous silicon-rich oxide (SiO_x, x<2) and stoichiometric silicon dioxide (SiO_2) by sputtering or plasma-enhanced chemical vapor deposition followed by annealing at a high temperature.[6,7] However, the results showed nonuniform dot size and dot spacing.

To address these problems, we have developed a sub-10nm-silicon-nano-disk (Si-ND) structure using the bio-template (ϕ7-nm-etching-mask) and damage-free chlorine (Cl) neutral beam (NB) etching.[8] The fabricated ND had a quantum effect, i.e. Coulomb staircase, at room temperature (RT). Two geometrical parameters of thickness and diameter in Si-ND can be independently controlled. Interestingly, the quantum effect of a single Si-ND is strongly dependent on its thickness, while almost independent of its diameter.[8] In this study, a 2D Si ND array with a high-density and well-ordered arrangement could be fabricated by using bio-template and an etching process combined with nitrogen trifluoride (NF_3) gas/hydrogen radical treatment (NF_3 treatment) and Cl NB etching. In this structure, the controllable band gap energy (from 2.2eV to 1.4eV) and high photon absorption coefficient ($>10^5$ cm^{-1}) could be obtained at RT by controlling the Si-ND structure.

Fabrication of high-density 2D array of Si-ND

The fabrication of a 2D Si-ND array using the bio-template and damage-free NB etching[8] is schematically shown in Fig. 1(a). The steps are as follows: multilayer films of 1.4-nm SiO_2, several nm-thick poly-Si and 3-nm SiO_2 (the 3-nm SiO_2 was fabricated by our developed neutral beam oxidation at a low temperature of 300 °C and is called NBO SiO_2 hereafter) were sequentially prepared on a Si wafer as shown in Fig.1(1), Fig.1(2), and Fig.1(3), respectively; (4) a 2D array of ferritin molecules (protein including iron oxide core (Fe-core) in the cavity) was placed through directed selforganization on the surface of NBO SiO_2; (5) ferritin protein shells were removed by heat treatment in oxygen atmosphere to obtain 2D Fe-core as a template; (6) etching was carried out using a NF_3 treatment and Cl NB etching to remove NBO SiO_2 and poly-Si, respectively; (7) and finally 2D Fe core was removed by using hydrochloric solution. The sample underwent NF_3 treatment for 30 min to remove NBO SiO_2 and NB etching for 90 seconds to remove 4-nm poly-Si. Figure 2 shows a SEM image of the top view of the sample after etching. We can see that the 2D Si-ND array has a high-density ($>7\times10^{11}$ cm^{-2}) and well-ordered arrangement. The 2D array is what remained after etching, proving that a good-quality 2D Si-ND array was successfully fabricated using the bio-template and Cl NB etching with NF_3 treatment. We performed NF_3 treatment to investigate the controllability of the ND diameter, i.e. the spacing between NDs. When the NF_3 treatment times were 15 and 30 min, the average gaps were about 1 and 3 nm (G_{ii} and G_{iii}), and the diameters were about 10 and 8 nm (D_{ii} and D_{iii}), respectively . These results suggest that the spacing between adjacent NDs can be controlled by changing the NF_3 treatment time, which also indicates that the formation of miniband in a 2D Si-ND array can be controlled. Although the spacing control by NF_3 treatment is accompanied by inevitable changes in diameter, as shown in Fig. 4, the diameter changes do not affect the quantum effect, which was proven in a previous work.[8]

978-1-4673-4840-9/13 $31.00 © 2013 IEEE

Optical Properties of 2D Si-ND array

The absorption properties of the structure were studied by measuring the transmission for samples by UV-vis-NIR. The absorption coefficient has been calculated in accordance with the equation below[9]

$$T = e^{-\alpha d}$$

α being the absorption coefficient, d the total thickness of the ND thickness and surface oxide thickness (3-nm thick), and T the transmittance of light passing through the structure. Figure 3(a) shows the results of an absorption coefficient of the structure as a function of ND thickness. We found that the absorption spectra strongly depend on the ND thickness and the absorption edge is blue-shifted when the ND thickness decreases due to the quantum size effect. Additionally, the absorption coefficient ($>10^5$ cm^{-1}) of 2D Si-ND array is extremely high, and therefore it is possible to obtain sufficient absorption if the NDs can be integrated into the 3rd dimension. To determine the optical band gap energy of the structure, the Tauc formula was used:

$$(\alpha h v)^{1/2} = A(h v - E_g),$$

where A is a constant, h is Planck/s constant, v is frequency, E_g is the band gap energy, and n is 1/2 in the case of indirect allowed and forbidden electronic transitions. The Tauc formulation as a function of ND thickness is plotted in Fig. 3(b). As the ND thickness changes from 2 to 12 nm, the E_g could be controlled from 2.2 to 1.4eV as shown in Fig. 4. From these results, we found that E_g could be certainly controlled by simply changing ND thickness by thin-film deposition in our proposed fabrication Based on the processes, all-Si tandem solar cells assembled with 3D ND array fabricated by stacking 2D Si-ND array as schematically shown in Fig. 5 could be constructed.

Conclusions

We created a 2D Si-ND array with a high-density and well-ordered arrangement using bio-template and an advanced etching process that included NF$_3$ treatment and damage-free Cl NB etching. The spacing between Si NDs can be controlled in the structure by changing NF$_3$ treatment time. The E_g can be easily controlled by changing the ND thickness during thin film deposition. The absorption coefficient of single layer 2D Si-ND is comparable to that of 3D QDSL. Our proposed processes for 2D Si-ND array and stacked ND are very feasible for the all-Si tandem solar cells comprising QDSL.

References

1) W. Shockley and H. J. Queisser, J. Appl. Phys. **32** (1961) 510.
2) M. A. Green et al, 20th EU-PVSEC (2005), 1AP.1.1.
3) A. Kongkanand et al., J. Am. Chem. Soc. **130** (2008) 4007.
4) G. Conibeer et al., Thin Solid Films **516** (2008) 6748.
5) A. J. Nozik: Physica E **14** (2002) 115.
6) E. C. Cho et al., Nanotechnology. **19** (2008) 245201.
7) Y. Kurokawa et al., Jpn. J. Appl. Phys. **45** (2007) L1064.
8) S. Samukawa et. al., Appl. Phys. Express 1 (2008) 074002.
9) B. Pejova and I. Grozdanov, Mater. Chem. Phys. 90 (2005) 35.

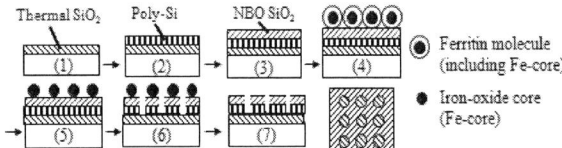

Figure 1. Fabrication flow of 2 dimensional Si nano-disk array by bio-template and chlorine neutral beam etching.

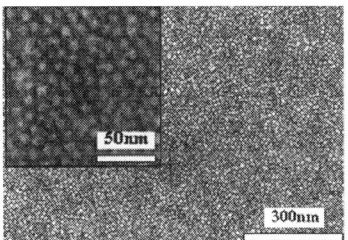

Figure 2. SEM images of 2 dimensional Si nano-disk array fabricated by Cl neutral beam etching with bio-template.

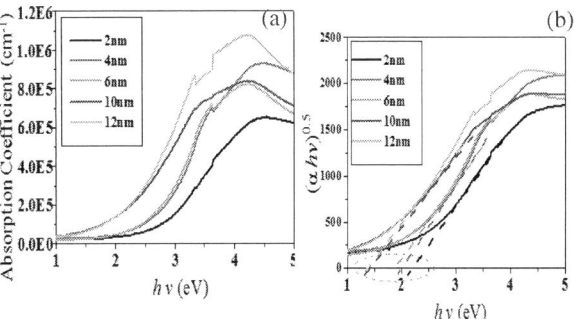

Figure 3. (a) Absorption coefficient (b) Tauc plot of 2 dimensional Si nano-disk array with different nano-disk thicknesses from 2 nm to 12 nm..

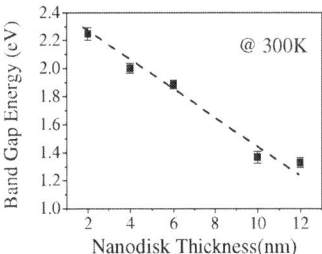

Figure 4. Band gap energy (E_g) of nano-disk with different Si nano-disk thicknesses by using UV-vis-NIR.

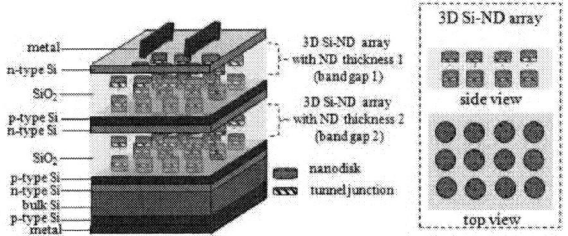

Figure 5. Scheme of all-silicon tandem solar cell assembled with 3 dimensional Si nano-disk array.

On Controlling EBL Parameters for Nanoelectromechanical Resonators Fabricated on Insulating/Semiconducting Structures

Ali B. Alamin Dow[1], H. Lin[1], C. Popov[2], U. Schmid[3], Nazir P. Kherani[1]

[1]Department of Electrical and Computer Engineering, University of Toronto, 10 King's College Road, ON M5S 3G4 Canada.
[2]Institute of Nanostructure Technologies and Analytics, University of Kassel, Heinrich-Plett-Str. 40, D-34132 Kassel, Germany
[3] Institute of Sensors and Actuator Systems, Vienna University of Technology, Floragasse 7, A-1040 Vienna, Austria.
Phone: +1- 416 -946- 7372, Fax: +1- 416 -946-0054 and E-mail: alamindow.alamindow@utoronto.ca

Abstract

The current work details the development and optimization of the fabrication processes for nanoelectromechanical resonators such as surface and bulk acoustic wave (SAW/BAW) devices that operate in GHz range, specifically based on nano-interdigitated transducers (n-IDTs). The method combines electron-beam lithography (EBL) and lift-off process to fabricate the n-IDTs with finger patterns having line widths of the order of 100 nm on AlN/UNCD(aluminum nitride/ultrananocrystalline diamond) combined structures deposited on crystalline silicon. The widespread availability of this method has led to the study of the combination of processing parameters for both EBL and lift-off for its application in the realization of n-IDTs. The fabricated devices exhibited high frequency range up to 15 GHz with minimum stop-band rejection of 25dB. Excellent filtering response of the devices and the compatibility of fabrication processes with existing manufacturing technologies pave the way towards advanced AlN/UNCD based nano-resonantors.

Introduction

Being an inseparable part in high volume data processing and emerging sensing complexity [1,2], it is essential to develop advanced and reliable fabrication techniques for the realization of high efficient devices such as nanoelectromechanical (NEMS) resonators [3, 4]. A comparison between conventional optical photolithography and electron beam lithography (EBL) confirms the advantages of EBL over conventional photolithography at minimum through its higher resolution, shorter wait time associated with preparation of photomasks, and cost effectiveness at the laboratory level. On the downside, this method leads to charge accumulation and dose variation effects. The positive and negative aspects of the method have led to the study of the combination of EBL and lift-off processes [4, 5, 6] and its applications in the fabrication of nano-interdigitated transducers (n-IDTs), used in surface acoustic wave (SAW) devices in the GHz range [7, 8, 9]. The study has been carried out on AlN thin films with varying the resolution of the structures. The fabrication of high resolution n-IDTs(<140 nm) is possible with EBL process, requiring no cleaning facilities and with good yield. Charge accumulation on AlN/UNCD layers was overcome by using a thin layer of chromium as an anti-static layer [5, 9, 10]. The charge accumulation, dose variations and beam current effects were investigated in detail as a function of the structure pitch. A minimum line width of 100 nm Ti-Pt was fabricated with good repeatability. The filtering response of the device was characterized and tested.

Process Development

The fabrication and optimization of SAW with various n-IDT widths on AlN/UNCD was studied and tested [5, 11, 12]. The lithography process was conducted using EBL tool (EBPG 5000+, Vistec Lithography Ltd.) with an acceleration voltage of 100 kV (Fig. 1).The exposure dose and the beam current were varied from 230-280 $\mu C/cm^2$ and from 5-15 nA, respectively. The tested structures were SAW nano resonators realized on AlN/UNCD by EBL and lift-off technique [5, 8]. The SAW nanoresonator was fabricated by transferring two aligned n-IDTs on top of the AlN surface using an optimized EBL process. The distance between the IDTs was fixed to 19 μm while their length was fixed at 90 μm. Each n-IDTs consists of 60 fingers. As shown in Fig. 2, the substrate was coated with 420 nm electro sensitive resist ZEP 520 (a). In order to overcome the charge accumulation on AlN surface, 20 nm chromium was thermally evaporated onto the ZEP520 resist (b) for discharging.

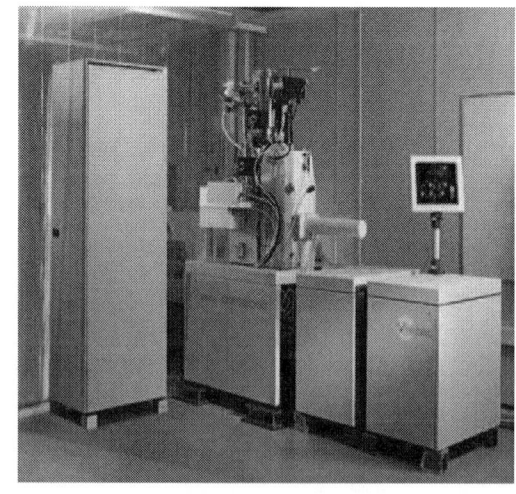

Fig.1. Electron beam lithography

978-1-4673-4840-9/13 $31.00 © 2013 IEEE

After writing the structures with EBL, the wafer was developed using ZEP- N50 developer for 60 s followed by dipping in a 9:1 mixture of methyl-isobutyl-ketone (MIBK): isopropanol (IPA) for 30 s (c). Subsequently, 20 s plasma descum was necessary to improve the adhesion of the n-IDTs. The n-IDTs were e-beam evaporated and made of 9 nm Ti and 110 nm thick Pt films (d). The fabrication process was completed by immersing the wafer in ZDMAC and thus realizing the n-IDTs (e). As shown in table 1, for the structures with 140~145 nm wide n-IDTs a dose of 240 μC/cm² at 15nA beam current was optimized for the realization of the device whereas the same current beam was optimized for the realization of 350 nm wide n-IDTs but at a dose of 275 μC/cm² (Fig. 3). However, for the 100 nm structures the optimization of the process resulted in dose and beam current of 230 μC/cm² and 5 nA respectively which differs from the previous trend. For the realization of structures with 100 nm wide n-IDTs, the plasma descum duration was reduced to 13 s to realize the structure (Fig. 4).

Table 1. EBL Process Parameters

Process	Device1	Device2	Device3	Device4
Accel. Voltage (Kev)	100	100	100	100
Current (nA)	5	15	15	15
Dose (μC/cm²)	230	240	240	275
Resolution (nm)	100	140	145	350

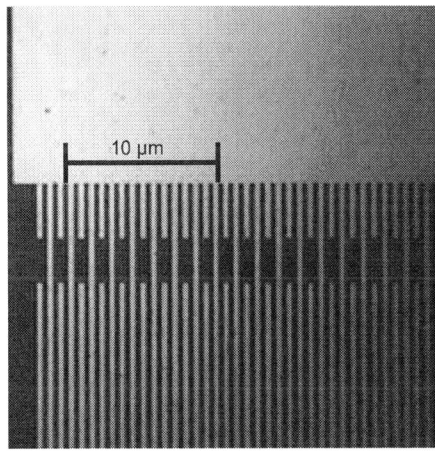

Fig.3. 350 nm wide n-IDTs

Further, from the testing of the fabricated devices, the plasma descum was found to be an essential step for the adhesion quality of the n-IDTs. Very good repeatability of the n-IDTs with widths of 140~350nm was obtained.

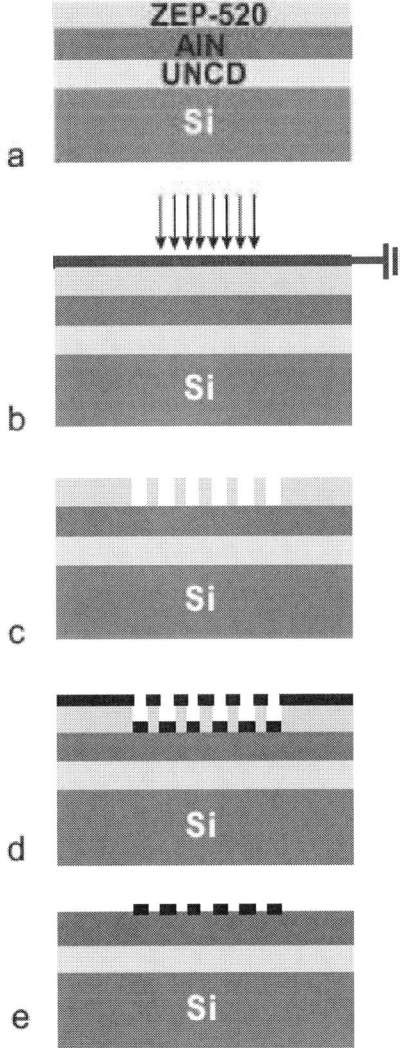

Fig.2. Fabrication process steps

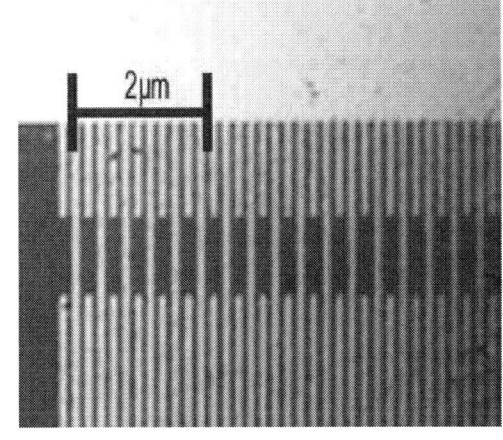

Fig.4. 100nm wide n-IDTs

Further optimization is necessary to improve the repeatability of the 100nm wide n-IDTs - which can be done using multiple resist coatings to facilitate the lift-off.

The advantages of using a single resist layer is its simplicity and lower cost compared to multiple coatings.

Device Characterization

In order to check the quality and repeatability of the fabricated devices, it is important to measure the frequency characteristics of the fabricated SAW nano resonators. Device testing was carried out using RF probe station (SussMicrotech) and Agilent 8361A network analyzer. As shown in Fig. 5, a fabricated SAW nano resonator with a finger width of 200 nm was tested.

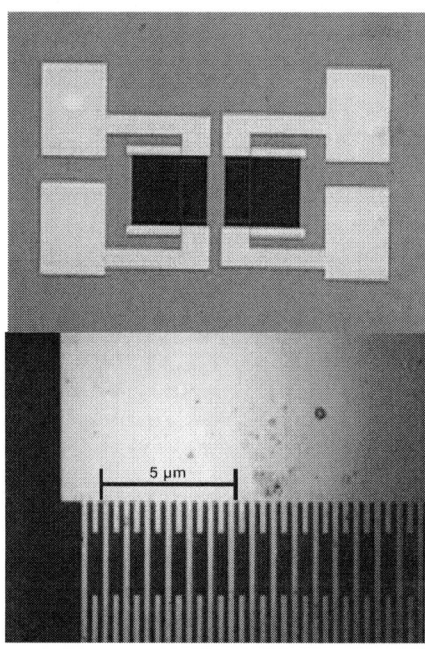

Fig.5. 200nm wide n-IDTs. The top figure shows the plan view including the contacts

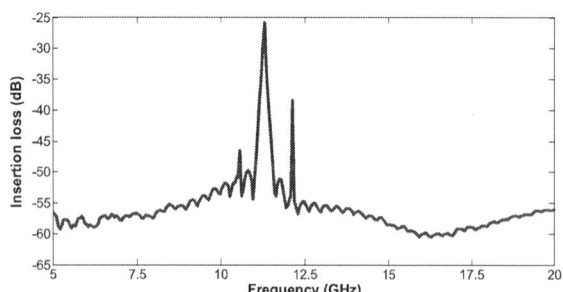

Fig.6. Frequency response of the fabricated device

The filtering response of the device was measured and the fabricated device exhibited a center frequency of 11.3 GHz (also called Sezawa mode) yielding a SAW velocity of 9040 m/s with an insertion loss of approximately 26 dB.

Conclusion

Based on the sensitivity of ZEP520 resist, it could be seen that n-IDTs with a line widths of 140 nm and higher could easily be achieved using a beam current of 15 nA with higher doses, whereas the structures with 100 nm wide n-IDTs were realized with a minimal current beam of 5 nA. From the measurement of the filtering response of the fabricated devices, the plasma descum was a very important step for metallization of the n-IDTs, however 13 s was the optimum duration for the 100 nm wide n-IDTs. A final remark, the fact that although structures below 140 nm wide need further optimization to increase the repeatability of the process, the performance of the other fabricated devices exhibited good frequency response with minimal insertion loss - thus, promising the realization of high performance SAW resonators for RF and sensing applications.

References

[1] D. Neculoiu, A. Muller, G. Deligeorgis, A. Dinescu, A. Stavrinidis, D. Vasilache, A.M. Cismaru, G.E. Stan, G. Konstantinidis: AlN on silicon based surface acoustic wave resonators operating at 5 GHz, Electronics Letters, Vol. 45, 2009.

[2] V. Cimalla, F. Niebelschütz, K. Tonisch, Ch. Foerster, K. Brueckner, I. Cimalla, T. Friedrich, J. Pezoldt, R. Stephan, M. Hein, O. Ambacher: Nano electromechanical devices for sensing applications, Sens. Actuators B, vol. 126, 2007.

[3] S. Petroni, G. Tripoli, C. Combi, B. Vigna, M. De Vittorio, M.T. Todaro, G. Epifani, R. Cingolani, A. Passaseo: GaN-based surface acoustic wave filters for wireless communications, Superlattices and Microstructures, Volume 36, Issue 4-6, p. 825-831, 2004.

[4] T. Palacios, F Calle, E. Monroy, E Munoz: Submicron technology for III-nitride semiconductors, J. Vac Sci. technol B, 20, pp. 2071–2074, 2002.

[5] Ali B. Alamin Dow, H. Lin, M. Schneider, Ch. Petkov, A. Bittner, A. Ahmed, C. Popov, U. Schmid, Nazir P. Kherani: Ultrananocrystalline Diamond-Based High-Velocity SAW Device Fabricated by Electron Beam Lithography, IEEE Trans. on Nanotechnology, Vol.11, pp. 979-984, 2012.

[6] H. Hatakeyama, T. Omori, K. Hashimoto and M. Yamaguchi: Fabrication of SAW devices using SEM-based electron beam lithographyand lift-off technique for Lab use, IEEE Ultrason. Symp..pp. 1896–1900,2004

[7] V. Mortet, O. Elmazria, M. Nesladek, M. B. Assouar, G. Vanhoyland, J. D'Haen, M. D'Olieslaeger, and P. Alnot: Surface acoustic wave propagation in aluminum nitride-unpolished freestanding diamond structures, Appl. Phys. Lett., vol. 81, pp. 1720–1722, 2002.

[8] Ali B. Alamin Dow, A. Ahmed, C. Popov, U. Schmid, Nazir P. Kherani: Nanocrystalline Diamond/AlN Structures for High Efficient SAW Nano-resonators, IEEE ISAF-PFM Symposium, Aveiro, 2012.

[9] O. Elmazria, M. El Hakiki, V. Mortet, B. M. Assouar, M. Nesladek, M. Vanecek, P. Bergonzo, O. Alnot: Effect of diamond nucleation process on propagation losses of AlN/diamond SAW filter', IEEE Trans. Ultrason. Ferroelectr. Freq. Control, 2004, 51, (12), pp. 1704–1709

[10] J. G. Rodriguez-Madrid, G. F. Iriarte, J. Pedros, O. A. Williams, D. Brink, F. Calle: Super-High-Frequency SAW Resonators on AlN/Diamond, IEEE Electron Device Letters, , Vol. 33 , pp. 495-497, 2012

[11] Kulisch, T. Sasaki, F. Rossi, C. Popov, C. Sippel, and D. Grambole: Hydrogen incorporation in ultrananocrystalline diamond/amorphous carbon films", Physica Status Solidi - rapid research letters, 2, 77-79, 2008.

[12] C. Popov, W. Kulisch, S. Boycheva, K. Yamamoto, G. Ceccone, Y. Koga: Structural investigation of nanocrystalline diamond/amorphouscarbon composite films, Diamond Relat. Mater., Vol. 13, 2071-2075, 2004.

Performance Analysis and Simulation of Two Different Architectures of (6:3) and (7:3) Compressors Based on Carbon Nano-Tube Field Effect Transistors

Shima Mehrabi[1], Keivan Navi[2,3], Omid Hashemipour[3]

[1]Department of Computer Engineering, Science and Research Branch, Islamic Azad University, Tehran, Iran.
[2]Department of Electrical and Computer Engineering, University of California, Irvine, USA
[3]Faculty of Electrical and Computer Engineering, Shahid Beheshti University, G.C., Tehran, Iran
[1]sh.mehrabi@srbiau.ac.ir, [2]navi@sbu.ac.ir, [2]hashemipour@sbu.ac.ir

Abstract

In this paper, two different architectures of (6-3) and (7-3) compressors, including the conventional topology based on full adder cells with the most interesting of those recently proposed and XOR/MUX based topology, are analyzed and compared for speed, power consumption and power-delay product at transistor-level in Carbon Nano-tube Field Effect Transistor (CNFET) technology. Simulations are carried out using Synopsys HSPICE with 32nm CNTFET technology. The results of simulation demonstrate the superiority of the XOR/MUX-based structures in terms of PDP and propagation delay around 9% and 14% respectively.

Keywords: CNFET, Compressor, Exclusive-OR (XOR), Full Adder, Multiplexer and Nanoelectronics.

Introduction

Since many studies have been accomplished on the implementation of fast and efficient Adders and Multipliers which are known as the arithmetic building blocks of microprocessors and digital signal processors (DSPs), choosing the appropriate implementation techniques and technologies are the two major approaches of today's VLSI circuit designs. Multipliers play an effective role in the performance of different practical circuits [1]. Fast multipliers are generally composed of three sub- functions: partial product generation, partial product accumulation, and carry-propagating addition [2, 3]. At the first step, Booth encodings are often used to reduce the number of partial products. A summation tree, which is called the Carry Save Adder (CSA), is used in the second sub-function to further reduce the partial products to two rows. The last step is normally fulfilled by a fast carry propagate adder, such as carry look-ahead adder or carry-skip adder [4].

To implement fast multipliers, many different architectures of Processing Elements (PEs) have been presented to perform arithmetic addition and multiplication. Compressors as one of the PEs are the fundamental building blocks which are being used for accumulating partial products during the multiplication process.

A compressor is a combinatorial device which is mostly used in multipliers to reduce the operands while adding terms of partial products. A typical (m:n) compressor takes *m* equally weighted input bits and produces *n*-bit binary number [5]. The simplest and the most widely used one is the (3:2) compressor (also known as a Full Adder cell) which has 3 inputs to be summed up and provides 2 outputs. Similarly, a (4:2) compressor can also be built from two cascaded (3:2) compressors [6].

In this paper, we analyze two different architectures of high-speed, low-power (6:3) and (7:3) compressors. The first circuit implementations designs are utilized of two recent CNTFET-based Full adder cells and the second implementation is composed of XOR/MUX gates circuits. To evaluate the performance of these two different architectural designs, they have been comprehensively compared at different voltages, load capacitors and Temperatures.

The rest of the paper is organized as follows: in section II, conventional design and architecture of (6:3) and (7:3) compressors are reviewed. Section III, include XOR/MUX implementation of these two compressors and in section IV, experimental results, analyses and comparisons are presented in section V and finally section V concludes the paper.

Architectural Design

A. Conventional (6:3) and (7:3) Compressors Architecture

At present, the most widely used compressors are (3:2) and (4:2) compressors. In other words, the simplest one is (3:2) compressor, known as a Full Adder cell. It has 3 inputs to be summed up and provides 2 outputs. The logic equations of full adder cell are shown in Eq.(1) and Eq.(2).

$$Sum = A \oplus B \oplus C \tag{1}$$

$$Carry = AB + C(A \oplus B) \tag{2}$$

Similarly, the (4:2) compressor is simply built by two cascaded (3:2) compressors. Moreover, as it shown in Fig.1, the critical path of (3:2) compressor equals 2 XOR gates delay, and there is no carry on the horizontal path. Also, a typical (4:2) compressor has a critical path delay of 3 XORs [7].

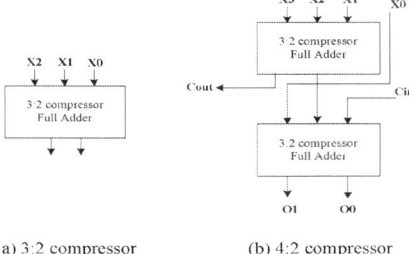

(a) 3:2 compressor   (b) 4:2 compressor

Fig. 1 Two fundamental Compressors Architectures

Although, both (3:2) and (4:2) compressors are ideal for constructing regularly structured Wallace tree with low complexity [14] but for the compression of a larger number of bits, higher order compressors are needed. Many researches show that the multipliers with high order compressors have better performance [7]. So, the (n:3) compressors together with Half Adders and Full Adders are utilized in order to achieve a fast multiplier [8-9]. A (6:3) compressor essentially comprises of a combinational logic circuit with six inputs and three outputs. Conventional architecture of high order inputs compressors such as (6:3) and (7:3) compressors based on the extended design of a conventional (3-2) compressor are shown in Fig. 2 (a) and Fig. 2 (b) respectively.

According to Fig.2, the conventional (6:3) and (7:3) compressors consist of three Full-Adder cells with one Half-Adder and four Full-Adder cells which are cascaded respectively. So the critical path of (6:3) consists of two Full-Adder and one Half-Adder cells. Obviously, the straightforward implementation of the (7:3) compressor entails for 6 gate delays.

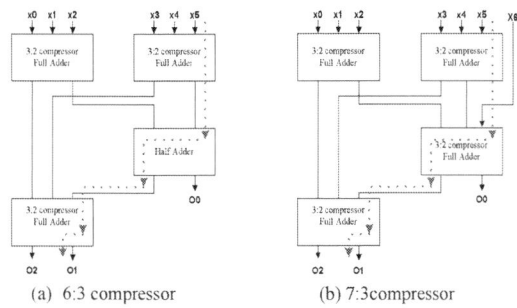

(a) 6:3 compressor   (b) 7:3compressor

Fig. 2 Block diagram of the conventional (6:3) and (7:3) compressors

B. (6:3) and (7:3) Compressors Architectures Based-on XOR/MUX Gates

The second compressor structure is focused on the design of XOR/MUX based (6-3) and (7:3) compressors, attempting to minimize the stage delays of the conventional structure which is designed using single bit full adder and half adder architectures.

By using Eq.(1) and (2) as general representations of the two outputs of 1-bit Full-Adder cell and suitably rewriting them, the architecture of the (6:3) compressor has been obtained as follows (Eq. (3), Eq.(4) and Eq.(5)):

$$h0 = \overline{(x_0 \oplus x_1)}x_0 + (x_0 \oplus x_1)x_2 \tag{3}$$

$$h1 = \overline{(x_3 \oplus x_4)}x_3 + (x_3 \oplus x_4)x_5 \tag{4}$$

$$h2 = (x_0 \oplus x_1 \oplus x_2)(x_3 \oplus x_4 \oplus x_5) \tag{5}$$

Based on Eq.(2) it can be found that the second design uses multiplexer to generate Carry output. Therefore, $h0$ and $h1$ indicate the outputs of multiplexers (MUX0 and MUX1) and $h2$ is the AND-gate output. With these 3 equations, the 3 outputs equations would be explored as follows (Eq. (4), (5) and (6)):

$$O0 = x_0 \oplus x_1 \oplus x_2 \oplus x_3 \oplus x_4 \oplus x_5 \tag{6}$$

$$O1 = h0 \oplus h1 \oplus h2 \tag{7}$$

$$O2 = \overline{(h1 \oplus h0)}ho + (h1 \oplus h0)h2 \tag{8}$$

Using the Eq. 6, Eq. 7 and Eq. 8, the architecture of the (6:3) compressor based on XOR/MUX gates is shown in Fig. 3.

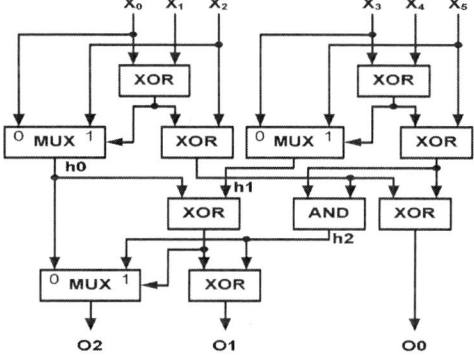

Fig. 3 XOR/MUX based (6:3) compressor Architecture

Similarly, these equations can be simply expandable for (7:3) compressors as follows to obtain the XOR/MUX based architecture:

$$h0 = \overline{(x_0 \oplus x_1)}x_{0+}(x_0 \oplus x_1)x_2 \tag{9}$$

$$h1 = \overline{(x_0 \oplus x_1 \oplus x_2 \oplus x_3)}x_3 + (x_3 \oplus x_4)x_5 \tag{10}$$

$$h2 = \overline{(x_0 \oplus x_1 \oplus x_2 \oplus x_3 \oplus x_4 \oplus x_5)}x_4 + (x_0 \oplus x_1 \oplus x_2 \oplus x_3 \oplus x_4 \oplus x_5)x_6 \tag{11}$$

where $h0$, $h1$ and $h2$ are the outputs of multiplexers (as the carry signals of Full-Adder cells). The outputs equations of XOR/MUX-based (7:3) compressors are the same as the XOR/MUX-based (6:3) compressor.(Eq.(12),Eq.(13) and Eq.(14))

$$O0 = x_0 \oplus x_1 \oplus x_2 \oplus x_3 \oplus x_4 \oplus x_5 \oplus x_6 \tag{12}$$

$$O1 = h0 \oplus h1 \oplus h2 \tag{13}$$

$$O2 = \overline{(h1 \oplus h0)}ho + (h1 \oplus h0)h2 \tag{14}$$

978-1-4673-4840-9/13 $31.00 © 2013 IEEE   323

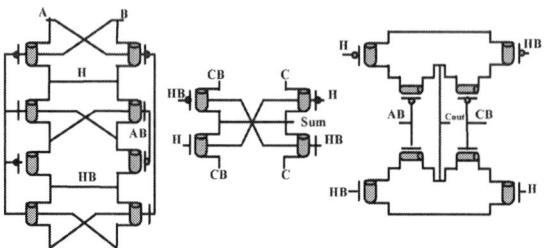

Fig. 5 Carbon Nano-Tube Full-Adder (CNTFA) cell design

The Bridge Full Adder cell [14](Fig.6) designed by 24 transistors is composed of two modules. The first module consists of two cascaded high-performance 2-input XOR circuit to generate the sum and the second one is the complementary bridge style circuit to generate C_{out} outputs with high driving capability.

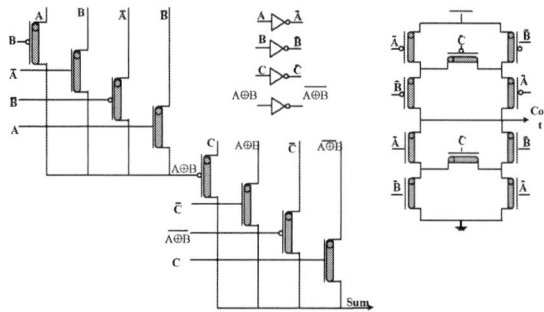

Fig.6 Bridge style Full-Adder cell design

.With 26-transistors Bridge full adder cell, the conventional architectures of (6:3) and (7:3) compressors require 92and 104 transistors respectively.

In this paper, we use the CNFET-based transmission gate multiplexer with only six transistors to design the full-swing structure of multiplexer and the full-swing CNFET-based pass-transistor XOR gate can be implemented with only eight transistors using pass transistor logic. The two full-swing circuit design of multiplexer and XOR gates are shown in Fig.7 (a) and (b) respectively. In order to have inverter gate, two additional transistors are needed to make an inverter signal.

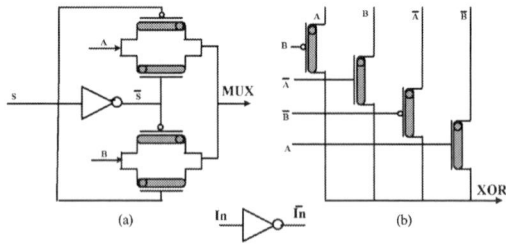

Fig. 7Circuit design of (a) Multiplexer Gate, (b) XOR Gate

Alternatively, the MUX/XOR-based design for (6:3) and (7:3) compressors with 78 and 88 transistors consumes less hardware with improvement of 8.3% in transistor count

---

Fig. 4 XOR/MUX based (7:3) compressor Architecture

As it is evident form Fig.5 and Fig.6, the output of the compressors require 4 (3 XOR and 1 MUX) and 5 (4 XOR and 1 MUX) gate delays respectively. So the XOR/MUX-based structures have shorter critical path compared to conventional structures. Moreover, according to Equation (1) and Equation (2), the (6:3) compressor requires the use of 7 XOR gates, 7 AND gates and 3 OR gates. Similarly, the (7:3) compressor requires the 8 XOR gates, 8 AND gates and 4 OR gates for the straightforward implementation.

The XOR/MUX-based (6:3) compressor requires the use of 7 XOR gates, 1 AND gate and 3 multiplexers, however the XOR/MUX-based (7:3) compressor requires the use of 8 XOR gates and 4 multiplexers.

IV. SIMULATION RESULTS

A: Performance Analysis and Comparison

Since silicon-based microchips will reach the physical limits of miniaturization in the early next decade, CNFET is one of the most promising devices among emerging technologies [10-12].

The 24 transistors Carbon Nano-Tube Full-Adder (CNTFA) cell is selected as the Full-Adder cell for the comparison (between conventional and XOR/MUX architecture) due to its high performance and novel structure [13].As it is depicted in Fig.5, it is composed of two modules and one sub-module. The sub-module reduces glitches and power consumption while the other two modules utilize the full-swing signals of sub-module to design full-swing low-power high performance circuits. The second module consists of four Carbon Nano-tube Transistors, which use no V_{DD} or GND to generate sum output and the third module consists of eight transistors, reusing sub-module signals and generating C_{out} output. With 24-transistors CNTFA, the conventional architectures of (6:3) and (7:3) compressors require 86 and 96 transistors respectively.

978-1-4673-4840-9/13 $31.00 © 2013 IEEE   324

compared to conventional designs. Moreover, the XOR/MUX based design offer minimum critical path.

Simulation results, shown in Table I, ease the comprehensive comparisons. In this experiment, circuits have been simulated at 0.65V supply voltage and 100MHz operation frequency. Simulation results demonstrate the superiority of the XOR/MUX-based designs in terms of power, delay and PDP, specifically at lower supply voltages.

TABLE I

SIMULATION RESULTS

V_{DD}	0.65V
Delay $\times$ 10^{-11} s	
6:3 Comp.(MUX/XOR)	3.48
7:3 Comp.(MUX/XOR)	5.03
6:3 Comp.(CNTFA)	5.2
7:3 Comp.(CNTFA)	23.69
6:3 Comp.(Bridge)	7.82
7:3 Comp.(Bridge)	9.73
Power $\times$ 10^{-7} W	
6:3 Comp.(MUX/XOR)	2.74
7:3 Comp.(MUX/XOR)	3.3
6:3 Comp.(CNTFA)	2.01
7:3 Comp.(CNTFA)	4.44
6:3 Comp.(Bridge)	1.8
7:3 Comp.(Bridge)	1.99
PDP $\times$ 10^{-18} J	
6:3 Comp.(MUX/XOR)	**9.55**
7:3 Comp.(MUX/XOR)	**16.61**
6:3 Comp.(CNTFA)	10.5
7:3 Comp.(CNTFA)	105.29
6:3 Comp.(Bridge)	14.16
7:3 Comp.(Bridge)	19.36

Conclusion

In this paper we have carried out a comprehensive analysis and comparison between two different architectures of (6:3) and (7:3) compressors and compare them at the transistor level, including recently XOR/MUX based of full adder cells and XOR/MUX based architecture. The obtained information is useful in the early design phases of an adder circuit, since architectural optimization techniques are based on the knowledge of the full adder cell used.

References

[1] N. Mehmood, "An Energy-Efficient 32-bit Multiplier Architecture in 90-nm CMOS". *Dissertation for receiving M. Sc. Degree in Electronic Devices*, Linkoping Institute of Technology, Sweden, 2006.

[2] K. Prasad and K.K. Parhi, "Low-Power 4-2 and 5-2 Compressors",*In Proc. of the 35th IEEE Asilomar Conf. on Signals, Systems and Computers*, vol. 1, pp. 129-133, 2001.

[3] A.M. Shams and M.A. Bayoumi,"A structured approach for designing low-power adders". *In Proc. of the 31st IEEE Asilomar Conf. on Signals, Systems and Computers*, vol. 1, pp. 757-761, 1997.

[4] J. Gu and C.H. Chang, , "Low voltage, Low power (5:2) compressor cell for fast arithmetic circuits",*In Proc. of IEEE*

Intl. Conf. on Acoustics, Speech and Signal Processing (ICASSP '03), vol. 2, pp. 661-664, 2003.

[5] S.B Sukhavasi, S.B Sukhasavi, V.B Madivada, H Khan, SR SastryKalavakolanu. "Implementation of Low Power Parallel Compressor for Multiplier using Self Resetting Logic", *International Journal of Computer Applications* (0975 – 888), Vol. 47,No.3, June 2012.

[6] P.D. Gopineedi, H. Thapliyal, M.B. Srinivas, and H.R. Arabnia,"Novel and Efficient 4: 2 and 5: 2 Compressors with Minimum Number of Transistors Designed for Low-Power Operations",In Proc., ESA, pp.160-168, 2006.

[7] W Ma, Sh Li, "A new high compression compressor for large multiplier", *9th International Conference on Solid-State and Integrated-Circuit Technology,pp.* 1877-1880, 2008.

[8] A. Dandapart, S. Ghosal, P. Sarkar, D. Mukhopadhyay, "A 1.2-ns 16×16-Bit Binary Multiplier Using High Speed Compressors", *International Journal of Electrical and Electronics Engineering*, 4:3,2010.

[9] S.R. Chowdhury, A. Banerjee, A. Roy and H. HiranmaySaha,"Design, Simulation and Testing of a High Speed Low Power 15-4 Compressor for High Speed Multiplication Applications",*In Proc. of IEEE 1st Intl. Conf. on Emerging Trends in Engineering and Technology (ICETET '08)*, pp. 434-438, 2008.

[10] H Iwai,"Roadmap for 22 nm and beyond",*Microelectronic Engineering, Elsevier.* Vol. 86, pp.1520-1528, 2009.

[12] K. Navi, M. Rashtian, A. Khatir, P. Keshavarzian and O. Hashemipour,"High Speed Capacitor-Inverter Based Carbon Nanotube Full Adder",*Nanoscale Research Letters, Springer*, vol. 5, no. 5, pp. 859-862, 2010.

[13] M. H. Ghadiry, A. AbdManaf, M. T. Ahmadi, H.Sadeghi, and M. NadiSenejani,"Design and Analysis of a New Carbon Nanotube Full Adder Cell", Journal of Nanomaterials,Vol. 2011, 2011.

[14] K. Navi, O. Kavehei, M. Rouholamini, A. Sahafi, S.Mehrabi and N. Dadkhahi, "Low-power and high-performance 1-bit CMOS full adder cell",*Journal of Computers*, vol. 3, no. 2, pp. 48-54, 2008.

[15] J. Deng, and H.-S.P, "A Compact SPICE Model for Carbon-Nanotube Field-Effect Transistors Including Nonidealities and Its Application—Part I: Model of the Intrinsic Channel Region",*IEEE Trans. on Electron Devices*, Vol. 54, No. 12, pp. 3186-3194, 2007.

[16] J. Deng, and H.-S.P,"A Compact SPICE Model for Carbon-Nanotube Field-Effect Transistors Including Nonidealities and Its Application—Part II: Full Device Model and Circuit Performance Benchmarking",*IEEE Trans. on Electron Devices*, Vol. 54, No. 12, pp. 3195-3205, 2007.

978-1-4673-4840-9/13 $31.00 © 2013 IEEE

One-step Formation of Atomic-layered Transistor by Selective Fluorination of Graphene Film

Kuan-I Ho[1], Jia-Hong Liao[1], Chi-Hsien Huang[1], Chang-Lung Hsu[2], Lain-Jong Li[2],
Chao-Sung Lai[1,*], Ching-Yuan Su[1,*]

[1]*Department of Electronic Engineering, Chang Gung University, Tao-Yuan 333, Taiwan*

[2]*Institute of Atomic and Molecular Sciences, Academia Sinica, Taipei 10617, Taiwan*

* *Corresponding author: chingyuansu@mail.cgu.edu.tw ; cslai@mail.cgu.edu.tw*

Abstract

In this work, the wafer scale fabrication of atomic layered transistors are demonstrated by selective fluorination of graphene with a remote CF_4 plasma, where the generated F-radicals preferentially fluorinated graphene surface at low temperature ($<200°C$) while this technique suppress the defect formation by screening out the ion damage effect. The resultant graphene shows electrical semiconducting and isolation after subjected to the fluorination for 5~20min, respectively. A back-gate transistor is then fabricated with a one-step fluorination of graphene film on SiO_2 substrate. The chemical structure, C-F bonds, is well correlated to the electrical properties in fluorinated graphene by XPS, Raman spectroscopy and electrical meter. This efficient method provide electrical semiconducting and insulator of graphene with a large area and selective pattering, where it turns out the potential for the integration of electronics down to atomic layered scale. (Example keywords: Graphene, CF_4 plasma and Transistor)

Introduction

Graphene,[1] an atom-thick graphite, has attracted intensive interest due to its two-dimensional and unique physical properties, such as high intrinsic carrier mobility (~200 000 cm2/V s),[2] excellent mechanical strength, and elasticity and superior thermal conductivity. Recently, it have been reported that graphene functionalized with hydrogen(H) or fluorine(F) can change the electronic structure from its pristine one. The fluorinated graphene, the so-called fluorographene,[3] showing good mechanical strength comparing to pristine graphene while the conductivity can be adjusted from semi-metallic to insulator by tuning the stoichiometry of carbon to fluorine ratio.[4] The theoretical calculation predict that partial fluorination of graphene from $C_{32}F$ to C_4F, is able to modify the energy bandgap from 0.8 to 2.9 eV, respectively.[4] This inspire the concept that the graphene can be patterned by selectively fluorination of graphene, making the conductor and insulator materials among the same graphene film. Lee et al., devised a method to selectively fluorine graphene by irradiating fluoropolymer-covered graphene with a laser. [5] F. Withers et al., demonstrated the nanopatterning of fluorinated graphene by the electron beam irradiation, making the insulating graphene to the conducting channel.[6] However, the integration of IC circuit requiring the large-scale and cost-effective process compatible to nowadays Si-based fabrication process. In this work, the selective fluorination of graphene was carried out by using the CF_4 plasma treatment in a plasma-enhanced chemical vapor deposition(PECVD) system. A filter was designed in this system to avoid the radical induced structure damage on graphene film. Moreover, the protection layer was selectively deposited on the channel region to attend the partial fluorination of graphene film through the diffusion process, which allows forming a transistor among the same graphene film. This method paves a way toward next-generation nano-electronic devices.

Excremental

A. Graphene film

For the CVD growth of graphene on Cu, high purity copper micro-wire (99.9% deom Nilaco Co.; 50 μ m in diameter) was put on a quartz plate and then loaded into the center of a tubular furnace (TF55030, Lindberg/Blue/M). The chamber was evacuated to ~ 5 mTorr and the temperature was increased from room temperature to 1000 ﹒ C. Prior to growth, a pretreatment step was performed by flowing a diluted hydrogen gas (H_2/Ar: 5 sccm/400 sccm at 500 Torr) through the chamber for 30 min. During the growth step, a gas mixture of methane and hydrogen (CH_4/H_2 = 20 sccm/30 sccm) was introduced and the pressure was controlled at 450 mTorr for 20 min. The system was then cooled down (cooling rate ~5 $°C\ s^{-1}$) to room temperature to complete the growth. To transfer the as-grown graphene onto the substrate(300 nm SiO_2/Si), the Cu substrate after the CVD growth was coated with a layer of Poly (methyl methacrylate) (PMMA) by spinning-coating method, followed by baking at 90°C for 1 minute. Then the PMMA-caped Cu substrate was immerged into a diluted $Fe(NO_3)_3 9H_2O$ to etch away the Cu thin layer. The PMMA-caped graphene film was floated on the solution surface, and then it was transferred to a deionized (DI) water to dilute and remove the etchant and residues. The PMMA/graphene was transferred to the receiving substrate and dried on a hot-plate. The PMMA was removed by warm acetone (90°C), and then the sample was rinsed with isopropyl alcohol and DI water.

B. Device favrication

The field-effect transistor device was fabricated by evaporating Au electrodes (30 nm thick) directly on top of

978-1-4673-4840-9/13 $31.00 © 2013 IEEE

the selected, regularly shaped graphene sheets using a copper grid (200 mesh, 20 μmspacing) as a hardmask. The typically obtained channel length between source and drain electrodes was around 20 μm. The electrical measurements were performed in ambient conditions using a Keithley semiconductor parameter analyzer, model 4200-SCS.

C. Characterizations.

The AFM images were performed in a Veeco Dimension-Icon system. Raman spectra were collected in a NT-MDT confocal Raman microscopic system (laser wavelength 473nmand laser spot-size is~0.5 μm). The Si peak at 520 cm^{-1} was used as reference for wavenumber calibration. STM analysis was carried out on a Veeco STM base in ambient condition. The UV-vis-NIR transmittance spectra were obtained using a Dynamica PR-10 spectrophotometer. XPS measurements were carried out by a Ulvac-PHI 1600 spectrometer with monochromatic Al KR X-ray radiation (1486.6 eV).

D. The Fluorination of Graphene Film

To form the fluorographene, the PECVD (the facility setup was published in our previous reports[7])chamber was evacuated to ~5 mTorr and the temperature was increased from room temperature to 200 ∘C. After the CF4 gas was inlet into the chamber, the gas flow and pressure was controlled. The degree of fluorination process is adjusted by controlling the exposure time when the plasma was ignited.

Results and discussion

To identify the degree of Fluorination of Graphene Film after exposed to CF$_x$ radicals induced from plasma. The X-ray photoelectron spectroscopy analysis was used to analysis the fluorine fraction in a graphene film. Figure 1(a) shows a typical fluorographene after exposed to CF4 plasma for 5min. Figure 1(b) shows the relation between the fluorine fraction and the exposed time. The result suggest that the fluorine fraction in a graphene film can adjusted from 0.2 to ~25 %, making the electrical properties of fluorographene from semiconducting to insulator.

Figure 2shows the The Raman spectrum of fluorographene during CF$_4$ plasma treatment with different exposure time. The Raman feature peaks is characterized. The disorder-induced peak(D peak) at 1350 cm^{-1} appears as fluorine chemisorbs on the graphene surface. The other feature peaks were located at the 2680 cm^{-1} (2D peak) and 1580 cm^{-1} (G peak). The ratio of 2D peak to G peak drop significantly. It is worthy to note that the Raman feature peak become unidentified after the exposed time upto 30 min, suggesting the fully fluorine functionalization.

Figure 3 shows the electrical characterization of the fluorographene. It is worthy to note that the electrical conductivity closed to the insulator after the exposure time attend 20min(few mega ohm). Based on these database, it allows us to make the transistor among the same graphene film as the illustration shown in Figure 4.

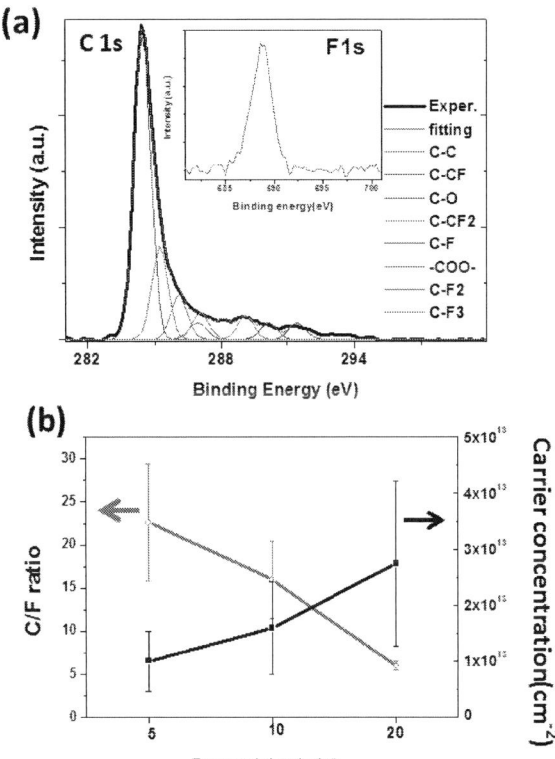

Figure 1. (a)The X-ray photoelectron spectroscopy analysis of fluorine functionalization during CF$_4$ plasma treatment. The inset shows the binding energy of F 1s state.(b) the fluorine fraction for single graphene to CF4 treatment for different exposure time.

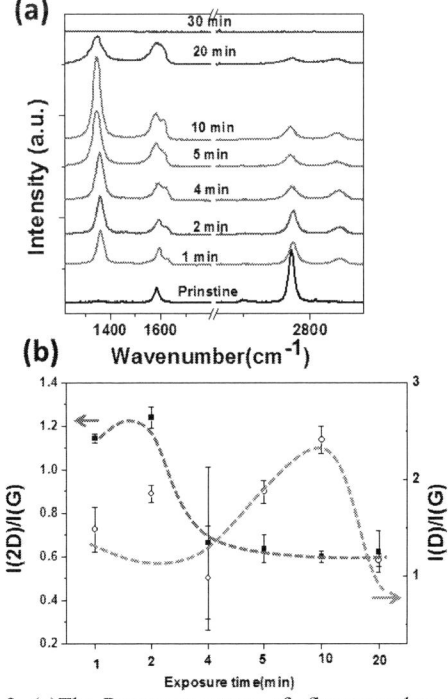

Figure 2. (a)The Raman spectrum of fluorographene during

CF_4 plasma treatment with different exposure time.(b) The relation between the intensity ratio of I(2D)/I(G) and I(D)/I(G) to exposure time.

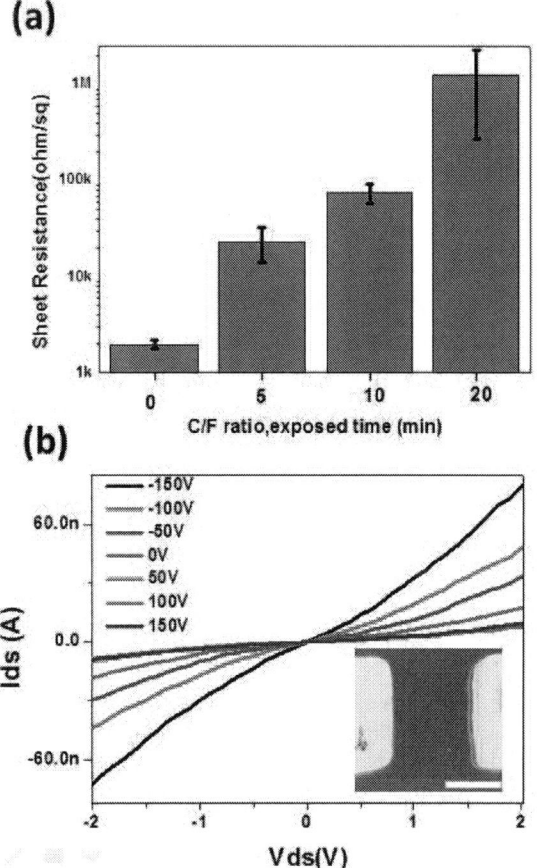

Figure 3. (a)The relation between the sheet resistance and the measured C/F ratio. (b) The Id/Vds curve of a back-gated graphene transistor subjected to the fluorine functionalization.

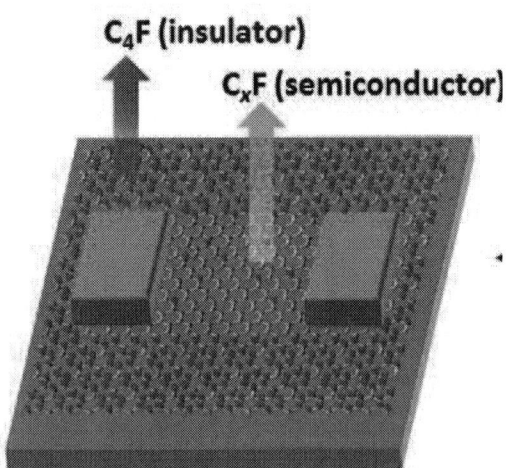

Figure 3. The illustration of a back-gate transistor is then fabricated with a one-step fluorination of graphene film on SiO2 substrate

Conclusion

In this work, we proposed the method to from the fluorine functionalized graphene by CF_4 plasma treatment. The result demonstrate that the C/F ratio can be adjusted by controlling the exposure time, from 5 to 20 min, during the plasma treatment. The A back-gate transistor is then fabricated with a one-step fluorination of graphene film on SiO_2 substrate. The chemical structure, C-F bonds, is well correlated to the electrical properties in fluorinated graphene by XPS, Raman spectroscopy and electrical meter. This efficient method provide electrical semiconducting and insulator of graphene with a large area and selective pattering, where it turns out the potential for the integration of electronics down to atomic layered scale.

Acknowledgement

This work is supported by the National Science Council of the Republic of China under the contract number of NSC 101-2218-E-182-003-MY2.

References

[1] K. S. Novoselov et al., Science, 306, p.666-669 2004.
[2] K. I. Bolotin et al., Solid State Commu., 146, 351-355, 2008.
[3] R. N. Rahul et al., Small, 2877, 2010.
[4] T. R. Jeremy et al., Nano Lett., 3001, 2010.
[5] W. H. Lee et al., Nano Lett., 2374, 2012.
[6] F. Withers et al., Nano Lett. 3912, 2011.
[7] C. H. Huang et al., Nanotech., 23, 2012.

Analysis of CNT Electronics Structure to Design CNTFET

Soheli Farhana, AHM Zahirul Alam, SMA Motakabber and Sheroz Khan

Department of Electrical and Computer Engineering, Faculty of Engineering
International Islamic University Malaysia
53100 Kuala Lumpur, Malaysia
Phone: +60361963384 and Fax: +60361964433 and E-mail: farhana_aub@yahoo.com

Abstract

In this paper, graphene electronic structure of carbon nanotubes has been analyzed and validate with past theoretical and experimental results. The energy dispersion relation, effective mass and intrinsic carrier concentration of graphene have been analyzed to build diverse carbon nanotubes. The electronics properties of graphene are subject to change with the different diameters and wrapping angles of carbon nanotubes. Different chiral vector (n,m) of a graphene allows to design carbon nanotube for a wide range of applications, which can be achieved from the analyzed carrier concentration calculation. The proposed calculations set a higher boundary for a wide range of applications including the integration of carbon nanotube internal architecture and carbon nanotube field effect transistors.

Keywords: Carbon nanotube, graphene, dos.

Introduction

Carbon nanotubes have especial characteristics that have made them especial, from electronic properties to mechanical properties. In fact carbon nanotubes are tiny tubes made of carbon atoms that are nanometer diameter and their length is larger and is about micrometers. 1D graphene sheets rolled into a tubular form to construct Carbon nanotubes [1] with electronic properties being predicted by their diameter and wrapping angle [2]. Scientists consider in carbon nanotubes to be the future aspirants to substitute CMOS technology [3]. Many applications have been proposed in earlier period researches for carbon nanotubes together with nano semiconductors devices [4-6]. Since carbon nanotubes are 1D sheets of graphite, theoretical calculations have predicted their electronic structure using a tight binding model [7,8] deduced from the electronic structure of graphite [9,10]. The density of states has also been calculated [11,12] and it has predicted a strict relationship to the carbon nanotube indices (n,m) [13] as expected.

This paper describes the Electronic properties, Energy dispersion relation, Effective mass and Intrinsic carrier concentration in the following sections.

Electronic Properties

Carbon nanotubes (CNTs) which have unique electronic properties are being considered for various next generation device applications like field effect transistors or sensing elements. Due to the tiny radius of carbon nanotubes, the quantization of wavevectors to be quantized in the circumferential direction. Moreover, the thinness of the nanotube's cylindrical shell obviously yields an even shorter length of confinement in the radial direction, thus making the material virtually one-dimensional as far as electron transport is concerned. Many published works to date have corroborated this claim with experimental evidence from device studies [14].

A. Reciprocal lattice

The primitive cell of a carbon nanotube can be described from the unit vectors [15,16]:

$$R_1 = \frac{a}{2}\left(\sqrt{3}\hat{x} + \hat{y}\right) \text{ and } R_2 = \frac{a}{2}\left(\sqrt{3}\hat{x} - \hat{y}\right) \quad (1)$$

where, a = 2.49 Å is the carbon to carbon atom distance between two carbon atoms.

The reciprocal lattice vectors are of the form [15,16]:

$$b_1 = \frac{2\pi}{a}\left(\frac{1}{\sqrt{3}}\hat{x} + \hat{y}\right) \text{ and } b_1 = \frac{2\pi}{a}\left(\frac{1}{\sqrt{3}}\hat{x} - \hat{y}\right)$$

Energy Dispersion Relation

In order to examine the conductivity properties of the nanotube, it is necessary to derive its k relation. This is done by starting from the equivalent relation of a two dimensional graphene lattice. The energy dispersion for carbon nanotubes can be calculated from the electronic structure of graphene. The energy dispersion of graphene is given by [17],

$$E_{2D}(K) = \pm V_{pp\pi}\{3 + 2Cos(KR_1) + 2Cos(KR_2) + 2Cos[K(R_1 - R_2)]\}^{1/2} \quad (2)$$

where Vpp, is the nearest neighbour transfer integral, R1 and R2 are the unit cell vectors of a carbon nanotube given in equation (1). The 3d view of energy dispersion relation is shown in fig. 1.

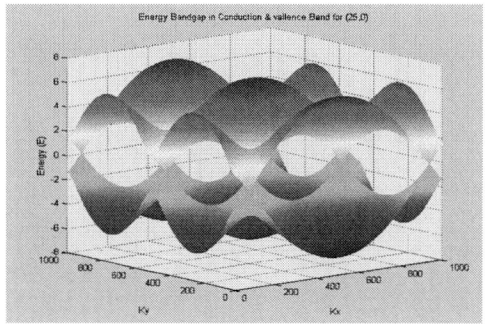

Fig. 1 Plot of the Energy dispersion relation between valence and conduction band for a CNT (25, 0)

978-1-4673-4840-9/13 $31.00 © 2013 IEEE

1D energy band calculates from equation (2) for single wall carbon nanotubes (SWNTs) as [18]:

$$E_{1D}(k) = \pm V_{pp\pi}\left[1 + 4\text{Cos}\left(\frac{\sqrt{3}K_x}{2}a\right)\text{Cos}\left(\frac{K_y}{2}\right)\right. \\ \left. + 4\text{Cos}^2\left(\frac{K_y}{2}\right)\right]^{1/2} \quad (3)$$

Where the wave vectors K_x and K_y are found using the relation

$$(K_x, K_y) = \left(k\frac{K_2}{|K_2|} + qK_1 \text{ for } \left(-\frac{\pi}{|T|} < k < \frac{\pi}{|T|}, \text{ and } q\right.\right. \\ \left.\left. = 1, \dots, N\right)\right.$$

Where k the wave vector along the nanotube axis, |T| is is the magnitude of the translational vector, and N is the number of hexagons within a unit cell. |T| and N are given by [16],

$$|T| = \frac{\sqrt{3}\pi d_t}{d_R} \text{ and } N = \frac{2(n^2 + nm + m^2)}{d_R}$$

where d_R is the greatest common divisor of (2n+m) and (2m+n). In addition, K1 and K2 denote the allowed reciprocal wave vectors along the tube and circumference axis given by [19],

$$K_1 = \frac{(2n + m)b_1 + (2m + n)b_2}{Nd_R} \text{ and } K_2 = \frac{mb_1 - nb_2}{N}$$

Density Of States

The allowed k-vectors in momentum space depend on vectors K_1 and K_2 as shown in fig. 2. It is possible to represent the area in momentum space for a single state as, $A_P^{1-state} = h^2|K_1||K_2|/2$ and a differential area as, $A_P = h^2|K_1|dk$, where dk is in the direction of K_2 and h is the Planck's constant divided by 27w. Therefore, the density of states per unit energy can be defined as follows [20]:

$$D(E)dE = 2\frac{dA}{A_p^{1-State}} = \frac{4}{h^2|K_1||K_2|}h^2|K_1|\frac{dk}{dE} \\ = \frac{2|T|}{\pi}\left(\frac{dE}{dk}\right)^{-1} \quad (4)$$

Equations (3) and (4) can be used to plot the density of states for any (n,m) carbon nanotube. Fig. 3 shows a plot for a CNT (25, 0).

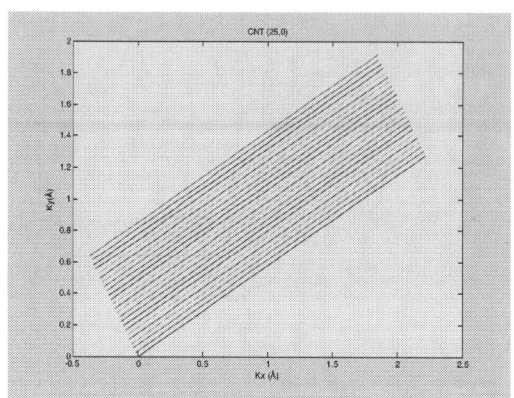

Fig. 2 Plot of the allowed wave vectors in K-space for a CNT (25, 0)

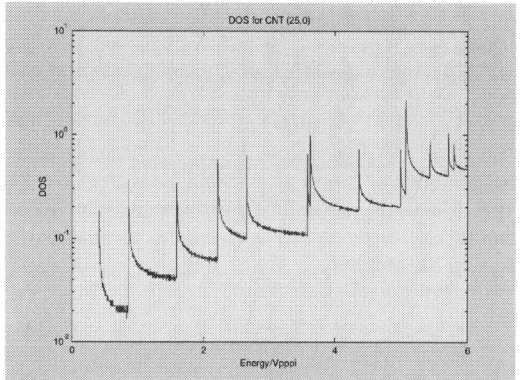

Fig. 3 Plot of the density of states for a CNT (25, 0)

Each peak in fig. 3 is called a Van Hove Singularity and its respective energy represents a conduction energy-band minimum value. The total number of Van Hove Singularities is the number of bands a CNT has. The density of states calculation shown in fig. 3 requires usage of mathematical software due to the complexity and relationship of the variables. However, using the approximated energy dispersion relation calculated in [21] an equation for the 1D density of states can be found as,

$$D(E)dE = \sum_{i}^{AllBand} \frac{4}{\pi V_{pp\pi}a\sqrt{3}}\frac{E}{\sqrt{E^2 - E_{c\,min_i}^2}}dE \quad (5)$$

where $E_{c\,min_i}$ is the conduction minimum for the given band. This minimum energy value is found using Figure 2. The density of states can be recalculated using equation (5), the approximated plot of density of states is shown in fig. 4.

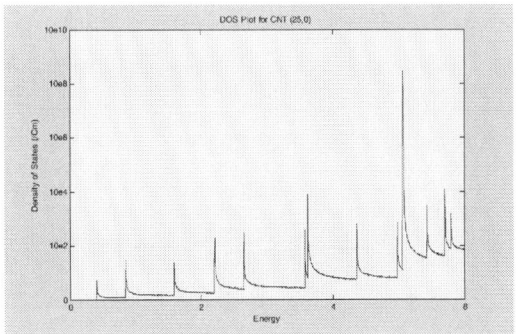

Fig. 4 Density of states plot for a CNT (25, 0) using approximated equation

Effective Mass

Given the complete description of the energy dispersion for carbon nanotubes, equation (3) can also be used to calculate the electron effective masses for each band. The effective mass for a semiconductor is calculated using the following equation [22]:

$$m_i^* = \frac{h^2}{\left(\frac{d^2 E}{dk^2}\right)} \qquad (6)$$

Using equation (6), the electron effective mass for various carbon nanotubes (n,m) have been calculated as shown in Table I.

TABLE I
EFFECTIVE MASSES FOR ELECTRONS IN SELECTED CARBON NANOTUBES

(n, m)	Effective mass for electrons
(3, 1)	$0.507\ m_o$
(3, 2)	$0.222\ m_o$
(4, 2)	$0.271\ m_o$
(5, 0)	$0.408\ m_o$
(5, 1)	$0.159\ m_o$
(6, 1)	$0.255\ m_o$
(7, 3)	$0.116\ m_o$
(9, 2)	$0.099\ m_o$
(25, 0)	$0.108\ m_o$

The effective masses in this table were calculated using equation (6). These effective masses of carbon nanotubes vary with the chiral vector inscribed in (n, m) indices. Compared these masses to the effective masses of Si (1.1 mo), and GaAs (0.067 mo) [22].

Carrier Concentration

With the density of states calculated, it is possible to calculate the carrier concentration of carbon nanotubes. The carrier concentration in any semiconductor is given by [23],

$$n_{cnt} = \int_{E_c}^{\infty} D(E)f(E)dE \qquad (7)$$

where D(E) is the density of states, f(E) is the Fermi level, and E_c is the conduction band minima.

Using equations (5), the carrier concentration becomes,

$$n_{cnt} = \sum_i^{\text{All Bands}} \left[\frac{4}{\pi V_{pp\pi} a\sqrt{3}} \cdot \int_{E_c}^{\infty} E(E^2 - E_{Cmin_i}^2)^{-1/2} \right.$$
$$\left. + \left(1 + e^{\frac{E - E_F}{KT}}\right)^{-1} dE \right] \qquad (8)$$

The equation (8) can be simplified using mathematical manipulations to arrive at the following expression:

$$n_{cnt} = \frac{4}{\pi V_{pp\pi} a\sqrt{3}} \int_0^{\infty} (E' $$
$$ + E_c)(E^2 + 2E_c E')^{-1/2} \left(1 \right.$$
$$\left. + e^{\frac{E' - E_F + E_c}{KT}}\right)^{-1} dE' \qquad (9)$$

The limits of integration have been changed by replacing the variable E in equation (8) with (E_c - E), where E_c is the lowest conduction energy. Furthermore, the summation has also been dropped as the Fermi function becomes negligible for conduction energy-minimums beyond the first band.

The integral in equation (9) is still very difficult to solve analytically, nevertheless, by letting x = E/KT and $\eta = \frac{E_F - E_c}{KT}$, equation (9) becomes,

$$n_{cnt} = \frac{4\sqrt{KT}}{\pi V_{pp\pi} a\sqrt{3}} \int_0^{\infty} (KTx $$
$$ + E_c)[x(KTx + 2E_c)]^{-1/2}(1 $$
$$ + e^{x-n})^{-1}dx \qquad (10)$$

The integral in this equation is similar to the complete Fermi integral [24] and can be approximately integrated [25] using two effective limits of the form,

Limit 1: $\eta \ll -1$

$$n_{cnt} = N_c I e^{\frac{E_F - E_c}{KT}}$$

Limit 2: $\eta \gg -1$, $n_{cnt} = N_c \frac{(E_F^2 - E_c^2)^{1/2}}{KT}$

where $N_c = \frac{4\sqrt{KT}}{KT}$

and, $I = \sqrt{\frac{E_c}{2KT}} + \frac{1}{\sqrt{KT}} \int_0^{\frac{6R_c}{KT}} \frac{(KT\,x + E_c)}{x^{1/2}(KT\,x + 2E_c)^{1/2}} e^{-x}dx$

This integral is found using any numerical integration method [26]. The upper limit has been replaced from infinity to $\frac{6R_c}{KT}$, as for values beyond this limit the function inside the integral approaches zero and can be neglected. The lower limit has also been changed to 0+, the value at x = 0 is found to be $\frac{E_c}{2KT}$ using the delta function integral [25]. Moreover, this integral is independent of the Fermi energy and characteristic of any particular carbon nanotube (n,m), it will remain constant at any bias voltage and the only physical dependency is temperature.

For intrinsic level calculation and regular doping, limit 1 is useful; limit 2 becomes important in degenerate level doping. Table II shows electron carrier concentration for different carbon nanotubes (n,m). An effective carbon nanotube wall thickness of 0.617Å [27] has been taken in these calculations. The intrinsic carrier concentrations of carbon nanotube (n,m) shown in Table II can be compared with the intrinsic carrier concentrations of Si (1.5 x 1010 cm-3), Ge (2 x 1013 cm-3), and GaAs (2.1 x 106 cm-3).

TABLE II
ELECTRON CARRIER CONCENTRATION FOR DIFFERENT
CARBON NANOTUBES

(n, m)	Intrinsic concentration (cm^{-3})
(3, 2)	1.187×10^5
(4, 2)	1.053×10^7
(5, 0)	4.037×10^5
(5, 1)	4.437×10^8
(6, 1)	2.734×10^9
(7, 3)	8.494×10^{12}
(9, 2)	7.921×10^{13}
(25, 0)	9.668×10^{14}

Conclusion

The energy dispersion relation, intrinsic carrier and effective mass for carbon nanotubes have been analyzed by derived equations from past theoretical and experimental results of carbon nanotubes. This results will helps the designer helps to construct in the electrical properties of carbon nanotubes as well as the conductivity of carbon nanotubes in the electronic band structure.

References

[1]. R. Martel, T. Schmidt, H. R. Shea, T. Hertel, and Ph. Avouris, "Single and multi wall carbon nanotube field effect transistors," *Appl. Phys. Letts.*, Vol. 73, No. 17, pp. 2447-2449, October 26, 1998.

[2]. J. Wildoer, L. Venema, A. Rinzler, R. Smalley, and C. Dekker, "Electronic structure of atomically resolved carbon nanotubes," *Nature*, Vol. 391, No 6662, pp. 59-62, January 1998.

[3]. S. J. Wind, J. Appenzeller, R. Martel, V. Derycke, and Ph. Avouris, "Vertical scaling of carbon nanotube field-effect transistors using top gate electrodes," *Appl. Phys. Letts.*, Vol. 80, Issue 20, 3817-3819, May 2002.

[4]. Y. Nosho, Y. Ohno, S. Kishimoto, and T. Mizutani, "N-type carbon nanotube field-effect transistors fabricated by using Ca

contact electrodes," *Appl. Phys. Letts.*, Vol. 86, No. 7, p. 073105, February 2005.

[5]. X. Zhou, J. Y. Park, S. Huan, J. Liu, and P. McEuen, "Band structure, phonon scattering, and the performance limit of single-walled carbon nanotube transistors," *Phys. Rev. Letts.*, Vol. 95, pp. 1468051, September 2005.

[6]. R. H. Baughman, A. A. Zakhidov, W. A. de Heer, "Carbon nanotubesthe route toward applications," *Science*, Vol. 297, No. 5582, pp. 787-792, August 2002.

[7]. R. Nizam, S. Mahdi A. Rizvi, Ameer Azam," Calculating Electronic Structure of Different Carbon Nanotubes and its Affect on Band Gap," *International Journal of Science and Technology*, Volume 1 No. 4, October 2011.

[8]. Ph. Avouris, "Molecular electronics with carbon nanotubes," *Acc. Chem. Res.*, Vol. 35, No. 12, pp. 1026-1034, December 2002.

[9]. P. R. Wallace, "The Band Theory of Graphite," *Phys. Rev. Letts.*, Vol. 71, pp. 622-634, May 1947.

[10]. R. Saito, M. Fujita, G. Dresselhaus, and M. S. Dresselhaus, "Electronic Band Structure of graphene tubules based on C60," *Phys. Rev. B.*, Vol. 46, No. 3, pp 1804-11, July 1992.

[11]. R. Martel, V. Derycke, J. Appenzeller, S. Wind, and Ph. Avouris, "Carbon Nanotube Field Effect Transistors and Logic Circuits," *39th Design Automation Conference (DAC 2002)*, pp. 94, 2002.

[12]. J.W. Mintmire and C.T. White, "Universal Density of States for Carbon Nanotubes," *Phys. Rev. Letts.*, Vol. 81, No. 21, pp. 2506-2509, September 1998.

[13]. C.T. White and J.W. Mintmire, "Density of states reflects diameter in nanotubes," *Nature (London)*, Vol. 394, pp. 29-30, 1998.

[14]. A. Javey, J. Guo, Q. Wang, M. Lundstrom, and H. Dai, "Ballistic carbon nanotube Field-effect transistors," *Nature*, 424, 654-657 (2003).

[15]. M.S. Dresselhaus, G. Dresselhaus, and Ph. Avouris, "Carbon Nanotube: Synthesis, Properties, Structure, and Applications," *Springer Verlag*, 2001.

[16]. R. Saito, M.S Dresselhaus, and G. Dresselhaus, "Physical Properties of Carbon nanotubes," *Imperial College Press*, London, July 1998.

[17]. A. Loiseau, P. Launois, P. Petit, S. Roche, J.P. Salvetat, Understanding Carbon Nanotubes, Springer, Berlin Heidelberg, 2006.

[18]. [18] R. Saito, G. Dresselhaus, M. S. Dresselhaus, "Trigonal warping effect of carbon nanotubes," Physical Review B, Vol. 61, No. 4, pp. 2981-90, January 2000.

[19]. R. Saito, G. Dresselhaus, M. S. Dresselhaus, "Trigonal warping effect of carbon nanotubes," *Physical Review B*, Vol. 61, No. 4, pp. 2981-90, January 2000.

[20]. D. Griffiths, "Introduction to Quantum Mechanics," *Prentice Hall, Inc.*, Upper Saddle River, NJ; 1995.

[21]. J. Guo, M. Lundstrom, and S. Datta, "Performance projections for ballistic carbon nanotube field-effect transistors," *Appl. Phys. Letts.*, Vol. 80, No. 17, 3192, April 29, 2002.

[22]. B. G. Streetman, *Solid State Electronics Devices*, Prentice Hall, 5th Edition, India, 2000.

[23]. M. Shur, *Physics ofSemiconductor Devices*, Prentice Hall, 1990.

[24]. M. Lundstrom, J. Guo, "Nanoscale Transistors, Device Physics: Modeling, and Simulation," *Springer Science Business Media, Inc.*, 2006.

[25]. J. Stewart, *Calculus early transcendentals, third edition*, Brooks/Cole Publishing Company, 1995.

[26]. R. and Education Association, *The Essentials of Numerical Analysis I, Research & Education Association (REA), Rev. Edition*, New Jersey, December 1992.

[27]. T. Vodenitcharova and L. C. Zhang, "Effective wall thickness of a single-walled carbon nanotube," Physical Review B, Vol. 68, p. 156401, 2003.

Long-wavelength III-V quantum-dot lasers monolithically grown on Si substrates

Qi Jiang, Andrew Lee, Mingchu Tang, Alwyn Seeds and Huiyun Liu*

Department of Electronic and Electrical Engineering, University College London,
London WC1E 7JE, United Kingdom,
*huiyun.liu@ucl.ac.uk

Abstract

Although great effort have been devoted for Si-based light generation and modulation technologies since the 1980s, monolithic growth of electrically pumped laser on Si remains the 'holy grail' for Si photonics. In this paper, room temperature lasing near the telecom wavelength of 1300 nm has been demonstrated at room temperature with low threshold current densities for InAs/GaAs quantum-dot lasers grown on both Si and Ge substrates.

Index Terms — **Quantum dots, Silicon photonics, III-V/Si integration, 1300 nm laser diodes, molecular beam epitaxy.**

Introduction

At present copper interconnects have been used everywhere from chip-to-chip to system-to-system interconnects since it is cheap, relatively manufacturable and good conductivity. However, limitations of copper interconnects are approaching, on both chip and system level. An attractive way to overcome this would be to integrate optical components and CMOS technology on one platform. This would allow optical electronic integrated circuits (OEIC) to be fabricated at low cost and high volume with high bandwidth and with virtually no crosstalk between channels. Other optical components such as integrated modulators, waveguides and photodetectors have been demonstrated on Si substrates {Michel, 2010 #249}[1–3] but because of the nature of indirect bandgap of group IV materials, it is very hard to use Si or Ge as the gain medium for lasers and very little reports have been published. [4][5]. On the other hand, III-V quantum dot (QD) lasers have been demonstrated with very low threshold current densities, temperature-insensitive operation above room temperature, and less sensitive to defects over traditional quantum well devices. And therefore III-V QD becomes one of the key components for producing high-performance OEIC.

In this paper, we studied the AlAs nucleation layer (NL) for the epitaxial growth directly grown on Si substrates. In comparison with the sample using a conventional GaAs NL, the photoluminescence (PL) intensity of 1.3-μm InAs/GaAs QDs is increased. A 5-layer InAs/GaAs laser structure on a Si substrate was fabricated with the use of AlAs NL. RT lasing occurs at ~1.29 μm with threshold current density (J_{th}) of ~650 A/cm^2. The growth of InAs/GaAs QD lasers on Ge and Ge/Si substrates is further studied. A J_{th} of ~200 mA/cm^2 has also been achieved for 1-mm long InAs/GaAs QD lasers on Ge substrates by using Ga pre-layer technique where the formation of antiphase domains and dislocations was significantly suppressed. By exploring the same approach to Ge/Si substrate, RT lasing at ~1.28 μm has been demonstrated with Jth of ~164 A/cm2 for InAs/GaAs QD device grown on Ge/Si substrates.

InAs/GaAs Quantum Dot Laser Diode Grown on Si Substrates

Due to the lattice mismatch between GaAs and Si, threading dislocations (TDs) can be nucleated at the interface between the GaAs buffer layer and Si substrate. Some of the TDs propagate into the III-V epitaxial layers built on GaAs/Si, leading to reduced optoelectronic conversion efficiency and lifetime of device. Therefore the density of TDs becomes one of most important parameter for the monolithic integration of GaAs on Si substrate. Hence studies on developing new NL to reduce the TD density within the GaAs buffer layer on Si substrates is critical for integrating GaAs-based photonic components with Si microelectronic circuits. Phosphorus-doped (100)-orientated Si substrates with 4° offcut towards the [110] planes were used for producing InAs/GaAs QDs directly grown on Si substrates by solid-source molecular beam epitaxy (MBE). Oxide desorption was performed by holding the Si substrate at a temperature of 900 °C for 10 minutes. The Si substrate was then cooled down to 400 °C for the growth of a 5-nm GaAs or a 5-nm AlAs NL, and a further 25 nm GaAs layer. An additional 970-nm GaAs layer was grown with a high growth rate of 0.7 ML/s at high temperature. InGaAs/GaAs dislocation filter layers, consisting of two repeats of a five-period (10-nm $In_{0.15}Ga_{0.85}As$/10-nm GaAs) superlattices (SPLs) and 400-nm GaAs, were used. Five InAs/InGaAs dots-in-a-well (DWELL) layers were then grown at optimized conditions as on GaAs substrates, with each layer consisting of 3.0 MLs of InAs grown on 2 nm of $In_{0.15}Ga_{0.85}As$ and capped by 6 nm of $In_{0.15}Ga_{0.85}As$.

The surface morphologies were investigated by using AFM measurement for both the GaAs [Fig. 1(a)] and AlAs NLs [Fig. 1(b)]. For the AFM measurements, the growth was terminated immediately after the deposition of a 5-nm AlAs layer or a 5-nm GaAs layer on Si substrates, followed by sample cool-down. For the sample with the GaAs NL, GaAs dots are observed with a density of over 5×10^{10}/cm^2. The height and width of the GaAs dots are ~ 4 nm and 20 nm, respectively. These indicate that the strain energy of 5-nm GaAs NL grown on a Si substrate is relaxed by the formation of coherent islands, which is similar to the In(Ga)As/GaAs system. In contrast, a lower density of ~ 4 × 10^9/cm^2 was obtained for AlAs dots grown on a Si substrate, with an irregular shape and larger average height and width of about 20 nm and 80 nm [Fig. 1(b)], respectively. The shape and size of AlAs dots indicate that the AlAs dots are defective. This suggests that the strain energy of 5-nm AlAs NL grown on a Si substrate is relaxed by the formation of both defects at the AlAs/Si interface and defective AlAs dots. Therefore, in comparison with the sample with the GaAs NL, more defects are confined at the III-V/Si interface and less defects propagate into the GaAs buffer layer are observed for the sample with the AlAs NL. The defect density in the III-V active region was estimated by EPD tests for both the sample with the AlAs NL and the one with the GaAs NL. The etchant used for the EPD delineation is a mixture of H$_3$PO$_4$, H$_2$O$_2$, and H$_2$O (in a 1:1:3 ratio). The defect densities of ~ 3 × 10^6/cm^2 and ~ 6 × 10^6/cm^2 were obtained for the samples with the AlAs NL and the one with the GaAs NL, respectively.

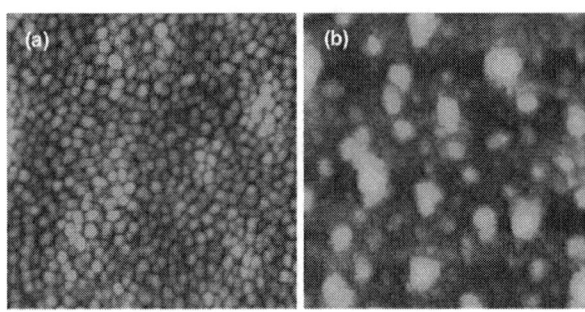

Fig. 1. 1 μm × 1 μm AFM images of the surface morphology of (a) a 5-nm GaAs nucleation layer and (b) a 5-nm AlAs nucleation layer, respectively.

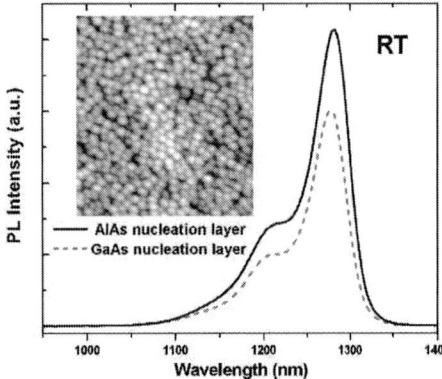

Fig. 2. Room-temperature PL spectra for Si-based InAs/GaAs quantum-dot samples with a GaAs and AlAs nucleation layers, respectively. The inset to shows a 1 × 1 μm² AFM image of InAs/GaAs QDs grown on a Si substrate with the AlAs nucleation layer.

Fig. 2 shows the RT PL spectra of InAs/GaAs QDs on Si substrates with the GaAs and AlAs NL. The InAs/GaAs QDs yield a RT emission at ~1.29 μm with a full width at half maximum (FWHM) of ~30 meV for both of the samples shown in Fig. 2. The increases of both peak intensity and

integrated intensity have been observed for AlAs NL sample. The inset of Fig. 2. shows uncapped InAs QDs on AlAs NL layer have a density of 4.3 × 10^{10} cm^{-2} where comparable to QD density grown on GaAs substrates.

Si-based InAs/GaAs QD laser diodes were consequently investigated with the incorporation of the AlAs NL. The laser

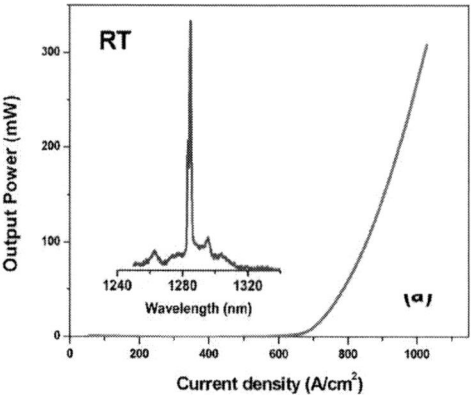

Fig. 3. Light output against current characteristic for a Si-based InAs/GaAs quantum-dot laser with cavity length of 3.5 mm. The inset shows the laser optical spectrum above threshold at room temperature.

was fabricated by standard lithography with wet etching techniques. The facets are cleaved without any HR coating. Fig. 3 shows the RT light output against drive current (L-I) characteristic for the device with a cavity length of 3.5 mm. The RT Jth is about 650 A/cm^2, which is lower that the previously reported values of III-V QD lasers monolithically grown on Si substrates, such as 900 A/cm2 for a 1.02 μm QD device and 725 A/cm^2 for a 1.3-μm QD laser. The inset of Fig. 3, demonstrated RT lasing at 1.29 μm.

InAs/GaAs Quantum Dot Laser Diode Grown on Ge Substrates

Because TDs are originated due to the lattice constant difference between GaAs and Si, an intermediate epitaxial layer with a similar lattice constant to GaAs and able to be grown on Si substrates with low defects density would be ideal to be used for reducing the density of TDs. Ge has been grown on Si substrates since the beginning of the Si photonics and this technology is very mature. At the same time the lattice constant between Ge and GaAs is almost identical (0.08% difference). III-V QD lasers grown on Ge-on-Si substrates were trialed.

Fig. 4 shows LI curve for a 1-mm long QD laser device. This laser started to lase at a threshold current of 100 mA at RT. The threshold current density of 200 mA/cm^2 for 1-mm long InAs/GaAs QD devices was obtained. This represents the similar quality growth of QDs compared to grown on GaAs substrates. Typical emission spectra has been found at ~1.29 μm at 100 mA in inset of Fig. 4. The behavior of ground and excited lasing has also been found on many InAs QD lasers grown on GaAs substrates.

978-1-4673-4840-9/13 $31.00 © 2013 IEEE

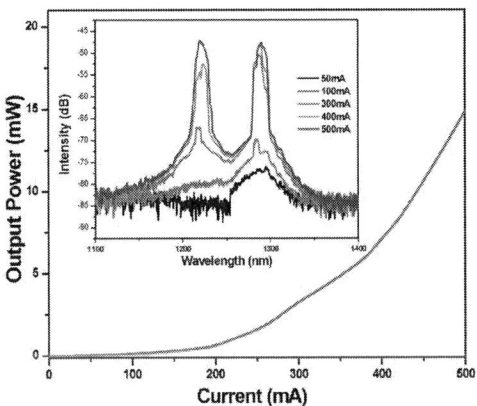

Fig. 4. Light output against current characteristic from 1-mm long InAs/GaAs quantum-dot laser diodes on Ge substrates at room temperature. The inset shows the emission spectra at different current.

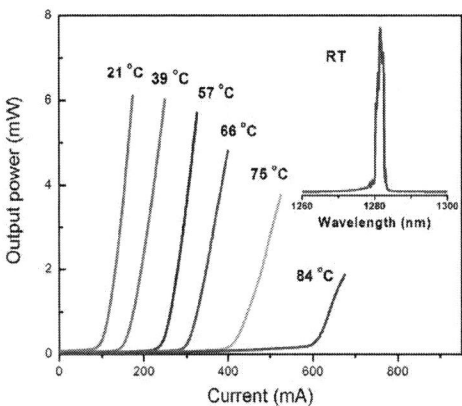

Fig. 5. Light output against current characteristic for 3-mm long InAs QD laser grown on Ge/Si substrate under pulsed condition with operating temperature from 20 to 84°C. The inset shows the laser optical above threshold at RT.

InAs/GaAs Quantum Dot Laser Diode Grown on Ge-on-Si Substrates

After our group successfully demonstrated the III-V QD lasers grown on Ge substrates, III-V QD lasers grown on Ge-on-Si substrates were investigated here. The knowledge and experience obtained on Ge substrates growth has been used on Ge-on-Si substrates growth. To form the Ge/Si virtual substrate, a 2-μm Ge layer was grown using chemical vapour deposition on phosphorus-doped (100)-oriented Si substrates with a 6o offcut towards the [111] planes. The dislocation density in the Ge epitaxial layer is about $5 \times 106/cm^2$. Similar laser structure to previous section was used for fabricating QD lasers in this section. The lasers were wet etched to 20-μm wide and cleaved to 3mm cavity length. The inset of Fig. 5 shows the spectra obtained for the InAs/GaAs QD laser above threshold current at room temperature and operating under pulsed conditions (1.0% and 0.1us). Fig 5. shows the light output versus current from 20 to 84 °C. RT threshold current density is about 164 A/cm². The characteristic temperature is ~50K in the temperature range from 20 to 84°C.

Conclusion

The use of an AlAs nucleation layer on Si substrates to reduce the density of threading dislocations has been demonstrated. A QD laser grown with AlAs nucleation layer has demonstrated better PL intensity and lower threshold current density. We also reported the InAs QD laser grown on Ge substrate with low J_{th} of 200 mA/cm² for 1-mm long device. The identical growth technique has been applied to Ge/Si substrates. And the InAs/GaAs QD laser grown on Ge/Si substrates was demonstrated with RT J_{th} of 164 A/cm² and operating at 84 °C for a 3-mm long device. These studies could represent an essential step toward the monolithic integration of 1.3-μm InAs/GaAs QD devices on a Si platform, as well as for the creation of other III-V photonic devices on a Si platform.

References

[1] J. Michel, J. Liu, and L. C. Kimerling, "High-performance Ge-on-Si photodetectors," *Nature Photonics*, vol. 4, no. 8, pp. 527–534, Jul. 2010.

[2] D. Liang and J. E. Bowers, "Recent progress in lasers on silicon," *Nature Photonics*, vol. 4, no. 8, pp. 511–517, Jul. 2010.

[3] J. Leuthold, C. Koos, and W. Freude, "Nonlinear silicon photonics," *Nature Photonics*, vol. 4, no. 8, pp. 535–544, Jul. 2010.

[4] H. Rong, A. Liu, R. Jones, O. Cohen, D. Hak, R. Nicolaescu, A. Fang, and M. Paniccia, "An all-silicon Raman laser," *Nature*, vol. 433, no. 7023, pp. 292–294, Jan. 2005.

[5] R. E. Camacho-Aguilera, Y. Cai, N. Patel, J. T. Bessette, M. Romagnoli, L. C. Kimerling, and J. Michel, "An electrically pumped germanium laser.," *Optics express*, vol. 20, no. 10, pp. 11316–20, May 2012.

Development of a high sensitivity photodetector using amorphous selenium and diamond cold cathode

K. Okano[1], T. Masuzawa[1], M. Onishi[1], I. Saito[1], A. T. T. Koh[2], D. H. C. Chua[2] and T. Yamada[3]

[1] Department of Material Science, International Christian University,
[2] Department of Materials Science and Engineering, National University of Singapore,
[3] Nanotube Research Centre, National Institute of Advanced Industrial Science and Technology,
3-10-2 Osawa, Mitaka
Tokyo, Japan
Tel/Fax: +81(0)422 33 3254, E-mail: kenokano@icu.ac.jp

Abstract

Amorphous-selenium (a-Se) based photodetector is a promising candidate for high sensitivity imaging device with high spatial resolution, high response speed, and a wide detection wave range. Although HARP (high-gain avalanche rushing amorphous photoconductor) is known to be the first a-Se based photodetector that reported carrier multiplication and achieved effective quantum efficiency of 10, physical and chemical properties of the material are yet poorly understood. In this study, we carefully considered the recipe of a-Se and as a result, the necessary condition for obtaining the carrier multiplication has been clarified. (Keywords: photodetector; amorphous selenium; diamond cold cathode)

Introduction

Amorphous-selenium (a-Se) based photodetectors are promising candidate for imaging devices, due to their high spatial resolution and response speed, as well as extremely high sensitivity enhanced by an internal carrier multiplication. HARP (high-gain avalanche rushing photoconductor) is the first a-Se based photodetector, which exploited the carrier multiplication and achieved effective quantum efficiency of 10 with real-time imaging capability [1]. Although the carrier multiplication phenomenon is analyzed using physical models such as the lucky drift model [2,3], the application of this multiplication effect can only be demonstrated in HARP films. Neither conditions nor physics of carrier multiplication are fully explained so far, mostly due to the lack in material characterization [4]. The conditions of the signal multiplication in a-Se films must be clarified in order to develop a novel imaging device with ultra high sensitivity and wide detectable wave range.

Here we present some example of fundamental conditions to induce the internal carrier multiplication. The conditions are refined by film characterization and photosensitivity measurement.

Experimental Details

The selenium-based amorphous films were deposited by rotational vacuum evaporation. The detailed recipe of the evaporation was described elsewhere [5]. Trigonal selenium powder and arsenic selenide (As_2Se_3) are used as evaporation source. The substrate was glass faceplate with a throw-hole. An indium-tin-oxide (ITO) thin film was deposited to serve as a transparent contact. The deposited films were characterized using Raman spectroscopy and time of flight secondary ion mass spectroscopy (TOF-SIMS). In Raman spectroscopy, an emission line of 647.1 nm from Kr^+ was used as excitation laser, and the spectra are recorded in a backscattering geometry [6]. The electronic properties of the films were evaluated in order to investigate a carrier multiplication effect. Current-voltage (I-V) characteristics of the films were measured and compared using the setup depicted in Figure 1. White halogen lamp (TECHNO LIGHT TKX-50R) with 2,000 lx was used to illuminate the film during the measurement.

Results and Discussions

Figure 2 illustrates the Raman spectra of the amorphous films, (a) without arsenic incorporation and (b) with arsenic incorporation. It was known from previous research that a-Se films without As_2Se_3 layers crystallize under laser exposure [6]. Raman spectra of the crystallized films showed a sharp peak at 235cm^{-1}, as illustrated in Figure 2(a). In contrast to the sharp peak in pure a-Se, a broad feature was observed with at 250 cm^{-1} for samples with arsenic incorporation. The broad feature in Figure 2(b) is attributed to a variation in bonding lengths, which is characteristic in amorphous materials. According to previous reports, the resistivity of such films was typically lower by orders of magnitude compared to the amorphous films [5]. In the crystallized films, it is difficult to obtain strong electric field of 80~100 V/μm, where carrier multiplication takes place. The Raman spectra clearly showed that the incorporation of arsenic was one of the key factors to cause a carrier multiplication.

The electronic properties of the films were evaluated based

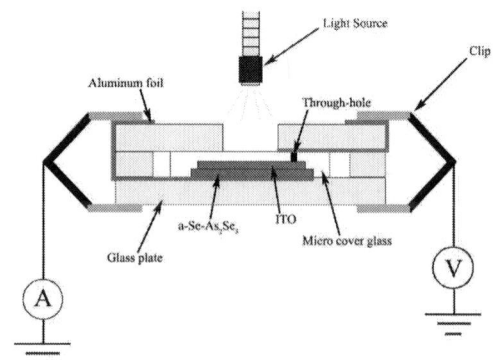

Figure 1. A setup apparatus of contact current-applied voltage measurement.

on I-V characteristics. Figure 3 illustrates a typical I-V curve of our films. The current increased linearly for voltages below 340 V. A drastic current increase was observed at 340 V, at which point I-V curve becomes exponential. This tendency was observed in each of the iterative measurement on the same sample, so that the sample was not irreversibly denatured. The resistance of the film at voltage lower than 340 V was as high as 1.6 TΩ, which is attributed to low carrier concentration [7]. No current rectification was observed in the measurement.

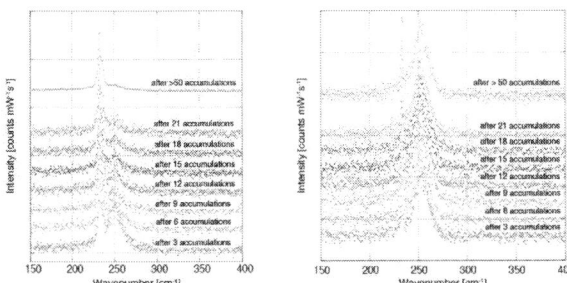

Figure 2. Raman spectra of a-Se films (a) without arsenic incorporation and (b) with arsenic incorporation [8]

It is widely known that a-Se contains high defect density; therefore, it can be regarded as an intrinsic semiconductor with many carrier traps. It is also known that a-Se shows p-type conduction by arsenic incorporation. Therefore, the structure of our films is assumed to be the alternative trapping layers and p type layers. The carrier multiplication process can be explained as follows;

1) Some carrier pairs are generated by incident light,
2) The carriers are accelerated in the p-type layer, because of the strong electric field in the film,
3) The accelerated carriers enter the trapping layer, which excite multiple carriers from the traps,
4) the excited carriers enters the p-type layer, where they are accelerated for the next carrier cascade.

The advantage of this "cascading multiplication" is that it can reduce the dark current even in signal multiplication mode. Since little or no incident photon is expected in the dark, most carriers are trapped in the trapping layer. This prevents carriers to enter the p-type layer, where they are accelerated

field reduction in the trapping layer. By choosing an optimum layer width, we can achieve both high multiplication gain and low dark current. A detailed analysis on the layer design and electronic properties is currently under investigation. The films will be then mounted as an anode on a diode-structured photodetector. The emission current-applied voltage characteristics of the photodetector will be evaluated in order to confirm the carrier multiplication in the device.

Conclusion

The conditions for the carrier multiplication have been clarified. From the Raman spectra, incorporation of arsenic was a necessary condition to maintain amorphous structure, which results in a high resistivity. Rapid current increase at the applied voltage of around 340 V in the electronic properties is also a necessary condition. When the film possesses these features, the carrier multiplication in photodetector should be observed in the emission current-applied voltage characteristics of the photodetector using a-Se and diamond cold cathode.

Acknowledgement

The present work is financially supported by MEXT-Support Program for the Strategic Research Foundation at Private Universities (S0801012) 2008-2012 and Grants-in-Aid for scientific research (21560053), both from the Ministry of Education, Culture, Sports, Science and Technology, Japan.

References

[1] K. Tanioka et al., IEEE Electron Device Letters, EDL-8,9, pp.392-394, 1987.
[2] A. Reznik et al., J. Non-Cryst. Solids 354, 2691–2696, (2008)
[3] O. Rubel et al., Phys. Status. Solidi. C 1, No. 5, 1186–1193 (2004)
[4] Reznik, A. et al., J. Appl. Phys. 102, 053711 (2007)
[5] I. Saito, et al., Jpn. J. Appl. Phys., Vol. 44, No. 11, L334-L337, 2005.
[6] K. Okano, et al., J. Non-Crystal. Sol. 353, 308–312, 2007.
[7] I. Saito, et al., "A transparent ultraviolet triggered amorphous selenium p-n junction," Appl. Phys. Lett. 98, 152102, 2011.
[8] I. Saito, et al., submitted to Sol. Stat. Comm.

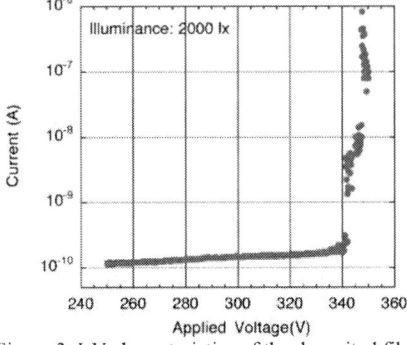

Figure 3. I-V characteristics of the deposited film

enough to cause carrier multiplication. For our films, the average electric field required for carrier multiplication was around 170 V/μm. This is roughly twice as large as previous reports. The higher threshold voltage could be explained by

978-1-4673-4840-9/13 $31.00 © 2013 IEEE

Fabrication and Characterization of Uni-Traveling-Carrier Photodetectors (UTC-PDs) with Dipole-Doped Structure at InGaAs/InP Interface

Q. Q. Meng[1*], C. Y. Liu[1], H. Wang[1,2], K. S Ang[1], K. Manoj[1], and T. X. Guo[1]

1.Temasek Laboratories@NTU (TL@NTU), Nanyang Technological University, 50 Nanyang Drive, Singapore 637553

2.School of Electrical and Electronic Engineering, Nanyang Technological University, Nanyang avenue, Singapore, 639798

*: Corresponding author, Phone: (65)97930516, E-mail: qqmeng@ntu.edu.sg

Abstract

This paper describes the fabrication and high speed characterization for uni-traveling-carrier photodetectors (UTC-PD) with dipole-doping layers. The dipole-doping InGaAs/InP layer has been introduced at the InGaAs (absorption layer) and InP (collection layer) interface to prevent current blocking in our UTC-PD devices. These UTC-PDs have achieved high photocurrent of 52 mA, high responsivity of 0.44 A/W as well as high 3dB bandwidth of ~17.5 GHz. Furthermore, a UTC-PD equivalent circuit model is used to simulate and interpret device results of our UTC-PDs.

Keywords: uni-traveling-carrier photodetector (UTC-PD), dipole doping, photocurrent, frequency response.

Introduction

In recent years, the uni-traveling-carrier photodetector (UTC-PD) has attracted much research attention. Compared to conventional photodiodes, such as PIN diodes [1,2], the UTC-PD has a unique mode of operation where only electrons are the active carriers [3-9]. The overshoot velocity of electrons is one order of magnitude larger than that of holes, and this helps shorten the carrier transit time, as well as reduce the space charge effect, which is a common issue in any surface illuminated photodiode. These factors result in high-speed and high saturation-output-powers simultaneously obtained from UTC-PDs[3].

Device structure and fabrication

In a typical UTC-PD layer structure, InGaAsP quaternary layers in the collection region were used to smooth out the abrupt conduction band barrier at the InGaAs–InP heterojunction interface. However, due to the growth constraints of the epi-wafer suppliers, structures with InGaAsP are currently unavailable. In order to minimize the current-blocking at InGaAs-InP interface, a dipole-doped interface structure is proposed for our devices [7]. The detailed layer structure is shown in TABLE 1. It can be seen that an InGaAs/InP region with p+-n+ is added between the InGaAs absorber and InP collector. 1-D energy band simulations suggest that the effective conduction band barrier at the InGaAs–InP heterojunction interface can be minimized, as in Fig.[7].

Top-illuminated UTC-PDs with cylindrical mesas of different diameters were fabricated using standard processing techniques. The mesa formation was based on wet etching processes to minimize the surface damage. The mesa was connected to the coplanar waveguide for RF probing with BCB planarization.

Testing Results of UTC-PD

In order to study the performance of our UTC-PD and to extract model parameters, the UTC-PD of different diameters have been tested both statically and dynamically. As shown in Fig.2, our UTC-PDs have achieved a high photocurrent of 52mA under a 200 mW optical power injection from a UTC-PD with a diameter of 36 μm. Fig.3 shows that a high responsivity of 0.44 A/W is achieved from a UTC-PD with a diameter of 44 μm. With the optical power of 44 mW, the photocurrent increases when V increases, which means the collection layer is more depleted at higher voltage.

Fig.4 presents the frequency response characteristics from a UTC-PD with a diameter of 15 μm under the constant optical power of 20 mW. The reverse bias is varied from 0.5 V to 2.5 V. The large 3dB bandwidth of this devices is ~17.5 GHz.

Equivalent Circuit Model of UTC-PD

The optical-to-microwave conversion frequency response in a UTC-PD is determined by both the carrier-transit time and the external parasitic RC time constant. Fig.5[10] shows the equivalent circuit model that considers both the carrier-transit time (Region 1) and the external parasitic RC time constant (Region 2) of our UTC-PD. In this model, the carrier transit-induced time delay is represented with a linear RC circuit (R_t, C_t in Region 1), which is combined in parallel to a voltage-controlled current source $g_m V_{RF}$. g_m is a constant adjusted according to the UTC-PD's optical-to-microwave conversion quantum efficiency.

The reason why we design the equivalent circuit model like this is that the 3-dB electrical bandwidth of UTC-PD is limited by both the carrier transit time and the external parasitic RC time, which can be expressed as in (1) [10]

978-1-4673-4840-9/13 $31.00 © 2013 IEEE

$$1/f_{3dB}^2 = 1/f_t^2 + 1/f_{RC}^2 \qquad (1)$$

In the proposed model, R_s represents the UTC-PD series resistance including bulk, sheet, and contact resistances; C_j is the junction capacitance; C_{dx} is the BCB capacitance; C_p is the pad capacitance; L is the total inductance of contact metal inductance and pad inductance. R_0 and R_L represent the source resistance and load characteristic resistance, and their values are both set at 50 Ω.

TABLE 1
LAYER STRUCTURE FOR THE UTC-PD WAFER.

Material	Thickness	Doping
$In_{0.6}Ga_{0.4}As$	30 nm	p-type 1×10^{19} cm^{-3}
InP	200nm	p-type 1×10^{19} cm^{-3}
$In_{0.53}Ga_{0.47}As$	50 nm	p-type 2×10^{18} cm^{-3}
$In_{0.53}Ga_{0.47}As$	100 nm	p-type 1×10^{18} cm^{-3}
$In_{0.53}Ga_{0.47}As$	300 nm	p-type 5×10^{17} cm^{-3}
$In_{0.53}Ga_{0.47}As$	22 nm	undoped
$In_{0.53}Ga_{0.47}As$	8 nm	p-type 1×10^{18} cm^{-3}
InP	8 nm	n-type 1×10^{18} cm^{-3}
InP	13 nm	n-type 1×10^{17} cm^{-3}
InP	200 nm	n-type 5×10^{16} cm^{-3}
$In_{0.53}Ga_{0.47}As$	30 nm	n-type 5×10^{18} cm^{-3}
InP	600 nm	n-type 5×10^{18} cm^{-3}
S. I. – InP substrate		

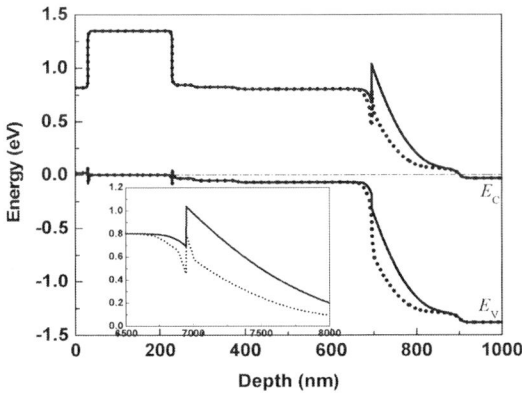

Fig.1 Effectiveness of using a dipole-doped interface to reduce the effective conduction band barrier at the InGaAs–InP heterojunction interface predicted by1-D energy band simulation. Inset: Conduction band energy diagram in the InGaAs/InP interface region.

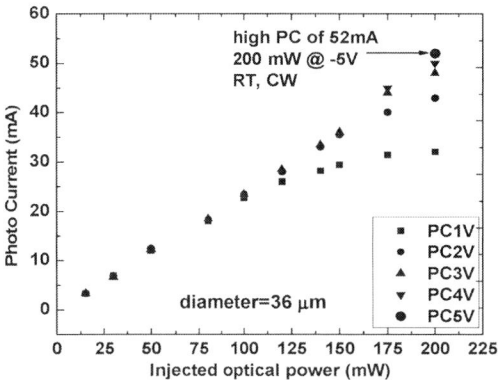

Fig.2 Photocurrent (PC) versus optical power of devices with a 36μm diameter at different reverse bias.

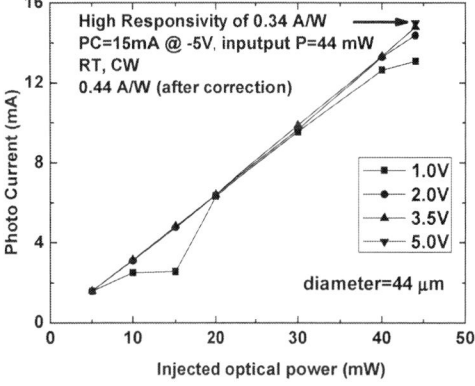

Fig.3 Photocurrent (PC) versus optical power of devices with a 44μm diameter at different reverse bias.

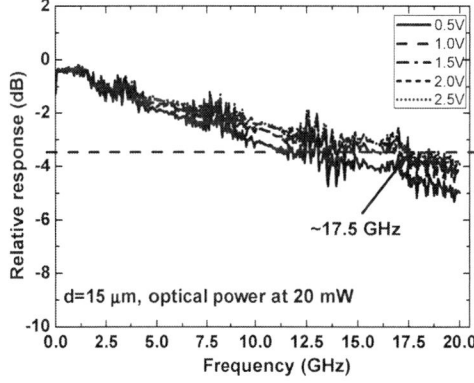

Fig.4 Frequency response of PD with a diameter of 15 μm

978-1-4673-4840-9/13 $31.00 © 2013 IEEE

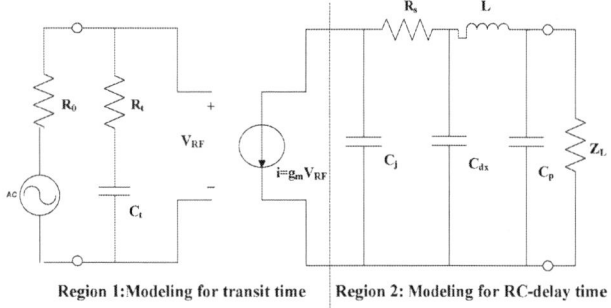

Region 1:Modeling for transit time | Region 2: Modeling for RC-delay time

Fig.5 Equivalent-circuit model of the UTC-PD.

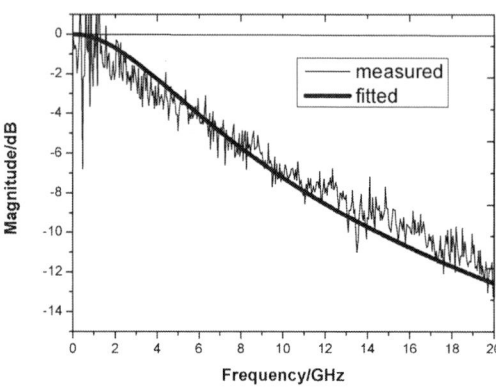

Fig.7 Measured and fitted S_{21} magnitude of the device with a diameter of 36 μm at a 5 V reverse bias. The thin solid line represents the measured parameters and the thick solid line represents the fitted parameters using proposed equivalent circuit model of UTC-PD.

Conclusion

In summary, we have designed and fabricated UTC-PD devices with dipole-doping structure. The measurement results have demonstrated reasonable good device performance. Meanwhile, a UTC-PD equivalent circuit model considering the space charge effect is used to simulate and predict the device performance. Simulation results show that the model produces a good matching with the measured S parameters in a frequency range from 10 MHz to 20 GHz. This model will be helpful to improve the design of high speed and high power UTC-PD in future work.

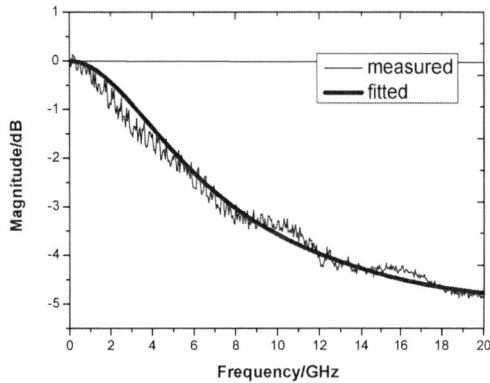

Fig.6 Measured and fitted S_{22} magnitude of the device with a diameter of 36 μm at a 5 V reverse bias. The thin solid line represents the measured parameters and the thick solid line represents the fitted parameters using proposed equivalent circuit model of UTC-PD.

TABLE.2
EXTRACTED MODEL PARAMETERS OF A UTC-PD

R_s	C_j	C_{dx}	L	C_p
81 Ω	0.88 pF	70 fF	0.16 nH	17 fF

For model parameter extraction, we can first extract the values for the RC time delay components in Region 2 by fitting the calculated S_{22} to measured S_{22}. Fig.6 shows the measured (thin line) and fitted S_{22} parameters (thick line) of the UTC-PD with the diameter of 36 μm at a 5 V reverse bias. The extracted parameters are summarized in Table 2. Then with the extracted parameters in Region 2, the calculated S_{21} can be best fitted with the measured S_{21} to get the values for the carrier transit time delay components (R_t, C_t, and g_m). Fig.7 shows the measured (thin line) and fitted (thick line) S_{21} magnitude of the same 36μm diameter device at a 5 V reverse bias using the extracted components parameters of S_{22}. The values of components in Region 1 are extracted to be R_t=90 Ω, C_t=10 fF and g_m=0.76 S.

As shown in Fig.6 and Fig.7 the fitted S parameters match pretty well with the measured ones in magnitude in the frequency range from 10 MHz to 20 GHz

References

[1] J. E. Bowers, C. A. Burrus, and R. J. McCoy, "InGaAs PIN photodetectors with modulation response to millimetre wavelengths," *Electronics Letters*, vol. 21, pp. 812-814, 1985.

[2] L. Y. Lin, M. C. Wu, T. Itoh, T. A. Vang, R. E. Muller, D. L. Sivco, and A. Y. Cho, "High-power high-speed photodetectors - Design, analysis, and experimental demonstration," *IEEE Transactions on Microwave Theory and Techniques*, vol. 45, pp. 1320-1331, Aug 1997.

[3] T. Ishibashi, T. Furuta, H. Fushimi, S. Kodama, H. Ito, T. Nagatsuma, N. Shimizu, and Y. Miyamoto, "InP/InGaAs uni-traveling-carrier photodiodes," *Ieice Transactions on Electronics*, vol. E83C, pp. 938-949, Jun 2000.

[4] N. Li, X. W. Li, S. Demiguel, X. G. Zheng, J. C. Campbell, D. A. Tulchinsky, K. J. Williams, T. D. Isshiki, G. S. Kinsey, and R. Sudharsansan, "High-saturation-current charge-compensated InGaAs-InP uni-traveling-carrier photodiode," *IEEE Photonics Technology Letters*, vol. 16, pp. 864-866, Mar 2004.

[5] D. A. Tulchinsky, X. W. Li, N. Li, S. Demiguel, J. C. Campbell, and K. J. Williams, "High-saturation current wide-bandwidth photodetectors," *IEEE Journal of Selected Topics in Quantum Electronics*, vol. 10, pp. 702-708, Jul-Aug 2004.

[6] H. Ito, S. Kodama, Y. Muramoto, T. Furuta, T. Nagatsuma, and T. Ishibashi, "High-speed and high-output InP-InGaAs unitraveling-carrier photodiodes," *IEEE Journal of Selected*

978-1-4673-4840-9/13 $31.00 © 2013 IEEE

Topics in Quantum Electronics, vol. 10, pp. 709-727, Jul-Aug 2004.

[7] H. Wang and S. Mao, "High speed InP/InGaAs uni-traveling-carrier photodiodes with dipole-doped InGaAs/InP absorber-collector interface," in *Compound Semiconductor Week (CSW/IPRM), 2011 and 23rd International Conference on Indium Phosphide and Related Materials*, 2011, pp. 1-3.

[8] X. Wang, N. Duan, H. Chen, and J. C. Campbell, "InGaAs-InP photodiodes with high responsivity and high saturation power," *IEEE Photonics Technology Letters*, vol. 19, pp. 1272-1274, Jul-Aug 2007.

[9] M. Chtioui, A. Enard, D. Carpentier, S. Bernard, B. Rousseau, E. Lelarge, F. Porrimereau, and M. Achouche, "High-power high-linearity uni-traveling-carrier photodiodes for analog photonic links," *IEEE Photonics Technology Letters*, vol. 20, pp. 202-204, Jan-Feb 2008.

[10] Gang Wang, Tsuneo Tokumitsu, Ikuo Hanawa, Yoshihiro Yoneda, Keiji Sato, and Masahiro Kobayashi, "A Time-Delay Equivalent-Circuit Model of Ultrafast p-i-n Photodiodes," *IEEE Transactions on Microwave Theory and Techniques*, vol. 51, No. 4, April 2003.

6.5 nm-thick Al₂O₃ Surface Passivated Layer Grown on Two Stacks of 10-Period InGaAs and GaAs-Capped InAs Quantum Dot Infrared Photodetector Focal Plane Arrays for High Temperature Operation

Shiang-Feng Tang[*], Tzu-Chiang Chen[+], Wen-Jen Lin[*] and Shih-Yen Lin[++]

*Materials and Electro-Optics Division, Chung-Shan Institute of Science and Technology, Lung-Tan, Taoyuan, Taiwan
+Department of Electrical and Electronic Engineering, National Defense University, Daxi, Taoyuan, Taiwan
++Research Center for Applied Sciences, Academia Sinica, Nankang, Taipei, Taiwan

Correspondent Author Email: shiangfengtang@yahoo.com.tw

Abstract

The study is to demonstrate a dual-band infrared images based on two stacked InGaAs and GaAs-capped InAs QDIP structure, with the use of nano-scale Al₂O₃ surface passivated layer by ALD system deposited to decrease the device shot noise and boost the operation temperature of the thermal imaging FPA to 180 K.

Keywords: Quantum Dot Infrared Photodetectors (QDIPs), Focal Plane Arrays (FPAs), Atomic Layer-Deposition (ALD), Molecular Beam Epitaxy (MBE), Cross-Sectional Transmission Electron Microscopy (XTEM). Photo-Enhanced Vapor Deposition (PVD)

Introduction

The three-dimensional confinement of quantum dots (QDs) offers the localization of carriers reducing the thermo-ionic emission for lowering the device dark current. The intersubband energy level spacing in the QDs is greater than the phonon energy and, therefore, reduces the phonon scattering, which is a dominant scattering mechanism in bulk and quantum wells (QWs). This is the main reason for long carrier relaxation times in QDIPs, which in turn increases the photoconductive gain. The responsivity and detectivity are also increased due to the increase in gain and photocurrent [1,2]. In addition, QDIPs are sensitive to normal incidence radiation, which is not possible in QWIPs, due to polarization selection rules, and requires specialized gratings to direct the radiation into the detector. The QDs are normally doped to about than 1-2 electrons per dot in order to prevent carriers from occupying the excited state which will increase the dark current.

The photogenerated current in QDIPs is generated due to absorption of photons. However, dark current is generated thermally or due to phonon interaction and is present when no light is incident on the detector. Once the background photocurrent is greater than the device dark current, the photodetector is background limited. It regards this operation condition as also background limited infrared performance, and it is a good measure of the operating temperature of the detector. Besides intrinsic three-dimensional confinement in QDIP being responsible for higher temperature operation, the decreases in leakage current of Al₂O₃/GaAs interfaces on mesas surrounding reveals indispensability to improve the QDIP FPA performance under higher operation. It is due to the removal of native oxides from the interface in the Al₂O₃/GaAs heterostructure and leads to relieve the Fermi-level pinning phenomenon.

Self-cleaning of interfacial As₂O₃ in atomic layer-deposition ALD-Al₂O₃ films on GaAs and InGaAs was found in previous studies, using Al(CH₃)₃ as the metal precursor. It provides an effective passivation of these GaAs-based substrates by Al₂O₃. Although the exact mechanism of the interfacial self-cleaning effect found in the ALD-Al₂O₃ on GaAs-based semiconductors is not known yet, C.-H. Chang, et al. [3] have proposed that it may involve ligand exchange (substitution) reactions between the trimethylaluminum [Al(CH₃)₃, TMA] and the native Al₂O₃, providing a pathway for the interfacial cleaning by forming Al₂O₃ and volatile trimethy-larsine. These reactions have been understood in alky-lating As₂O₃ with aluminum alkyls to form trialky-larsines. In this paper, it is demonstrated firstly that the dislocation-free interfaces of ALD-Al₂O₃ on InGaAs and GaAs-capped InAs QDIP FPA under extremely high operation temperature of 180K. The cross-sectional transmission electron microscopy (XTEM) of interfaces of ALD-Al₂O₃ and GaAs and the spectral profile for photoresponse of the proposed QDIP sample with ALD-Al₂O₃ and PVD are also measured and investigated.

Device Fabrication and Experimental Methodology

The hetero-structured epi-wafer consists of two stacks of 10-period InGaAs and GaAs-Capped InAs quantum dot Infrared photodetector by solid-source MBE [2]. The schematic of QDIP hetero-structure are shown in Fig.1. The structure is designed to detect the infrared spectra through two atmospheric spectral windows of 3~5 (mid-wavelength) and 8~14 μm (long-wavelength). The upper stack of dot-in-a-well (DWELL) QDIP and bottom one of regular QDIP facilitate the infrared dual-band detections for mid- and long-wavelength, respectively.

The epi-wafer is defined for device mesa by photolithography and then cleaned up for the growth of passivated layer of Al₂O₃ and SiO2, respectively. The ALD-Al₂O₃ thin films were grown at 200°C in vertical-type ALD reactor and the gas pressure of the reactor was kept at 1 torr during the deposition process. Trimethyaluminum, and H₂O were used as precursors, and argon (99.9995%) was employed as the carrier, dilute and purge gas. Both TMA and water precursors were kept in a stainless tank at room temperatures.

All the injection pipes were heated to 100°C, and the mixing box to 150°C. The flow rate of argon was fixed at 200 sccm, using alternating pulse of ALD-Al_2O_3 in the sequence of precursor/ purging gas/ oxidizing gas/ purging gas in period at 3/3/3/3s for each self-limiting cycle, respectively. An Al_2O_3 deposition rate of 0.078 nm/ cycle was measured by spectroscopic ellipsometer. The 210 nm-thick SiO_2 thin film is grown on the same QDIP structure by PVD under substrate temperature lowering to 160 °C for comparison to that of ALD- Al_2O_3. The single device and FPA were fabricated using ALD reactor to deposit 6.5 nm-thick Al_2O_3 passivated layer and followed by III-V compound semiconductor photolithography to define via-contact and mesa with wet chemical etching recipes of buffer HF and H_3PO_4: H_2O_2: H_2O solutions, respectively. Finally, the standard metal deposition and rapid thermal annealing (RTA) processes are conducted, respectively. In addition, the FPA using the same device structure was mechanically thinned down to 50 µm using a combination of lapping and polishing. And then the etching-stop process (recipe: $C_6H_8O_7$: $K_3C_6H_5O_7 \cdot H_2O$: H_2O_2) is done for backside removal fully to increases the infrared illuminated absorption and remove the dislocations and pinholes induced by delaminating and image deterioration [4].

Fig.2 shows the 320×256 pixel-numbered FPA hybridized with snapshot-mode ROIC mounted in a 68 pin leadless ceramic chip carrier, which was put in the temperature-controlling cryogenic chamber with optical cold infrared spectral window.

The ROIC chip based on complimentary metal–oxide–semiconductor circuit has the advantages of low power consumption and high noise rejection while providing sophisticated multiplexing and control logic for an IR QDIP FPA module. The integral and readout of the signal is based on the snapshot mode in contrast to the rolling mode for the suppressing of thermal image trail while the tracking target moves in a high-speed condition. Due to the non-uniformity calibrations of gain and offset for this QDIP FPA, a quasi two-point temperature corrected process and replacement of bad sensitive pixels by neighboring good ones are applied for real-time video image with the frame rate of 30 Hz [2]. The operability of the QDIP FPA would be measured. The thermal images are taken by the home-made IR video demonstrated systems with adjusting suitable integrated time (T_{it}), gain and offset voltages under setting operation temperature and the collecting layout of video evaluating system in detail shown in Fig.3 (a) and (b), respectively.

Results and Discussions

Fig.4 shows the high-resolution XTEM image for exhibiting the dislocation-free interface of top GaAs and Al_2O_3 passivated layer. The transition region (the Al_2O_3 /GaAs interface) is clearly shown and illustrates the Al_2O_3 /GaAs interface is sharp [5].

Under higher reverse bias of -4 V and 30 K, it is dominated in the mid-wavelength infrared spectrum which is attributed to bottom stack QDIP detection. And the calibrated photoresponses of test samples with ALD-Al_2O_3 and PVD-SiO_2 passivated layers are shown in Fig.5, respectively. Except for the different pattern exhibited in response spectra with ALD- Al_2O_3 and PVD-SiO_2, the peak wavelength almost

is not change, i.e., ~4.3µm.

It can be deduced that the profile of the photoresponse for the sample with ALD is obviously smoothing than that with PVD process due to interface leakage current suppression effectively and resulting in lower device shot-noise. It may be reasonably explained that the relieves of the Fermi-level pinning on the interfaces of Al_2O_3 and top GaAs layer. Fig. 6 shows the uncorrected noise equivalent temperature difference (NEDT) histogram of FPA with detector pixels of 320×256 and infrared optics of F/3 under the temperature of 30 K and bias of -4V. This NEDT value for F/3 optics was 832 mK and comparable to that for F/1 optics (3–5 µm) previously proposed from Sanjay Krishna. et.al,. [6] The non-corrected uniformity is about 6.7%.

Fig.7 shows the infrared face-to-face head image of two men taken by FPA with 320×256 pixels under 180 K almost to approach thermoelectric cooling temperature. The operability of the QDIP FPA has reached 98%. The inoperative pixels are attributed to the defects of indium bumps.

Conclusions

Based on merely thick 6.5nm ALD-Al_2O_3 passivated layer on two stacks of 10-period InGaAs and GaAs-Capped InAs QDIP structure, a 320×256 FPA without light-coupling grating, and anti-reflected coating process is observed. Using the home-made IR video demonstration systems, the thermal image can be obtained at high temperatures up to 180 K. Under extreme high operation temperature, it is one of best imaging demonstration for the QDIP FPA.

Acknowledgement

The authors wish to thank the Department of Natural Sciences of the National Science Council and Department of Industrial Technology of Ministry of Economic Affairs, Taiwan, R.O.C. for supporting this work.

References

[1] Shiang-Feng Tang, et al., Appl. Phys. Lett. 78, pp. 2428-2430, 2001.

[2] Shiang-Feng Tang, et al., IEEE Photonics Tech. Lett., 18, pp.986-988, 2006.

[3] C.-H. Chang, et al., Appl. Phys. Lett. 89, pp. 242911-1-242911-3, 2001.

[4] Shiang-Feng Tang, et al., IEEE Sensors 2012 Conf., Taiwan, pp. 1396, 2012..

[5] Cheng-Wei Cheng and Eugene A. Fitzgerald, Appl. Phys. Lett. 96, pp. 202101-2-202101-3, 2010.

[6] Sanjay Krishna et al., Appl. Phys. Lett. 86, pp. 193501-1-193501-3, 2005.

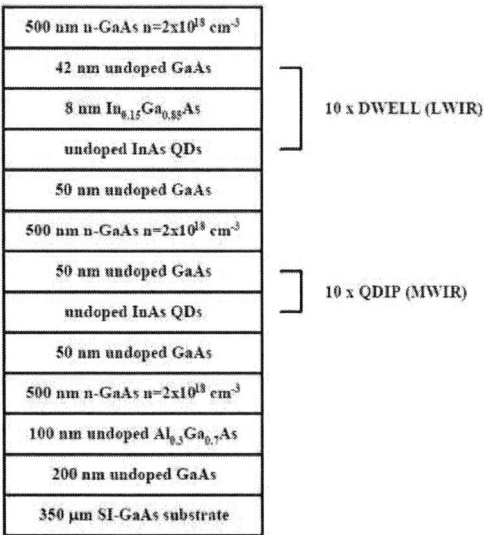

| 500 nm n-GaAs n=2x10^{18} cm^{-3} |
| 42 nm undoped GaAs |
| 8 nm In$_{0.15}$Ga$_{0.88}$As |
| undoped InAs QDs |
| 50 nm undoped GaAs |
| 500 nm n-GaAs n=2x10^{18} cm^{-3} |
| 50 nm undoped GaAs |
| undoped InAs QDs |
| 50 nm undoped GaAs |
| 500 nm n-GaAs n=2x10^{18} cm^{-3} |
| 100 nm undoped Al$_{0.3}$Ga$_{0.7}$As |
| 200 nm undoped GaAs |
| 350 μm SI-GaAs substrate |

10 x DWELL (LWIR)

10 x QDIP (MWIR)

Fig. 1 Schematic of two stacks of InGaAs and GaAs-capped InAs QDIP structure grown by solid-source MBE system for dual-band infrared detection.

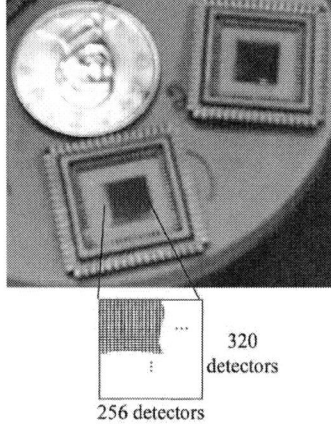

320 detectors

256 detectors

Fig. 2   The 320×256 pixel-numbered FPA is hybridized with snapshot-mode ROIC mounted in a 68 pin leadless ceramic chip carrier and its dimension with a comparison of NT$10.

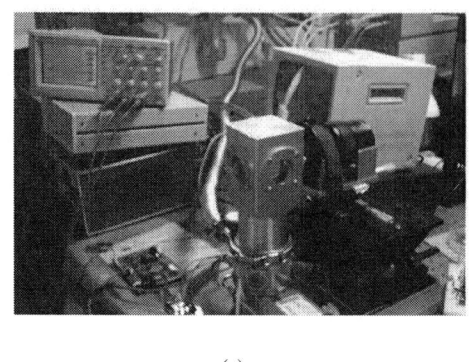

(a)

(b)

Fig. 3 (a) The thermal images are taken by the home-made IR video demonstrated system, (b) the collecting layout of video evaluating system.

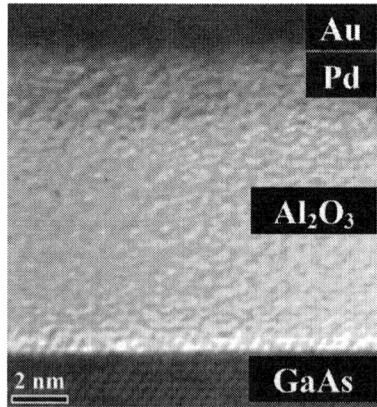

Fig. 4 High-resolution XTEM image of interfaces among GaAs, Al$_2$O$_3$, Pd and Au, respectively.

978-1-4673-4840-9/13 $31.00 © 2013 IEEE   344

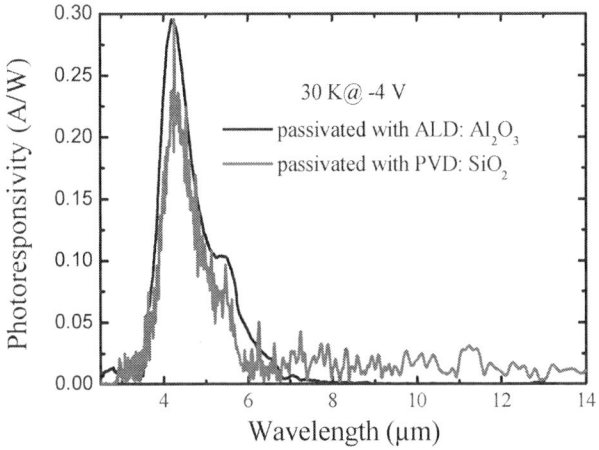

Fig.5 Calibrated spectral photoresponses of test samples with ALD-Al$_2$O$_3$ and PVD-SiO$_2$ passivated layers, respectively.

Fig.7 The infrared face-to-face head image is taken by FPA with 320×256 pixels under the extreme high operation temperature of 180K.

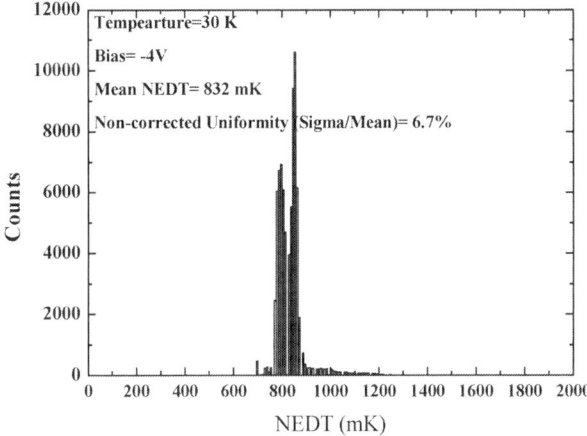

Fig. 6 Uncorrected noise equivalent temperature difference (NEDT) histogram of FPA with detector pixels of 320×256 is measured with infrared optics of F/3 under the temperature of 30K and the bias of -4V.

978-1-4673-4840-9/13 $31.00 © 2013 IEEE   345

Influence of Trap Depth on Charge Transport in Inverted Bulk Heterojunction Solar Cells employing ZnO as electron transport layer

Naveen Kumar Elumalai, [1,2] Chellappan Vijila,*[1] Arthi Sridhar, [2] and Seeram Ramakrishna [2]

[1]Institute of Materials Research and Engineering, A*STAR (Agency for Science, Technology and Research),
3 Research Link, Singapore 117602, e-mail: c-vijila@imre.a-star.edu.sg
[2]National University of Singapore, Singapore, 117576

Abstract

Inverted organic solar cells with device structure ITO/ZnO/ poly(3- hexylthiophene) (P3HT):[6,6]-phenyl C61 butyric acid methyl ester (PCBM) /MoO₃/Ag were fabricated employing low temperature solution processed ZnO as electron selective layer. Devices with varying film thickness of ZnO interlayer were investigated. The optimum film thickness was determined from photovoltaic parameters obtained from current-voltage measurements. The type of charge transport process, distribution of trap states and the ohmicity of the contacts in the optimized device were evaluated using the temperature and illumination intensity dependent study. The results demonstrate the effect of trap depth on device performance and its distribution on the stability of contacts.

Keywords: Zinc Oxide, Trap depth, Temperature dependence, Electron selective layer, Solution processed.

Introduction

Organic solar cells (OSCs) are kind of excitonic solar cells which gained lot of research interests in the past two decades. It has numerous potential advantages including low cost, light weight, compatibility with flexible materials and ease of fabrication [1]. Bulk heterojunction (BHJ) OSCs employ interpenetrating networks of donor and acceptor phases for exciton dissociation and appropriate electrodes with suitable work function for collecting the charges [2]. Owing to issues with respect to stability/device degradation, inverted device architecture has become a favorable choice [3]. In this configuration the ITO is made as cathode by incorporating suitable electron transporting layers (ETL), which reduces the work function of ITO as well as blocks the holes from reaching it. ETLs such as Ca, Al, ZnO, and TiO₂ are used for such purpose [4, 5]. Among them ZnO is widely used owing to its stability, high electron mobility, high degree of transparency to visible light and solution process-ability [6, 7]. On the other hand, hole transporting layers (HTL) such as poly (3, 4-ethylene dioxythiophene) :(polystyrene sulfonic acid) (PEDOT: PSS) Pedot: PSS, Molybdenum oxide (MoO₃) etc. have been widely used [5]. MoO₃ is found to be much resistant to degradation and has become an alternative to PEDOT: PSS especially in inverted devices [8]. The morphology and thickness of these buffer layers play a major role in determining device performance and the corresponding charge transport process affecting the same [2].

In this work, we fabricated inverted BHJ devices with

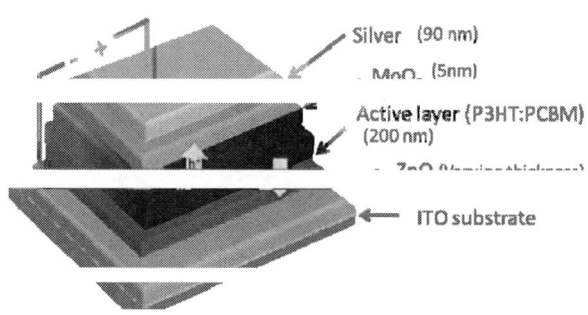

Fig. 1 Inverted device architecture employing ZnO as electron selective layer.

structure ITO/ZnO/P3HT:PCBM/MoO₃/Ag employing solution processed ZnO as electron selective layer. The optimum thickness of the ZnO film is determined from the photovoltaic parameters observed. The efficacy of the optimized ZnO film in charge transport processes, its influence on charge generation, trap states distribution and ohmicity as the electron transporting contact is evaluated using temperature and illumination intensity study.

Experimental Methods

The ITO-coated glass substrates were first cleaned with detergent, ultrasonicated in water, acetone and isopropyl alcohol, and subsequently dried overnight in an oven. The ZnO precursor was prepared by dissolving zinc acetate di-hydrate

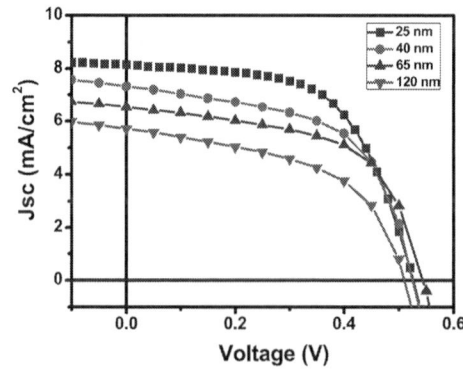

Fig. 2 Current-Voltage (J-V) characteristics of the devices with different ZnO film thickness.

Table.1 Photovoltaic parameters obtained from the J-V characteristics for different ZnO film thickness.

ZnO film thickness	Voc (V)	Jsc (mA/cm^2)	FF (%)	PCE (%)
25 nm	0.52	8.15	59.2	2.54
40 nm	0.52	7.33	57.57	2.21
65 nm	0.54	6.54	57.45	2.04
120nm	0.51	5.72	51.47	1.50

(Zn (CH$_3$COO)$_2$·2H$_2$O, Aldrich, 99.9%, 1 g) and ethanolamine (NH$_2$CH$_2$CH$_2$OH,Aldrich, 99.5%, 0.35g) in 2-methoxyethanol (CH$_3$OCH$_2$CH$_2$OH, Aldrich, 99.8%, 15 mL) under vigorous stirring for 8 h for the hydrolysis reaction in air. The ZnO precursor solution was then spin cast atop the cleaned ITO substrates at different rpm (revolutions per minute) to obtain different film thickness. Film thickness of about 25 nm, 40nm, 65nm and 120nm is obtained for rpm speeds of about 3500 rpm, 2000 rpm, 1000 rpm and 500 rpm respectively. After spin coating the coated substrates are sintered at 140 °C for 2 hrs. Then the blend solution of P3HT and PCBM dissolved in the ratio of 1:0.8 in 1, 2-dichlorobenzene solution was spin coated on the ITO/ZnO layer. The thickness of the blend layer is about 200 nm. The film was then annealed at 110 °C for 10 minutes. The MoO$_3$ layer of thickness ~5 nm was then thermally evaporated in a vacuum chamber with a base pressure of ~10^{-7} mbar. Finally, silver (~90 nm) electrode was deposited onto the MoO$_3$ layer at a pressure ~10^{-5} mbar. The optimised device with ZnO film thickness ~25nm is then subjected to temperature and illumination intensity dependence study. The sample was mounted in a continuous helium flow cryostat (Janis) under the vacuum of about 10^{-5} mbar. The device was illuminated with white light whose intensity is calibrated to maximum of 100 mW cm^{-2} inside the cryostat. The J-V characteristics were measured by varying the light intensities (5 mw cm^{-2} to 100 mw cm^{-2}) and by varying the temperatures from 123 K to 343 K.

Results and discussion

The inverted device architecture employing solution processed ZnO as electron buffer layer and MoO$_3$ as hole transport layer is shown in Figure 1. Film thickness of the ZnO layer is varied keeping all the other components of the device constant. The effect of film thickness on photovoltaic parameters is evaluated from current voltage (J-V) characteristics as shown in figure 2. The device with the ZnO film thickness of about 25 nm exhibited maximum power conversion efficiency of ~2.54%. This device exhibited short circuit current density (Jsc) of about 8.15mA/cm^2 and fill factor of ~ 59.2% which is also the highest compared with the other counterparts with different ZnO film thickness. The comparison is listed in table 1. The Voc of the devices did not show large variation with the ZnO film thickness. The order of difference in Voc among them is just 0.02-0.03V whereas the Jsc is greatly influenced by film thickness. The device with

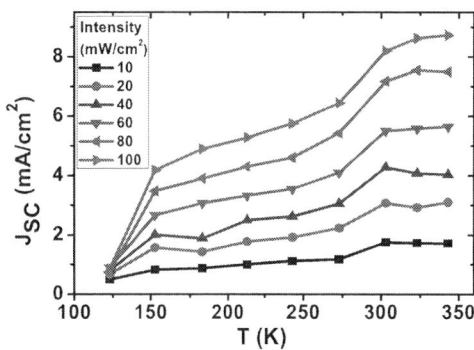

Fig. 3 Variation of short circuit current density as a function of temperature under different light intensities.

ZnO film thickness of about 120nm showed the lowest performance efficiency ~1.5% owing to low Jsc ~5.72 mA/cm^2 and poor fill factor ~51.5% comparatively. This lowering of Jsc and fill factor can be attributed to decreased light transmission and increased series resistance respectively. Reduction of film thickness from 65nm to 40nm improves the Jsc from 6.54mA/cm^2 to 7.33 mA/cm^2 respectively. The PCE is improved by 68%, Jsc is improved by 42% and fill factor is improved by 15% with reduction in film thickness from 120nm to 25nm. Further reduction in film thickness < 25nm, leads to inconsistent ZnO film, leading to device leakage resulting in poor performance as a consequence of incomplete inversion of the device. The optimum thickness of the ZnO film in this case is found to be 25nm (D4). In order to improve the PCE of the device further, a better understanding of the optimized device is needed. Its efficacy with respect to charge transport processes; including charge generation and recombination has to be understood. Therefore, the optimized device is subjected to the light and temperature dependent study in order to evaluate the controlling parameters. Variation of Jsc as a function of temperature for different illumination intensities is shown in figure 3. The Jsc is found to be increasing with temperature under any light conditions, depicting the trend of positive temperature co-efficient. Such trend is observed when the charge mobility is increased with temperature. In the case of organic semiconductors namely polymer and

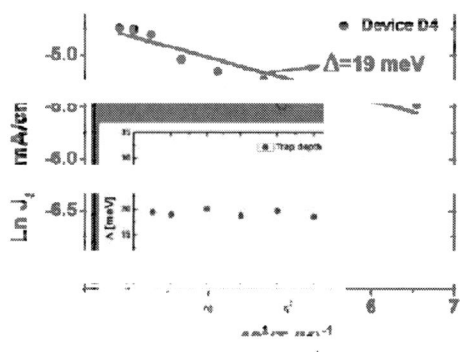

Fig. 4 Trap depth for the optimized device. (Inset) Variation of the trap depth with illumination intensity.

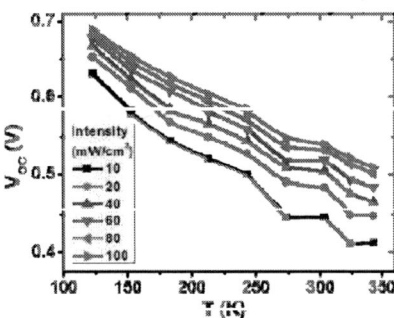

Fig. 5 Variation of open circuit voltage as a function of temperature under different illumination intensities.

methanofullerene, charge mobility is thermally activated. Temperature enables charge transport to occur via localized states by means of hopping [9, 10]. Apart from trap states from active layer, owing to the nature of ZnO nano-crystallinity and low sintering temperature, deep and shallow trap states are incumbent within the ZnO ETL both at surface level and in the bulk. These trap states distribution in the device as a whole largely influences the charge extraction. Therefore, the observed positive temperature co-efficient of Jsc with temperature as shown in figure 3 is a consequence of these processes. These trap states play a major role in affecting the device performance by acting as charge recombination centers having a detrimental effect [11]. The calculated trap depth or the activation energy in the optimized device under 1 sun condition is found to be 19 meV as shown in figure 4. The trap depth is determined using the equation as follows

$$J_{sc}(T, P_L) = J_0(P_L) \exp\left(-\frac{\Delta}{kT}\right) \quad (1)$$

J_0 (P_L) is a pre-exponential factor which includes the photogenerated charge-carrier density, the electric field and their mobility; T is the temperature; k is the Boltzmann constant and Δ is the trap depth with respect to the corresponding band minimum [12]. Figure 4 (inset) shows that the trap states are distributed narrowly within the range of 19 – 20 meV even under different illumination intensities, indicating that activation energy is independent of carrier generation.

The variation of Voc with temperature under different photon flux is shown in figure 5. The Voc increases linearly with increasing temperature which on extrapolation (T=0K) intercepts the y-axis at 0.78 - 0.82V, which is higher than the work function difference between the contact electrodes ~0.65-0.7V. This indicates that the Voc is determined by the effective band gap (E_g) of the absorber layer, depicting that the contacts are ohmic [13, 14]. Even at low temperature T<200 K, the Voc showed steady increase without any saturation at any point, indicating that the ohmic contact formed by the ZnO film serving as ETL, is highly stable. The interface between the ZnO and blend layer also plays a major role in device performance. Further improvement in device PCE can be achieved by optimizing this interface as well as by reducing the trap depth. It can be accomplished by increasing the

crystallinity of ZnO as well as from surface treatment, increasing its surface density of states (DOS) [7]. Such surface states alter the fermi level pinning and improve diffusion co-efficient enabling effective charge extraction at the interface [15].

Conclusion

Devices with structure ITO/ZnO/P3HT:PCBM/MoO$_3$/Ag utilizing solution processed ZnO as electron buffer layer is fabricated. The effect of ZnO interlayer thickness on photovoltaic parameters is evaluated from J-V characteristics. The optimum thickness of ZnO interlayer is found to be 25nm owing to improved Jsc and FF due to increase in incident photons and decreased charge recombination respectively. The charge mobility is thermally activated and follows hopping transport resulting in positive temperature co-efficient of Jsc with temperature. The trap depth in the optimized device is found to be ~19 meV and independent of the incident photon flux. The Voc is determined by the effective band gap (E_g) of the absorber layer indicating that the ZnO ETL employed is ohmic. The linearly increase in Voc without saturation even at lower temperature denotes that the contact formed using ZnO interlayer is highly stable. Decreasing the trap depth by improving the ZnO nano-crystallinity will result in better device performance.

References

[1] P. Kumar and S. Chand, "Recent progress and future aspects of organic solar cells," *Prog. Photovoltaics,* pp. 377-415, 2011.

[2] S. K. Hau, H.-L. Yip, and A. K. Y. Jen, "A Review on the Development of the Inverted Polymer Solar Cell Architecture," *Polymer Reviews,* vol. 50, pp. 474-510, 2010.

[3] Y.-H. Lin, P.-C. Yang, J.-S. Huang, G.-D. Huang, I.-J. Wang, W.-H. Wu, M.-Y. Lin, W.-F. Su, and C.-F. Lin, "High-efficiency inverted polymer solar cells with solution-processed metal oxides," *Sol Energ Mat Sol C,* vol. 95, pp. 2511-2515, 2011.

[4] R. Po, C. Carbonera, A. Bernardi, and N. Camaioni, "The role of buffer layers in polymer solar cells," *Energy Environ. Sci.,* vol. 4, p. 285, 2011.

[5] E. L. Ratcliff, B. Zacher, and N. R. Armstrong, "Selective Interlayers and Contacts in Organic Photovoltaic Cells," *J. Phys. Chem. Lett.,* vol. 2, pp. 1337-1350, 2011.

[6] Z. Hu, J. Zhang, Y. Liu, Z. Hao, X. Zhang, and Y. Zhao, "Influence of ZnO interlayer on the performance of inverted organic photovoltaic device," *Sol. Energ. Mat. Sol. C,* vol. 95, pp. 2126-2130, 2011.

[7] Y.-J. Kang, K. Lim, S. Jung, D.-G. Kim, J.-K. Kim, C.-S. Kim, S. H. Kim, and J.-W. Kang, "Spray-coated ZnO electron transport layer for air-stable inverted organic solar cells," *Sol. Energ. Mat. Sol. C,* vol. 96, pp. 137-140, 2012.

[8] D. Y. Kim, J. Subbiah, G. Sarasqueta, F. So, H. Ding, and Y. Gao, "The effect of molybdenum oxide interlayer on organic photovoltaic cells," *Appl. Phys. Lett.,* vol. 95, pp. 093304-093304, 2009.

[9] H. Bässler, "Charge Transport in Disordered Organic Photoconductors a Monte Carlo Simulation Study," *Phys. Status Solidi B,* vol. 175, pp. 15-56, 1993.

[10] V. D. Mihailetchi, J. K. J. van Duren, P. W. M. Blom, J. C. Hummelen, R. a. J. Janssen, J. M. Kroon, M. T. Rispens, W. J. H. Verhees, and M. M. Wienk, "Electron Transport in a Methanofullerene," *Adv. Funct. Mater.,* vol. 13, pp. 43-46, 2003.

[11] H. Antoniadis, L. J. Rothberg, F. Papadimitrakopoulos, M. Yan, M. E. Galvin, and M. A. Abkowitz, "Enhanced carrier photogeneration by defects in conjugated polymers and its mechanism," *Phys. Rev. B,* vol. 50, pp. 14911-14915, 1994.

[12] I. Riedel, J. Parisi, V. Dyakonov, L. Lutsen, D. Vanderzande, and J. C. Hummelen, "Effect of temperature and illumination on the electrical characteristics of polymer-fullerene bulk-heterojunction solar cells," *Adv. Funct. Mater.,* vol. 14, pp. 38-44, 2004.

[13] M. Kemerink, J. M. Kramer, H. H. P. Gommans, and R. a. J. Janssen, "Temperature-dependent built-in potential in organic semiconductor devices," *Appl. Phys. Lett.,* vol. 88, pp. 192108-192108, 2006.

[14] Y. Shen, A. R. Hosseini, M. H. Wong, and G. G. Malliaras, "How to make ohmic contacts to organic semiconductors," *ChemPhysChem,* vol. 5, pp. 16-25, 2004.

[15] P. P. Boix, J. Ajuria, I. Etxebarria, and R. Pacios, "Role of ZnO Electron-Selective Layers in Regular and Inverted Bulk Heterojunction Solar Cells," *J. Phys. Chem. Lett.,* pp. 407-411, 2011.

A Dual-silicon-nanowire Based Nanoelectromechanical Switch

You Qian, [1,2] Liang Lou, [1] Vincent Pott, [2] Minglin Julius Tsai, [2] and Chengkuo Lee[1*]

[1]Department of Electrical and Computer Engineering, National University of Singapore,
4 Engineering Drive 3, Singapore
[2]Institute of Microelectronics, A*STAR (Agency for Science, Technology and Research),
11 Science Park Road, Singapore Science Park II, Singapore
*Email: elelc@nus.edu.sg

Abstract

A dual-silicon-nanowires based nanoelectromechanical (NEMS) switch is fabricated using standard complementary metal-oxide-semiconductor (CMOS) compatible process. The switch comprises a capacitive paddle and two silicon nanowires both connect with the paddle, form a U-shape structure. The high electrostatic force generated from the large capacitive paddle and high flexible structure favor of silicon nanowires result to ultra-low pull-in voltage. The pull-in voltage is measured at 0.9V. According to the preliminary results, this switch demonstrates great potential in decreasing pull-in voltage.

Introduction

Over the past 50 years, the evolution of integrated circuit (IC) technology has affected many aspects of our lives. As the basic building block of IC, the metal–oxide–semiconductor field-effect transistor (MOSFET) provides dramatic improvements in speed, miniature, cost and functionality of IC. The scalable manufacturing processes and reliable performance also keep the MOSFET as the core of semiconductor industry. However, when semiconductor devices become smaller and applications become increasingly demanding and diverse, complementary approaches will be needed.

Alternative design has been proposed to break through the fundamental limitation of subthreshold swing such as tunnel transistor [1]. Still these designs cannot achieve zero off-state current. In contrast, a mechanical switch has essentially zero off-state current and abrupt switching behavior. These advantages allow the mechanical switches dramatically decrease the static power consumption, and maintain high driving current when scale down the supply voltage. There is even no limitation for supply voltage to further scale down as long as the pull-in voltage of mechanical switch can be designed to be lower.

Among the different actuation mechanisms, which can be realized by CMOS process, electrostatic actuation is favored for its low power consumption and fast switching speed [2, 3]. But one of the major problems with electrostatic switch is the high actuation voltage [4-6]. We report a NEMS switch using silicon nanowire (SiNW) to effectively lower down the pull-in voltage.

Design

A dual-silicon-nanowire based NEMS switch comprises a capacitive paddle with dimension of 2 μm by 4 μm and two silicon nanowires both connect with the paddle, form a U-shape structure, which is shown in Fig 1. The nanowires act as fix-free beams and hold the capacitive paddle to suspend on top of the substrate with a gap of 145 nm. The nanowires are 5 μm long with cross-section of 90 nm by 90 nm. When sufficient voltage applied between capacitive paddle and the substrate, the nanowire will bend down and form an electrical contact through the paddle and substrate. The device could turn on and off of the current through two terminals by applying a sufficient voltage between them.

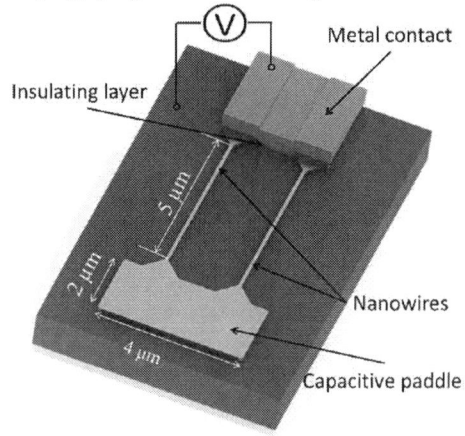

Fig. 1 A schematic illustration of the U-shape NEMS switch.

Simulation

A COMSOL simulation is performed to estimate the effect on the variation of SiNW length. The simulation fixes the size of the capacitive paddle and the gap where the length of SiNW is varied to find the suitable parameter. The result can be found on table 1. The pull-in voltage of pure cantilever switch without SiNW is as high as 26.5 V. When a very short SiNW is added in simulation, the pull-in voltage dramatically decreases until the SiNW length increasing to 5 μm. Therefore, a better approach to further reduce the pull-in voltage will be to enlarge the capacitive paddle. From the simulation result, 5 μm SiNW is chosen for real fabrication process.

978-1-4673-4840-9/13 $31.00 © 2013 IEEE

TABLE I. The SiNW length versus pull-in voltage in COMSOL simulation

SiNW length(μm)	Pull-in voltage(V)
0	26.5
0.2	6.97
1	2.86
4	1.3
5	1.04
6	0.95
7	0.92
10	0.88

Fabrication

The fabrication process of the U-shape NEMS switch starts with a silicon-on-insulator (100) wafer, with device layer of 117 nm and buried oxide (BOX) layer of 145 nm. After photolithography, the width of photoresist nanowire pattern is about 160 nm. The width of the photoresist pattern is further reduced to 110 nm by using the plasma trimming. It induced the feeding gas He/O2 + N2, where the ratio of He/O2 is 7/3. The He/O2 gases are deployed to oxidize the photoresist while the N2 gas is used to smoothen the surface of the photoresist. This photoresist trimming process achieves a critical dimension of around 110 nm. The SiNW is then patterned along (110) directions by reactive ion etching (RIE). To further shrink down the dimension of SiNW, thermal oxidation is conducted with the final cross section of the SiNW being 90 nm x 90 nm (Figure 2). The capacitive paddle is also oxidized on top and has a thickness of 90 nm. To increase the conductivity of SiNW and capacitive paddle, a p-type implantation process using BF2+ with a dosage of 1×10^{14} ion/cm2 is done, followed by annealing for dopant activation.

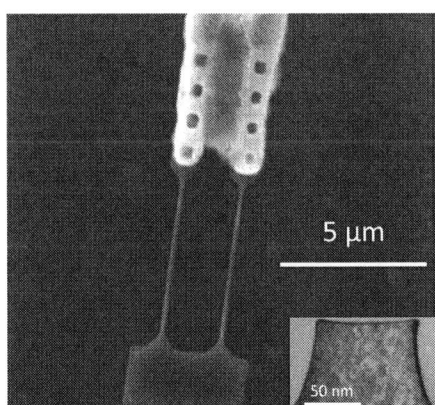

Fig. 2 SEM photo of a U-shape NEMS switch after HF vapor releasing, Inset: TEM image of a SiNW cross-section.

Measurement result shows that the SiNW has a resistivity of 0.021 Ω•cm, and the resistance per SiNW is 130 kΩ. Next, an extra SiO2 layer of 400 nm is deposited for protection. Further RIE process opened via on the protection layer. Lastly, Al PVD deposition and patterning is conducted for wiring and probe pad.

The device is released by vapor hydrofluoric (VHF), which etching away the BOX layer between U-shape NEMS switch and silicon substrate. VHF is generated by 49% hydrofluoric acid (HF) at room temperature. To eliminate the stiction caused by liquid in the experiment, the chip would be pre-heated to 140 ℃ and put into the VHF in room temperature for 1 minute. After the releasing, the device is still above 100 ℃.

After the release process, the NEMS switch is checked under the scanning electron microscope (SEM). It shows a clean release process and the residues could be considered as acceptable. No stiction or deflection could be found for SiNWs while the paddle, which is made from single crystal Si, remains free standing above the substrate.

Results and Discussion

Fig 3 shows the I-V plots of the fabricated U-shape NEM switch. Since it operates as a two-terminal switch, the voltage applied is between the U-shape structure and the substrate. The voltage is swept from 0–2 V, with no compliance been set for the current.

The pull-in voltage is measured at 0.9V. The measured off-state current is around 10 pA, which is equivalent to the noise level of the testing setup. And the on state current increases rapidly to 0.1 μA after pull-in. Based on the applied voltage step of 5 mV, this subthreshold slope is less than 7 mV/decade. This represents an ideal on/off current characteristic, where the sub-threshold slope is lower the theoretical limit of CMOS devices (60 mV/decade).

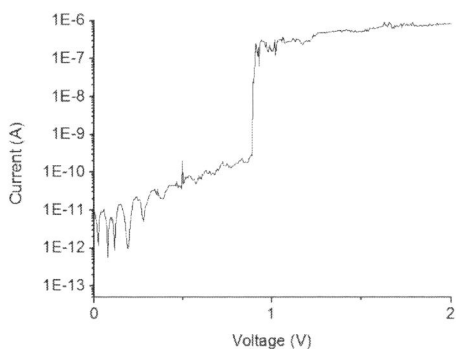

Fig 3. I-V characteristic of the NEMS switch.

Leakage current can be observed in off state, probably due to degraded oxide layer, which caused by localize joule heating. The nanowire suffers from accumulative Joule heating effect after taking into consideration of its extremely small cross-section area. The oxide insulation layer just behind the SiNW would be physically. And we also observe SiNW become welded in ambient air from SEM. In Fig.4, we can clearly see that the SiNW is melted and touched substrate but eh paddle together with the contact area still remind suspended above the substrate.

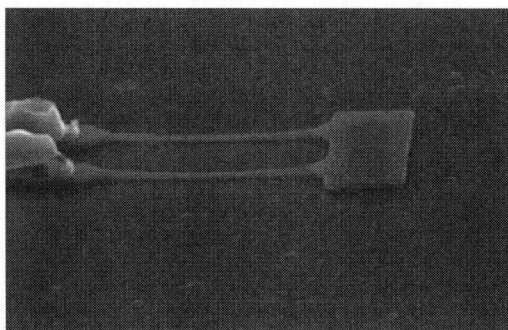

Fig. 4 SEM photo SINW welded on the substrate

Conclusion

The dual SiNWs based U-shape NEMS switch shows a low actuation voltage in electrical test. It takes advantage of the high electrostatic force generated from the large capacitive paddle and high flexibility from the single crystal SiNW. The device shows great potential on the voltage compatibility with modern CMOS circuit.

Acknowledgment

This work was supported by the Science and Engineering Research Council, Agency for Science, Technology and Research, Singapore, under Grant Nos. 102-101-0022 and 102-165-0084.

References

[1] K. Boucart and A. M. Ionescu, "Double-gate tunnel FET with high-kappa gate dielectric," IEEE Transactions on Electron Devices, vol. 54, pp. 1725-1733, Jul 2007.

[2] S. W. Lee, S. J. Park, E. E. B. Campbell, and Y. W. Park, "A fast and low-power microelectromechanical system-based non-volatile memory device," Nature Communications, vol. 2, Mar 2011.

[3] D. N. Guerra, M. Imboden, and P. Mohanty, "Electrostatically actuated silicon-based nanomechanical switch at room temperature," Applied Physics Letters, vol. 93, Jul 21 2008.

[4] J. Rubin, R. Sundararaman, M. Kim, and S. Tiwari, "A Low-Voltage Torsion Nanorelay," IEEE Electron Device Letters, vol. 32, pp. 414-416, Mar 2011.

[5] Y. Hayamizu, T. Yamada, K. Mizuno, R. C. Davis, D. N. Futaba, M. Yumura, and K. Hata, "Integrated three-dimensional microelectromechanical devices from processable carbon nanotube wafers," Nature Nanotechnology, vol. 3, pp. 289-294, May 2008.

[6] M. Y. Liao, S. Hishita, E. Watanabe, S. Koizumi, and Y. Koide, "Suspended Single-Crystal Diamond Nanowires for High-Performance Nanoelectromechanical Switches," Advanced Materials, vol. 22, pp. 5393, Dec 14 2010.

978-1-4673-4840-9/13 $31.00 © 2013 IEEE

Device modeling and optimization of high- performance thin film CIGS solar cell with $Mg_xZn_{1-x}O$ buffer layer

Saurabh Kumar Pandey and Shaibal Mukherjee

Hybrid Nanodevice Research Group (HNRG), Discipline of Electrical Engineering, Indian Institute of Technology, Indore-453441, India
Phone: +91-732-4240715, Fax: +91-731-2361482, +1-484-2317876
Email: shaibal@iiti.ac.in

ABSTRACT

A comprehensive device modeling for thin film CIGS-based solar cell with MgZnO buffer layer has been performed. The effects of thickness, doping, and alloy composition of various device constituent layers are extensively studied while optimizing device performance of the solar cell at room temperature. In this study, a maximum power conversion efficiency of 21.4% is achieved with performance parameters of 1.2V for open circuit voltage (V_{oc}), 34.8 mA/cm^2 of short circuit current density (J_{sc}) and a fill factor of 85.7%. Different aspects of constituent layer parameters thickness, doping, and alloy composition calibrations has been considered to identify the optimization criteria for the design of CIGS based solar cell.

Keywords: CIGS, conversion efficiency, deep defect, modeling, Photovoltaic (PV) cell

I. Introduction

Polycrystalline thin-film solar cells have reached a high level of interest on photovoltaic market because of their performance as well as low production costs [1]. Among various thin film solar cells, Cu (In, Ga) Se$_2$ (CIGS) solar cells seems to be most popular because of low material consumption and high efficiency [2]. CIGS is one of the most promising absorber materials for thin film photovoltaic's (PV) because it is a direct band gap (E$_g$) semiconductor with a high absorption coefficient [3]. Moreover its band gap can be tuned over a wide range (1.0 to 1.68 eV) by varying the Ga concentration in the film. In a CIS/CIGS model, the absorber acts as a p-type doped region with typical thickness of 2 to 3 μm [4-5].Device performance of thin film CIGS solar cell is very sensitive to variations in buffer layer. Changes in this layer results in a notable impact on the device energy-conversion efficiency. Two types of defect mentioned may be responsible for high Ga-content CIGS. That is, a shallow one of 0.3 eV is widely acknowledged but found to have little impact on the decrease of the performance [5], while a deep one lying in the upper part of the energy gap was found and believed to have a more significant influence [6].

In this present work, ATLAS device simulation software has been used to simulate the CuIn$_{1-x}$Ga$_x$Se$_2$ cell module structure. The aim of the simulation of CIGS solar cell structure was to check the device performance by varying the alloy composition, doping concentration and thickness of the CIGS absorber layer and buffer layer. The device performance is mainly based on the material parameters, optical parameters, and electrical parameters of each layers used in the structure. This paper starts with a device structure (Section II), followed by the results and discussion (Section III), followed by the conclusions (Section IV).

II. Device structure

Analytical investigations were performed using commercial device simulation software. The cell behavior in the dark can be described by the Poisson, electron and hole continuity, and drift-diffusion equations. Unlike most simulations found in the literature, our model of the illuminated cell (standard AM1.5D solar spectrum) features a transfer matrix approach [8] for the calculation of light propagation: this approach accounts for the effects of reflection at the interfaces between adjacent layers, a commonly overlooked effect that can significantly impact the cell efficiency. The characteristics of the CIGS absorber can be found in [9] and here we have focused on the device optimization and defect analysis of cell module.

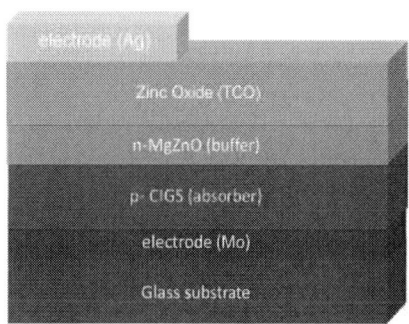

Fig.1 Schematic diagram of the simulated thin film CIGS solar cell.

978-1-4673-4840-9/13 $31.00 © 2013 IEEE

Permittivity	9	10.5	13.6
Band gap, eV	3.37	4.1	1.15
Electron mobility, cm^2 V^{-1}s^{-1}	300	300	100
Hole mobility, cm^2 V^{-1}s^{-1}	30	20	25
Defect density	N_{DG}= 10^{17} cm^{-3}	N_{AG} = 10^{18} cm^{-3}	N_{DG} =10^{14} cm^{-3}

DOS*: Density of States

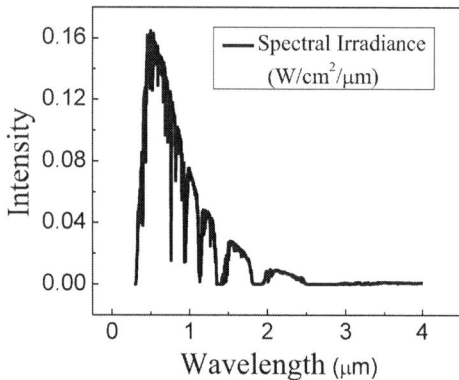

Fig.2 AM1.5 spectrum used for the illumination of thin film CIGS solar cell.

Fig.1 presents the cross-sectional view of the device studied in this paper. In this paper, the device modeling and numerical simulation for the CIGS solar cell in terms of constituent layer structure, doping and composition and their effect on light emission from the device and its threshold behavior is thoroughly examined. Here, we have chosen transfer matrix method to simulate AM1.5 conditions sampled between 0.3 microns and 1.2 microns at 50 samples. Comprehensive study on the influence of various device parameters was performed to confirm the validity of the model and improve the physical understanding of CIGS cells.

The Table I shows the description of material parameters required by the carrier transport equations, including the base parameters and other additional parameters for CIGS based quaternary alloys. In this device optimization, the intensity of light illumination has been taken AM1.5.

Table I *Material parameters used in the simulation study for CIGS solar cell. All parameters were taken at room temperature.*

Parameters	i-ZnO	n-MgZnO	p-CIGS
Recombination lifetime			
Electrons, sec	1e-7	6.5e-9	1e-7
Holes, sec	1e-7	1e-9	1e-7
Electron Affinity	4.0	4.05	4.1
Effective DOS* in conduction band	2.2×10^{18}	2.2×10^{18}	2.2×10^{17}
Effective DOS in valence band	1.8×10^{19}	1.8×10^{19}	1.5×10^{19}

III. Results and Discussion

The optimal design conditions for the buffer and absorber layer has been considered in order to match the best device technology. From the simulation results, it has been found that the CuIn$_{1-x}$Ga$_x$Se$_2$ cell has higher V_{oc} and smaller J_{sc} due to higher absorber coefficient which leads to improved fill factor thus increased power conversion efficiency. However, the device performance was degraded as the Ga alloy composition exceeds 30%.

A. Absorber layer

(a) Doping concentration

It can be observed from Fig.3 that as we increase the doping concentration from 5×10^{16} to 5×10^{18} cm^{-3}, Cu deficiency increases the majority carrier concentration by increasing the number of Cu vacancies. This reduces the device open circuit voltage (V_{oc}) and increases short circuit current density which leads to degradation of power conversion efficiency from 20.8% to 17%.

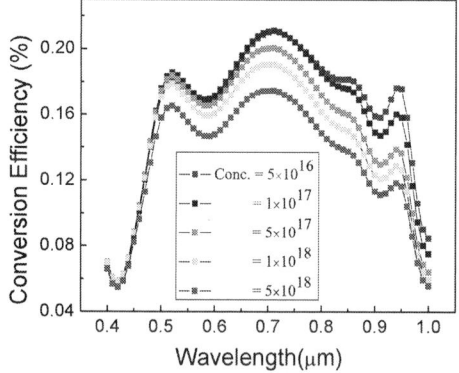

Fig.3 Simulated power conversion efficiency of the CIGS solar cell as a function of acceptor concentration of absorber layer.

As we increase the dopant concentration from 2×10^{14} cm^{-3} to 2×10^{18} cm^{-3}, the short circuit current density decreases from 39 mA/cm^2 to 22 mA/cm^2 thus leads to increase in V_{oc} as shown in Fig.4.

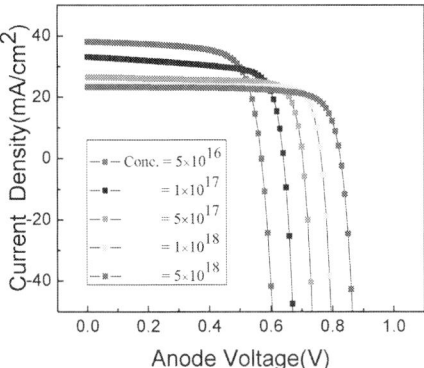

Fig.4 J-V characteristic of solar cell with variable CIGS absorber layer doping concentration.

(b) Alloy composition

It can be observed from Fig.5 that as we increase the Ga composition from 10% to 30%, there is slight increment in emission intensity due to effective carrier confinement in the wide band gap region.

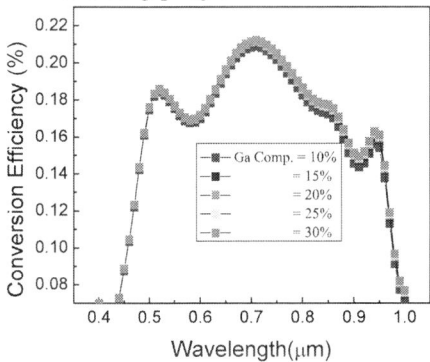

Fig.5 Simulated power conversion efficiency of the CIGS solar cell as a function of Ga composition of absorber layer.

Similarly the current density reduces on increasing the indium composition from 10 % to 30% and then after it reduces as shown in Fig.6.

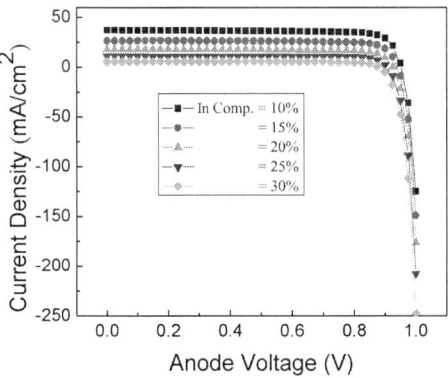

Fig.6 J-V characteristic of solar cell with variable Ga composition of absorber layer.

B. Buffer layer

(a) Doping concentration

It can be observed from Fig.7 on increasing the doping concentration from $5x10^{16}$ to $5x10^{18}$ cm^{-3} in buffer layer, the power conversion efficiency increases linearly thus leads to better device performance.

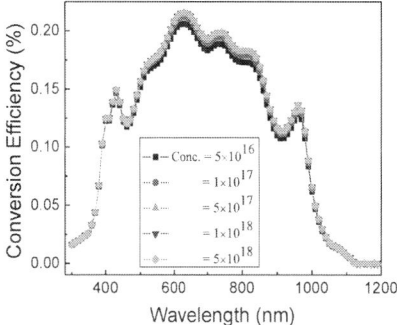

Fig.7 Simulated power conversion efficiency of the CIGS solar cell as a function of buffer layer concentration.

Similarly the short circuit current density reduces from 35 mA/cm^2 to 22 mA/cm^2 on increasing the doping concentration in buffer layer as shown in Fig.8.

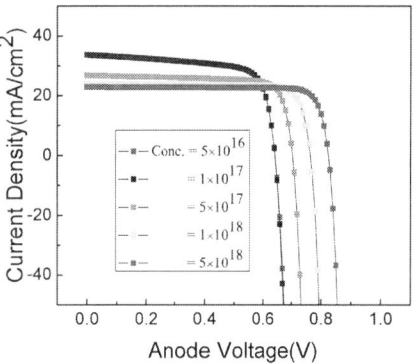

Fig.8 J-V characteristic of solar cell with variable doping concentration of buffer layer.

C. Defect analysis

The output parameters depend strongly on the number of defects and their capture cross-sections. The peak energy and distribution of deep defects, however, have relatively minor influence on performance. Extreme values for width or peak energy do lead to small changes in the lower quality factor; this is attributed to a change in occupancy of the defect states close to the Fermi level and a resulting change in the band structure.

It can be observed from fig.9 that on inclusion of surface defect density, the conversion efficiency degraded from 21% to 17%.

978-1-4673-4840-9/13 $31.00 © 2013 IEEE   355

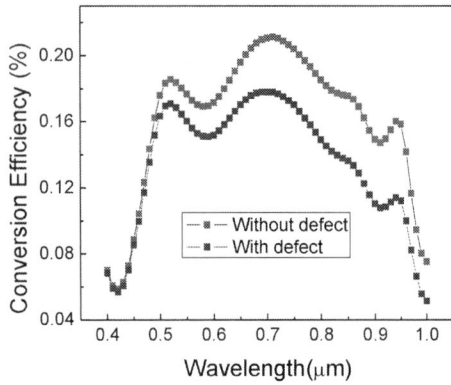

Fig.9 Simulated power conversion efficiency as a function of defect states of CIGS solar cell

Similarly the short circuit current density also reduces from 41 mA/cm^2 to 30 mA/cm^2 when defects are considered into the device structure as shown in fig.10.

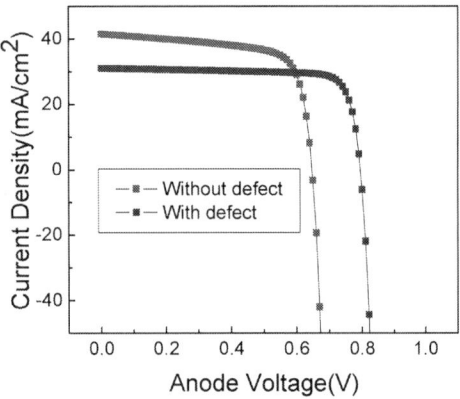

Fig.10 Simulated J-V characteristic as a function of defect density of the CIGS solar cell.

IV. Conclusion

We have discussed theoretical investigation of optimized device structure of thin film CIGS solar cell performance under illuminated behavior. The presence of surface defect density at the interfaces is a specific feature of these cells that must be accounted for in any physics-based model. We found that the amount of Ga composition and the doping concentration significantly alter the CIGS device performance. From the simulation results, it has been found that the increasing thickness of the absorber layer results in higher efficiency of solar cell, where optimum thickness is around 1μm. Furthermore, thickness of MgZnO buffer layer has been found in the range of 300 nm as the optimum value.

References

[1] H.Komaki, Y.Kamikawa-Shimizu, T. Yoshiyama, et al. "Fabrication of integrated CIGS modules using the in-line three-stage process," 33rd IEEE Photovoltaic Specialists Conf., 2008.

[2] J.S. Britt, S.Wiedemann, U.Schoop, D.Verebelyi, "High-volume manufacturing of flexible and lightweight CIGS solar cells," 33rd IEEE Photovoltaic Specialists Conf., 2008.

[3] M. Gloeckler, J. R. Sites, "Band gap grading in Cu (In. Ga) Se2 solar cells", Department of Physics, Colorado State University, pp 4, September 2004.

[4] Clas Persson, "Thin-film ZnO/CdS/CuIn1-xGaxSe$_2$ solar cells: anomalous physical properties of the CuIn1-xGaxSe2 absorber", Brazilian Journal of Physics. V.36 sep. 2006.

[5] G. Hanna, A. Jasenek, U. Rau, H.W. Schock, "Influence of the Ga-content on the bulk defect densities of Cu(In,Ga)Sez." Thin Solid Films 387(2001)71-73.

[6] A. Rockett , D. Liaoa, J.T. Heath. J.D. Cohen. Y.M. Strzhemechny. LJ. Brillson, K.Rarnanathan, W.N.Shafannan. "Near-surface defect distributions in Cu(In,Ga)Se2," Thin Solid Films 431 -432 (2003) 301-306.

[7] ATLAS User's Manual: "Device Simulation Software", Santa Clara, CA: SILVACO International, (2009).

[8] C. C. Katsidis and D. I. Siapkas, "General transfer-matrix method for optical multilayer systems with coherent, partially coherent, and incoherent interference", *Applied Optics*, 41, pp. 3978-398 (2002).

[9] G. Sozzi et al., "Numerical Analysis of the Effect of Grain Size and Defects on the Performance of CIGS Solar Cells", *Proc. CS-ManTech* (2010).

A one-way terahertz plasmonic waveguide based on surface magneto plasmons in a metal-dielectric-semiconductor structure

Bin Hu,[1] Qi Jie Wang,[1,2,*] and Ying Zhang[3]

[1]Division of Microelectronics, School of Electrical & Electronic Engineering, Nanyang Technological University, 50 Nanyang Ave., Singapore, 639798
[2]Division of Physics and Applied Physics, School of Physical and Mathematical Sciences, Nanyang Technological University, Singapore, 637371
[3]Singapore Institute of Manufacturing Technology, 71 Nanyang Drive, Singapore, 638075
qjwang@ntu.edu.sg

Abstract

A novel one-way-propagating terahertz plasmonic waveguide at a subwavelength scale is proposed based on a metal-dielectric-semiconductor structure. Unlike one-way plasmonic devices based on interference effects of surface plasmons, the proposed one-way device is based on nonreciprocal surface magneto plasmons under an external magnetic field. Theoretical and simulation results demonstrate that the waveguide has a broadly tunable one-way-propagating frequency band by the external magnetic fields. The proposed one-way device can be used to realize various high performance tunable plasmonic devices such as isolators, switches and splitters for ultracompact integrated plasmonic circuits.

Keywords: one way, terahertz, and surface magneto plasmons

Introduction

In recent years, terahertz (THz) science and technology has been rapidly developed for its huge potentials in imaging, spectroscopy, biomedical sciences, and integrated circuits [1]. For those applications, THz plasmonic waveguides, based on either surface plasmons (SPs) or spoof surface plasmons have been developed due to the sub-wavelength confinement for miniaturized devices [2]. However, all of these plasmonic waveguides are two-way waveguides, i.e. THz waves propagate in both the forward and the backward directions. One-way-propagating waveguides are of particular importance for functional devices such as isolators, switches and splitters.

So far, most of the proposed one-way-propagating plasmonic devices are based on the idea of SPs interference [3, 4]. This effect is strongly depends on the fabrication. Another approach is to use the nonreciprocal effect of SPs under an external magnetic field (MF) (also called surface magneto plasmons (SMPs) [5-9]). The nonreciprocal effect of SMPs refers to the different dispersions of the forward and backward propagating SMPs. They terminate at different cut-off frequencies, making it possible to realize an absolute one-way plasmonic waveguide.

In this letter, we propose a THz one-way sub-wavelength plasmonic waveguide that needs only 1 Tesla. By tuning the applied MF, we find through both theoretical analyses and numerical simulations that the central frequency of the one-way-propagating frequency band can be broadly tuned from 1.5 to 0.36 THz when the MF increases from 0.5 to 5 Tesla. In addition, the bandwidth of the one-way-propagating frequency band can be broadened by using a dielectric material with a higher permittivity. We also demonstrate a one-way THz plasmonic isolator with an extinction ratio of 100% and a tunable THz plasmonic switch.

Theory

The schematic structure of the one-way waveguide is composed of a metal-dielectric-semiconductor structure as shown in Fig. 1 (a). The metal and the semiconductor layers are half-infinite and the thickness of the dielectric layer is denoted by w. A TM-polarized SMP (with the MF component parallel to the y-axis) propagates along the z-direction at the dielectric/metal and dielectric/semiconductor layers. The external static MF is applied uniformly on the whole structure along the y-axis (as indicated by B), forming a Voigt configuration. The dielectric constants of both the metal and the semiconductor can be expressed by a tensor [10, 6]:

$$\vec{\varepsilon} = \varepsilon_\infty \begin{bmatrix} \varepsilon_{xx} & 0 & \varepsilon_{xz} \\ 0 & \varepsilon_{yy} & 0 \\ -\varepsilon_{xz} & 0 & \varepsilon_{xx} \end{bmatrix} \quad (1)$$

where in the lossless case, the parameters in Eq. (1) can be expressed as $\varepsilon_{xx} = 1 - \omega_p^2/(\omega^2 - \omega_c^2)$, $\varepsilon_{xz} = -i\omega_p^2\omega_c/[\omega(\omega^2 - \omega_c^2)]$, and $\varepsilon_{yy} = 1-\omega_p^2/\omega^2$. Here ω is the angular frequency of the incident wave, ω_p is the plasma frequency of metal or semiconductor (in the following, the plasma frequencies of the metal and the semiconductor are denoted by ω_{pm} and ω_{ps}, respectively), ε_∞ is the high-frequency permittivity, and $\omega_c = eB/m^*$ is the cyclotron frequency. e and m^* are the charge and the effective mass of electrons, respectively. It should be noted that ω_{ps} is in the order of ~10^{13} Hz. When B is 1 Tesla, the cyclotron frequency (ω_c ~10^{12} Hz) is comparable with ω_{ps}, thus the effect of the B can be observable. However, ω_{pm} is in the order of ~10^{16} Hz, making B to be very large to observe the effect of the MF. Therefore, using semiconductors instead of metals can dramatically weaken the desired MFs. In THz frequency, due to $\omega_{pm} \gg \omega$, metals resemble perfect conductors [11], and can be regarded as isotropic, with a permittivity of $\varepsilon_m = \varepsilon_{m,xx}$. From the Maxwell equations and Eq. (1), the electromagnetic (EM) fields in the three layers can be expressed as

$$H_y^{(II)} = C e^{-\kappa_2(x-w/2)}, \quad (2a)$$
$$E_z^{(II)} = -\frac{i\kappa_2}{\omega\varepsilon_0\varepsilon_m} C e^{-\kappa_2(x-w/2)}$$

978-1-4673-4840-9/13 $31.00 © 2013 IEEE   357

$$H_y^{(I)} = A e^{-\kappa_1 x} + B e^{\kappa_1 x}, \tag{2b}$$

$$E_z^{(I)} = -\frac{i\kappa_1}{\omega\varepsilon_0\varepsilon_d}(A e^{-\kappa_1 x} - B e^{\kappa_1 x})$$

$$H_y^{(III)} = D e^{\kappa_3(x+w/2)}, \tag{2c}$$

$$E_z^{(III)} = \frac{i\beta\varepsilon_{s,xz}D e^{\kappa_3(x+w/2)} - \varepsilon_{s,xx}\kappa_3 D e^{\kappa_3(x+w/2)}}{i\omega\varepsilon_0(\varepsilon_{s,xx}^2 + \varepsilon_{s,xz}^2)}$$

where $\varepsilon_{s,ij}$ are the corresponding elements of permittivity tensor of the semiconductor; ε_d is the dielectric constant of the dielectric layer; A, B, C, and D are the amplitudes of the EM fields; β is the propagation constant of the waveguide. κ_1, κ_2, and κ_3 are defined by $\kappa_1^2 = \beta^2 - k_0^2\varepsilon_d$, $\kappa_2^2 = \beta^2 - k_0^2\varepsilon_m$, $\kappa_3^2 = \beta^2 - k_0^2\varepsilon_s$, where $\varepsilon_s = \varepsilon_{s,xx} + \varepsilon_{s,xx}^2/\varepsilon_{s,xz}$ is the Voigt dielectric constant of the semiconductor [8]. Considering the continuity of H_y and E_z at the interfaces of the metal/dielectric and semiconductor /dielectric, we have the dispersion relation of the waveguide from Eq. (2):

$$\{\frac{\kappa_2\kappa_3}{\kappa_1^2}\frac{1}{\varepsilon_m\varepsilon_s} + \frac{1}{\varepsilon_d^2} - i\frac{\beta\kappa_2}{\kappa_1^2}\frac{\varepsilon_{s,xz}}{\varepsilon_m\varepsilon_s\varepsilon_{s,xx}}\}\tanh(\kappa_1 w) \tag{3}$$
$$+[\frac{\kappa_3}{\kappa_1}\frac{1}{\varepsilon_d\varepsilon_s} + \frac{\kappa_2}{\kappa_1}\frac{1}{\varepsilon_m\varepsilon_d} - i\frac{\beta}{\kappa_1}\frac{\varepsilon_{s,xz}}{\varepsilon_d\varepsilon_s\varepsilon_{s,xx}}] = 0$$

The nonreciprocal effect can be seen clearly from Eq. (3) that the dispersion relation curves are different for $\beta > 0$ (SMPs propagating along the z direction) and $\beta < 0$ (SMPs propagating along the –z direction). Without loss of generality, here we consider InSb, gold, and air as the semiconductor, metal, and dielectric materials in the calculation. The corresponding parameters of InSb at room temperature are given by $m^* = 0.014m_0$ (m_0 is the free electron mass in vacuum), $\omega_{ps} = 1.26\times10^{13}$ Hz, and $\varepsilon_\infty = 15.68$ [12]. $\omega_{pm} = 1.37\times10^{16}$ Hz [13]. The width of the waveguide is $w = 0.1\lambda_{ps}$, where λ_{ps} is the InSb plasma wavelength defined as $2\pi c/\omega_{ps}$. Fig. 1(b) shows the dispersion curves of the forward and backward propagating THz SMPs waves with ($B = 1$ Tesla) and without the external MFs in the frequency region $\omega/\omega_{ps} = [0, 1]$. It can be seen that the dispersion curves are symmetric for the forward and backward propagating waves when no magnetic field exists (shown as the black dotted lines). However, when an MF is applied, the dispersion curves of the two propagating waves become different. The forward-propagating mode vanishes at a frequency of $\omega/\omega_{ps} = 0.59$, while the backward propagating mode vanishes at a higher frequency of $\omega/\omega_{ps} = 0.61$. This means that the THz waves in the frequency region of $\omega/\omega_{ps} = [0.59, 0.61]$ (corresponding to $f = [1.18, 1.23]$ THz) can only propagate backwards.

Simulation Results

To verify the theoretical calculations, we do the finite element method (FEM) simulations, the results of which are plotted in Figs. 2 (a)-(d). Without the external MF, the field distributions of the forward and backward propagating waves are the same (see (a) and (b), respectively). However, when 1 Tesla MF is applied, the forward-propagating THz wave is blocked (see Fig. 2(c)), while the backward-propagating wave can still propagate through the slit (see Fig. 2(d)). The reason for the results in Fig. 2 (c) is that the forward-propagating SMPs mode is blocked under the MF, thus SMPs cannot be excited by the edge diffractions [3], when the THz wave impinges on the slit. The disappearance of the spots in the slit

of Fig. 2 (d), caused by the SMPs interference, also proves that there is no forward-propagating THz waves. The transmitted intensity curves of the forward and the backward propagating waves under the MF are depicted in Fig. 2 (e). It can be seen the one way region is $\omega = [0.592, 0.612]$ ω_{ps}, which agrees very well with the theoretical results. In the simulations, we consider a slit with finite length of $L = 300\mu m$. However, the one-way effect can also be observed for other slit lengths.

It is important to study the factors that affect the performance of the one-way-propagating frequency band. According to Eq. (3), we calculate the cutoff frequencies of the forward (ω_{Vf}) and the backward (ω_{Vb}) propagating modes versus the applied MFs, and versus the permittivity of the dielectric layer, shown in Fig. 3 (a) and (b), respectively. It is found that the external MF affects the cutoff frequencies of both the forward and backward propagating modes (see Fig. 3(a)). The cutoff frequencies of the two modes decrease dramatically (from 0.76 ω_{ps} to 0.18 ω_{ps} corresponding to from 1.5 THz to 0.36 THz) with the increase of the MF (from 0.5 Tesla to 5 Tesla). It is interesting to notice that the bandwidth of the one-way-propagating band $\Delta\omega$ depends little on the MF. On the other hand, it is mainly determined by the permittivity of the dielectric layer ε_d as seen in Fig. 3(b), where $\Delta\omega/\omega_{ps}$ increases from 0.02 to 0.075 (corresponding to from 0.04 THz to 0.15 THz) as ε_d increases from 1 to 3.

The effects of B and ε_d on the one-way-propagating frequency band can also be explained analytically. By applying the non-retardation limit [14], i.e. $\beta >> k_0$, into Eqs. (3), we have $\kappa_1 \approx \kappa_2 \approx \kappa_3 \approx \beta$, and $\beta w >> 1$. Thus the cutoff frequencies of the forward and the backward propagating modes ω_{Vf} and ω_{Vb} can be expressed as

$$\omega_{Vf} = \frac{1}{2}[\sqrt{\omega_c^2 + 4\omega_{ps}^2\varepsilon_\infty/(\varepsilon_d + \varepsilon_\infty)} - \omega_c] \tag{4}$$

$$\omega_{Vb} = \frac{1}{2}[\sqrt{\omega_c^2 + 4\omega_{ps}^2} - \omega_c]$$

respectively. It can be seen that the one-way-propagating bandwidth can be obtained analytically as $\Delta\omega = \omega_{Vf} - \omega_{Vb}$, which increases with ε_d. ω_{Vf} rather than ω_{Vb} is dependent on ε_d is owing to the asymmetry of the mode distribution caused by the MF. From simulations, the forward-propagating wave is mostly confined on the insulator/ semiconductor interface, while the backward propagating wave is confined on the insulator/ metal interface. Because the permittivity of metal is very large ($|\varepsilon_m| >> \varepsilon_d$), the impact of varying ε_d on the insulator/metal interface is much weaker than on the insulator/ semiconductor interface.

Last, we present an application example of this one-way THz waveguide for realizing a THz plasmonic switch. It should be noted that if we change the propagation direction of the applied MF, i.e. from B to –B, the dielectric tensor matrix in Eq. (1) will be transposed. As a result, the dispersion relation of the forward and backward propagating waves is interchanged, which means the one-way-propagating band will not block the forward wave, but the backward wave. According to this principle, we design a T-shape waveguide, as shown in Fig. 4. The width of the waveguide is $0.1\lambda_{ps}$. The MF intensity is 1 Tesla, and the incident angular frequency is $\omega = 0.6\omega_{ps}$ within the one-way-propagating band. It is clearly seen that when the direction of the external MF is changed, a tunable THz plasmonic switch can be realized.

Conclusion

In conclusion, we present a one-way-propagating THz plasmonic waveguide and a tunable THz plasmonic switch based on the nonreciprocal effect of SMPs. The proposed waveguide structure and the concept may open a new avenue of realizing various tunable THz plasmonic devices.

References

[1] Duling and D. Zimdars, "Terahertz imaging: Revealing hidden defects," *Nat. Photonics* **3**, 630–632 (2009).

[2] K. Wang and D. M. Mittleman, "Metal wires for terahertz wave guiding," *Nature* **432**, 376–379 (2004).

[3] T. Xu, Y. Zhao, D. Gan, C. Wang, C. Du, and X. Luo, "Directional excitation of surface plasmons with subwavelength slits," *Appl. Phys. Lett.* **92**, 101501 (2008).

[4] J. Chen, Z. Li, S. Yue, and Q. Gong, "Efficient unidirectional generation of surface plasmon polaritons with asymmetric single-nanoslit," *Appl. Phys. Lett.* **97**, 041113 (2010).

[5] J. J. Brion, R. F. Wallis, A. Hartstein, and E. Burstein, "Theory of surface magnetoplasmons in semiconductors," *Phys. Rev. Lett.* **28**, 1455–1458 (1972).

[6] B. Hu, Q. J. Wang, S. W. Kok, and Y. Zhang, "Active Focal Length Control of Terahertz Slitted Plane Lenses by Magnetoplasmons," *Plasmonics* **7**, 191–199 (2011).

[7] B. Hu, Q. J. Wang, and Y. Zhang, "Broadly tunable one-way terahertz plasmonic waveguide based on nonreciprocal surface magneto plasmons," *Opt. Lett.* **37**, 1895 (2012).

[8] B. Hu, Q. J. Wang, and Y. Zhang, "Slowing down terahertz waves with tunable group velocities in a broad frequency range by surface magneto plasmons," *Opt. Express* **20**, 10071–10076 (2012).

[9] B. Hu, Q. J. Wang, and Y. Zhang, "Voigt Airy surface magneto plasmons," *Opt. Express* **20**, 21187 (2012).

[10] E. D. Palik and J. K. Furdyna, "Infrared and microwave magnetoplasma effects in semiconductors," *Rep. Prog. Phys.* **33**, 1193–1322 (1970).

[11] C. R. Williams, S. R. Andrews, S. a. Maier, a. I. Fernández-Domínguez, L. Martín-Moreno, and F. J. García-Vidal, "Highly confined guiding of terahertz surface plasmon polaritons on structured metal surfaces," *Nat. Photonics* **2**, 175–179 (2008).

[12] J. Gomez Rivas, C. Janke, P. H. Bolivar, and H. Kurz, "Transmission of THz radiation through InSb gratings of subwavelength apertures," Optics Express **13**, 847 (2005). S. Linden, C. Enkrich, M. Wegener, J. Zhou, T. Koschny, and C. M. Soukoulis, *Science* (New York, N.Y.) **306**, 1351 (2004).

[13] M. S. Kushwaha and P. Halevi, "Magnetoplasmons in thin films in the Voigt configuration," *Phys. Rev. B* **36**, 5960–5967 (1987).

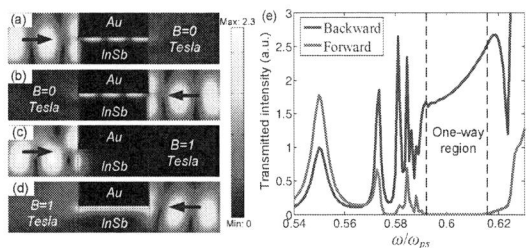

Fig. 2 Fig. 2 (a)-(d) Field distribution $|H_y|^2$ of the forward and backward propagating waves in a sub-wavelength. The width and length of the slit are $0.1\lambda_p$ and $300\mu m$, respectively. The incident frequency is $\omega = 0.6\omega_{ps}$ (in the one-way-propagating frequency band). The arrows indicate the incident directions. (e) Transmitted intensities of the forward (corresponding to (c)) and backward propagating (corresponding to (d)) waves when 1 Tesla MF is applied.

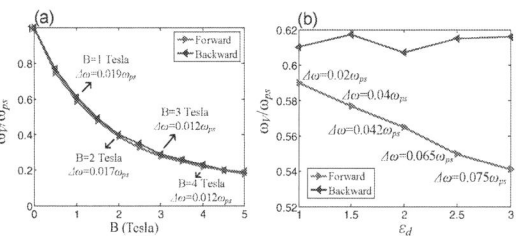

Fig. 3 Effects of (a) the external MFs B and (b) the permittivity of the dielectric layer ε_d on the one-way-propagating frequency band. ω_V represents the cutoff frequency. $\Delta\omega = \omega_{Vb} -\omega_{Vf}$ is the bandwidth of the one-way-propagating band.

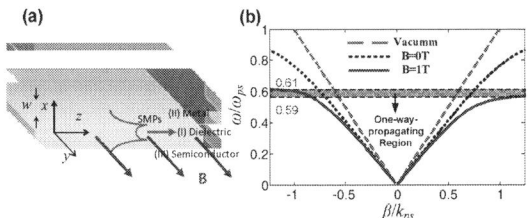

Fig. 1 (a) Schematic structure of the one-way-propagating THz plasmonic waveguide. It is composed of metal (II), dielectric (I), and semiconductor (III) layers. (b) Dispersion relations of the THz surface magneto plasmons without and with an external MF.

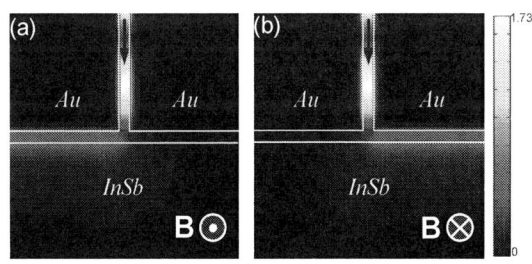

Fig. 4 $|H_y|^2$ distributions of a designed THz plasmonic T-shape switch tuned by an external MF. (a) The MF direction is along the y-axis. (b) The MF direction is along the $-y$-axis.

High Frequency SAW Nanotransducer Utilizing Ultrananocrystalline Diamond/ AlN Bimorph Architecture

Ali B. Alamin Dow[1], H. Lin[1], C. Popov[2], U. Schmid[3], Nazir P. Kherani[1]

[1]Department of Electrical and Computer Engineering, University of Toronto, 10 King's College Road, ON M5S 3G4 Canada.
[2]Institute of Nanostructure Technologies and Analytics, University of Kassel, Heinrich-Plett-Str. 40, D-34132 Kassel, Germany
[3] Institute of Sensors and Actuator Systems, Vienna University of Technology, Floragasse 7, A-1040 Vienna, Austria.
Phone: +1- 416 -946- 7372, Fax: +1- 416 -946-0054 and E-mail: alamindow.alamindow@utoronto.ca

Abstract

This article reports on the development and realization of a super high frequency surface acoustic wave (SAW) nanotranducer architecture based on a bi-layer of aluminum nitride (AlN) and ultrananocrystalline diamond (UNCD). The SAW nanotransducer was fabricated on an AlN/UNCD structure using electron beam lithography and lift off processes. The SAW device consists of nano inter-digitated transducers (n-IDTs) at both the input and output ports. The fabricated devices exhibited response over a high frequency range, as high as 18.4 GHz, with minimal insertion losses. The good frequency characteristics of the fabricated devices and compatibility with existing fabrication technologies open the way for the realization of advanced AlN/UNCD based transducers.

Introduction

Over the last decade surface acoustic wave devices (SAW) have attracted much research and development interest from both industry and academia for a variety of applications including high volume data processing, satellite communications [1, 2], high sensitivity chemical, physical and biological sensors [3, 4] and mobile communications. Moreover, SAW devices are considered potentially viable for harsh environment applications such as data and sensing processing in high temperature combustion ambient. Hence, the increase in sensing complexity and high volume data transmission and/or receiving demand highly efficient SAW devices that can be fabricated with minimal processing restrictions so as to address high frequency and complex sensing applications [2, 3, 5]. Classically, there are a number of platform materials that can be coupled with microfabrication techniques to synthesize SAW devices, for example, quartz and lithium substrates. However, the drawback of classical SAW materials is their incompatibility with existing semiconductor fabrication technologies and in particular the complementary metal-oxide-semiconductor (CMOS) technology [1, 3, 5, 6]. Also, the high demand for SAW devices in high frequency applications, such as advanced sensors and communication systems, and calls for the development of advanced functional SAW thin films on platform materials such aluminum nitride (AlN) [6, 7]. Moreover, the combination of AlN with nanocrystalline diamond can offer many advantages such as high Young's modulus that provides high acoustic velocity. Further, both AlN and nanodiamond can be realized using existing fabrication processing such as CMOS technology. The combination of nanointerdigitated transducers (n-IDTs) that are fabricated at the nanometer scale together with AlN/nanodiamond architecture could improve the filtering response of SAW devices [8, 9, 10]. Recently, AlN has been identified as being among candidate piezoelectric materials that can be developed and integrated using existing fabrication technologies, and together with ultra-nanocrystalline diamond (UNCD) can drastically increase the operating frequency of SAW devices [9, 10]. Here, we present the development, fabrication and testing of super high frequency surface acoustic wave (SAW) nanotransducers on an ultrananocrystalline diamond (UNCD)/AlN bimorph architecture. The incorporation of UNCD was motivated by its high acoustic wave properties to enhance the frequency characteristics of SAW devices, albeit the attainment of smooth UNCD is critical for low loss propagation.

Material Synthesis

Ultrananocrystalline diamond (UNCD) films were prepared by microwave plasma chemical vapor deposition using precursor gases containing 17% CH_4 in N_2. Monocrystalline (100) 75 mm diameter silicon wafers were used as substrates, which were pretreated ultrasonically in a suspension in order to enhance the nucleation density. The suspension contained a mixture of 80 mg ultradisperse diamond powder (5 nm average crystallite size) and 50 mg nanocrystalline diamond powder (250 nm average crystallite size) in 75 ml n-pentane. The achieved nucleation density was on the order of 1×10^{10} cm^{-2} [11, 12]. The UNCD films were deposited at a substrate temperature of 600°C, a microwave power of 800 W and a working pressure of 23 mbar. The deposition rate under these conditions was 3 nm/min. These deposition conditions were used to deposit 2μm thick UNCD/a-C films, which were comprehensively characterized with respect to their crystallinity, composition, topography and bonding structure [9, 12]. X-ray diffraction (XRD) and selected area electron diffraction revealed patterns characteristic for the diamond phase (Fig. 1). Furthermore, the size of the diamond crystallites was determined from the XRD peaks to be on the order of 3-5 nm. These nanocrystallites are embedded in an amorphous carbon matrix (a-C) with a grain boundary width of 1-1.5 nm, as shown by transmission electron microscopy (Fig. 2) [9, 11]. The ratio of the volume fractions of the two phases (crystalline and amorphous) estimated from the density of the coatings and from the total crystallinity is close to unity.

978-1-4673-4840-9/13 $31.00 © 2013 IEEE   360

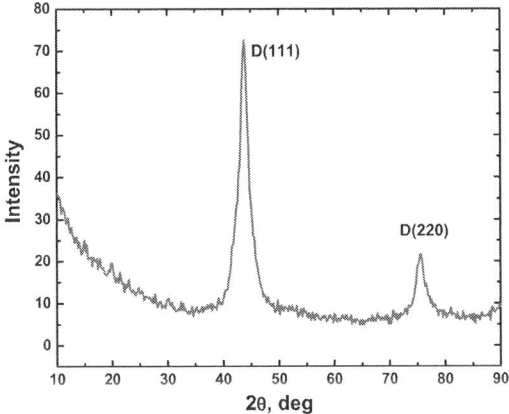

Fig. 1. XRD pattern of UNCD films.

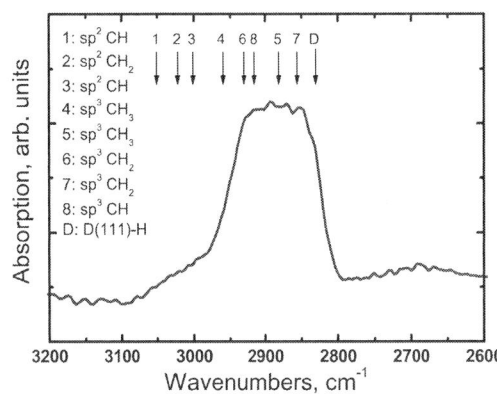

Fig. 3. FTIR spectrum of UNCD showing the C-H stretching modes

Investigations of the UNCD films with Raman spectroscopy, X-ray photoelectron spectroscopy (XPS), electron energy loss spectroscopy (EELS) and Auger electron spectroscopy (AES) showed the presence of sp^2-bonded carbon atoms (up to 15 at%) localized in the amorphous matrix [11, 12]. The bulk films are composed mostly of carbon; the concentration of nitrogen and oxygen is below 1 at% each. Although no H_2 was added to the precursor gas mixture, the UNCD films contain about 8-10 at% H in the bulk, as revealed by elastic recoil detection (ERD) analysis, originating from the CH_4 molecules. Fourier transform infrared (FTIR) spectroscopy showed that the hydrogen is bonded predominantly in sp^3-CH_x groups (Fig. 3). Atomic force microscopy (AFM) studies exhibited the characteristic topography of UNCD films, composed of structures with diameters of several hundred nanometers, which themselves possess a substructure. The rms surface roughness values are in the narrow range of 10 – 14 nm [9, 11].

The surface composition of as-grown UNCD/a-C films was investigated by XPS showing that the as-grown surfaces are very clean, with oxygen and nitrogen concentrations of about 1 at%. From the depth profiles of hydrogen concentration in the UNCD films, obtained by nuclear reaction analysis (NRA), surface H concentration of 12-14 at% is inferred which hints hydrogen termination of the surface [9, 11].

The AlN thin films were deposited on UNCD in a production-type DC sputtering machine from "Von Ardenne" (LS 730) with a target thickness of about 1.7 μm. Prior to deposition the substrate was cleaned with deionised water, acetone and isopropanol. The deposition chamber was filled with N_2 with a purity of 99.9999%. The aluminium target with purity 99.999% and 150mm diameter was positioned at a distance of 65mm above the substrate. Prior to deposition, the chamber was evacuated below 2×10^{-6} mbar and the target was pre-sputtered in argon atmosphere of 6×10^{-3} mbar for purification purposes. During deposition the sputtering pressure was 6×10^{-3} mbar and the DC power was 500W [9].

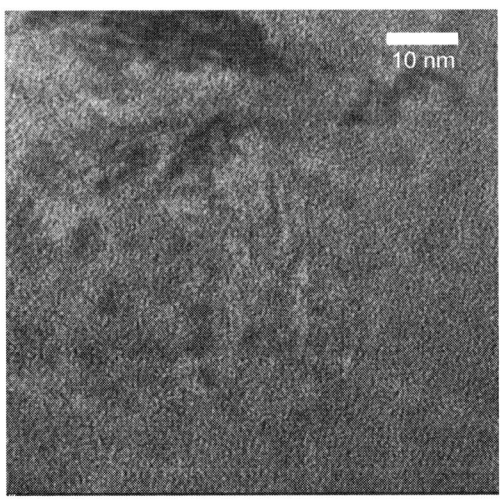

Fig. 2. Bright field TEM image of the UNCD film

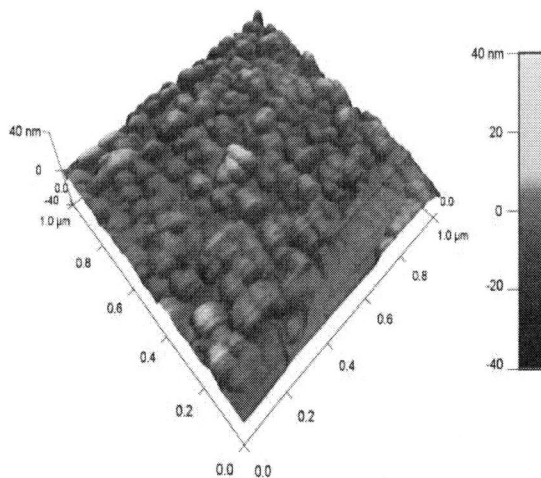

Fig. 4. AFM image of AlN grown on UNCD

The sputtering rate was measured to be 12.25 nm/min. The substrate temperature increased during processing due to plasma heating from room temperature up to 150°C. AFM measurements were performed to evaluate the surface roughness of the aluminium nitride thin film due to the importance of that property for SAW devices [9]. The roughness was found to be 5.07 nm (Fig. 4).

Device Fabrication and Characterization

The device designs were written on top of AlN using optimized electron beam lithography (EBL) and subsequently realized by lift-off technique. The fabrication process involves aligning n-IDTs on the wafer using EBL. In the design, the length of the n-IDTs and the distance between the input and output transducers were fixed to 90 µm and 19 µm, respectively. Each n-IDT consists of 120 reflectors and 60 fingers. Two finger widths have been realized at 140 nm (device 1) and 400 nm, corresponding to periods of 560 nm (device 2) and 1600 nm). Electro sensitive resist ZEP 520 was spun on the wafer with a resulting thickness of 420 nm and subsequently 20 nm chromium was thermally evaporated to overcome charge accumulation on AlN/UNCD bi-layer. The EBL process was carried out using EBL tool (EBPG 5000+, Vistec Lithography Ltd.) with an acceleration voltage of 100 kV. The process parameters were optimized and an exposure dose of 240~275 µC/cm² at an optimal current of 15 nA were used to write the structures. The exposed structures were developed in the ZEP-N50 developer for 60 s followed by 30 s in a 9:1 mixture of methyl-isobutyl-ketone (MIBK): isopropanol (IPA). Subsequently, a plasma descum for 20 s was carried out for adhesion purposes and electron beam evaporation of 120 nm of platinum with an interleaved adhesion layer of 9 nm titanium was performed to realize the device n-IDTs. Removing the ZEP 520 resist in ZDMAC (Fig. 5) completes the fabrication process. The filtering response of the device was characterized using an RF probe station (SussMicrotech) and a network analyzer (Agilent 8361A). After positioning the device on the probe station, an RF-signal was directly applied on n-IDTs at the input port thus giving rise to a SAW which is subsequently converted to an RF signal at

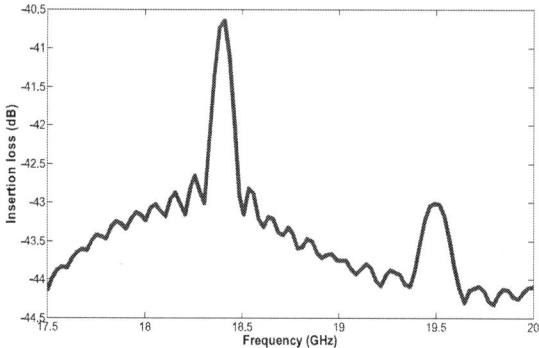

Fig. 6. Frequency response of device 1

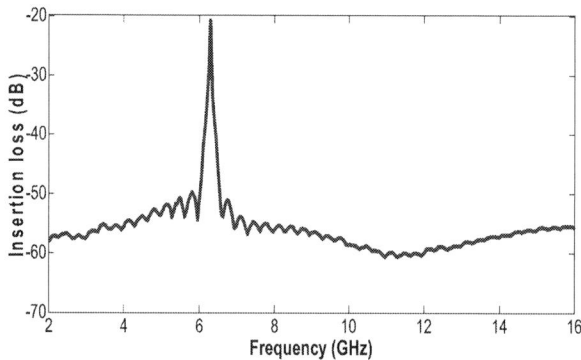

Fig. 7. Frequency response of device 2

the output port which in turn can be measured by a network analyzer. As shown, the frequency response of the fabricated devices (Figs. 6 and 7) exhibit a centre resonance frequencies of 18.4 GHz for the device with a 560 nm period (device 1) and 6.8 GHz for the device with a 1.6 µm period (device 2) with a minimal insertion loss of approximately 40.5 dB and 21 dB, respectively, and corresponding acoustic phase velocity v of 8316 m/s and 10064 m/s, respectively. The electromechanical coupling coefficient (K^2) of the devices was evaluated using $K^2 = 2(1 - v_m/v_0)$ where v_0 is the free SAW velocity and v_m is the SAW velocity with a shorting metallic surface. Thus, electromechanical coupling coefficients of 0.74 and 0.96 for devices 1 and 2, respectively, were obtained.

Conclusion

In conclusion, we have presented the realization and testing of SAW nano transducers with n-IDT dimensions in the sub-µm range synthesized on AlN/UNCD bi-layer. The AlN/UNCD thin films synthesis was presented and discussed. The fabricated devices exhibited good filtering performance with an operating frequency up to 18.4 GHz with low insertion loss

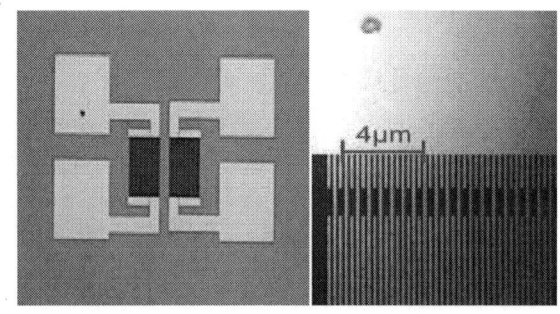

Fig. 5. Micrographs of the fabricated device

characteristics. The resulting frequency response together with the fabrication technology pave the way towards the realization of high frequency SAW devices that can meet future challenges in complex sensing and high volume data processing applications.

Acknowledgment

This research was supported by the Natural Sciences and Engineering Research Council (NSERC) Canada Discovery grants and the Ontario Research Fund – Research Excellence program.

References

[1] David Morgan,"Surface Acoustic Wave Filters With application to communications and signal processing", Elsevier, Academic Press, 2d Edition, 2007.

[2] V. Cimalla, F. Niebelschütz, K. Tonisch, Ch. Foerster, K. Brueckner, I. Cimalla, T. Friedrich, J. Pezoldt, R. Stephan, M. Hein, O. Ambacher: Nano electromechanical devices for sensing applications, Sens. Actuators B, vol. 126, 2007.

[3] C S Lam, C Y J Wang, S M Wang: A review of the recent development of temperature stable cuts of quartz for SAW applications, Symp. on Piezoelectricity, AcousticWaves, and Device Applications, 2004.

[4] I Ingrosso, S Petroni, D Altamura, M Devittorio, C Combi, A Passaseo: Fabrication of AlN/Si SAW delay lines with very low RF signal noise, Microelectronic Engineering,Volume: 84, Issue: 5-8, Pages: 1320-1324, 2007.

[5] D. Neculoiu, A. Muller, G. Deligeorgis, A. Dinescu, A. Stavrinidis, D. Vasilache, A.M. Cismaru, G.E. Stan, G. Konstantinidis: AlN on silicon based surface acoustic wave resonators operating at 5 GHz, Electronics Letters, Vol. 45, 2009.

[6] U. Schmid, J. L. Sanchez Rojas, "Piezoelectric Properties of Sputtered AlN Thin Films and Their Applications", Advances in Science and Technology, Vol. 54, pp. 41-49, 2008.

[7] T. Palacios, F Calle, E. Monroy, E Munoz: Submicron technology for III-nitride semiconductors, J. Vac Sci. technol B, 20, pp. 2071–2074, 2002.

[8] Y.C. Lee , S.J. Lin, V. Buck, R. Kunze, H. Schmidt, C.Y. Lin, W.L. Fang, I.N. Lin," Surface acoustic wave properties of natural smooth ultra-nanocrystalline diamond characterized by laser-induced SAW pulse technique", Diamond & Related Materials, vol. 17, 446–450,2008.

[9] Ali B. Alamin Dow, H. Lin, M. Schneider, Ch. Petkov, A. Bittner, A. Ahmed, C. Popov, U. Schmid, Nazir P. Kherani: Ultrananocrystalline Diamond-Based High-Velocity SAW Device Fabricated by Electron Beam Lithography, IEEE Trans. on Nanotechnology, Vol.11, pp. 979-984, 2012.

[10] H. Hatakeyama, T. Omori, K. Hashimoto and M. Yamaguchi: Fabrication of SAW devices using SEM-based electron beam lithographyand lift-off technique for Lab use, IEEE Ultrason. Symp.,pp. 1896–1900,2004

[11] C. Popov, W. Kulisch, S. Boycheva, K. Yamamoto, G. Ceccone, Y. Koga: Structural investigation of nanocrystalline diamond/amorphouscarbon composite films, Diamond Relat. Mater., Vol. 13, 2071-2075, 2004.

[12] W. Kulisch, T. Sasaki, F. Rossi, C. Popov, C. Sippel, and D. Grambole: Hydrogen incorporation in ultrananocrystalline diamond/amorphous carbon films", Physica Status Solidi - rapid research letters 2, 77-79, 2008.

978-1-4673-4840-9/13 $31.00 © 2013 IEEE

Hump Phenomenon in Transfer Characteristics of Double-Gated Thin-body Tunneling Field-Effect Transistor (TFET) with Gate/Source Overlap

Hyun Woo Kim[1], Min-Chul Sun[1,2], Sang Wan Kim[1] and Byung-Gook Park[1]

[1]Inter-University Semiconductor Research Center (ISRC) and Department of Electrical Engineering and Computer Science, Seoul National University, 1 Gwanak-ro, Gwanak-ku, Seoul 151-742, Republic of Korea
[2]TD Team, System LSI Business, Samsung Electronics Co., Ltd. Yongin 446-711, Republic of Korea
Tel.: +82-2-880-7279, E-mail address: hyunoo1218@snu.ac.kr

Abstract

Most of research groups have studied on Tunneling Field-Effect Transistors (TFETs) with assumption that there are no gate/source overlap and abrupt source/channel junction. In this work, we study the electrical characteristics of double-gated thin-body TFET with gate/source overlap and no abrupt source/channel junction. From transfer characteristics, hump phenomenon occurring with increasing gate bias is observed. This phenomenon affects the threshold voltage (V_{TH}), which worsens device matching and also makes it difficult to design logic circuits. The reason why tunneling current is suddenly increased is due to tunneling components in a direction normal to the channel. Theses hump phenomena are not seen in the previous TFET study using unidirectional nonlocal band-to-band tunneling model.

Keywords: TFET, hump characteristics, and band-to-band tunneling

Introduction

TFETs have been extensively researched as one of the candidates for switching applications in low-power electronic circuits because it is possible to obtain subthreshold swing (SS) value less than 60 mV/dec [1-4]. However, most of research groups have studied on TFET with assumption that there are no gate/source overlap and abrupt source/channel junction. In this work, double-gated thin-body TFET is used for obtaining high current drivability using field-coupling effect and it can be fabricated as FinFET structures. And we study the electrical characteristics of double-gated thin-body TFET with gate/source overlap and no abrupt source/channel junction. From transfer characteristics, hump phenomenon occurring with increasing gate bias is observed. This phenomenon affects the threshold voltage (V_{TH}), which worsens device matching and also makes it difficult to design logic circuits. Hence we scrutinize the cause of hump phenomenon in TFET with gate/source overlap.

Results and Discussion

1. Device Sturcture

The model device structure studied in this work is shown in Fig. 1. It is a TFET with double-gated thin-body channel and 20 nm of offset between gate and drain. In order to simulate the dopant diffusion of real fabricated devices, each source and drain edges are designed to have 20 nm of diffused region (Fig. 1(b)). This makes the simulation results more realistic. Other details of design parameters are summarized in Fig. 1(c). For the

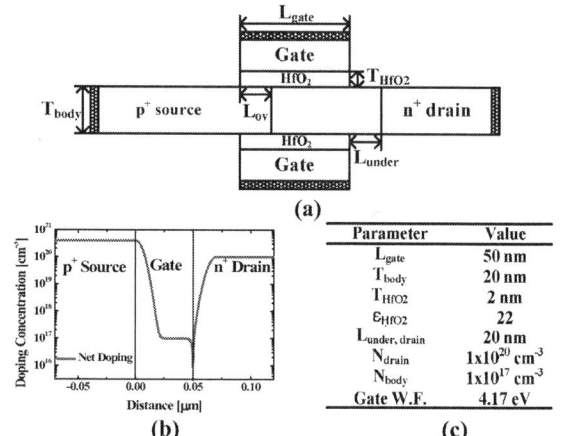

Parameter	Value
L_{gate}	50 nm
T_{body}	20 nm
T_{HfO2}	2 nm
ε_{HfO2}	22
$L_{under, drain}$	20 nm
N_{drain}	1×10^{20} cm^{-3}
N_{body}	1×10^{17} cm^{-3}
Gate W.F.	4.17 eV

Fig. 1. (a) The scheme of TFET, (b) doping profile and (c) device parameters used this work.

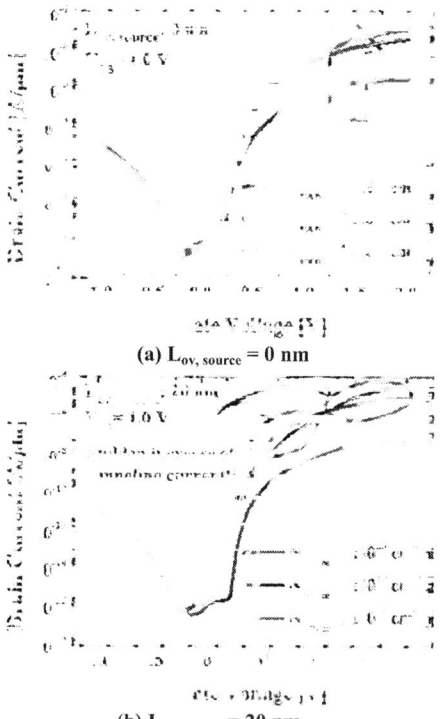

(a) $L_{ov, source} = 0$ nm

(b) $L_{ov, source} = 20$ nm

Fig. 2. Transfer characteristics with increasing N_{source} at $L_{ov, source} = 0$ and 20 nm.

accurate calculation of the device characteristics, the nonlocal band-to-band tunneling model of Sentaurus™ TCAD simulation of Synopsys Inc. (ver. D-2010.12) is used. In this

978-1-4673-4840-9/13 $31.00 © 2013 IEEE

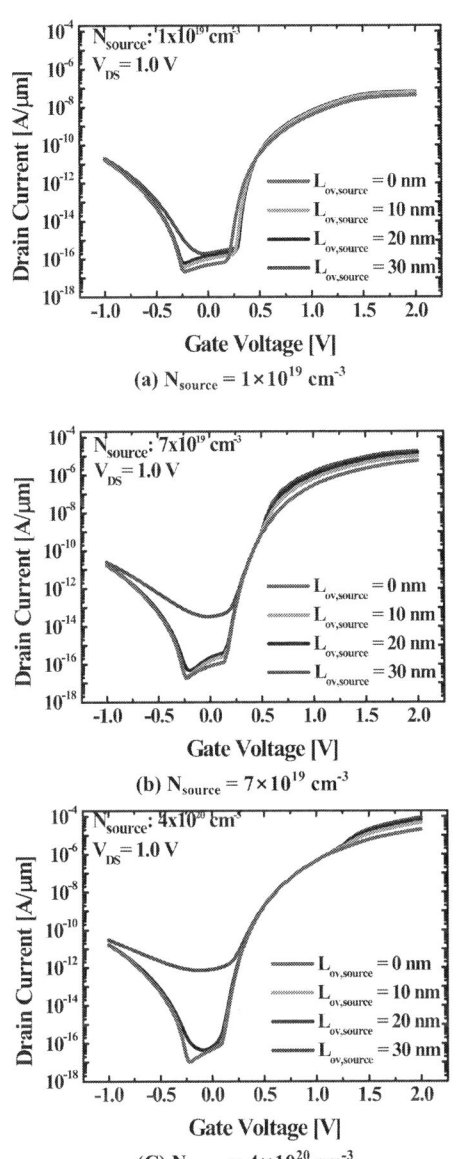

(a) $N_{source} = 1 \times 10^{19}$ cm^{-3}

(b) $N_{source} = 7 \times 10^{19}$ cm^{-3}

(C) $N_{source} = 4 \times 10^{20}$ cm^{-3}

Fig. 3. Transfer characteristics of TFET with increasing $L_{ov, source}$

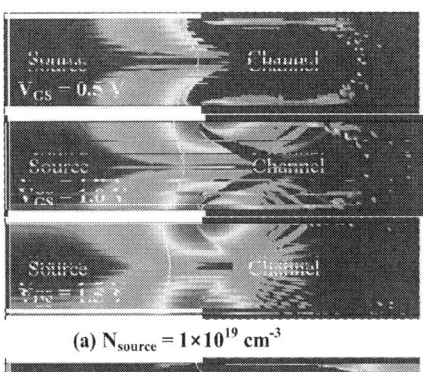

(a) $N_{source} = 1 \times 10^{19}$ cm^{-3}

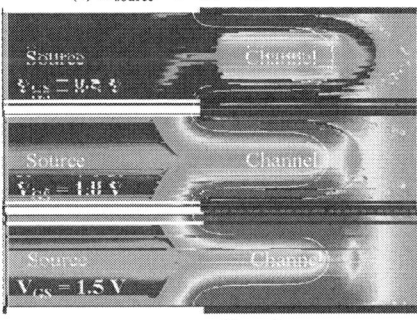

(b) $N_{source} = 7 \times 10^{19}$ cm^{-3}

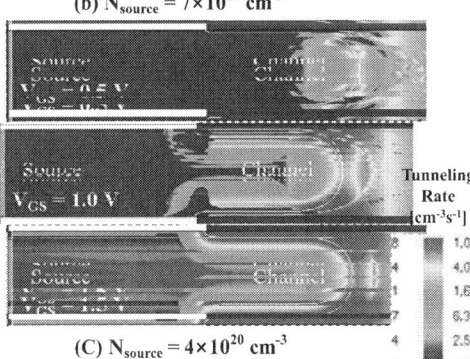

(C) $N_{source} = 4 \times 10^{20}$ cm^{-3}

Fig. 4. Tunneling rates at $V_{GS} = 0.5$, 1.0 and 1.5 V with increasing N_{source} in case of $L_{ov, source} = 20$ nm.

tool, automatic calculation of tunneling direction from valence band gradient completely gets rid of uncertainty of the results due to inappropriate assumption of the tunneling direction [5, 6].

2. Analysis on Hump phenomena

Firstly, the change of transfer characteristics with gate/source overlap is studied shown in Fig. 2. For a device with zero gate/source overlap, the drain current gradually increases as source doping concentration (N_{source}) increases due to thinning of tunneling barrier. However, a device with 20 nm of the overlap shows a sudden increase of drain current with the N_{source} becomes larger. This hump phenomenon is not seen in the previous TFET study which relies on unidirectional nonlocal band-to-band tunneling model [7].

Next, gate/source overlap length is varied from 0 to 30 nm to understand more on the hump characteristics and N_{source} are varied to maximum of 4×10^{20} cm^{-3} to consider boron solubility in silicon (Fig. 3). While the overlap length does not make any change in the phenomenon when N_{source} is as low as 1×10^{19} cm^{-3}, the larger overlap increases the hump behavior. Also, the onset point where the hump starts to appear looks dependent on N_{source}. Finally, in order to understand the physics of the hump phenomena, tunneling rate maps in the semiconductor regions are studied with the gate bias changed. In case of $N_{source} = 1 \times 10^{19}$ cm^{-3}, tunneling mostly occurs in a direction parallel to the channel because the overlapped source region is fully depleted regardless of the gate bias as shown in Fig. 4(a).

In case of $N_{source} = 7 \times 10^{19}$ cm^{-3}, tunneling components in a direction parallel to the channel is dominant during low gate bias. Then, as gate bias is increased, tunneling components in a direction normal to the channel starts to appear as shown in Fig. 4(b). These components become more than that in a direction parallel to the channel with high gate bias. As a result, tunneling current is suddenly increased due to an additional tunneling path. Also, while $L_{ov, source}$ is increased, the hump phenomena get worse because tunneling area of a direction normal to the

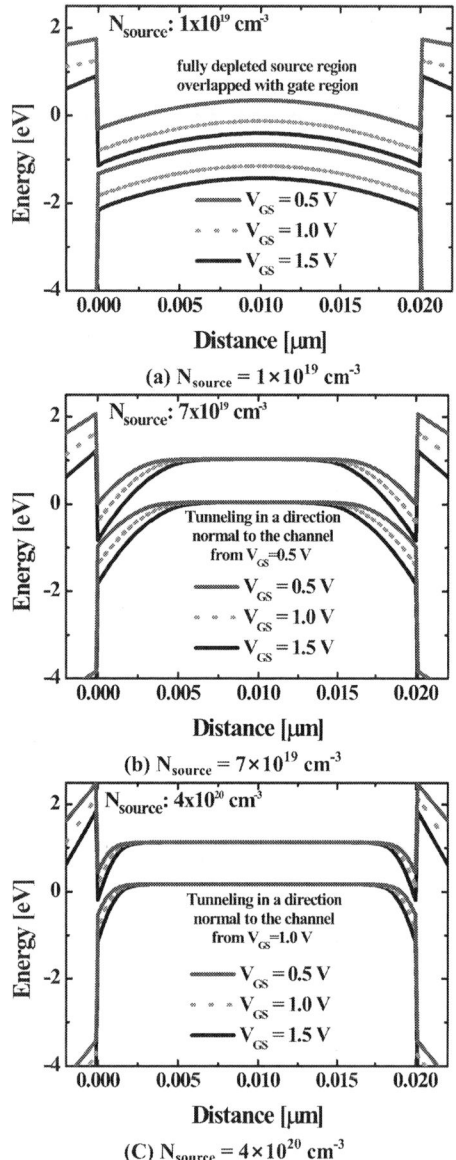

(a) $N_{source} = 1 \times 10^{19}$ cm^{-3}

(b) $N_{source} = 7 \times 10^{19}$ cm^{-3}

(C) $N_{source} = 4 \times 10^{20}$ cm^{-3}

Fig. 5. Energy band diagrams along a direction normal to the channel at 10 nm from gate edge in case of $L_{ov, source}$ = 20 nm.

channel becomes wider more. In case of $N_{source} = 4 \times 10^{20}$ cm^{-3}, the gate bias where the hump phenomenon begins is higher than that of $N_{source} = 7 \times 10^{19}$ cm^{-3}(Fig. 3). It is because of the overlapped source region with high N_{source} requires higher gate bias to be thin tunneling barrier. Thus, tunneling in a direction normal to the channel starts to occur as gate bias is increased by more than 1.0 V as shown in Fig. 4(c). Fig. 5 shows the energy band diagram along a direction normal to the channel at 10 nm from gate edge in TFET with $L_{ov, source}$ = 20 nm. In order to take place the tunneling in a direction normal to the channel, valence band in source must be aligned with conduction band in channel. When N_{source} is low, the overlapped source region is fully depleted and band alignment can not be occurred. However, as N_{source} is increased, tunneling barrier can be reduced enough to occur tunneling by gate bias because the overlapped source region is not more fully depleted.

Conclusion

From the electrical characteristics of double-gated thin-body TFET with gate/source overlap, hump phenomenon occurring with increasing gate bias is observed. The reason why tunneling current is suddenly increased is due to tunneling components in a direction normal to the channel. Theses hump phenomena are not seen in the previous TFET study using unidirectional nonlocal band-to-band tunneling model. From these results, in real fabrication, these hump phenomena will be a critical issue for logic circuit design using TFET.

Acknowledgements

This work was supported by the Smart IT Convergence System Research Center funded by the Ministry of Education, Science and Technology as Global Frontier Project (SIRC-2011-0031845).

References

[1] W.-Y. Choi et al., "Tunneling field-effect transistors (TFETs) with subthreshold swing (SS) less than 60 mV/dec," *IEEE Electron Device Letters,* vol. 28, no. 8, pp. 743-745, 2007.

[2] C. Hu et al., "Prospect of Tunneling Green Transistor for 0.1V CMOS," in *IEDM Techn. Dig.,* 2005.

[3] S. H. Kim et al., "Germanium-source tunnel field effect transistors with record high I_{ON}/I_{OFF}," in *VLSI Symp. Tech.,* 2009.

[4] K. Jeon et al., "Si tunnel transistors with a novel silicided source and 46mV/dec swing," in *VLSI Symp. Tech.,* 2010.

[5] Sentaurus User's Manual ver. D-2010.12.

[6] M.-C. Sun et al., "Design of Thin-Body Double-Gated Vertical-Channel Tunneling Field-Effect Transistors for Ultralow-Power Logic Circuits," *Japanese Journal of Applied Physics,* vol. 51, pp. 04DC031-04DC03-5, 2012.

[7] A. Chatopadhyay et al., "Impact of a Spacer Dielectric and a Gate Overlap/Underlap on the Device Performance of a Tunnel Field-Effect Transistor," *IEEE Transactions on Electron Devices,* vol. 58, no. 3, pp. 677-683, 2011.

Epi Defined (ED) FinFET: An alternate device architecture for high mobility Ge channel integration in PMOSFET

S. Mittal[1,*], S. Gupta[2], A. Nainani[2], M.C. Abraham[2], K. Schuegraf[2], S. Lodha[1], U. Ganguly[1]

[1]CEN, Department of EE, Indian Institute of Technology Bombay, 400076 India, [2]Applied Materials Inc. Santa Clara, USA, 94085

*smittal@iitb.ac.in

Abstract

Band to band tunneling (BTBT) is a major challenge in Ge FinFETs due to its smaller band gap. Narrow fin widths reduce BTBT due to quantum confinement (QC). However, Line Edge Roughness (LER) on narrower fins causes large V_T variability. Previously, we have proposed an architecture named Epitaxially Defined (ED) FinFET to reduce V_T variability due to LER wherein channel depletion is defined by low doped highly uniform epitaxy (thus named Epi Defined FinFET) (epi-thickness non uniformity<2%) over a thick highly doped Si fin instead of lithography based patterning subject to LER (non-uniformity<50% i.e. 2nm LER on a 4nm fin). In the present work, we propose integration of Ge into EDFinFET architecture in which Ge (or SiGe) is grown on top of Si fin. Proposed structure shows 10× reduction in LER based V_T variability in comparison to FinFETs. Valence band QC in gate oxide/Ge/Si stack is used to control BTBT. Biaxial stress in thin Ge epitaxially grown on Si results in 27% higher I_{ON}. Thin Ge film required is lower than critical defect free thickness of Ge epitaxy on Si. Hence defect free Ge integration into FinFET architecture in enabled. We also show that EDFinFET can enable multiple V_T just by the application of a bias at the body terminal.

Introduction

FinFET has been introduced at 22 nm node recently. For high performance, integration of high mobility channel material instead of Si is quite desirable. Ge is an exciting candidate because of its highest reported hole mobility and CMOS compatibility. In this work, we examine integration of Ge into FinFET architecture for PMOS device. However, the fundamental problem with Ge and other high mobility materials is their narrow band gap which leads to high Band to Band tunneling (BTBT) leakage current. Therefore, in order to enable their integration into FinFET architecture, BTBT based leakage needs to be minimized. Reduction of Wfin has been proposed as an attractive solution since it increases quantum confinement which leads to higher band gap and thus BTBT leakage reduces [1]. Device variability has become a challenging issue with scaling [2]. V_T Variability due to Random Dopant Fluctuations (RDF) is no longer a problem in FinFET architecture due to undoped fins [3]. However FinFETs are susceptible to V_T variability due to Line Edge Roughness (LER) [4] due to requirement of very narrow fin widths (~ $L_G/3$ [5]) for required electrostatic control of channel. For Ge, QC required to control BTBT leakage produces stronger constraint for fin width definition. Previously, we have proposed an architecture named Epitaxially Defined (ED) FinFET [6] to reduce V_T variability due to LER in Si FinFETs. In this paper, we first demonstrate the challenges of V_T variability in Ge FinFETs. Second we propose the integration of Ge in EDFinFET. We demonstrate key advantages in

performance, variability and multiple V_T compared to conventional Ge FinFETs.

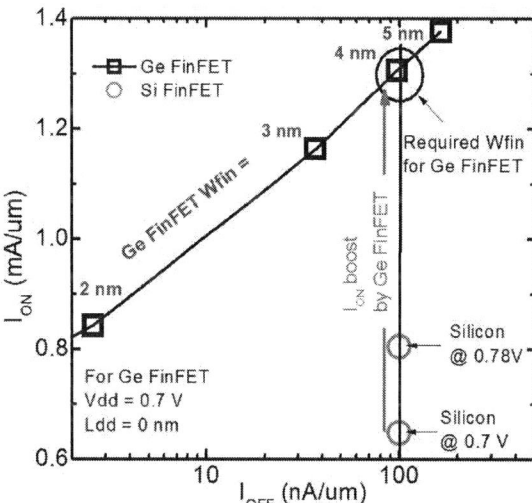

Fig.1: I_{ON} vs. I_{OFF} for Ge FinFET for different Wfin. Wfin of 4 nm is required to meet I_{OFF} specification for Ge FinFET. Also shows I_{ON} boost by Ge FinFET over Si FinFET.

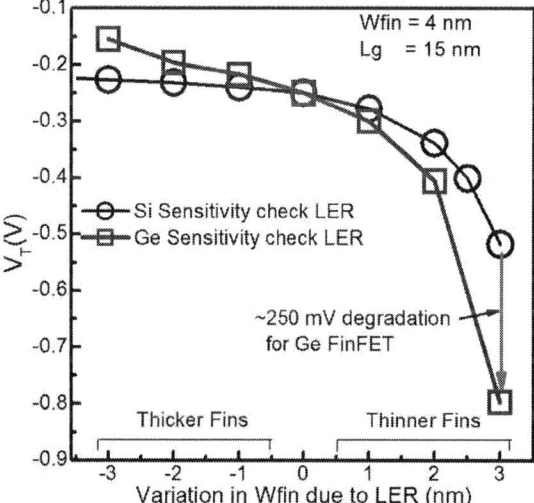

Fig.2: Change in V_T with variation in fin width for Si and Ge FinFET. Ge FinFETs owing to more quantum confinement are more susceptible to LER based variability. Typical value of variation in fin widths is $3\sigma = 3$nm [11-13].

Description of Approach

Sentaurus[TM] [7] non-local tunneling model has been well calibrated to data in literature [8] to account for BTBT leakage in Ge. TAT, SRH recombination/generation models and

978-1-4673-4840-9/13 $31.00 © 2013 IEEE   367

mobility models are also calibrated for excellent match with literature [8]. To simulate LER based variability, structures were generated by a Gaussian autocorrelation model [9] with the RMS amplitude $3\sigma = 1.5$ nm (roughness on edge and not on width) and correlation length $\Lambda = 30$ nm and is modeled by a sine function (e.g. LER=$3\sigma \sin(2\pi x/\Lambda) \pm W/2$) where W is the fin width (figure 3b) [7]. To get I_{ON} for EDFinFET, Monte Carlo simulations with the inclusion of biaxial stress were done in Sentaurus. Biaxial stress was calculated using Sentaurus Sband [7].

Challenges with Ge FinFETs

Figure 1 shows I_{ON} and I_{OFF} achieved for Ge PMOS FinFET for different fin widths. A 4 nm of fin width is required to meet 100 nA/μm I_{OFF} specifications [10]. On the other hand, Si FinFET needs 5 nm of fin width based on electrostatic control requirements. Figure 1 also shows I_{ON} boost achieved by Ge FinFET over Si FinFET, which highlights the advantage of Ge as a high mobility channel material.

V_T variation due to change in fin width, for nominal fin width of 4nm in Ge FinFET, is shown in Figure 2. We have used 3σ variation of 3 nm for fin width in FinFET. This is representative of various lithography capabilities like SADP and EUV [11-13]. Comparison with Si FinFET of the same fin width shows that Ge FinFETs are more susceptible to V_T variability. For thinner fins, Ge FinFETs are 250 mV poorer in variability with respect to Si FinFET. This is due to the fact that Ge has a deeper hole quantum well and therefore subjected to more quantum confinement than Si, resulting into large variation in V_T with Wfin. This shows the critical nature of LER based variability for Ge FinFETs.

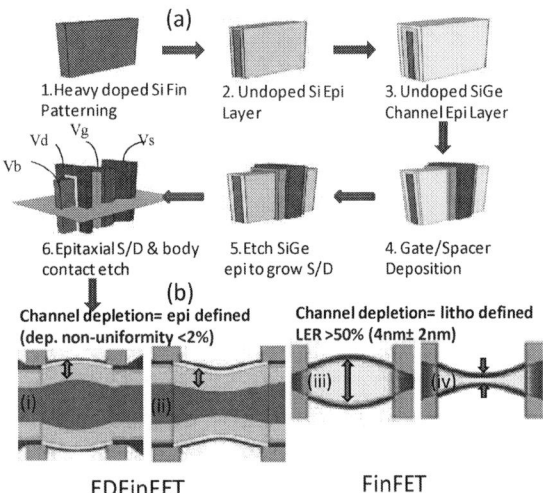

Fig. 3 (a) Fabrication steps for SiGe Epi Defined (ED) FinFET depicting simplicity of fabrication. (b) EDFinFET and FinFET structure subjected to LER variation of $3\sigma=1.5$ nm (fin width variation of 3 nm) and correlation length of 30 nm. In EDFinFET, channel depletion is epitaxy defined and in FinFET it is litho defined.

Epitaxially Defined (ED) FinFET for Ge PMOS

A. Device Structure and V_T variability comparison

As discussed above, large V_T variability for Ge FinFET is due to extreme quantum confinement effects. In response, we propose Ge channel integration into EDFinFET [6] architecture for PMOS device. The fabrication steps are shown in Figure 3 which is similar to as given in [6]. We start with heavy n-doped fin (doping density $\sim 10^{20}$ cm^{-3}) with fin width greater than 2-3 times the fin width of conventional FinFET (around > 10 nm) on a p-well. The fin is fabricated like any other conventional FinFET fin. It is therefore susceptible to the same extent of LER. A low-doped Si is conformally grown by epitaxy on top of it with abrupt doping (~ 1 nm/decade [14]). After which, the Ge channel epi is grown conformally on top of low doped Si. The gate is patterned next followed by spacer deposition. Following which, SiGe epi is etched out in a self aligned way to be able to grow source-drain epitaxially. Body contact to the initial heavy doped fin can be taken by etching a part of source or drain. In EDFinFET, channel depletion is defined by low doped highly uniform epitaxy (thickness non uniformity<2%) (Figure 3b- (i) & (ii)) over a thick highly doped Si Fin instead of lithography based patterning subject to LER (non-uniformity<50% i.e. 2nm LER on a 4nm fin) (Figure 3b –(iii) & (iv)). This configuration therefore provides uniform quantum confinement by epitaxial control (c.f. LER based quantum confinement non-uniformity) and thus reduces V_T fluctuation. The improvement in LER based variability thus obtained is shown in Figure 4. EDFinFET improves V_T variability by around 500 mV over Ge FinFETs due to reasons mentioned above. So EDFinFET shows better V_T variability resistance compared to Ge FinFET.

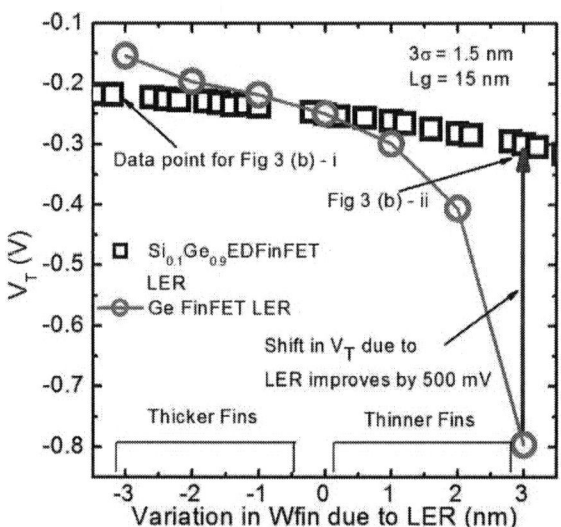

Fig. 4: V_T variation due to LER for SiGe EDFinFET and Ge FinFET. EDFinFET shows huge 500 mV improvements over sensitivity check of Ge FinFET.

B. Performance Comparison

In Ge EDFinFET, strained Ge (or SiGe) channel is confined from two sides by SiO$_2$ and Si. The hole quantum well of strained Ge on Si produces quantum confinement effect. This gives required quantum confinement to reduce BTBT leakage in SiGe EDFinFET. Net effective band gap thus obtained is sufficient to reduce BTBT leakage below specification. This is shown in Figure 5 (a) where minimum current through SiGe EDFinFET for different SiGe epi layer thicknesses is plotted, for three different mole fraction of Ge in SiGe. Below 1.5 nm, I_{OFF} below 100 nA/um can be obtained by all three mole fractions. The structure for which I_{OFF} criteria was met, I_{ON} at I_{OFF} = 100 nA/um is shown in Figure 5 (b). Inclusion of Si into SiGe increases alloy scattering and thus reduces the mobility. Therefore as shown, higher the Si percentage, lower is the I_{ON}. Highest I_{ON} is achieved by 1.5 nm of 100 % Ge grown on top of Si. Such thin defect free Ge is below critical thickness limit and demonstrated elsewhere [1]. So EDFinFET enables defect free integration of Ge into FinFET architecture.

has 27% higher I_{ON} than FinFET. Also EDFinFET is a surface inversion device while FinFET is a bulk inversion device. So there is an inherent EOT benefit for EDFinFET as discussed in [6]. This also enhances I_{ON} for EDFinFET. EDFinFET has poorer subthreshold slope (being a single gate device compared to FinFETs). Dynamic Threshold (DT) MOS configuration [15] can make subthreshold slope same as that of FinFET as shown on log scale plot. In DTMOS, body is ramped along with gate resulting into enhanced gate control and therefore results in better subthreshold slope.

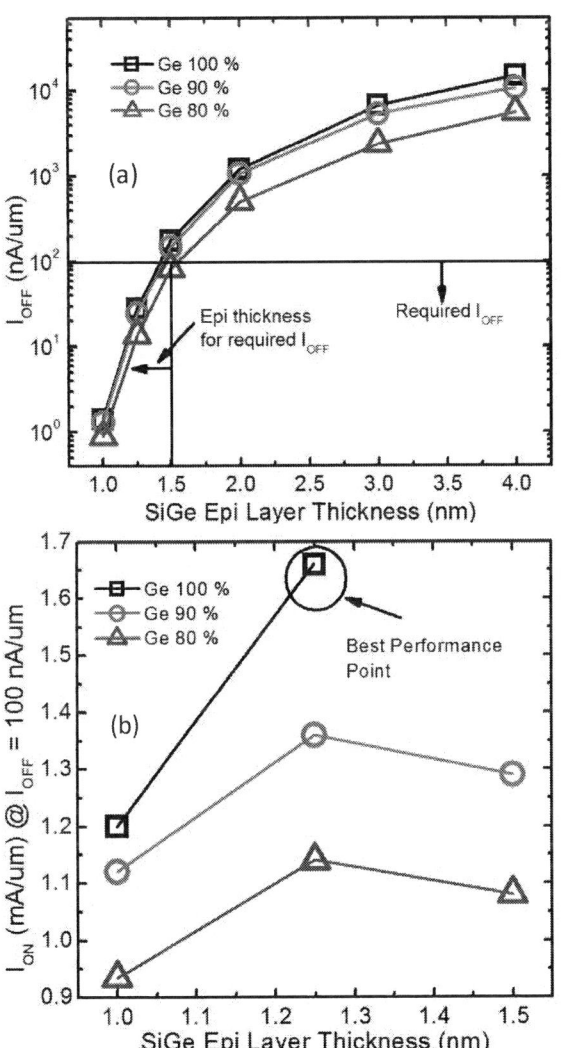

Fig. 5 (a): I_{OFF} vs SiGe Epi Layer thickness. For 1.5 nm or below, I_{OFF} below 100 nA/um can be achieved for all three Ge mole fraction cases. (b) Monte Carlo I_{ON} (boost due to biaxial stress included), for data points for which I_{OFF} can be below 100 nA/μm. I_{ON} is calculated here at I_{OFF} = 100 nA/μm. 1.5 nm of 100 % Ge gives best performance.

Id-Vg comparison is done in Figure 6 where drift-diffusion plots are shown on log scale to compare sub-threshold slope and monte-carlo plots are shown on linear scale to assess I_{ON} accurately. It can be observed in linear scale plot that EDFinFET, due to being biaxially stressed from body beneath,

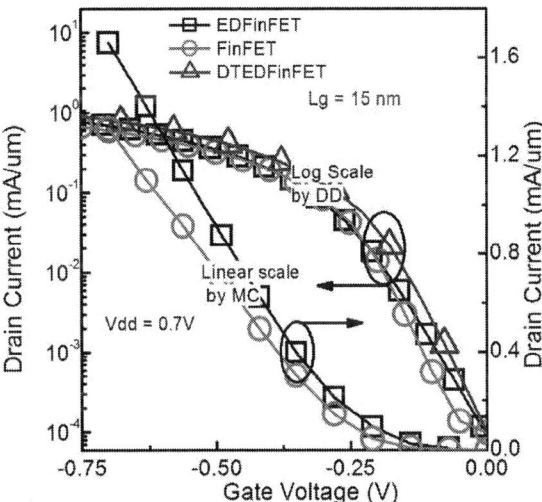

Fig.6: I_D-V_G comparison for FinFET and EDFinFET. EDFinFET due to biaxial stress and EOT benefit [6] shows higher I_{ON} than FinFET. EDFinFET has poor SS which can be restored by using EDFinFET in DTMOS configuration (referred as DTEDFinFET).

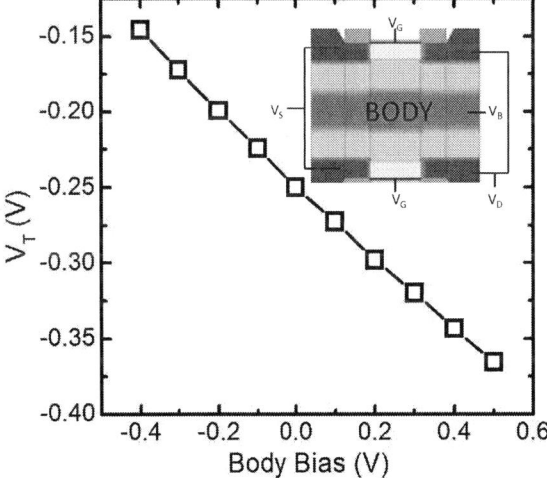

Fig.7: V_T for EDFinFET with body bias. Multiple V_T can be enabled for EDFinFET with just an application of body bias post fabrication.

C. Multiple V_T capability:

Multiple V_T is a much desired capability so as to be able to enable various technologies like High Performance (HP), Low Operating Power (LOP) and Low standby power (LSTP). In planar architecture, this capability is achieved by various implant steps. This capability is not available for FinFETs due to undoped fins. In literature, complex fabrication procedures like gate multi-work function and gate-source/drain overlap engineering are proposed [16] which are technically very challenging and expensive. However EDFinFET can achieve multiple V_T by varying body bias. Range of V_T obtained by the variation of centre body bias is shown in Figure 7.

D. I_{ON} per unit area

EDFinFET fins are 33 nm wide in comparison to 4nm wide FinFET. EDFinFET fins can be grown taller as the thicker fin can support taller aspect ratios due to enhanced mechanical stability. By doing the stable fin calculation using [17], it can be shown that EDFinFET fins are 2.4 times taller for same area. Which after multiplying with inherent I_{ON} benefits of EDFinFET, can give upto 3X higher I_{ON} per foot print area. Similarly the inherent I_{ON} benefit for Dynamic Threshold EDFinFET (DTEDFinFET) is higher than that of EDFinFET. Therefore overall benefit in on current per unit foot print that can be achieved by EDFinFET technology is 3-6 times. Thus thicker fins of EDFinFET provide an area scaling advantage in terms of higher current drive per unit area of silicon.

Conclusions

A modified Epi Defined FinFET (EDFinFET) [6] proposed here shows several advantages over conventional Ge FinFETs. Firstly, defining channel depletion through epitaxy instead of litho leads to 10X improvement in LER based variability. Secondly, very thin defect free Ge can be epitaxially grown on Si. This helps reduce I_{OFF} by reducing BTBT through QC and increase I_{ON} by biaxial strain due to mismatch in lattice of Ge and Si. Thirdly, thicker fin enhances mechanical stability and enables taller fin height that provides 4-6x higher drive current per unit footprint area. Finally, multiple V_T capability with body bias terminal can be realized that enables circuit level power optimization. Overall EDFinFET can enable high mobility Ge material integration into FinFET architecture with variety of benefits which are otherwise challenging for conventional FinFETs.

References

[1] T. Krishnamohan et.al., *VLSI Technology, 2005. Digest of Technical Papers.* , pp. 82- 83, 14-16 June 2005

[2] X. Wang et.al., *IEEE Transactions on Electron Devices*, Vol. 58, No. 8, pp. 2293–2301 (2011)

[3] E. Baravelli et.al., *IEEE Transactions on Electron Devices*, Vol. 54, No. 9, pp. 2466–2474 (2007)

[4] X. Wang et.al., *IEEE International Electron Devices Meeting (IEDM)* , Dec 2011, pp. 541-544.

[5] J. Kedzierski, et al., *IEEE Transactions on Electron Devices* , vol. 50, no. 4, pp. 952-958, Apr. 2003.

[6] S. Mittal, et al., *70th Annual Device Research Conference (DRC), 2012*, pp. 127-128.

[7] Sentaurus TCAD Design Suite. [Online]. http://www.synopsys.com

[8] G. Hellings et.al., *IEEE Transactions on Electron Devices*, Vol. 57, No. 10, pp. 2539–2346 (2010).

[9] A. Asenov, *IEEE Transactions on Electron Devices*, vol.50, no.5, pp. 1254- 1260 (2003)

[10] ITRS 2010.

[11] P. Kedar, PhD Thesis, EECS Department, University of California, Berkeley, 2010.

[12] M. C. Chiu, et al., *Proc. SPIE 7140, Lithography Asia 2008, 714021* , 2008.

[13] J. V. Hermans, et al., "Performance of the ASML EUV Alpha Demo Tool," in *Proc. SPIE 7636, Extreme Ultraviolet (EUV) Lithography*, 2010, pp. 76361L-76361L-12.

[14] K.-W. Ang, et al., *IEEE International Electron Devices Meeting (IEDM)*, Dec 2011, pp. 3551-3554.

[15] F. Assaderaghi et.al., *IEEE Transactions on Electron Devices* ,vol. 44, no.3, pp.414-422 (1997)

[16] S.A. Tawfik et.al., *IEEE Transactions on VLSI Systems* , vol.19, no.1, pp.151-156, (2011)

[17] J D Choi, *International Memory Workshop Short Course*, 2010.

FinFET Device Capacitances: Impact of Input Transition Time and Output Load

Archana Pandey, Swati Raycha, S. Maheshwaram, S. K. Manhas, S. Dasgupta, A. K. Saxena, Bulusu Anand

Department of Electronics and Computer Engineering, Indian Institute of Technology Roorkee

Roorkee, Uttarakhand - 247667 India

Tel: +91-1332-285347, Fax: +91-1332-273560, Email: anandfec@iitr.ernet.in

Abstract

FinFET devices with source drain underlaps are attractive due to their high I_{on}/I_{off} ratios [1]. However, a thorough understanding of the device parasitics on underlap FinFET circuit performance is yet to be attained. In this paper, we report a new Extension Transistor Induced Capacitance Shielding (ETICS) phenomenon. Due to this phenomenon, the effective values of a FinFET logic gate's input and parasitic capacitances depend strongly on transition times of its terminal voltages. We show that understanding of this phenomenon is essential for circuit design.

Keywords: Underlap FinFET, parasitic capacitance, three transistor equivalent circuit, capacitance shielding and inverter delay.

Introduction

FinFET is the most attractive choice amongst the novel device architectures proposed to overcome the issues in nanometer regime such as diminishing circuit performance, exponentially increasing leakage power, lower reliability and issues in scaling supply voltage etc. due to its higher gate control and relatively planar-compatible fabrication process. Researchers have worked on several ideas for improving FinFET device design and circuit topologies utilizing device's double gate structure [1]. However, a serious consideration of device parasitic effects is necessary to maximize FinFET circuit performance. This is because the relationships between device capacitances and drive current must be quantitatively known for formulating a circuit design methodology. Conventional digital circuit design methodologies [2] are based on the assumption that the values of input capacitance (C_{in}) and parasitic capacitance (C_p) of an inverter (or any logic gate) are unique for a given input voltage transition ΔV_{in} (and are very weak functions of load capacitance C_L). However, we observe that the values of C_{in} and C_p of an inverter significantly decrease and increase with an increase in the value of C_L and an increase in the value of t_{rin}, respectively due to ETICS phenomenon. We explain this phenomenon by observing that the drain extension region forms a transistor which shields a part of the device gate fringing field capacitances.

Simulation Setup

We use 2D Sentaurus TCAD [3] device and mixed-mode simulations in this work (Fig. 1). Our FinFET device structure and current values confirm to ITRS [4]. The FinFET device drawn gate length is 16nm, the length of extension regions is 20-30nm (variable), underlap length varies within the range 0-4nm (variable) and the gate oxide thickness is 1.1nm with nitride spacers. We assume that the top gate of the device has been disabled by increasing the top gate oxide thickness. The device parameters such as drive current and threshold voltage are as per ITRS roadmap targeted values. The simulation setup includes quantum mechanical carrier confinement and lattice scattering, surface scattering, dopant ion scattering, longitudinal and transverse electric field effects in its mobility models. We adjust the doping profile in the extension region to vary underlap length. In this, we assume a Gaussian reduction of doping density from drain/source pad to the channel region such that the value of standard deviation is at least 3nm/dec, as suggested in [5]. We calibrate this setup with fabricated devices [6] in Fig. 2. The value of C_{in} is extracted by integrating the current entering the input node of an inverter within 20%-80% transition of input voltage. The value of C_p is extracted using the integral of the difference of the currents through the sources of PFinFET and NFinFET for 80%-20% output transition.

Three Transistor Equivalent Circuit

A FinFET device can be considered as a series stack of three transistors (as discussed in [7] by two of the co-authors), as shown in Fig. 3(a). In this stack, we call the one formed by the fin-channel, gate oxide and the gate as the main transistor M_M. The other transistors in the stack M_{pd}/M_{ps} are formed by the drain/source underlap region within the fin, drain/source extension spacer material and the gate. Inversion electrons appear in the low-doped underlap regions of the fin extension (transistors M_{pd} and M_{ps}) due to the fringing electric field from gate to underlap regions through the spacer dielectric. In Fig. 3(b), we observe that along with the channel region ($-0.01\mu m \leq X \leq 0.0\mu m$), the electron density in underlap regions ($X < -0.01\mu m$ and $X > 0.01\mu m$) also increases by orders of magnitude when V_g is increased from $V_g = 0$ to $V_g = 0.4V$. The transistors M_{pd}/M_{ps} have values of turn-on voltages higher than that of M_M. This can be seen from Fig. 3(b) where M_{pd}/M_{ps} regions are seen to have much lower values of inversion electron density compared to M_M for a given value of gate voltage V_g. We use this equivalent circuit to explain the ETICS phenomenon.

Results and Discussion

We observe using TCAD simulations that for underlap FinFET inverters (Fig. 3(c)), the values of C_{in} and C_p for a given ΔV_{in} are strong functions of input node's transition time (t_{rin}) and C_L, as seen in Fig. 4(a) and 4(b).

A. Increase in the value of C_{in} and C_p with increasing t_{rin}

Consider an input rising transition where the value of t_{rin} is increased while keeping the value of C_L constant. If the value of t_{rin} in Fig. 5(a) is increased, the capacitor C_{gd}' of Fig. 3(a) is connected through transistor M_{pd} to the output node for a longer time. Here, transistor M_{pd} turns-off at times when input

978-1-4673-4840-9/13 $31.00 © 2013 IEEE   371

voltage attains a value V_i. If the value of C_L is kept constant, the output transition time is almost unaffected, which we also observe in Fig. 5(a). Therefore, for a given value of ΔV_{in}, the miller capacitance C_{gd}' finds a larger value of output voltage transition (V2 > V1, as marked in Fig. 5(a)) before it is disconnected from the output node. This can also be observed when we compare the change in potential within the P-device's fin at the two points, 1nm and 19nm away from gate edge respectively, for a change in input voltage ΔV_{in} from $0.2V_{dd}$ to $0.8V_{dd}$. In Fig. 7 we compare the reduction in respective potentials at these points normalized with their respective values for trin=2ps and a load FO1. We observe in the figure that for a constant FO (FO1), change in potential at 19nm away from gate edge is almost independent of changes in t_{rin}, signifying that the Miller capacitance introduced by C_{gd} (Fig. 3(a)) is unaffected by changes in t_{rin}. Whereas, the change in potential at 1nm away from gate edge increases as t_{rin} is increased resulting in an increased value of Miller capacitance due to C'_{gd} as t_{rin} is increased. This is the reason behind the phenomenon that we observe in Fig. 4. This increases the effective values of C_{in} and C_p.

B. Decrease in the values of C_{in} and C_p with increasing Fan-Out FO (or C_L):

Consider the case (Fig. 5(b)) where the value of C_L (or FO) is increased while keeping the value of t_{rin} constant. If the value of C_L in Fig. 5(b) is increased the output transition time increases, as we observe in Fig. 5(b). However, if the value of t_{rin} is kept constant, the time for which the capacitance C_{gd}' of Fig. 3(a) is connected through transistor M_{pd} to the inverter's output node does not increase. Therefore, for a given value of ΔV_{in}, the miller capacitance C_{gd}' finds a smaller value of output voltage transition (V2 < V1, as marked in Fig. 5(b)) before it is disconnected from the output node. We observe in Fig. 6 that the change in voltage at the point 1nm away from gate edge is

larger than the change in voltage at the point 19nm from gate edge due to the ETICS phenomenon. This decreases the effective values of C_{in} and C_p.

We perform the same experiment for various fin widths. The impact of this phenomenon increases with reducing fin thickness W_{fin} (Fig. 7) due to an increase of the effective resistance of the extension regions.

Conclusion

The values of C_{in} and C_p of an inverter significantly decrease and increase with an increase in the value of C_L and an increase in the value of t_{rin}, respectively (Fig. 4). Hence, the actual and effective loading will differ for various FO and t_{rin} (Table 1). The large difference between the actual FO and effective extracted FO is due to the large parasitic capacitance in FinFETs and varies significantly with t_{rin} and actual FO. The effective values of FO thus obtained are significant for determining the number of fins of a driver inverter where the delay is pre-specified. This would not allow a use of existing delay estimation methods. Hence an efficient FinFET circuit design methodology would need quantitative modeling of ETICS phenomenon.

Acknowledgement

Authors acknowledge Department of Science and Technology, Government of India, for its financial support.

References

[1]A. B. Sachid et al, IEDM Tech. Dig., p. 1, (2008).
[2]I. E. Sutherland et al, *Logical effort: Designing fast CMOS circuits*, Morgan Kaufmann Publisher, 1998.
[3]*Sentaurus TCAD Manuals, Synopsys* Inc., Mountain View, CA, 2009, (ver.2009.06).
[4] The International Technology Roadmap for Semiconductors, 2010.
[5]V. Trivedi, J. G. Fossum, , M. M. Chodhury, "Nanoscale FinFETs with Gate-Source/Drain Underlap," *IEEE Trans. on Electron Devices*,vol.52, no.1,pp. 56-62, Jan. 2005.
[6]K. von Arnim et al, Symp. on VLSIT Dig, p.106, (2007).
[7]S. Nema et al, Proc. ICCCD 2010, IIT Kharagpur.

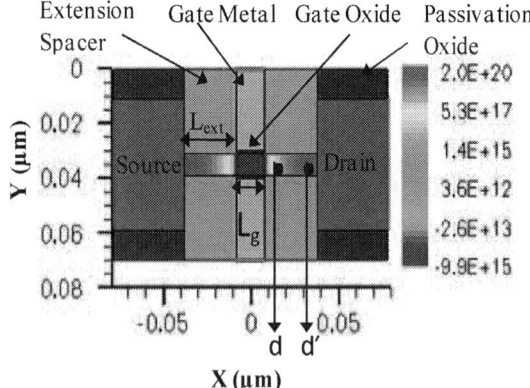

Fig. 1. Underlap NFinFET device structure

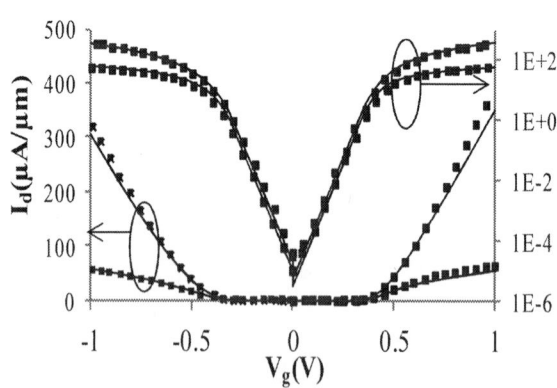

Fig. 2. Calibration of TCAD models with experimental data at L_g=75nm[6]. Solid line and symbols denote data from our simulations and the experiments report, respectively.

978-1-4673-4840-9/13 $31.00 © 2013 IEEE

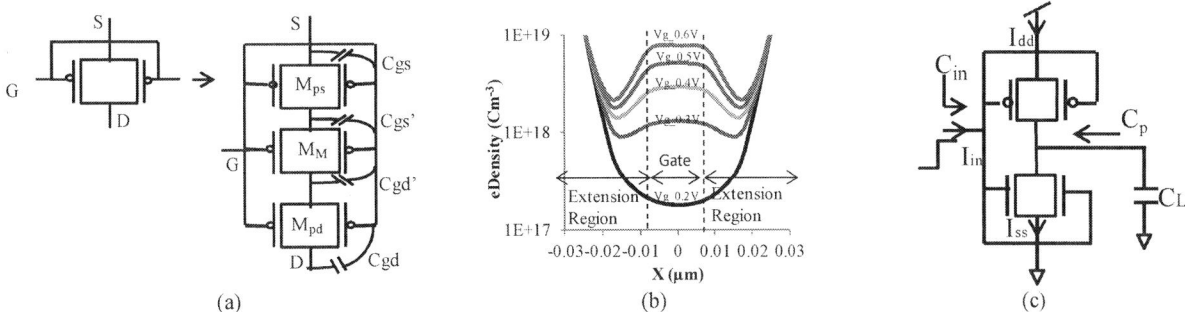

Fig. 3. (a) Schematic of an underlap PFinFET and its three transistor equivalent circuit. (b) Electron density variation with gate voltage in the main and parasitic transistors. (c) FinFET inverter circuit schematic.

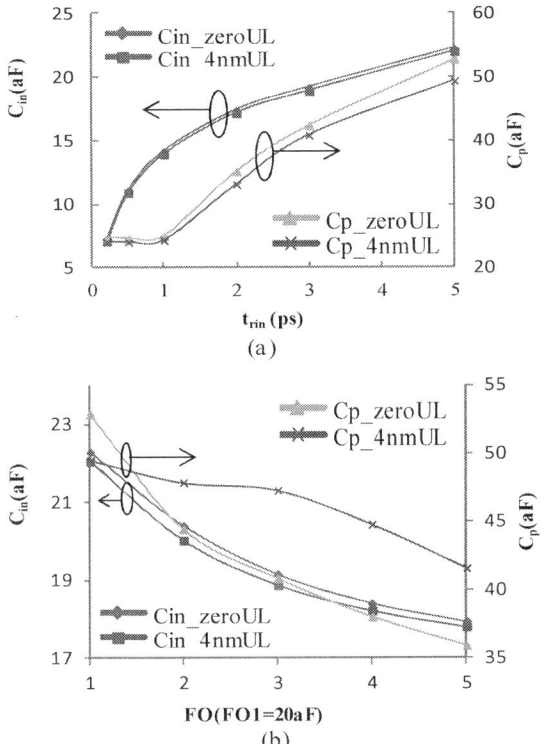

Fig. 4. Variation of C_{in} and C_p with (a) Trin (C_L=20aF) (b) FanOut (Trin=5ps) (UL=underlap).

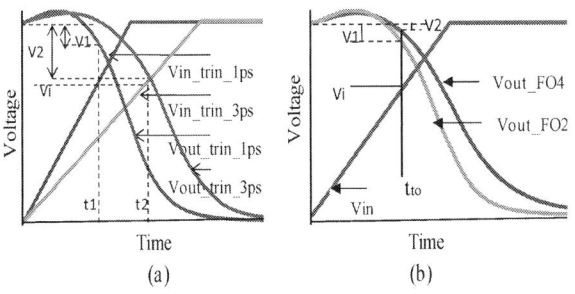

Fig. 5. Inverter input and output voltage transitions obtained using simulations (a) Increasing t_{rin} keeping FO constant. Output transition time varies by a negligible 3%. (b) Increasing FO keeping t_{rin} constant.

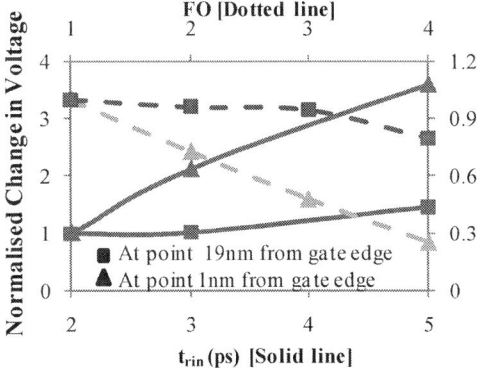

Fig. 6. Variation of the change in voltage at points d and d' (shown in Fig. 1) with increasing C_L and with t_{rin} for a constant ΔV_{in}. (UL=4nm). The change has been normalized with respect to its value at t_{in}=2ps/FO1 (solid/dotted) at the respective point.

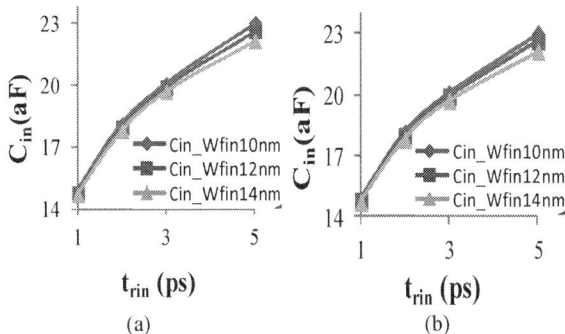

Fig. 7. (a) C_{in} versus t_{rin} (b) C_{in} versus FO. for several values of fin widths.

TABLE I

EFFECTIVE FAN OUT VALUES OF A FINFET INVERTER CONSIDERING CP VS CL AND CP VS T_{RIN} VARIATION (UNDERLAP LENGTH = 4NM)

Actual FO	Effective FO (t_{rin} = 3 ps)	Trin (ps)	Effective FO (Actual FO=4)
1	3.04		
2	3.89	1	5.21
3	4.74	2	5.31
4	5.63	3	5.63
5	6.55	5	6.24

978-1-4673-4840-9/13 $31.00 © 2013 IEEE

Droplet based lab-on-chip microfluidic microsystems integrated nanostructured surfaces for high sensitive mass spectrometry analysis

Guillaume Perry[1,2], Florian Lapierre[1,2], Yannick Coffinier[2], Vincent Thomy[1], Rabah Boukherroub[2], CongXiang Lu[3], Siu Hon Tsang[3,4] Beng Kang Tay[3,4] and Philippe Coquet[1,4]

[1] Institut d'Electronique, de Microélectronique et de Nanotechnologie (IEMN, UMR CNRS 8520), Université Lille 1, 59652 Villeneuve d'Ascq, France
[2] Institut de Recherche Interdisciplinaire (IRI, CNRS-USR 3078), Université Lille 1, 59658 Villeneuve d'Ascq, France
[3] School of Electrical and Electronics Engineering, Nanyang Technological University, Singapore 639798
[4] CINTRA CNRS/NTU/THALES, UMI 3288, Research Techno Plaza, 50 Nanyang Drive, Singapore 637553

Abstract

We present in this paper a microsystem coupling electrowetting on digital microfluidic and nanostructured surfaces for matrix-free Laser Desorption/Ionization Mass Spectrometry (LDI-MS) analysis of small biomolecules. Silicon nanowires are processed to form highly sensitive pads for LDI analysis and also to produce superhydrophobic surfaces for enhanced transfer of droplets containing the analytes to the analyzing pads. By this way, analysis of low molecular weight compounds with high sensitivity can be achieved. In addition, wetting properties of carbon nanotubes surfaces are investigated in the perspective of further increasing the detection performances.

Keywords: Silicon Nanowires, Carbon Nanotubes, Superhydrophobic Surfaces, EWOD, Mass Spectrometry Analysis.

Introduction

Since the emergence of Lab-On-Chip applications for biomedical protocols, electrowetting on dielectric-digital microfluidic (EWOD-DMF) systems have been implemented in many domains such as enzyme assays, immunoassays, DNA-based applications, cell-based assays, tissue engineering and proteomics [1,2]. Meanwhile, Digital MicroFluidic (DMF) Lab-On-Chip devices have been coupled with off-line Matrix-Assisted Laser Desorption/Ionization Mass Spectrometry (MALDI-MS) analysis as an alternative to optical methods [3]. MALDI-MS allows the analysis of a wide variety of compounds, including polymers, peptides and proteins. Even though this is a pioneering system in LoC applications for MS analysis, several drawbacks can be underlined: protein or peptide concentrations remain are still relatively high (> nmol/µL) and the analysis protocol is somewhat time consuming and complicated due to the manipulation of viscous matrix and the drying of proteins on the surface. Due to these drawbacks, a lot of work has been devoted to the development of matrix-free laser desorption–ionization techniques through

replacing the matrix with nanoparticles or micro/nanostructured surfaces.

In this paper, we present an original study consisting of coupling EWOD-DMF and matrix-free Laser Desorption/Ionization Mass Spectrometry (LDI-MS) analysis using superhydrophobic nanostructured silicon substrates. Compared to a classical hydrophobic surface, it leads to an improvement of the microfluidic actuation (both in terms of minimal applied voltage threshold and droplet speed) and it allows a matrix-free LDI-MS analysis of very low concentrations of peptide samples (down to fm/µL) through a rapid and simple protocol (without addition of any organic matrix) [3-7]. Then we describe our preliminary results dealing with wetting and electrowetting properties of nanostructured carbon nanotubes surfaces.

Material and Methods

A. ElectroWetting

Among the different techniques used for liquid motion, droplet-based displacement has generated a huge interest due to its convenient and efficient properties compared to the classical single-phase continuous flow. The most promising strategy to displace droplets on a programmable platform is to use ElectroHydroDynamics (EHD) force, in particular, ElectroWetting-On-Dielectric (EWOD). It relies on the modification of a liquid droplet–solid surface contact angle (CA) by applying an electrical potential between the droplet and the substrate. Through the use of a simple electrode network, EWOD permits the control of the trajectory of conductive liquid droplets in real time. Allowing fluidic basic operations such as merging, coalescing and mixing, which pave the way to Lab-on-Chip applications. The EWOD system is comprised of two parts: a hydrophobic base consists of an electrode network and a conductive counter-electrode, as shown in Fig.1. The gap between these two plates is 300µm so that a 1µL droplet can be displaced at 100 V_{TRMS} @1kHz.

The device was fabricated using standard micro-fabrication

978-1-4673-4840-9/13 $31.00 © 2013 IEEE

techniques. The base plate consists of an array of independently addressable control electrodes (1 mm²) patterned in a 10 nm thin layer of nickel, which is further coated with a SU-8 layer (2 μm) as a dielectric. Electrical contacts are SU-8 free to facilitate voltage supply towards the electrodes. Finally, a thin hydrophobic layer of Teflon (AF 1600, 20 nm) is spin coated to complete the base. It results in a contact angle and a contact angle hysteresis of 115° and 11°, respectively.

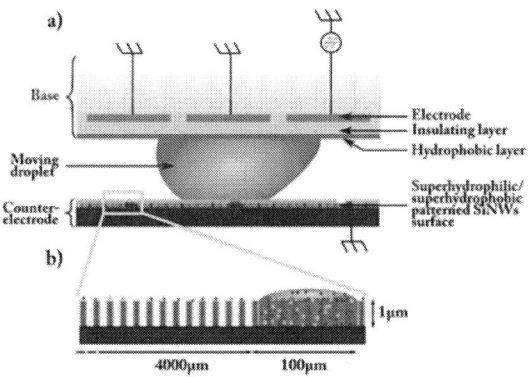

Fig. 1 a) EWOD set up integrating superhydrophobic counter electrode with superhydrophilic pads. b) Illustration of the liquid impregnation in the superhydrophilic areas.

B. Counter Electrode - Superhydrophobic surfaces

The originality of our system relies on the use of superhydrophobic surfaces with a contact angle hysteresis close to 0° as counter-electrode. It reduces the drag friction, leading to low voltage displacement and self-cleaning properties, thus limiting non-specific adsorption of biomolecules along the droplet pathway. The superhydrophobic surfaces consist of silicon nanostructures chemically modified with octadecyltrichlorosilane (OTS). First, the silicon wafer was degreased in acetone and isopropyl alcohol, rinsed with deionized water and then cleaned in a piranha solution. The clean substrate was immersed in HF (5% i.e. 2.625 M)/AgNO₃(0.005 M) aqueous solution at 54°C for 30 min, leading to a wet etching of the Si wafer. The resulting surface was rinsed copiously with deionized water and immersed in an aqueous solution of HCl/HNO₃/H₂O (1:1:1, v/v/v) at room temperature overnight to remove the silver nanoparticles and dendrites deposited during the chemical etching. The nanowires have a diameter in the range of 10-200 nm and are 1 μm in height, as shown in Fig. 2. This surface morphology gives optimum LDI-MS performance [4]. To achieve the superhydrophobicity, the silicon nanowire surface was chemically modified with OTS. First, the substrate was UV/ozone-treated for 30 min to remove any organic contaminants on the surface and to generate surface hydroxyl groups. The surface was then reacted with a 10⁻³ M OTS solution in hexane for 4 h at room temperature in a dry nitrogen purged glovebox. The resulting surface was rinsed with CH₂Cl₂, i-PrOH and dried under a gentle stream of nitrogen. The combination of the roughness with hydrophobic coating leads to a superhydrophobic character with 160° contact angle and 0° contact angle hysteresis.

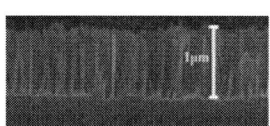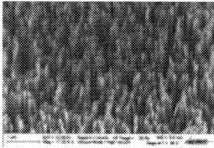

Fig. 2 SEM images of silicon nanowire substrate prepared by chemical etching of Si (100) in HF-AgNO₃ aqueous solution. Left: cross view. Right: tilted view.

C. Counter electrode – Superhydrophilic pads

To allow analyte deposition on localized plots, UV-ozone through a physical mask, containing 150 μm diameter apertures, was performed to remove locally the OTS on the SiNWs surface. By this way, we can control locally the wetting properties of a macroscopic superhydrophobic surface. We denoted the inside zone or area as the superhydrophilic patterns and the outside zone or area as the superhydrophobic part of the cover plate. The dimensions of the superhydrophilic area allow droplet displacement and transfer of a small amount of liquid for the analysis.

D. Operating principle

A 1 μL droplet containing the analyte is encapsulated between the superhydrophobic counter-electrode, connected to the ground, and the base electrode. The droplet is then displaced along the microsystem by EWOD-DMF. The displacements are carried out using a LabView program that multiplexes an electrical square voltage (100 VTRMS, 1 kHz) to an array of 96 electrodes (4 mm² in size). During the droplet actuation, a small amount of the solution is adsorbed on the superhydrophilic areas, Fig. 1b. A rinsing step with deionized water droplet is performed after the sample deposition. This procedure enhances the Signal to Noise (S/N) ratio, via a re-suspension/re-deposition step. The system is then opened to remove the droplet and the counter electrode is taken to perform off-line LDI-MS analysis.

Mass Spectrometry Analysis

LDI-MS analysis was performed using a Voyager-DE-STR time-of-flight (ToF) mass spectrometer (Applied Biosystem) with delayed extraction, operating with a pulsed N2 laser at 337 nm (3 ns pulse). Superhydrophobic cover substrates were attached to the usual MALDI-MS target using conductive double-side carbon tape. Positive ion mass spectra were acquired in a reflector mode of operation, an accelerating potential of 20 kV and a grid voltage at 73%. Each spectrum is the result of 10 laser pulses.

Results

The sample droplet is a 1 μL mixture of peptides Mix1, obtained from a sequazyme peptide mass standard kit of AppliedBiosystems AB) (Part number P2-3143-00). The characteristics of the peptides are detailed in Table 1. It is to be noted that the neurotensin concentration is 5 times lower than the other peptides. These peptide solutions were then diluted in a 1 mM ammonium citrate aqueous solution to 50, 25 and 10 fmol mL^{-1} concentrations.

Molecule	m/z	pI	Charges	Sequence	pH
Des-Arg-Bradykinin	904.5	10	1 (+)	PPGFSPFR	5.5
Angiotensin I	1296	7.4	2 (+)	DRVYIHPFHL	
Fibrinopeptide B	1570	4.0	3 (−)	EGVNDNEEGFFSAR	
Neurotensin	1672.9	8.7	1 (+)	ELYENKPRRPYIL	

Table 1. Properties of the investigated peptides. The amino-acid colors in the peptide sequence correspond to (red) acidic residues, (blue) basic residues, (green) hydrophobic uncharged residues and (black) to other residues.

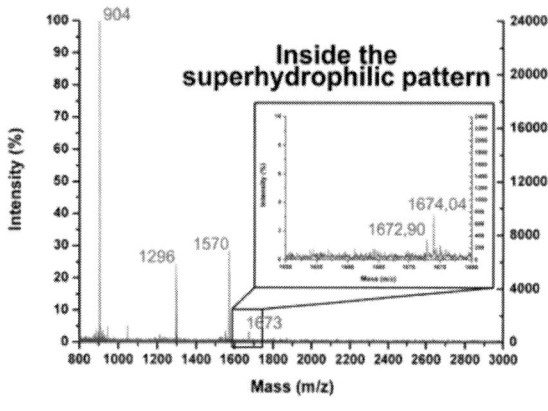

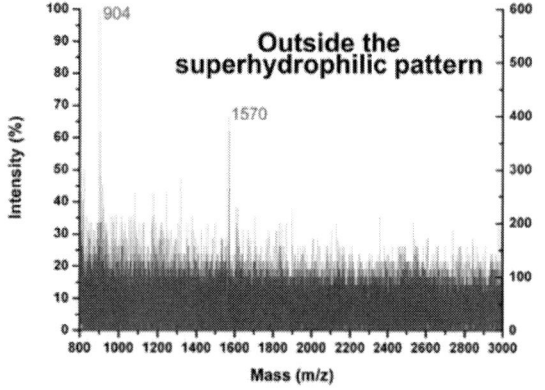

Fig. 3. Mass spectra of a peptide mixture (25 fmol mL^{-1}) deposited after 1 displacement by EWOD on 5 electrodes, (a) inside the superhydrophilic and (b) on the superhydrophobic surface. The interaction time of the droplet mixture with the superhydrophilic pattern is set at 2 ms.

Fig. 3 displays the LDI-MS spectra for the experiment with the displacement of Mix1 droplet inside and outside of the superhydrohilic pattern. The noise floor is dramatically reduced inside the superhydrophilic pattern. Inside this pattern, all peptides are detected, including those present at a low concentration. This illustrates the high sensitivity that can be obtained with this system. Compared to those data obtained without rinsing step (data not shown), rinsing with a water droplet can remove a significant amount of the peptides adsorbed on the superhydrophobic surface. These results are a

proof of concept of a rapid, easy to handle and highly sensitive MS analysis technique. To further increase the sensitivity of detection, we are now investigating the use of carbon nanotubes (CNT) as an alternative to silicon nanowires.

Nanostructured Carbon Nanotubes surfaces

The surface morphology and thermal properties of the nanostructures are found to be essential features contributing to the Desorption/Ionization performances [4]. In this context, CNTs show superior advantages among other nanostructure materials, such as large aspect ratio, excellent thermal and electrical conductivity. Hence, by controlling the diameter and the density of the CNT, a much wider spectrum of tunable surface energy can be achieved [8]. Furthermore, the ease of chemical functionalization of CNT allows almost any desired chemical species to be attached and enhance the molecular selectivity and biocompatibility for a larger range of applications. Fig. 4 shows the results of the contact angle measurement on CNT surfaces obtained with 3 different approaches: low pressure CVD (LPCVD) method, plasma assisted CVD (PECVD) method and a newly developed method that produces 3D Network CNT [9]. The surface energy is strongly dependent on the morphology of CNT structure; the contact angle varies from 7.6° (PECVD mode) to 144.6° (LPCVD mode). It demonstrates a full range of tunability and precision control of the surface energy as a foundation for the future integration of these CNTs based surfaces in the EWOD microsystem for LDI-MS experiments.

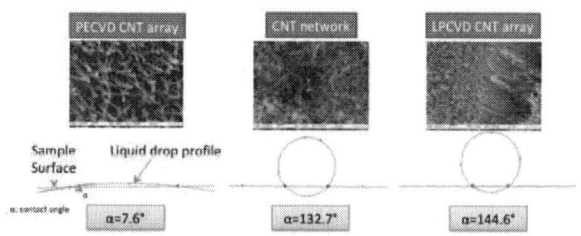

Fig. 4. Results of the contact angle measurement for (a) PECVD CNTs, (b) CNT network and (c) LPCVD CNTs.

Conclusion

This proof of concept of a rapid, easy to handle and highly sensitive MS analysis technique for Lab-on-Chip has now to be pursued. To reach a lower limit of detection, EWOD could be used to induce micromixing, enhance the mass transfer toward the superhydrophilic areas. Furthermore, to improve the device performance, we can introduce a high level of selectivity and thus a specific recognition toward biological targets: i.e. selecting one specific biomolecule among complex biological fluids via an appropriate surface chemistry inside the apertures. Another approach will be replacing the silicon nanowires with nanostructured carbon nanotubes surfaces. Preliminary results have shown the possibility of easily controlling their wetting properties from

978-1-4673-4840-9/13 $31.00 © 2013 IEEE

superhydrophilic to superhydrophobic. Furthermore, the excellent thermal properties of the CNTs could be of great interest to enhance the desorption/ionization performances.

References

[1] L. Malic, D. Brassard, T. Veres and M. Tabrizian, *Lab Chip* 10 (2010) 418-431.

[2] V. Srinivasan, V. Pamula, and R. Fair, *Lab Chip* 4 (2004) 310-315.

[3] A. R. Wheeler, H. Moon, C. A. Bird, R. R. O. Loo, C. J. Kim, J. A. Loo, R. L. Garrell, *Anal. Chem.* **77** (2005) 534

[4] G. Piret, H. Drobecq, Y. Coffinier, O. Melnyk and R. Boukherroub, Langmuir, 2010, 26, 1354–1361.

[5] F. Lapierre, G. Piret, H. Drobecq, O. Melnyk, Y. Coffinier, V. Thomy and R. Boukherroub, *Lab Chip* 11 (2011) 1620-1628.

[6] M. Jönsson-Niedziolka, F. Lapierre, Y. Coffinier, S. J. Parry, F. Zoueshtiagh, T. Foat, V. Thomy, R. Boukherroub, *Lab Chip* 11 (2011) 490-496

[7] G. Perry, V. Thomy, M. R. Das, Y. Coffinier and R. Boukherroub, *Lab Chip* 12 (2012) 1601-1604

[8] C.L. Choong, W.I. Milne, K.B.K. Teo, *Int. J. Mater. Form*, 1 (2008) 117-125

[9] J. Wei, K. P. Yung, and B. K. Tay, *Method of Forming A Self Assembled, Three Dimensional Carbon Nanotube Network and Networks So Formed*, PCT/SG2009/000274.

Impact of metal contact on the performance of cupric oxide based thin film solar cells

[1,2]S. Masudy-Panah, [1]V. Kumar, [2]C. C. Tan, [1]K Radhakrishnan, [2]D. Z. Chi, and [2]G. K. Dalapati[*],

[1]School of Electrical and Electronic Engineering, Nanyang Technological University, Nanyang Avenue, Singapore 639798
[2]Institute of Materials Research and Engineering, A*STAR (Agency for Science, Technology and Research), 3 Research Link, Singapore 117602
*Corresponding author: dalapatig@imre.a-star.edu.sg

Abstract

In this paper we have studied the effects of the contact material on the photovoltaic characteristics of p-CuO/n-Si solar cells fabricated by using the RF sputtering method and annealed by rapid thermal process. We have fabricated three types of solar cells with different front contact. Type (I) is the p-CuO/n-Si solar cells with Cu front contact, Type (II) is the p-CuO/n-Si solar cells with Al front contact and Type (III) is the p-CuO/n-Si solar cells with Cu followed by Al front contact. It has been shown that the Cu/Al as a front contact to the p-CuO can improve the performance of p-CuO/n-Si solar cells. It has been shown that by using the Cu/Al as the front contact the conversion efficiency increases by 70% compared to the Cu or Al contacts.

Key words— CuO thin film solar cells, front contact

Introduction

Cupric oxide (CuO) is an attractive material for photovoltaic devices and has created centre of attention for solar cell applications due to its high optical absorption, well matched band gap with the solar spectrum, abundance in the earth crust, low cost fabrication and nontoxic constitute [1]-[4]. Theoretically, the maximum energy conversion efficiency of a cupric oxide solar cell is 20% [4]. Although various type of homojunction and heterojunction solar cells have been fabricated using cupric oxide [1–4], efficiency of them is very poor. The main reasons for low efficiency of CuO solar cells are the poor quality CuO, interface defects between active coper oxide absorption layer and substrate as well as the copper oxide layer and contact [4]-[5].

In this paper we have studied the effects of contact material on the efficiency of cupric oxide based solar cells and the stability of them. We have fabricated the CuO based solar cells on the Si substrate to address the effects of contact material on the efficiency of our fabricated solar cells.

Experiment

Si substrate were ultrasonicated with deionized water using the standard RCA cleaner for 5 min and dried with nitrogen gas blowing. Then Si substrate were rinsed with 2% dilute HF for 2 min to remove native silicon dioxide and dried with nitrogen gas blowing. Subsequently, substrate has been immediately loaded into the sputtering chamber. The evacuation pressure was 3×10^{-7} Torr.

Poly crystalline CuO thin films were sputtered on the single crystalline n-type Si (100) substrates (5Ωcm) by RF magnetron sputtering. Sputtering was performed at room temperature.

Deposition of CuO was performed at base pressure of 3.3×10^{-3} Torr. Pure argon gas has been used as ambient gas and its flow rate was 25sccm. The profilometer has been used to measure the thickness of deposited CuO and the thickness of deposited CuO was around 200nm. All the samples were annealed in nitrogen ambient using the rapid thermal annealing (RTP) technique for 1min at the temperature 300°C. Then the aluminium back ohmic contact for the silicon substrate was deposited by DC magnetron sputtering through the mask. The evacuation pressure for contact deposition was 2×10^{-7} Torr. Sputtering was performed at room temperature, argon ambient and DC power of 200w.

Results and Discussions

The commonly used material for the front contact of copper oxide solar cells that has been utilized by researcher is copper. [3]. The procedure that has been used for fabricating the front contact of our samples is similar as Al back contact. In Table I we have compared the circuit current, open circuit voltage, fill factor and efficiency of our fabricated solar cells with the Al and Cu front contact.

As shown at this table Al shows better open circuit voltage and short circuit current when we compare it with copper.

Using the Al front contact can improve the efficiency of the solar cells by almost 30%.

Table I. I_{sc}, V_{oc}, FF and η of the CuO based solar cells with Al and Cu front contact

Front contact	Cu	Al
V_{oc}(V)	0.27	0.3
I_{sc}(mA/cm^2)	0.7	1.1
FF	0.3	0.35
η	0.1%	0.13%

The main problem of Al front contact for the CuO based solar cells is its stability. In Table II we have compared the current voltage characteristics of the same devices after two days. As shown at this table the short circuit current, open circuit voltage and consequently the efficiency of solar cells with the Al front contact drastically degraded. The efficiency degradation of solar cells is mainly because of an increase in the series resistance [5] and recombination [6].

Table II. I_{sc}, V_{oc}, FF and η of the CuO based solar cells with Al and Cu front contact two days after contact forming

Front contact	Cu	Al
V_{oc}(V)	0.26	0.2
I_{sc}(mA/cm^2)	0.69	0.6
FF	0.3	0.3
η	0.099%	0.08%

To solve this problem and enjoy the higher short circuit current of the solar cells with Al front contact we need to do metal contact engineering. To do it we have used both Al and Cu. For this propose we sputtered copper followed by Al as front contact.

The experimental procedure of forming front contact with the two different materials is as follow;

Deposition of front contact material was performed at evacuation pressure of 2×10^{-7} Torr and base pressure of 5.3×10^{-3} Torr. Pure argon gas has been used as ambient gas and its flow rate was 25sccm. Then the copper was deposited by DC magnetron sputtering through the mask. The evacuation pressure was 2×10^{-7} Torr. Sputtering was performed at room temperature, argon ambient and DC power of 250W. After that Al was deposited by DC magnetron sputtering through the same mask. Sputtering was performed at room temperature, argon ambient and DC power of 200W.

In Table III we have compared the photocurrent of the solar cells with Cu, Al and Cu followed by Al. All the characteristics have been measured two days after fabrication. As shown at this table by using the copper as the interfacial layer between the Al and copper oxide not only the stability of this kind of contact improves but also this kind of contact shows better short circuit current and consequently better efficiency. As shown in this table using the copper followed by Al front contact can improve the efficiency of fabricated solar cells by almost 70%.

Table III. I_{sc}, V_{oc}, FF and η of the CuO based solar cells with Al and Cu and Cu/Al front contact two days after contact forming

Front contact	Cu	Al	Cu/Al
V_{oc}(V)	0.26	0.2	0.32
I_{sc}(mA/cm^2)	0.69	0.6	1.4
FF	0.3	0.3	0.38
η	0.099%	0.08%	0.17%

Conclusion

In this paper we have studied the effects of the contact material on the photovoltaic characteristics of p-CuO/n-Si solar cells fabricated by using the RF sputtering method. It has been shown that the front contact material has the great impact on the performance and efficiency of CuO based solar cells. By optimizing the structure of the front contact we could improve the efficiency of the p-CuO/n-Si heterojunction solar cell by 70%.

REFERENCES

[1] H. Kidowaki, T. Oku and T. Akiyama, "Fabrication and characterization of CuO/ZnO solar cells," *J of Phys: Conference Series.*, vol. 352, pp. 1-5, 2012.

[2] Yee-Fun Lim, Joshua J. Choi, and Tobias Hanrath. "Facile Synthesis of Colloidal CuO Nanocrystals for Light-Harvesting Applications," *Journal of Nanomaterials.*, vol. 2012, pp. 1-5, 2012.

[3] F. Gao, X. Liu, J. Zhang, M. Song, and N. Li, "Photovoltaic properties of the p-CuO/n-Si heterojunction prepared through reactive magnetron sputtering," *Journal of Applied Physics.*, vol. 111, pp. 084507- 084510, 2012.

[4] T. Oku, R. Motoyoshi, K. Fujimoto, T. Akiyama, B. Jeyadevan, J. Cuya, "Structures and photovoltaic properties of copper oxides/fullerene solar cells," *Journal of Physics and Chemistry of Solids.*, vol. 72, pp. 1206-1211, 2011.

[5] A. Mittiga, E. Salza, F. Sarto, M. Tucci, and R. Vasanthi, "Heterojunction solar cell with 2% efficiency based on a Cu$_2$O substrate," *Applied Physics Letters*, vol. 88, pp.163502-163503,2006.

[6] Steven S. Hegedus, Brian E. McCandless, "CdTe contacts for CdTe/CdS solar cells: effect of Cu thickness, surface preparation and recontacting on device performance and stability," *Solar Energy Materials & Solar Cells*, vol. 88, pp.75- 95, 2005.

Through-silicon via Fabrication with Pulse-reverse Electroplating for High Density Nanoelectronics

Nay Lin and Jianmin Miao

School of Mechanical and Aerospace Engineering, Nanyang Technological University
50 Nanyang Avenue, 639798 Singapore
mjmmiao@ntu.edu.sg

Abstract

In this paper, fabrication of through-silicon vias (TSV) with different diameters ranging from 60 to 150 μm is reported. It was observed that at the low current density of 20 mA/cm^2, all the through-holes with different diameters are filled with copper without voids and pores. At higher current density of 40 mA/cm^2, however, the pillars with diameters bigger than 100 μm tend to have voids at the middle portion of pillars. Focused ion beam (FIB) examination of the copper pillars fabricated with low current density reveals the difference in grain size and internal structure of the grain along the length of the pillar. Current-potential characters of solution were studied for the electrolyte bath used in the process. It shows the limiting current density around 40-60mA/cm^2. The microstructures of TSV fabricated at low and high current densities are investigated and it shows that high current density produces porous copper with void at the core of TSV.

(keywords: TSV, copper, pulse reverse electroplating)

Introduction

With the current rate of integration towards smaller and more powerful devices, the through-silicon via (TSV) interconnection technology is one of the most promising candidates for a new generation of electronic and integrated micro-systems [1]. Due to its superiority in electrical performance such as low electrical resistivity [2], high resistance to electromigration [3], copper has replaced the aluminum interconnection technology for many years for the single device layer chips. The three-dimension integration or chip stacking technology [4] will continue to deploy copper as a mainstream material for wafer level packaging due to its high performance, low cost and existing manufacturing technologies. The development of three-dimensional integration also provides opportunities for device and system designers to develop new device architectures such as three-dimensional radio frequency RF MEMS [5]. In order to accurately provide the designer with appropriate fabrication related parameters, the characterization of both electrical and mechanical properties of the through-silicon vias is extremely important. The developments of three dimension hanging structures require the mechanical properties of TSV to be well characterized and controlled in order to meet the stringent mechanical reliability requirements of the MEMS devices. The mechanical properties of the copper deposit are directly related to the microstructure which can be controlled by varying the process parameters during the electrodeposition. Recent research in the pulse and pulse reverse electrodeposition shows the exciting development of nano-twinned copper with high

mechanical strength and ultra-low electrical resistivity [6]. The components and devices such as electrical interconnection and three dimensional RF inductors require materials with such properties for their excellent performance. Low resistivity of the material provides low resistive losses dissipated in the form of thermal energy. In previous studies, pulse-reverse plating with aspect ratio dependent current modulation technique was shown to be able to fabricate high mechanical strength through-silicon via interconnections [7]. In this technique, low current density was used at the beginning of the plating with increasing forward current density as the plating continues. Ab-initio total energy calculation with supercell approach shows a thermodynamically favorable configuration of nanotwinned crystal during electrodeposition [8,9]. The formation of nanotwinned structures in copper grains can improve the mechanical properties of copper without sacrificing the electrical resistivity. In this paper, microstructures of copper TSV fabricated below and above the limiting current density are reported.

Experimental Details

Fabrication steps- Schematics of fabrication of through wafer interconnect are shown in Fig. 1. Silicon wafers of 100 mm diameter and 450 μm thickness were patterned with photolithography to form the holes of diameters ranging from 80 to 150 μm followed by temporary bonding to a support wafer to perform deep etching process. Through-vias with diameters ranging from 80 to 150 μm were etched through by DRIE process, followed by 2 μm silicon oxide growth by wet thermal oxidation process. Device wafer was temporally bonded to a contact wafer by positive photoresist AZ9260. Photoresist underneath the device wafer was cleared by combination of UV exposure, developer and oxygen RIE plasma. Bottom-up electroplating was performed in a home-made electroplating setup as reported in the previous article [4]. The electrolyte chemistry consists of copper 40 g/L, sulfuric acid 150 g/L, chloride 50 mg/L, leveler 15 ml/L and brightener 10 ml/L. Pulse-reverse current cycle was used throughout the plating process. Duration of forward and backward current is 2 seconds and 0.1 second, respectively as shown in Fig. 2 and the total electroplating time was 10 hours to fill the through holes. Since the agitation or forced convection has strong influence on the electroplating uniformity, the agitation by rotation at 20 rpm was applied. However, it should be noted that the actual intensity of agitation in the deep trenches and vias differ significantly from the applied agitation. Therefore, the limiting plating current density would be much lower at the bottom of deep aspect ratio via due to its mass transfer limitation or lack of convection

978-1-4673-4840-9/13 $31.00 © 2013 IEEE

which comes with agitation.

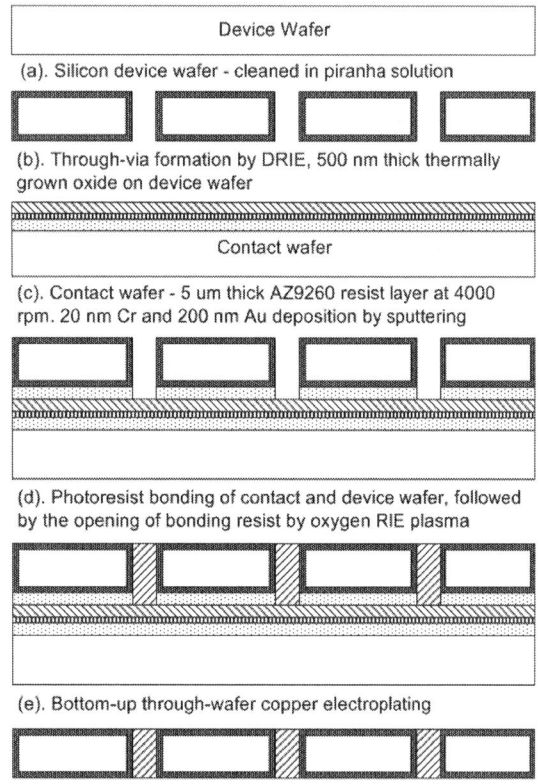

(a). Silicon device wafer - cleaned in piranha solution

(b). Through-via formation by DRIE, 500 nm thick thermally grown oxide on device wafer

(c). Contact wafer - 5 um thick AZ9260 resist layer at 4000 rpm. 20 nm Cr and 200 nm Au deposition by sputtering

(d). Photoresist bonding of contact and device wafer, followed by the opening of bonding resist by oxygen RIE plasma

(e). Bottom-up through-wafer copper electroplating

(f). Wafer separation by removing bottom bonding resist in acetone, followed by seed layer etching and polishing

Fig. 1 Bottom-up fabrication process for through-silicon via.

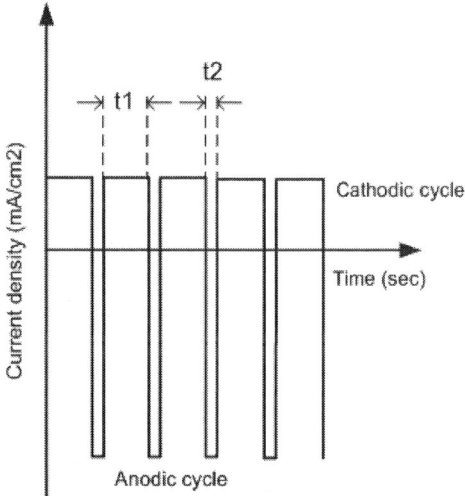

Fig. 2 Current modulation applied for fabrication of TSV.

Fabrication results and discussion

Fig. 3 shows the SEM picture of cross sections of the TSV with varying diameters from 60 to 150 μm fabricated with current density of 20 mA/cm². It can be observed from the figure that the all the through-holes are properly filled with copper.

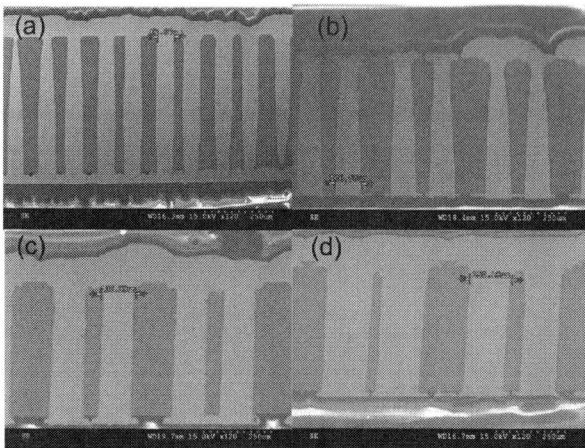

Fig. 3 SEM picture of electroplated TSV of (a) 60 (b) 100 (c)130 (d) 150 μm diameter.

As can be seen from the figures, the over-plating of the copper is relatively constant for the all portions of the wafer with various diameters of TSV. Therefore, the chemical mechanical polishing (CMP) step which usually follows after the plating process could be achieved easily.

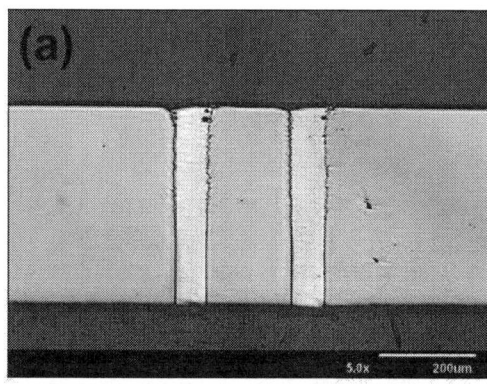

Fig. 4 Optical picture of (a) fully filled TSV of 60 μm diameter (b) voided 130 μm diameter.

Fig. 4 shows the fabricated TSVs of the 60 and 120 μm diameter. Although the pillars with smaller diameters such as 60 μm are completely filled with copper, the larger diameter through-holes have a higher chance to have voids at the middle of the TSV at higher current densities.

978-1-4673-4840-9/13 $31.00 © 2013 IEEE

Electrochemical studies

In order to study the underlying electrochemical nature of plating, electrochemical current potential-curves are measured in the copper sulfate solution with increasing concentrations of brightener and leveler (Two organic additives were used to achieve void free filling of TSV). The measurements were done with three electrode system using Ag/AgCl electrode as reference. Commercially pure copper plate is used as the anode. The measurements were done under the well agitated small scale electrochemical cell. Magnetic agitation was used to ensure the well stirred plating condition during the experiments. The scanning was done at the rate of 0.2 V/sec.

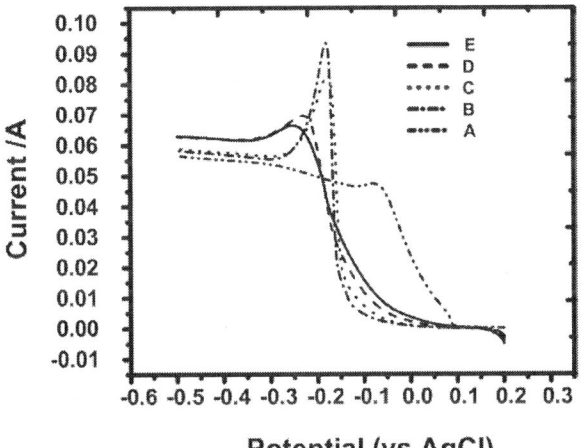

Fig. 5 Current-potential curves for various concentrations of brightener and leveler.

Fig. 5 shows the current-potential curves of the copper sulfate solution with increasing concentrations of brightener and leveler. Curve A represents the solution without any additives. Curves B and C show the current-potential curves with the brightener solution at the concentration of 5 and 10 ml/l respectively. Curve D and E show the characteristics of solutions with 7.5 and 15 ml/l of leveler concentration respectively. Abrupt increase in the current density in case B and C can be seen in Fig 3. However, such increase in the current density is suppressed in the solution with the concentration used in TSV electroplating. As can be seen in the figure, the current saturates around 40 to 60 mA/cm^2.

Microstructure of copper TSV

Fig. 6 shows the FIB image of the microstructure of copper at the bottom of the 60 μm pillar with low current density of 20 mA/cm^2. It can be clearly observed from the figure that the grain sizes are of the order of hundreds of nanometer to 1 μm. The parallel structures in each individual grain show the existence of nanotwins in copper. Smaller grain sizes at the bottom indicate the higher frequency of adatom nucleation during the electroplating process. Existence of higher density of twinned crystals represents the stress/strain relaxation during grain growth leading toward the twinned orientation of copper atoms. Although experimental and theoretical studies of nanotwins progressed recently, the exact description of

formation of nanotwins is far from complete understanding due to its combined electrochemical, material and mass transport phenomena. The use of reverse current is also an important factor that contributes to the smaller grain size due to removal from electroplated copper [9] in anodic cycle.

Fig. 6 FIB image of the copper TSV pillars with 60 μm diameter at the bottom.

Fig. 7 FIB image of the copper TSV pillars with 60 um diameter at the top.

Fig. 7 shows the FIB image of the copper TSV with 60 μm diameter. The grain size is about few micrometers. Almost none of the grains have copper nanotwinned structure. The variation of the grain structure may be due to the local variation of plating conditions at different positions along the length of the pillar. But in both top and bottom cross sections, no void or pore is observed. Near to the opening side of the through-hole electroplating, grain sizes are observed to be larger. The mode of electrodeposition is grain growth dominant rather than the nucleation of new grains. The fluid convection at the opening of the through-holes may have played a key role in the process by excess supply of copper ions by ion migration in addition to diffusion which is the only ion transport mechanism in deep trenches and vias [10]. Plating at high current densities involves the gas evolution from the electrochemical reaction. The trapping of gas bubbles in the deposit forms the porous copper deposit as shown in the

Fig. 8.

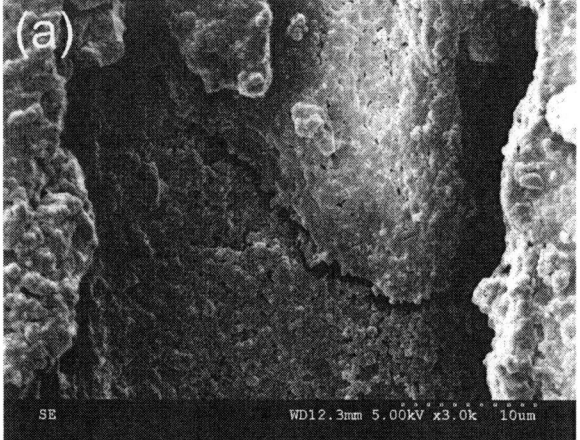

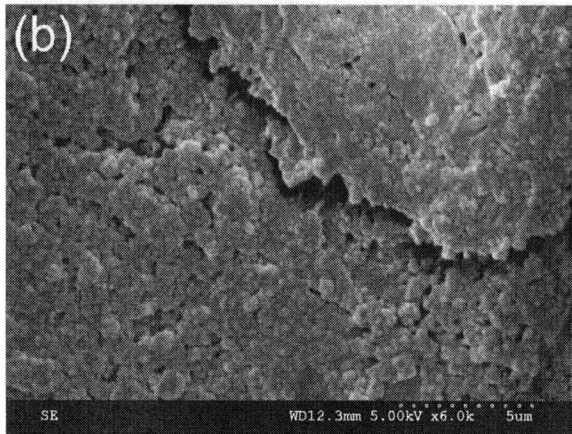

Fig. 8 SEM examination of surface morphology of the electroplated copper in the void (current density 40 mA/cm^2).

The Fig. 8(a) shows the microstructure/surface morphology of the inner surface of the void. Fig. 8(b) shows detailed surface morphology of the inner surface of the void formed during electroplating. It shows the porous granular structure at the core of the void as can be seen in Fig. 8b. It shows that plating performed at the higher current densities tends to give porous copper and void at the core of TSV.

Conclusion

The copper through-silicon vias with varying diameters were fabricated with the pulse-reverse constant current modulation. The experimental results show that all the through-holes with different diameters are filled with copper without voids or pores. But at the higher current density, the copper pillars with diameters larger than 100 μm have the higher chance of having a void at the middle of the pillar. It may be due to the variation of electrical field distribution inside the deep trenches although the exact reasons of voiding at higher current density needs to be studied in more details due to the complicated nature of copper electrodeposition in the deep high-aspect ratio trenches. Electrochemical studies show the point around 40 mA/cm2 beyond which the current saturates. The FIB examination on 60 μm diameter with low current density reveals the difference in grain size and internal structure of the grains. At the bottom

of the pillars, the grains sizes are small at the order of 1 μm with significant density of nano-scale twins. At the top of the pillars, the grain sizes of the copper are bigger as compared to that of the grains at the bottom. TSV fabricated with higher current densities shows a void at the core with porous microstructure and surface morphology.

References

[1] S. F. Alsarawi, D. Abbott, and P. D. Franzon, A review of 3-D packaging technology, IEEE Trans. Compon. Packng. Manufg. Technol., Part B 21 (1998) 2-14.

[2] J. R. Lloyd, J. Clemens and R. Snede, Copper metallization reliability, Microelectron. Reliab. 39 (1999) 1595-1602

[3] D. G. Pierce and P. G. Brusius, Electromigration: A review, Microelectron. Reliab. 37 (1997), 1053-72

[4] P. Dixit, J. M. Miao and R. Preisser, Fabrication of high aspect ratio 35 um pitch through-wafer copper interconnects by electroplating for 3-D wafer stacking, Electrochem Solid-State Lett. 9 (2006) G305-8.

[5] K. X. Ma, J. K. Ma, J. B. Sun, J. M. Miao, M. A. Do and K. S. Yeo, A miniaturized silicon-based ground ring guarded patch resonator and filter, IEEE Microwave and Wireless Comp. Let. 15 (2005), 478-480.

[6] L. Lu, Y. F. Shen, X. H. Chen, L. Qian and K. Lu, Ultrahigh strength and high electrical conductivity in copper, Science 304 (2004), 422

[7] L. Xu, P. Dixit, J. M. Miao, J. H. Pang, X. Zhang, K. N. Tu and R. Preisser, Through-wafer electroplated copper interconnect with ultrafine grains and high density of nanotwins, Appl. Phy. Let. 90 (2007) 033111-13.

[8] D. Xu, W. L. Kwan, K. Chen, X. Zhang, V. Ozolin and K. N. Tu, Nanotwin formation in copper thin films by stress/strain relaxation in pulse electrodeposition. Appl. Phy. Let. 91 (2007) 254105-07.

[9] T. R. Anthony, Forming electrical interconnections through semiconductor wafers, J. Appl. Phy., 52 (1981) 5340-49

[10] K. M. Akahashi, M. E. Gross, Transport phenomena that control electroplated copper filling of submicron vias and trenches, J. Electrochem. Soc. 146 (1999) 4499-4503

Designing a Display Unit to Drive the 8x8 LED Dot-Matrix Displays

Wan-Fu Huang

National Kaohsiung University of Applied Sciences
415, Chien-Kung Road,
Kaohsiung, Taiwan
phone: +886-975-259-378, fax: +886-7-381-1182, email: edhuang@kuas.edu.tw

Abstract

This paper presents a field programmable gate array (FPGA) prototype of a display unit to drive the eight-by-eight LED dot-matrix displays of two colors. The circuit design was Verilog-based and downloaded to the Spartan-3 FPGA chip of the Spartan-3 Starter Board. The display unit was able to generate four character patterns to rotate in four directions successfully. The design itself suggests a wide variety of the small-sized display application domains with a single red-green eight-by-eight LED dot-matrix display.
Keywords: FPGA, LED Dot-matrix Display, Verilog, matrix transposition

Introduction

When a small, simple display device is required on a machinery surface or in a building, a light-emitting-diode (LED) matrix display can be an adequate choice. Although humans had been using the 5x7 LED matrices to create dot matrix patterns for characters of different languages [1-3], nowadays the small symbols and characters displayed on the 8x8 LED dot-matrix displays are seen almost everywhere in the world. This paper suggests a display unit to light a popular red-green eight-by-eight LED dot-matrix display [4] for the similar display purpose with four characters rotating in sequence.

The display unit of this paper allows four alphanumeric characters to rotate on an eight-by-eight LED dot-matrix display in order in one of the four directions, either leftward, rightward, upward, or downward. The circuit design was based on the Verilog [5] hardware description language, synthesized by the Xilinx Synthesis Technology (XST) synthesizer [6] and implemented with the ISE design flow [7]. The bit stream file after the ISE design flow was downloaded onto the Spartan-3 xc3s200-4ft256 FPGA chip [8] on an FPGA lab board—the Spartan-3 Starter Board [9]. The lab board housed the prototype of the display unit.

The Hardware System

A. The Complete System

The whole display unit system was composed of a red-green two-color eight-by-eight LED dot-matrix display, inserted onto the sockets soldered on a printed circuit board, and an FPGA lab board—the Spartan-3 Starter Board. The printed circuit board is connected to an expansion connector of the FPGA lab board by a 34-pin cable, as shown in Fig. 1.

B. The Red-Green Eight-by-Eight LED Dot-Matrix Display and the LED On-Off Control

Fig. 1 The complete display unit system

The CSM-88261DF LED dot-matrix module [4] in Fig. 1 is a red-green two-color eight-by-eight LED dot-matrix display with a column-cathode row-anode structure. The internal equivalent circuit of the LED dot matrix and the external on-off control of the LEDs are shown in Fig. 2. The 8x8 LED matrix comprises eight light dot rows by eight light dot columns. Each light dot is composed of one red LED and one green LED. Totally there are sixty-four light dots, sixty-four red LEDs, and sixty-four green LEDs. The anode terminals of the sixteen LEDs in a single row are connected together and designated c1, c2, ..., and c8 from the top to the bottom in Fig. 2. The connected cathode terminals of the eight red LEDs in a column are designated r1, r2, ..., and r8 from the left to the right. The connected cathode terminals of the eight green LEDs in a column are designated g1, g2, ..., and g8 from the left to the right in the figure.

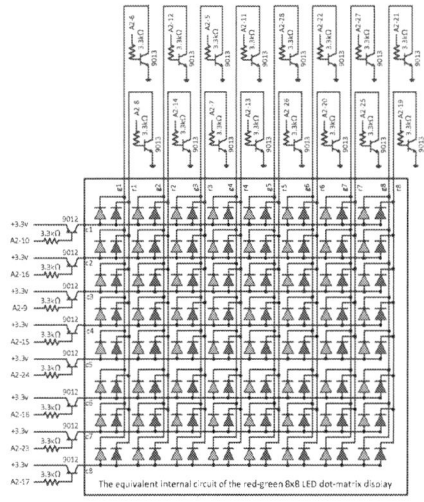

Fig. 2 The 8x8 LED matrix and the LED on-off control

A PNP 9012 [10] transistor and a NPN 9013 transistor [11] were used to control the current flow in the LED. To light the red LED at the upper-left corner, a logic 0 was applied to the base terminal of the 9012 transistor and a logic 1 was applied to the base terminal of the 9013 transistor simultaneously, as shown in Fig. 3. A 3.3kΩ resistor was connected between the logic input and the transistor base terminal in either transistor.

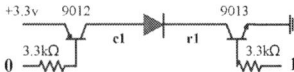

Fig. 3 The 9012/9013 transistor pair to turn on a red LED.

C. The Function Components of the FPGA Lab Board

All the input signals of the display unit were provided by the components on the Spartan-3 Starter Board. The 50 MHz clock source set the system clock. A push-button switch provided the system reset. Another push-button switch started or halted the system operation. The combination of a third push-button switch toggle signal and a slide switch selection signal determined the rotation direction, either leftward, rightward, upward, or downward. A second slide switch selected the character rotation speed, either normal, one rotation step per period of 0.2 sec, or fast, one rotation step per period of 0.1 sec.

An on-board discrete LED indicated the rotation speed, fast or normal. The display unit function block was housed in the Spartan-3 xc3s200-4ft256 FPGA on the lab board. Three 10-bit output signals of the display unit controlled the operation of the 8x8 LED dot-matrix display by way of the on-board A2 expansion connector and the 34-pin connection cable.

Rotating the Character Dot-Matrix Pattern

A. Representing the 8x8 Dot patterns in the 10x10 Matrices

Four ASCII [12] alphanumeric characters were designed into the display unit. English letters K, U, A and S are the test example of this paper. They were designed to display on the eight-by-eight LED dot-matrix display in the way that characters K and A were displayed in red color and U and S in green color. The rotating character order was red K, green U, red A, and green S in any one of the four rotating directions. Each character was mapped onto a single eight-by-eight dot matrix as shown in Fig. 4.

Fig. 4 The 8x8 matrix representations of any four characters.

In order to distinguish the vertically or horizontally adjacent characters amid the rotating moments, two conceptual all-dark rows and two conceptual all-dark columns were inserted below the bottom row and at the right of the right-most column in the matrix respectively. Thus each visible 8x8 character dot pattern was represented by a ten-by-ten dot matrix in the circuit description, as depicted in the case of K, Fig. 5. The

two always-dark rows or columns rotated with the eight regular ones. When the dark rows or columns came inside the

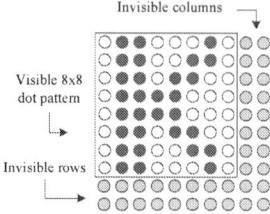

Fig. 5 Representing an 8x8 visible dot pattern in a 10x10 dot matrix.

visible LED matrix, they were dark dot patterns and when they were rotated beyond the bottom edge or the right-most edge, they were invisible and unnoticeable. The truly visible character pattern was still in the dimension of eight by eight and displayed only on the eight-by-eight LED dot-matrix display as shown in Fig. 4.

With the 10x10 matrix representations of the 8x8 character dot patterns, a character rotating in any direction took eleven steps to rotate from its starting position back to its starting position as shown in the case of the left-rotating K, Fig. 6. Accordingly, it took 11*4=44 steps for four characters to finish a complete rotation cycle in any one of the four rotation directions.

B. Exploiting the Matrix Transposition

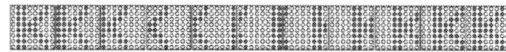

Fig. 6 A character pattern takes eleven steps to rotate one full cycle from the starting position back to the starting position.

Since each character was represented in a ten-by-ten matrix, each row vector carried ten elements, i.e., ten character dot information in a row and ten row vectors represented the whole character dot matrix. Simply rotating the row vectors leftward or rightward would achieve the goal of rotating the whole character dot pattern leftward or rightward. But to rotate the whole character dot pattern upward or downward, this paper transposes [13] the character dot matrix twice.

The first matrix transposition converted the column vectors to the new or transposed row vectors so that the very basic horizontal rotation operations could be applied to the transposed row vectors to rotate the transposed row dot vectors leftward or rightward. The second matrix transposition converted the transposed row vectors back to the original column vectors and restored the true shape of the original character dot pattern.

This second matrix transposition also converted the leftward rotation of the transposed row dot pattern onto the upward rotation of the original column dot pattern and the rightward rotation of the transposed row dot pattern onto the downward rotation of the original column dot pattern. The matrix transposition and its effect are depicted in Fig. 7 and Fig. 8.

From a viewer's perception, the true character dot movement was in a column by column per step manner during the leftward or rightward rotation and in a row by row per step manner during the upward or downward rotation.

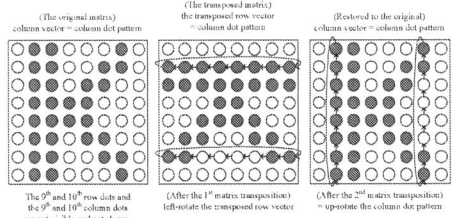

Fig. 7 Up-rotating the column dot patterns by way of left-rotating the transposed row dot patterns.

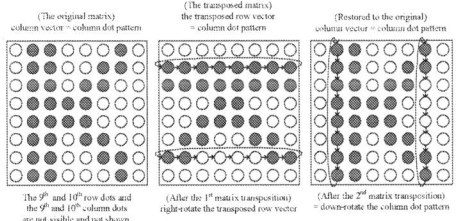

Fig. 8 Down-rotating the column dot patterns by way of right-rotating the transposed row dot patterns.

The Circuit Design

The display unit circuitry is shown in the circuit block diagram of Fig. 9. The system clock was an on-board 50MHz clock. The display unit got a system reset for 300 ms after power-on. Pressing the reset push-button also reset the system. After the reset, the character pattern on the LED matrix started to rotate either leftward or upward depending on the vh slide-switch position. A down vh slide-switch position assumed a horizontal rotation, either leftward or rightward. An up vh slide-switch position assumed a vertical rotation, upward or downward. Pressing the udlr push-button switch once reversed the leftward-rightward or upward-downward rotation direction. Pressing the paus push-button halted the character pattern rotation. A second press on the paus push-button resumed the system operation. When the sel slide switch was in the down position, the rotation speed was normal, one rotation step per period of 0.2 sec, and the speed LED was off. When the sel slide switch was in the up position, the rotation speed was fast, one rotation step per period of 0.1 sec, and the speed indication LED was on.

The circuit function block residing in the FPGA outputted 10-bit crow signals to control eight 9012-transistor on-offs to

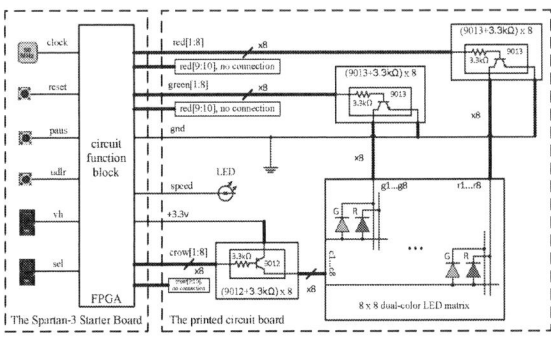

Fig. 9 The circuit block diagram of the display unit.

designate which LED row to be displayed. Two output bits, the 9th and 10th crow bits, were purposely connected to some irrelevant pins because they did not truly light any LED row. The 10-bit red output signals controlled eight 9013-transistor on-offs to choose the red LED columns for display. Another 10-bit green output signals controlled eight 9013-transistor on-offs to choose the green LED columns for display. The output bits including the 9th and 10th red and the 9th and 10th green bits were purposely connected to some irrelevant pins too. The character dot pattern was rendered by a row-scan display approach [14], meaning that the LED dot pattern was displayed row by row repetitiously. The character pattern refreshing rate was set at 100 Hz with a row display period of 1 ms.

The simplified system function block diagram of the display unit is shown in Fig. 10.

The Resource Utilization and the Test Results

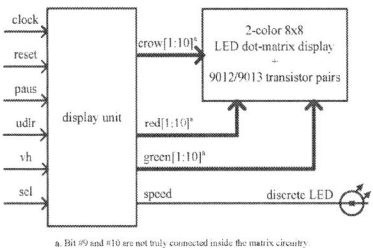

Fig. 10 The simplified system block diagram of the display unit.

The Verilog-based design of the display unit was synthesized by the XST synthesizer and implemented with the ISE design flow. The digital resources utilization reported in the Map Report and the Place and Route Report [7] after the design flow is summarized in Table I. Table I shows that this display unit circuit consumed 99% slices of the Spartan-3 xc3s200-4ft256 FPGA.

The final design was tested to drive the LED dot-matrix

TABLE I
SPARTAN-3 FPGA DIGITAL RESOURCES UTILIZATION

Map Report:			Utilization	Resource Total	Percentage
Design Summary					
	Number of Errors		0		
	Number of Warnings		0		
Logic Utilization					
	Number of Slice Flip Flops		2574	3840	67%
	Number of 4 input LUTs		3095	3840	80%
Logic Distribution					
	Number of Occupied Slices		1918	1920	99%
	Total Number of 4 input LUTs:		3159	3840	82%
		Number used as logic	3095		
		Number used as a route-thru	64		
	Number of bonded IOBs		37	173	21%
	Number of BUFGMUXs		1	8	12%
Average Fanout of Non-Clock Nets:			3.51		
Place and Route Report:					
	Clock Fanout	BUFGMUX0	1298		

display. The four-direction rotating patterns were recorded. The leftward rotating pattern recording photos are shown in Fig. 11, in which all the lit K and A dots are red and all the lit U and S dots are green. The upward rotating pattern recording

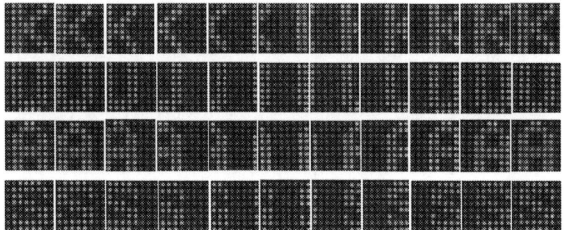

Fig. 11 The rotating pattern recordings for the leftward rotation.

photos are shown in Fig. 12 in which all the lit K and A dots are red and all the lit U and S dots are green. In either direction, the characters were rotated in the order of red K, green U, red A, and green S and each single character took eleven steps to finish its turn of rotation. The rotating pattern recordings for the rightward rotation and the downward rotation were exactly the same as those shown in Fig. 11 and Fig. 12 except that the pattern pictures of each character were in reverse order.

As for the rotation speed test, the display unit took 88 seconds to finish ten complete rotation cycles of four characters when the display unit operated in the normal speed mode in either one of the four rotation directions. Dividing 88 seconds by 44 steps in a complete rotation cycle and by 10 cycles got a 88/(11*4*10)=0.2 sec time period, meaning that one normal rotation step consumed 0.2 sec. When the display unit operated in the fast speed mode, the system took 44 seconds to finish 10 complete rotation cycles of four characters in either one of the four rotation directions. This meant that the fast rotation speed was 44/(11*4*10) = 0.1 sec per rotation step. Both normal and fast rotation speeds met the circuit design specification exactly.

Fig. 12 The rotating pattern recordings for the upward rotation.

Conclusions

This paper suggests a display unit prototype to drive a two-color eight-by-eight LED dot-matrix display for four-direction rotation with two rotation speed modes. The character dot patterns of the four alphabets rotated in the order of red K, green U, red A, and green S in either one of the leftward, rightward, upward, and downward rotation directions. The rotation was in a column by column manner during the leftward or rightward rotation and in a row by row manner during the upward or downward rotation. In the normal speed mode the character patterns on the LED matrix moved one row or column step forward per period of 0.2 seconds. In the fast rotation speed mode one row or column was moved forward one step per period of 0.1 seconds. The test results matched the design specification exactly in terms of the rotation speed, the character pattern, and the color appearance order.

The suggested display unit can rotate any four display characters in the same manner as that demonstrated in this paper as long as the dot patterns of K, U, A, and S are replaced by those of the new four-character set in the Verilog code.

References

[1] M. S. Beg and W. Ahmad. "Dot Matrix Alphanumeric Display System for Arabic," *IEEE Transactions on Consumer Electronics,* Vol. CE-33, No. 1, February 1987, pp. 47-50.

[2] W. L. Goh and K. T. Lau, "A Microprocessor-based Dot Matrix Display System for Japanese Hiragana Syllables," *IEEE Transactions on Consumer Electronics,* Vol. 35, No. 1, February 1989, pp. 32-36.

[3] W. L. Goh and K. T. Lau, "Dot Matrix Display System for Korean Numerals," *IEEE Transactions on Consumer Electronics,* Vol. 37, No. 4, November 1991, pp. 892-896.

[4] *Dot Matrix Display CSM-88251DF/88261DF,* China Semiconductor Corporation. http://www.datasheetarchive.com/CSM-88261DF-datasheet.html #contextual

[5] *IEEE Standard for Verilog Hardware Description Language,* IEEE Std 1364-2005, April 7, 2006.

[6] *XST User Guide for Virtex-4,Virtex-5, Spartan-3, and Newer CPLD Devices,* UG627 (v 12.4), Xilinx Inc., San Jose, CA, December 14, 2010.

[7] *ISE Design Suite Software Manuals – PDF Collection,* (V 13.4), Xilinx Inc., San Jose, CA, January 18, 2012.

[8] *Spartan-3 FPGA Family Data Sheet,* DS099, Xilinx Inc., San Jose, CA, December 4, 2009.

[9] *Spartan-3 Starter Kit Board User Guide,* UG130 (v1.1), Xilinx Inc., San Jose, CA, May 13, 2005.

[10] *SS9012 PNP Epitaxial Transistor, Rev. A4.* Fairchild Semiconductor Corporation, November 2002.

[11] *SS9013 NPN Epitaxial Transistor, Rev. A4.* Fairchild Semiconductor Corporation, November 2002.

[12] *American National Standard for Information Systems—Coded Character Sets—7-Bit American National Standard Code for Information Interchange (7-Bit ASCII),* ANSI X3.4-1986 (R2007), American National Standards Institute, Inc., June 14, 2007.

[13] H. Anton, *Elementary Linear Algebra,* 9th ed., USA: John Wiley & Sons, Inc., 2005, PP.33.

[14] W. F. Huang, "Three Approaches to Light an 8x8 LED Dot-Matrix Display," unpublished.

Physical/Process Parameter Dependence of Gate Capacitance and Ballistic Performance of InAs$_y$Sb$_{1-y}$ Quantum Well Field Effect Transistors

Iftikhar Ahmad Niaz[*1], Md. Hasibul Alam[1], Imtiaz Ahmed[1,2], Zubair Al Azim[1,2], Nadim Chowdhury[1], Quazi Deen Mohd Khosru[1,2]

[1]Dept. of Electrical and Electronic Engineering, Bangladesh University of Engineering and Technology, Dhaka-1000, Bangladesh
[2]Dept. of Electrical and Electronic Engineering, Green University of Bangladesh, Section # 6, Mirpur-2, Dhaka-1216, Bangladesh
Phone: 88029345988 , Fax: (8802) 9668054, E-mail: *iftikhar.oni@gmail.com

Abstract

This paper reports complete Capacitance-Voltage (CV) characterization of InAs$_y$Sb$_{1-y}$ Quantum Well Field Effect Transistor (QWFET) along with an analysis of ballistic transport performance. 1-D coupled Schrodinger-Poisson equations are solved for electrostatic performance analysis of QWFET considering wave function penetration and strain effects. Dependence of CV characteristics on some important process and physical parameters like oxide thickness, channel composition, top barrier composition, channel thickness and temperature is studied in this work. We observed that before cross-over point the capacitance increases rapidly and gradually reaches saturation after that point. This is because the slope of the sheet charge density gradually increases and finally reaches almost a constant value. The pattern of variation of simulation results are consistent with the results of experimentally grown device by Ali et al. Ballistic current is also reported to improve from compressive to tensile strained channel.

Keywords: Buried channel device, high κ dielectric, quantum well FET, strain and wave function penetration.

Introduction

In recent times, there has been great interest in exploring the substitution of the Si channel in scaled logic field-effect transistors with a III-V compound semiconductor [1]. In this quest for a III-V CMOS technology mixed anion InAs$_y$Sb$_{1-y}$ quantum wells are emerging as a potential candidate with their high electron mobility as well as high hole mobility [2]. Though there is extensive research on this device [3]-[7], a complete Capacitance-Voltage (CV) characterization is still missing in literature to the best of our knowledge. The effects of parameter variation on electrostatic performance are important to have an insight on complete CV characterization. In this work, we carried out a complete study of gate capacitance using self consistent Schrödinger-Poisson method in InAs$_y$Sb$_{1-y}$ QWFETs by varying different process parameters as well as physical parameters like oxide thickness, channel composition, top barrier composition, channel thickness and temperature. We also investigated ballistic performance of the device.

Simulation Methodology

We analyzed CV characteristics and ballistic transport performance of the structure in Fig. 1. We obtained the results for the device solving 1-D coupled Poisson and Schrödinger equations [8]. We used Finite Difference Method for solving Poisson equation [9] and Hamiltonian Matrix Formalism [10] for solving Schrödinger equation numerically. Strain effect was also incorporated in our simulator [11]. The ballistic current was calculated with the model shown in [12] which is based on the theory developed by Natori [13]. The band diagram of the device along with carrier profile is drawn in Fig. 2. A rectangular well is formed in the channel region in the band diagram. The carrier profile shifts in the right along z axis as gate voltage becomes more negative. After a certain negative voltage the carrier density in the channel region becomes almost zero which

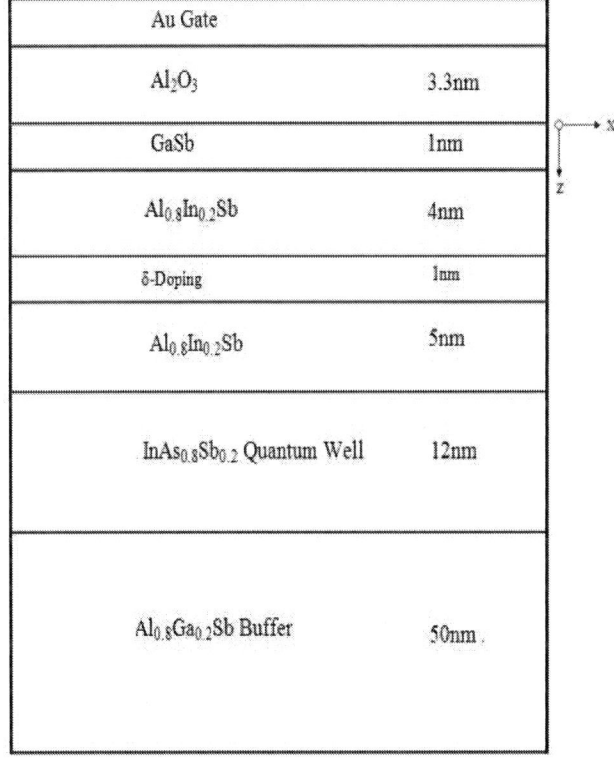

Fig. 1 Basic device structure of Buried-Channel AlInSb/InAsSb QWFET.

978-1-4673-4840-9/13 $31.00 © 2013 IEEE   389

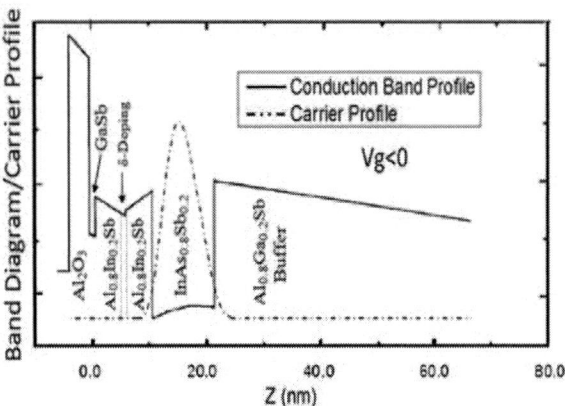

Fig. 2 Conduction band profile with carrier concentration (not drawn to scale.

corresponds to the pinch-off condition.

Results and Discussions

A. Effect of oxide thickness

We observed the effect of oxide thickness variation in channel sheet carrier density and gate capacitance in Fig. 3 and Fig. 4 respectively and found that after cross-over point both sheet carrier density and capacitance increase as oxide thickness decreases. This can be explained by the domination of oxide capacitance after cross-over since oxide capacitance is higher for lower oxide thickness.

Fig. 5 shows the gate capacitances for three different dielectric materials (κ_{SiO2}=3.9, κ_{Al2O3}=6.86, κ_{HfO2}=20). We observe that the capacitances do not vary significantly for the same Equivalent Oxide Thickness (EOT). For high-κ dielectric gate leakage current reduces. Hence we can infer that high-κ dielectric yields better electrostatic and gate leakage performance. A thorough

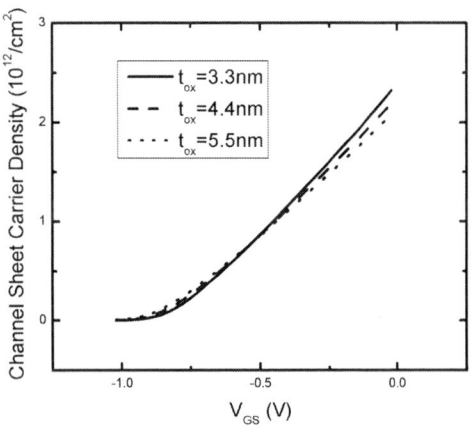

Fig. 3 Channel sheet carrier density as a function of gate voltage for different oxide thickness.

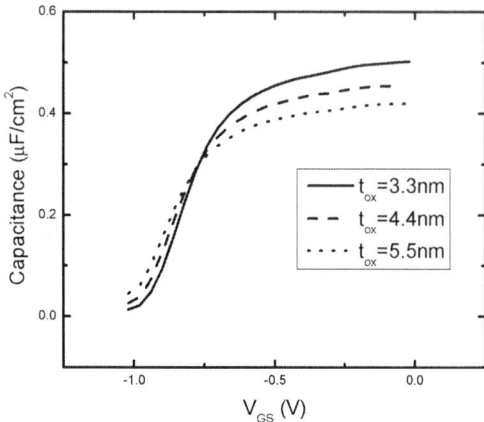

Fig. 4 Gate Capacitance as a function of gate voltage for different oxide thickness for top barrier of $Al_{0.8}In_{0.2}Sb$ and channel of $InAs_{0.8}Sb_{0.2}$ at T=300K.

research has to be performed for the suitability of using high-κ dielectrics without significant interface traps and surface oxidation.

B. Strain Effects

Effect of strain between channel and top barrier layer is presented in Fig. 6 which clearly shows that gate capacitance near zero volt increases gradually from tensile to compressive strained channel i.e. with higher InSb composition. The reason is

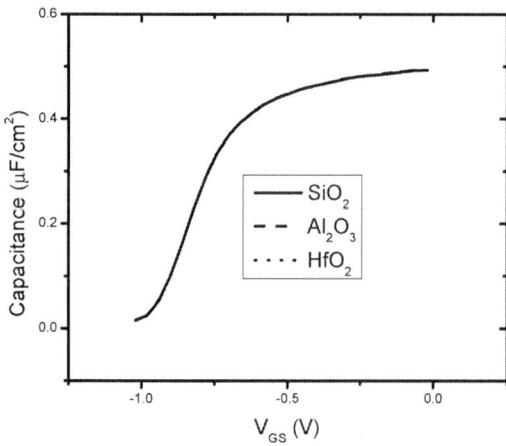

Fig. 5 Gate Capacitance as a function of gate voltage for different dielectric materials for EOT=2nm.

that with lower bandgap of InSb there is a higher band offset with top barrier layer and as such higher carrier confinement in InSb rich channels.

It is found from Fig. 7 that there is minimal effect of top barrier composition variation on CV characteristics. The gate capacitance in the ON region is slightly higher for top barrier

978-1-4673-4840-9/13 $31.00 © 2013 IEEE

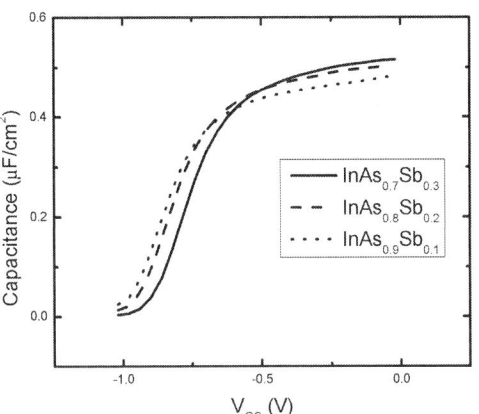

Fig. 6 Gate Capacitance as a function of gate voltage for different compositions of InAs and InSb at channel layer.

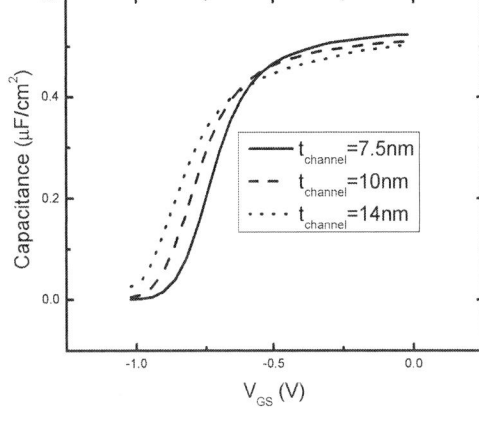

Fig. 8 Gate Capacitance as a function of gate voltage for different channel thickness.

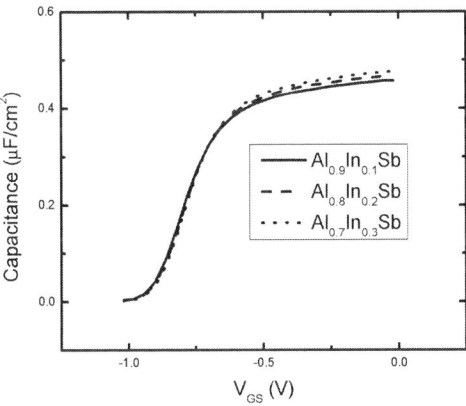

Fig. 7 Gate Capacitance as a function of gate voltage for different compositions of AlSb and InSb at top barrier with channel of $InAs_{0.8}Sb_{0.2}$ and Al_2O_3 as gate dielectric, t_{ox}=3.3nm and $t_{channel}$=12nm.

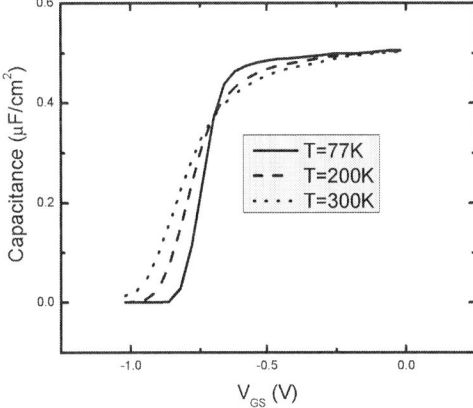

Fig. 9 Gate Capacitance as a function of gate voltage for different temperatures.

with higher InSb composition.

C. Effect of channel thickness variation

Fig. 8 shows the change of gate capacitance for three different channel thicknesses. We observe the highest capacitance with the lowest channel thickness after the cross over point. This can be explained by the highest rate of increase of channel sheet carrier density with increasing gate voltage in case of lowest channel thickness.

D. Effect of temperature

We investigated CV characteristics for three different temperatures of 77K, 200K and 300K in Fig. 9. Bandgap of top barrier and back barrier decreases as temperature increases. As a result carrier confinement in channel region is decreased. Hence capacitance decreases as temperature increases.

E. Ballistic transport performance

We inspected ballistic current with the variation of channel composition which is illustrated in Fig. 10. It is found that with increasing InAs composition i.e. from compressive to tensile strained channel ballistic performance improves significantly.

Conclusions

We investigated the effects of parameter variation on electrostatic and ballistic performances of $InAs_ySb_{1-y}$ QWFET. Process and physical parameters: oxide thickness, channel composition, top barrier composition, channel thickness and temperature were varied. We observed that all these parameters except top barrier composition and different dielectric materials with same EOT have strong effects on CV characteristics. We also observed that ballistic performance improves from compressive to tensile strained channel. Our simulation work provides consistent pattern of variation with experiments on

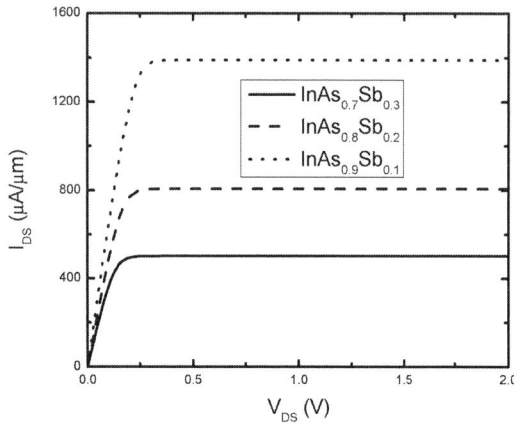

Fig. 10 Drain Current as a function of drain voltage for different compositions of InAs and InSb at channel layer with top barrier of $Al_{0.8}In_{0.2}Sb$ at T=300K, $t_{channel}$=12nm, V_{GS}= -0.02V, t_{ox}=3.3nm.

$InAs_ySb_{1-y}$ QWFETs [3] and gives an insight of performance enhancement with varying appropiate process/physical parameters.

Acknowledgment

The authors would like to express their gratitude to Mr. Raisul Islam for his guidance regarding simulation procedure.

References

[1] D.-H. Kim and J. A. del Alamo, *IEDM Tech. Dig.* ,2009, 861

[2] J. B. Boos et al., *IEICE Trans.* vol. E85-A/B/C/D, no. 7, July (2008)

[3] A. Ali et al, *IEDM Tech. Dig.*,2010, p: 6.3.1 - 6.3.4

[4] A. Ali et al, *IEDM Tech. Dig.*, 2011, p: 1397 – 1403

[5] A. Ali et al, *IEEE Electron Device Lett*, vol. 32, no. 12, pp. 1689-1691, Dec, 2011

[6] Suman Datta, Indium Phosphide and Related Materials Conference Proceedings, 2006 p: 174 – 176

[7] Bennett et al, *Journal of Electronic Materials*, Vol. 36, No. 2, 2007

[8] F. Stern, "Self-consistent results for n-type si inversion layers," *Phys. Rev. B*, vol. 5, pp. 4891 –4899, 1972

[9] J. Kisulaas, "Numerical Methods in Engineering with MATLAB," Cambridge University Press, 2nd Ed., 2009.

[10] S. Datta, Quantum Transport: Atom to Transistor. Cambridge University Press,2005.

[11] Joachim Pipreck, Semiconductor Optoelectronic Devices: Introduction to Physics and Simulation. Academic Press: San Diego, California, 2003, pp. 27–32.

[12] F. Assad, et al., *IEEE Trans. Electron Devices*, vol. 47, no. 1, pp. 232 –240, Jan. 2000.

[13] K. Natori, *J. Appl. Phys.* , vol. 76, pp. 4879–4890, Oct. 1994.

Metal-polymer nanocomposite films with ordered vertically-aligned metal cylinders for optical application

Linda Y.L. Wu*[1], B. Leng[1], W. He[1], A. Bisht[2], C.C. Wong[2]

*[1] *Singapore Institute of Manufacturing Technology, 71 Nanyang Drive, Singapore 638075*
Fax: +65-67916377, E-mail: ylwu@SIMTech.a-star.edu.sg
[2] *School of Materials Science and Engineering, Nanyang Technological University, 50 Nanyang Avenue, Singapore 639798*

Abstract

Ordered nanostructures of conventional plasmonic metals (Ag, Au etc.) in a dielectric matrix are candidate materials for fabrication of metamaterials. Control over the dimensions of these building blocks offers tunability in the metamaterial design. To that effect we demonstrated the large area fabrication of poly(styrene-b-4-vinyl pyridine) (PS-b-P4VP) diblock copolymer films wherein the self assembly of the PS and P4VP phases creates monodisperse hexagonal cylindrical nanotemplates whose structure can be controlled by varying the molecular weights, layer thicknesses and layer stacking. Silver and gold precursors were infused into the P4VP micro domains and chemically reduced within. We investigated the PS-b-P4VP template film synthesis and successfully obtained micro domain sizes from 30nm to 100nm and template thicknesses up to 700 nm (5 layers) was achieved. The P4VP micro domains were selectively removed by acetic acid etching which was further confirmed by Atomic Force Microscopy (AFM) using the high aspect ratio tips. Using oxygen plasma etching we have shown the effective loading of gold and silver into the P4VP micro domains. Finite Difference Time Domain (FDTD) simulation of a proposed meta-lens with typical material dimensions demonstrates negative refraction and super-resolution imaging.

Introduction

Artificially designed materials with an array of complex metallic subwavelength components, giving rise to extraordinary optical properties which do not exist in nature are referred to as metamaterials. These materials have been previously shown to exhibit negative refraction in the wavelengths spanning a few microns using the now ubiquitous split ring resonators and metallic wires.[1] The working wavelength was brought down towards the NIR (near infra-red) using the hexagonal fishnet structures.[2] These metamaterial designs suffer from intrinsic losses as they are based on resonance mechanisms and as a consequence they are also operational only at certain wavelengths in the spectrum.[3] A sub-class of metamaterials referred to as indefinite metamaterial was introduced by some groups [4-6], as these materials work in the regions where imaginary part of permittivity $\varepsilon''\approx0$, the losses are negligible. An indefinite metamaterial has an anisotropic permittivity such that $\varepsilon_{xy}>0$ and $\varepsilon_z<0$. These materials do not have any cutoff in their equifrequency contour plots as they have a hyperbolic dispersion relation as compared to a spherical one for air. This allows them to support high-k propagating states and makes them capable of collecting the high frequency components

which carry sub-wavelength information about the object.[7] These materials have highly tunable working wavelengths, which depend mostly on the filling factor of the metal in the dielectric.[8] Therefore, it becomes imperative to control the structural dimensions of the nanostructures through fabrication.

Diblock copolymers have attracted great attention [9-11,18] for being used as templates for fabrication of nanostructured materials. Their popularity lies in the fact that a wide range of morphologies (lamellae, cylinders and spheres) can be produced via a microphase separation process. These self-assembled microphase domain patterns can be tuned by changing the relative lengths of blocks or interconnected network morphologies[12] which can be further used as hosts/templates for the incorporation of various inorganic nanomaterials.[13] Therefore, diblock copolymers have become promising candidates for the preparation of many functional nanostructures, which may be used in surface patterning[14-15] and lithography[16-17] for the fabrication of organic-inorganic hybrid devices.[18]

In this work, poly(styrene-b-4-vinyl pyridine) (PS-b-P4VP) diblock copolymers with cylindrical micro domains were used as the matrices to prepare nanocomposite films infused with gold or silver. Different sizes of the micro domains were achieved by varying the molecular weights of PS (polystyrene) and P4VP (poly4-vinyl pyridine) blocks. A stacked fabrication method was adopted to create multi layered 3D structure with higher film thickness.

Experiments

PS-b-P4VP diblock copolymers with different molecular weights: PS(24k)-b-P4VP(9.5k), PS(48k)-b-P4VP(20.3k), PS(93k)-b-P4VP(35k), PS(330k)-b-P4VP(125k) were purchased from Polymer Source, Inc. Toluene (C_7H_8), chloroform ($CHCl_3$), gold chloride trihydrate (99.9+%, $HAuCl_4 \cdot 3H_2O$), silver nitrate (99+%, $AgNO_3$) and sodium borohydride (99%, $NaBH_4$) were all purchased from Sigma-Aldrich and used as received. Silicon wafers and quartz pieces were used as the substrates with proper cleaning.

Given amounts of PS-b-P4VP diblock copolymers (with different molecular weights) were dissolved in toluene/ and chloroform mixtures (toluene/chloroform = 80/20 vol%). The solutions were stirred for at least 4 h and then filtered through a 0.2 μm-sized microfilter before being spin-coated onto pre-cleaned substrates. The polymer solution concentration is kept at 20 mg/ml or 40 mg/ml, and different spin ramp and speed parameters were programmed to achieve the desired film thickness. The films were then solvent annealed in a saturated

978-1-4673-4840-9/13 $31.00 © 2013 IEEE

toluene/chloroform environment (toluene/chloroform = 20/80 vol%) for 60 minutes.

After the solvent annealing, the PS-b-P4VP films were immersed in either HAuCl₄ or AgNO₃ solution for 2 minutes to complex gold or silver ions with the 4-vinylpyridine units, followed by being rinsed several times with DI water. NaBH₄ was used to reduce the gold or silver salt into gold or silver nanoparticles.

Multilayers of alternating cylindrically structured PS-b-P4VP and pure PS were prepared for thicker layers. Firstly, a single layer cylindrical structured PS-b-P4VP and metal loading was performed and UV exposure for a few second was applied to crosslink the polymer. Secondly, given amounts of PS polymers (MW: 22,200 g/mol) were dissolved in toluene, and spin coated on the UV-crosslinked PS-b-P4VP layer. The PS coated layer was also UV-crosslinked. Then step1 and step 2 were repeated to obtain alternating stake of the layers.

Results and Discussion

Self-assembled cylindrically structured PS-b-P4VP

With different molecular weight compositions of PS-b-P4VP diblock copolymer, the micro domain size of the self-assembled layer varies. Figure 1 shows the AFM top-view images of the PS-b-P4VP cylindrical structures of various P4VP micro domain sizes with the size distribution graphs obtained by AFM phase analysis. It can bee seen that with increasing molecular weight (both phases), the micro domain size increases from about 25 nm to 70 nm.

Table 1 summarizes the micro domain sizes with respective molecular weight and polydispersity index (PDI) of the four types of PS-b-P4VP diblock copolymer self assembled layers. Due to the chemically dissimilar PS and P4VP monomers, the entropy that drives these two blocks towards mixing is small. However, the thermodynamic forces driving separation is counterbalanced by the covalent linkages between the two blocks. Therefore, during the solvent annealing, the energetically preferable arrangement allows the longer PS blocks to reside on the convex side of PS-P4VP interface, resulting in a micelle like formation as cylindrical structure in a top down manner.[4] Increasing the length of the P4VP block causes the formation of larger micelles, while the spacing in-between is mainly dependent on the length of the PS block. It provides a simple way of tuning the diameter and distance of metallic nanoparticles when deposited onto a substrate.

Metal loading into cylindrically structured PS-b-P4VP template

In order to load metal into one phase of the diblock copolymer, a proper removal of the phase without distortion of the cylindrical structure is essential. We used concentrated acetic acid due to the preferred solubility of the P4VP phase in this acid. Figure 2 shows a porous PS template produced from the PS(330k)-b-P4VP(125k) thin film with removed P4VP phase. A high aspect ratio tip was mounted onto AFM and scanned

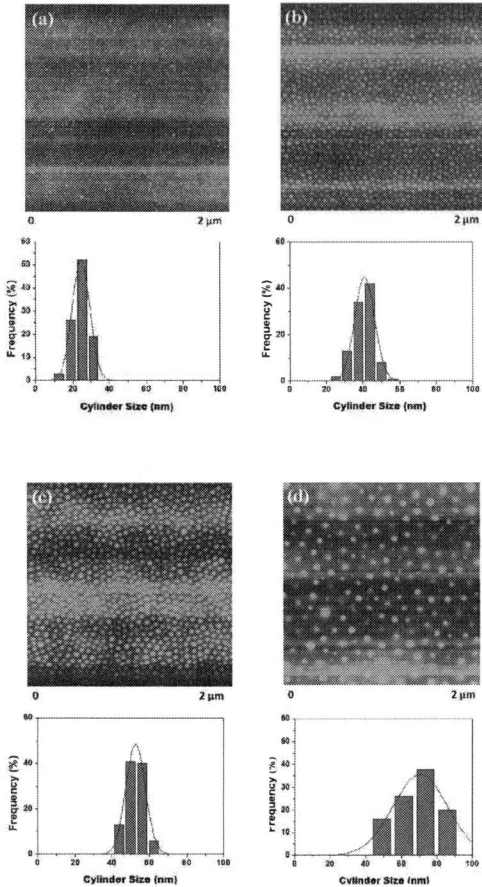

Fig. 1. AFM images of hexagonal micelle patterns of solvent annealed block copolymer thin film: (a) PS(24k)-b-P4VP(9.5k), (b) PS(48k)-b-P4VP(20.3k), (c) PS(93k)-b-P4VP(35k) and (d) PS(330k)-b-P4VP(125k). Corresponding Gaussian-fitted histogram of particle size distribution is indicated.

Table 1. Characteristics of PS-b-P4VP diblock copolymers, and the mean size, standard deviation of gold nanoparticles from the fitting to each histogram in Fig. 1.

	PS(24k)-b-P4VP(9.5k)	PS(48k)-b-P4VP(20.3k)	PS(93k)-b-P4VP(35k)	PS(330k)-b-P4VP(125k)
PS (g/mol)	24,000	48,000	93,000	330,000
P4VP (g/mol)	9500	20,300	35,000	125,000
PS/P4VP	2.53	2.36	2.66	2.64
PDI	1.10	1.13	1.15	1.18
Structure	cylindrical	cylindrical	cylindrical	cylindrical
d (nm)	24.4 ± 4.8	40.9 ± 5.4	52.7 ± 5.2	70.6 ± 15.3
R_q (nm)	1.9	2.3	3.4	3.5

across the porous template generating the depth profile of the porous structure. The match depth of the pore with the film thickness proved complete removal of the P4VP phase, and the pores are vertical standing and perpendicular to the substrate surface. This conclusion also applies to lower

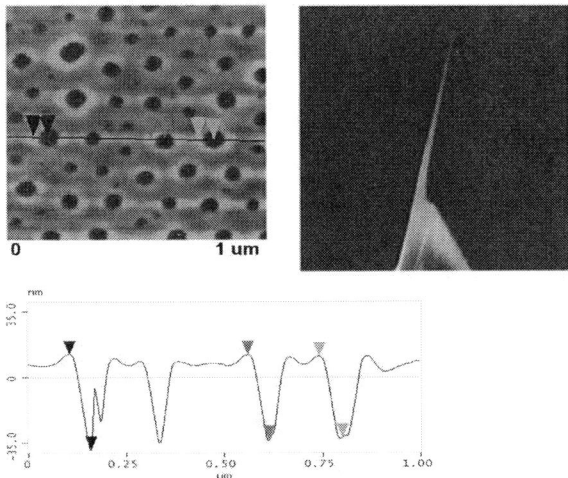

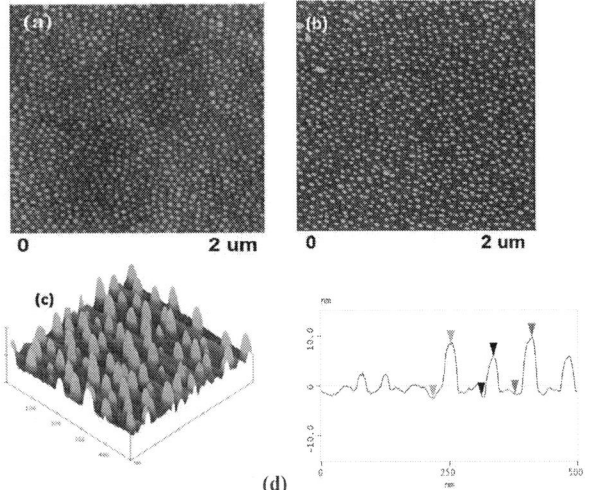

Fig. 2 PS porous template after removal of P4VP cylindrical micelles (a), high aspect ratio AFM tip (b) and depth profile of the cylindrical micelle (c), average depth 50 nm.

Fig. 3. AFM images of PS(48k)-b-P4VP(20.3k) thin film: (a) cylindrical patterns of Ag filled structure, (b) top view and (c) 3D view of Ag nano-clusters remnant after O_2 plasma etching, (d) cross section profile of the Ag nano-clusters.

molecular weight PS-b-P4VP block copolymers, as shorter blocks with high chain mobility are easier to self assemble into this entropic preference arrangement in a saturated solvent vapour condition.

Metal loading was performed by infusing $HAuCl_4$ or $AgNO_3$ salt solution into the P4VP domains pores, and then reduced to form Au or Ag nano-clusters. To prove the effective loading of metal, a Ag loaded PS(48k)-b-P4VP(20.3k) thin film (80 nm thick) was subjected to an oxygen plasma etching to remove the PS matrix. The topography images by AFM in 2D and 3D with depth profile are shown in Fig. 3. Nano clusters array with the same period and diameter with that of P4VP micro domains before the etching process are observed. Highly ordered vertically aligned nanorods are seen. From the depth profile, the average height of the nano-clusters is shown to be 8.15 nm (+/- 1.44 nm) for 20 nanorods sampled. The collapse of the some of the Ag-loaded cylinders may be due to the removal of the supporting polymer matrix during the etching, which could cause partial reduction of the height of the nano rod-like clusters. Multi-layers of alternating cylindrically structured PS-b-P4VP and pure PS were prepared. Coating thickness up to 700nm was obtained with a 5-layer structure.

Numerical simulation and discussion

In order to understand the applicability of the metal-polymer nanocomposite film as meta-lens, we perform full-wave simulations taking into consideration the cylindrical structure and the material properties. We use FDTD Lumerical™ 8.0.4, a commercial solver for performing electromagnetic calculations. The model consists of an object in the form of a very small slit, with width w ≪ λ, etched into a thin mask which sits on top of our meta-lens as shown in Fig. 4. A TM plane wave is incident from the top of the mask as it is obvious that a TE plane wave would not couple to the slit. We take a particular structure of our fabricated sample wherein

w=50nm, H=500nm, d=20nm, t=50nm, a=40nm. Periodic boundary conditions are used along the x-axis. Figure 5 shows the negative refraction capabilities of these meta-lenses. A gaussian wave at 466nm incident at an angle of 60° to these lenses is shown to refract negatively in the lens, as is evident by the Poynting vector(energy flow) plot. Since the characteristic lengths of the meta-lens are small compared to the wavelength of interest, we remain in the domain where effective media description of these nanorods is applicable[19,20].

The distribution of electric field intensities in the meta-lens is shown in Fig. 6 at three different wavelengths (λ=391.3nm, 435.5nm and 491nm). After exiting the slit object, the beam splits into two and then are refocused by the lens on the other side. The image/focal spot seem to move closer to the lens with an increase in the source wavelength. Further simulation for 3D structure using periodic boundary conditions both in x-axis and z-axis also showed beam splitting, which will be reported separately. From these simulation results, we confirmed the capability of our ordered vertically-aligned metal cylinders in polymer matrix could be using for meta-lens application. Further imaging experiments will be performed.

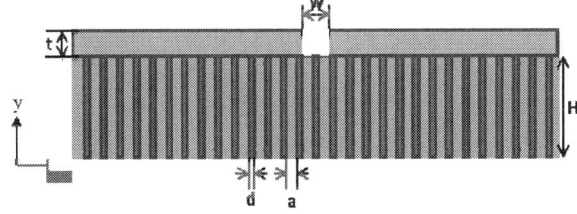

Fig. 4. Schematic of the nanowire metalens structure with an object on top. Various dimensions indicated in the structure are assumed to be < λ. w : width of the slits, a : centre to centre distance between nanorods, d: diameter of the rods, H : height of the rods and t : thickness of the object/mask.

978-1-4673-4840-9/13 $31.00 © 2013 IEEE

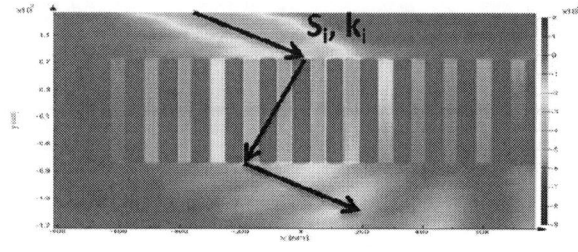

Fig. 5. Poynting vector (S) plot showing the negative refraction property for a gaussian wave incident at an angle to the meta-lens at 466nm. S_i and k_i indicate the incident Poynting vector and wave vectors. Arrows mark the direction of energy flow in the lens.

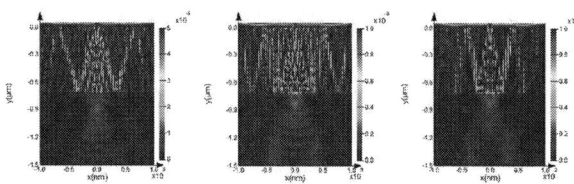

Fig. 6. Electric field intensity distributions in the meta-lenses. From left λ = 391.3nm, 435.5nm and 491nm. H in all three cases is 700nm. The focus of the object is clearly seen forming on the other side of the meta-lens. These are 2D simulations.

Conclusions

Metal-polymer nanocomposite films were self-assembled into nano-cylindrical array structures. After etching away the P4VP phase, ordered vertically-aligned metal cylinders were formed by infusion Ag or Au salt solution into the pores and reduced to metal nano-clusters. The negative refraction capability of the ordered vertically-aligned metal cylinders in polymer matrix has been confirmed by FDTD full wave simulation using a slit object. Negative refraction was detected in the electric field intensities distribution at three different wavelengths (λ=391.3nm, 435.5nm and 491nm). The beam splitting and refocusing phenomena indicated that the metal-polymer ordered structure can be potentially used as meta-lens for optical imaging application.

Acknowledgments

We acknowledge the financial support by the Metamaterials Programme: Superlens (092 154 0099) funded by Agency for Science, Technology and Research of Singapore (A*Star).

References

[1] R. A. Shelby, D. R. Smith, S. Schultz, *Science* **292**, 77 (2001).

[2] G. Dolling, M. Wegener, C. M. Soukoulis and S. Linden, *Opt. Lett.* **32**, 53-55 (2007).

[3] V. A. Podolskiy and E. E. Narimanov, *Opt. Lett.* **30**, 75 (2005).

[4] D. R. Smith and D. Schurig, *Phys. Rev. Lett.* **90**, 077405 (2003).

[5] B. D. F. Casse, W. T. Lu, Y. J. Huang, E. Gultepe, L. Menon, and S. Sridhar, *Appl. Phys. Lett.* **96**, 023114 (2010).

[6] R. Wangberg, J. Elser, E. E. Narimanov, and V. A. Podolskiy, *J. Opt. Soc. Am. B.* **23**, 498 (2006).

[7] Z. Jacob, I. I. Smolyaninov, and E. E. Narimanov, *Appl. Phys. Lett.* **100**, 181105 (2012).

[8] Y. Liu, G. Bartal and X. Zhang, *Opt. Express* **16**, 15439 (2008).

[9] P. A. Mistark, S. park, S. E. Yalcin, D. H. Lee, O. Yavuzcetin, M. T. Tuominen, T. P. Russell, and M. Achermann, *ACS Nano*, **2009**, Vol. 3, No. 12, 3987-3992

[10] S. Park, B. Kim, O. Yavuzcetin, M, T. Tuominen, and T. P. Russell, *ACS Nano*, **2008**, Vol. 2, No. 7, 1363-1370

[11] S. Park, J. Wang, B. Kim, W. Chen, and T. P Russell, *Macromolecules*, **2007**, 40, 9059-9063

[12] F. S. Bates and G. H. Fredrickson, *Physics Today*, Feb **1999**, 32

[13] A. Fahmi, T. Pietsch, C. Mendoza, N. Cheval, *Materials Today* **2009**, *12*, 44-50.

[14] L, Song, and Y. M. Lam, *Nanotechnology*, **2007**, Vol. 18, No. 7, 075304

[15] B. H. Kim, D. O. Shin, S. Jeong, C. M. Koo, S. C. Jeon, W. H. Hwang, S. Lee, M. G. Lee, and S. O. Kim, *Adv. Mater.*, **2008**, 20, 2303-2307

[16] B. H. Kim, J. Y. Kim, S. Jeong, J. O. Hwang, D. H. Lee, D. O. Shin, S. Choi, S. O. Kim, **2010**, *ACS Nano*, Vol. 4, No. 9, 5464-5470

[17] S. O. Kim, H. H. Solak, M. P. Stoykovich, N. J. Ferrier, J. J de Pablo, and P. F. Nealey, *Nature*, **2003**, Vol. 424, 411-414

[18] J. Bang, U. Jeong, D. Y. Ryu, T. P. Russell, and C. J. Hawker, *Adv. Mater.*, **2009**, 21, 4769-4792.

[19] J. Elser, V. A. Podolskiy, I. Salakhutdinov, and E. E. Narimanov, *Appl. Phys. Lett.* **89**, 261102 (2006).

[20] A. Sihvola, Electromagnetic Mixing Formulas and Applications, Institute of Electrical Engineers (1999).

978-1-4673-4840-9/13 $31.00 © 2013 IEEE

Ag-doped SiO₂/TiO₂ hybrid optical sensitive thin films with visible absorption enhancement for diffractive optical element application

P. Junlabhut[1*], S. Boonruang[2], and W. Pecharapa[1,3]

[1] College of Nanotechnology, King Mongkut's Institute of Technology Ladkrabang, Bangkok, Thailand
[2] Photonics Technology Laboratories, National Electronics and Computer Technology Centre, NSTDA, Pathumthani, Thailand
[3] Thailand and Center of Excellence in Physics (ThEP Center), CHE, 328 SiAyutthaya RD, Bangkok, Thailand
*Corresponding author, e-mail: pjunlabhut@gmail.com

Abstract

This paper reports the synthesis of Ag-doped SiO₂/TiO₂ hybrid optical sensitive thin films deposited by sol-gel spin coating technique. The structural properties of thin films were characterized by XRD, TEM and EDX. The optical absorption of the films in visible region measured by UV-VIS can be enhanced by Ag nanoparticles due to surface plasmon resonance effect. Photosensitive film can be utilized as effective photosensitive material for diffractive optical element with controllable period by interference angle using interference lithography technique. AFM was employed to investigate the fabricated patterns. The diffraction pattern highly correlated to the performance of diffractive optical element is scrutinized.

Keywords: Ag- doped SiO₂/TiO₂ hybrid film, photosensitive, absorption enhancement and interference lithography technique.

Introduction

Optical elements (OE) have been extensively utilized in various applications including optoelectronics, optical communication, optical computing, optical data storage, imaging system and optical sensors. Optical elements such as gratings, waveguide, micro lenses and diffractive elements have been widely used to improve the ability of optical device system. The optical function of optical elements strongly depends on its structure. So far, several processes have been introduced for the fabrication of functional OE, including using X-ray, UV light and electron beam [1-3]. Typically, the fabrication techniques are complicated process including deposition of photosensitive films on the substrate. The photo patterns were obtained by fabrication and transferring process. Essentially, the transfer of surface profiles to substrates was obtained by etching process.

Photosensitive materials mainly contain functional polymer and photo-induced material. During illumination, cross-linking of the materials is initiated and the formation of organic network structure can be strengthened via photo polymerization process [4]. Along with other research works, the photosensitive materials have been improved using organic and inorganic materials which are suitable for photo patterning process [5-7]. The photo patterning using organic and inorganic materials can directly transfer the surface relief profiles into substrate without further etching process. Silicate

based organic-inorganic hybrid sol-gel materials have been extensively investigated due to their considerably advantageous properties. Organic-inorganic hybrid materials based on silicate were extensively investigated due to their distinguished properties. Methacryloxy propyl trimethoxysilane (MAPTMS) Silane coupling agent was considered for practical usage due to its unique chemical and mechanical properties such as good optical transparency, high porosity and ease of processing at low temperature [8-9]. Hybrid materials structure is typically composed of organic and inorganic phase. Titania (TiO₂) is an inorganic compound that could be employed to improve the properties of optical hybrid material owing to exceptional properties such as high refractive index, hardness, transparency and good chemical and thermal stability [10]. The organic and inorganic networks SiO₂/TiO₂ are promised as potential material for wide range of optical applications since their optical properties can be tuned by controlling amount of composition between each material [11]. Photo patterns of hybrid photosensitive material are typically obtained under UV illumination. The improvement of hybrid photosensitive materials processes in longer wavelength is therefore in focus by changing the structure of material. Photosensitive materials in longer wavelength are selected to modify the structure of hybrid photosensitive materials to get the suitable wavelength sensitivity. In addition, the absorption enhancement in suitable wavelength was achieved using noble metal nanostructures. Pt, Pd, Au and Ag nanoparticles can effectively modify the absorptivity of the hybrid photosensitive films in visible wavelength due to their strong surface plasmon resonance phenomena [12-13]. However, there are still few of research works emphasizing on the utilization of Ag-doped SiO₂/TiO₂ hybrid optical sensitive thin films for the fabrication of optical elements by specific optical interference technique.

In this present work, the preparation of photosensitive Ag-doped SiO₂/TiO₂ hybrid films was carried out by sol-gel spin coating on glass substrate. The enhancement of optical absorption of the hybrid films was achieved using Ag doping. The fabrication of optical diffractive element utilizing as-prepared was conducted by interference technique.

Experiment

A. Sample preparation

The synthesis of Ag-doped SiO₂/TiO₂ hybrid optical sensitive thin films was carried out by hydrolysis silica and titania precursor. Silicon network was prepared by dissolving silane coupling agent methacryloxy propyl trimethoxysilane (MAPTMS) in absolute ethanol and hydrolyzing with De-ionization water (DI water) and 0.2 M HCl. silver nitrate (AgNO₃) used as Ag source was added in silica solution with designated Ag concentration of 0%, 2%, 5% and 10%. The solution was homogeneously mixed and stirred for 30 min at room temperature, producing the silica network solution. Similarly, Titania network was prepared by dissolving titanium isopropoxide in acetyl acetone following by homogeneously stirred for 30 min at room temperature, resulting to titania solution. After 30 min of stirring to induce a solution network, both solutions were mixed together and continuously stirred for another 10 min. The addition of titania was used to modify the structural and optical properties of the hybrid films. Typically, photo initiator was introduced to initiate photo-induced reaction. Upon the light exposure, photo polymerization reactions were induced by photo initiator. 10% wt. of bis (.eta.5-2, 4-cyclopentadien-1-yl) bis [2, 6-difluoro-3-(1H-pyrrol-1-yl) phenyl] titanium or irgacure 784 was added to the mixture under continuous stirring for another 20 min. After vigorous homogenization via magnetic stirrer, the mixture silica and titania solution was changed to bright yellow color due to photo initiator additive. The hybrid solution was sealed and aging in darkness for 24 hr at room temperature. The sol-gel was coated on borosilicate glass slide by spin coater. Prior to coating process, glass substrates were washed using washing-up liquid, DI water, acetone and isopropyl alcohol, respectively. The precursor was spun onto borosilicate glass at 2500 rpm for 45 s. As-prepared hybrid films were baked at 80 °C for 20 min to remove all solvents. After that, the Ag-doped SiO₂/TiO₂ hybrid films were obtained.

B. Material characterization

The optical absorption of the hybrid films was investigated by UV-VIS spectroscopy (Themo Electron Corporation Helios α). The optical absorption of hybrid films was measured in visible region at 280-800 nm. The structural properties of Ag-doped SiO₂/TiO₂ were characterized by X-ray diffraction: XRD (Rigaku Mini Flex) using Cu-Kα radiation operated at 40 kV and 15 mA over 2θ range of 20-80°. For XRD investigation, the hybrid material was synthesized in powder form. The diffraction patterns were recorded with a scanning rate 2°/min. The morphologies of hybrid films and element analysis were examined by Transmission electron micrograph: TEM (FEI, TECNAI G2 20). The preparation of the samples was conducted by diffusing a detached piece of hybrid film in NaOH and droped onto copper grids.

C. Fabrication of diffractive optical element

Photo patterning was fabricated on Ag-doped SiO₂/TiO₂ hybrid optical sensitive thin films using interference pattern method. Fig. 1 shows the schematic diagram of interference photo patterning set up. A He-Cd laser source (λ=442 nm; 35 mW) is used to create the interference pattern. The laser beam was collimated using lens LA 1509 to minimize the interference due to the multiple surfaces reflection. Then the laser beam is split into two paths by 3:1 beam splitter. The optical paths are set to be equal before interfering and expanded by concave lens LA 1582, then collimated using the convex lens LA 1443, respectively. This set up is used for

fabrication of diffraction optical elements whose grating period is presented by the following equation;

$$v = \frac{\lambda}{2 \sin\theta}$$

Where λ is the wavelength of the He-Cd laser and θ is the incident beam angle on the Ag-doped SiO₂/TiO₂ films. This formula suggests that the spacing of the fringe can be assigned by changing the incident angle of two beams.

D. Diffractive optical element characterization

The surface profile and period of the diffraction optical element was examined by Atomic force microscopy: AFM (Seiko Instruments, SPA400). The diffraction patterns of diffraction optical elements are characterized using green laser diode.

Results and discussions

X-ray diffraction spectra of as-synthesized Ag-doped SiO₂/TiO₂ powders are shown in Fig. 2. The XRD patterns of sample can be indexed to the silicon titanium oxide (Si-O-Ti) phase. The peak observed for Si-O-Ti phase is accorded with data standards JCPDS (ICSD 89-8099). The XRD study indicated the formation of silver (Ag) crystalline phase. Five distinct peaks at 2θ value of 38.1°, 44.3°, 51.6°, 64.4° and 77.3° correspond to cubic Ag orientation (111), (200), (104), (220) and (311), respectively [14-15]. One can see the titanium oxide (Ti-O) phase at 52.01°. These results suggested that the silver, silica and titania are mixed on an atomic scale.

TEM image of Ag-doped SiO₂/TiO₂ thin film is shown in the Fig. 3(a). The silver nano particles are spherical in shape with a smooth surface morphology and size of Ag nanoparticles varies in the range of 5-15 nm. These result indicated the Ag nanoparticles are well-dispersed on SiO₂/TiO₂ film. The result is also agreeable with the X-ray diffraction spectra. The element analysis of the sample was performed using EDX spectroscopy as shown in Fig. 3(b). The corresponding result shows the existence of Ag, Si, Ti and O in as-prepared thin film. The formation of the Ag-doped SiO₂/TiO₂ thin film was affirmed by EDX result.

Fig. 4 shows the optical absorption spectra of the Ag-doped SiO₂/TiO₂ thin films with various Ag content at 0%, 2% and 10%. The broad peak in UV region at about 300-350 nm is associated to the typical TiO₂ absorption [16].

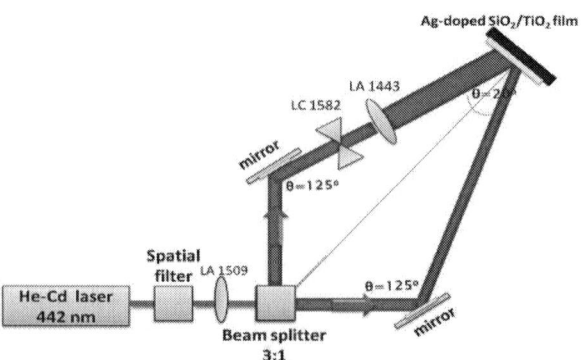

Fig. 1 schematic of interference photo patterning set up.

978-1-4673-4840-9/13 $31.00 © 2013 IEEE

The transparent film becomes bright yellow and shows the prominent absorption band at 410 nm due to the surface plasmon resonance (SPR) of the embedded silver nano particles. SPR peak additionally confirms the reduction of silver oxide to silver nanoparticles. The SPR peak might be related to the shape and size of silver nano particles on the surface of the films [17-18]. This result shows SPR phenomena which become stronger and more intense as the silver doping content increase to 2%. On the other hand, greater amount of silver content at 10% results to the decreasing SPR effect, that may due to agglomeration of silver particles. The result implies the good formation of silver nanoparticles, accompanying the strong SPR-induced absorption band in visible region.

The fabrication of diffraction patterns on the Ag-doped SiO$_2$/TiO$_2$ thin film was investigated by AFM. The photo patterns were obtain after developing process and post baking. The diffraction optical elements are shown in Fig. 5. This result shows the perfect sinusoidal profile with fringe spacing width of 1.7 µm and depth of 15 nm.

The diffraction patterns of diffraction optical elements are demonstrated using 532-nm green laser diode. The measurement set up is illustrated in Fig. 6. A collimated laser beam was used for reconstruction. The result shows the diffraction patterns in the first order [19]. The calculation of the first order diffracted beam is found to be ~30°, which is in well agreeable to the fringe spacing monitored by AFM measurement.

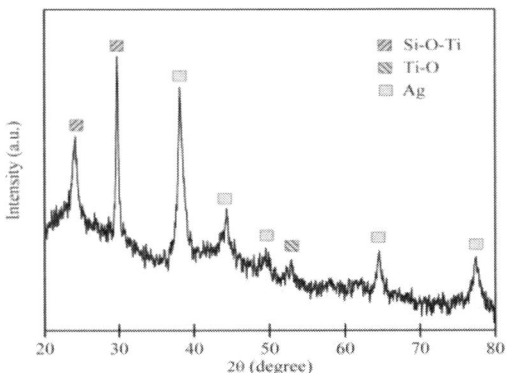

Fig. 2 Diffraction spectra of as-synthesized Ag-doped SiO$_2$/TiO$_2$ powder.

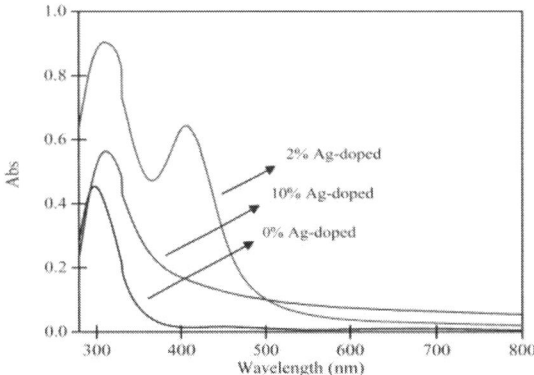

Fig. 4 Optical absorption of Ag-doped SiO$_2$/TiO$_2$ thin films with various Ag content at 0%, 2% and 10%.

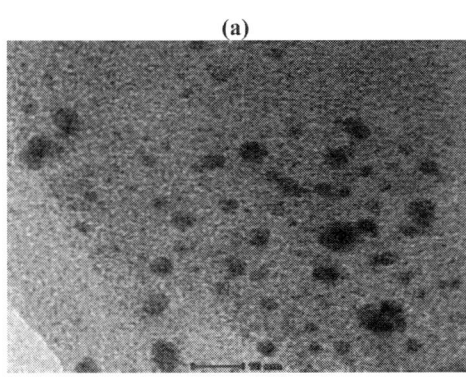

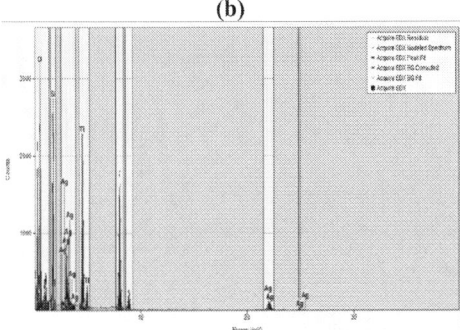

Fig. 3 TEM image of Ag-doped SiO$_2$/TiO$_2$ thin film (a) morphology (b) EDX profile.

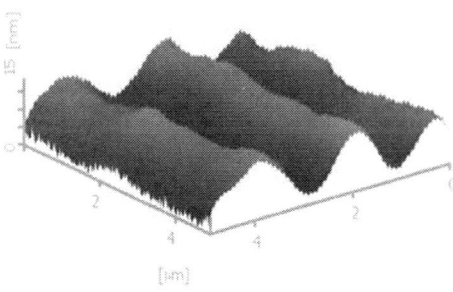

Fig. 5 AFM image of diffractive optical element using Ag-doped SiO$_2$/TiO$_2$ thin films fabricated by the interference photo patterning system.

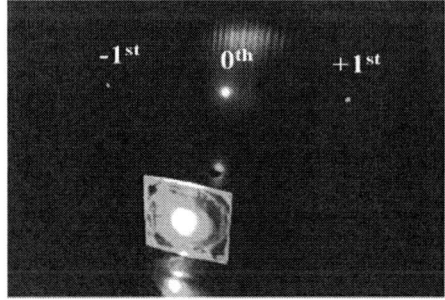

Fig. 6 Measured diffraction patterns of the fabricated diffractive optical elements using laser diode 532 nm at normal incidence.

978-1-4673-4840-9/13 $31.00 © 2013 IEEE

Conclusion

The Ag-doped SiO_2/TiO_2 hybrid optical sensitive thin films were deposited on glass substrate by sol-gel spin coating technique. The structural properties of hybrid films are confirmed by XRD analysis and TEM image. The SPR effect of silver nanoparticle on optical absorption in longer wavelength of the hybrid film is revealed. The hybrid optical sensitive films exhibited better photosensitivity in visible regime. The interference photo patterns were fabricated by simply interference lithography technique. The diffraction optical elements were obtained, which was confirmed by AFM and diffraction pattern. The fabricated diffraction optical elements on as-prepared hybrid film exhibit such good diffraction performance.

Acknowledgments

This work has partially been supported by National Nanotechnology Center (NANOTEC), NSTDA, Ministry of Science and Technology, Thailand, through its program of Center of Excellence Network. Authors would like to thank Thailand Graduate Institute of Science and Technology (TGIST), NSTDA, for student fund supported no. TG-44-22-53-069M.

References

[1] H. J. Jiang, et al., *Opt. Commun,* 185, pp. 19-24, November 2000.

[2] X. C. Yuan, et al., *J. Phys. D: Appl. Phys,* 34, pp. L125-L128, November 2001.

[3] G. Della Giustina., *Microelectron eng*, 86, pp. 745-748, December 2008.

[4] I. Fortunati, et al., *Proc. of SPIE,* 6988, pp. 69881 J-1-69881 J-9.

[5] B. S. Bae, J. Sol-gel. Sci. Techn, 31, pp. 309-315, 2004.

[6] D. J. Kang, et al., *Opt. Express*, 12, pp. 3947-3953, August 2004.

[7] J. Yang, et al., *Chinese Optics Letters*, 3, pp. 399-401, July 2005.

[8] D. D. Le, et al., *Adv.Nat.Sci.:Nanosci.Nanotechnol*, 1, pp. 015007 (5pp), May 2010.

[9] Z. Hongxi, et al., *Appl. Phys. Lett*, 84, pp. 1064 – 1066, Febuary 2004.

[10] K.Balachandran, et al., *Int. J. Eng. Sci*, 2, pp. 3695-3700, 2010.

[11] Y. Yu, et al., *Mater. Chem. Phys,*126, pp. 962-972. 2011.

[12] A. Ulatowska-Jaeza, et al., *B. Pol. Acad. Sci-chem*, 59, pp. 253-261, 2011.

[13] M. Abdul Majeed Khan, et al., *Nanoscale. Res. Lett*, 6, pp. 434 (6pp), 2011.

[14] G. V. Krylova, et al., *J. Sol-Gel. Sci. Technol*, 50, pp. 216-228. March 2009.

[15] A. Amarjargal, et al., *Ceram. Int*, 38, pp. 6365-6375, December 2012.

[16] L. Yang, et al., *J. Mater. Res*, 20, pp. 3141-3149, November 2005.

[17] F.H. Scholes, et al., *NSTI-nanotech*, 2, pp. 460-462, 2005.

[18] R. Bryaskova, et al., *J. Colloid. Interf. Sci,*349, pp. 77-85, May 2010.

[19] P. Junlabhut, et al., *Proceedings of SPIE*, 7743, May 2010.

Study of Optical Radiation Efficiency of nanoparticles

Hasan Sarwar, Md. Mydul Islam

United International University
Hs80, Rd8/A, Sat Masjid Road, Dhanmondi
Dhaka 1209, Bangladesh
hsarwar@cse.uiu.ac.bd,mmi8000@yahoo.com

Abstract

Optical radiation efficiency is the ratio of scattering cross section to extinction cross section. Optical radiation efficiency has been calculated, using classical electrostatics, for spherical nanoparticles of Ag, Au, Cu, Co, Cr, Ni, Pd, Pt, Sn, and Ti. The particle diameter was varied and calculated for the wavelengths in the range of 400-1200nm. The change in radiation efficiency in glass medium with respect to the size of the particle has been observed. The validation of calculation is confirmed by finding and comparing the efficiencies for Ag, Au, Cu in Dhumale 2012. This work has considered all the materials considered by Tanabe2007.

Keywords: Radiation Efficiency, Nanoparticles.

Introduction

Nanoscale Noble metals, while interacting with light, produce interesting phenomena through enhancement of optical and photothermal properties, known as Surface Plasmon Resonance (SPR). SPR refers to the fact that an incident photon on a metal surface creates a resonance with the collective excitation of conduction electrons. The use of SPR has opened up promising opportunity in enhancing performance of thin film solar cell [1], in the areas of imaging, sensing, biology, and medicine [2].

Optical radiation efficiency η is an indicator of re-radiation capacity of metal nanoparticles. It is the ratio of the scattering cross section to the extinction cross section. Tanabe (2007) calculated η for 11 different spherical (symmetric) nanoparticles for air medium only [3]. Dhumale (2012) calculated η for 4 different materials using air, glass and water medium [4]. In this paper, we calculate η for Ag, Au, Cu, Ni, Pt, Cr, Co, Pd, Ti, Sn in glass medium at a free space wavelength of 400nm, 700nm, and 1200nm.

Theory

Optical radiation efficiency is the ratio of scattering cross section (C_{sca}) and extinction cross section. Extinction cross section is the sum of scattering cross section (C_{sca}) and absorption cross section (C_{abs}). η is expressed as

$$\eta = \frac{Cscat}{Cabs + Cscat}$$

The scattering and absorption cross sections are given by

$$Cscat = \frac{1}{6\pi}\left(\frac{2\pi}{\lambda}\right)^4 |\alpha|^2 \text{, and}$$

$$Cabs = \frac{2\pi}{\lambda}\text{Im}[\alpha]$$

Where,

$$\alpha = 3V\left[\frac{\varepsilon_p/\varepsilon_m - 1}{\varepsilon_p/\varepsilon_m + 2}\right]$$

Here, α is the polarizability of the particle, V is particle volume, ε_p is dielectric function of the particle, ε_m is dielectric function of the surrounding medium. ε_m is assumed 3.7 for all considered wavelength. ε_p is given by $\text{Re}(\varepsilon_p)+\text{Im}(\varepsilon_p)$, where $\text{Re}(\varepsilon_p) = n^2$-$k^2$ and $\text{Im}(\varepsilon_p) = 2nk$ where n is the refractive index and k is the extinction coefficient. The values of n and k for respective wavelengths are assumed from optical constant data provided in [5].

Results and Discussion

Optical radiation efficiency is calculated for Au, Ag, Cu, Ni, Ti, Co, Pt, Cr, and Sn. Fig 1-3 shows radiation efficiency of the above materials surrounded by glass medium. Observation reveals Ag provides best efficiency in 400nm, 800nm and 1200nm wavelengths. Calculation shows as the diameter is increased from 1nm to 100nm, radiation efficiency increases. Ag and Au are promising candidates. Au shows poor performance at 400nm wavelength. It should be noted that particle dimension cannot be increased indefinitely, as the diameter should be substantially smaller compared to the wavelength to allow absorption and scattering. Ti can reach a maximum of 25% efficiency at 400nm. At 400nm, only Ag can reach upto 70% efficiency in glass medium, all other materials remain below 45%. At 700nm and for 100nm diameter particle, except Ag, Au, and Cu, all other materials have efficiency below 35%. At 1200nm and for 100nm diameter, except Ag, Au, and Cu, all other materials show efficiency below 10%.

978-1-4673-4840-9/13 $31.00 © 2013 IEEE   401

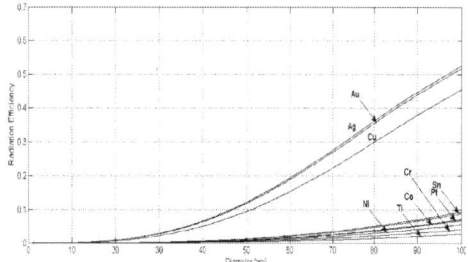

Fig. 1 Radiation efficiency for different particle sizes at 400nm wavelength surrounded with glass medium.

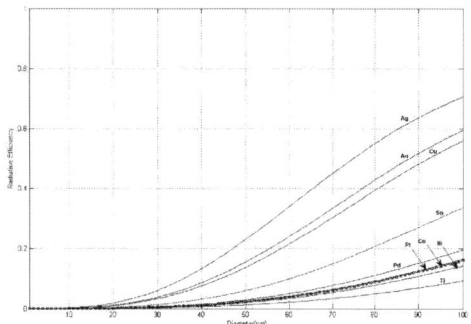

Fig. 2 Radiation efficiency for different particle sizes at 800nm wavelength surrounded with glass medium.

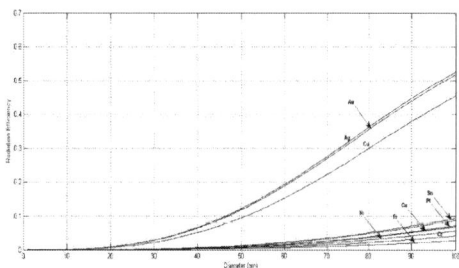

Fig. 3 Radiation efficiency for different particle sizes at 1200nm wavelength surrounded with glass medium.

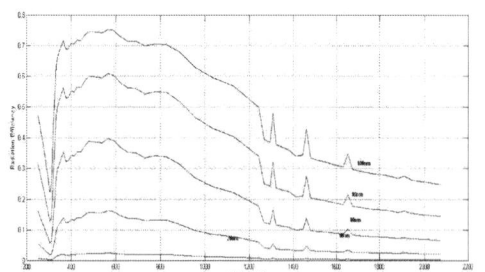

Fig. 4 Radiation efficiency of Ag nanoparticls at varying wavelength.

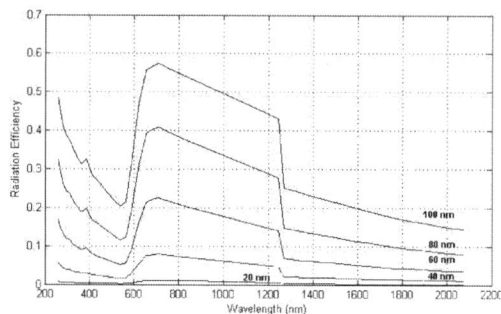

Fig. 5 Radiation efficiency of Cu nanoparticls at varying wavelength..

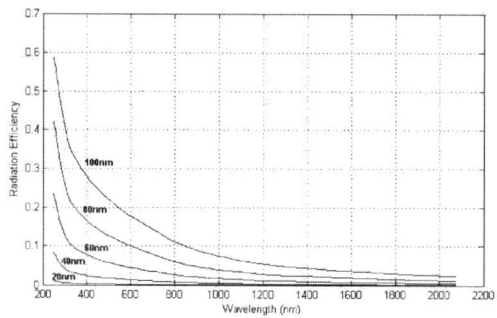

Fig. 6 Radiation efficiency of Ni nanoparticls at varying wavelength.

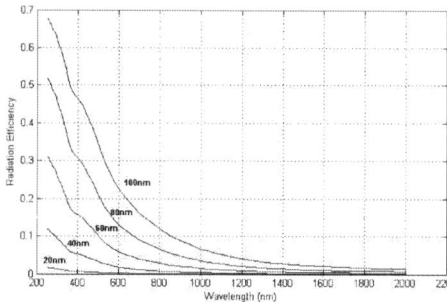

Fig. 7 Radiation efficiency of Cr nanoparticls at varying wavelength..

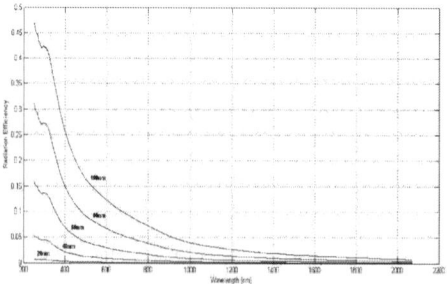

Fig. 8 Radiation efficiency of Ti nanoparticls at varying wavelength..

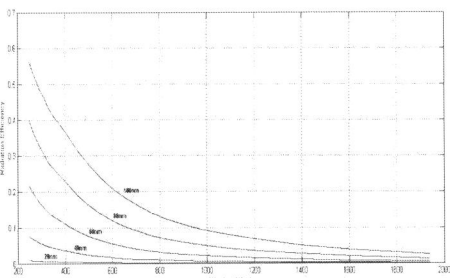

Fig. 9 Radiation efficiency of Co nanoparticls at varying wavelength..

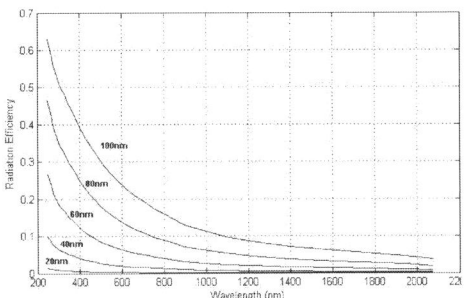

Fig. 10 Radiation efficiency of Pd nanoparticls at varying wavelength..

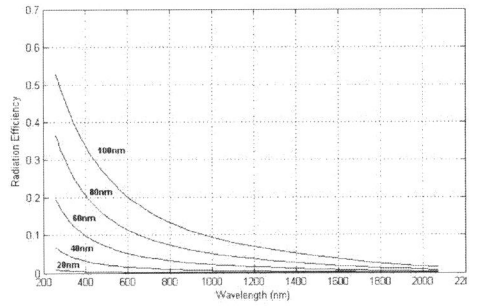

Fig. 11 Radiation efficiency of Pt nanoparticls at varying wavelength..

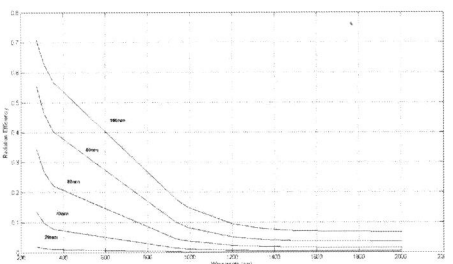

Fig. 12 Radiation efficiency of Sn nanoparticls at varying wavelength..

TABLE I
COMPARISON WITH DHUMALE [4] FOR GLASS MEDIUM

	400nm d=60nm		700nm d=60nm		1200nm d=60nm	
	Dhu male	This paper	Dhu male	This paper	Dhu male	This paper
Ag	34	34	35	34	19	19
Au	06	5	27	24	18	18
Cu	09	8	23	21	14	15
	400nm d=100nm		700nm d=100nm		1200nm d=100nm	
	Dhu male	This paper	Dhu male	This paper	Dhu male	This paper
Ag	70	70	71	70	52	52
Au	22	22	63	59	51	51
Cu	29	30	58	56	44	45

Fig.4-Fig.11 showing efficiency variation within wavelength ranging between 200nm to 2000nm surrounded with glass medium. From Fig. 4 and Fig. 5, we see Ag and Cu show better efficiency compared to Ni, Cr, Ti, Co, Pd, Pt, and Sn as shown in Fig. 6 – Fig. 12. Cu shows consistent efficiency within 600-1200nm wavelength. Above 1200nm wavelength, radiation efficiency falls below 30% for 100nm diameter particle. Efficiency falls below 40% for 100nm Ag particle above 1200nm radiation. On the contrary, Ni, Pd, Pt, Ti's efficiency falls below 30% above 400nm wavelength for 100nm diameter particle, Sn's efficiency falls below 40% above 600nm wavelength for 100nm diameter particle. The dips seen in Ag and Cu nanoparticles may be attributed to plasma resonance [3,6]. It is seen that the positions of the dips do not depend on the particle size, which is consistent with the theory of resonant frequencies and experimental results [3,6]. A comparison of efficiency values calculate in [4] is shown with the similar values we get from our calculation. It proves great similarity for those values of Ag, Au and Cu.

Conclusion

Optical radiation efficiency is calculated for Au, Ag, Cu, Ni, Ti, Co, Pt, Cr, and Sn surrounded with glass medium. Of them, Ag, Au, and Cu shows higher efficiency within 600-1200nm wavelength ranges. Other materials could not produce efficiency of more than 30% within that region. These data might be useful to understand the behavior of these materials while surrounded with glass medium.

Acknowledgement

We would like to acknowledge Prof. Dr. M. Rezwan Khan, Vice Chancellor of United International University for his helpful suggestions and support.

References

[1] K. R. Catchpole, et al., *Optics Express,* 16, pp. 21793-21800, December 2008.

[2] K. J. Prashant, et al., *Accounts of Chemical Research*, 41, pp. 1578-1586, December 2008.

[3] K. Tanabe, et al., *Materials Letters*, 61, pp. 4573-4575, March 2007.

[4] V. A. Dhumale, et al., *Bul. Mater. Sci.*, pp. 143-149, April 2012.

[5] E.D. Palik (Ed.), *Handbook of Optical Constants of Solids*, Orland, Academic Press, 1985

[6] H.R. Stuart, et al., *Appl. Phys. Lett.*, 73, pp. 3815-3817, December, 1998.

Nano-needle Pressure Sensor Integrated with Printed Organic Transistors

Jiseok Kim[1], Tse Nga Ng[2], Woo Soo Kim[1]

1. School of Engineering Science, Simon Fraser University, Surrey, BC Canda, V3T 0A3
2. Palo Alto Research Center, 3333 Coyote Hill Road, Palo Alto, CA 94304, USA
Email: tnng@parc.com, woosook@sfu.ca

Abstract

This paper presents a highly sensitive capacitive pressure sensor composed of a polymer dielectric film with nano-needle structures made from the breath figure method. The pressure sensitivity of the sensor reached 1.76 kPa^{-1} in the low pressure range (<1 kPa), which is comparable to the sensitivity of human skin. The pressure sensors were integrated with inkjet-printed organic thin film transistors to enable flexible, large-area tactile sensing applications. (Keywords: pressure sensor, organic transistors and printed flexible electronics)

Introduction

Mimicking the sense of touch has been very challenging, because the human body is capable of detecting touch with a mass-loading sensitivity better than 1 kPa (the pressure of 10 grams on an area of 1 cm^2). A highly sensitive pressure sensor is needed for the development of artificial skin [1-4]. Ideally the pressure sensors should be conformal and allow large-area coverage. Processes using solution materials, lamination, and inkjet printing are advantageous in allowing additive fabrication over a variety of different surfaces. Inkjet printed organic thin-film transistors (OTFTs) have been used as large-area switching matrices in photosensor and memory as shown in Fig. 1 [5, 6], and here OTFTs are combined with pressure sensors to enable current read-out and device structures that will be useful for integration into sensor matrix arrays.

Experimental Procedure

The breath figure method was applied to fabricate nano-needle arrays [7] in Fig. 2(a). In brief, a PS-b-P2VP co-polymer solution was drop-cast onto a glass substrate inside a humid chamber (>90 RH%) at room temperature, to allow condensation of water droplets on the polymer film. A honeycomb array was obtained after evaporation of water and organic solvent. Then adhesive tape was used to remove the top portion of the polymer film and expose a nano-needle surface. Tips of the nano-needle array are very sharp with a radius of curvature around 10 nm. A capacitive pressure sensor was fabricated by placing the nano-needle dielectric film between two aluminum electrodes. The sensor capacitance was measured to be 0.05 pFcm^{-2} at 1 kHz without external pressure. Here the OTFTs [Fig. 2(b)] were fabricated by inkjet printing [6,8]. The nano-needle capacitor was integrated with an OTFT by placing the capacitor over the OTFT gate electrode, as shown in Fig. 3.

Results and Discussions

A. Sensitivity of nano-needle pressure sensor

Fig. 4 compares the pressure response of dielectric films with solid, hemispherical, or nano-needle structures. In the low pressure range below 1kPa, the sensitivity of the nano-needle sensor reached 1.76 kPa^{-1}, which exceeds the typical human-skin pressure sensitivity. For pressure over 1.5 kPa, the sensitivity of the nano-needle sensor decreases to the level similar to that of a solid film, because the base portion of the nano-needle film is being compressed. Likewise, a pressure sensor with the hemispherical filler shows higher sensitivity than the solid film in the low pressure range, and then the sensitivity of the two structures converge at higher pressure.

B. OTFT characteristics

The printed OTFTs are top-gate structures, with a p-type semiconductor and a fluoropolymer dielectric with capacitance C_g = 5 nF/cm^2. The transistor channel length is 35 ±5 μm. The effective mobility of the devices is 0.1 cm^2 V^{-1} s^{-1} at a gate voltage of 20 V. The typical OTFT transfer and output characteristics are shown in Fig. 5.

C. Current response of the sensor integrated with an OTFT

To demonstrate direct current read-out, the nano-needle pressure sensors were integrated with organic transistors. When pressure was applied, the nano-needle dielectric decreased in thickness and thus the total capacitance between the semiconductor layer and the electrode above the structured film increased. This increase in capacitance led to increased charge accumulation in the OTFT channel, which in turn increased the OTFT drain current.

Since a typical OTFT had a small loading area, it was challenging with conventional tools to apply a force small enough to stay below the pressure range of less than 1 kPa. To overcome this difficulty, a paper force gauge [9] was introduced for characterization at low pressure. By varying the input voltage to the paper force gauge, its piezoresistive element bent to apply pressure; the corresponding movement was also tracked by measuring the change in current through the gauge. Thus, with the help of the force gauge, we were able to obtain relative measurements for pressure below 1 kPa. While the force gauge was applying pressure to the integrated OTFT-sensor, the drain current of the OTFT was measured by a Keithley 2400 source meter, with source-drain voltage at -10V and gate voltage at -50 V. The output current of the gauge and the drain current of the OTFT were simultaneously measured under increasing pressure, as shown in Fig. 6. The OTFT-sensor with a nano-needle dielectric film shows about 3 times higher sensitivity compared to one with a solid unpatterned film; the slopes are -51.44 and -16.31, respectively. The OTFT nano-needle sensors provide linear current response to pressure, and this result is promising for incorporating such OTFT pressure sensors into a matrix to improve spatial resolution.

978-1-4673-4840-9/13 $31.00 © 2013 IEEE

Conclusions

A capacitive pressure sensor with a nano-needle dielectric has been fabricated by a facile breath figures method and demonstrated a record sensitivity (up to 1.76 kPa⁻¹) among polymer tactile sensors. The capacitor with the nano-needle dielectric was integrated with an inkjet printed OTFT. The current of OTFTs increased with pressure due to an increase in capacitance, which in turn enhanced charge accumulation in the conduction channel. The structural advantage of the nano-needle film facilitated very high sensitivity for the capacitive pressure sensors. These pressure sensors integrated with printed OTFTs will be enabling for many flexible, large-area pressure sensing applications.

References

[1] J. J. Boland., *Nat. Mater.* **9**, 790 (2010).

[2] T. Someya, et al., *Proc. Natl. Acad. Sci. U.S.A.* **102**, 12321 (2005).

[3] S. C. B. Mannsfeld, et al., *Nat. Mater.* **9**, 895 (2010).

[4] J. Kim, T. N. Ng, and W. S. Kim, *Appl. Phys. Lett.* **101**, 103308 (2012).

[5] T. N. Ng, et al., *Adv. Mater.* **21**, 1855 (2009).

[6] T. N. Ng, et al., *Org. Electron.* **12**, 2012 (2011).

[7] J. Kim, B. Lew, and W. S. Kim., *Nano. Research Lett.* **6**, 616 (2011).

[8] T. N. Ng, et al., *Sci. Rep.* **2**, 585 (2012).

[9] T. Aktar, J. Joseph, and W. S. Kim, *IEEE Electron Dev. Lett.* **33**, 902 (2012).

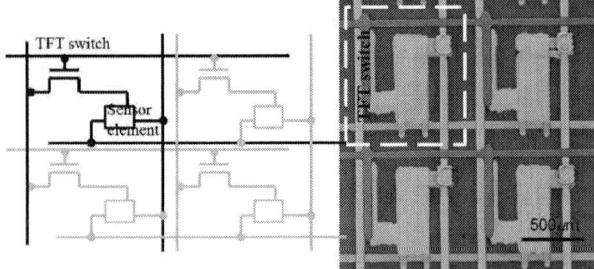

Fig. 1 (a) Schematics of an active-matrix TFT array. (b) Optical micrograph of a printed active-matrix TFT array.

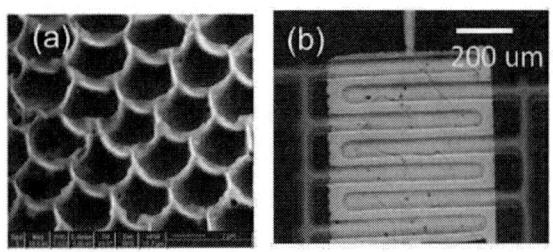

Fig. 2 (a) SEM image of a nano-needle film. (b) Optical micrograph of an inkjet printed OTFT with inter-digitated source/drain electrodes.

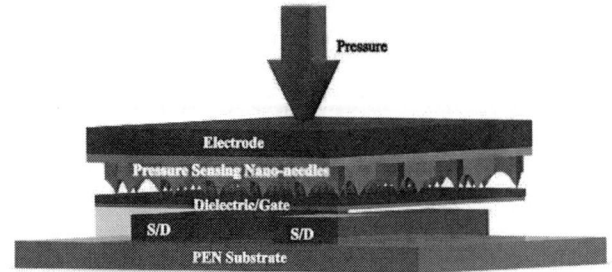

Fig. 3 Schematics of the pressure sensor integrated on a top-gate OTFT.

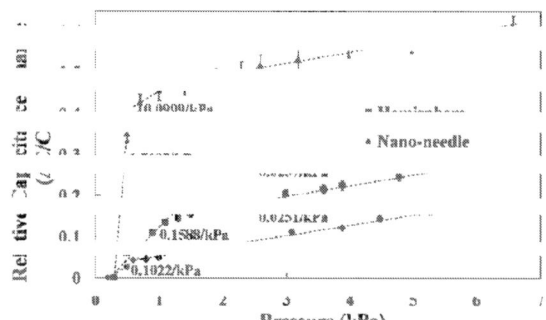

Fig. 4 Relative change in capacitance versus pressure for dielectric films with solid, hemispherical, or nano-needle structures.

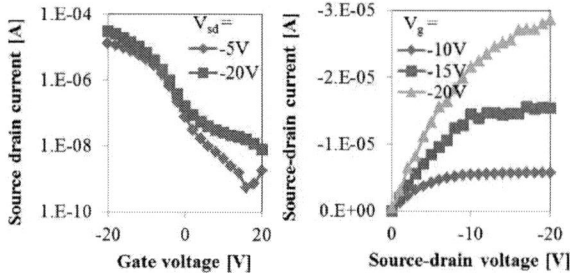

Fig. 5 Transfer and output characteristics of the printed OTFT, with W=12mm and L=35μm.

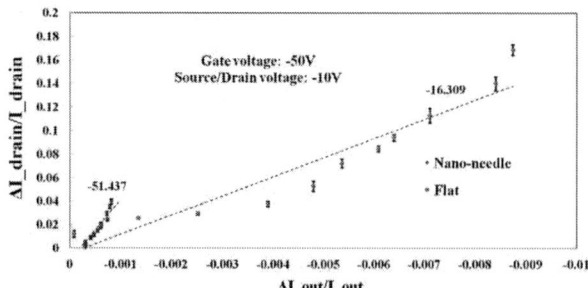

Fig. 6 Relative change in the drain current I_{drain} of an OTFT versus the force-gauge current I_{out}. The absolute current through the force gauge is tuned to apply pressure less than 1 kPa.

Resonant Cavity Far Infrared Photo-detector based on Self-Assembled InAs/GaAs Quantum Dots

C.M. S. Negi[1*], Dharmendra Kumar[2], Saral K. Gupta[3] and Jitendra Kumar[2]

[1] Department of Electronics, School of Physical Sciences, Banasthali University, Rajasthan-304022, India, [2] Department of Electronics Engineering, Indian School of Mines, Dhanbad, Jharkhand- 826 004, India, [3]Department of Physics, School of Physical Sciences, Banasthali University, Rajasthan-304022, India

[1*] nchandra@banasthali.in

Abstract

In this paper, a novel design for far infrared resonant cavity enhanced quantum dot photo-detector (RCE QDP) based on bound-to-bound intervalence subband transitions in self assembled InAs/GaAs quantum dots (QDs) is presented. The device structure consists of QDs active layers inserted between two distributed Bragg reflectors (DBRs). Quantum efficiency is calculated as a function of wavelength for different reflectivity of mirrors. Dark current is calculated as a function of the voltage for different value of temperature and QD radius.

Keywords: Resonant cavity, quantum dot photodetector, quantum efficiency, dark current, valance subband structure, intersubband transitions.

Introduction

The far infrared radiation (FIR) is of particular interest for various applications including, biomedical sensing, studies of lattice vibrations in crystals, anti-ferromagnetic resonance, energy gap measurements in superconductors [1] and imaging through fog and imaging through dust. Photo-detectors currently operating in this range are based on intersubband transitions in semiconductor quantum wells [2], quantum coherence in coupled quantum wells [3], germanium or silicon doped with shallow impurities [4] and quantum dot infrared photodetectors (QDIP) [5]. The most dramatic result of using the QDIP to detect mid and far-infrared signals was its high-temperature performance [6], and insensitivity to incident light polarizations. So far, however, most of the QDIP devices reported in the literature are based on intersubband transitions in the conduction band of semiconductor QDs [7]. Due to the complexities related to the valence band, hole intersubband transitions have not been explored as intensively as their electronic counterparts.

In this paper we purpose a far infrared photodetector based on bound-to-bound intervalence subband transitions in InAs/GaAs quantum dots placed inside a Fabry-Perot resonant microcavity. Such resonant cavity enhanced (RCE) devices has advantage of wavelength selectivity and the huge increase of the resonant optical field introduced by the cavity. The increased optical field allows RCE photodetector structures to be thinner and therefore faster, while simultaneously increasing the quantum efficiency at the resonant wavelengths. Off-resonance wavelengths are rejected by the cavity making RCE photodetectors promising for low crosstalk applications. Self assembled InAs/GaAs QD is modeled by anisotropic parabolic potential along x-y plane and finite well potential along growth (z-direction). Luttinger Kohn Hamiltonian is numerically solved to calculate the Eigen energies, Eigen vectors and wave functions of the QDs. Calculated energy levels and wave functions are subsequently used to calculate dipole matrix elements, absorption coefficients, quantum efficiency and dark current of resonant QDP.

Device Model

The purposed RCE QDP general structure is shown in figure 1. The structure consists of InAs/ GaAs QDs layer placed in a Fabry-Perot resonant microcavity. The mirrors of resonant cavity consist of 5 pairs of GaAs/Al$_x$Ga$_{1-x}$As DBR, top mirror has reflectivity R$_1$ and bottom mirror has reflectivity R$_2$.

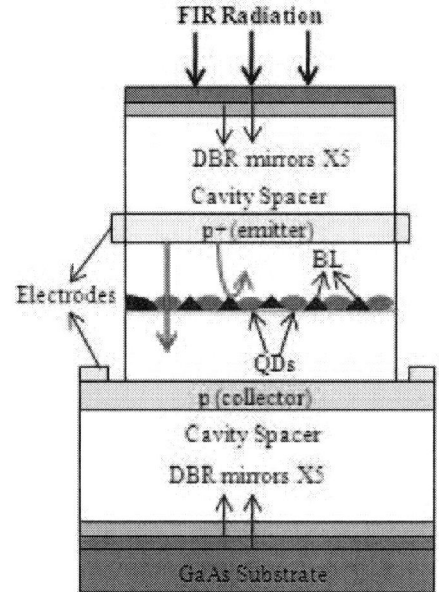

Fig. 1 Structure of RCE QDP

The active region consists of layer of AlGaAs in which array of QDs is buried. The area between the QDs is covered by AlGaAs material, due to its large band gap it acts as a blocking layer (BL). We assume that barrier height formed by BL and its thicknesses are sufficiently large that both tunneling currents and thermionic currents through the area covered by BL are negligible. The p$^+$ and p regions are made-up of GaAs material and act as QDP emitter and collector respectively.

To describe the valance band structure of QD, We have numerically solved 4X4 Luttinger Hamiltonian as given by equation (1) [8],

$$H = \begin{bmatrix} H_{hh} & b & c & 0 \\ b^{\dagger} & H_{lh} & 0 & -c \\ c^{\dagger} & 0 & H_{lh} & b \\ 0 & -c^{\dagger} & b^{\dagger} & H_{hh} \end{bmatrix} \quad (1)$$

978-1-4673-4840-9/13 $31.00 © 2013 IEEE

Where,

$$H_{hh} = -\frac{\hbar^2}{2m_0}\left[(\gamma_1 - 2\gamma_2)\frac{d^2}{dz^2} + (\gamma_1 + \gamma_2)\left(\frac{d^2}{dx^2} + \frac{d^2}{dy^2}\right)\right] \quad (2)$$

$$H_{lh} = -\frac{\hbar^2}{2m_0}\left[(\gamma_1 + 2\gamma_2)\frac{d^2}{dz^2} + (\gamma_1 - \gamma_2)\left(\frac{d^2}{dx^2} + \frac{d^2}{dy^2}\right)\right], \quad (3)$$

$$b = \frac{\sqrt{3}\hbar^2}{2m_0}\gamma_2\frac{d}{dz}\left(\frac{d}{dx} - i\frac{d}{dy}\right), \quad (4)$$

$$c = \frac{\sqrt{3}\hbar^2}{2m_0}\gamma_3\left(\frac{d}{dx} - i\frac{d}{dy}\right)^2. \quad (5)$$

In the above equations γ_1, γ_2, γ_3 are the Luttinger parameters. Solution of the in-plane parabolic confinement potential would be the harmonic oscillator wave function. Using these wave functions as our basis, the matrix element can be written as

$$H_{hh} = [E_z^{hh} + \hbar\omega_x^{hh}(n_x + \frac{1}{2}) + \hbar\omega_y^{hh}(n_y + \frac{1}{2})]\delta_{n_x,n_x'}\delta_{n_y,n_y'} \quad (6)$$

$$H_{lh} = [E_z^{lh} + \hbar\omega_x^{lh}(n_x + \frac{1}{2}) + \hbar\omega_y^{lh}(n_y + \frac{1}{2})]\delta_{n_x,n_x'}\delta_{n_y,n_y'} \quad (7)$$

The analytical expression for the absorption coefficient can be obtained by using density matrix approach [9] and can be written as

$$\alpha(\omega) = \omega\sqrt{\frac{\mu}{\varepsilon_r}}\frac{\sigma\hbar\Gamma|M_{fi}|^2}{(E_f - E_i - \hbar\omega)^2 + (\hbar\Gamma)^2} \quad (8)$$

Where, σ is carrier density in quantum dot, ε_r is the dielectric permittivity of semiconductor, $\hbar\omega$ is incident photon energy, $\hbar\Gamma$ is spectral line width, and E_i is valence sub band energy of initial states and E_f is valence sub band energy of final states. M_{fi} is the dipole matrix element of the transition between the valence sub-bands and is defined as

$$M_{fi} = e\int\psi_f x\psi_i dx \quad (9)$$

Where, ψ_f and ψ_i are wave function of initial and final states of valence sub band in quantum dot.

The Quantum efficiency (η) of a photodetector placed inside a resonant cavity at wavelength (λ) is given by [10]:

$$\eta = \left\{\frac{[1 + R_2 e^{-\alpha(\omega)d}]}{1 - 2\sqrt{R_1 R_2}e^{-\alpha(\omega)d}\cos(4\pi nL/\lambda + \phi_1 + \phi_2) + R_1 R_2 e^{-2\alpha(\omega)d}}\right\} X$$
$$(1 - R_1)(1 - e^{-\alpha(\omega)d}) \quad (10)$$

Where, L is the cavity Length, n is refractive index, ϕ_1 and ϕ_2 are phase shifts of the mirrors.

Dark Current Model

A critical parameter in the operation of FIR detectors is the dark current, because it determines the appropriateness of the device to be used in high temperature operation. In this paper we used a model which includes the thermionic emission of holes from QDs and field assisted tunneling of holes from QDs to describe the dark current.

Fig 2 schematically illustrates the capture, thermionic emission and field assisted tunneling through the potential barrier.

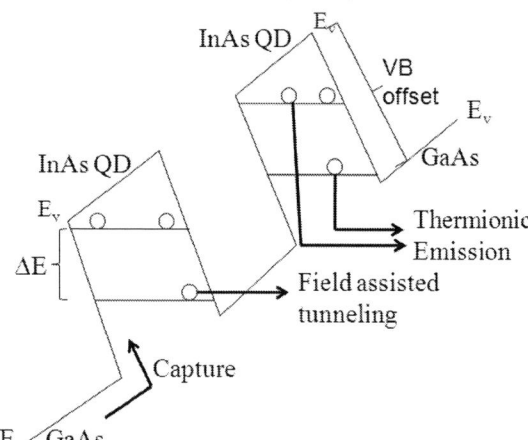

Fig. 2 Valance band schematic of InAs–GaAs QDP

The detailed balanced relation used to equate a rate of hole capture into and hole emission from QDs can be expressed as

$$\frac{J(dark)p_c}{e\sigma_{QD}} = G_{th} + G_{tun} \quad (11)$$

Where σ_{QD} is the QD sheet density in each QD layer, p_c is the capture probability of a hole crossing a QD having BL and can be presented as [11]

$$p_c = \frac{p_{oc}(N_{QD} - N)}{N_{QD}}, \quad (12)$$

Where, p_{oc} is the capture probability for uncharged QDs, N_{QD} is the maximum no of holes which can occupy each QD, G_{th} is the rate of hole emission from QDs associated with thermionic emission and can be written as

$$G_{th} = G_o \exp(-E_i/kT)\exp\left[\frac{\pi\hbar^2 N}{m_{lh}^* kTr_{QD}^2}\right], \quad (13)$$

Where G_o is the thermionic emission rate constant, E_i is the ionization energy of the ground state of QD, m_{lh}^* is the effective mass of light hole, r_{QD} is QD radius, h is the Plank constant and k is the Boltzmann constant and T is the absolute temperature.

The field assisted tunneling for one dimensional triangular barrier and can be expressed as [12]

$$G_{tun} = G_{ot}\exp\left(-\frac{4}{3}\frac{\sqrt{2m_{lh}^* e}\phi_B^{3/2}}{hE}\right)\exp\left(-\frac{\Delta E}{kT}\right)$$
$$X\exp\left[\frac{\pi\hbar^2 N}{m_{lh}^* kTr_{QD}^2}\right], \quad (14)$$

Where, G_{ot} is the field-assisted tunneling emission rate constant, ΔE is the energy difference from the quantum dot ground state to the highest filled quantum dot energy level, e is the electronic charge, E is the electric field across the device, ϕ_B is the potential barrier height from the uppermost filled quantum dot energy level, and is expressed as

$$\Phi_B = \frac{(VB_{os} - E_{nh} - 0.5eEL_z)}{e}, \quad (15)$$

978-1-4673-4840-9/13 $31.00 © 2013 IEEE

Where VB_{os} is the valance band offset between QD and barrier, E_{nh} is the energy level of the hole in the valance sub-band and L_Z is the confinement length along transverse direction.

The analytical expression of N by assuming $N < N_{QD}$ can be obtained from a detailed balance relation and is given by

$$N = \frac{\ln\left[\dfrac{j_{max} r_{QD}^2 \exp\left(\dfrac{L_e}{L_e + L_c}\dfrac{(v_d + v)e}{kT}\right)}{e[G_o \exp(-E_i/kT) + G_{ot1}]}[1 + \exp(-ev/kT)]\right]}{\left(\dfrac{e^2}{kTC} + \dfrac{\pi \hbar^2 N}{m_{lh}^* kT r_{QD}^2}\right)} \quad (16)$$

Where j_{max} is the maximum current density that can be extracted from the emitter, Le and Lc are the distance between emitter and collector from the QD layer, respectively, $vd = 2\pi e N_A L_e (Le + Lc)/\varepsilon_{QD}$, N_A is the acceptor concentration, ε_{QD} is permittivity, $C = \dfrac{2\varepsilon_{QD} r_{QD}}{\pi\sqrt{\pi}}$ is the QD capacitance, and

$$G_{ot1} = G_o \exp(-E_i/kT) + G_{ot} \exp\left(-\frac{4}{3}\frac{\sqrt{2m_{lh}^* e}\phi_B^{3/2}}{hE}\right)\exp\left(-\frac{\Delta E}{kT}\right)$$

One can obtain the equation of dark current as

$$J(dark) = e/2\sigma_{QD} \frac{[1 - \exp(-ev/kT)]}{[1 + \exp(-ev/kT)]}(G_{th} + G_{tun}) \quad (17)$$

Results and Discussions

In this section we present results for FIR QDP structure, we have used following parameters: L_e=50nm, L_c=500nm, L=30μm, σ_{QD}=10^{14} meter^{-2}, m_{lh}=0.027m_o. Luttinger Hamiltonian in the presence of strain is numerically solved for different values of r_{QD} and L_z= 5nm to calculate various parameters used in equation (13-14) and the evaluated parameter is given in table 1.

Table1

r_{QD} (nm)	E_{nh} (eV)	E_i (eV)	ΔE (eV)	ϕ_B (eV)
3	0.0592	0.418	0.1153	0.410
6	-0.0683	0.5383	0.0518	0.5158

G_o, G_{ot} and j_{max} are used in fitting parameters in the model and are based on [11, 13].

Fig. 3 (a) shows quantum efficiency (η) of FIR RCE QDP of QD radius 3 nm and L_z= 5nm as a function of wavelength for different values of R_1 for fixed value of R_2=0.98. Figure shows oscillatory behavior as a result of resonant cavity structure. For The wavelength corresponding to the maximum η the waves confined between reflectors interfere constructively and for wavelength corresponding to the η minimum waves interfere destructively. Fig. 3(a) also indicates the maximum attainable η for conventional QDP for the same active layer thickness, η for conventional QDP is multiplied 5 times in the graph to make out the comparison. It is noticeable that η of RCE is exceptionally larger than the η of conventional QDP. It is also observe from the figure that the conventional QDP provides roughly constant η over a wide wavelength range while RCE QDP can be design to have significantly enhanced η at specific wavelength. This wavelength selectivity property of RCE QDP can used a background noise reduction factor, which blocks unwanted background noise outside the pass-band of resonant cavity. Figure 3(b) shows η of RCE QDIP of QD radius of 6nm as a function wavelength. It is observe from the figure (3) that increase in QD radius shows the shift in the peak wavelength, larger QD radius decreases the separation between energy levels in the valance subband that results in the shift in the value of peak wavelength towards higher values.

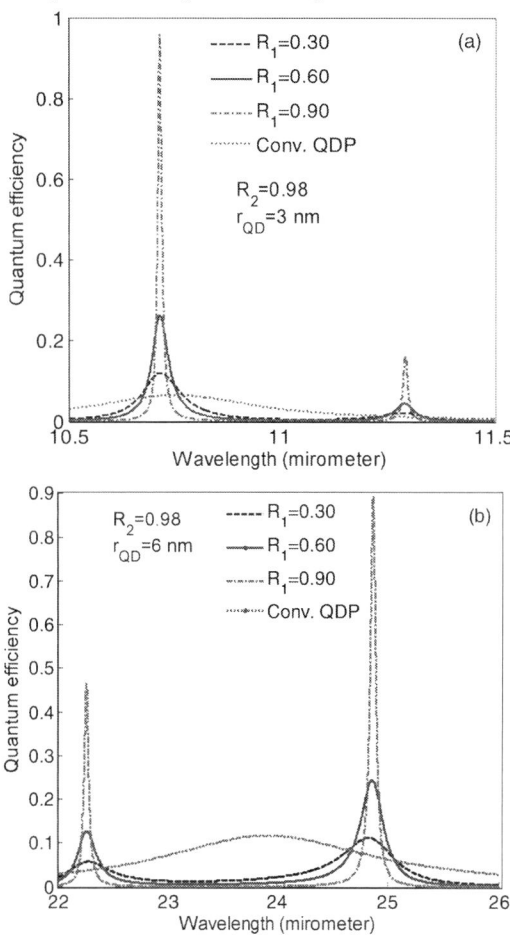

Fig. 3 Quantum efficiency as a function of wavelength for different values of R_1 (a) for r_{QD}=3 nm (b) for r_{QD}=6nm.

Figure 4 display temperature dependent dark current-voltage characteristics of RCE QDP from 180-300K. It is observe that as expected the dark current increases with the increase in temperature and the bias voltage. Figure shows that at 0.5 V as the temperature increased from 180 K to 300 K the dark current density increases 3 orders of magnitude from 1.4×10^{-5} A/cm^2 to .01A/cm^2, this rapid increase of dark current density with temperature at lower applied voltage is due to its exponential dependence on temperature. The main mechanism producing the dark current in this region is thermionic emission of hole confined in the QD heterostructure supported by its exponential dependence on temperature. At 5 V as the temperature increased from 180 K to 300 K the dark current density increases only 20 times in magnitude from 0.33 A/cm^2 to 6.56 A/cm^2 , in high applied bas region the dark current density mainly depends on applied voltage this is due to the field assisted tunneling process of the holes in the QDs.

Figure 5 shows dark current density as a function of applied voltage for two different QD radius at T=300k. Figure display large dark current density for small r_{QD}, this can be attributed to the small value of ionization energy for r_{QD}=3 in comparison with the r_{QD}=6 as given in table 1.

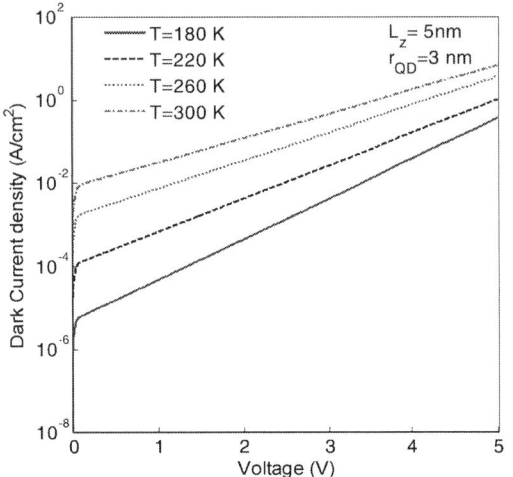

Fig. 4 Dark current density as a function of applied voltage at different temperature.

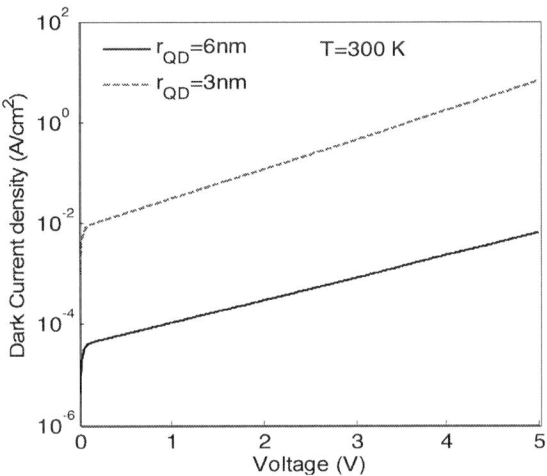

Fig. 5 Dark current density as a function of applied voltage at for different QD radius.

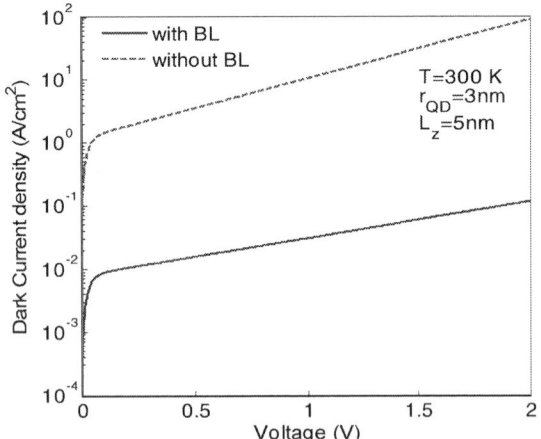

Fig. 6 Dark current density as a function of applied voltage at for different QD radius.

Figure 6 shows comparison of the dark current characteristics

for RCE QDP with BL and without BL. This figure clearly indicates the AlGaAs layer used in the active region of RCE QDP acts as blocking layer for the dark current and reduces dark current by a greater extent in comparison with the RCE QDP without BL.

Conclusion:

We have investigated QE and characteristics of dark current in FIR RCE QDP. Valance subband structure of QDs embedded in the device is evaluated by numerical digonalization of Luttinger Hamiltonian. QD radius has the effect of shifting the peak wavelength of the photodetector. We find prominent effect of QD radius on dark current density. Blocking layer inserted between QDs in the active region of the device significantly reduces dark current density of QDP. To the best of our knowledge, no experimental and theoretical studies related to the QDIP based on the valance subband transitions in QDs are available for comparison of our results. We believe that this study will make a significant contribution to the development of FIR photodetectors.

References

[1] M. F. Kimmitt, *"Far-Infrared Techniques"*, Pion Limited, London, 1970.

[2] A. G. U. Perera, S. G. Matsik, H. C. Liu, M. Gao, M. Buchanan, W. J.Schaff, and W. Yeo, "GaAs/InGaAs quantum well infrared photodetector with a cutoff wavelength at 35 μm" *Appl. Phys. Lett.,* Vol. 77, No.5, pp.741-743, 2000.

[3] A. K. Wojcika, F. Xiea, V.R. Chagantia, A. Belyanina an J. Konom, "Mid/far-infrared photodetectors based on quantum coherence in coupled quantum wells", *Journal of Modern Optics* Vol. 55, No. 19–20, pp. 3305–3317, 2008.

[4] J. Wolf and D. Lemke, *Infrared Phys. 25, 327,* 1985.

[5] S. J. Xu, S. J. Chua, T. Mei, X. C. Wang, X. H. Zhang, G. Karunasiri, W. J. Fan, C. H. Wang, J. Jiang, S. Wang, and X. G. Xie, "Characteristics of InGaAs quantum dot infrared photodetectors", *Appl. Phys. Lett.,* Vol. 73, No. 21, pp 3155-3177, 1998.

[6] J. W. Kim, J. E. Oh, S. C. Hong, C. H. Park, and T. K. Yoo, "Room temperature far infrared (8-10μm) photodetectors using self-assembled InAs quantum dots with high detectivity, *IEEE Electron Device Lett.,* Vol. 21, pp. 329-331, 2000.

[7] V. Ryzhii1, I. Khmyrova, M. Ryzhii and V. Mitin, "Comparison of dark current, responsivity and detectivity in different intersubband infrared photodetectors", *Semicond. Sci. Technol,* Vol. 19, pp. 8–16 2004.

[8] J. Kumar, S. Kapoor, S.K. Gupta and P.K. Sen, Theoretical investigation of the effect of asymmetry on optical anisotropy and electronic structure of Stranski-Krastnov quantum dots, *Phys. Rev. B,* Vol. 74, pp. 115326(1-10), 2006.

[9] G. H. Wang, and Q. Guo, and K. X. Guo, "Refractive Index Changes Induced by the Incident Optical Intensity in Semiparabolic Quantum Wells", *Chin. J. Phys.,* Vol 41, pp. 296-306, 2003.

[10] M. Selim lhiiia and S. Strite, "Resonant cavity enhanced photonic devices" *J. Appl. Phys., Vol. 78, No. 2, 1995,* pp. 607-639.

[11] V. Ryzhii, "Physical model and analysis of quantum dot infrared photodetectors with blocking layer", *J. Appl. Phys.,* Vol. 89, No. 9, pp. 5117-5124, 2001.

[12] P. Martyniuk and A. Rogalski, "Insight into performance of quantum dot infrared photodetectors", *Bulletin of the Polish Academy of Sciences Technical Sciences,*Vol. 57, No. 1, , pp. 103-116, 2009.

[13] A. D. Stiff-Roberts, X. H. Su, S. Chakrabarti, and P. Bhattacharya, "Contribution of field-assisted tunneling emission to dark current in InAs–GaAs quantum dot infrared photodetectors" , *IEEE Photonics Technology Letters,* Vol. 16, No.3, pp.867-869. 2004.

Enhanced Conversion Efficiency of Cu(In,Ga)Se$_2$ Solar Cells with Periodic Nanosphere Arrays

Ming-Yang Hsieh[1], Shou-Yi Kuo[1*], Fang-I Lai[2], Hau-Vei Han [3], Tsung-Yeh Chuang[4,5], and Hao-Chung Kuo[3]

[1]Department of Electronic Engineering, Chang Gung University,Taiwan
[2]Department of Photonics Engineering, Yuan-Ze University,Taiwan
[3] Institute of Electro-Optical Engineering, National Chiao Tung University,Taiwan
[4]Department of Electrophysics, National Chiao Tung University,Taiwan
[5]Green Energy and Environment Research Laboratories, Industrial Technology Research Institute,Taiwan

Presenting author e-mail address: sykuo@mail.cgu.edu.tw

Abstract

Enhanced photoelectric conversion is demonstrated in a Cu(In,Ga)Se$_2$ solar cell with the polystyrene (PS) nanospheres, which is one of the simplest and the fastest methods to build a two-dimensional closely packed periodic structure. Owing to the scattering of the PS nanospheres, the light path length can be increased when compared to a bare CIGS solar cell without PS nanospheres on surface. The dimensions of the PS nanospheres are analyzed by using rigorous coupled-wave analysis (RCWA) method, and the optimized efficiency increased from 15.03 to 15.78%.

Introduction

Solar cells are renewable and secure energy compared with existing energy sources. From the structural difference, solar cells can be roughly divided into bulk and thin film devices. Among all thin film solar cells, Cu(In,Ga)Se$_2$ (CIGS) solar cell have great potential as a new source with low cost and high efficiency solar electricity, which had achieved efficiencies 20.3%[1].

In order to improving the solar cells conversion efficiency, the antireflection layer plays an important role. Due to the gradual refractive index between air to the material surface, the undesirable loss from Fresnel reflection can be suppressed efficiently [2-4]. The various antireflective surface structures on solar cells have been investigated, such as imprinting, colloidal arrays, and layer by layer. Recently, Chen et al. used self-assembled nanospheres to increase effectively for InGaP solar cells by extra scattering induced by nanospheres [5]. The self-assembled nanospheres structure is one of the simplest and fastest methods to fabricate period structures.

In this work, we use the rigorous coupled-wave analysis (RCWA) method to simulate the antireflection characteristics that nanospheres on AZO/ZnO/CdS/CIGS/Mo/glass, as shown in Figure.1. The electrical and optical properties of devices covered by nanospheres with various diameters have been studied. Also, the optimum of size and period of nanospheres used for the CIGS solar cell has been examined.

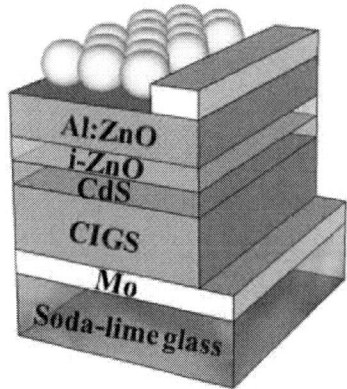

Fig.1. Schematic illustration of CIGS solar cell with PS nanospheres

TABLE I
BASE PARAMETERS FOR CIGS

	AZO	i-ZnO	CdS	CIGS
thickness(nm)	300	50	50	2000
ε	9	9	10	13.6
μ_n (cm^2/V s)	50	50	10	300
μ_p (cm^2/V s)	5	5	1	30
N_A (cm^{-3})	0	0	0	8×10^{16}
N_D (cm^{-3})	3×10^{20}	5×10^{17}	1×10^{17}	0
E_g (eV)	3.3	3.3	2.4	1.2
χ (eV)	4	4	3.75	3.89

Modeling process

In this study, we analyze the optical characteristic of CIGS solar cells with Polystyrene (PS) nanospheres in a broad spectral range by using RCWA method. In the simulations, several layers including 300nm AZO transparent conductive oxide layer, 50nm intrinsic ZnO layer, 50nm CdS buffer layer, 2000nm CIGS absorber layer, 500nm Mo bottom contact and glass substrate construct the CIGS solar-cell structures. Tables 1 shown the description of the base parameters that have been

978-1-4673-4840-9/13 $31.00 © 2013 IEEE

used in this study (*e*, dielectric constant; m_n, electron mobility; m_p, hole mobility; N_A, effective density of states in conduction band; N_D, effective density of states in valence band; E_g, band gap energy and *c*, electron affinity). Simulation of the diameter of PS nanospheres was varied from 100 to 2000 nm and to find out the optimization CIGS solar cell. In this work, all simulations have been performed under an AM 1.5G light spectrum.

Results and discussion

Figure 2 shows the contour plots for reflectance as a function of the wavelength and nanospheres diameters of CIGS solar cells. The resulting short-current density is calculated as shown below for eq. (1).

$$J_{sc} = \frac{e}{hc} \int_{400\,nm}^{1000\,nm} \lambda [1 - R(\lambda)] I_{AM1.5G} d\lambda \quad (1)$$

where e is the electron charge, h is Planck's constant, $I_{AM1.5G}$ is

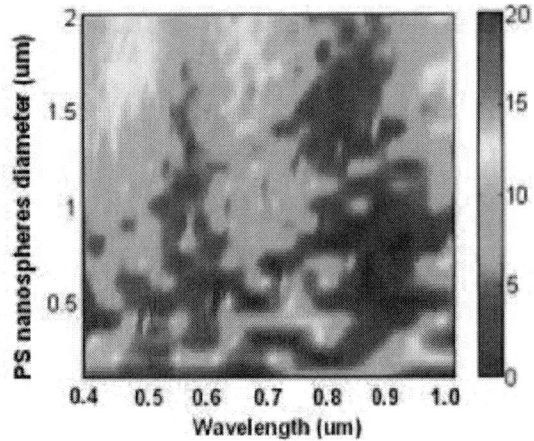

Fig.2. Contour plots for reflectance as a function of the wavelength and nanospheres diameters of CIGS solar cells

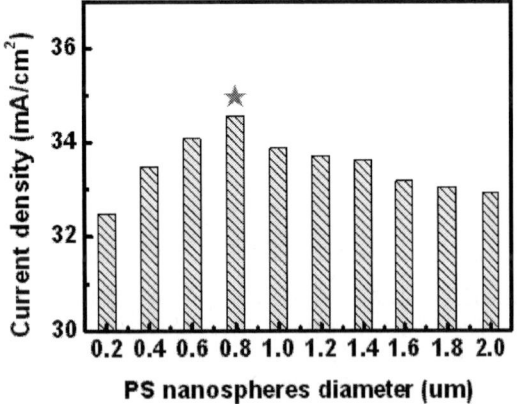

Fig.3. Calculated of short-circuit current density for different diameters of PS nanospheres.

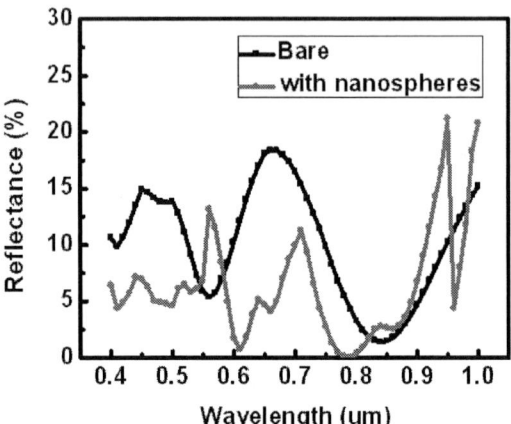

Fig.4. Reflectance spectra of a bare CIGS solar cell and a CIGS solar cell with PS nanospheres

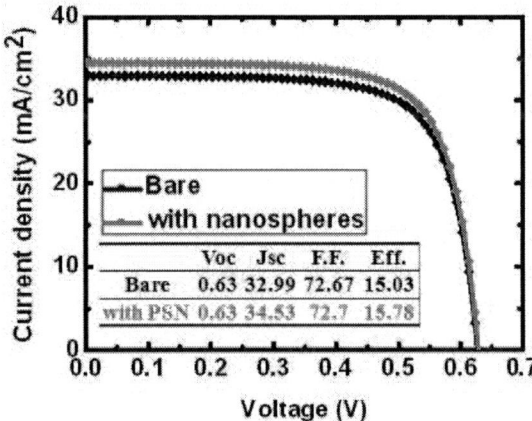

Fig.5. J-V characteristics of the bare CIGS solar cell and the CIGS solar cell with nanospheres

the intensity of the AM 1.5G solar spectrum, and the $R(\lambda)$ is the simulated reflectance of the CIGS cell. As shown in Figure 3, the optimal Jsc is obtained for a PS nanospheres diameter is 800nm. Figure 4 shows the reflectance of the CIGS solar cells with PS nanospheres compared with bare CIGS solar cell. The average reflectance of a bare CIGS solar cell of 10.04 % at incident wavelength between in 400nm to 1000nm, the reflectance of the nanospheres was determined to be 6.34%. Figure 5 shows the photovoltaic current-voltage (*J-V*) curves of CIGS solar cells with flat and with nanospheres. From the results, the bare CIGS solar cell and the cell with the 800 nm-diameter PS nanospheres represented average conversion efficiencies (η) of 15.03 and 15.78%. The extra gain (G_p) in photocurrent is calculated as shown below for eq. (2).

$$G_P = \frac{J_{sc}(with\ nanospheres) - J_{sc}(bare)}{J_{sc}(bare)} \quad (2)$$

Owing to the extra scattering cause to path length be increased, the extra gains in photocurrent Gp for the CIGS solar cell with nanospheres is 4.7%.

Conclusions

In summary, the close-packed pattern of Polystyrene nanospheres simulated by using the RCWA method and optimized their diameter. In 800 nm-diameter, the reflectance of CIGS solar cell can be improved to get the lowest and short-circuit current density will get the highest. From the results of photovoltaic performances, the efficiency increased from 15.03 to 15.78%

References

[1] O. Lupan, S. Shishiyanu, V. Ursaki, H. Khallaf, L. Chow, T. Shishiyanu, V. Sontea, E. Monaico, S.Railean, Sol. Energy Mater. Sol. Cells 93 (2009) 1417.

[2] Y. Li, J. Zhang, S. Zhu, H. Dong, Z. Wang, Z. Sun, J. Guo, and B. Yang, J. Mater. Chem. **19**(13), (2009)1806–1810

[3] Y. M. Song, S. J. Jang, J. S. Yu, and Y. T. Lee, Small 6(9), (2010) 984–987.

[4] S. Wang, X. Z. Yu, and H. T. Fan, Appl. Phys. Lett. **91**(6), (2007) 061105

[5] H. C. Chen, C. C. Lin, H. W. Wang, M. A. Tsai, P. C. Tseng, Y. L. Tsai, H. W. Han, Z. Y. Li, Y. A. Chang, Hao-Chung Kuo, P. Yu, and S. H. Lin, IEEE Photonics Technology Letters 23 (2011) 691.

Sensitivity Improved Surface Plasmon Resonance Sensor
Based on Graphene and Gold Nanorods

Shuwen Zeng[1,2,3*], Mathieu Sylvain Bergont[2], Aurelien Olivier[2], Xuan-Quyen Dinh[2], Xia Yu[3], Ken-Tye Yong[1*]

[1]School of Electrical and Electronic Engineering, Nanyang Technological University, Singapore, 639798

[2]CINTRA CNRS/NTU/THALES, UMI 3288, Research Techno Plaza, 50 Nanyang Drive, Border X Block, Singapore, 637553

[3]Singapore Institute of Manufacturing Technology, 71 Nanyang Drive, Singapore, 638075

[*]Email: szeng2@ntu.edu.sg; ktyong@ntu.edu.sg

Abstract

In this study, we proposed a new sensing configuration for enhancing the surface plasmon resonance (SPR) based on graphene and gold nanorods (Au NRs). Both analytical modeling based on Fresnel equations and numerical analyses through finite element method (FEM) are performed to optimize the number of graphene layers and the aspect ratio (AR) of Au NRs. The improved sensitivity up to 10^6 degree/ RIU can be achieved by adding a monolayer of graphene onto the Au sensing film with coupling of localized SPR of Au NR (AR=2).

Keywords: Surface plasmon resonance, sensor, graphene, gold nanorods

Introduction

Surface Plasmon resonance (SPR) has been widely studied in the past decade for their potential as a highly sensitive probe to detect biological and chemical interactions in real time [1-4]. This optical phenomenon is associated with electromagnetic waves that consist of collective oscillations of free electron gas propagating along the interface between metal and dielectric. A well-known method to excite SPR is proposed by Kretschmann [5]. The configuration is based on a prism coupler, the light energy will transfer from photons to surface plasmons when the incident angle equals to the resonance angle (θ_{SPR}). This SPR resonance angle is always larger than θ_c, thus total light reflection would occur. The reflective light signal (intensity and phase) can be measured. For example, the reflectivity (R) would decrease to almost zero when SPR excites. However, conventional SPR sensors are limited for sensing of tiny molecules such as cytokine and TNT. There are two main strategies that can further improve the sensitivity: (i) using phase measurement method – sharper signal change than that of intensity and wavelength (ii) exploring new sensing structures such as functionalization the sensing film with nanoparticles and graphene[6-9].

Recently, graphene-based photonics have attracted great attention in various applications such as optical modulators, photovoltaic devices, saturable absorbers and ultrafast lasers, etc[10]. Another important application is to enhance the sensitivity of SPR [11-15]. It has been demonstrated that monolayer graphene only absorbs 2.3% of transmitted light and

there will be a charge transfer when graphene makes contact with metals [16-17]. Furthermore, the structure of hexagonal cells existing in graphene is similar to the carbon ring in most of chemical and biological molecules. Thus, the π-stacking interactions will help the sample molecules adsorb more tightly to the sensing film of SPR. But many researchers only focus on the interactions between graphene and the metallic sensing film, few of them studied the sensitivity enhancement by coupling of functionalized metallic nanoparticles on the graphene-modified sensing film, even the metallic nanoparticles such as gold nanorods (Au NRs), have a high electron density on their surface and the localized SPR (LSPR) of Au NRs can strongly couple with the surface plasmon wave (SPW) on the sensing film. Compared to spherical nanoparticles, nanorods are more flexible in tuning the LSPR peak by changing the aspect ratio (AR) and also more suitable for longer wavelength light excitation source (785 – 1064 nm). In this paper, we propose a new sensing configuration for enhancing the sensitivity of SPR based on graphene and gold nanorods. A sensitivity up to 10^6 degree/ RIU is achieved by adding a monolayer of graphene onto the Au sensing film with coupling of LSPR of Au NR (AR=2).

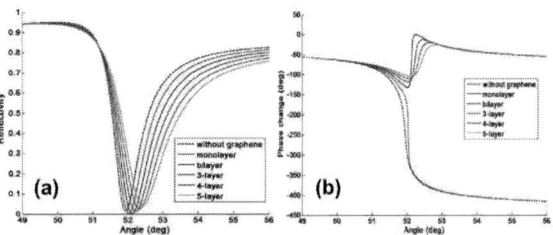

Fig.1. The effects of different number of graphene layers on (a) SPR angle and (b) phase of *p*-polarization.

Analytical Modeling

In our Kretschmann configuration, the incident light passes through an equilateral SF11 prism onto a graphene-coated Au thin film evaporated on a BK7 glass slide and then reflected out through the other edge of the prism. First, we use Fresnel equation and transfer matrix method (TMM) to optimize the number of graphene layers that are transferred onto the Au thin film[18]. The refractive indices for each layer at excitation wavelength 785nm are 1.76652 for SF11 prism, 1.51108 for BK7 glass slide, $0.14891 + i\,4.7830$ for 50nm Au thin film and $2.90 + i\,1.50$ for graphene (0.34nm for monolayer)[19-21]. As it is reported that SPR curves are not affected by the thickness of glass slide, we set it as 100nm for simplification [22-23]. Fig.1 shows the effects of different number of graphene layers on the SPR dip angle and on the phase of p-polarization. One can see that both monolayer and bilayer graphene exhibit more pronounced SPR signal changes than that of without graphene. For angular measurement, the monolayer graphene-coated Au thin film experience a much lower reflectivity at SPR angle, which means that more light energy has been transferred to surface plasmons. For phase measurement, it also has a sharper phase change at SPR angle which indicates larger signal change when the concentration of sensing medium changes. Here, we suggested a differential phase measurement method for detection of biological and chemical molecules[6]. Since SPR only affect the intensity and phase of p-polarization light (ϕ_p), the phase of s-polarization light (ϕ_s) can serve as a reference channel to effectively eliminate the environmental noise (See Fig. 2). According to the results shown in Fig. 2, the sensitivity for monolayer graphene-coated Au thin film is estimated to be 2.387×10^4 °/RIU ($S = \Delta\phi_d / \Delta n$, $\Delta\phi_d = |\phi_p - \phi_s|$).

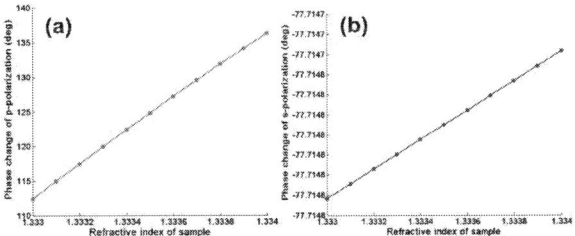

Fig.2. The relations between sample refractive index and phase of (a) p-polarization(ϕ_p) and (b) s-polarization(ϕ_s).

Numerical Analysis

To further study the perturbation of electric field when SPR excites, we employed finite element method (FEM) to have close observations on the field intensity distribution inside the multi-layer sensing structure. The solution space is set as 1.5 $\mu m \times 1.2\ \mu m$ which is larger than the wavelength of incident light 785nm in order to avoid diffraction effects. In Fig. 3(a), the media in the modeling from left to right are respectively, SF11 prism, BK7 glass slide, Au thin film and water. One can see that when the incident angle is at SPR angle, clear evanescent waves are excited and penetrate to the sensing

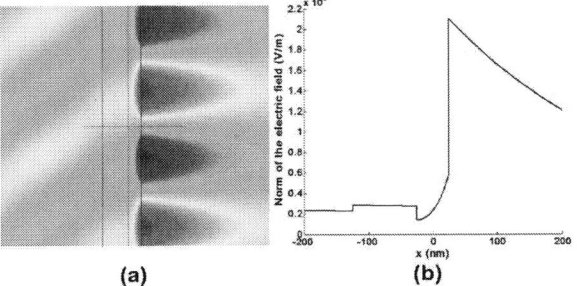

Fig.3. (a) Electric field in y component (b) Cross-section plot of total electric field, in the case of without graphene and Au NR.

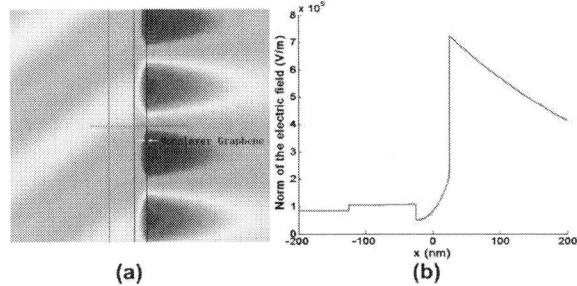

Fig.4. (a) Electric field in y component (b) Cross-section plot of total electric field, in the case of with monolayer graphene but without Au NR.

medium. The red line in Fig. 3(a) shows the cross section range of the total electric field that we are interested in. The center of the line is in the middle of 50nm thick-Au thin film. When a monolayer graphene is coated on the Au thin film, the total electric field is enhanced by almost 4 times compared to that of without graphene as shown in Fig. 4(b). It has been demonstrated that the field intensity is linearly related to the sensitivity of the whole SPR system [24].

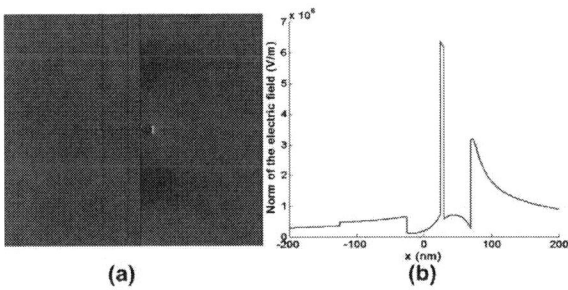

Fig.5. (a) Electric field in x component (b) Cross-section plot of total electric field, in the case of without monolayer graphene but with Au NR (20nm in width, 40nm in length, AR=2).

Then we coupled Au NRs with AR=2 (20nm in width and 40nm in length) with a gap of 5nm to the sensing film and it shows that even without graphene-coated on the Au thin film, the maximum electric field is enhanced by 30 times compared to that of without Au NRs (See Fig. 5). More importantly, when the Au NR is coupled with monolayer graphene-coated Au thin film, the electric field is further improved by 2 more times as shown in Fig. 6. These inspiring results indicate that the total sensitivity of our proposed SPR system can enhanced by 60 times compared to conventional SPR setups, up to 1.43×10^6 °/RIU.

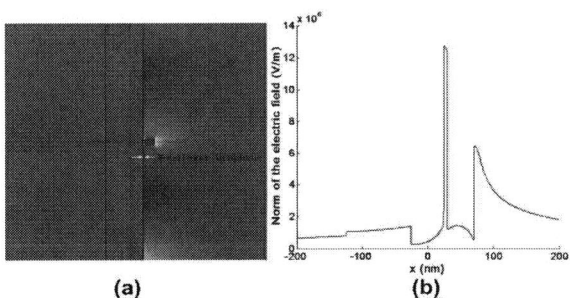

Fig.6. (a) Electric field in x component (b) Cross-section plot of total electric field, in the case of with both monolayer graphene and Au NR (20nm in width, 40nm in length, AR=2).

Conclusions

In summary, a novel SPR sensing structure based on graphene and Au NRs has been proposed and demonstrated to be capable for significantly improving the sensitivity. And it is worth noting that the value can be further enhanced depending on the specific resolution of experimental SPR configurations. The flexible fabrications of uniform and large area graphene-coated metallic thin films promise the realization of this ultrahigh sensitive structure [25-26]. The optimum number of graphene layers is 1 and the aspect ratio for Au NR is 2. Here, the large field enhancement due to strong coupling between LSPR of Au NR and SPW of the graphene-modified Au sensing film plays a key role. Typically, a SPR detection process is that the targeted biological analytes are first flowed onto the functionalized sensing film followed by Au NRs functionalized with capture-analytes. In this way, sandwich structures with targeted analytes between the Au NRs and the sensing film are formed [4, 24]. In our case, the sensing film is monolayer graphene (MG)-coated Au thin film with functional groups such as $-SH$ and $-NH_2$ that are suitable for binding to most of biological analytes.

References

[1] Raether, H., *Surface plasmons on smooth and rough surfaces and on gratings.* 1988, Berlin: Springer-Verlag.

[2] Shuwen Zeng, Ken-Tye Yong, Indrajit Roy, Xuan-Quyen Dinh, Xia Yu, Feng Luan, *A review on functionalized gold nanoparticles for biosensing applications.* Plasmonics, 2011. **6**(3): p. 491-506.

[3] R. Schasfoort and A. Tudos, *Handbook of surface plasmon resonance.* 2008, Cambridge, UK: RSC.

[4] Shuwen Zeng, Xia Yu, Wing-Cheung Law, Yating Zhang, Rui Hu, Xuan-Quyen Dinh, Ho-Pui Ho, Ken-Tye Yong, *Size dependence of Au NP-enhanced surface plasmon resonance based on differential phase measurement.* Sensors and Actuators B: Chemical, 2012.

[5] Kretschmann, E. and H. Raether, *Radiative decay of nonradiative surface plasmons excited by light.* Z. Naturforsch. A, 1968. **23**: p. 2135.

[6] Wu, S.Y., et al., *Highly sensitive differential phase-sensitive surface plasmon resonance biosensor based on the Mach-Zehnder configuration.* Optics Letters, 2004. **29**(20): p. 2378-2380.

[7] Wang, T.J. and C.W. Hsieh, *Surface plasmon resonance biosensor based on electro-optically modulated phase detection.* Optics Letters, 2007. **32**(19): p. 2834-2836.

[8] Law, W.C., et al., *Nanoparticle enhanced surface plasmon resonance biosensing: Application of gold nanorods.* Optics Express, 2009. **17**(21): p. 19041-19046.

[9] Moirangthem, R.S., Y.C. Chang, and P.K. Wei, *Investigation of surface plasmon biosensing using gold nanoparticles enhanced ellipsometry.* Optics Letters, 2011. **36**(5): p. 775-777.

[10] Kandammathe Valiyaveedu Sreekanth, Shuwen Zeng, Jingzhi Shang, Ken-Tye Yong, Ting Yu, *Excitation of surface electromagnetic waves in a graphene-based Bragg grating.* Scientific Reports, 2012.

[11] Wu, L., et al., *Highly sensitive graphene biosensors based on surface plasmon resonance.* Optics Express, 2010. **18**(14): p. 14395-14400.

[12] Choi, S.H., Y.L. Kim, and K.M. Byun, *Graphene-on-silver substrates for sensitive surface plasmon resonance imaging biosensors.* Optics Express, 2011. **19**(2): p. 458-466.

[13] Verma, R., B.D. Gupta, and R. Jha, *Sensitivity enhancement of a surface plasmon resonance based biomolecules sensor using graphene and silicon layers.* Sensors and Actuators B-Chemical, 2011. **160**(1): p. 623-631.

[14] Maharana, P.K. and R. Jha, *Chalcogenide prism and graphene multilayer based surface plasmon resonance affinity biosensor for high performance.* Sensors and Actuators B-Chemical, 2012. **169**: p. 161-166.

[15] Salihoglu, O., S. Balci, and C. Kocabas, *Plasmon-polaritons on graphene-metal surface and their use in biosensors.* Applied Physics Letters, 2012. **100**(21).

[16] Nair, R.R., et al., *Fine structure constant defines visual transparency of graphene.* Science, 2008. **320**(5881): p. 1308-1308.

[17] Giovannetti, G., et al., *Doping graphene with metal contacts.* Physical Review Letters, 2008. **101**(2).

[18] Hansen, W.N., *Electric Fields Produced by the Propagation of Plane Coherent Electromagnetic Radiation in a Stratified Medium.* J. Opt. Soc. Am., 1968. **58**(3): p. 380-388.

[19] Weast, R.C., *CRC Handbook of chemistry and physics.* 68th ed. 1987, Boca Raton, Fla: CRC Press.

[20] Johnson, P.B. and R.W. Christy, *Optical constants of the noble metals.* Physical Review B, 1972. **6**(12): p. 4370-4379.

[21] Weber, J.W., V.E. Calado, and M.C.M. van de Sanden, *Optical constants of graphene measured by spectroscopic ellipsometry.* Applied Physics Letters, 2010. **97**(9).

[22] Xiao, C.D. and S.F. Sui, *Numerical simulations of surface plasmon resonance system for monitoring DNA hybridization and detecting protein lipid film interactions.* European Biophysics Journal with Biophysics Letters, 1999. **28**(2): p. 151-157.

[23] Xiao, C.D. and S.F. Sui, *Characterization of surface plasmon resonance biosensor.* Sensors and Actuators B-Chemical, 2000. **66**(1-3): p. 174-177.

[24] Law, W.C., et al., *Sensitivity improved surface plasmon resonance biosensor for cancer biomarker detection based on plasmonic enhancement.* Acs Nano, 2011. **5**(6): p. 4858-4864.

[25] Eda, G., G. Fanchini, and M. Chhowalla, *Large-area ultrathin films of reduced graphene oxide as a transparent and flexible electronic material.* Nature Nanotechnology, 2008. **3**(5): p. 270-274.

[26] Song, B., et al., *Graphene on Au(111): A highly conductive material with excellent adsorption properties for high-resolution bio/nanodetection and identification.* Chemphyschem, 2010. **11**(3): p. 585-589.

Thickness Effect of Sputtered ZnO Seed Layer on the Electrical Properties of Li-doped ZnO Nanorods and Application on the UV Photodetector

C. Y. Kung[a,*], S. L. Young[b,*], M. C. Kao[b], H. Z. Chen[b], J. H. Lin[c], H. H. Lin[c], Lance Horng[d], Y. T. Shih[d]

[a]Department of Electrical Engineering, National Chung Hsing University, Taichung, Taiwan
[b]Department of Electronic Engineering, Hsiuping University of Technology, Taichung, Taiwan
[c]Department of Electrical Engineering, Hsiuping University of Technology, Taichung, Taiwan
[d]Department of Physics, Changhua University of Education, Changhua, Taiwan
Phone: +886-4-22850359, E-mail: cykung@nchu.edu.tw and slyoung@mail.hust.edu.tw

Abstract

Well-defined $Zn_{0.9}Li_{0.1}O$ nanorods were hydrothermally synthesized at the low temperature of 90°C on the sputtered ZnO seed layer in different thickness, 20 nm and 50 nm. The x-ray diffraction patterns of both samples with a single diffraction peak (002) showed the same wurtzite hexagonal structure. The length and diameter of the nanorods were both increased from 1.12 μm to 1.47 μm and 95 nm to 113 nm, respectively, while increasing the sputtered ZnO seed layer thickness from 20 nm to 50 nm. The optical properties were influenced by the presence of oxygen vacancies induced by Li doping in the $Zn_{0.9}Li_{0.1}O$ nanorods. Besides, p-type ZnO-based semiconductor nanorods were obtained by the doping of Li into ZnO. Finally, the UV photodetectors with (p-$Zn_{0.9}Li_{0.1}O$ nanorods)/(n-Si subatrate) structure were also achieved. The I-V characteristics of the UV photodetector with p-ZnO:Li nanorod/n-Si structure were also measured. The UV illumination/dark current ratio of the photodetectors with different seeded layer, 20nm and 50nm, were also measured as 201% and 364%, respectively.

Keywords: ZnO, seed layer, nanorod and UV photodetector.

Introduction

The II-IV zinc oxide (ZnO) semiconductor with a direct band-gap of 3.37 eV and a large exciton binding energy of 60 meV has attracted great attention due to the need to investigate the versatile physical properties [1-2]. Although ZnO had been studied for the past decades, the renewed interests are now focused on the potential applications on the optoelectronics [3] and spintronics [4]. As to the characteristic of direct wide band-gap, ZnO is potentially used in the low cost production of green, blue, and white visible light emitting devices [5-7]. As for another characteristic of the large exciton binding energy, ZnO could lead to the lasing action based on exciton recombination above room temperature for laser device applications [8-9]. The n-type doped ZnO is known to be easily synthesized. However, the fabrication of the ZnO based optoelectronic devices has met the challenges of both low reproducibility and the lack of low-resistivity p-type ZnO [10]. The p-type ZnO can be achieved by the doping of group-I elements, such as Li, Na and K. However, due to the low solubility of the dopants and high self-compensation of Na and K, Li is the most ideal acceptor for p-type ZnO films [11-13].

Progressive studies on the improvement of good photoluminescence performance for optoelectronic device applications, such as the photoconductors for electro-photography, the photo-anodes for solar cells, and the sensors for gas detection have been continuously proceeded [14-15]. There have also been reported that highly conductive and optical transparent ZnO films are suitable for transparent electrodes in solar cell [16] and liquid crystal display applications [17]. In this work, the smaller element Li, compared with Zn, is designed to be doped into the ZnO nanorods to synthesis the p-type ZnO:Li nanorods grown on the different thickness (20 nm and 50 nm) of sputtered ZnO seed layer. Thus, the nanorods have grown on different thickness of sputtered ZnO seed layer on the n-Si substrate to observe the morphological effect on the characteristic of p-ZnO:Li/n-Si photodetector.

Experimental Procedure

The $Zn_{0.9}Li_{0.1}O$ (denoted as ZnO:Li) nanorods were fabricated on seeded n-type silicon substrates by hydrothermal method. After the substrate cleaning processes, sputtered seed layers (20 nm and 50 nm) of ZnO were deposited on n-type silicon substrates by RF magnetron sputtering system in Ar+N$_2$ at 300 °C. The source solutions for ZnO and ZnO:Li nanorods growth were prepared with the precursors, zinc acetate dehydrate $Zn(C_2H_3O_2)_2 \cdot 2H_2O$, lithium acetate dehydrate $LiCH_3COO \cdot 2H_2O$ and $(CH_2)_6N_4$ in stoichiometric proportions dissolving in de-ionized water. The seeded substrates were placed upside down into the solutions contained in different closed vials heated at 90 °C for 3 h to grow the ZnO:Li nanorods. Then, the metal electrodes were deposited via the thermal evaporation on both p-type ZnO:Li nanorods and n-type Si substrate for the characterization measurement of UV photodetector.

The crystal structure and grain orientation of ZnO films were determined by the x-ray diffraction (XRD) patterns using a Rigaku D/max 2200 x-ray diffractometer with Cu-Kα radiation. The XRD data were recorded at room temperature under the 2θ range from 20° to 60° with a step width of 0.01° and a scan speed of 0.5°/min. Morphological characterization was observed using a field emission scanning electron microscopy

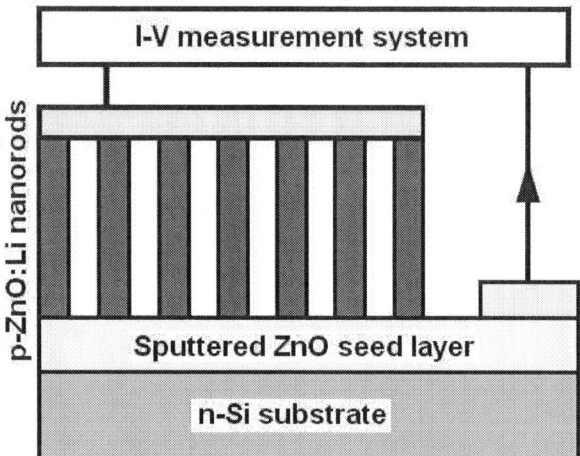

Fig. 1 Structure diagram of the designed p-ZnO:Li/n-Si photodetector.

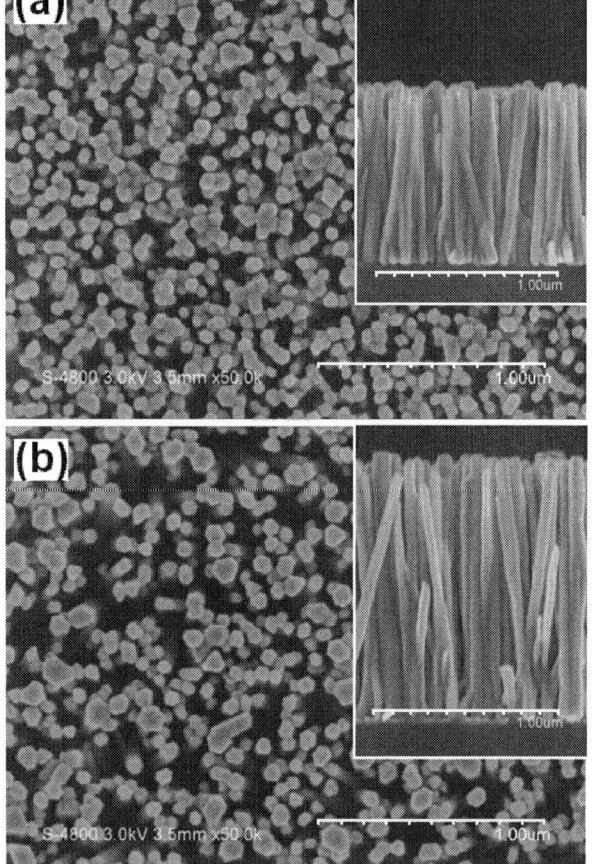

Fig. 2 FE-SEM surface morphology and cross-section images of ZnO:Li nanorods. (a), (b) for 20 nm and (c), (d) for 50 nm sputtered seed layer samples.

(FE-SEM, JEOL JSM-6700F) at 3.0 kV. The electrical properties including carrier type, concentration, mobility and resistivity were measured by Ecopia HMS-3000 Hall-effect measurement system. Finally, the current-voltage (I-V) curves

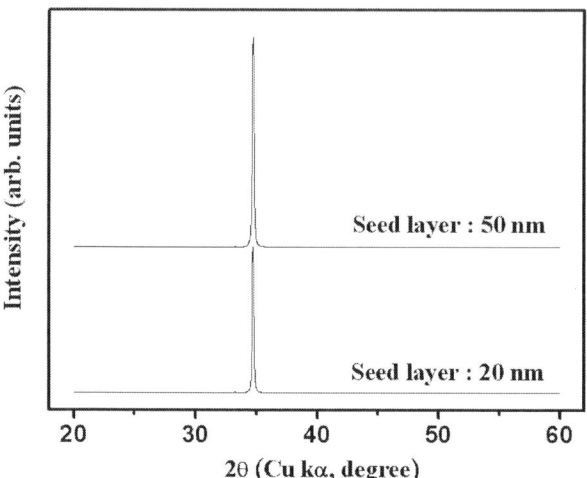

Fig. 3 X-ray diffraction patterns of the ZnO nanorods with sputtered ZnO seed layer of thickness 20 nm and 50nm.

Table 1 Electrical properties of ZnO:Li nanorods grown on the ZnO seed layer with a layer thickness of 20 nm and 50 nm.

Seed layer thickness (nm)	Concentration (cm^3)	Mobility (cm^2/V-s)	Resistivity (Ω-cm)
20	1.48×10^{17}	10.1	5.48
50	1.62×10^{17}	35.2	1.47

were measured by the HP 4145 semiconductor parameter analyzer with the applied voltage varied from $-5V$ to $+5V$ under both darkness and illumination. The structure diagram of the (p-$Zn_{0.9}Li_{0.1}O$ nanorods)/(n-Si subatrate) UV photodetector was shown as Fig. 1.

Results and Dissusion

Fig. 2 shows the FE-SEM surface morphology and cross-section images of the ZnO:Li nanorods grown on the different thickness of seed layers, 20nm and 50 nm. The length and diameter of the nanorods are both increased from 1.12 μm to 1.47 μm and 95 nm to 113 nm, respectively, while increasing the sputtered ZnO seed layer thickness from 20 nm to 50 nm. The result indicates thicker seed layer with larger grain size will enhance the growth rate of ZnO crystal structure.

The XRD patterns of the ZnO:Li nanorods grown on the different ZnO seed layer, 20 nm and 50 nm, are shown as Fig. 3. As shown in the figure, both of the XRD patterns can be indexed as the same hexagonal wurtzite structures. By comparing the both samples, it can be found that the intensity of (002) diffraction peaks is increasing with the increase of seed layer thickness due to the enhancement of larger (002) diffraction plane and dominant (002) crystallization. The result is consistent with longer length and larger diameter of the nanorods grown on 50 nm seed layer compared with that of 20 nm seed layer as indicated in Fig. 2.

Table 1 shows the electrical properties of ZnO:Li nanorods grown on the ZnO seed layer with a thickness of 20 nm and 50 nm, respectively, measured by Hall-effect measurement system. From the measured results, both samples show p-type

978-1-4673-4840-9/13 $31.00 © 2013 IEEE

conductivity and the electrical properties are obviously

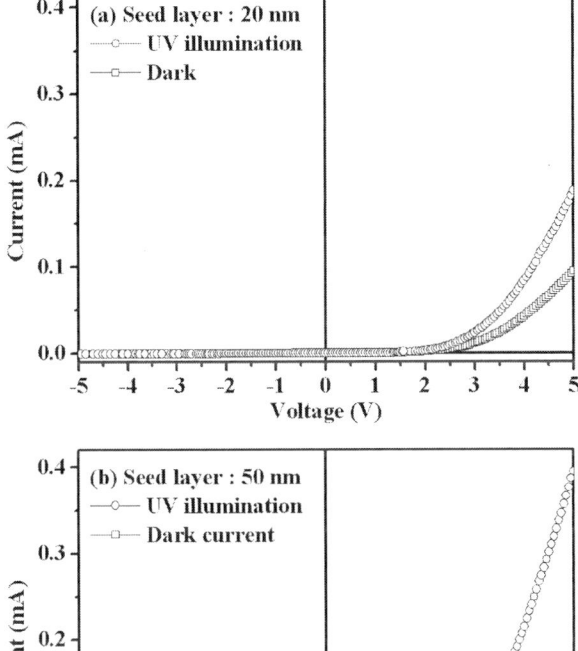

Fig. 4 I-V characteristic curves of the UV photodetectors measured under UV illumination and in dark of the UV photodetectors with seed layer (a) 20 nm and (b) 50nm.

enhanced with the seed layer thickness. The better results including hole concentration of 1.62×10^{17}, carrier mobility of 35.2 cm^2/V-s and resistivity of 1.47 Ω cm are obtained for the ZnO:Li nanorods grown on the seed layer with a thickness of 50 nm compared with that of nanorods grown on the seed layer with a thickness of 20 nm, with a, as shown in Table 1. The result of the improved electrical property with seed layer thickness is also consistent with the results revealed in Fig. 1 and Fig. 2.

Fig. 4 shows the I-V characteristic curves of the (p-Zn$_{0.9}$Li$_{0.1}$O nanorods)/(n-Si subatrate) structure which was utilized to measure the photocurrent (I_{UV}) and dark current (I_{dark}) under UV illumination and in the dark, respectively, with the bias voltage varied from -5V to $+5$V. The photocurrent was measured using a 30 W Xe lamp with an incident wavelength 385 nm as the irradiation source. Fig. 4 (a) and (b) show the results measured with different seed layer thickness, 20 and 50 nm, respectively, and show a good rectifying behavior for both devices. In a dark environment, the turn-on voltage is 2.5 V and 1.8V for devices with 20 and 50 nm seed layer, respectively. The I_{dark} and I_{UV} of the photodetectors are 0.1084 and 0.3946 mA at +5 V for devices with 20 nm seed layer and 0.1084 and 0.3946 mA at +5 V for devices with 50

nm seed layer, respectively. In other words, we obtain an obviously increased photo-to-dark current ratio (I_{UV}/I_{dark}) from 201% to 363.94% as shown in Fig. 4 (b) which illustrates the possibility for photodetector application.

Conclusions

The influence of sputtered ZnO seed layer thickness on the electrical properties of Li-doped ZnO Nanorods and corresponding application on the UV Photodetector are investigated in this study. The length and diameter of the nanorods are both increased while increasing the thickness of sputtered ZnO seed layer. Both of the XRD patterns can be indexed as the same hexagonal wurtzite structures. From the measured results, both samples show p-type conductivity and electrical properties including carrier concentration, mobility and resistivity of the nanorods are obviously enhanced with the seed layer thickness. Finally, the UV photodetectors with (p-Zn$_{0.9}$Li$_{0.1}$O nanorods)/(n-Si subatrate) structure are also achieved. The UV dark/illumination current ratio of the photodetectors with different seeded layer, 20nm and 50nm, are also obtained as 201% and 364%, respectively, which shows the possibility for photodetector application.

Acknowledgment

This work was sponsored by the National Science Council of the Republic of China under the grants No. NSC 101-2221-E-164-004 and NSC 101-2221-E-005-014.

References

[1] Ü. Özgür, Y.I. Alivov, C. Liu, A. Teke, M.A. Reshchikov, S. Doğan, V. Avrutin, S.J. Cho, H. Morkoç, J. Appl. Phys. 98 (2005) 041301.

[2] D.P. Norton, Y.W. Heo, M.P. Ivill, K.A. Ip, S.J. Pearton, M.F. Chisholm, T. Steiner, Mater. Today 7 (2004) 34.

[3] M. Godlewski, E. Guziewicz, K. Kopalko, G. Łuka, M. I. Łukasiewicz, T. Krajewski, B. S. Witkowski, S. Gierałtowska, Low Temp. Phys. 37 (2011) 235.

[4] S.J. Pearton, D.P. Norton, Y.W. Heo, L.C. Tien, M.P. Ivill, Y. Li, B.S. Kang, F. Ren, J. Kelly, A.F. Hebard, J. Electron. Mater. 35 (2006) 862.

[5] H. Zeng, G. Duan, Y. Li, S. Yang, X. Xu, W. Cai, Adv. Funct. Mater. 20 (2010) 561.

[6] Tsukazaki, M. Kubota, A. Ohtomo, T. Onuma, K. Ohtani, H. Ohno, S.F. Chichibu, M. Kawasaki, Jpn. J. Appl. Phys. 44 (2005) L643.

[7] Y.S. Choi, J.W. Kang, D.K. Hwang, S.J. Park, Trans. Electron. Devices 57 (2010) 26.

[8] S. Chu, M. Olmedo, Z. Yang, J. Kong, J. Liu, Appl. Phys. Lett. 93 (2008) 181106.

[9] C. Zhang, F. Zhang, T. Xia, N. Kumar, J.I. Hahm, J. Liu, Z.L. Wang, J. Xu, Optics Express 17 (2009) 7893.

[10] Ü. Özgür, D. Hofstetter, H. Morkoç, Proc. IEEE 98 (2010) 1255.

[11] S.-Y. Tsai, M.-H. Hon, Y.-M. Lu, J. Cryst. Growth 326 (2011) 85.

[12] N.R. Yogamalar, A.C. Bose, Appl. Phys. A 103 (2011) 33.

[13] S.N. Das, J.-H. Choi, J.P. Kar, T.I. Lee, J.-M. Myoung, Mater. Chem. Phys. 121 (2010) 472.

[14] J. Chen, C. Li, D.W. Zhao, W. Lei, Y. Zhang, M.T. Cole, D.P. Chu, B.P. Wang, Y.P. Cui, X.W. Sun, W.I. Milne, Electrochem.

978-1-4673-4840-9/13 $31.00 © 2013 IEEE

Commun. 12 (2010) 1432.

[15] M.W. Ahn, K.S. Park, J.H. Heo, J.G. Park, D.W. Kim, K.J. Choi, J.H. Lee, S.H. Hong, Appl. Phys. Lett. 93 (2008) 263103.

[16] M.C. Kao, H.Z. Chen, S.L. Young, Appl. Phys. A 98 (2010) 595.

[17] B.Y. Oh, M.C. Jeong, T.H. Moon, W. Lee, J.M. Myoung, J.Y. Hwang, D.S. Seo, J. Appl. Phys. 99 (2006) 124505.

Selective enhancement of red upconvesion luminescence of Er^{3+} by doping with Mn^{2+} ions

En-hai Song, Fen Xiao, Shi Ye, Qin-yuan Zhang*

State Key Laboratory of Luminescent Materials and Devices, and Institute of Optical
Communication Materials, South China University of Technology
Wushan RD.,Tianhe District,Guangzhou,P.R.China,510641
Tel: +86-02087114204; Email:qyzhang@scut.edu.cn

Abstract

Well-defined $KZnF_3$:Yb^{3+},Er^{3+},Mn^{2+} nanocubes were successfully prepared by a facile solvothermal method. The XRD and SEM measurements results show that the cubes have good crystallinity and the average cubic size is about 40 nm. The incorporation of Mn^{2+} ions into $KZnF_3$:Yb^{3+},Er^{3+} leads to the great increase in the red to green luminescence intensity ratios from 10:1 to 214:1 and the red emission was enhanced about 20 times. The selective enhancement of red upconversion luminescence of Er^{3+} can be ascribed to the exchange energy transfer interactions between Er^{3+} ions and Yb^{3+}-Mn^{2+} dimers. The case of $KZnF_3$:Yb^{3+},Er^{3+},Mn^{2+} provides a new strategy to get the pure red upconversion luminescence, which show potential applications in the fields of lighting, displays and biological nanolabels.

Key Words: Selective enhancement; exchange energy; Er^{3+}/Mn^{2+}; upconversion luminescence.

Introduction

Since the Photon upconversion (UC) light emitting mechanism has been reported in 1960s [1], numerous attentions have been focused on UC process and UC luminescent materials because of their potential applications in many realms, such as compact solid-state lasers, solar cell, muti-color displays, biological imaging and lighting [2-6]. As is well known, the UC process converts long wavelength radiation into shorter wavelength emitting light via a two or multi-photon absorption mechanism. Recently, an increasing interest has been devoted on rare earth (RE) ions doped UC nanocrystals for applications in biological labeling [6-8] due to their unique advantages, such as little photodamage and high excitation penetration depths in tissues [9]. However, the low emission penetration depth is still an obstacle for deep tissue applications [10, 11]. As is well known, the NIR spectral range of 700-100 nm and the red region of 600-700 nm are considered to be the "optical window" of cells [9, 10, 12]. Therefore, tuning both the emission and excitation bands into the "optical window" is very important for deep tissue applications of fluorescent labels [12]. Among the UC nanocrystals, the most efficient systems are Yb^{3+}/Er^{3+} doped nanocrystals [13-15]. However, the Er^{3+} simultaneously provides green and red emissions that are not good for the deep

tissue applications [10]. Although the red to green emission ratio can be enhanced by increasing the concentration of Yb^{3+} in Yb^{3+}/Er^{3+} codoped system, the UC emission intensity was usually greatly decreased [16]. Moreover, it is difficult to obtain pure red upconverison luminescence via this strategy [17]. Therefore, the greatest primary task currently is to develop a more effective method to solve this problem.

In this work, the Mn^{2+} ions doped $KZnF_3$:1%Yb^{3+},1%Er^{3+} nanocystals were successfully synthesized by a solvothermal method. The selected enhancement of red upconversion luminescence of Er^{3+} by doping with Mn^{2+} and strong pure red emission of Er^{3+} has been observed in this system. The UC luminescence properties and the selective enhancement mechanism have been carefully investigated.

Experimental

The raw materials $Zn(NO_3)_2 \cdot 6H_2O$ (AR), KOH (AR), KF(99%), $Mn(CH_3COO)_2$ (99.9%), Yb_2O_3 (99.998%), Er_2O_3(99.99%) and Oleic acid (AR) were all purchased from Guangzhou Chemical Reagent Co., Ltd (China) without further purification.

The samples $KZnF_3$:1%Yb^{3+},1%Er^{3+},xMn^{2+}(x=0, 1%, 3%, 5%, 10% and 15%) and $KZnF_3$:1%Yb^{3+},5%Mn^{2+} were synthesized by a facile solvothermal method using oleic acid as a stabilizing agent. In a typical synthesis, 4 mmol KOH (AR), 2 mL distilled water, 15 mL ethanol and 5 mL oleic acid were mixed together under magnetic stirring to form homogeneous solution. Then, 5 mL of an aqueous solution containing (1-x) mmol $Zn(NO_3)_2 \cdot 6H_2O$ (AR), x mmol $Mn(CH_3COO)_2$, 0.01 mmol $Yb(NO_3)_3$, 0.01 mmol $Er(NO_3)_3$ were added into the solution and vigorous stirring for 10 minutes. Last, 8 mmol KF was added into the complex under vigorous stirring. The mixture was agitated for 30 minutes and then transferred into the 50 mL autoclave, sealed and hydrothermal treated at 220 °C for 24 h. After the reaction, the system was cooled to room temperature naturally, and the products were colleted and centrifuged several times with the distilled water and absolute ethanol to remove the surplus oleic acid and other remnant substances, and finally dried at 60 °C for 10 hours.

The crystal structure of the products were identified by a Philips Model PW 1830 X-ray powder diffractometer with Cu-K_{α} radiation (λ=1.5406 Å) at 40 kV tube voltage and 40 mA tube

978-1-4673-4840-9/13 $31.00 © 2013 IEEE

current. The size and morphology of the products were measured by a field emission scanning microscope (SEM, JEOL JEM-1010). The UC luminescence spectra were recorded on a TRIAX320 fluorescence spectrofluorometer (Jobin-Yvon Co., France) upon continuous wave excitation of a 976 nm Laser Diode with the powder density 10 W cm^{-2} (Coherent corp., USA). Emitted light was focused onto the monochromator and was monitored at the exit slit R928 photomultiplier tube (PMT). The Photoluminescence (PL) and Photoluminescence excitation (PLE) spectra in visible and NIR regions were determined on a FL920 combined with time resolved and steady state fluorescence spectrophotometer (Edinburgh Instruments LTD) fitted with a 450 W xenon lamp as the excitation source. All measurements are performed at room temperature.

Results and discussion

Fig. 1 (a) shows the typical XRD patterns of KZnF$_3$:1%Yb^{3+},1%Er^{3+},xMn^{2+} (x=1% and 15%) together with the Joint committee on Power Diffraction Standards (JCPDS) card No. 06-0439. It can be observed that all diffraction peaks are in good agreement with the JCPDS card No. 06-0439, indicating that the crystal structure of the products are identical to that of KZnF$_3$ and the incorporation of Yb^{3+}, Er^{3+} and Mn^{2+} ions into KZnF$_3$ do not cause any significant change in the host structure. Moreover, the increase concentration of Mn^{2+} ions resulted in the main peaks of KZnF$_3$ shifted towards the smaller angles, as shown in Fig. 1(b), indicating the substitute of Zn^{2+} ions by Mn^{2+} with the larger ion radius caused the host lattice to expand. Fig. 1(c) shows the typical SEM image of KZnF$_3$:1%Yb^{3+},1%Er^{3+},5%Mn^{2+}. It can be seen that the sample consists of good nanocubes which have smooth surfaces and the average cubic size is about 40 nm.

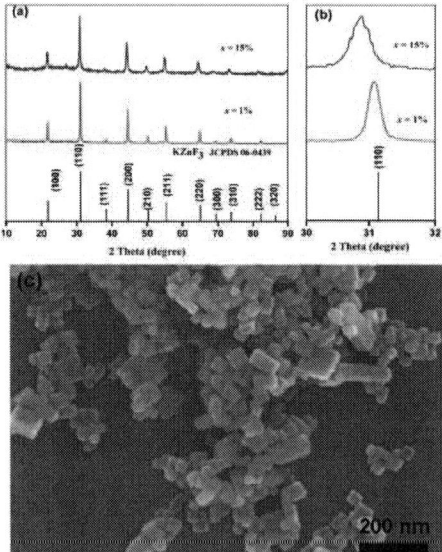

Fig. 1 (a) XRD patterns of KZnF$_3$:1%Yb^{3+},1%Er^{3+},xMn^{2+} (x=1% and 15%); (b) XRD patterns of samples from 30° to 32° and (c) SEM image of KZnF$_3$:1%Yb^{3+},1%Er^{3+},5%Mn^{2+}.

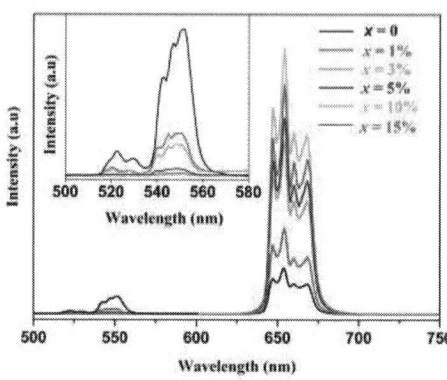

Fig .2 UC emission spectra of KZnF$_3$:Yb^{3+},Er^{3+},xMn^{2+} (x=0, 1%, 3%, 5%, 10% and 15%) under 976nm LD excitation. The inset shows the UC emission spectra of samples from 500 to 580 nm.

Fig. 2 shows the UC emission spectra of KZnF$_3$:1%Yb^{3+}, 1%Er^{3+},xMn^{2+}. The UC spectrum of KZnF$_3$:1%Yb^{3+},1%Er^{3+} consists of three emission peaks located at 524, 548 and 654 nm, corresponding to the $^2H_{11/2} \rightarrow {}^4I_{15/2}$, $^4S_{3/2} \rightarrow {}^4I_{15/2}$ and $^4F_{9/2} \rightarrow {}^4I_{15/2}$ transitions of Er^{3+} ions, respectively. And two photons absorption processes are responsible for the green and red UC emissions, respectively [18]. Since these transitions of Er^{3+} belong to the f-f transitions which are well shielded by outer shells, the UC emissions of all Yb^{3+}/Er^{3+} doped systems are almost the same except the difference on the ratio of red to green emission intensity. For instances, in NaYbF$_4$:Er^{3+}, the red UC emission is much stronger than that of the green UC emission[19]. While, the green UC emission is much stronger than that of the red UC emission in CaMoO$_4$:Yb^{3+},Er^{3+}[20]. For the sample KZnF$_3$:1%Yb^{3+},1%Er^{3+}, it can be clearly seen that the red emission peak (654 nm) is stronger than that of the green (524 and 548 nm) emission and the red to green emission ratio was calculated to about 10. The intense red UC emission can be ascribed to the efficient cross-relaxation (CR) process ($^4F_{7/2}$ + $^4I_{11/2} \rightarrow {}^4F_{9/2}$), which consequently makes the $^4F_{9/2} \rightarrow {}^4I_{15/2}$ transition of Er^{3+} more efficient[17].

To further increase the red to green emission ratio and eventually get the bright pure red UC luminescence, the KZnF$_3$:1%Yb^{3+},1%Er^{3+},xMn^{2+} (x = 1%-15%) samples were prepared. It can be seen from Fig. 2 that the green emission of Er^{3+} gradually decreases with increasing Mn^{2+} concentration (x) and completely disappeared when x>10%. This result indicates that the Mn^{2+} has significant effect on tuning the red to green emission ratio of Er^{3+}. Fig. 3 shows the dependence of the red to green emission ratio (I$_R$/I$_G$) on Mn^{2+} concentration (x) and the red UC emission intensity of Er^{3+}. It can be observed in this figure that the I$_R$/I$_G$ value monotonously increases with increasing the Mn^{2+} concentration and the larger I$_R$/I$_G$ value was determined to be 194 and 214 for KZnF$_3$:1%Yb^{3+},1%Er^{3+}, xMn^{2+} with x =10% and 15%, respectively. It is worth noting that the red UC emission intensity of Er^{3+} was obviously enhanced gradually with increasing the Mn^{2+} concentration. The red UC emission intensity of Er^{3+} first increases with increasing Mn^{2+} concentration and reaches a maximum at x = 10%. After the concentration of Mn^{2+} is over the critical value,

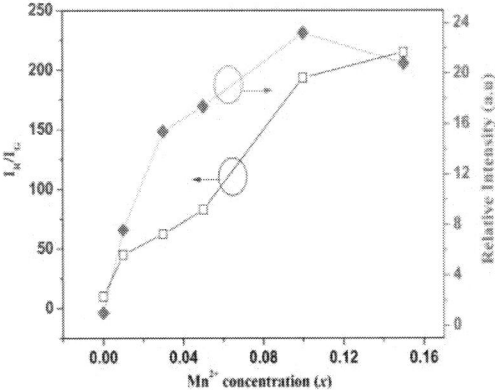

Fig. 3 Red to green UC emission intensity ratio (I_R/I_G) and Relative red UC emission intensity of KZnF$_3$:1%Yb^{3+}, 1%Er^{3+}, xMn^{2+} versus Mn^{2+} doping concentration (x).

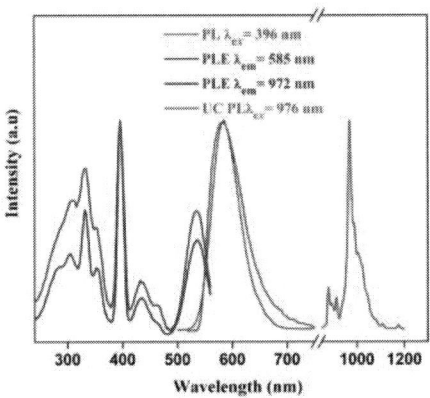

Fig. 4 UC emission spectrum, Photoluminescence and Photoluminescence excitation spectra of KZnF$_3$:1%Yb^{3+}, 5%Mn^{2+}.

the emission intensity of Er^{3+} sharply decreases because of the concentration quenching effect, which is mainly caused by nonradiative energy transfer between the UC emission centers and the defects or Mn^{2+} centers. Furthermore, the sample KMnF$_3$:1%Yb^{3+},1%Er^{3+} was also prepared in this work. Under 976 nm LD excitation, KMnF$_3$:1%Yb^{3+},1%Er^{3+} shows a single red UC band centered at 654 nm which is quite similar to that of the samples KZnF$_3$:1%Yb^{3+},1%Er^{3+},xMn^{2+} with x = 10% and 15%. And the red emission intensity of KZnF$_3$:1%Yb^{3+},1%Er^{3+}, 10%Mn^{2+} is about eight times stronger than that of KMnF$_3$:1%Yb^{3+},1%Er^{3+}. Therefore, the KZnF$_3$:1%Yb^{3+}, 1%Er^{3+},10%Mn^{2+} nanocrystals show great potential application in the fields of biological labeling and lighting.

Fig. 4 shows the UC emission spectrum, photoluminescence and photoluminescence excitation spectra of KZnF$_3$:1%Yb^{3+}, 5%Mn^{2+}. It can be observed that KZnF$_3$:1%Yb^{3+},5%Mn^{2+} shows a broad yellow band centered at 585 nm, which might be attributed to the ^{4}T$_{1g}$(G)→^{6}A$_1$(S) transition of Mn^{2+} because its spectral profile is similar to that of KZnF$_3$:Mn^{2+},Yb^{3+} under UV excitation and the Yb^{3+} ion has no emission in this spectral region. Upon 396 nm, the spectrum consists of a broad band

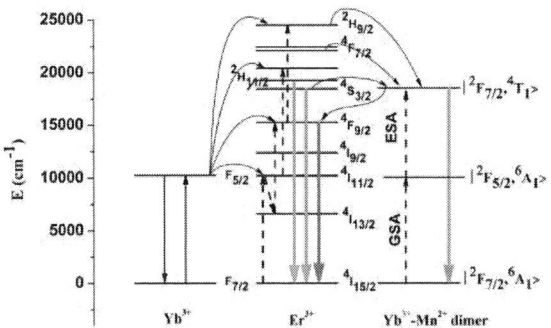

Fig. 5 Schematic diagram of the UC mechanism of Yb^{3+}-Er^{3+}-Mn^{2+} co-doped system.

centered at 585 nm and an asymmetric band centered at 972 nm, corresponding to the ^{4}T$_1$(G)→^{6}A$_1$(S) transition of Mn^{2+} and the ^{2}F$_{5/2}$→^{2}F$_{7/2}$ transition of Yb^{3+}, respectively. The excitation spectrum of Mn^{2+} consists of six sub-bands located at 308, 333, 353, 396, 435 and 535 nm, corresponding to the transitions of Mn^{2+} from the ground state ^{6}A$_1$(^{6}S) to the ^{4}T$_{1g}$(P), ^{4}E$_g$(D), ^{4}T$_{2g}$(^{4}D), [^{4}A$_{1g}$(^{4}G), ^{4}E$_g$(^{4}G)], ^{4}T$_{2g}$(^{4}G) and ^{4}T$_{1g}$(^{4}G) energy levels[21], respectively. Meanwhile, the excitation spectrum of Yb^{3+} shows quite similar to that of Mn^{2+} except a little difference in relative intensity, indicating there are presence of energy transfer from Mn^{2+} to Yb^{3+} in KZnF$_3$:Mn^{2+},Yb^{3+}[22]. These facts demonstrated that the room temperature Mn^{2+} UC luminescence occurs in this system, and which is the first observation of Mn^{2+} UC luminescence in fluorides. The UC mechanism of Mn^{2+} can be ascribed to the ground stated absorption and excited state absorption steps (GAS/ESA) based on Yb^{3+}-Mn^{2+} dimers[23], and which has been demonstrated in other Yb^{3+}/Mn^{2+} doped crystals, such as CsMnBr$_3$:Yb^{3+}[24], MnCl$_2$:Yb^{3+}[25],Zn$_2$SiO$_4$:Yb^{3+},Mn^{2+}[26] and CsMnCl$_3$:Yb^{3+}[27].

Comparing Fig. 2 and Fig. 4, a significant spectral overlap between the green UC emission band of Er^{3+} and the excitation spectrum of Mn^{2+} was observed. Therefore, the nonradiative resonance energy transfer from Er^{3+} to Mn^{2+} can be expected in KZnF$_3$:Yb^{3+},Er^{3+},Mn^{2+}. Moreover, the ^{4}I$_{15/2}$→^{4}I$_{9/2}$ absorption (about 654 nm) band of Er^{3+} and the UC emission band of Mn^{2+} have a large spectral overlap, the nonradiative resonance energy transfer from Mn^{2+} to Er^{3+} can be also expected in this system. So, it is reasonable to suggest that the two steps energy transfer resulted in the green UC emission and red UC emission of Er^{3+} showed the opposite changing trends with increasing Mn^{2+} concentration in KZnF$_3$:Yb^{3+},Er^{3+},xMn^{2+} as shown in Fig. 2. Fig.5 shows UC mechanism of Yb^{3+},Er^{3+},Mn^{2+} doped system. For the Yb^{3+}-Mn^{2+} dimer, the $|^2$F$_{7/2}$,^{6}A$_{1g}$(S)>, ^{2}F$_{5/2}$,^{6}A$_{1g}$(S)> and $|^2$F$_{7/2}$,^{4}T$_{1g}$> represent the ground state intermediate excited state and the higher excited state (emitting state), respectively. Therefore, the selective enhancement mechanism of Er^{3+} can be ascribed to the nonradiative energy transfer from ^{4}F$_{7/2}$, ^{2}H$_{11/2}$ and ^{4}S$_{3/2}$ levels of Er^{3+} to the $|^2$F$_{7/2}$,^{4}T$_{1g}$> state of Yb^{3+}-Mn^{2+} dimer, followed by back energy transfer to the ^{4}F$_{9/2}$ level of Er^{3+}. Meanwhile, the energy absorbed by the Yb^{3+}-Mn^{2+} dimer was also transferred to the ^{4}F$_{9/2}$ level of Er^{3+} through resonance energy transfer and the $|^2$F$_{7/2}$,^{4}T$_{1g}$>→$|^2$F$_{5/2}$,^{6}A$_{1g}$(S)> radiative

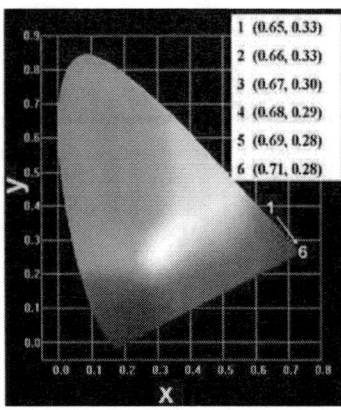

Fig .6 CIE Chromaticity coordinates of $KZnF_3$:1%Yb^{3+},1%Er^{3+}, xMn^{2+} (1 for $x=0$; 2 for $x=1$%; 3 for $x=3$%; 4 for $x=5$%; 5 for $x=10$% and 6 for $x=15$%) under 976 nm LD excitation.

transition of the dimer was quenched. This is the reason why the UC luminescence of Mn^{2+} can not be observed in Er^{3+} co-doped $KZnF_3$:Yb^{3+},Mn^{2+}. As a result of the exchange energy transfer between the Er^{3+} and Yb^{3+}-Mn^{2+} dimers, the bright pure red UC luminescence of Er^{3+} has been obtained in $KZnF_3$:1%Yb^{3+}, 1%Er^{3+},xMn^{2+} when $x>10$%.

The chromaticity coordinates of $KZnF_3$:Yb^{3+},Er^{3+},xMn^{2+} under 976 nm LD excitation are shown in Fig.6. With increasing the Mn^{2+} concentration, the chromaticity coordinates obviously changed from (0.65, 0.33) to (0.71, 0.28), corresponding to the orange red to deep red color, respectively. In other words, a wide range of the red UC emissions of Er^{3+} can be achieved by controlling the relative intensities of the red to green emissions depending on the Mn^{2+} doping concentration.

Conclusions

In conclusion, we have synthesized and investigated the selective enhancement upconversion luminescence of $KZnF_3$:Yb^{3+},Er^{3+},Mn^{2+} nanocubes with NIR (976 nm) excitation. The ratios of red to green UC emission as well as the intensity of the red emission can be enhanced by varying the Mn^{2+} concentrations. And the selective enhancement mechanism of Er^{3+} has been demonstrated to be the exchange energy transfer between Er^{3+} and Yb^{3+}-Mn^{2+} dimer. A pure red UC luminescence with coordinates of (0.69, 0.28) has been obtained in $KZnF_3$:1%Yb^{3+},1%Er^{3+},10%Mn^{2+}, suggesting that the successfully doping of Mn^{2+} in $KZnF_3$:Yb^{3+},Er^{3+} nanocubes exhibit potential applications as biological label or display materials.

Acknowledgments

This work is financially joint supported by the NSFC (Grant Nos. 50872036, 21101065 and 51125005), the Fundamental Research Funds for the Central Universities, SCUT, and China Postdoctoral Science Foundation funded project.

References

[1] F. Auzel, Che. Rev., 104, pp. 139-174, 2004.

[2] J. Suyver, A. Aebischer, D. Biner, P. Gerner, J. Grimm, S. Heer, K. Krämer, C. Reinhard, H. Güdel, Opt. Mater., 27, pp. 1111-1130, 2005

[3] F. Wang, D. Banerjee, Y. Liu, X. Chen, X. Liu, Analyst, 135, pp. 1839-1854, 2010.

[4] M. Haase, H. Schafer, Angew. Chem. Int. Ed., 50, pp. 5808-5829, 2011.

[5] M. Wang, G. Abbineni, A. Clevenger, C. Mao, S. Xu, Nanomed.-Nanotech. Bio. Med., 7, pp. 710-729, 2011.

[6] J. Zhou, Z. Liu, F. Li, Chem. Soc. Rev, 41, pp. 1323-1349, 2012.

[7] C. Wang, L. Cheng, H. Xu, Z. Liu, Biomaterials, 33, PP. 4872-4881, 2012.

[8] P. Yuan, Y.H. Lee, M.K. Gnanasammandham, Z.P. Guan, Y. Zhang, Q.H. Xu, Nanoscale, 4, PP. 5132-5137, 2012.

[9] M.Y. Xie, X.N. Peng, X.F. Fu, J.J. Zhang, G.L. Li, X.F. Yu, Scripta mater., 60, pp. 190-193, 2009.

[10] G. Tian, Z. Gu, L. Zhou, W. Yin, X. Liu, L. Yan, S. Jin, W. Ren, G. Xing, S. Li, Y. Zhao, Adv. Mater., 24, pp. 1226-1231, 2012.

[11] J. Wang, F. Wang, C. Wang, Z. Liu, X. Liu, Angew. Chem. Int. Ed., 50, pp. 10369-10372, 2011.

[12] M. Nyk, R. Kumar, T.Y. Ohulchanskyy, E.J. Bergey, P.N. Prasad, Nano lett, 8, pp. 3834-3838, 2008.

[13] Y. Dai, P. Ma, Z. Cheng, X. Kang, X. Zhang, Z. Hou, C. Li, D. Yang, X. Zhai, J. Lin, ACS Nano, 6, pp. 3327-3338, 2012.

[14] B. Dong, S. Xu, J. Sun, S. Bi, D. Li, X. Bai, Y. Wang, L. Wang, H. Song, J. Mater. Chem., 21, pp. 6193-6200, 2011.

[15] J. Shan, M. Uddi, N. Yao, Y. Ju, Adv.Funct. Mater., 20, pp. 3530-3537, 2010.

[16] Y. Song, Y. Huang, L. Zhang, Y. Zheng, N. Guo, H. You, RSC Advances, 4, pp. 4777-4781, 2012.

[17] F. Vetrone, J.C. Boyer, J.A. Capobianco, A. Speghini, M. Bettinelli, J. Appl. Phys., 96, pp. 661-667, 2004.

[18] S. Zhao, Y. Hou, X. Pei, Z. Xu, X. Xu, J. Alloy Compd., 368, pp. 298-303, 2004.

[19] S. Zeng, G. Ren, Q. Yang, J. Mater. Chem., 20, pp. 2152-2156, 2010.

[20] J.H. Chung, J.H. Ryu, J.W. Eun, J.H. Lee, S.Y. Lee, T.H. Heo, K.B. Shim, Mater. Chem. Phys., 134, pp. 695-699, 2012.

[21] F. Rodriguez, M. Moreno, J. Phys. C: Solid State Phys., 19, pp. L513-L517, 1986.

[22] S. Ye, Y. Li, D. Yu, G. Dong, Q.Y. Zhang, J. Mater. Chem., 21, pp. 3735-3739, 2011.

[23] C. Reinhard, R. Valiente, H.U. Güdel, J. Phys. Chem. B, 106, pp. 10051-10057, 2002.

[24] P. Gerner, O.S. Wenger, R. Valiente, H.U. Güdel, Inorg. Chem., 40, pp. 4534-4542, 2001.

[25] P. Gerner, C. Reinhard, H.U. Güdel, Chem. Eur. J., 10, pp. 4753-4741, 2004.

[26] P. Gerner, C. Fuhrer, C. Reinhard, H.U. Güdel, J. Alloys Compd., 380, pp. 39-44, 2004.

[27] R. Valiente, O. Wenger, H.U. Güdel, Chem. Phys. Lett., 320, pp. 639-644, 2000.

Indium Phosphide (InP) Colloidal Quantum Dot based Light-Emitting Diodes Designed on Flexible PEN Substrate

Yohan Kim[†], Tonino Greco[‡], Christian Ippen[‡], Armin Wedel[‡] and Jiwan Kim[*†]

[†] Flexible Display Research Center, Korea Electronics Technology Institute, Seongnam-si, Gyeonggi-do, 463-816, Korea
[‡] Fraunhofer Institute for Applied Polymer Research, Geiselbergstrasse 69, Potsdam-Golm, 14476, Germany

[*] jiwank@keti.re.kr, +82-31-789-7418

Abstract

Quantum dot light-emitting diodes (QD-LEDs) using InP/ZnSe/ZnS muitishell colloidal quantum dots (QDs) which were prepared by simple heating-up method were designed on polyethylene naphthalate (PEN) substrate for rugged optoelectronic device. The synthesized InP/ZnSe/ZnS multishell QDs exhibited an emission peak at 545 nm for clear green color with a full-width at half-maximum (FWHM) of 50 nm, and photoluminescent (PL) quantum yield (QY) of 45 %. The maximum luminance and current efficiency of InP based QD-LEDs fabricated on PEN substrate reached 640 cd/m^2 and 1.0 cd/A.

Introduction

Colloidal quantum dots (QDs) are considered as the rising candidate for the next-generation light-emitting materials because of its superb properties such as tunable color spectra, narrow emission bandwidth and cost-effective solution based processing [1,2]. These unique optical/electrical properties of colloidal QDs with core/shell hetero structure are determined by changing the particle size due to the quantum confinement effect [3]. Since the first report on colloidal QDs based light-emitting diodes (QD-LEDs) in 1994, many research groups have investigated to enhance the device performances based on the various colloidal QDs compositions and efficient carrier transport layers [4-6]. While the performance of QD-LEDs has been upgraded dramatically, the future task is the substitution of Cd-based QDs by less toxic materials. Until now, limited studies about QD-LEDs using Cd-free colloidal QDs were reported so far and the most investigated devices have been fabricated on rigid substrates [7,8]. Flexibility is one of the key elements for future information display, but the optoelectronic properties of QD-LEDs on the flexible substrate are not fully investigated yet [9-11].

In this work, unique Cd-free QD-LEDs were designed on a plastic substrate using green indium phosphide (InP) based QDs and their optoelectronic characteristics were investigated.

Experimental Details

A. Synthesis of colloidal quantum dots

The InP/ZnSe/ZnS muitishell QDs were synthesized by heating-up method. The heating-up synthesis is simple and reproducible comparing conventional hot-injection method due to removing manual injection process. The detailed heating-up synthesis procedure for InP based multishell QDs is described in previous literatures [11]. The raw solution of multishell QDs in ODE was purified several times and then it was redispersed in nonane for fabrication of QD-LED device.

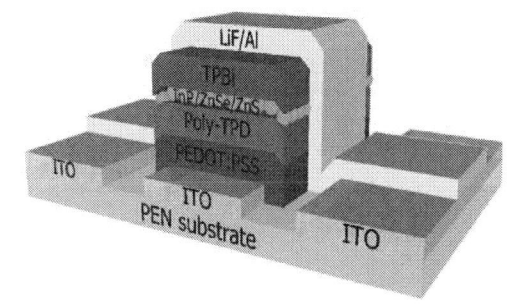

Fig. 1 A cross-section schematic of InP/ZnSe/ZnS QD-LEDs designed on PEN substrate.

B. Fabrication of QD-LEDs

Using muitishell InP QDs, we fabricated the devices on indium-tin oxide (ITO) coated polyethylene naphthalate (PEN) flexible substrate. ITO/PEN substrate was chosen because of lower sheet resistance and better thermal stability than other plastic substrate. Fig. 1 presents a schematic of the InP/ZnSe/ZnS QD-LEDs. ITO/glass (Samsung Corning Precision Materials, Korea) and ITO/PEN (OIKE & Co., Ltd) were used as rigid or flexible substrates for QD-LED fabrication. The completed device consists of ITO as the anode, PEDOT:PSS as the hole injection layer (HIL), poly-TPD as the hole transport layer (HTL), colloidal InP/ZnSe/ZnS QDs as the light emission layer, TPBi as the electron transport layer (ETL) and LiF/Al as a metallic cathode layer.

C. Characterizations

UV-vis spectra were recorded with a PerkinElmer Lambda 19 spectrometer. Photoluminescent (PL) spectra and quantum yield (QY) were measured using a Hamamatsu C9920-02. Energy level of QDs was determined by ultraviolet photoelectron spectroscopy (Riken Keiki AC-2) and UV-vis measurement. A spectroradiometer (Minolta CS1000) was employed for measurement of the electroluminescence (EL) spectrum and J-V-L was measured with an experimental set-up consisting of a Keithley 2400 source meter and calibrated with fast silicon photodiode at ambient condition.

978-1-4673-4840-9/13 $31.00 © 2013 IEEE

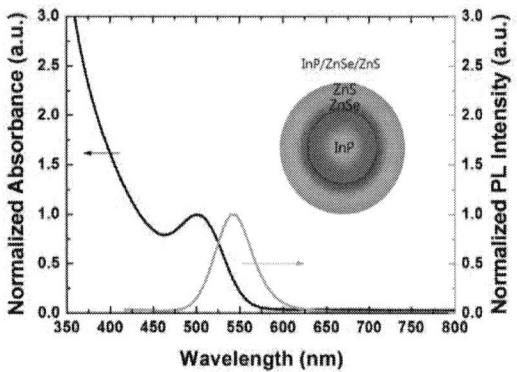

Fig. 2 UV-vis absorption and PL spectra of colloidal InP/ZnSe/ZnS QDs. The inset shows a structure of mutishell QDs.

Results and Discussions

Fig. 2 shows the UV-vis absorption, PL spectrum and a multishell structure of green colloidal InP/ZnSe/ZnS QDs. As the ZnSe/ZnS shells were formed, the QY of InP QDs was improved upto 45% [12]. The InP/ZnSe/ZnS multishell QDs exhibited an emission peak at 545 nm for clear green color with a FWHM of 50 nm. Fig. 3 presents the performance of InP based QD-LEDs fabricated on two different substrate. For both devices, the normalized EL spectra of InP/ZnSe/ZnS QD-LED in Fig. 3(a) show bright and narrow green emission with 555 nm of peak wavelength. There were specific parasitic emissions which were identified as an EL from poly-TPD layer exhibits at 425 nm in both devices. This emission peak from HTL is usually generated from unbalanced transport of carriers and leakage current through various defects in the QD layer [13]. However, the device fabricated on PEN substrate shows a small increment of this poly-TPD peak. We assume that the exciton recombination region shifted slightly from QD to HTL layer due to higher sheet resistance of ITO/PEN substrate. Fig. 3(b) presents luminance-voltage characteristics of InP based QD-LEDs. The maximum luminance reaches 640 cd/m^2 (maximum current efficiency; 1.0 cd/A) for the ITO/PEN substrate and 1440 cd/m^2 (maximum current efficiency; 1.98 cd/A) for the ITO/glass substrate. Current density of the device with ITO/PEN substrate is lower than the device with ITO/glass substrate at the same applied voltage. These performances were strongly related with the characteristics of each ITO substrate, which are summarized in Table I. The higher sheet resistance of ITO/PEN substrate resulted in lower performance of QD-LEDs. Additionally, lower transmittance of ITO/PEN substrate induced to decrease the intensity of emitting light from the QD layer.

The performance of our device is not matched with traditional Cd based QD-LEDs yet, but it shows the great potential for various flexible and eco-friendly information displays. The developed technology for designed device structures with uniform QD layer, efficient transport materials and flexible substrates can be also applicable in various fields. With further investigation about highly efficient and stable InP colloidal QDs, optimization of device structure, and high quality of flexible ITO substrate, the QD-LED device on flexible substrate can be one of major flexible display applications in near future.

(a)
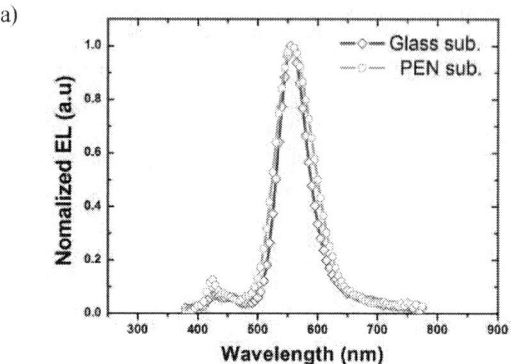

(b)
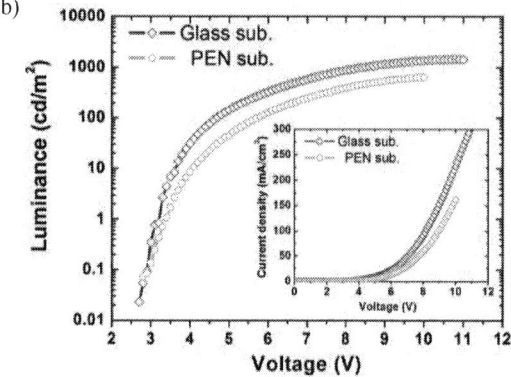

Fig. 3 (a) Normalized EL spectra, (b) Luminance-voltage characteristics (inset: current density-voltage characteristics) of InP/ZnSe/ZnS QD-LEDs designed on ITO/glass and ITO/PEN substrate.

TABLE I
The Characteristics of ITO/glass and ITO/PEN substrate

Substrate	Sheet resistance ($\Omega/\square$)	Thickness of ITO (nm)	Transmittances (% @ 550 nm)
ITO/glass	9.9	130	88.77
ITO/PEN	15.3	430	80.52

Conclusions

In conclusion, we demonstrated the Cd-free QD-LEDs designed on flexible PEN substrate using InP/ZnSe/ZnS muitishell colloidal QDs (FWHM: 50 nm, QY: 45%). The maximum luminance and maximum current efficiency of InP QD-LEDs reached 640 cd/m^2 and 1.0 cd/A as for the device with flexible substrate, which were comparable with the performance of device designed on rigid substrate (1440 cd/m^2, 1.98 cd/A). Although the current device performance has not been fully optimized yet, further research on ITO/substrate and InP based QDs should lead to additional improvements in the device performance. We believe that materials and device structure presented in this study can be applicable to various flexible information displays in the near future.

978-1-4673-4840-9/13 $31.00 © 2013 IEEE

Acknowledgment

This research was supported by QD-LED Project of International Cooperation Program funded by the Ministry of Knowledge and Economy

References

[1] V. Wood and V. Bulović, "Colloidal quantum dot light-emitting devices," Nano Reviews, 1, 5202-1–7, 2010.

[2] P. Peiess, M. Protiere and L. Li, "Core/Shell semiconductor nanocrystals," Small, 5, 154–168, 2009.

[3] D. Bera, L. Qian, T. K. Tseng and P. H. Holloway, "Quantum dots and Their Multimodal Applications: A Review," Materials, 3, 2260–2345, 2010.

[4] V. L. Colvin, M. C. Schlamp and A. P.Alivisatos, "Light-emitting diodes made from cadmium selenide nanocrystals and a semiconducting polymer," Nature, 370, 354–357, 1994.

[5] W. K. Bae, J. Kwak, J. W. Park, K. Char, C. Lee and S. Lee, "Highly efficient green-light-emitting diodes based on CdSe@ZnS duantum dots with a chemical-composition gradient," Adv. Mater., 21, 1690–1694, 2009.

[6] T.H. Kim, K. S. Cho, E. K. Lee, S. J. Lee, J. Chae, J. W. Kim, D. H. Kim, J. Y. Kwon, G. Amaratunga, S. Y. Lee, B. L. Choil, Y. Kuk, J. M. Kim and K. Kim, "Full-colour quantum dot displays fabricated by transfer printing," Nature Photon., 5, 176–182, 2011.

[7] J. Lim, W. K. Bae, D. Lee, M. K. Nam, J. Jung, C. Lee, K. Char and S. Lee, "InP@ZnSeS, core@composition gradient shell quantum dots with enhanced stability," Chem. Mater., 23, 4459–4463, 2011.

[8] X. Yang, D. Zhao, K. S. Leck, S. T. Tan, Y. X. Tang, J. Zhao, H. V. Demir and X. W. Sun, "Full visible range covering InP/ZnS nanocrystals with high photometric performance and their application to white quantum dot light-emitting diodes," Adv. Mater., 24, 4180–4185, 2012.

[9] Z. Tan and J. Xu, "Colloidal nanocrystal-based light-emitting diodes fabricated on plastic toward flexible quantum dot optoelectronics," J. Appl. Phys., 105, 034312-1–5, 2009.

[10] Y. Kim, S. M. Kim, and, J. Kang and C. J. Han, "Colloidal Quantum Dot LED Transparent Display on Flexible Substrate," SID Digest, 42, 1505–1508, 2011.

[11] G. P. Crawford, "Flexible flat panel display technology," Wiley, 1, 1–9, 2005.

[12] C. Ippen, T. Greco and A. Wedel, "InP/ZnSe/ZnS: A novel multishell system for InP quantum dots for improved luminescence efficiency and its application in a light-emitting device" J. Soc. Inf. Display, 13, 91–95, 2012.

[13] S. Coe, W. K. Woo, J. S. Steckel, M. Bawendi and V. Bulovic, "Tuning the performance of hybrid organic/inorganic quantum dot light-emitting devices," Organic Electronics, 4, 123–130, 2003.

Modeling of the nipip HIT structure with the hole thermionic emission mechanism

H.-T. Hsiao, T.-Y. Kuo, C.-H. Lin*

Department of Opto-Electronic Engineering, National Dong Hwa University,

No. 1, Sec. 2, Da-Hsueh Rd.,
Shoufeng, Hualien 97401, Taiwan
Tel: 886-3-8634188 / Fax: 886-3-8634180 / chlin0109@mail.ndhu.edu.tw

Abstract

We will show the current-voltage behaviors and band diagrams of HIT cells with and without hole thermionic emission model. When the hole thermionic emission model is not included, only drift diffusion model is used. In such a case, the calculated current would be much larger than the practical value.

Introduction

By considering the thermionic emission mechanism, the modeling of nano-scale heterojunction device can obtain a more accurate result. In the past, it is popular to consider the III-V heterojunction with including the mechanism of the electron thermionic emission [1]. We would like to investigate the effect of including hole thermionic emission mechanism for the heterojunction with intrinsic thin layer (HIT) solar cells. There is a large valence band offset between a-Si and c-Si in the HIT cell. It is worth to study the heterojunction with the hole thermionic emission mechanism. We will show the current-voltage behaviors and band diagrams of HIT cells with and without hole thermionic emission model. When the hole thermionic emission model is not included, only drift diffusion model is used. In such a case, the calculated current would be much larger than the practical value.

Model

We start from the current density formula considering the electron thermionic emission from reference [1], and we rewrite it with considering the hole thermionic emission instead. The current density considering hole thermionic emission can be expressed as:

$$J = A_2^* T^2 \left(\exp\left(\frac{\eta_{12} - \phi_{B2}}{kT}\right) - \exp\left(\frac{\eta_{12} - \phi_{B2} - \phi_q}{kT}\right) \right) \quad (1)$$

The label of 1 and 2 refers to semiconductor 1 and semiconductor 2. A_2^* is the effective Richardson constant of region 2 in unit of A/K-cm^2, and the hole effective mass difference between region 1 and region 2 is included in a coefficient, η [1]. This formula will be included to simulate HIT solar cells.

Results and Discussion

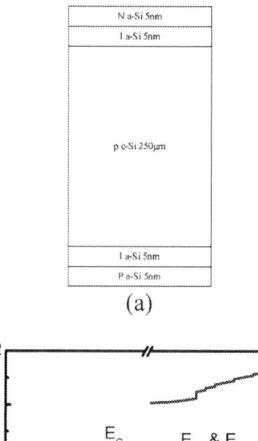

(a)

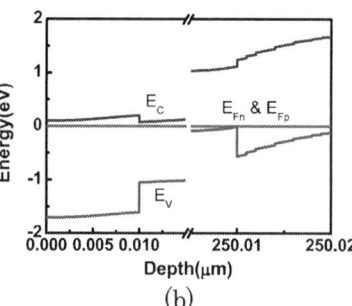

(b)

Fig. 1 (a) The schematic structure of a nipip HIT solar cell and (b)the thermal equilibrium band diagram.

Fig.1 (a) shows the studied structure of a typical nipip HIT solar cell and Fig.1 (b) shows the thermal equilibrium band diagram without considering hole thermionic emission mechanism at heterojunction. This band diagram is exactly the same with that considering hole thermionic emission mechanism, and we only show one of them here. The J-V curve of the nipip HIT solar cell is show in Fig.2.

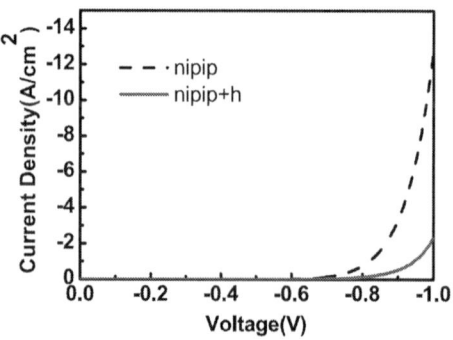

Fig. 2 The J-V curve of the nipip HIT solar cell. The symbol "+h" means the thermionic emission model is included.

978-1-4673-4840-9/13 $31.00 © 2013 IEEE   428

Fig.3(a) and Fig.3(b) show the 1-V band diagrams of both models. The hole thermionic emission model results in a much larger discontinuity of hole quasi-Fermi level (E_{Fp}) as compared to the drift diffusion model. However, it should be noted that there is still a small discontinuity of E_{Fp} existing in drift diffusion model. This small discontinuity is due to the appearance of the intrinsic a-Si layer, and the details will be studied as shown in Fig.4(c) and Fig.5(c).

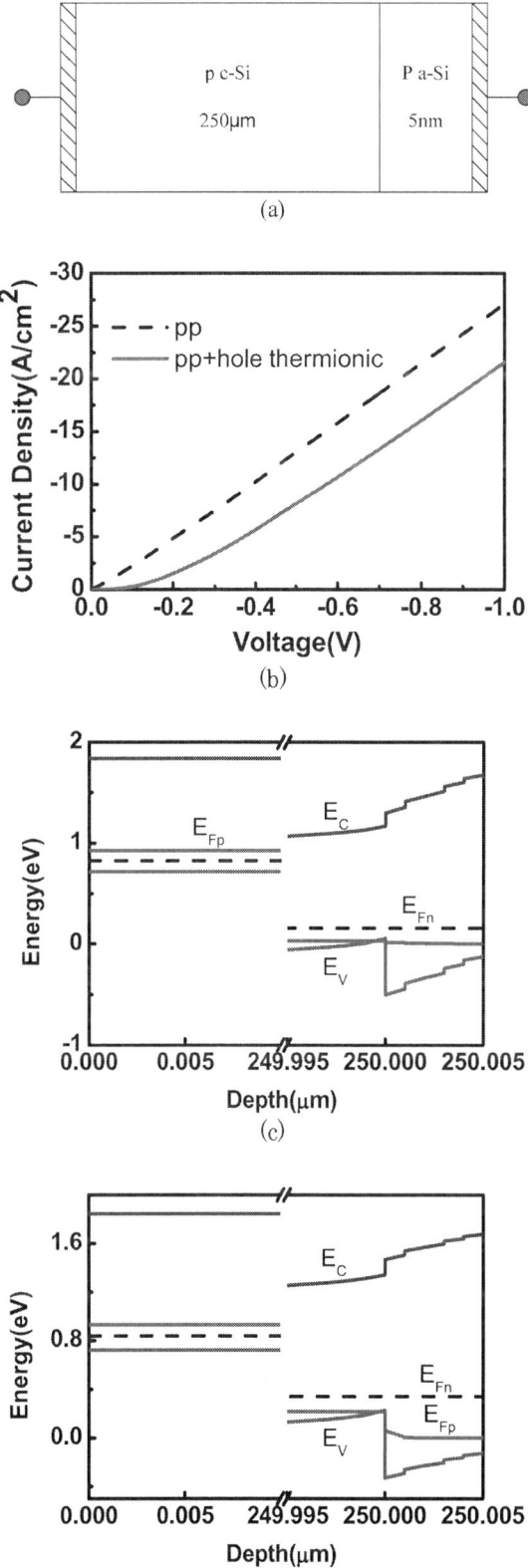

(a)

(b)

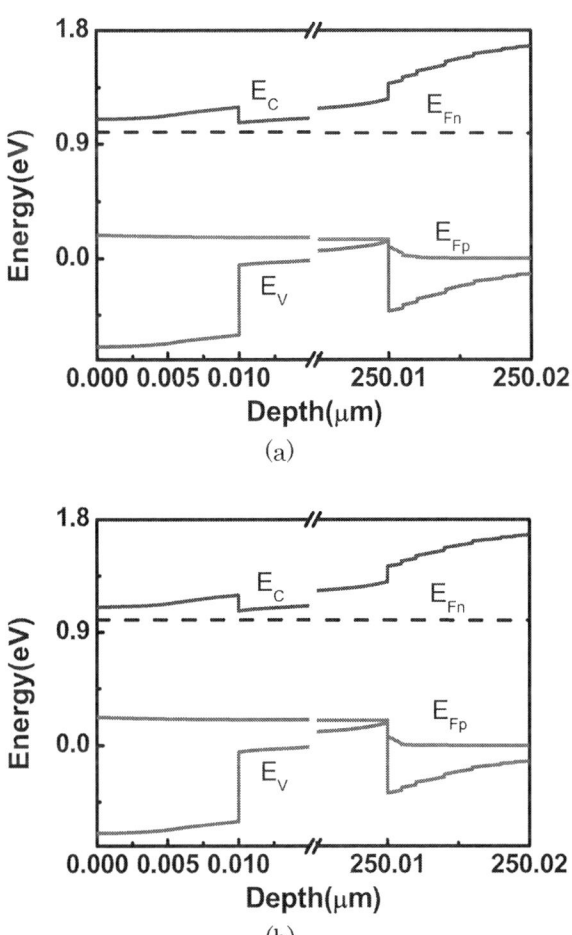

(a)

(b)

Fig. 3 (a) the 1-V band diagram of drift diffusion model. (b) the 1-V band diagram of hole thermionic emission model.

Fig.4 shows the structure of a p a-Si/p c-Si device (Fig.4(a)), J-V curve (Fig.4(b)) and band diagram with drift diffusion model (Fig.4(c)) and hole thermionic emission model (Fig.4(d)) at 1 V. As shown in Fig.4(c) and Fig.4(d), the E_{Fp} offset at the heterojunction of the drift diffusion model differs from the hole thermionic emission model. The current across the p a-Si/p c-Si heterojunction calculated by hole thermionic emission model is smaller than the current calculated by drift diffusion model due to the E_{Fp} offset, and this smaller current will limit the overall current in the nipip HIT cell. Then, we also simulated the p a-Si/i a-Si/p c-Si structure (Fig.5). As shown in Fig.5(b), adding an intrinsic a-Si layer to the interface between n type a-Si and p type c-Si indeed results in a small discontinuity of E_{Fp} even for the case without hole thermionic emission model.

(c)

(d)

Fig. 4 (a) the structure of a p a-Si/p c-Si device. (b) J-V curve without (drift diffusion model) and with hole thermionic emission model. (c) band diagram without hole thermionic emission model. (d) band diagram with hole thermionic emission model.

978-1-4673-4840-9/13 $31.00 © 2013 IEEE   429

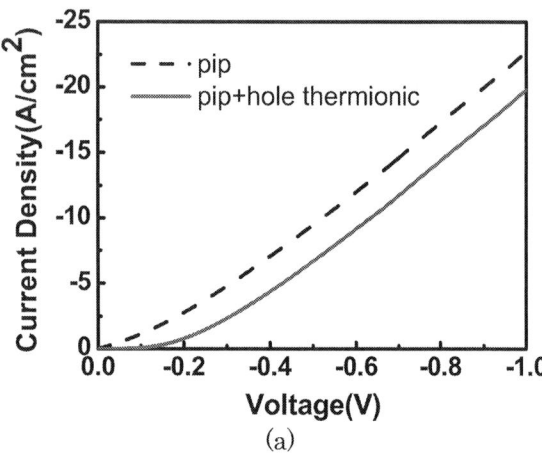

(a)

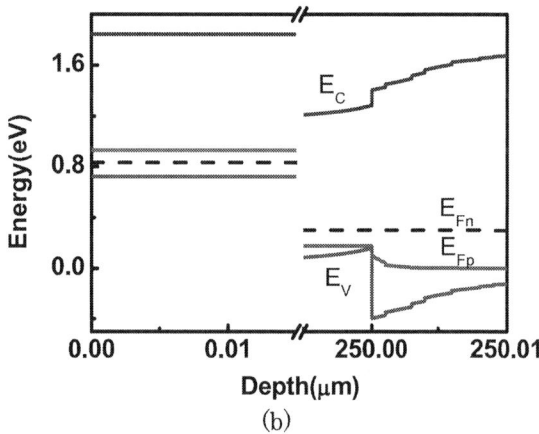

(b)

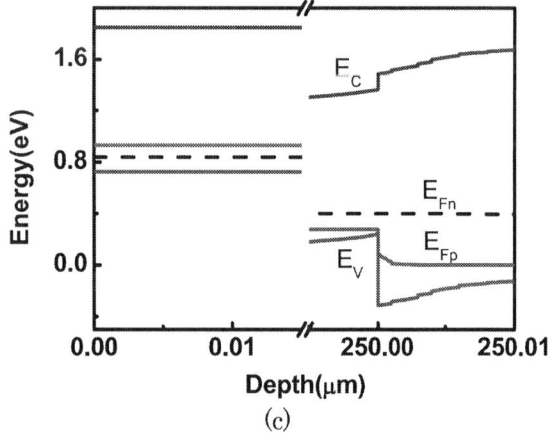

(c)

Fig. 5 (a) J-V curve without and with hole thermionic emission model. (b) band diagram without hole thermionic emission model. (c) band diagram with hole thermionic emission model.

Summary

We have successfully included the hole thermionic emission mechanism for HIT solar cells. From the band diagram at 1V, we found that the difference mainly occurs at the p a-Si/p c-Si heterojunction. The discontinuity of hole quasi-Fermi level results in the smaller current and limit the overall current in the pinin HIT solar cell.

References

[1] K. Horio and H. Yanai, "Numerical Modeling of Heterojunctions Including the Thermionic Emission Mechanism at the Heterojunction Interface," *IEEE Trans. Electron Devices*, Vol. 37, 1990, pp.1093-1098.

[2] C. M. Wu and E. S. Yang, "Carrier transport access heterojunction interface," *Solid-State Electron.*, Vol. 22, 1979, pp. 241-248.

Design guidelines for (111) Si inclined nanohole arrays in thin film solar cells

Lei Hong, [1,2] Rusli, [*1] Xincai Wang, [2] Hongyu Zheng, [2] Hao Wang, [1] HongYu Yu[3]

[1]School of Electrical and Electronic Engineering, Nanyang Technological University, 50 Nanyang Avenue, Singapore
[2]Singapore Institute of Manufacturing Technology, A*STAR (Agency for Science, Technology
and Research), 71 Nanyang Drive, Singapore
[3]South University of Science and Technology of China, Shenzhen, China
*email addresses: erusli@ntu.edu.sg

Abstract

In this paper, a systematic design and analysis of slanting silicon nanohole structure is simulated using the finite element method. The slanting nanohole structure is based on the Si (111) wafer. The impact of the hole diameter and structural periodicity has been investigated. It is found that the absorption is significantly enhanced due to the strong light trapping ability of slanting nanohole structure. The optimal structural parameters are achieved when the periodicity is 700 nm and the diameter to periodicity ratio is 0.85. The highest ultimate efficiency achieved is 32.9 %, higher than that of vertical nanohole structure with a value of 29.7%.

Keywords: silicon nanohole, light absorption, ultimate efficiency and scattering

Introduction

Silicon nanostructure based thin film solar cell has attracted tremendous research interest to boost the light harvesting ability and facilitate the carrier diffusion in order to improve the power conversion efficiency[1-5]. However, fabrication of these periodic nanostructures with low cost implementation remains a challenge. Recently, a simple and low cost maskless approach for fabricating large scale Si nanohole (SiNH) arrays has been proposed, which can be potentially applied to c-Si solar cells to improve their efficiency, while reduce the cost [6]. It is also found that the metal nanoparticles catalyst electroless etching technique shows a strong preference along the (100) crystallographic orientation of Si [7-9]. In addition, it has been demonstrated that the Si (111) demonstrate a better power conversion efficiency than Si (100) wafer, especially in the Si based hybrid solar cell [10].

Therefore, in this paper, we performed a systematical design and analysis of the inclined SiNH structure simulated using the Finite Element Method (FEM) method. The inclined angle chosen is 40° for the Si (111) NH structure [9]. The influence of the SiNH diameter and structure periodicity on light absorption has been examined to obtain the optimal structural parameters measured by the ultimate efficiency. It is found that the absorption is significantly enhanced due to a strong light harvesting ability of SiNH, which prolongs the optical path length of sunlight significantly within the structure. In addition, comparison of the light absorption spectra for the inclined SiNH and the vertically aligned SiNH structure is also investigated.

Simulation Methodology

Figure 1 shows the schematic of the three dimensional,
top-view and cross-sectional view of the slanting nanohole structure for simulation. The height (H) of the SiNH is fixed as 1μm and the underlying Si thin film is also 1 μm. The diameter (D) of the NHs and periodicity (P) of the structure are varied to achieve the optimal condition. The slanting angle of the NHs is fixed at 40° as reported [9]. The incident light is normally shone on the periodic SiNH structure. The incident light wavelength is ranging from 300 nm to 1100 nm, covering the main part of solar spectrum of interest. The interaction between incident light and the SiNH structure is solved using the finite element method [11]. The optical characteristics can then be obtained from spatial distribution of the electric field. In order to facilitate the performance comparison, the absorption spectra is weighed by the AM 1.5G solar spectrum to get the ultimate efficiency (η) [12, 13]. The optical constant of the Si material is taken from the literature [14].

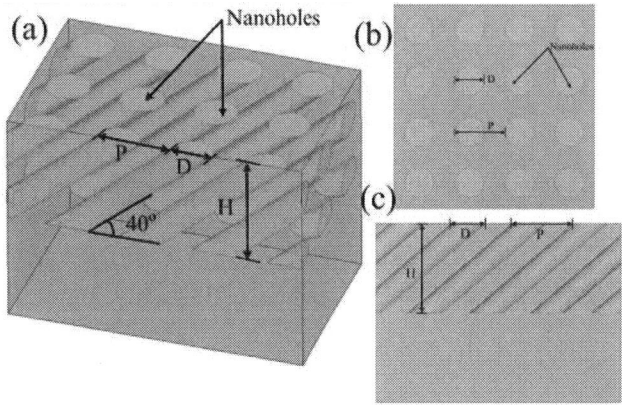

Fig. 1 Schematic illustration of the slanting SiNH structure of (a) 3-dimensional view, (b) top view and (c) cross-sectional view.

Results and Discussion

Figure 2 depicts the absorption (A) spectra for fixed structural periodicity of 700 nm with a varying D/P ratios. For small D/P ratio, the absorption is relatively lower especially for light below 700 nm. As D/P ratio increases, the light absorption improves over the solar spectrum from 300 nm to 1100 nm. However, further increase the D/P ratio to 0.9, although the absorption for short wavelength light is enhanced, the absorption degrades for longer wavelength.

Figure 3 shows the absorption spectra of fixed D/P ratio of 0.8 for different structural periodicities. From the figure we can see that the change of P has little impact for light below 500 nm. When P is 500 nm, the absorption of light ranging from 500 nm to 600 nm is higher. This is because that when P is 500 nm, light with the wavelength comparable with the nanohole P will scatter more efficiently around the SiNH structure, inducing stronger light absorption. As P increases to 600 nm, the absorption for light ranging from 600 to 700 nm is enhanced. At P of 700 nm, a high and broadband light absorption is achieved. When P further increases to 800 nm, the performance degrades.

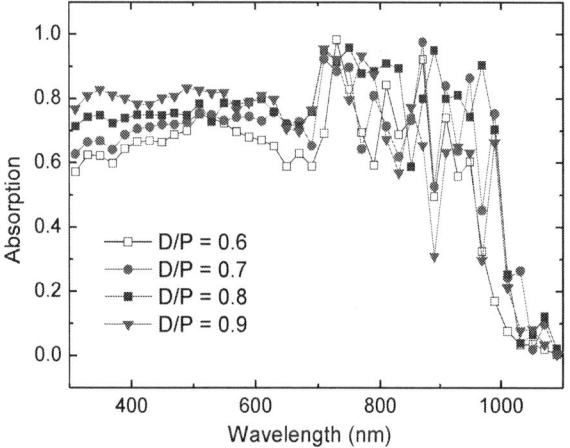

Fig. 2 Absorption spectra of the slanting SiNH structure with different D/P ratios at a fixed P of 700 nm.

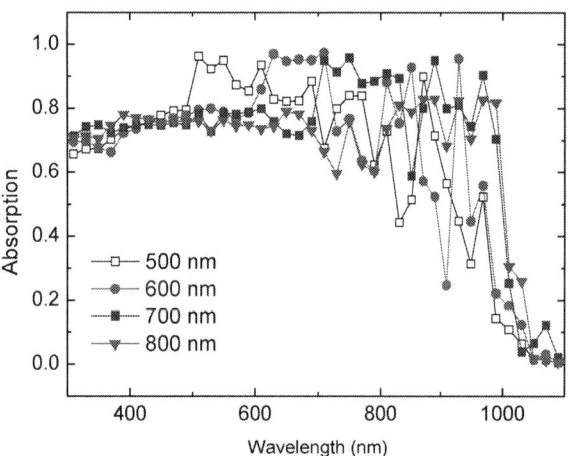

Fig. 3 Absorption spectra of the slanting SiNH structure with different periodicities at a fixed D/P ratio of 0.8.

In order to determine the optimized geometric configuration, the ultimate efficiency (η) is calculated to facilitate the performance comparison by assuming that the internal quantum efficiency is 1. Figure 4 shows the η for different P as the D/P ratio changes. It can be seen that the ultimate efficiency generally increases and then decrease as the D/P ratio change from 0.5 to 0.95. The optimized D/P ratios for SiNH structure with different P are located around 0.75 to 0.85. For small P of 300 nm, η is low. This is because, at this small P, the feature size of the nanostructure is much smaller than the longer wavelength light, hence it may not generate strong scattering for the long wavelength light, which leads to a low η value. As P increases, the optimized η increase. The optimized η of 32.9 % is achieved when the P value is 700 nm and D/P ratio is

0.85. It is also higher than that of its vertical counterpart SiNH structure, which has a highest efficiency of 29.7 %. Therefore, it indicates that the slanting SiNH structure has better light absorption ability than the vertical SiNH structure.

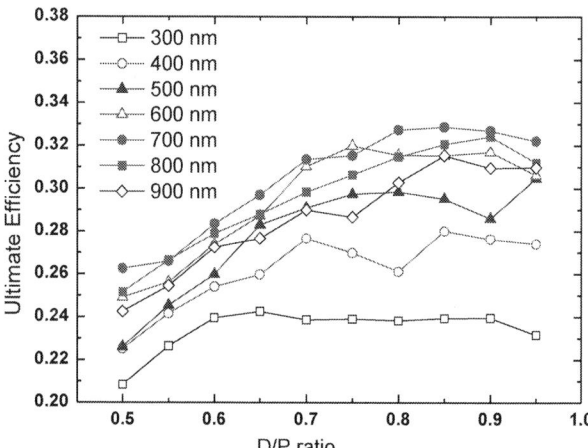

Fig. 4 Ultimate efficiency of slanting SiNH structure with varying D/P ratio for different structural periodicity.

Conclusions

In summary, simulation was used to design efficient slanting nanohole structure. The effect of structural periodicity and hole diameter have been investigated. It is found that the optimal condition is achieved when the periodicity is 700 nm and diameter/periodicity ratio is 0.85. The highest ultimate efficiency achieved is 32.9 %, higher than the vertical counterpart of 29.7 %.

References

[1] Z. Fan, H. Razavi, J.-W. Do, A. Moriwaki, O. Ergen, Y.-L. Chueh, P.W. Leu, J.C. Ho, T. Takahashi, L.A. Reichertz, S. Neale, K. Yu, M. Wu, J.W. Ager, A. Javey, Nature Materials, 8 (2009) 648-653.

[2] E.C. Garnett, M.L. Brongersma, Y. Cui, M.D. McGehee, Annual Review of Materials Research, 41 (2011) 269-295.

[3] H. Lu, C. Gang, Nano Letters, 7 (2007) 3249-3252.

[4] J. Zhu, Z. Yu, G.F. Burkhart, C.-M. Hsu, S.T. Connor, Y. Xu, Q. Wang, M. McGehee, S. Fan, Y. Cui, Nano Letters, 9 (2009) 279-282.

[5] J. Zhu, C.-M. Hsu, Z. Yu, S. Fan, Y. Cui, Nano Letters, 10 (2010) 1979-1984.

[6] F. Wang, H.Y. Yu, X. Wang, J. Li, X. Sun, M. Yang, S.M. Wong, H. Zheng, Journal of Applied Physics, 108 (2010).

[7] P. Kuiqing, Z. Mingliang, L. Aijiang, W. Ning-Bew, Z. Ruiqin, L. Shuit-Tong, Applied Physics Letters, 90 (2007) 163123-163121.

[8] X. Li, Current Opinion in Solid State and Materials Science, 16 (2012) 71-81.

[9] Z. Ming-Liang, P. Kui-Qing, F. Xia, J. Jian-Sheng, Z. Rui-Qin, L. Shuit-Tong, W. Ning-Bew, Journal of Physical Chemistry C, 112 (2008) 4444-4450.

[10] L. He, C. Jiang, H. Wang, D. Lai, Rusli, Applied Physics Letters, 100 (2012).

[11] J.-M. Jin, D.J. Riley, Finite Element Mesh Truncation, in: Finite Element Analysis of Antennas and Arrays, John Wiley & Sons, Inc., 2008, pp. 55-99.

[12] L. Junshuai, Y. HongYu, L. Yali, W. Fei, Y. Mingfei, W. She

Mein, Applied Physics Letters, 98 (2011) 021905 (021903 pp.).

[13] W. Shockley, H.J. Queisser, Journal of Applied Physics, 32 (1961) 510-519.

[14] Handbook of Optical Constants of Solids, Academic Press, Orlando, FL, USA, 1985.

Design Guidelines for Periodic Nanowire Arrays in Thin-Film Silicon/Organic Hybrid Solar Cell

Hao Wang,[1,2] Lei Hong,[1] Lining He,[1,2] and Rusli[1,2]

1School of Electrical and Electronic Engineering, Nanyang Technological University, Singapore 639798
2CINTRA UMI CNRS/NTU/THALES 3288, Research Techno Plaza, 50 Nanyang Drive, Singapore 637553
Nanyang Technological University, School of Electrical and Electronic Engineering, 50 Nanyang Avenue, Singapore 639798
E-mail: hwang2@e.ntu.edu.sg

Abstract

In this work, we perform optical simulation and investigate the effect of the periodicity (P) and diameter (D) of SiNW arrays (SiNWs) on light harvesting in thin-film silicon/PEDOT:PSS hybrid solar cells. The hybrid cell structure comprises 2 µm thick silicon thin film with SiNWs textured surface and 50 nm thick PEDOT:PSS top layer. It is found that the light harvesting in the SiNWs is greatly enhanced compared to the planar counterpart.

Keywords: hybrid solar cell, SiNWs, PEDOT:PSS, simulation, optical properties

Introduction

There has been extensive research work done on clean and renewable solar energy to make it more cost-effective. Hybrid solar cells based on SiNWs and organic materials have drawn much attention for their low-cost and low-temperature process. Besides poly (3-hexylthiophene) (P3HT) [1] and 2,2′,7,7′-Tetrakis-(N,N-di-4-methoxyphenylamino)-9,9′-spiro bifluorene (Spiro-OMeTAD) [2, 3], a transparent p-type conductive polymer, poly (3,4-ethylene dioxythiophene):polystyrenesulfonate (PEDOT:PSS) [2, 4-9] has been used to form heterojunction hybrid solar cell with Si and achieved a power conversion efficiency of ~10% [7, 9]. Recently, our group has demonstrated 2.2 µm thin-film crystalline SiNWs/PEDOT:PSS hybrid solar cells with enhanced efficiency compared to planar hybrid solar cells of the same thickness [10]. Silicon nanostructures are commonly employed in solar cells because of their capability in enhancing light harvesting and carrier extraction [11-13]. Several simulation work have been done on the optimization of SiNWs, and it was found that their properties, including periodicity, length and diameter, have great impact on the light harvesting [14-16].

In this paper, we simulated optical absorption in hybrid solar cells that are formed between a 2 µm thick Si with a thin layer of 50 nm PEDOT:PSS. The Si layer comprises a 1 µm thick SiNWs on top of a 1 µm thick Si thin film. It is found that the incorporation of SiNWs in the Si/PEDOT:PSS hybrid solar cells can significantly enhance the light harvesting. When the diameter to periodicity (D/P) ratio of the SiNWs is fixed at 0.5, it is predicted that an ultimate efficiency of ~22.3% is achievable for the aforementioned structure. An even higher ultimate efficiency of ~24.5% is realized for the same structure when P is set at 500 nm.

Simulation Approach

The simulation is performed using a full wave finite element method (FEM) [17]. The unit cell has a total thickness of 2.05 µm, which is a sum of the thickness of the underlying Si film ($T1$ = 1 µm), the length of the SiNWs (L = 1 µm) and the thickness of the PEDOT:PSS top layer ($T2$ = 50 nm), as indicated in Fig. 1. The lateral dimension of the unit cell is equal to the periodicity P of SiNWs/PEDOT:PSS hybrid structure. The interaction of incident light field and the hybrid structure was realized by applying periodical boundary condition to the unit cell to simulate the periodic structure as shown in Fig. 1. The incident light has an energy range from 1 to 4 eV, and is incident normal to the PEDOT:PSS top layer. The optical constants of PEDOT:PSS were obtained from the literature [18]. The simulated spatial distribution of the energy flux within the hybrid structure was used to extract its optical characteristics. For comparison, planar hybrid cell with only a 2 µm thick planar Si and the PEDOT:PSS top layer has also been simulated.

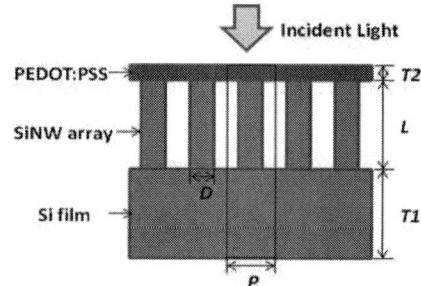

Fig. 1 Schematic of the cross-sectional view of the SiNWs/PEDOT:PSS hybrid structure. The thickness of the Si film $T1$ is 1 µm. The SiNWs length L is 1 µm. The PEDOT:PSS thickness $T2$ is 50 nm.

Results and Discussion

The incident light is trapped inside the hybrid structure, and absorbed in the PEDOT:PSS top layer and SiNWs. We focus on light absorption in Si, which includes the SiNWs and the underneath planar Si, as they are the main absorber layer of the hybrid solar cell.

Fig. 2(a) illustrates the absorption spectra in Si with a fixed D/P ratio of 0.5 and different P of 100 nm, 500 nm, and 800 nm. The corresponding results for planar cell are also plotted for comparison. As can be seen, the cell with P = 100 nm shows a lower absorption compared to the planar cell in the lower energy region (E < 2.7 eV) except for an absorption peak centered at ~2.5 eV. For P = 500 nm, we observe higher light absorption compared to the planar cell for E < ~2.3 eV, and an oscillating light absorption behavior for higher energy range. When P is further increased to 800 nm, there is a decrease in light absorption over a broad band compared to the cell with P = 500 nm, except for a peak centered at $E \sim 1.4$ eV.

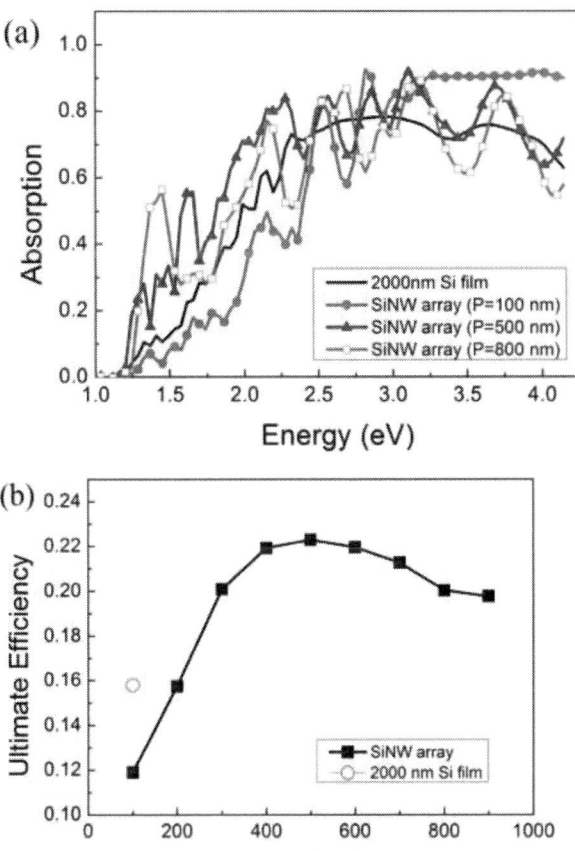

Fig. 2 (a) Absorption spectra of planar cell and SiNWs cells with P = 100, 500 and 800 nm and D/P ratio = 0.5. (b) The ultimate efficiency of planar cell and hybrid SiNWs cells as a function of P when the D/P ratio is fixed

To evaluate the light harvesting capability of the simulated structure for solar cell application, we have calculated the ultimate efficiency η under the standard Air Mass 1.5 spectrum [19] using the equation [14-16, 20]:

$$\eta = \frac{\int_{Eg}^{\infty} \frac{I(E) \times \alpha(E) \times Eg}{E} dE}{\int_{0}^{\infty} I(E) dE} \tag{1}$$

where $I(E)$ is the spectral solar intensity corresponding to the Air Mass 1.5 direct normal and circumsolar spectrum [19], E is the photon energy, E_g is the energy corresponding to the silicon band gap of ~ 1.1 eV, and $\alpha(E)$ is the absorption spectrum of the hybrid cells. This formula assumes any photon trapped inside the solar cell with energy greater than the band gap would generate only one electron-hole pair with energy equal to E_g, which can be extracted as electric current without loss [20].

Fig.2 (b) shows the ultimate efficiency η calculated based on the light absorption in Si as a function of P from 100 nm to 900

nm with a fixed D/P ratio of 0.5. The η of the planar cell is also shown for comparison. It can be seen that η first increases with P until it reaches a maximum value of 22.3% at P = 500 nm, and thereafter a further increase in P would lead to a decrease in η. This maximum efficiency achieved is $\sim 41\%$ higher than the 15.8% efficiency of the planar cell. It is also noted that η of the SiNWs cells is greater than that of the planar cell only for $P >$ 200 nm. For smaller P = 100 nm, the light trapping capability is actually worse than the planar counterpart.

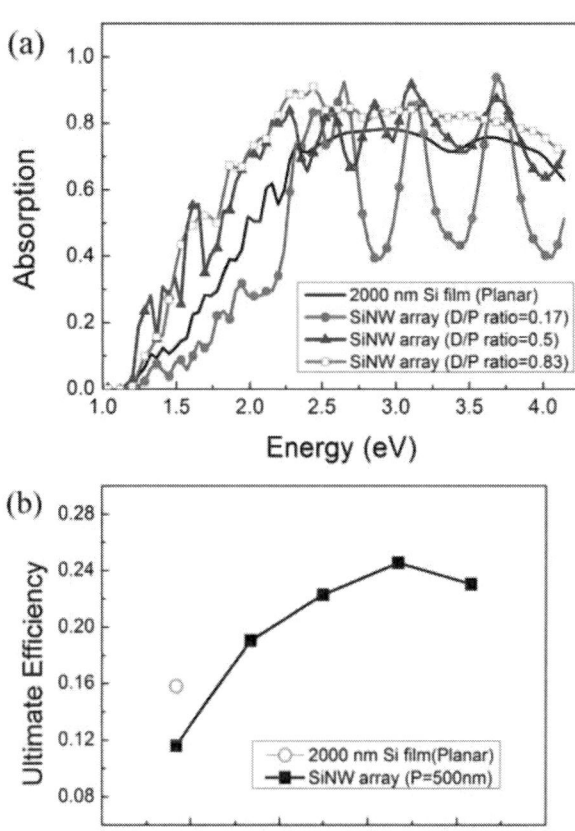

Fig. 3 (a) Absorption spectra and (b) ultimate efficiency of planar cell and SiNWs cells with P = 500 nm and D/P ratio of 0.17, 0.5 and 0.83.

Fig. 3(a) illustrates the absorption spectra in Si of the planar and SiNWs hybrid cells, with a fixed P of 500 nm and 3 different D/P ratios of 0.17, 0.5, and 0.83. As shown, the SiNWs cell with a D/P ratio of 0.17 exhibits an oscillating behavior and a lower absorption compared to the planar cell over a broad band. For cell with D/P = 0.5, we observe light absorption with oscillatory behavior and it is higher compared to the planar cell. When D/P ratio is further increased to 0.83, the light absorption further enhances and the spectrum becomes smoother compared to cells with lower D/P ratios.

Fig.3 (b) shows η calculated based on the light absorption in Si for the planar cell and SiNWs cells with a fixed P of 500 nm and varying D/P ratios from 0.17 to 0.83. It can be seen that η first increases with the D/P ratio until it reaches a maximum value of $\sim 24.5\%$ at D/P = 0.67. Subsequently a further increase in the D/P ratio leads to a decrease in η. This maximum efficiency is ~55% higher than the 15.8% efficiency calculated for the planar cell. It is noted that the light absorption in the SiNWs cells is greater than that of the planar cell for $D/P > \sim 0.3$. For $D/P = 0.17$, the light trapping capability is actually worse than the planar counterpart.

Conclusions

In conclusion, we have studied the optical absorption of Si thin film/PEDOT:PSS hybrid solar cells. It is found that the light absorption can be greatly enhanced by incorporating SiNWs in the Si/PEDOT:PSS hybrid solar cells. An ultimate efficiency of ~22.3% is predicated to be achievable when the periodicity and diameter to periodicity ratio are set at 500 nm and 0.5 respectively. A maximum ultimate efficiency of ~24.5% is achieved for a diameter to periodicity ratio of 0.67 when P is fixed at 500 nm. This work suggests how the structural parameters of hybrid solar cells based on silicon thin films with nanowire arrays textured surface and PEDOT:PSS can be optimized to achieve high efficiency.

References

[1]   S. Avasthi, *et al.*, "Role of majority and minority carrier barriers silicon/organic hybrid heterojunction solar cells," *Advanced Materials*, vol. 23, pp. 5762-5766, 2011.

[2]   L. He, *et al.*, "Highly efficient Si-nanorods/organic hybrid core-sheath heterojunction solar cells," *Applied Physics Letters*, vol. 99, 2011.

[3]   X. Shen, *et al.*, "Hybrid heterojunction solar cell based on organic-inorganic silicon nanowire array architecture," *Journal of the American Chemical Society*, vol. 133, pp. 19408-19415, 2011.

[4]   E. C. Garnett, *et al.*, "Silicon nanowire hybrid photovoltaics," in *35th IEEE Photovoltaic Specialists Conference, PVSC 2010, June 20, 2010 - June 25, 2010*, Honolulu, HI, United states, 2010, pp. 934-938.

[5]   S.-C. Shiu, *et al.*, "Morphology dependence of silicon nanowire/poly(3,4-ethylenedioxythiophene): poly(styrenesulfonate) heterojunction solar cells," *Chemistry of Materials*, vol. 22, pp. 3108-3113, 2010.

[6]   L. Wenhui, *et al.*, "Si/PEDOT:PSS core/shell nanowire arrays for efficient hybrid solar cells," *Nanoscale*, vol. 3, pp. 3631-4, 2011.

[7]   L. He, *et al.*, "High efficiency planar Si/organic heterojunction hybrid solar cells," *Applied Physics Letters*, vol. 100, 2012.

[8]   H. Lining, *et al.*, "Simple Approach of Fabricating High Efficiency Si Nanowire/Conductive Polymer Hybrid Solar Cells," *IEEE Electron Device Letters*, vol. 32, pp. 1406-8, 2011.

[9]   L. He, *et al.*, "Si nanowires organic semiconductor hybrid heterojunction solar cells toward 10% efficiency," *ACS Applied Materials and Interfaces*, vol. 4, pp. 1704-1708, 2012.

[10]   L. He, *et al.*, "Effects of nanowire texturing on the performance of Si/organic hybrid solar cells fabricated with a 2.2 m thin-film Si absorber," *Applied Physics Letters*, vol. 100, 2012.

[11]   B. M. Kayes, *et al.*, "Comparison of the device physics principles of planar and radial p-n junction nanorod solar cells," *Journal of Applied Physics*, vol. 97, 2005.

[12]   J. Zhu, *et al.*, "Optical absorption enhancement in amorphous silicon nanowire and nanocone arrays," *Nano Letters*, vol. 9, pp. 279-282, 2009.

[13]   K. Peng, *et al.*, "Aligned single-crystalline Si nanowire arrays for photovoltaic applications," *Small*, vol. 1, pp. 1062-1067, 2005.

[14]   L. Junshuai, *et al.*, "Design guidelines of periodic Si nanowire arrays for solar cell application," *Applied Physics Letters*, vol. 95, p. 243113 (3 pp.), 2009.

[15]   L. Junshuai, *et al.*, "Si nanopillar array optimization on Si thin films for solar energy harvesting," *Applied Physics Letters*, vol. 95, p. 033102 (3 pp.), 2009.

[16]   H. Lu and C. Gang, "Analysis of optical absorption in silicon nanowire arrays for photovoltaic applications," *Nano Letters*, vol. 7, pp. 3249-3252, 2007.

[17]   J.-M. Jin, *The Finite Element Method in Electromagnetics, 2nd Edition*, 2nd ed.: Wiley, 2002.

[18]   L. A. A. Pettersson, *et al.*, "Optical anisotropy in thin films of poly(3,4-ethylenedioxythiophene)-poly(4-styrenesulfonate)," *Organic Electronics: physics, materials, applications*, vol. 3, pp. 143-148, 2002.

[19]   Air Mass 1.5 Spectra, American Society for Testing and Materials [Online]. Available: http://rredc.nrel.gov/solar/spectra/am1.5/

[20]   W. Shockley and H. J. Queisser, "Detailed balance limit of efficiency of p-n junction solar cells," *Journal of Applied Physics*, vol. 32, pp. 510-519, 1961.

Electronic structure of Ge/Si$_x$Sn$_y$Ge$_{1-x-y}$ quantum dots

J. Chen, W. J. Fan, D. H. Zhang, Q. Xu, X. W. Zhang

School of Electrical and Electronic Engineering,
Nanyang Technological University, Singapore 639798

Abstract

The electronic band structures and optical gains of Ge/Si$_x$Sn$_y$Ge$_{1-x-y}$ truncated pyramid-shaped quantum dots (QDs) are calculated using the 8-band **k.p** model. The large bowing factors in the calculation of band gaps of both Γ-conduction and L-conduction valley in the barrier are considered so that the band gaps of the barrier are small. The strains are calculated by constant strain method and valence force field (VFF) method. By constant strain method, strains are uniformly distributed in the QD, and there is no strain in the barrier. Due to the small conduction band offset and strong quantum confinement effect, it requires greater tensile strain to make Ge QD a direct-band material by setting Si and Sn compositions at 11% and 26% respectively. By VFF method which considers the four nearest-neighbour interactions for each atom, the calculated strains are not uniform and the strains exist in the barrier. These barrier strains have raised the valence band edge of the barrier and changed the potential profile of the structure, so that the holes may not be confined in the QD. The VFF method should be more accurate to reflect the real situation.

Key words: Band structure, **k.p** method, quantum dots, Germanium, optical gain

PACS: 73.21.La, 73.22.-f, 78.67.Hc

Introduction

Group-IV semiconductors such as germanium (Ge) are considered as interesting material for future optoelectronics semiconductor devices, because they are compatible with silicon technology and their optical properties can be applied by using different growth patterns. The experimental and theoretical investigations have shown that these alloys have a wide tunability of band gap, indicating a huge potential for optoelectronic applications such as laser diodes and photodetectors [1-5]. The main difficulties of their application are their large lattice difference of the constituents, and their indirect band gaps. To create a direct-band group-IV semiconductor, some research has been done to grow tensile strained Ge layers sandwiched between SiGeSn barriers [6-8]. Under tensile strain, the Γ-conduction band minimum of Ge decreases faster than the L-conduction band minimum. When the

tensile strain is greater than 0.0161, the band edge of the Γ-conduction valley is lower than L-conduction valley, and the optical gains of Ge/SiSnGe quantum wells have done theoretically studied [9]. However, there is experimentally evidence that in many cases the shape the QDs more closely approximate to truncated pyramids [10-11], and the calculated energies provide the best fit to photoluminescence experiments for truncated pyramids [12]. In this paper, we simulate the tensile strained Ge/SiSnGe quantum dots (QDs) to investigate its electronic band structure and optical properties. We apply 8-band **k.p** model which includes the coupled conduction and valence bands, and spin-orbit splitting to study the electronic band structure.

Theoretical Model

In the simulation, the top of the pyramid is truncated, and the pyramid axis is along the direction of positive z direction. Every surface of the pyramid is a simple crystal plane, and the full pyramid has its base width (b) twice of its height (x). After the top part is truncated, the effective height h of truncated QDs is about 1.71 nm, and the base width is about 11.4 nm. The schematic diagrams of truncated QDs are shown in Fig. 1. The QD is in the centre of the unit cell. The width and the height of the unit cell are 4 and 3 times of the height of QD respectively. The zero point is taken at the centre of the unit cell. The assumed temperature is T=300 K. The reference level of energy is the unstrained valence band edge of the Ge QD. Because the QDs are assumed to be along the [001] direction, we do not include piezoelectric effects in our simulation.

By constant strain method, the strains only exist in the QD region by $\varepsilon_{xx}=\varepsilon_{yy}=(a_0-a)/a$ and ε_{zz} = $-2\varepsilon_{xx}$ C12/C11, where C11 and C12 are the elastic stiffness constants, and a and a_0 are the lattice constants of Ge QD and SiGeSn barrier respectively. By VFF method which includes the four nearest-neighbour interactions, the atoms in the barrier are considered as one material virtually whose properties and parameters are calculated by linear interpolation. The lattice constant of the barrier is great than Ge, and it induces tensile strain in the QDs. The presence of strain significantly affects the electronic band structure of the QDs. Applying the first principle, our research group has calculated the parameters used in VFF calculation listed in

Table I, and the detailed VFF processes follow Refs. 13, 14, and 15.

Table I: Parameters used in VFF calculation.

Materials	Si	Ge	Sn
Bond distance d^0_{ij} (Å)	2.34000	2.44522	2.80549
α (10^3 dyne/cm)	48.271	37.411	27.682
β(10^3 dyne/cm)	13.037	10.306	5.264

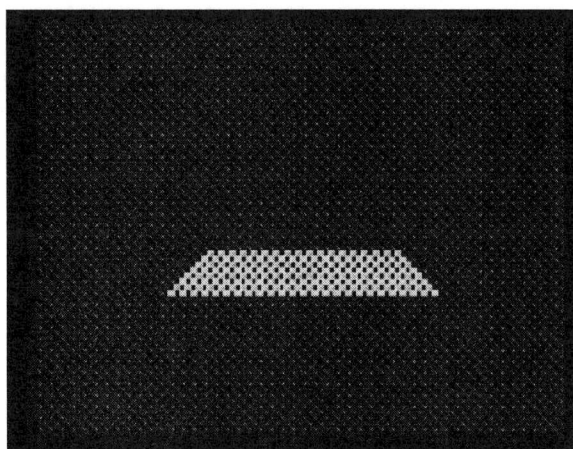

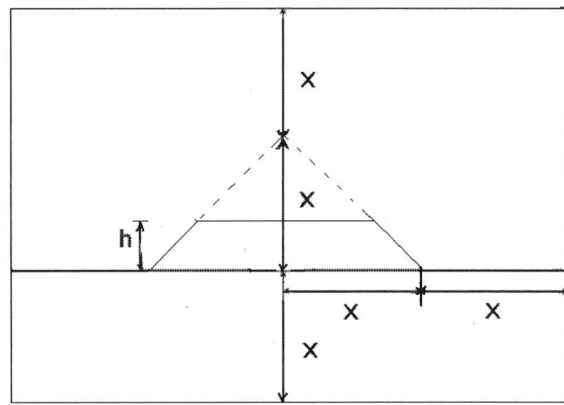

Fig. 1. Schematic diagrams of truncated pyramid-shaped Ge quantum dot with side views.

We apply 8-band k.p model which includes the coupled conduction and valence bands, and spin-orbit splitting. The detailed eight-band Hamiltonian follows Ref. 16. In this calculation, most of the parameters of the barrier can only be obtained by interpolation among Si, Ge and α-Sn. The parameters are listed in Table II [17-21]. For the band gap of SiSnGe at Γ-conduction valley at 300K, the bowing

factors are b_{SiGe} = 0.14 [22], b_{GeSn} = 2.55 [23], b_{SiSn} = 3.05 [22], and at L-conduction valley, the bowing factors are b_{SiGe} = 0.335 [24], b_{GeSn} = 0.89 [23], b_{SiSn} = 3.123 [24].

Table II: Parameters used in 8-band k.p calculation.

Materials	Si	Ge	Sn
E_g,Γ (eV)	4.185 [a]	0.7985 [a]	-0.413 [a]
m_c (m_0)	0.528 [a]	0.038 [a]	-0.058 [a]
γ_1	4.22 [a]	13.38 [a]	-15 [a]
γ_2	0.39 [a]	4.24 [a]	-11.45 [a]
γ_3	1.44 [a]	5.69 [a]	-8.55 [a]
ΔE_{vav} (eV)	-0.47 [b]	0	0.69 [c]
Δ(eV)	0.044 [a]	0.29 [a]	0.8 [a]
E_p (eV)	21.6 [a]	26.3 [a]	24.0 [a]
a_c (eV)	1.98 [d]	-8.24 [d]	-8.714 [e]
a_v (eV)	2.46 [d]	1.24 [a]	1.62 [e]
b (eV)	-2.1 [a]	-2.9 [a]	-2.01 [e]
d (eV)	-4.8 [e]	-5.3 [e]	-0.39 [e]
n_r		4.03249	

[a] Reference 17.
[b] Reference 18.
[c] Reference 19.
[d] Reference 20.
[e] Reference 21.

Theoretical Results and Discussion

A. By Constant Strain Method

The constant strain method was applied in Ref. 9, and we are different from Ref. 9 that we have considered the bowing factors in the calculation of band gap of the barrier SiGeSn. Because these bowing factors are quite large, the band gap of the barrier is much smaller than those in Ref. 9. Correspondingly the Si composition should be set higher than Ref. 9, at 11% to obtain appropriate band gap. (The Si composition cannot be too high neither due to the large bowing factors involving Si composition, especially for L-valley).

Fig. 2 shows the potential profiles of various bands for Sn composition of (a) 20% and (b) 26%. In the QD region (Ge), the light hole (LH) band edge is lifted while the heavy hole (HH) band edge is lowered due to the tensile strain. Due to the larger magnitude of the deformation potential a_c of the Γ-conduction valley, the band edge of the Γ-conduction valley of Ge is lower than the L-conduction valley for both Figs. 2(a) and 2(b). For electrons to accumulate at the Γ-conduction valley, we have to consider the factor of the small band gap of the barrier. In Fig. 2(a), the Sn composition is only 20%, and the calculated energy levels C1 (the ground state of the Γ-conduction valley) and LC1 (the ground state of the

978-1-4673-4840-9/13 $31.00 © 2013 IEEE

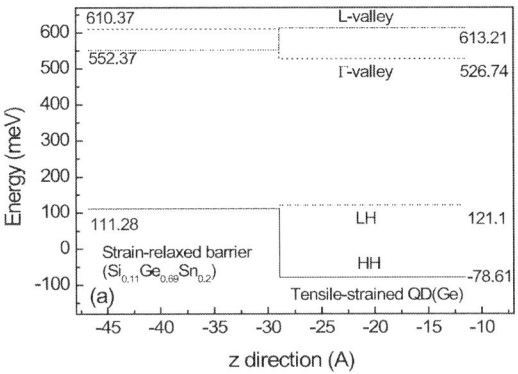

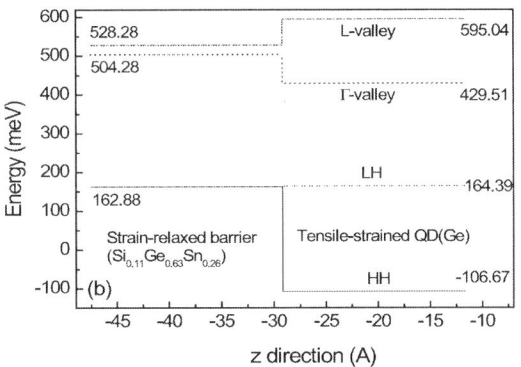

Fig. 2 Potential profiles of various bands for Sn composition of (a) 20% and (b) 26%. In the QD region (Ge), the LH band edge is lifted while the HH band edge is lowered due to the tensile strain.

L-conduction valley) are 608.24 meV and 552.82 meV respectively. It shows C1 is still higher than LC1 although the band edge of the Γ-conduction valley of Ge is lower than the L-conduction valley of Ge. The reasons are that the Γ-conduction band offset is rather small and the quantum confinement effect of QD is much stronger than quantum well (QW), so the C1 is quite close to the band edge of the Γ-conduction valley of the barrier. But for the L-conduction valley, the electrons may accumulate in the barrier due to the lower potential. The quantum confinement effect in the barrier is much smaller than QD, so the LC1 is just a little higher than the band edge of the L-conduction valley of the barrier. In Fig. 2(b), the Sn composition increases to 26%, and the band edge of the Γ-conduction valley of the barrier is lower than that of the L-conduction valley of the barrier. At this time, electrons can accumulate in the QD region, and then we can conclude that Ge becomes a direct-band material. Comparing Figs. 2(a) and 2(b), we suggest that dislike QW,

the condition that the band edge of the Γ-conduction valley of Ge QD is lower than the L-conduction valley is not sufficient to conclude Ge becomes a direct-band material. When the Sn composition is 20%, the lattice mismatch is (5.78667-5.647)/5.647 = 0.0247 which may not be sufficient. Now we focus on the valence band, when the Sn composition increases from 20% to 26%, the valence band offset decreases from about 10 meV to less than 2 meV. If the Sn composition increases further, holes may not confined in the Ge QD, but in the barrier due to barrier's higher potential. From this analysis of potential profile, it shows the compositions of the barrier must be delicately selected to make Ge QD a direct-band material.

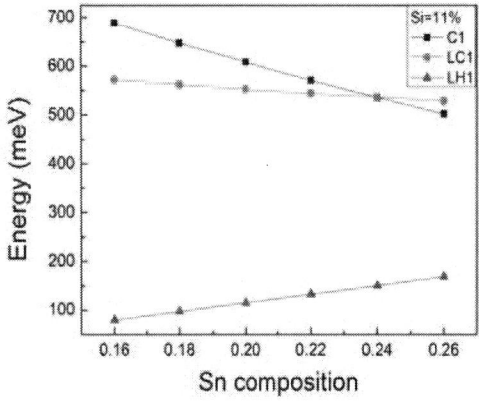

Fig. 3 By constant strain method, the energy levels of C1 (the ground state of the Γ-conduction valley), LC1 (the ground state of the L-conduction valley) and LH1 (the ground state of light hole band) as a function of the Sn composition in the SiGeSn barrier. The Si composition is fixed at 11%. The reference level is the unstrained valence band edge of the Ge QD.

Fig. 3 shows by constant strain method, the energy levels of C1 (the ground state of the Γ-conduction valley), LC1 (the ground state of the L-conduction valley) and LH1 (the ground state of light hole band) as a function of the Sn composition in the SiGeSn barrier. The Si composition is fixed at 11%. The reference level is the unstrained valence band edge of the Ge QD. When the Sn composition increases, because C1 and H1 are in the QD region, C1 (H1) deceases (increases) is due to the increased tensile strain. From Fig. 2, we know LC1 is in the barrier, so LC1 decreases mainly due to the decreased band gap of the L-conduction valley of the barrier. At about 24% of Sn, C1 is lower than LC1, so the desirable range of Sn is from 24% to 26% in which Ge is a direct-band material.

978-1-4673-4840-9/13 $31.00 © 2013 IEEE

The optical transition matrix elements (TME) measure the momentum of the transition between the hole subbands and the electron subbands. The QD is under tensile strain which raises the energy levels of light hole above heavy hole. The TME for the transverse electric (TE) mode is the sum of the major contribution from the electron to heavy hole transition and the minor contribution from the electron to the hybrid state of light hole and spin-orbit split-off hole transition. For the transverse magnetic (TM) mode, the contribution comes from the electron to the hybrid state of light hole and spin-orbit split-off hole transition, so the numerical values of TME for the TM mode are much greater than the TE mode, and we only focus on TM mode to investigate the optical properties. When the Sn composition increases from 24% to 26%, the TME decreases from 0.531 to 0.496 because the valence band offset drops significantly, and the holes are less confined in the QD region resulting in a less efficient overlapping of the electron-hole wave functions. Fig. 4 shows the maximum optical gains for TM mode as a function of carrier density for Sn compositions at 24% and 26% in the barrier. When the carrier density increases, the maximum optical gains for both cases increases steadily, and the structure with greater Sn composition in the barrier has the greater maximum optical gain despite its smaller value of TME. Possible reason is that the volumes of the QDs are very small, and the structures with greater Sn composition have greater lattice constant and greater volume. It suggests that the TME is not the dominant factor for optical gains. Additionally the magnitude of the maximum gain is desirable although it requires to supply a large amount of carrier density.

uniform and it is dependant on the shape of the structure and the surrounding environment of each atom. In this case, the strains exist in the barrier especially the part adjacent to the QDs, although the magnitude is smaller than the strains in the QDs. These strains in the barrier can change the potential profile of various bands, and then significantly affect the electronic band structure and optical properties of QDs. Fig. 5 shows the VFF strain distribution of components ε_{xx} and ε_{zz} in y-z plane (x=0) with the barrier of 11% of Si and 20% of Sn. The red (blue) region represents the greatest (smallest) value. The red region is QD, and ε_{xx} is positive meaning tensile strains exist in the QD because the lattice constant of Ge is smaller than that of the barrier. Fig. 5(b) shows that the component ε_{zz} in QD is negative, and the smallest value locates in the centre of QD, and the greatest values are in the barrier just adjacent to the QD region. Because the strains are not uniform, we cannot plot the potential profile similar as Fig. 2(a). The valence bands concern us more because of its very small band offset, so Fig. 6 shows the potential profiles of (a) heavy hole and (b) light hole in the y-z plane (x=0) with the barrier of 11% of Si and 20% of Sn. The HH potential is higher in the barrier region because a small positive strains exist in the barrier, and the LH potential is higher in the QDs. It is not so obvious to indicate where the holes will accumulate. We should investigate the squared wave functions of holes to have a better understanding.

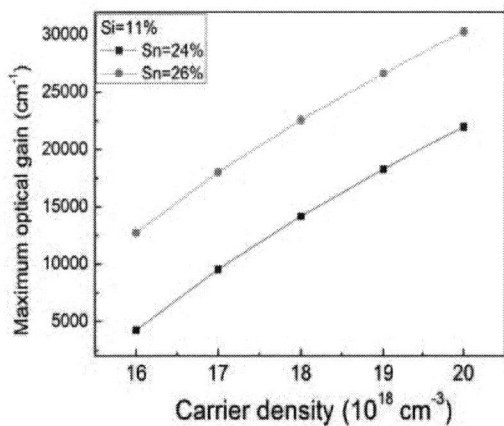

Fig. 4 The maximum optical gain for TM mode as a function of carrier density for Sn compositions at 24% and 26% in the barrier.

B. By VFF Method

The VFF is different from the constant strain method in such a way that the VFF strains in the structure are not

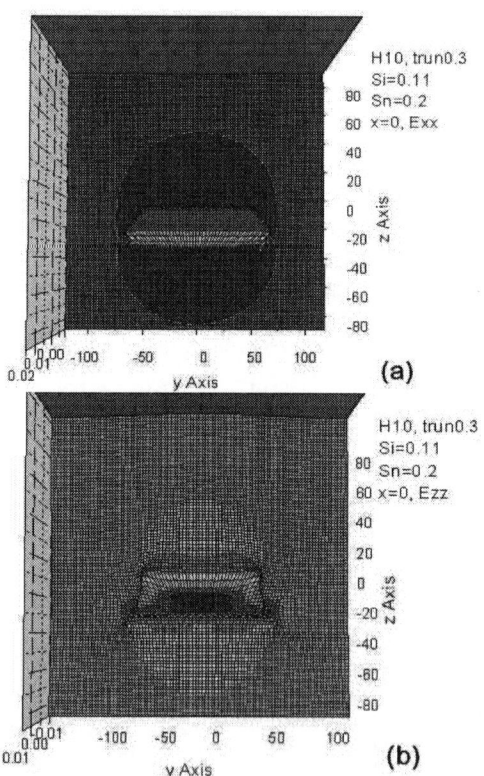

Fig. 5 The VFF strain components (a) ε_{xx} and (b) ε_{zz} distribution in the y-z plane (x=0) with the barrier of 11% of Si and 20% of Sn.

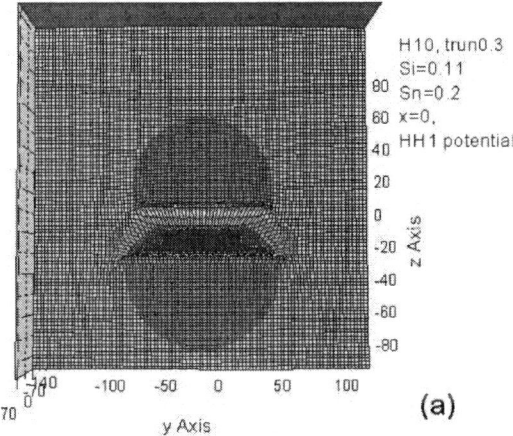

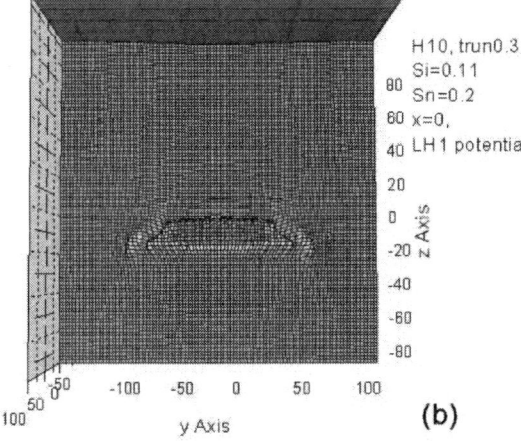

Fig. 6 By VFF method, the potential profiles of (a) heavy hole and (b) light hole in the y-z plane (x=0) with the barrier of 11% of Si and 20% of Sn.

Fig. 7 shows the squared wave function of C1 and H1 in the y-z plane (x=0) with the barrier of 11% of Si and 20% of Sn for both constant strain method and VFF method. For constant strain method, both electrons and holes are confined in the QD region, and this is consistent with the potential profiles shown in Fig. 2. But for VFF method, electrons are confined but holes are not confined in the barrier region. The main reasons for this are that the valence band offset is very small and the strains existing in the barrier raise the HH band edge of the barrier higher than the LH band edge of the QD.

Fig. 8 shows the energy levels of C1, LC1 and H1 as a function of the Sn composition in the barrier. Fig. 8 is very similar to Fig. 3 that both C1 and LC1 decrease, and C1 decreases faster due to the larger magnitude of the conduction band deformation potential. And LC1 is above C1 if the Sn composition in the barrier is greater than 24%. The difference between Fig. 8 and Fig. 3 is that holes are not confined in Fig. 8, so that H1 is the ground state of HH

in the barrier. When the Sn composition increases, H1 increases because the strains in the barrier increase, and the HH band edge is lifted higher. As electrons and holes are separated in the QD and the barrier respectively, their wave functions may not overlap efficiently, and the result is not desirable.

Si=11%, Sn=20%	C1, the ground state of Γ-conduction band	H1, the ground state of valence band
By constant strain method		
By valence force field method (VFF)		

Fig. 7 The squared wave function of C1 and H1 in the y-z plane (x=0) with the barrier of 11% of Si and 20% of Sn for both constant strain method and VFF method.

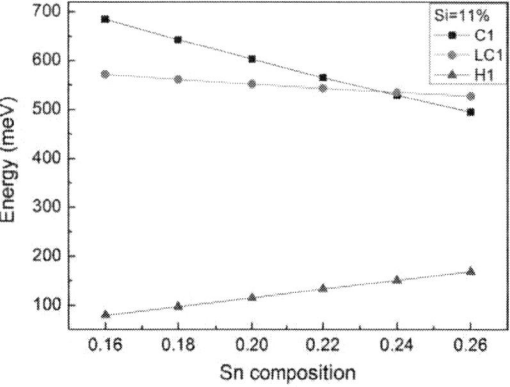

Fig. 8 By VFF method, the energy levels of C1 (the ground state of the Γ-conduction valley), LC1 (the ground state of the L-conduction valley) and LH1 (the ground state of light hole band) as a function of the Sn composition in the SiGeSn barrier. The Si composition is fixed at 11%. The reference level is the unstrained valence band edge of the Ge QD.

Conclusions

We have considered the large bowing factors in the calculation of the barrier band gap. This narrow barrier band gap is the limiting factor for us to choose the compositions of the elements Si, Ge and Sn, and it also affects the electronic structure and optical properties of $Ge/Si_xSn_yGe_{1-x-y}$ QD. The constant strain method is often applied in quantum well in which strains only exist in the QD region. For the tensile-strained Ge QD, due to the small Γ-conduction band offset and strong quantum confinement effect, the C1 is quite close to the band edge of the Γ-conduction of the barrier. The L-conduction band edge is lower in the barrier. So to make Ge become a direct-band material, it requires greater tensile strain not only to force the band edge of the Γ-conduction valley lower than the L-conduction valley in the QD, but also reduce the band gap of Γ-conduction valley lower than the L-conduction valley in the barrier. If the Si composition is fixed at 11%, when the Sn composition varies from 24% to 26%, Ge becomes a direct-band material. When the Sn composition increases, the valence band offset decreases, and the wave function of valence band is less confined in the QD, resulting in a lower value of TME for TM mode. This limiting factor prevents us from increasing the Sn composition higher than 26%. The optical gains are more affected by the volume of the QD, so the structure with higher Sn composition has the greater maximum gain despite its lower value of TME. The VFF method is more accurate to reflect real situation. The VFF method is different from constant strain method in terms of that the strain distribution of VFF is non-uniform in the system and strains exist in the barrier. The strains in the barrier raise the valence band edge of the barrier, and the valence band offset is small, so that holes may not be confined in the QD. Unless there is updated data in the calculation, the results shows $Ge/Si_xSn_yGe_{1-x-y}$ quantum dots are difficult to become a direct-band material by the VFF method.

References

[1] G. Sun, H. H. Cheng, J. Menendez, J. B. Khurgin, and R. A. Soref, Appl. Phys. Lett. **90**, 11992 (1990).

[2] G. He and H. A. Atwater, Phys. Rev. Lett. **79**, 1937 (1997).

[3] P. Moontragoon, Z. Ikonic, and P. Harrison, Semicond. Sci. Technol. **22**, 742 (2007).

[4] J. Kouvetakis, J. Menendez, and A. G. V. Chizmeshya, Annu. Rev. Mater. Res. **36**, 497 (2006).

[5] J. Kouvetakis and A. G. V. Chizmeshya, J. Mater. Chem. **17**, 1649 (2007).

[6] R. A. Soref and L. Friedman, Superlatt. Microstruct. **14**, 189-194 (1993).

[7] J. Taraci, J. Tolle, J. Kouvetakis, M. R. McCartney, D. J. Smith, J. Menendez, and M. A. Santana, Appl. Phys. Lett. **78**, 3607-3609 (2001).

[8] J. Menendez and J. Kouvetakis, Appl. Phys. Lett. **85**, 1175-1177 (2004).

[9] S. W. Chang and S. L. Chuang, IEEE J. Quantum Electron. vol. **43**, no. 3, pp 249-256 (2007).

[10] D. S. L. Mui, D. Leonard, L. A. Coldren, and P. M. Petroff, *Appl. Phys. Lett.* **66** (1995), p. 1620.

[11] P. W. Fry, I. E. Itskevich, D. J. Mowbray, M. S. Skolnick, J. J. Finley, J. A. Barker, E. P. O'Reilly, L. R. Wilson, I. A. Larkin, P. A. Maksym, M. Hopkinson, M. Al-khafaji, J. P. R. David, A. G. Cullis, G.. Hill, and J. C. Clark, *Phys. Rev. Lett.* **84** (2000), p. 733.

[12] H. Lee, W. Yang, R. Lowe-Webb, and P. C. Sercel, in *Proceedings of the 24th International Conference on the physics of Semiconductors*, edited by D. Gershoni *(World Scientific, Singapore, 1998)*.

[13] Hongtao Jiang and Jasprit Singh, Phys. Rev. B **56**, 4696 (1997).

[14] W. H. Press, B. P. Flanner, S. A. Teukolsky, and W. T. Veteering, Numerical Recipes (Cambridge University Press, New York, 1989).

[15] C. Pryor, J. Kim, L. W. Wang, A. J. Williamson, and A. Zunger, J. Appl. Phys. **83**, 2548 (1998).

[16] J. Chen, W. J. Fan, Q. Xu, X. W. Zhang, J. W. Luo, S. S. Li and J. B. Xia, J. Appl. Phys. **105**, 123705(2009).

[17] O. Madelung, M. Schultz, and H. Weiss, Eds. Physics of Group IV Elements and III-V Compounds, 1st ed. New York: Springer-Verlag, 1982, vol. 17a.

[18] M. M. Rieger and P. Vogl, Phys. Rev. B, vol. **48**, no. 19, pp. 14276-14287 (1993).

[19] M. Jaros, Phys. Rev. B, vol. 37, no. 12, pp. 7112-7114 (1988).

[20] C. G. Van de Walle, Phys. Rev. B, vol. 39, no. 3, pp. 1871-1883 (1989).

[21] P. Moontragoon, N. Vukmirovic, Z. Ikonic, and P. Harrison, J. Appl. Phys. **103**, 103712 (2008).

[22] V. R. D'Costa, C. S. Cook, J. Menendez, J. Tolle, J. Kouvetakis, S. Zollner, Solid State Communi. **138**, 309-313 (2006).

[23] W. J. Yin, X. G. Gong, S. H. Wei, Phys. Rev. B **78**, 161203 (2008).

[24] P. Moontragoon, Z. Ikoni'c and P. Harrison, Semicon. Sci. Technol. 22 (2007) 742-748.

Copper oxide based low cost thin film solar cells

[1,2]Vinay. Kumar, [1,2]S.Masudy-Panah, [2]C. C. Tan, [1]T. K. S. Wong, [2]D. Z. Chi,
and[2]G. K. Dalapati[*],

[1]School of Electrical and Electronic Engineering, Nanyang Technological University, Nanyang Avenue,
Singapore 639798

[2]Institute of Materials Research and Engineering, A*STAR (Agency for Science, Technology and Research), 3
Research Link, Singapore 117602
*Corresponding author: dalapatig@imre.a-star.edu.sg

Abstract

Copper oxide is one of the earliest semiconductor materials investigated for solar cells in the early 1900's before silicon cells became widespread. It is environmentally friendly, nontoxic and furthermore copper is an abundant metal. In spite of having low power conversion efficiencies when compared to theoretical values, there is much scope to further improve the efficiency. Copper oxide exists in two stable forms namely, CuO and Cu_2O with a direct band gap in each case. The band gap can be tuned between 1.6 eV (CuO) to 2.3 eV (Cu_2O).

In the present work, semiconducting copper oxide have been deposited on glass and (100) silicon substrates by using radio frequency (RF) sputtering technique with a CuO target in argon ambient. After deposition, thermal annealing treatment was carried out at different temperatures ranging from 300°C to 550°C in rapid thermal annealing (RTA) system. The structural properties and composition of the deposited films have been studied by using X-ray diffraction and X-ray photoelectron spectroscopy analysis. Optical and electrical properties were studied by using UV-Vis spectrophotometer and current-voltage (I-V) characteristics. The band gap of ~1.6 eV was obtained for sputtered CuO oxide after annealing at 300°C. Hetero-junction solar cells was fabricated using p-type CuO and n-type Si(100) substrates. The I-V characteristics of hetero-junction solar cell under sunlight of air mass 1.5 and 100 mW/cm^2 illumination shows open circuit voltage of ~380 mV and short-circuit current of ~ 1 mA/cm^2.

Key words—CuO, Hetero-junction solar cells, Low cost thin film solar cells,

Introduction

Semiconducting behavior of copper oxide is well understood in early 1900's, and its potential for the design solar cells application[1]. However, with the advent of strong semiconducting nature in silicon and germanium materials, research in copper oxide had slowed down. After reaching the advanced research in these materials and due to the global rise in energy demand, now there is a necessity to develop low cost solar cells. Recently, transition metal oxides attracted researchers due to their attractiveoptical and electrical properties. Amongst the various oxides, copper oxide semiconductor is quite promising for solar cells application. Copper Oxide exists in two forms: cuprous Oxide (Cu_2O) andcupric Oxide(CuO). The attractive features of copper oxide for solar cells applications are nontoxicity, highly abundance in earth's crust and easily reproducible[1]. Many research groupsworked on cuprous oxide based cells, achieving an efficiency of about 3.8%[2-4]. Cupric oxide is equally important to consider, however, because it has the ideal material properties required for photovoltaic conversion applications with a direct band gap. The direct band gap of the CuO can be tuned from 1.2-1.7ev by properly selecting the deposition method and conditions. Generally, CuOis a p-type semiconductor due to copper vacancies [5]. Understanding the oxidation states of the material is important to improve the junction quality. Cuprous oxide based films prepared by thermal oxidation process on ZnO film reported a conversion efficiency of 3.8%[2].

The purpose of the present study was to fabricate p-type CuO/n-Si(100) by sputter deposition method and to investigate the electrical and optical properties. The junction and film properties were studied, inorder to improve the film quality.

Experiment

Copper Oxide thin films were deposited on glass and n-Si(100) substrates by RF sputtering technique in an argon atmosphere at 25sccm flow rate. The base and working pressures of the chamber were maintained at4×10^{-7}Torr and 3.3×10^{-3}Torr respectively. The thickness of the film is about 230nm. The sputtered films wereannealed innitrogen ambientat different temperatures ranging from 300-500°C to improve the film quality.The crystalline structure and surface morphology of deposited films were characterized by X-ray diffraction (XRD) and atomic force microscopy(AFM) analysis. The surface roughness of the film is ~0.67 nm.

978-1-4673-4840-9/13 $31.00 © 2013 IEEE

Optical characteristics to determine the band gap of the deposited films were obtained from Uv-vis Spectrometer.

Results and Discussions

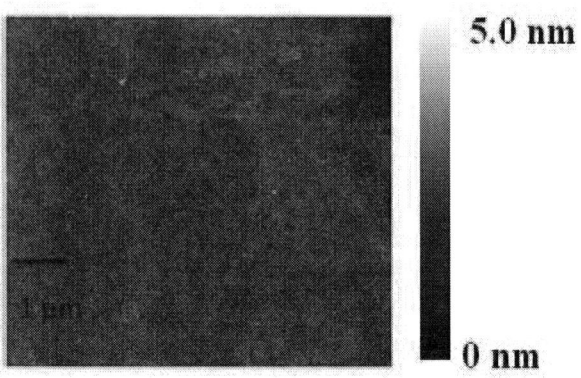

Fig. 1: The RMS surface roughness of the film is 0.67nm, which is a very smooth film. The films are pin-hole free and continuous.

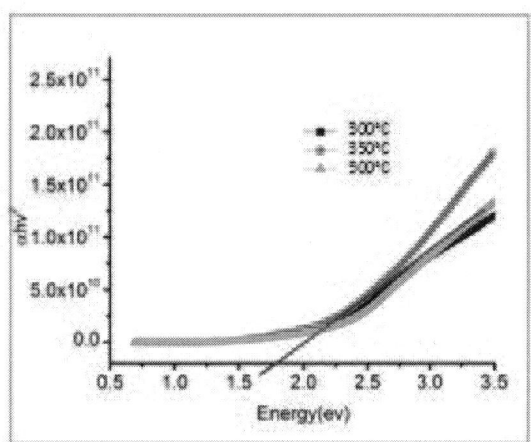

Fig. 2: Band of the films was determined from UV-vis spectrometer. The sputter deposited film shows a direct band gap of 1.6 eV for CuO film.

The optical band gap(E_g) of the semiconductor is related to the optical absorption coefficient (α) and the incident photon energy. The relation can be given as:

$$\alpha = (E_g - h\nu)^n$$

Where $h\nu$ is the incident photon energy, α is the absorption coefficient, E_g is the band gap, n= ½, 2, 3/2 and 3 corresponds to direct allowed, indirect allowed, direct forbidden, indirect forbidden respectively. The intersection of the extrapolation of the linear portion of the curve gives the band gap.According to Fig. 2, the sputter deposited CuO film is found to be direct-allowed semiconductor after annealing at 300°C. The band

gap, observed here to be around 1.6ev, is promising for solar cell application. The band gap of the CuO semiconductor can be varied by introducing the oxygen gas in to the chamber during deposition.

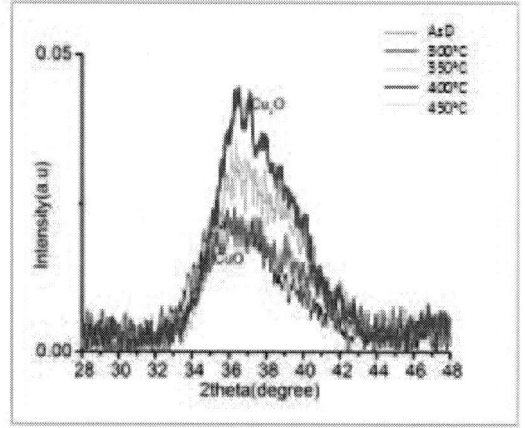

Fig. 3: The XRD patterns showing the presence of copper oxide (I) and (II) at different annealed temperatures.

Fig. 3 shows the broad XRD pattern of the copper oxide films that were annealed at different temperatures;show the mixture of CuO and Cu$_2$O phase of the oxide.It is observed that films annealed on glass at 300°C appear to be reddish brown and the XRD diffraction peaks match closely with the CuO phase.With the increase of temperature of RTA, the films turned more likely in to Cu$_2$O phase. Three distinct nature of the diffraction peak is observed for the films annealed at 300°C, however increasing the RTA anneal temperature the peak can be resolved into two peaks due to the Cu$_2$O phase. This study has clearly shown that the RTA annealing temperature and conditions are very important to get a CuO rich phase, the optimum annealing temperature of CuO phase is around 300°C.

Fig. 4 shows the I-V characteristics of the p-CuO/n-Si hetero-junction cell in the dark and under illumination conditions. Fig. 4(a) shows the variation of dark current with different annealing conditions. According to Fig. 5, the forward current is 0.1A/cm^2 at applied bias of 2V. The reverse current of the hetero-junction is ~10^{-5} A/cm^2 for the applied bias of -3V for the sample annealed at 300°C. This shows the high quality rectifying property of the p-CuO/n-Si hetero-junction solar cells. The dark current in the reverse bias increases under illumination, as shown in Fig. 4(b). This is due to the generation of electron and holes under illumination. According to the Fig. 4(b), the open circuit voltage (V$_{oc}$) and short circuit current density (I_{sc}) is decreasing with the increase of annealing temperatures. The best performance of the solar cells with V$_{oc}$340mV andI_{sc}, 1mA/cm^{2}were obtained at 300°C.

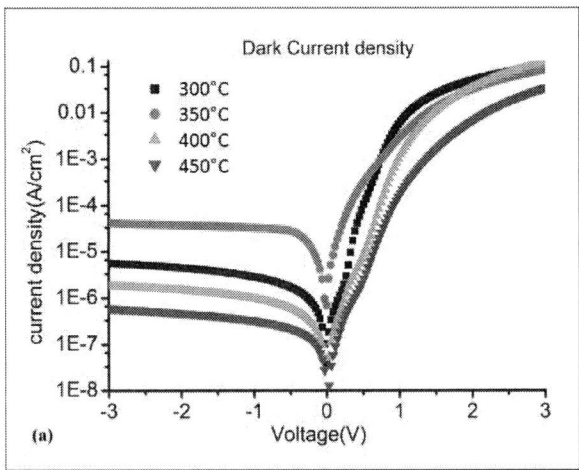

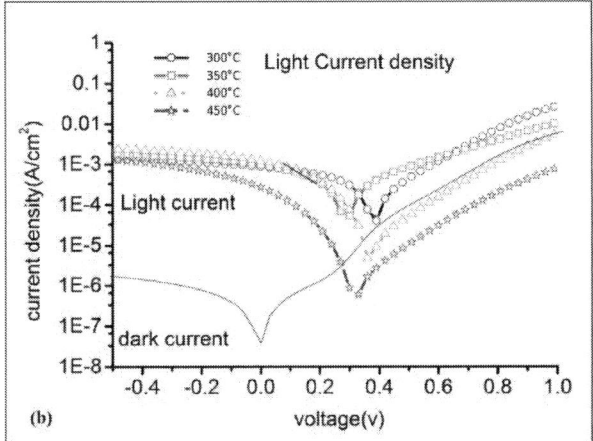

Fig. 4:(a) Current-voltage characteristics p-CuO/n-Si(100) hetero-junction solar cells. The p-CuO thin film is showing rectifying characteristics with n-Si (100). (b) Photo current-voltage characteristics p-CuO/n-Si(100) hetero-junction solar cells.

Conclusion

p-CuO/n-Si heterojunction were prepared by RF magnetron sputtering. The effect of annealing temperature on the electrical and optical properties of CuO films were studied and optimum annealing temperature for sputter deposited CuO film is 300°C. The solar cell with p-CuO and n-Si exhibits anopen-circuit voltage of 340mV and short-circuit current of 1mA/cm². By optimizing the deposition conditions and thermal treatment under different ambient it is possible to improving the film quality further, as a result the efficiency of the cell can be improved. Present work shows the feasibility of CuO based solar cells at low temperature for the application of low cost thin film solar cells.

REFERENCES

[1] A. Goetzberger and V.H. Hoff man, "Photovoltaic Solar energy Generation," New York:springer, 2007.

[2] T. Miami, "High efficiency oxide solar cells with ZnO/Cu₂Oheterojunction fabricated on thermally oxidized Cu₂O sheets," *Applied physics express*, Volume 4, Issue 6, pp. 062301-062304, 2011.

[3] Hideke Tanaka, Takahiro shimakawa, "Electrical and Optical Properties of TCO-Cu₂Oheterojunction devices," *Thin Solid films*, Vol 469–470, pp 80-85, 2004.

[4] A. Mittiga, E. Salza, F. Sarto, M. Tucci, and R. Vasanthi, "Heterojunction solar cell with 2% efficiency based on a Cu₂O substrate," *Applied Physics Letters*, vol. 88, pp.163502-163503,2006.

[5] Takeo Oku, RyosukeMotoyoshi, Kazuya Fujimoto, Tsuyoshi Akiyama, BalachandranJeyadevan, John Cuya, "Structures and photovoltaic properties of copper oxides/fullerene solar cells", *Journal of Physics and Chemistry of Solids*, Vol 72, pp 1206-1211, 2011.

Manipulating Surface Plasmon Polaritons on the meta-surface

Zhengji Xu[1], Dao Hua Zhang[1]*, Tao Li[2], Changchun Yan[3], Dongdong Li[1], Yueke Wang[1] and Fei Qin[1]

[1]School of Electrical & Electronic Engineering, Nanyang Technological University, Nanyang Avenue, 639798, Singapore
[2]National Laboratory of Solid State Microstructures, College of Physics, College of Engineering and Applied Sciences, Nanjing University, Nanjing 210093, China
[3]School of Physics and Electronic Engineering, Jiangsu Normal University, 221116, China
* Corresponding author: edhzhang@ntu.edu.sg

Abstract

We proposed a concept of meta-surface composed of sub-wavelength groove arrays in gold film on chromium(Cr) layer with designable parameters, which allows for flexible manipulations of the Surface Plasmon Polaritons (SPPs). The SPP propagation and dispersion properties were systematically studied in the theoretical meta-surface scheme by numerical simulations. For details, a conventional SPP wave on metal surface was launched by a proper grating and then incident onto the meta-surface. Adopting periodic structural parameters, we find the surface waves can be split into two beams within the structure, which depends on the specific designs. This phenomenon is valid in visible wavelength of 633nm. Our meta-surface design would open a new avenue in manipulating the surface wave and hold the promise to develop new kinds of nanophotonic elements and interconnects.

Introduction

The SPP propagation property is well been studied in the last decades. Usually, SPP propagation is investigated on the noble metal thin films such as gold and silver. SPP on gold films and silver films with nanostructures is widely and deeply researched by several research groups[1, 2]. The metamaterials suggest a way to design and modulate optical properties of natural materials by nanofabrication techniques. The metamaterials with negative refraction can be used for beam splitting[3-5], super-resolution imaging[6, 7] and generating dark hollow light cone(DHLC)[8, 9]. By introducing gain materials into metamaterials, the lasing spaser[10, 11] had been proposed.

The most challenging part to realize the functions of metamaterials is fabrication. The meta-surface we proposed is easy to fabricate and characterize. In this paper, we numerically demonstrate that by utilizing meta-surface composed of sub-wavelength groove arrays in gold film with designable parameters, the effective permittivity of the gold film surface can be tuned.

Structure and Modeling

The structure of the meta-surface is shown in Fig 1(a). It composed of sub-wavelength groove arrays in 150nm thick gold film on 60nm Cr layer whose top surface is placed by 3 gratings opening. The whole structure is 5um times 10 um and on which the gratings are 1um long, 150nm wide and 210nm deep, the distance between each gratings' middle axis is 610nm. SPP is launched by the light penetrate through the gratings. The SPP will propagate to both directions. The sub-wavelength groove arrays are 150nm deep. Each groove is 50nm wide and 3um long. The period of groove arrays is a variable P. P is varied from 110nm, 130nm to 150nm. Fig 1(b) is illustrating the perspective view of our model. The 633nm linearly polarized incidence is illuminated at the back of the substrate.

We do our simulation using a finite element method. In the simulations, the incidence wave is with an electric field of 1 V m^{-1}. It is assumed to be the normal incident source. The permittivities of gold and Cr used are -9.586 + 1.18i and -13.2 + 14.6i, respectively, at this incident wavelength[12, 13]. The performance of the meta-surface investigated as P is varied. The generalized minimal residual method (GMRES) is selected for our simulations because of its rapid convergence and short calculation time.

Theory and simulation results

A. Theory of beam splitting mechanism

From the effective medium theory(EMT)[14], the effective permittivity of layered structure is[9]:

$$\varepsilon_x = \frac{N\varepsilon_m Q + (1-N)\varepsilon_d}{NQ + (1-N)}, \quad (1)$$

$$\varepsilon_y = N\varepsilon_m + (1-N)\varepsilon_d, \quad (2)$$

where $Q = 2\varepsilon_d/(\varepsilon_m + \varepsilon_d)$ and N is the metal filling ratio. When $\varepsilon_y < 0$, $\varepsilon_x > 0$ and $k_0 = \omega/c$, The dispersion equation can be expressed as:

$$k_y^2 + \frac{\varepsilon_x}{\varepsilon_y}k_x^2 = k_0^2\varepsilon_x, \quad (3)$$

Where k_y and k_x are wave vectors along the y and x directions. Based on the EMT and dispersion relation we calculated the effective permittivity. It is shown the maximum beam divergence angle in the beam propagation direction is related to the angle between the asymptotes of the hyperbola and k_x/k_0 axis.

978-1-4673-4840-9/13 $31.00 © 2013 IEEE   446

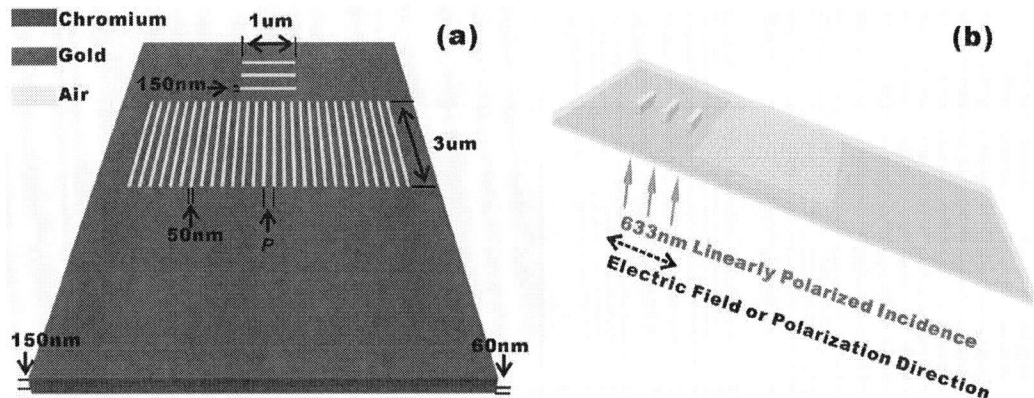

Fig. 1 (a)Schematic of the meta-surface composed of sub-wavelength groove arrays in gold film on chromium. The period of groove arrays is denoted as P. (b)Perspective view of the meta-surface model. The 633nm linearly polarized incidence is illuminated at the back of the substrate.

B. Analysis of the maximum beam divergence angle

The maximum SPPs beam divergence angle on the meta-surface can be expressed as[15]:

$$\tan \theta = \sqrt{\frac{\varepsilon_x}{|\varepsilon_y|}} , \qquad (4)$$

Using (4), we calculated the divergence angles for dielectrics permittivity of P =110 nm, P =130 nm and P = 150nm are $45.96°, 48.97°, 51.89°$, respectively. As shown in Fig. 2, the divergence angle follows the trend of the period of groove arrays and the theoretical calculated result are matched with the simulation result very well. The calculation detail is listed in the Table I.

TABLE I
MAXIMUM BEAM DIVERGENCE ANGLE CALCULATION

P(nm)	N	ε_x	ε_y	$\tan \theta$	θ
110	0.545	5.098	-4.769	1.034	45.96
130	0.615	7.276	-5.51	1.149	48.97
150	0.667	9.74	-5.987	1.275	51.89

When the SPP start propagate on the grooves, the edge of grooves bridge the momentum gap for the SPPs modes, high frequency components are created, and those high frequency components are supported by the anisotropic gold dielectric multi-grooves structure. The calculation results showing that dispersion relation of the meta-surface structure is similar with stratified metal-dielectric composite metamaterials which is hyperbolic shape. That is why excited SPP beam diverge to two directions inside the designed meta-surface structure.

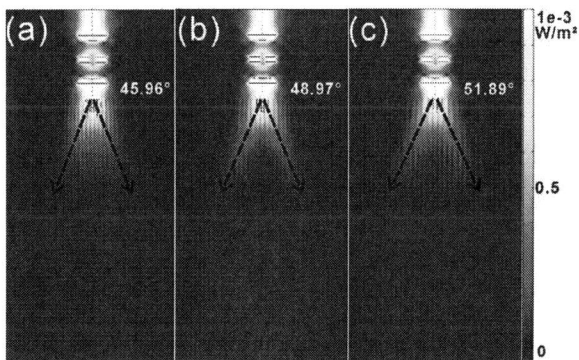

Fig. 2 Power flow distribution at the surface of the meta-surface. The calculated divergence angles shown by black arrows are based on (4): 45.96, 48.97 and 51.89 degree for period of groove arrays of (a) P =110nm,(b) P =130nm and (c) P =150nm, respectively.

Conclusion

In conclusion, we proposed meta-surface composed of sub-wavelength groove arrays in Au film on Cr layer with designable parameters, which allows for Surface Plasmon Polaritons propagating on the meta-surface. By simulating the power flow distributions of the meta-surface, we found the splitting angle is following the trend of varying the period of the groove arrays. We also calculated the maximum beam divergence angle of the beam splitting. Using effective medium theory we can explain the beam splitting phenomenon very well. It is believed that the meta-surface will have potential applications in near-field optics and nanophotonics devices.

Acknowledgements

The project is supported by the National Research Foundation (NRF-G-CRP 2007-01), A*Star (092154009), Singapore, the National Natural Science Foundation of China (50975187 and 61078019), and the Natural Science Foundation (BK2011203) of Jiangsu Province, China.

References

[1] L. Yin, V. K. Vlasko-Vlasov, J. Pearson, J. M. Hiller, J. Hua, U. Welp, D. E. Brown, and C. W. Kimball, "Subwavelength Focusing and Guiding of Surface Plasmons," *Nano Letters,* vol. 5, pp. 1399-1402, 2005/07/01 2005.

[2] X. Liu, Y. Wang, and E. O. Potma, "A dual-color plasmonic focus for surface-selective four-wave mixing," *Applied Physics Letters,* vol. 101, pp. 081116-4, 2012.

[3] C. C. Yan, D. H. Zhang, Y. Zhang, D. D. Li, and M. A. Fiddy, "Metal-dielectric composites for beam splitting and far-field deep sub-wavelength resolution for visible wavelengths," *Opt. Express,* vol. 18, pp. 14794-14801, 2010.

[4] Y. K. Wang, D. H. Zhang, J. Wang, X. F. Yang, D. D. Li, and Z. J. Xu, "Waveguide devices with homogeneous complementary media," *Opt. Lett.,* vol. 36, pp. 3855-3857, 2011.

[5] Y. K. Wang, D. H. Zhang, J. Wang, M. Yang, D. D. Li, and Z. J. Xu, "Beam splitting with subwavelength resolution using combined metallodielectric films," *Journal of Optics,* vol. 14, p. 015103, 2012.

[6] J. B. Pendry, "Negative Refraction Makes a Perfect Lens," *Physical Review Letters,* vol. 85, pp. 3966-3969, 2000.

[7] Y. K. Wang, D. H. Zhang, C. C. Yan, D. D. Li, and Z. J. Xu, "Efficient and wide spectrum half-cylindrical hyperlens with symmetrical metallodielectric structure," *Applied Physics A,* vol. 107, pp. 31-34, 2012/04/01 2012.

[8] C. C. Yan, D. H. Zhang, D. D. Li, H. J. Bian, Z. J. Xu, and Y. K. Wang, "Metal nanorod-based metamaterials for beam splitting and a subdiffraction-limited dark hollow light cone," *Journal of Optics,* vol. 13, p. 085102, 2011.

[9] Z. J. Xu, D. H. Zhang, C. C. Yan, D. D. Li, and Y. K. Wang, "Concentric cylindrical metamaterials for subwavelength dark hollow light cones," *Journal of Optics,* vol. 14, p. 114014, 2012.

[10] N. I. Zheludev, S. L. Prosvirnin, N. Papasimakis, and V. A. Fedotov, "Lasing spaser," *Nat Photon,* vol. 2, pp. 351-354, 2008.

[11] C. C. Yan, D. H. Zhang, Z. J. Xu, D. D. Li, and A. Yang, "A multi-layered split ring metamaterial for a multiwavelength and tunable lasing spaser," *Journal of Optics,* vol. 14, p. 045101, 2012.

[12] E. D. Palik, "Handbook of Optical Constants of Solids," ed: Elsevier, 1985.

[13] P. B. Johnson and R. W. Christy, "Optical Constants of the Noble Metals," *Physical Review B,* vol. 6, p. 4370, 1972.

[14] R. Wangberg, J. Elser, E. E. Narimanov, and V. A. Podolskiy, "Nonmagnetic nanocomposites for optical and infrared negative-refractive-index media," *J. Opt. Soc. Am. B,* vol. 23, pp. 498-505, 2006.

[15] Z. Jacob, L. V. Alekseyev, and E. Narimanov, "Semiclassical theory of the hyperlens," *J. Opt. Soc. Am. A,* vol. 24, pp. A52-A59, 2007.

Beam focusing by an anisotropic metal-dielectric multilayer structure

Dongdong Li, Dao Hua Zhang*, Yueke Wang, Zhengji Xu,
Jun Wang, Fei Qin and Wenjuan Wang

School of Electrical and Electronic Engineering
Nanyang Technological University
Singapore, 639798
Corresponding author: edhzhang@ntu.edu.sg

Abstract

we demonstrate subwavelength beam focusing by a slab of anisotropic material. The proposed device consists of a slab of anisotropic material on top of which a Chromium layer with a small slit is used as the mask. By carefully designing the thickness of the anisotropic layer, we can selectively modify the phase of different wave components exiting the output surface, so that a subwavelength-sized light spot can be generated as results of near-field constructive interference. Our study showed that a light spot with FWHM of 0.36λ can be generated at optical frequencies.

keywords: beam focusing; metal-dielectric; anisotropic

Introduction

As one of the widely studied metamaterial, the anisotropic metamaterial have shown potentials in realizing several breakthrough applications such as subwavelength imaging, waveguiding and beam focusing [1-13]. Compare with the double negative metamaterials, they are easier to be realized and have lower losses [10,11]. Different schemes have been proposed to realize anisotropic metamaterials, such as the metal-dielectric multilayer and metallic nanowire array structures. Both theoretical and experimental studies showed that negative refraction and subwavelength imaging can be achieved by such anisotropic metamaterials. In our previous work, we have shown that by configuring the metal-dielectric multilayer structure as a scanning probe, the near-field information of an object can be obtained in the far-field [6]. The electromagnetic energy emitted from the object first spread inside the metal-dielectric multilayer and then refocus at the output due to negative refraction, and the size of the refocused beam is about the half of the incident wavelength. In this work, we further show that, the sharpness of the refocused beam can be increased by tuning the thickness of the metal-dielectric multilayer and a subwavelength-sized light spot can be generated as results of near-field constructive interference. Compare with other beam focusing devices such as the Fresnel zone plates and plasmonic lenses which rely on specially designed concentric annular rings to realize their functionalities, the proposed structure can be easily realized by alternating metal-dielectric layers. In addition, the size of the structure is much smaller than the Fresnel zone plate and the plasmonic lens. Thus it can be easily integrated with nano-scale photonic devices.

Structure

The structure with labeled dimensions is illustrated in figure 1. It consists of an anisotropic material covered by a Chromium (Cr) layer on which a 30 nm slit is opened. It is immersed in air with a permittivity of 1. For simplicity, the effective permittivity of the anisotropic material are taken as ε_x=5.36-0.08i and ε_y=-5-0.42i, respectively. While the permittivity of the Cr is extracted from experimental data [14]. In our studies, an 570 nm transverse magnetic-polarized (i.e., $H\|z$ axis) plane wave at normal incidence onto the interface between air and the metamaterial is considered. The structure is investigated numerically by a finite element method (FEM) using the Radio Frequency (RF) Module of COMSOL Multiphysics 3.5. The boundaries of the simulation domains except the incident boundary are set with perfectly matched layers to reduce the interference of the back scattered waves. The size of the simulation domain is 3000 nm×3000 nm, and the maximum mesh size of the entire structure is 1 nm. The Cr slit serves as a subwavelength scatter and ensure that the incident wave propagates only through the slit. Figure 1 (b) shows the structure constructed by alternating silver (Ag) and silicon dioxide (SiO$_2$) multilayer with equivalent permittivity along the x-axis and y-axis. At λ=570 nm, the complex permittivity of Ag and SiO$_2$ are ε_{Ag}=-12.2-0.84i and 2.2, respectively. When the thickness of the Ag and SiO$_2$ are equal to each other (both of them are 15 nm), the effective permittivities are ε_x=5.36-0.08i and ε_y=-5-0.42i, respectively.

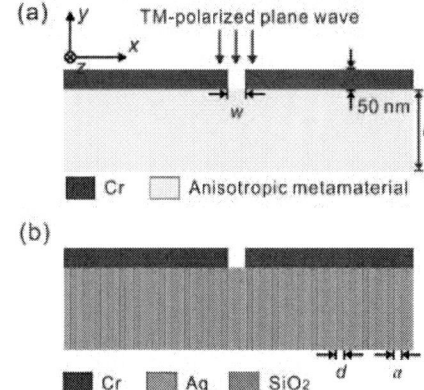

Fig. 1. Cross-sectional schematic of (a) an anisotropic metamaterial covered by a Cr layer with one small slit. The thickness of the Cr layer is 50 nm. The thickness of the anisotropic metamaterial layer is a variable. (b) the structure constructed by alternating Ag-SiO$_2$ multilayer with equivalent permittivity along the x-axis and y-axis. The thicknesses of each individual Ag and SiO$_2$ layers are d=15 nm and a=15 nm, respectively.

978-1-4673-4840-9/13 $31.00 © 2013 IEEE

Simulations and discussion

The structure illustrated in Fig. 1(a) with different thickness h=460 nm, h=600 nm, h=800 nm and h=1200 nm are numerically simulated. The time averaged power flow (or the time averaged Poynting vector) distribution is used to evaluate the intensity of the beam inside the anisotropic material and the Air region. The simulation results are shown in Fig. 2(a) to Fig. 2(d). It can be seen that the beams first spread and form a cone shape inside the anisotropic material layer. Depending on the thickness of the anisotropic material layer, then they refocus at different locations in the air region. The focal point decreases as the thickness of the anisotropic material layer decreases. For the structure with h=460 nm, the refocusing point approaches to the output surface of the anisotropic layer and the beam spot is much sharper than the rest three structures. To qualitatively measure the sharpness of the refocused beams, the full width at half maximum (FWHM) around the corresponding focus points are measured. It is found that the FWHM of the structure with h=1200 nm is 280 nm, which is about the half wavelength of the incident light. In contrast, the FWHM of the structure with h=460 nm has a sub diffracted size of 205 nm (about 0.36λ).

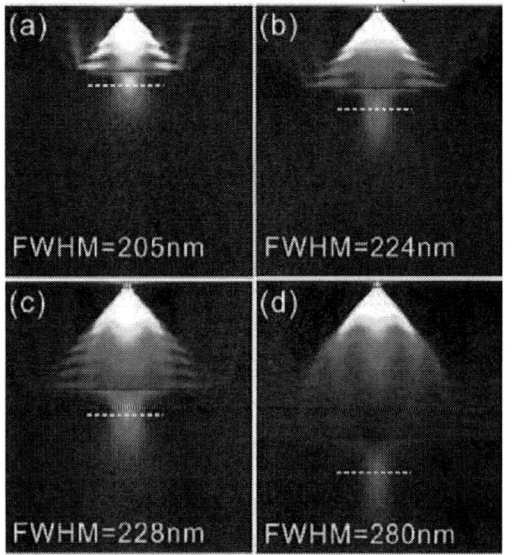

Fig. 2. Energy flow distributions for different anisotropic layer thickness h with a 570 nm incidence. (a) h=460 nm; (b) h=600 nm; (c) h=800 nm; (d) h=1200 nm.

To understand the physics, we investigate the dispersion relation of the anisotropic layer. The dispersion relation of the anisotropic material for TM waves is given by:

$$\frac{k_x^2}{\varepsilon_y} + \frac{k_y^2}{\varepsilon_x} = k_0^2 \tag{1}$$

where k_x and k_y are the wave vectors of the anisotropic metamaterial along transverse and longitudinal directions, and k_0 is the wave vector in air. ε_x and ε_y are the effective permittivities of the anisotropic metamaterial along transverse and longitudinal directions. In the case of ε_x>0 and ε_y<0, the dispersion relation is in hyperbolic shape, as illustrated in Fig. 3.

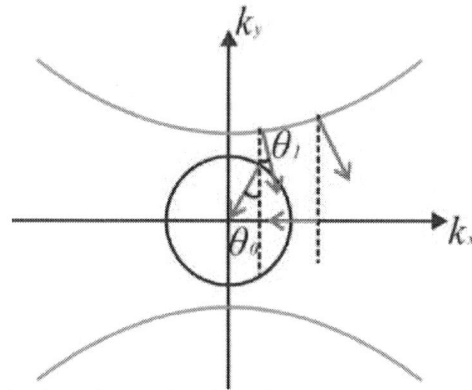

Figure 3. Dispersion relation of the anisotropic material with ε_x>0 and ε_y<0 and air. The red arrows denote the power flow direction (poynting vector direction). θ_1 denotes the refraction angle in the anisotropic material, while θ_0 denotes the refraction angle of the propagating waves in air.

The refraction angle for the poynting vector in the anisotropic medium can be calculated by:

$$\theta_1 = \tan^{-1}\left(\frac{S_x}{S_y}\right) = \tan^{-1}\left(-\frac{k_x/\varepsilon_y}{k_y/\varepsilon_x}\right) \tag{2}$$

It can be seen that the refraction angle in the anisotropic material layer is dispersive, which indicates that waves carrying different wave vector components will have different refraction angles. This explains the formation of the cone shaped energy distribution inside the anisotropic material layer. Next when the beams inside the anisotropic material layer encounter the air region at the output surface, the beam undergo negative refraction as illustrated in Fig. 3. For propagating wave components, the refraction angle in the air region can be calculated by:

$$\theta_0 = \tan^{-1}\left(\frac{S_x}{S_y}\right) = \tan^{-1}\left(\frac{k_x}{k_y}\right) \tag{3}$$

while for the evanescent components, the refraction angle in the air region becomes 90°, as illustrated in Fig. 3.

For the structures with large h, the beam refocus at a further location from the output surface. The contribution of the evanescent wave components to the focused beam spot is small. While for the case of smaller h, the evanescent wave components plays an important role. Due to the contribution of the evanescent wave components, it is possible to realize a beam spot with dimension smaller than the diffraction limit. It should also be noted that, due to the dispersive nature of the refraction angle, waves with different transverses component will refocus at different locations. This will cause aberration and decrease the sharpness of the focused spot. By tuning the thickness of the anisotropic layer, the focal point can be controlled. When a appropriate thickness is used, a focus point with FWHM smaller than the incident wavelength can be realized. Thus in order to achieve a sharp beam spot at the output, the thickness of the indefinite medium should be optimized. In our simulations, the optimized thickness of the anisotropic material layer is 460 nm. With this thickness, a beam spot with FWHM of 0.36λ can be achieved.

978-1-4673-4840-9/13 $31.00 © 2013 IEEE

Conclusion

In this work, the near-field beam focusing properties of the anisotropic material consisting of alternating metal-dielectric layers are investigated. Beam spot with a size below the diffraction limit can be realized in the near-field by optimizing the thickness of the anisotropic material layer. The size of the proposed structure can be smaller than 1 um, which is much smaller than the Fresnel zone plate and the plasmonic lens, indicating that it can be easily integrated with nano-scale photonic devices.

Acknowledgement

This project is supported by National Research Foundation (NRF-G-CRP 2007-01) and A*Star (092154009), Singapore.

References

[1] J. B. Pendry, "Negative refraction makes a perfect lens," Phys. Rev. Lett. 85, 3966-3969 (2000).

[2] N. Fang, H. Lee, C. Sun, and X. Zhang, "Sub–diffraction-limited optical imaging with a silver superlens," Science 308, 534-537 (2005).

[3] A. Houck, J. B. Brock, and I. L. Chuang, "Experimental observations of a left-handed material that obeys Snell's law," Phys. Rev. Lett. 90, 137401 (2003).

[4] X. H. Hu and C. T. Chan, "Photonic crystals with silver nanowires as a near-infrared superlens," Appl. Phys. Lett. 85, 1520-1522 (2004).

[5] H. Shin and S. H. Fan, "All-angle negative refraction for surface plasmon waves using a metal-dielectricmetal structure," Phys. Rev. Lett. 96, 073907 (2006).

[6] C. Yan, D. H. Zhang, D. Li, H. Bian, Z. Xu and Y. Wang, "Metal nanorod-based metamaterials for beam splitting and a subdiffraction-limited dark hollow light cone," Journal of Optics, 13, 8 (2011).

[7] C. Yan, D. H. Zhang, D. Li, and Y. Zhang, "Dual refractions in metal nanorod-based metamaterials," Journal of Optics, 12, 6 (2010).

[8] C. Yan, D. H. Zhang, D. Li, Y. Zhang, D. Li and M.A Fiddy, "Metal-dielectric composites for beam splitting and far-field deep sub-wavelength resolution for visible wavelengths," Optics Express, 18(14), 14553-14567 (2010).

[9] Li, D. H. Zhang, C. Yan,, and Y.Wang, "Two-dimensional subwavelength imaging from a hemispherical hyperlens," Appl. Opt. 50(31), G86-G90 (2011).

[10] J. Elser, V. A. Podolskiy, I. Salakhutdinov, and E. E. Narimanov, "Nanowire metamaterials with extreme optical anisotropy," Appl. Phys. Lett. 89, 261102 (2006).

[11] V. A. Podolskiy and E. E. Narimanov, "Strongly anisotropic waveguide as a nonmagnetic left-handed system," Phys. Rev. B 71, 201101 (2005).

[12] T. Dumelow, J. A. P. da Costa, and V. N. Freire, "Slab lenses from simple anisotropic media," Phys. Rev. B 72, 235115 (2005).

[13] G. Shvets, S. Trendafilov, J. B. Pendry, and A. Sarychev, "Guiding, focusing, and sensing on the subwavelength scale using metallic wire arrays," Phys. Rev. Lett. 99, 053909 (2007).

[14] E. D. Palik and G. Ghosh, The electronic handbook of optical constants of solids (Academic, New York, 1999).

Performance Improvement of Triple-Junctions GaAs-Based Solar Cell using SiO₂-Nanopillars/SiO₂/TiO₂ Graded-Index Anti-Reflection Coating

Jheng-Jie Liu[1], Wen-Jeng Ho[1*], Jhih-Kai Syu[1], Yi-Yu Lee[1], Ching-Fuh Lin[2], and Hung-Pin Shiao[3]

[1]Department of Electro-Optical Engineering, National Taipei University of Technology

1, Sec. 3, Chung-Hsiao E. Rd., Taipei 106, Taiwan

[2]Graduate Institute of Photonics and Optoelectronics, National Taiwan University, Taipei, Taiwan.

[3]Win Semiconductor Corp., Taoyuan, Taiwan.

* wjho@ntut.edu.tw

Abstract

This paper demonstrate experimentally the top-cell external quantum efficiency (EQE) improving of the triple-junction GaAs-based solar cell using a SiO₂-nanopillars/SiO₂/TiO₂ graded-index anti-reflection coating (GI-ARC). The reflectance and EQE of the solar cells with SiO₂/TiO₂ double-layer ARC and with SiO₂-nanopillars/SiO₂/TiO₂ GI-ARC are measured and compared. The reflectance of cell with SiO₂-nanopillar/SiO₂/TiO₂ GI-ARC had most favorable low reflectance between 550-700 nm wavelengths. Therefore, the EQE enhancement of 5.88% on top-cell and -1.56% on middle-cell and the best current matching were achieved.

Keywords: SiO₂-nanopillars, graded-index anti-reflection coating, triple-junction solar cell, external quantum efficiency, current matching.

Introduction

High conversion efficiency of GaAs-based solar cells under one-sun and/or concentrated condition has been demonstrated for used in space application and in the terrestrial solar power plants. GaAs-based multi-junction solar cells (MJ-SCs) can provide wide range absorption in the solar spectrum from visible to infrared band to generate higher conversion efficiency than single-junction one through epitaxial growth technique and device structure optimization [1]. Therefore, MJ-SCs require a broadband antireflection coating (ARC) to prevent larger losses of incident light due to surface reflection [2,3]. The broadband effectiveness can be obtained using a sub-wavelength surface structure ARC [4,5] and a different transparent multilayer ARC [2]. However, the fabrication of ARC for MJ-SCs is very challenging owing to the broadband absorption and the additional need for current matching between the subcells, simultaneously. Because the total photocurrent generated by the MJ-SC was limited by the subcell that has the lowest current-generating capability one. Thus, the ARC for MJ-SCs must be also designed to minimize the light losses in the absorption range of the limiting subcell in order to achieve higher total currents. In fact, the GaAs/Ge based triple-junction solar cells (3J-SCs) are normally current limited by either the top or the middle subcell, the spectral range required for the specific ARC to meet the top and middle subcells can be reduced to 300-950 nm.

In this paper, we demonstrate experimentally the performance enhancement of a top subcell-limited triple-junction GaAs-based solar cell using a SiO₂-nanopillars/SiO₂/TiO₂ graded-index anti-reflection coating (GI-ARC). The sub-wavelength nano-sized SiO₂ nanopillars formed on SiO₂/TiO₂ layer using a reactive ion etching (RIE) process with CF₄ are described and presented by the images of scan electron microscope (SEM). The reflectance spectra are examined and compared before and after surface AR-coatings. The dark current-voltage (I-V) and photo I-V under AM1.5G illumination are measured. The measured external quantum efficiency (EQE) response of the solar cell with SiO₂-nanopillar/SiO₂/TiO₂ GI-ARC indicates that the EQE enhancement of 5.88% on top-cell and -1.56% on middle-cell are obtained and the best current matching between the top and the middle subcells can be achieved, compared to the solar cell coated with SiO₂/TiO₂ DL-ARC one.

Experiment

The epitaxial structure of the triple-junction GaAs-based solar cell is shown in the device schematic of Fig. 1. It is consisted of an n^+/p GaInP top cell with a bandgap energy of 2.0 eV and an n^+/p GaInAs middle cell with a bandgap energy of 1.2 eV. The subcells are connected in series by means of a tunnel junction. The Ge bottom cell junction was created during metal-organic chemical vapor deposition (MOCVD) growth of GaInAs middle cell and GaInP top cell by diffusion of an n-type dopant (As) into a p-type Ge substrate. The growth conditions were 700 ℃ and 100 mbar reactor pressure. The growth rate was 5 um/h for GaInAs and 2 um/h for GaInP. After MOCVD growth, Ni/Ge/Au/Ag/Au front- and Au/Ag/Au back contacts were evaporated and defined by an image-reversal photolithographic lift-off process. The bare solar cell had a chip size of 5×5 mm² after isolated etching and sawed dicing.

The front side of the bare cell was firstly deposited with a 59.4 nm–thick TiO₂, 394 nm-thick SiO₂, and 35 nm-thick Ag filmrs by a thermal evaporation. Then Ag film was annealed at 200 ℃ for 20 min in N₂ ambient to form Ag nanoparticles The sub-wavelength SiO₂ nanopillars were formed using a reactive ion etching (RIE) process with CF₄ to etch off a 300 nm depth of SiO₂ top-layer that involved using Ag nanoparticles as the etching masks and then removed residual Ag films [6]. Therefore, SiO₂-nanopillars with a dimension of 120-300 nm and a height of 300 nm and a period of about 300 nm were

978-1-4673-4840-9/13 $31.00 © 2013 IEEE

formed on surface of the solar cell, the cross-section SEM image of the sub-wavelength SiO_2-nanopillars as shown in Fig. 2.

The reflectance of the fabricated triple-junction GaAs-based solar cell sample was measured and compared using UV/VIS/NIR spectrophotometer before and after surface AR-coatings. The current-voltage (I-V) characteristics of solar cell device was characterized both in the dark and photo-illumination conditions. In this work, the photovoltaic performance was obtained under one sun AM 1.5G (100 mW cm^{-2}) normally illumination at 25 □ and the solar simulator system has been calibrated using a standard solar cell before sample measurement. In addition, the external quantum efficiency (EQE) response of the bare solar cell (without any surface coating), solar cell coated with a SiO_2/TiO_2 double layer anti-reflection coating (DL-ARC), and solar cell using a SiO_2-nanopillars/SiO_2/TiO_2 GI-ARC are also measured and examined, respectively, for the wavelength range from 300 nm to 1100 nm.

Results and Discussion

The ideality factor (n) of 4.35 and the reverse saturation current (I_0) of 1.5×10^{-9} A/cm^2 of the bare-type top-cell current-limited triple-junction GaAs-based solar cell were obtained from the dark I-V measurement at 25 ℃. A small reverse saturation current generated and a small average value of ideality factor of 1.45 for each diode (3.45/3) can be attributed to the excellent epitaxial film quality grown by MOCVD. The measured reflectance spectrum of the triple-junction GaAs-based solar cell before and after surface coating is shown in Fig. 3. The reflectance of the solar cell with a SiO_2-nanopillar/SiO_2/TiO_2 GI-ARC had most favorable low reflectance than the cell coated with a SiO_2/TiO_2 DL-ARC one, particularly, at the wavelengths between 550-700 nm. On the other hand, the photovoltaic performance of the solar cell with a SiO_2-nanopillars/SiO_2/TiO_2 GI-ARC are clearly exhibited that the short-circuit current (I_{sc}) is increased from 3.61 mA to 3.79 mA and the conversion efficiency (η) is increased from 29.54 % to 30.77%, compared to the cell coated with a SiO_2/TiO_2 DL-ARC one. The increasing in I_{sc} and η is attributed to the profited of adding a SiO_2-nanopillars layer on the SiO_2/TiO_2 DL-ARC, which having lower reflectance with a broad range of wavelength than DL-ARC one. Furthermore, the measured EQE response of the solar cell with different surface coating is presented in Fig. 4. The average EQE enhancements (△EQE) of top-cell by 5.88% and -1.56% of middle-cell are achieved, due to the low reflection between the wavelength 550 nm and 700 nm, respected to the solar cell coated with SiO_2/TiO_2 DL-ARC one. Because the photocurrent generated is in proportion to the values of EQE. Therefore, the EOE value increasing in the top cell and decreasing in the middle cell can be applied to improve the top cell current limited, in this work, to achieve the best current matched between these sub-cells as the solar cell using the proposed SiO_2-nanopillars/SiO_2/TiO_2 GI-ARC. Furthermore, the interesting enhanced mechanism of the nanopillars graded-index anti-reflection coating for using to enhance the top-cell or/and middle-cell performance is undergoing in-depth study in our laboratory.

Conclusion

We demonstrated the top-cell external quantum efficiency (EQE) improvement of the top-cell current-limited triple-junction GaAs-based solar cell using a SiO_2-nanopillars/SiO_2/TiO_2 GI-ARC. The reflectance of the solar cell with SiO_2-nanopillar/SiO_2/TiO_2 GI-ARC had most favorable low reflectance with a broad range of wavelength, especially at short wavelength-range. The average EQE enhancements of top-cell by 5.88% and -1.56% of middle-cell were achieved and the best currents matched are obtained between the top-cell and middle-cell, compared to the solar cell coated with SiO_2/TiO_2 DL-ARC one.

Acknowledgment

The authors would like to thank the financial support from the National Science Council of Republic of China under Grant NSC-100-2221-E-027- 053-MY3.

References

[1] W. Guter, et al., Current matched triple-junction solar cell reaching 41.1% conversion efficiency under concentrated sunlight, *Appl. Phys. Lett.*, Vol. 94, pp. 223504-1-223504-3, 2009.

[2] D. J. Aiken, Antireflection coating design for series interconnected multi-junction solar cells, *Prog. Photovolt: Res. Appl.*, Vol. 8, No. 6, pp. 563-570, 2000.

[3] R. Homier, et al., Antireflection coating design for triple-junction III-V/Ge high-efficiency solar cells using low absorption PECVD silicon nitride, *IEEE J. of Photovoltaics*, Vol. 2, pp. 393-397, 2012.

[4] J. Tommila, et al., Nanostructure broadband antireflection coatings on AlInP fabricated by nanoimprint lithography, *Solar Energy Mater. Solar Cells*, Vol. 94, No. 10, pp. 1845-1848, 2010.

[5] S. L. Diedenhofen, et al., Broadband and omnidirectional anti-reflection layer for III/V multi-junction solar cells, *Solar Energy Mater. Solar Cells*, Vol. 101, pp. 308-314, 2012.

[6] W. J. Ho, et al., Broadband wavelength and wide-acceptance angle of the SiO_2 sub-wavelength surface structure for solar cells using CF$_4$ reactive ion etching, *Thin Solid Films*, in press. DOI:10.1016/j.tfs.2012.09.028

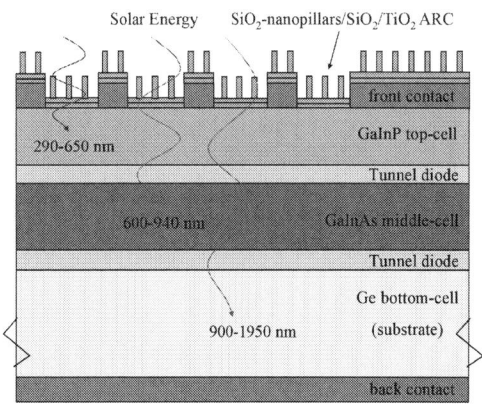

Fig. 1 Device schematic of triple-junction GaAs-based solar cell with a SiO_2-nanopillars/SiO_2/TiO_2 GI-ARC.

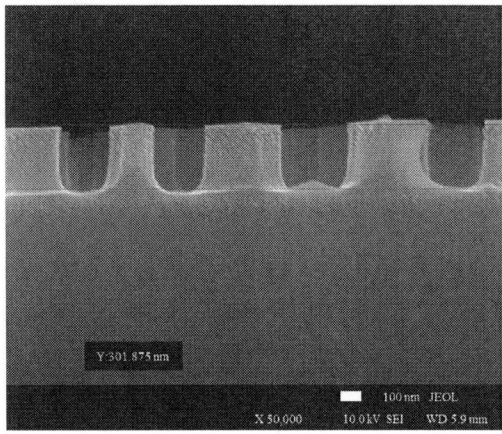

Fig. 2 Cross-section SEM image of the sub-wavelength SiO_2 nanopillars obtained by a reactive ion etching (RIE) process with CF_4.

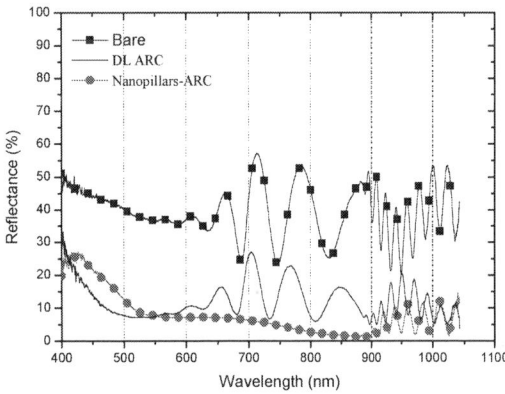

Fig. 3 The measured reflectance spectrum of the triple-junction GaAs-based solar cell before and after surface coatings.

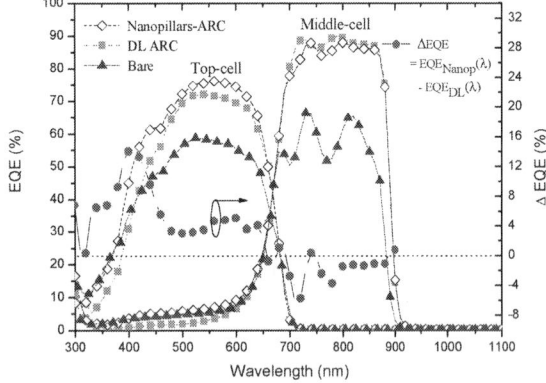

Fig. 4 The measured EQE response and EQE enhancement factor of the triple-junction GaAs-based solar cells with different surface coatings.

Tunable Subwavelength Terahertz Plasmonic Stub Waveguide Filters

Jin Tao[1], Qi Jie Wang[1,2*], Bin Hu[1], Xiao Yong He[1], Ying Zhang[3]

[1] Division of Microelectronics, School of Electrical and Electronic Engineering, Nanyang Technological University, 639798, Singapore

[2] Division of Physics and Applied Physics, School of Physical and Mathematical Sciences, Nanyang Technological University, 639798, Singapore

[3] Singapore Institute of Manufacturing Technology, 638075, Singapore

* qjwang@ntu.edu.sg

Abstract

Tunable subwavelength terahertz plasmonic stub waveguide filters based on indium antimonide are proposed and numerically investigated. The transmission line theory and the Finite Different Time Domain simulation results reveal that the single-stub waveguide structure can realize a stop-band filtering function and the central wavelength of the notch is linearly dependent on the stub length while nonlinearly dependent on the stub width. The central wavelength of the notch can be actively controlled by tuning the temperature. The proposed filters may have applications in THz highly-integrated plasmonic circuits.

Introduction

Terahertz (THz) technology is now drawing extensive attention because of its potential applications in biochemical sensing, spectroscopy, and high-speed communication. With the rapid development of THz sources [1, 2] and detectors [3], there is a high demand for THz components, such as waveguides, polarizers, filters, and collimators [4]. Waveguide filter is probably the most basic component among various components. Various waveguide filters have been demonstrated within the last few years. Most of these are based on conventional guiding structures, such as metal tubes [5], metal wires [6], and dielectric waveguides [7]. Confining and manipulating electromagnetic waves at dimensions much smaller than the wavelength are of great importance for miniaturized integrated photonic circuits. The minimum confinement of a guided mode in conventional waveguides is limited by the diffraction limit.

Plasmonic devices, based on surface plasmon-polaritons (SPPs) propagating at metal-dielectric interface, have shown great potential to guide and manipulate light by metallic nanostructures at sub-wavelength scales [8]. Most of the efforts in plasmonic devices have been concentrated at optical and near-infrared frequencies [9]. In the THz region, both the real and imaginary values of the metal permittivity are huge, with typical absolute values of the order 10^5. These values of permittivity give rise to weak coupling of the electromagnetic field to the electrons in the metal. Some semiconductors have a permittivity at THz range close to that of metal at optical range. The permittivity of semiconductors can be modified [10] by varying temperature and/or dopant concentrations which can be used to realize the tunability of plasmonic devices.

In this paper, we propose planar tunable subwavelength plasmonic stub waveguide filters in terahertz range using indium antimonide (InSb) as the "metal" material. Characteristics of deep-subwavelength InSb plasmonic slot waveguide are first investigated. The single-stub slot InSb stub waveguide filters are proposed and analyzed by transmission line theory and Finite Different Time Domain (FDTD) simulation [11]. The results obtained from both methods agree well. Our structures may have great potential for ultra-compact THz integrated circuits.

Characteristics of subwavelength InSb plasmonic slot waveguide

The inset of Fig. 1 shows the InSb slot waveguide structure composed of two parallel InSb plates with a dielectric core. The dispersion relation of the fundamental surface plasmon (SP) mode in the InSb slot waveguide is given by [12]

$$\tanh\left(\sqrt{n_{eff}^2 - \varepsilon_d}\, k_0 w/2\right) = \frac{-\varepsilon_d \sqrt{n_{eff}^2 - \varepsilon_{insb}(\omega, T)}}{\varepsilon_{insb}(\omega, T)\sqrt{n_{eff}^2 - \varepsilon_d}}, \quad (1)$$

where n_{eff} is the effective refractive index of the SP wave in the waveguide and $k_0 = 2\pi/\lambda_0$ is the free-space wave vector. ε_d denotes the permittivity of medium in the slot guide region and the permittivity $\varepsilon_{insb}(\omega, T)$ of InSb can be described by the Drude model [13] approximation,

$$\varepsilon_{insb}(\omega, T) = \varepsilon_\infty - \omega_p^2(T)/\omega[\omega + i\Gamma(T)],$$

where ω is the angular frequency of the incident electromagnetic radiation, and $\varepsilon_\infty = 15.75$ is the high-frequency permittivity; $\omega_p(T)$ and $\Gamma(T)$ are the plasma frequency and collision rate of the charge carriers respectively. The plasma frequency $\omega_p(T) = \sqrt{Ne^2/\varepsilon_\infty\varepsilon_0 m^*}$ depends on the density of carriers, where e is the effective charge unit, ε_0 is the permittivity in vacuum, and m^* is the effective mass of electrons. Different from metal, the plasma frequency of the InSb significantly depends on the temperature T. The intrinsic concentration in the InSb is described by the formula $N = 5.76\times10^{14}T^{1.5}\exp(-0.129/K_B T)\mathrm{cm}^{-3}$ [14], where K_B and T are the Boltzmann constant and temperature, respectively. At 295 K, the permittivity of InSb at 1 THz is -44.19+15.58i, which is similar to that of metals at optical frequencies. The InSb permittivity is changed to -26.69+10.18i at temperature of 280 K. The dielectric constant $\varepsilon(\omega, T)$ of

InSb material is affected by the variation of the temperature in the terahertz range because of plasma frequency and collision rate of the charge carriers are both dependent on the temperature. Therefore, the tunable plasmonic filters based on thermal tuning can be expected.

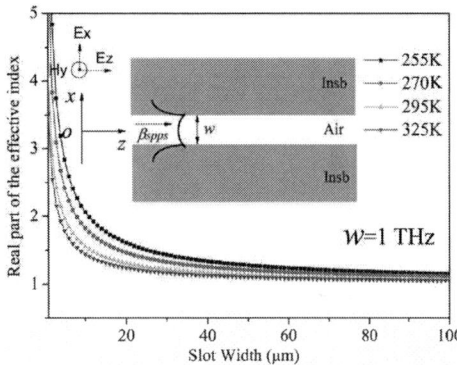

Fig. 1. Real part of the effective index of InSb slot waveguide as a function of slot widths at different temperatures for incident frequency at 1 THz. The inset: schematic structure of an InSb slot waveguide.

In InSb slot waveguide, the fundamental transverse magnetic mode (TM$_0$) always exists regardless of the waveguide width, while other high-order modes have a cutoff waveguide width. To satisfy the single mode propagation condition [15], the width of the waveguide must be smaller than $\lambda_0 \mathrm{atc}\tan(\sqrt{-\mathrm{Re}(\varepsilon_{insb})\varepsilon_d}\,/\,\pi\sqrt{\varepsilon_d})$. In this paper, the dielectric in the guided region is assumed to be air with a permittivity $\varepsilon_d = 1$. Figures 1 shows the dependence of the real part and imaginary part, respectively, of the effective refractive index of InSb slot waveguide on the slot widths at different temperatures for an incident light at 1 THz. One can see that the real part effective index of the waveguide decreases with increasing w at a certain temperature, and for a given width it decreases as the temperature rises.

Analysis of InSb plasmonic slot stub waveguide filters

The InSb plasmonic waveguide filter with a single-stub structure is shown in Fig. 2(a). The plasmonic waveguide can be modeled as a transmission line with characteristic impedance. Figure 2(b) shows the transmission line equivalent circuit of the single-stub structure. Z_0 is the characteristic impedance of the waveguide. The characteristic susceptance the of the single-stub structure is given by $Y_s = -jY_0 \tan\left(\beta(w_s)L\right)$, where $j = \sqrt{-1}$, $Y_0 = 1/Z_0$, $\beta(w_s)$ and $\beta(w)$ are the propagation constant of the stub and input waveguide[16]. The transmittance of the single-stub structure can be expressed as

$$T_s = \frac{4}{4 + \left(\dfrac{\beta(w)w}{\beta(w_s)w_s}\right)^2 \tan^2\left(\dfrac{2\pi n_{eff}L}{\lambda}\right)} \exp\left(-\frac{D}{L_{\mathrm{SPP}}}\right), (2)$$

where $L_{\mathrm{SPP}} = (2\mathrm{Im}\beta)^{-1}$ is the characteristic propagation length of the SP mode. The equation above exhibits a clear physical

process: the first part accounts for interference between the incident wave and the wave reflected from the stub. The exponential factor describes the attenuation of the SP wave. When the phase delay satisfies $2\pi n_{eff}L/\lambda = (m+1/2)\pi$ ($m = 0, 1, 2, ...$), the transmittance T_s reaches the minimum. Therefore, the wavelength λ_m of the notch of the transmittance is determined as

$$\lambda_m = 2n_{eff}L/(m+1/2). \qquad (3)$$

It can be seen that the wavelength λ_m is linear to the stub length L and the effective refractive index of the stub.

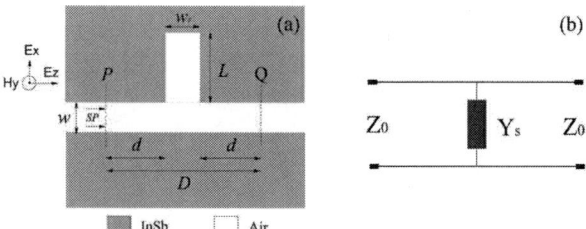

Fig. 2. (a) Schematic structure of an InSb plasmonic slot waveguide filter with the single-stub structure. (b) Transmission line equivalent circuit of the single-stub structure.

In the following part, FDTD method is used to investigate the transmission response of the stub filters. The perfectly matched layer absorbing boundary conditions is set at all boundaries of the simulation domain. Since the width of the InSb slot waveguide is much smaller than the operating wavelength in the structure, only fundamental waveguide mode is supported. The incident wave for excitation of SP mode is a TM-polarized (the magnetic field is parallel to y axis) fundamental mode. The width w of the incident waveguide is set to be 50 μm and the distance L is fixed to 120 μm. Two power monitors are, respectively, set at the positions of P and Q to detect the power of the incident and the transmission field.

The transmission spectra of the InSb single-stub filter are shown in Fig. 3(a), which are obtained by the transmission line model (red dash curve) and FDTD (black solid curve). One can see that there are two transmission notches at the frequency of 0.213 THz (corresponding to m=0) and 0.642 THz (corresponding to m=1), with minimum transmittance of 2% and 3% respectively. The maximum transmittance at the frequency of 0.42 THz is nearly 80% and the full width half maximum (FWHM) are 0.14 THz and 0.125 THz, respectively. It is found that the FDTD results almost agree well with those calculated by the transmission line model. There is a small deviation at the positions of the transmission notch because the phase changes at the junction of the stub and incident waveguide are not considered in the transmission line model. Submitting ω=0.213 THz into Eq. (3) gives a total phase change $\Delta\varphi$=0.019π for the first notch with the stub length L=300 μm, and the effective index n_{eff}=1.1302. The second resonance frequency can be approximately calculated as 0.661 THz by the formula neglecting the frequency dependence on the phase change. The magnetic fields (H_y) at the frequencies of 0.213 THz and 0.43 THz corresponding to the transmission

978-1-4673-4840-9/13 $31.00 © 2013 IEEE

notch and peak in Fig. 3(a) are displayed in Fig. 3(b).

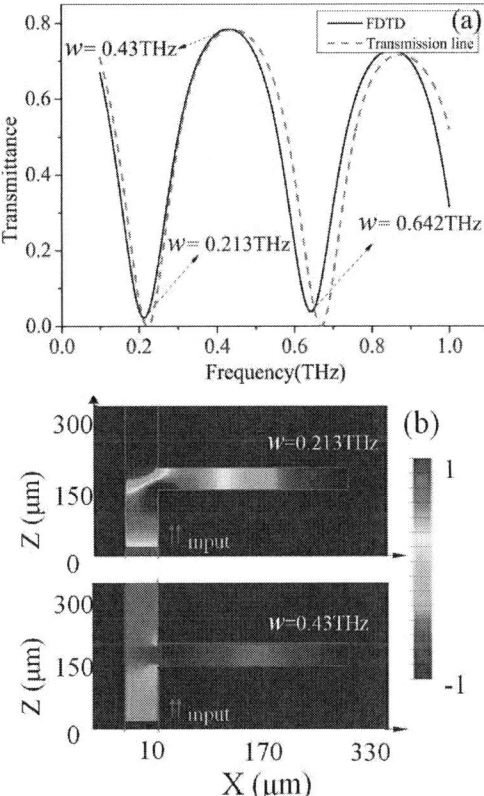

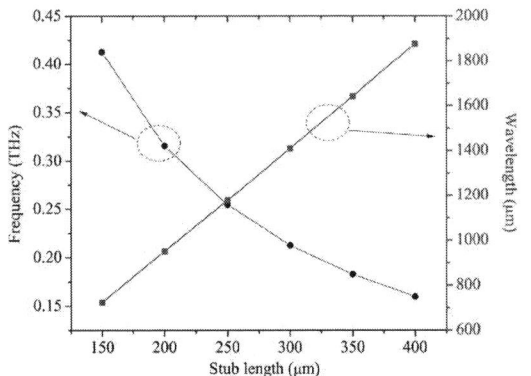

Fig. 3. (a) Transmission spectra of the InSb single-stub filter with the waveguide width w=50 μm, the stub width w_s=50 μm, the length L=300 μm at temperature of 295 K. (b) The corresponding simulated magnetic field (H_y) distributions for frequencies at 0.213 THz (one notch in Fig. 3(a)) and 0.43 THz (one peak in Fig. 3(a)), respectively.

Figure 4 shows that the central wavelength moves to a longer wavelength with the increase of the stub length, as expected from Eq. (3) that central wavelength of the notch has a linear relationship with the length of the stub.

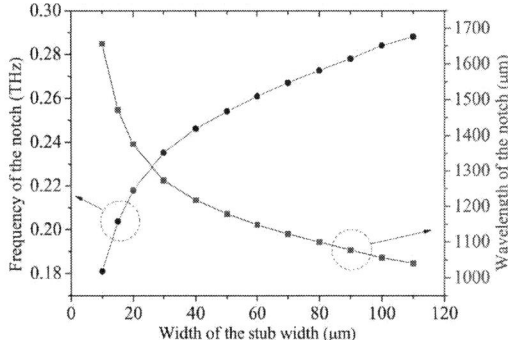

Fig. 4. Relationship between the frequency (wavelength) of the transmission notch and the stub length. The width of the waveguide is w=50 μm, and the stub width is w_s=50 μm at temperature of 295 K.

Figure 5 shows the central wavelength (frequency) of the first transmission notch versus the stub width of w_s for stub length of 250 μm at temperature of 295 K. As revealed in Eq.

(3), the above relationship between the notch position and w_s is mainly due to the contribution of inverse-proportion-like dependence of the effective index n_{eff} on the stub width w_t as shown in Fig. 1. Therefore, one can realize filtering at various required frequencies by properly choosing the width and/or the length of the stub structure.

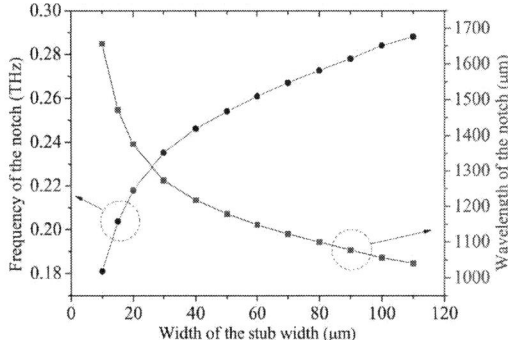

Fig. 5. Relationship between the frequency (wavelength) of the first transmission notch and the stub width. The width of the waveguide is w=50 μm, and the stub length L=250 μm at temperature of 295 K.

As discussed in the above section, the permittivity of the InSb and the effective index of the InSb slot waveguide are affected by the variation of the temperature at the terahertz range. One can expect that the filtering characteristics of the InSb single-stub waveguide filter can be controlled by tuning the temperature. Figure 6(a) shows the transmission spectra of the InSb single-stub waveguide filter at temperatures of 225 K and 325 K, respectively. The other parameters of the structure are set as: slot waveguide width w=50 μm, stub width w_s=30 μm, stub length L=200 μm. From Fig. 6(a), it can be seen that the central frequency of the transmission notch shifts from 0.231 THz to 0.305 THz as temperature increases from 225 K to 325 K. The transmission of the pass band at 225 K is lower than that at 325 K because the imaginary part of the effective index the InSb slot waveguide is higher at lower temperatures which causes higher transmission losses in the waveguide. Figure 6(b) shows the relationship between the central frequency of the transmission notch and the temperature. One can see that the central frequency of the transmission notch increases with the increase of the temperature. This is because the wavelength of the transmission notch has a linear relationship with the effective index of the stub which decreases with the increase of the temperature (shown in Fig. 1). Therefore, active control of SP wave in InSb slot stub waveguide structure can be realized by tuning the temperature.

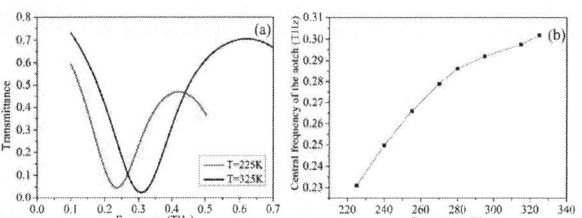

Fig. 6. (a) Transmission spectra of the InSb slot single stub waveguide filter at the temperatures of 225 K and 325 K with w=50 μm, w_s=30 μm, and L=200 μm. (b) Relationship between the central frequency of the transmission notch and temperature.

Conclusions

In summary, we propose planar tunable subwavelength plasmonic stub waveguide filters at terahertz range. The InSb plasmonic slot waveguide has a strong confinement and guides THz waves beyond the diffraction limit. The single-stub InSb plasmonic slot waveguide structure can operate as a notch filter with its central wavelength designed by choosing the width and/or the length of the stub structure. These filtering characteristics of the stub structures can be actively controlled through temperature tuning. The proposed structures open a possibility for future applications in ultra-compact THz integration circuits.

References

[1] R. Köhler, A. Tredicucci, F. Beltram, et al., Nature, 417, pp.156-159, 2002.

[2] M. A. Belkin, Q. J. Wang, C. Pflugl, et al. IEEE, J. Sel. Top. Quant. Electron. 15, pp. 952-967, 2009.

[3] S. Komiyama, O. Astafiev, et al. Nature, 403, pp. 405-407, 2000.

[4] N. Yu, Q. J. Wang, M. A. Kats, et al. Nat. Mater. 9, pp. 730-735, 2010.

[5] G. Gallot, S. P. Jamison, R. W. McGowan, et al. J. Opt. Soc. Am. B, 17, pp. 851-863, 2000.

[6] X. Y. He, J. Opt. Soc. Am. B, 26, pp. A23-A28, 2009.

[7] S. P. Jamison, R. W. McGown, and D. Grischkowsky, Appl. Phys. Lett. 76, pp. 1987-1989, 2000.

[8] D. K. Gramotnev, S. I. Bozhevolnyi, Nature Photon. 4, pp. 83-91, 2010.

[9] J. Tao, X. G. Huang, X. Lin, Q. Zhang, and X. Jin Opt. Express, 17, pp. 13989-13994, 2009.

[10] J. G. Rivas, M. Kuttge, H. Kurz, P. H. Bolivar, and J. A. Sánchez-Gil, Appl. Phys. Lett., 88, pp 082106-082108, 2006.

[11] Lumerical FDTD Solution, http://www. Lumerical.com/

[12] R. Gordon and A. Brolo , Opt. Express, 13, pp. 1933-1938, 2005.

[13] J. A. Sánchez-Gil, J. G. Rivas Phys. Rev. B, 73, pp. 205410, 2006.

[14] M. Oszwaldowski and M. Zimpel J. Phys. Chem. Solids, 49, pp. 1179-1185, 1998.

[15] I. P. Kaminow, W. L. Mammel, and H. P. Weber Appl. Opt. 13, pp. 396-405, 1974.

[16] Y. Matsuzaki, T. Okamoto, M. Haraguchi, M. Fukui, and M. Nakagaki Opt. Express 16, pp. 16314-16325, 2008.

Luminescence properties of cerium doped silicon nitride with MgO additive

Y. Y. Ma, F. Xiao, S. Ye, Q. Y. Zhang[*]

State Key Lab of Luminescence Materials and Devices
Institute of Optical Communication Materials
South China University of Technology
Guangzhou 510641, P. R. China

Abstract

This work focuses on the luminescent materials $Si_3N_4:Ce^{3+}$ with magnesium oxide additive prepared by the solid-state reaction method. The detailed phase formation was investigated based on X-ray diffraction profiles. The photoluminescence (PL) and photoluminescence excitation (PLE) spectra, the decay lifetime depend on the concentration of doped-Ce^{3+} ion and MgO additive were investigated. The emission intensity was effectively enhanced approximately 18.7 times ascribing to MgO additive. $Si_3N_4:Ce^{3+}$ phosphor shows good absorption in ultraviolet (UV) region and emits broad blue emission, which can be used as a potential candidate for application in phosphor-converted light emitting diodes (LEDs).

Keywords: White LEDs, silicon nitride, additive, photoluminescence.

Introduction

White light emitting diodes (WLED) have been developed for a potential solid-state light source and display device applications as a replacement for traditional incandescent and halogen lamps due to their several excellent features[1]. The advantages of white LEDs application include reliability, long lifetime, low power consumption, extraordinary luminous efficiency and environmental friendliness[2]. Basically, there are two methods for generating white-light emitting LEDs. First is combining different LED chips which exhibit red, green and blue (RGB) radiation. The second way is formed by a blue or ultraviolet (UV) LED chip and the down-convertion phosphors. Since the superiority of stabilization, convenient machining, low cost and high color rendering, white LEDs consisting of a high performance LED chip and high efficiency inorganic phosphor have been used in commercial applications extensively. Currently, a InGaN-based blue chip combining with a yellow phosphor ($YAG:Ce^{3+}$) are considered to be the most common approach to produce the WLEDs. This general approach faces serious problems, such as low thermal quenching, poor color rendering index and narrow visible range. However, these problems can be solved by mixing the RGB phosphors with a UV LED chip. There are several benefits, including high color rendering index, high luminous efficiency and stable light color, which are almost independent of the changed current. Consequently, the quest for phosphors with blue emitting is one of the important tasks.

In recent years, (oxy)nitride compounds have gained increasing interest in the investigations of new luminescent materials applied in white LEDs, which show excellent thermal and chemical stability and are considered to be efficient host materials. All these phosphor materials were synthesized using Si_3N_4 powder as a starting material. Hence, there are many researchers focusing on the synthesis of rare-earth activated Si_3N_4. Recently, photoluminescence properties of Eu^{2+}, Ce^{3+} and Tb^{3+} doped α-Si_3N_4 have been reported[3]. However, owing to high absorption of the α-Si_3N_4 host and photoionization of Ce^{3+}, the emission intensity of Ce^{3+} is much lower in α-Si_3N_4 than that in Ca-α-SiAlON. It is reasonable to increase their brightness by introducing some additive, which is valuable for phosphors used in commercial products. In addition, Deeley *et al.* has found that MgO was one of the most effective additives for hot-pressing Si_3N_4, which was often added with silicon nitride in sintering process as a sintering aid[4].

In present work, we successfully synthesized $Si_3N_4:Ce^{3+}$ phosphors and investigated the effects of MgO additives on the structure and luminescence performance. The relations between spectral changes and content of additives and activators were examined. The luminescence decay characteristics of the phosphors were also studied.

Experimental Section

The nominal compositions of $Si_3N_4:Ce^{3+}$ with varied MgO concentrations ($x = 0$, 3, 8, 14, 20, 26 and 30 wt%) are synthesized, in which the concentration of the Ce^{3+} ions was fixed at 1 mol% in the ternary systems of Si_3N_4–CeO_2–MgO. The activator Ce^{3+} was doped in concentration value of y against total amount is 0.5, 1, 3, 5 and 7 mol%, respectively. The raw materials silicon nitride (Si_3N_4), light magnesia powder (MgO) and cerium oxide (CeO_2) were then weighed out and thoroughly mixed in an agate mortar, subsequently the mixture powder was converted in to an Al_2O_3 crucible. The samples were fired at 1450 °C for 4 h under N_2–H_2 (5%) atmosphere in a horizontal tube furnace. The heating rate of the

---

[*] Author to whom correspondence should be addressed; Email: gyzhang@scut.edu.cn

samples was 3 °C/min and cooled down naturally to room temperature after the firing processing.

The crystalline structure of the prepared samples was characterized by X-ray powder diffraction (XRD) with a PANalytical Model X'Pert Pro X-ray diffractometer using a monochromatized Cu Kα radiation. All these XRD measurements were performed at room temperature. The particle morphologies of samples were investigated with a scanning electron microscope (Zeiss Model EVO 18). To investigate their luminescent properties, PL and PLE were measured at room temperature with a fluorescence spectrometer (Edinburgh Instruments Ltd. Model FLS920) equipped with a 450-W xenon lamp as an excitation source, and a nanosecond flash lamp was used as an excitation source for the luminescence decay curves. CIE color coordinates X and Y were calculated using the PL spectral data.

Results and Discussion

Fig. 1 shows XRD patterns of the prepared $Si_3N_4:1\%Ce^{3+}$ samples with varying MgO addition concentration (a, $x = 0$; b, $x = 8$ wt%; c, $x = 20$ wt% and d, $x = 30$ wt%). when the MgO concentration is less than $x = 20$ wt%, it reveals that the prepared samples consist of α-Si_3N_4 and β-Si_3N_4 crystalline phases (JCPDS card no. 01-076-1409 and no. 01-082-0708), respectively[5, 6]. That can be attributed to the commercial α-Si_3N_4 powders are always α/β mixtures with a certain amount of the β phase due to different heat-treatment history in the fabrications[7]. α-Si_3N_4 has a hexagonal crystal structure with the space group $P31c$ (No. 159), its unit cell parameters are found to be $a = b = 7.7696$ Å, $c = 5.6318$ Å and the cell volume $V = 294.43$ Å^3. Compared with α-Si_3N_4, β-Si_3N_4 also has a hexagonal crystal structure with the space group $P6_3$ (No.173), which parameters are as follows: $a = b = 7.6277$ Å, $c = 2.9187$ Å and $V = 147.07$ Å^3. However, other impure phases are detected in samples with a composition of $x \geq 20$ wt%. Therefore, the results indicate that MgO adding into a single Si_3N_4 phase is limited in a range of $x \leq 20$ wt%. There may be many reasons for the absence of MgO. Firstly, a loss of MgO attributed to volatilization is easily explained by the high vapor pressure of MgO at high temperatures[4]. Secondly, the liquid phase may be formed which strongly ascribed to MgO additive at low temperature. After the formation of a liquid phase, MgO dissolved into it, and existed as amorphous phase[8]. Finally, the content of the MgO additives is not much. Consequently, the magnesium oxide additives are not detected by XRD.

The XRD patterns are exhibited for different Ce^{3+}-doped Si_3N_4 compounds ($x = 14$ wt%; e, $y = 0.5$ % and f, $y = 7$ %, respectively) are presented in Fig. 1. The remains of the Si_3N_4 phase in the highly Ce^{3+}-doped samples can be explained by the high solubility of doped rare-earth ions.

The morphologies of the samples were obtained by scanning electron microscopy. Fig. 2 shows the SEM images of $Si_3N_4:Ce^{3+}$ powders synthesized with MgO additive. The calcined phosphor show a severe agglomeration of particles with an irregularly morphology. Moreover, the distributions of estimated particles range from 800 nm to 2 μm.

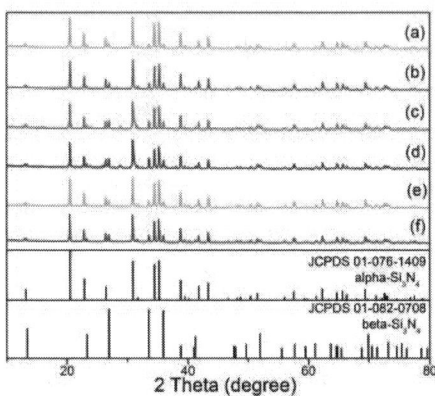

Fig. 1. The XRD patterns of $Si_3N_4:1\%Ce^{3+}$ compounds with different MgO content (a, $x = 0$; b, $x = 8$ wt%; c, $x = 20$ wt% and d, $x = 30$ wt%) and $Si_3N_4:yCe^{3+}$ with 14 wt% MgO addition (e, $y = 0.5$ % and f, $y = 7$ %), respectively.

Fig. 2. A typical SEM image of the $Si_3N_4:1\%Ce^{3+}$ samples with 14 wt% MgO additive.

Fig. 3a shows typical PLE spectra, measured at room temperature, of the particles with different concentration of MgO addition ($x = 0, 3, 8, 14, 20, 26$ and 30 wt%). The broad excitation band ranged from 260 to 380 nm consists of two excitation bands centered at 280 and 340 nm which can be assigned to $4f \rightarrow^2D_{3/2}$ and $4f \rightarrow^2D_{5/2}$ of Ce^{3+}, respectively[9]. The broad absorption band indicates that the phosphor matched well with the UV-LED chips, which is essential for improving the efficiency of WLEDs. Furthermore, it is observed clearly that the excitation intensity is enhanced obviously by the increase of MgO concentration from 0 to 26 wt%.

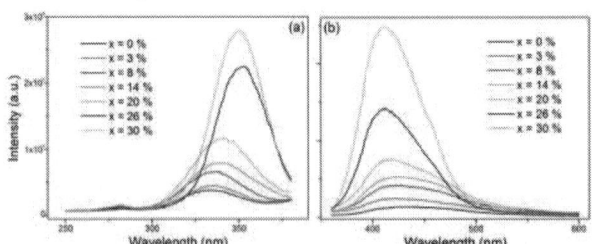

Fig. 3. PLE (a, monitor at peak wavelength of emission band) and PL (b, $\lambda_{ex} = 340$ nm) spectra of $Si_3N_4:1\%Ce^{3+}$ with varied MgO content ($x = 0, 3, 8, 14, 20, 26$ and 30 wt%).

In Fig. 3b, the PL spectrum shows a single and broad emission band centered at 415 nm, which is assigned to the 5d → 4f transition of Ce^{3+}. The full width at half maximum of the emission band is about 98 nm. Compared with PLE spectra, a similar tendency of luminescent intensity is observed at emission spectra. The Si_3N_4:Ce^{3+} phosphor with high concentration MgO additive ($x \geq 0$) emits strong blue luminescence. When $x = 30$ wt%, the emission intensity of Si_3N_4:Ce^{3+} is approximately 18.7 times higher than the sample without MgO additive. These evidences indicated that MgO makes strong contributions to Ce^{3+} emitting. The enhancement of emission in phosphors with additive are readily accomplished, which can be demonstrated by various mechanisms. It includes the effect of flux agent[10] and charge compensation[11, 12], distortion or dislocation of crystal structure[13, 14], change of coordination surrounding the Ce^{3+} ions[15, 16]. In this work, magnesium oxide is selected in terms of sintering aid which function equates with flux. Moreover, when Ce^{3+} ions are incorporated into a host lattice, the charge balance is necessarily required. Thus, MgO additive may act as charge compensator contributing to the electric neutrality. Additionally, as the emission peak of the phosphors was almost unchanged, suggesting that the crystallographic environment of Ce^{3+} has not significant change induced by adding MgO. All careful examination of the results suggested that MgO likely acting as flux and charge compensation play a role of enhancing luminescence intensity.

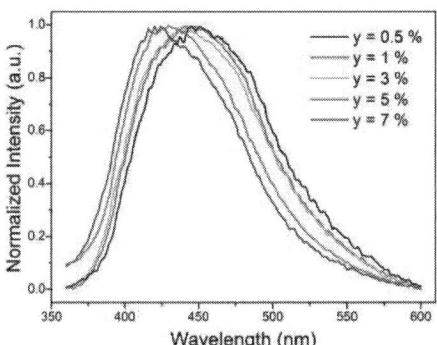

Fig. 4. PL spectra of Si_3N_4:yCe^{3+} phosphors with 14 wt% MgO additive ($y = 0.5, 1, 3, 5$ and 7 mol%).

The normalized emission spectra of Si_3N_4:yCe^{3+} samples with varied Ce^{3+}-doping concentration ($y = 0.5, 1, 3, 5$ and 7 mol%) presents a broad band under the excitation of the 340 nm ultraviolet light, as shown in Fig. 4. The emission band of Si_3N_4:yCe^{3+} shows an asymmetric shape in visible region, which indicates the existence of different luminescent centers. With increasing of Ce^{3+}-doping content, the main peaks in the emission spectrum of phosphor shifted to the short wavelength. The Ce^{3+} ions concentration has some influences on the position of the emission spectra, indicating that a few abrupt disturbances to the local structure is incurred by increasing Ce^{3+}-doping level. Since there is no suitable occupation for the rare-earth ions in Si_3N_4, relatively low Ce^{3+} concentration can be accommodated in this lattice. Here, the larger Ce^{3+} most

probably coordinating on specific atomic sites in two closed interstices of the unit cell within the SiN_4 tetrahedron framework[3]. With increasing Ce^{3+} content, more disorder will be introduced into the crystal lattice accordingly and cause some changes in the crystal field of the activator, then resulting in the smaller centroid shift. So the blue-shift emission can be attributed to the abnormal crystal field strength acting on Ce^{3+} ions. Owing to the existence of two types of phase in Si_3N_4:Ce^{3+} system, it is indistinct to determinate the influence of different crystallographic sites. Further studies are required to discuss the relationship between Ce^{3+} doped concentration and crystal field environments.

Further investigating the effect of MgO, decay characteristics of the phosphors are examined, as shown in Fig. 5. The decay curve of the phosphor with MgO obeyed a second-order exponential equation: $I = A_1\exp(-t/\tau_1) + A_2\exp(-t/\tau_2)$, where I is the luminescence intensity, A_1 and A_2 are constants, t is the time, τ_1 and τ_2 are rapid and slow lifetimes for exponential components, respectively. The average luminescence lifetime for Ce^{3+} can be determined by the formula as $\tau = (A_1\tau_1^2 + A_2\tau_2^2)/(A_1\tau_1 + A_2\tau_2)$, which is determined to be 28.59, 30.22, 31.42, 30.09, 32.49, 28.62 and 29.52 ns monitored at 415 nm ($x = 0, 3, 8, 14, 20, 26$ and 30 wt%). These results were corresponded to the value typically observed for the 5d → 4f allowed transition of Ce^{3+} ions. It is known that the radiative lifetime of the Ce^{3+}-excited state depends on the crystal field at the ion site. As the increasing of additive, the environment around Ce^{3+} will change, which influence on the lifetime of Si_3N_4:Ce^{3+} phosphors. Owing to the variety and uncertain occupation of Ce^{3+} ions, the value of lifetime exhibit irregular change.

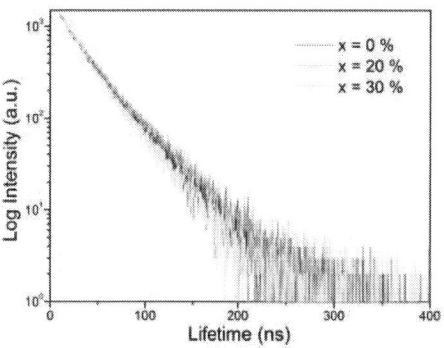

Fig. 5. The luminescence decay curves of Si_3N_4:$1\%Ce^{3+}$ phosphors with varied concentration of MgO ($x = 0, 20$ and 30 wt%).

The luminescence decay lifetime of the Ce^{3+}-doping phosphors also can be best fitted with second-order exponential equation above, which are found to be 31.69, 30.09, 28.59, 27.94 and 31.12 ns. The decay curves of Si_3N_4:$0.5Ce^{3+}$ with 14 wt% MgO additive are measured and represented as shown in Fig. 6.

Chromaticity analyses are performed as well to further investigate the PL properties. A CIE 1931 chromaticity diagram of the Si_3N_4:yCe^{3+} phosphor is shown in Fig. 7. The color located at the blue region in the chromaticity diagram.

978-1-4673-4840-9/13 $31.00 © 2013 IEEE

The chromaticity coordinates of $Si_3N_4:yCe^{3+}$ phosphors were $X = 0.182$, $Y = 0.208$; $X = 0.178$, $Y = 0.192$; $X = 0.175$, $Y = 0.186$; $X = 0.172$, $Y = 0.149$ and $X = 0.165$, $Y = 0.142$ ($y = 0.5$, 1, 3, 5 and 7 mol%, respectively). The change in chromaticity coordinate associated with the shift of Ce^{3+} emission.

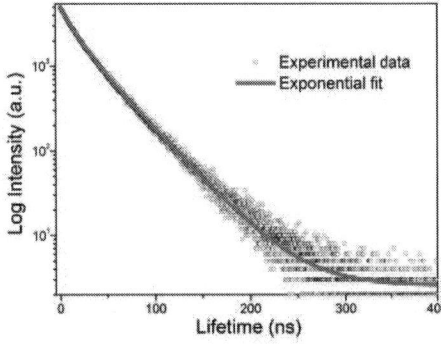

Fig. 6. The decay curves of $Si_3N_4:0.5Ce^{3+}$ samples with 14 wt% MgO additive.

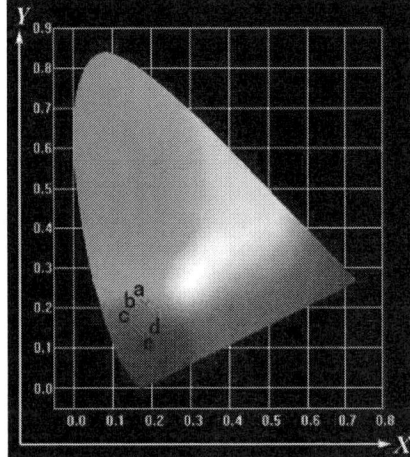

Fig. 7. CIE chromaticity diagram for $Si_3N_4:yCe^{3+}$ phosphors (a, $y = 0.5$ %; b, $y = 1$ %; c, $y = 3$ %; d, $y = 5$ % and e, $y = 7$ %).

Conclusions

In summary, we have synthesized a series of $Si_3N_4:Ce^{3+}$ phosphors by a solid-state reaction method. A certain amount of magnesium oxide was acted as a flux and charge compensator for synthesizing phosphor. The $Si_3N_4:Ce^{3+}$ phosphors exhibited blue-emitting light by a strong excitation band, centering at about 340 nm, which matches well with UV LED chips. By systematically controlling the MgO addition, the strong blue emission can be achieved. All the experiment results show that these phosphors can serve as a potential blue emitting phosphor for UV LEDs. Furthermore, the results provide valuable insight for further development of MgO additive, which effects will be unraveled until more researches are done.

Acknowledgements

This work is financially joint supported by the NSFC (Grant Nos. 50872036, 21101065 and 51125005), the Fundamental Research Funds for the Central Universities, SCUT, and China Postdoctoral Science Foundation funded project.

References

[1] S. Ye, F. Xiao, Y.X. Pan, Y.Y. Ma, and Q.Y. Zhang, Mater. Sci. Eng. R-Reports. **71**(1), pp. 1-34, 2010.

[2] S.J. Ho, B.I. Won, C.L. Dong, Y.J. Duk, and S.K. Shi, J. Lumin. **126**, pp. 371-377, 2007.

[3] Y.Q. Li, N. Hirosaki, R.J. Xie, T. Takeda, and M. Mitomo, J. Lumin. **130**, pp. 1147-1153, 2010.

[4] A. Glachello, P.C. Martinengo, and G. Tommasini, J. Mater. Sci. **14**, pp. 2825-2830, 1979.

[5] M. Billy, J.C. Labbe, and A. Selvaraj, Mater. Res. Bull. **18**(8), pp. 921-934, 1983.

[6] J. Schneider, F. Frey, N. Johnson, and K. Laschke, Z. Kristallogr. **209**, pp. 328-333 1994.

[7] G. Petzow and R. Sersale, Pure Appl. Chem. **59**, pp. 1673-1680, 1987.

[8] T. Yano, J. Yamane, K. Yoshida, S. Miwa, and M. Ohsaka, Mater. Sci. Eng. **9**, pp. 012024, 2010.

[9] H.S. Jang and D.Y. Jeon, Appl. Phys. Lett. **90**, pp. 041906, 2007.

[10] Y.X. Pan, M.M. Wu, and Q. Su, J. Phys. Chem. Solids. **65**, pp. 845-850, 2004.

[11] C. Jiang, L. Fang, M.R. Shen, F.G. Zheng, and X.L. Wu, Appl. Phys. Lett. **94**, pp. 071110-071113, 2009.

[12] Y.G. Su, L.P. Li, and G.S. Li, Chem. Mater. **20**, pp. 6060-6067, 2008.

[13] H. Yamamoto and S. Okamoto, Displays. **21**, pp. 93-98, 2000.

[14] S. Okamoto, H. Kobayashi, and H. Yamamoto, J. Appl. Phys. **82**, pp. 5594-5597, 1999.

[15] S. Okamoto, Yamamoto, H. , Appl. Phys. Lett. **78**, pp. 655-657, 2001.

[16] S.H. Shin, J.H. Kang, D.Y. Jeon, and D.S. Zang, J. Lumin. **114**, pp. 275-280, 2005.

Field Effect Transport Properties of Electrochemically prepared Graphene Quantum Dots

Hemen Kalita[#], Harikrishnan V[#], M. Aslam*

Department of Physics
Indian Institute of Technology Bombay, Powai, Mumbai- 400076 India.
*Corresponding author: m.aslam@iitb.ac.in, Phone:022-2576-7585
# The authors contributed equally to this work

Abstract

Herein we report the field effect properties of lithographically fabricated FET with graphene quantum dots (GQDs) as channel. GQDs are synthesized via an electrochemical avenue using multiwall carbon nanotubes. As-prepared dots of 4.5±0.55 nm average diameter are found to be p-type in nature under ambient conditions. Field effect measurements yield hole mobility of 0.01 cm^2 V^{-1}s^{-1} and I_{on}/I_{off} ratio of 45. After annealing of devices in Argon atmosphere at 300^0C for 20 min, the channel is found to show ambipolar transport with significant increase in resistance.

Keywords: Graphene, quantum dots, p-type, mobility, hole dopant, adsorbates

Introduction

Graphene, a single atom sheet of sp^2-bonded carbon atoms densely packed in a honeycomb crystal lattice due to its high carrier mobility and ultrathin body has been receiving tremendous interest in recent years [1]-[5]. It also has outstanding intrinsic properties like high electrical and thermal conductivity, increased mechanical stiffness, good optical transmittance etc [6]-[8]. Graphene based devices have shown excellent promise in numerous applications like energy storage [9], fuel cells [10], transparent conducting electrodes [11] and memory systems [12]. And it has also attracted attention as a channel material for future high speed nanoelectronics devices [1], [13], [14]. The atomic level thickness and high surface area of graphene makes it an especially attractive material for high performance chemical and gas sensors [15]. But, despite its high intrinsic carrier mobility, the zero bandgap limits its usage for electronic and opto-electronic applications [1]. However, the introduction of a finite bandgap in graphene nanoribbons (GNRs) and graphene quantum dots (GQDs) due to quantum confinement and edge effects can be seen as a way to overcome this problem. Earlier works on GQDs are done on theoretical prediction but recently more focus is on experimental synthesis and characterization. This new material provides an opportunity to perform systematic studies on transport properties and leads to potential use in VLSI Technologies.

Here we present the field effect properties of lithographically fabricated Field Effect Transistor (FET) with as-prepared and annealed GQDs as channel.

Experimental Section

Graphene quantum dots used for this study were prepared electrochemically using multiwall carbon nanotubes (MWCNT) as precursor material [16].

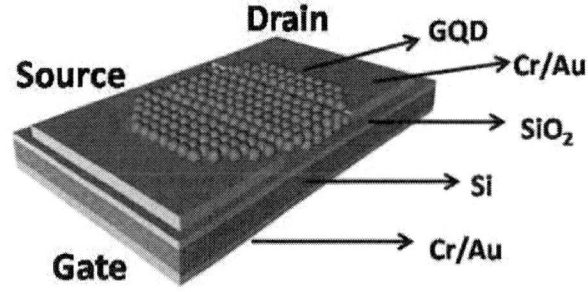

Fig. 1 The schematic view of the device prepared for the electrical measurements.

For the device fabrication Si (100) (0.001-0.005 Ω-cm) with a thermally grown 100 nm thick SiO$_2$layer were used as substrates. Source and drain contact pads (Au(150nm)/Cr(50nm)) were patterned on Si/SiO$_2$ by optical lithography and lift-off process. Here chromium is serving as a buffer layer. The electrodes are squares of dimension 100 x 100 μm and are separated by a 10 μm channel, where the prepared graphene quantum dots were dropcasted and dried in the ambient environment. Au (100nm) was thermally deposited on the back side of the wafer for gate contacts. Fig. 1 shows a schematic view of the prepared device.

Crystallographic measurements were performed on Xpert PANAlytic X-ray diffractometer using Cu Kα radiation (λ = 0.15404 nm). HRTEM (JEOL 2100F, field emission gun transmission electron microscope is used to confirm the particle size distribution of the GQDs. Atomic force microscope (AFM) measurements are performed using digital instruments Nanoscope IV Multimode scanning probe microscopy. Images were taken in tapping mode using etched silicon probes. Raman spectra recorded at room temperature using confocal micro-Raman spectrometer equipped with Ar$^+$ laser (514.5 nm). Current-voltage measurements were carried out at room temperature in air using a Keithley 4200SCS sourcemeter.

978-1-4673-4840-9/13 $31.00 © 2013 IEEE

Results and Discussion

High resolution transmission electron microscopy shows that the average size of the prepared quantum dots is 4.5 ± 0.5 nm (Fig. 2). Fig. 3 shows an AFM image of the as-prepared quantum dots along with its sectional analysis. The height profile of synthesized quantum dots is determined to be around one nm. Fig. 4 shows the Raman spectrum of the as prepared dots with a G band at 1590 cm^{-1} and D band at 1350 cm^{-1}. The Raman spectra show that defects are less in case of the as-prepared GQD. The X-ray diffraction spectrum (Fig. 5) of the prepared quantum dots shows a broader (002) peak at 26.8^{O}, as compared to multiwall carbon nanotubes, indicating a smaller crystalline size. The d spacing is calculated to be 3.30 Å, a value close to that of bulk graphite.

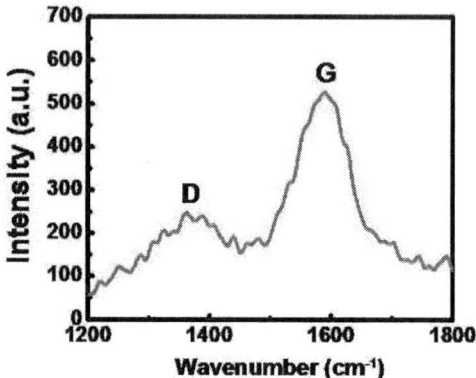

Fig. 4 Raman spectrum of the as-prepared GQDs.

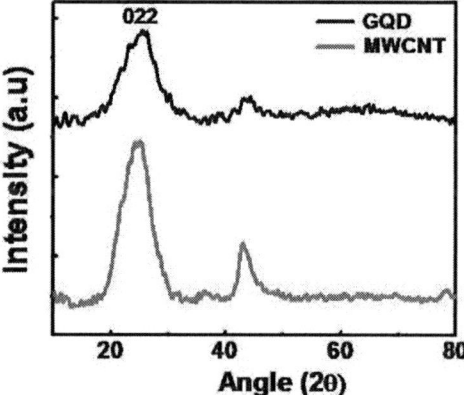

Fig. 5 X-ray diffraction spectra of multiwall carbon nanotubes and as-prepared graphene quantum dots.

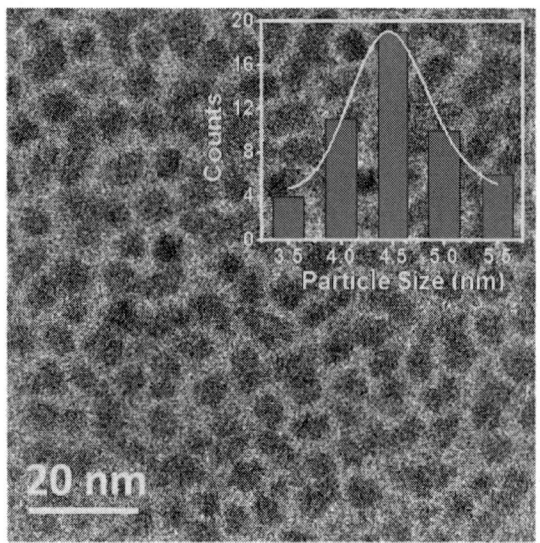

Fig. 2 HRTEM image of as-prepared GQDs. The inset shows the size distribution of GQDs. Average diameter of the dots is calculated to be 4.5 ± 0.55 nm.

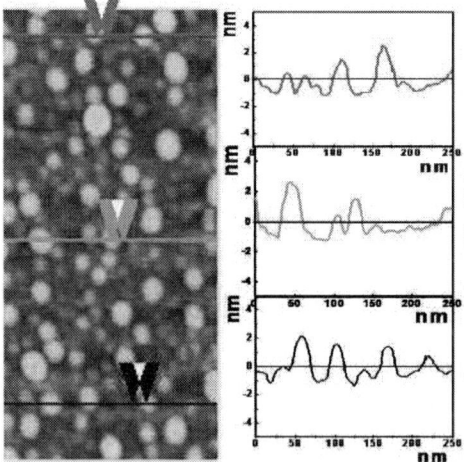

Fig. 3 Atomic force microscopy image of the dots with height profiles of GQDs at different positions.

Fig. 6 shows the transfer characteristics of the FET device with as-prepared GQD as the channel material. Increase in I_{ds} with increase in negative gate voltage bias and nearly invariant I_{ds} for positive gate bias indicate that the total current is dominated by hole current, suggesting the p-type nature of GQDs. This p-type nature obtained is attributed to the physisorption of water molecules. I_{ds}-V_{ds} characteristics are recorded with drain voltage (V_{ds}) sweeps from -20V to +20V. The curve is found to be non-linear and asymmetric. The hole mobility of the GQDs is calculated from (1) and it is found to be 0.01 cm^2 V^{-1}s^{-1} and the I_{on}/I_{off} ratio is found to be 45.

Electron and hole mobility can be extracted from the linear regime of the transfer characteristics using

$$\mu = [(\Delta I_{ds}/V_{ds}) \times (L/W)]/C_{ox}\Delta V_{g}, \qquad (1)$$

where L and W are channel length and width, respectively, C_{ox} is silicon oxide gate capacitance and I_{ds}, V_{ds} and V_{g} are drain-source current, drain-source voltage and gate voltage respectively.

The prepared devices are annealed at 300^{0}C for 20 min in Ar atmosphere to remove any adsorbates and contaminants present. After annealing the device shows ambipolar characteristics with significant increase in resistance as shown in Fig. 7. Graphene quantum dots, due to their high surface area to volume ratio are extremely sensitive to external adsorbates. This leads to rapid adsorption of ambient water molecules present in their environment, hence rendering them p-type. The removal of these adsorbed water molecules during annealing leads to weakened coupling among dots leading to higher interdot tunneling resistance. These effects culminate in the higher resistance and ambipolar transport observed.

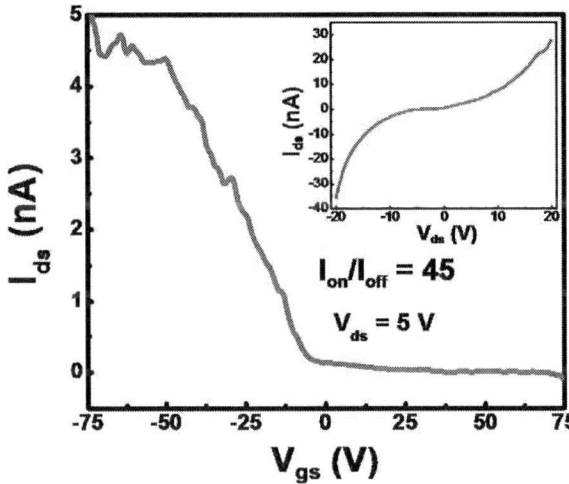

Fig. 6 Transfer characteristics I_{ds} vs V_{gs} of as-prepared GQDs at fixed bias at V_{ds}=5V showing p-type characteristics. The inset shows the I_{ds} vs V_{ds} characteristics.

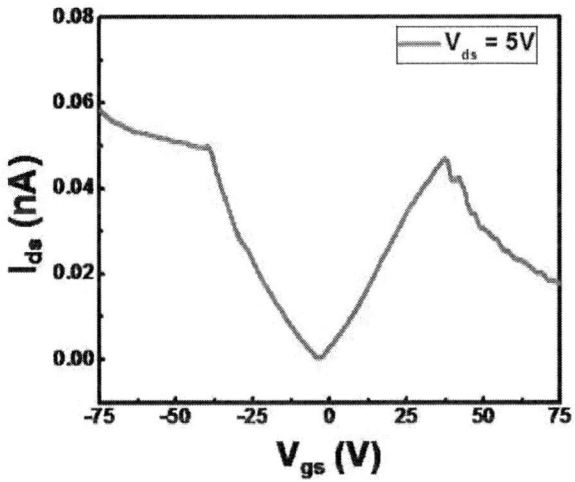

Fig. 7 Transfer characteristics I_{ds} vs V_{gs} of annealed GQDs at fixed bias at V_{ds}=5V showing ambipolar characteristics.

Conclusions

Field effect measurements of GQD-FET yield hole mobility of 0.01 cm^2 V^{-1}s^{-1} and I_{on}/I_{off} ratio of 45. After annealing of devices in Argon atmosphere at 300^{0}C for 20 min, the channel is found to show ambipolar transport with significant increase in resistance. This increase, which is found to be of the order of 3 is determined to occur due to desorption of adsorbed water molecules. Increase in conductance and reappearance of hole doping is observed to occur when the device is reintroduced to the ambient environment, suggesting reversible physisorption. Hence, modifying graphene

quantum dot surface by external adsorbates provides a mean to control doping and strength of interdot coupling in the quantum dot array, offering interesting possibilities to develop size tunable solid state quantum dot electronics.

References

[1] K. S. Novoselov, A. K. Geim, S. V. Morozov, D. Jiang, Y. Zhang, S. V. Dubonos, I. V. Grigorieva and A. A. Firsov, *Science*, vol. 306, pp. 666–669, Oct. 2004.

[2] F. Schedin, A. K. Geim, S. V. Morozov, E. W. Hill, P. Blake, M. I. Katsnelson and K. S. Novoselov, *Nat. Mater.*, vol. 6, pp. 652–655, Sep. 2007.

[3] X. C. Dong, Y. M. Shi, Y. Zhao, D. M. Chen, J. Ye, Y. G. Yao, F. Gao, Z. H. Ni, T. Yu, Z. X. Shen, Y. X. Huang, P. Chen and L. J. Li, *Phys. Rev. Lett.*, vol. 102, pp. 135501-1–135501-4, Mar. 2009.

[4] N. Tombros, C. Jozsa, M. Popinciuc, H. T. Jonkman and B. J. Van Wees, *Nature*, vol. 448, pp. 571–574, Aug. 2007.

[5] X. C. Dong, Y. M. Shi, W. Huang, P. Chen and L. J. Li, *Adv. Mater.*, vol. 22, pp. 1–5, 2010. DOI: 10.1002/adma.200903645.

[6] A. K. Geim and K. S. Novoselov, *Nat. Mater.*, vol.6, pp183-191, 2007.

[7] K. I. Bolotin, K. J. Sikes, Z. Jiang, M. Klima, G. Fudenberg, J. Hone, P. Kim, H. L. Stormer, *Solid State Commun.*, vol.146 , pp.351-355, 2008.

[8] S. V. Morozov, K. S. Novoselov, M. I. Katsnelson, F. Schedin, D. C. Elias, J. A. Jaszczak and A. K. Geim, *Phys. Rev. Lett.*, vol.100 ,pp. 016602-016605, 2008.

[9] D. Stoller, S. J. Park, Y. W. Zhu, J. H. An and R. S. Ruoff, *Nano Lett.* vol.8, pp.3498-3502, 2008.

[10] B. Seger and P. V. Kamat, *J. Phys. Chem. C.*, vol.113, pp.7990-7995, 2009.

[11] Xuesong Li, Yanwu Zhu, Weiwei Cai, Mark Borysiak, Boyang Han, David Chen, Richard D. Piner, Luigi Colombo and Rodney S. Ruoff, *Nano Lett.*, vol.9, pp.4359-4363, 2009.

[12] Augustin J. Hong, Emil B. Song, Hyung Suk Yu, Matthew J. Allen, Jiyoung Kim, Jesse D. Fowler, Jonathan K. Wassei, Youngju Park, Yong Wang, Jin Zou, Richard B. Kaner, Bruce H. Weiller and Kang L. Wang, *ACS Nano.*, vol.5, pp.7812-7817, 2011.

[13] B. Huard, J. Sulpizio, N. Stander, K. Todd, B. Yang, and D. Goldhaber-Gordon, *Phys. Rev. Lett.*, vol. 98, no. 23, p. 236 803, Jun. 2007.

[14] M. C. Lemme, T. Echtermeyer, M. Baus and H. Kurz, *IEEE Electron Device Lett.*, vol. 28, no. 4, pp. 282–284, Apr. 2007.

[15] Jesse D. Fowler, Matthew J. Allen, Vincent C. Tung, Yang Yang, Richard B. Kaner and Bruce H. Weiller, *ACS Nano*, vol. 3, pp.301-306, 2009.

[16] D. B. Shinde and V`. K. Pillai, *Chem. Eur. J.* 18,12522 (2012).

In-Situ Observation of Temperature Distribution of Microheaters Using Near-Infrared CCD Imaging System

Takanari Saito[1], Weichih Lin[2], Ibuki Atsumo[1], and Jun-ichi Shirakashi[1*]

[1]Department of Electrical and Electronic Engineering, Tokyo University of Agriculture & Technology,
2-24-16 Nakacho, Koganei, Tokyo 184-8588, JAPAN.
[2] Nanoscience Centre, University of Cambridge, UNITED KINGDOM.
*Phone/Fax: +81-042-388-7919, E-mail: shrakash@cc.tuat.ac.jp

Abstract

Temperature distribution of microheaters during electromigration (EM) was observed by near-infrared (NIR) charge coupled device (CCD) imaging system. The temperature of the hot spot, located at the tip of the microheater, was well controlled with varying the applied voltage. Microheaters were broken during EM processes, and breakdown region clearly corresponded to the position of the hot spot. These results imply that NIR CCD imaging system is a useful tool for the investigation of the temperature distribution of microheaters.
Keywords: in-situ observation, thermal distribution, microheater, NIR, CCD and EM

Introduction

Recently, electromigration (EM) has been used as a common technique for the fabrication of nanometer-scale gaps between conducting electrodes [1-3]. Furthermore, EM processes in metal wires have been widely analyzed using scanning electron microscopy (SEM) [4] and transmission electron microscopy (TEM) [5]. In EM phenomena, local Joule heating assists EM by enhancing atomic diffusion [6]. Therefore, experimental studies of thermal distribution during EM are strongly required. In particular, heating properties in Al [7, 8] and Pt [9] wires during EM have been investigated using thermal emission microscopy. Although these methods for the investigation of EM provide high spatial resolution, they require high voltages and/or high vacuum environments, as well as complicated experimental procedures and/or expensive facilities. In addition, in order to obtain direct thermal cartography, the thermoreflectance allows us to detect the temperature profile variation on a large thermal range [10]. However, the method requires precise calibration procedures of the sample because of the optical sensitivity. Furthermore, infrared (IR) thermography is also currently used for testing micro-devices [11], but the wavelength of IR radiation is greater than 3 µm, resulting in low spatial resolution due to longer wavelength.

Previously, we have reported that the heating process of the metal wires such as graphite and W under current flow was simply and easily observed using a hand-made, in-situ, near-infrared (NIR) charge coupled device (CCD) imaging system [12]. In this report, we study temperature distribution of microheaters [13, 14] under current flow using our NIR CCD imaging system.

Experimental Details

The schematic of experimental setup is shown in Fig. 1(a), which consists of NIR microscopy equipped with objective 100x and source measure units (SMUs). The experiments were carried out in obscurity and ambient air. Microheater devices are fabricated by repeated lift-off processes [14]. Optical image of microheaters is shown in Fig. 1(b). The inset also exhibits an expanded image around the tip of the microheater. The microheater is composed of a pair of conductive Au wires and an electroresistive Pt layer. A Zn layer is then deposited on the surface. The thickness of the Au, Pt, and Zn layers is 100 nm, 200 nm, and 300 nm, respectively. The narrowest part of the devices, which corresponds to the tip, is approximately 5 µm. The detailed information of the microheaters has been reported elsewhere [13, 14].

First, for the calibration of gray-scale value of the NIR image versus temperature, electromagnetic radiation emitted from the microheater was measured using a spectrometer equipped with a multichannel analyzer (MCA). The MCA was aligned over the center of the device heated by applying controlled emission voltages. The temperature of the

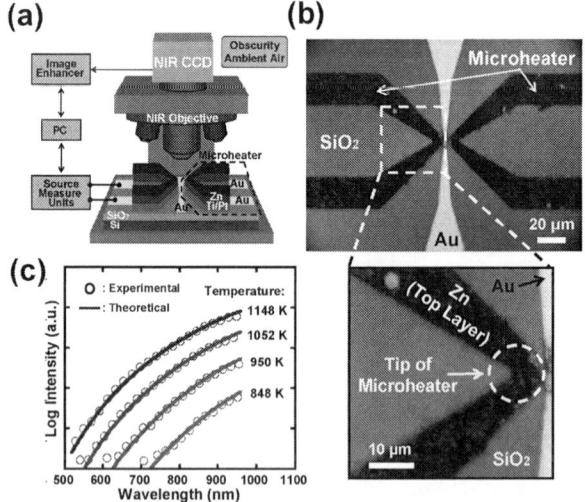

Fig. 1 (a) Schematic illustrations of experimental setup for in-situ observation of temperature distribution of microheaters. (b) Optical image of microheaters. The inset shows enlarged image around the tip of the microheater. (c) Typical measured spectrum from microheaters (circles). Theoretical blackbody curves (solid lines) show good ageement with the experimental data.

978-1-4673-4840-9/13 $31.00 © 2013 IEEE

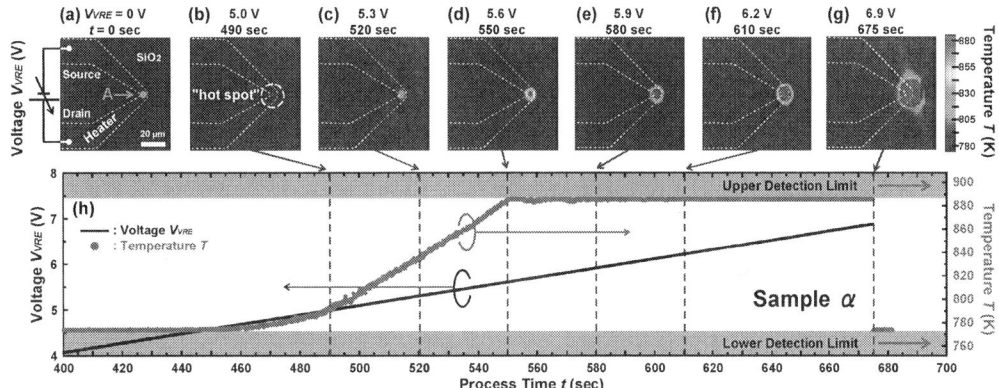

Fig. 2 (a)-(g) Time evolution of VRE in the microheater (sample α) with a series of temperature distributions. (h) Voltage and temperature as function of time. The temperature was obtained from point A, as shown in Fig. 2 (a).

microheater was determined by fitting the measured spectrum to a perfect blackbody spectrum, as shown in Fig. 1(c). Then, NIR images of the microheater heated by Joule effect with an applied bias voltage were taken using in-situ NIR CCD imaging system. We determined the gray-scale values of the pixels on the microheater, as a function of the applied bias voltage. The combined blackbody/NIR measurements allow us to estimate the relationship between gray-scale value of the NIR image and temperature of the microheater.

Then, we performed two different EM procedures, one is voltage-ramped EM (VRE) [1] and the other is feedback-controlled EM (FCE) [2, 3]. In both procedures, applied bias voltages (V_{VRE} and V_{FCE}) were increased in steps of 1 mV. Especially for FCE scheme, the voltages V_{FCE} were actively controlled and adjusted in response to the changing resistance of the microheaters. In this paper, temperature distribution of the microheaters during both EM procedures was observed by NIR CCD imaging system.

Results and Discussion

A. In-Situ Observation of Temperature Distribution of Microheaters during VRE

Figs. 2(a)-2(g) exhibit the temperature distribution of a microheater (Sample α) during VRE. Dashed lines also represent the boundary of the microheater. The generated "hot spot" was located almost at the tip of the microheater, as indicated by the dashed circle in Fig. 2(b). It is generally observed that the temperature of the hot spot was increased with increasing the applied voltages V_{VRE}. Fig. 2(h) represents the voltage V_{VRE} and temperature T as function of the process time t. The temperature T was obtained from point A, as shown in Fig. 2(a). At the process time t of 460 sec after starting the VRE, the temperature T of the point A was detected as about 760 K. Then, the temperature T increased with increasing the voltage V_{VRE} and reached the upper detection limit of the system, approximately 880 K. It should be noted that the temperature of the samples, which was determined by fitting the spectra obtained with the MCA to the theoretical blackbody emittance, was seen to be around 1000 K as shown in Fig. 1 (c). Therefore, it is expected that the temperature of the "hot spot" is easily increased above 1000 K, especially in the final stage of the heating (Fig. 2(g)).

B. In-Situ Observation of Temperature Distribution of Microheaters during FCE

We also investigated the temperature distribution of a microheater (Sample β) during FCE, as shown in Figs.

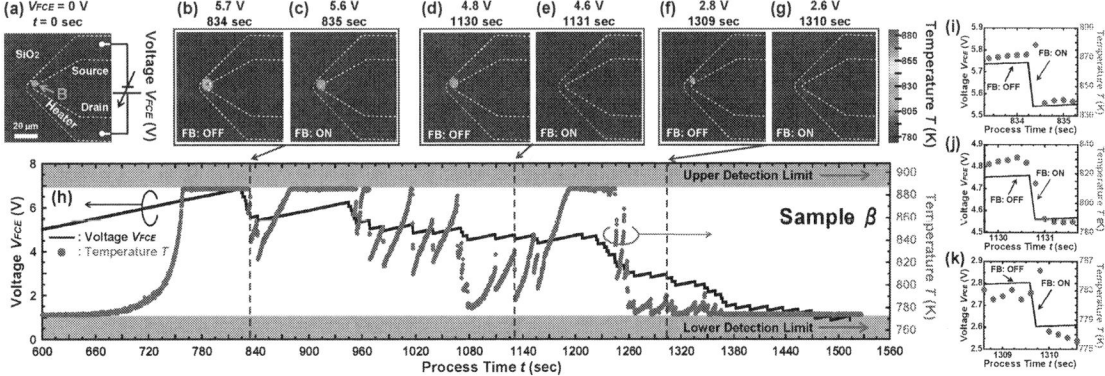

Fig. 3 (a)-(g) Time evolution of the temperature distribution of the microheater (sample β) during FCE. (h) Voltage and temperature obtained from point B (Fig. 3(a)) as function of process time. (i)-(k) Time enlarged scale of Fig. 3(h) around 834 sec, 1130 sec, and 1309 sec, respectively.

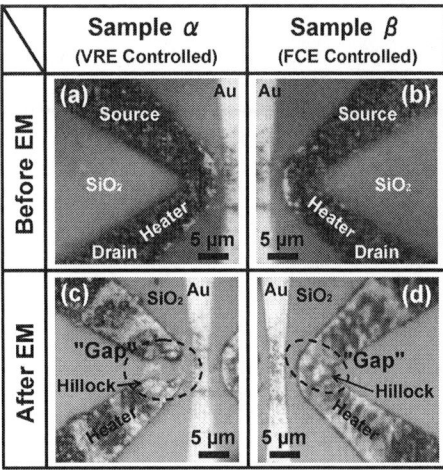

	Sample α (VRE Controlled)	Sample β (FCE Controlled)

Before EM: (a) Source, Au, SiO₂, Heater, Drain, 5 μm | (b) Au, Source, SiO₂, Heater, Drain, 5 μm

After EM: (c) SiO₂, Au, "Gap", Hillock, Heater, 5 μm | (d) Au, SiO₂, "Gap", Hillock, Heater, 5 μm

Fig. 4 Optical images of the samples α (a) and β (b) before VRE and FCE processes, respectively. Samples α (c) and β (d) after VRE and FCE processes, respectively. Clear gap formation is observed in the broken circles.

3(a)-3(g). The voltage V_{FCE} and temperature T obtained from point B (Fig. 3(a)) as function of the process time t are also shown in Fig. 3(h). Similarly as in the case of VRE, "hot spot" was generated at the tip of the microheater, and the temperature obtained from point B ("hot spot") was increased with increasing the applied voltages V_{FCE} until the voltage feedback process occurred. Figs. 3(c), 3(e), and 3(g) show the temperature distribution images of the microheater with voltage feedback "ON". The images with feedback "OFF" are also exhibited in Figs. 3(b), 3(d), and 3(f). Figs. 3(i)-3(k) represent the voltage V_{FCE} and temperature T with enlarged time scale around 834 sec, 1130 sec, and 1309 sec, respectively. From the figures, the temperatures T obtained at feedback "ON" are clearly lower than those at feedback "OFF". These results strongly suggest that the temperature of the microheater is successfully controlled and suppressed by adjusting the applied voltages V_{FCE} through the FCE process, as compared with VRE.

C. Threshold Temperature for Detection in NIR CCD Imaging System

Previously, we have reported the threshold temperature (lower detection limit of the temperature) for detection in the system is approximately 600 K in the graphite wire [12]. In the case of the microheaters, here, the value is determined to be approximately 760 K (Figs. 2(h) and 3(h)). It is considered that the difference between these threshold temperatures is due to the difference of emissivity between the graphite wire and the microheater. In general, it is known that graphite has a high emissivity of 0.8 [15, 16] and 0.8-0.9 [17]. On the other hand, the emissivity of Zn, which is the top layer of the microheater, is determined to be 0.02-0.06 [18-21] and 0.05-0.5 [22]. Thus, the radiation magnitude of the microheater tends to be lower than that of the graphite wire in our NIR CCD imaging system. As expected, the detection range of the system depends on the emissivity of the materials.

D. Gap Formation in Microheaters before and after EM Processes

Figs. 4(a) and 4(b) show optical images of the samples α and β before EM processes, respectively. In both of the samples, the electron flow is from source to drain. As a result of the breaking of the microheaters during both VRE and FCE processes, the gaps were clearly formed and corresponded to the position of the "hot spot", as shown in Figs. 2(b)-2(g) and Figs. 3(b)-3(g). Moreover, the hillocks are also observed on the surface of the microheaters, especially in the drain side, as shown in Figs. 4(c) and 4(d). Furthermore, heating process of the microheater was clearly regulated using FCE procedures, as shown in Figs. 3(a)-3(h). Consequently, it is considered that the gap width formed by the FCE process was more uniform and narrower than that of the VRE, as shown in Fig. 4(d). These results imply that NIR CCD imaging system can easily investigate the heating process of microheaters during EM.

Conclusions

We describe the temperature distribution of microheaters during EM investigated by NIR CCD imaging system. Generated hot spots are located almost at the tip of the microheaters and are well regulated with varying the applied voltages. Heating process of the microheaters controlled using FCE procedure is observed by in-situ NIR microscopy. The gaps are clearly formed by both EM processes (VRE and FCE) and correspond to the position of the hot spot. The uniform and narrow gap is fabricated using FCE, which is due to that thermal heating is well controlled and regulated by FCE process. These results suggest that the simple NIR CCD imaging system easily investigates the heating process during EM and is useful for the controlled formation of gaps with nanometer-scale.

References

[1] H. Park, et al., *Appl. Phys. Lett.*, 75, pp. 301-303, July 1999.

[2] D. R. Strachan, et al., *Appl. Phys. Lett.*, 86, pp. 043109-1-3, January 2005.

[3] S. Itami, et al., *J. Nanosci. Nanotechnol.*, 10, pp. 7464-7468, November 2010.

[4] T. Taychatanapat, et al., *Nano Lett.*, 7, pp. 652-656, February 2007.

[5] D. R. Strachan, et al., *Phys. Rev. Lett.*, 100, pp. 056805-1-4, February 2008.

[6] M. L. Trouwborst, et al., *J. Appl. Phys.*, 99, pp. 114316-1-7, June 2006.

[7] S. Kondo, et al., *J. Appl. Phys.*, 79, pp. 736-741, January 1996.

[8] S. Kondo and K. Hinode, *Appl. Phys. Lett.*, 67, pp. 1606-1608, Septembre 1995.

[9] D. R. Ward, et al., *Appl. Phys. Lett.*, 93, pp. 213108-1-3, November 2008.

[10] G. Tessier, et al., *Appl. Phys. Lett.*, 78, pp. 2267-2269, April 2001.

[11] R. Furstenberg, et al., *Rev. Sci. Instrum.*, 78, pp. 064903-1-5, June 2007.

[12] Y. Kuwabara, et al., *IEEE-NEMS 2011*, pp. 681-684, 20-23 February 2011.

[13] W. Lin, et al., *IEEE-NEMS 2011*, pp. 897-900, 20-23 February 2011.

[14] W. Lin, et al., *IEEE-MEMS 2012*, pp. 452-455, January 29-February 2 2012.

978-1-4673-4840-9/13 $31.00 © 2013 IEEE

[15] X. J. Chen, et al., *Cryst. Res. Technol.,* 42, pp. 971-975, August 2007.

[16] V. V. Kalaev, et al., *J. Cryst. Growth.,* 249, pp. 87-99, February 2003.

[17] R. Pugno, et al., *J. Nucl. Mater.,* 363-365, pp. 1277-1282, June 2007.

[18] Available from: http://calex.co.uk/downloads/application_guidance/emissivity_tables.pdf [Accessed November 9, 2012].

[19] Available from: http://www.coleparmer.com/TechLibraryArticle/254#anchor76 [Accessed November 9, 2012].

[20] Available from: http://www.omega.com/literature/transactions/volume1/emissivitya.html [Accessed November 9, 2012].

[21] Available from: http://www.omega.com/temperature/z/pdf/z088-089.pdf [Accessed November 9, 2012].

[22] Available from: http://www.raytek.com/Raytek/en-r0/IREducation/EmissivityTableMetals.htm [Accessed November 9, 2012].

Characterization of a-Se p-n junction fabricated using electrolysis in NaCl *aq.*

M. Onishi[1], K. Komiyama[1], K. Takeno[1], I. Saito[1], W. Miyazaki[1], T. Masuzawa[1],
A.T. T. Koh[2], D. H. C. Chua[2], T. Yamada[3], N. Sano[4], K. Okano[1]

[1]Department of Material Science, International Christian University,
[2]Department of Materials Science and Engineering, National University of Singapore,
[3]Nanotube Research Centre, National Institute of Advanced Industrial Science and Technology,
[4]Department of Chemical Engineering, Kyoto University
3-10-2 Osawa, Mitaka
Tokyo, Japan
Tel/Fax: +81(0)422 33 3254, E-mail: kenokano@icu.ac.jp

Abstract

In this paper, we introduce an electro-chemical doping method of amorphous selenium (a-Se) using NaCl*aq.* Recently, an a-Se photovoltaic device fabricated using this method [1], has been announced and opened up the potential of a new impurity doping method. This study will further explore its possibilities by doping chlorine (Cl) and sodium (Na) and aim to fabricate a p-n junction by reversing the applied voltage during the electrolysis. The device is characterized through photoelectric measurements. The I-V characteristics show rectification under light illumination. (keywords: amorphous, electrolysis and photoconductivity)

Introduction

Studies of photovoltaic devices, especially for solar cell applications have reached a certain saturation point since efficiency and cost balance is at trade-off. Practically speaking, the efficiency of solar cells matters greatly due to the limited amount of space available for cell installation. On the other hand, devices that may be implemented onto formerly unavailable spaces, such as windows, can still be a valuable option.

This paper will introduce one simple yet maybe a revolutionary method of utilizing electrolysis of salt water for the fabrication of an amorphous selenium (a-Se) p-n diode. Although it is known that chlorine is introduced into a-Se films for the compensation hole traps introduced by the inclusion of arsenic by mixing a chloride compound into the evaporation source [2,3], to the best of the author's knowledge, these methods do not introduce chlorine into lattice of the a-Se network such that the material exhibits n-type conduction. Combined with the results in the former literature, this method would help inexpensive fabrication of transparent

photovoltaic devices using amorphous materials in the near future.

Experimental

A. Thin film fabrication

Figure 1(a) illustrates the film fabrication process. The substrate is made of borosilicate glass that has two thin Au films with a 1mm gap (Fig. 2) that function as an electrolysis electrode. Metallic selenium was charged on a molybdenum boat inside the evaporator. The turntable rotation speed was kept on 80 rpm. 200 nm of the a-Se film was deposited onto the substrate at 20 nm per second.

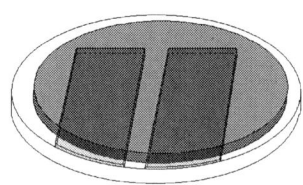

Figure 2. Illustration of an a-Se thin film, deposited on top of two Au electrodes on a glass plate.

B. Preparation of the electrolyte

29.0 g of Pure NaCl was dissolved into 75 cm^3 of deionized water at 27-30 °C to form saturated NaCl*aq.* in a glass beaker. The setup was kept at 27-30 °C using a heater and was monitored by a thermometer dipped into the electrolyte near the edge of the beaker.

C. Bi-directional electrolysis

Electrolysis of salt water (NaCl*aq.*) is performed aimed to dope Cl$^-$ ions and Na$^+$ ions into a-Se. The significance of this method is that both ions can be doped into the films by just switching the polarity of the electrolysis.

As illustrated in Fig. 1(b) using sodium chloride solution was used for the electrolyte, the Au films under the a-Se, and the Cu strip that are placed 4 cm away facing the a-Se film functions as the electrodes. Each Au electrode was named A and B for notation. Au electrode A was positively biased relative to the Cu electrode, where in this case, the Au electrode functions as the anode and the Cu strip as the cathode. The anode will attract negative ions, in this case Cl$^-$ and OH$^-$. Consequently, Au electrode B was negatively biased relative to the Cu electrode, which implies that this time, B functions as the cathode and Cu as the anode. The cathode attracts positive ions such as Na$^+$ and H$^+$.

As illustrated in Fig.3, the electric current that determines

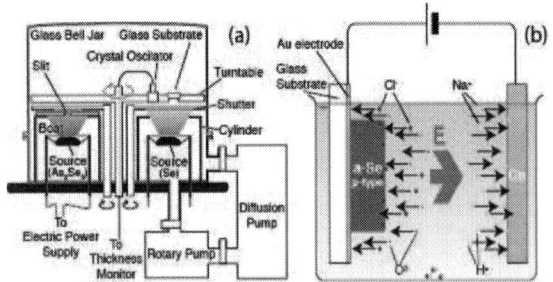

Figure 1. Schematic diagram of (a) rotational thermal evaporator for arsenic doping and (b) the electrolysis procedure for chlorine doping

978-1-4673-4840-9/13 $31.00 © 2013 IEEE

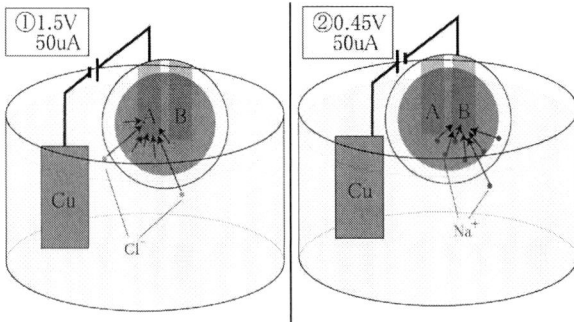

Figure 3. A schematic of (1) Cl and (2) Na electrolysis in NaCL*aq*.

the ion incorporation was kept at 50 μA for both Au(A)-Cu and Au(B)-Cu electrolysis where the bias was 1.5 V and 0.45 V respectively.

D. Time-of-flight Secondary Ion Mass Spectroscopy (ToF-SIMS)

The incorporation of the dopants were characterised by ToF-SIMS measurements. Note that a correlation sample cannot be prepared at this point, thus the relative signal intensity before and after the electrolysis is examined in order to qualitatively determine the depth profile of the impurities.

ION-TOF ToF-SIMS was used for the measurement. Ga was used for negative mode analysis in order to detect Cl ions, and Bi was used for positive mode to detect Na ions. Cs and Ar were used to etch the thin film respectively. The parameters for the analysis are shown in Table 1.

TABLE I
ToF SIMS measurement parameters

Polarity	Analysis parameters	Sputter Parameters
Negative	Ga / 25 keV/ 1.50 pA	Cs / 1 keV / 4.00 nA
	Area: 100 x 100 μm^2	Area: 200 x 200 μm^2
Positive	Bi / 25 eV / 0.80 pA	Ar / 1 keV / 16.00 nA
	Area: 100 x 100 μm^2	Area: 200 x 200 μm^2

E. Current–Voltage (I-V) characteristics and characterisation of the photocurrent.

Figure 4 is the schematic diagram of the I-V measurement

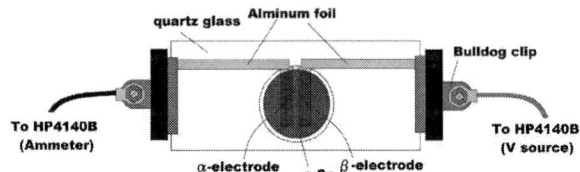

Figure 4. Top view of the I-V setup

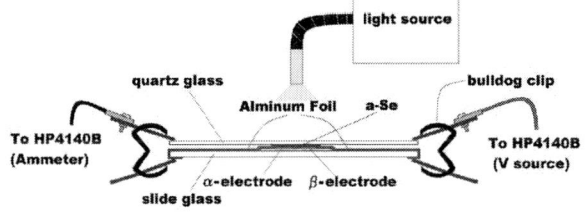

Figure 5. Side view of the I-V setup and the light source

setup. The a-Se sample is placed on a slide glass. Then an Al foil is placed onto the Au electrode of which are exposed out of the a-Se thin film as illustrated in the figure. A quartz slide glass of the same size is then placed on top of the setting to pressurize the Al film housing on to the electrode. The side view of the complete setup is shown on Fig. 5. Note that light is illuminated from the quartz glass side so that the whole spectrum passes through and reach the a-Se. The reason for the quartz glass is to reduce the amount of heat reaching the a-Se to prevent unnecessary degradation. HP4140B was used for the voltage source and the pico ammeter, and a Techno KTX-50R with a xenon lamp was used for the light source. The distance between the top of the quartz and the light guide was set to 3 cm, of which the luminance during the measurement was fixed at 2.6 x 10^4 ± 0.5 lx. For notation, the direction of the applied bias was from Au (A) towards Au (B).

Results and Discussions

Former studies show that electrolysis is effective for chlorine incorporation and the chlorine ions submerges for more than 60 % of the film thickness [1]. A diode that activates on UV light illumination was successfully fabricated by using an As incorporated film on one side and the chlorine incorporated layer on the other. In accordance to this result, a reverse biased electrolysis was performed in order to incorporate Na into the film. The ToF-SIMS results before and after the electrolysis is shows that the intensity of Na is approximately one and a half order of a magnitude higher after the electrolysis, which confirms that the ions have successfully incorporated into the film [1].

The I-V characteristics before and after the electrolysis has been measured but the linear characteristics that the non-doped a-Se has indicated [4] was preserved for the Na doped sample. The Cl doped sample showed the same results, which implies that a) Cl doped a-Se–Au(A) is ohmic and b) Na doped a-Se – Au(B) is also ohmic. It was found that the ratio between the photocurrent and the dark current has become smaller after the electrolysis. This is either because the amorphous network has been suppressed by the dopants (same effect as As$_2$Se$_3$ doped a-Se), or because it caused many defects. This is currently under investigation.

On the other hand, Cl doping using Au(A) and Na doping using Au(B) were performed sequentially in order to obtain a p-n or p-i-n diode of Cl doped and Na doped a-Se. The dark current for this dual doped a-Se was below detection limit, however the photocurrent showed rectifying characteristics. As it is written before, the I-V characteristics for the Cl doped a-Se and Na doped a-Se was linear, which eliminates the possibility of Schottky rectification. Non-linear I-V characteristics for the photocurrent have been reported in the past. In case of a-Se, the numerous densities of states in the wide bandgap traps the donor electrons and acceptor holes at the junction disabling it from forming a p-n (p-i-n) junction at thermal equilibrium. It is assumed that photo-assisted excitation triggers the donor and acceptor to excite to mobile states (*i.e.* conduction band and valence band) and form a p-n (p-i-n) junction. Although this assumption must be carefully investigated through activation energy analysis, however the rectification clearly indicates that a diode structure is formed within the film due to dual doping through electrolysis.

978-1-4673-4840-9/13 $31.00 © 2013 IEEE

Conclusion

The effect of Na doping using electrolysis has been confirmed through ToF-SIMS measurements. Sequential bi-directional doping of a-Se on Cl and Na was performed and a rectifying characteristics were confirmed under visible light illumination. Since Cl doped a-Se and Na doped a-Se itself indicated linear I-V characteristics, the diode characteristics is not Schottky Therefore it is speculated that a p-n or p-i-n a-Se diode has been fabricated by just switching the polarity of the electrolysis.

Acknowledgement

The present work is financially supported by MEXT-Support Program for the Strategic Research Foundation at Private Universities (S0801012) 2008-2012 and Grants-in-Aid for scientific research (21560053), both from the Ministry of Education, Culture, Sports, Science and Technology, Japan.

References

[1] I. Saito, et al., Appl. Phys. Lett. **98**, 152102, (2011)
[2] G. Belev, et al., J. Non-crys. Sol. **345&346**, 484 (2004)
[3] S.O. Kasap et al., J. Phys. D: Appl. Phys., **18**, 703 (1985)
[2] S.Touihri, et al., physica status solidi, **159**, p.569 (1997)

Investigation of work function and surface energy of aluminum: an ab-initio study

Shuguang Cheng[1], Cher Ming Tan[2], Tianqi Deng[1], Feifei He[2], Shuai Zhang[2] and Haibin Su[1]

1 Division of Materials Science, Nanyang Technological University, 50 Nanyang Avenue, Singapore 639798

2 School of Electrical & Electronic Engineering, Nanyang Technological University, 50 Nanyang Avenue, Singapore 639798

0065 83527639 sgcheng@ntu.edu.sg

Abstract

The work function and surface energy of aluminum with different orientations are investigated by employing the DFT simulation. We mainly focus on two situations: pure aluminum surface and aluminum surface with impurities. The numerical results indicate that the work function of Al (100) is larger than Al (110). With the introduction of the impurities (carbon atoms), the work function increases because of the extra electric dipoles on the surface. We also find that the surface energy of Al (100) is smaller than that of Al (110) indicating that Al (100) surface is more stable. When there are impurities on the surface, the surface energy decreases for silicon impurity and increases for calcium impurity. The magnitude of the increase is related to the orientation of the surface.

Keywords: DFT method, Atomic simulation, surface energy, work function and aluminum

Introduction

Metal is the most popular kind of material used in our daily life, and in most cases, we are dealing with materials usually terminated by surfaces, i.e. finite size samples. In a metallic bulk sample, since there is fundamental difference between atoms inside the sample and atoms on the surface, there are peculiar properties attributed by the material surfaces. The atoms arrangement on the surface is pivotal to the physical and chemical properties of materials, and many phenomena can occur at the interface between the sample and surrounding environment such as fatigue fracture, flaw and crack. Metal crack is always dangerous and unpredictable, and although the fatigue fracture rate could be reduced by several methods, such as adding rare earth elements, annealing etc., it is important also to estimate the remaining life of the parts before fracture occur after remanuafacturing of parts in a non- destructive manner.

Theoretical investigations of surface properties of metal are proved to be rather successful since we can perform surface modeling in the microscopic regime and provide fatigue sensitive information such as work function and surface energy[1,2]. In this work, we report the theoretical investigation of surface properties of aluminum. By employing the Density Function Theory (DFT) method [3], the work function and surface energy of aluminum is calculated under different conditions, specifically with different orientations and impurity. We mainly focus on two type of aluminum surface, namely Al (110) and Al (100). For the discussion of the influence of impurities, we use carbon for work function discussion and we use silicon and calcium for surface energy discussion. The results are explained from the viewpoint of microscopic physics and the mechanism is discussed as well [4].

Method, Simulation and discussion

The work function and surface energy of metal Aluminum is computed with the VASP (Vienna ab initio simulation package) program. By using the Perdew–Burke–Ernzerhof (PBE) functional for electronic exchange and correlation, the projector augmented wave method is used to represent the electron wave functions. The valence wave functions are expanded in a basis set consisting of plane waves. All plane waves up to a kinetic energy cut-off of 500 meV are included to guarantee the accuracy of calculation in the process of convergence. The system we considered possess the displacement invariance, thus a supercell is adopted in the numerical calculation. The convergence with respect to number of k-points in the Brillouin zone was checked. The parameters of the calculation are: E_{vacuum} =5.88 eV and E_{Fermi} = 1.60 eV. The size of the supercell for Al (100) in the plane is 14.3nm x14.3nm. For Al (110) in the plane the size of the supercell is 17.0x17.0nm.

With respect to the work function calculation, the valence bands are filled up to the Fermi energy. The electron distribution is described by the Fermi-Dirac distribution. The discrepancy between the Fermi energy and vacuum level gives the work function. Physically the work function is the minimum amount of energy needed to remove an electron from the metal to the vacuum. The work function of a metal surface is strongly related to the details of the surface. Boundary orientation, the presence of oxide layer at the surface or metal (aluminum for instance), the impurity atoms absorbed at the surface and the roughness of the metal surface can change the work function substantially [5].

The work function variation should be attributed to the formation of electric dipoles on the surface. With the electrons redistribution and the appearance of electric dipoles, the electric field is changed and the energy that an electron needs to leave the sample is varied accordingly. Since the work

TABLE I
Work function of aluminum (eV)

	Pure Al	With C impurities
(100)	4.27	4.31
(110)	3.92	3.97

TABLE II
Surface energy of aluminum (J/m^2)

	Pure Al	With Si impurities	With Ca impurities
(100)	0.95	0.805	0.965
(110)	1.02	0.938	1.034

function is sensitive to the surfaces details, its measurement can give valuable information of the surface, including the presence of vacancies, fatigue and micro-cracks. Hence it is possible to evaluate and predict the lifetime of metal parts through the time variation of the work function.

Fig. 1 and Fig. 2 show the electric potential energy along the direction perpendicular to the surface in pure aluminum (100) face without (Fig.1) and with C impurities (Fig.2). In the actual calculation, the supercell we adopt is $5 \times 5 \times 6$ layers plus vacuum with the height of 15Å. The abrupt chang from the oscillatory part to the flat part is the interface between solid and the vacuum. The peaks are the large values of potential in between atoms layers and the valleys are at the vicinity of atoms.

The work function can be obtained by calculating the discrepancy between the vacuum energy and the Fermi level we have set before. Our main results of work function are given in Table I. We can see that the work function of Al (100) is larger than that of Al (110). For Al (110), we use an orthorhombic unit cell and there are dipoles created by the spreading of electron and thus the work function is smaller. While for Al (100) surface, the orientation of high density electrons is almost un-modified and consequently a smaller reverse dipole is induced and a higher work function is accompanied [7].

As it is rather common to find impurities in the aluminum we consider carbon impurity as an example, and we calculate the work functions of Al (100) and (110) surfaces with a carbon atom in the center. We find that the work function increases for both Al (100) and (110) surfaces when there are impurities on the surface as shown in Table I.

Next we investigate the properties of surface energy. If we break a solid into two pieces, energy is needed and the total surface of the solid is increased. Thus the surface energy is defined as half of the energy needed to break a bulk solid into two [6]. Similar to work function, surface is related to the surface details of the material, such as boundary orientation, impurities and roughness. However, the calculation of surface energy is much more sensitive than that of work function. This is because the surface energy comes from the discrepancy between the total energy of two bulks and the energy of a single bulk. So the sample size we considered should be large enough to eliminate the boundary effect. Using the linear fitting method, the total energy of a bulk solid is calculated by adding extra N layers as follows:

$$E = \varepsilon V + 2\sigma S \qquad (1)$$

Here V is the bulk volume, ε is the bulk energy density, S is the cross-sectional areas and σ is the surface energy. We can rewrite Eqn (1) as $\sigma = E / 2S - \varepsilon Ndx / 2$.

By employing this method, the surface energy is obtained

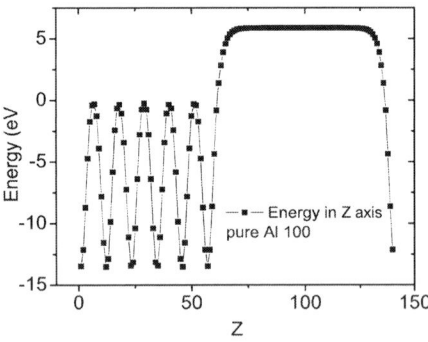

Fig. 1 Electric potential energy along z axis (z axis is the coordinate perpendicular to the surface) in pure Al 100. The interval from 6 Al layers in the bulk to the vacuum was plot with 210 points.

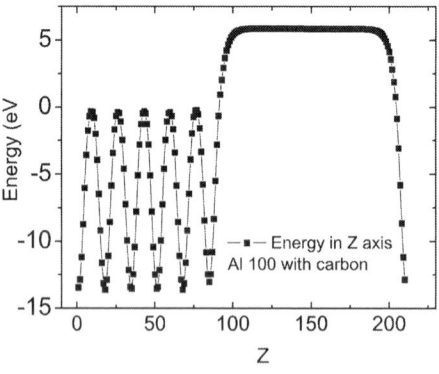

Fig. 2 Electric potential energy along z axis in Al 100 surface with carbon impurities.

978-1-4673-4840-9/13 $31.00 © 2013 IEEE

and the main results are shown in Table II. The numerical calculation of surface energy indicates that the surface energy of Al (100) is smaller than that of Al (110). As smaller surface energy indicates that the material is more stable, Al (100) is more stable than Al (110), i.e. Al (100) is more robust again surface contamination as compared to Al(110).

When there are impurities atoms on the surface, electron density of states (DOS) are changed accordingly. The surface energy is determined by the DOS of orbital s and p electrons at several atomic layers. The higher the DOS is, the lower the surface energy and vice versa. When the impurities are silicon, the surface energy is decreased in both cases (Al (100) and Al (110)). However, for Al (100), the surface energy is about 15 percent smaller than that of pure aluminum. For Al (110), the surface energy is only 8 percent smaller as compared to its pure counterpart. When the impurities are calcium, the situation is different. The surface energy increases slightly for both cases. The results indicate that the presence of different impurity can either make Al surface more stable or less.

Conclusion

The work function and the surface energy of aluminum are investigated by using the DFT method. We focus on the influence of aluminum orientation and impurities at the surface on the work function and surface energy. The work function of Al (100) is larger than the work function of Al (110). Also the work function increases when there are carbon impurities at the surface for both two aluminum surface orientations. On the other hand, the surface energy of Al (110) is larger than that of Al (100), indicating that Al (100) surface is more stable than Al (100) surface.

When there are impurities, we find the surface energy decreases when the impurity is silicon and the surface energy increases when the impurity is calcium. It indicates that absorption of silicon impurity at the aluminum surface makes the surface more stable.

References

[1] N. D. Lang and W. Kohn, *Theory of metal surfaces: Work function*, Phys. Rev. B 3 pp. 1215-1223, 1971.

[2] H. L. Skriver, N. M. Rosengaard, *Surface energy and work function of elemental metals*, Phys Rev. B, 46 pp. 7157-7168, 1992.

[3] J. P. Perdew, J. A. Chevary, S. H. Vosko, K. A. Jackson, M. R. Pedersen, D. J. Singh, C. Fiolhais, *Atoms, molecules, solids, and surfaces: Applications of the generalized gradient approximation for exchange and correlation*, Phys. Rev. B, 46, 6671, 1992

[4] B. Mutasa, D. Farkas, *Atomistic structure of high-index surfaces in metals and alloys*, Surface Science, 415(3): pp. 312−319. 1998

[5] C. J. Fall, N. Binggeli, and A. Baldereschi, *Anomaly in the anisotropy of the aluminum work function.* Phys. Rev. B 58, R7544-7547, 1998.

[6] F. Y. Zhang, Y. Y. Teng, M. X. Zhang, S. L. Zhu, *Density*

functional theory study of surface energiesof Al(001), (110) and (111), Corrosion Science and Protection Technology, 17(1): 47−49, 2005

[7] J. Schöchlin, K. P. Bohnen, K. M. Ho. Surface, *Structure and dynamical at the Al(111)-surface.* Surf Sci, 324(2/3): 113−121 1995.

Multicolored Cell Imaging with Bioconjugated Fluorescent Quantum Dots

YuchengWang[#], Rui Hu[#], Guimiao Lin[#], Ken-Tye Yong[#]

[#]School of Electrical and Electronic Engineering, Nanyang Technological University
Singapore 639798, Singapore
ywang2@e.ntu.edu.sg

Abstract

Photoluminescent quantum dots (QDs) have been extensively used in biomedical research for sensing, imaging and disease therapy. In this contribution, we demonstrate the use of bioconjugated cadmium telluride (CdTe) QDs as fluorescent optical probes for targeted in vitro cell imaging. A co-surfactant system of mercaptopropionic acid (MPA) and Cysteine (Cys) mixture were introduced for the synthesis of CdTe QDs in aqueous phase. The QDs as synthesized possess narrow emission profile and the colour is size tunable over the visible range. Multichannel imaging of cancer cells using different QD-conjugates were demonstrated. Specificity of cancer cell labelling was obtained by conjugating the QDs with bio-functional molecules such as folic acid and transferrin. This work suggests that bioconjugated CdTe QDs can be used for high-contrast in vitro cell imaging and cancer cell detection.

Key words: CdTe quantum dots, cancer cell labelling, multichannel imaging

Introduction

Application of semiconductor QDs in biological and biomedical fields has become an attractive research area during the past few decades. Due to their unique optical and chemical properties including bright and narrow fluorescence emission, sized-tunable colour, long lifetime, high resistance to photo-bleaching and relatively large surface area ready for functionalization [1], these nanosized materials have been intensively studied as promising alternatives to organic dyes and fluorescent proteins and to be used as luminous probes in biosensing, cellular imaging and potential novel applications such as multiplex/multimodal imaging, cellular tracking and in vivo tumor imaging/therapy [2-5]. So far, although II-VI QDs (such as CdSe, CdTe) with high crystallity and high quality, namely high fluorescence efficiency and narrow emission bandwidth, have been successfully synthesized by organometallic method [6], methods to fabricate QDs in aqueous phase have also drawn great attentions, since the dots obtained from this method are initially water-dispersible, ready for bioconjugation and thus can be directly applied in biological environment [7-9] without time-consuming phase transfer steps. In aqueous phase method, thiolated molecules such as mercaptopropionic acid (MPA) and thioglycolic acid (TGA) are frequently used as surfactants to stabilize QDs suspension in polar solvent, passivate and protect the particle surface [10]. More recently, there are also reports using amino acid Cysteine (Cys) as surfactant to prepare dots. In this case, the Cys-capped QDs display both carboxylic and amine functional groups on their surface for bioconjugation [11, 12]. However, unstable performance of Cys-capped QDs has been reported, such as loss of fluorescence and cytotoxic effects due to hydrolysis and oxidation of the Cys capping layer [13, 14].

In this paper, we report targeted cancer cell imaging using aqueous phase synthesized CdTe QDs as optical probes. According to our previous work, a mixed capping layer containing Cys and MPA shows great advantage over Cys-only cap in protecting the QDs from photo-degradation [15]. Therefore, in order to achieve a capping layer displaying carboxylic and amine functional groups, meanwhile, maintain the stable fluorescence of the QDs, a co-surfactant system containing both MPA and Cys was employed in the synthesis reaction. After purification and characterization, folic acid (FA) and transferrin (Tf) are linked to the surface of the as-prepared dots using carbodiimide chemistry method to produce QD-FA and QD-Tf conjugates. Considering the high binding affinity between FA/Tf and the folate receptors (FR)/transferrin receptors (TfR) overexpressed on many type cancer cell membrane [16, 17], these bio-functionalized QDs are investigated as potential optical probes for targeted cancer cell imaging. In vitro cell imaging study was performed using human pancreatic cancer cell (PANC-1). Application of aqueous phase CdTe QDs for targeted cancer cell labelling as well as the capability for multi-colour imaging was demonstrated by using fluorescent microscopy technique.

Experimental

Materials: Cadmium perchlorate hydrate ($Cd(ClO_4)_2 \cdot 6H_2O$), sodium borohydride ($NaBH_4$), tellurium (Te) powder (99.8%, 200 mesh), L-cysteine (99.5%, Cys), 3-mecaptopropionic acid (MPA), folic acid, apo-transferrin and1-ethyl-3-(3-dimethylaminopropyl)carbodiimide (EDC) were purchased from Sigma-Aldrich. All the chemicals are used as received. Ultrapure water with 18.2 MΩ/cm was used for sample preparation in all steps.

Preparation and Characterization of CdTe QDs: 48mg of tellurium powder and 30mg of sodium borohydride ($NaBH_4$) were mixed with 5ml nitrogen-saturated DI water to prepare precursor NaHTe. The mixture was stirred for 3~8 hours to until the final solution turned to colourless or light pink. 672 mg $Cd(ClO_4)_2 \cdot 6H_2O$, Cys and MPA ($n_{Cys\&MPA} : n_{Cd}=2$, $n_{Cys} : n_{MPA}=1:3$) were mixed with 100ml DI water to prepare Cd precursor. pH value of the mixture was adjusted to 12 by dropwise titration with 2M NaOH solution. The synthesis reaction was carried out in a three-necked flask. After injection of fresh NaHTe, the reaction solution was vigorously stirred and heated to 100°C. It took 10min~15h to obtain QDs with different emission wavelength, varying from 500 to 700nm. The resulted QDs were purified by addition of ethanol and centrifugation (6000rpm, 10min) to remove unreacted Cd precursor, excess surfactants and by products. After three times wash with ethanol, the precipitates were re-

978-1-4673-4840-9/13 $31.00 © 2013 IEEE

dispersed in DI water for further characterization and in vitro cell experiments. UV-visible absorption spectra were obtained from a spectrophotometer (Shimadzu UV-2450). Photoluminescence (PL) spectra were measured at room temperature using a spectrofluorometer (Fluorolog-3 spectrofluorometer) with excitation of 470nm.

Bioconjugation: In order to prepare optical probes for targeted cancer cell imaging, the CdTe QDs capped with Cys and MPA were conjugated with FA or Tf using EDC as coupling reagent. As for QD-FA conjugates, 800µl solution of FA (0.5mg/ml, in DMSO) and 200µl EDC solution (5mg/ml, in water) were mixed for 1h at room temperature. 100µl QD stock solution (5mg/ml, in water) was then added and the reaction was further incubated for 3h. After that, the conjugates were separated from the mixture by centrifugation (8000rpm for 5min) and redispersed in PBS (pH=7.4) for cellular experiments. In case of QD-Tf preparation, 100µl QD stock solution was mixed with 150ul EDC solution (5mg/ml, in water) for 2mins. After that, 250µl Tf solution (2mg/ml, in PBS) was added and the mixture was stirred for 3h at room temperature. Purification by centrifugation (6000rpm for 5min) was performed and the precipitates were redispersed in PBS for use.

Cell Culture and Live Cell Imaging: Human pancreatic cancer cells, PANC-1 (CRL-1469, American Type Culture Collection), were cultured in Dulbecco's Modified Eagle's Medium (DMEM, Hyclone) supplemented with 10% fetal bovine serum (FBS,Hyclone), 100 µg/mL penicillin (Gibco) and 100 µg/mL streptomycin (Gibco) at 37 °C in a humidified atmosphere containing 5% CO_2. After being seeded and cultured for 24 hours, the PANC-1 cells were gently washed with PBS and treated with the QD-FA/Tf conjugates (50ppm in DMEM) for 2h at 37°C with 5% CO_2. Subsequently, the cells were washed with PBS three times before imaging.

Result and Discussion

CdTe QDs with tunable emission wavelength ranging from 500 to 700nm were successfully synthesized by using a co-surfactant system containing surfactant Cys and MPA. Figure.1 shows the absorption and photoluminescence profiles of the QDs with different colours. These profiles suggest that the colour of the dots can be tuned by simply manipulating the time of particle growth, and narrow emission of the dots can be excited over a wide range of absorption wavelength.

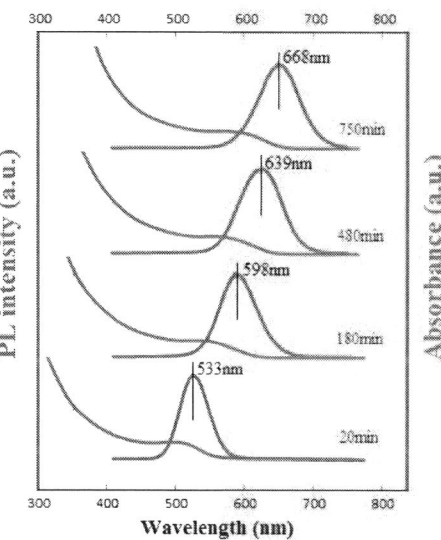

Fig.1 Spectral profile of the as-prepared QDs: absorption (blue) and PL emission (red).

Since the CdTe QDs were synthesized in a co-surfactant system containing Cys and MPA, a mixed capping layer of Cys and MPA was formed on the as-prepared dots due to the coordination between the thiol groups and surface Cd atoms. In this case, both carboxylic and amine groups are available on QDs surface to couple with biomolecules. Considering FR and TfR are found to be overexpressed in most type of cancer cells, in this work, the QDs were functionalized with FA and TF, which act as recognition moieties to target the QDs for human pancreatic cancer cell imaging. Figure.2 is the schematic illustration of the QD-FA/Tf coupling process, where crosslink agent EDC was introduced to catalyze condensation reaction as commonly used. Specifically, in case of QD-Tf preparation, stable amide bond was formed between carboxylic groups on QDs surface and amine groups of Tf. As for QD-FA coupling, the amine groups on the surface of QDs can be directly utilized to link with FA. Attribute to the presence of Cys on QD surface, usage of linker molecules, such as 2,20-(ethylenedioxy)-bis-ethylamine [18], can be saved in preparing QD-FA conjugates.

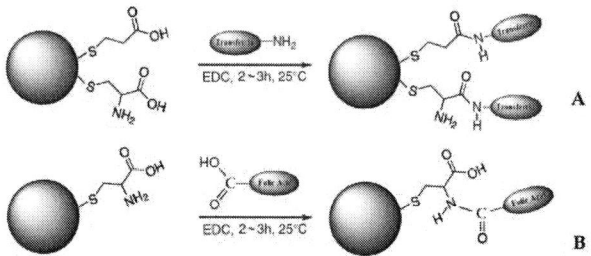

Fig. 2 Crosslink CdTe QDs with (A)FA and (B)Tf using EDC as coupling agent.

For cell imaging study, mono- and multi-colour imaging of PANC-1 cells were demonstrated. Figure.3 shows the fluorescent images of PANC-1 cells treated with QD_{533}-FA or QD_{639}-Tf conjugates. As can be seem from these images, robust optical labelling of the cancer can be achieved by using the bio-functionalized QDs. Due to the high binding affinity between FA/Tf and their receptors on cancer cell membrane, the dots are expected to be captured on the membrane or internalized via receptor-mediated uptake [19, 20]. However, results from confocal microscope are required to confirm the cellular uptake. Furthermore, in order to confirm the QD conjugates specifically stain the cancer cell, cancer cells were incubated with non-conjugated QDs in control experiments. As expected, minimal staining of the cells was observed, which demonstrate the specificity of cancer cell labelling using QD-FA and QD-Tf conjugates as optical probes.

We also demonstrate the feasibility of multiplex cellular labelling using QDs with different emission colours. Figure.4 shows PANC-1 cells stained with both 533nm- and 639nm-emitting QDs (QD_{533} and QD_{639}), which are coupled with FA and Tf respectively. Due to the narrow emission bandwidth, these two populations of QDs are well spectrally separated and thus can be differentiated by filter system easily. Based on these results, we can foresee the application of bio-functionalized CdTe QDs for multiple cellular targets labelling.

Fig. 3 Imaging of live pancreatic cancer cells treated with (A) folic acid-conjugated 533nm-emitting CdTe QDs (QD_{533}), (B) non-conjugated QD_{533}, (C) transferrin-conjugated QD_{639} and (D) non-conjugated QD_{639}. Left panel correspond to transmission images, fluorescence images are shown in the middle, and the overlay images of transmission and fluorescence are on the right.

Fig. 4 Image of live pancreatic cancer cells treated with QD_{533}-FA and QD_{639}-Tf. Left images correspond to transmission image, overlay image of the luminescence from QD_{553} (green) and QD_{639} (red) is shown in the middle and the right one is the overlay of fluorescence and transmission.

Conclusion

In summary, we have demonstrated the synthesis of aqueous CdTe QDs using a co-surfactant system containing Cys and MPA. The obtained QDs are stable in fluorescence and display both carboxylic and amine ends on their surface to link with biomolecules, therefore shows advantages in the procedure of bio-functionalization. QDs were coupled with FA and Tf via EDC catalyzed reaction and, specific pancreatic cancer cell labelling was successfully achieved using these bio-functionalized QDs. Multiplex imaging was also demonstrated, indicating the dots prepared in our co-surfactant system have the potential for multiple cellular targets labelling.

Acknowledgements

The study was supported by the Start-up grant (M4080141.040) from Nanyang Technological University, and Tier1 Academic Research Funds (M4010360.040 RG29/10) from Singapore Ministry of Education and partially from the Singapore Ministry of Education under Tier 2 Research Grant MOE2010-T2-2-010 (4020020.040 ARC2/11).

Reference

[1] I. L. Medintz, H. T. Uyeda, E. R. Goldman, and H. Mattoussi, "Quantum dot bioconjugates for imaging, labelling and sensing," *Nature Materials,* vol. 4, pp. 435-446, Jun 2005.

[2] S. Kim, Y. T. Lim, E. G. Soltesz, A. M. De Grand, J. Lee, A. Nakayama, J. A. Parker, T. Mihaljevic, R. G. Laurence, D. M. Dor, L. H. Cohn, M. G. Bawendi, and J. V. Frangioni, "Near-infrared fluorescent type II quantum dots for sentinel lymph node mapping," *Nature Biotechnology,* vol. 22, pp. 93-97, Jan 2004.

[3] D. R. Larson, W. R. Zipfel, R. M. Williams, S. W. Clark, M. P. Bruchez, F. W. Wise, and W. W. Webb, "Water-soluble quantum dots for multiphoton fluorescence imaging in vivo," *Science,* vol. 300, pp. 1434-1436, May 2003.

[4] I. L. Medintz, A. R. Clapp, H. Mattoussi, E. R. Goldman, B. Fisher, and J. M. Mauro, "Self-assembled nanoscale biosensors based on quantum dot FRET donors," *Nature Materials,* vol. 2, pp. 630-638, Sep 2003.

[5] J. K. Jaiswal, H. Mattoussi, J. M. Mauro, and S. M. Simon, "Long-term multiple color imaging of live cells using quantum dot bioconjugates," *Nature Biotechnology,* vol. 21, pp. 47-51, Jan 2003.

[6] C. B. Murray, D. J. Norris, and M. G. Bawendi, "SYNTHESIS AND CHARACTERIZATION OF NEARLY MONODISPERSE CDE (E = S, SE, TE) SEMICONDUCTOR NANOCRYSTALLITES," *Journal of the American Chemical Society,* vol. 115, pp. 8706-8715, Sep 1993.

[7] N. Gaponik, D. V. Talapin, A. L. Rogach, K. Hoppe, E. V. Shevchenko, A. Kornowski, A. Eychmuller, and H. Weller, "Thiol-capping of CdTe nanocrystals: An alternative to organometallic synthetic routes," *Journal of Physical Chemistry B,* vol. 106, pp. 7177-7185, Jul 2002.

[8] A. M. Smith, H. W. Duan, A. M. Mohs, and S. M. Nie, "Bioconjugated quantum dots for in vivo molecular and cellular imaging," *Advanced Drug Delivery Reviews,* vol. 60, pp. 1226-1240, Aug 2008.

[9] Y. G. Zheng, S. J. Gao, and J. Y. Ying, "Synthesis and cell-imaging applications of glutathione-capped CdTe quantum dots," *Advanced Materials,* vol. 19, pp. 376-+, Feb 2007.

[10] W. B. Cai, A. R. Hsu, Z. B. Li, and X. Y. Chen, "Are quantum dots ready for in vivo imaging in human subjects?," *Nanoscale Research Letters,* vol. 2, pp. 265-281, Jun 2007.

[11] N. N. Mamedova, N. A. Kotov, A. L. Rogach, and J. Studer, "Albumin-CdTe nanoparticle bioconjugates: Preparation, structure, and interunit energy transfer with antenna effect," *Nano Letters,* vol. 1, pp. 281-286, Jun 2001.

[12] W. C. Law, K. T. Yong, I. Roy, H. Ding, R. Hu, W. W. Zhao, and P. N. Prasad, "Aqueous-Phase Synthesis of Highly Luminescent CdTe/ZnTe Core/Shell Quantum Dots Optimized for Targeted Bioimaging," *Small,* vol. 5, pp. 1302-1310, Jun 2009.

[13] H. Zhang, P. Sun, C. Liu, H. Gao, L. Xu, J. Fang, M. Wang, J. Liu, and S. Xu, "L-Cysteine capped CdTe-CdS core-shell quantum dots: preparation, characterization and immuno-labeling of HeLa cells," *Luminescence,* vol. 26, pp. 86-92, Mar-Apr 2011.

[14] C. Kirchner, T. Liedl, S. Kudera, T. Pellegrino, A. M. Javier, H. E. Gaub, S. Stolzle, N. Fertig, and W. J. Parak, "Cytotoxicity of colloidal CdSe and CdSe/ZnS nanoparticles," *Nano Letters,* vol. 5, pp. 331-338, Feb 2005.

[15] R. H. Yucheng Wang, Guimiao Lin, Indrajit Roy, Ken-Tye Yong, "Optimizing the Aqueous Phase Synthesis of CdTe Quantum Dots using Mixed-ligands System for Bioimaging Applications," *unpublished.*

[16] R. J. Lee and P. S. Low, "FOLATE-MEDIATED TUMOR-CELL TARGETING OF LIPOSOME-ENTRAPPED DOXORUBICIN IN-VITRO," *Biochimica Et Biophysica Acta-Biomembranes,* vol. 1233, pp. 134-144, Feb 1995.

[17] F. Erogbogbo, K. T. Yong, I. Roy, G. X. Xu, P. N. Prasad, and M. T. Swihart, "Biocompatible luminescent silicon quantum dots for imaging of cancer cells," *Acs Nano,* vol. 2, pp. 873-878, May 2008.

[18] M. Geszke, M. Murias, L. Balan, G. Medjandi, J. Korczynski, M. Moritz, J. Lulek, and R. Schneider, "Folic acid-conjugated core/shell ZnS:Mn/ZnS quantum dots as targeted probes for two photon fluorescence imaging of cancer cells," *Acta Biomaterialia,* vol. 7, pp. 1327-1338, Mar 2011.

[19] D. S. Lidke, P. Nagy, R. Heintzmann, D. J. Arndt-Jovin, J. N. Post, H. E. Grecco, E. A. Jares-Erijman, and T. M. Jovin, "Quantum dot ligands provide new insights into erbB/HER receptor-mediated signal transduction," *Nature Biotechnology,* vol. 22, pp. 198-203, Feb 2004.

[20] A. M. Derfus, W. C. W. Chan, and S. N. Bhatia, "Intracellular delivery of quantum dots for live cell labeling and organelle tracking," *Advanced Materials,* vol. 16, pp. 961-+, Jun 2004.

Nano-IGZO layer for EGFET in pH sensing characteristics

Chia-Ming Yang[1,2,3,4], Jer-Chyi Wang[1,2,3], Tzu-Wen Chiang[1], Yi-Ting Lin[1], Teng-Wei Juan[1], Tsung-Cheng Chen[1], Ming-Yang Shih[1], Cheng-En Lue[4], Chao-Sung Lai[1,2,3,*]

[1] Department of Electronic Engineering, Chang Gung University, Taoyuan, Taiwan
Phone: +886-3-2118800 ext. 5960 E-mail: cslai@mail.cgu.edu.tw
[2] Healthy Aging Research Center, Chang Gung University, Taoyuan, Taiwan
[3] Center for Biomedical Engineering, Chang Gung University, Taoyuan, Taiwan
[4] Department of Device Engineering, Inotera Memories Inc., Taoyuan, Taiwan

Abstract

In-Ga-Zn-O (IGZO) was widely applied in the substrate of TFT to replace a-Si in recent year. In this study, IGZO layer with thickness of 70 nm is first proposed as a pH sensing membrane directly on P-type Si substrate acting as an extended gate of conventional extended-gate field-effect Transistor (EGFET). Material criteria of extended gate electrode are low resistance and high capacitance. Therefore, Ar/O_2 ratio was modified in the rf sputtering with IGZO target. Post deposition anneal was also performed to check the sheet resistance and pH sensing performance. EGFETs were measured in standard pH buffer solution by using B1500A and constant voltage constant current (CVCC) circuit. Similar I_{DS}-V_{GS} curves including transconductance (G_m) and substrate swing (S.S.) are obtained in various sputtering conditions of IGZO compared to commercial NMOSFET in CD4007. pH application range is only between pH 2 to pH 10. IGZO-EGFET prepared by Ar/O_2 ambience of 24/1 in sputtering can have a sensitivity of 59.5 mV/pH. Lower sensitivity and linearity can be observed in the samples with RTA treament at higher temp and in O_2 ambience. N_2 anneal at 500°C can be used to improve pH sensing performance for IGZO-EGFET prepared by Ar/O_2 ambience of 20/5 in sputtering. Nano-IGZO layer is verified to be the sensing membrane in EGFET to have a high sensitivity of 59.5 mV/pH for the first time. More studies on enlargement pH application range and minimization of non-ideal effect still need to be investigated.

Introduction

Amorphous indium-gallium-zinc oxide (α-IGZO) was firstly proposed as the channel material of thin-film transistors (TFTs) on polyethylene terephthalate (PET) substrate with a Hall mobility exceeding 10 $cm^2V^{-1}S^{-1}$. [1] It can be a candidate to replace amorphous silicon (α-Si) in TFT applications with advantages of high mobility, wide band gap and high transmittance rate on flexible substrate. [2-3] In the meantime, α-Si was proposed as the sensing material of ion-sensitive field-effect transistor (ISFET) [4-5] with narrow application range and high drift. Another field effect sensor platform which called extended-gate field-effect transistor (EGFET) was derived from ISFET, and its configuration is composited of a sensor electrode and a conventional metal-oxide-semiconductor field-effect transistor (MOSFET). The sensor electrode is connected on the gate of MOSFET and extended from MOSFET by a metal signal line, which is an easy and reliable method in encapsulation due to most electrical signals are away from the measurement environment. Therefore, interference from the environment including light-induced drift on ISFET can be also minimized with easy process and low cost [6]. Although α-Si can be used as the material for the ISFET, the sensing performance of α-Si can be performed in pH 1 to 7, which is not accetable for real applications. To study the pH sensing properties of IGZO, a systematic investigation of preparation of IGZO eletrode by rf reactive sputtering is carried out based on EGFET and corresponding measurement platform.

Experimental

IGZO electrode based on the electrolyte-insulator-semiconductor (EIS) structures with a single-layer sensing membrane were fabricated as the extended gate of EGFETs to investigate hydrogen ion sensing properties. IGZO layer was deposited by radio frequency (rf) sputtering directly on p-type (100) silicon wafer after standard RCA cleaning. The IGZO target with atomic ratio of 1:1:1:4 and 99.9% purity was used in reactive rf sputtering with the power at 150 W. The flow rate of Ar/O_2 was set as 25/0, 24/1 and 20/5 in sccm, respectively. The thickness of IGZO sensing membrane was controlled to 70 nm by deposition time and then verified by ellipsometer measurement. To improve the sensing performance, a post deposition anneal was performed in N_2 and O_2 ambient at temperature of 500, 700, and 900°C for 1 min for the sample with Ar/O_2 ratio of 20/5, respectively. Backside contact of IGZO electrode was deposited by Al evaporation with thickness of 300 nm. SU8-2005 photo-sensitive epoxy was used to define sensing area. The encapsulated electrode was attached on a printed-circuit board (PCB) by using a silver paste and a hand-made epoxy was used to keep Cu line from electrolyte in measurements. Detail process flows of all groups are shown in Fig. 1(a). A commercial product of n-MOSFET in CD4007 was chosen as the FET device of the EGFET, and the gate electrode of nMOSFET was connected to IGZO electrode. pH sensitivities of sensing membranes were all extracted by drain-to-source current - gate-to-source voltage (I_{DS}-V_{GS}) curves of EGFET measured in various pH buffer solution of Merck Inc. through Ag/AgCl reference electrode by B1500A semiconductor parameter analyzer. Schematic of measurement setup is shown in Fig. 1(b). A constant voltage constant current circuit was used to monitor the time-dependent output voltage change with fixed I_{DS} of 1 μA. [7]

978-1-4673-4840-9/13 $31.00 © 2013 IEEE

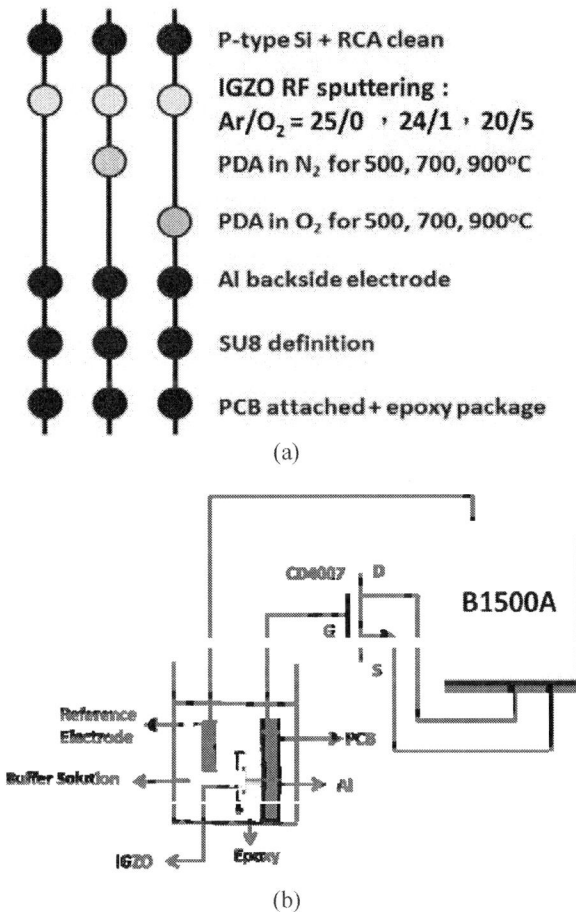

(a)

(b)

Fig. 1 (a) Process flow of IGZO electrode, and (b) measurement setup of IGZO-EGFET.

Results and Discussion

As shown in the Fig. 2, I_{DS}-V_{GS} curves of NMOSFET in CD4007 and EGFETs with IGZO-electrode prepared by different Ar/O_2 ratio had similar behaviors. A NMOSFET had lower threshold voltage (V_T) and EGFETs with more O_2 flow had higher V_T which could be explained by more O composition and more dielectric-like properties. The coupling capacitance of extended gate can be higher and then higher V_T is presented. However this V_T shift will not affect the pH sensing performance. Fig. 2 shows a typical pH-dependent response of I_{DS}-V_{GS} curves of IGZO-EGFET prepared by Ar/O_2 of 24/1. V_{GS} in same I_{DS} increases with pH increases, which means the hydrogen ions binding on the IGZO surface makes the surface potential difference and corresponding V_T shift. Output voltage (V_{out}) is calculated by interpolation of I_{DS}-V_{GS} curves for I_{DS} = 1 uA as shown in the inset of Fig. 2. Sensitivity and linearity can be calculated by linear fitting between V_{out} and corresponding pH of buffer solution used in measurement. Fig. 4 shows the pH-dependent V_{out} response for IGZO-EGFET prepared by different Ar/O_2 ratio. Ideal pH response is 59.6 mV/pH from Nernst equation as shown in Fig. 3. As shown in Fig. 4, higher or lower Ar/O_2 flow makes lower sensitivity and non-linear response, especially in basic buffer solution. The mechanism is still waiting more material analysis

to prove including XPS and AFM. Calculated sensitivity and linearity of IGZO-EGFET of different Ar/O_2 ratio are shown in Fig. 5. The highest sensitivity of 59.5 mV/pH and linearity of 99.7% are observed in the group with Ar/O_2 ratio of 24/1 by a linear fitting from pH 2 to 10. However, sensitivity and linearity are both decreased with O_2 increase and decrease. Because the low sensitivity and linearity in the group with Ar/O_2 ratio of 20/5, additional post deposition anneal was used to improve the pH sensing performance. As shown in Fig. 6, sensitivity and linearity can be increased by anneal, but too high temperature makes lower sensitivity and linearity. Although sensitivity and linearity were both improved, final performance is still far from real applications. In addition, time-dependent output response of IGZO-EGFET with Ar/O_2 ratio of 24/1 for pH 4 and 6 are shown in Fig. 7. Drift coefficient is calculated by linear fitting V_{out} from 120 to 300 min. Drift coefficient of IGZO-EGFET with Ar/O_2 ratio of 24/1 are 0.75 and 0.21 for pH 4 and 6, which is comparable to other sensing material. Since the pH sensing performance can be changed by minor adjustments in rf sputtering process and RTA didn't help to improve to an acceptable criterion, further investigation on material analysis for mechanism study is also suggested.

Conclusions

IGZO layer is verified to be the sensing membrane in EGFET to have a high sensitivity of 59.5 mV/pH for the first time. IGZO electrodes prepared by Ar/O_2 ambience of 24/1 in sputtering can have a sensitivity of 59.5 mV/pH. However, sensitivity and linearity could decrease a lot even only a minor modification of Ar/O_2 ratio. With RTA treatment, lower sensitivity and linearity can be observed in the samples with higher temp and O_2 ambience. Drift coefficient of IGZO-EGFET with Ar/O_2 ratio of 24/1 are 0.75 and 0.21 for pH 4 and 6, respectively. More studies on enlargement pH application range and process stability still need to be investigated.

Acknowledgements

This work is supported by National Science Concil of the Republic of China under contract. (NSC 98-2221-E-182 -057 -MY3 and NSC 101-2218-E-182-004)

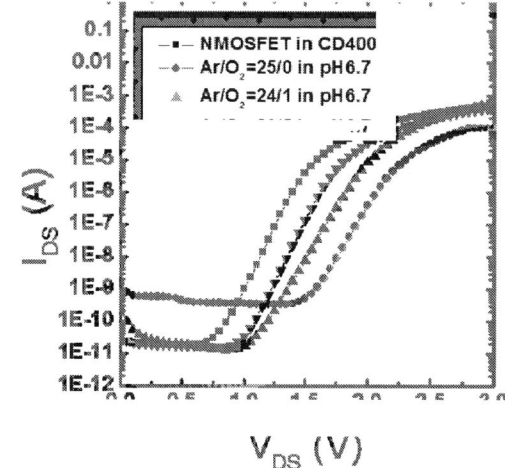

Fig. 2 I_{DS}-V_{GS} curves of NMOSFET and EGFETs with IGZO-electrode with different Ar/O_2 ratio tested in pH 6.7.

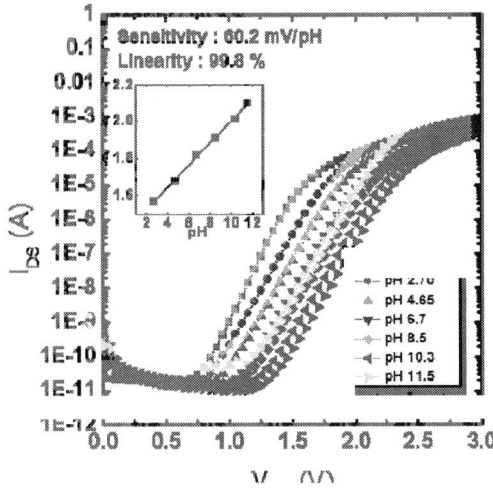

Fig. 3 I_{DS}-V_{GS} curves of EGFETs with IGZO-electrodes with Ar/O$_2$ ratio of 24/1 measured in various pH buffer solutions.

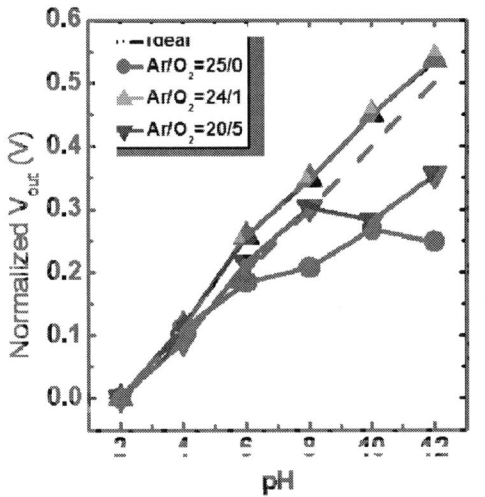

Fig. 4 pH dependent output voltage of EGFETs with IGZO-electrodes with different Ar/O$_2$ ratio.

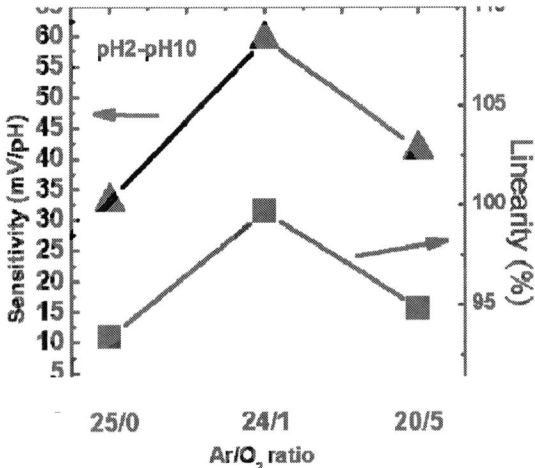

Fig. 5 The calculated pH sensitivity and linearity of EGFETs with IGZO-electrodes with different Ar/O$_2$ ratio.

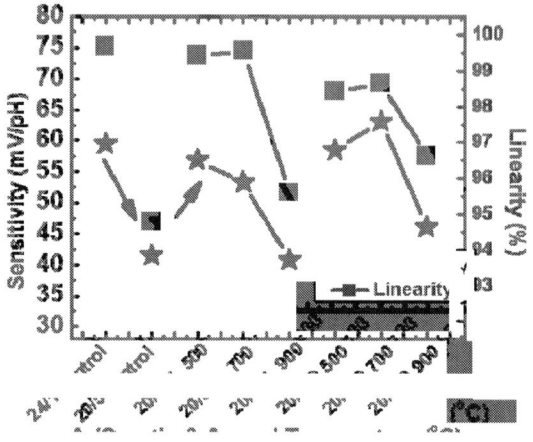

Fig. 6 The calculated pH sensitivity and linearity of EGFETs with IGZO-electrodes with Ar/O$_2$ ratio of 24/1 and 20/5 annealed with N$_2$ and O$_2$ ambience at different temperature.

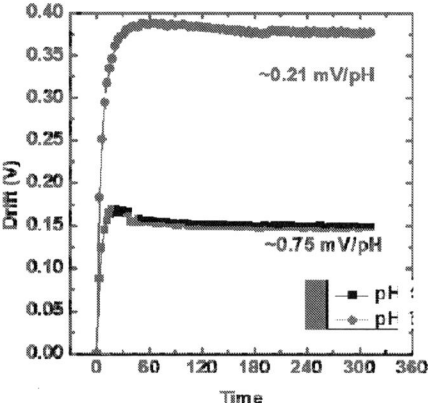

Fig. 7 Time-dependent output voltage of EGFET with IGZO-electrode with Ar/O$_2$ ratio of 24/1 measured in pH 4.

References

[1] K. Nomura, H. Ohta, A. Takagi, T. Kamiya, M. Hirano, and H. Hosono, "Room-temperature fabrication of transparent flexible thin-film transistors using amorphous oxide semiconductors," *Nature*, vol . 432, 2004, pp. 488-492

[2] H. Yabuta, M. Sano, K. Abe, T. Aiba, T. Den, and H. Kumoni, "High-mobility thin film transistor with amorphous InGaZnO$_4$ channel fabricated by room temperature rf-magnetron sputtering," *Appl. Phys. Lett.*, vol. 89, 2006, pp. 112123-1-3.

[3] K. Nomura, A. Takagi, T. Kamiya, H. Ohta, and H. Hosono, "Amorphous oxide semiconductors for high-performance flexible thin-film transistors," *Jpn. J. Appl. Phys.*, vol. 45, 2006, pp. 4303-4308.

[4] J.-C. Chou, Y.-F. Wang and J.-S. Lin, "Temperature effect of a-Si:H pH-ISFET", *Sens. Actuators B*, vol. 62, 2000, pp. 92-96.

[5] J.-C. Chou, Y.-F. Wang and J.-S. Lin, "Study and simulation of the drift behavior of hydrogenated amorphous silicon gate pH-ISFET", *Sens. Actuators B*, vol. 62, 2000, pp. 97-101.

[6] J.-C. Chen, J.-C. Chou, T.-P. Sun, and S.-K. Hsiung, "Portable urea biosensor based on the extended-gate field effect transistor sensor," *Sens. Actuators B*, vol. 91, 2003, pp.180-186.

[7] C.-S. Lai, T.-F. Lu, C.-M. Yang, Y.-C. Lin, D. G. Pijanowska, and B. Jaroszewicz, "Body effect minimization using single layer structure for pH-ISFET applications," *Sens. Actuators B*, vol. 143, 2010, pp.494-499.

The Side Effects on N-Type FinFET Devices

Hsin-Chia Yang[*,1], Chong-Kuan Du[1], Wen-Shiang Liao[1], Jing-Zong Jhang[1], Yi-Hong Lee[1], Tsao-Yeh Chen[1], Ko-Fan Liao[1], Mu-Chun Wang[1], Sungching Chi[1], Shea-Jue Wang[2]

[1]Dept. of Electronic Engineering, Minghsin University of Science and Technology, Taiwan

[2]Dept. of Materials and Resources Engineering, National Taipei University of Technology, Taipei, Taiwan

*Corresponding email: hsinchia_yang@yahoo.com.tw

Abstract

Fin-FET is so expected because it protects I_{off} current from outrageously leaky as the channel length gets shorten continuously. It thus keeps the threshold voltage and the swing from rolling-up. Those good characteristics are manifested by the fully depleted region and the lack of leaky body as the gate is biased. In this study, the fin-thickness effect is to be noticeably discussed. The correlated swings are to be determined and compared between the two kinds. The P-well Vt adjustment at two different energies, 10KeV and 6KeV, are also put into split. (Keywords: FinFET Devices, 3-D Fin Structure, Leakage current, Over Exposure, Over Etching)

Introduction

Of all the options for developing the next-generation devices, 3-D fin-like FinFET ones are thought to be noticeably distinguishing. It is designed be thin fin structure with gate poly-silicon crossing over as Gate [1-3]. The common but unique foundation of fins is silicon on insulator (SOI) making the fin-like channels stand individually. Those channels are thus fully depleted as the gates are biased. The two ends of the channels are assigned to be Sources and Drains. To make them so, selective dry etching techniques are imposed followed by the grown oxide of intrinsically excellent conformality and then the poly-silicon chemical deposition of good step coverage. Somehow the effective widths of the devices may be large enough depending on how high the aspect-ratio is, while the channel length is to be determined relying on the gate poly-silicon etching. One thus counts on how reliable the processing is and how neat the design of the FinFET structure is. Nevertheless, the body-free structure protects the devices from being leaky resulting in the possibility of lower I_{off} current even in the deep sub-nanometer regime.

In this paper, a <100> p-type Silicon on Insulator (SOI) wafer is used for fabrication of FinFET devices. Channel length and width less than 90nm might even be achieved by using 90nm photo mask with over-exposure and over-etching technique. Fin thickness (channel width) effects are taken into account and the differences of electric characteristics at different implant energies for the adjustment of threshold voltage are paid attentions to. The threshold voltage [5-6], and swing [7] are to be compared.

II. Experiments and Preparation of FinFET Devices

An SOI wafer with the structure of silicon on insulator is put to Furnace to grow thin sacrificing oxide followed by the dispensed photo resist (PR). The exposure of PR to ultra-violet rays define the "I" shape devices. After the development, there leave I's where the beneath silicon gets protected from being dug out as a sequence of ion dry etching processes are used. The high aspect-ratio thin fins can be constructed and get p-typified by P-well Vt-adjustment implants at different energies for comparison. A layer of high quality 14-angstrom nitrided oxide as gate dielectric is then grown followed by 150nm heavy-doped poly-silicon as Gate, as shown in Fig.1.

III. Results and Discussions

The characteristic curves of I_D-V_D of NFinFET devices are so conclusive for the two different channel width/length sizes, (W/L=0.11μm/0.1μm, 0.12μm/0.1 μm), as shown in Fig. 2 (A) at 10KeV for P-well Vt implant. The channel width (the fin thickness), 0.11 micron, always gives much better I_D at I_G=1.0V over 80% than the one of channel 0.12 micron. The superior electric performances stand at 6KeV for P-well Vt implant as shown in Fig.3 (A). As referred to different Vt implant energy, I_D-V_D of Fig. 3 (A) is superior to that of Fig. 2 (A) because of less implant energy taking boron ions to shallower extent from the surface of the fins such that boron ions get out-diffused much more easier during high temperature activation. The resulted threshold voltage is thus lowered down making the I_D higher in the case of lower implant energy just as expected.

Moreover, Swing is determined according to the following formula:

$$S \equiv \frac{dV_G}{d \log I_D}\bigg|_{subthreshol}$$ (1)

and found to be 80mV/decade and 110mV/decade corresponding to the slopes of I_D-V_G in sub-threshold regions in Fig.2 (B) and Fig.3 (B). In addition, I_{off} (V_D=0.05V)= 1.23×10^{-6}A in Fig.2 (B) and I_{off} (V_D=0.05V)= 2.28×10^{-6}A in Fig.3 (B) both give satisfactory leakage currents at V_G=0.0V.

The maximum trans-conductance g_{m_max} is 3×10^{-6} (A/V) in Fig.4 (B) which is superior to 1×10^{-7} (A/V) in Fig.5 (B). Therefore, it is acceptable that the nFinFET devices with channel width/length (0.11μm/0.1μm) using P-well Vt adjustment implant at 10KeV are comparably promising.

In addition, many of them have negative linear threshold voltages. That is very reminding, and one shall immediately go back to the long lasting model, in which the threshold voltage is presented as follows:

978-1-4673-4840-9/13 $31.00 © 2013 IEEE

$$V_{th} = [-0.56 - \frac{k_B T}{e}\ln(\frac{p}{n_i})] + 2\frac{kT}{e}\ln(\frac{p}{n_i}) + \frac{Q_{dep}}{C_{ox}^{(1)}} \quad (2),$$

In equation (2), k_B is Boltzmann constant (1.38×10^{-23} J/K), e is carrier charge (1.6×10^{-19}C), p is p-type dopant concentration, n_i is the intrinsic dopant concentration, and Q_{dep} is the depleted space charge that causes the burden to strongly inverse p-type into n-type. Somehow, Q_{dep} becomes much less as the channel width is so thin making the depletion region contribute less space charge and lightening the burden of strong inversion. Both the large negative flat band voltage in the parenthesis and the eliminated positively compensated term (Q_{dep}/C_{ox}) make Vt tends to be negative for n-channel FinFET devices. In some sense, the I-V curves may, more or less, get improved.

V. Conclusion

At channel length 0.1μm, FinFET devices reasonably suppress the leakage current and promote the drain current since the body is thin enough and the effective width is lengthened. Swing is determined to be in the satisfying region. One is then intrigued in the variation of threshold voltage as the channel length is shortened less than 0.1μm in the near future. Moreover, It fits the theoretical plots to explore the fin thickness effects. It is then concluded that the whole fin is to be fully depleted as the width gets thinner. But how thin it is supposed to be? To make the fin thickness thin may also depend on the process capability, repeatability and reliability. One may wonder if the thinner the fin thickness is, the better the electric performance is as long as the fin is thicker than a couple of hundreds angstroms. Is the silicon material strong enough to stand firmly without being collapsed even though the processes are reliable? Those interesting questions wait to be answered in the near future

VI. Acknowledgement

All the authors sincerely thank National Nano Device Laboratory in Taiwan for being so fully supportive.

Reference

[1] Xuejue Huang, Wen-Chin Lee, C. Kuo, D. Hisamoto, Leland Chang, J. Kedzierski, E. Anderson, H. Takeuchi, Yang-Kyu Choi, K. Asano, V. Subramanian, Tsu-Jae King, J. Bokor, Chenming Hu, "Sub-50 nm P-channel FinFET," *IEEE Transactions on Electron Devices,* vol.48, no.5, pp.880-886, May. 2001.

[2] M. Masahara, K. Endo, Y.X. Liu, S. O'uchi, T. Matsukawa, R. Surdeanu, L. Witters, G. Doornbos, V.H. Nguyen, G. Van den bosch, C. Vrancken, M. Jurczak, S. Biesemans, E. Suzuki, "Four-Terminal FinFET Device Technology," *2007, IEEE International Conference on Integrated Circuit Design and Technology (ICICDT)* , pp.1-4, May. 30 2007-Jun. 2007.

[3] Feng Wang, Yuan Xie, K. Bernstein, Yan Luo, "Dependability analysis of nano-scale FinFET circuits," *2006,IEEE Computer Society Annual Symposium on Emerging VLSI Technologies and Architecture,* vol, pp.6 pp., 2-3 Mar. 2006.

[4] M. Saitoh, N. Yasutake, Y. Nakabayashi, K. Uchida, T. Numata,"Understanding of strain effects on high-field carrier velocity in (100) and (110) CMOSFETs under quasi-ballistic transport, " *2009 IEEE International Electron Devices Meeting (IEDM),* pp. 1-4, 7-9 Dec. 2009.

[5] J.J. Barnes, K. Shimohigashi, R.W. Dutton, "Short-channel MOSFETs in the punchthrough current mode," *IEEE Journal of Solid-State Circuits,* vol. 14, no. 2, pp. 368- 375, Apr. 1979.

[6] T. Rudenko, V. Kilchytska, M.K.M. Arshad, J.-P. Raskin, A. Nazarov, D. Flandre; "On the MOSFET Threshold Voltage Extraction by Transconductance and Transconductance-to-Current Ratio Change Methods: Part II—Effect of Drain Voltage," *IEEE Transactions on Electron Devices,* vol.58, no.12, pp.4180-4188, Dec. 2011.

[7] P. Chotimanus, N. Cooharojananone, S. Phimoltares, "Real swing extraction for video indexing in golf practice video, " 2012 Computing, Communications and Applications Conference (ComComAp), pp.420-425, 11-13 Jan. 2012.

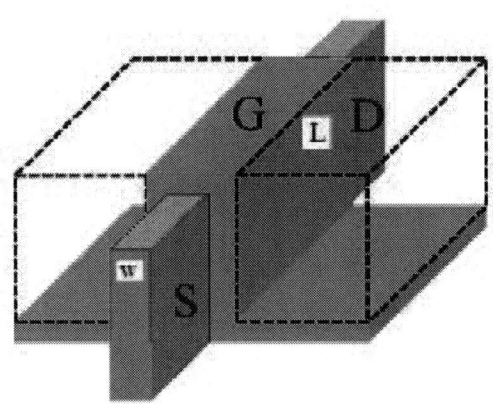

Fig. 1 The top view of the 3-D FinFET device

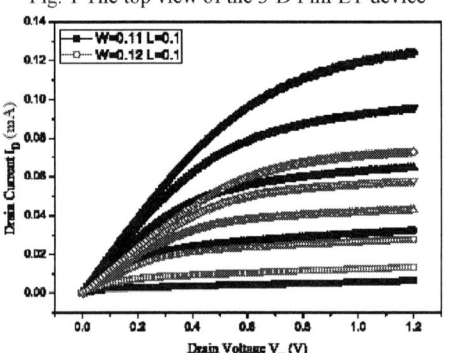

(A)

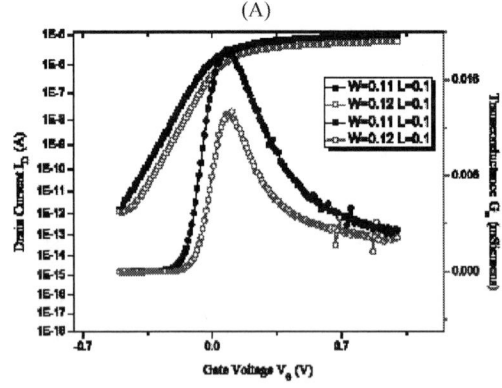

(B)

Fig.2 The characteristic curves with channel width/length (W/L = 0.11μm/0.1μm) and (W/L = 0.12μm/0.1μm) at P-well Vt implant energy 10KeV. (A) I_D-V_D curves (B) I_D-V_G and trans-conductance

978-1-4673-4840-9/13 $31.00 © 2013 IEEE

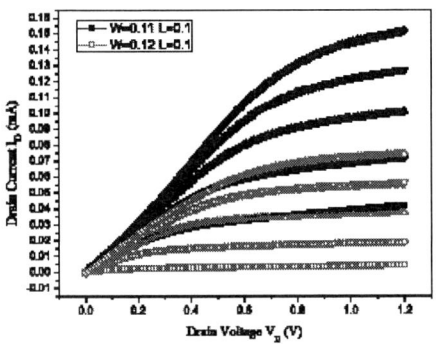

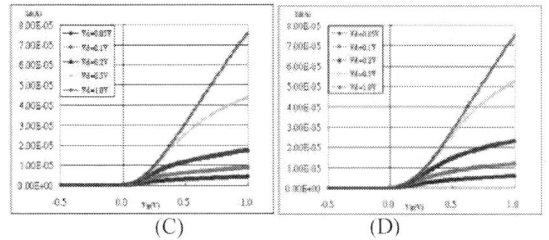

(C)   (D)

Fig.5 I_D-V_D curves at P-Well Vt implant energy 6KeV. (A) W/L = 0.11μm/0.1μm (B) W/L = 0.115μm/0.1μm. (C) W/L = 0.11μm/0.24μm (D) W/L = 0.115μm/0.24μm.

(A)

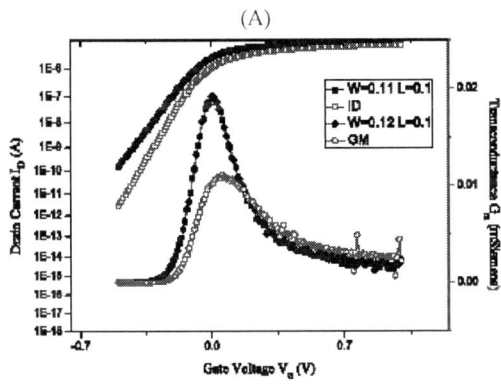

(B)

Fig.3 The characteristic curves with channel width/length (W/L = 0.11μm/0.1μm) and (W/L = 0.12μm/0.1μm) at P-well Vt implant energy 6KeV. (A) I_D-V_D curves (B) I_D-V_G and trans-conductance

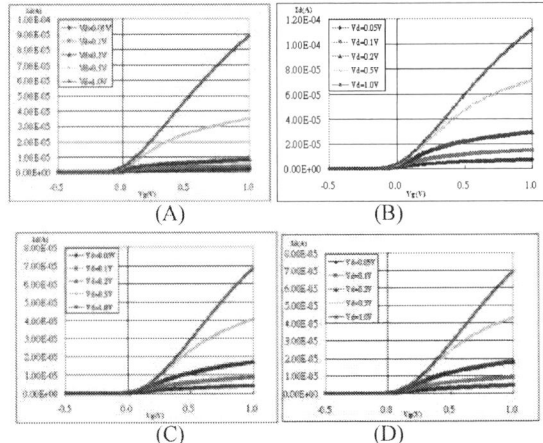

Fig.4 I_D-V_D curves at P-Well Vt implant energy 10KeV. (A) W/L = 0.11μm/0.1μm (B) W/L = 0.115μm/0.1μm. (C) W/L = 0.11μm/0.24μm (D) W/L = 0.115μm/0.24μm.

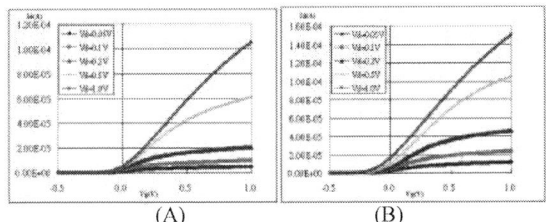

978-1-4673-4840-9/13 $31.00 © 2013 IEEE   485

Next Promising P-Type FinFET Devices without or with Cobalt-Silicide Applied to the Gate

Hsin-Chia Yang*[1], Guo-Wei Wu[1], Wen-Shiang Liao[1], Wei-Yen Peng[1], Sung-Ching Chi[1], Mu-Chun Wang[1], Shea-Jue Wang[2]

[1]Dept. of Electronic Engineering, Minghsin University of Science and Technology, Taiwan

[2]Dept. of Materials and Resources Engineering, National Taipei University of Technology, Taipei, Taiwan

*Corresponding email: hsinchia_yang@yahoo.com.tw

Abstract

The leakage current is suppressed on 3-D structural fin-like channels of FinFET as the sizes of devices get substantially shrunk. The devices with channel width/length (0.12μm/ 0.10μm) are focused on and the baseline device is taken to be 15KeV phosphorous ions (the precursor PH₃) for N-well Vt implant and heavily doped poly-silicon for the gate. One is thus intrigued in what if the Vt implant energy is changed to 20KeV or the gate poly-silicon is fully replaced with cobalt silicide. (Keywords: fully silicide, CoSi₂ silicide, work function, threshold voltage)

Introduction

Fin-FET is one of the next-generation promising devices, which successfully offers fully carrier-free depletion region. With this, drain induced barrier lowering (DIBL) is to be reduced to the minimum or, even better, DIBL may never happen. Devices without DIBL or with un-dominant DIBL enjoy the controllability of switching on or off. Therefore, Fin-FET devices [1,2,3] are expected to provide lower I_{off} current as the channel length gets shorten continuously.

In this study, to maintain a high driving current as stated, one considers PMOS devices using Fin-FET structure. And one expects that the lithography accompanied with processes makes the thickness of source-drain fin thin enough. The improved threshold voltage (Vt) can be resulted because of the fin structure. In addition, the different work function of cobalt-silicide is put to tests and into splits. The N-well Vt adjustment implanted dosage corresponding to different energies, 20KeV and 15KeV phosphorus, is also taken into account. Those electrical characteristics may be demonstrated by the fully depleted region and the lack of leaky body as the gate gets biased.

Experiment and Preparation of FinFET Devices

The orientation <100> of silicon on the oxide dielectric as an insulator (SOI) is mainly the key substrate for fabricating the devices. Careful consecutive photolithography and etching processes are imposed to gain the high aspect ration and 3-D fin-like channels [4, 5], which is deliberately made in the shape of "I", followed by phosphorous N-Well Vt implants at different energies, 15KeV and 20KeV. The 14 angstrom nitrided oxide as gate dielectric is then grown. The two ends of

"I" are, of course, named Source and Drain. Then a layer of 150nm heavy-doped CVD poly-silicon or cobalt silicide is chemically deposited onto the grown gate oxide of the thin high wall as shown in Fig.1.

Results and Discussions

The baseline PFinFET device with width/length sizes, (W/L=0.12μm/0.1μm), is to be evaluated as compared to different Vt implant energy or different gate materials. As shown in Fig. 2, Fig.3, and Fig.4, the device with N-Well Vt implant energy 15KeV always performs the best among them. It is speculated that, compared to implant energy 20KeV, less energies, such as 15KeV, makes the phosphorous ions penetrate through the silicon shallower from the surface, causing more doped ions to out-diffuse during high temperature activation. Lower concentrations of phosphorous for body result in lower threshold voltage and bring up more I_D. As for the gate materials, work functions also play the crucial role on threshold voltages as the traditional poly-silicon is replaced with cobalt silicide. By referring to Fig.2 to Fig.4, the work function of cobalt silicide can be determined to be lower than that of poly-silicon such that the electrical performances look inferior to those of the tradition ones.

Furthermore, Vt_linear can be found by examining I_D-V_G curves near $V_G =0$. And it is quite normal that p-channel devices with poly-silicon gates usually have negative threshold voltages, as referred to the following equation with $\Phi_{work} = -0.56V$.

$$V_{th} = [\Phi_{work} + \frac{k_B T}{e} \ln(\frac{n}{n_i})] - 2\frac{kT}{e}\ln(\frac{n}{n_i}) - \frac{Q_{dep}}{C_{ox}^{(1)}} \qquad (1),$$

where Φ_{work} is work-function related value, k_B is Boltzmann constant (1.38×10^{-23} J/K), e is carrier charge and equal to (1.6×10^{-19}C), n is n-type dopant concentration, n_i is the intrinsic dopant concentration, and Q_{dep} is the depleted space charge. As the gate is biased crossing over the threshold voltage, n-type channels thus get strongly inversed into p-type channels resulting in routes between conductive Sources and Drains. For PFinFET devices, the depleted region is thin enough to make the threshold voltage less negative, and thus the devices get enhanced on the I-V characteristics corresponding to certain boundary conditions. Somehow, some crucial changes may happen as the gate poly-silicon is replaced with fully-covered metal silicide, e.g., cobalt silicide.

In the end, one is interested in Swing (mV / decade) in the

sub-threshold current according to the slope formula as follows:

$$S \equiv \frac{dV_G}{d \log I_D}\bigg|_{V_d V_{th}} \qquad (2)$$

It may further help check with the performances of promising devices in Fig.2 (A), e.g., $S_{0.12}$=84mV/decade and I_{off} (V_D=0.05V)=3.63×10⁻⁶A.

Conclusion

The PFinFET devices are expected to well control the leakage current, especially for the shorter channel length. In this paper, the channel length is fixed to 0.1µm and many other concerning characteristics are shown to be promising.

Acknowledgement

All the authors sincerely thank National Nano Device Laboratory in Taiwan for being so fully supportive.

References

[1] W.S. Hsin, *et al.*, "Drain-induced-barrier lowering and subthreshold swing fluctuations in 16-nm-gate bulk FinFET devices induced by random discrete dopants," *Annual Device Research Conference (DRC)*, pp.109-110, 18-20 Jun. 2012.

[2] S.H. Rasouli, *et al.*, "Variability analysis of FinFET-based devices and circuits considering electrical confinement and width quantization," *IEEE/ACM International Conference on Computer-Aided Design - Digest of Technical Paper ICCAD*, pp.505-512, 2-5 Nov. 2009.

[3] Bin Yu, *et al.*, "FinFET scaling to 10 nm gate length," *IEDM, International Electron Devices Meeting*, pp.251-254, 8-11 Dec. 2002.

[4] S. Inaba, *et al.*, "FinFET: the prospective multi-gate device for future SoC applications," *European Solid-State Circuits Conference(ESSCIRC)* pp.50-53, Sept. 2006.

[5] R.A. Vega, *et al.*, "Three-Dimensional FinFET Source/Drain and Contact Design Optimization Study," *IEEE Transactions on Electron Devices*, vol.56, no.7, pp.1483-1492, Jul. 2009.

[6] W. Xiong, *et al.*, "Impact of strained-silicon-on-insulator (sSOI) substrate on FinFET mobility," *IEEE , Electron Device Letters*, vol.27, no.7, pp. 612- 614, Jul. 2006.

[7] M. Poljak, *et al.*, "Technological constrains of bulk FinFET structure in comparison with SOI FinFET," *International Semiconductor Device Research Symposium*, pp.1-2, 12-14 Dec. 2007.

[8] F. Peijie, *et al.*, "Comparison of silicon-on-insulator and Body-on-Insulator FinFET based digital circuits with consideration on self-heating effects," *International Semiconductor Device Research Symposium (ISDRS)*, pp.1-2, 7-9 Dec. 2011.

[9] Y.T.Bing, *et al.*, "A novel 25-nm modified Schottky-barrier FinFET with high performance," *IEEE, Electron Device Letters*, vol.25, no.6, pp. 430- 432, Jun. 2004.

[10] S.Kumar, *et al.*, "Leakage Analysis for FinFET Devices using Self-Consistent Electro-Thermal Modeling," *IEEE International Conference on Integrated Circuit Design and Technology, 2007. (ICICDT)*, pp.1-4, May 30 2007-Jun. 2007.

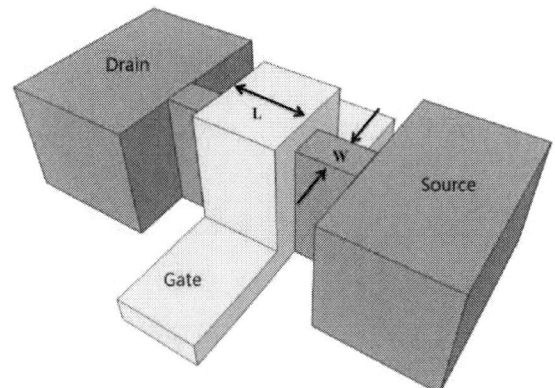

Fig. 1 The cross section of the Fin-FET

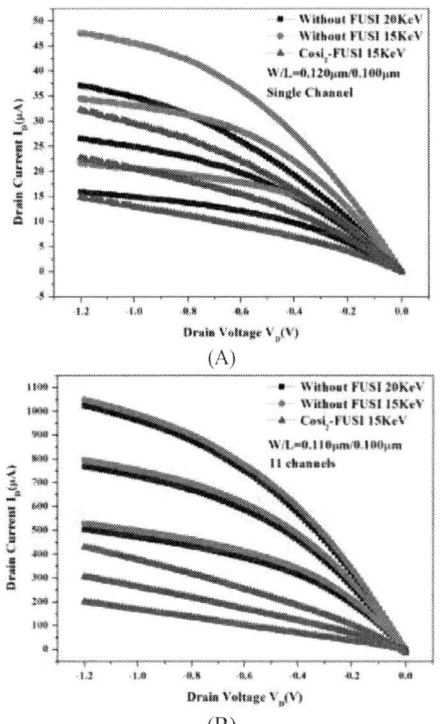

(A)

(B)

Fig.2 I_D-V_D characteristic curves with channel width/length (A) Single Channel (W/L = 0.120µm/0.100µm) and (B) 11 Channels (W/L = 0.11µm/0.10µm).

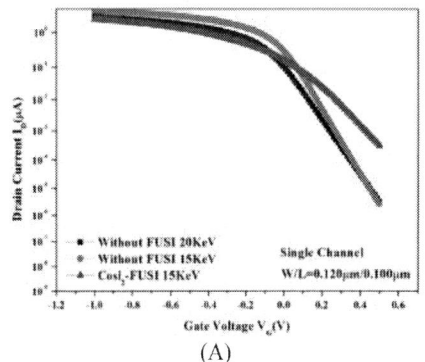

(A)

978-1-4673-4840-9/13 $31.00 © 2013 IEEE

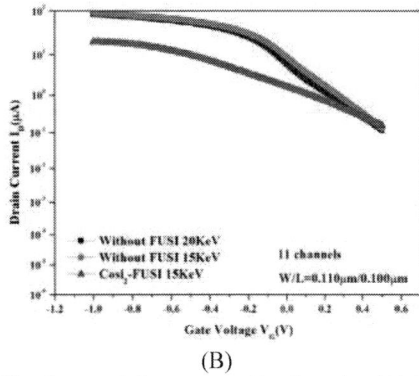

(B)

Fig.3 I_D-V_G characteristic curves with channel width/length (A) Single Channel (W/L = 0.120μm/0.100μm) and (B) 11 Channels (W/L = 0.11μm/0.10μm).

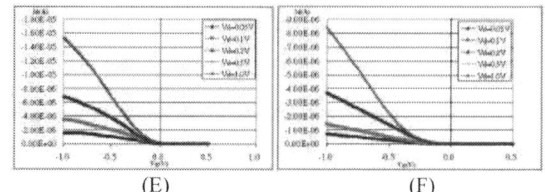

(E)　　　　　　　　　(F)

Fig.5 I_D-V_D curves at N-Well Vt implant energy 20KeV. (A) W/L = 0.11μm/0.1μm (B) W/L = 0.115μm/0.1μm. (C) W/L = 0.11μm/0.24μm (D) W/L = 0.115μm/0.24μm. (E) W/L = 0.12μm/0.1μm (F) W/L = 0.12μm/0.24μm.

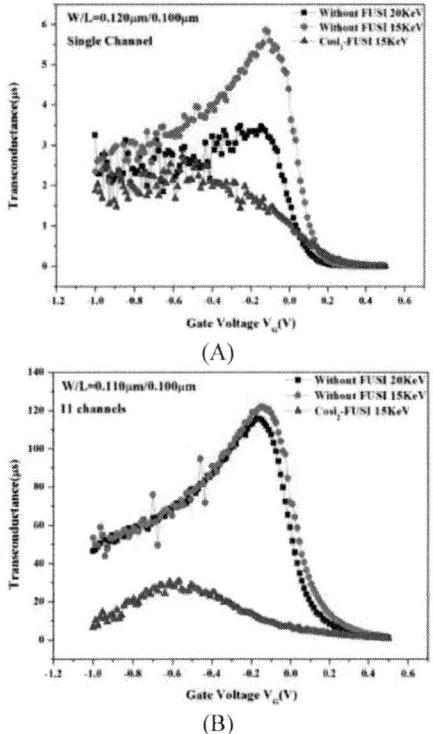

(A)

(B)

Fig.4 G_M characteristic curves with channel width/length (A) Single Channel (W/L = 0.120μm/0.100μm) and (B) 11 Channels (W/L = 0.11μm/0.10μm).

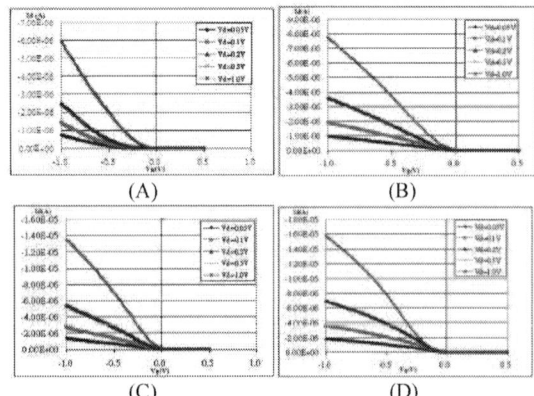

978-1-4673-4840-9/13 $31.00 © 2013 IEEE　　　　　488

The Improvement of MOSFET Electric Characteristics through Strain Engineering by Refilled SiGe as Source and Drain

Hsin-Chia Yang[1], Jie-Min Yang[1], Wen-Shiang Liao[1], Mu-Chun Wang[1]*, Shea-Jue Wang[2]*,
Chun-Wei Lian[1], Chao-Wang Li[1], Chong-Kuan Du[1]

[1]Dept. of Electronic Engineering, Minghsin University of Science & Technology, Taiwan
[2]Dept. of Materials and Resources Engineering, National Taipei University of Technology, Taipei, Taiwan
No. 1 Hsin-Hsing Road, Hsin-Fong, Hsin-Chu 304, Taiwan
*Corresponding email: hsinchia_yang@yahoo.com.tw

Abstract

Strained Engineering including both global and local strains effectively enhances the mobility of carriers, in which global strains are generated by the mismatching of lattice constants at the junction of Si and $Si_{0.775}Ge_{0.225}$ and local strains are aroused by Source/Drain refilled with SiGe. In this paper, junction breakdown voltage, punch-through voltage, and the variation of threshold voltages are to be determined. Strained and non-strained devices are also put in comparison. It is then concluded that the strained engineering technique works promisingly for meeting the requirements of next generation devices. (Keywords: Mobility, Strained Engineering, SiGe-Refilled Source/Drain)

Introduction

The intriguing stain engineering technique make shrunk MOSFET devices regain stages to lead the roles in IC industry. The enhancement of the electrical performances is to be recognized by both global and uni-axial strain engineering [1]. In this paper, mismatched lattice constants between SiGe [3]and Si are deliberately implemented for the devices to be fabricated. As SiGe is inserted between single crystal silicon and epitaxial silicon cap, the overall environment is subjected to a global strain improving the mobility of carriers [2,4]. In addition, the uni-axial strain is to be created as the silicon of Source and Drain are dug out and the holes are refilled with silicon germanium. I_D-V_D characteristics curves are thus shown to be improved over 30% to 50% enhancement. The superior I_D-V_G characteristics curves and the maximum trans-conductance (g_m) using strain engineering are discussed. The threshold voltages are moderately rolled up as the channel length gets less. Unfortunately, the breakdown voltages with strain engineering are observed to be worse than without. Somehow, even though the trade-off can not be avoidable, the whole devices are capable of tolerating the less robust breakdown voltage and proven to the promising candidates for the next generation applications.

In this paper, NMOSFET [5] devices fabricated on the <110> substrate are taken into account. Channel lengths that are less than 90nm are formed by 90nm photo masks with over exposure and over etching technique. In addition to carrier mobility, one also watches the junction breakdown voltage [6,7], the punch-through voltage, and the variation of threshold voltages. In addition, Swing (mV / decade) is to be used for checking up the performance and to be determined by the

formula as follows:

$$S \equiv \frac{dV_G}{d \log I_D}\bigg|_{V < V_{th}} \qquad (1)$$

Experiments and Preparation of NMOSFET Devices

Unto <110> substrate, a 5nm epi-silicon as a buffer layer is grown for adjusting the concentration profile followed by 10nm epitaxial Silicone Germanium (SiGe) and a 1.5nm epi-silicon cap layer for channels and 1.4nm grown nitrided oxide, respectively. 150nm heavy-doped CVD poly-silicon is chemically deposited on the gate oxide and the side wall is to be formed by using TEOS deposition and etching processes. As shown in Fig.1, the as-formed source and drain are etched in with 50nm in depth and refilled with 90nm silicon germanium in which there exists 30% germanium. One uses Agilent Instruments 4156C accompanied with a personal computer having ICS software for the measurements. The chuck temperatures are to be sustained at $25^{\circ}C \pm 1 ^{\circ}C$.

Results and Discussion

NMOSFET devices with three different channel width/length sizes, (W/L=10 m/10 m, 10 m/1 m, 10 m/0.24 m), are to be evaluated. As shown in Fig. 3, the threshold voltages rolled up at less channel length as expected. But it is encouraging that the strain engineering suppresses the rolling up trend.

Junction breakdown voltage is determined by the limit of 1 A as the bias is to be applied to the drain, as shown in Fig.4 (A). It is found that the junction breakdown voltages are quite consistent and predictable. With the buffer layer, the refilled and strained devices enjoy higher sustaining breakdown voltage and show more robust capability no matter what local strain is applied.

The leakage current for determining punch-through voltage is also set to 1 A. The leakage current test can be found in Fig. 4 (B). In general, the devices with shorter channel lengths suffered higher leakage current because the depletion region easier penetrates the shorter path and touches the depletion region on the side of the source. It is so promising that the strained devices are less leaky than non-strained devices.

In addition, no apparent degraded I_D-V_D performances are found as referred to Fig.5 after breakdown and punch through tests.. In Fig.6 (A), I_D-V_G curves of refilled NMOS show almost

978-1-4673-4840-9/13 $31.00 © 2013 IEEE

the same swing as unstrained ones. Unfortunately just as demonstrated in Fig.6 (B), the g_m performance is degraded after the strain engineering is applied. As confirmed in Fig.6 (A), the higher leakage currents at V_G=0V along with the more resistive junction dominantly cause the degradation. Somehow, the improved mobility associated with the swing (~80mV/decade) because of the strain engineering is to be recognized.

Conclusion

The strained devices do show promising feature upon the electrical performances, such as better threshold voltages rolling trend, higher breakdown voltages, higher punch-through voltages, and somewhat good swing even though there exist junction un-matched problems and higher leakage currents. All those encourage ones to promote the strain engineering techniques for the future uses.

Acknowledgement

All the authors sincerely thank National Nano Device Laboratory in Taiwan for being so fully supportive

References

[1] Ming Mao Chu, June-Hua Chou, "Physical yield improvement for SiGe Selective Epitaxial Growth fabrication process on nano scale pMOS strain engineering," *Nanotechnology Materials and Devices Conference, 2009. NMDC '09. IEEE* , pp.42-45, 2-5 June 2009.

[2] Fischetti, M.V.Laux, S.E, "Band structure, deformation potentials, and carrier mobility in strained Si, Ge , and SiGe alloys," *Journal of Applied Physics* , vol.80, no.4, pp.2234-2252, Aug 1996.

[3] Zhang, J. J.Hrauda, N.Groiss, H. Rastelli, A.Stangl, J. Schaffler, F. Schmidt, O .G .Bauer, G, "Strain engineering in Si via closely stacked, site-controlled SiGe islands," *Applied Physics Letters* , vol.96, no.19, pp.193101-193101-3, May 2010.

[4] Currie, M.T. Leitz, C. W. Langdo, T. Λ. Taraschi, G. Fitzgerald, E. A. Antoniadis, D. A, "Carrier mobilities and process stability of strained Si n- and p-MOSFETs on SiGe virtual substrates," *Journal of Vacuum Science & Technology B: Microelectronics and Nanometer Structures* , vol.19, no.6, pp.2268-2279, Nov 2001.

[5] T.S.Perova, Lyutovich, K.Potapova, D.Parry, C.P.Kasper, E.Moore, R. A."Strain and composition in thin SiGe buffer layers with high Ge content studied by micro-Raman spectroscopy" *The Fourth International Conference on Advanced Semiconductor Devices and Microsystems*, pp. 191- 194, 14-16 Oct. 2002.

[6] Eng-Huat Toh,Grace Huiqi Wang,Lap Chan, Guo-Qiang Lo, Sylvester, D.Chun-Huat Heng, Samudra, G. Yee-Chia Yeo, "A complementary-I-MOS technology featuring SiGe channel and i-region for enhancement of impact-ionization, breakdown voltage, and performance," *Solid State Device Research Conference, 2007. ESSDERC 2007. 37th European* , pp.295-298, 11-13 Sept. 2007.

[7] Jiahui Yuan, Cressler, J.D. "A novel superjunction collector design for improving breakdown voltage in high-speed SiGe HBTs," *Bipolar/BiCMOS Circuits and Technology Meeting, 2009. BCTM 2009. IEEE* , pp.75-78, 12-14 Oct. 2009.

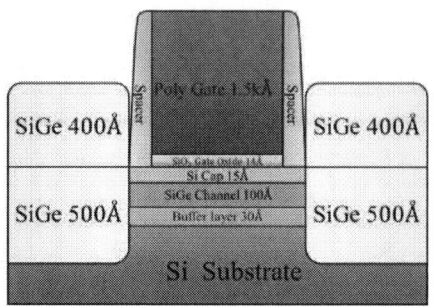

Fig. 1 Cross section of the device with refilled SiGe as Source and

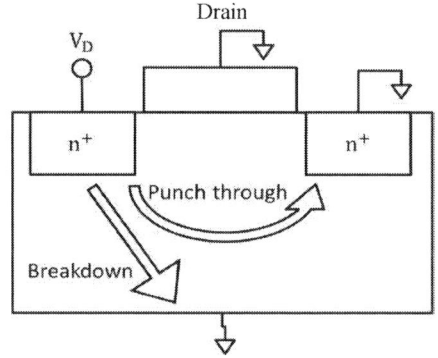

Fig. 2 Punch through and junction breakdown tests

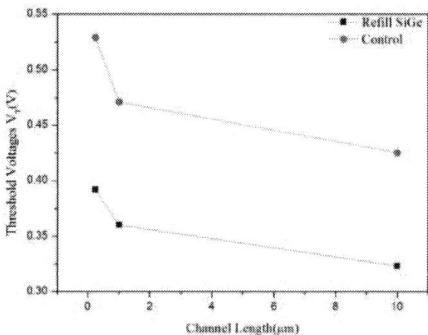

Fig. 3 Threshold voltages of three different lengths

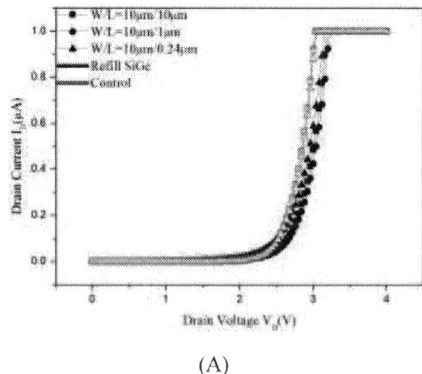

(A)

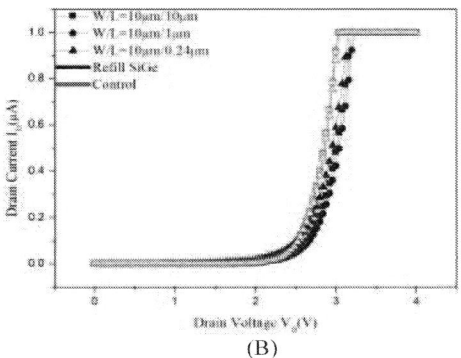

(B)

Fig. 4 (A) Punch through tests (B) Breakdown tests

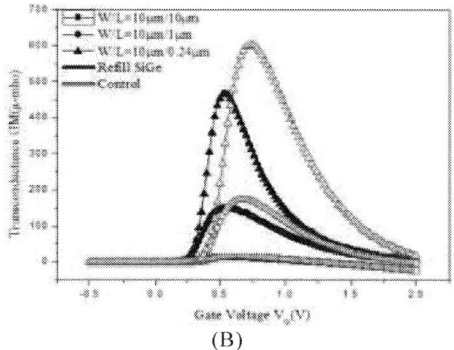

(B)

Fig. 6 (A) I_D-V_G curves (B) Trans-conductance g_m

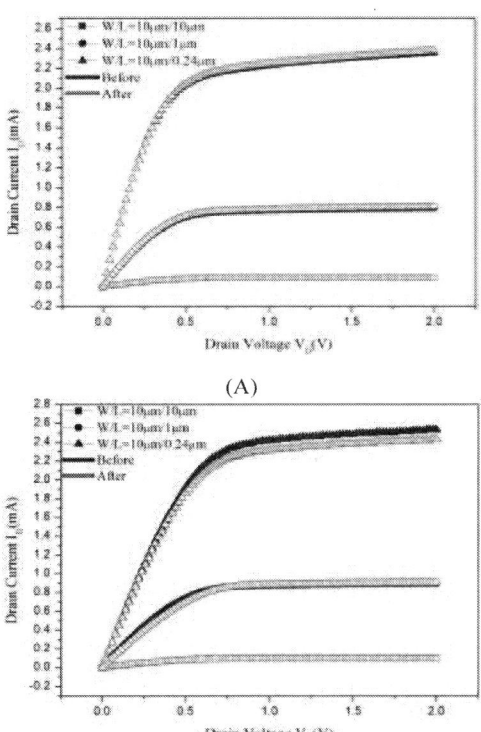

(A)

(B)

Fig. 5 I_D-V_D curves before and after punch-through and breakdown voltage tests (A) Un-strained (B) Strained

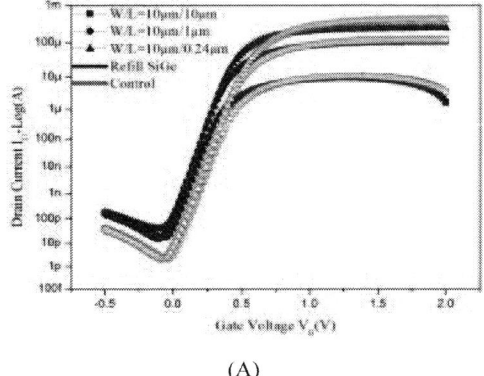

(A)

978-1-4673-4840-9/13 $31.00 © 2013 IEEE   491

Study of Surfactant Modified MWNT/Polyimide Composites by In-Situ Polymerization

Hung-Han Ko, Yao-Yi Cheng[*], Ching-Wei Wang

Institute of Organic and Polymeric Materials, National Taipei University of Technology
1, Sec. 3, Chung-hsiao E. Rd.
Taipei,10608,Taiwan, R.O.C
Tel: (886-2) 2771-2171 #2433; E-mail:qqq751008@gmail.com

Abstract

In this study, we attempt to enhance the electrical conductivity of polyimide (PI) by non-covalently modified multiwall carbon nanotubes (MWNTs). The multiwall carbon nanotubes were first oxidized (o-CNT) and modified with nitric acid (n-CNT). The o-CNT and n-CNT were then modified with anionic surfactant (Sodium dodecylbenzene sulfonate) (SDBS) to improve their dispersion in the PI matrix. We also compare the properties of the surfactant modified MWNTs prepared by filtration or centrifuge. We confirmed the successful preparation of the surfactant modified MWNTs by Fourier transform infrared (FTIR), scanning electron microscope (SEM) and UV visible spectroscopy. Furthermore, the thermal stability, electrical and mechanical properties of PI/MWNT nanocomposites can be increased through in-situ polymerization.

Keywords: *carbon nanotube; polyimide; surfactant*

Introduction

Carbon Nanotubes (CNTs) with excellent electrical conductivity, mechanical and thermal properties [1] can be potentially used as nano-reinforcements in polymer composite [2], which has attracted much attention recently [3-5]. However, the lack of interfacial interactions between CNTs and polymer matrix causes facile aggregation of CNTs when loading high weight percent CNTs. Therefore, it is necessary to modify CNT to enhance the compatibility within a polymer matrix.

Polyimide (PI) has been widely applied in aircraft structure, aerospace and semiconductor because of their high thermal stability, fabulous electrical insulation, mechanical properties and radiation resistance properties [6]. In this research, surfactant was used to assist uniform dispersion of CNT in solutions. In addition, we fabricated the well dispersed PI/CNTs nanocomposites by in situ polymerization method.

Experimental

2.1 Materials

Pristine multiwall carbon nanotubes with a diameter 10-20nm and length 5-10 μm were fabricated by chemical vapor deposition with purity of 95%. Sodium dodecylbenzene sulfonate (SDBS, Acros, 99%), Nitric acid (HNO₃, Scharlau 65%), 3, 3', 4, 4'- biphenyl tetracarboxylic dianhydride (BPDA, Chriskev, 98%), p-phenylenediamine (p-PDA, Acros, 99%) and 1-methyl-2-pyrrolodinone (NMP, Acros, 99%) were used as received without further purification.

2.2 Preparation of oxidized and acid modified MWNTs

Pristine CNTs was purified by oxidation in air, at 550℃ for 45 minutes, to remove amorphous carbon and residual metal catalyst to obtain the oxidized CNTs (o-CNT). The o-CNT was mixed with 6M nitric acid using ultra-sonication at 50℃ for 30 minutes. The dispersed mixture was refluxed at 120℃ for 2 hours and then washed and filtered with distilled water to obtain nitric modified CNTs (n-CNT). The process scheme of preparing multiwall carbon nanotubes is shown in Fig.1.

2.3 Preparation of surfactant modified MWNTs

For the preparation of surfactant modified CNTs, we investigate the effect of the filtration and centrifuge. The sodium dodecylbenzene sulfonate (SDBS) was sonicated with o-CNTs or n-CNTs at 50℃ for 1 hours in order to make surfactant treated CNTs. SDBS: CNT of 0.5:1 and 1:1 by weight was denoted with SD-CNTs 0.5:1and SD-CNT 1:1, respectively. After sonication, the dispersed mixture was filtered and dried in the oven to receive surfactant modified n-CNT (SD-nCNTf). Alternatively, the dispersed mixture was centrifuged at 2500 rpm for 40 min and 3000 rpm for 20min to obtain the precipitated CNTs, which was then dried in an oven to receive surfactant modified o-CNT (SD-oCNT) or n-CNT (SD-nCNT).

2.4 Preparation of PI/CNTs composites by in-situ polymerization

Surfactants-CNTs were pre-treated with NMP under ultra-sonication for 6 hours to obtain uniformly dispersed CNTs suspension. Subsequently, the CNTs suspension was mixed with *p*-PDA in the flask. After 10 minutes, BPDA was added while the mixture solution was stirred for 12 hours to get PAA-CNTs mixture. Then, the PAA precursor was coated on glass plates for multi-step thermal curing (at 100℃, 150℃, 200℃, 250℃, 350℃, each temperature held for 1 hour) to form PI/ CNTs films.

Results and Discussion

Fig. 2 and Fig. 3 compares the FTIR spectra of pure SDBS, SDBS modified MWNTs by filtration (SD-nCNTf) and SDBS modified CNTs by centrifuge (SD-nCNT). As shown in Fig.1, adsorption bands were observed at 2959, 2926 and 2855 cm⁻¹ for C–H stretching vibration, 1466 cm⁻¹ for C–H bending vibration, 1378cm⁻¹ for CH₃ stretching vibration and 1188, 1133, 1046, 1013and 832 cm⁻¹ for Sulfonate group. In Fig. 3, the adsorption peaks of SDBS can be observed, which reveals

978-1-4673-4840-9/13 $31.00 © 2013 IEEE

that CNTs was successfully treated with SDBS.

Fig. 4 shows SEM images of oxidized and surfactant CNTs. As observed in Fig. 4(b) and 4(c), the nanotubes present higher degree of separation with the treatment of SDBS. This result suggests that the aggregation of CNTs can be minimized through the treatment of SDBS surfactant.

UV-visible spectroscopy can be used to examine the dispersed condition of CNTs in NMP solution. Using UV-vis spectra, the carbon nanotube absorption peaks at 253 nm can be detected [8]. Fig.5 shows the absorbance spectra of solutions of o-CNT, n-CNT with various SDBS treatments, in which a characteristic peak exist around 260 nm. SD-nCNT1:1 and SD-nCNT0.5:1 presents the highest absorbance, which suggests that centrifuge process can keep more surfactant with CNTs; therefore, the CNTs could be uniformly dispersed in the NMP solutions.

Table 1 shows the intensity and G/D ratio obtained from Raman spectra of CNTs. The values of G/D ratio were practically unchanged after the treatment of SDBS treated with CNTs, because the surfactant functionalization of CNTs is a non-covalently treatment without damaging the surface of CNTs [9].

Table 2 shows the electrical conductivity of PI and PI loaded with 2 wt% of various CNTs. The criterion for electrostatic charge mitigation (10^6- 10^8 Ω/cm^2) can be met when PI is prepared by in-situ polymerization with SD-nCNT1:1 loading of 2 wt%. The surface resistance and volume resistance of SD-nCNT1:1 significantly decrease to 5.25x10^6 (Ω/cm^2) and 2.45x10^6 (Ω-cm), respectively.

Fig.6 presents the TGA curves of PI/CNTs nanocomposites. As summarized in Table 2, the thermal decomposition temperature of the PI/CNTs composites can be slightly increased with loading of 2 wt% CNTs. T_d reaches a maximum temperature of 602.3°C, when 2 wt% SD-oCNT0.5:1 is added to the PI composite. Compared with SD-nCNTf1:1, SD-nCNT1:1 has more effect on the enhancement thermal stability of the polyimide nanocomposites.

Table 3 summarizes the tensile strength and elongation of PI composite films with various modified CNTs loading. The results also demonstrate that SDBS could effectively de-bundle CNTs as observed in results of UV-visible and SEM. SD-nCNTf1:1 increased about 1.3 times to 153.3MPa. However, excess SDBS will reduce the mechanical properties of polyimide. The tensile strength of the polyimide loading 2 wt% SD-nCNT1:1 is lower than 2 wt% SD-nCNT0.5:1, and the polyimide loading SD-oCNT1:1lower than SD-oCNT0.5:1.

Reference

[1] S. Iijima, *Nature*, Vol. 354, 1991, pp. 56-58.
[2] S.-M Yuen et al., *Composites Science and Tech.*, 68, 2008, pp2842-2848.
[3] X. Chen et al., *Materials Science and Engineering*, A492, pp. 236–242, March 2008.
[4] J. Shen et al., *Composites Science and Tech.*, 67, 2007, pp. 3041- 3050.
[5] P.C. Ma et al., Carbon, 44, 2006, pp. 3232–3238
[6] Q.-Y. Tang et al., *Polym. Int.*, Vol. 59, 2010, pp. 1240-1245.
[7] K.-L Wu et al., *Journal of Applied Poly. Science*, Vol. 116, 2010, pp. 3111-3117.
[8] J. Rausch et al., *Composites, A41*, 2010, pp. 1038-1046.
[9] I. Madni et al., Colloids and Surface A: Physicochem. Eng. Aspects, 358, 2010, pp. 101-107.

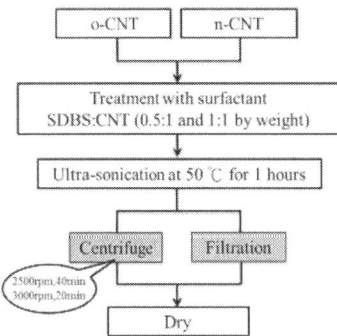

Fig. 1 Reaction scheme for preparing surfactant-CNTs.

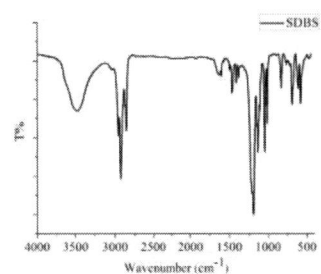

Fig. 2 FTIR spectrum of pure Sodium dodecylbenzene sulfonate.

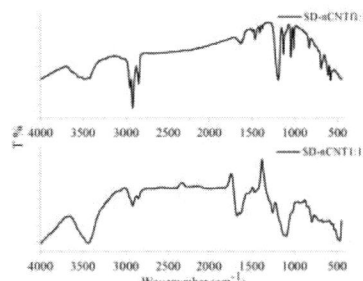

Fig. 3 FTIR spectrum of SD-nCNT and SD-nCNTf.

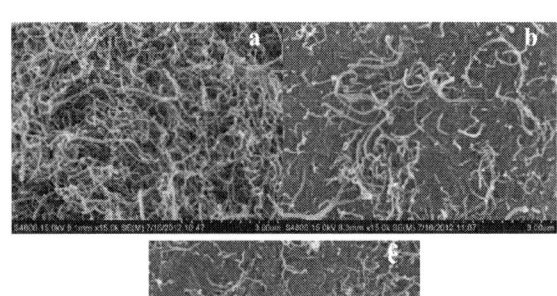

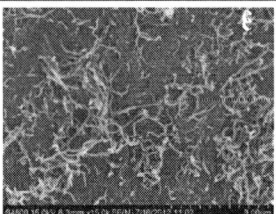

Fig 4 SEM images of (a) o-CNT, (b) SD-nCNT1:1 and (c) SD-nCNTf1:1.

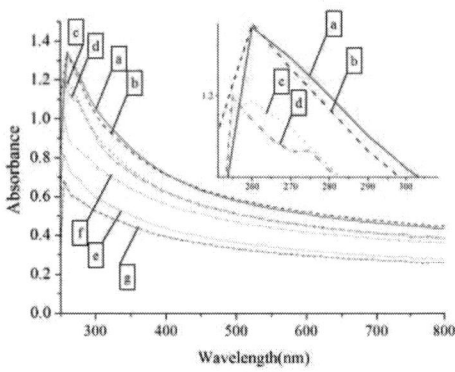

Fig 5 UV- visible absorption spectra of (a) SD-nCNT1:1,
(b) SD-nCNT0.5:1, (c) SD-nCNTf1:1, (d) SD-oCNT1:1,
(e) SD-oCNT0.5:1, (f) n-CNT and (g) o-CNT in NMP.

Table 1 Raman results of various CNTs.

Carbon nanotubes	I_{D-band}	I_{G-band}	I_G/I_D
o-CNT	20670	22460	1.08
n-CNT	47370	50640	1.07
SD-oCNT1:1	19420	20970	1.08
SD-nCNT1:1	40750	47900	1.17
SD-nCNTf1:1	22410	26390	1.18

Table 2 Electrical and thermal stability properties of PI and
PI/SD-CNTs composites.

Sample	Content (wt%)	Surface resistance (Ω/cm^2)	Volume resistance (Ω-cm)	[a]Td (℃)
PI	0	2.22×10^{14}	1.67×10^{14}	594.2
PI/SD-nCNTf1:1	2	2.77×10^9	9.46×10^8	594.2
PI/SD-nCNT1:1	2	5×10^6	2.45×10^6	598.7
PI/SD-nCNT0.5:1	2	6.91×10^9	7.47×10^9	590.7
PI/SD-oCNT1:1	2	7.49×10^9	6.25×10^{10}	597.6
PI/SD-oCNT0.5:1	2	7.85×10^9	7.33×10^{10}	602.7

[a] Decomposition temperature at 10% weight loss.

Table 3 Mechanical properties of PI and PI/SD-CNTs composites.

Sample	Content(wt%)	Tensile Strength(Mpa)	Elongation(%)
PI	0	114.7 ± 16.7	3.7 ± 0.5
PI/SD-nCNTf1:1	2	153.3 ± 6.6	6.1 ± 0.5
PI/SD-nCNT1:1	2	130.4 ± 6.7	5.3 ± 1.7
PI/SD-nCNT0.5:1	2	134.2 ± 5.6	4.5 ± 1.1
PI/SD-oCNT1:1	2	120.2 ± 7.0	5.8 ± 1.6
PI/SD-oCNT0.5:1	2	129.0 ± 11.9	5.8 ± 0.7

A Novel InGaAs Photodiode Fabrication and Its Application

Chii-Wen Chen[1], Wen-Chin Lee[1*], Meng-Chyi Wu[2], Chong-Long Ho[2],
Chia-Hao Chuang[2], Dong-Ying Hsieh[1]

[1] Dept. of Electronic Engineering, Minghsin University of Science & Technology, Hsinchu, Taiwan
[2] Institute of Photonics Technologies, National Tsing Hua University, Hsinchu, Taiwan
No. 1, Hsinhsing Road, Hsinfong, Hsinchu 30401, Taiwan
* Corresponding email: chise520@gmail.com ; Phone:+886-921-892-115; FAX: +886-3-559-1402

Abstract

In this study, a two-dimension array of InGaAs photodiode with 0.9~1.7nm absorption wavelength was presented. By using the characteristic of wide wavelength range in InGaAs material, we designed several active regions in our novel devices for inducing expected photocurrent. While the distances between light source and these several equal-area active regions increases, we found the current generated in photodiode decreases non-linearly. Furthermore, this device was applied on detecting skin physiology, when the tungsten bulk light source penetrated through skin and was absorbed by photodiode, there had individual different currents in each equal-area active regions, and the relationship between photocurrent and moisture in skin will be interpreted.*(Keyword: Photonics, InGaAs, Photodiode, Large Size, Skin Moisture).*

Introduction

In the recent, many research of photodiode (PD) were focused on their high-speed and light-power applications such as fiber-optic communications and military telescope. On the other, large area photodiodes were applied on green energy technology. PD has both photovoltaic and photoconductive mechanisms. For instance, solar cells could generate current by light in photovoltaic mode, that is the most popular topic on green energy development[1]. There are many designs modules in PD for example, RF system design has four active regions in the PD to enhance the link gain, and sensitivity. However, this enlarge approach would lead to increase capacitance, drop photoreceiver's the bandwidth, and the current noise. We must find a balance between noise and sensitivity for special PD application[2]. Someone produces a RF output signal by using a trams-impedance amplifier to generate the required RF power directly in traditional RF system. But it won't need RF power amplifier and will reduce phase noise in this system if high power photoreceiver's were applied to handling PD [3]. In this study, we fabricated PD with large area and applied them on skin physiology. Owing to better absorption efficiency at wavelength range from 0.9nm to 1.8nm in InGaAs PD, these devices have variable currents at different distance from skin under tungsten bulb light[4]~[7]. We measured the skin humidity by using skin hygrometer in order to find the correlation between distance, light-induced current and skin humility.

Device Fabrication

The photodiode structure in this study were presented as shown in Fig.1. We deposited 2000Å SiO_2 on substrate firstly, then applied O_2 plasma diffusion-on source wafer after patterning and wet-etching, and deposited 1500Å SiO_2 passivation for secondary diffusion[8]~[10]. As shown in Fig.2, the dark current rose exponentially, and depletion region and potential barrier increased abruptly while the reverse bias stepped up linearly. Where the steeper potential barrier, then was higher electric field, it will cause tunnel breakdown after reverse bias excessing critical condition of InGaAs. This last step could increase the concentration in active region and reduce the dark current effectively. We used Cr / Zn / Au material for P-metal layer because of good adhesion with SiO_2, excellent electrical conductivity and easy package bonding. We found they had lower leakage current at 400°C after alloy annealing for reduced defects between lattices[11]~[13]. Finally, we selected silicon oxide and aluminum oxide as AR coating material, Their combination would effectively enhance light absorption and avoid light loss at wavelength range form 1.2nm to 1.6nm as Fig.3 shown.

Results and Discussions

Experiment were set up by using 3mm diameter tungsten bulb as light source and a PD moved step by step in 2.5mm pitch on skin surface, as illustrated in Fig.4. The skin humidity by using hygrometer and light-induced currents with different distance between PD and skin surface of five specimens were averaged by measured three times respectively. In this experiment, we measured photocurrents under different distance or humidity to skin by applied a 5mm diameter incandescent bulb as light source. Curves in Fig.5 illustrated that photocurrents had a significant change with in 10mm from skin surface. Although light transmitted from 5mm in candescent bulb couldn't penetrate through well from the skin to PD. As Fig.6 shown, although specimen A had moister skin then B, the deleted current has no obvious change. This trend was repeated in specimen E that had the highest moisture on skin. It means skin humidity has no relationship with photo current. However, we found there had a significant change in photo detected with different distance between light and PD. Especially at a distance of 7.5mm to 10mm, they decayed rapidly with comparison with distance over 1mm.

978-1-4673-4840-9/13 $31.00 © 2013 IEEE

Conclusions

The relationship between skin moisture and distance from PD measured by hygrometer and photo current were presented in this paper. We found photo current showed a downward trend while distance increased, but it not had fierce change to do with humidity.

Acknowledgement

We would like to thank group members of Institute of Photonics Technologies, National Tsing Hua University for their supports in equipments, and thank Mr.L.C. Yan of Minghsin University of Science & Technology for his helps in this work.

References

[1] J.C. Campbell, A.G. Dentai, G.J. Qua, J. Long, V.G. Riggs, "Planar InGaAs PIN photodiode with a semi-insulating InP cap layer," *Electronics Letters* , vol.21, no.10, pp.447-448, May 9 1985.

[2] A. Joshi, J. Rue, S. Datta , "Low-Noise Large-Area Quad Photoreceivers Based on Low-Capacitance Quad InGaAs Photodiodes " *Photonics Technology Letters, IEEE* , vol.21, no.21, pp.1585-1587, Nov. 2009.

[3] S. Datta, A. Joshi, D. Becker, R. Howard, "High phase linearity, high power handling, InGaAs photodiodes for precise timing applications" *Optical Fiber Communication - incudes post deadline papers, Conference on OFC 2009*, pp.1-3, 22-26 March 2009.

[4] L.B. Karlina, B.Ya. Ber, P.A. Blagnov, A.M. Boiko, M.M. Kulagina, A.S. Vlasov, "Influence of phosphorus on undoped and zinc doped InGaAs " *Indium Phosphide and Related Materials Conference. IPRM. 14th* , pp. 213- 215, 2002.

[5] Eun Soo Nam, M.S. Oh, S.E. Hong, H.S. Kim, "Two Dimensional 32x32 InGaAs/InP Photodiode Arrays and Dark Current Characteristics Limited by the Diffusion and Generation-Recombination " *The Joint International Conference on Optical Internet and Next Generation Network,* pp.78-80, 9-13 July 2006.

[6] K. Ohnaka, M. Kubo, J. Shibata, "A low dark current InGaAs/InP p-i-n photodiode with covered mesa structure " *IEEE Transactions on Electron Devices*, vol.34, no.2, pp. 199-204, Feb 1987.

[7] C.L. Ho, M.C. Wu, W.J. Ho, J.W. Liaw, "Bandwidth enhancement for p-end-illuminated InP/InGaAs/InP p-i-n photodiodes by utilizing symmetrical doping profiles" *Journal of Lightwave Technology*, vol.17, no.5, pp.912-917, May 1999.

[8] P. Maigne, J.A. Beraldin, T.M. Vanderwel, M. Buchanan, D. Landheer, " An InGaAs-InP position-sensing photodetector " *IEEE Journal of Quantum Electronics,* vol.26, no.5, pp.820-823, May 1990.

[9] K.R. Linga, G.H. Olsen, V.S. Ban, A.N. Joshi, W.F. Kosonocky, "Dark current analysis and characterization of InxGa1-xAs/InAsyP1-y graded photodiodes with x>0.53 for response to longer wavelengths (>1.7 µm) " *Journal of Lightwave Technology*, vol.10, no.8, pp.1050-1055, Aug 1992.

[10] Y.W. Chen, W.C. Hsu, Y.J. Chen, "Low dark current InGaAs(P)/InP p-i-n photodiodes " *Optoelectronics, Proceedings of the Sixth Chinese Symposium* , pp. 95- 98, 12-14 Sept. 2003.

[11] M. Purica, E. Budianu, I. Grozescu, E. Rusu, S.V. Slobodchikov, " Design and optimization of InGaAs/InP photodetector for coordinate sensitive detection systems " *Electron Devices for Microwave and Optoelectronic Applications, 2001 International Symposium on* ,pp.113-118, 2001.

[12] R.D. Dupuis, J.C. Campbell, J.R. Velebir, "Planar InGaAs PIN photodetectors grown by metalorganic chemical vapour deposition " *Electronics Letters* , vol.22, no.1, pp.48-50, January 2 1986.

[13] S.H. Chang, Y.K. Fang, S.F. Ting, S.F. Chen, C.Y. Lin, C.S. Lin, C.Y. Wu, "Fabrication of very high quantum efficiency planar InGaAs PIN photodiodes through prebake process " *Circuits, Devices and Systems, IEE Proceedings* , pp. 637- 640, 9 Dec. 2005.

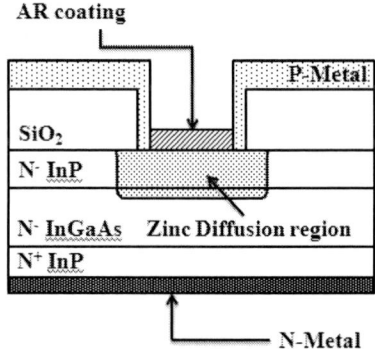

Fig.1 Cross section of Photodiode

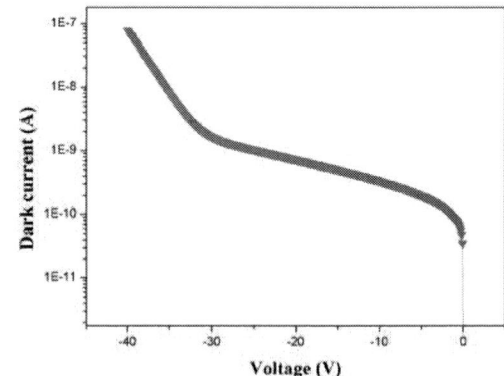

Fig.2 Dark current Cuive

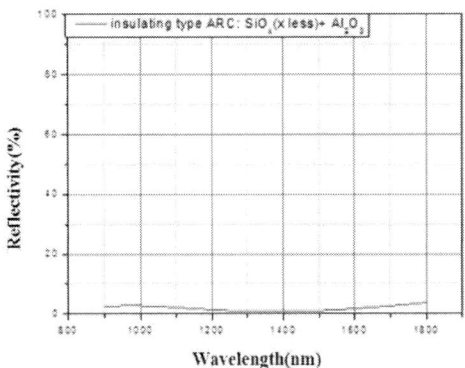

Fig.3 The reflectivity ARC on the top layer of InP.

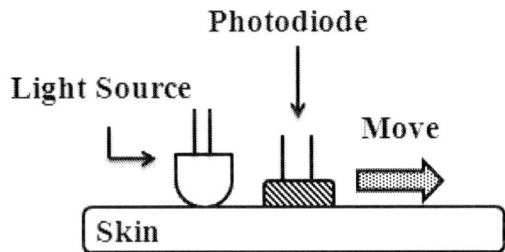

Fig.4 Measurement Diagram

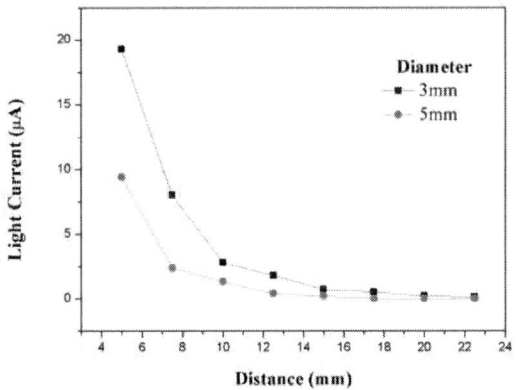

Fig.5 Different size tungsten bulb for skin

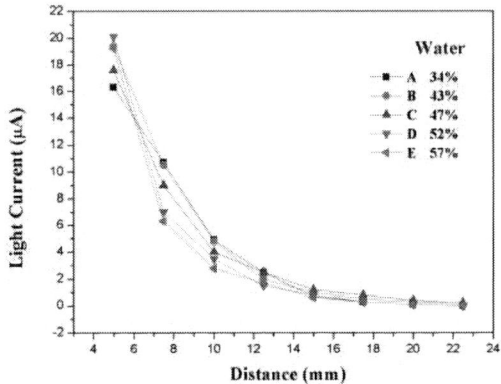

Fig.6 Humidity and photocurrent chart

Study on the Characterizations and Applications of the pH-Sensor with GZO/Glass Extended-Gate FET

*Jung-Lung Chiang, Chia-Yu Kuo

Department of Electronic Engineering
Chung Chou University of Science and Technology
No. 6, Lane 2, Sec. 3, Shanjiao Rd., Yuanlin Township, Changhua County 510, Taiwan, R.O.C.
Tel: 886-48359000-2200, *E-mail: lung@dragon.ccut.edu.tw

Abstract

The GZO thin films of the thickness about 200~250nm were deposited on the glass which used as a pH sensor head based on the extended-gate field-effect transistor (EGFET) by r. f. sputtering system in this study. The pH sensing characteristics of the GZO/glass EGFET sensing structure were investigated and a semiconductor parameter analyzer (Keithley 4200) was used to measure the drain-source current versus gate-source voltage curves in various buffer solutions. In the experimental results, it can be obtained that the various pH sensitivities of the GZO pH-EGFET at different measuring temperature ambiance. Furthermore we also found that the pH sensitivity is increasing with temperature increased, and the pH response is a good linearity in this study. In addition, the GZO thin film was found that the resistance was about $8.37 \times 10^{-3}\Omega$-cm and the average of transmittance was about 80% in the visible range.

Kywords: pH-sensor, GZO, EGFET, Sensitivity.

Introduction

Twenty-first century, thanks for scientific and technological development, the human life has increased. The treatments of the chronic diseases and preventive medicine are playing an important role in modern and future studies. Various diseases detection and real-time monitoring will be attention. And biosensors will become to maintain the quality of life. Additionally, the development of high sensitivity, stability, miniaturization and a number of physiological parameters detection of biological sensing system, will be a big challenge in medical testing and business opportunities.

Presently, In_2O_3, and ZnO are the main materials used for transparent conducting of oxide thin films (TCO). Recently, ZnO-based TCO thin film has attracted much attention as a transparent and conductive film material because it is very low cost, non-toxic, and the process is simple [1-2]. Transparent and conductive oxide thin film has very high development potential, and may not be easily affected by the plasma process of hydrogen [3].

Recently, the development of pH-sensor is tending toward low cost, disposable, and real-time measurement, owing to apply in human blood [4], urine [5-7], and saliva [8] in the clinic as point-of-care testing (POCT) micro-system [9]. The pH value plays a role to detect healthy level of the human body. In general, commercial pH-glass electrode is difficult to integrate with electrical circuit, for that purpose, ion-sensitive

field-effect transistor (ISFET) technology was developed by Bergveld in 1970 [10], but the sensitive membrane and metal-oxide-semiconductor field effect transistor (MOSFET) were immersed in solution, which caused poor thermal stability of the transistor, On the contrary, the extended-gate field-effect transistor (EGFET)[11] is separated the H^+ sensitive membrane and MOSFET, which can avoid temperature affection.

Generally, the stability and reliability of the sensors are into the commercialization of critical. Temperature effect is one of the factors, which affect the accuracy of measurement. The temperature effect of factor is divided into: (1) reference electrode, (2) the test solution, (3) the threshold voltage of the device, (4) the temperature effect of the surface potential.

In this study, the GZO thin films have been deposited on the glass as a pH sensing material. The resistance and transmittance of GZO thin films were measured. And temperature effect of the GZO pH-EGFET at different measuring temperature ambiance were also measured and discussed.

Experimental

A. Deposition of GZO Thin Film

In this study, the GZO thin film was deposited on the commercial glass by radio frequency(rf) sputtering system using the compound GZO target material(ZnO: 97 wt% and Ga_2O_3: 3 wt%) in an Ar gas atmosphere at 300°C substrate temperature. The deposition conditions were carried out under a background pressure of $\sim 1 \times 10^{-6}$ Torr, with sputtering at a total operating pressure of 10×10^{-3} Torr in the 20sccm flowing rate of Ar gas for 15min. The rf power was set at 100 W and operated at 13.56 MHz. The detail deposition conditions are shown in Table I.

TABLE I
DEPOSITION PARAMETERS OF GZO THIN FILMS

Substrate	Commercial glass
Substrate-to-Target Distance	~6 cm
GZO Target	(ZnO: Ga_2O_3=97 : 3 wt%)
Gas (Ar)	20 sccm
RF Power	100 W
Sputtering Pressure	10 mTorr
Substrate Temperature	300°C

B. Analysis of Surface Properties of GZO/Glass

After deposition, The crystallographic structure of the GZO thin film was analyzed using an X-ray diffractometer (XRD, Shimadzu Oceania-XRD-6000), and optical transmittance was analyzed using a UV-VIS spectrophotometer (SHIMADZU, UV-1700) in the wavelength range from 200nm to 1100nm. The influence of the surface morphologies and cross-section were observed using a field emission scanning electron microscope (FE-SEM, JEOL-JSM-6700F). The electrical properties, such as resistance, were studied through a four-point probe flat station measurement (QUATEK, QTI-QT-50).

C. Package of GZO/Glass Sensing Structure and I-V Measurement Set-Up

In this study, the GZO thin film was deposited on the glass and the GZO/glass structure was encapsulated as sensor head using the epoxy resin. The pH-EGFET device was encapsulated as shown in Fig. 1. The sensing window with the exposed gate region is close to 2x2 mm². Afterwards, the sensor head connected with the gate of MOSFET. The extended-gate field-effect transistor based on the GZO/glass sensing structure was finished and the pH-sensing properties were studied using a semiconductor parameter analysis measurement (Keithley 4200) in the pH 7, 9 and 11 different buffer solutions, respectively. The set-up of measurement was show in Fig. 2 and procedures described as follows[12]:

Step 1: The drain-to-source voltage (V_{DS}) was set at 0.2 V. The scanning voltage (gate-to-source voltage, V_{GS}) was set to the region (V_{GS} step = 0.1 V).

Step 2: To decrease the deviation or error response, the reference electrode and a pH sensing head were immersed in pH buffer solution for 10s to maintain the temperature balance. All measurements were carried out in a black box.

Step 3: The I-V characteristic curves were obtained at the setting temperature in the H^+ concentration range between pH=7, 9 and 11, respectively.

Step 4: Similar measuring procedures from steps (1) through (3) were repeated for temperatures of 15, 25, 35 and 45 °C (±0.1 °C).

Results and Discussion

A. XRD Analysis

In this study, figure 3 shows the XRD analysis of GZO film at as-deposition. The pattern in this figure shows that the only diffraction peak for the as-deposition GZO thin film is the ZnO (002) peaks located at 2θ= 34.2°. The GZO film had adequate (002) crystal plane diffraction peak, which indicated that the film is polycrystalline with a hexagonal wurzite structure.

B. Transmittance Analysis

In this study, the optical transmittance was studied using a UV-VIS spectrophotometer. According to the experimental result, the average of transmittance of GZO film was about 80% in the visible range.

C. Surface Morphology and Electrical Properties Analysis

In this study, the surface morphology of GZO thin film deposited onto the glass substrate using the sputtering method was examined by a field emission scanning electron microscope (FE-SEM), as shown in Fig. 4(a) and 4(b). The surface morphology in Fig. 4(a) shows that the GZO film has a irregular island structure surface and is well-distributed without any cracks. The resulting GZO thin film thickness shows in Fig.4(b) approximately 200-230nm was estimated. Additionally, the electrical characteristics of the GZO conductive oxide thin film was obtained by a four-point probe flat system. The lowest resistivities of GZO thin films are $8.37 \times 10^{-3} \Omega$-cm.

D. pH Sensitivity and Temperature Effect

In this study, the current-voltage of pH response curves were shown in Fig.5. By the experimental results, the GZO film is soluble in an aqueous solution of acidic (pH1~6), having a poor stability of the output potential, and can not be rendered in the acidic solution preferred sensing capability.

According to the mentioned above, the GZO thin films were deposited on the glass which used as a pH sensor head based on the extended-gate field-effect transistor (EGFET) by r. f. sputtering system was measured in basic buffer solutions in this study. The temperature effects of the EGFET based on the GZO/glass sensing structure were investigated and the drain-source current (I_{DS}) versus gate-source voltage (V_{GS}) curves were measured in pH=7,9,11 buffer solutions at different temperature (15-45°C).

In Fig.6, the current-voltage and transconductance curves of GZO pH-EGFET were measured in pH=7,9,11 buffer solutions at 25°C in this study. According to the experimental results, the highest value of transconductance curves was fixed at 0.29 μA/V, meaning that the ambient temperature was maintained at constant and the temperature effect of sensing head could be ignored temporarily during the measuring process. Moreover, when the drain current was set at 0.2 mA, various gate voltages could be extracted, and the I_{DS}-V_{GS} curves were shifted in parallel with pH7 to pH11 buffer solutions, and the threshold voltage shifted towards positive with increasing concentration of pH buffer solutions. This means that the threshold voltage shift ΔV_T depends upon the concentration of H^+ ions; namely, the threshold voltage shifts variation with respect to the gate-source voltage (Sensitivity= $\Delta V_T/\Delta pH = \Delta V_{GS}/\Delta pH$). The voltage sensitivity of H^+ ions could be extracted from the change in V_T or V_{GS}, the tendency of the voltage variation was near 99 % linearity and were plotted as shown in Fig. 7.

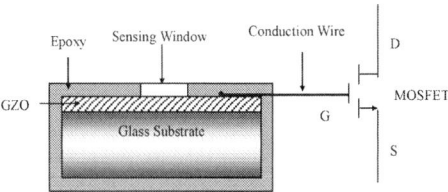

Fig. 1 The extended-gate sensing structure of pH sensor.

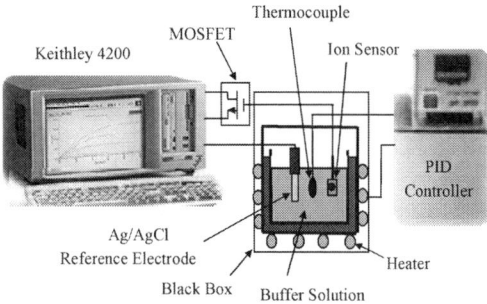

Fig. 2 The measurement structure of pH sensor.

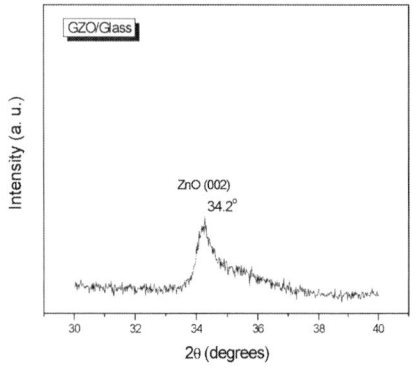

Fig. 3. XRD pattern of GZO film.

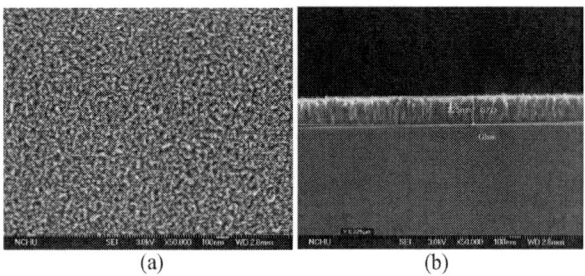

Fig. 4 FE-SEM of GZO film (a) surface morphology and (b) cross section.

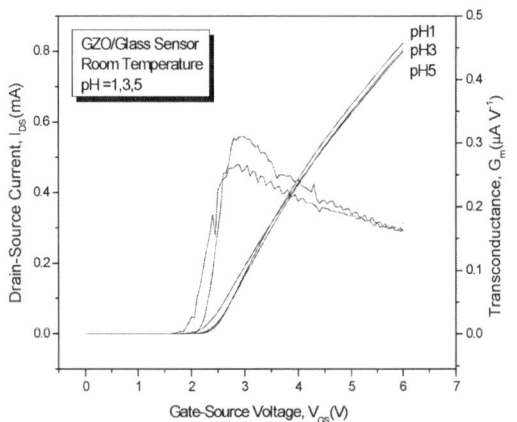

Fig. 5 Current-voltage curves of GZO pH-EGFET in acidic buffer solution (pH=1,3,5) at 25°C.

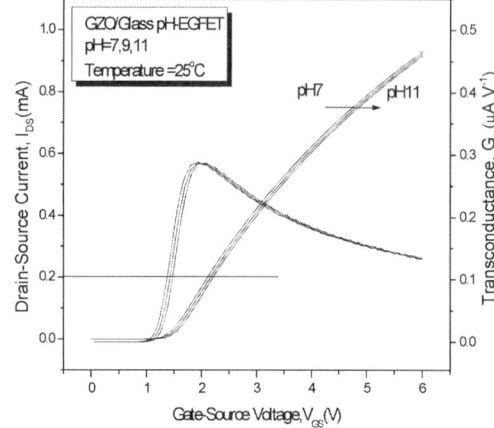

Fig. 6 Current-voltage curves of GZO pH-EGFET in pH=7,9 and 11 buffer solution.

At the same experimental measuring processes, the results can also shown in Fig. 7. The pH sensitivities of the GZO pH-EGFET for 15, 25, 35 and 45 °C are obtained 19.77, 28.99, 31.19 and 32.41mV/pH, respectively. Furthermore we also found that the pH sensitivity is increasing with temperature increased, and the pH response is a good linearity in the basic solutions in this study. These measurement tendencies and results are similar to those in references 13, and 14.

In addition, the relationship between the pH sensitivity and temperature of the GZO pH-EGFET is shown in Fig. 8, by the temperature coefficient of the sensitivity (T.C.S.), which can be approached to be 0.0952 mV/pH°C for linear-fitting and 0.136 mV/pH°C was obtained for 2-order polynomial fitting. This T.C.S. value is comparable to those obtained for other materials [15] as listed in Table II. The experimental results found that the GZO films can be as a sensing material for a pH sensor in the alkaline buffer solutions is suitable and feasible. And the temperature effect of GZO-pH-EGFET is superior behavior.

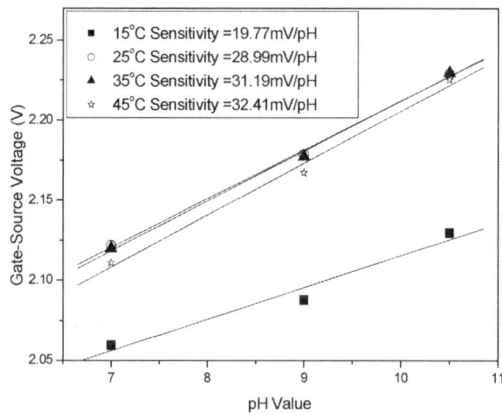

Fig. 7 pH sensitivities of GZO EGFET at 15, 25, 35 and 45°C.

978-1-4673-4840-9/13 $31.00 © 2013 IEEE

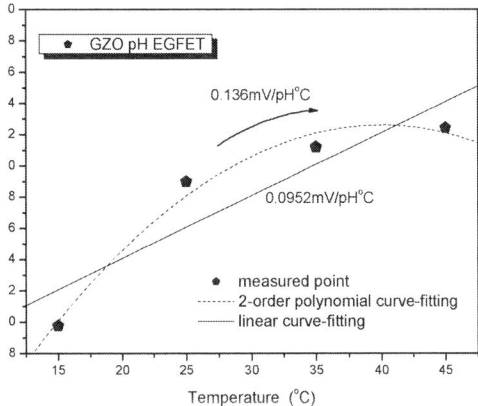

Fig. 8 Temperature coefficient of pH sensitivity of GZO pH-EGFET

TABLE II
COMPARISON OF THE TEMPERATURE COEFFICIENT FOR
VARIOUS SENSING MATERIALS

Membrane	Test range	Temperature coefficient of sensitivity (mV/pH°C)
GZO[this paper]	7-11	0.0952
SnO$_2$[15]	2 ~ 10	0.166
a-WO$_3$[15]	1~7	0.268
Ta$_2$O$_5$[15]	2 ~ 12	0.134
AlN[15]	1 ~ 11	0.130
PbTiO$_3$[15]	2 ~ 12	0.101

Conclusions

In this study, the thickness of GZO thin films about 200~250nm have been deposited on the glass substrate by r.f. sputtering system and first proposed as the pH sensing material based on the EGFET. The sensing properties are examined by the current-voltage(I-V) measurement in the alkaline buffer solutions, which reveals that the pH response of approximately 29mV/pH at 25°C can be obtained and the pH sensitivity is increasing with increased temperature at 15~45°C. In addition, the resistance of GZO film was about $8.37 \times 10^{-3} \Omega$-cm and the average of transmittance was about 80% in the visible range.

Acknowledgements

This work has been supported by the National Science Council, The Republic of China, under contract No. NSC101-2221-235-001, and NSC101-2120-S-005-001.

References

[1] S. Major, et al., "Effect of hydrogen plasma treatment on transparent conducting oxides", *Appl. Phys. Lett.* 49, pp. 394-396, 1986.

[2] T. Minami, et al., "Group III impurity doped zinc oxide thin films prepared by rf magnetron sputtering", *Jpn. J. Appl. Phys.* 24, pp. L781-L784, 1985.

[3] K. Ellmer, " Magnetron sputtering of transparent conductive zinc oxide: relation between the sputtering parameters and the electronic properties", *J. Phys., D-Appl. Phys.* 33, pp. R17- R32, 2000.

[4] O.A. Boubriak, et al., "Determination of urea in blood serum by a urease biosensor based on an ion-sensitive field-effect transistor", *Sens Actuators B* 26–27, pp. 429-431, 1995.

[5] L.-T. Yin, et al., "Enzyme immobilization on nitrocellulose film for pH-EGFET type biosensors", *Sens. Actuators B* 148, pp. 207–213, 2010.

[6] J.-Q. Wang, et al., "pH-based potentiometrical flow injection biosensor for urea", *Sens. Actuators B* 91, pp. 5-10, 2003.

[7] J.-C. Chen, et al., "Portable urea biosensor based on the extended-gate field effect transistor", *Sens. Actuators B* 91, pp. 180-186, 2003.

[8] M. Moritsuka, et al., "The pH change after HCl titration into resting and stimulated saliva for a buffering capacity test", *Aust. Dent. J.* 51, pp. 170–174, 2006.

[9] C.S. Lee, et al., "Ion-sensitive field-effect transistor for biological sensing", *Sensors* 9, pp. 7111-7131, 2009

[10] P. Bergveld, "Development of an ion-sensitive solid-state device for neurophysiological measurements", *IEEE Trans. Biomed. Eng.* 17, pp.70–71, 1970.

[11] J. Van der Spiegel, et al., "The extended gate chemical sensitive field effect transistor as multi-species microprobe", *Sens. Actuators B* 4, pp. 291-298, 1983.

[12] J. L. Chiang, et al., "Hydrogen ion sensors based on indium tin oxide thin film using radio frequency sputtering system", *Thin Solid Films*, 517, pp. 4805-4809, 2009.

[13] J. L. Chiang, et al., "Temperature effect on AlN/SiO$_2$ gate pH-ion-sensitive field-effect transistor devices", *Japn. J. Appl. Phys.*, 41, pp. 541-545, 2002.

[14] J. L. Chiang, et al., "Study on the temperature effect, hysteresis and drift of pH-ISFET devices based on amorphous tungsten oxide". *Sens. Actuators B*: Chemical 76, pp. 624-628, 2001.

[15] J. L. Chiang, et al., "Study of the pH-ISFET and EnFET for biosensor applications", *J. Medical and Biological Engineering*, 21, pp. 135-146, 2001

Characteristics of Al-doped ZnO Nanorods
Synthesized by the Hydrothermal Process at Low Temperature

Jung-Lung Chiang[1],*, Sui-Chu Tsai[2]

[1]Department of Electronic Engineering, Chung Chou University of Science and Technology
[2]Graduate School of Engineering Technology, Chung Chou University of Science and Technology
No. 6, Lane 2, Sec. 3, Shanjiao Rd., Yuanlin Township, Changhua County 510, Taiwan, R.O.C.
Tel: 886-48359000-2200, *E-mail: lung@dragon.ccut.edu.tw

Abstract

In this study, the Al-doped ZnO (AZO) nanorods were synthesized on silicon substrate by the hydrothermal process. The SiO_2/silicon substrate was submerged horizontally in 0.06M mixture solution of zinc nitrate hexahydrate ($Zn(NO_3)_2 \cdot 6H_2O$), diethylenetriamine (DETA), and doped the solution of aluminium nitrate, nonahydrate ($Al(NO_3)_3 \cdot 9H_2O$) at about 90 °C for 4h. It has been shown that the ZnO nanorods are of single hexagonal phase of wurtzite structure. SEM image revealed that the range of diameter of ZnO nanorods about 83 ~ 425 nm. In addition, the ZnO nanorods/Si_2O/Si structure was utilized as a pH sensor head to immerse into various pH buffer solutions. Experimental results were found that the Al-doped ZnO nanorods have a pH response. Afterwards, the sensor head will be designed the extended-gate field-effect transistor (EGFET) to study the pH-sensing characteristics in the different pH buffer solutions.
keywords: Al-doped ZnO, nanorods, synthesized, pH-sensing.

Introduction

Zinc oxide (ZnO) is II - VI semiconductor materials and has a wide bandwith (~3.37eV) at room temperature and is a direct energy gap material, electron binding energy of 60 meV, and has anti-oxidation, high temperature and good chemical stability and so superior characteristics. In recent years, the nano zinc oxide materials (nano-wires, nano-rods, nano-sheets and nano-tube) have been extensive researched. Especially, in high-tech applications domain, such as: the solar-cells, sensors, photocatalytic, photovoltaic elements and surface acoustic wave devices [1-3]

Zinc-containing volume is rich, and its price is cheaper than ITO materials. The zinc oxide materials have no toxicity, and in a hydrogen plasma with high chemical stability and low growth temperature characteristics. According to the mentioned above, the zinc oxide transparent conductive electrode was investigated in a wide range [4]. In the recent years, the resistivity of ZnO was approximate 1~100 Ω-cm without doped. In order to obtain lower resistivity of ZnO materials, the growth process has been doped other material. Such as the zinc oxide doped with aluminum, gallium and Indium materials, and the AZO[5], GZO[6] and IZO[7] were been formed.

Several process and methods for ZnO doped were proposed and investigated, such as thermal evaporation [8], hydrothermal system process [9], chemical vapor deposition (CVD) [10], sputtering[5], pulsed laser deposition[11] etc. In this study, the low cost, convenient and low growth temperature hydrothermal method was utilized to growth ZnO with doping aluminum materials. Accordingly, the characteristics of surface morphologies were observed using a scanning electron microscope(SEM), and the ZnO doped Al were deposited on the Si_2O/Si substrate for pH sensor based on the extended-gate field-effect transistor (EGFET) structure.

Experimental

A. Al-Doped ZnO (AZO) of Synthesize

In this study, a size of about 5 x 5 mm^2 SiO_2/silicon wafer was used as a substrate. The substrate cleaned using acetone and ethanol respectively for 15 minutes, and finally rinse with deionized water and dried with nitrogen. The substrate cleaning is completed, the Al-doped zinc oxide nanorods using a simple two-stage approach to growth: (1) spin-coated thin film layer of zinc acetate on the SiO_2/Si substrate; (2) the hydrothermal synthesis method was used to product Al-doped zinc oxide nanorods materials.

First, pre-treatment procedures were described as follows:

Step 1: Zinc acetate powder is added to the alcohol, formulated into 0.0075M zinc acetate solution.

Step 2: Spin-coated the zinc acetate solution on the Si_2O/Si substrate.

Step 3: Baking the samples at 135°C.

Secondary, the SiO_2/silicon substrate was submerged horizontally in 0.06M mixture solution of zinc nitrate hexahydrate ($Zn(NO_3)_2 \cdot 6H_2O$), diethylenetriamine (DETA), and doped the solution of aluminium nitrate, nonahydrate ($Al(NO_3)_3 \cdot 9H_2O$) at about 90 °C for 4h.

B. Surface Morphology and element Analysis

In this study, the surface morphology of Al-doped zinc oxide nanorods deposited onto the SiO_2/Si substrate using the hydrothermal synthesis was examined by a field emission scanning electron microscope (FE-SEM). In order to study the size of diameter of AZO nanorods, the pre-treatment procedures were between 6 and 12 times. In addition, the energy dispersive X-ray spectroscopy (EDS) was used to examine chemical composition.

978-1-4673-4840-9/13 $31.00 © 2013 IEEE

C. Electrical and pH sensing characteristics Analysis

In this study, The Al-doped ZnO nanorods/SiO₂/Si structure was utilized as a pH sensor head and connected with the MOSFET, that the extended-gate field-effect transistor was finished. Afterwards, the semiconductor parameter analysis instrument (Keithley 4200) was used to measure the current-voltage characteristics curves in the different pH buffer solutions, respectively.

Results and Discussion

Fig. 1(a) and Fig. 1(b) show the typical SEM image of Al-doped ZnO nanostructures growth on the SiO₂/Si substrate for 6 and 12 times pretreatment, respectively. It was observed that the morphology of Al-doped ZnO is that of a rod clusters and the hexagonal phase of wurtzite structures. In Fig. 1(a) is the high magnification image of Al-doped ZnO nanorods. It was shown that the nanorods have an average diameter of about 425 nm and an average length exceed 1μm. Fig. 1(b) shows that the average diameter of nanorods is about 83 nm.

Accordingly, the experimental results of the diameter of Al-doped ZnO nanorods by pretreatment with different times as listed in Table I. It can be found that the average diameter decreased with the pretreatment increasing. The maximum diameter of AZO nanorods is 477 nm and the minimum is about 68 nm in this study.

TABLE I DIAMETER OF AL-DOPED ZNO NANORODS BY DIFFERENT PRE-TREATMENT TIMES

Pre-treatment times	6	9	12
Max. (nm)	477	290	99
Min. (nm)	323	152	68
Average (nm)	425	220	83

In addition, the AZO nanorods structure was also examined chemical composition by the energy dispersive X-ray spectroscopy (EDS). The EDS spectra include Al, Zn and O elements which shown in Fig. 2. The experimental result confirmed the truth of Al-doped into the ZnO nanorods structure.

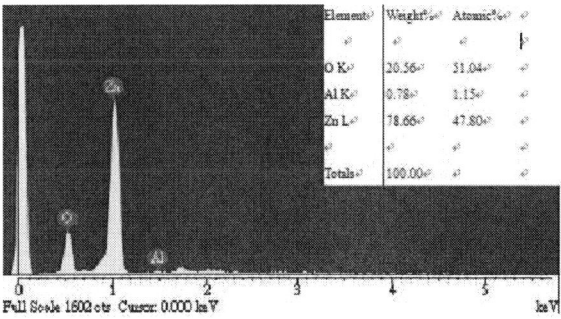

Fig. 2. EDS spectra of the Al-doped ZnO nanorods.

In Fig. 3 shows that the current-voltage characteristics curve by different pretreatment. It can be found that the current increased with the pretreatment increasing. The 12 times pretreatment has a superior current behavior and electronic characteristics.

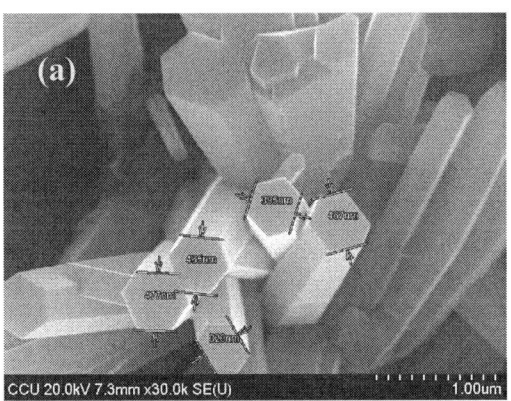

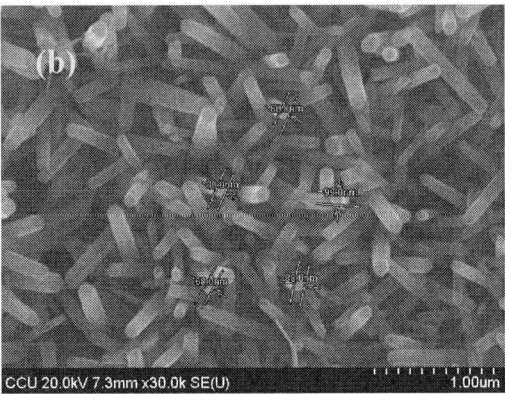

Fig. 1 The Al-doped ZnO nanorods of surface structure (a) 6 times pre-treatment, (b) 12 times pre-treatment.

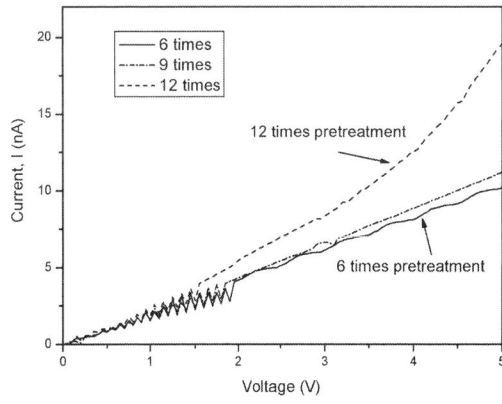

Fig. 3. Current-voltage characteristics curve for 6, 9 and 12 times pretreatment.

Fig. 4 shows that the pH response curves of Al-doped ZnO nanorods based on the extended-gate field-effect transistor(EGFET) were examined in pH 6~10 buffer solutions. When the drain current was set at 0.2 mA, various gate voltages could be extracted, and the I_{DS}-V_{GS} curves were shifted in parallel with pH6 to pH10 buffer solutions, and the threshold voltage shifted towards positive with increasing concentration of pH buffer solutions. This means that the threshold voltage shift ΔV_T depends upon the concentration of H^+ ions; namely, the threshold voltage shifts variation with respect to the gate-source voltage. The voltage sensitivity of H^+ ions could be extracted from the change in V_T or V_{GS}, the tendency of the voltage variation was near linearity. It was found that the Al-doped ZnO nanorods have the good pH characteristics and can be apply for biosensor to detect other ion. The experimental results and pH measurement tendencies are similar to the ref. 12.

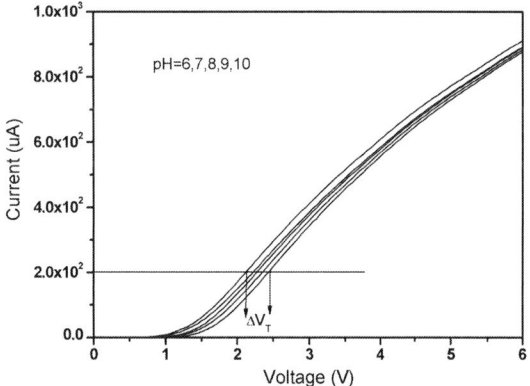

Fig. 4 pH response curves of Al-doped ZnO nanorods based on the extended-gate field-effect transistor(EGFET).

Conclusions

In this study, we successfully synthesized the Al-doped ZnO nanorods on the SiO2/Si substrate by the hydrothermal process. The average diameter of nanorods was about 425 nm for 6 times pretreatment and decreased in size 83 nm for 12 times pretreatment. According to the experimental results, the current increased with the pretreatment increasing. The 12 times pretreatment has a superior current behavior and electronic characteristics. In addition, the pH-sensing characteristics of Al-doped ZnO nanorods based on the EGFET was shown linear response.

Acknowledgements

This work has been supported by the National Science Council, The Republic of China, under contract No. NSC101-2221-235-001, and NSC101-2120-S-005-001.

References

[1] K. Hara, et al., "Highly efficient photon-to-electron conversion with mercurochrome-sensitized nanoporous oxide semiconductor solar cells", *Solar Energy Materials and Solar Cells*, Vol 64, 2000, pp. 115-134.

[2]J.A. Rodriguez, et al., "Reaction of NO₂ with Zn and ZnO: photoemission, XANES, and density functional studies on the formation of NO₃", *Journal of Physical Chemistry B*, Vol. 104, 2000, pp. 319-328.

[3]H. Yumoto, et al., "Application of ITO films to photocatalysis", *Thin Solid Films*, Vol. 345, 1999, pp. 38-41.

[4]D. L. Raimondi, et al., "High resistivity transparent ZnO thin films", *J. Vac. Sci. Technol*, Vol. 7, 1969, pp. 96-99.

[5]J. Oda, et al., "Improvements of spatial resistivity distribution in transparent conducting Al-doped ZnO thin films deposited by DC magnetron sputtering", *Thin Solid Films*, Vol. 518, 2010, pp. 2984-2987.

[6]W. T. Yen, et al., "Influences on optoelectronic properties of damp heat stability of AZO and GZO for thin film solar cells", *Advanced Materials Research: Multi-Functional Materials and Structures II*, Vols. 79 - 82, 2009, pp. 923-926.

[7]H. C. Pan, et al., "Influence of sputtering parameter on the optical and electrical properties of zinc-doped indium oxide thin films", *Journal of Vacuum science technology A*, vol. 23, pp. 1187-1191, 2005.

[8]H. Huang, et al., "Catalytic growth of zinc oxide nanowires by vapor transport", *Advanced Materials*, Vol. 13, 2001, pp. 113 -116.

[9]Y Chen, et al., "Hydrothermal synthesis of hexagonal ZnO clusters", *Materials Letter*, Vol. 61, 2007, pp. 4438-4441.

[10]C. L. Wu, et al., "Growth and characterization of chemical-vapor-deposited zinc oxide nanorods", *Thin Solid Films*, Vol. 498, 2006, pp.137-141.

[11]H. Kumarakuru, et al., "The growth of Al-doped ZnO nanorods on c-axis sapphire by pulsed laser deposition", *Surface & Coatings Technology*, Vol. 205, 2011, pp. 5083-5087.

[12]J. L. Chiang, et al., "Ion sensitivity of the flowerlike ZnO nanorods synthesized by the hydrothermal process", *Journal of Vacuum Science & Technology B: Microelectronics and Nanometer Structures*, Vol. 27, 2009, pp. 1462-1465.

Inspecting the effects of post-annealing on ZnO nanorods by optical second harmonic generation

Chung-Wei Liu,[1] Shoou-Jinn Chang,[1,2] Chun-Chu Liu,[3] Ruei-Jie Huang,[4] Yan-Shen Lin,[4] Min-Chia Su,[4]

Peng-Han Wang,[4] and Kuang-Yao Lo[2,3*]

[1]Institute of Microelectronics & Department of Electrical Engineering and Center for Micro/Nano Science and Technology,
National Cheng Kung University, Tainan 701, Taiwan.
[2]Advanced Optoelectronic Technology Center, National Cheng Kung University, Tainan 701, Taiwan.
[3]Department of Physics, National Cheng Kung University, Tainan 701, Taiwan
[4]Department of Electrophysics, National Chia Yi University, Chia Yi 600, Taiwan.

[*]Tel.: +886-6-275-7575 ext. 65261; fax: 886-6-74-7995; Email: kuanglo@mail.ncku.edu.tw

Abstract

We report on the effects of post-annealing on ZnO nanorods array using optical second harmonic generation (SHG). The high-quality and nearly vertically aligned ZnO nanorod arrays were deposited on ZnO-seeded glass substrates by aqueous solution method. The SHG result indicated the transformation of oxygen deficiency on ZnO nanorods surface after post-annealing treatment process. The chemical composition of the nanorods was investigated by X-ray photoelectron spectroscopy. The corresponding structural properties of the nanorod arrays were investigated by scanning electron microscopy and X-ray diffraction.

Keywords: ZnO, nanorod, annealing and SHG.

Introduction

ZnO-based semiconductors optoelectronic devices have attracted much attention because of their wide direct band gap ($Eg = 3.4$ eV), high excitation binding energy (60 meV), which is larger than that of GaN (26 meV), low-cost materials, and ease of manufacturing [1.2]. With a high surface-to-volume ratio, one-dimensional (1D) semiconductor nanostructures exhibit unique properties that make them desirable for novel device applications [3].

Numerous methods, such as solution method [4], chemical vapor deposition [5], and physical vapor deposition [6], have been used to synthesize 1D ZnO nanostructures. The deposition of a ZnO seed layer on a substrate is essential for these techniques, and it influences the quality of as-grown ZnO nanorods [7]. The spin coating method is popular among the various techniques for ZnO seed layer deposition because it is cheap and can be widely used in industry fabrication [8].

Oxygen-deficient defects are generated during the growth of ZnO nanorods. These defects reduce the optical properties, such as the suppression of UV emission efficiency in photoluminescence (PL) property and UV detection efficiency in photodetector devices [9, 10].

Several studies have recently reported that oxygen-deficient defects could be remarkably reduced through post-annealing process, and optical properties can be enhanced [11, 12]. However, the structural property is difficult to investigate because of the high surface-to-volume ratio of a ZnO. Inspection with high sensitivity on the surface structure should be developed to detect the quality transformation of ZnO nanorods.

Second harmonic generation (SHG) proved to be a sensitive technology for the surface structure [13]. For the symmetry of the crystalline structure or the surface structure, polarized SHG can analyze the polar structure with a resolution of several tens of nanometers, which is more sensitive than X-ray diffraction (XRD) [14, 15]. In addition, SHG can also detect the bulk quality of these materials with high surface-to-volume ratio by acquiring the nonlinear susceptibility.

In this study, as-grown ZnO nanorods were treated by post-annealing in an oxygen-rich chamber to reduce O vacancy defects during post-annealing. The quality transformation of ZnO nanorods was analyzed by reflective SHG (RSHG), compared using X-ray diffraction (XRD), X-ray photoelectron spectroscopy (XPS), and a scanning electron microscope (SEM). Abundant information would be revealed from RSHG results because RSHG is highly sensitive on surface structure.

Experiment

ZnO-seeded glass substrate was prepared by the spin coating method. A 5 mM solution was prepared by dissolving zinc acetate [$Zn(CH_3COO)_2 \cdot 2H_2O$] in ethanol, and the solution was repeatedly dropped on the glass substrate with a spin rate of 4500 rpm. Then, the solution-coated glass substrate was thermally treated to volatilize the solution and crystallize the ZnO seed layer. ZnO nanorod arrays were grown on the ZnO-seeded layer by aqueous solution method. The mixed aqueous solution consisted of 25 mM zinc nitrate hexahydrate [$Zn(NO_3) \cdot 2H_2O$, 99.0%] and 25 mM hexamethylenetetramine [$C_6H_{12}N_4$, 99.0%]. Subsequently, the ZnO-seeded glass substrate was dipped in the mixed aqueous solution at 90 °C for 1.5 h to grow ZnO nanorods. The post-annealing process was performed to reduce the oxygen-deficient defect in as-grown ZnO nanorod. The annealing condition was 800 °C for 20 min at oxygen-rich ambient.

The surface morphology of the samples was characterized by an SEM operated at 15 kV. The Bruker D8 Discover XRD system with Cu-K_a radiation was applied to analyze the structural properties of the ZnO nanorod arrays. The composition of the ZnO nanorod surface was characterized by XPS measurement.

SHG was performed by transmission type with varied incident angles. The laser light source was an Ytterbium femtosecond laser with 1044 μm wavelength and 280 fs pulse duration. The laser spot was not tightly focused and the pulse energy was well controlled to avoid the surface annealing effect. SHG signal was acquired by a photomultiplier tube and collected with a lock-in amplifier system.

Results and discussion

SEM observation was performed to confirm the morphological change in the ZnO nanorods as a result of the annealing process. The cross section and the in-plane SEM images for the as-grown and annealed ZnO nanorods are shown in Fig. 1. The ZnO nanorods were approximately 1 μm long and 100 nm wide. No evident change in the morphology was found after the annealing treatment. This phenomenon indicates that the annealing treatment has no influence on the outward shape and morphology of ZnO nanorods.

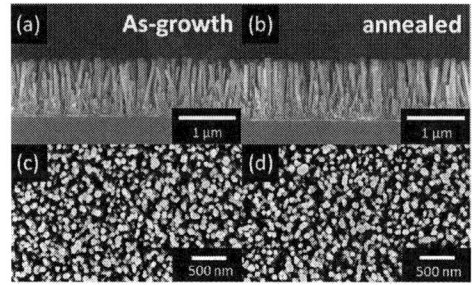

Figure 1 SEM images for as-growth and annealed ZnO nanorods. Upper are the cross section of SEM; Bottom are the in-plane of SEM.

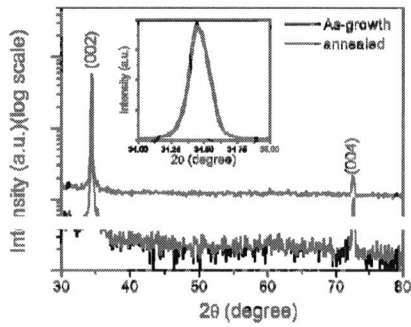

Figure 2 XRD pattern for as-growth and annealed ZnO nanorods. The inset shows peak of (002).

Bulk quality was inspected through XRD (Fig. 2). The XRD spectra of the ZnO nanorods consisted of (002) and (004) peaks, which belong to the ZnO wurtzite structure. Only (002) and (004) peaks can be detected, indicating that the nanorods are aligned perpendicular to the glass substrate. To investigate the crystal quality of the nanorod structure, the full-width at half-maximum (FWHM), d-spacing, and intensity of the (002) peak were considered. The inset in Fig. 2 shows the peak of (002). FWHM values of 0.181° and 0.176° were obtained for the as-grown and annealed ZnO nanorods, respectively, and their corresponding d-spacing values were 0.2600 and 0.2601 nm. XRD results indicate that the bulk quality was improved through post-annealing process, but the enhancement was not obvious.

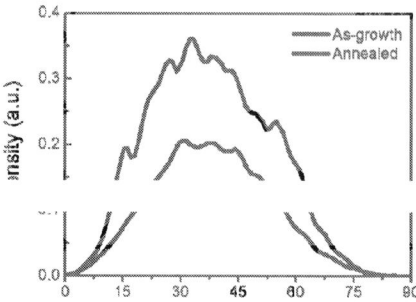

Figure 3 SHG for as-growth and annealed ZnO nanorods.

The SHG intensity of annealed ZnO nanorods is significantly larger than that of the as-grown ZnO nanorods. From the reduction of susceptibility from SHG, shown as Eq. (1) [16].

$$I_{pp} = [\sin(\cos\theta) \times [d_{15}\cos\theta\sin 2\theta + d_{31}\cos^2\theta\sin\theta + d_{33}\sin^2\theta\sin\theta]]^2 \quad (1)$$

where d15, d31 and d33 are nonlinear susceptibility coefficient, and θ are incident angle.

The SHG data was fitted by Eq. 1, and the d_{33} value for the as-grown and annealed sample was obtained. The d_{33} values were 0.4 and 0.65 for the as-grown and annealed ZnO nanorods, respectively. A higher d_{33} means well-crystallized ZnO in the ZnO nanorods. Except for the enhancement of the bulk quality of ZnO nanorods by annealing treatment to acquire sufficient activation energy, oxygen vacancy defect did not increase under oxygen-rich annealed chamber.

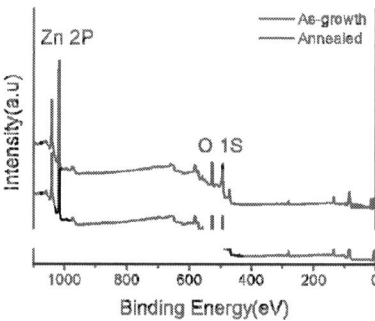

Figure 4 XPS for as-growth and annealed ZnO nanorods.

To check further the surface composition of the ZnO nanorods, XPS was performed to analyze the actual composition after annealing. As shown in Fig. 4, the XPS spectra of the ZnO nanorods consists of Zn 2P and O 1S peaks, which belong to the ZnO composition. The atomic percentage of Zn and O was constant at approximately 50% and 50%, respectively. Annealing treatment with oxygen-rich environment can provide enough activation energy to restructure ZnO nanorods without the generation of additional oxygen vacancy defects.

Conclusion

Annealing treatment can enhance the quality of ZnO nanorods by providing sufficient activation energy to remove defects, resulting in well-crystallized ZnO nanorods. Although the morphology of the ZnO nanorods was not evidently changed by post-annealing, XRD results indicate that the bulk quality was improved. However, FWHM and d-spacing only slightly changed. ZnO nanorods have high surface-to-volume ratio. Therefore, the bulk inspection method is not sufficient to analyze the quality change. SHG results revealed that the surface quality of ZnO nanorods was enhanced by the annealing treatment in an oxygen-rich chamber. The SHG experiment with high surface sensitivity can provide quality inspection in analyzing ZnO nanorod fabrication.

Acknowledgement

The authors would like to thank the National Science Council (NSC) of the Republic of China (Taiwan) for financially supporting this research under Contract No. NSC 99-2112-M-415-006-MY3

References

[1] S. J. Pearton, et al., *Superlattices Microstruct.,* 34, pp. 3-32, 2003.

[2] U. Ozgur, et al., *J. Appl. Phys.,* 98, pp. 041301, 2005.

[3] Z. L. Wang., *J. Phys.: Condens. Matter,* 16, pp. R829-R858, 2004.

[4] Lionel Vayssieres., *Adv. Mater.,* 15, pp. 464-466, 2003.

[5] D. Barreca, et al., *Sen. Actuators, B,* 149, pp. 1-7, 2010.

[6] D. Yuvaraj, et al., *Appl. Mater. Interfaces,* 2, pp. 1019-1024, 2010.

[7] P. K. Das, et al., *Mater. Chem. Phys.,* 122, pp. 18-22, 2010.

[8] L. E. Greene, et al., *Nano Lett.,* 5, pp. 1231-1236, 2005.

[9] X. L. Wu, et al., Appl. Phys. Lett., 78, pp. 2285-2287, 2001.

[10] C. Soci, et al., *Nano Lett.,* 7, pp. 1003-1009, 2007.

[11] L. Wu, et al., *Opt. Mater.,* 28, pp.418-422, 2006.

[12] C. C. Lin, et al., *Appl. Phys. Lett.,* 86, pp. 183103, 2005.

[13] K. Y. Lo, et al., *Appl. Phys. Lett.,* 90, pp. 161904, 2007.

[14] K. Y. Lo, et al., *Appl. Phys. Lett.,* 92, pp. 091909, 2008.

[15] Y. J. Huang, et al., *Appl. Phys. Lett.,* 95, pp. 091904, 2009.

[16] S. W. Chan, et al., *Appl. Phys. B,* 84, pp. 351-355, 2006.

Development of Networked Electronic Nose Based on Multi-walled Carbon Nanotubes/Polymer Composite Gas Sensor Array

Mario Lutz[1], Chatchawal Wongchoosuk[2,*], Adisorn Tuantranont[3], Supab Choopun[4], Pisith Singjai[4], Teerakiat Kerdcharoen[5,6,*]

[1]Materials Science and Engineering Programme, Faculty of Science, Mahidol University, Bangkok, Thailand
[2]Department of Physics, Faculty of Science, Kasetsart University, Bangkok, Thailand, Email: chatchawal@ku.ac.th
[3]Nanoelectronic and MEMS Lab, National Electronic and Computer Technology Center, Pathumthani, Thailand
[4]Department of Physics and Materials Science, Faculty of Science, Chiang Mai University, Chiang Mai, Thailand
[5]Department of Physics and Center of Nanoscience and Nanotechnology, Faculty of Science, Mahidol University, Bangkok, Thailand
[6]NANOTEC Center of Excellence at Mahidol University, National Nanotechnology Center, Email: sctkc@mahidol.ac.th

Abstract

In this paper, we have presented the design and invention of an electronic nose (E-nose) that can work and analyze the results via a network system such as LAN or WiFi. The MWCNTs/polymer composites were used as a gas sensor array. The constant current circuit was newly designed for gas sensor measurement. The fabricated E-nose showed a high performance for indoor air monitoring with very low noise.
Keywords: E-nose, polymer sensor, indoor air monitoring, E-nose circuits, PS/CNT sensor, gas sensor array

Introduction

Recent decades have observed significantly increasing interest in the applications of electronic nose (E-nose) ranging from quality control of foods [1,2] and beverages [3,4], environment protection [5], medical applications [6] to human identification [7,8]. Because E-nose offers many advantages in the analysis of volatiles, i.e. a low detection limit, cost- and time effectiveness, robustness, simplicity, and operator independence [9]. In principles, an E-nose consists of three main parts: (i) air flow system, (ii) detection system, and (iii) control and data analysis system [10]. For air flow system there are two main types of flow systems including static and dynamic flow systems. In detection system, sensor array with measurement circuit is used to detect the aroma molecules. For data analysis system, pattern recognition and machine learning such as artificial neural networks (ANN), linear discriminant analysis (LDA), support vector machine (SVM), and principal components analysis (PCA) are typically used in data analysis on a computer. Although E-nose has been in the market for several years, development of E-nose is still necessary to provide opportunities for new applications.

In this work, we have presented the development of an E-nose that can work and analyze the results via a network system such as LAN and WiFi. The fabrication of room temperature gas sensor array will be proposed. The measurement circuits of the networked E-nose will be highlighted. The constructed networked E-nose will also be demonstrated for uses in real world applications such as indoor air monitoring.

Experimental Details

A. Treatment of Carbon Nanotubes

Multi-walled carbon nanotubes (MWCNTs) were dispersed in a solution of ethanol and NaOH/ethanol using ultrasonication for 30 min. The NaOH/ethanol solution contained 150 g/l of NaOH in 80 vol% ethanol and 20 vol% water. The dispersion was filtered by using a syringe micro-filter system with a nylon micro-filter and then washed with ethanol until the pH value of the filtrate was about 7. After this treatment, the remaining filtered and dried MWNTs can be easily dispersed in different solvents.

B. Preparation of MWCNTs/Polymer Composite Sensor array

The sensor elements consist of interdigitated gold electrodes deposited on aluminum oxide substrate (2x2.5 mm). In order to prepare MWCNTs/polymer composite, polystyrene (PS) and polystyrene-co-maleic acid (PS-Maleic AC) were employed as the polymer matrices while dimethylformamide (DMF) and tetrahydrofuran (THF) were used as solvents. Four different conditions of CNTs/polymer composites were produced as summarized in Table I. The deposition of the CNTs/polymer composites on the sensor element was achieved by drop coating using a micro pipette, Gilson Pipetman P20.

TABLE I
DEPOSITION CONDITIONS FOR THE MWCNTS/POLYMER COMPOSITE COATING

Sensor Number	Polymer / MWCNT	Solvent	Drop amount	Substrate Temp
1	PS	THF	5 μl	80 °C
2	PS	THF	3 μl	80 °C
3	PS-Maleic Ac	THF	3 μl	25 °C
4	PS	DMF	3 μl	80 °C

C. Fabrication of Networked E-nose

The networked E-nose consists of three main parts including

(i) air flow system, (ii) detection system, and (iii) control and data analysis system. The air flow system refers to the way to deliver aroma molecules into the detection system. In this networked E-nose, a simple dynamic flow system was employed. The gas sensor array was installed at the bottom of the rectangular chamber. The air carrying the odor molecules can be introduced into the sensor chamber through a hole with an area of 4 cm². In the detection system, a constant current circuit as shown in Fig. 1 was used to measure the resistance of the gas sensor array. The constant current measurement circuit consists of an operational amplifier in a constant current configuration of 500 µA with the sensor array to be measured in the feedback loop (see Fig. 1a). Each side of the sensor was connected to a discrete buffer circuit consisting of an operational amplifier in the voltage follower configuration as shown in Fig. 1b. The buffer circuit separates the constant current circuit from the differential amplifier to prevent a disturbance of the current source. The differential amplifier calculates the difference between the two input voltages and multiplies the result by a gain factor. If R3=R4 and R2=R5, the gain factor was R2/R3. In Fig. 1c, R3 and R4 were 1 kΩ while R2 and R5 were 2 kΩ, so the gain factor was 2. For example, if the resistance of the sensor was 1 kΩ, the current of 500 µA produced a voltage difference of 0.5V. Therefore, this voltage difference of 0.5 V was amplified by a factor of 2, so the resulting output voltage was 1V.

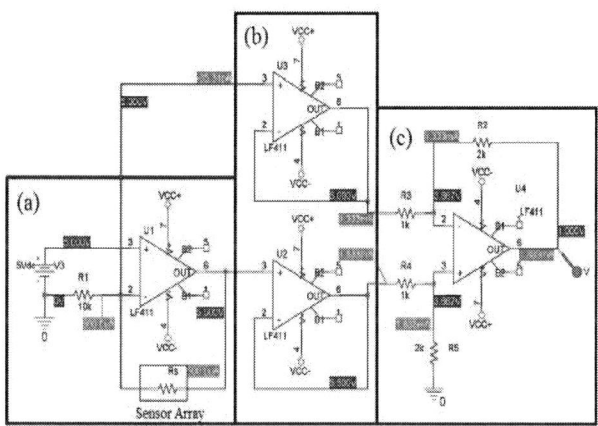

Fig. 1 Measurement circuit with an output of 0-10 V (1 mV/Ω) including (a) 500 µA constant current source, (b) buffer circuit and (c) differential amplifier having gain of 2.

Instead of using a single sensor, an analog multiplexer can be used to select one out of many sensors to be measured as shown in Fig. 2. In such a case, each sensor was connected to one input of the analog 8-channel multiplexer. The multiplexer internally connected only one input to its output. The input-to-output connection was selected by a digital input.

In the control and data analysis system, microchip PIC24FJ 48GA002 microcontroller was used to control the measurement systems. The CS5526 was used as the A/D convertor. The measurement software was written by LabVIEW. The data from networked E-nose was stored and transferred into a Laptop every second via a small WiFi-Router for subsequent analysis.

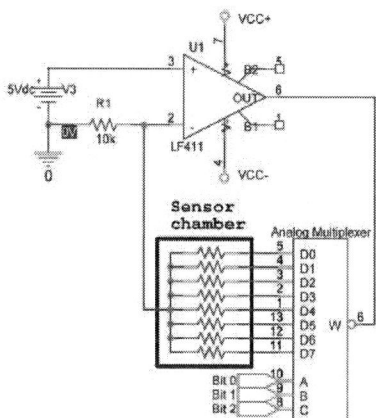

Fig. 2 Connection of gas sensor array with 8-channel multiplexer circuit.

Results and Discussion

A. Electronic Circuit Test

As the design measurement circuits, the sampling rate can be programmed ranging from 3.76 samples per second up to 202 samples per second. Note that the noise level and resolution depends on the sampling frequency.

TABLE II
RMS NOISE OF DESIGN MEASUREMENT CIRCUIT

Output Rate	-3 dB Filter	Input Range (Bipolar/Unipolar Mode)					
(Hz)	Frequency	25 mV	55 mV	100 mV	1 V	2.5 V	5 V
3.76	3.27	90 nV	90 nV	130 nV	1 µV	2 µV	4 µV
7.51	6.55	110 nV	130 nV	190 nV	1.5 µV	3 µV	7 µV
15.0	12.7	170 nV	200 nV	250 nV	2 µV	5 µV	10 µV
30.1	25.4	250 nV	300 nV	500 nV	4 µV	10 µV	15 µV
60.0	50.4	500 nV	1 µV	1.5 µV	15 µV	45 µV	85 µV
123.2	103.6	2 µV	4 µV	8 µV	72 µV	190 µV	350 µV
168.9	141.3	10 µV	20 µV	30 µV	340 µV	900 µV	2 mV
202.3	169.2	30 µV	55 µV	105 µV	1.1 mV	2.4 mV	5.3 mV

TABLE III
RESISTANCE RESOLUTION OF MEASUREMENT CIRCUIT

Output Rate	Resistance Range					
(Hz)	2.5 kΩ	5.5 kΩ	10 kΩ	100 kΩ	250 kΩ	500 kΩ
3.76	0.06 Ω	0.06 Ω	0.09 Ω	0.66 Ω	1.32 Ω	2.64 Ω
7.51	0.07 Ω	0.09 Ω	0.13 Ω	0.99 Ω	1.98 Ω	4.62 Ω
15.0	0.11 Ω	0.13 Ω	0.17 Ω	1.32 Ω	3.3 Ω	6.6 Ω
30.1	0.17 Ω	0.2 Ω	0.33 Ω	2.64 Ω	6.6 Ω	9.9 Ω
60.0	0.33 Ω	0.66 Ω	0.99 Ω	9.9 Ω	29.7 Ω	56.1 Ω
123.2	1.32 Ω	2.64 Ω	5.28 Ω	47.52 Ω	125.4 Ω	231 Ω
168.9	6.6 Ω	13.2 Ω	19.8 Ω	224.4 Ω	594 Ω	1320 Ω
202.3	19.8 Ω	36.3 Ω	69.3 Ω	726 Ω	1584 Ω	3498 Ω

Table II shows the dependence of voltage noise by the sampling rate while Table III shows the resistance resolution for 10 µA constant current depending on the sampling rate. To calculate the resistance resolution, the peak-to-peak noise voltage was calculated first by multiplying the RMS noise voltage by 6.6 and then divided by 10 μA.

The best ratio of performance and speed of measurement was achieved by using 15 Hz sampling rate. From the Tables, at an input range of 25 mV and 15 Hz sampling rate, the peak–to–peak noise of the resistance, using 10 μA constant current, was only about 0.11 Ω. This design circuit with low noise can also support the measurement of resistance from many gas sensor elements up to 64 gas sensors.

B. MWCNTs/Polymer Composite Characterization

The pure dispersed MWCNTs and MWCNTs/polymer composites were investigated by using a scanning electron microscope (SEM) as displayed in Fig. 3 and 3b, respectively. Using chemical treatment, the MWCNTs were quite well dispersed in solvents (see Fig. 3a). The diameters of MWCNTs are within a range of 60-180 nm. After the deposition of the MWCNTs/polymer composites on the sensor element, SEM image (see Fig. 3b) showed the insertion of MWCNTs inside the hole of polymer. Creation of these nanochannels can be expected to enhance the surface-to-volume ratio and reactivity between gas molecule and sensing material [11,12] leading to an increase in the sensitivity of typical polymer gas sensor.

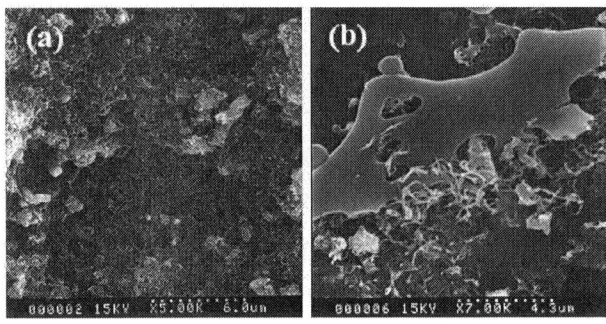

Fig. 3 SEM images of (a) pure dispersed MWCNTs and (b) MWCNTs/PS composite surfaces.

C. Indoor Air Monitoring

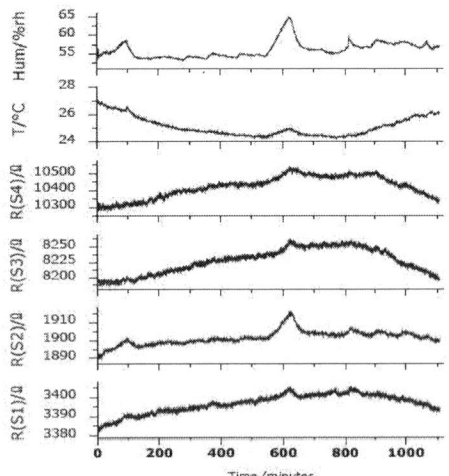

Fig. 4 Gas sensor signals from long time measurement in a room.

The sensor elements showed a highly selective sensitivity to acetic acid vapors but as good as no response to ethanol, methanol and water vapor. The long-term stability of the sensor elements was tested over a period of 18.5 hours in normal air in the laboratory. It can prove that this networked E-Nose based on MWCNTs/polymer composite gas sensor array can be used to monitor the environment with low noise (See Fig. 4).

Conclusion

The design and implementation of a networked E-nose has been presented for online detection of indoor air. The MWCNTs/polymer composite gas sensors as prepared by drop coating technique were used as a gas sensing array. The new designed constant current circuit was employed to measure the resistance of gas sensor array. The results showed the designed measurement circuit with very low noise (at sampling rate < 100 Hz, the relative noise < 0.02%) and insertion of MWCNTs inside the hole of polymer that can enhance the sensitivity of typical polymer gas sensor. It is hoped that this work will have many uses for new development of low cost networked E-nose system for indoor air monitoring.

Acknowledgments

This research was supported by Mahidol University and Kasetsart University. C.W. acknowledges TRF-CHE-KU Research Grant for New Scholar (MRG 5580229).

References

[1] K. Tikk, et al., Meat Science, Vol. 80, pp. 1254-1263, 2008.

[2] N. Barié, et al., Sens. Actuat. B, Vol. 114, pp. 482–488, 2006.

[3] C. Wongchoosuk, et al., Mater. Res. Innovat., Vol. 13, pp. 185-188.

[4] P. Lorwongtragool et al., Proceeding of ECTI-CON 2011, pp. 163 – 166, 2011.

[5] R.E. Baby, et al., Sens. Actuat. B, Vol. 69, pp. 214–218, 2000.

[6] H. P. Chan, et al., Lung Cancer, Vol. 63, pp. 164–168, 2009.

[7] C. Wongchoosuk, et al., Sensors, Vol. 9, pp. 7234-7249, 2009.

[8] C. Wongchoosuk, et al., Proceeding of DSR 2011, pp. 1-4, 2011.

[9] R. Fend, et al., J. Clin. Microbiol. Vol. 44, pp 2039-2045, 2006.

[10] C. Wongchoosuk, et al., Malodor detection based on electronic nose, Chapter 3 in air quality monitoring, assessment and management, ISBN: 978-953-307-317-0, pp. 41-74, 2011.

[11] C. Wongchoosuk, et al., Sens. Actuat. B, Vol. 147, pp. 392-399, 2010.

[12] C. Wongchoosuk, et al., Sensors, Vol. 10, pp. 7705-7715, 2010.

Current Matched Improving of Triple-Junctions GaAs-Based Solar Cell using Periodic Patterns Incorporated with Indium Nanoparticle Plasmonics

Yi-Yu Lee[1], Wen-Jeng Ho[1*], Cheng-Ming Yu [1], Jheng-Jie Liu[1], Ching-Fuh Lin [2], and Hung-Pin Shiao[3]

[1]Department of Electro-Optical Engineering, National Taipei University of Technology

1, Sec. 3, Chung-Hsiao E. Rd., Taipei 106, Taiwan

[2]Graduate Institute of Photonics and Optoelectronics, National Taiwan University, Taipei, Taiwan.

[3]Win Semiconductor Corp., Taoyuan, Taiwan.

* wjho@ntut.edu.tw

Abstract

We demonstrate experimentally the improvement of the external quantum efficiency (EQE) and current-matching of the triple-junction GaAs-based solar cell using periodic-patterns incorporated with random nano-sized indium-nanoparticles plasmonics. The average EQE decreased of -9.8% in top-cell and increased of +2.4% in middle-cell were obtained as the solar cell with periodic-patterns random nano-size indium-nanoparticles. Thus, photocurrent generated in the current-limited middle-cell was improved of ~3.4% and the conversion-efficiency was enhanced of ~3.0%, which current matching was also achieved under 1-sun AM1.5G illumination.

Keywords: External quantum efficiency (EQE), nanoparticles, plasmonics, triple-junction solar cell, current matched.

Introduction

Recently, high conversion efficiency of GaAs-based solar cells under one-sun and/or concentrated condition has been demonstrated for used in space applications and in the terrestrial solar power plants. GaAs-based multiple-junction (MJ) solar cells can be further provided a wide range absorption in the solar energy spectrum from visible-band to near-infrared-band and obtained higher conversion efficiency [1]. The optimization of the device structure, epitaxial layer quality, and device processing as well as using broadband anti-reflective coating for a high-efficiency MJ solar cells have been proposed in recent years [2,3]. In addition, the resonant excitation of plasmon in metallic nanoparticles causing a strong absorption in the Si-based photodiode was proposed by Schaadt et al. [4]. Therefore, deposited metallic nanoparticles such as gold (Au) and silver (Ag) nanoparticles on the surface of solar cells to enhance the photovoltaic performance have been successfully demonstrated by some researchers [5-8]. However, there is rarely research on using Indium material for nanoparticles to enhance the conversion efficiency of solar cells, especially, for GaAs-based multiple-junction solar cells.

In this paper, we demonstrate experimentally the performance enhancement of a middle subcell-limited triple-junction GaAs-based solar cell using periodic-patterns incorporated with random nano-sized indium-nanoparticles plasmonics. The periodic-patterns random nano-sized indium-nanoparticles are formed by evaporated the indium films through a 500-mesh screen-mask on the TiO_2 layer and subsequently annealed at low temperature. We characterized the periodic-patterns random nano-sized indium-nanoparticles by using the images of scan electron microscope (SEM). The reflectance spectra are examined and compared before and after surface AR-coatings. The dark current-voltage (I-V) and photo I-V under AM1.5G illumination are also measured and analyzed. Moreover, the measured external quantum efficiency (EQE) response is used to confirm the contribution of the reflection and transmission and to indicate the EQE enhanced in the middle-cell when the triple-junction GaAs-based solar cell using periodic-patterns incorporated with random nano-sized indium-nanoparticles plasmonics.

Experiment

The schematic structure of the triple-junction GaAs-based solar cell consisted of a GaInP top cell with an absorbed range from 300 nm to 650 nm, a GaInAs middle cell with an absorbed range from 650 nm to 900 nm and a Ge bottom cell with absorbed range from 990 nm to 1850 nm. The Ge bottom cell junction was created during metal-organic chemical vapor deposition (MOCVD) growth of the middle cell and top cell by diffusion of an n-type dopant (As) into a p-type Ge substrate. The subcells are connected in series by means of the tunnel diodes. After MOCVD growth, the Ni/Ge/Au/Ag/Au and Au/Ag/Au for ohmic-contacts were deposited on front-side and back-side by e-beam evaporation, respectively. The solar cell had a chip size of 5×5 mm^2 after isolated etching and sawed dicing.

For the indium nanoparticles plasmonics characterization, the glass-substrate and the front side of bare solar cell were firstly deposited the TiO_2 space layer with a thickness of 15-75 nm and then a fixed thickness of 3.8 nm indium films was deposited on the TiO_2 layer by a thermal evaporation through a 500-mesh screen-mask. The periodic patterns were obtained after the indium films evaporation. Next the samples were annealed at 200 ℃ for 20 min in an H_2 ambience which can transfer the deposited indium films to the indium nanoparticles. The top-view scanning electron microscope (SEM) image of the proposed periodic-patterns incorporated with random nano-sized indium-nanoparticles is illustrated in Fig. 1. The dimension and the period of the periodic-patterns are about 40 micrometer and 60 micrometer, respectively. The inset in Fig.1 is the distribution profile of the nano-sized indium-nanoparticles with the dimensions from 29 nm to 76 nm within each pattern. The optical reflectance and the transmission of the samples with a TiO_2 layer and with periodic-patterns indium-nanoparticles on TiO_2 layer were characterized using UV/VIS/NIR spectrophotometer. Worthy of note in this study, the used triple-junction GaAs-based solar cell was a middle-cell current-limited device. The current-voltage (I-V) characteristics of the triple-junction GaAs-based solar cell were measured under dark and one sun AM 1.5G (100 mW cm^{-2}) normally illumination at 25 □. The solar simulator system has been calibrated using a standard solar cell before sample measurement. The external quantum efficiency (EQE) response of the bare solar cell (without any surface coating), solar cell coated with a TiO_2 space layer, and solar cell with periodic-patterns indium-nanoparticles/TiO_2 space layer are examined and compared for the wavelength range from 350 nm to 1100 nm, respectively.

Results and Discussion

The measured transmission spectrum of the glass-substrate, the substrate with a 55 nm–thick TiO_2 layer, and the substrate with periodic patterns random nano-size indium nanoparticles on TiO_2 layer is shown in Fig. 2. The minimum transmission of 75% was observed at the wavelength of 650 nm. The magnitudes of transmission of the samples with TiO_2 layer and with indium-nanoparticles on TiO_2 layer are same between the wavelengths of 350 nm to 750 nm. Beyond 750 nm wavelength, however, the magnitude of transmission of the sample with indium-nanoparticles on TiO_2 layer was slightly less than the sample with TiO_2 layer one. Due to the shading of the metal-particles of 31%, the transmission performance of the sample with indium-nanoparticles on TiO_2 layer was not significantly reduced, however, which can be attributed to the nanoparticles surface plasmonics effect. The measured reflectance spectrum of the triple-junction GaAs-based solar cell before and after surface coating is shown in Fig. 3. The average reflectance of 17.89%, 8.85% and 12.83%, corresponding to the wavelength range of the top-cell, middle-cell and bottom-cell, were obtained as the solar cell with a 55 nm–thick TiO_2 layer. On the other hand, the average reflectance of 19.65%, 8.83% and 12.82%, corresponding to the wavelength range of the top-cell, middle-cell and bottom-cell, were obtained when the solar cell with periodic-patterns indium-nanoparticles/TiO_2 space layer. The reflection increased in the top-cell and slightly decreased in the middle cell and the transmission maintained the same in both top and middle subcells, the combination of the reflective and transmitted optical properties can be enhanced light absorption as well as photocurrent generated in the middle subcell and achieved current matched between the subcells.

From dark I-V measurement, the ideality factor (n) and the reverse saturation current (I_0) of the bare-type triple-junction GaAs-based solar cell of 4.97 and 5.95×10^{-12} A/cm^2 were obtained at 25 ℃. However, the improving in ideality factor (n = 4.71) and reducing in reverse saturation current ($I_0 = 5.95\times10^{-12}$ A/cm^2) are indicated that using a passivation layer can be suppressed the surface recombination when the solar cell coated with a 55 nm-thick TiO_2 space layer. In addition, the average value of ideality factor of about 1.6 for each diode can be attributed to a good epitaxial film quality grown by MOCVD. The photovoltaic performances of the bare-type

solar cell, solar cell with a 55 nm-thick TiO_2 space layer, and solar cell with periodic-patterns indium-nanoparticles on TiO_2 space layer in the subsequently processing are presented in the Fig. 4. The short-circuit current (I_{sc}) of the solar cell with periodic-patterns indium-nanoparticles on TiO_2 space layer increasing by 15.54% (increased from 2.96 mA to 3.42 mA) and by 3.32% (increased from 3.31 mA to 3.42 mA), compared to the bare-type solar cell and the solar cell with a 55 nm-thick TiO_2 space layer. Similarly, the conversion efficiency (η) of the solar cell with periodic-patterns indium-nanoparticles on TiO_2 space layer increasing by 12.44% (increased from 21.7% to 24.4%) and by 2.95% (increased from 23.7% to 24.4%), compared to the bare-type solar cell and the solar cell with a 55 nm-thick TiO_2 space layer. From the above-mentioned results, the increasing in I_{sc} of 3.32% and in η of 2.95% can be attributed to the contribution of the indium-nanoparticles plasmonics when the solar cell with periodic-patterns indium-nanoparticles on TiO_2 space layer. Furthermore, the measured EQE response of the solar cells with different surface coating is presented in Fig. 5. The average EQE enhancements ($\triangle$EQE) of the solar cell with periodic-patterns indium-nanoparticles on TiO_2 space layer decreased of -9.8% in top-cell and increased of +2.4% in middle-cell were obtained, respected to the solar cell with a 55 nm-thick TiO_2 space layer one. The obtained EOE value decreasing in the top-cell and increasing in the middle-cell was agreement with the reflection increased in the top-cell and decreased in the middle cell. Because the photocurrent generated in solar cell is in proportion to the values of EQE. Therefore, the EOE value increasing in the middle cell can be applied to improve the middle-cell current limited to achieve the best current matched between these sub-cells [9]. Furthermore, the I_{sc} increasing by 3.32% was also confirmed by the EQE response measurement in this experiment.

Conclusion

We demonstrated the external quantum efficiency (EQE) improvement and current matching enhancement of the middle-cell current-limited triple-junction GaAs-based solar cell using periodic-patterns incorporated with random nano-sized indium-nanoparticles plasmonics. The reflection increased in the top-cell and slightly decreased in the middle cell and the transmission maintained the same in both subcells were obtained. The obtained EOE value decreasing in the top-cell and increasing in the middle-cell was agreement with the reflection increased in the top-cell and decreased in the middle cell. The short-circuit current increasing by 3.32% and the conversion efficiency increasing by 2.95% as well as to reach the best current matching were achieved as the solar cell using the proposed periodic-patterns incorporated with random nano-sized indium-nanoparticles plasmonics.

Acknowledgment

The authors would like to thank the financial support from the National Science Council of Republic of China under Grant NSC-100-2221-E-027- 053-MY3.

References

978-1-4673-4840-9/13 $31.00 © 2013 IEEE

[1] W. Guter, et al., Current matched triple-junction solar cell reaching 41.1% conversion efficiency under concentrated sunlight, *Appl. Phys. Lett.*, Vol. 94, pp. 223504-1-223504-3, 2009.

[2] D. J. Aiken, Antireflection coating design for series interconnected multi-junction solar cells, *Prog. Photovolt: Res. Appl.*, Vol. 8, No. 6, pp. 563-570, 2000.

[3] R. R. king, et al., Solar cell generations over 40% efficiency, *Prog. Photovolt: Res. Appl*, Vol. 20, pp. 801-815, 2012.

[4] D. M. Schaadt, et al., Enhance semiconductor optical absorption via surface Plasmon excitation in metal nanoparticles, *Appl. Phys. Lett.*, Vol. 86, pp. 063106-063108, 2005.

[5] H. Atwater, et al., Plasmonics for improved photovoltaic devices, *Nature Materials*, Vol. 9, pp. 205-213, 2010.

[6] S. Pillai, et al., Surface Plasmon enhanced silicon solar cells, *Journal of Appl. Phys.*, Vol. 101, pp. 093105-093112, 2007.

[7] N. Kalfagiannis, et al., Plasmonics for orgnic solar cells, *Solar Energy and Solar Cells*, Vol. 104, pp. 165-174, 2012.

[8] K. Nakayama, et al., Plasmonic nanoparticles enhanced light absorption in GaAs solar cells, *Appl. Phys. Lett.*, Vol. 93, pp. 121904-1-121904-3, 2008.

[9] M. –Y. Chiu, et al., Improved optical transmission and current matching of a triple-junction solar cell utilizing sub-wavelength structures, *Optics Express*, Vol. 18, pp. S308-A313, 2010..

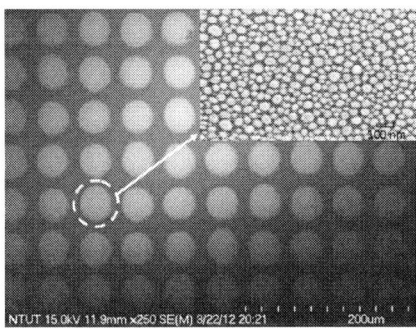

Fig. 1 Top-view scanning electron microscope (SEM) image of the proposed periodic-patterns incorporated with random nano-sized indium-nanoparticles. The inset is the distribution profile of the nano-sized indium-nanoparticles.

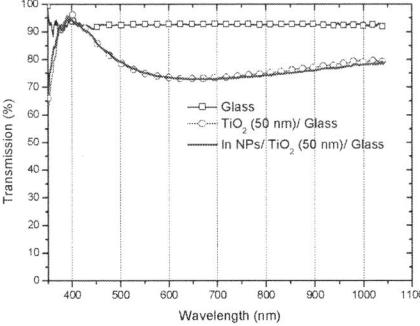

Fig. 2 Measured transmission spectrum of the glass-substrate, the substrate with a 55 nm–thick TiO_2 layer, and the substrate with periodic patterns random nano-size indium nanoparticles on TiO_2 layer.

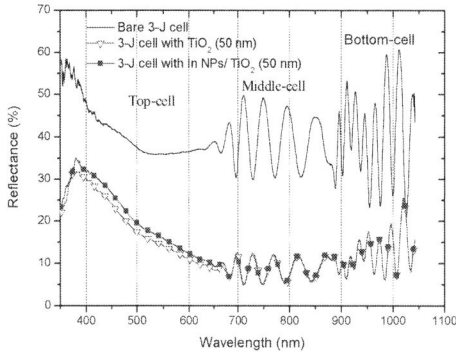

Fig. 3 Measured reflectance spectrum of the bare-type triple-junction GaAs-based solar cell, the solar cell with a 55 nm-thick TiO_2 space layer, and the solar cell with periodic-patterns indium-nanoparticles on TiO_2 space layer.

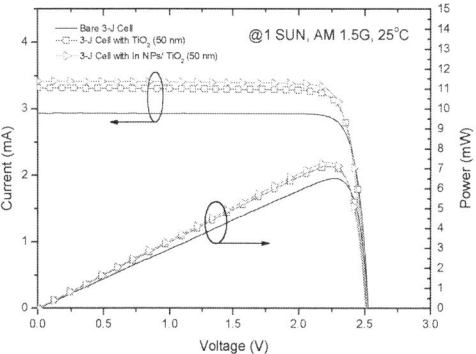

Fig. 4 Photovoltaic performances of the bare-type triple-junction GaAs-based solar cell, the solar cell with a 55 nm-thick TiO_2 space layer, and the solar cell with periodic-patterns indium-nanoparticles on TiO_2 space layer.

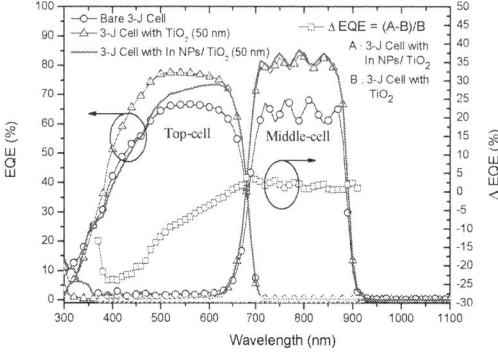

Fig. 5 Measured EQE response of the bare-type triple-junction GaAs-based solar cell, the solar cell with a 55 nm-thick TiO_2 space layer, and the solar cell with periodic-patterns indium-nanoparticles on TiO_2 space layer.

Horizontally Suspended Carbon Nanotube Bundles Patterned on Silicon Trench Sidewalls

Jingyu Lu and Jianmin Miao*

School of Mechanical and Aerospace Engineering, Nanyang Technological University, Singapore
Email: mjmmiao@ntu.edu.sg; Tel.: +65-67906038

Abstract

Arrays of horizontally suspended carbon nanotube (HSCNT) bundles were grown from selected silicon trench sidewalls with help of micromachining technologies. The key process is the patterning of an iron (Fe) catalyst layer onto selected silicon trench sidewalls through a shadow mask placed on top of silicon trenches by the tilted electron beam evaporation technique. These HSCNTs bundle arrays are promising in micro/nano electronics applications.

Keywords: carbon nanotube, horizontal growth

Introduction

Carbon nanotubes (CNTs) have drawn great attention from both industries and academic communities for the past two decades due to their unique electrical[1], mechanical[2], thermal[3] and chemical[4] properties. Great efforts have been invested on their application explorations, especially in micro/nano electronics. However, the progress has long been hurdled by the strong interaction between CNTs and the supporting substrate[5], the majority of horizontally aligned CNTs obtained so far are lying on the substrate. Therefore, great efforts have been invested to obtain HSCNTs.

Generally, there are two types of solutions to obtain HSCNTs. The post-growth assembly method, including the gel chapping[6], dielectrophoresis[7], contact printing[8], or the assembly with microsystem technologies[9], usually brings organic functional groups or defects to CNTs, which shadows the benefits of suspending CNTs. Another solution is to grow HSCNT directly, and in most cases, the buoyancy effect of feeding gas flow is used to align isolated CNTs horizontally across trenches on the substrate to obtain sparse HSCNTs[10], and the van der Waals effect is employed to obtain densely packed HSCNTs by growing CNTs from trench sidewalls[11]. Previously some groups have the substrate with almost all external surfaces grown with dense CNTs[12], some of which are HSCNTs, but the too much substrate surface coverage limits the CNT applications. Recently, we demonstrated the growth of HSCNT mats and random bundles from selected silicon trench sidewalls[11, 13]. In this paper we report the first fabrication of HSCNT bundle arrays grown on selected silicon trench sidewalls with the help of microsystem technologies, these CNT bundle arrays display great potential in microelectronic interconnects applications.

Experiment

The fabrication process of HSCNT bundle arrays is sketched in Fig. 1. First, trench structures were patterned on a 4" silicon wafer by the traditional photolithography, and then were fabricated with the depth of about 100 μm by the deep reactive ion etching (DRIE) process. The coating of a 10 nm thick Al_2O_3 layer on the substrate was followed via the atomic layer deposition technique, so as to provide the buffer layer between the silicon substrate and the Fe catalyst layer, which was deposited onto the selected trench sidewalls with the thickness of about 2 nm by the tilted electron beam evaporation technique. Note that, during the deposition process, a shadow mask was placed on top of the trenches, so that after removal of the photoresist layer with acetone solution, the patterned Fe catalyst layer was only left on selected trench sidewalls. The CNT growth was performed in a conventional CVD furnace, and the details can be found in our previous report[11].

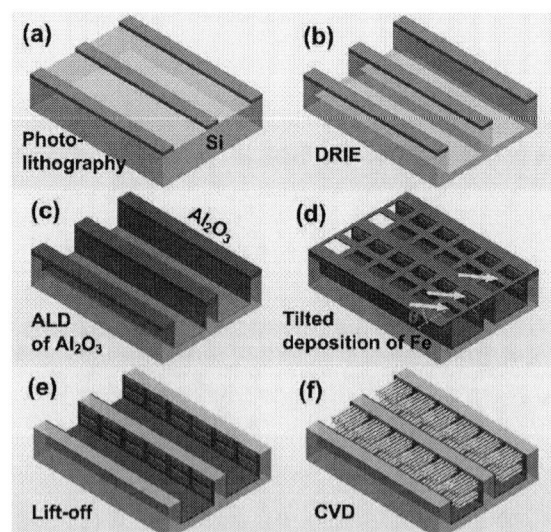

Fig.1. Fabrication process of horizontally suspended CNT bundle arrays.

Results and discussions

The scanning electron microscopy (SEM, by Hitachi S-3500

978-1-4673-4840-9/13 $31.00 © 2013 IEEE

N) images of typical as-grown CNT bundle arrays are shown in Fig. 2. The overall view in Fig. 2(a) shows that these CNT bundles are horizontally suspended from selected trench sidewalls, and they are well patterned as displayed in the top view in the insert. The magnified view of an HSCNT bundle array in Fig. 2(b) reveals that these HSCNT bundles are densely packed with the width of 58 μm, which is determined by the dimension of the open windows of the shadow mask.

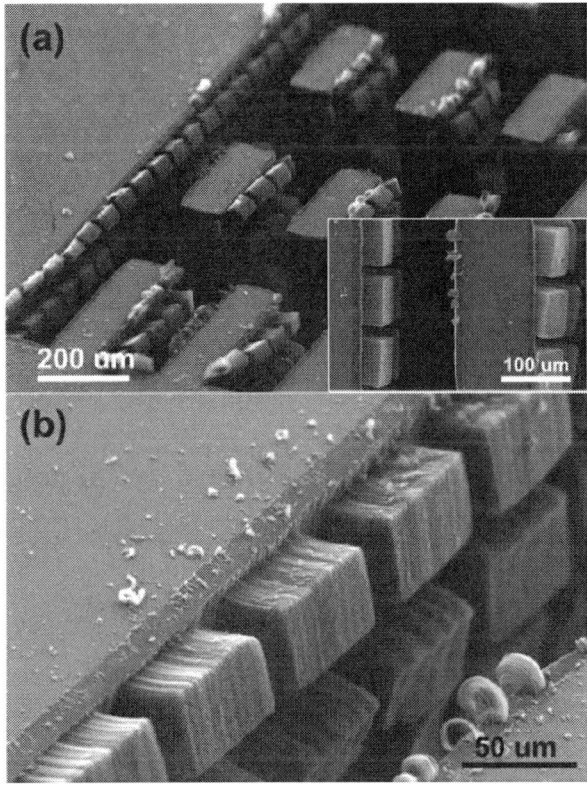

Fig. 2. (a) the SEM image of typical HSCNT bundle arrays grown from the silicon trench sidewalls, the insert is the local top view of HSCNT bundles, (b) the tilted magnified view of HSCNT bundles.

Fig. 3. (a) A 50 μm long HSCNT bundle array, (b) 6 μm long HSCNT bundle arrays, (c) local zoom-in view, and (d) HSCNT bundle arrays with smaller vertical coverage depth than those shown in (b).

The length of these HSCNT bundles can be controlled by changing the growth duration as compared in the 50 μm long HSCNT bundles in Fig. 3(a) and 6 μm long HSCNT bundles in Fig. 3(b). While Fig. 3(c) is the magnified view of a single HSCNT bundle appeared in Fig. 3(b). The vertical coverage depth of HSCNT bundles can also be brought under control, simply by changing the deposition direction of the Fe catalyst layer relative to the substrate top surface as indicated in Fig. 1(d). The effect can be shown by comparing Fig. 3(b) and (d); and more details can be found in a previous report[13].

The high resolution transmission electron microscopy (HRTEM, by FEI Titan) images of these CNTs in Fig. 4 and the insert reveal that these CNTs are multiwalled CNTs (MWCNTs), and the outmost diameters of these CNTs are 10-30 nm. Compared with the statistically 1/3 of metallic single walled CNTs, MWCNTs always conduct electricity in a metallic way, which is benefit for the interconnect applications.

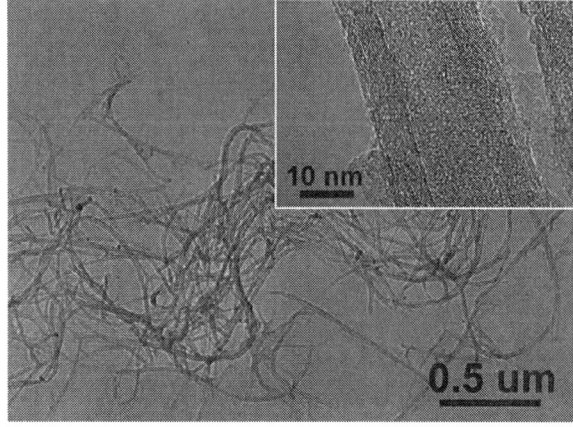

Fig. 4. HRTEM image of CNTs with the magnified view of a MWCNT segment in the insert.

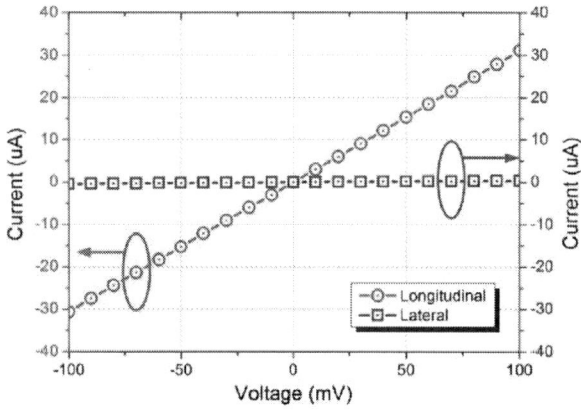

Fig. 5. I-V curves of a HSCNT bundle measured in longitudinal and lateral direction.

To characterize the electrical performance of these CNT bundles, a pair of tungsten probes were brought into contact with both ends of a HSCNT bundle, and the probes were connected to the semiconductor analyzer HP 4156B. The resultant I-V curves taken along the longitudinal and lateral directions are shown in Fig. 5. At the same applied bias, the current along the longitudinal direction of the bundle is about one order of magnitude larger than that taken in the transverse direction, which indicates the strong anisotropic nature of the CNT bundle. The longitudinal resistivity is estimated to be 10.1 Ω·mm, which is about 5 times that of vertically aligned CNT (VACNT) carpets[14], because all VACNTs in the carpet contribute to the electron conduction with both ends in good contact to the electrodes, while only a few HSCNTs here dominate the contribution. The resistivity can be greatly decreased by the liquid-induced densification of CNT bundles[15]. These HSCNT bundle arrays will be promising for interconnects and biosensor applications.

Conclusions

The fabrication of well aligned HSCNT bundle arrays was demonstrated without any post-growth assembly procedures. Micromachining technologies were applied to fabricate the trench structures, and the titled electron beam evaporation technique was used to deposit and pattern the Fe catalyst layer onto desired silicon trench sidewalls. The dimension of the HSCNT bundles can be controlled with the operational parameters, and the resultant CNTs are MWCNTs, with the bundles highly anisotropic in electrical conduction. These CNT bundles are well patterned with very good horizontal alignment and free of interaction to the trench bottom, and the fabrication is easy to process with high efficiency. Thus they have great potential in horizontal interconnects and biosensor applications.

References

[1] H. Dai, "Carbon nanotubes: synthesis, integration, and properties," *Acc. Chem. Res.,* vol. 35, pp. 1035-1044, 2002.

[2] J. P. Salvetat, *et al.*, "Mechanical properties of carbon nanotubes," *Appl. Phys. A: Mater. Sci. .Proc.,* vol. 69, pp. 255-260, 1999.

[3] R. S. Ruoff and D. C. Lorents, "Mechanical and thermal properties of carbon nanotubes," *Carbon,* vol. 33, pp. 925-930, 1995.

[4] D. Tasis, *et al.*, "Chemistry of carbon nanotubes," *Chem. Rev.,* vol. 106, pp. 1105-1136, 2006.

[5] R. Czerw, *et al.*, "Substrate-interface interactions between carbon nanotubes and the supporting substrate," *Phys. Rev. B,* vol. 66, p. 033408, 2002.

[6] C. Ji, *et al.*, "Suspended, straightened carbon nanotube arrays by gel chapping," *ACS Nano,* vol. 5, pp. 5656–5661, 2011.

[7] S. Shekhar, *et al.*, "Ultrahigh density alignment of carbon nanotube arrays by dielectrophoresis," *ACS Nano,* vol. 5, pp. 1739-1746, 2011.

[8] H. Liu, *et al.*, "Transfer and alignment of random single-walled carbon nanotube films by contact printing," *ACS Nano,* vol. 4, pp. 933-938, 2010.

[9] Y. Hayamizu, *et al.*, "Integrated three-dimensional microelectromechanical devices from processable carbon nanotube wafers," *Nat. Nanotechnol.,* vol. 3, pp. 289-294, 2008.

[10] Z. Jin, *et al.*, "Ultralow feeding gas flow guiding growth of large-scale horizontally aligned single-walled carbon nanotube arrays," *Nano Lett.,* vol. 7, pp. 2073-2079, 2007.

[11] J. Lu, *et al.*, "Growth of horizontally aligned dense carbon nanotubes from trench sidewalls," *Nanotechnology,* vol. 22, p. 265614, 2011.

[12] TalapatraS, *et al.*, "Direct growth of aligned carbon nanotubes on bulk metals," *Nat. Nanotechnol.,* vol. 1, pp. 112-116, 2006.

[13] J. Lu, *et al.*, "Facile growth of horizontally suspended carbon nanotubes," *Mater. Lett.,* vol.81, pp. 165–168, 2012

[14] C. T. Lin, *et al.*, "Anisotropic electrical conduction of vertically-aligned single-walled carbon nanotube films," *Carbon,* vol. 49, pp. 1446-1452, 2011.

[15] T. Wang, *et al.*, "Through-silicon vias filled with densified and transferred carbon nanotube forests," *IEEE Electron Dev. Lett.,* vol. 33, pp. 420-422, 2012.